OCEAN YEARBOOK 17

**OCEAN YEARBOOK 17
BOARD OF EDITORS**

Alexandru Bologa
(Romania)
Elisabeth Mann Borgese
(Canada)
Mary Brooks
(Canada)
Aldo Chircop
(Canada)
Gao Zhiguo
(China)
Edgar Gold
(Australia)
Alejandro Gutierrez
(Costa Rica)
Hou Wenfeng
(China)
Peter Serracino Inglott
(Malta)
Kuria Kairu
(Kenya)
Hugh Kindred
(Canada)
Moira McConnell
(Canada)
Gary Newkirk
(Canada)
Masako Otsuka
(Japan)
Kim Prochazka
(South Africa)
Robert Race
(Canada)
R. Rajagopalan
(India)
Dawn Russell
(Canada)
Phillip Saunders
(Canada)
Robin South
(Fiji)
Diafara Toure
(Senegal)
John Vandermuelen
(Canada)
David VanderZwaag
(Canada)
Christian Wiktor
(Canada)

OCEAN
YEARBOOK 17

Pacem in Maribus

Sponsored by the
International
Ocean Institute

Edited by
Elisabeth Mann Borgese,
Aldo Chircop, and
Moira McConnell

Assistant Editor: Scott Coffen-Smout

Supported by the Marine and Environmental Law
Programme of Dalhousie Law School

The University of Chicago Press

Chicago and London

The University of Chicago Press, Chicago 60637
The University of Chicago Press, Ltd., London

©2003 by The University of Chicago
All rights reserved. Published 2002
Printed 2003 in the United States of America

International Standard Book Number: 0-226-06620-7
Library of Congress Catalog Card Number: 79:642855

COPYING BEYOND FAIR USE. The code on the first page of an article in this volume indicates the copyright owner's consent that copies of the article may be made beyond those permitted by Sections 107 or 108 of the U.S. Copyright Law provided that copies are made only for personal or internal use, or for the personal or internal use of specific clients and provided that the copier pay the stated per-copy fee through the Copyright Clearance Center Inc. To request permission for other kinds of copying, such as copying for general distribution, for advertising or promotional purposes, for creating new collective works, or for resale, kindly write to Permissions Department, The University of Chicago Press, 1427 E. 60th St., Chicago, IL 60637. If no code appears on the first page of an article, permission to reprint may be obtained only from the author.

The paper used in this publication meets the minimum requirements of American National Standard for Information Sciences—Permanence of Paper for Printed Library Materials. ANSI Z39.48-1984.∞

Contents

The International Ocean Institute   ix
Marine and Environmental Law Programme, Dalhousie Law School   xvii
Acknowledgments   xv

Issues and Prospects

Non-European Sources of Law of the Sea,   *R. P. Anand*   1
Professionalization of the Field of Marine Affairs: Towards Standard-Setting in Education, Training and Practice,   *Aldo Chircop*   19
The International Tribunal for the Law of the Sea and Marine Environmental Protection: Expanding the Horizons of International Oceans Governance,   *Donald R. Rothwell*   26
From the Rhodian Sea Law to UNCLOS III,   *Wolfgang Graf Vitzthum*   56

Living Resources

Solving the Tuna-Dolphin Problem in the Eastern Pacific Purse-Seine Fishery,   *Martin Hall, Marcela Campa and Martha Gómez*   60
Ecolabelling in the Fisheries Sector,   *Christine LeBlanc*   93
Aquaculture in the Fiji Islands: The Need for a Comprehensive Statutory Framework,   *Timothy Pickering*   142

Transportation and Communications

Twilight of Flag State Control,   *Awni Behnam and Peter Faust*   167
Marine Biosecurity Issues in the World Oceans: Global Activities and Australian Directions,   *Chad Hewitt*   193
Ballast and Biosecurity: The Legal, Economic and Safety Implications of the Developing International Regime to Prevent the Spread of Harmful Aquatic Organisms and Pathogens in Ships' Ballast Water,   *Moira McConnell*   213

Environment and Coastal Management

Beyond Fiduciary Duty: Aboriginal Rights and Integrated Coastal Zone Management in New Zealand and Canada,   *Matthew Heemskeerk*   256
Integrating Risk Assessment and Management and Disaster Mitigation in Tourism Development Planning in Jamaica,   *Bevon Morrison*   290

The Prince Edward Islands: Southern Ocean Oasis, *Evgeny A. Pakhomov and Steven L. Chown* 348

Integrated Coastal Management in Developing, Decentralizing Countries: The General Paradigm, the U.S. Model and the Indonesian Example, *Jason M. Patlis, Maurice Knight, and Jeff Benoit* 380

Communicating the Harmful Impact of TBT: What can Scientists Contribute to EU Environmental Policy Planning in a Global Context?, *Cato C. ten Hallers-Tjabbes, Jan P. Boon, José Luis Gómez Ariza, and John F. Kemp* 417

Empowering Local Communities: Case Study of Votua, Ba, Fiji, *Joeli Veitayaki, William Aalbersberg and Alifereti Tawake* 449

Security and Military Activities

Technology Cooperation and Transfer, Piracy and Armed Robbery at Sea: A Discussion Paper in Two Parts for UNICPOLOS II, *Elisabeth Mann Borgese and François N. Bailet* 464

Abortions on the High Seas: Can the Coastal State invoke its Criminal Jurisdiction to Stop Them? *Adam Newman* 512

Military Activities in the Exclusive Economic Zone: The Case of Aerial Surveillance, *Ivan Shearer* 548

Climate Change and Arctic Shipping

Global Warming and Canada's Shipping Lanes: An Oceanographer's View, *Carl Anderson* 563

Climate Change and Canada's Shipping Lanes: The Background Science, *Henry Hengeveld* 580

The Future of the Arctic Ocean: Competing Domains of International Public Policy, *Douglas M. Johnston* 596

Appendices

A. Annual Report of the International Ocean Institute 625
   1. Report of the International Ocean Institute, 2000–2001 626

B. Selected Documents and Proceedings 632
   1. Oceans and the Law of the Sea: Report of the Secretary-General, 2001 633
   2. Oceans and the Law of the Sea: Report of the Secretary-General, 2001, Addendum 782

3. United Nations Convention on the Law of the Sea, Report of the Eleventh Meeting of States Parties, New York, 14–18 May 2001   813
4. Report on the Work of the United Nations Open-ended Informal Consultative Process Established by the General Assembly in its Resolution 54/33 in Order to Facilitate the Annual Review by the Assembly of Developments in Ocean Affairs at its Second Meeting   835
5. Report of the Fourth Global Meeting of Regional Seas Conventions and Action Plans   927
6. Review of Accomplishments in the Implementation of the Global Programme of Action, 1995–2001   949
7. Montreal Declaration on the Protection of the Marine Environment from Land-Based Activities   962
8. Conclusions of the Co-Chairs from the First Intergovernmental Review Meeting on the Implementation of the Global Programme of Action for the Protection of the Marine Environment from Land-Based Activities, Montreal, 26–30 November 2001   966
9. Oceans and Coasts at Rio+10: Toward the 2002 World Summit on Sustainable Development, Johannesburg: Concluding Statement by Conference Co-Chairs, Paris, 3–7 December 2001   973
C. Directory of Ocean-related Organizations   978

Contributors   1109

Index   000

The International Ocean Institute

Pacem in Maribus

The International Ocean Institute (IOI) was created to promote education, capacity building, and research as a means to enhance the peaceful and sustainable use and management of ocean and coastal spaces and their resources, as well as the protection and conservation of the marine environment, guided by the principle of the Common Heritage of Mankind. Professor Elisabeth Mann Borgese founded the IOI in 1972 as an independent, non-profit, non-governmental organisation headquartered at the University of Malta.

For 30 years the IOI has stood at the forefront of organisations in addressing these issues with the concern of future generations through an interdisciplinary and comprehensive approach. The IOI has also prepared working papers for the Third United Nations Conference on the Law of the Sea (UNCLOS III: 1973–1981), the Preparatory Commission for the International Seabed Authority, and for the International Tribunal for the Law of the Sea (1982–1994), as well as for various governments. It has provided consultants to UNEP, the World Bank, the United Nations Industrial Development Organisation (UNIDO), and the Asian-African Legal Consultative Committee (AALCC). It contributed to the formulation of recommendations of the World Summits in Rio de Janeiro (1992) and Johannesburg (2002) and of the World Water Forum in Japan (2003) on oceans, coasts and islands, as well as to the review of the developments in ocean affairs through the United Nations open-ended informal consultative process.

The goals of the IOI are to:

- Enhance the ability of developing countries to develop and manage their own resources sustainably for their own benefit, to establish self-reliant development, and help with education and eradication of poverty from community to national levels;
- Enhance abilities for self-reliant development at the community level, taking into account the diversity in developing as well as developed countries, including control and protection of natural resources for future generations; the eradication of poverty in coastal areas; and mitigation of and adaptation to natural hazards;
- Enhance participation of people, in particular women, in development projects which take into account environmental issues;

- Establish sustainable mechanisms able to tackle inter-related social, environmental and economic issues in an integrated fashion.

The approach by which the IOI gradually achieves its goals include:

- Strengthening of institutions through capacity building, sharing and dissemination of information, and generating incentives and contact between local and national authorities;
- Establishing partnerships and networks with the IOI Operational Centres, other NGOs, donors and between authorities and communities;
- Increasing awareness and understanding of the sensitivity and the importance of the coastal zone and the marine environment of sustainable development, through demonstrations, training, provision of educational material and information to local NGOs, schools and authorities;
- Encouraging self-reliant development of sustainable livelihoods by means of aquaculture, farming, value added processing of resources, protection of water resources and application of traditional and new technology;
- Emphasising decentralised decision making to local authorities and communities, and implementation of agreements, regulations, and development projects with the involvement of the private sector; and Increasing the abilities at local and national levels to transfer and apply scientific (social and natural sciences) knowledge and information, from generators to users, through hands-on training, case studies, and demonstration sites; and providing incentives through linkages to other sites, and to international agreements and commitments.

The IOI's activities include training projects, information dissemination, conferences, research and publications:

- Training of hundreds of decision-makers and professionals, mainly from developing countries, through short and long duration interdisciplinary courses in ocean and coastal management;
- Development work among coastal communities with the objective of improving their livelihood while restoring and preserving coastal ecology;
- Information dissemination to international organizations and national institutions through the global IOI networks and the IOI websites;
- Organisation of the annual *Pacem in Maribus* Conference and other seminars and workshops;

- Research on a variety of ocean-related areas such as international and regional agreement and policies on oceans and the coastal zone; on regional and sub-regional co-operation and on scientific and technological approaches to sustainable management of living and nonliving marine resources;
- Education and awareness-creation about ocean resources, marine and coastal environments, and the need to care for them;
- Technology evaluation, transfer, and evaluation of the effects thereof;
- Publication of the *Ocean Yearbook* in collaboration with the Dalhousie University Law School, Canada; *Across the Oceans*, the IOI's Newsletter; as well as directories of experts, funding opportunities, and potential clients for IOI services. IOI's web site (www.ioinst.org) is now under modification and updating. Regional operational centres also publish their own newsletters, research papers and reports; and
- Services include advice, consultancy, and information regarding ocean and coastal environments.

The IOI gained a worldwide respect and reputation through launching such projects as the coastal and eco-villages projects, dedicated women and youth programmes, training in ocean governance, risk assessment and others. The IOI Virtual University is a new initiative. It will be based on a network of education, training and research centres with expertise in ocean, coastal and marine-related affairs and governance. The Centres will be joined together in a partnership so as to provide for all interdisciplinary and comprehensive coverage of the subject areas. The overall objective will be to enhance the abilities of developing countries to develop and govern their own marine and coastal resources and environment sustainably.

The IOI scope and presence is truly international with 25 Operational Centres around the globe and with several new Centres or affiliates in the development stage. The IOI network provides a flexible mechanism with a governing and co-ordinating structure that generates synergism and strategic planning of the network of semiautonomous nodes. This cohesive and comprehensive mechanism is capable of co-operating equally well with other intergovernmental systems and the private sector. The current centres and their host institutions are:

> IOI – Canada, Dalhousie University, Halifax, Canada; IOI – China, National Marine Data and Information Service, State Oceanic Administration, China; IOI – Costa Rica, Universidad Nacional, Costa Rica; IOI – Pacific Islands, University of the South Pacific; IOI – India, Indian Institute of Technology, Madras, India; IOI – Japan, Yokohama City University, Japan; IOI – Malta, University of Malta, Malta; IOI – Black Sea, National Institute for Marine Research and

Development 'Grigore Antipa', Romania; IOI – Senegal, Centre de Recherches Oceanographiques de Dakar – Thiaroye, Senegal; IOI – Southern Africa, University of Western Cape, South Africa; IOI – Eastern Africa, Kenya Marine and Fisheries Research Institute, Mombasa, Kenya; IOI – Ukraine, Institute of Biology of the Southern Seas, Sevastopol, Ukraine; IOI – Russia, P.P. Shirshov Institute of Oceanology; IOI – Western Africa, Nigerian Institute for Oceanography and Marine Research, Lagos, Nigeria; IOI – Thailand, Office of Thai Marine Policy and Restoration Committee, Bangkok, Thailand; IOI – Caspian Sea, Astrakhan State Technical University, Astrakhan, Russia; IOI – Volga River Basin, Nizhny Novgorod State University of Architecture and Civil Engineering, Russia; IOI – Indonesia, Centre for Marine Studies, University of Indonesia; IOI – Western South Atlantic, Centre for Marine Studies, Foundation of the Federal University of Panama, Brazil; IOI – Germany, Centre for Tropical Marine Ecology, Germany; IOI – South American Cove, Catholic University of Valparaiso, Chile; IOI – Regional Centre for Australia and the Western Pacific, International Marine Project Activities Centre, Australia; IOI – Baltic, University of Kalmar, Sweden; IOI – Islamic Republic of Iran, Iranian National Centre for Oceanography (INCO), Tehran, Iran; IOI – Slovenia, Marine Biology Station at the Institute of Biology, Piran, Slovenia.

Each Operational Centre is autonomous, uniquely identifying its own regional priorities for research, capacity building and development, while benefiting from the support of the overall IOI network. Regional approaches to research and capacity building enable the Institute to draw upon the different strengths of the Operational Centres to cater to the needs identified within each region. A director, generally supported by a small staff with a large number of experts and volunteers on call, runs each Centre. The directors are members of the IOI's Planning Council, which meets annually. The IOI is governed by a Board that takes decisions on policy, programme and budget matters.

Growing steadily and responding to global changes, the IOI network is now aiming at a multiplier effect to its spectrum of activities. It plans to move from direct training to training-the-trainers; from direct implementation of projects to offering advisory and consultative services; from a network of centres to a network of clusters and affiliates. IOI is also developing online and distance education systems.

INTERNATIONAL OCEAN INSTITUTE
GOVERNING BOARD MEMBERS

François Bailet (Ex-officio)
(Canada)
Awni Behnam *(President)*
(Switzerland)
Noel Brown
(Jamaica)
Salvino Busuttil
(Malta)
Anita Coady (*Treasurer*)
(Canada)
Nikolaus Gelpke
(Switzerland)
Derek W. Keats (Ex-officio)
(South Africa)
Federico Mayor
(Spain)

Iouri Oliounine (Ex-officio)
(Russia)
Alain Piquemal
(France)
Russell E. Reichelt
(Australia)
Peter Serracino Inglott (Ex-officio)
(Malta)
Bhagwat Singh (Ex-officio)
(India)
Tuiloma Neroni Slade
(Samoa)
Layashi Yaker
(Algeria)
Alexander Yankov (*Vice President*)
(Bulgaria)

INTERNATIONAL OCEAN INSTITUTE
PLANNING COUNCIL MEMBERS

François Bailet (Ex-officio)
(Canada)
Alexandru Bologa
(Romania)
Anita Coady (*Treasurer*)
(Canada)
Alejandro Gutierrez
(Costa Rica)
Derek W. Keats (Chairman)
(South Africa)
Kuria Kairu
(Kenya)
Qin Li
(China)
Iouri Oliounine (Ex-officio)
(Russia)
Masako Bannai Otsuka
(Japan)
Olusegun Oyewo
(Nigeria)

Kim Prochazka
(South Africa)
Victoria Radchenko
(Ukraine)
R. Rajagopalan
(India)
Peter Serracino Inglott
(Malta)
Robin South
(Australia)
Diafara Toure
(Senegal)
Joeli Veitayaki
(Fiji)
Hou Wenfeng
(China)
Viatcheslav Zaitsev
(Russia)

Marine and Environmental Law Programme
Dalhousie Law School

Established in 1883, Dalhousie Law School is the oldest common law school in Canada. As a leading law school, Dalhousie has traditionally played a critical role in the development of national legal education in Canada, in servicing the needs of the Atlantic region and as a focal point for graduate level education globally. The Law School offers the full breadth of undergraduate and graduate level education and is home to the Marine and Environmental Law Programme (MELP), the Indigenous Blacks and Mi'kmaq Programme, the Health Law Institute, the Commonwealth Judicial Institute and the Law and Technology Institute.

Dalhousie Law School is pleased and honored to support and provide a home for the Ocean Yearbook through the Marine and Environmental Law Programme. In 1974, MELP was established as a center of excellence for students wishing to pursue research, practice, and teaching in the marine and environmental law fields. The core programme faculty teach courses and conduct research on business and environmental law, coastal zone management, environmental law, international environmental law, land-use planning, law of marine environmental protection, law of the sea, maritime law and practice, maritime law and policy (carriage of goods by sea), ocean law and policy (international fisheries), and oil and gas law. The Programme has attracted students from numerous countries from all regions. Its principal offerings include a Certification in Marine or Environmental Law for Bachelor of Laws (LL.B.) students and specializations in these fields at the Master of Laws (LL.M.) and Doctorate in the Science of Law (J.S.D.) levels. MELP is further supported by the Sir James Dunn Law Library, which houses one of the largest collections in marine and environmental law and policy in North America. In 1980, the law librarian initiated the *Marine Affairs Bibliography*.

The MELP faculty and students have had a prolific research and publication record in most major periodicals in these fields. Further, faculty are active in legal and policy development projects for governments in Canada, Europe, in various developing regions and countries, including Southeast Asia, South Asia, Latin America and the Caribbean. The Halifax metropolitan area boasts one of the largest concentrations of expertise in coasts and oceans in the world. The city is home to the Bedford Institute of Oceanography, six universities, various federal and provincial government agencies with coastal and ocean management capabilities, and a strong private sector base in marine science, technology, and consulting. This wealth of expertise has enabled Dalhousie to identify ocean studies as an area of focus in its mission statement. As a result, most faculties and departments have developed unique strengths in coastal and ocean studies, contributing in turn to a rich multi- and interdisciplinary environment. MELP interacts with a wide variety of departments and programs dealing with ocean studies at Dalhousie. In particular, the Marine Affairs Programme, the School for Resource and Environmental Studies, the Department of Biology, the Department of Sociology, and the Department of Oceanography have strengths in various aspects of ocean studies.

Acknowledgments

Many individuals have contributed to this volume of the *Ocean Yearbook*, and only a few can be specifically acknowledged here. We especially acknowledge the contribution of time and expertise from numerous peer reviewers. We are grateful for the ongoing administrative and substantive support provided by the Marine and Environmental Law Programme and the Sir James Dunn Law Library at Dalhousie University Law School. The Editors wish to particularly acknowledge the continuing key role played by the Assistant Editor, Scott Coffen-Smout. The International Ocean Institute provided kind assistance from its Malta headquarters and its Halifax Operational Centre. We extend thanks to Jacqui Shaw for her research and editing assistance, and to Susan Rolston for compiling the volume index and for revising the directory of oceans-related organizations. Warm thanks also go to the Board of Editors for numerous formal and informal contributions and to the University of Chicago Press for its editorial and infrastructural support. Special thanks are extended to Teresa Mullen and Marsha Ross at the University of Chicago Press.

THE EDITORS

Issues and Prospects

Non-European Sources of Law of the Sea†

R. P. Anand
Jawaharlal Nehru University, New Delhi

LAW OF THE SEA: THE PRODUCT OF EUROPEAN CIVILIZATION

There is a widespread belief amongst Western, especially European, scholars that the law of the sea, like other rules of interstate conduct of modern international law, is a product of Western European Christian civilization to which non-European countries have contributed practically little or nothing. It is asserted with a sense of pride that international law is a "product of the conscious activity of the European mind" and "European beliefs" and is based on European state practices that were developed and consolidated during the last three centuries.[1] Thus, relating the story of the development of international law, Professor J.H.W. Verzijl states:

> The body of positive international law once called into being by the concordant practice and express agreement of European states, has since the end of the eighteenth century onwards, spread over the rest of the world as a modern *ratio scripta*, to which extra-European states have contributed extraordinarily little. International law as it now stands is essentially the product of the European mind and has practically been 'received' . . . lock, stock and barrel by American and Asiatic states.[2]

Relying entirely and almost exclusively on European history and European sources, with rare exceptions,[3] most of the Western scholars affirm or

---

†EDITORS' NOTE.—An earlier version of this article was presented as a keynote address at the Pacem in Maribus XXVIII Conference—The European Challenge, Hamburg, Germany, December 3–6, 2000.

1. J. H. W. Verzijl, "Western European Influence on the Foundations of International Law," in *International Law in Historical Perspective*, ed. J. H. W. Verzijl. (Leyden: A. W. Sijthoff, 1968), pp. 435–36.
2. *Ibid*, p. 442.
3. See C. H. Alexandrowicz, *An Introduction to the History of the Law of Nations in the East Indies (16th, 17th and 18th Centuries)* (Oxford: Clarendon Press, 1967).

© 2003 by The University of Chicago. All rights reserved.
0-226-06620-7/03/0017-0001 $0.1.00 *Ocean Yearbook* 17:1–18

confirm this opinion. As Professor B.V.A. Roling asserts: "There is no doubt about it: the traditional law of nations is a law of European lineage."[4]

Kunz confirms: "Our international law is a law of Christian Europe. It has its roots in the *Republica Christiana* of medieval Europe."[5] Practically every study on the history of international law in Europe expresses and confirms this opinion.[6] In fact it is noted "with a certain amount of amusement" how the Asian States grasp "as the highest and, indeed, as universal values certain fundamental ideas created and elaborated by the West." This offers curious but clear,

> [E]vidence of the lasting dependence of non-western nations in the conduct of their international affairs upon fundamental concepts of the western world from which their political leaders nevertheless so ardently crave to liberate their states without, however, being able either to derive any different workable principle of international law from data of their own national history or to develop independent legal principles susceptible of replacing the traditional standard principles of existing international law.[7]

Although some of the ancient countries, like China, India, Egypt, and Assyria, with quite advanced forms of civilizations, might have had certain generally accepted principles and rules of interstate conduct, the western jurists feel that these practices "reveal little that could, even in the broader sense of the word, be considered as international law."[8]

FREEDOM OF THE SEAS: THE PARAMOUNT PRINCIPLE

The bulk and essence of maritime law during the last more than two centuries can be summed up in the simple phrase, "Freedom of the Seas." What it meant was that beyond a limited area of territorial sea where the coastal state exercised sovereign jurisdiction, an area that was deemed essential for its security and protection of its other vital interests, the vast areas of the

---

4. B. V. A. Roling, *International Law in an Expanded World* (Amsterdam: Djambaton, 1960), p. 10.
5. J. L. Kunz, "Pluralism of legal and value systems and International Law," in *The Changing Law of Nations*, ed. J. L. Kunz (Columbus, Ohio: Ohio State University Press, 1968), p. 48.
6. See A. Nussbaum, *A Concise History of the Law of Nations* (New York: Macmillan, 1962), p. 191; See also several writers quoted in R. P. Anand, *New States and International Law*, (Delhi: Vikas Pub. House, 1972), p. 7.
7. Verzijl (n. 1 above), p. 445.
8. Nussbaum (n. 6 above), p. 10.

ocean were open and free and could not be appropriated and must not be controlled by anyone. In these areas of what were called the "high seas," all states enjoyed—or at least until recently were supposed to enjoy—as Article 2 of the 1958 Convention on the High Seas[9] declared: freedoms of unobstructed navigation, uncontrolled fishing, right to lay down and maintain submarine cables and pipelines, and freedom to fly over, and such other undefined freedoms as they might like to exercise with due regard to the similar rights and freedoms of others.

The history of the law of the sea is, to a large extent, the story of the development of the freedom of the seas doctrine and the vicissitudes through which it has passed through the centuries. For the last nearly 200 years, it had been accepted as an undisputed principle, almost a dogma, that no one could dare challenge. Recognized and referred to as *jus cogens* (universally accepted principles), it was supposed to be in the interests of all mankind. It expressed in a sense the essence and substance of the law of the sea. All other rules relating to interstate conduct more or less revolved around this doctrine and their validity or otherwise was to be judged and depended on the touchstone of this incontrovertible principle. Thus, even when coastal state's jurisdiction in a part of the sea close to its coastline came to be recognized as territorial sea for the protection of its security and other interests, its limits were always sought to be kept as narrow as absolutely essential to maintain this freedom in wide areas. In any case, beyond the narrow limits of the territorial sea, even limited jurisdiction for the protection of coastal fisheries was totally denied until the end of the World War II. Contiguous zones for the protection of coastal economic, health, and financial interest were either refused or merely tolerated, in the name of the freedom of the seas, by the biggest maritime power, Great Britain, which ruled the waves for more than 200 years.[10]

Origin of the Principle

It is generally assumed and widely asserted that it was the seventeenth century Dutch jurist, Hugo de Groot or Hugo Grotius, who propounded the doctrine of the freedom of the seas for the first time in the modern period. Although it is believed that the principle was clearly accepted under Roman law and had been reduced to a legal formula according to which the sea was recognized as *"commune omnium,"* or common property of all, after the

9. Convention on Fishing and Conservation of Living Resources of the High Seas. Done at Geneva, 29 May 1958. Entered into force, 20 March 1966.
10. See T. W. Fulton, *The Sovereignty of the Sea* (London: W. Blackwood, 1911) (Reprinted Millwood, New York: Krauss Reprint Co., 1976), p. 593–603; J. L. Brierly, *The Law of Nations,* (6th Ed. by Sir H. Waldock) (Oxford: Oxford University Press, 1963), p. 205–206.

disintegration of the Roman Empire it had been lost and forgotten through the centuries.[11] The "reawakening" of the principle was brought about by Hugo Grotius. As Meurer put it: "Up to modern times the freedom of the seas slumbered the sleep of the Sleeping Beauty until there appeared from Netherlands the knight whose kiss awakened her once more."[12]

It is well-known that Grotius enunciated and elaborated his thesis relating to the freedom of the seas in his famous book *Mare Liberum (Free Seas)* published anonymously in 1609.[13] Few works of such small size have gained such great reputation as *Mare Liberum*. It is said to be "the first and the classic exposition of the doctrine of the freedom of the seas."[14] Grotius wrote this remarkable book, which has earned him the title of the "founder" or "father" of international law, in order to defend his country's right to navigate in the Indian Ocean and Eastern seas and to trade with India and the East Indies (Southeast Asian Islands), over which Spain and Portugal asserted a commercial monopoly as well as political domination. In fact, *Mare Liberum* was merely one chapter (Chapter XII) of a bigger work, *De Jure Praedae (On the Law of Spoils)* which Grotius, as advocate of the Dutch East India Company, had prepared as a legal brief but which he had refrained from publishing.[15]

This was a period of keenest international commercial rivalry between Spain, Portugal, Holland, and England, all of whom were struggling to gather the riches of the East. Ever since Rome made Eastern products fashionable and its Egyptian subjects went out to seek them in the Indies, the European world had been possessed by the splendour of the East. Aromatic spices from India and the East Indies were in the greatest demand and yielded the largest profit. Spice trade with the East, especially pepper, then became a great motivating factor of history. As a recent writer points out: "Pepper may not mean much to us, but in that age it ranked with the precious stones. Men risked the perils of the deep and fought and died for pepper."[16] Spain and Portugal, the two Iberian Powers who were the first to look for a sea route to India and the Spice Islands, claimed a legal title to half the non-Christian world each under a Papal Bull of 4 May 1493, by which Pope Alexander VI divided the world between the two and defined

---

11. See C. Meurer, *The Program of the Freedom of the Sea* (Tr. from German by L. J. Franchenberg) (Washington: Govt. Print. Off., 1919), p. 4–7.

12. *Ibid,* p. 7.

13. H. Grotius, *The Freedom of the Seas or The Right which belongs to the Dutch to take part in the East Indies Trade* (Tr. by R. Van Deman Magoffin and Edited with an Introduction by J. B. Scott) (New York: Oxford University Press, 1916).

14. W. S. M. Knight, "Seraphin de Freitas: Critic of *Mare Liberum*," *Transactions of Grotius Society,* Vol. 11 (1926), p. 1.

15. See W. S. M. Knight, *The Life and Works of Hugo Grotius,* (London: Sweet and Maxwell, 1925), p. 79.

16. G. F. Hudson, quoted by K. M. Panikkar, *Asia and Western Dominance* (London: Allen & Unwin, 1953), p. 25.

a line of demarcation running 100 leagues west of Azores and Cape Verde Islands and granted to Spain all lands west of it, and to Portugal all lands to the east. By a bilateral treaty of 1494, the two Powers fortified their title.[17]

ASIAN TRADITIONS IGNORED

It is submitted that the contribution of Asian, African, and other extra-European countries toward the development of modern international law, or their attitude, outlook, and behaviour toward its rules in their international relations, is more often than not based on ignorance of their history and lack of information or understanding of their cultures and cultural traditions. Europeans generally do not want to look beyond European history, written during the colonial period, to acknowledge that when European adventurers arrived in Asia in the fifteenth century, "they found themselves in the middle of a network of states and interstate relations based on traditions which were more ancient than their own and in no way inferior to notions of European civilization."[18] These rules of interstate conduct might have differed, and in fact did differ, from the European state practice; but there is no doubt about their widespread acceptance amongst Asian States. Thanks to their liberal traditions of freedoms of peaceful navigation and international maritime trade—and permission to foreign merchants to establish themselves by their own laws—the Europeans got an easy foothold in Asia.[19] Whether expressed in the form of a doctrine or not, there is no doubt that the unobstructed freedoms of navigation and commercial shipping were accepted by all countries in the Indian Ocean and other Asian seas centuries before history was ever recorded, long before Grotius was ever heard of, or Europe emerged as a formidable force on the international stage. Besides historical records, numerous travelers' memoirs testify to this state of affairs.[20] Freedom of the seas was also a recognized rule in the Rhodian Maritime Code and was unequivocally adopted in Roman law. From the first century A.D., regular maritime commercial relations were established between Rome and several states in India and the Indian Ocean region, and they continued for nearly 300 years.[21]

 17. See Panikkar (n. 16 above), p. 31–32.
 18. Alexandrowicz (n. 3 above), p. 224.
 19. Alexandrowicz, *Ibid.*
 20. See *The Travels of Marco Polo* (W. Marsden ed. and tr. 1948); I. Batutta, *Travels in Asia and Africa (1325–54)* (H. A. R. Gibb tr.); Narrative and Journey of Abd-er-Razak, A Persian Traveler and Ambassador of Shah Rukh (1442), *India in the Fifteenth Century* (R. H. Major tr. & ed.)
 21. See H. G. Rawlinson, *Intercourse between India and the Western World from the Earliest Times to the Fall of Rome* (1926), p. 9–12; See also E. H. Warmington, *The Commerce between the Roman Empire and India* (Delhi: Vikas Pub. House, 1974), p. 35 ff.

6   Issues and Prospects

On the eve of European penetration into the Indian Ocean, not only was the principle of freedom of the seas and trade well recognized in customary law of Asia, but also in some states this principle was codified and well-publicized. Examples include the maritime codes of Macassar and Malacca, which were compiled at the end of the thirteenth century, based on customary practices.[22] Resisting the Dutch attempts to monopolize the maritime trade of the Spice Islands, the ruler of Macassar is reported to have said in 1615 that sea was common to all and that "it is a thing unheard of that any one should be forbidden to sail the seas."[23]

FREEDOM OF THE SEAS: A CASUALTY IN EUROPE

While the salutary practices of freedoms of navigation and unobstructed maritime trade continued to prevail and prosper in Asia, in Europe the Rhodian and Roman traditions of the freedom of the seas foundered in the turbulent waters of disputes and conflicts of numerous smaller states which emerged from the ruins of Rome, each vying with the other. Maritime commerce died in a "state of wild anarchy" in Europe, and even the memory of Rhodian law did not last beyond the thirteenth century. By this time, all European seas came to be more or less appropriated by European states, leading to numerous disputes and almost continuous warfare. Thus, in addition to the wide claims of Spain and Portugal, Venice claimed sovereignty over the Adriatic Sea, Genoa occupied the Liguarian Sea, England dominated the undefined British seas, and Denmark closed the Baltic by closing the Sound and extended control over the northern seas.[24]

Portugal Disturbs Peaceful Navigation in the Indian Ocean

When the Portuguese arrived in India by the end of the fifteenth century, they found no maritime powers, no warships, and no arms in the sea. The Indian Ocean had never been a theatre of any serious naval conflicts. Asians were not peaceful peoples but felt no need to fight for the sea that was but of limited use for navigation, maritime trade, and catching small quantities of fish. They were essentially land powers. The hub of Asian activities and

---

22. For a translation of both codes, see J. M. Pardessus, *Collection de Lois Maritimes* (1895), p. 6; See also Sir S. Raffles, "The Maritime Code of the Malays," *Journal of the Royal Asiatic Society (Straits Branch)*, vol. 2 (Dec. 1879), p. 1–20.

23. Quoted in G. J. Resink, *Indonesia: History between the Myths* (The Hague: W. van Hoeve, 1968), p. 45.

24. See Fulton (n. 10 above), p. 3–5; P. B. Potter, *Freedom of the Seas in History, Law and Politics* (New York: Longmans, Green and Co., 1924), p. 36–38.

relations, their struggles and conflicts, related to the vast and fertile land on the largest continent of the world. The absence of armed shipping in the Indian Ocean helped tiny Portugal to control vast areas of the ocean. The Europeans were sea powers trained in the rough waters of the Atlantic and the North Sea, whose challenges hardened them into expert navigators and naval warriors. Portugal sought to apply European custom to control the vast Indian Ocean and enforce its control by its armed carracks and galleons against the unarmed Indian Ocean ships engaged in peaceful trade. Although Portugal was fairly successful in gaining a share of the Asian spice market and in disturbing peaceful navigation in the Indian Ocean, it could not wipe out the Asian maritime trade.[25] But the Portuguese monopoly of the Eastern spice trade and its huge profits aroused the jealousy of other European powers which began to challenge Portugal's authority in the late sixteenth century.

Contest of Wits and Arms in Europe

It was to contest the Portuguese monopoly, as we have noted earlier, that Grotius, taking his cue from the Asian maritime practices of free navigation and trade, propounded his doctrine in a brief he prepared for the Dutch East India Company. The company asked Grotius, who was associated with it as a lawyer, to defend the company's capture of a Portuguese vessel laden with Eastern spices in the Straits of Malacca in 1604. Learning as much as he could about India and the East Indies, their traditions of free trade and commerce throughout history, and the Portuguese attempts to stultify the traditional freedom of navigation to these countries, Grotius wrote *De Jure Pradae* in 1605 to defend the action. He tried to "show that war might rightly be waged against, and prize taken from the Portuguese, who had wrongfully tried to exclude the Dutch (and others) from [trade with eastern countries].''[26] His greatness lies in keenly observing the maritime customs of Asian countries, presenting them in the form of a doctrine supported by logical arguments, Christian theology, and the authority of the venerable Roman law, and recommending these views to European countries. This fact of history has been generally ignored by historians of international law. There is little doubt, as Professor Alexandrowicz said, "that Grotius either conceived or perfected his doctrine of the freedom of the seas under the influence of the maritime traditions of the East.''[27]

Besides Asian traditions, Grotius relied on logic. He tried to establish

---

25. See M. A. P. Meilink-Roelofsz, *Asian Trade and European Influence* (Gravenhage: Nijhoff, 1962), p. 136–72.
26. Knight (n. 14 above), p. 80.
27. Alexandrowicz (n. 3 above), p. 229; See also *Ibid*, p. 44.

8   Issues and Prospects

two propositions: first, "that which cannot be occupied, or which never has been occupied cannot be the property of any one, because all property has arisen from occupation"; and second, "that which has been so constituted by nature that although serving some one person it still suffices for the use of all other persons, is today and ought in perpetuity to remain in the same condition as when it was first created by nature."[28] The air belongs to this class of things, and so does the sea. Therefore, argued Grotius with disarming logic of the time: "The seas is common to all because it is so limitless that it cannot become a possession of one, and because it is adapted for the use of all, whether we consider it from the point of view of navigation or of fisheries."[29]

It must be pointed out, however, that in spite of all this learning and logic, neither Grotius nor Holland were in favour of freedom of the seas as a principle. As the Dutch defeated the Portuguese and seized the profitable trade of the Spice Islands, they sought to create their own monopoly. Grotius conveniently forgot the freedom of the seas principle he had propounded with such fervour and went to England with a Dutch delegation four years later in 1613 to argue in favour of a Dutch monopoly of trade with the Spice Islands. In fact he was surprised to find that his own book, published anonymously, was being quoted by the British against him.[30] Successive attempts by each European state to demand freedom of the lucrative spice trade of the East Indies, and later attempts by each of them to try to create a monopoly for itself, along with a similar game being played in the Atlantic, led to a spate of books by numerous scholars in Europe. Most or all of these works were nothing more than apologies by these writers for their countries' policies and interests. In this battle of books and wits, which continued in the din of actual war, it was not Grotius, it must be pointed out, who won, as is generally assumed. The real victor was John Selden, British scholar and statesman, whose *Mare Clausum, sen de Domino Maris Libri Duo (The Closed Sea or Two books concerning the Rule over the Sea)* written at the behest of the English Crown, remained the most authoritative work on maritime law in Europe for the next 200 years.[31] Although several other publicists countered Selden's arguments, all the European countries continued to follow his prescription in controlling as much ocean as their power would permit. Selden won this protracted "battle" not by the brilliance of his arguments, but by the "louder language" of the powerful British navy.[32]

28. Grotius (n. 13 above), p. 28 ff.
29. Grotius, *Ibid.*
30. See G. N. Clark, "Grotius' East India Mission to England," *Transactions of the Grotius Society*, Vol. 20 (1934), p. 79; also Knight (n. 14 above), p. 136–43.
31. In England, "*Mare Clausum* became in a sense a law book." Fulton (n. 10 above), p. 374.
32. See Potter (n. 24 above), p. 61.

RESURGENCE OF THE FREEDOM OF THE SEAS

It was only in the late 18th or really early 19th century that freedom of the seas came to be revived under the patronage of Great Britain, which had emerged as the greatest power of the world. The needs and demands of the industrial revolution in Europe—larger markets, sources of raw material and surplus capital that could not be invested in Europe—led to huge colonial empires in Asia and Africa. As Europeans got more interested in commercial prosperity and free trade, and ever more Europeans started traveling to these widespread colonies, Selden's *Mare Clausum* became an anachronism which was no longer necessary. It was more useful for them to have open and free seas in order to exploit vast unexplored areas of the world that no one nation could reach alone. Pretensions to sovereignty over the sea and monopoly of trade slowly died their natural death and England became not only the strongest champion of the freedom of the seas, but also its policeman.[33] Grotius, a dejected and rejected man in his life, and a false prophet for 200 years, was acclaimed and proclaimed a hero and his, in some respects illogical, arguments came to be accepted without any questions.

LAW VAGUE AND UNCERTAIN

In any case, the freedom of the seas principle accepted by the Europeans had nothing in common with Asian maritime practices. Unlike Asians, who had maintained these freedoms for centuries for peaceful commercial relations, the chief purpose for their revival in nineteenth century Europe was joint exploitation of Asia and Africa to satisfy the needs of European industries. It may also be mentioned that, but for general agreement on vague freedom of the seas, implying freedom of peaceful navigation with a few agreed "rules of the road," which benefited all Europeans, there was little agreement on other rules. Freedom of fisheries, which England came to accept only after three wars with Holland and other conflicts with neighbours, continued to be a subject of serious disputes among Europeans. There was no agreement on a uniform limit of territorial sea, or freedom of navigation through the maritime belt or straits, especially for warships. The same was true of the contiguous zone, and England, ever since the repeal of its *Hovering Act* in 1876, continued to question the legality of such jurisdiction exercised by other states.[34] Moreover, a large part of the law of

---

33. See Sir G. Butler and S. Maccoby, *The Development of International Law* (London: Longmans, Green and Co., 1928), p. 53.

34. See Fulton (n. 10 above), p. 593–603; J. L. Brierly, "The doctrine of the contiguous zone and the dicta in *Croft vs. Dunphy*," *British Yearbook of International Law*, vol. XIV (1933), p. 156.

the sea relating to war, contraband, blockade and rights of neutrals was always at the mercy of belligerents stretching their rights according to their needs and the contingencies of war. Thus, during the two World Wars, the belligerents outstretched their legal jurisdiction on the sea on the basis of controversial doctrines they propounded, like "ultimate enemy destination" and "long distance blockades," and enforced them, over the strong protest of the neutrals, through navicerts systems of their own.[35] Thus it is important to note that, apart from a few general principles, much of the maritime law, as it developed in the nineteenth and the first half of the twentieth centuries, was controversial, uncertain, and in several respects nothing more than a panorama of conflicting rules.

LEGAL VACUUM

Even more important is the fact that, beyond a limited maritime belt, the vast areas of the ocean—more than 70 percent of the globe—remained a legal vacuum, an area of "no law" beyond what are referred to as a few "rules of the road." Freedom of the seas meant essentially non-regulation and *laissez-faire* which was in the interest of the big maritime powers. This law, or rather lack of law under the freedom of the seas doctrine, was often used in the nineteenth century by European powers to threaten small states, to get concessions from them, or simply to subjugate them.[36] Even later, it gave them a license to use the freedom in furtherance of their immediate interests—whether for navigation, fisheries or military maneuvers—irrespective of the rights of others. The protracted and sometimes bitter fishery disputes between smaller European countries—Holland, Denmark, Norway, and Iceland—on the one hand, and Great Britain, on the other, numerous such disputes on the American continent, and almost continuous protests by neutral states against violation of their freedoms of navigation and trade by belligerent maritime powers, were constant reminders of the dissatisfaction of the smaller coastal states. The situation became even more serious during and after the World War II when the maritime powers took the liberty to further stretch this freedom and enclose even wider areas of the ocean either for defeating the enemy, or for conducting nuclear and missile tests, threatening the life and liberty of all peaceful users of the seas. Protests by

35. See C. J. Colombos, *International Law of the Sea*, 6th ed. (New York: D. Mckay Co., 1967), p. 62, 748–52; J. Stone, *Legal Controls of International Conflict* (New York: Rinehart, 1959), p. 484 ff., 500 ff.

36. There is no dearth of cases of trigger-happy western naval commanders using naval ordnance against "backward" peoples of Asia and Africa on the smallest excuse, or no excuse at all. It was the classic age of punitive or minatory bombardments. See for details of numerous cases R. R. Palmer and J. Colton, *A History of the Modern World,* 3rd ed. (New York: Knopf, 1965), p. 548 ff.; 615 ff.

smaller states to such uses of the sea were almost always rejected on the ground that what was not prohibited in law was permitted, and that these were "reasonable" measures of security and self-defense.[37]

Most of the rules of modern maritime law were based on the practice of a few dominant maritime powers. Many a time their interest differed and their practices were not uniform. The situation was tolerated not only because of the over-bearing influence of the European maritime powers, especially Great Britain, along with France, Germany, and Russia, as well as extra-European powers like USA and Japan, which were all helped by this undefined and wide freedom of the seas, but also because the sea was of only limited importance and use. But the law even for these limited purposes was imprecise and not beyond doubt. An attempt was made to codify the law under the auspices of the League of Nations in 1930, but it failed because the big maritime powers, especially Great Britain, insisted on a narrow three-mile limit of the territorial sea, and the smaller states were deeply concerned about protecting their fisheries and other coastal interests in wider zones.[38]

POST–1945 ERA: A NEW WORLD

By the end of World War II, the whole balance of forces had changed. The West European powers, which had dominated the world scene for nearly 300 years, were no longer at the center of the world stage. Out of the ruins of World War II emerged the United States and the Soviet Union with enough strength to dominate the world, and to challenge each other seriously. The world, divided into two power blocs, plunged into a bitter Cold War that affected all aspects of international relations and law.

With the weakening of Europe, colonialism collapsed and there emerged numerous Asian and African states which for a long time had no formal legal status and no recognized role in the formulation of international law. Comprising a majority of the new extended world society, the Asian-African states, along with the thus far equally neglected and disgruntled Latin American states—the so-called Third World as they came to be called—acquired a new influence in the divided post-war world society. Non-aligned with any of the power blocs, as most of these countries were, they became allies to take concerted action and play an important role in international legal and political structures in pursuance of their interests.

There was another development. So far the uses of the sea were few

---

37. See for such defense by both the US and British Governments, M. Whiteman, *Digest of International Law,* Vol. 4 (Washington, DC: U.S. Dept. of State, 1965), p. 585 ff. 600 ff.
38. See J. S. Reeves, " The codification of the law of territorial waters," *American Journal of International Law,* Vol. 24 (1930): 493.

and the coastal states were mostly concerned about their security, protection of their nearshore areas for fisheries, and their commercial fleets. The tremendous advances in marine technology after World War II revealed a new world with nine times as much vegetation available in the sea as was cultivated on land. Even more important, it came to be found that natural resources and minerals in quantities beyond anyone's imagination were present not only in the water of the sea but also on the ocean floor and in the underlying layers. By 1945, geologists had confirmed that huge quantities of sorely needed oil and gas resources lay buried under the seabed off the shores of various countries, outside the territorial sea, and technology was making them accessible. These invaluable resources could not be left there or risked to be exploited by other distant water states, as had been the case with fisheries for centuries.

The development of technology also revolutionized commercial fishing operations. Significant technological breakthroughs in the ability to detect, concentrate, and harvest fish in the high seas increased the capacity of a few technologically advanced countries to indulge in overfishing, threatening entire fishery resources near the coasts of other states. The need to protect coastal resources—both living and nonliving—had become all the more evident.

FREEDOM OF THE SEAS NOT IMMUTABLE

Law could not remain unaffected by all these changes. Unlimited freedom of the seas, which had served the interests of a few maritime powers in an age with limited uses of the sea, could no longer remain unchallenged or unchanged. As Professor Gidel said as early as 1950:

> The expression 'freedom of the high seas' is in reality a purely negative, worn-out concept, nothing more; it has no meaning for us, except as the anti-thesis of another, a positive concept, which has long since disappeared. The idea of the freedom of the high seas is, paradoxically, a survival of the idea—long since dead—that the high seas are subject to dominion and sovereignty, just like any territorial dominion.[39]

Europe had largely lost its control and hold over the law of the sea. It was no longer a law to be made by and for the European countries. Once it came to be realized that the sea was much more than a navigation route or a storehouse of fisheries which could be freely exploited under the old

---

39. See United Nations' *Memorandum on the Regime of the High Seas* (reputed to have been prepared by the French jurist Gidel) *UN General Assembly Document No. A/CN.4/38*, 14 July 1950, pp. 2–3.

freedom of the seas doctrine, the old law lost its charm and sanctity. Most of the initiative and calls for change in the law came from extra-European countries. The first and most important challenge to the traditional freedom of the seas doctrine in the period following World War II came from the United States, which had emerged as the strongest maritime power after the war. The twin proclamations by President Harry Truman on 28 September 1945, referred to developments in technology as necessitating the extension of US coastal jurisdiction to establish conservation zones in contiguous high seas areas to protect fisheries and the right to exclusive exploitation of the mineral resources of the continental shelf.[40] In both proclamations, the littoral state extended its limited jurisdiction to areas of the high seas close to its coasts, without any claim to an extension of territorial waters, and specifically declared unaffected the high seas character of the areas and the right to free and unimpeded navigation in those waters. In spite of this disclaimer, the Truman Proclamations were certainly novel claims that modified, if not grossly violated, the freedom of the seas doctrine.

The United States' proclamations led to numerous claims by other states not only for continental shelf jurisdiction but also for protection of their fisheries. By 1958, nearly a score of countries had made such continental shelf claims. Some Latin American countries went even further. Argentina, Chile, Peru, Ecuador, Costa Rica, El Salvador, and Honduras all extended their jurisdiction or sovereignty to 200 miles to protect their fisheries from depredations by outsiders. Practically every proclamation claiming special rights to the continental shelf or fisheries contained the statement that freedom of the high seas was fully recognized and maintained. But as the 1950 UN Memorandum on the Regime of the High Seas suggested, these disclaimers could not be taken seriously.[41]

CONFLICTING AND DIVERSE CLAIMS

There was a lot of confusion during this period about the legal validity or otherwise of all these claims about continental shelf and fisheries jurisdiction. The confusion was confounded by widening claims relating to the territorial sea. By 1958, at least 27 of the 73 independent coastal states claimed specific breadths of territorial sea in excess of the so-called "traditional" three-mile limit. These claims ranged between 5, 6, 12 and 200 miles. Six others, while rejecting the three-mile rule, did not specify their limits.[42]

    40. See Proclamation No. 2667, 10 Fed.Reg. (1945), p. 12, 303.
    41. See UN Memorandum (n. 39 above), p. 2–3.
    42. See Draft synoptical table prepared by the UN Secretariat in pursuance of the Resolution of First Committee (Territorial Sea and Contiguous Zone) at its 14th meeting (March 13, 1958) UN Document A/Conf. 13/C.11; March 20, 1958; *Ibid,* Rev 1, April 1, 3, 1958; *Ibid,* Rev. 1, Corr. 2, Apr. 22, 1958.

14   Issues and Prospects

Some countries sought to achieve the same purpose without extending their territorial waters or fisheries jurisdiction by adopting straight baselines for measuring the territorial sea joining outermost islands, islets, or rocks off their coasts. Thus, Norway essentially extended its territorial seas by redrawing its baselines and enclosing vast bodies of waters, large and small bays, and countless arms of the sea making them internal waters subject to the absolute sovereignty of Norway. This method for protection of coastal fisheries from outsiders was upheld by the International Court of Justice in the *Anglo-Norwegian Fisheries* case in 1951.[43]

UNITED NATIONS EFFORTS TO CODIFY THE LAW

The divergent standpoints adopted by different states since the World War II on the territorial sea, fisheries jurisdiction, continental shelf, and other issues of the law of the sea made the already ambiguous and uncertain situation "a confused medley of conflicting solutions."[44] To bring order in this confusing situation, the United Nations (UN) organized two conferences in 1958 and 1960 to develop and codify the law in a systematic manner. Four conventions[45] were concluded in 1958 which, on the whole, reasserted the traditional freedoms of the sea and accepted coastal state sovereign jurisdiction over its continental shelf and exclusive right to exploit its resources up to a depth of 200 meters or, beyond that limit to wherever the depth of the superjacent waters admitted of exploitation of the natural resources. Although coastal states were permitted to extend maritime zones and adopt fish conservation measures over adjacent waters, no agreement could be reached about the extent of territorial sea or fisheries jurisdiction, and the agreement on the definition of continental shelf was vague and controversial. Another attempt was made in 1960 to reach agreement on the territorial sea, but it also failed.[46]

Many coastal states still wished, and some claimed, wider territorial sea, but were unable to move the entrenched powers, or successfully challenge their historic "rights" and change the traditional law. During the two conferences, there was a continuous struggle between numerically strong but poor, newly independent Asian-African nations and their allies in Latin America, supported by the Soviet Group, on the one hand, and politically

43. *I.C.J. Reports*, 1951, p. 132.
44. UN Memorandum (n. 39 above), p. 112.
45. Convention on the Territorial Sea; Convention on the High Seas; Convention on Fishing and Living Resources of the High Seas; and Convention on the Continental Shelf.
46. See R. P. Anand, "Winds of Change in the Law of the Sea," in R. P. Anand (ed.), *Law of the Sea: Caracas and Beyond* (New Delhi: Radiant Publishers, 1978), p. 41 ff.

dominating, rich, satisfied, European and North American maritime powers, and some other small Asian-African countries under their influence, on the other.[47] While the maritime powers recounted and reassured the virtues of the freedom of the seas as a "time-honoured" principle, the dissatisfied states of the Third World thought that it was a "time-worn" old doctrine which could still serve and be useful but only if modified and adapted according to changed needs of the changed international society. Rejecting the three-mile rule for territorial sea as a "fallen idol," the new members of the international community said that "agreement among maritime powers alone was not law" and that "rules should be based on general state practice, not on that of a handful of states that had repeatedly been challenged and now finally rejected."[48] The Western powers were still strong enough to enforce the traditional law of *laissez-faire* which favoured them. The developing countries did not like this law, but could not help it.

RENEWED CHALLENGE TO THE FREEDOM OF THE SEAS

In a sense, the 1958 Conventions had become outmoded by the time they were written. Since then the pressure to change the old freedom of the seas increased even more with a further widening of the international society. Moreover, technology soon made it feasible to exploit the vast resources of the seabed and ocean floor, especially oil and gas, at depths beyond geological continental shelf. Indeed, exploitation became possible at any depth and countries started stretching their continental shelf jurisdiction to include the whole continental margin extending to a depth of 2,500 meters. It also came to be known that beyond the continental margin, generally referred as the deep seabed, there lay extensive deposits of manganese nodules containing manganese, nickel, copper, and cobalt, all metals essential for industrial economies.

In 1967, a perceptive representative of a very small country, Arvid Pardo of Malta, informed the UN General Assembly about the inadequacies of the current international law and freedom of the seas, which could and would encourage appropriation of vast areas of the sea which were suddenly found to contain untold wealth by those who had the technological competence to exploit them. To avoid a potentially disastrous scramble for sovereign

---

47. See A. H. Dean, "The Second Conference on the Law of the Sea: Fight for the Freedom of the Seas," *American Journal of International Law*, Vol. 54 (1960), p. 752; R. L. Friedheim, "The Satisfied and Dissatisfied States negotiate International Law," *World Politics*, Vol. 18 (1965), p. 20–41.

48. See U. Mya Sein (Burma), Shukairy (Saudi Arabia) and Hassan (UAR), *Second UN Conference on the Law of the Sea: Official Records, Summary Records of Plenary Meetings and Meetings of the Committee of the Whole*, UN Doc. A/CONF.1918 Geneva, March 17 to April 26, 1960, p. 58, 74, 102.

16   Issues and Prospects

rights over the seabed, he suggested the creation of an effective international regime for the seabed and ocean floor beyond a clearly defined national jurisdiction, and acceptance of that area as a "common heritage of mankind" that would not be "subject to national appropriation in any manner whatsoever, to be used and exploited for exclusive benefit of mankind as a whole."[49]

Pardo's essentially internationalist approach was heralded by many as an idea whose time had come. The General Assembly not only accepted Pardo's suggestion but also established a Seabed Committee to prepare for the Third United Nations Conference on the Law of the Sea (UNCLOS III). In 1970, it unanimously adopted a Declaration of Principles Governing the Seabed and Ocean Floor. The Assembly declared that the seabed beyond the limits of national jurisdiction was not subject to national appropriation or sovereignty but was "the common heritage of mankind" and must be exploited for the benefit of humanity as a whole, "taking into particular consideration the interests and needs of the developing countries."[50]

Although the maritime powers sometimes denied the legal force of these declarations of the General Assembly, there was clear indication that the new majority had started asserting itself. At UNCLOS III, organized to regulate new uses of the sea for the vastly extended international society, the new states were determined to play a more vigorous role. Over the objections of "old guards" and defenders of the traditional law, who preferred a conference only for formulation of law for the exploitation of the seabed beyond the limits of national jurisdiction, these states wanted a comprehensive conference to review the whole international law of the sea. They wanted to be able to "analyze, question and remold, destroy if need be, and create a new equitable and rational regime for the world's ocean and deep ocean."[51]

FURTHER EROSION OF THE FREEDOM OF THE SEAS

In the meantime, the trend to curb the freedom of the seas by extending coastal state jurisdiction for the protection of security and economic interests of the coastal states continued or even increased after 1960. By the end of 1973, nearly 35 percent of the ocean, an area equal to the landmass of the planet, was claimed by the coastal states. Deploring this trend, some well-

49. A. Pardo, UN Document A/C.1/PV.1515, Nov. 1, 1967, p. 6.
50. G.A. Res. 2749 (XXV) 25 UN GAOR Supp. No. 28, 24, UN Doc. A/8028 (1970)
51. C. W. Pinto, "Problems of developing states and their effects on the law of the sea," in L. M. Alexander (ed.), *Needs and Interests of the Developing Countries* (Kingston, Rhode Island: The University Press, 1973), p. 4; See also Lusaka Declaration of Third Conference of Heads of State or Government of Non-aligned Countries, September 1970, UN DOC. A/AC 134/34, 30 April 1971, p. 5.

meaning jurists regretfully felt that the era of *mare liberum* "may now be drawing to a close."[52] But others, like Sir Hersch Lauterpacht, pointed out that "in so far as the original conception of the freedom of the seas, as it came to full fruition in the nineteenth century, acquired a rigidity impervious to needs of the international community and to a regime of an effective order on the high seas, "the loss of paramountcy" provides no occasion for anxiety."[53]

Third United Nations Conference on the Law of the Sea

At UNCLOS III, which met at its substantive session in Caracas, Venezuela, in 1974, the new majority of the developing countries made it clear that it was only the strong maritime powers "that profited from these undefined freedoms" of the traditional law.[54] The continuing *laissez-faire* on the high seas had ceased to serve the interest of international justice.[55] In seeking to establish a new legal order, the developing countries said, they would be "seeking not charity but justice based on the equality of rights of sovereign countries with respect to the sea."[56] Only a new international law could establish this new order, because "between the strong and the weak, it is freedom which oppresses and law which protects."[57] The developing countries, in short, were determined, as the President of Venezuela said in opening the conference, that the sea could not be permitted to "be used in such a way that a few countries benefited from it while the rest lived in poverty, as had been done with the riches of the land."[58]

On April 30, 1982, after nine years of intense, arduous, sometimes bitter, and protracted negotiations, the UN Conference adopted "a comprehensive constitution for oceans,"[59] a Convention that was said to be the most significant international agreement since the Charter of the United Nations. Without going into the details of this Convention it may be pointed out that for the first time there was an agreement on a wide range of issues. For the

52. W. Friedman, "Selden Redivivus: Towards a Partition of the Seas," *American Journal of International Law* (1971), p. 763.
53. H. Lauterpacht, "Sovereignty over submarine areas," *British Year Book of International Law*, Vol. XXVII (1950), p. 198, 403–407.
54. See Vratusa (Yugoslavia), *Third UN Conference on the Law of the Sea, Official Records*, Vol 1, p. 92, UN Sales No. E. 75, V.3 (1975).
55. Warioba (United Republic of Tanzania), *Ibid*, p. 92.
56. H. S. Amersinghe (Sri Lanka), *Ibid*, p. 218.
57. Raharijaona (Madagascar), *Ibid*, p. 106.
58. C. A. Peres, *Ibid*, p. 36.
59. T. B. Koh, President of the UNCLOS III quoted in R. P. Anand, "Odd Man Out: The United States and UN Convention on the Law of the Sea," in J. M. Van Dyke, *Consensus and Confrontation: The United States and the Law of the Sea Convention* (Honolulu: Law of the Sea Institute, University of Hawaii, 1985), p. 73.

18   Issues and Prospects

first time in history there emerged a consensus in favour of agreed limits of territorial sea of 12 nautical miles, 24 miles of contiguous zone, a new exclusive economic zone (EEZ) extending up to 200 miles, and legal continental shelf extending to the end of the continental margin up to a depth of 2,500 meters or even beyond. An almost "revolutionary" effect of the acceptance of coastal stated jurisdiction over a 200-mile EEZ has been the "elimination of freedom of fishing and the substitution of coastal state sovereign rights over the exploration, exploitation, conservation, and management of living resources."[60] Moreover, the seabed beyond the limits of national jurisdiction came to be reaffirmed and accepted as the "common heritage of mankind." Although the exact meaning and content of "common heritage" might be somewhat vague, like numerous other concepts of international law, an international machinery for the exploitation of the oceans' resources has come to be devised and accepted by an overwhelming majority of states.

While in the beginning some of the Western powers, led by the United States, refused to sign the 1982 Convention, and Chapter XI of the Convention relating to exploitation of the deep seabed resources had to be modified to accommodate their interests by an agreement concluded in 1994, practically all the states have come around to accepting it in its modified form.[61] The basic premise of the consensus reached at UNCLOS III is clear and beyond doubt, namely, that in future the sea must be used for the benefit of all and not merely for the interests of a few great powers.

For the first time in centuries, freedom of the seas has lost its charm and stranglehold. It has come to be modified and adapted to fulfil new needs of the new international society. Although navigation is vitally important, the sea is not merely a navigation route, as it has been for centuries, but is a new area of wealth still largely unexplored, which will be the scene of the next adventure and expansion of humanity. While Europe is still extremely important, international law is no longer confined to Europe and must, therefore, serve the interests of the worldwide community of states. Freedom of the seas will still be a relevant concept, but this freedom will not be unlimited. It will be the same kind of freedom that individuals enjoy in a national society, namely, freedom under generally agreed and widely accepted legal principles as adopted by the worldwide community of states.

---

60. See J. Stevenson and B. Oxman, "Third United Nations Conference on the Law of the Sea: The 1975 Geneva Session," *American Journal of International Law*, 69 (1975): 774–81.

61. See B. Oxman, "Law of the Sea Forum: the 1994 Agreement on Implementation of the Seabed Provisions of the Convention on the Law of the Sea," *American Journal of International Law*, 88 (1994), p. 687 ff; L. B. Sohn, "International Law Implications of the 1994 Agreement," *American Journal of International Law*, 88 (1994): 696.

Issues and Prospects

Professionalization of the Field of Marine Affairs: Towards Standard-setting in Education, Training and Practice†

Aldo Chircop
Marine Affairs Program and Marine and Environmental Law Programme, Dalhousie University

INTRODUCTION

The purpose of this article is to put forward two propositions for the promotion of interuniversity academic and professional standards for degree programmes in integrated coastal and ocean management, and the need for standard-setting professional associations in marine affairs. These institutional propositions are interrelated. Two observations are offered in support of these propositions: there is precious little interuniversity cooperation in curriculum development and, although there exist national associations for professionals in the field of marine affairs broadly defined, to date none has ventured into standard-setting and membership services in the truest sense of a professional association.

Growth of Marine Affairs Professions

The marine affairs professions may be loosely defined with reference to policy, planning, and management services offered by practitioners to a wide range of sectoral marine activities and integrated planning and management functions performed by institutions and groups as part of a public or self-imposed mandate. Practitioners provide services to, among others, industry and regulators in maritime transport, living resource activities, aquaculture, tourism, mineral development, and sector-specific environment protection. Integrated planning and management functions include integrated coastal management, response to pollution of the marine environment from land-based activities, economic and environmental planning generally, multiple-

†EDITORS' NOTE.—An earlier version of this article was presented to the International Conference on Coastal and Ocean Space Utilization (COSU 2000), Cancun, Mexico, 1–4 November 2000.

© 2003 by the University of Chicago. All rights reserved.
0-226-06620-7/03/0017-0002 $0.100 *Ocean Yearbook* 17:19–25

use conflict avoidance and management, community development, and conflict management and resolution generally.

The institutional tools vary from governmental (at various levels) to non-governmental, private sector and community based. The practice does not necessarily carry distinct job titles in marine affairs, but job terms of reference would indicate expected functions and services. There are commonalities of interests and practices, but hardly any common professional titles or designations that serve as uniform categories for those in private, public or community service practice. Defining the marine affairs, or even more specifically the coastal management practitioner, is not easy.

Proliferation of Marine Affairs Education and Training

The growth in the field of marine affairs over the last three decades has been accompanied by development of accredited academic and professional education and training programmes by learning institutions, mainly at the tertiary level. Undergraduate and graduate degree, diploma, and certificate programmes are offered by numerous European and North American universities. In a North American context, the Marine Affairs and Policy Association (MAPA) has developed a directory of the largest number of such programmes in any one country, i.e., the US. Like their US counterparts, Canadian universities in British Columbia, Newfoundland, Nova Scotia, and Quebec offer graduate degrees or diplomas in the field. Likewise in Mexico, universities in Baja and Jalisco offer similar programmes. There are a number of European countries, notably the UK, with several university programmes. In addition to these programs, which by and large tend to adopt a macroscopic view of integration, there are also marine sector-specific degree and diploma programmes. For instance, the World Maritime University in Sweden has long offered degree programs focussing on the needs of maritime transport. There are also innumerable training programmes offered by nonaccredited institutions, mainly nongovernmental organizations.

The growth curve of new programmes is steep and is likely to remain so for several years. Over the last five years, for instance, several new marine affairs programmes have been initiated in, among other countries, Australia, Canada, Costa Rica, Cuba, Indonesia, Mexico, Philippines, Thailand, and the UK.

There is great diversity in programme institutional home, objectives, curriculum, structure, duration of programme, and pedagogy/andragogy. There are also differences in evaluation and assessment requirements. Diversity per se is not necessarily negative, as each educational institution is expected to address the needs of its constituency or market share in its own context and with its own philosophical and cultural perspectives. There are also significant similarities amidst the diversity. For instance, the majority of these universities are producing graduates with a mission of integration and interdisciplinarity. The graduates of these programmes are expected to take up important work positions that will involve decision making or decision support on matters of economic, environmental and human well-being.

With the exception of some sector-specific areas (e.g., maritime transport) and generic associations such as the Coastal Society (CS), the Marine Affairs and Policy Association (MAPA), and Coastal Zone Canada Association (CZCA), there are precious few professional associations cutting across the field of marine affairs and those that exist have limitations. Although valuable in their area of operation, sector-specific associations (e.g., Chartered Institute of Transport) are limited to specific subject areas. Equally valuable are multisector organizations (such as CS, MAPA, CZCA) which organize important biennial conferencing, website services and opportunities for networking. However, these organizations are not engaged in standard setting (e.g., requirements for practice, ethical practices, disciplinary procedures, insurability of practitioners), accreditation (e.g., providing professional certification, accreditation of education and training programmes), and membership services (e.g., continuing education, except in an implied manner through conference participation). MAPA comes closest to the latter by virtue of its workshops on the teaching of integrated coastal management. But then the driving force behind MAPA is US universities providing a marine affairs education.

There is no pressure on universities to respond to practitioners' needs as advanced by professional associations. Field and practitioner responsiveness in curricula tends to result from individual university efforts to monitor the field, market surveys, regular programme reviews and feedback from alumni/ae. Outside of what each individual university determines to be desirable standards, a legitimate question is what practice-wide academic and professional standards should be expected of universities offering marine affairs education?

In many ways, standards for sector-specific programmes are less challenging because of more focussed field and practitioner definition, and the presence of sector-specific professional associations that go beyond generalized services, such as conferencing, by providing for some control in recruitment and membership categories. True integrated management programmes in marine affairs are in a different situation. Standards tend to be generic tertiary university standards (not necessarily field-specific) and different universities are left to their own devices in deciding on curricular content, unless those universities operate under a national or university association standard monitoring system.

This is the case with integrated coastal management. Integrated management programmes may be initiated through diverse disciplinary entry points and consequently there are likely to be significant differences in content and methodology. Programmes initiated in the natural sciences and engineering disciplines can be expected to have a higher quantitative content than programmes initiated in the humanities and social sciences, where qualitative content is considered very important in addressing and influencing human behaviour. There are differences between programmes on what an integrated management education should consist of.

In older and discipline-specific professions, such as law, medicine, accounting, engineering and many health professions, the existence of a standard-setting professional association is a critical element. The professional association tends to have a say in curriculum content and control in ultimate certification for practice. Standards are consequently a consistent mixture of minimum academic and professional norms. Professional certification means something and has to be maintained. In a malpractice scenario, the malfeasant does not lose the degree, but may well be censured and possibly lose the practising license. The license is both for practice and regulatory purposes. This type of standard-setting naturally goes farther than might be possible in an integrated coastal and ocean management context, particularly because of the chartered character of traditional discipline-based professional associations and the lack of likelihood of exclusive professional regulatory prerogatives in the diverse marine affairs context. However, it is often overlooked that the marine affairs professions, like their traditional counterparts, perform public services in pursuit of social values and goals. If society has a right to expect high standards of practice from professionals dedicated to public order, health and education, should there not be parallel expectations from professions that affect economic, environmental and human well-being in a coastal and ocean management context?

The Need for Standardization in Marine Affairs Programmes

But then, is there a need for standardization in marine affairs programmes? This article argues in the affirmative for programmes focussing on integrated planning and management, such as integrated coastal and ocean management. Although diversity will, and should, remain important, there is a need for balance by a common core. Although this article does not have the space for discussion on content of a common core, the benefits of a degree of academic standardization are proposed as follows:

- Better definition of integrated coastal and ocean management education on a global basis. A common core of knowledge, skills and attitudes should be identified cooperatively by educational and professional associations. Although employers have a role to play, the process of identifying a core component should be above the exclusive interests of any individual employer. For universities that are part

of or monitored by government, there must be readiness for an inclusive approach in their particular context.
- Easier, intelligent and rational new programme development without the need of re-inventing the wheel or adopting a foreign programme "model" lock, stock, and barrel. The common core could then be socialized into the local operational environment of the university concerned.
- Readability among similar university programmes, irrespective of their location or market reach. The expectations of an integrated coastal and ocean manager would be universal and explicit, rather than local and implicit.
- Enhanced recognition and market mobility for marine affairs programme graduates. Core knowledge, skills and attitudes can be expected of any graduate, in any work force. Employers will have clearer expectations, which in turn serve to guide universities in curriculum development.
- Curricular advances at any university are easier to measure against a prevailing norm.
- Prospects for the development of widely recognized and international faculty training programmes, set against international standards.
- Enhanced interuniversity cooperation in marine affairs. It is suggested that the existence of a common core would facilitate student exchanges and the academic requirements that normally accompany these.

If this is a valuable proposition, then it would need a forum to initiate what will be a long-term process aimed at systemic change. Perhaps this could initially be a national university forum that has already attained a certain momentum, and which is generous enough to open its ranks to meaningful international participation and truly international discourse. It is essential for such a forum to be apolitical and open enough to any educational institution from any country. The participants must be accredited education and training institutions since these institutions would already have achieved an acceptable educational standard according to their own national or regional quality control standards (e.g., associations of universities). The purpose here is to enhance quality and relevant education in institutions already bound by quality standards.

The Need for Standard-setting Professional Associations

Following graduation, it is unusual for a marine affairs graduate to find and become a member of an association that provides ongoing professional guidance and support. As indicated earlier, the most that exists in North America

24 Issues and Prospects

is a regular conference circuit, which naturally is a valuable component of the post-graduation experience. The purpose of a professional association is to enable a career choice starting with quality education and continuing with a life-long commitment to quality professional service (responsible, efficient and effective practice). Excellence in professional service should be expected of all graduates given their impact on society.

This is where standard-setting professional associations are needed for post-educational support. The likely lack of a chartered status (for reasons given above) means that compliance with standards will be primarily voluntary. However, if over the long-term professional certification is maintained, it is suggested that this in itself could attain market value for the individual practitioner ("good standing").

In addition to existing services such as conferencing, information and website services, and networking, existing professional associations and new ones that may emerge should be challenged into performing the following services:

- Standard setting: professional associations could work more closely with education and training institutions and perhaps even perform the function of accreditation for purely professional practice purposes. A code of conduct regarding professional ethics developed by a professional association would complement research and intellectual property ethics increasingly promoted in tertiary education. Such a code would respect and promote values of the society where practice occurs. It is suggested that subscription to such standards would render would-be practitioners more attractive in hiring processes and career advancement generally. Membership would provide "added value."
- Better definition of the profession: this should lead to better designation of management positions by employers.
- Membership services: continuing education and lifelong learning (guided by principles of adult education)
- Neutral ground for sensitive discourse: a truly independent/impartial professional association with a code of conduct would facilitate open discussion of sensitive and controversial issues.

The argument for national professional association standards is easier than for international standards. However, with the growth of ISO standards in relation to a wide variety of goods and services, there is an opportunity to explore a longer-term process for integrated coastal and ocean management standard-development on the basis of international cooperation under ISO auspices. The strength of this vision is quality control of a planning and management service that ultimately has global implications.

Is the field of marine affairs ready for the development of this type of standard-setting professional association? This article argues that some countries might be in a position to experiment and test the proposition. There is a need for an initiative aimed precisely at developing such a professional association. The trigger for such an initiative is an explosion of coastal and ocean management issues in just about every marine sector and challenging all levels of government and society in this country. There are issues of socio-economic and cultural equity cutting across aboriginal peoples, established resource users and government agencies. A process of fiscal austerity and cost-recovery services, including privatization have tested governance structures. There is increasing demand by the public for principled, transparent, efficient, equitable and cost-effective decision making. At the same time, there is a significant body of professional expertise in coastal and ocean management that is missing an ongoing institutional framework for much needed open dialogue among marine affairs professionals and interested groups.

At this early stage, the discourse on the need of an association of marine affairs professionals taking an integrated approach has more questions than answers: how should the field of marine affairs be defined? What professions ought to be included? How would such an association complement existing sector or issue-specific and other generic associations already in place? What benefits would such an association provide to make it attractive for the membership? How should standards be developed and according to what and whose criteria? How should such an association be organized in a federal state? What models of existing professional associations are useful to learn from?

CONCLUSION

This article has attempted to highlight the need for and benefits of greater interuniversity cooperation in education and training and the development of standard-setting professional associations. The intent is to propose ideas for increasing the relevance and effectiveness of the professional service providers our universities are producing on a cooperative basis. It is suggested that the marine affairs professions in today's globalized world and greater efforts at integration cannot really claim to be completely independent of issues and standards developed in other professions and individual disciplines. Ultimately, these two propositions strike at the essence of quality management and quality assurance in marine affairs.

Issues and Prospects

The International Tribunal for the Law of the Sea and Marine Environmental Protection: Expanding the Horizons of International Oceans Governance

Donald R. Rothwell*
Faculty of Law, University of Sydney, Australia

INTRODUCTION

The 1982 UN Convention on the Law of the Sea (UNCLOS)[1] ushered in a new era of oceans governance to which the international community has been adjusting for the past two decades. In addition to the significant development of a range of maritime zones plus comprehensive provisions for the protection of the marine environment and marine scientific research, and the recognition of new regimes for archipelagic States and the deep seabed, the Convention also addressed dispute resolution. This in itself is not extraordinary as nearly all multilateral and bilateral treaties contain some provisions addressing dispute resolution. What made UNCLOS distinctive was that it sought to make the peaceful settlement of all disputes under the Convention compulsory.[2] This action did not only refer to the provisions of Article 33 of the UN Charter,[3] but went further through the establishment of complex dispute resolution mechanisms. These mechanisms not only relied upon existing institutions such as the International Court of Justice (ICJ), but also developed new institutions, including the following:

- The International Tribunal for the Law of the Sea (ITLOS);
- An Arbitral Tribunal under Annex VII;
- A Special Arbitral Tribunal under Annex VIII.

---

*Research support for this article was provided by a University of Sydney Sesqui R&D Grant. The author also acknowledges the invaluable research assistance of Kristen Daglish, however, all errors and omissions remain his responsibility.

1. 1833 UNTS 396.
2. UNCLOS, Art. 279.
3. 1 UNTS xvi.

When it is considered that UNCLOS also created additional institutions such as the Commission on the Limits of the Continental Shelf (CLCS) and the International Sea-Bed Authority (ISBA), which have the capacity to be engaged in both formal and informal dispute settlement, it will be appreciated that the Convention has significantly revised how law of the sea disputes will be settled in the future.

However, the law of the sea is not contained within a single international instrument. While UNCLOS may be classified as the primary instrument, other multilateral and bilateral conventions and treaties give effect to the principles of the law of the sea that have developed throughout the past 50 years. In some instances, such as those treaties dealing with maritime boundaries, the instruments themselves seek to resolve disputes between the parties.[4] In other instances, however, these additional instruments foster the development of subregimes in the law of the sea that can generate their own specific disputes. This has especially been the case with respect to fisheries, which while subject to overall management in framework provisions dealing with the exclusive economic zone (EEZ) and high seas in UNCLOS, have increasingly been the subject of regional and subregional management throughout the world's oceans.[5] These management arrangements may cause numerous disputes between the parties concerning scientific data, quotas and catch limits, fishing practices, and the sovereign rights of coastal states. This in turn raises questions concerning how these disputes may be resolved within the framework of the specific fishing instrument, or within the wider UNCLOS.

The purpose of this article is to explore some of these issues with particular reference to environmental disputes. While it may be possible to characterise UNCLOS distinctively as a regulatory instrument for the oceans, in this article it is argued that the Convention has such a significant environmental impact that it can also legitimately be characterised as a multilateral environmental instrument for the oceans.[6] Accordingly, particular emphasis will be given to law of the sea disputes with an environmental focus. A review will be undertaken of the dispute resolution mechanisms of UNCLOS, and the recent work of ITLOS.

---

4. See e.g., the 1978 Treaty between Australia and the independent State of Papua New Guinea concerning Sovereignty and Maritime Boundaries in the Area between the Two Countries, including the Area known as Torres Strait, and Related Matters [1985] ATS 4.

5. For a discussion of the CCAMLR area as an RFMO see E. J. Molenaar, "Southern Ocean Fisheries and the CCAMLR Regime," in A. G. Oude Elferink and D. R. Rothwell, eds., *The Law of the Sea and Polar Maritime Delimitation and Jurisdiction* (The Hague: Kluwer Law International, 2001), pp. 293–315.

6. On how UNCLOS should be interpreted in the context of environmental disputes, see E. Hey, *Reflections on an International Environmental Court* (The Hague: Kluwer Law International, 2000), p. 5.

OCEANS GOVERNANCE AND DISPUTE RESOLUTION

Following the entry into force of UNCLOS on 16 November 1994 there has been growing interest in both oceans governance and dispute resolution. It has only been since the realisation that law of the sea and international environmental law concepts and principles could be usefully blended, that a vision of integrated oceans governance began to emerge.[7] There has been growing interest in marine and oceans policies that promote sustainability and other environmental values at both the national and regional level. Interest in dispute resolution has been further heightened with the commencement of the work of ITLOS and the inevitable scrutiny the early decisions have received. Indeed it could be argued that, on the basis of the literature that has been generated on this topic in the past 10 years and especially since 1996, dispute resolution has become a new subspecialty of the law of the sea. However, the point should be reinforced that one of the more prominent areas of dispute resolution practice throughout the 20th century was in the law of the sea. It is possible to go back even farther into the late 19th century when it is recalled that the Behring Sea Fur Seals Arbitration of 1893[8] was perhaps the first of the modern law of the sea disputes involving not only issues concerning maritime zones and high seas rights, but also resource management and environmental considerations.[9]

Law of the sea disputes have a strong record of peaceful settlement through a range of mechanisms, some established through international institutions such as the UN Charter or Part XV of UNCLOS. Some disputes have been successfully settled through ad hoc measures, and others have been resolved via negotiation. This is not to suggest that all law of the sea disputes have been simply resolved through peaceful means. Some of the most prominent disputes have involved armed force, as in the case of the Corfu Channel Case,[10] the Fisheries Jurisdiction Case,[11] and the Red Cru-

---

7. See discussion in J. M. Pureza "International law and ocean governance: audacity and modesty," *Review of European Community and International Environmental Law* 8 (1999): 73–77; R. W. Knecht "A perspective on recent developments that could affect the nature of ocean governance regimes," *Law of the Sea Institute Proceedings* 21(1995): 177–90; W. T. Burke "State practice, new ocean uses, and ocean governance under UNCLOS," *Law of the Sea Institute Proceedings* 28 (1994): 219–34; L. K. Kriwoken et al., eds., *Oceans Law and Policy in the Post-UNCED Era: Australian and Canadian Perspectives* (London: Kluwer Law International, 1996).

8. Behring Sea Arbitration Award between Great Britain and the United States Given at Paris, 15 August 1893, 179 Consolidated Treaty Series 97.

9. For discussion see C. P. R. Romano, *The Peaceful Settlement of International Environmental Disputes: A Pragmatic Approach* (The Hague: Kluwer Law International, 2000), pp. 133–50.

10. International Court of Justice (ICJ), *Corfu Channel Case* (United Kingdom v. Albania). ICJ Reports 3. The Hague: ICJ, 1949, p. 3.

11. ICJ, *Fisheries Jurisdiction Case* (Merits) (United Kingdom v. Iceland). ICJ Reports 3. The Hague: ICJ, 1974, p. 3.

sader.[12] Nevertheless, even in these instances the parties to the dispute eventually utilised peaceful means of dispute settlement to resolve outstanding matters.

The most fundamental of all means of peaceful dispute settlement is negotiation. The importance of negotiation is reinforced throughout the UN Charter and the institutions created by the UN that facilitate negotiated settlement of disputes, and also in the provisions of UNCLOS. Not only in Part XV, but elsewhere in the Convention reference is made to the obligation on parties to settle their disputes by negotiation. The clearest example of the reliance upon negotiation and the confidence attached to its capacity to resolve significant disputes can be found in Articles 74 and 83, which both anticipate the settlement of maritime boundary disputes between opposite and adjacent States by way of "agreement on the basis of international law."[13] This is to be contrasted with the earlier equivalent provisions found in both the 1958 Convention on the Territorial Sea and Contiguous Zone[14] and the 1958 Convention on the Continental Shelf.[15] These conventions suggested that States should seek to resolve their maritime boundary disputes by agreement and created a technical framework to be used in achieving this result. Under UNCLOS, parties are given much freer scope to apply whatever delimitation methods may be appropriate to their circumstances, provided the outcome is effected "by agreement on the basis of international law . . . in order to achieve an equitable solution."[16] Parties have taken the opportunity provided by this broad framework that encourages innovative dispute settlement to adopt a variety of approaches in their maritime boundary delimitations that often go beyond the more traditional approaches of equidistance or median lines.[17] There is, likewise, considerable evidence proving that negotiation is a successful dispute settlement method with respect to fisheries management and conservation, resulting in the finalisation of numerous regional, subregional, and bilateral fisheries instruments applying to EEZ fish stocks and highly migratory and straddling fish stocks. Once again, UNCLOS encourages the State parties to explore resolution of these "disputes" via negotiation so as to reach agreement.[18]

12. Commission of Enquiry (Denmark – United Kingdom), *The Red Crusader Inquiry*, International Law Reports 35 (1967): 485.
13. While UNCLOS, Art. 15 is not expressed in such clear terms, it is self-evident that in the case of delimitation of the territorial sea, it is anticipated that coastal States will initially seek to reach agreement on their maritime boundaries.
14. 516 UNTS 205.
15. 499 UNTS 82.
16. UNCLOS, Arts. 74 (1), 83 (1).
17. See for example Australia's practice in maritime boundary delimitation, especially with Papua New Guinea, Indonesia and UNTAET/East Timor; see further S. Kaye, *Australia's Maritime Boundaries* (Wollongong: Centre for Maritime Policy, University of Wollongong, 1995).
18. UNCLOS, Art 64.

Notwithstanding the success of negotiation as a dispute resolution tool in matters pertaining to the law of the sea, third party interventions such as arbitration or adjudication have also been successfully used. One law of the sea phenomenon since 1945 has been the expansion in both the customary international law and the conventional definition of maritime zones. As these zones have crept gradually farther seaward there has been greater need for maritime boundary resolution between opposite and adjacent States. While this discussion would suggest that negotiation may be a favoured method of settlement of such disputes, there are clearly sensitive issues at stake that at times demand a more formal dispute settlement mechanism. Accordingly, maritime boundary disputes are often resolved through arbitration or adjudication. From the 1909 Grisbådarna[19] case involving Norway and Sweden to the 1977 Anglo-French Continental Shelf Arbitration,[20] arbitration remains a popular method of law of the sea dispute settlement. With States having the capacity to determine the arbitrators, the issue for resolution, and even the law upon which the decision should be made, ad hoc arbitration through nonpermanent tribunals remains attractive because of its flexibility.

The International Court of Justice (ICJ) has also played a prominent role in maritime boundary delimitation beginning with the important North Sea Continental Shelf Cases.[21] Indeed, the Court's jurisprudence in this area is so extensive that it now extends to maritime boundary disputes involving nearly all continents.[22] Such has been the role of the Court in maritime boundary delimitation, that its distinctive jurisprudence is clearly recognised as one of the most significant contributions of the court to the law of the sea. However, this has not been the only area where the ICJ has been utilised in adjudicating law of the sea disputes. As previously mentioned, the Corfu Channel Case in 1949 was an important early indicator of the capacity of the Court to adjudicate in cases concerning law of the sea issues. This decision remains important today for the impact it had upon navigation rights through straits. The Nuclear Test Cases[23] were important in focussing attention on marine environmental protection and developing some early principles of international environmental law,[24] while in both the Icelandic Fisheries Case and the more recent Turbot Case[25] between Spain and Canada the

19. (1909) XI RIAA 147.
20. (1979) 18 ILM 397.
21. [1969] ICJ Reports 1.
22. The Court has been involved in maritime boundary dispute adjudication involving States from Europe, Africa, Australia, Asia, and North America.
23. ICJ, *Nuclear Test Cases* (Australia v. France; New Zealand v. France). ICJ Reports 253. The Hague: ICJ, 1974, p. 253.
24. See discussion in Romano (n. 9 above), pp. 279–306.
25. ICJ, *Fisheries Jurisdiction Case* (Spain v. Canada). ICJ Reports 87. The Hague: ICJ, 1995, p. 87; ICJ *Fisheries Jurisdiction Case* (Spain v. Canada). ICJ Reports 58. The Hague: ICJ, 1996, p. 58.

ICJ has had the opportunity to explore, and in some instances develop, the law dealing with fisheries management and conservation.

It is clear even from this relatively brief survey of dispute resolution in the law of the sea that there has been considerable practice developed throughout the 20th century. The peaceful settlement of law of the sea disputes has therefore been a characteristic of much of the past 100 years, and this has been further reinforced in the period since 1958 and 1982 following the respective negotiations of the four Geneva Conventions on the Law of the Sea and UNCLOS.

UNCLOS AND THE SETTLEMENT OF DISPUTES

Part XV of UNCLOS has three sections that create the framework for dispute settlement under the Convention. The first outlines the general provisions, the second the compulsory procedures entailing binding decisions, and the third outlines limitations and exceptions to the jurisdiction of the specific dispute resolution bodies.

General Provisions for Dispute Settlement

Not surprisingly, UNCLOS reaffirms the peaceful settlement of disputes consistent with Article 33 of the United Nations Charter.[26] However, this provision does not bar the use of compatible peaceful means of dispute settlement established under UNCLOS itself.[27] To further emphasize the flexible nature of the peaceful means of dispute settlement open to the parties, the Convention further provides that nothing should prevent the parties from settling a dispute by any "peaceful means of their own choice."[28] This attempt by the Convention to encourage all means available to the parties for peaceful dispute settlement is not at the exclusion of the specific measures established under the Convention, and there is a residual capacity for Part XV to be activated to resolve a dispute.[29] UNCLOS also recognizes that parties to a dispute may be parties to other multilateral conventions, whether of a general, regional, or bilateral nature that may in turn contain their own dispute resolution mechanisms. In this case, and if the parties so choose, disputes may be referred to those procedures entailing binding decisions in

---

26. UNCLOS, Art. 279.
27. See 1959 Antarctic Treaty 402 UNTS 71, Art XI, which merely restates the peaceful means of dispute settlement that the parties should utilize in resolving their disputes, including reference to the International Court of Justice.
28. UNCLOS, Art. 280.
29. Ibid, Art. 281.

lieu of the provisions contained within Part XV.[30] The only specific dispute resolution mechanism Section 1 of Part XV refers to is the obligation to exchange views regarding settlement by negotiation or other peaceful means[31] and conciliation through procedures established under Annex V of UNCLOS.[32]

Compulsory Means of Dispute Settlement

The principal provision of Part XV is Article 287, which outlines the various procedures available to parties for the peaceful settlement of disputes via the compulsory mechanisms established under the Convention.[33] From the time of signature or accession or any time thereafter, State parties have the capacity to declare which means of dispute settlement they prefer.[34] The suite of dispute settlement bodies include the following:

- ITLOS;
- ICJ;
- Annex VII Arbitration; or
- Annex VIII Special Arbitration.

Anticipating variable practice amongst State parties concerning their acceptance of the compulsory means of dispute settlement, Article 287 also outlines various procedures where a declaration has not been made, and if two disputing parties have not selected the same methods of dispute settlement.[35] Where the parties have selected the same procedure for settlement of the dispute, only that dispute settlement mechanism may be utilised unless the parties otherwise agree.[36] As is the case with similar declarations indicating acceptance of the jurisdiction of the ICJ, declarations made under Article 287 may be subject to variation or revocation by the parties.[37]

One technical difficulty may arise in the interpretation of Article 287 as both Paragraphs 4 and 5 refer to "the same procedure" when anticipating State declarations accepting jurisdiction. The issue is whether "the same

30. Ibid, Art. 282.
31. Ibid, Art. 283.
32. Ibid, Art. 284.
33. See discussion on this provision in R. R. Churchill and A. V. Lowe, *The Law of the Sea*, 3rd ed. (Manchester: Manchester University Press, 1999) pp. 455–59
34. See Table 5 listing the current Declarations made by State Parties to UNCLOS reflecting an election of choice of procedures under UNCLOS, Art. 287 and Art 298 Optional exceptions.
35. UNCLOS, Art. 287 (3), (5).
36. Ibid, Art. 287 (4).
37. Ibid, Art. 287 (6), (7). For discussion on these opting procedures see Romano (n. 9 above), pp. 95–97.

procedure" should be taken literally to extend to precisely the same declaration, or whether it is interpreted more flexibly? In this regard, Eiriksson has noted the following:

> It is unwise to give much thought to situations which may never arise in practice, but the Tribunal would be well advised to regard the expression of preferences as something which a potential applicant should take into account in the spirit of comity. The Tribunal should not in any event entertain the development of conditional choices beyond the letter of article 287.[38]

As there have been so few disputes in which the compulsory dispute settlement provisions of UNCLOS have to date been activated it is premature to make any definitive assessment on this question, though clearly it has the potential if interpreted strictly to limit the operation of Part XV.

Each of the four courts or tribunals referred to in Article 287 are given broad jurisdiction under Article 288 to address the following issues:

- Any dispute concerning the application or interpretation of the Convention submitted consistently with Part XV;
- Any dispute concerning the interpretation or application of an international agreement related to the purposes of the Convention submitted consistently with that agreement;
- The specific jurisdiction of the Sea-Bed Disputes Chamber of ITLOS and other tribunals established under Part XI consistently with any matter submitted to them.

It is clear then that the jurisdiction conferred on the Article 287 Courts and Tribunals is very wide and extends to a multitude of law of the sea issues arising not only under UNCLOS but also related instruments.[39]

Section 2 of Part XV also contains a number of provisions that address procedural issues. The applicable law is UNCLOS and other "rules of international law" that are not "incompatible" with the Convention.[40] This, however, does not prejudice the court or tribunal from having jurisdiction to determine a matter *ex aequo et bono*.[41] Local remedies must also have been exhausted consistent with international law prior to parties activating the mechanisms under Part XV.[42] This rule has already been the subject of dis-

---

38. G. Eiriksson, *The International Tribunal for the Law of the Sea* (The Hague: Martinus Nijhoff, 2000), p. 117.
39. Note the discussion in J. G. Merrills, *International Dispute Settlement*, 3rd ed. (Cambridge: Cambridge University Press, 1998), pp. 174–75.
40. UNCLOS, Art. 293 (1).
41. Ibid, Art. 293 (2).
42. Ibid, Art. 295.

cussion before ITLOS in the M/A "Saiga" (No. 2) Case.[43] It particularly has potential application in matters involving the release of vessels pursuant to the municipal law of the arresting State under which there will exist local remedies that can be explored by the parties. In the "Camouco" Case,[44] however, ITLOS ruled that Article 292 dealing with the prompt release of vessels provided complete relief in these categories of cases, and accordingly the exhaustion of local remedies rule did not apply in these instances.[45]

Decisions rendered by a Part XV Court or Tribunal are also considered final, and parties to the dispute are required to comply with the decision.[46] The decisions do not, however, have any binding force beyond the parties to the dispute,[47] though they may of course be seen as significant for the development of the law of the sea in general and accordingly be influential in the future interpretation of the law.[48] In this regard, it is important to note that there is no avenue of appeal from decisions of ITLOS or either the Annex VII or Annex VIII Tribunals to the ICJ.

Limitations on Jurisdiction[49]

Notwithstanding the general establishment of compulsory means of dispute settlement, UNCLOS does place some limitations on the jurisdiction of courts, ITLOS, and arbitral tribunals. When signing or at any other stage becoming a party to the Convention, a State may declare that it does not accept the procedures under Section 2 for compulsory dispute settlement with respect to a number of matters. These can include the following:

- Certain disputes concerning maritime boundary delimitations or those involving historic bays or titles;

---

43. "*The M/V 'Saiga'* (Saint Vincent and the Grenadines v. Guinea) Provisional Measures Order of 11 March 1998," *International Legal Materials (ILM)* 37 (1998): 1202; see also discussion in Eiriksson (n. 38 above), pp. 163–64.

44. *The Camouco Case* (Panama v. France) Application for Prompt Release, Judgment, 7 February 2000.

45. See discussion in Eiriksson (n. 38 above), p. 164.

46. UNCLOS, Art. 296 (1).

47. Ibid, Art. 296 (2).

48. The ITLOS decision in the Southern Bluefin Tuna Cases (Australia v. Japan, New Zealand v. Japan) Request for Provisional Measures (Cases No. 3 and No. 4) are seen as important in the development of the international law interpretation of the "precautionary approach/principle"; see the discussion in A. Trouwborst, *Evolution and Status of the Precautionary Principle in International Law* (The Hague: Kluwer Law International, 2002), pp. 169–74.

49. See the discussion in Merrills (n. 39 above), pp. 176–78; 193–96; Churchill and Lowe (n. 33 above), pp. 455–456; J. Collier and V. Lowe, *The Settlement of Disputes in International Law* (Oxford: Oxford University Press, 1999), pp. 92–93.

- Disputes concerning military activities; or
- Disputes in respect of which the United Nations Security Council is exercising its functions under the United Nations Charter.[50]

Jurisdiction is more precisely defined for disputes involving coastal State sovereign rights, marine scientific research, and fisheries. Article 297 particularly provides for the application of jurisdiction under Section 2, Part XV in the case of the following:

Coastal State sovereign rights or jurisdiction:

- Where a coastal State has acted in contravention of freedoms of navigation, overflight, or laying of submarine cables;
- Where a State in exercising freedoms of navigation, overflight, or the laying of submarine cables has acted in contravention of laws and regulations adopted by a coastal State; or
- Where a coastal State has acted in contravention of specified rules and procedures for protection and preservation of the marine environment.

Marine Scientific Research:

- Jurisdiction generally resides over disputes concerning marine scientific research with the exception of cases concerning the exercise of a coastal State right to refuse permission to conduct such research in either the exclusive economic zone or continental shelf, or where authorised research has been suspended; or
- A specific provision also exists permitting the referral of some disputes concerning marine scientific research to conciliation.

Fisheries Management and Conservation:

- Fisheries disputes generally shall be submitted for resolution excepting that a coastal State is not obliged to accept settlement of a dispute in matters with respect to management and conservation of living resources within the EEZ; or
- Annex V Conciliation is envisaged in instances where it is alleged that a coastal State has failed to comply with proper conservation and management measures, or determine total allowable catch, or where there has been a refusal to allow access to surplus catch.

Notwithstanding these limitations, disputes that have been excluded under Article 297 or Article 298 may with the agreement of the parties be submitted

---

50. UNCLOS, Art. 298.

36   ITLOS and Marine Environmental Protection

for resolution before any of the Courts or Tribunals identified in Section 2, Part XV.[51]

Miscellaneous Provisions

In addition to these mechanisms, Part XV also deals with important questions such as Provisional Measures. Article 290 makes it clear that both the ICJ and ITLOS have the capacity to issue provisional measures if *prima facie* jurisdiction is considered to exist, and if issuing provisional measures is appropriate in terms of preserving the rights of the parties or preventing serious harm to the marine environment.[52] The capacity to issue provisional measures even extends to ITLOS or the ICJ in instances where an Annex VII or Annex VIII Arbitral Tribunal is selected as the preferred dispute resolution body. In these instances, all the ICJ or ITLOS must determine is whether that arbitral tribunal would have *prima facie* jurisdiction and that the urgency of the matter requires that provisional measures be issued.[53]

An important additional provision is that dealing with the prompt release of vessels and crews. Article 292 provides that in instances where a dispute arises over prompt release of vessels between the detaining State and the flag State, the matter may be submitted to a court or tribunal for prompt release of the vessel and crew within 10 days from the time of detention. In such instances, the court or tribunal is required to deal with the matter without delay and to address only the issue of the release, without prejudice to the merits of the case. To further emphasise the urgency of the matter, when an order for release has been made the State concerned is required to comply promptly with the order.[54]

Finally, it should be noted that there is also scope under Part XV for experts to be engaged by a court or tribunal where a dispute involves scientific or technical matters. Such experts can sit with the court or tribunal but will not have the right to vote.[55]

THE INTERNATIONAL TRIBUNAL FOR THE LAW OF THE SEA

While UNCLOS was responsible for the establishment of a number of new institutions for the law of the sea, the International Tribunal for the Law of the Sea (ITLOS) was the most anticipated. This reflected the significance

   51. Ibid, Art. 299.
   52. Ibid, Art. 290 (1).
   53. Ibid, Art. 290 (5). These issues arose in both the Southern Bluefin Tuna case and the MOX Plant case and are discussed further below.
   54. Ibid, Art. 292 (4).
   55. Ibid, Art. 289.

TABLE 1.—ADJUDICATION AND ARBITRATION OPTIONS UNDER PART XV OF UNCLOS

Adjudication	Arbitration
ICJ	Annex VII Tribunal
ITLOS	Annex VII Special Tribunal

given to dispute settlement in the Convention, and also a curiosity as to how the law of the sea would respond to its own distinctive tribunal capable of settling and resolving disputes arising under the new regime ushered in by UNCLOS. There was also uncertainty as to how ITLOS would interact with other dispute settlement bodies, especially the ICJ,[56] and speculation on whether given the prominent jurisprudence of the ICJ in law of the sea matters there was the potential for conflicting interpretations arising.

With the Tribunal having only been formally established in 1996, it remains too early to assess conclusively its impact.[57] It is possible, however, to assess the work of the Tribunal to date and make some observations of the role that it will play in law of the sea dispute settlement and more generally in oceans governance. Under the framework created by Part XV of UNCLOS, ITLOS is one of the four specialist courts and tribunals having competence to compulsorily resolve law of the sea disputes. The framework is outlined in Table 1. The ICJ is a creature of the UN Charter and has its own Statute and Rules. Accordingly, all that UNCLOS provides for the ICJ is additional jurisdiction to resolve law of the sea disputes. Given the prominent role the Court has already taken in resolving such disputes, any matters arising directly under Part XV will not be a novelty for the Court. The inclusion of the ICJ within the Part XV, Section 2 framework is however an important reinforcement of the role of the Court in compulsory dispute settlement.

The creation of ITLOS as a counterpart to the ICJ in Part XV is, however, of real significance. ITLOS is not an *ad hoc* creation, but rather it is a permanent body with a permanent seat, judges, registry and staff. In many

---

56. For discussion on how the decisions of ITLOS may interact with the ICJ, see A. E. Boyle, "Dispute settlement and the Law of the Sea Convention: problem of fragmentation and jurisdiction," *International and Comparative Law Quarterly* 46 (1997): 37–54.

57. For an ongoing review of the work of the Tribunal, see a series of articles by S. Rosenne, "International Tribunal for the Law of the Sea: 1996–97 Survey," *International Journal of Marine and Coastal Law* 13 (1998): 487–514; "International Tribunal for the Law of the Sea: 1998 Survey," *International Journal of Marine and Coastal Law* 14 (1999): 453–65; "International Tribunal for the Law of the Sea: Survey for 1999," *International Journal of Marine and Coastal Law* 15 (2000): 443–74.

respects both in terms of appearance and in its procedure it mirrors the ICJ. ITLOS is therefore one of two options for State parties who elect to have their disputes determined by adjudication. It also has the important capacity to issue provisional measures[58] and play a specialist role with respect to seabed disputes.

The Tribunal is located in the Free and Hanseatic City of Hamburg in Germany and is comprised of 21 members. Following the initial elections in 1996, one third of the members served a 3-year term, another third, a 6-year term, and the remaining number, a 9-year term. All members subsequently appointed will fill 9-year terms with the next elections due in 2005.[59] The President and Vice-President of the Tribunal are elected for 3 years. ITLOS may also appoint a Registrar and such other officers as it deems necessary.[60] The Tribunal is organised as follows:

- The Tribunal in plenary—comprising all 21 members with a quorum of 11;
- A Sea-Bed Disputes Chamber comprised of 11 members, which may itself form an *ad hoc* chamber composed of three members; and
- Special Chambers comprised of three or more members to deal with particular categories of disputes.

To date, in addition to sitting in plenary, ITLOS has also elected members to form the Sea-Bed Disputes Chamber and also two Special Chambers: a Chamber for Marine Environmental Disputes, and a Chamber for Fisheries Disputes.

The Tribunal's jurisdiction, provided for in Part XV, is further reinforced in Annex VI, Article 21, which provides the following:

> The jurisdiction of the Tribunal comprises all disputes and all applications submitted to it in accordance with this Convention and all matters specifically provided for in any other agreement that confers jurisdiction on the Tribunal.

However, in recognition that UNCLOS and ITLOS may post-date some general, regional, or bilateral instruments dealing with law of the sea, Article 22 of Annex VI leaves it open to the parties to those instruments to also refer disputes concerning the interpretation or application of that instrument to ITLOS. This potentially provides an independent source of ITLOS jurisdic-

---

58. UNCLOS, Art. 290.
59. The most recent election of ITLOS judges took place on 19 April 2002, at which seven judges were elected for a term of nine years commencing from 1 October 2002.
60. UNCLOS, Annex VI, Art. 12.

Issues and Prospects 39

TABLE 2.—ITLOS CASES: 1996–2002

Cases No 1 and No 2	The M/V Saiga
Cases No 3 and No 4	Southern Bluefin Tuna (New Zealand v. Japan; Australia v. Japan)
Case No 5	The Camouco
Case No 6	The Monte Confurco
Case No 7	The "Swordfish Stocks"
Case No 8	The "Grand Prince"
Case No 9	The "Chaisiri Reefer 2"
Case No 10	The MOX Plant Case (Ireland v. United Kingdom)

tion beyond the technical scope of Part XV. In all other respects Annex VI parallels the framework created by Part XV, though specifically elaborated by the Tribunal's Rules of Procedure.

The Tribunal has now been operating for 6 years and during that time it has considered ten cases,[61] involving eight distinctive disputes brought before the Tribunal (table 2).

To date the principal focus of the Tribunal's work has been with respect to prompt release cases, though it has also had occasion to consider fisheries disputes. Not surprisingly the Court's jurisprudence has attracted considerable interest from law of the sea commentators eager to assess how the new institution is developing.[62] It is not the intention here to specifically comment on all ITLOS decisions to date,[63] other than to note that the Tribunal has seemed anxious to fulfil the role conferred upon it in UNCLOS and to develop a distinctive approach to its work. It is of interest to note that ITLOS has yet to hear a matter relating to maritime boundary delimitation, in contrast to the work of the ICJ that has concentrated in this area of law of the sea disputes. What does seem clear, however, is that from the cases in which ITLOS has been involved in there is considerable potential for it to play

61. See the list in Table 2.
62. See e.g., M. D. Evans, "Bonded reason: *the Camouco*," *Lloyd's Maritime and Commercial Law Quarterly* (2000): 315–22; B. H. Oxman and V. P. Bantz, "The 'Camouco' (Panama v. France) (Judgment) ITLOS Case No. 5, available online: <http://www.un.org/Dept/los/ITLOS/JudgmentCamouco.htm>. International Tribunal for the Law of the Sea, February 7, 2000," *American Journal of International Law* 94 (2000): 713–21; V. Lowe, "The M/V Saiga: The first case in the International Tribunal for the Law of the Sea," *International and Comparative Law Quarterly* 48 (1999): 187–99.
63. For some commentary, however, see V. Lowe, "The International Tribunal for the Law of the Sea: survey for 2000," *International Journal of Marine and Coastal Law* 16 (2001): 549–70; R. O'Keefe, "ITLOS flags its intent," *Cambridge Law Journal* 59 (2000): 428–31; B. Chigara, "The International Tribunal for the Law of the Sea and customary international law," *Loyola of Los Angeles International and Comparative Law Journal* 22 (2000): 433–52.

an important role in oceans governance, not only through its resolution of disputes, but also through a contribution to the development and understanding of the law.

The Southern Bluefin Tuna Cases

The Southern Bluefin Tuna Cases between Australia/New Zealand (A/NZ) and Japan are an interesting study in dispute resolution under the law of the sea principally because they seek to utilise a number of the Part XV mechanisms of UNCLOS described earlier. They are also somewhat distinctive from other law of the sea disputes in that they involve fisheries rather than maritime boundaries and concern the application of not only UNCLOS but also other international instruments including those dealing with the environment. Both the 1999 ITLOS decision and the 2000 Annex VII Arbitration Tribunal decision have been the subject of exhaustive comment in the academic literature.[64] It is not the intention here to revisit the substantive issues raised by those decisions or the subsequent commentaries, but rather to look at how dispute resolution processes were applied in this case and the final outcome of those processes.

In 1993, Australia, New Zealand, and Japan concluded the Convention for the Conservation of Southern Bluefin Tuna (CCSBT).[65] The Convention, which generally entered into force on 20 May 1994, establishes a regime for the conservation and optimum utilisation of southern bluefin tuna,[66] a fish stock that is highly migratory but that is found in significant quantities in the Australian and New Zealand EEZs.[67] Japan, as a principal market for the

64. See for example T. Stephens, "A paper umbrella which dissolves in the rain? Implications of the Southern Bluefin Tuna Case for the compulsory resolution of disputes concerning the marine environment under the 1982 LOS Convention," *Asia Pacific Journal of Environmental Law* 6 (2001): 297–318; B. Kwiatkowska, "The Southern Bluefin Tuna (New Zealand v. Japan; Australia v. Japan) Cases," *International Journal of Marine and Coastal Law* 15 (2000): 1–36; M. Hayashi, "The Southern Bluefin Tuna Cases: Prescription of provisional measures by the International Tribunal for the Law of the Sea," *Tulane Environmental Law Journal* 13 (Summer 2000): 361–85; S. Marr, "The Southern Bluefin Tuna Cases: The precautionary approach and conservation and management of fish resources," *European Journal of International Law* 11 (2000): 815–31; L. Sturtz, "Southern Bluefin Tuna Case: Australia and New Zealand v. Japan," *Ecology Law Quarterly* 28 (2001): 455–86; B. Kwiatkowska, "The Australia and New Zealand v. Japan Southern Bluefin Tuna (Jurisdiction and Admissibility) Award of the First Law of the Sea Convention Annex VII Arbitral Tribunal," *International Journal of Marine and Coastal Law* 16 (2001): 239–93.
65. [1994] Australian Treaty Series No. 16.
66. CCSBT, Art. 3.
67. For background on the CCSBT, see A. Bergin and M. Haward, "Southern Bluefin Tuna Fishery: recent developments in international management," *Marine Policy* 18 (1994): 47.

stock, has long had an interest in exploiting the fishery, and accordingly the three States working within the UNCLOS framework negotiated this regional instrument to regulate access to the fishery. The Convention does not exclude other States acceding,[68] however, it has only been since 2001 that other States have begun to express an active interest in becoming a party to the Convention. South Korea formally acceded in October 2001.[69] The Convention establishes a Commission that is responsible for oversight of the Convention, including determinations made at annual meetings regarding catch quotas.

Between 1997–1998 there emerged different views within the Commission over the acceptable quota for the take of southern bluefin tuna and this eventually resulted in Japan establishing an experimental fishery program (EFP) partly in an effort to establish appropriate scientific data on the status of the stocks and the viability of continued commercial fishing. Australia and New Zealand disagreed with the Japanese action and notwithstanding discussions within the Commission, resolution of the differing views proved impossible. In 1999, A/NZ began to move towards the commencement of formal dispute resolution under the CCSBT, Article 16 of which essentially mirrors reliance upon traditional peaceful means of dispute settlement.[70] Following the failure of the parties to reach agreement over the dispute resolution process, A/NZ in August 1999 commenced an application for provisional measure before ITLOS[71] (table 3).

---

68. CCSBT, Art. 18.

69. At the most recent meeting of the CCSBT it was agreed that "pressure should be applied" by the CCSBT for Indonesia to also accede: Commission for the Conservation of Southern Bluefin Tuna *Report of the Eighth Meeting* (15–19 October, 2001, Miyako, Japan), para. 14.

70. CCSBT, Art. 16 provides as follows: If any dispute arises between two or more of the Parties concerning the interpretation or implementation of this Convention, those Parties shall consult among themselves with a view to having the dispute resolved by negotiation, inquiry, mediation, conciliation, arbitration, judicial settlement or other peaceful means of their own choice. Any dispute of this character not so resolved shall, with the consent in each case of all parties to the dispute, be referred for settlement to the International Court of Justice or to arbitration; but failure to reach agreement on reference to the International Court of Justice or to arbitration shall not absolve parties to the dispute from the responsibility of continuing to seek to resolve it by any of the various peaceful means referred to in paragraph 1 above. In cases where the dispute is referred to arbitration, the arbitral tribunal shall be constituted as provided in the Annex to this Convention. The Annex forms an integral part of this Convention.

71. As neither A/NZ had a national acting as an ITLOS Judge, they were entitled to nominate a judge *ad hoc:* UNCLOS, Annex VI, Art. 17 (2). Professor I. Shearer of the Faculty of Law, University of Sydney, was jointly nominated as the Judge ad hoc in this instance.

TABLE 3.—SOUTHERN BLUEFIN TUNA DISPUTE: KEY DATES

1993	Convention for the Conservation of Southern Bluefin Tuna (i.f. 20/5/94)
1997–1998	Dispute arises over quota for Japanese take of SBT
1997–1998	Japan seeks to continue experimental fishing
1999	Australia/New Zealand request dispute settlement
1999	23 June—Japan accepts mediation under SBT Convention; Australia accepts mediation providing Japan cease experimental fishing
1999	14 July—Japan rejects cessation of experimental fishing arguing it is consistent with SBT Convention; Japan accepts arbitration under SBT Convention; Australia views Japan's response as a rejection of mediation
1999	30 July—Australia/New Zealand file a request for provisional measures with ITLOS
1999	27 August—ITLOS issues provisional measures
2000	May—Arbitration before an UNCLOS, Annex VII Arbitration Tribunal
2000	4 August—Arbitration Tribunal rejects jurisdiction by 4/1 majority; Sir Kenneth Keith in a Separate Opinion in dissent
2000	5 August—Australia supports continuing efforts for amicable resolution of the dispute, including by negotiation
2000	22 December—Australia announces continuing ban on port access by Japanese tuna boats
2001	29 May—Australia/New Zealand/Japan negotiate new quota on a one-off basis
2001	October—CCSBT VIII fails to reach agreement on national allocation of quota, though there is broad consensus on total allowable catch

ITLOS Provisional Measures

The A/NZ request for provisional measures against Japan to halt the EFP for southern bluefin tuna was granted by ITLOS on 27 August 1999. In doing so, ITLOS sought to exercise jurisdiction under UNCLOS, Article 290 (5) by which the measures were issued pending the constitution of an Annex VII Arbitral Tribunal which A/NZ had both indicated was their eventual intention. By a majority, ITLOS agreed that the *prima facie* jurisdiction of the Annex VII Tribunal was established, and accordingly under Article 290 (5) ITLOS had a capacity to issue provisional measures. This was an important decision given the context of the dispute, which was one that had primarily arisen under the CCSBT and not UNCLOS, though clearly important principles of oceans governance under UNCLOS were at stake.

The decision to issue provisional measures had particular ramifications given that Part XV of UNCLOS acknowledges in Article 282 that in the case of a "general, regional or bilateral agreement," procedures that entail a binding decision "shall apply in lieu of the procedures" provided for in Part XV. A principal question therefore in the Southern Bluefin Tuna Cases was characterisation of the dispute under one of a number of relevant international instruments. Was it a dispute that arose only under the CCSBT, in which case the parties had open to them the dispute resolution provisions provided for under that Convention, or rather was this a dispute that ultimately raised such fundamental issues of fisheries management and conservation within EEZ areas that the dispute was more properly characterised as one that fell under UNCLOS? While some consideration was given to this question by ITLOS,[72] it was not the subject of exhaustive consideration. In addition, there was little detailed discussion as to the urgency of the provisional measures sought notwithstanding that this too is a matter that arises for consideration under Article 290 (5).[73]

The Tribunal issued the provisional measures sought and ordered the parties to refrain from continuing with the EFP and to not exceed their annual allocations of total allowable catch, as agreed to under the CCSBT. Notwithstanding that the parties were moving towards final resolution of the dispute before an Annex VII Arbitral Tribunal, ITLOS also encouraged the parties to resume negotiations to reach agreement on measures for the conservation and management of the stock.

Annex VII Arbitral Tribunal

As part of the dispute resolution mechanisms that A/NZ had activated in this case, they were obliged to also establish an Annex VII Arbitral Tribunal under UNCLOS, and the ITLOS provisional measures had been founded on this. Accordingly, in early 2000 mechanisms were put in place allowing for written pleadings to be submitted to an Annex VII Arbitral Tribunal and for oral hearings to commence in May 2000. A hearing on jurisdiction was held before the Tribunal sitting at the International Centre for Settlement of Investment Disputes (ICSID) in Washington, DC from 7–11 May 2000 and the decision was handed down on 4 August 2000. By 4/1,[74] the Tribunal found that in this case the dispute was substantially founded under the CCSBT and that Article 16 of that instrument did not anticipate the parties resorting to any other method of dispute settlement, including those available under UNCLOS. Here the key provisions in Article 16 were that if the

---

72. See the discussion in the judgments by Judge *ad hoc* Shearer.
73. See the discussion in the judgments by Judge Vukas and Judge Treves, and comment by Kwiatkowska (2000) (n. 64 above), pp. 21–23.
74. The majority members were Judge Schwebel (President), Judge Feliciano, Judge Tresselt, and Judge Yamada; Judge Keith was in dissent.

parties had been unsuccessful in reaching a resolution of the dispute by way of negotiation, inquiry, conciliation, arbitration, adjudication, or any other forms of peaceful settlement, and had also rejected formal settlement of the dispute by the ICJ or through the CCSBT Arbitral Tribunal, they had "the responsibility of continuing to seek to resolve it by any of the various peaceful means"[75] provided for. While this provision has ultimately been described as being circular,[76] the majority of the members of the Annex VII Arbitral Tribunal did find that these means of dispute settlement were sufficiently comprehensive to activate the provisions of UNCLOS, Articles 281–283.

However, notwithstanding the decision of the Tribunal that it lacked jurisdiction, this was not seen to absolve the parties of their obligation to settle the dispute by peaceful means. Here the Tribunal noted:

> The Tribunal recalls that Article 16 (2) prescribes that failure to reach agreement on reference to arbitration shall not absolve the parties to the dispute from the responsibility of continuing to seek to resolve it by any of the various peaceful means referred to in paragraph 1.
> ... Whatever the mode or modes of peaceful settlement chosen by the Parties, the Tribunal emphasizes that the prospects for a successful settlement of their dispute will be promoted by the Parties abstaining from any unilateral act that may aggravate the dispute while its solution has not been achieved.[77]

2001 Meeting of the CCSBT Commission

Since the international litigation in 1999–2000 the parties have returned to the negotiating table, principally through the CCSBT Commission, in which they have sought to resolve their continuing differences regarding the total allowable catch permitted and the Japanese EFP. However, while the most recent meeting of the Commission in Miyako from 15–19 October 2001 did achieve broad consensus on global catch levels for southern bluefin tuna, there was no agreement reached on national allocations. This led both A/NZ to brand the meeting as a "failure,"[78] a view that was rejected by Japan.[79] It would seem that there is more work ahead of the CCSBT parties before this dispute is finally resolved.

75. CCSBT, Art. 16 (2).
76. See the comment by Judge ad hoc Shearer, Southern Bluefin Tuna Case.
77. Annex VII, para. 70.
78. CCSBT VIII, Attachment N-1 (Australia), Attachment N-4 (New Zealand).
79. Ibid, Attachment N-2 (Japan); Korea at Attachment N-3 noted "Regarding the failure to reach consensus, it believes more time is required to reach full cooperation in the Commission."

The MOX Plant Case

The MOX Plant Case is the most recent decision of ITLOS and like the Southern Bluefin Tuna Cases not only raises issues concerning provisional measures but also addresses environmental issues. The case deals with the building by the United Kingdom of a mixed oxide (MOX) plant at Sellafield adjacent to the Irish Sea and concerns raised by Ireland as to the potential environmental impact of the facility. British Nuclear Fuels (BNFL) (which has existed in its present form since 1971) is responsible for most of the activities carried out at the Sellafield site. The reprocessing of nuclear waste fuel and discharges began at Sellafield in the 1950s (at that time it was known as Windscale). Sellafield lies in Cumbria, in the Northwest of England, on the coast of the Irish Sea about 112 miles from the Irish coast. The U.K. government has recognised that Ireland has a "legitimate interest" in the activities carried out at Sellafield, given its proximity and potential impact on the Irish Sea. Ireland has long been pressing Britain to stop discharging radioactive waste into the Irish Sea, claiming it contaminates its shores and affects its citizens' health. It claims that routine and accidental discharges of artificial radionuclides into the Irish Sea from Sellafield have occurred since the early 1950s, making the Irish Sea one of the most radioactively polluted seas in the world. Irish protests have accelerated over the past decade, however, for the most part it seems that Britain has ignored Irish concerns.[80]

Following final approval for the commencement of operations at the Sellafield MOX Plant facility in mid-2001, Ireland began to seriously explore options for the commencement of compulsory dispute resolution proceedings against the United Kingdom under UNCLOS. On 25 October 2001 this resulted in a request by Ireland for Provisional Measures from ITLOS and also a notification that the dispute would finally go to an Annex VII Arbitral Tribunal. ITLOS handed down its orders regarding Provisional Measures on 3 December 2001, which assumed the parties would proceed to an eventual 2002 hearing before an Annex VII Tribunal[81] (table 4). Like the Southern Bluefin Tuna Cases, a key issue for ITLOS in the MOX Plant Case was establishing *prima facie* jurisdiction and to that end much attention was given in the orders to relevant provisions of UNCLOS, Part XV. The United Kingdom rejected the primacy of UNCLOS and argued that the parties were obliged to settle their disputes under a range of other related instruments, including the 1992 Convention for the Protection of the North East Atlantic (OSPAR Convention),[82] and the 1957 Treaty Establishing the European Atomic Energy Community (EURATOM),[83] and that these obligations trig-

---

80. *Independent*, 25 November, 2001; *Guardian*, 26 November, 2001.
81. "MOX plant dispute," *Environmental Policy and Law* 32 (2002): 25–26.
82. (1993) 32 ILM 1068.
83. 298 UNTS 167.

TABLE 4.—MOX PLANT CASE: KEY DATES

1991	British Nuclear Fuels (BNFL) decides to build a mixed oxide (MOX) plant
1993	A MOX production facility begins producing small quantities of MOX fuel; Ireland expresses concerns about the proposed MOX plant.
1994	Planning permission is granted for the construction of the MOX plant.
1996	Construction of a MOX plant at Sellafield is completed.
1997	First public consultation exercise on the operation of the Sellafield MOX plant begins. A total of five will be undertaken between 1997 and 2001.
1998	Britain signs an international agreement to cut radioactive waste discharges into the Irish Sea to almost nothing by 2020.
1999	30 July—Ireland raises specific concerns with the United Kingdom over irregularities in United Kingdom practice under UNCLOS
1999	23 December—Ireland again writes to the United Kingdom detailing its concerns about the MOX plant by reference to clearly identified provisions of UNCLOS.
2000	18 February—Nuclear Installations Inspectorate (NII) sets 15 recommendations before it will allow the restart of MOX production at Sellafield.
2001	14 June—Report in the *Guardian* that Ireland is going to take the British government to an international tribunal over proposals to open the new MOX facility for violating freedom of information rules, prompted by the British refusal to release details of the development.
2001	July—The final public consultation invites comments on the findings of a report that concluded the Sellafield MOX plant would result in a financial benefit of more than £150 million to the United Kingdom over its lifetime.
2001	3 October—British government authorises the operation of the MOX plant on the basis that it is economically justifiable. Ireland claims this is despite not all NII recommendations having been met.
2001	5–24 October—Exchange of letters between the United Kingdom and Ireland. Ireland essentially requests the suspension of the operation of the plant while the British government seems strongly in favour of its early operation.
2001	25 October—Ireland requests by notification to the United Kingdom that the dispute be submitted to an arbitral tribunal to be established under Annex VII of the UNCLOS and requests provisional measures in the interim.
2001	6 November—Two environmental groups, Friends of the Earth and Greenpeace, file papers asking for a judicial review of the government decision authorising the commencement of production of MOX fuel.
2001	9 November—Ireland submits a request for the prescription of provisional measures under article 290 paragraph 5 of the UNCLOS to the ITLOS pending the constitution of the arbitral tribunal.
2001	3 December—ITLOS issues adjusted provisional measures

gered the application of Article 282, UNCLOS, which deprived ITLOS of jurisdiction. However, ITLOS noted that even if those instruments "contain rights or obligations similar or identical with the rights or obligations set out in the Convention, the rights and obligations under those agreements have a separate existence from those under the [LOS] Convention."[84] Accordingly, the existence of other international or regional instruments under which Ireland may have been able to obtain relief was not fatal in this instance to a finding that ITLOS had *prima facie* jurisdiction, as the claim made by Ireland was one founded on UNCLOS.[85]

ITLOS then began an assessment of the scope of its jurisdictional capacity under UNCLOS, Article 290 to issue provisional measures and in particular considered the "urgency of the situation."[86] Following a review of the situation and evidence presented by Ireland and the United Kingdom, the Tribunal took the view that this ground had not been made out sufficiently to justify "the prescription of the provisional measures requested by Ireland, in the short period before the constitution of the Annex VII arbitral tribunal."[87] Nevertheless, the Tribunal was persuaded that given the importance of cooperation to prevent marine pollution and under general international law, and considering its capacity to issue orders at variance with those that had originally been requested, it would be appropriate to issue certain provisional measures in this instance.[88] To that end, ITLOS ordered that Ireland and the United Kingdom would do the following:

- Exchange information with respect to possible consequences for the Irish Sea arising out of the commissioning of the MOX plant;
- Monitor risks or the effects of the operation of the MOX plant; and
- Devise measures to prevent marine pollution that might result from the operation of the MOX plant.[89]

In late 2001, the MOX plant was formally commissioned and became operational. In addition to taking the matter to Annex VII Arbitration, Ireland has kept open the prospect of pursuing dispute resolution mechanisms under the OSPAR Convention.[90]

At one level the decision in the MOX Plant Case is disappointing in that the Tribunal refused to issue the substantive orders sought and did not outline in sufficient detail the grounds for its refusal to do so. This of course

84. The MOX Plant Case (Ireland v. United Kingdom) Request for Provisional Measures (Case No. 10) (3 December 2001), para. 50.
85. Ibid, para. 53–62; Ireland had founded its claim on UNCLOS, Arts. 192, 193, 194, 207, 211, 213.
86. UNCLOS, Art. 290 (5).
87. MOX Plant Case, para. 81.
88. Ibid, para. 82–88.
89. Ibid, para. 89.
90. "MOX Plant Dispute" (n. 81 above).

depends upon the quality of the evidence presented to the Tribunal and it also goes to the issue of the 'urgency' of the situation at hand. Nevertheless, it was clear that there existed a potential environmental threat from the commencement of the operations of the Sellafield site and given the relatively short period of time until proceedings were to commence before the Annex VII Arbitral Tribunal, it would not have seemed inappropriate in this instance to issue the orders that were sought. In that respect it should be noted that in the Southern Bluefin Tuna Cases interim measures were issued notwithstanding the very short period of time remaining until the expiry of the existing fishing season and the next meeting of the Southern Bluefin Tuna Commission.

The promising aspect of the MOX Plant Case was the willingness of the Tribunal to find *prima facie* jurisdiction notwithstanding the potential claims that Ireland may have had available to it under a range of related international instruments. Here, ITLOS seemed to reaffirm the broad view that it took on the issue of *prima facie* jurisdiction in the Southern Bluefin Tuna Cases, while also reinforcing that environmental and marine regimes create distinctive rights and obligations and that though there is clearly a strong connection between these instruments the remedies available under them are in each case separate. This suggests some emerging distinction between the position of ITLOS in the interpretation of UNCLOS, Articles 281 and 282, and that of *ad hoc* Annex VII Tribunals.

Finally, the case clearly demonstrates the potential of ITLOS in marine environmental disputes. To date the Tribunal has had little opportunity in this area, but it is clear from this decision and also from Southern Bluefin Tuna that ITLOS does have a real role to play in international environmental marine disputes and to that end it can be anticipated that the Tribunal may have more opportunities to develop its jurisprudence in this area in the future.

MARINE ENVIRONMENTAL DISPUTES AND THE LAW OF THE SEA[91]

One of the phenomena of international law in the past 50 years has been the development of international environmental law, and in particular the nearly parallel development of modern international environmental law alongside the law of the sea as reflected in UNCLOS in the period from the late 1960s through to the early 1990s. It was accordingly not surprising that UNCLOS, negotiated as it was at a time of growing international environmental awareness, contained significant provisions dealing with the marine environment, in particular Part XII. While Part XII represented a significant

---

91. See A. E. Boyle, "UNCLOS, the Marine Environment and the Settlement of Disputes," in *Competing Norms in the Law of Marine Environmental Protection*, ed. H. Ringbom (London: Kluwer Law International, 1997), pp. 241–56; and generally D. R. Rothwell, "Reassessing international environmental dispute resolution," *Asia Pacific Journal of Environmental Law* 6 (2001): 201–14.

Issues and Prospects 49

step forward in the development of marine environmental protection law, it creates more of an overarching framework than detailed rules for the protection of the marine environment.[92] The content of these rules that have been developed consistently with international environmental law and the law of the sea are more often to be found in conventions concluded by the IMO dealing with pollution from shipping, regional marine environmental instruments adopted in Europe and the Southwest Pacific, and in actual state practice that has given effect to the precautionary principle and sustainable development.[93] In light of these developments, it is difficult to argue that no clear connection exists between the law of the sea and the marine environment and accordingly the dispute resolution provisions of Part XV of UNCLOS apply with equal force to marine environmental disputes.[94]

This is confirmed in Part XV where specific reference is made in Article 297 (1)(c) to disputes arising where a coastal State has acted "in contravention of specified international rules and standards for the protection and preservation of the marine environment." Not only can such rules have been recognised in UNCLOS, but the Article goes on to provide that those rules "established . . . through a competent international organization or diplomatic conference in accordance with this Convention" will also be recognised. The practice of ITLOS to date has also recognised the potential for environmental disputes to arise and the need for it to respond to the special challenges such disputes may raise. In this regard it is significant that the Tribunal has established a 'Chamber for Marine Environmental Disputes' under the 'Special Chambers' mechanism of Annex VI. Under the resolution adopting the Chamber on 8 October 1999, ITLOS provided that the Chamber would be available to deal with disputes concerning the interpretation or application of any provision of the following:

- UNCLOS concerning the protection and preservation of the marine environment;
- Special conventions and agreements relating to the protection and preservation of the marine environment referred to in Article 237 of the Convention; and
- Any agreement relating to the protection and preservation of the marine environment that confers jurisdiction on the Tribunal.[95]

While Judge Mensah, the inaugural President of ITLOS, described the organization of this Chamber as evidence of the Tribunal being " 'user-friendly'

92. That framework has been further enhanced by the 1992 Rio Declaration (1992) 31 ILM 818, and Agenda 21, especially Chapter 17.
93. For an assessment of the author's views on these developments in the Asia Pacific, see B. Boer, R. Ramsay and D. R. Rothwell, *International Environmental Law in the Asia Pacific* (The Hague: Kluwer Law International, 1998), 133–44.
94. This point is reinforced by Boyle (n. 91 above), pp. 241–56.
95. ITLOS Resolution on the Chamber for Marine Environmental Disputes (8 October 1999), reproduced in Eiriksson (n. 38 above), p. 366.

in respect of cases relating to the protection and preservation of the marine environment,"[96] the real evidence of such an approach will come with how the Chamber deals with its first case.

While the Southern Bluefin Tuna Cases clearly raised important issues of environmental law, especially the application of the precautionary approach/principle, the cases were not brought on classical environmental grounds. In addition, as A/NZ sought to utilise an Annex VII Arbitral Tribunal for final determination of the dispute, there was never an intention to have ITLOS finally rule on the merits of the environmental issues raised in the cases. The MOX Plant Case squarely raised issues of marine environmental law, however, the ITLOS refusal to grant the Provisional Measures sought resulted in little attention being given to the substantive issues raised. While the two decisions are therefore ultimately disappointing in terms of their contribution to an understanding of marine environmental protection, there is enough content in both to indicate the capacity of ITLOS to play a significant role in this area in the future.

A further aspect of the interaction of the law of the sea with international environmental law and its consequences for dispute resolution arises from the 1992 Convention on Biological Diversity.[97] The Convention's overarching objective of the conservation of biological diversity, which extends to marine ecosystems, is clearly compatible with UNCLOS, Part XII. However, its provisions dealing with biotechnology and genetic resources have considerable potential to assist in the regulation of marine biotechnology in ways never fully anticipated in UNCLOS. While Article 22 of the Convention usefully provides that it shall not affect existing rights and obligations under other international agreements, and in particular indicates that it shall be implemented "with respect to the marine environment consistently with the rights and obligations of States under the law of the sea," the decision of the Southern Bluefin Tuna Cases suggests that notwithstanding the complementary interaction of international instruments unforeseen problems can arise in dispute resolution.[98] How a dispute concerning marine biodiversity would be characterised is difficult to judge, however, it is noteworthy that while the Convention on Biological Diversity does contain dispute resolution provisions including the reference of disputes to arbitration,[99] these mechanisms are not as extensive as those under UNCLOS nor are they compulsory.[100]

96. T. A. Mensah, "The International Tribunal for the Law of the Sea and the protection and preservation of the marine environment," *Review of European Community and International Environmental Law* 8 (1999): 1 at 5.
97. (1992) 31 ILM 818.
98. P. Birnie and A. Boyle, *International Law and the Environment*, 2nd ed. (Oxford: Oxford University Press, 2002), p. 588 note of the effect of Art. 22 that "[t]his raises interesting possibilities concerning application of the Law of the Sea Convention's provisions on enforcement, compliance and dispute settlement to biodiversity issues."
99. Convention on Biological Diversity, Art. 27.
100. Birnie and Boyle (n. 98 above), p. 589.

CONCLUSION

Oceans governance is beginning to have an increased influence on the contemporary law of the sea with the result that the legal and institutional frameworks created by UNCLOS and reflected in bodies such as ITLOS will increasingly be impacted upon by this development.[101] As coastal States begin to place greater emphasis on integrated oceans management, taking into account not only obligations under the law of the sea but also environmental responsibilities for sustainable development and planning, a greater mixing will take place of traditional law of the sea obligations and international environmental law. Part of the challenge here will be to ensure some degree of integration between the various multilateral, regional, subregional, and bilateral instruments that have been negotiated over the past 30 years. It will also be important to ensure that the instruments and the regimes that they create are flexible enough to respond to change.

The Southern Bluefin Tuna Cases and the MOX Plant Case represent initial attempts by ITLOS to address some of these issues. In Southern Bluefin Tuna the Tribunal was confronted with resolving some of the issues that arise when global and subregional instruments interact. Notwithstanding the ultimate finding in this case by the Annex VII Arbitral Tribunal, ITLOS was certainly prepared to take an expansive view of its jurisdictional competence, but also importantly there was extensive discussion of important international environmental principles such as the precautionary approach. It is clear that the application of precaution was influential for a number of the judges in their ultimate decision to grant provisional measures for a fishery under threat. Likewise, in the MOX Plant Case the Tribunal also took an expansive view as to its jurisdictional competence. While ultimately, the Provisional Measures sought were not granted primarily as a result of concerns over a lack of evidence over the urgency of the matter, the case does demonstrate the potential of ITLOS to be utilised in marine environmental disputes notwithstanding the broad obligations created by UNCLOS with respect to land-based marine pollution.

The lesson from these early ITLOS decisions is that coastal States seeking to rely upon existing and developing marine environmental principles do have open to them clear means by which they can seek some resolution from the new UNCLOS institutions. ITLOS has already signalled its interest in hearing marine environmental cases through the establishment of a "Special Chamber," the judges have certainly shown their individual interest in such disputes through their judgements and willingness to go beyond a narrow assessment of the law and take into account environmental principles. It is possible to conclude therefore, that subject to overarching technical limitations such as founding jurisdiction, ITLOS does have a clear potential

---

101. For general comments, see Birnie and Boyle, ibid, pp. 390–91.

to play an important role in marine environmental disputes whether it be through the speedy issuing of provisional measures or resolution of substantive issues. This is not only an important institutional development for marine environmental protection, but also significant for the further development of the law. As coastal States and the international maritime community generally grapple with the challenges of effective oceans governance and implementation of and compliance with international obligations, the emergence of ITLOS as an international institution capable of contributing to that process is to be welcomed (table 5).

TABLE 5.—DECLARATIONS MADE UNDER PART XV OF UNCLOS

State	Choice of Procedure [Article 287, Paragraph 1]	Optional Exceptions to Application of Part XV, Section 2 [Article 298, Paragraph 1]
Algeria	(a) — (b) Accepts jurisdiction only with prior agreement between all parties (c) — (d) —	—
Argentina	(a) First preference (b) Second preference (c) (d) Second preference for questions concerning fisheries, protection and preservation of the marine environment, marine scientific research and navigation	(a) Does not accept (b) Does not accept (c) Does not accept
Australia	(a) First preference (b) Second preference (c) — (d)	Does not accept with respect to disputes concerning application or interpretation of Arts. 15, 74, or 83 relating to sea boundary delimitations, historic bays or titles.
Austria (EC)	(a) First preference (b) Third preference (c) — (d) Second preference	
Belarus	(a) Accepts for questions concerning the prompt release of detailed vessels or their crews (b) — (c) Accepts (d) Accepts for questions concerning fisheries, protection and preservation of the marine environment, marine scientific research and navigation, including pollution from vessels and by dumping	(a) Does not accept (b) Does not accept (c) Does not accept

TABLE 5.—(*Continued*)

State	Choice of Procedure [Article 287, Paragraph 1]	Optional Exceptions to Application of Part XV, Section 2 [Article 298, Paragraph 1]
Belgium (EC)	(a) Accepts (b) Accepts (c) — (d) —	—
Cape Verde	(a) First preference (b) Second preference (c) — (d) —	(a) — (b) Does not accept (c) — Does not accept procedures for disputes concerning law enforcement activities under Article 297 paras 2 and 3
Chile	(a) First preference (b) — (c) — (d) Second preference	(a) Does not accept (b) Does not accept (c) Does not accept
Cuba	(a) — (b) Does not accept jurisdiction (c) — (d) —	
Egypt	(a) — (b) — (c) Accepts (d) —	(a) Does not accept (b) Does not accept (c) Does not accept Excludes disputes under Article 297
Finland (EC)	(a) Accepts (b) Accepts (c) — (d) —	—
France (EC)	—	(a) Does not accept (b) Does not accept (c) Does not accept
Germany (EC)	(a) First preference (b) Third preference (c) Second preference (d) Fourth preference for questions concerning fisheries, protection and preservation of the marine environment, marine scientific research and navigation, including pollution from vessels and by dumping	—
Greece	(a) Accepts (b) — (c) — (d) —	—
Guinea-Bissau	(a) — (b) Does not accept jurisdiction (c) — (d) —	

Issues and Prospects 53

TABLE 5.—(Continued)

State	Choice of Procedure [Article 287, Paragraph 1]	Optional Exceptions to Application of Part XV, Section 2 [Article 298, Paragraph 1]
Iceland	—	Under Article 298 reserves right that any interpretation of Article 83 be submitted to conciliation under Annex V, Section 2
Italy (EC)	(a) Accepts (b) Accepts (c) — (d) —	(a) Does not accept (b) — (c) —
Netherlands (EC)	(a) — (b) Accepts (c) — (d) —	
Nicaragua	(a) Does not accept (b) Accepts (c) Does not accept (d) Does not accept	Accepts only in the ICJ
Norway	(a) — (b) Accepts (c) — (d) —	(a) Does not accept (b) Does not accept (c) Does not accept
Oman	(a) Accepts (b) Accepts (c) — (d) —	—
Portugal (EC)	Will choose from (a), (b), (c) or (d), the latter preferred for questions relating to fisheries, protection and preservation of the marine environment, marine scientific research, navigation and marine pollution	(a) Does not accept (b) Does not accept (c) Does not accept
Russian Federation	(a) Accepts for questions concerning the prompt release of detained vessels and crews (b) — (c) Accepts (d) Accepts for questions concerning fisheries, protection and preservation of the marine environment, marine scientific research and navigation, including pollution from vessels and dumping	(a) Does not accept (b) Does not accept (c) Does not accept
Slovenia	(a) — (b) — (c) Accepts (d) —	(a) Does not accept (b) Does not accept (c) Does not accept

TABLE 5.—(*Continued*)

State	Choice of Procedure [Article 287, Paragraph 1]	Optional Exceptions to Application of Part XV, Section 2 [Article 298, Paragraph 1]
Spain	(a) — (b) Accepts (c) — (d) —	—
Sweden	(a) — (b) Accepts (c) — (d) —	—
Tunisia	(a) First preference (b) — (c) Second preference (d) —	(a) Does not accept (b) Does not accept (c) Does not accept
Ukraine	(a) Accepts for questions concerning the prompt release of detained vessels and crews (b) — (c) Accepts (a) Accepts for questions concerning fisheries, protection and preservation of the marine environment, marine scientific research and navigation, including pollution from vessels and dumping	(a) Does not accept (b) Does not accept (c) —
United Kingdom of Great Britain and Northern Ireland	(a) Will consider on case by case basis (b) Accepts (c) — (d) —	—
United Republic of Tanzania	(a) Accepts (b) — (c) — (d) —	—
Uruguay	(a) Accepts without prejudice to recognition of jurisdiction of ICJ (b) — (c) — (d) —	Does not accept procedures for disputes concerning law enforcement activities

Issues and Prospects

From the Rhodian Sea Law to UNCLOS III[†]

Wolfgang Graf Vitzthum
University of Tübingen, Germany

This essay deals with the historical development of the law of the sea, taking the Lex Rhodia[1] as a starting point and finishing with the latest developments in the progressive development and the codification of this area of law, the 1982 United Nations Convention on the Law of the Sea[2] and related agreements, thus covering a period of roughly 2000 years. Of special interest are the changes that the law of the sea went through over the centuries, as well as the continuing influence of those historical developments on today's law of the sea and on general international law.

Ancient and medieval law of the sea contained few or no rules that would be considered public international law from today's viewpoint. The important and often cited collections of the law of the sea, the Lex Rhodia, the Rôles d'Oleron,[3] or the Visby Rules[4] (to name only a few) were primarily concerned with the legal regulation of sea trade. They thus constituted mainly what we would call rules of maritime law. Those rules emanated from the customs of seafaring merchants and others who participated in maritime trade and filled the gap that was left by a lack of state-sponsored legislation. Such rules managed to gain widespread acceptance over the centuries. The Roman Empire did nothing more than incorporate the law of the sea that had been developed since about 800 B.C. on the island of Rhodes into its own set of laws. The Lex Rhodia also influenced the development of maritime law during the Middle Ages. Some rules from the Lex Rhodia, such as those

---

[†] EDITORS' NOTE.—An earlier version of the essay was presented as a keynote address at the Pacem in Maribus XXVIII Conference—The European Challenge, Hamburg, Germany, 3–6 December 2000. The text was subsequently updated and revised by the author to reflect the entry into force of the 1995 United Nations Fish Stocks Agreement on 11 December 2001.

1. W. Ashburner, *The Rhodian Sea Law* (Clarendon Press: Oxford, 1909), (reprinted Scientia: Aalen, 1976), p. 1 et seq.; cf. also Digests 14, 2, 1.

2. United Nations Convention on the Law of the Sea, 1982. A/Conf. 63/122. Entered into force, 16 November 1994. Accessed 15 April, 2002 on the World Wide Web: http://www.un.org/depts/los/.

3. J. M. Pardessus, *Collection de Lois Maritimes autérieurs au XVII siècle*, vol. 1 (Paris: 1828) (reprinted Bottega d'Erasmo: Torino, 1959), p. 323 et seq.

4. Pardessus, p. 463 et seq.

© 2003 by The University of Chicago. All rights reserved.
0-226-06620-7/03/0017-0004$01.00 *Ocean Yearbook* 17:56–59

concerning general average (*große Haverei*), can still be found in the maritime law of many jurisdictions.

The fact that the law of the sea of Antiquity and the early Middle Ages contained only private law rules can be taken as an indication that the States at the time were not especially interested in extending their jurisdiction over the seas. Instead, the sea was left to merchants and their trade. However, it remains doubtful if this general behaviour amounted to a freedom of the seas in a legal sense, as it is difficult to trace whether the States considered themselves obliged to respect this freedom. For example, although Roman law contained rules according to which the sea was to be considered as free and not capable of being appropriated, it was also the Romans who considered themselves to be masters of the enitre Mediterranean. In the Roman conception this sea was *mare nostrum*. It has to be asked, however, if this idea was more than the taking over of a policing responsibility by the Roman Empire as it did in Pompeius' successful fight against piracy in 67 B.C.

Real challenges to the freedom of the seas—irrespective of whether they be legal or only factual—emerged with the rise of new seafaring powers in the Middle Ages. Venice, being of paramount influence in the south, and the Hanseatic League, Denmark, and Sweden in the north, tried to establish a position of predominance in their respective seas. The Age of Discoveries saw the rise of other countries to sea power: Spain and Portugal came first, followed by the Dutch and the English. While earlier quests for dominance were aimed at sea areas close to the territory (like the Adriatic Sea in the Venetian case), now the sea powers were struggling for global dominance. Thus, Portugal and Spain divided the Atlantic Ocean between themselves in the 1494 Treaty of Tordesillas. This soon met with opposition by other seafaring nations that realized both the potential of overseas trade and the effect of the dominance of the ocean by one or two powers.

As the English and the Dutch extended their overseas operations, conflicts arose between the two countries and with the Iberian powers. England is an example of how maritime interests influenced a State's position on the freedom of the seas. As long as England opposed foreign efforts to achieve sea dominance, it was a fervent advocate of the freedom of the seas. However, as England developed more and more into the pre-eminent sea power, its attitude changed. The acquisition of sovereignty over the sea was now considered possible, with the consequence that the sovereign could exclude others from the use of the sea. In this situation the Dutch authorities published in 1609 the book *Mare Liberum*[5] that had been written a few years earlier by the Dutch lawyer Hugo Grotius. Grotius argued that the sea was not capable of being subjected to the sovereignty of any State. The book provoked England to publish in 1635 a counter-study by John Selden with

---

5. H. Grotius, *Mare Liberum: sive, De iure quod Batavis competit ad Indicana commercia dissertatio* (New York: Carnegie Endowment for International Peace, 1952).

58   Issues and Prospects

the title *Mare Clausum*[6] in which Grotius's views were opposed. After having reached a predominant position at sea England lost its interest in safeguarding the concept of *mare clausum*. In times of increasing trade, it considered it beneficial to its interests if the use of the sea was as unrestricted as possible. By supporting the freedom of the seas again, England promoted the general acceptance of this principle at the dawn of the 19th century.

With the recognition of the freedom of the seas, coastal States—seeking protection of their territory—started to claim territorial waters around their coastlines for their sovereignty. This marked the beginning of the establishment of an abundance of zones covering wide parts of the sea. These zones provide for preferential or exclusive coastal State rights, or even full sovereignty in their respective areas. Thus, with the general recognition of the freedom of the high seas a process went hand and hand that widely reduced the areas to which that principle applied.

The history of the law of the sea in the 20th century, especially after World War II, saw an immense acceleration of this process. It has been described as "terraneisation" of the oceans, as sea areas were now more and more subjected to regimes similar to those in force on the land territories. The discovery of vast reserves of natural resources in the seabed and the increasing importance of fisheries led to the seas being perceived mainly as a supply of resources. Thus, States wanted to gain exclusive rights over as much of the resource-rich areas as possible. The proclamations of continental shelves by a number of States from 1945 onwards were only the beginning. The 1958 Geneva Conventions on the Law of the Sea did not effectively slow the process of coastal States' rights being constantly extended seawards. The "creeping jurisdiction" had just begun. Exclusive fishery zones were being called for and the seaward limits of the continental shelves were extended farther and farther.

These tendencies, together with the idea of establishing a New International Economic Order that should lead to a more equitable global distribution of wealth, were the incentives for convoking a further international conference, the Third United Nations Conference on the Law of the Sea (UNCLOS III), which took place from 1974 to 1982. One of the aims of UNCLOS III was to find a compromise between the interests of individual States on the one hand and those of the world community as a whole on the other. The deep seabed was declared the "common heritage of mankind" even before the conference had started, and the validity of the freedom of the high seas was emphasized once more. The concept of "common heritage"—which basically accords rights and responsibilities in respect of certain resources or areas to the world community as such—has meanwhile found acceptance in fields outside the law of the sea as well.

6. J. Selden, *Of the Dominion; or, Ownership of the Sea* (New York: Arno, 1972). Translation of *Mare Clausum*, Reprint of the 1652 ed.

The outcome of UNCLOS III, the 1982 UN Convention on the Law of the Sea, considerably extended the rights of coastal States to the detriment of the global commons, although the legal position of the commons in theory was reinforced. On the other hand, the 1982 Convention also signified a remarkable development of the law of the sea, with the introduction of a sophisticated system for the settlement of disputes and the establishment of the first international court dealing exclusively with law of the sea matters, the International Tribunal for the Law of the Sea. Of further importance are the provisions of the 1982 Convention enabling an improved protection of the marine environment.

Because some major industrial powers had strong reservations about certain rules contained in the 1982 Convention, especially the deep-seabed regime, it was only in 1994 that the Convention reached the necessary number of ratifications for its entry into force. This was mainly achieved by the Agreement on the Implementation of Part XI of the 1982 Law of the Sea Convention, which had brought considerable changes to this regime by favouring a free market approach and thus facilitating accession of industrial countries to the 1982 Convention.

In 1995, an additional agreement implementing the 1982 Convention was concluded, the UN Fish Stocks Agreement that just recently entered into force. Its aim is to overcome problems with the management of fish stocks that cross borders of jurisdiction. The 1982 Convention had left many of the related problems unsolved. Although regulation was necessary to ensure the sustainable development of such fish stocks, the UN Fish Stocks Agreement in practice means a further increase of coastal States' rights in relation to fisheries, one of the uses guaranteed, in principle, to all States by the freedom of the high seas.

Living Resources

Solving the Tuna-Dolphin Problem in the Eastern Pacific Purse-Seine Fishery

Martin A. Hall, Marcela Campa and Martha Gómez
Inter-American Tropical Tuna Commission, La Jolla, California

INTRODUCTION

One of the longest and most controversial by-catch problems, that of the incidental mortality of dolphins in the purse-seine fishery for tuna in the eastern Pacific Ocean, serves as a model to illustrate the development of a program to reduce by-catches in a fishery. In spite of some of its unique characteristics, there are some phases in the identification of the problem, its quantification, and the approaches to its solution that are common to most, if not all, fisheries. After briefly reviewing the historical development of the problem, this article describes five phases that produced (a) a valid statistical description of the problem, including the identification of the causes of mortality and of the specific solutions to each one of these; (b) upgrades and dissemination of technology to avoid or mitigate the problem; (c) a measure of the impact of crew skill and motivation on dolphin mortality, the "human factor"; (d) a training program for captains and crews; and (e) an international management program to consolidate and expand the achievements of the program. Dolphin mortality has been reduced by more than 98 percent, without eliminating a fishery that provides significant economic and social benefits to the region. Many factors that have contributed to the success of this program are identified.

BRIEF HISTORY OF THE PROBLEM

The purse-seine revolution in the eastern Pacific Ocean (EPO) tuna fisheries took place in the late 1950s.[1] This "revolution" was brought about by the

1. F. G. Alverson, "Distribution of fishing effort and resulting tuna catches from the eastern tropical Pacific by quarters of the year, 1951–1958," *Inter-American Tropical Tuna Commission Bulletin* 4 (1960): 321–446; G. C. Broadhead, "Recent changes in the efficiency of vessels fishing for yellowfin tuna in the eastern Pacific Ocean," *Inter-American Tropical Tuna Commission Bulletin* 6 (1962): 283–332; R. E. Green, W. F. Perrin, and B. P. Petrich, "The American tuna purse-seine fishery,"

development of improved freezing methods and the use of nylon webbing for the nets and a new device to retrieve them (the power block). These developments resulted in an almost complete replacement of a less productive baitboat fishery by a purse-seine fleet that currently takes practically all the catch of yellowfin tuna (*Thunnus albacares*) and skipjack tuna (*Katsuwonus pelamis*), the main targets of the surface fishery. With better material and more power on the boats, the nets increased in size to reach about 1.5 km long by 200 m deep. Improvements in vessel technology and design, together with an increasing demand for tuna, led to expansion of the fishery. Another species of tuna, big-eye tuna (*Thunnus obesus*), has also become a target for purse seiners in recent years. This species was traditionally taken by long-liners.

The detection of the tuna was also greatly influenced by technological progress. From sightings with the naked eye, with simple 7X binoculars, with the more powerful 20X to 25X binoculars, and later with the addition of helicopters and "bird radar," the detection technology developed at a fast pace, and new ideas were quickly adopted. The helicopters greatly extended the area searched and provided accurate information on tuna sightings. Bird radar, capable of detecting a single seabird at a distance of more than 10 miles, proved very useful, given the frequent association of bird flocks with tuna schools. Acoustic devices (e.g., sonar) have not played a role as significant as in other fisheries.

After detection, the capture of the tuna is the result of an operation called "purse seining." The net encircles the school of fish, and when the circle is complete a cable at the bottom edge of the net is pulled, closing the opening at the bottom of the net. Each deployment of the net is called a "set." There are three types of sets that correspond to three ways of detecting the tuna schools:

School Sets.—A tuna school is detected by evidence of its presence on the surface of the ocean, that is, the sea surface is disturbed, and it appears to boil or to be disturbed by a local breeze. Frequently, birds associated with the tuna school are detected from the vessel with the radar mentioned earlier. The operation is called "school fishing," and the sets are called "school sets." This technique usually produces small yellowfin (modal size of 50 cm, or 2.5 kg, for the 1976–1995 period) and skipjack tuna.

Log Sets.—Tuna schools tend to associate with floating objects during the night, and then leave them early in the morning. When the fishermen find a floating object with tuna around it, they surround it with the net shortly before sunrise, capturing the fish associated with the object. Because

---

in *Modern Fishing Gear of the World*, vol. 3, ed. H. Kristjonsson (London: Fishing News (Books) Ltd., 1971), pp. 182–94; J. S. Cole, "Synopsis of biological data on the yellowfin tuna, *Thunnus albacares* (Bonnaterre, 1788) in the Pacific Ocean," *Inter-American Tropical Tuna Commission Special Report* 2 (1980): 71–150.

the most common objects are tree trunks and branches, this method of fishing is called "log fishing," and the sets are called "log sets." This technique catches very small yellowfin (modal size of 40 cm, or 1.2 kg, for the 1976–1995 period) and big-eye tuna, but the main catch is skipjack tuna. In recent years a variant of this fishery has developed in which a vessel deploys a number of fish-aggregating devices (FADs), allows them to drift with radio buoys, and fishes on them later.

Dolphin Sets.—In the EPO yellowfin are frequently found associated with groups of dolphins. It is not known why they associate with one another, but most of the hypotheses proposed to explain the association are based on feeding or predator protection,[2] although energy-management reasons have also been proposed.[3] Several dolphin species (especially spotted dolphin, *Stenella attenuata,* and spinner dolphin, *S. longirostris*) are known to associate with tuna. After encircling the tuna and dolphins, the net is "pursed," and both are captured. The technique is called "dolphin fishing," and the sets are called "dolphin sets." This technique produces almost exclusively yellowfin, and these are larger (modal size of 70–80 cm, or 6.9–10.4 kg, for the 1976–1995 period) than those caught by other methods of purse-seine fishing.

During the baitboat era, fishers became aware that one of the most prized tuna species, the yellowfin tuna, traveled in association with herds of dolphins, and they used the easier-to-detect groups of dolphins as indicators of the presence of tuna. Dolphins did not take the hooks, so the direct impact of the fishery on the dolphins was probably minimal. When the purse-seine fishery developed, the tuna were encircled by the net instead of hooked, and the dolphins became an incidental, and unwanted, capture. Entanglements in the net and other problems created by the inability of the fishers to release them resulted in very high dolphin mortalities in the 1960s. The technology that increased the productivity of the fishery by an order of magnitude or more also caused a major ecological impact on other species by the loss in the selectivity of the fishing gear compared to the hook-and-line method.

---

2. W. F. Perrin, R. W. Warner, C. L. Fiscus and D. B. Holts, "Stomach contents of porpoise, *Stenella* spp., and yellowfin tuna *Thunnus albacares,* in mixed species aggregation," *Fishery Bulletin U.S.* 71 (1973), pp. 1077–92; W. E. Stuntz, The Tuna-Dolphin Bond: a Discussion of Current Hypotheses, Southwest Fisheries Center, NMFS (National Marine Fisheries Service [U.S.]), Administrative Report no. LJ-81-19, p. 9, August 1981; Anonymous, Annual Report of the Inter-American Tropical Tuna Commission for 1993, p. 316, 1994.).

3. E. F. Edwards, "Energetics of associated tunas and dolphins in the eastern tropical Pacific Ocean: A basis for the bond," *Fishery Bulletin U.S.* 90 (1992): 678–90; E. F. Edwards, Separation/attraction Research on the Tuna-dolphin Bond: Review and Criteria for Future Proposals, National Marine Fisheries Service, Southwest Fisheries Science Center, Administrative Report. LJ-96-17, p. 25, (1996).

During the first period of the purse-seine fishery (1959–1972), the levels of dolphin mortality were very high, but the figures are very inaccurate because of the scarcity of reliable data.[4] Most of the fleet flew the U.S. flag and landed the catch in California ports to be processed in the canneries there. A biologist taken on a fishing trip saw and reported the high mortality levels.[5] The strong public response to this information was one of the driving forces leading to the passage of the Marine Mammal Protection Act (MMPA) of 1972 that made mandatory an observer program, gear, and procedures, and set limits to total mortality. As a result of these actions, and of the increasing awareness on the part of captains and crews, the mortality decreased considerably, first to a level of around 100,000, and then to less than half of that figure. But in the early 1980s, many coastal nations of the Pacific began to develop their national fleets. The opening of the European and other markets, the oil crisis that favored oil-producing nations (such as Ecuador, Mexico, and Venezuela), and the lower processing costs in the region contributed to this development. Some of the new markets required large tuna and paid higher prices for a product with a higher yield. As these sizes were normally found in association with dolphins, this led to increases in fishing effort directed toward dolphins that reached record levels in 1986–1990. Many of the new captains and crews were inexperienced in the use of dolphin release maneuvers and gear, and dolphin mortality went up again. It reached a peak of 133,000 dolphins in 1986.[6] From then on, a series of actions and programs began to reduce that mortality to reach the current level of around 1,700 dolphins in 2000. Currently, the fishery covers an area that goes roughly from California to northern Chile, and extends offshore to about 150° W longitude. The main fleets fly the flags of Ecuador, Mexico, Venezuela, Vanuatu, Colombia, Unites States, and Panama.

What we present in this article is the strategy followed, and the sequence of actions developed to implement it, to achieve the success of the international program coordinated by the IATTC to reduce dolphin mortality. There were different phases in this program, with well-defined objectives. These phases overlapped in time, but there is a sequence of steps, many of

---

4. T. D. Smith and N. C. H. Lo, Some Data on Dolphin Mortality in the Eastern Tropical Pacific Tuna Purse Seine Fishery Prior to 1970, U.S. Department of Commerce, NOAA Tech. Memo. NOAA-TM-NMFS-SWFC-34, p. 26, 1983; R. C. Francis, F. T. Awbrey, C. L. Goudey, M. A. Hall, D. M. King, H. Medina, K. S. Norris, M. K. Orbach, R. Payne and E. Pikitch, *Dolphins and the Tuna Industry*, xii (Washington, D.C.: National Academy Press, 1992), p. 176.

5. W. F. Perrin, "The porpoise and the tuna," *Sea Frontiers* 14, no. 3 (1968): 166–174; W. F. Perrin, "Using porpoise to catch tuna," *World Fishing* 18 (1969): 42–45.

6. M. A. Hall and S. D. Boyer, "Incidental mortality of dolphins in the eastern tropical Pacific tuna fishery in 1986," *Report of the International Whaling Commission* 38 (1988): 439–41.

which are pre-requisites for others. They will show the roles of technology, education, regulation, public awareness, and motivation in the solution of a complex problem of the type we will be facing more and more frequently in the future. Even though the characteristics of this fishery are rather unique, it is clear that many of the lessons that we learned here can be applied to other fisheries. Other contributions address the issue of by-catches in general, (classification, reduction strategies, ecological implications)[7] and make an ecological comparison of the impacts of the different forms of fishing for tuna in the region, especially the three types of purse-seine sets.[8]

The period covered by this article is mainly 1984 to the present, the tenure of the first author as the chief scientist of the Tuna-dolphin Program of the Inter-American Tropical Tuna Commission. A brief chronology of the main events in the fishery is included in Appendix 1.

THE SITUATION PRIOR TO 1985

Two major factors had a deep influence on the size and composition of the fleet by flag in the early 1980s. After many years in which the U.S. fleet had been the dominant one in the region and in which the vast majority of the catches were unloaded in U.S. ports and processed in U.S.-based canneries, the early 1980s saw major change. A period of growth overfishing in the late 1970s and early 1980s was followed by the strong El Niño event of 1982–1983, and the result was a major failure for the vessels fishing in the EPO. The low production led many of them to leave the area, and many of those, especially U.S.-flagged vessels, opted for the Western Pacific fishing grounds. The other process that took place in those early years of the decade was a gradual, but steady, development of the fleets of several Latin American countries. Oil-producing nations such as Ecuador, Mexico, and Venezuela had a major advantage by being able to provide one of the major needs of the vessels at a competitive, or even subsidized price. In addition, in some cases they were closer to the fishing grounds than other fleets. Other economic factors that affected the fishery were the gaps in the cost of labor and in the taxation system between California-based canneries and those other locations. Many U.S. canners moved their operations to Puerto Rico, driving

---

7. M. A. Hall, "Strategies to reduce the incidental mortality of marine mammals and other species in fisheries," in *Developments in Marine Biology*, vol. IV of *Whales, Seals, Fish and Man*, ed. A. Schytte Bix et al. (Tromso, Norway: Elsevier Publ., 1995), pp. 537–44; M. A. Hall, "On by-catches," *Reviews in Fish Biology and Fisheries* 6 (1996): 319–52.

8. M. A. Hall, "An ecological view of the tuna dolphin problem: impacts and trade-offs," *Reviews in Fish Biology and Fisheries* 8 (1998): 1–34.

the center of gravity of the fishery toward the South. Another driving force for this development was the adoption by most nations of a new Law of the Sea that caused conflicts between coastal nations and the United States concerning access to their Exclusive Economic Zone (EEZs). Extended jurisdiction gave greater and exclusive access to coastal states.

When the fleet was almost completely composed of U.S.-flag vessels, the protection of the dolphins (monitoring, mandatory equipment, and limits on dolphin mortality, etc.) was mostly a U.S. issue, ruled by the MMPA of 1972 and controlled by the National Marine Fisheries Service, an organ of the U.S. Department of Commerce. The Act defined a series of requirements for vessels fishing in the region.

The main objective of the MMPA of 1972 was to maintain marine mammals at "optimum sustainable populations," understood to be "the number of animals which will result in the maximum productivity of the population or the species, keeping in mind the carrying capacity of the habitat and the health of the ecosystem of which they form a constituent element." The Act established a moratorium on the taking of marine mammals, but permitted certain exceptions to the moratorium, including the incidental take of marine mammals in the course of commercial fishing operations. In addition, the MMPA requires carrying observers in a significant proportion of the trips to obtain estimates of dolphin mortality and verify compliance with the regulations. Furthermore, it calls for a ban on the importation of fish or products from fish that have been caught with commercial fishing technology that results in the incidental kill or incidental serious injury of ocean mammals in excess of the U.S. standards.

The MMPA was amended, among other times, in April 1976, through Public Law 94-265; October 1981, through Public Law 97-58; July 1984, through Public Law 98-364; November 1988, through Public Law 100-711; November 1990, through Public Law 101-627; and in September 1994, through Public Law 103-238 (see Appendix 1).

When the fishery became more international the new countries joining in had to develop their own programs and regulations. In 1976 the IATTC asked the Commission staff to undertake a program concerning the issue of dolphin mortality in the fishery. The objectives of the IATTC were extended to include the following: "it should strive to maintain a high level of tuna production and also to maintain [dolphin] stocks at or above levels that assure their survival in perpetuity, with every reasonable effort being made to avoid needless or careless killing of [dolphins]." Some countries with fleets in the region, but that were not members of the IATTC, decided to join the Program later on, in view of the fact that they shared the objectives, had the incentive of trying to comply with the requirements of the MMPA to export to the United States, and could benefit from the combined effort. This produced the international program that started with very modest observer coverage in 1979. By 1983, the observer program had a coverage rate of 34.0 percent of the trips by U.S. vessels, but only 2.7 percent for vessels of other flags.

The actions started in 1984 and subsequent years can be better interpreted if we identify the different phases of the program, with the specific objectives sought after in each one of them and the periods in which they took place. These phases had no formal existence; they were only operational divisions that helped us focus our attention and concentrate our efforts on some specific need or angle of the problem. The duration of these phases was not determined by planning, but by the availability of personnel, information, and other resources. The time periods mentioned should not be considered as the only ones in which some actions took place, but only as references to a sequence of objectives, which in some cases were prerequisites for others.

PHASE 1 (1984–1986)

Objectives.—Producing a better estimate of mortality and increasing our understanding of the factors that affect dolphin mortality.

Two ratio estimates have been used to estimate dolphin mortality.[9] The mortality per set, calculated from observer data, was multiplied by the total number of dolphin sets obtained from either observed trips, or logbook data from the vessels. The mortality per ton of tuna caught was calculated from the observer records and extrapolated to the tonnage caught by all vessels. In both cases the data had been stratified into two groups, the U.S. fleet and all other vessels, to account for the fact that the regulations and the experience of the different crews were different. By 1984 it was obvious that the coverage of the non-U.S. fleet was insufficient, and the first step was to increase observer coverage to a level that would produce a better estimate of mortality. To determine that level, a series of simulations were run[10] using pooled data from previous years. The simulations produced two main results: at low coverage the ratios tended to overestimate the mortality, and with coverages of 33–40 percent, the biases were very low and the reductions in variance achieved by increasing the sample size over those levels were not cost-effective, that is, a large increase in sampling effort yielded a low reduction in the variance. Accordingly, the first goal was to bring all fleets to that level or higher. Between 1983 and 1986, the coverage of the non-U.S. fleet

9. N. C. H. Lo and T. D. Smith, "Incidental mortality of dolphins in the eastern tropical Pacific, 1959–1972," *Fishery Bulletin U.S.* 84 (1986): 27–34.
10. Anonymous, Tuna-dolphin workshop. "Incidental Mortality of Dolphins in the Eastern Tropical Pacific Tuna Fishery, 1979–1988." Working Document no. 2. San Jose, Costa Rica, March 14–16, 1989.

increased from 2.7 to 24.4 percent. With the addition of Mexico to the program in 1986, the vessels of all nations fishing in the region were sampled, and the issue of representativeness was addressed.

Besides the question of variance, there was the question of a potential "observer effect"[11] that could not be addressed with simulations. The idea was that vessels with observers on board behaved differently from those without them, so the extrapolation of ratios from observed vessels to the others provided a biased estimate. The issue of the "observer effect," from the point of view of modifying the behavior of the fishers, was addressed when the nations boosted observer coverage of their fleets to 100 percent.

After some initial resistance in 1984–1985, the industry changed its position, and became much more agreeable even to increases in observer coverage. Two reasons were behind this change: 1) a perception by the industry that the better the coverage the lower the estimates, because of the elimination of the biases and of the extrapolation procedures that they distrusted; and 2) the idea that a better monitoring system for the performance of vessels and crews led to improved performance. Both of these perceptions were probably right, at least in part.

From the scientific point of view, the confidence intervals around the mortality estimates shrank considerably, but perhaps the main gain in the increased coverage was the acceleration of the data collection rates. To reduce the incidental mortality of dolphins, we had to find which were the factors contributing to it. Environmental factors (currents, visibility, time of day), catch factors (tonnage of tuna in the net, species composition), dolphin factors (group size and composition, behavior in the net), gear factors (availability, and condition of the dolphin safety gear, deployment conditions, malfunctions), and skipper and crew factors (experience, training, and motivation) may contribute to the occurrence and level of dolphin mortality. Because of the multiplicity of factors, any analyses attempting to identify those affecting the mortality levels would require a large sample size to eliminate the effect of other covariates, or at least to fully explore the range of variability. A large database was needed for this, and with greater coverage it soon became available. The pay-off was very high. Many factors affecting dolphin mortality rates were identified; some were known, but many others were identified for the first time in these analyses. This list of environmental, catch-related, gear-related, and other factors mentioned earlier provided a list of potential targets for technological changes, education of fishers, and regulation. We now had the tools to reduce dolphin mortality without shutting down the fishery.

11. B. E. Wahlen and T. D. Smith, "Observer effect on incidental dolphin mortality in the eastern tropical Pacific tuna fishery," *Fishery Bulletin U.S.* 83 no. 4 (1985): 521–30.

PHASE 2 (1986–1989)

Objective.—Reducing mortality through adoption of available technological changes in fishing gear and auxiliary equipment by the international fleet.

When the fishers started fishing with purse seines, encircling mixed groups of tuna and dolphins, it became apparent to them that many dolphins were going to die in their fishing operations. It was also obvious to them that there were no benefits from this incidental mortality. Sets with high mortality took much longer than those with low, or no mortality; there was no possible utilization of the dolphins; and, if the dolphin populations decreased, the fishers' way of life would be threatened because the dolphins were the best indicator of the presence of tuna (easy to detect, always on the surface, and leading to the best-priced tuna, etc.) It was to the fishers' advantage to fish on dolphins, but to release them alive.

In the early years of the fishery, several ingenious approaches to the solution originated in the fishers themselves, or in programs organized by the National Marine Fisheries Service.[12] They included some major changes, and some small innovations or additions to the gear and techniques used to prevent the entanglement or facilitate the release of dolphins. Together, all of these produced a fairly complete solution, but many of them had yet to be included in the new sector of the fleet. They included the following:

a. The backdown maneuver: after the group is encircled, the vessel moves in reverse, pulling the net. As a result of this, and when the maneuver is executed at the right speed, and with the right arc, the net takes an elongated shape, and it sinks at the end opposite the vessel. The dolphins that tend to remain close to the surface can now swim out of the net, or in some cases the net simply moves under them until they are out of the area encircled. The tuna, swimming deeper, are retained in the net.

b. The Medina panel:[13] In the area of the net where dolphins are entangled more frequently the mesh size (usually 10 cm in mesh size) was replaced by a much smaller mesh (around 3 cm) to keep the dolphins from getting their snouts caught in the webbing.

c. Hand rescue: A raft with a rescuer, and sometimes additional swimmers in the water, enter the net to help in the manual release of entangled animals. As the crews become more efficient in their han-

---

12. J. M. Coe and G. Sousa, "Removing porpoises from a tuna purse seine," *Marine Fisheries Review* 34 nos. 11–12 (1972): 15–19; J. M. Coe, D. B. Holts and R. W. Butler, "The "tuna-porpoise" problem: NMFS dolphin mortality reduction research, 1970–81," *Marine Fisheries Review* 46, no. 3 (1984): 18–33.

13. E. G. Baker, W. K. Taguchi and S. B. Reilly, "Porpoise rescue methods in the yellowfin tuna purse seine fisheries and the importance of Medina panel mesh size," *Marine Fisheries Review* 39, no. 5 (1977): 1–10.

dling of the animals they have a greater impact in reducing dolphin mortality. More recently, divers with breathing hoses have been added to rescue animals from deeper areas of the net.
d. Floodlight: A very powerful floodlight is used in sets that are prolonged into the hours of darkness. It is not clear whether the floodlight helps the rescue efforts of the crew, or facilitates the dolphin escape, but it is effective in reducing mortality.
e. Towing speedboats: Some of the speedboats remain close to the net and use a towing bridle to keep the net open. In some cases, currents, malfunctions, or poor maneuvers cause the net to collapse. This collapse may result in many animals being trapped inside the undulations or billowing of the net.

All these pieces of equipment and procedures became mandatory for the U.S. fleet first. The other fishing nations matched the U.S. regulations as a result of the "comparability" requirements of the MMPA, and also because of the effectiveness of these equipment and procedures that were already available to them. During the 1986–1989 period a major focus of the IATTC program was to make sure that all vessels carried the necessary equipment, and that their crews had mastered the maneuvers needed to rescue dolphins. A major part of this task was accomplished through "trial sets." A "trial set" is a deployment of the net made only to see the condition of the net, the way it is connected to the boat, and the way the backdown maneuver is carried on by the crew. Employees of the IATTC and of national programs have performed this procedure many times in the different ports of the region and have produced reports containing the recommendations to improve the performance of the rescue maneuvers and to avoid entanglement caused by net deficiencies. In addition, the observers at sea report on the rescue maneuvers and equipment, providing best diagnostic tools to the problems noticed. Diagrams of the net at different stages of development allowed the experts to identify patterns caused by a maneuver made at a too low or too high speed, and to separate equipment problems from the others caused by currents, malfunctions, and so forth.

Another equipment-related action was the emphasis on maintenance of the components of the vessel systems (hydraulic, electrical, mechanical) whose malfunctions contributed to increased mortality by preventing the execution of the backdown maneuver, slowing down the sets and increasing the chances of the net collapsing and trapping a number of animals, and so forth.

This was the main focus of the mortality reduction activities during the period 1986–1989. It had started earlier, and continued after 1989, but it was our thinking that bringing all the vessels to an acceptable standard on the condition and availability of gear and equipment could make major gains. As a result of the information obtained in Phase 1, the fishers knew which were the major factors causing mortality, so they could avoid them.

But if they failed to avoid the problems, Phase 2 gave them the technical tools to mitigate their impact. A list of recommended steps to reduce dolphin mortality is given in Appendix 2, simply to illustrate the variety of causal factors considered.

PHASE 3 (1989–1991)

Objective.—Reducing mortality through monitoring the human components: skill and motivation of captains and crews.

Increased observer coverage made it possible for the IATTC staff to study the causes of mortality of dolphins in more detail. Cursory examination of the data indicated that the mortalities per set or per ton of yellowfin caught in dolphin sets differed among vessel captains. The data were stratified into various categories, for example, areas, seasons, stocks of dolphins set on, and vessel flag, and the performances of the captains were evaluated. The mortality rates were greatest for sets during which malfunctions occurred, so it is important that the number of such sets be minimized. Because malfunctions are chance events, the sets where they occurred were eliminated from the analyses.

What these studies showed was that the "human factor" was even more significant than the equipment. Most of the mortality was the result of the actions of a relatively small number of captains. This information allowed industries and governments to realize that the monitoring of individual performances was a crucial step in reaching a solution to the problem.

PHASE 4 (1989–PRESENT)

Objective.—Training and education of the fishers.

Having identified the high priority of the crew's skill and motivation, the IATTC developed a program to interact with them more directly, rather than through the governments or the vessel owners. This idea was aided by an initiative of an industry leader who proposed that the IATTC bring together all the captains of the vessels under his charge for an educational meeting. In the past, the idea of communicating with the captains had always been hampered by the fact that they spent most of the time at sea, and when they finish a trip they are, understandably, eager to return to their home towns (in America or Europe) for the few days they have between trips. Trips normally last 40–60 days. His initiative was a costly one for him, but an excellent opportunity for the IATTC. The first seminar took place in Cartagena, Colombia, and it lasted 2 full days. The seminar included the following topics:

- Activities of IATTC's international tuna-dolphin program.
- Historical review of the purse-seine fishery in the EPO.

- Historical review of incidental dolphin mortality, including the high mortality levels in the early years, and the current statistics.
- The public perception of the problem (illustrated with videotaped materials, magazine and newspaper articles, etc.): this showed the fishers, in many cases isolated from society for most of the year, the way in which their fishery has been portrayed, the charges made, and so forth.
- The observer program: duties and responsibilities of observers, observer-crew interactions.
- Factors contributing to dolphin mortality and mitigation actions for each; a discussion of the technical and other problems identified in our studies; how the different factors affected the average mortality per set; responsibilities of captains and boat owners.
- Dolphin-safety gear and procedures.
- Services provided by the IATTC concerning gear.
- New ideas and innovations in the fleet; suggestions and experiences from the participants.
- Laws and regulations.
- Performances by the different fleets.
- Performances of the different captains; comparative statistics such as proportion of sets with zero mortality, without backdown, in darkness, and so forth.
- Individual record of performance: At the end of the seminar the individual statistics of each captain were discussed with him or her in private to avoid embarrassment in some cases, emphasizing the reasons why his or her performance was below average, when that was the case, and discussing specific problems of each vessel when needed. Many of these discussions led to suggestions made to vessel owners about improvements in the operation of the vessels. This phase was especially important during 1989–1991.

Over the years, the fishers achieved the reductions in mortality in two ways: better decision making and better preparedness. There was a steady decline in the proportion of sets in adverse conditions, showing an increasing awareness of those conditions on the part of the fishers, and also the development of a risk-adverse attitude that made them pass up fishing opportunities when the circumstances indicated that there is a high probability of a problem set. But when those conditions occur their impact (mortality per set) is much lower.

Figures 1 and 2 show the evolution in the control of some of the causes of mortality over the 1986–1997 period. In both cases, the upper panel shows the proportion of sets with the problem in question, and the lower one, the change in the mortality per set occurring in those sets with the problem. Malfunctions (fig. 1) cannot be eliminated, but better mainte-

72 Solving the Tuna Dolphin Problem

Sets with major malfunctions

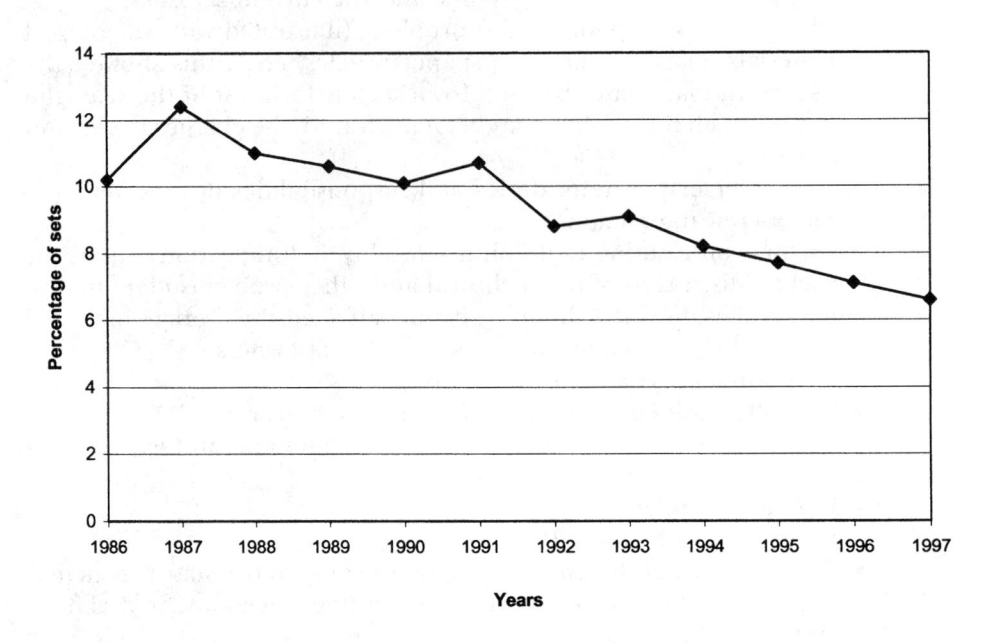

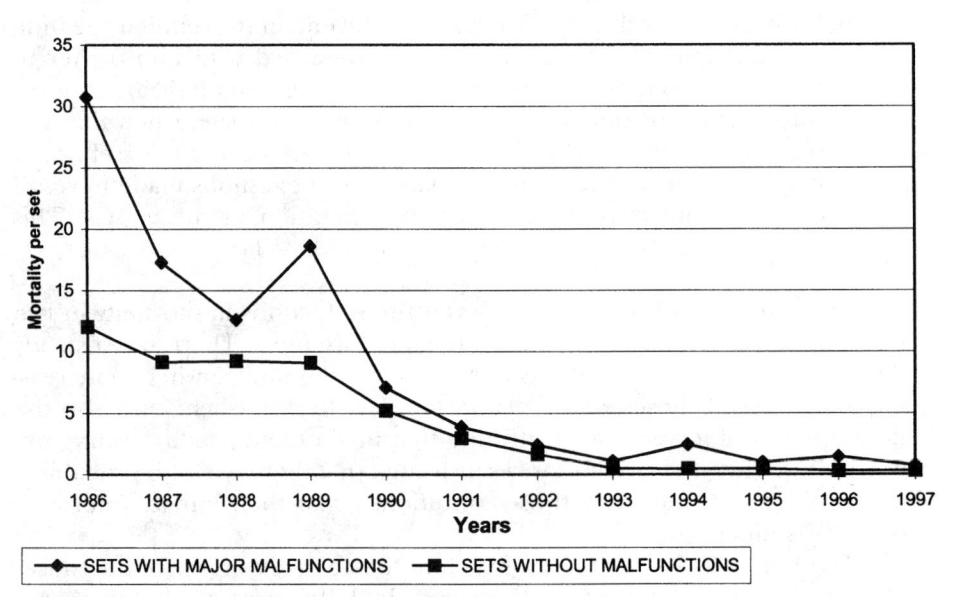

FIG. 1.—Upper panel, percentages of sets with major malfunctions for 1986–1997 period; lower panel, average mortality per set in sets with and without malfunctions for the same period (source: IATTC Observer Database).

Sets with net collapse

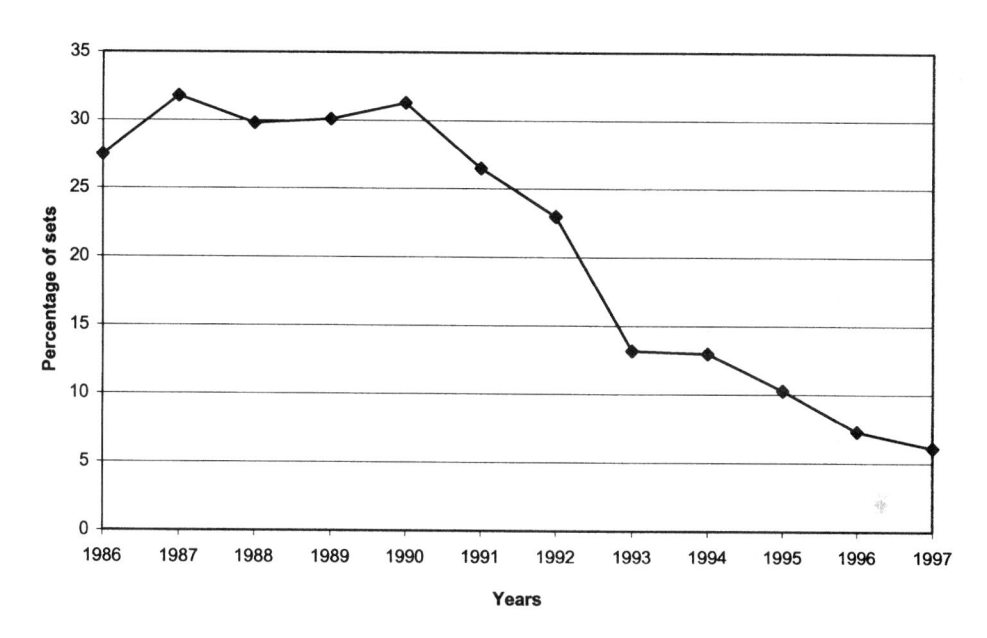

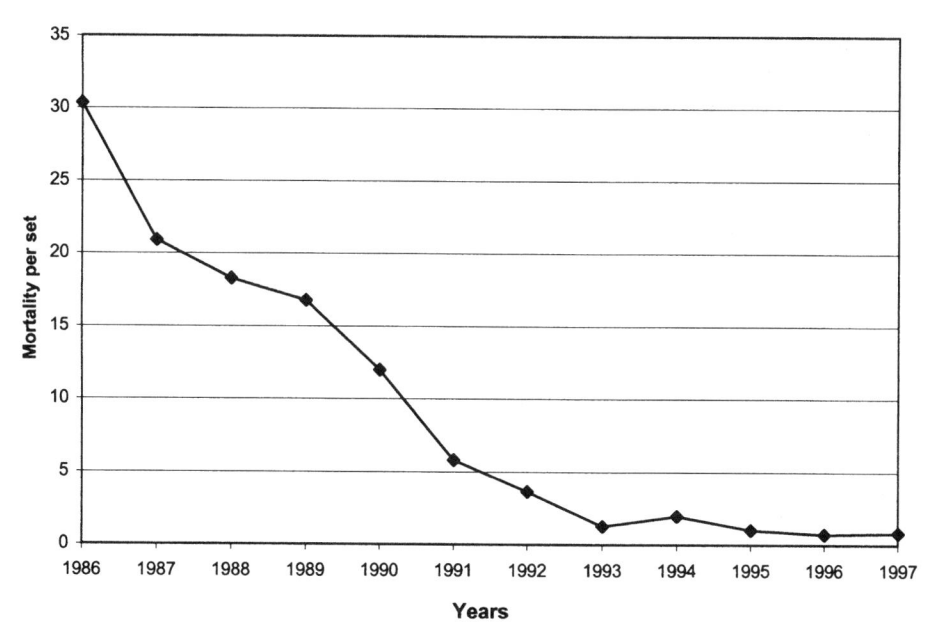

FIG. 2.—Upper panel, percentages of sets with net collapses for 1986–1997 period; lower panel, average mortality per set in sets with net collapses for the same period (source: IATTC Observer Database).

nance can reduce them, and better preparedness can reduce their impact when they occur (lower mortality per set). Net collapses (fig. 2) can be caused by setting in areas with strong subsurface currents and are a major cause of mortality. Avoiding these areas results in a decrease of the proportion of sets with the problem, and better equipment and training of the crews mitigates the impact when net collapse cannot be avoided. The variables shown are only selected examples to illustrate the evolution of the learning process. The combination of improved equipment, training, regulation, and motivation, moves the process in a gradual but inexorable way.

PHASE 5 (1992–PRESENT)

Objective.—Reducing mortality through international management.

During the development of Phases 1 to 4, dolphin mortality declined considerably. However, many believed that an international agreement was needed to consolidate the process in a multilateral way. Beginning in the 1980s, a series of embargoes had been placed by the United States on most tuna-producing nations operating in the EPO, as a result of the provisions in the MMPA. Joseph discusses these embargoes.[14] Most countries affected by the embargoes denounced them as unilateral actions and in some cases questioned the basic motivation behind them, alluding to protectionism.

In addition to the embargoes, another factor was affecting the interactions among countries participating in one way or another in the fishery. In 1990, at the urging of some environmental groups, several of the major canneries in the United States adopted a policy of buying only dolphin-safe tuna. The definition of dolphin-safe used by the U.S. canneries for the purse-seine fleet meant that the total catch from a trip in which dolphins had been intentionally encircled one time or more could not be sold to any cannery participating in this policy. That restricted the market and resulted in lower prices. Some vessels switched to fishing on fish-aggregating devices (FADs) and logs, others moved to other ocean areas, and others continued to fish on dolphins, selling their catches to canneries in countries that had no restrictions regarding tuna caught in association with dolphins.

The verification of this label was carried on by an environmental group. The task was daunting because of the paucity of personnel to carry it out. Tracking the flow of tuna throughout the world was very difficult, and it was criticized as ineffectual. In addition to the doubts about the veracity of the label, there were ecological concerns. The alternative ways of fishing that

14. J. Joseph, "The tuna-dolphin controversy in the eastern Pacific Ocean: biological, economic, and political impacts," *Ocean Development and International Law* 25 (1994): 1–30.

did not involve dolphins had their own problems,[15] and pushing the whole fleet toward them would have resulted in massive increases in by-catches of undersized tuna and many other species, and in a fishery with a much lower productivity. Dolphin-safe as used by the U.S. canneries was not ecologically sound. Many environmental groups that initially had supported the policy realized its full ecological implications and the misleading arbitrariness of the label. Tuna caught in other fisheries (i.e., in gill nets) were automatically classified as dolphin-safe even when it was known that dolphin mortalities were occurring in those fisheries. This group of "ecologically-minded" environmental organizations still wanted to see a continued reduction in dolphin mortality, but did not approve of the approach taken. Other approaches were needed. It had to be a multilateral approach, involving all the countries participating in the fishery. It had to consolidate the gains achieved between 1986 and 1991 (dolphin mortality reduced from over 130,000 to close to 15,000), and to provide incentives to continue the reductions in dolphin mortality. It had to be transparent and developed with all the sectors involved. It had to be ecologically sound.

The main obstacle to develop this international program was that most systems dealing with dolphin mortality could be considered as indirect forms of resource allocation. Based on our experience with individual captain performance, it appeared that the best system would be to divide a total dolphin limit by the number of vessels participating in the program and to assign to each vessel an equal or very similar share. This approach, placing the individual responsibility for each vessel in the hands of its captain and crew, creates a major incentive to improve technology and performance and facilitates the "natural selection" of the fishers. This proposal was adopted. The Dolphin Mortality Limits (DMLs) were the core of the Agreement for the Conservation of Dolphins (La Jolla Agreement), signed in 1992, and continued to be a centerpiece of the subsequent agreements signed. Each vessel receives a DML at the beginning of the year, and if it reaches its limit it must stop fishing on dolphins, and switch to other, usually less desirable, ways of fishing for tuna, or stop altogether.

By placing the fate of each vessel on the shoulders of captain and crew, individual responsibility is emphasized rather than taking actions against flags that do not discriminate between good and bad operators. The unit of management is the unit of fishing. Even though the nations must enforce the provisions of the agreement, the fishers are individually identified and the economic fate of the companies depends on their performance. Vessels are treated fairly, and the allocation issue is minimized. By reducing the total limits every year, the DMLs are almost always reduced too, so the captains and crews must continuously work to keep reducing the mortality to

15. Hall (n. 8 above).

stay within their limits. The incentive for research and innovation is always present. Indirectly, these DMLs may serve to curtail the expansion of the fleet fishing on dolphins by making it economically unfeasible to fish under a very low limit, as would be the case with many new vessels entering the fishery.

An international structure was needed to oversee the new program. It was obvious that all countries with fleets operating in the area, or involved in the fishery in other ways, should participate in a panel reviewing the development of the new program and should vote on its decisions. An innovation of the new agreements was the fact that most sectors interested in the issue (i.e., industries, environmental groups, etc.) had had a large role in structuring them and had participated in their evolution. To retain this participation, a major asset of the program, we proposed to give transparency to the process by allowing both the industry and the environmental community to be present during the consideration of infractions, compliance, and assignment of limits, and so forth.

This panel, the International Review Panel (IRP), has been functioning since 1992 and has served to show to all parties the benefits of openness and cooperation. The IRP reviews all the observer reports that show the possible occurrence of an infraction and determines, after examining the circumstances surrounding the event, whether they believe that an infraction was committed. The information is then passed to the nations to investigate the issue, and if appropriate, to take legal action. To avoid "national feelings" from getting in the way of the judgement of the participants, we proposed that potential infractions be "judged" in a blind way, without identifying the name or the flag of the vessel in question.

Interested governments adopted the "Declaration of Panama" in 1995 that lead to the 1998 Agreement on the International Dolphin Conservation Program (AIDCP). This Agreement came into force in 1999, consolidating the process started by the La Jolla Agreement of 1992. The AIDCP and the 1992 La Jolla Agreement contain similar objectives concerning the reduction of dolphin mortality, but add limits by stock, consideration of by-catches, overcapacity, and other issues.

The AIDCP was adopted as a legally binding instrument, representing a stronger expression of commitment on the part of the signatories than the previous agreement. By continuing the system of individual dolphin mortality limits, it allows the process of "natural selection" of captains started in 1992 to continue. Captains with poor records of dolphin mortality become a liability to the company that employs them, and eventually they are replaced by others who can produce tuna without causing high dolphin mortality. The AIDCP also strengthened provisions regarding dolphin stock management, establishing a per-stock, per-year dolphin mortality cap for each stock of dolphins between 0.2 percent and 0.1 percent of the minimum estimated abundance (N_{min}) for the years up to 2001. The minimum estimated abundance is a very conservative figure, related to the Potential Biological Removal concept used in the US domestic laws. It replaces the best point estimate of abundance of a stock by a much lower figure, determined using a percentile of the distribution around that value that takes into consideration the imperfections and uncertainties of the estimation process. In 2001 and thereafter the per-stock per-year cap will be 0.1 percent N_{min}. If the annual mark cap is exceeded for a particular year, all sets on that stock (including mixed herds) will cease for that year. However, in no event will the total mortality exceed 5,000 dolphins.

The result of the process is that the "ecological costs," in terms of dolphins, of producing tuna in the region is coming down, both in absolute numbers, and in terms of dolphin mortality per ton of tuna produced. Since 1986 dolphin mortality has dropped by 98 percent, from 133,000 to close to 1,700, while the average mortality per set has gone from over 12 dolphins/set, to under 0.2 dolphin/set today.

In recent years, however, other ecological costs have gone up as a result of the U.S. dolphin-safe policy that leads to an increase in the number of sets on FADs with a concomitant increase in by-catches of many species. Sets on floating objects, including FADs and on school sets take smaller tuna (i.e., big-eye tuna caught in the surface fishery going from an average weight of 20 kg in 1991, to around 5 kg in 1997). While it is not an issue with big-eye, sets on floating objects contribute as much to the reduction in yield per recruit of yellowfin as sets on FADs. Within the provisions of the AIDCP is a new definition of dolphin-safe, which is tuna taken in sets in which there is no mortality or serious injury to dolphins. This definition recognized the value of taking yellowfin associated with dolphins while providing strong protection for dolphins.

TECHNOLOGY AND THE FUTURE OF THE FISHERY

Technology created the tuna-dolphin problem. By allowing the capture of massive aggregations of tuna and dolphins, it increased the productivity of the fishery, but also increased the ecological costs of that production. Dolphins and other species are caught in the very large nets, designed only to encircle the tuna. Fishers had to come up with ideas and develop modifications to mitigate those impacts. Are there other ideas for gear and methods that could completely eliminate the mortality, or even the encirclement of dolphins? We will briefly explore some of the options available for further improvement. These and others are discussed by DeMaster.[16]

16. D. P. DeMaster, "Strategic Plan to Develop and Evaluate "Dolphin-safe" Methods of Fishing for Yellowfin Tuna in the Eastern Tropical Pacific," National Marine Fisheries Service, Southwest Fisheries Science Center, Administrative Report LJ-92-16 (1992), 21 pp.

Detecting tuna without using dolphins.—Some experiments have been carried out testing the potential use of laser detection systems[17] to find schools of tuna not associated with dolphins. These have shown some promise, but are still limited by the restricted area swept and the relatively low penetration they achieve, especially away from the perpendicular. There is also considerable interest in the potential use of acoustic detection systems (carried by the vessels, fixed, anchored, etc.). Satellite technology is currently in use that can help locate areas with the appropriate temperatures or other ecological variables adequate for tuna, and thus facilitate the search for areas with fish. Perhaps in the future satellites can help locate tuna schools many meters below the surface.

Separating the tuna from the dolphins.—Once the tuna school is detected with a dolphin herd it may be possible to separate the tuna and dolphins. Either attraction or repulsion systems can be used (noises, visual stimuli, etc.). The bond between the tuna and dolphins is strong, so it is not a trivial matter to generate the "force" needed to counter the attraction. Ideally, these stimuli would create a spatial separation (vertical or horizontal) that could be used to limit the encirclement to tuna, or facilitate the release of the dolphins. Some differences in the sensory systems of the species could help achieve this goal: dolphins have a sophisticated acoustic sensory system, while tuna have a powerful and highly discriminating sense of smell.

Improving the current release systems.—Many ideas have been proposed to improve the efficiency of the backdown procedure and to keep the net from collapsing. They include inflatable corklines, cables to lower the net in the backdown area, systems to increase the rigidity of the corkline (thus keeping the net open), use of acoustic Doppler profilers to detect subsurface currents that cause collapses, use of remotely operated videocameras to monitor the net, and so forth. One of the most imaginative ideas is the Freitas panel. It consists of a set of plastic strips that would replace the webbing in the area of potential dolphin entanglement. These strips would generate a visual barrier that should keep the dolphins and the tuna inside, but if they came into contact with the barrier they would be able to swim through it. Experiments are required to test the effectiveness of these ideas. With dolphin mortality well under control, some of the emphasis in the search for technological improvements has shifted toward the development of more selective gear that could reduce not only dolphin mortality, but also the by-catches of other species, such as billfishes, sharks, rays, mahi-mahi, and of undersized tuna of several species of commercial and/or ecological significance. Some of the modifications discussed are the following:

17. C. W. Oliver, W. A. Armstrong and J. A. Young, "Development of an Airborne LIDAR System to Detect Tunas in the Eastern Tropical Pacific Purse-seine Fishery," U.S. Department of Commerce, NOAA Technical Memo, NOAA-TM-NMFS-SWFSC-204, p. 67, (1994).

The Misund-Beltestad sorting grid.—A rigid grid that is placed on the side of the seine, and that allows the release of small individuals (of a width determined by the spacing of the bars of the grid.) The grid has been tested with mackerel and saithe in the North Sea.[18] A recent pilot experiment carried out by V. Scholey, following a design by R. Olson and M. Scott at an IATTC laboratory tank showed that tuna crossed the grid when crowded, and that most fish survived contact with the bars. The result is encouraging, but more tests at sea are needed. This system could reduce the by-catch of small fish, including juvenile tuna, but it won't solve the problem of the by-catch of large sharks, rays, billfishes, and so forth.

Partitioning of the catch and selection in a sorting system.—If the catch could be transferred in small amounts to a shallow pen where the fishers could sort even at the individual level, it may be possible to address also the by-catches of other large species. The floating pens currently used for bluefin sea ranching, and the technology for their transfer from the net to the pen, could serve as the basis for this development. Basically, a floating pen is towed, or deployed in any other way, adjacent to the net. The net is opened to allow the transfer of part of the catch still alive to the pen. Fishers located on the sides of the floating pen select either the fraction to be retained, or that to be released. The maneuver imagined appears slow and cumbersome, but the creativity of the fishers may make it feasible.

Changes in mesh size or type.—The mesh size is commonly used to select the sizes in the catch. However, the need to avoid dolphin entanglement has resulted in a mesh size of 3.175 cm. This is too small to be an effective selection tool for tuna. It may be necessary to dedicate vessels to different types of sets, and use different gear accordingly. Vessels dedicated to log or school fishing may be required to use a larger mesh size, or even change the mesh type. Purse seines used in the EPO have rhomboidal openings, but in other areas a hexagonal mesh is used in the nets because the openings do not close under tension, as happens in the current nets.

Acoustic systems to aid in decision making.—Frequently, sets are made especially on floating objects that result in large proportions of waste.[19] Besides causing an unwanted ecological impact, the fishers may end up with catches that were not worth their time. Acoustic systems showing the fishers the size and species composition of the school before encirclement may lead to better decisions concerning when to deploy the net and when to pass up the current school so as to look for a more profitable one. They may also help manage a mixed-species fishery, where the catch of one of the target species, but not of the others, may have reached its limit, or where a school may contain the "good size" of one species or undersized individuals of another species.

18. O. A. Misund and A. Beltestad, "Size-selection of mackerel and saithe in purse seines," I.C.E.S., C.M. 1994/B: 28, Ref. G.H., pp. 1–12, (1994).
19. Hall (n. 8 above).

80   Solving the Tuna Dolphin Problem

DISCUSSION AND CONCLUSIONS

The complexity of the issues surrounding the tuna-dolphin problem makes it impossible to address all its angles in a single document. The picture painted in the article is therefore incomplete. Many concepts have been explored or tested on this problem. The fact that the "perfect" solution for the dolphins—cessation of encircling dolphin-associated tuna—had a major ecological cost forced environmental groups to question their objectives. Some pursued an ecosystem approach and rejected the single-species focus. Others, more interested in the charismatic species, would have preferred another solution.

Over the years, the tuna industry, and the fishers themselves, went from an attitude of denial/resistance to one of acceptance of the problem, and now are becoming involved in the solution. When they acknowledged that the problem had to be tackled they took the first step toward survival of the fishery. The nations went from a harsh period of unilateral actions and accusations of protectionism, to a truly multilateral atmosphere, where understanding prevailed at the most difficult times. We hope that this understanding will still be the guiding force when we address other major issues affecting the fishery, such as fleet overcapacity, by-catches of other species, gear competition, and so forth. The sequence of requirements that had to be met to achieve a solution is very long, but these are some of the important ones:

   a. Public pressure originating in environmental groups brought the subject to the limelight, and through political and economic expressions, kept the process moving. Even efforts not supported by science, such as the old dolphin-safe policy, may have helped generate the alliance of forces that produced the international agreements.
   b. The participation of the fishers in the process was crucial. They contributed their experience, their ideas, and especially their motivation. They were responsible for most of the major innovations. The training seminars circulated information on causes of mortality and solutions, shared experiences, and so forth.
   c. The transition from global mortality limits used by some nations to individual vessel limits for all vessels of the international fleet was very important. By putting their fate in their own hands, the fishers were given a chance to defend their interests and the survival of their fishery. Individual responsibility resulted in the evolution, by "natural selection," of a group of more competent fishers in terms of producing tuna with lower impacts on dolphins. Those less intelligent in making decisions, less skilled in performing the dolphin-saving maneuvers, or less motivated, were replaced.
   d. Public education is a requisite for public pressure to be exerted in

Living Resources 81

the right direction. Over the years the public saw that the solutions were not as simple as brief slogans made them look. Their understanding of the trade-offs involved allowed the process to move forward.

e. Much of the scientific effort was directed toward identifying solutions, rather than just quantifying the problem. When there was a commitment to reduce mortality the solutions were available. The scientists worked closely with the fishers; the regulations and the reasons behind them were explained to the fishers.

f. The solutions proposed were enacted in a gradual way. The fishers were not asked to produce miracles overnight. Instead, our experience with the fishers suggested that it would be more effective if the international program "moved the bar" in small increments, trying to keep it within reach. It was important that the fishers perceived that they could achieve the expected mortality reduction goals.

g. The scientists were involved in forging a solution, contributing not only to research, but also to other aspects such as management, education, and so forth. Because the scientists are more detached from the interests at stake, they could assist in the search for fair solutions.

h. Many individuals were important for the process. In general, the leadership of the different organizations evolved and matured intellectually. More realistic attitudes and positions were taken. Social and economic impacts were not ignored. Extreme demands were replaced by reasonable ones. Consolidating the process was given high priority.

i. The interactions among nations, environmental groups, and the fishing industry went from antagonism to open-minded discussion. Even though many differences still exist, the perception of the "other side" as being unable to understand or insensitive, has changed. When all the dust settled, it emerged clearly that the interests of an intelligent fisher and an intelligent environmentalist were not too far apart.

APPENDIX 1: CHRONOLOGY OF IMPORTANT EVENTS THAT TOOK PLACE IN THE EASTERN PACIFIC OCEAN TUNA FISHERY

Most of the events in this appendix relate to the United States. Because the United States was the most important market in the Americas, events there, particularly those related to trade, have had a major influence on the region. After the U.S. portion of the international fleet decreased, however, international conservation efforts increased and even events within the United States occurred largely in the context of international agreements and management efforts.

1949. The Inter-American Tropical Tuna Commission (IATTC)[20] was established to study and recommend conservation measures for the tuna resources of the eastern Pacific Ocean (EPO). More specifically, its objective was to maintain tuna stocks at levels capable of producing maximum yields on a sustainable basis. Over time, the IATTC's role in the EPO has expanded.

1960–1970. The U.S. fleet dominated the fishery and was responsible for the majority of dolphin deaths.[21]

1968. The first data regarding the tuna-dolphin problem were collected by a scientist from the Department of Commerce's National Marine Fisheries Services (NMFS) on board a U.S. purse seiner.[22]

1972. The U.S. Congress passed the Marine Mammal Protection Act (MMPA)[23] as a result of the public awareness regarding, among other things, high dolphin mortality that occurred during fishing operations for tuna in the EPO.

The Act went on to allow certain exceptions to the moratorium. First, permits could be issued for taking and importing, for purposes of scientific research, public display, photography, or enhancing the survival of a species or stock.[24] Second, incidental taking of marine mammals in the course of commercial fishing operations was permitted.[25] Third, taking of marine mammals by "any Indian, Aleut, or Eskimo who resides in Alaska and who dwells on the coast of the North Pacific Ocean or the Arctic Ocean,"[26] as long as it is for subsistence purposes, or done for purposes of creating and selling authentic native articles of handicrafts and clothing, was permitted. These exceptions would not apply if the species in question were considered to be endangered. With respect to commercial fishing operations, the Act states, "it shall be the immediate goal that the incidental kill or incidental serious injury of marine mammals permitted in the course of commercial fishing operations be reduced to insignificant levels approaching zero mortality and serious injury rate."[27] Furthermore, the Secretary of the Treasury

---

20. Convention for the Establishment of an Inter-American Tropical Tuna Commission (IATTC), Washington, 1949, 1 UST 230, TIAS 2044.
21. See Minutes of the 52nd Meeting of the IATTC, La Jolla, California, October 1993, Dr. Joseph's Comments.
22. See Joseph (n. 14 above).
23. Marine Mammal Protection Act of 1972 (MMPA), 16 U.S.C. 1361–1407, Public Law 92-522, 1972.
24. MMPA, §1371 Sec 101. (a) (1).
25. Ibid. (a) (2).
26. Ibid. (b).
27. Ibid. (a) (2).

Living Resources 83

calls for a ban on "the importation of commercial fish or products from fish that have been caught with commercial fishing technology which results in the incidental kill or incidental serious injury of ocean mammals in excess of United States standards."[28] Moreover, the Act states that it is unlawful to import into the United States any marine mammal taken in violation of Title I of the Act (Title I—Conservation and Protection of Marine Mammals); any marine mammal product if any fish (fresh or frozen) was caught in a manner proscribed by the Secretary, even if the marine mammal was caught incidental to the catching of the fish.[29]

1974. The U.S. government instituted an observer program for its vessels, and the NMFS was put in charge of it.

1976. The Magnuson Fisheries Conservation and Management Act was enacted.[30] It established a 200-mile fisheries conservation zone, but excluded tuna from national management jurisdiction, given the fact that they are considered to be highly migratory species. The Act went on to amend the MMPA.[31] At the 33rd meeting of the IATTC, it was noted that porpoise populations are highly migratory and interact with international tuna resources. Hence, anything affecting porpoise stocks also affects the potential yield from tuna stocks.

1979. The IATTC initiated an international observer program. The program was intended to cover trips made by the international purse-seine fleet setting on dolphins in the EPO. The IATTC began the program so it could obtain reliable data on dolphin mortality.

1980. The United States imposed an embargo on Mexican tuna imports as a reaction to a seizure of a U.S. vessel by the Mexican government within Mexico's Exclusive Economic Zone (EEZ).

1981. Amendments to the MMPA set forth the conditions under which permits could be granted for the incidental take of marine mammals in the course of commercial fishing.[32] The United States imposed a second embargo on Mexico under Section 101 of the MMPA.

1984. The MMPA was amended once again. Congress reauthorized the general permit for an indefinite period, and continued with the mortality quota

    28. Ibid.
    29. Ibid. §1372 Sec 102 (c).
    30. The Magnuson Fishery Conservation and Management Act of 1976 (MFCMA), 16 U.S.C. 1801 *et seq.*, Public Law 94-265.
    31. MMPA of 1972 as amended by Public Law 94-265 (1976).
    32. MMPA of 1972 as amended by Public Law 97-58 (1981).

of 20,500 dolphins. In addition, individual quotas were established for eastern spinner and coastal spotted dolphins. Furthermore, the amendments require that foreign nations that wanted to export yellowfin tuna or yellowfin tuna products to the United States would have to implement a program comparable to that of the United States to address the issue of dolphin mortality in the EPO. In addition, a foreign nation's dolphin mortality rate would have to be comparable to that of the U.S. fleet. If nations did not comply with the U.S. standards, the 1984 amendments would allow import restrictions to be placed against those nations not in compliance with such standards.[33]

1986. Mexico established a voluntary export limit of 20,000 tons, which led to a withdrawal of both U.S. embargoes. Subsequently, Mexico joined the IATTC observer program.

1988. Congress amended the MMPA yet again. The amendments specified that backdown procedures for dolphin sets must be completed no later than 30 minutes after sundown. Other restrictions required placement of an observer on every fishing trip made by U.S. vessels with carrying capacity greater than 400 short tons and prohibited the use of explosives while setting on tuna.

Furthermore, a 5-year interim exemption was set for commercial fisheries from the MMPA's general prohibitions against taking marine mammals. During that time commercial fishermen would not be penalized by the MMPA for reported incidental taking of marine mammals. However, the MMPA would still penalize unreported or intentional taking of marine mammals during fishing operations.[34]

With regard to the U.S. standards required from foreign nations that wish to export yellowfin tuna or yellowfin tuna products into the United States, the amendments provide specific guidelines for what would be considered an "acceptable foreign program" and "comparable mortality rates." For a foreign nation's program governing the incidental taking of marine mammals to be considered comparable to that of the United States, the program would have to prohibit the "encircling of single species schools of certain dolphin stocks (particularly the eastern spinner dolphin) and making "sundown sets,"[35] have an observer coverage equal to that of the United States, and have its tuna fishing operations monitored by the IATTC.[36] With respect to the requirement that a foreign nation's fleet maintain a comparable average rate of incidental taking of marine mammals with

33. MMPA of 1972 as amended by Public Law 98-364 (1984).
34. MMPA of 1972 as amended by Public Law 100-711 (1988).
35. Ibid. §1371 Sec 101. (a) (2) (B) (I).
36. Ibid. (IV).

that of the United States, the amendments required that the average rate of incidental taking by foreign fleets be no more than 2 times that of the U.S. vessels for the same period ending in 1989, and no more than 1.25 times for the same period ending in 1990 and thereafter.[37] The Act further provided that the average rate for the incidental taking of marine mammals never exceed the average rate of the preceding year and, moreover, that the mortality rate be reduced by significant amounts during the successive years.

The amendments also required that, of the total number of marine mammals incidentally taken by vessels in a year, the total number of eastern spinner dolphins and coastal spotted dolphins taken incidentally by a foreign fleet must not exceed 15 percent and 2 percent, respectively.

Any nation failing to meet these requirements would be subject to a primary embargo within 90 days, and any nation that traded yellowfin tuna or yellowfin tuna products with the embargoed nation would be subject to a secondary embargo, that is, if the intermediary nation trading yellowfin tuna or yellowfin tuna products with the embargoed nation did not place a ban on such nation's products within 60 days of the effective day of the primary embargo. A secondary embargo might not be imposed if the intermediary nation acted to prohibit the importation of yellowfin tuna or tuna products that were subject to a primary embargo by the United States.[38] It should be stated that a secondary embargo applied to all yellowfin tuna from the intermediary nation, regardless of where or how the tuna was harvested.[39] Furthermore, failure to adopt a parallel import ban within 6 months required certification of the intermediary nation under the Fishermen's Protective Act of 1967, and might result in additional import restrictions.[40] Consequently, the United States imposed secondary embargoes on all yellowfin tuna and tuna products on approximately 20 intermediary nations, some were subsequently lifted.[41]

1990. The U.S. Congress enacted the Dolphin Protection Consumer Information Act.[42] This Act allowed a "dolphin safe" label to be used on cans of tuna not caught in association with dolphins. For tuna caught in the EPO to qualify as dolphin safe, the tuna had to be caught by vessels under 400 short tons, or accompanied by a certificate from a qualified observer that

37. Ibid (II).
38. Ibid. (a) (2) (C).
39. Earth Island Institute vs. Mosbacher, 746 F. Supp. 964, 975 (N.D. Cal. 1990), *aff'd*, 929 F. 2d. 1449 (9th Cir. 1991).
40. MMPA (n. 34 above), §1371 Sec 101(a)(2)(D).
41. Marine Mammal Commission, Annual Report to Congress (1992).
42. The Dolphin Consumer Protection Information Act, Public Law 101-627 (1990).

no dolphin sets were made during the entire trip during which the tuna was caught. The Act amended the MMPA Act of 1972.[43]

U.S. Courts ordered embargoes on yellowfin tuna and yellowfin tuna products from Ecuador, Mexico, Panama, Vanuatu, and Venezuela. This was also the result of the lawsuit filed by Earth Island Institute against the U.S. Department of Commerce,[44] arguing that as of January 1990 only tuna from countries whose dolphin kill rate was comparable to that of the United States, and whose take of eastern spinner and coastal spotted dolphins during 1989 did not exceed the established quotas, could be imported. The Court ruled in favor of Earth Island Institute, and further ruled that the MMPA does not require that the comparison between foreign and U.S. overall dolphin mortality rates be based upon data for an entire calendar year, but only for "the same period." However, with respect to the mortality rates of eastern spinner and coastal spotted dolphins, the comparison between foreign and U.S. mortality rates had to be based upon data for an entire calendar year.

Within months, Ecuador's embargo was lifted as a result of implementing legislation by which it banned its nationals from setting on dolphins. The same thing happened to Panama for 1991, but the embargo was reinstated for 1992 because a Panamanian vessel intentionally set on dolphins. Colombia was also embargoed in 1992 because its observer coverage did not satisfy the 75 percent observer coverage required by the United States.

1992. The NMFS redefined the concept of intermediary nations. The term "intermediary nation" meant a nation that exports yellowfin tuna or yellowfin tuna products to the United States and that imports tuna or yellowfin tuna products that are subject to a direct ban on importation into the United States. This new definition amended the MMPA.[45] However, the concept of "intermediary nations" did not include nations that certified and provided reasonable proof that they had not, within the preceding 6 months, exported any yellowfin tuna or tuna products subject to a direct ban on importation into the United States.[46] Under this new more favorable definition of "intermediary nation" more secondary embargoes were lifted.

The U.S. Congress passed the International Dolphin Conservation Act.[47] The Act established a 5-year moratorium on the encirclement of dol-

---

43. Marine Mammal Protection Act of 1972 as amended by Public Law 101, 627 (1990).
44. MMPA (n. 40 above).
45. MMPA §1362 (5).
46. (n. 38 above).
47. The amendments enacted by the International Dolphin Conservation Act of 1992, codified at 16 U.S.C. §§1411–1418.

phins with purse-seine nets. The moratorium would have started on March 1994, contingent upon another major fishing country agreeing to it. During that time any country that committed to the moratorium and whose tuna or tuna products had been embargoed by the United States under Section 101 of the MMPA would be free to export such products to the United States. However, if the countries for which the embargoes had been lifted failed to further enter into a binding agreement regarding the aforementioned moratorium the embargoes would be reinstated.

The Act was an intent on the U.S. side to resolve the tuna-dolphin conflict, especially with those nations on which the United States had implemented tuna embargoes, but given the fact that no country agreed to the moratorium to begin with, little, if any, improvement on the tuna-dolphin conflict was reached. As a result, the American Tunaboat Association's general permit of 20,500 annual mortality quota continued in effect throughout 1999. However, the incidental mortality rate for each successive year had to be reduced by statistically significant amounts each year, as previously mentioned.

In 1992, the Agreement for the Conservation of Dolphins (the "La Jolla Agreement"), which established the International Dolphin Conservation Program under the auspices of the IATTC, was adopted.

Separately, at an Intergovernmental Meeting of the IATTC[48] the parties agreed to adopt a resolution in which they reaffirm their commitment to adopt an International Dolphin Conservation Program, under the exact guidelines as the La Jolla Agreement with the difference that such resolution would apply to all members of the IATTC, and the La Jolla Agreement would apply to its signing members.

1994. The MMPA was once again amended. The most significant change was the establishment of a new regime to govern the taking of marine mammals incidental to commercial fishing, replacing the 5-year interim exemption in place since 1988.[49]

1995. At the 29th Intergovernmental Meeting of the IATTC six parties to the La Jolla Agreement (Colombia, Costa Rica, Ecuador, Mexico, Panama, and Venezuela), taking into consideration that by that time most of the nations that fished for tuna in the EPO were embargoed by the United States, issued a joint statement urging the United States to lift the primary and secondary tuna embargoes that were currently in effect, and also to redefine

---

48. "Minutes of the 50th Meeting of the Inter-American Tropical Tuna Commission," La Jolla, California, June 1992, Appendix 7.

49. Marine Mammal Protection Act of 1972 as amended by Public Law 103-238 (1994).

the concept of "dolphin safe" to include tuna products caught in accordance with the La Jolla Agreement.[50]

At the 30th Intergovernmental Meeting the Declaration of Panama was approved.[51] In the Declaration the governments reaffirmed their commitment to the La Jolla Agreement and announced their intention to formalize it as a legally binding instrument. It was agreed that such a legally binding instrument would establish per-stock, per-year annual mortality limits, would further address the problem of by-catch as a whole rather than just dolphins, and would provide strengthened enforcement measures through national and international channels. Also there would be commitment on the part of the governments to the conservation of ecosystems and the sustainable use of living marine resources related to the tuna fishery within the EPO. However, the Declaration clearly stated that the adoption of a legally binding instrument would be contingent upon the United States making changes to certain provisions established under its laws (lifting primary and secondary embargoes, redefining the term "dolphin safe," and opening the U.S. market to tuna caught in the EPO in compliance with the La Jolla Agreement). The Declaration was also endorsed by the Center for Marine Conservation, the Environmental Defense Fund, Greenpeace International, the National Wildlife Federation, and the World Wildlife Fund.

1997. The International Dolphin Conservation Program Act was enacted by the U.S. Congress.[52] The Act starts by amending Section 2 of the MMPA (Findings and Declaration of Policy) by clearly stating that the purposes of the MMPA are, among other things, to place into effect the Declaration of Panama, including the establishment of the International Dolphin Conservation Program,[53] and to eliminate the ban on imports of tuna from those nations that are in compliance with the International Dolphin Conservation Program.[54] The Act defines the term "International Dolphin Conservation Program" as the international program established by the La Jolla Agreement, and also defines the term "Declaration of Panama," as the declaration, which was signed in Panama on 4 October 1995.[55] The Act amends the Dolphin Protection Consumer Information Act by redefining the term dolphin safe with respect to tuna fished in the EPO:

    50. "Minutes of the 29th Intergovernmental Meeting on the Conservation of Tunas and Dolphins in the Eastern Pacific Ocean," of the IATTC, La Jolla California, June 1995, Appendix 5.
    51. "Minutes of the 30th Intergovernmental Meeting on the Conservation of Tunas and Dolphins in the Eastern Pacific Ocean," of the IATTC, Panama City, Panama, October 1995, Appendix 4.
    52. International Dolphin Conservation Program Act, Public Law 105-42 (1997).
    53. Ibid. Sec. 2 (a) (1).
    54. Ibid. (3).
    55. Ibid. Sec. 3 (28) and (29).

"a tuna product that contains tuna harvested in the eastern tropical Pacific Ocean by a vessel using purse seine nets is dolphin safe if
    (a) the vessel is of a type and size that the Secretary has determined, . . . is not capable of deploying its purse seine nets on or to encircle dolphins; or
    (b) the product is accompanied by a written statement executed by the captain;
    (ii) (I) the Secretary or a designee; (II) a representative of the IATTC; or . . .
which states that there was an observer approved by the International Dolphin Conservation Program on board the vessel during the entire trip and that such observer provided the certification required."[56]

The certification of the captain and observer can be in two different ways. It can certify that the tuna is dolphin safe because no dolphins were killed or seriously injured during the sets in which the tuna was caught, or it can certify that the tuna is dolphin safe for a certain trip because no tuna was caught during the trip that was harvested by use of a purse-seine net intentionally deployed on or used to encircle dolphins, and that no dolphins were killed or seriously injured during the sets in which the tuna was caught.[57] Certification (in one way or the other) and enactment of the new definition for dolphin safe, will depend on the initial and final findings of the study required by Section 304 of the MMPA as amended,[58] by which the Secretary of Commerce, in consultation with the Marine Mammal Commission and the IATTC, must conduct a study on the effect of intentional encirclement of dolphins and dolphin stocks incidentally taken in the course of purse-seine fishing for yellowfin tuna in the EPO. The study was to commence 1 October 1997, and an initial finding was to be made by the Secretary of Commerce during March 1999. A final finding is to be made between 1 July 2001 and 31 December 2002.

The Act further amends the Marine Mammal Protection Act by eliminating Title III—Global Moratorium to Prohibit Certain Tuna Harvesting Practices, and replacing it with Title III—International Dolphin Conservation Program.[59] The Act also states that the Secretary of State, in consultation with the Secretary of Commerce, shall seek to secure a binding international agreement to establish an International Dolphin Conservation Program with certain guidelines. This constitutes a contingency for the International Dolphin Conservation Program Act to go into effect, as does the requirement that there be a certification by the Secretary of Commerce regarding the

---

56. Ibid. Sec. 5 (d) (2).
57. Ibid. Sec. 5 (h).
58. Ibid.
59. Ibid. Sec. 6.

availability of sufficient funds to complete the first year of the aforementioned study, and that the study has commenced.

1998. The Agreement on the International Dolphin Conservation Program was signed. The Agreement limits total incidental dolphin mortality in the purse-seine tuna fishery to no more than 5,000 animals annually, and it commits to establish, among other things, incentives for vessel captains to reduce dolphin mortality, a system of technical training and certification for fishing captains and crews, research to improve gear, equipment, and fishing techniques, a system for the assignment per year-per stock DMLs, a system for the tracking and verification of tuna.[60] The Agreement contains an important provision concerning the implementation of the Agreement at national levels, for which it requires each party to adopt the necessary measures (laws and regulations) to ensure the implementation and compliance.[61] It also requires that the parties meet periodically for purposes of monitoring implementation of the Agreement.[62] It calls for the establishment of a Scientific Advisory Board and, an International Review Panel under the guidelines of the La Jolla Agreement, National Scientific Advisory Committees, an On-Board Observer Program, also under the guidelines of the La Jolla Agreement, that covers 100 percent of the fishing trips made by vessels of greater than 400 short tons of carrying capacity.[63] The Agreement emphasizes the importance of "transparency" within the Agreement, and calls for public participation by means of intergovernmental organizations and nongovernmental organizations taking part in meetings as observers.[64] With respect to settlement of disputes, the Agreement calls for consultation among the parties as the best way to resolve a dispute, but if this method does not prove successful the parties shall settle the dispute through any peaceful means in accordance with international law.[65] Furthermore, the Agreement also recognizes the sovereign rights of states in accordance with international law.[66] The Agreement was open for signature at Washington from 21 May 1998 until 14 May 1999,[67] but shall remain open to accession by any state or regional economic integration organization that meets the requirements.[68] The Agreement shall go into force once four nations have ratified it.[69] In 1998 the United States and Panama ratified the Agreement.

60. Ibid. Art. V, 1, 2, 3.
61. Ibid. Art. VII.
62. Ibid. Art. VIII.
63. Ibid. Art. X, XI, XII, and XIII respectively.
64. Ibid. Art. XVII.
65. Ibid. Art. XX.
66. Ibid. Art. XXI.
67. Ibid. Art. XXIV.
68. Ibid. Art. XXVI.
69. Ibid. Art. XXVII.

1999. The International Agreement on the IDCP became effective on 15 February 1999 with the ratification of Ecuador and Mexico. In March, according to Section 304 of the MMPA as amended by the IDCPA of 1997, the National Marine Fisheries Service reported to Congress the initial finding regarding whether the intentional deployment on or encirclement of dolphins with purse seine nets is having a significant adverse impact on any depleted dolphin stock in the Eastern Tropical Pacific Ocean. NMFS concluded there is not enough data to confirm that encircling dolphins to catch tuna causes a significant adverse impact on three depleted dolphin stocks of the eight major dolphin stocks found in the ETP.[70] The NMFS will continue the study until completing the final finding by the end of 2002.

On 29 April, the U.S. Department of Commerce according to the initial finding of the NMFS study adopted a new dolphin-safe label standard for tuna caught by the encirclement of dolphins in the Eastern Tropical Pacific Ocean. The new dolphin-safe standards under the IDCPA allow processors and canners to use the dolphin-safe label on tuna caught in the presence of dolphins, provided that no dolphins were killed or seriously injured.

2000. Since 1 January the Agreement on the International Dolphin Conservation Program substituted the La Jolla Agreement. Early in 2000 Venezuela initiated its own observer program called Programa Nacional de Observadores de Venezuela (PNOV), and at the end of 2000, Ecuador started its own observer program, Programa Nacional de Observadores Pesqueros de Ecuador (PROBECUADOR). In April, the United States Department of Commerce lifted the 9-year embargo on Mexican tuna. In the same month, several environmental groups successfully asked a federal court to stop the application of the new dolphin-safe label standards arguing that the Department of Commerce had not made sufficient progress in conducting certain scientific research required by the Congress in 1997. The U.S. Administration appealed the Court's verdict and a decision is awaited at the time of writing. In August, the Assistant Administrator for Fisheries removed the embargo as intermediary nation to Costa Rica, which had been embargoed since 1992.

APPENDIX 2: RECOMMENDED STEPS TO PREVENT UNNECESSARY DOLPHIN MORTALITY DURING PURSE-SEINING OPERATIONS

1) Maintenance of dolphin safety gear:

- Repair all damage to the dolphin safety panel frequently.
- Keep spaces between the fine mesh, webbing, and the corks and handhold spaces at a minimum.

70. Report to Congress prepared by SFSC/NMFS/NOAA/U.S. Department of Commerce of March 25, 1999.

- Maintain inflatable raft, speedboats (equipped with towing bridles), vessel hydraulics, bow thruster (if fitted), and skiff in good repair.

2) Dolphin safety and release procedures:

- Set the net with the wind blowing toward the port side of the boat.
- Use the speedboats to tow on the net in areas of potential collapse.
- Have a speedboat circling constantly outside the net just off the stern during net roll.
- Carry out the backdown maneuver in all sets in which dolphins are captured, even if there are only a few.
- Tow the bow *ortza* away from the vessel to make on opening through which the dolphins can escape if a malfunction makes backdown impossible.
- Use the raft as a platform for dolphin herding and rescue during and after backdown. Man the raft with an athletic, conscientious crewmember, equipped with a mask and snorkel.
- Keep the backdown channel open as long as possible during backdown by using a suitable turning radius, using the bow thruster, and towing with speedboats on the sides of the channel if necessary.
- Release as many live animals as possible during backdown.
- Hand-release any remaining live animals after backdown from the raft and/or speedboats, or with the aid of swimmers if it is safe for crewmembers to enter the water.
- Do not roll live animals toward or through the power block during net roll. Lower them to the deck and release them over the side.
- Do not sack up live animals with the intention of brailing them aboard and releasing them from the deck. Make every reasonable effort to release them from the backdown channel during or shortly after backdown.
- Use the speedboats to tow on the corkline to clear canopies that are trapping dolphins.

3) Sets and conditions to avoid:

- Setting on pure schools of common dolphins (whitebellies).
- Sundown sets. If backdown must be carried out in darkness, illuminate the channel with a high-intensity sodium-vapor floodlight.
- Setting in areas of strong currents or during rough weather.

Living Resources

Ecolabelling in the Fisheries Sector[†]

Christine LeBlanc
Osgoode Hall Law School, York University, Toronto

INTRODUCTION

The Marine Stewardship Council's (MSC) unique global certification program is the first ecolabelling program specifically geared towards using market forces to promote sustainably harvested products in the fisheries sector. The MSC claim is that the fish product is harvested from "a well-managed and sustainable fishery."[1] Though similar "sustainably harvested" ecolabels are known in the forestry sector, no such labelling scheme, until now, has been made applicable to marine resources.[2]

The MSC certified its first fishery in March 2000.[3] As of October 2001, the Marine Stewardship Council has certified six fisheries as being sustainable: the Western Australia Rock Lobster, the Thames-Blackwater Driftnet Herring, the Alaska Salmon, the New Zealand Hoki, the Burry Inlet Cockle, and England's South West Mackerel Handline fisheries. Several other fisheries are presently undergoing a full certification assessment (Alaska Pollock, British Columbia Salmon, South Georgia Toothfish, Banco Chinchorro Lobster, Mexican Baja California Spiny Lobster) South African Hake

---

[†] EDITORS' NOTE.—This article was the runner-up in the 2001 *Ocean Yearbook* Student Paper Competition. An earlier version of this paper was presented at the Conference on Communication and the Environment at the University of Cincinnati in July 2001. The author would like to thank Ronald LeBlanc, Rosemonde LeBlanc and Michelle Poirier for all their encouragement and support, as well as their generous efforts in reviewing this text. The author would also like to thank Professor J. Stepan Wood for his comments on an earlier draft and staff at the *Ocean Yearbook* for their assistance. Any shortcomings or errors in this article are attributable solely to the author.

1. (March 2001). *MSC Certification Methodology* (Issue 3). www.msc.org.
2. Other fisheries-related labelling programs are dolphin-friendly and organic seafood labels. The Marine Aquarium Council has suggested the creation of a label to ensure that ornamental fish taken from coral reefs are adequately protected. See C. Deere, *Eco-labelling and Sustainable Fisheries* (IUCN: Washington, D.C. and FAO: Rome, 1999), pp. 10–12.
3. World Wildlife Federation, Press Release, "WWF welcomes fisheries ecolabel," 3 March 2000. Accessed 9 September 2002 on the World Wide Web: http://www.panda.org/news/press/news.cfm?id=1872.

© 2003 by the University of Chicago. All rights reserved.
0-226-06620-7/03/0017-0006$01.00 *Ocean Yearbook* 17:93–141

94   Living Resources

and the Loch Torridor Nephrop. A total of nearly thirty fisheries are at some stage of certification.[4]

Although the South West Mackerel Handline Fishery met the MSC standard for certification in September 2001, the public report was not available at the time of writing and will not be discussed in this article.[5] Another aspect that will not be discussed in this article is the MSC chain of custody certification. The chain of custody certificate essentially affirms that a given product can carry the MSC logo because it has been independently verified that it originated from a certified fishery. For example, one could buy frozen herring fillets carrying the MSC logo if it could be proven that the fillets originate from a MSC-certified fishery, and that the fillets have not been mixed with fish that do not originate from the MSC fishery.

The objective of this article is to examine ecolabelling for sustainably harvested fish products—yet the scope of this article is limited to studying the new MSC ecolabelling initiative. I will not discuss the activities of other certification organizations that may soon operate in the fisheries sector, such as the Global Aquaculture Alliance, which will eventually grant ecolabels to responsibly harvested products in the aquaculture sector,[6] nor the Nordic Technical Working Group on Fisheries Ecolabelling Criteria that has developed criteria for fisheries specifically located in the Northeast Atlantic.[7]

I will argue that although the MSC Principles and Criteria for Sustainable Fishing embody the concept of a sustainable fishery, the way in which these Principles and Criteria are applied to real-life scenarios sometimes illustrate a nonnegligible departure from the set of guidelines that the MSC uses to define sustainability in the fisheries sector. By this I do not intend to second guess the findings of scientific experts, but rather, to center my argument on the fundamental elements of trust and transparency that are required from any label or claim that relies on consumers' confidence to effectively influence market forces and encourage consumers to purchase "green" products.

Part I will be dedicated to discussing the concept of ecolabelling generally. In Part II, I will explain how the MSC is structured, what it aims to accomplish, and how a fishery becomes certified through the MSC. In Part

4. (October 2001). Fisheries news. *Fish 4 Thought—The MSC Quarterly Newsletter* (Issue 1). Accessed 9 September 2002 on the World Wide Web: http://www.msc.org.

5. MSC, News Release, "South West Mackerel Handliners Get MSC Sustainability Seal of Approval," 4 September 2001. Accessed 9 September 2002 on the World Wide Web: http://www.msc.org

6. Information was accessed 9 September 2002 at the Global Aquaculture Alliance Web site on the World Wide Web: http://www.gaalliance.org

7. See the Technical Working Group on Fisheries Ecolabelling Criteria. Accessed 9 September 2002 on the World Wide Web: http://www.norden.org/fisk/sk/criteria.asp.

III, I will analyze how the standards set out by the MSC have been applied to the fisheries certified thus far.

PART I—ECOLABELLING

Ecolabels Defined

Although there is no commonly accepted definition for the term "ecolabel," it is a label that tells the consumer something about the environmental characteristics of the product. Many would claim that it is also a label "that reflects the results of life-cycle analyses,"[8] meaning that it takes into account a product's impact on the environment from the beginning of its existence to its final disposal. The approach taken by the International Organization for Standardization (ISO) in recent years is useful: type I labels (referred to as ecolabels) are viewed as "voluntary schemes that utilize preset criteria established by third parties to evaluate a product's environmental characteristics throughout its life-cycle."[9] Although some authors would classify "sustainably harvested" as a single-issue label not necessarily reflecting a complete life-cycle analysis, notable organizations that are in the business of certifying sustainably harvested products such as the Marine Stewardship Council and the Forest Stewardship Council have appropriated for themselves the use of the term "ecolabel." For the purposes of this paper, the term ecolabel will be used in the sense used by those organizations who award these sustainably harvested ecolabels: it is a label that informs the consumer about a specific environmental characteristic and reveals how the product may be considered superior to other competitive products.

Types of Ecolabels

It is important to note that there are as many ways to classify labelling schemes as there are authors writing about them. I have classified ecolabels as mandatory, voluntary self-claims, or voluntary third-party labels. Labelling

---

8. A. E. Appleton, *Environmental Labelling Programmes: International Trade Law Implications* (London: Kluwer Law International, 1997), p. 2. For information on the development of the concepts of ecolabels and environmental labelling generally, see pp. 1–11.

9. Ibid., p. 4, referring to *Environmental Labelling—Guiding Principles, Practices and Criteria for Multiple Criteria-Based Practitioner Programmes—Guide for Certification Procedures, 3.1,* Working Draft—ISO/WD 14024.2, ISO Document ISO/TC 207/SC 3/WG1 N 43, (Sept. 1995).

schemes can be considered either "single attribute" or "multi-criteria,"[10] both of which can be mandatory or voluntary.

Single attribute labels are common and examples are numerous: "recyclable," "contains recycled materials," "biodegradable," "ozone friendly," and "dolphin friendly." These labels focus on one positive environmental attribute of the product. Multi-criteria labelling schemes generally rely on a life-cycle analysis. Two major types of labelling schemes provide an example of a multi-criteria scheme. Firstly, the environmental report card displays information and allows the consumer to compare products and make his or her own purchasing decisions without being prompted in any direct way, for example, the kilograms of carbon dioxide produced for a given product.[11] Secondly, the widespread "seal of approval" type of label is based on expert opinion that attempts to persuade the consumer to buy a specific product, for example, Germany's Blue Angel Programme.[12]

Mandatory Ecolabels

Mandatory labels are government-mandated programs. These types of labels may be required where a product contains a harmful substance or was manufactured using a harmful process, such as "Contains CFCs." This type of label may also be required to disclose specific information about the product. The EnerGuide program in Canada, which displays energy efficient ratings such as city and highway fuel consumption of a new vehicle and energy efficiency for home appliances is an example of this type of label.[13] These types of labels inform the consumer about the product but do not suggest whether the product is good or bad for the environment. Mandatory programs tend to focus on a single characteristic of a product.

Voluntary Ecolabels

Some voluntary programs are government sponsored while others are sponsored by the private sector.

10. "Single-issue" and "multi-criteria" are terms used by E. Staffin, "Trade barrier or boom? A critical evaluation of environmental labeling and its role in the greening of world trade," *Colombia Journal of Environmental Law* 21 (1996): 205–86 at 215–19.
11. Ibid., pp. 232–33.
12. Ibid., pp. 219–20.
13. Accessed 9 September 2002 at the EnerGuide website on the World Wide Web: http://energuide.nrcan.gc.ca.

Voluntary Self-claim Labelling Scheme
The plethora of self-claim labels is perhaps the biggest culprit for the existing confusion among consumers regarding ecolabels. The self-claim label is basically a characteristic the manufacturer chooses to emphasize about its product, without an independent verification system to ensure that the claim is accurate. Although legislation in some countries aims to prevent the use of empty or misleading claims, these labels have sometimes been deceptive.[14]

Voluntary Third-party Claims
In a voluntary third-party scheme, an independent organization verifies the claim made on behalf of a product and determines whether the product meets the requirements of a specific ecolabelling program. Generally the process involves evaluating a product against predetermined criteria. This type of scheme promotes transparency in the ecolabelling process.

Existing Ecolabelling Programs

The oldest ecolabelling program is Germany's Blue Angel Programme. In existence since 1977, this voluntary, government-sponsored program has served as a base model for numerous labelling schemes in a number of countries.[15] The system is based on product categories for which specific minimum criteria are developed. Manufacturers must apply to have a product qualify to receive a Blue Angel label.[16]

In 1992, the European Community adopted a uniform ecolabelling program that did not hinder national programs. Indeed, labelling programs now exist in numerous European countries and in other countries around the world.[17] Such schemes include "White Swan" in Norway, Finland, and Sweden, Japan's "Eco Mark" launched in 1989, and Canada's "Environmental Choice" launched in 1988. The Environmental Choice program was also implemented in Australia and New Zealand. These schemes are generally considered seal of approval programs.

Ecolabels designed specifically for sustainably harvested products are recent and began as a forestry initiative. Although the only "sustainably harvested" label found thus far in the fisheries sector is the one granted by the Marine Stewardship Council, its Principles and Criteria, as well as various

---

14. S. Dawson and N. Gunningham, "The more dolphins there are the less I trust what they're saying: can green labelling work?" *Adelaide Law Review* 18 (1996): 1–33, 2–3.
15. R. Wynne, "The emperor's new eco-logos? A critical review of the Scientific Certification Systems Environmental Report Card and the Green Seal Certification Mark programs," *Virginia Environmental Law* 14 (1994): 51–149 at 60–62.
16. Ibid., p. 60.
17. Ibid., pp. 60–63.

other certification methodologies, bear some resemblance with the ecolabelling program designed by the Forest Stewardship Council.[18]

Rationale Behind Ecolabelling

Driven by Consumer Demand
Ecolabelling encourages competition between goods on the open market by using environmental characteristics to entice consumers to differentiate between products. It is hoped that consumers will cause an increase in the market share of these "greener" products, thus encouraging competitors to be innovative so that they may offer similar products. Hence, ecolabelling represents supply-demand market behaviour; the consumers' demands for environmentally superior products are met by an increase in the supply of these products.[19]

Environmentally harmful activities are treated as externalities, and ecolabelling aims to compensate for this. Manufacturers do not assume the cost of ozone depletion, unsustainable fishing, or the disappearance of threatened species—and these costs are left for society to absorb.[20] Without some form of intervention, there are no short-term market incentives for manufacturers to consider these factors, thus "producers may lack compelling reasons to 'internalize' the cost of externalities."[21]

The gain in popularity of ecolabelling is perhaps due to the very nature of the world system that has encouraged its initial development. Because of the predominance of international trade rules, which essentially claim that environmental measures must be consistent with the provisions set out in the General Agreement on Tariffs and Trade (GATT), ecolabelling is a way to sidestep the trade-centered regime.

As ecolabels are market-based instruments, "the underlying premise of a labelling program is that the strong environmental values of consumers can be used as a market force to leverage environmental improvement."[22] Furthermore, because life-cycle ecolabelling encourages environmental change at its source, it provides an incentive to create products that are better for the environment.[23] An environmental labelling program can change consumer behaviour, raise consumer awareness about the environ-

---

18. For more information on the Forest Stewardship Council initiation, visit the FSC Web site on the World Wide Web: http://www.fscus.org. (Accessed 9 September 2002).
19. Appleton (n. 8 above), p. 14.
20. Ibid., pp. 11–12.
21. Ibid., p. 12.
22. A. Okubo, "Environmental labeling programs and the GATT/WTO regime," *Georgetown International Environmental Law Review* 11 (1999): 599–646 at 602.
23. Appleton (n. 8 above), p. 15.

mental effects of products, and increase the production of environmentally superior products and technologies.[24]

The effectiveness of any ecolabelling scheme depends both on the producer and the consumer, as well as the consumer's willingness (and ability) to pay for these products.[25] Some studies indicate that the concept is proving to be a promising mechanism to educate consumers and cause changes in the market. A survey conducted by Environmental Research Associates concluded that 41 percent of consumers demonstrated awareness towards labels, as they always or usually checked environmental labels while making purchases and claimed that environmental reasons played a prominent role in their decisions to purchase certain categories of products.[26] American consumers seem interested in purchasing ecolabeled seafood.[27]

A particular risk is involved in this type of marketing. Unlike an obvious product characteristic such as price, consumers may be incapable or unwilling to test the truth about the environmental claims by verifying the information for themselves. Although people may desire to make environmentally responsible choices, they do not always have the knowledge, time, or interest to dedicate a significant amount of effort to do so. Like any other product characteristic that is highlighted with marketing techniques, consumers will make purchasing decisions based on what labels tell them—or rather, what consumers believe the labels are telling them. For an ecolabel to be truly effective, it must encourage the consumer to make the best environmental choice available to them.

Putting relatively little trust in the consumer's abilities may seem paternalistic, yet claiming the opposite would imply that consumers who are preoccupied with ecolabels are informed consumers. This may be the case, but the argument leaves little room for the role labels can play in promoting consumer awareness and education. Consumers may care about the environment, yet not be well informed about the implications of their particular purchases. Some have argued that "Report Card" type labels that simply display information without judgment have proved to be somewhat ineffective because consumers may not possess adequate knowledge to understand the label's information.[28]

24. Okubo (n. 22 above), p. 601.
25. Appleton (n. 8 above), p. 16.
26. K. Forstbauer and J. Parker, "The role of ecolabeling in sustainable forest management," *Journal of Environmental Law Review* 17(8) (1994): 165–90 at 170. Similar statistics reported in the UK and Australia: see Dawson and Gunningham (n. 14 above), pp. 1–2.
27. C. Wessels, H. Donath and R. Johnston, "U.S. Consumer Preferences for Ecolabeled Seafood: Results of a Consumer Survey," (Kingston, Rhode Island: University of Rhode Island, Department of Environmental and Natural Resource Economics, September 1999). Accessed 9 September 2002 on the World Wide Web: http://www.riaes.org, at p. 53 (under "Publications").
28. Generally, see Wynne (n. 15 above), pp. 95–102.

The effects of an ecolabelling scheme can have significant positive impacts on the environment. An ecolabel allows a consumer to know about a particular environmental characteristic of a product. As consumers become more aware, manufacturers who lag behind in environmentally friendly technologies risk losing out on their share of the market by not keeping up with their competitors. Hence, ecolabelling can provide the motivation for new, more environmentally conscious technologies.[29] But the most obvious benefit of an ecolabelling program is its potential reduction of negative environmental impacts. Manufacturers raise the standard of their products to bring them in line with more environmentally responsible processes and technologies. As competitors change to keep up, the result is an upward movement of environmental consciousness on the market.

The success of an ecolabelling scheme in the fisheries sector is ultimately measured by its ability to promote consumer-driven responsible fishing practices. Ecolabelling in the fisheries sector can provide information about the environmental consequences of certain fish products and allow consumers to make informed decisions about their purchases. It also allows consumers to express their environmental concerns through their purchasing behavior, which could in turn encourage retailers and other consumers to buy sustainable fish products. Ecolabelling could raise the standards in fish harvesting practices and provide a competitive advantage to products derived from sustainable sources. Best of all, changes could be seen on a global scale, as ecolabels could educate and promote awareness in consumers around the world and hence have far-reaching effects in other countries.[30]

Manufacturers' Incentives
Consumer demand is not the only factor that plays a role in the increasing popularity of ecolabelling. One must not forget that corporations themselves use ecolabels for their own benefit, albeit sometimes by misleading consumers.[31]

According to Wynne, the green consumer market is "dysfunctional" because producers trying to create a 'green' image for themsleves have taken advantage of consumer trust.[32] Consumer trust and confidence in the label is "a prerequisite for eco-labelling to succeed in stimulating environmental concern."[33] In fact, it is easy to understand how consumers can be confused when faced with a myriad of complicated claims—some without accuracy

---

29. G. Richards, "Environmental labeling of consumer products: The need for international harmonization of standards governing third-party certification programs," *Georgetown International Environmental Law Review* 7 (1994): 235–76 at 247–48.
30. Deere (n. 2 above), p. 7.
31. Dawson and Gunningham (n. 14 above), p. 2.
32. Wynne (n. 15 above), p. 54.
33. Appleton (n. 8 above), p. 18.

or clear meaning—and how it can be difficult to differenciate between mere marketing strategies and products that truly are environmentally superior.[34] Indeed, if ecolabels "are to invigorate the green consumer market, they must be able to direct consumers to the 'greenest' products."[35]

Ecolabels are not only beneficial to the environment, they also serve the interest of environmentally conscious manufacturers by providing them with a niche that allows them to seek out the market share of green consumers and offsetting competitors who have enriched themselves by ignoring environmental considerations. Environmentally conscious manufacturers highlight an environmental quality about their product in the hopes that it will entice consumers to buy the product. For "green" consumers, the presence of an ecolabel, even on a more expensive product, serves as a reminder that there are differences between products that may not be apparent in its price or quality.

Although corporations may sometimes be pushed into certain labelling initiatives through government regulations, it is unlikely that many companies would voluntarily pay for the certification of their products unless they believed consumers would respond to the labels in a positive manner and that the labels would benefit the company in some way. Several advantages may be conferred to companies using an ecolabel, including an "improved corporate image, premium prices and expanded consumer base."[36] To these advantages one can also add that corporations using ecolabels could gain a competitive advantage and an increase in market share which could lead to an increase in exports, especially to countries considered to be greener markets.

Greening of the Corporate Image.—Companies may have a very strong incentive to certify their products if it allows them to boast the environmentally conscious attitude of the corporation and give it a competitive edge (or simply allow the company to keep up with fellow competitors). To illustrate this in the realm of sustainably harvested products, there is perhaps no better example than Home Depot's efforts to render the company more "green" in the eyes of consumers, especially after being the object of numerous visible public protests and demonstrations by environmental organizations such as the Rainforest Action Network. Home Depot has pledged to stop selling timber products from endangered forest areas by the end of 2002.[37]

---

34. Wynne (n. 15 above), p. 54.
35. Ibid., p. 64.
36. K. Kloven, "Eco-labeling of sustainably harvested wood under the Forest Stewardship Council: Seeing the forest for the trees," *Colorado Journal of International Environmental Law and Policy* (1998): 48–55 at 53.
37. ENN. (15 March 1999). Activists blast Home Depot despite move to go green. Accessed 9 September 2002 on the World Wide Web: http://www.cnn.com/NATURE/9903/15/home.depot.enn/. Home Depot carries products from ancient temperate rainforests of British Columbia, old growth lauan and ramin from

Obtaining Premium Prices.—Corporations may have an incentive to certify their products and use an ecolabel as this may allow them to charge higher prices for their products. Although the company would incur expenses to certify their product, it is reasonable to assume that this cost could be recovered in the long term, as consumers may be willing to pay more for what they consider to be a higher quality product.

Expanding the Consumer Base or Market Share.—The use of an ecolabel may help companies break into an established market.[38] Some have argued that certain consumers may be hesitant to buy fish products as they know several fish stocks are under threat. Consequently, an ecolabel may act as an assurance for these consumers to continue buying certified fish products, thus ensuring that manufacturing and processing companies do not lose out on that market share.[39]

To sum up, it is unlikely that the existence of ecolabels is the result of a process that is solely driven by the consumer. As the rationale of ecolabels is rooted in market forces, it is logical that ecolabels resemble a two-way street: providing a benefit to both the consumer and the manufacturer.

What Makes a Good Ecolabel for the Fisheries Sector?

Accuracy and Trustworthiness

An ecolabel should represent what is truly being claimed about the product. For example, if the label is a type of seal of approval and claims the product is "environmentally superior," the product should indeed be environmentally superior when compared to competing products. Being otherwise could very well undermine the trust of consumers and contribute to their confusion, which is a very real phenomenon.[40]

In the fisheries context, a label claiming that the fish come from sustainable, well managed sources should mean that the product being purchased was harvested according to sustainable fishing practices. It is insufficient that the product was fished without a significant amount of by-catch and that it respected the total allowable catch limits. The product must have been fished *sustainably*, taking all relevant factors into consideration. Substantial environmental requirements must be satisfied to meet the consumers' expectations. Accuracy and substantive environmental expectations are neces-

---

Southeast Asia, and bigleaf mahogany from the Amazon. See: Rainforest Action Network, Press Release, "Home Depot Announces Commitment to Stop selling Old Growth Wood" (26 August 1999). Accessed 9 September 2002 on the World Wide Web: http://www.ran.org/news/newsitem.php?id=74.

38. Wynne (n. 15 above), p. 131.
39. Deere (n. 2 above), p. 9.
40. See generally Richards (n. 29 above), p. 253.

sary for the certification scheme to be a credible and trustworthy one in the eyes of the consumer.

Independence
Third-party certification schemes are more desirable than self-claim schemes because independent, third-party certification brings in "greater consumer acceptance, reduction in political oversight, and reduction in cost to the government."[41] Although the government should play a role in ensuring truthful advertising, it could be argued that governments should refrain from having direct involvement in the independent third-party certifications to ensure consumers do not believe the label was awarded to a product because of political lobbying.

Partiality and promotion of self-interest become evident when a manufacturer grants itself a label. Corporations should be prevented from espousing responsible environmental values as a lure to convince people to buy that product where it is not warranted. On the other hand, under an independent scheme, consumers are likely to have more trust in the label granted by an organization having no interest in the final outcome of certification for that particular product. Indeed, if ecolabelling standards are developed independently of corporate interest, the requirements to qualify for the label are much more likely to be transparent and trustworthy.

Transparency and Clarity
The consumer must be able to trust the process by which the product has earned certification and this process should be described in ways accessible to the interested consumer. Transparency is crucial for the certification process to be credible (i.e., public reports). For the certification process to be transparent, it must be readily understood and verifiable by the consumer. In other words, the consumer must understand the reasons behind the claim on the label. When public reports are available, they must be written in such a way that a curious consumer is able to understand the certification process and the reasons why the product was certified. Leaving too many questions unanswered may mean a loss of credibility for the entire certification process.

Flexibility and Consistency
The process must be consistent from one product to the next to ensure uniformity and homogeneity in the methodology, yet the certification process should be flexible. In the fisheries context, the process should be uniform in the sense that important and relevant elements should always be considered when determining if a fishery is sustainable, yet the criteria should be flexible enough to recognize that fisheries differ greatly from each

41. Ibid.

other and criteria must be manipulated to best suit the reality of the fishery at hand.

The remainder of this article will discuss the voluntary, third-party certification scheme adopted by the MSC. The approach taken is that of a curious consumer trying to understand the MSC certification process by using data and information publicly available. Hence, much of the following information has been obtained from the MSC Web site.[42] My rationale for taking this approach is to find out just how credible the MSC scheme is to the consumer contemplating purchasing an ecolabelled product.

PART II—SUSTAINABLE FISHERIES AND THE MARINE STEWARDSHIP COUNCIL

Although the Marine Stewardship Council began as a joint initiative between the World Wildlife Fund (WWF) and Unilever, a leading buyer of frozen fish products, the MSC is now an independent, nonprofit international body and aims to reverse the decline in world fisheries "by seeking to harness consumer purchasing power to generate change and promote environmentally responsible stewardship of the world's most important renewable food source."[43] To accomplish this, the MSC ensures that a fishery meets the Principles and Criteria for Sustainable Fishing.[44] If a fishery meets these Principles and Criteria, it is deemed to have met the MSC standard. Products that originate from such a fishery are eligible to display the MSC logo.

The Role of Accreditation Bodies

The MSC itself does not directly assess whether a fishery meets the standard for sustainability, but rather, an independent certification body carries out the assessment. The role of the MSC is to approve and accredit these groups, and this is done in accordance with certain requirements specifically set out in the Guidelines for Certifiers[45] and Accreditation Manual.[46]

---

42. The Marine Stewardship Council's Web site is located on the World Wide Web at http://www.msc.org.
43. About MSC. Accessed 9 September 2002 on the World Wide Web: http://www.msc.org.
44. Principles and criteria for sustainable fishing. Accessed 9 September 2002 on the World Wide Web: http://www.msc.org.
45. (August 2000). *MSC Guidelines for Certifiers 3*. Accessed 9 September 2002 on the World Wide Web: http://www.msc.org.
46. (March 2001). Accreditation manual (requirements for certifiers). *MSC Accreditation Manual Part 3* (Issue 4). Accessed 9 September 2002 on the World Wide Web: http://www.msc.org.

Organizations apply to become certifiers and once accredited, carry out the 'hands-on' mandate of the MSC as independent, third-party certification bodies. These organizations will then assess whether or not a fishery meets the MSC Principles and Criteria for sustainable fishing. The MSC has thus far accredited the following five organizations for fishery and chain of custody evaluations: Scientific Certification Services (California, USA), SGS Product and Process Certification (Netherlands), Moody Marine Ltd. (Birkenhead, UK), Tavel Ltd. (Halifax, Canada), TQCSI Marine, South Australia).[47] Additional organizations have been certified to carry out only the chain of custody evaluation.[48]

There are several categories in which potential accreditation bodies must prove themselves before they are accredited by the MSC. This is to "ensure that the accredited certification bodies are committed to maintaining standards throughout their organization."[49] Importantly, the MSC specifies certain methodologies to ensure that fishery certifications are consistent, controlled and transparent, thus allowing consumers to rely on the MSC claim regardless of which certification body has carried out the evaluation.[50]

Thus, the Accreditation Manual outlines the necessary and recommended ways an accredited certifier should operate. For example, the manual highlights the necessity of having formal dispute resolution procedures, disclosing potential conflicts of interest, the funding of accreditation bodies, et cetera. These requirements display how the MSC exercises control over the daily operations of certification bodies. It is important to remember that the MSC is the body that will be recognized by consumers. In fact, consumers will have no idea who specifically carried out the certification of the fishery or fishery product when they purchase MSC-labeled products. For consumers, the identifiable body remains the MSC. They must visit the MSC Web site and read a lengthy public report to find out which certifier carried out the certification process.

It is critical that the MSC ensure consistency in its procedures. Consumers will purchase products displaying the MSC logo regardless of which specific certification body completed the assessment. For these reasons, it is crucial for the MSC to take certain steps to ensure the smooth functioning of the bodies that will be carrying out the operational aspect of the MSC mandate. By compelling the potential certification body to disclose certain

---

47. Information accessed 9 September 2002 on the World Wide Web at the MSC Web site: http://www.msc.org.

48. Organizations accredited to carry out only the chain of custody evaluation are as follows: Integra Food Secure Ltd. (Exeter, UK), McAlister Elliott & Partners Limited (Lymington, UK) and Surefish (Seattle, USA). Accessed 9 September 2002 on the World Wide Web at the MSC Web site: http://www.msc.org.

49. *MSC Accreditation Manual* (n. 46 above), Part 3.2, Section 1 (1.1).

50. *MSC Certification Methodology* (n. 1 above), p. 6.

information, the MSC is ensuring that its mandate will be carried out by reliable organizations.

But to what extent do the various requirements imposed by the MSC encumber the accreditation body? The certification body is constrained in this sense: numerous guidelines and requirements concerning several aspects of the certification body's operations are outlined. The potential certification body will be evaluated according to these guidelines. This is done to ensure that certification bodies conform to all MSC accreditation requirements and to allow the MSC to monitor compliance with such requirements.[51] The accredited body is re-evaluated every 5 years.[52] However, the obligatory nature of these requirements is not always clear. Although the MSC specifies certain requirements for many criteria, in many cases the implementation of these requirements is left to the discretion of the certifier.

In the following section I will delineate some of the requirements that must be met by potential certification bodies. Because the requirements are lengthy and numerous, the following is not meant to be comprehensive; rather, it is a cursory outline meant to give the reader a general sense of how the MSC exerts control over the certification process.

Types of Requirements that Must Be Met by Potential Certification Bodies

MSC's Requirements Concerning Organizational Structure, Financial Stability, and Compliance with National Legislation[53]
Certifiers must comply with national and international laws including legislation concerning legal status and financial operations.[54] The certification body must also provide documents describing its organizational structure and ownership, lines of authority, evidence of reliable funding, et cetera.[55]

Independence, Conflict of Interest, Confidentiality, and Dispute Resolution Procedures[56]
The credibility of third-party certifiers depends on their ability to remain objective and free from outside influence. To accomplish this objective, the MSC requires that the certification body disclose all potential conflicts of interest, financial or otherwise.[57] This requirement also applies to the em-

---

51. Overview. *MSC Accreditation Manual* (n. 46 above).
52. Information on the MSC accreditation process is available on the World Wide Web at http://www.msc.org. See also: *MSC Accreditation Manual* (n. 46 above), Part 3.1, Section 7, p. 14.
53. MSC Accreditation Manual (n. 46 above), Part 3.1, Sections 1, 2 and 3.
54. Ibid., Part 3.1, Section 1, p. 5.
55. Ibid., Part 3.1, Sections 2 and 3.
56. Ibid., Part 3.1, Sections 4, 5, 6 and 10.
57. Ibid., Part 3.1, Section 4.

ployees who will conduct the certification operations.[58] A dispute resolution mechanism must be accessible to stakeholders and the certifiers must keep records of all complaints, appeals, and disputes.[59] The certifying body must define the entity (group, person) responsible for making a certification decision, which in turn must follow clearly defined decision-making procedures.[60] Certification bodies must also have policies and procedures to ensure confidentiality[61] in addition to having written policies about conflict avoidance.[62]

Credibility and Transparency through Appropriate Documentation, Competence, and Public Information
Document control is important to ensure that the certification body conforms to MSC requirements.[63] All assessment procedures must be documented and available for review by the MSC.[64] This includes all documents and records that contribute to the certification process and all documents destined for public distribution.[65] To ensure the competency of permanent and contract personnel, certifying bodies must provide their records of education, training, and experience.[66] As for information that is made available to the public, several requirements are outlined by the MSC and these will be discussed at length in Part III of this article.

Fairness and Integrity of Claims (maintenance and extension of certification, changes to certification requirements, suspension and withdrawal of certification, information on certificates, control of claims and logos)
Certifiers are responsible for instructing their client of exactly what they can and cannot claim about the fish product so that the MSC logo is not misappropriated. To ensure this, certifiers must have a written contract with their client.[67] This contract dictates the terms of use of the MSC logo, claim, certificates, and so forth. Both parties must be aware of the conditions under which a certificate may be suspended or withdrawn.[68] The contract must also include procedures for dealing with the incorrect use of MSC claims.[69]

Obligations to be fulfilled by the certified operation must be clear to all parties and are contractually binding.[70] The contract must specify a period of

58. Ibid., Part 3.1, Section 4, p. 9.
59. Ibid., Part 3.1, Section 6, p. 12.
60. Ibid., Part 3.1, Section 5, p. 10.
61. Ibid., Part 3.1, Section 10, p. 17.
62. Ibid., Part 3.1, Section 4, p. 9.
63. Ibid., Part 3.1, Section 12, p. 20.
64. *MSC Guidelines for Certifiers* (n. 45 above), Item 3.1.
65. Ibid., Items 4.1 and 4.2.
66. *MSC Accreditation Manual* (n. 46 above), Part 3.1, Section 11, p. 18.
67. *MSC Guidelines for Certifiers* (n. 45 above), Item 12.
68. *MSC Accreditation Manual* (n. 46 above), Part 3.1, Sections 8 and 9, pp. 15, 16.
69. Ibid., Part 3.1, Section 18, p. 26.
70. Ibid., Part 3.1, Section 7, p. 14.

validity and must provide the certification body with clear rights to revise certification requirements and withdraw the certificate in case of noncompliance.[71] The MSC or the certification body reviews all public claims for accuracy before they are released.[72] In the end, although it is the accredited certification body that carries out all assessments, it must do so by adhering to firm MSC requirements. Although the certification body signs a contract with the client fishery, the MSC maintains control of certain elements within this contract.[73]

In summary, requirements set out by the MSC, if properly met, will ensure credibility, transparency, and competence of those certification bodies undertaking the assigned task of certifying a fishery.

How a Fishery Is Certified

Regardless of size or location, the MSC certification process is available to any client who wishes to certify a fishery.[74] Certification is awarded upon completion of an in-depth assessment by the certification body. Before describing how this assessment is carried out, I will explain the three MSC Principles and Criteria for Sustainable Fishing, which are used to define what constitutes a sustainable fishery. These principles form the backbone of the entire certification ideology and shape the foundation for the methodology used to measure sustainability in the fishery. Although the first step is to certify a specific fishery, it is then possible to certify fish products (such as frozen fillets, lobster tails, etc.) if manufacturers can demonstrate that the products are the same as those that have been deemed sustainable by certification. Although very interesting, as noted in the introduction, Chain of Custody certification is not discussed in this article.[75]

MSC Principles and Criteria for Sustainable Fishing

The MSC Principles and Criteria help determine if a fishery is sustainable by evaluating if it "allows target fish populations to recover to healthy levels

---

71. Ibid., Part 3.1, Section 8, p. 15.
72. *MSC Certification Methodology* (n. 1 above), Item 16.4.3, p. 30.
73. *MSC Guidelines for Certifiers* (n. 45 above), Item 12.
74. "About MSC," accessed 9 September 2002 on the World Wide Web: http://www.msc.org—Vision, Mission and Values.
75. More information is available on Chain of Custody certification as accessed 9 September 2002 on the World Wide Web: http://www.msc.org. Processors with chain of custody exist for the Western Australia Rock Lobster, Thames-Blackwater Herring, Alaska Salmon, New Zealand Hoki, South West Handline Mackerel fisheries and Burry Inlet cockle fisheries.

where they have been depleted in the past. Such a fishery will ensure that there is a future for the industry."[76]

According to the MSC, a sustainable fishery is one that is "conducted in such a way that: it can be continued indefinitely at a reasonable level; it maintains and seeks to maximize ecological health and abundance; it maintains the diversity, structure and function of the ecosystem on which it depends as well as the quality of its habitat, minimizing the adverse effects that it causes; it is managed and operated in a responsible manner, in conformity with local, national and international laws and regulations; it maintains present and future economic and social options and benefits; it is conducted in a socially and economically fair and responsible manner."[77]

Having articulated the MSC's definition of a sustainable fishery, I will now state and discuss each one of the three distinct Principles that identify such a fishery. It is important to understand that each individual Principle is further defined by several Criteria. Because the Criteria are numerous (17 Criteria for Principle 3), for the sake of brevity only the Principles themselves will be cited in the following discussion.

Principle 1
   A fishery must be conducted in a manner that does not lead to overfishing or the depletion of the exploited populations and of those populations that are depleted, the fishery must be conducted in a manner that demonstrably leads to their recovery.[78]

To determine if this Principle is followed, three Criteria must be satisfied: i) the fishery must maintain a high level of productivity within the target population; ii) fishing must not alter the age or genetic structure and the sex composition, and most intriguingly, iii) where the target population is one that is depleted, the fishery must be conducted in such a way to permit the target population to recover.

Regarding the last criterion, it is thus possible to obtain MSC certification for a fishery that is depleted, provided that one can demonstrate that the depleted population is fished in a way that encourages the recovery of the stocks. In fact, a depleted population is usually less robust than a stock that exists in healthy numbers, and a depleted population is sometimes more susceptible to environmental harms because of low numbers. With healthy, populous fish stocks, there is more room for error, a necessary element in

---

76. "Fisheries Certification," accessed 9 September 2002 on the World Wide Web: http://www.msc.org.
77. Preamble. *MSC Principles and Criteria for Sustainable Fishing.* (n. 44 above).
78. Ibid.

fisheries management—this is not the case when considering low-abundance fish populations.[79]

A recovering or nonabundant fish population can be sustainably harvested—sustainability is not defined by simple abundance. More accurately, a scenario in which a certified fishery collapses could potentially destroy the credibility of the MSC certification process in the eyes of the consumer and this is a more likely occurrence in a depleted fish stock. To illustrate this point by means of a fictitious example: if consumers purchasing a MSC-certified cod became aware that the fish stock was plummeting dangerously or that the stock has crashed, they may wonder how this was possible given the MSC's assurance that they were purchasing sustainably harvested fish products. Sustainable fish stocks are meant to be able to adapt to reasonable changes in the environment, not to be in danger of disappearing.

The counter-argument then becomes obvious: given the application of such stringent criteria to depleted stocks, one must also remember that some unpredictable calamity could render a large, healthy fish stock into a dwindling one very rapidly. Obviously, these potential stock failures do not prevent us from certifying sustainable products at present.

The difference lies in the *present* state of the fish stocks—large, healthy, sturdy stocks are naturally more capable of surviving pressure than smaller, more fragile and vulnerable stocks recovering from a state of depletion. Because labelling schemes use market forces to allow certain producers to gain a competitive advantage over others by convincing consumers to purchase certain products, truthful representation to the consumer is imperative. When consumers purchase seafood from "sustainable and well-managed sources," they logically expect that the products originate from healthy sources.

This discussion begs the question: does it make sense to allow certification of severely depleted fisheries? In my opinion, the MSC should not be certifying fisheries that are severely depleted. My reasoning stems from the potential effect that certification may have on consumers who may not understand why a depleted fishery is receiving MSC certification, while that fishery is perceived as being at risk. Without a serious public education campaign, a scenario such as the fictitious one referred to earlier could seriously undermine the credibility of the MSC label. Any labelling scheme should preoccupy itself with the reaction consumers will have towards its label. This problem will be revisited in the case analysis of the Alaska Salmon fishery in Part III.

79. R. Hilborn, "Uncertainty, Risk and the Precautionary Principle," in *Global Trends: Fisheries Management*, ed. E. Pikitch et al. (Seattle: American Fisheries Society, 1997), pp. 100–106 at p. 100. (Proceedings of the American Fisheries Society Symposium 20).

Principle 2
    Fishing operations should allow for the maintenance of the structure, production, function and diversity of the ecosystem (including habitat and associated dependent and ecologically related species) on which the fishery depends.[80]

Clarified by another three Criteria, this Principle is met when the trophic levels (the balance of predator-prey relationships in the ecosystem) are not modified. In addition to this, functional relationships between species are maintained: biodiversity is preserved at the genetic, species, and population levels and harm to threatened and endangered species is minimized.

Principle 2 accurately exemplifies the ecosystem approach that encompasses the understanding that no system is isolated and that "fisheries contain a system of systems which [. . .] are contained in one another, and the operational limits of the manageable systems are not easily drawn."[81] In this approach, not only is the habitat of the resource seen as an absolute necessity to the health of the stock, but also the Principle incorporates the need to look at the impact that the fishery may have on both dependant and ecologically related species. By looking at dependant species, the MSC is ensuring that the natural, delicate balance of predator-prey relationships in the ecosystem is not substantially altered. As a result of including all ecologically related species, the assertion is broad enough to include the ecosystem as a whole, as all species are interconnected. Theoretically, Principle 2 appears all encompassing—in practice, only time will tell if MSC certifiers will adequately be able to respect ecosystem integrity, as put forward by this Principle.

It was specified at the first meeting of the MSC Standards Council, held in Vancouver on 24–25 June 2000 that the drafters of this Principle did not intend that the fishery have no impact whatsoever on the marine ecosystem. This would have been unrealistic and unfeasible. Rather, it was intended that a fishery "should not have any unacceptable impacts on the marine ecosystem."[82]

Principle 3
    The fishery is subject to an effective management system that respects local and international laws and standards and incorporates institu-

---

80. MSC Principles and Criteria (n. 44 above).
81. S. M. Garcia and R. J. R. Grainger, "Fisheries Management and Sustainability: A new perspective of an old problem?," in *Developing and Sustaining World Fisheries Resources,* ed. D. A. Hancock et al. (Collingwood, Australia: CSIRO Publishing, 1997), pp. 631–54 at p. 648. (The State of Science and Management, Proceedings of the 2nd World Fisheries Congress)
82. Summary of First MSC Standards Council Meeting 24th and 25th June 2000. (On file with the author).

tional and operational frameworks that require use of the resource to be responsible and sustainable.[83]

The goal of this Principle is to ensure that the management and operational framework are such that the fishery is able to effectively comply with the MSC's mandate of continued sustainable practices. This Principle is unique in the sense that it is subdivided into two Criteria: management system Criteria (11 elements), and operational Criteria (6 elements), which are listed:[84]

The management system shall

i) ensure that the fishery is not conducted under a controversial unilateral exemption to an international agreement;
ii) ensure that long-term objectives are consistent with MSC Principles and Criteria, that there is a transparent consultative process that involves all interested parties,
iii) consider the cultural context, scale, and intensity of the fishery;
iv) observe legal rights (including customary rights) of people who depend on fishing for food and livelihood;
v) have an appropriate dispute resolution mechanism;
vi) provide incentives (social and economic) to support sustainable fishing while ensuring that fishing operations shall not be driven by subsidies that contribute to unsustainable fishing;
vii) use a precautionary approach when dealing with scientific uncertainty and acting on the basis of best available information;
viii) incorporate a research plan, and provides for the dissemination of information;
ix) require periodical assessments of the biological resource and the impact of the fishery;
x) set catch levels that maintain a high productivity of the target fish population, take into account the productivity of the ecological community, including non-target species, identify appropriate fishing methods to minimize the impact on the ecosystem, provide for the recovery of depleted fish populations; and ensure that no-trade zones and mechanisms to close the fishery once the catch limit has been reached are established;
xi) ensure that compliance, monitoring, control, surveillance, and enforcement procedures exist.

The fishing operation shall:

i) avoid using gear that captures non-target species;
ii) implement fishing methods that minimize the impact on habitat;

83. MSC Principles and Criteria (n. 44 above).
84. Ibid., Principle 3.

iii) not use destructive fishing methods such as explosives or poisons;
iv) minimize lost gear, oil spills, et cetera;
v) respect legal and administrative requirements and comply with the fishery management system; and
vi) co-operate with management authorities in the collection of catch, discard, and other information.

To recapitulate, the MSC Principles and Criteria are used to measure whether a fishery is considered sustainable or not. By means of third-party certifiers, experts are hired to determine if the fishery meets the MSC Principles and Criteria. If a fishery meets the Principles and Criteria set out by the MSC, it will have proved itself to merit MSC certification. Before I point out how these principles and criteria were put into practice by the MSC's certification of three fisheries, I will explain how a fishery becomes certified.

The Process of Fishery Certification

Initial Certification Feasibility Review
The purpose of this initial review is for the accredited certification body to familiarize the client (the agency that is seeking to have a fishery certified) with MSC requirements and to ensure all parties understand the nature of the MSC objectives. It is at the initial review stage that issues meriting further investigation are identified so that they may be examined later during the pre-assessment visit.[85]

Assessment Contract
The certification body must satisfy the MSC's requirement that basic information has been acquired, such as defining the scope and area of the client's activities, analyzing fishery issues, identifying fish stock issues as well as the MSC Principles and Criteria issues that arise from those activities. The assessment contract also describes the competencies needed to certify the particular fishery.[86]

Pre-assessment Evaluation
The purpose of the pre-assessment evaluation is to determine the scope of the certification project and to evaluate the readiness of the fishery to undergo certification. The pre-assessment evaluation is largely based on document review. Ultimately, a report is produced that gives an overview of the fishery and its management practices, highlights key issues, and identifies potential problems in the certification process. It is at this stage that prospec-

85. *MSC Certification Methodology* (n. 1 above) Section 2, p. 12.
86. Ibid., Section 3, p. 13.

tive stakeholders and eventual field sites are identified. A decision is made as to whether the fishery should move to the assessment stage, and a budget estimate is produced.[87]

Stakeholder Consultation
The concerns of various stakeholders are taken into account so that the certification body can be made fully aware of all issues relevant to the fishery. Stakeholders include government agencies, management authorities, fishery organizations, nongovernmental organizations, conservation groups, other commercial fishers that impact the certified fishery but are not within the scope of the certification, research organizations, et cetera. Interestingly, as much as the MSC certification scheme emphasizes stakeholder consultation, none of the documents inform the reader about what weight is given to stakeholder input.[88] If some stakeholders have firm reasons to believe the certification should not go ahead, is the MSC obliged to take their concern into consideration at all? This issue will be discussed at length in the analysis of the New Zealand Hoki Fishery, in Part III.

Assessment Planning and Team Selection
It is at this stage that the accredited certifier hires a multidisciplinary assessment team made up of members with expertise in the fishery undergoing certification.[89] Even though the MSC sets out the guidelines defining how a fishery attains a well-managed and sustainable status, it is the expert, multidisciplinary assessment team members who decide if the fishery meets the MSC Principles and Criteria. Given the assessment team's important mission, it is not surprising that the MSC has rigid criteria regarding the selection of experts by the certification body.[90]

Matching MSC Principles and Criteria and Their Weighting Prior to the Assessment Visit
It is worth noting that the three MSC Principles and Criteria described earlier cannot of themselves constitute the standard for evaluating the fishery as they are too broad and require further interpretation specific to the fishery at hand. Thus each fishery has Scoring Guideposts, which are essentially a list of indicators associated to one of the three Principles. Each Principle will have many such indicators listed in the Scoring Guideposts designed specifically for a given fishery. Although the MSC does provide a generic list of Scoring Guideposts, this list must usually be amended to adjust to the fishery at hand.[91]

    87. Ibid., Section 4, p. 13–14.
    88. Ibid., Section 5, p. 14–16.
    89. Ibid., Sections 6, 6.4, 6.5, p. 17–19.
    90. Ibid., Section 6.3, p. 17.
    91. Ibid., Section 7.1, p. 19.

TABLE 1.—DESCRIPTION OF THE 80 PERCENT AND 100 PERCENT SCORING GUIDEPOSTS USED TO ASCERTAIN IF THE STANDARD DESCRIBED IN SCORING INDICATOR 2A FOR THE ALASKA SALMON FISHERY HAS BEEN MET

Scoring Criteria 2A	Scoring Guidepost 80%	Scoring Guidepost 100%
Are the non-target species in the fishery known?	Identities of all significant discards are known and some information is collected and available on the numbers caught and removed from the fishery.	All significant discard species, including salmon, are known and estimates of the quantity caught/removed are available.
Are the levels of catch/mortality for discards known?		Estimates of discard mortalities available.

SOURCE.—Data reproduced from the Performance Criteria & Scoring Guideposts Against the Marine Stewardship Council Principles & Criteria—Alaskan Salmon Fishery Issue.[206]

For example, to ascertain if the second Principle (and its accompanying Criteria) is met for the Alaska Salmon fishery, the Principle contains four indicators. Indicator 2A is shown in Table 1. The assessment team also defines the appropriate performance level of each indicator, and these are referred to as Scoring Guideposts of 80 percent or 100 percent. The performance level required to satisfy Indicator 2A in Table 1 is determined before any data are collected or evaluated. For the indicator to be met, it must score at least 80 percent.

Assessment Visit and Evaluation of the Data
During the assessment visit, the Scoring Guideposts performance levels that were selected as appropriate in the pre-assessment stage are now specifically quantified and evaluated for the fishery. Evaluating the data allows the assessment team to determine if the minimal performance levels are attained for each Scoring Guidepost indicator and this in turn will allow the team to determine if each of the three Principles and Criteria are satisfied.

Determining if a fishery satisfies each of the three Principles and Criteria is done by prioritizing, weighting, and setting scores for each individual Scoring Guidepost indicator. An example of how criteria are weighted and scored is illustrated in Table 2. Two distinct calculations are achieved as follows: first, each indicator is weighted between 0 and 1, hence attributing a sense of importance or priority; second, the assessment team evaluates the data and assigns each indicator a score between 0 and 100. Finally, the weight of each indicator is multiplied by its score, giving a final normalized,

TABLE 2.—EXAMPLE OF WEIGHTED SCORING CRITERIA FROM THE BURRY INLET COCKLE FISHERY

Scoring Criteria	Weight	Score
1B.1 Is the fishery related mortality recorded/estimated?	25	95
1E.1 Are assessment models used?	20	80
1F.1 Is the stock(s) at or above the reference levels?	100	100

SOURCE.—Marine Stewardship Council and Moody Marine Ltd., Certification Report for Burry Inlet Cockle Fishery (March 2001), Scoring Criteria, pp. 19, 23 and 25. Accessed 9 September 2002 on the World Wide Web: http:www.msc.org.

weighted measure of performance, again on a scale from 0 to 100.[92] The indicators themselves are then summarized to give a final performance measure of the Principle. This calculation is performed for each of the three MSC Principles.

Each Scoring Guidepost indicator must obtain a minimum value of 80 percent to be considered a pass, while a score of 100 percent would describe the ideal fishery.[93] Indicator 2A considered in Table 1 would have to meet the minimal requirement described in the 80 percent column to be considered a pass.

To illustrate this concept, one may consider that the weighting method used by the MSC is not unlike writing an examination: the pupil cannot hope to successfully complete an exam if he or she fails too many questions that are individually worth few points, nor will the pupil pass if he or she answers all questions correctly except one that is worth many points.

What happens if a fishery scored below the required 80 percent threshold? Does the entire process fail? No, it would not. What is important is that each of the three distinct Principles must independently pass. No fishery will be certified if even one of these principles does not score a pass. But a fishery can obtain certification while failing one or several indicators (although a fishery will fail entirely if any individual indicator scored lower than 60).[94] For example, it would be possible for the Alaska salmon fishery to not meet the 80 percent requirement described in Table 1 and still receive certification and be allowed to display the MSC logo. In fact, all fisheries certified thus far have failed on numerous indicators; nonetheless, all have been deemed to pass each of the three individual Principles and Criteria.

Conditions and Recommendations
Failing to meet a minimal 80 percent threshold may result in a conditional certification.[95] The assessment team will simply identify criterion-specific

92. Ibid., Section 10.1, p. 21–22.
93. Ibid., Section 10.4, p. 22.
94. Ibid., Section 10.3, p. 22.
95. Ibid., Section 11, p. 23.

conditions, which are designated as Minor or Major Corrective Actions. These corrective actions are designed to increase the performance of the deficient indicator to a score of 80.[96]

The consumer, however, will not be aware that the certification is conditional simply by looking at the fish product being purchased. In fact, a curious consumer will only discover whether corrective actions have been stipulated by reading the public reports available on the MSC Web site. Thus, the consumer will believe that the fishery was deemed sustainable and satisfied the three MSC Principles and Criteria—which is, in fact, quite true. However, conditional certifications can potentially be misleading, as the fish product perhaps did not originate from a fishery that is as sustainable as what is portrayed to the consumer.

As previously mentioned, a fishery can be given minor or major corrective actions. A major corrective action can potentially prevent a fishery's certification, but a minor corrective action will not. In the case of a minor corrective action, the condition specified by the assessment team must be satisfied by the client-fishery within a specified timeframe. The condition is a contractual obligation binding on the client. In the case where such a condition is not adequately addressed, the certificate is revoked.[97]

Peer Review, Final Report, and Certification
Once the fishery evaluation has been made, the assessment team prepares a final report and it is submitted to the client for feedback. All client comments must be documented.[98] The report is then sent to another assessment team for external peer review. This peer review panel is comprised of experts whose task is to review and comment upon the certification methodology as well as the results of the assessment. Any questions the peer review panel members raise must be addressed by the assessment team.[99]

When all the conditions have been met, the final report is issued along with the certification. When appropriate, the client has the right to bear the MSC logo on its certified product. The certificate is valid for 5 years, although re-evaluations are carried out annually. Finally, reports on the certification process must be made available to the public.[100]

Conclusion for Part II

The MSC's third-party, independent certification scheme was developed with great attention to detail and is thoroughly documented. Despite lengthy

   96. Ibid., Section 11.2, p. 23.
   97. Ibid., Section 11.4, p. 23.
   98. Ibid., Section 12.5, p. 24.
   99. Ibid., Section 13, p. 24.
   100. Ibid., Section 15, p. 28.

documents, consumers can conclude several things about the certification process: the credibility of the scheme is enhanced by the fact that independent experts carry out very specific evaluations of fisheries; the process is very well documented thus rendering it quite transparent and constancy is ensured within the MSC certification process as requirements are imposed upon the independent certifier.

The MSC scheme allows flexibility as expert assessment teams interpret the MSC Principles and Criteria for each individual fishery. Expert members of the assessment teams are not obliged to try to classify the fishery into generic, predetermined categories. On the contrary, the MSC principles are adaptable to any fishery in the world. The expert assessment teams determine whether the fishery passes or fails the test; hence the decision making is left to those who know the subject matter best: fisheries management personnel and expert biologists with specializations in the field.

PART III—HOW THE MSC PRINCIPLES AND CRITERIA HAVE BEEN APPLIED TO CERTIFIED FISHERIES

The Western Australia Rock Lobster: How Empty Public Reports could Undermine Consumers' Trust in the MSC Label

In March 2000, the Western Australia Rock Lobster became the first seafood product in the world to be granted a Marine Stewardship Council ecolabel.[101] This fishery is a significant one for Australia as it represents 20 percent of the total value of Australian fisheries, and much of the lobster is destined for export.[102] The desired level of catch is achieved by controlling the number of traps (pots) in use. Minimum size requirements prevent the capture of young lobsters, thus ensuring a future breeding biomass. Well over a dozen other fisheries (commercial, recreational and aboriginal) take place in the waters where Rock Lobster is fished. Two of these, notably the trawl fisheries in the Abrolhos Island and the commercial crab fishery, were cited as potentially conflicting with the Rock Lobster fishery.[103] Both the trawling operations and the crab fishery have since been limited to sandy areas, where there is little risk of disrupting the lobsters and their habitat.

101. World Wildlife Fund, Press Release, "First seafood eco-label receives warm welcome at Europe's largest trade show" (Brussels, Belgium) 11 May 2001. Accessed 9 September 2002 on the World Wide Web: http://www.panda.org/endangeredseas/seafood.cfm.

102. Commercial fisheries of western Australia. Department of Fisheries Web site, accessed 9 September 2002 on the World Wide Web: http://www.wa.gov.au/westfish/comm/broc/lobster/index.html.

103. Marine Stewardship Council. (April, 2000). Western Australia Rock Lobster Fishery public summary. Item 1.4.1 and 1.4.2, pp. 9–10. Accessed 9 September 2002 on the World Wide Web: http://www.msc.org.

Weaknesses of the Rock Lobster Fishery

Although the fishery earned a passing mark on all 3 individual principles and was recommended for certification, there were lacunae referred to in the public report, especially regarding the second Principle. The most significant observation made about the Rock Lobster public report is its failure to provide substantive information. In fact, the paucity of detail in the Rock Lobster public report is its most noteworthy attribute.

Incomplete Ecological Risk Assessment
For the Rock Lobster fishery Scoring Guideposts, Criterion C of Principle 2 (2C) assumes the completion of adequate studies and/or assessments of the impacts of the fishery on the environment based on existing information. It is also assumed that such an evaluation is based, at least in part, on information from fished versus unfished areas. Also, 2C expects that there have been studies to address specific identified impact issues, and that scientifically robust methods have been used to evaluate ecological risk assessments.[104] The Rock Lobster fishery failed to attain the passing mark of 80 percent needed to meet the earlier-mentioned criteria. However, the reasons why the fishery failed on this point are not included in the public report. As is the case for most of the fishery's flaws, the reader is unable to draw any substantive conclusions regarding the weaknesses of the fishery operation, as they are simply not discussed in the report. Exceptions are the description of the elements needed to meet the 80 percent scoring guidepost and the statement that the appropriate performance level for this indicator has not been met. Describing the elements needed to meet the 80 percent scoring guidepost and then stating that the appropriate performance level for this indicator has not been met do not provide any guidance to the potential consumer.

Lack of Attention Given to the Production of the Public Report
The assessment team called for increased participation from the environmental community within 24 months of certification.[105] Interestingly, the report quotes this requirement as stemming from Principle 2, Criteria D (2D), but surprisingly, in the Performance Criteria and Scoring Guideposts designed specifically for the Rock Lobster fishery, 2D has no mention of transparency in the decision-making process. Rather, 2D centers on unacceptable impacts on the ecosystem, populations, or habitats. Faced with this, interested consumers might wonder how the Guideposts and the public re-

---

104. Performance criteria and scoring guideposts against the Marine Stewardship Council principles and criteria—*Western Australia Rock Lobster Fishery Issue* 1, p. 9.
105. MSC rock lobster report (n. 103 above), p. 26.

port fit together, and how much attention is dedicated to producing the public report. This may simply be an oversight in the publication of the report. But with all the MSC requirements regarding the format that must be followed by the certification body to provide information to the public,[106] it is worrisome that the MSC does not appear to be concerned about such details. Moreover, it would appear that fisheries are announced as being newly certified even before the MSC itself receives the public report.[107] It seems odd that these public reports—the crucial link that establishes trust and transparency between the MSC and the potential consumer—are left to the sole care of certification bodies, seemingly without any input or verification from the MSC.

No Data on By-catch

The assessment team required that a formal system for recording by-catch data be formalized within 12 months.[108] The report does point out that numerous environmental groups raised concerns about the incidence of by-catch, damage to habitats related to fishing gear use, and harmful effects on endangered and threatened species.[109] The public report does not describe any of these potential issues, nor the reasons why the assessment team felt these issues were not significant enough to prevent certification.

The Implications of Public Reports that Lack Substance

Even though guidelines were specified in the MSC Certification Methodology document, it is surprising that such significant differences in quality exist between the Rock Lobster and other certification public reports. According to both the Accreditation Manual and the Certification Methodology, the public certification summary must include, inter alia, the following elements: general information about the fishery and its management system, a general description of environmental and socioeconomic concerns, details on the assessment team and assessment process (i.e., stakeholder consultation), main strengths and weaknesses of the fishery, conclusions, conditions, and recommendations.[110] Interestingly, all these elements were present in

---

106. *MSC Certification Methodology* (n. 1 above), Section 15, pp. 28–29.
107. David Bell, MSC Fisheries Research Assistant, Personal Communication, 18 October 2000. Communication via e-mail concerning the Alaska Salmon Public Report: by 18 October 2000; it had not been made available to the MSC although it had received certification on 5 September 2000. This potential scenario appears to have been corrected by section 16.1 of the *MSC Certification Methodology*. Ibid.
108. MSC rock lobster report (n. 103 above), p. 26.
109. Ibid., p. 22.
110. MSC Accreditation Manual (n. 46 above), Part 3.3, Section 3, pp. 54–55. See also *MSC Certification Methodology* (n. 1 above), Section 15, pp. 28–29.

the Rock Lobster report. Yet so few details were provided that a reader comes away with much less knowledge and confidence in the process than with the other certification reports, even though formal MSC requirements were followed. Given that consumers must trust the value of the ecolabel for it to be an effective marketing tool, public documents should at least provide enough information to convince the consumer to purchase that product if consumers choose to investigate beyond the label.

Conclusion on the Rock Lobster Fishery: Readers Aren't Given Sufficient Details

The fact that the assessment team gives potential readers so few details pertaining to the incomplete risk assessment may potentially undermine the trustworthiness of the MSC label: if the product is sustainable and well managed, why are readers kept in the dark? If consumers invest the time to read lengthy public reports to gain more knowledge about a given product, they should not be left guessing at the reasons why the assessment team recommended the correction of certain weak points: problems and solutions should be clearly spelled out. If information is to be made public, then it should be substantive. Otherwise, transparency is damaged if the public reports are void of any substance and serve more as a public relations tool.

In this fishery certification, we see the emergence of a problem that could potentially plague the MSC: although basic requirements have been met, there is a world of difference between the quality of the Rock Lobster public report and the public certification reports of other MSC certified fisheries. Such inconsistency impresses upon the consumer the appearance that the MSC lacks interest over constancy in the process, at least as far as the public reports are concerned. A consumer may then be left to wonder if the expert evaluation is also lacking in constancy. Unfortunately, finding an answer to this question is beyond the reach of most citizens. Although it is important to allow some flexibility to those carrying out the evaluation, the MSC should realize that such variability in the quality of public reports could cause consumers to doubt the ability (or concern) of the MSC to oversee and ensure that certifiers are carrying out assessments appropriately. After all, much of this scheme's success ultimately depends on whether consumers trust the MSC claim enough to purchase certified products.

The Thames-Blackwater Herring Driftnet Fishery: How a Fishery Can Be Certified Before Meeting the MSC Principles and Criteria

The Thames Blackwater Herring is a small fishery located in the United Kingdom that received MSC certification in March 2000. Like the Western

Australia Rock Lobster and the Alaska Salmon, the Thames-Blackwater Herring fishery was once closed due to low levels of catch.[111]

This herring fishery sought the MSC ecolabel to get a competitive advantage over the larger and less expensive North Sea Herring.[112] Since certification, fishery officers claim they have received up to 50 percent more for their catch.[113]

Weaknesses of the Herring Fishery

Authorities Do Not Have the Ability to Close the Fishery Once the TAC Is Reached
Within the Regulatory Area, all vessels participate in the driftnet fishery, but just outside this area, a pair-trawl fishery catches significant amounts of Blackwater-Thames Herring. These trawlermen, who believe their gear produces a better quality of catch, disagree with the reasons for their exclusion from the Regulatory Area and do not support the MSC certification scheme.[114]

In the 1997–1998 season, 50 percent of the herring catch was obtained outside the Regulatory Area.[115] Presently, the Ministry of Agriculture, Fisheries and Food (MAFF, which determines the total allowable catch) and the Kent Essex Sea Fisheries Committee (KESFC, which is responsible for enforcing bylaws) are unable to fully close the fishery once the total allowable catch is reached.[116] The public report does not clearly explain why this problem occurs, but it would appear this is caused by "administrative inconsistencies" that remain unknown to the reader.[117] Once the total allowable catch (TAC) is met inside the estuarine area, the fishery closes, yet the trawlers off the coast are allowed to continue to fish until the TAC for the North Sea herring stock is met.[118] This poses a particular threat to the fishery not only because it has been subject to poor recruitment for the past 6 years,[119] but also because driftnet fishermen can potentially see all their efforts to

---

111. Fisheries certification public summary. *Marine Stewardship Council and SGS AgorControl* (March 2000), 1. (Thames-Blackwater Herring Driftnet). Accessed 9 September 2002 on the World Wide Web: http://www.msc.org.
112. Marine Stewardship Council (undated). MSC Certification of the Thames Herring Driftnet Fishery: Frequently Asked Questions (FAQs). Accessed 9 September 2002 on the World Wide Web: http://www.msc.org.
113. MSC, News Release, "First Proof of MSC Label Benefits to UK Fishing Industry—Thames Herring Prices up by 50%," 2 November 2000, Accessed 9 September 2002 on the World Wide Web: http://www.msc.org.
114. MSC and SGS AgorControl Herring Report (n. 111 above), p. 2.
115. Ibid., p. 7.
116. Ibid., p. 19.
117. MSC Herring FAQs (n. 112 above).
118. MSC and SGS AgorControl Herring Report (n. 111 above), pp. 7 and 17.
119. MSC Herring FAQs (n. 112 above).

fish responsibly being undermined by the trawl fishermen who continue to fish according to a different TAC, just outside the Regulatory Area.

The assessment team actually designated this issue as a major corrective action: a problem deemed significant enough to prevent certification.[120] Although herring catches have not exceeded the established TAC in the past 3 years, the KESFC was obliged to develop a plan to rectify this problem within a 2-year timeframe.[121] As a consequence, this major corrective action was downgraded to a minor corrective action, allowing the fishery to be certified.[122] Indeed, the inability to close a fishery once the TAC has been met encourages potential overfishing and is contrary to the fundamental concept of sustainable fishing.

A consumer purchasing Thames-Blackwater herring may do so because of the belief that it is a sustainably harvested and environmentally superior fish product. Even if consumers are educated about the substance and reasoning behind the three MSC principles, it is unlikely that they will take the time to sift through the public report and other documents to discover if these principles were strictly adhered to during a specific fishery's certification process. Consumers could assume that the Principles and Criteria have been met and would undoubtedly be very surprised to hear that a pair-trawl fishery may be involved in potential unsustainable fishing practices, just outside the Regulatory Area.

This scenario also leaves us to wonder if the pair-trawl Thames herring fishery will benefit from its association with the MSC-certified driftnet herring fishery. Although the fish caught in the pair-trawl fishery will not display the MSC logo, it is feasible that consumers may purchase it because they mistakenly believe that all Thames-Blackwater herring has been certified by the MSC. Only future marketing and consumer behaviour studies will shed light on the severity of this potentially dangerous association.

Even if this herring fishery has been deemed as sustainable and well managed, one can note that the conclusions of the report, although generally favorable, demonstrate significant departures from the Principles and Criteria used to define a sustainable fishery. This problem was circumvented by granting the herring fishery a conditional certification.

A conditional certification imposes an obligation upon the client fishery to provide a solution to a specified problem within a certain time frame. In this case, KESFC has agreed to "develop an appropriate mechanism following a timetable"[123] and readers are unable to discover just what that entails. Before conditional certification is granted, the client must agree in writing

---

120. MSC and SGS AgorControl Herring Report (n. 111 above), p. 19.
121. This plan is not available on the MSC Web site, nor was it available on KESFC related Web sites.
122. MSC and SGS AgorControl Herring Report (n. 111 above), pp. 19–20.
123. Ibid.

to abide by certain conditions. The consumer, however, will not be aware that any conditions are attached to the certification unless the public report is read, as the logo does not differ between a conditional fishery and an unconditional one. Ultimately, the certification is withdrawn if conditions are not met within the specified timeframe.[124]

So the problem endures. Consumers today are purchasing what they believe to be certified as sustainable products. The product believed to be sustainable is undergoing an assessment that will only be resolved within the next 2 years. Although the assessment team agreed to certify the fishery and was of the opinion that it merited certification, the MSC has allowed its label to be displayed on a product that is presently weak on certain basic elements of the MSC Principles and Criteria.

Data Are of Questionable Quality
MAFF sets the TAC by "largely" following scientific advice of the government research agency Centre for Environment, Fisheries and Aquaculture Science (CEFAS).[125] Yet CEFAS was unable to provide documentation on the methodology used to calculate both the TAC and the Sustainable Spawning Biomass.[126] This resulted in a minor corrective action. Also, the stock assessment itself may be flawed because CEFAS conducts its annual survey (on which stock assessment is based) in October at a time when conditions in the estuary differ greatly from conditions out at sea. Doing so may result in inadequate estimates.[127]

CEFAS makes another stock assessment (on landings) once recruitment data become available and may then modify the seasonal TAC. Yet the quality of this recruitment data is also questionable as fishermen estimate the weight of their landings and submit a catch form to MAFF. These fish landings are not actually weighed nor are there any cross-correlations of data undertaken, although some samples are sent to CEFAS for analysis.[128] The absence of any cross-checking of data also resulted in a minor corrective action. These weaknesses amount to management decisions being based on unverified estimates of available fish quantities and numbers caught—a far cry from the MSC's requirement embodied in the first Principle that fishery data must be well known. In addition, there is absolutely no record of bycatch and discards.[129] Nonetheless, the assessment team claimed the data,

---

124. MSC Accreditation Manual (n. 46 above), Part 3.1, Section 9.
125. The public report states that the TAC is based "largely" on the scientific advice of CEFAS, leading the reader to assume that the TAC is perhaps not solely based on scientific advice. See MSC and SGS AgorControl Herring Report (n. 111 above), p. 5.
126. Ibid., p. 19.
127. Ibid., pp. 17–18.
128. Ibid., pp. 6 and 19.
129. Ibid., p. 3.

although dependant on voluntary fishermen's contributions, appeared to be of good quality and fishery stock assessments were extensive given the small size of this fishery.[130]

In addition to the two minor corrective action descriptions noted earlier, the following elements merited minor corrective action at the time of certification: there was no formal management plan; some fishermen (trawlers) had not been consulted;[131] there is a systematic potential for economic incentives to cause nonsustainability of the fishery as it is essentially open access and is easily accessible to a greater number of vessels than are presently exploiting it.[132] In addition, because large nets were used, there was an obvious danger of "ghost fishing" when the nets are accidentally lost. The report claimed that ghost fishing is inconsequential in this fishery, thus it was spared a corrective action although no data are available on the gravity of the problem.[133]

Strong Points of the Herring Fishery

Driftnets adequately target herring and have very little impact on the ecosystem because other species are rarely caught in the nets.[134] Moreover, there is no evidence that trophic links and relationships are adversely affected.[135]

In some cases, endangered and threatened species were accidentally caught in the herring driftnets, notably Twait shad and Allis shad. These fish are threatened mostly by activities other than fishing and their by-catch is negligible.[136] English Nature, an environmental nongovernmental organization, claims the incidental catches of these two species are sustainable.[137] Fishers further explain that these fish do not make their way into the nets for two reasons: because they only enter the estuary long after the driftnet fishery is over and because they are too big for the nets.[138]

The local community is generally supportive of the labelling initiative, particularly fish merchants and those directly involved in the certified fishery. Scepticism stems from persons indirectly involved who are pessimistic of the benefits of the certification and ecolabelling and fear interference by the MSC in local fisheries management.[139]

130. Ibid., p. 17.
131. Ibid., p. 19.
132. Ibid., pp. 8–22.
133. Ibid., p. 11.
134. Ibid., p. 15.
135. Ibid., p. 15.
136. Ibid., p. 11.
137. Ibid., p. 15.
138. Ibid., p. 15.
139. MSC Herring FAQs (n. 112 above).

Oddly enough, unlike other certified fisheries, there is no Performance Criteria and Scoring Guideposts document for the herring fishery that is available to the general public. Hence, interested consumers cannot have a clear idea of the criteria used to evaluate this fishery. This seems to be at odds with the information provided in the Certification Methodology document, which states that the MSC must publish performance indicators and scoring guidelines on their Web site to allow stakeholders to comment.

Conclusion for the Thames-Blackwater Herring Fishery: Implications of Conditional Certifications

The public report on the Thames-Blackwater fishery clearly indicates the strengths and weaknesses of the fishery, as well as which weaknesses resulted in conditional certification. The reader immediately notices that the Herring report is much more thorough than the Rock Lobster report.

One can readily observe some faults in the certification process. I would argue that this illustrates what is perhaps the biggest defect in the MSC's certification process: the herring fishery was certified as sustainable even though there were major problems within that fishery. The MSC logo will not inform consumers that certification is conditional. The most significant implication in this certification is that a certificate has been awarded *prior to* the fishery meeting the MSC's Principles and Criteria. It is the author's opinion that awarding certification prior to meeting all the criteria may undermine the confidence and trust of consumers. Unless consumers are very diligent about reading the lengthy public report, they will never know that the MSC is certifying fishery products that will only 'pass the test' in a few years time. The danger in this is that consumers may lose trust in the MSC scheme. Consumers' reaction to this certification will depend on their level of awareness.

Alaska Salmon: Fish Management that Is So Flexible, the Fishery Can Be Opened or Closed on a Daily Basis

In September 2000, the MSC announced that the Alaskan Salmon fishery would qualify for the MSC ecolabel.[140] As a result, the Alaskan Salmon is the first fishery to be certified in North America.

As salmon is in demand all over the world, it was hoped that the certification would help salmon products gain a greater foothold in European

140. ENN Press Release, "Alaska salmon earn landmark seal of approval," 7 September 2000. Accessed 9 September 2002 on the World Wide Web: http://www.enn.com/enn-news-archive/2000/09/09072000/asalmon_31195.asp.

markets where "green" products tend to have a bigger impact on consumers than they do in North America. Already, some corporations in Europe have claimed they are eager to purchase certified salmon.[141] Indeed, perhaps the popularity and high demand of Alaskan salmon will make the whole notion of sustainably harvested certified fish products surge forward. Consumers around the world will have an opportunity to gain awareness and hopefully enhance the movement to discriminate between products that are fished responsibly from those that are not.

In Alaska, salmon are harvested by driftnets, gillnets, purse seine, and trawling. The Alaska Salmon fishery boasts a strong, healthy fishery, but this was not always the case: overfishing caused the salmon fishery to plummet after 1940, resulting in salmon runs becoming so low that the regions of Alaska were declared a national disaster area in 1953.[142] Since then, the Alaskan salmon has arguably been on an uphill journey to success, and salmon hatcheries have existed since the late 1970s.

Uniqueness and Strengths of the Alaskan Salmon Fishery

One unique feature about the Alaskan salmon fishery is that unlike other certified fisheries, it is a multispecies fishery that has been certified as a whole. In other words, five species of salmon have been included in the certification rather than just the one.[143]

Another unique feature, from a management perspective, is that the authority to open or close a fishery was delegated directly to area resource managers, those who arguably possess the best knowledge for the management of the resource. Because authority is delegated to those overseeing the daily salmon fishery, this allows for management that is immediate in its effect. Those who monitor the fish have the authority to close the fishery, or open it, on a daily basis. This authority is vested in them under an Emergency Order that "gives the department field staff the authority to make regulatory announcements based on management decisions that can be placed into effect immediately and carry the full force of the law."[144] Measuring salmon abundance frequently gives resource managers a better grasp of the number of fish in the rivers at any given time. This method of verifying

141. Ibid.
142. Alaska Department of Fish and Game. Alaska Salmon Management: A story of success. Accessed 9 September 2002 on the World Wide Web: http://www.state.ak.us/adfg/ under "Publications."
143. The five species are Chinook, Sockeye, Coho, Chum, and Pink Salmon.
144. Marine Stewardship Council and Scientific Certification Systems Inc. (September 2000). The summary report on certification of commercial salmon fisheries in Alaska, p. 5. Accessed 9 September 2002 on the World Wide Web: http://www.msc.org.

128   *Living Resources*

the quantity of salmon returning to their spawning grounds is a safer way to monitor the fishery than merely providing an estimate at the beginning of the season.

Stakeholder consultation was extensive for this fishery. Stakeholders in both the United States and Canada had to be contacted. This process is well documented in the public report. Although the efforts in contacting various stakeholders were substantial, their comments were minimal yet positive as general satisfaction was expressed with the Alaska Salmon fishery as a whole.[145]

Weaknesses

It is difficult for the reader to determine exactly what the weaknesses of the salmon fishery are, as these were not discussed at length in the report. In fact, only the last three pages of the report point out which Performance Indicators called for specific requirements for continued certification.

To maintain certification over time, the Alaska Department of Fish and Game (ADF&G) must comply with four indicators. To begin, the ADF&G must determine the number of salmon managed on different types of escapement goals and categorize the various stocks. The ADF&G must provide for the continued management of the fishery in the event of a substantial change in the ocean survival rate, as observed in the 1950–1970s. Also, the ADF&G must provide, inter alia, details on (a) the types of analysis employed, (b) an assessment of what catch and socioeconomic impacts may ensue if lower ocean survival rates should occur, and (c) how the ADF&G would respond to these conditions. Also, a sampling program to identify fish, bird, and marine mammal by-catch must be implemented within 3 years. Finally, within 2 years, the ADF&G is obligated to report on the progress made concerning both the reduction of the number of permits and research conducted on the interactions between hatchery programs and wild stocks.[146]

Although the report points out which actions are essential to have the certification upheld, the reasons why these actions are necessary are neglected in the document. Similarly to the Rock Lobster public report, the consumer is informed of improvements required in the salmon fishery, but is unapprised of the gravity of these difficulties.

As will be discussed later, the New Zealand Hoki fishery has received a minor corrective action because fishery management responsibilities were improperly defined. Yet the assessment team for the Alaska Salmon fishery did not determine that their statement to the effect that "authorities are

145. Ibid., pp. 18–19 and 21–27.
146. Ibid., pp. 30–32.

overlapping and there may be contradictory or confusing regulations"[147] was an obstacle warranting a corrective action. Thus, consumers may wonder if certain fisheries are assessed more rigorously than others are.

What Is the Effect of the Declaration of a State Disaster Emergency?

Although scientists are aware that cyclical variations are a normal occurrence in salmon populations, recent salmon declines have surprised many fishers.[148] Yet a potential glitch in the MSC certification became apparent on 19 July 2000, when Alaska State Governor Tom Knowles declared an economic disaster in numerous commercial salmon fishing communities in Alaska due to extremely low runs of salmon.[149] The area affected covered 240,000 square miles, or 41 percent of the surface area of the entire state, which included 30,000 residents, 80 percent of which were Native Americans.[150] As a result of this disaster, many individuals required assistance to secure fuel, food, and basic necessities. In fact, the Declaration of Disaster states that biologists doubt the future viability of the resource in the affected area. It is noteworthy that the salmon fishery is open only in areas not affected by the declaration of disaster.

How will this recent news of poor salmon runs in portions of Alaska affect the MSC certification of Alaska salmon? In the MSC's "Frequently Asked Questions" concerning the Alaska Salmon certification, their response to the above question was notably vague. However, the MSC stressed that fishery closures ensure sustainability and prevent disaster.[151]

Only time will tell how the disaster may affect, if at all, the MSC certified salmon fishery. The author acknowledges that salmon fisheries can exist sustainably in some geographic areas while not in others. Yet, there is a certain irony in a scenario where consumers on the one hand are purchasing sustainably harvested salmon products and on the other knowing that some Alaskans are in a dismal economic situation because of low salmon numbers.

147. Ibid., p. 8.
148. (30 July 1998). Scientists: Alaska's salmon bust may be Northwest bounty. *Seattle Times Company*. Accessed 9 September 2002 on the World Wide Web: http://seattletimes.nwsource.com/news/local/html98/fish_073098.html. Available in archives.
149. Governor Tony Knowles (19 July 2000). *Declaration of Disaster Emergency*. Office of the Governor, State of Alaska. Anchorage, Alaska: 24 August 2001. Accessed 9 September 2002 on the World Wide Web: http://www.gov.state.ak.us/press/01-08-24_disaster.html.
150. Accessed 9 September 2002 on the World Wide Web: http://www.ak-prepared.com/ykn/images/fisheriesmapweb.jpg.
151. Marine Stewardship Council (undated). Alaskan Salmon Certification (Frequently Asked Questions). Accessed 9 September 2002 on the World Wide Web: http://www.msc.org.

It is easy to see that without a public education program, some consumers may be confused by the situation.

Conclusion on the Alaska Salmon Fishery: The MSC Must Take Notice of Consumers' Perception of the MSC Claim

Once again the consumer is left wondering what kind of deficiencies arose from this MSC-certified fishery. It should be a source of concern that consumers are unable to draw substantive conclusions from a report they have read. Moreover, consumers could be quite perplexed at the varying degrees of thoroughness given to various fishery reports. For the process to be trustworthy, it is imperative that the MSC insist on the publication of thorough and in-depth public reports. This would enable consumers to judge for themselves whether or not the labelled product is worth purchasing. In the end, the success of the MSC labelling scheme will ultimately depend on its ability to entice consumers to purchase certified products.

The Alaska Salmon fishery seems to have raised little concern among stakeholders. All seem to agree that this fishery is sustainable and well managed and worthy of such recognition. Although only time will tell if the Declaration of Disaster confuses potential consumers, the MSC should give much thought as to whether or not consumers will perceive this fishery to be sustainable. Ultimately, if consumers (especially Alaskan consumers) have difficulty understanding that the MSC can label salmon as being sustainable while having a large number of people affected by low salmon runs, people may not have much faith in the accuracy or trustworthiness of the MSC claim.

New Zealand Hoki: MSC Stamp of Approval Angers Seal Protectionists

The New Zealand Hoki public report is clear, concise and thorough—any reader should start with this report as the MSC process is plainly described. It does not suffer from the lack of information and explanation found in other public reports. The thoroughness of this particular report is evident by the fact that a significant proportion of the document is dedicated to an explanation of each criterion and indicator and then discloses the assessment team's findings. By reading the lengthy report, readers can gain precise knowledge on the scores earned for each indicator.

Weaknesses with the Hoki Fishery

There are two substocks of hoki in New Zealand: the western and eastern. Although not genetically distinct, the stocks have different biological characteristics and can be fished at different rates. Each stock must therefore be considered and modeled separately.[152] Although the eastern stock is about one-third the size of the western stock, it accounts for 40 percent of the hoki catch.[153]

The result of assessing the western hoki stock is inconclusive because of the existence of two data sets and two modeling techniques. However, both sets of data confirm that the western stock biomass is above the levels required for sustainable yield, and is hence a healthy stock in this respect.[154]

As for the eastern hoki stock, the situation is entirely different. The threat of overfishing was identified during stakeholder consultations.[155] If current fishing levels continue, there is a high probability that the eastern hoki will fall below the maximum sustainable yield level needed to ensure the stock can be fished on a long-term basis. In fact, the eastern hoki stock biomass may fall below the maximum sustainable yield level within the next 5 years.[156]

Obviously the risk to the eastern stock would decline if fishing efforts were decreased. However, a problem arises from the present management regime that does not recognize the existence of two very distinct substocks of hoki. Present management is based on Quota Management Area boundaries, rather than the biological boundaries of the two substocks.[157]

It is the author's submission that a fishery cannot be certified while there are serious doubts concerning the sustainability of one of its substocks. Potential nonsustainable fishing pressure of the eastern hoki substock is a clear violation of the first MSC Principle, which states that a fishery must be conducted in a manner that prevents overfishing. This has resulted in a minor corrective action requiring the Hoki Fishery Management Company to implement an action plan.[158] Yet it is hard to believe that although the long-term sustainability of a substock of the targeted New Zealand hoki is dubious, this fishery nonetheless qualified for the MSC logo.

Impact of Hoki Fishery on Benthic and Pelagic Environments, Seals, and Seabirds

Benthic and pelagic mid-water trawls are used in the hoki fishery. The impact of bottom trawling on the benthic biodiversity is unknown.[159] This is

152. *SGS Products and Process Certification.* Marine Stewardship Council. MSC Fisheries Certification. New Zealand Hoki: March 2001, p. 6. Accessed 9 September 2002 on the World Wide Web: http://www.msc.org.
153. Ibid., p. 7.
154. Ibid., p. 8.
155. Ibid., p. 59.
156. Ibid., p. 37.
157. Ibid., p. 16.
158. Ibid., p. 64.
159. Ibid., p. 22.

132   Living Resources

further exacerbated by the fact that knowledge of these environments and natural functional relationships are limited.[160] The report even states that if deepwater benthic habitats are damaged by the hoki fishery, "recovery within a reasonable time scale is unlikely."[161] As well, there has not been a full ecological assessment for the hoki fishery, and impacts on non-target species are poorly understood.[162]

The public report points out that more knowledge is needed regarding the impacts of the fishery on seals and seabirds, and that very minimal attention has been dedicated to other key species, including sharks.[163] In fact, the very approach used to determine the impacts on seal and seabird populations is limited and more advanced population dynamic models were not used (the report is silent as to why this has occurred).[164] The report also states that the assessment of risks to seals is inadequate.[165]

Seals

There was considerable evidence that seals and seabirds were caught in the process of fishing hoki.[166] Although seals are not considered to be a vulnerable or threatened population, the size of the New Zealand fur seal population is unknown.[167] On this issue, a reader may note a serious clash between the authors of the public report and New Zealand environmentalists. The public report states that "there is an occasional, incidental by-catch of New Zealand fur seals and seabirds in the hoki fishery."[168] Yet according to Cath Wallace, Marine Co-ordinator for Environment and Conservation Organisations of New Zealand (a network of groups that share a concern for the environment), the New Zealand hoki fishery "drowns over 1000 fur seals each year. A multitude of other animals are crushed by trawl nets when these scrape across the bottom."[169] Environmentalists allege that the hoki fishery may have been responsible for over 5600 seal deaths between 1989 and 1998,[170] and that industry has been slow at implementing seal excluder devices.[171]

  160. Ibid., p. 39.
  161. Ibid., p. 41.
  162. Ibid., pp. 41–42.
  163. Ibid., pp. 44–45.
  164. Ibid., p. 44.
  165. Ibid., p. 45.
  166. Ibid., p. 46.
  167. Ibid., p. 12.
  168. Ibid.
  169. B. Burton, "Stamp of approval for NZ fishery angers seal protectionists," *Environment News Service* (20 March 2001). Accessed 9 September 2002 on the World Wide Web: http://ens.lycos.com/ens/mar2001/2001L-03-20-02.html.
  170. Ibid.
  171. MSC and SGS Products and Process Certification Hoki Report (n. 152 above), p. 60.

Seabirds

Stakeholders raised the concern that seabird data were inadequate as dead seabirds would not always be included in the catch and would usually remain unobserved.[172] Moreover, Barry Weeber, speaking for the Royal Forest and Bird Protection Society, claims that albatrosses make up 60 percent of the 1100 seabirds caught while fishing for hoki.[173] The species most affected by the hoki fishery include the Buller's and Salvin's albatrosses,[174] both considered vulnerable by the Species Survival Commission of the International Union for the Conservation of Nature (IUCN).[175]

The authors of the report assert that the hoki industry is developing measures to reduce the mortality of seabirds.[176] The report also alleges that the fishery does not pose threats to any endangered or threatened species. This may be true, as the albatross species mentioned are only considered vulnerable, but the reader is left to wonder how threatened the albatross may become in the future in light of hoki fishing activities.

Conclusion for the New Zealand Hoki Fishery: The MSC Cannot Afford Bad Press on Behalf of Environmentalists

After surveying the New Zealand Hoki public report, the reader may be left to wonder just how many minor corrective actions a fishery can receive and still be granted certification. In the New Zealand Hoki fishery, of the 31 indicators measured, 10 necessitated minor corrective actions and arguably, some of these are not so 'minor.'

After the hoki fishery received the MSC certification, environmentalists were quick to point out the flaws that affect seal and seabird populations. If environmentalists are not giving their blessings to a certification program that aims to help conserve fish stocks and promote sustainable fishing prac-

---

172. Ibid.
173. Burton (n. 169 above).
174. The Buller's Albatross is classified as vulnerable as it is restricted to a very small breeding area. Major threats are accidental mortality and alien invasive species. If a decline is observed, it may be upgraded to endangered status. For the Salvin's Albatross, major threats include accidental mortality, atmospheric pollution and natural disasters. It is also classified as vulnerable because breeding is largely restricted to one tiny group of islands. The 2000 IUCN *Red List of Threatened Species,* accessed 9 September 2002 on the World Wide Web: http://www.redlist.org/.
175. Burton (n. 169 above). Weeber referred to a third species of vulnerable albatross, the "white-capped" Albatross. This species was not found in the IUCN database under that name. The scientific name of the species was not provided in the reporting, or in the Hoki Public Report (n. 152 above). The Hoki public report did, however, also refer to three un-named albatross species as being vulnerable.
176. MSC and SGS Products and Process Certification Hoki Report (n. 152 above), p. 12.

tices, consumers will likely start asking themselves questions. It is important to note that environmentalists did not condemn the MSC certification process, but rather heavily criticized the fact that the hoki fishery was awarded certification notwithstanding it arguably has such adverse consequences on wildlife.

Once again, this potential problem hinges on public perception of the MSC certification scheme. In the Hoki Public Report, it is hard to determine how stakeholder comments and concerns were considered. Were the concerns of organizations such as Environment and Conservation Organisations of New Zealand and the Royal Forest and Bird Protection Society ignored? Are the environmentalists simply wrong? The reader doesn't know. Either way, after reading conflicting opinions between the public report and the press releases issued by environmental groups, readers are undoubtedly left without the knowledge of who has a better grasp on levels of wildlife mortality—the assessment team or the environmentalists? It is the author's opinion that to maintain trust in the eyes of potential consumers, the MSC cannot afford to take the accuracy of impacts to wildlife very lightly. It is also the author's opinion that if any further such mishaps be reported by the media, the MSC scheme will not be able to gain the trust of consumers by having to release rebuttals to environmentalists' claims, and this will affect MSC credibility no matter how legitimate and deserving the certification program may be.[177]

To sum up, the reader will undoubtedly be left to wonder how the hoki fishery was granted MSC certification because the impact on seals is largely unknown and vulnerable albatrosses are caught in the nets. The underlying criticism that stems from this certified fishery is this: although minor corrective actions require that weaknesses be addressed to ensure that the hoki fishery keeps its MSC seal of approval, the obstacles plaguing the New Zealand fishery are not so minor after all.[178]

The Burry Inlet Cockle Fishery: On the Road to Small-scale Perfection

The Burry Inlet Cockle Fishery was certified in March 2001. It is located in southwest Wales and is licensed and regulated by the South Wales Sea Fisher-

---

177. B. May, "MSC fights back over 'greenwash claims'," *Worldcatch News Network* (20 March 2001). Article appeared on the Worldcatch News Network Web site on 20 March 2001. Accessed 10 December 2001 on the World Wide Web: http://www.msc.org under "Certified Fisheries." Author's note: In July 2002, the MSC provided an Objections Procedure by which organizations may lodge an objection to a determination that a fishery meets or does not meet the MSC standard.

178. Numerous minor corrective actions have since been addressed. See Fishing Surveillance Audit Report, available on World Wide Web www.msc.org. Accessed 9 September 2002.

ies Committee (SWSFC). Cockles are small bivalve mollusks that live in the intertidal zone of sandy or muddy beaches. Cockle license holders gather cockles by hand raking and sieving them.[179] In 2000, this Burry Inlet fishery yielded nearly 7200 tons of cockles to 55 full-time and 35 temporary license holders.[180] Cockles are collected in especially designed sacks that reveal a 'fill-line' and are labelled with the licensee's number. Quotas are established so that the fishery is not overexploited and most of the catch is sold in local markets.[181]

Interestingly, the Scoring Guideposts for the cockle fishery are the only ones to show numerical values. Hence, the reader is able to see the precise score for each indicator rather than just a pass/fail rating.[182]

Some Points of Minor Concern

Young mussels found on rocks (termed 'crumble') prevent certain areas from being hand raked for cockles. These young mussels, while being a food source for marine birds, are removed periodically with dredgers. The Countryside Council for Wales (CCW) approves of the removal of crumble.[183] Yet in spite of this, some crumble areas are maintained as a food source for birds.[184] Although the removal of crumble is thought to be within acceptable limits, further research is needed to study the relationship between cockle and mussel biomass.[185]

The ramifications of hand gathering cockles on non-target species have not been studied,[186] but by-catch is deemed to be negligible. Although incidental mortality has never been measured, it is not considered to be a problem because cockles are robust and a high survival rate is assumed.[187] Areas of crumble removal have not been studied specifically, but it is thought that the impact on mussels is negligible as the crumble continues to reappear.[188] There is no evidence of unacceptable impacts on oystercatcher shorebirds, which feed on cockle beds.[189] Landings and fishing practices are randomly

---

179. A. Hough and T. J. Holt (March 2001). Marine Stewardship Council and Moody Marine Ltd. certification report for Burry Inlet Cockle Fishery. p. 5. Accessed 9 September 2002 on the World Wide Web: http://www.msc.org.
180. Marine Stewardship Council (undated). Burry Inlet Cockles FAQ. Accessed 9 September 2002 on the World Wide Web: http://www.msc.org.
181. Ibid.
182. MSC and Moody Marine Ltd., Cockle Report (n. 179 above), pp. 17–44.
183. Ibid., p. 8.
184. Ibid.
185. Ibid., pp. 31–32.
186. Ibid., p. 26.
187. Ibid., p. 28.
188. Ibid., p. 27.
189. Ibid., p. 29.

136   *Living Resources*

inspected on a regular basis.[190] Curiously, the SWSFC is unauthorized to prosecute someone for illegally fishing cockles, but can demand that the cockles be returned to the beds.[191]

Interestingly, although many of the lower-scoring indicators were rated below 80 due to lack of information, for the most part, the risks were deemed to be totally acceptable. For example, the indicator 2D1 scored 70 on the indicator measuring whether there was adequate knowledge of the physical impacts on the habitat due to use of gear. The assessment attributed a score of 70 because "there is indirect evidence that the effects of hand gathering are transient and of low impact on habitat. No direct evidence exists."[192]

One would think that hand gathering cockles buried in mud is the fishing method that would be least intrusive on the ecosystem. However, the assessment team gave this indicator a score of 70 because the negligible environmental impacts of hand gathering cockles were unproved. It is the author's opinion that the explanation advanced for this low score enhances consumer trust in the MSC scheme because the assessment team refused to accept the theory that the impact of hand gathering is likely negligible. The assessment team pointed out that direct environmental impact information was lacking and assigned a low score accordingly. Indeed, the cockle Scoring Criteria illustrates numerous indicators that have scored relatively low for similar reasons pertaining to lack of direct observation.

It was decided that only three conditions were to be imposed on this fishery: (1) formalized fishery management objectives to meet short- and long-term resource and environmental objectives must be clearly stated (although such goals are already implicitly recognized); (2) details of cockle growth, mortality, and so forth, must be integrated in the calculation of fishable biomass, and be clearly documented; and (3) short- and long-term objectives will soon be set in terms of ecosystem impacts and impact minimization as the estuary is a Special Protection Area and a candidate Special Area of Conservation. These objectives must be set for fishery management purposes. In this regard, the SWSFC will have to cooperate with the CCW and other authorities in formulating these objectives.[193]

Conclusion for Burry Inlet Cockle Fishery: Giving Readers the Ability to Understand the Assessment Team's Evaluation

In this most recently certified fishery, we see that the assessment team's report has a new and different format, one that focuses on listing all the crite-

   190. Ibid., p. 43.
   191. Ibid., p. 43.
   192. Scoring Criteria 2D.1 for the Burry Inlet Cockle Fishery, Ibid., p. 30.
   193. Ibid., p. 47.

ria evaluated and showing the reader the numerical score and the justification for it. It is the author's submission that pointing out numerical scores can satisfy readers as the numbers help make the certification much more transparent and accessible.

It is easy to see how this small-scale fishery could serve as a benchmark for similar sustainable operations around the world. Indeed, a community can gain much when its fishing operations are sustainable. There is the possibility that the MSC logo could bring with it an increased market for cockles accompanied by higher demand and higher prices. Should this happen, the SWSFC would have to carefully assess the increased pressure on the cockle fishery.

Ultimately, the Burry Inlet cockle fishery appears to be a prime example of how a fishery can be sustainable, have a low impact on its environment, provide employment, and contribute to the local fish markets while maintaining conservation objectives in an expressly protected estuarine habitat.

CONCLUSION

Will the MSC Scheme Be Available to Developing Nations?

Presently, the MSC has nearly thirty fisheries in some stage of the certification process. Interestingly, two of these fisheries are located in Mexico, a developing nation. Moreover, several corporations and groups from the South (developing nations) have shown their interest and support in the MSC scheme by becoming MSC signatories.[194]

For the MSC scheme to be truly unbiased, fisheries in both rich and poor countries should be able to have their sustainable fisheries qualify under the MSC. Doing otherwise would imply that only fisheries with monetary resources can qualify for the sustainable and well-managed designation, while the cost of certification may be unaffordable by poor fisheries.

It is the certification bodies themselves that set a price to carry out the assessment. Curiously, MSC documents are silent on the topic of how (and if) they have any control over what amounts can be charged for the assessment. If the MSC has no such guidelines, it should endeavour to create them quickly. Otherwise, corporations and fisheries in poorer countries could be excluded from the certification scheme even though their fisheries may be sustainable. As most fish caught in countries of the South are sold to wealthy nations,[195] the promotion of sustainable fishing practices is obviously a global issue of far-reaching consequences.

194. MSC Signatory Programme. Accessed 9 September 2002 on the World Wide Web: http://www.msc.org.
195. FAO. Focus: fisheries and food security. Accessed 9 September 2002 on the World Wide Web: http://www.fao.org/FOCUS/E/fisheries/trade.htm.

International Trade Concerns

The international trade implications of the MSC scheme is another issue that may be of concern, especially to countries of the South. The MSC certification is entirely based on how the fish was produced, which is referred to as a processing and production method (PPM). This feature is crucial as WTO/GATT panel report have stated that PPMs cannot serve as a reason to differentiate between goods.[196] One can consider only the physical characteristics of products. Hence for trade purposes, a fish harvested from a sustainable operation is the same as a fish harvested by dynamiting coral reefs or using cyanide.

The Technical Barriers to Trade (TBT) Agreement[197] arguably applies to the MSC's labelling scheme because labelling is considered a non-product-related PPM (labelling that does not affect the physical qualities of the product per se). The TBT states that all technical regulations and standards must not serve as disguised trade protectionism. Non-product-related PPMs fall under the category of technical standards if they are voluntary, as is the case with the MSC scheme.

Commentators from the South have questioned the legality of voluntary schemes, as it results in de facto discrimination. Although the GATT traditionally applies to government actions, this trend has brought up the issue of the applicability of the GATT/WTO regime to trade distortions caused by private groups.[198] In private ecolabelling schemes, from a trade point of view, the consequences on imports and consumer choices may be the same, regardless if the scheme is controlled by government or by a private party.[199] The TBT is applicable to nongovernment bodies, although these groups are directly regulated by governments rather than the TBT Agreement itself, hence members must prevent such private parties from creating unnecessary obstacles to trade.[200]

The WTO recognized these ambiguities in the potential applications of the TBT and created the Committee on Trade and Environment (CTE) to look into the problem. The report produced by the CTE in 1996 canvasses the wide scope of diametrically opposing opinions as to the applicability of TBT on various forms of ecolabelling schemes. On one hand, "the negotiating history of the TBT Agreement indicates clearly that there was no inten-

---

196. United States Restriction on Imports of Tuna, GATT DOC. DS21/R (3 Sept. 1991). (This was not adopted.)
197. Agreement on Technical Barriers to Trade (TBT), 15 April 1994. Marakesh Agreement Establishing the World Trade Organization, Annex 1A, I.L.M. 33 (1994), 81. Accessed 9 September 2002 on the World Wide Web: http://www.wto.org.
198. Okubo (n. 22 above), p. 630.
199. Ibid., pp. 631–633.
200. TBT Agreement (n. 197 above), Articles 3 and 4 generally.

tion of legitimizing the use of measures based on non-product-related PPMs under the TBT Agreement, and that voluntary standards based on such PPMs are inconsistent with the provisions of the Agreement as well as with other provisions of the GATT."[201] On the other extreme, "all forms of ecolabelling, including eco-labels that involve non-product-related PPMs, are covered by the TBT Agreement and that the inclusion of non-product-related PPM-based elements in an eco-labelling regime is not per se a violation of WTO rules."[202] The CTE offered no clarifications or solutions in its report, leaving us to wonder if the Agreement covers standards based on non-product-related PPMs.[203]

Needless to say, how trade rules will affect the MSC scheme will only be clarified when (and if) this unique ecolabelling initiative will be put to the test.

Can the MSC Keep Up with Future Growth?

Seeing that nearly thirty fisheries are presently considering the MSC certification, one must wonder how the MSC will manage so many fisheries around the world. Indeed, the six fisheries certified so far represent but a tiny fraction of fisheries in the world. To date, the MSC seems able to meet the task: the organization has come a long way since its beginning and has offered new documents, better-explained procedures, and constant updates over time. Also, the fact that a satellite office is now open in Seattle may lead us to believe the MSC may adopt some type of regional management approach when certified fisheries become numerous and widespread. But how the MSC will handle the management of so many fisheries remains to be seen. Although it may seem daunting for a relatively small and inexperienced operation like the MSC to control a proliferation of fisheries, it must be recalled that other labelling programs such as the Blue Angel Programme or Japan's Eco Mark have certified thousands of products. The MSC certification will serve as a test case to evaluate whether a third-party scheme can manage a fisheries certification programme on a global scale—and if the scheme will satisfy any technical guidelines that may eventually be issued by the Food and Agriculture Organization of the United Nations (FAO).[204]

---

201. World Trade Organization. *Report (1996) of the Committee on Trade and Environment*, at para. 70, p. 18. Accessed 9 September 2002 on the World Wide Web: http://www.wto.org.
202. Ibid., para. 73.
203. Ibid., para. 74-81.
204. The FAO has discussed the feasibility of developing guidelines for ecolabelling in the fisheries sector. The FAO stated that any agreement would have to establish clear accountability for the promoters of the scheme. The Nordic Council of Ministers took part in this consultation. See FAO. *Report of the Technical Consulta-*

In reality, it would appear that much of the significant labelling programs operating in the realm of sustainably harvested products lie in the hands of independent third-party initiatives, such as the Forestry Stewardship Council.

On a Positive Note . . .

An examination of the five fisheries allows us to draw some concrete conclusions about the MSC certification process and some of its deficiencies. Although I have outlined several flaws in both the certification process and how that process was carried out in the fisheries certified thus far, I would like to point out that these difficulties are not insurmountable and are far outnumbered by the positive influences the MSC labelling scheme can have on the development of sustainable fishing practices in the world.

It is the author's opinion that the most important flaw in the MSC certification process is the granting of a conditional certification. One could argue that, to be a completely trustworthy organisation, the MSC should delay certification until the certification body could wholeheartedly grant the "sustainably harvested" MSC logo. Such is the case for the Blue Angel Programme, where clients reapply for the label if they do not meet all the criteria. Although the MSC never directly mentions the rationale behind its preference for conditional certification as opposed to delayed certification, elements of the answer are fairly obvious. First, one of MSC's primary mandates is to increase awareness about sustainable fisheries among consumers. By delaying certification, consumers may not be aware that sustainability can play a role in fish products for several years, after all MSC requirements are met. Conditional certification allows consumers to become more aware of sustainability issues, at the expense of potential loss of trust if consumers learn that "sustainably harvested" fish products have actually been certified prior to meeting all MSC requirements.

Secondly, conditional certification may simplify financial matters for both the certification body and the client as the expert assessment team is paid only once (the consumer cannot determine if re-assessment visits are paid for separately). If the fishery certification was delayed until all MSC requirements were met, additional funds may have to be disbursed to re-hire an assessment team to ensure conditions are met. Delaying certification

---

tion on the Feasibility of Developing Non-Discriminatory Technical Guidelines for Eco-Labelling of Products from Marine Capture Fisheries. The Fisheries Department of the Food and Agriculture Organization. Rome, Italy: 21–23 October 1998. Accessed 9 September 2002 on the World Wide Web: http://www.fao.org/fi/faocons/ecolab/r594e.asp.

may also discourage potential clients from applying for certification, as it could be a lengthy process before they qualified for the MSC label.

Thirdly, there is merit in rewarding clients who manage their fishery responsibly, and they should benefit from their environmentally responsible behaviour. Given the negative outlook for world fisheries, time is of the essence and sustainable harvesting must make ground before too many fisheries collapse.

One must keep in mind that although corrective actions were issued for all fisheries and that some of these, arguably, departed from the MSC Principles and Criteria, all fishery clients have agreed to develop action plans to remedy these shortcomings within a specified timeframe. This in itself is a tremendous achievement, as surely no fishery in the world can be deemed perfect. As Brendan May, chief executive of the MSC points out:

> How, other than through the MSC, could the environmental community have persuaded a fishery publicly to agree to a series of corrective actions which must be undertaken in a specific time frame in order to retain the right to use our logo? Without the MSC process, and the fishery's willingness to participate in it, many of these issues could still be unresolved with no agreement on how or when to make much needed improvements.[205]

Though the response to this global initiative has thus far been very promising, only time will tell if the scheme will prove to be successful. We will have to wait to see how consumers react to the MSC label, and if they seek to buy certified fish products. The success of the scheme will ultimately depend on its ability to convince consumers that these certified fish products are the products they should be buying. Success will also depend on the industry's reaction to the initiative—though apparently favourable thus far, success will be achieved if there is a shift in the market share for sustainably harvested certified fish products, encouraging other fisheries to adopt (and market) responsible fishing practices.

Lastly, it is important that a scheme such as the MSC's certification process be open to public scrutiny. Obviously, the MSC aims to mandate such scrutiny as public reports and other information are accessible on its Web site. It is the author's opinion that the MSC would serve the public well by ensuring that its public reports are constant as between the fisheries, and by making it clearer to consumers why certain certifications scored poorly on specific indicators.[206]

---

205. May (n. 177 above).
206. Marine Stewardship Council (April 2000). *Performance Criteria and Scoring Guideposts Against the Marine Stewardship Council Principles & Criteria—Alaskan Salmon Fishery Issue.* On file with author.

Living Resources

Aquaculture in the Fiji Islands: The Need for a Comprehensive Statutory Framework

Timothy Pickering
Marine Studies Programme, University of the South Pacific, Suva, Fiji

INTRODUCTION

Aquaculture is a new industry to many parts of the Pacific, including Fiji. In times past it has been traditionally practised in some of the Pacific Island countries, but not in others. Nowadays many so-called "micro-states" in the Pacific, wishing to stimulate economic development but with meagre natural resources, have identified aquaculture along with fishing and tourism as ways to promote economic growth based upon their natural advantages: clean water and high marine biodiversity.[1]

Progress to develop aquaculture in most Pacific Island nations has been disappointingly slow,[2] and Fiji is no exception despite its relatively large size and its many infrastructure advantages. It is still very much a fledgling industry; however, many years of work to develop aquaculture in Fiji now appear to be on the verge of paying off, with some projects already operating commercially and with others in the pipeline. Fiji along with Cook Islands now leads the Pacific Island developing countries in terms of aquaculture development and diversification.

This article reviews the current institutional arrangements for aquaculture in Fiji and examines how these might be further developed so that this new sector of the economy may become better established. Definitions of "aquaculture" and its property-rights components are discussed, then Fiji's laws are analyzed to find out whether these components are present or absent within current institutional arrangements. Any missing elements will have to be provided for in legislation if the sustainable development of aquaculture industries in Fiji is to be successfully fostered.

---

1. J. Bell, *Aquaculture: A Development Opportunity for Pacific Islands*, (Noumea: Information Paper 19, 1st SPC Heads of Fisheries Meeting, Noumea, New Caledonia, 9–13 August 1999, Secretariat for the Pacific Community, 1999).
2. K. R. Uwate, *Aquaculture Development in the Pacific Islands Region*, (Honolulu: Pacific Islands Development Programme, 1984).

FIJI GOVERNMENT OBJECTIVES IN AQUACULTURE

The size of the economies of Pacific Island developing countries is small, and a large proportion of the population is part of the subsistence economy rather than the cash economy. The pool of domestic savings available for investment is therefore small, so Pacific Island developing countries rely heavily on foreign investment or external economic assistance to achieve economic growth.[3]

In past media statements, the government of Fiji has made it clear that it places a high priority upon economic growth, and sees attraction of foreign investment as vital to achieve this. Regarding its stance on aquaculture specifically, the government is supportive of appropriate and sustainable aquaculture development and is actively involved in such development. In doing so, the government of Fiji has four main objectives:[4]

1. *Food security*—improvement of human nutrition and increase in availability of protein at the subsistence level of the economy;
2. *Rural development*—provide an additional source of income and employment for artisanal fishers, reduce urban drift, and ease pressure on capture fisheries;
3. *Import substitution*—provide for local demand using locally produced seafood and avoid loss of foreign exchange on imported goods;
4. *Export earnings*—bring in overseas dollars and contribute to export-led economic growth.

CURRENT ECONOMIC VALUE OF FIJI AQUACULTURE

Food and Agriculture Organization (FAO) statistics for worldwide aquaculture production by region show that Oceania always comes last in terms of both tonnage and value.[5] Even then, the Oceania figures are dominated by Australia and New Zealand production of pearl oyster, edible oyster, salmon, and mussels, with Pacific Island aquaculture production being globally insignificant. Still, aquaculture has the potential to be regionally very significant in these smaller economies, as export earners, for import substitution (particularly to support tourist industries), and for food security.

---

3. R. Thistlethwaite and G. Votaw, *Environment and Development: A Pacific Island Perspective*, (Manila: Asian Development Bank, 1992).

4. M. Lagibalavu, "Fiji country report," in T. Adams, (ed.), *Meeting Report: First Heads of Fisheries Meeting, 9–13 August 1999* (Noumea: Secretariat for the Pacific Community, 1999).

5. FAO, *The State of World Fisheries and Aquaculture* (Rome: FAO Fisheries Department, Food and Agricultural Organization of the United Nations, 1999).

In Fiji, accurate economic statistics on aquaculture production and value are difficult to obtain. Fiji Fisheries Division staff collect statistics on municipal market sales of all fish products and imports, which are published in the Fiji Fisheries Division Annual Report series. The problem is that their database makes no distinction about the mode of production for the species being recorded, that is, aquaculture in comparison with capture fisheries. Presumably this is because, until very recently, Fiji fish production was almost entirely from capture fisheries; however, it makes identification of the aquaculture component difficult. In order to track the growth of the aquaculture sector and provide a measure of progress, the Fisheries Division will need to add aquaculture to the list of other fisheries sectors (artisanal, longline, pole and line, etc.) for which it collects statistics.

Similarly, estimates of the aquacultured component of fish exports were not available at the time of writing. In fact, on the value of all fish exports from Fiji, the Fisheries Division Annual Report 1997 states that; "It is difficult to give [an] exact estimate of the product due to the nature of [the] export market (Japan, Taiwan, USA—auction market). However estimated, the export value could well [be] above $F135 million. This contributes 5% to the nation[']s Gross Domestic Product." Since export licenses are required for fish products, it should be possible to extract this information from government statistics, although the mode of production may not be identifiable. It is still the case, however, that almost all of the edible aquaculture products are sold locally, while the exports are made up of non-edible products like dried seaweed, aquarium-market giant clams, and pearls.

With the available statistics it is possible to identify aquacultured production for the main commercial species, if they are species for which there is little capture-fishery production. For example, in 1996 the Fisheries Division recorded production of 122 T of tilapia in 1997, with a value of F$366,000, and sales of aquarium-sized giant clams of F$746. In 1998, tilapia production was 242 T valued at F$726,000, and virtually all was disposed of locally either for subsistence or through market sales. For others, such as *ura* (prawns), the statistics collected are for more than one type (*Macrobrachium* and *Penaeus* spp. combined) and from more than one source, so a breakdown of this market is hard to determine, however aquacultured prawns are currently estimated to be worth F$500,000 per year, of which almost all is sold locally. *Kappaphycus* seaweed production in 1999 was 120 T worth F$100,000, of which almost all was exported to Denmark. In 2000, seaweed exports totalled 515 T worth F$247,200 F.O.B.[6] No information is available about the value of blacklip pearls from aquaculture production in Fiji.

In summary, the aquaculture ventures that currently make money in Fiji (penaeid shrimp, tilapia, *Kappaphycus* seaweed, and blacklip pearl) are

6. S. Mario, personal communication. 2001.

currently worth about F$1.45 million per annum. This can be broken down roughly into a domestic value of F$1.2 million (shrimp, and tilapia) and exports of F$250,000 (seaweed), plus blacklip pearl for which the value is not known.

Aquaculture production in Fiji is therefore still very small at less than 10 percent of the value of all fishery exports. It can, however, be expected to increase. With a range of additional projects that may soon come on stream, Fiji is poised for diversification into additional species, and for rapid growth of its aquaculture sector.

INSTITUTIONAL ARRANGEMENTS FOR AQUACULTURE

"Institutional arrangements" means the governance arrangements for aquaculture, which can exist at the national governmental, regional, and nongovernmental, and private-sector levels. Generally, the role of government in aquaculture is threefold:

1. Government regulates aquaculture to ensure sustainability;
2. Government enables aquaculture by providing in legislation for necessary property rights to be acquired by private-sector investors (with procedures for conflict resolution); and
3. Government develops aquaculture by taking a lead in research and development.

These can be elaborated in the following potential areas where governments are typically involved:

1. Management of environmental effects of aquaculture ("regulate");
2. Protection of public health ("regulate");
3. Biosecurity and border protection ("regulate");
4. Provide efficient means for people to acquire the necessary property rights for aquaculture, through creation and allocation of aquaculture rights and through provision of efficient mechanisms for transactions with resource owners, plus conflict resolution mechanisms for allocation of scarce resources ("enable");
5. Research and development, especially pre-commercial phases, environmental impacts, environmental monitoring ("develop");
6. Human resource development (training) and technology transfer ("develop");
7. Provide financial incentives for aquaculture development where appropriate ("develop");
8. Legal support; legislation to empower or control all of the above ("enable," "regulate");

9. National aquaculture planning, to provide a coordinated approach to all of the above ("regulate," "enable," "develop").

Fiji Fisheries Division

The lead government agency for aquaculture in Fiji is the Fisheries Division of the Ministry of Agriculture, Fisheries, Forests (MAFF), and the Agricultural Landlords and Tenants Act (ALTA). This department administers the Fisheries Act,[7] under which Fiji fisheries waters are defined, and which provides for licensing and regulation of fisheries activity. The budget and staffing levels of the Division have remained almost constant in recent years, while the demands upon its resources have been increasing owing to the rapid expansion of commercial fisheries in Fiji.[8]

Fiji's fisheries statute is based upon British legal concepts, as a legacy of its British colonial history. In many respects this legislation is outmoded and in need of review. For example, no provision is made for aquaculture in the Fisheries Act. Customary laws and traditional practices also apply to Fiji's coastal fisheries. As just one example, fisheries in some areas may be closed during the mourning period following the death of a high chief. In brief, Fiji's fisheries law is a mixture of a British-derived statute and traditional and customary laws.

Since 1997, MAFF has had available to it government funding for projects under a Commodity Development Framework (CDF), which is aimed at diversification of the primary sector and generation of new export products. Because of hardships in other economic sectors in Fiji (for example, sugar), there has been increasing pressure to develop marine resources. The Fisheries Division decided to redirect its activities away from its traditional service-oriented functions, to focus only on selected commodities that can become full-fledged industries within a 3–4-year timeframe. The aim is to "jump-start" particular industries through government action to initiate development projects and provide infrastructure, without waiting for donors or private sector investment.[9]

Most of the marine commodities selected by the Fisheries Division of MAFF under the CDF have been aquacultured commodities, for example, *Kappaphycus* seaweed and milkfish *Chanos chanos*. The aim of each CDF proj-

7. Cap. 158 of the Laws of Fiji.
8. G. R. South and J. Veitayaki, "Fisheries in Fiji," in B. Lal and T. R. Vakatora (eds.), *Research Papers of the Fiji Constitution Review, Vol. 1. Fiji in Transition.* (Suva: School of Social and Economic Development, The University of the South Pacific, 1998), pp. 291–311.
9. Lagibalavu (n. 4 above); MAFF, *Fiji Fisheries Division—Annual Report 1997.* (Suva: Fiji Fisheries Division, Ministry of Agriculture, Fisheries, Forests and ALTA, 1997).

ect was to start from the market and work backwards to processing and production, to develop a total project package that can later be devolved to the private sector as a viable "going concern." In this way, government hoped that CDF support would overcome the "fledgling industry syndrome" that is often faced by the private sector in bridging the gap between technical success and commercial success.

To make best use of the CDF funding, a new Aquaculture group was created within the Fisheries Division, with three sub-groups; a Freshwater Programme (tilapia, *Macrobrachium* prawns, Asian carps), a Brackishwater Programme (milkfish, penaeid shrimps), and a Mariculture Programme (giant clam, trochus, blacklip pearl oyster, and seaweeds). A range of other new projects is also being looked at. The Fisheries Division currently has 24 scientific/technical staff and 40 support staff working in aquaculture.

So far, it is fair to say that the government of Fiji, through the Fisheries Division of MAFF, has concentrated mainly upon the "development" role of government, especially numbers 5 and 6 above.

Lands and Survey Department

The Lands and Survey Department administers the Crown Lands Act,[10] which includes seabed lands below the high tide mark. The Act provides for the granting of foreshore leases and licenses, after an application process that includes consultation with traditional fishing rights owners and other affected coastal users. Only two applications for aquaculture have ever been made, and these are still being processed.

Environment Department

As part of efforts to meet international obligations toward the environment stemming from the United Nations Conference on Environment and Development (UNCED) and Agenda 21,[11] an Environment Department has been set up by the Fiji government to oversee environmental issues and infuse an environmental dimension into the development-oriented activities of other government ministries. Various new statutory powers needed by this department are not yet a reality, because the draft Sustainable Development Bill[12] was still awaiting a slot in Parliament's legislative program at the time of the

---

10. Cap. 132 of the Laws of Fiji.
11. United Nations Environment Programme, 1992. Available online: <http://www.unep.org>.
12. Fiji's Parliamentary processes have since resumed; however, the Sustainable Development Bill still awaits introduction to Parliament.

attempted coup in May 2000 when Fiji's Parliament was suspended. Once in law, this Bill will formalize environmental requirements for development projects such as environmental impact assessment (EIA), which presently can only be implemented at a policy level under current legislation.

Fiji Trade and Investment Board (FTIB)

The Fiji Trade and Investment Board (FTIB) facilitates foreign investment in Fiji and provides a "one-stop shop" for overseas investors who want to establish businesses in Fiji. A single application can be made to FTIB, which will then liaise with other government departments to gain all necessary statutory approvals (work permits, customs, or finance approvals, etc.) and then communicate the government's decision to the investor. This streamlines the investment process for outsiders who may be unfamiliar with the workings of government in Fiji.

The Private Sector

The role of the private sector in aquaculture includes:

1. Responsible and sustainable use of natural resources
2. Maximize profit and contribute toward export-led economic growth
3. Organize into a professional association to represent their industry's interests in a coherent and sensible fashion
4. Provide employment
5. Research and development, especially pilot production scale and market research

Being very much a fledgling industry in Fiji, it is perhaps to be expected that the business part of the private sector has so far concentrated upon numbers 2 and 5, above—maximizing profit and research and development. It is important to note that the private sector for aquaculture in Fiji can have two components—the business level and the community or subsistence level. For example, tilapia-farming development has mainly occurred at the community or subsistence level in Fiji, with the prime objective being food security rather than monetary gain.

REGIONAL/INTERNATIONAL-LEVEL INSTITUTIONS

Fiji has assistance available to aquaculture under a variety of bilateral and multilateral arrangements. Bilateral assistance in fisheries and aquaculture has been provided to Fiji mainly by the government of Japan through its overseas development assistance agencies, the Japan International Cooperation Agency (JICA), and the Overseas Fishery Cooperation Foundation (OFCF). There has also been assistance from the government of Australia through AusAID (for example, the joint MAFF/ACIAR project on tilapia genetic improvement) and the government of New Zealand (assistance with seaweed farming development during the 1980s).

Regional multilateral initiatives that can support aquaculture in Fiji include the following:

FAO South Pacific Aquaculture Development Project Phase Two (SPADP-II)

Based in Suva, Fiji, and serving the Forum countries of Australia, Cook Islands, Federated States of Micronesia, Fiji, Kiribati, Nauru, New Zealand, Niue, Palau, Papua New Guinea, Republic of the Marshall Islands, Samoa, Solomon Islands, Tonga, Tuvalu, and Vanuatu, this project reached the end of its second 5-year phase in 1999 and has not been extended. The aim of the project was to provide technical assistance to countries, to identify aquaculture priorities, provide information, and assist with implementation of aquaculture trials.

Secretariat for the Pacific Community

Participating countries want to see continued regional support for aquaculture after the expiration of SPADP II. Discussions have been underway to find out whether the Secretariat for the Pacific Community (SPC)—formerly known as the South Pacific Commission) could take on a coordination role for aquaculture under a new arrangement that involves the SPC and other regional organisations. The SPC has no current direct involvement in aquaculture, so it must find support for a new staff position to make this possible. Following these discussions, in 2001 the SPC was able to secure AUSAid support to establish a Regional Aquaculture Programme, the major outputs of which will be a regional aquaculture strategy, identifying and meeting training needs, exchange of aquaculture information, and a technical advisory service.

International Center for Living Aquatic Resources Management

The International Center for Living Aquatic Resources Management (ICLARM) is an international aquaculture research institute with a South

150   Living Resources

Pacific facility at Aruligo near Honiara in the Solomon Islands. Its work has been on giant clam, bêche-de-mer, pearl oyster, and reef finfish. The aim of ICLARM is to develop methodologies for target species that encompass both technical and economic aspects, to demonstrate economic viability, and hand over to Pacific Island governments and the private sector a complete and workable aquaculture development package for that species. Political upheavals in the Solomon Islands led to the destruction of the ICLARM facilities at Aruligo, however, they are now in the process of establishing a new project in New Caledonia.

The University of the South Pacific

A new F$25 million facility for the Marine Studies Programme was completed in 1998 at USP's Laucala Campus in Suva. A seawater tank-room is part of the new complex to enable staff and post-graduate research and short-course aquaculture training to take place in Fiji on a range of species. A new undergraduate course on aquaculture has been developed for the BSc (Biology) and BSc (Marine Science) degree programmes and has been offered since 2001.

The University of the South Pacific's Institute of Marine Resources (IMR), the applied research and consultancy arm of the Marine Studies Programme, was relocated to the Solomon Islands and a Director was appointed in 1999 with a strong background in aquaculture. Phase I of IMR was officially opened in May 1999 on land next to the ICLARM facility at Aruligo. However, following the destruction of that facility during the Solomon Island's political unrest from 2000 to the present, the IMR has been relocated to Fiji for the medium term.

Forum Secretariat

The Forum Secretariat, based in Suva, administers various aid budgets on a regional basis. A focus of recent assistance has been private sector development, which can include aquaculture development. Their various sources of funding can be used to obtain technical assistance, consultants, training, and marketing opportunities for applicants from the private sector. The permanent Chair of the Council of Regional Organisations in the Pacific (CROP) is also based within the Forum Secretariat. A Marine Sector Working Group was appointed for CROP, which ensures that the activities of various CROP organisations' activities in the marine sector are integrated and avoid either gaps or overlaps.

Proposed Regional Coordination of Aquaculture

In 1999, it was agreed that, in the post-SPADP-II era, SPC, ICLARM, USP, and SPC-member countries should elaborate a Regional Aquaculture Strategy to coordinate the roles of the various regional institutions.[13] The various institutions would have complementary roles. The ICLARM can undertake long-term research and development of new species and technologies from baseline research through to marketing. The USP can carry out short-term research on specific topics and focus on capacity building (education and training of undergraduate students, regional governmental staff and private sector personnel in aquaculture). The SPC and member governments can implement aquaculture development projects at the grassroots level and provide assistance to the private sector. Such coordination would help to provide a better institutional environment for developing aquaculture industries in the region. Fiji, as an SPC-member country, will be one beneficiary of this regional strategy.

A DEFINITION OF AQUACULTURE

In any discussion of the need in Fiji for a comprehensive statutory framework for aquaculture, it is important to bear in mind what aquaculture is, and what it is not. Aquaculture legislation invariably contains a legal definition of "aquaculture" or some equivalent term like "marine farming"; however, efforts to come up with a workable definition can be problematic and this has been the subject of many debates. Before analysis of a Fiji aquaculture framework, attention will first be paid to the vexed question of a definition for "aquaculture."

There are three reasons why a definition of aquaculture is sought. First, a biological definition is necessary because scientists need to know what processes lie inside or outside their chosen field of study. Second, an economic definition is needed because economists need to know what data or statistics to collect and analyze, for example, to compare aquaculture production with fisheries production. Third, a legal definition is required because resource managers and resource users need to know what rules will or will not apply to aquaculture activity.

A precise yet workable definition of aquaculture has proved elusive, because of the wide range of activities that can fall within it. Most people have a vague notion that aquaculture is the rearing or cultivation of aquatic organisms, much the same as for terrestrial agriculture but in an aquatic medium.

---

13. T. Adams, "Outputs of the Meeting," in T. Adams (ed.), *Meeting Report: First Heads of Fisheries Meeting, 9–13 August 1999*. (Noumea: Secretariat for the Pacific Community, 1999).

Aquaculture is thus usually thought of as being separate from fishing (for example, university courses are often entitled "Fisheries and Aquaculture") and more similar to farming on land. Under British-based law, however, the property rights regime applying to aquatic (especially marine) environments is often very different from that applying to agriculture on land.

A biological definition is provided by Reay: "Aquaculture is man's attempt, through inputs of labour and energy, to improve the yield of useful aquatic organisms by deliberate manipulation of their rates of growth, mortality and reproduction."[14]

In biological terms, aquaculture is human intervention to manipulate the terms of the classic Russell equation for production of biomass from a fish population:

$$\text{Biomass} = \text{Recruitment} + \text{Growth} - \text{Natural Mortality} - \text{Harvest}$$

Through investment in an aquaculture system (tanks, ponds, enclosures, etc.) and in appropriate husbandry, the aquaculturist attempts to maximize *recruitment* and *growth*, and minimize *mortality*. By contrast, fisheries management is usually an attempt to control only *harvest*.

Bardach et al.[15] showed the wide possible range of aquaculture activities by describing the following seven general categories:

1. Transplantation of aquatic organisms from a poor natural environment to a better environment (for example, oysters or other molluscs);
2. Release into the wild of hatchery-reared juveniles (for example, giant clam restocking);
3. Trapping of naturally occurring juveniles in or on openwater structures until ready for harvest (for example, pearl oyster or Pacific oyster cultivation on rafts);
4. Trapping of naturally occurring juveniles in or on open-water structures and cultivating them (for example, sea-cage farming of groupers or tuna);
5. Release of hatchery-reared juveniles into or onto open-water structures and cultivating them (for example, sea-cage farming of snappers or groupers);
6. Trapping of naturally occurring juveniles in closed waters and cultivating them (for example, milkfish in ponds);

14. P. J. Reay, *Aquaculture*. (London: Studies in Biology No. 106, Edward Arnold, 1979).
15. J. E. Bardach, J. H. Ryther and W. O. McLarney, *Aquaculture: The Farming and Husbandry of Freshwater and Marine Organisms*. (New York: Wiley-Interscience, 1972).

7. Release of hatchery-reared juveniles into closed waters and cultivating them (for example, penaeid shrimps in ponds).

These categories, which span extensive through intensive types of aquaculture, suggest that aquaculture, as an activity, is in fact a continuum of activities that merge from farming into capture fisheries. This is a definition of aquaculture that includes any human intervention in the natural lifecycle of an aquatic organism.

A biological definition, however, says nothing about the flow of benefits from investment in an aquaculture system. This is a crucial point for an investor, who must be able to capture most of the benefits of their husbandry or else go bankrupt. Accordingly, Reay[16] goes on to give an economic definition: "Aquaculture is production of aquatic organisms from the basis of site leasehold or stock ownership."

This narrows aquaculture down to situations where there is exclusive ownership of or access to fish, so that benefits flow chiefly to the investor rather than to any noninvestors. It rules out restocking of common property fishery resources even though, in biological terms, the same activities and processes are involved, and includes pure capture fisheries operating in areas that may be exclusively owned by an individual or corporation. This is an issue that FAO has struggled with in its attempts to record aquaculture statistics separately from capture fishery statistics.

Legal definitions of aquaculture are also necessary, because the scope of laws applying to aquaculture need to be set out. For example, in many countries the harvest of fish from a marine farm often has a less restrictive compliance regime (such as no size or tonnage limits) compared with harvesting the same species from a wild fishery.

However, legal definitions of aquaculture activity are very difficult to narrow down without causing anomalies. Biological definitions (like those above) are unsuitable as legal definitions because they are very broad and overlap with pure capture fisheries, or say nothing about peoples' intent. For example, a person throwing food scraps into a pond could be defined as "fish-farming," when really they are just littering. A fisher could apply for a "farm" license to escape size or tonnage limits, without any intention of investing in the various inputs that make aquaculture more productive than fishing. Such inputs are a matter of degree; at what point do we consider that an activity is "farming" rather than "fishing"?

Economic definitions are also unsatisfactory from a legal viewpoint, as they do not cover non-profit aquaculture (where objectives may be social or environmental, such as restoring degraded coral reefs), nor do they cover aquaculture where fish harvests are not exclusively controlled (such as restocking a common property fishery).

16. Reay (n. 14 above).

The best attempt to devise a comprehensive definition for "aquaculture" is the new FAO official definition,[17] which reads thus:

> Aquaculture is the farming of aquatic organisms including fish, molluscs, crustaceans and aquatic plants. Farming implies some sort of intervention in the rearing process to enhance production, such as regular stocking, feeding, protection from predators, etc. Farming also implies individual or corporate ownership of the stock being cultivated. For statistical purposes, aquatic organisms which are harvested by an individual or corporate body which has owned them throughout their rearing period contribute to aquaculture while organisms which are exploitable by the public as a common property resource, with or without appropriate licences, are the harvest of fisheries.

It can be observed, however, that this new FAO definition of aquaculture is really no more robust than any of the earlier attempts, since all it has done is pool several different definitions, each taken from a different perspective (scientific, economic, or legal). It does not provide any conceptual relief to the difficulty of keeping aquaculture legally distinct from capture fishing, and applying a different set of rules to it.

These definitional issues have not yet been any cause for worry in Fiji or other Pacific Islands where aquaculture is still in its infancy, where there is little difference in compliance regimes between aquaculture and fisheries, and where the compliance regimes attached to capture fisheries are not very strict anyway. However, they will become relevant in the future if, as in Australia, New Zealand, Japan and other countries, aquaculture takes its place alongside capture fisheries as a major economic activity that shares the same resources but is managed under a separate legislative or property rights regime. For example, in New Zealand for a time, application was being made under marine farming legislation to establish scallop "farms" covering large areas of seabed, as a way to circumvent the cost of entry and compliance regime attached to the scallop capture fisheries being managed and re-stocked under the Fisheries Act's Quota Management System.

When the time comes for Pacific Island governments to legislate for aquaculture, they will need to put some thought into this issue. If they simply adopt the definitions used in the aquaculture legislation of other states outside the region, they will sooner or later encounter the same problems that

---

17. FAO, *FAO Technical Guidelines for Responsible Fisheries 5: Aquaculture Development* (Rome: FAO Fisheries Department, Food and Agricultural Organization of the United Nations, 1997), at p. 6.

these other states, for example New Zealand,[18] have been grappling with in trying to integrate aquaculture management with stock enhancement and fisheries management.

The best way to define aquaculture from a legislative perspective may be to focus on the *property rights* that are involved in it. To focus on the activity of aquaculture is useful only to paint a general picture to a layperson of the things that aquaculture might include. The range of all possible permutations for aquaculture activity defies definition in terms of what aquaculture excludes, because in practice aquaculture activities form a continuum with capture fisheries.

Unlike activities, property rights can be clearly defined in law. The principles of land law have been around for centuries. Aquaculture can be legally provided for in terms of a particular set of property right elements that need to be acquired in order to carry out the activity. These fall into three categories:

(1) A right to harvest fish. A right to harvest fish may stem either from ownership of land and its resources, or from the granting of a statutory license (such as a fishing permit) that conveys a right to use a resource found in a defined area (sometimes referred to as a usufruct right).

(2) A right to occupy space. All aquaculture farms (and even a part of fishery restocking processes) need to be located on a site. The right to occupy a site may be obtained by purchase of freehold or leasehold land, or by lease or license for public land vested with the state.

(3) A right to cause environmental impacts within acceptable limits. If statutory permissions are required for any aspect of the aquaculture operation (for example, to discharge effluent from the farm), then permission will need to be granted through a process to determine that any environmental impacts are within acceptable limits.

Because aquaculture property rights are a bundle of rights that merge with fishery property rights, and because the difference is a matter of degree, a robust legal definition of aquaculture (as distinct from fishing) based on property rights will still remain elusive. By following this approach, however, it becomes possible to legally provide for aquaculture in a generic way that enables it to take place and enables it to be managed, while avoiding any need for a precise legal definition of it. One simply defines in legislation the property rights elements and compliance regimes that are available for the establishment of aquaculture enterprises.

18. Fisheries Task Force, *Fisheries Act Review: Public Discussion Document* (Wellington: MAF Policy, Ministry of Agriculture and Fisheries, Government of New Zealand, 1991).

As far as a rights-based definition itself is concerned, the most that can be said about aquaculture is that: An aquaculture right is the right to establish a more exclusive type of fishery with fewer legal restrictions on catch.

AQUACULTURE PROPERTY RIGHTS IN FIJI

A crucial part of private sector involvement in aquaculture development is obtaining secure property rights to provide the basis for investment. Secure property rights are an important part of business confidence in any sector of the economy. As mentioned earlier, aquaculture requires rights to harvest fish, rights to occupy space, and permission for any environmental impacts. Government needs to ensure that a statutory mechanism is in place for allocation of property rights for aquaculture in a way that fosters private sector (both commercial and traditional/subsistence) investment yet safeguards the public interest in marine resources.

Sources of aquaculture property rights for the coastal marine environment are not very clear in Fiji at present. Fiji, in common with all other Pacific Island developing countries, has no aquaculture legislation that provides explicitly for it to be either enabled or regulated (although Tonga is in the process of developing aquaculture legislation). It is necessary to acquire aquaculture rights in the marine environment either under generic legislation which does not explicitly contemplate aquaculture, or else follow the strategy of obtaining freehold or leasehold land and concentrate upon land-based aquaculture. So far, commercially successful private sector involvement in aquaculture in Fiji has been almost entirely land-based (penaeid shrimp, tilapia). It is timely that we now discuss what, if any, generic institutional arrangements may enable aquaculture in Fiji's marine environment.

Bearing in mind that a marine aquaculture right is essentially a right to subdivide a fishery area and establish a more exclusive claim over the rights to harvest fish in the farm area, it is useful to examine the situation for ownership of fishing rights for inshore areas in Fiji.

Another legal strategy to obtain exclusivity for protection of a farmed fish stock is to acquire an exclusive occupation right by obtaining a lease over the seabed and water column of the farm area so that others may not enter the area. It is therefore also useful to examine ownership of the seabed in Fiji.

Ownership of Rights to Harvest Fish in Fiji

In pre-Cession days, customary fishing rights in Fiji were traditionally owned from land out to the outer edge of barrier reefs, enclosing the reef edge, reef flat, and lagoon area bounded by the reef. The situation today is that traditional fishing rights areas (*qoliqoli*) are being codified by the Native Lands and Fisheries Commission through demarcation of boundaries, with negotiations to resolve any disputes. Four hundred and eleven such *qoliqoli* have been surveyed and are now listed in the Register of Fijian Customary Fishing Rights. The owners of traditional fishing rights in a particular *qoliqoli* have an automatic right to fish there for noncommercial purposes (as do all other people for noncommercial purposes, except if using a net). Rights to allocate commercially harvested fish within all Fiji waters are vested with the State. Traditional fishing rights owners must obtain a fishing license from the government if they are fishing commercially within their *qoliqoli*. Fishing licenses may also be granted by the State to nontraditional rights holders, but only after consultation with the traditional rights holders who may entirely refuse to have any commercial license allocated or may apply any restrictions that they see fit. In practice, this customary right of control over commercial fishing is respected in all cases by the state, and there was formerly only one exception—tuna pole and line baitfishing.

Traditional fishing rights in Fiji do not extend beyond the barrier-reef edge. Rights to harvest fish in the open sea areas within Fiji's Exclusive Economic Zone and Territorial Waters outside the reef, which with advancing technology may someday be of interest for aquaculture, are vested with the state and the process to grant a license does not legally require consultation with traditional fishing rights owners.

Ownership of the Seabed in Fiji

The situation of seabed ownership has recently been reviewed by South and Veitayaki.[19] In brief, under the 1874 Deed of Cession whereby Great Britain gained sovereignty over Fiji, traditional owners lost ownership of fishing rights and ownership of seabed lands to the British Crown. Usage rights ("usufructory rights") for fishing were subsequently returned to traditional owners, but not ownership of seabed lands below the high tide mark that remained vested with the state. Since cession, Fiji has thus followed the British system of marine tenure where rights to land, to water, and to harvest fish are divisible from each other in terms of ownership. This is in contrast to the traditional Pacific Island view that the sea is an integral part of the land.

Theoretically, there is a right of public access to coastal lands and waters below the high tide mark, while more permanent and enforceable property rights can be obtained by applying to the Lands Department for a foreshore lease or license. In practice though, traditional rights extend to more than

19. South and Veitayaki (n. 8 above).

just fishing rights in the minds of many, and this has been a source of misunderstanding and, in some cases, conflict. One example has been the famous Tavarua Cloudbreak surf spot in Fiji, where surf tour parties who followed the British system of public access encountered opposition and even violence from villagers who viewed it as a customary right to control access to the surf spot and receive cash payments for granting permission. Such misunderstandings can only have an adverse effect upon investor confidence in marine tourism ventures, or any other business like aquaculture for which a marine location is essential.

Possible Sources of Property Rights for Aquaculture

South and Veitayaki[20] have examined these ownership issues mainly from the standpoint of fostering customary marine tenure for fisheries. They did not explicitly address the issue of aquaculture rights, but these are implicitly a part of fishing rights so the discussion in their paper has implications for aquaculture. We will now analyze their description of existing marine tenure arrangements in Fiji, to find possible sources of (i) a right to take fish for an aquaculture activity, that provides a "shield" against any statutory prohibition on the taking of fish, (ii) a right to exclusively take fish in the farm area, that provides a "sword" to protect farmed fish from being taken by others, and (iii) a right to occupy seabed lands more or less exclusively, as a "sword" to exclude trespassers from entering the area.

A Right to Harvest Fish for Aquaculture (A "Shield")

The starting point is the Fisheries Act[21] that prohibits the taking of fish in Fijian fisheries waters by way of trade, business, or other commercial purposes without a license. There is no statutory exemption for aquaculture (be it harvesting, or spat collecting), and there is no other source of authorization to take fish commercially except under a fishing license. Fijian waters are defined as all internal waters, territorial waters, archipelagic waters, and waters of the Exclusive Economic Zone. Internal waters include freshwater ponds, or even the water in somebody's goldfish bowl. Technically, all aquaculture in Fiji, whether land-based or marine, requires the authorization of a fishing license before fish can be "taken" by way of trade or business. This is irrespective of the ownership of the fish, about which the Fisheries Act makes no distinction. The prohibition on taking fish without a license also applies to aquaculture by traditional fishing rights owners, since they too

20. *Ibid.*
21. n. 7 above.

are required to hold a license if the fish are being taken for commercial purposes.

Nobody seems to be very worried about the fact that all aquaculturists in Fiji are technically committing offences against the Fisheries Act. The Fisheries Division of MAFF has not (in practical terms, quite reasonably) followed a policy of mounting any prosecutions against aquaculture operators. This legal situation can be tolerated when aquaculture is still in its infancy. But when aquaculture achieves the higher economic profile in Fiji that is being hoped for, then the position of aquaculturists in terms of the Fisheries Act will need to be validated. No one will want to make million-dollar investments knowing that, technically, they are at the mercy of a benevolent bureaucracy. If anyone who objected to aquaculture development were to lay a complaint, the Fisheries Division would be forced to act according to the law as written.

There are two options to validate the position of people who want to "take" fish from an aquaculture operation. The first is to create for aquaculture a statutory exemption from the prohibition on taking fish without a license, for example, by providing for a separate aquaculture license, as is allowed under the Fisheries Act, Section 5(3)(b). The problem with this is that aquaculture would need to be defined in a way that makes it different from fishing, and returning to the discussion above about legal definitions of aquaculture will show why this option should not be preferred. Alternatively, the Minister could name specific people and conditions for exemptions under this Section for the case of aquaculture. This, however, is not good administrative practice as it can lead to inconsistency and ministerial whims in the absence of overarching policy principles set in legislation. It could provide a temporary "fix" only if policy on granting exemptions is sufficiently well-defined and transparent.

The second option is to expand the scope of a fishing license by amending the Fisheries Act so that the bundle of rights encapsulated in these licenses can enable aquaculture operations as well. The types of rights being granted in "Fishing" in comparison with "Aquaculture" licenses is essentially the same. The main difference is that the conditions attached to these rights for environmental or other purposes need to be flexible enough so that aquaculture operations will not be unduly penalized compared to the more strict compliance regime usually attached to capture fishing. This second option is recommended as a better legislative "fix" for validation of aquaculture operations in terms of the Fisheries Act. Note, however, that neither option provides exclusivity to ensure security of investment in aquaculture.

Currently, anybody who wants to take fish for commercial gain from a traditional fishing rights area (including members of traditional fishing rights-owning groups) must apply to the Fisheries Division for a fishing license. This may be granted after consultation with, and a favorable response from, the traditional fishing right owners for the area. Granting a fishing license will make the harvest of fish from a marine farm "legal," but such a right will not be exclusive; it will be shared with all others who hold fishing rights in the area unless fishing rights owners specify that it does not. They would all be within their rights if they helped themselves to farmed fish. There is currently no mechanism to obtain an exclusive fishing license for an area, except by negotiation and trade with all others who hold rights in the area to cease their fishing, or by decision of the registered owners of the customary fishing rights (and even they do not have the power to restrict noncommercial fishing in the area except that involving nets). It is not expected that this mechanism of negotiation and trade will be practicable, given the diffuse nature and large number of individual stakeholders in Fiji's inshore fisheries.

To summarize, relatively simple amendments to current fisheries law in Fiji could make it possible for aquaculturists to acquire a fishing right that acts as a "shield" against statutory prohibition on the commercial taking of fish. However, it would still not be an exclusive fishing right that acts as a "sword" to defend farmed fish against others.

An Exclusive Right to Harvest Fish (A "Sword")

There is currently no statutory provision for granting an exclusive fishing right within a designated area to only one license-holder. A statutory mechanism to create this type of "aquaculture license" would need to be created as an amendment to the Fisheries Act. The circumstances under which such a potentially powerful fishing right is granted (size of area, location, rights of other stakeholders, conflict resolution procedures, etc.) would need to be carefully circumscribed, along similar lines to marine farming legislation in other countries such as the Part on Marine Farming in New Zealand's Fisheries Act[22] or their former Marine Farming Act.[23]

In practice, customary control of *qoliqoli* by traditional owners through their Chiefs is quite strong in many parts of Fiji. Negotiation and trade by aquaculture investors directly with traditional owners can potentially result in acquisition of the very type of exclusivity and control over farmed fish that such investors are seeking. Although *vakavanua* ("in accordance with custom"), such arrangements would have no status under statutory law in Fiji, nor would they carry any weight with those "trespassers" who do not feel obligated to respect the customs of that area. Fiji's first commercial pearl

22. Fisheries Act, 1983 014. Commenced: 1 October 1983. Available online: <http://rangi.knowledge-basket.co.nz/gpacts/public/text/1983/an/014.html>. The new Part IVA on Marine Farming was inserted by the Fisheries Amendment Act 1993 067. Commenced: 7 July 1993. Available online: <http://rangi.knowledge-basket.co.nz/gpacts/public/text/1993/an/067.html>.

23. Marine Farming Act, 1971 029. Commenced: 1 January 1972.

farm was established under this type of arrangement, however the farmer eventually applied for a Foreshore Lease under the Crown Lands Act in order to gain legal protection for the farm structures and stock.

A Right to Occupy Space and Protect Aquacultured Fish (A "Sword")

An aquaculture operation often also needs to occupy space in the marine environment, and lease arrangements for that space can provide for the occupation to be exclusive so that farmed fish may be legally protected from interference by "trespassers."

A right to occupy (more or less exclusively) seabed lands may be obtained by applying to the Lands Department for a foreshore lease or license under the Crown Lands Act, which may be granted after consultation with traditional fishing rights owners (if the application falls within a traditional fishing rights area) and with the public at large. If all goes smoothly, the application process currently takes about one year, since determining the extent and value of loss or diminution of customary fishing rights takes some time. There is a long history of such valuation for loss of traditional fishing rights in Fiji, mainly in the case of foreshore reclamation projects.

At the time of this writing, only two applications have ever been made for a lease or license for aquaculture in Fiji. These applications are still being processed. One is an application for a foreshore license covering an area of 20 ha in Vanua Levu, with the application made by Pearls Fiji Ltd., the only commercial pearl farm, and currently the sole private marine farm in Fiji. Their purpose for making application was to gain an exclusive area where they could prevent boats or people entering the farm, to avoid damage to structures or theft of farmed stock. A second application has been received for a coral reef area in the Mamanuca Group of the Fiji Islands, to provide legal protection for restocking of giant clams, bêche-de-mer, pearl oysters, and other sedentary species. These applications are setting the precedent for acquisition of enforceable rights over marine farm areas in Fiji.

The property rights contained in a lease are usually quite strong, far stronger than are actually needed for most aquaculture operations. Since acquisition of strong rights by one stakeholder means loss of such rights by other stakeholders, it means that lease applications are only likely to successfully pass through the "objections" part of the application process if the farm area is very small, or if the locality is very isolated. The application process itself is not an easy one, and would most likely involve payment of compensation to traditional fishing rights owners. In general, strong property rights are costly to obtain.

What is really needed for most aquaculture operations is a less draconian property right, that gives sufficient exclusivity over the farmed fish species, but which continues to allow public access and which leaves intact any existing fishing rights for other species. An aquaculturist is interested in possession of the farmed fish, rather than possession of the seabed land. Provided the fish are left alone and farm property is not damaged (fisheries legislation can provide separate offences for these), trespass rights are not always necessary. Currently, however, the long, tortuous, costly, and still untested avenue of acquiring a foreshore lease or license is the only way to gain exclusive rights over farmed fish in Fiji that are enforceable in law.

Conditions Attached to Aquaculture Rights

Regarding the third component of aquaculture property rights, statutory requirements regarding environmental impacts of activities in Fiji are still quite minimal in practice. Pending passage through Parliament of the Sustainable Development Bill, there are as yet no statutory requirements for EIA or water-quality standards. EIAs can, however, be implemented at the policy level when operating the application procedures for statutory permissions under various existing pieces of legislation. Investors in major aquaculture projects can therefore expect to have to carry out some sort of an EIA. It will be difficult for an investor to tell in advance just what environmental management regime will be attached to their project, as this will be implemented through license or lease conditions on a case-by-case basis in the absence of any generic statutes or regulations.

The Present Climate for Aquaculture Investment

The current legislative situation will make it difficult to reach the full potential of aquaculture in Fiji. Aquaculture appears to have been not contemplated at all by the drafters of fisheries or marine spaces legislation, presumably because this legislation predates the post–World War II era of global aquaculture expansion, so was simply not an issue at that time. Technically, anyone who harvests farmed fish for commercial sale without a license is committing an offence under the Fisheries Act. Rights to harvest farmed fish cannot be provided under present legislation except through granting fishing licenses, which cannot be guaranteed completely exclusive. Statutory rights to protect farmed fish from harvest by others can only be provided laboriously through trespass rights conferred by seabed leases under the Crown Lands Act. Such leases are much stronger property rights than are actually needed for many forms of aquaculture. Rights can potentially be acquired through customary means, but the nature of such rights is less clear and less enforceable. The means to attach a compliance regime to aquaculture activity is also unclear, and is presently done on an *ad hoc* basis.

For marine aquaculture to reach its potential in Fiji, the situation regarding property rights and environmental management regimes needs to be clarified and improved through amendments to the relevant pieces of legislation.

The Position of Traditional Fishing Rights Owners

Traditional fishing rights owners, both as existing users of coastal resources who may be affected by aquaculture and also as potential aquaculturists themselves, have at times been unhappy with these present institutional arrangements. Over the years, mention has been made in various meetings of provincial councils and the Great Council of Chiefs (GCC) that the ownership of marine spaces and resources needs to be clarified, preferably by return of ownership of seabed lands within fishing rights areas to traditional owners.[24]

Additionally, fishing rights owners in some places are frustrated that they themselves cannot easily establish more exclusive and enforceable harvesting rights over areas that they want to use for aquaculture activities such as coral aquaculture or giant clam restocking, except through the mechanism of marine protected areas whose primary aim is the cessation of all fish harvesting.[25] They are empowered under the Fisheries Act to finely control commercial harvesting within their *qoliqoli*, as well as to enforce any restrictions, but it is true that very few registered fishing rights owners use these provisions. Provision in legislation of a more exclusive kind of fishing right to cater for aquaculture would therefore benefit both traditional fishing rights owners with aquaculture aspirations and aquaculture investors from the commercial part of the private sector.

The issue of seabed lands was raised during the two constitutional reviews that took place in Fiji during 1990 and 1997. Not long before the 1999 election, the Cabinet of the governing SVT-Party decided that legislation should be drafted to return to traditional fishing rights owners the ownership of seabed lands within their demarcated fishing rights areas.[26] The outcome of the 1999 election, however, was a change of government in Fiji, and the new coalition government did not make any pronouncements on this issue.

Following the attempted coup of 19 May 2000, a new interim government was established in Fiji, which promulgated a "Blueprint for the Protection of Fijian & Rotuman Rights and Interests, and the Advancement of their Development."[27] A list of proposed legislative actions by decree included

24. South and Veitayaki (n. 8 above).
25. Muaikaba Fishing Cooperative (pers. comm.). (1997).
26. J. Hicks, "Fishing laws to change," *The Fiji Times* (28 April 1999).
27. Full-page paid advertisement, "Blueprint for the Protection of Fijian & Rotuman Rights and Interests, and the Advancement of Their Development," *The Daily Post* (14 July 2000).

the following item: "Ownership rights to Customary *Qoliqoli* – The conferment of ownership rights, similar to customary ownership of land, on all traditional qoliqoli, as requested by the GCC and the NLTB. [This will take some time as the survey and demarcation of boundaries by the Native Lands and Fisheries Commission needs to be completed. Appropriate safeguards will be included in the legislation on the right of public access and the protection of the interests of investors.]"

The legality of the interim government was successfully challenged in the locally famous Chandrika Prasad case, and the government's February 2001 appeal against that High Court ruling was lost.[28] At the time of writing, there is uncertainty about the validity of interim government decisions such as the Blueprint and any decrees that stemmed from it, and uncertainty about the nature of any future government. Issues relating to lease of communal lands are at the core of these political developments, and a wide divergence of views is apparent. Accordingly, it is still open to speculation how any new arrangements for leasehold or other "renting" arrangements of seabed for aquaculture might operate.

One option is to follow the existing model administered by the Native Land Trust Board for agricultural land leases. Under this system, investors seeking aquaculture sites would presumably apply to the Board, who would then negotiate with landowners, fix rentals and other conditions, and collect lease money for payment to landowners. The new situation regarding compensation for any loss of fishing rights would also need to be made clear.

Alternatively, the government may wish to take a fresh approach to marine tenure, rather than transfer existing land tenure problems into the marine environment. It must also be remembered that areas outside of *qoliqoli* remain vested with the state, so a dual system (with dual administrations within government) will need to be operated in any case.

An Opportunity for a Legislative "Fix" for Aquaculture

Because government's decision in the Blueprint would treat traditional fishing rights holders as "owners" of the coastal seabed rather than as just one of several groups of "users," the proposed return to them of seabed land ownership within *qoliqoli* has the potential to clarify the marine tenure situation and provide more certainty and security for all parties concerned. If aquaculture policy can be included in the same legislative agenda and be "fixed up" at the same time, it could also open the way to real expansion of commercial marine aquaculture in Fiji on a sound legal footing.

28. *Republic of Fiji and Attorney-General of Fiji v. Chandrika Prasad*, in the Court of Appeal, Fiji Islands on appeal from the High Court of Fiji Islands, (1 March 2001) Civil Appeal No. ABU0078 of 2000S. Judgement available online: <http://ccf.org.fj/extras/COA/content.htm> (accessed 15 May 2002).

In other words, there is a legislative opportunity to design a seabed tenure system that provides in statute for at least some of the property rights needed for aquaculture, in a way that treats it generically along with other marine activities and in a way that avoids having to legally define "aquaculture," if government follows through with the proposed transfer of seabed ownership within *qoliqoli*. For this to become reality, Fisheries Division officials and private sector aquaculturists will need to ensure that they have input to the policy-making process for implementation of the government's seabed lands decisions. It must be remembered that, as yet, there has been no explicit mention of aquaculture in any of the discussion about return of ownership of seabed lands. Without specialist policy input from the aquaculture sector both inside and outside government, the opportunity may be missed.

It would be timely to also review other legislative requirements for aquaculture, to find out if any aspects of property-rights allocation or environmental management need to be provided specifically for the aquaculture sector. First, all aquaculturists need their position to be validated in terms of the statutory prohibition in the Fisheries Act on taking fish without a licence. The fishing licence provisions of that Act could be amended to make them "aquaculture-friendly" and, if necessary, to also provide for allocation of more exclusive fishing rights in small areas for aquaculture so that the need for seabed leases can be avoided altogether. Second, there may be particular environmental issues (such as powers to control the spread of fish diseases) which need special-purpose aquaculture legislation, rather than just be dealt with generically. This could be done by amending the Fisheries Act to include a new part on the management of aquaculture.

CONCLUSION

Fiji, along with the Cook Islands, now leads the Pacific Island developing countries in terms of aquaculture production, and Fiji can be regarded as being at a consolidation and diversification phase in its development of aquaculture industries. Efforts by the Fisheries Division of MAFF have so far mainly focused upon the development aspect (especially research and training) of governmental roles in aquaculture, and will need to widen its focus to include the other two main roles for government; to enable aquaculture (for example, provide legislation for aquaculture licensing) and regulate aquaculture (provide for environmental management).

There are legislative impediments to expansion and diversification of coastal marine aquaculture in Fiji either by private sector investors or by traditional fishing rights owners. These impediments are the result of gaps in relevant statutes, the drafting of which simply did not contemplate aquaculture.

Steps that still need to be taken to further strengthen institutional arrangements for aquaculture in Fiji, to provide for its consolidation and expansion in a sustainable manner, include:

- Ensure that any new seabed tenure system for coastal waters can also smoothly provide for allocation of space-occupation and trespass rights over seabed in a way that is appropriate for all sectoral interests in aquaculture;
- Amend the Fisheries Act to validate the position of aquaculture operators in terms of the statutory prohibition on taking fish without a licence, by expanding the scope of a fishing licence (or providing a special category of fishing licence) to include aquaculture;
- Consider amending the Fisheries Act so that a fishing licence granted for harvest of fish from an aquaculture operation can provide a degree of exclusivity sufficient to protect investment in the farmed fish;
- Amend the Fisheries Act to include a new part on the management of aquaculture, to provide for management of the environmental and other stakeholder aspects of aquaculture rights allocation; and
- At a policy level, add aquaculture to the list of other fisheries sectors (artisanal, longline, pole and line, etc.) for which the government collects statistics.

Transportation and Communications

Twilight of Flag State Control

Awni Behnam and Peter Faust*
United Nations Conference on Trade and Development, Geneva

PREFACE

Shipping continues to represent the most valuable use of the oceans. More than 90 percent of world trade moves by sea. In 2001, seaborn trade approached 6 billion tons and the world fleet reached 808 million dead-weight tons (d.w.t.).[1] However, hidden beneath this apparently healthy global growth and progress lies a very disturbing reality—that at the start of the new millennium half of the world fleet is under flags of convenience (FOC). This phenomenon is spreading out of control. Efficient world shipping, with generally low freight rates and sophisticated use of advanced technology is only one aspect of the equation. The other darker aspect of the equation is the unchecked market forces, alongside a great myth called flag State control.

Hugo Grotius in his celebrated 1604 dissertation *Mare Liberum* states, "Every nation is free to travel to every other nation and to trade with it."[2] This came to be known as the doctrine of "Freedom of the Seas." Under the Freedom of the Seas doctrine, the use of the oceans for trade among nations provides for the extension of the nation State from one coast to another by the ship carrying its nationality and flag. The ship deemed an extension of the sovereign State whose flag it flies is symbolic of its jurisdiction, control, responsibility, and liability in law. Nevertheless, shipping for the most part throughout history has been concentrated under a few flags. The post-World War II expansion of trade changed the qualitative and quantitative distribution of shipping and flags, apart from the economic implications of the engagement of nations in shipping. The technology revolution ushered in the vast increase of ship size and the ability to carry unconven-

---

*The authors acknowledge the technical and editorial clearances made by John Pappas.

1. UNCTAD. *Review of Maritime Transport 2001* UNCTAD/RMT/2001.Geneva, 2001.
2. H. Grotius, "The Freedom of the Seas or the Right Which Belongs to the Dutch to Take Part in the East Indian Trade." Dissertation, Oxford University, 1916 (New York, Oxford University Press, 1916), p. 7.

tional cargoes, such as dry bulk, oil, gas, toxic chemicals, and radioactive materials. These cargoes brought into the limelight both the environmental and safety issues that had to be addressed in terms of flag State control.

Churchill and Lowe explain that "the ascription of nationality to ships is one of the most important means by which public order is maintained at sea. As well as indicating what rights a ship enjoys and to what obligations it is subject, the nationality of a vessel indicates which State is to exercise flag State jurisdiction over the vessel."[3] Nationality is granted when a ship is entered on the national register of the country authorizing it to fly its flag. The exercise of control, however, is not automatic; it is generally dependent on the ability and willingness "vigour." The first is dependent on the existence or lack of a genuine link between the vessel and the flag State relating to the existence of the flags of convenience. The latter is applicable to all flag registries. The 1982 United Nations Law of the Sea Convention[4] (UNCLOS) in Articles 91–94 recognized the obligations of flag States in the exercise of jurisdiction and control, but did not impart any solution to the basic problem of enforcing the obligation for a genuine link. Consequently, UNCLOS did not, in essence, strengthen the genuine link between a ship and the flag State.

This became evident when major international conventions relating to safety standards, pollution, and social conditions were being elaborated in the 1970s and 1980s in the Intergovernmental Maritime Consultative Organization (IMCO) (precursor to the International Maritime Organization (IMO)) and other international organizations. Enforcement lay squarely and largely with the flag States, however, in the late 1980s and early 1990s, it became apparent that the root cause of the problem had not been addressed. Consequently, the role of port State control increased in maritime governance.

Edgar Gold captures this in his article entitled "Learning from Disaster."[5] He explained that as international organizations had no enforcement power (which had traditionally been left to flag States), acceptance or adherence to international codes and conventions did not entail that the accepting State was willing or able to enforce such codes. Moreover, in the case of the flag of convenience, flag States had little or no actual controls, which resulted in some open-registry States not being involved in any enforcement at all. Gold agrees that the governance deficit, which he terms as "enforcement leakage," was the root cause of the major disasters and

3. R. R. Churchill and A. V. Lowe, *The Law of the Sea*, 3d ed. (Manchester: Manchester University Press, 1999), p. 205.
4. Accessed January 2002 on the World Wide Web: <http://www.un.org/depts/los/>.
5. E. Gold, "Learning from disaster: Lessons in regulatory enforcement in the maritime sector," *Review of European Community and International Environmental Law* 8(1) (1999): 18.

substandard shipping in the 1980s and was a driving force for alternative methods of control.

The authors of this article believe that to explore the lacuna in the effective exercise of jurisdiction and control by flag States and to understand more fully the evolving role of alternative governance tools, including port State control and coastal State control as a priori may not be sufficient. In this regard, at least in the 1980s, it was thought that the final frontier for restoring an acceptable level of governance would be through international action, giving effect to the prime responsibility of flag State. As such, hopes were high for the elaboration of the United Nations Convention on Conditions for Registration of Ships[6] under the auspices of the United Nations Conference on Trade and Development (UNCTAD). While this Convention failed in its prime objective, the authors will nevertheless discuss, in detail, aspects of the Convention as it contains the elements of effective flag State control, without which the twilight of flag State control becomes a genuine reality.

WHAT ARE FLAGS OF CONVENIENCE?

The flag of convenience issue became one of the most critical in international shipping debate, for a number of reasons.

First of all, flag of convenience fleets (or open-registry fleets) expanded at a faster rate than any other flags of the world merchant fleet. In fact, these vessels now account for one-half of the world's dead-weight tonnage, consisting mainly of tankers and bulk carriers. This, in itself, is a cause for serious concern, as the system permits owners of vessels to be replaced by "faceless operators" who conceal their identities behind a veil of nominal holding companies. Trading nations dependent upon shipping cannot fail to be alarmed at the prospect of the shipping industry falling increasingly into the hands of shipowners who cannot be identified.

Second, there appears little doubt that the flag of convenience phenomenon lies at the root of irrational and erratic development of the world fleet over the years, during which the amount of money invested in shipbuilding, and especially tanker building, far exceeded demand. This appears to be due, to a large degree, to the fact that the profits and cash flow from flag of convenience operations are constantly recycled into new shipbuilding so as to keep the money outside normal fiscal controls. The result has been disruption, not only to the shipping industry, but to the shipbuilding industry as well.

Third, and most significant in the controversy, is the occurrence in re-

---

6. The United Nations Convention on Conditions for Registration of Ships was adopted in February 1986 under the auspices of UNCTAD, TD/RS/CONF/23, Geneva: 1986.

cent decades of a number of alarming incidents involving shipwrecks, scuttling of vessels, maritime fraud, breaking of the United Nations embargoes, and environmental disasters. Those incidents provide clear illustrations of the problems of enforcing the law when flag States have no more than a nominal connection with the shipowners who fly their flags. This means half of the world fleet is de facto an international anomaly, "stateless" and not subject to the jurisdiction of the State of the flag they fly.

In general terms, it can be said that a vessel flies a flag of convenience when it has no real economic connection (or no genuine economic link) with the country whose flag it flies. From the viewpoint of the countries of registration, an open-registry country is one that accepts vessels on its shipping register with which it has no genuine economic link. According to UNCTAD's classifications, there are six main open registries—those of Liberia, Panama, Bahamas, Cyprus, Bermuda, and Vanuatu—the first four being much more important than the last two. There are also multiple numbers of small developing island countries offering flag of convenience facilities in the international market.

Although the Convention on the High Seas of 1958[7] and, later, UNCLOS states that there must be a genuine link between a vessel and its country of registration, it does not define what is meant by genuine link. Over the years a number of bodies, including the Rochdale Committee in the United Kingdom, defined what they regarded as features of an open registry.[8] The first body with official international standing to pronounce itself on the matter of the genuine link was the Ad Hoc Intergovernmental

---

7. Convention on the High Seas, *United Nations Treaty Series*, Vol. 450, pp. 84 and 86.

8. (i) The country of registry allows ownership and/or control of its merchant vessels by noncitizens;

(ii) Access to the registry is easy. A ship may usually be registered at a Consul's Office abroad. Equally important, transfer from the registry at the owner's option is not restricted;

(iii) Taxes on the income from the ships are not levied locally or are low. A registry fee and an annual fee, based on tonnage, are normally the only charges made. A guarantee or acceptable understanding regarding future freedom from taxation may also be given;

(iv) The country of registry is a small power with non-national requirements under any foreseeable circumstances for all the shipping registered (but receipts from very small charges on a large tonnage may produce a substantial effect on its national income and balance of payments);

(v) Manning of ships by non-nationals is freely permitted;

(vi) The country of registry has neither the power nor the administrative machinery to effectively impose any Government or international regulations, nor has the country the wish or the power to control the companies themselves.

United Kingdom Committee of Inquiry into Shipping. Report 51 Cmnd. 4337. London: HMSO, 1970. The Committee became known as the Rochdale Committee after its chair.

Working Group, which met under the auspices of UNCTAD in February 1978,[9] and concluded unanimously that the following elements are normally relevant in determining whether a genuine link exists:

a. the merchant fleets contributes to the national economy of the country;
b. revenues and expenditures of shipping, as well as purchases and sales of vessels, are treated in the national balance-of-payments accounts;
c. the employment of nationals on vessels; and
d. the beneficial ownership of the vessel.

REASONS FOR USING FLAGS OF CONVENIENCE

There are numerous reasons why shipowners register vessels under flags of convenience—including evasion of taxes, avoidance of various governmental regulations, and freedom from restrictions on the use of cash flows. Some of the less reputable shipowners undoubtedly use these flags with the specific aim of concealing their identities and escaping the responsibilities and law enforcement procedures relating to maritime safety, pollution prevention, and so forth, that apply under normal flags. However, there are also numerous transnational corporations of good standing involved in open-registry operations. The reasons why these companies choose flags of convenience relate principally to crew costs.

Although crew costs represent a comparatively small proportion of the total costs of operating tankers and bulk carriers, these represent the only costs, other things being equal, that can be substantially varied by changing the flag of a vessel. Capital costs, which represent the major cost item, remain more or less the same regardless of the flag of a vessel, but the transnationals can achieve substantial cost savings by using a flag of convenience and recruiting a crew from a low-cost labour supplying country.[10]

CONSEQUENCES OF OPEN-REGISTRY OPERATIONS: THE ISSUE OF JURISDICTION AND CONTROL

One of the main consequences of open registration is that it enables the traditional maritime operators (countries) to maintain their domination

9. UNCTAD. *Report of the Ad Hoc Intergovernmental Working Group on the Economic Consequences of the Existence or Lack of a Genuine Link between Vessel and Flag of Registry.* UNCTAD. TD/B/C.4/177, Annex to the Report, p. 1.Geneva: United Nations, February 1978.
10. A. D. Couper, C. J. Walsh, B. A. Stanberry and G. L. Boerne, *Voyages of Abuse: Seafarers, Human Rights and International Shipping* (London: Pluto Press, 1999), p. 11. Also "Comparative Labour Costs," UNCTAD TD/222/Sup.4, (May 1979).

over world shipping. This is despite their increasing inability to operate under their own country flags because of high labour costs and scarcity of seafarers.[11] At the beginning of the 1990s, these traditional maritime operators (countries) owned 86 percent of the total world dead-weight tonnage, and although this amount has been reduced, they still own a very high percentage—72 percent in 2001. Moreover, whereas national flag fleets have declined from 65 to 25.2 percent of the world total, the percentage of fleets operating under flags of convenience has increased almost correspondingly.[12] If the market forces had, in theory, been allowed to operate freely, much of this tonnage would have been transferred to developing countries, which operate economically under their own flags. In other sectors of the world economy, particularly in the 1980s and early 1990s, transnationals had been forced to set up subsidiaries in low-cost countries to benefit from cheaper labour, but in shipping the open-registry phenomenon constitutes an artificial barrier to such a process.[13]

However, another consequence that has aroused the most public discussion over the years has been the amount of misconduct and irresponsible conduct associated with the operation of open-registry vessels. There are, of course, a large number of companies that operate FOC vessels in a responsible manner, not because of the open-registry system, but because they believe that safety and respect of international standards is good business. However, the same system that permits these companies to operate with more freedom than they could achieve under their home flags also permits the operation of vessels by irresponsible owners. The problem arises from the fact that whereas a country with a normal registry can exercise authority over the owners, crew, and vessel, an open-registry country in practice can only exercise authority over the nominal owners listed on its register book. Because these countries, unlike the normal registry countries, do not impose taxes, they do not have an incentive to identify the real owners; the key crew members of these vessels are non-nationals and consequently the only real remedy that an open-registry country can apply in the event of misconduct is to de-register a vessel. However, de-registering is not an effective measure, because nominal owners can circumvent this by changing their company

11. Couper addressed another aspect of the phenomenon stating the existence of globalized shipping with its substandard sector clearly represents unfair competition to decent shipping companies. But it is more socially pervasive than that. Because of competitive pressures, some FOC ships (in the absence of an ITF Blue Certificate) can induce a downward leveling in the conditions of seafarers under *all* flags by undermining the economic viability of socially responsible owners. Couper et al. (n. 10 above), p. 173.
12. See UNCTAD. *Review of Maritime Transport 1979.* TD/B/C.4/198. Geneva: UNCTAD, May 1980, Chapter II, p. 8 and UNCTAD TD/B/C.4/289 for the 1984 Review, and also *Review of Maritime Transport 2001* (n. 1 above).
13. See UNCTAD. *The Repercussions of Phasing Out Open Registries.* TD/B/C.4/AC.1/5.Geneva: UNCTAD, September 1979, p. 14, para. 43.

and ship name and re-registering. Some of the International Labour Organization and IMO conventions are enforceable by port States, but what action can be taken by a port State? A port State can take action against the crew or a vessel, but the owners are effectively outside its jurisdiction. Some of the traditional maritime countries were urging action to correct the abuses of the open-registry system without altering the system itself. However, the abuses are an inherent part of the system, stemming from the fact that the real owners live outside the jurisdiction of the flag State. Consequently, it is simply not possible to take effective action to correct abuse without first ensuring that there is a genuine economic link between owners and countries of registration.

As ships for the most purposes operate only under the jurisdiction of the flag State, in the high seas only international rules and regulations are applicable, provided they bind the flag State either directly (ratification) or indirectly (common use). Under these conditions, if there is no genuine link (according to the standard terminology) between the ship and the State, serious questions arise:

> How can the State effectively exercise its jurisdiction and control to ensure that the ship flying its flag will meet its national and international obligations in maritime navigation?
> How can the flag State conceive, define, and effectively implement a maritime policy, of which the ship is a part?
> How can one define the responsibility of the ship, its owner, or even the flag State with insufficient links, mainly in the case of damage to third parties in operating the vessel?

For decades, the international community has faced a challenge. The proper exercise of jurisdiction and control over vessels by a flag State was not only of national concern, but above all, of international concern. It would be futile to expect a flag State to exercise such control unless minimum requirements were laid down in an international convention on the registration of vessels.

It should be recalled that at a meeting of European and Japanese Ministers of Transport held in Tokyo in February 1971, a number of decisions were taken in regard to flags of convenience. Through those decisions, the governments of the countries concerned, that is, 11 European countries, took note of the growing tendency of shipowners in some of their countries to register under a flag of convenience and expressed certain concern regarding that practice. They also recalled, "that registration of ships under flags of convenience was not in conformity with the principle reflected in the Geneva Convention on the High Seas, 1958,[14] which required that a

14. UNCTAD. *Economic Consequences of the Existence or Lack of a Genuine Link between Vessel and Flag of Registry*. TD/B/C.4/168. Geneva: UNCTAD, 1977, p. 4.

State must effectively exercise its jurisdiction and control in administrative, technical, and social matters over ships flying its flag."

It will be noted of course that the latter principle was also embodied in the adopted 1982 United Nations Convention on the Law of the Sea, in particular with respect to the provisions relating to the nationality of vessels and the duties of flag States (Articles 91 to 97). Article 91 states that each State shall fix the conditions for the grant of its nationality to ships, for the registration of ships in its territory and for the right to fly its flag. However, there must exist a genuine link between the State and the ship. Article 94 sets forth various duties of the flag State concerning the exercise of its jurisdiction and control in administrative, technical, safety, and social matters.

At this juncture, one may reflect on some arguments that have been made with regard to the law of the sea. Statements highlighted in the media often remind us that there was no need for an international agreement on conditions for registration of ships as that had been adequately covered by UNCLOS.

It is clear from an analysis of the relevant provisions of UNCLOS that although individual States have certain flexibilities to establish conditions concerning the grant of their nationality, registration, and right to fly their flags, there must exist as a very minimum a genuine link between the State and the ship. The Convention provides no definition of this genuine link or in any other way gives guidance as to the conditions for the registration of ships to meet the genuine link requirement. Article 94 deals with the separate question of what the duties of a State are vis-à-vis vessels flying its flag. In this respect the Convention sets forth international norms for the exercise of jurisdiction and control by flag States.

When this issue was first raised in the meetings of the Committee on Shipping in UNCTAD in 1978, it had been the position of the majority of developing countries that the absence of a genuine link between open-registry vessels and the flag State makes it impossible for the flag State to fulfil its international obligations to exercise jurisdiction and control over vessels flying its flag. However, the UNCLOS left unresolved the issue as to what elements constitute the genuine link between State and vessel and enable a State to effectively exercise its jurisdiction and control set forth in Article 94. In this context it is important to note that the statement in Article 91 that, "Each State shall fix the conditions for the registration of ships" is a statement of *obligation* and not a statement of freedom. It does not say, "Each State shall *remain free* to fix the conditions for the registration of ships." Thus, developing countries contended that the text of UNCLOS does not preclude, and is not inconsistent with, establishing by another international agreement minimum conditions for the registration of ships.[15] In fact, they

---

15. "In spite of the fact that the 'genuine link' requirement appears to have had little influence on State practice since the High Seas Convention came into

argue that such an international agreement would be complementary to UNCLOS by, in effect, establishing the minimum conditions for registering ships without affecting the right of individual States to actually fix conditions for the grant of nationality, registration, and right to fly its flag that are more stringent than the stated minimum. Such a definition, in the form of establishing minimum conditions upon which vessels may be registered in a State would result in fulfilment of the duties of flag States, as set forth in Article 94.

AN INTERNATIONAL ATTEMPT TO REGULATE FLAGS OF CONVENIENCE

The concerns about the genuine link and open registry did not just arise in the 1970s. The open-registry issue was first raised at an international forum at the ILO meeting in 1933 when attention was drawn to the appalling employment and labour conditions on open-registry vessels. The shipowners of developed countries at a meeting of ILO in December 1937 deplored the system, particularly when it resulted in the avoidance of allegiance and obligations to any country.[16]

The International Law Commission, which did the preparatory work for the Convention on the High Seas, adopted at its seventh session a draft provision in which the minimum national element between the flag State and the ship was stated in terms of the nationality of the owners of the ship, namely that a vessel should be either the property of the State or more than one-half owned by nationals domiciled in the flag State.

In 1958, the United Nations Conference on the Law of the Sea considered this issue. Because it could not agree on the definition of the constitutive elements of registry, it adopted the concept of the genuine link,[17] and very significantly added the phrase that "in particular the State must effectively exercise jurisdiction and control in administrative, technical and social matters over ships flying its flag."

In 1959, major maritime countries argued at a meeting of IMCO (now

---

force, the requirement is repeated in UNCLOS (Art. 91), although the requirement is not linked to the effective exercise of jurisdiction by the flag State, as it is in Article 5 of the High Seas Convention. (The effective exercise of flag State jurisdiction is dealt with by Article 94, discussed later.) There seems little reason for supposing that Article 91 will have any more influence on State practice than Article 5 of the High Seas Convention. The direct attack mounted on flags of convenience in the past few years by UNCTAD may prove more effective." Churchill and Lowe (n. 3 above).

16. See UNCTAD, *The Repercussions of Phasing Out Open Registries* (n. 13 above).

17. The term "genuine link" came to the attention of the international community in the 1955 I.C.J. case, The Nottebohm Case (Liechtenstein v. Guatemala) ICJ Rep.4, 23, (1955) dealing with the nationality of an individual.

IMO), that there must be a genuine link between the vessel and the flag State. Initially, they also led a campaign to keep open-registry countries out of membership of the IMCO maritime safety council.

Contrary to popular belief, this issue was first raised in UNCTAD not by the developing countries or by the UNCTAD secretariat, but by some developed market economy countries. These countries raised the issue at meetings of the Committee on Shipping and urged the UNCTAD secretariat to expedite the work in investigating open registries and to give top priority to the work.

The first meeting to devote itself to a full discussion of the subject was the Ad Hoc Intergovernmental Working Group on the Economic Consequences of the Existence or Lack of a Genuine Link between Vessel and Flag of Registry, which was convened in February 1978 at the request of the UNCTAD Committee on Shipping.[18]

At its first session this Intergovernmental Working Group, which consisted of representatives of 44 countries, unanimously adopted a Resolution that declared, among other things, that "the expansion of open-registry fleets has adversely affected the development and competitiveness of fleets of countries which do not offer open-registry facilities, including those of developing countries."[19] The Group went on to identify the elements, mentioned earlier, that it considered normally relevant when establishing whether a genuine link exists between a vessel and its country of registry and asked the UNCTAD secretariat to undertake further studies on various aspects such as the beneficial ownership of open-registry fleets, and the trade routes on which open-registry vessels are employed. It declared that the subject of open-registry shipping should be kept under continual review within UNCTAD.

Following that session, the UNCTAD Secretariat proceeded to identify, for the first time, the countries of beneficial ownership of the open-registry fleets and found a high degree of concentration. No less than 81 percent of the dead-weight tonnage of these vessels were traced to beneficial owners in a small group of only six countries (the United States of America, Greece, Hong Kong, China, Japan, and the Federal Republic of Germany).[20] The Secretariat's study also revealed the lack of knowledge on the part of the governments of countries of beneficial ownership: many of these had no knowledge at all of the activities of their citizens under flags of convenience, while others had only partial knowledge. This finding was important, because it indicated that the open-registry fleets were operating outside the

18. The Committee on Shipping, which was established in UNCTAD in 1968, was abolished in 1996 due, to a large extent, to the pressure brought on the organization to stop UNCTAD from work on shipping and related issues, particularly flags of convenience.
19. UNCTAD, TD/B/C.4/177 (n. 9 above).
20. See UNCTAD, Secretariat. *Beneficial Ownership of Open Registry Fleets.* UNCTAD V. TD/222/Supp.1. Report. Geneva: UNCTAD, May 1979.

control, and even influence, of governments, of "home" countries. In the absence of information from governments the secretariat successfully sought data from commercial sources. At the same time, the UNCTAD secretariat ascertained that a large percentage of the open-registry fleets were engaged, as mentioned earlier, in the trade with developing countries.[21]

In January 1980, the Group of 77 developing countries tabled a proposal for phasing out[22] flags of convenience within a reasonable period of time but subsequently Liberia and Panama dissociated themselves from that proposal. The majority of developed market economy countries, with the notable exception of France, tabled a proposal calling for action to eliminate the adverse effects without agreeing to phasing out.[23] Despite this disagreement, however, it was noteworthy that there was complete agreement between the Group of 77 and developed market economy countries on one point: the need for greater knowledge regarding the owners of flag of convenience vessels.

The United Nations Conference on Conditions for Registration of Ships held under the auspices of UNCTAD was opened on 16 July 1984. The second and third parts of the Conference took place in January and July 1985, respectively. The third part ended on 19 July 1985 after achieving agreement on the three key elements—management, ownership, and manning—of an international instrument setting out the conditions under which vessels should be accepted on national shipping registers. The question of participation by nationals in the management, manning, and ownership of vessels had constituted one of the most difficult areas of negotiation for the Conference. The consensus agreement on those issues meant that the quasi-totality of substantive issues had then been resolved, and by the end of the meeting on 7 February a new international agreement would be adopted by the international community to address open-registry shipping.

LIMITED SUCCESS: THE UN CONVENTION ON CONDITIONS FOR REGISTRATION OF SHIPS

Prior to the decision of the General Assembly and at the first session of the United Nations Conference it was quite clear that the majority of developing

21. See UNCTAD. *Report of the Ad Hoc Intergovernmental Working Group on the Economic Consequences of the Existence or Lack of a Genuine Link between Vessel and Flag of Registry*, at its Second Session. TD/B/C.4/191. Geneva: UNCTAD, January 1980.

22. "Phasing out" refers to the gradual elimination of the practice of registering ships in countries with which the ships have no genuine link. In this context, the existence or lack of a genuine link implies some form of economic linkage. Phasing out does not imply the immediate "abolition" of the shipping registries of countries presently offering open-registry facilities, but rather a gradual tightening of the conditions on which those countries will accept new registrations or retain vessels on their registers, as well as simultaneous restrictions on the establishment of any new open-registry facilities.

23. See UNCTAD. TD/B/C.4/191 (n. 21 above), Annexes II and III.

countries, the Eastern European countries, and China wished to achieve a gradual phasing out of open registries. This objective became unattainable for the simple reason that the majority of developed market economy countries were opposed to such action in the face of the staunch defence of open registries by the developed countries, in particular, the United States. The developing countries had no option but to concede and reduce their aims, hoping to achieve such aims indirectly through a change in their options and focus by the start of the first round of negotiations.

The economic and social consequences of open registries for international shipping had been studied and debated at considerable length in a number of reports by the UNCTAD Secretariat. Government representatives listened to the pros and cons of the open-registry system. Participants were constantly reminded of the virtues of the open-registry system by those who exalted the open-registry system and were told that it was the last bastion of free enterprise. This, while neglecting that other free enterprises existed but within the boundary of national and international norms and controls as was the case with land-based free market industries.

The Secretariat held the view that the fundamental problems in world shipping caused by open registration all stemmed from situations in which the flag State was entirely absent and lacking in the ownership, management, or manning of a ship. In fact, this situation afforded an owner the greatest chance of escaping responsibility for the enforcement of standards that were so meticulously elaborated by ILO and IMO—and this remained the case, notwithstanding attempts by certain States, to make such owners somewhat more accountable by means of port State controls. The continuation of this situation was the primary cause for the high degree of unfair competition in world shipping today, caused by owners who could cut costs by ignoring the standards for which they could not readily be held accountable. Moreover, there was no need to point out that type of unfair competition that affected all maritime countries in their endeavour to develop their national merchant fleets.

It was understandable that countries that failed to achieve a greater involvement in shipping wanted to draw up proposals to tighten conditions of registration in such a way as to maximize the involvement of the flag State. However, it was necessary to reflect that any set of conditions that would maximize the flag State's involvement in ownership, management, and manning, without any flexibility on the three aspects, would inevitably conflict with the maritime laws of the vast majority of maritime countries. Such extreme proposals inevitably led to extreme counter proposals, as the maritime countries felt obliged to defend legislation and national practices that they had developed over the years.

The vast majority of maritime countries had laws that enabled them to identify the shipowners and operators of ships on their registers and to make such owners and operators accountable, even though their legislation did not stipulate involvement of nationals or residents in each and every aspect of ownership, management, and manning. Taking this into account, it seemed that the most fruitful course was not to draw up a set of maximum conditions, but rather to examine what would be the minimum conditions needed to ensure a genuine link between a vessel and a flag State, as distinct from the so-called "links" that exist only on paper, or in the form of "brass-plate" companies. There was every reason to believe that it would be possible to draw up a set of minimum conditions that would both put an end to the practice of permitting artificial links between vessels and flag States and be consistent with the laws of most, if not all the responsible maritime countries. This, particularly if coupled with the flexibility allowing for the fact that many countries placed different degrees of emphasis on the separate aspects of ownership, management, and manning.

In 1984, the Conference on Conditions for Registration had before it a number of alternative texts for consideration including one submitted by the developed market economy countries. Nevertheless, the substantial debate highlighted that the divergencies of views were centred on the following fundamental issues. First, was an international agreement necessary to fill a gap in present international legislation to enable all flag States to exercise control over ships on their register? Second, if an international agreement was needed what should be the nature of that agreement? Should it be mandatory, recommendatory, or a mixture of both? Third, what were the elements of the genuine link between a State and ships flying its flag? Were economic elements needed for the exercise of effective jurisdiction and control or was perhaps the establishment of an efficient maritime administration sufficient? With respect to the economic elements, would it be necessary for nationals of the flag State to have a participation in the equity capital, manning, and management of national flagships and shipowning companies? That proposal became the central issue before the Conference.

The decision to proceed was agreed when an understanding on the following points was reached:

1. An international agreement on conditions for registration of ships was necessary and would fill a gap in present international legislation;
2. With respect to the genuine link, all groups agreed that:
    a. Vessels have the nationality of the country whose flag they fly;
    b. There must exist a genuine link between a State and ships flying its flag;
3. Flag States must effectively exercise their jurisdiction and control in administrative, technical, and social matters over ships flying its flag;
4. A flag State should assure itself that nationals participate in the manning of vessels as key officers and crew, the level thereof to be fixed in accordance with the national laws and regulations;

5. There should be a management or representative office of the shipowning company in the flag State (the nature and extent of the responsibilities of such office was still subject to negotiation);
6. There was a need for the flag State to have a competent and efficient maritime administration to ensure that ships under the national flag comply with the relevant legislative provisions and regulations regarding technical matters such as safety of navigation and prevention of pollution, as well as social standards of seafarers;
7. The flag State must have a register of ships that should contain all necessary information regarding shipowners and operators, as well as technical matters;
8. A provision should be included that ensured that interested parties have easy access to the relevant information from the register of ships, in particular, the names and other information on the real owners and operators of the ship. The flag State should also ensure that shipowners were accountable for their actions;
9. Countries agreed that provision should be included that vessels on bareboat charter may be registered in the charterer's State for the duration of the charter where national legislation so permits;
10. Some measures were required to protect the interests of labour-supplying countries.

It was also generally agreed that points 2 to 10 should be incorporated in the international agreement. In addition, it was stipulated that there should be a time frame for the implementation of the agreement and, moreover, the agreement should include measures to ensure its implementation.

At the July 1985 session of the Conference, a turning point had been reached when developing countries (the Group of 77), after much soul-searching due to the pressures exerted on them, made a proposal that would allow States to opt for either the manning or the ownership articles in addition to the management link article. On 7 February 1986 the Conference completed the negotiations and adopted the International Convention on Registration of Ships.

The United Nations Convention on Conditions for Registration of Ships[24] introduced new standards of responsibility and accountability for the world shipping industry. For the first time an international instrument existed that defined the elements of the genuine link that should exist between a ship and the State whose flag it flies. The Convention filled a major gap in international maritime jurisprudence, as the components of the genuine link had never been identified.

Articles 8, 9, and 10—the heart of the Convention—provided for partic-

24. United Nations Convention on Conditions for Registration of Ships, TD/RS/CONF/23 (1986).

ipation by nationals of the flag State in the ownership, manning, and management of ships, thus establishing key economic links between a ship and the flag State that are often missing in present practice. A distinctive feature was that States had an option between the two mandatory articles on ownership and manning. This element of flexibility was introduced to take account of the different conditions prevailing in flag States. Some flag States might lack sufficient manpower among their nationals or "persons domiciled or lawfully in permanent residence" within their territory to provide for significant participation by nationals in the crews of ships flying their flag, while other flag States might not have sufficient capital to participate effectively in ship ownership. Among the important provisions, Article 9 on manning provides that the State of registration shall ensure that the manning of its ships "is of such a level and competence as to ensure compliance with applicable international rules and standards, in particular those regarding safety at sea." Article 9 also stipulates that the State of registration shall ensure that the terms and conditions of employment "are in conformity with applicable international rules and standards" and that "adequate procedures exist for the settlement of civil disputes between seafarers employed on ships flying its flag and their employers."

A balanced approach is evident in the Article on management. On the one hand, the principle is set out that before entering a ship on its register of ships, a registration State would ensure that the shipowning company or its subsidiary was established and/or had "its principal place of business within its territory." On the other hand, where this was not the case, the flag State was expected to ensure that there was "a representative or management person who shall be a national of the flag State or be domiciled therein." The article on management is also significant in that it makes the State of registration responsible for ensuring that persons accountable for the management and operation of ships are in a position to meet the financial obligations that may arise from the operation of such ships and to cover risks that are normally insured in international maritime transportation in respect of damage to third parties.

Another important article (Article 5) in the instrument provides for the establishment by a flag State of a "competent and adequate national maritime administration which shall be subject to its jurisdiction and control," and that is charged with a number of specific mandatory tasks such as ensuring that a ship flying its flag complies with a State's "laws and regulations concerning registration of ships and with applicable international rules and standards concerning, in particular, the safety of ships and persons on board and the prevention of pollution of the marine environment" and ensuring that it carries on board documents, "in particular, those evidencing the right to fly its flag and other valid relevant documents." At present, a number of States have no such national maritime administrations.

Article 6 on identification and accountability provides that a State shall take the necessary measures to ensure that owners and operators of a ship on its register are "adequately identifiable for the purposes of ensuring their full accountability." This provision is of particular importance to identifying and punishing perpetrators of maritime fraud.

The Convention induces greater transparency in the operations of open-registry vessels through Articles 6 and 11. It also provides the legal basis for registration of bareboat chartered vessels in Article 12.

Article 5 of the Convention deals with national maritime administration and consequently with safety standards. It states, "The flag State shall have a competent adequate maritime administration which shall be subject to its jurisdiction and control."

As Sturmey[25] admits, this is one of the strongest provisions of the Convention. Moreover, Sturmey has also made a sound proposal for the improvement of this Article, namely for an appropriate United Nations agency to be set up to review the structure and performance of the maritime administration of States, upon their request, and to recommend what measures might be taken to enable them to meet the standards set in the Convention.

One could agree with him that if the Convention were to be enforced, it would not only be the present open-registry countries that would need to act, because plenty of "normal" registers are defective with respect to some of the provisions of the Convention. Moreover, the Convention would render an international service if it were responsible for an improvement in maritime administrations throughout the world, including parastatal and private bodies, for example, the classification societies and surveyors that work in conjunction with administrations in so many ways.

The Convention did not please all concerned. It was a product of a compromise necessitated by the divergent positions held by States on flags of convenience. Some criticized the Convention, claiming that it did not change the status quo. Others accepted realities and saw the Convention as a step forward in the struggle to limit flags of convenience and their undesirable side effects. Some claimed that open-registry countries were the real beneficiaries, as they would not need to make any changes. The truth was thought to be somewhere in between. If it had come into force the Convention would have provided an international legal instrument for States to take sanctions at the national level against the more undesirable aspects of the phenomenon of flags of convenience. It also provided a policy platform for developing countries for their future planning. Perhaps most important it provided those countries who supplied labour and wished to attract vessels

---

25. Discussion with the author. Professor S. G. Sturmey directed the shipping activities of UNCTAD, 1968–1976 and later was the Principal Adviser to the Minister of Merchant Marine of the Côte d'Ivoire. See also S. G. Sturmey, "The United Nations Convention on Conditions for Registration of Ships," *Lloyds Maritime and Commercial Law Quarterly* 1987, no. 1 (February 1987): 97-117.

to their national register a means to adapt their standard of registration without compromising the integrity of their registers and thus have their national labour manning vessels under their own flags. Thus, it opened the door to competition that the traditional open-registry countries perhaps could not afford to ignore as they would have to tighten their control over vessels under their register in accordance with the minimum conditions set out in the Convention.

Unfortunately, the Convention on Ship Registration still awaits ratification and is unlikely to enter into force. The die-hard believers of the Convention hoped that it would provide, nevertheless, sufficient guidance to national maritime administrations and set a standard for new registries.

Since the adoption of the Convention, the number of flags of convenience (open registries) have expanded even further. There appear, however, to have been some qualitative changes. The Convention may be said to have been the catalyst for Norwegian and Danish international ship registries to be opened and modelled along the lines the Convention proposes for a genuine link, thus allowing Norwegian and Danish vessels to be registered under their national flags but on an offshore tax haven concept. At the same time, some developing countries decided to follow the concept "if you cannot beat then join them." Operators from such countries as China (with Hong Kong and mainland China), Taiwan (Province of China), Saudi Arabia, Republic of Korea, and Singapore became fully involved in open-registry operations by registering some of their vessels under open-registry flags.

Clearly, the attempt to bring flags of convenience under a new international regime of jurisdiction and control has totally failed.

Professor Alastair Couper has poignantly documented at length the continuing and tragic predicament and suffering of seafarers in his recent book, *Voyages of Abuse*,[26] which shows how the human rights of seafarers are flagrantly violated on a daily basis. The book shows the failure of the international legal framework to provide seafarers with even the minimum protection. It describes cases of their mistreatment outside of territorial seas, their servitude on substandard vessels, their mounting loss of life and disenfranchisement as FOC States abandon them in a flagrant dereliction of duty, which contravenes international law and minimum norms of decent behaviour. Clearly, there exists today a major lacuna between the intention of the drafters of UNCLOS and the applicability of the law as it concerns the duties and responsibilities of the flag States. This lacuna, as demonstrated by the extensive details in the elaboration of the International Convention on Registration of Ships, remains the greatest challenge to the international community.

---

26. Couper et al. (n. 10 above).

THE RESPONSE TO LACK OF FLAG STATE CONTROL

Port State Control

As mentioned earlier, the harsh reality of shipping is that flag States are not universally doing their duty to regulate and effectively police tonnage included in their registers. There is often little identifiability and even less accountability, particularly in relation to the owners and operators of substandard ships. The UN Convention on the Conditions for Registration of Ships addresses these issues by calling for flag States to install competent and adequate maritime administrations ensuring that ships flying their flags comply with applicable international rules and standards. As noted earlier, the Convention has never entered into force and the shipping scene under some flags continues to be characterized by unidentifiable and consequently unaccountable one-ship companies.

While the international community has not been able to reach agreement on an instrument defining the genuine link and setting minimum obligations for flag States, it has recognized that the situation of substandard shipping has become untenable. Consequently, those directly and adversely affected by these operations have either unilaterally or plurilaterally taken considered necessary measures to protect their environment and maritime safety from such substandard operations.

Port State control is not meant to be a substitute for flag State obligations, but rather a complementary instrument. It is clear that port State control, at the scale observed today, would not be necessary if flag States were to dispose of their duties in a more responsible manner.

The concept of port State control is not new, but already contained in the SOLAS Convention[27] and later on in UNCLOS. It is equally recognized that the coastal State has full authority to determine the conditions of and prescribe the policy for the access to and use of its port. In fact, a number of modern international instruments recognize not only the power but also the duties of port States to undertake inspections of vessels to ensure compliance with international rules and regulations.[28]

Port State control, at its inception, was an instrument used by individual governments to ensure compliance with selective obligations placed on foreign vessels. A major shortcoming of this approach is the possibility to avoid controls through shifting operational patterns and the lack of coordination among port States. To overcome these problems, port States have formed

27. IMO, International Convention for the Safety of Life at Sea (SOLAS), (1974). Entered into force 25 May 1980 and accessed November 2001 on the World Wide Web: <http://www.imo.org/>.

28. For a discussion of the roles of flag States, coastal States and port States, see J. Hare, "Flag, coastal and port state control—Closing the net on unseaworthy ships and their unscrupulous owners," *Sea Changes* 16 (1994): 57.

regional groupings to coordinate and cooperate in the exercise of port State control. The first such grouping was the Paris Memorandum of Understanding (Paris MOU) of 1982, concluded by European States and later joined by Canada, thus extending its regional scope and common inspection system to Europe and the North Atlantic. Today, all major trading and shipping areas of the world are covered by regional MOUs, the major ones being listed below:

- Tokyo MOU (Asia and Pacific region);
- Acuerdo de Viña del Mar (Latin America region);
- Caribbean MOU;
- Mediterranean MOU;
- Indian Ocean MOU;
- Abuja MOU (West and Central Africa region);
- Black Sea MOU.

The various MOUs provide for elements of regional standardization of inspection criteria and procedures. Consequently, these agreements introduce standards in the implementation of safety and prevention of pollution measures and equally improve predictability of inspection measures for shipowners. Apart from the element of standardization, the major benefit of regional cooperation can be seen in the exchange of information in a regular and systematic fashion. The agreements have established databases such as the Asia-Pacific Port State Computer Information System (APCIS). In addition, information is also publicized through detention lists and/or lists of banned ships. The disclosure of information is considered to be an effective way to reduce substandard shipping.

Apart from general inspection for compliance with internationally agreed instruments and standards, MOU members, either individually or collectively, undertake focused inspection campaigns reacting to specific threats to marine safety and the environment. An example of such focused inspection campaigns is the campaign on oil tankers undertaken by the Paris MOU members following the *Erika* disaster in November 1999. This campaign was carried out from 1 September to 30 November 2000 and produced an above-average detention rate of 11 percent with some interesting detailed results. It was shown that, for instance, members of the International Association of Classification Societies (IACS) had surveyed all detained ships and that five of the detentions involved items for which the class society is directly responsible.[29]

---

29. For details of the results of the inspection campaign on oil tankers, see Paris MOU. Campaign highlights detention rate and structural defects. Accessed November 2001 on the World Wide Web: <http://www.parismou.org/whatsnew/2001-0424.html>.

Another example of a focused inspection campaign carried out at the national level is that of the Australian Maritime Safety Authority (AMSA). This campaign is to run in phases over 2 years. It started in 1 December 2000 with phase 1 focusing on Collision Avoidance and with phase 2 on Requirements of SOLAS relating to global marine distress and safety systems to follow. These focused inspection campaigns are announced in advance and conducted in addition to the regular port State control activities.[30]

Up to now, it is fair to state that port State control has been a relatively efficient instrument of eliminating or at least reducing substandard shipping. Nevertheless, the general level of maritime safety and pollution prevention could only be raised to standards acceptable to the global community at large if all flag States would live up to their obligations and implement mechanisms to ensure identifiability and accountability of owners and improve general standards of maritime operations.

Coastal State Control

In contrast to port State control, coastal State control is not clearly defined or integrated into a coherent concept of protecting marine safety and the environment by combating substandard shipping. While the coastal States' authority with regard to the prevention of pollution is established in UNCLOS, it is not clear to what extent measures to prevent, reduce, and control pollution may interfere with the concept of innocent passage of foreign vessels.[31]

The issue of coastal State control has, however, become more topical in the context of increasing problems related to illegal, unreported, and unregulated (IUU) fisheries. Fishing in areas under national jurisdiction without the authorization of the coastal State has become a major offence of IUU fishing requiring active cooperation between flag States, coastal States, and port States in the combat of these malpractices.[32]

ISM Code and Quality Management

Another instrument to combat substandard shipping that is closely linked to flag State control is the International Safety Management (ISM) Code.[33]

30. See AMSA. Port State Control—Focused Inspection Campaign. Accessed November 2001 on the World Wide Web: <http://www.amsa.gov.au/sp/psc/fic/fic.htm>.
31. See Hare (n. 28 above).
32. The problem of IUU fishing and potential solutions were debated at length at the first meeting of the United Nations Open-ended Informal Consultative Process on Oceans and the Law of the Sea and the proceedings are contained in the report thereof: Report no. A/55/274. New York: 2000.
33. IMO. International Management Code for the Safe Operation of Ships and for Pollution Prevention (ISM Code) (1993). Entered into force 1 July 1998, accessed November 2001 on the World Wide Web: <http://www.imo.org/>.

While the ISM Code is based on an intergovernmental instrument, the approach chosen is somewhat different from flag State and port State control mechanisms, as it prescribes a certain type of behaviour that is directed at the company level and is to be implemented by ship operators. In fact, the philosophy behind the ISM Code, both with regard to its objectives and the implementation mechanisms, is similar to that of quality assurance as reflected in the provisions of the ISO Standard (ISO 9000). The linkage between the two approaches was particularly clear in the beginning of the discussions of ISM, when no decision had been taken as to whether or not the Code would become compulsory. It is primarily the focus that differs. While ISM focuses on safety and environment, ISO 9000-based quality assurance focuses on the customer and service. Both elements are integral parts of a quality/safety management system. Equally, those companies that had chosen to follow a path of quality assurance through ISO 9000 have generally also faced fewer difficulties in adapting to ISM Code requirements.

The introduction of the ISM Code primarily involved a cultural change that had gradually impacted on the way shipping companies were being managed. This change involved an integration of ship- and shore-based management in pursuit of the common goals of improving maritime safety, protection of the environment, and consequent improving the profitability of shipping operations. In doing so, it was necessary to redirect behavioural patterns from those based on activity toward those based on results. Thus, companies should change from an "inspection culture" widely accepted in the context of flag State and port State control mechanisms to a "safety culture" where the systematic improvement of operations safety becomes the underlying principle of the companies' activities. This changeover is to be achieved through a disciplined approach to ship operations.

The introduction of ISM in the context of embracing quality management systems seems to be a logical continuation of the general quest for the improvement of maritime safety and the eradication of substandard shipping. The potentially positive impact of mandatory introduction of higher quality tonnage in major markets such as, for instance, double-hull tankers, will certainly be amplified by the concurrent introduction of quality management systems. For the shipping companies themselves it seems to be clear that, apart from the mandatory aspects of certain technical features and the ISM Code, the introduction of such systems will have a positive impact on their market position, and consequently their profitability. Pressure on charterers to refrain from chartering low-cost substandard tonnage, will further help to provide the necessary economic incentives for shipping companies to engage in quality management programmes.

ISM, due to its mandatory nature, and quality management in line with ISO 9000, are important elements of a commercial strategy for developing countries' shipowners, as they provide for predictable standards enabling them to compete in international markets, particularly, but by no means exclusively, in those cases where employment possibilities are increasingly being sought in third country trades as exporters of shipping services. Management improvements will not only be imposed by mandatory regulation, but equally by commercial necessity.

Certification of management system improvements will be increasingly requested by charterers and shippers, particularly to the extent that they themselves are subjected to measurable quality standards. Yet, it has to be noted that certification is not very popular in developing countries and is so far concentrated in countries of the Far East and Southeast Asia. The reasons for this are many and extend equally to similar problems faced by developing countries in the implementation of the ISM Code. Only a few of these countries, particularly of those in sub-Saharan Africa, have the basic ingredients for successful implementation of any such codes at their disposal. The lack of certification bodies and accreditation mechanisms, including the lack of maritime administrations, is particularly endemic, and resulting costs are an important element in the final cost equation determining the competitiveness of developing countries' carriers. This, however, is a problem of certification only, and should not be a deterrent to the introduction and application of quality management procedures, which can be expected to significantly reduce operating costs. Certification is thus considered by a number of companies to be a threat rather than an opportunity, particularly in light of the relative cost thereof. Economies of scale are attainable in certification, thus tending to place small shipping companies operating small fleets at a competitive disadvantage. However, while these cost considerations may be paramount in the short term, in the long term they have to be checked against the benefits of quality management, both with regard to safety operations (ISM Code certification) and the impact on the marketability of shipping services through ISO 9000 certification. At any rate, the issue of cost of implementing ISM and ISO 9000 is often exaggerated by ship owners/operators. The actual cost is minimal when compared to the operating costs and many companies combine the ISM certification with their class—a sort of one-stop shopping.

Certification of any type requires the existence of a "competent authority," either as a national authority or as an institution entrusted with the power of certification. Depending on the type of certification—mandatory, safety, or voluntary management system certification—the need for administrative action arises. This is not necessarily new. The UN Convention on Conditions for Registration of Ships already calls for the establishment of a "competent and adequate national maritime administration" to ensure effective flag State control. Mandatory ISM Code certification has put new pressure on countries and their registries to provide for appropriately equipped administrative structures capable of addressing and fulfilling flag State obligations in a meaningful manner, even if the actual audits of compliance will be subcontracted to classification societies or similar third-party institutions.

It might still be too early to pass a definite judgement on the success or failure of the ISM Code as an instrument designed to ". . . provide an international standard for the safe management and operation of ships and for pollution prevention."[34] The basic conceptional problem that arises is whether a change in operational philosophy can be administratively imposed on operators and whether implementation mechanisms can be sufficiently stringent to ensure compliance also by those who might not show the necessary responsiveness to new approaches to vessel operations. Preliminary indications appear to cast serious doubts, though, on whether the ISM objectives can be met.

Some of the problems that have emerged relate, for instance, to the validity of certificates, both "documents of compliance" (DOC) and "safety management certificates" (SMC).[35] The ISM Code requires ships to undergo certification inspection (audit) by class or third party every 2.5 years, whereas ISO 9000 requires audits (internal) every year to ensure compliance. External ISO 9000 audits use representative sampling techniques and may not audit ships every year. Substandard tonnage only has to pass the "test" once every 2.5 years and there seems to be a general perception that this is insufficient policing to make the ISM Code an effective instrument. This concern is also reflected in recent moves by major classification societies. Lloyds Register, American Bureau of Shipping, and Det Norske Veritas are apparently reconsidering their approach to ISM certification due to an adverse record of detentions/inspection by port State control of vessels with their ISM certification, when in fact many of the vessels are not classed by the same classification societies. As a result, LR, ABS, and DNV require ISM audits of vessels every 12 months when they are not responsible for class matters. These problems will certainly be aggravated once all vessels, including the large number of general cargo vessels, will be subjected to the ISM Code.

Another potential problem of the ISM Code relates to the scope of application. It only covers passenger ships and cargo vessels above 500 grt. Thus, for instance, fishing vessels and coastal vessels engaged in cabotage trades—vessels typically below 500 grt—are not required to conform to the ISM Code standards. Consequently, particularly in coastal zones there is the potentially dangerous situation of substandard vessels trading alongside international tonnage.

34. ISM Code, Preamble (n. 33 above).
35. The DOC is issued to a company and the SMC to a vessel following initial verification of compliance with the requirements of the ISM Code.

CONCLUSION: THE MYTH OF FLAG STATE CONTROL

A cursory review of the sections of the Reports of the Secretary-General of the United Nations[36] to the fifty-fourth and the fifty-fifth sessions of the General Assembly that deal with the shipping industry, navigation, and fisheries, revealed the problems relating to safety of navigation, safety of ships, welfare of crews, certification, illegal unreported and unregulated (IUU) fishing, and related re-flagging issues. The reports consistently placed the responsibility on the flag State and the exercise of its jurisdiction and duties. The reports stressed that the primary responsibility in the enforcement of IMO and other regulations lie with the flag State. However, the truth was, and remains, that flags of convenience were created with the very intention that the flag State, even if it had the will, would not, and could not, exercise such duties ipso facto.[37] To go any further and pretend that flags of convenience States can abide by the self-imposed rules or model self-imposed codes is tantamount to placing arsonists in charge of the fire brigade.

Port State control has been a natural response by coastal/port states that are most directly and adversely affected by safety and environmental consequences of irresponsible operations. Nevertheless, it can only bring a partial solution to the problem as it relies too heavily on ad-hoc measures and regionalizes the issue of substandard shipping. Similarly, it must be seen that the concept of ISM and the related Code that attempt to set operational standards and disciplines stop short of addressing the real flag State issues. They do little, if anything, to create the genuine link called for in UNCLOS with the legal, economic, and operational consequences thereof. The link, which could be created through certification mechanisms, is too weak to be effective. The Code undoubtedly presents a step in the direction of improving ship management, but its implementation is still flawed and existing problems might well be amplified after July 2002 when the bulk of the world fleet becomes subject to the provisions of the ISM Code.

It is also clear that while a flag of convenience State does not have the leverage of the genuine link to exercise control, other flag States that have a genuine link to the vessel do not exercise that control with the required vigour for well-known reasons, including the unfair competition that flag of convenience operators enjoy. Consequently, it is not only a regulatory issue

---

36. United Nations. Oceans and the Law of the Sea Reports of the Secretary General, A/55/61, A/54/429/Corr1, New York. Accessed January 2002 on the World Wide Web: <http://www.un.org/depts/los>.

37. "The State has sovereign rights over the ships and seafarers as part of its territory. This is now a form of political sophistry in the global industry. The flag State is in control de jure, but only theoretically; in practice, the de facto authority lies with the foreign-based beneficial owner (via the master), who can dictate procedures and conditions on board regardless of the flag of the ship" Couper et al. (n. 10 above).

but also an issue that permeates the whole structure of the industry, the technology change, the concentration of capitals, economic performance, and so forth.

The issue, therefore, is one of governance as part of the global governance of the oceans. As Gold indicates, there are a multiplicity of actions by governments, civil society, international organizations, port States, coastal States, the business community, and the media, among others, that have to provide the global community with "safer ships and cleaner seas."

The authors' theses is that flag State control is no longer a viable proposition. Furthermore, the authors did not intend to propose alternatives or to resolve the lacuna, however, they point out certain needs that may be addressed by the international community, inter alia, the need to

- Elaborate an international instrument designed to deal with the problems of a variety of illegal maritime acts and fraud, and specifically the problem of jurisdiction and extradition. Such a convention could expand the jurisdiction of States and list the illegal maritime acts to be covered. This expansion of jurisdictional capabilities of States should be linked to extradition requirements, so that a State could either prosecute offenders in its custody or extradite them to a requesting State. So far, governments have not found any existing international legal instruments to govern illegal maritime acts or offences appropriately. Crimes that lead to destruction of living resources, endanger safety of life, and result in pollution are not "extraditable crimes," governed by international treaty. It seems that the time has come for the international community to address these issues seriously, in the same way as it rose to the challenge of air piracy, drug trafficking, and terrorism;
- Ensure and promote coherence among all involved institutions relative to their work on the oceans and UNCLOS and Agenda 21;
- Promote increasing political accountability as regards the various uses of the oceans through a vigorous approach to regional cooperation and coordination;
- Involve all stakeholders of the oceans, particularly civil society and nongovernmental organizations, in the evolving framework of governance;
- Pay greater attention to the recent proposal on levying charges for the use of the "global commons in high seas"[38]; and

38. The German Advisory Council on Global Change presented a paper at the International Conference on Financing for Development (Monterrey, March 2002) in which it described the use of the high seas for transportation as an example of global open access where the high seas are not subject to the legal sovereignty of any State. It argued in this context for levying user charges to create incentives to reduce shipping-induced marine pollution and to close the prevailing regulatory gap. See WBGU. *Special Report, Charging the use of the Global Commons.* Berlin: 2002.

- Set higher goals and prepare the ground for further progress in promoting a system of global governance of the oceans that is comprehensive and interdisciplinary; that is democratic, inclusive and transparent; and that can contribute to addressing universal concerns. It may be necessary to establish an "Oceans Senate." This could be a quasi-legislative or deliberative body composed of elected members from member States, heads of multilateral institutions, eminent persons who have made significant contributions to the oceans, and representatives of civil society and industry. The deliberations or recommendations of the Oceans Senate could be transmitted to the General Assembly for endorsement and legitimacy.

It is important to recall that Professor Elisabeth Mann Borgese expressed in 1976 at the Algiers meeting on Reshaping the International Economic Order a vision of world shipping sailing under a single international flag. Perhaps the time has come to consider a single international registry for ships.

Transportation and Communications

Marine Biosecurity Issues in the World Oceans: Global Activities and Australian Directions

Chad L. Hewitt
Centre for Research on Introduced Marine Pests
CSIRO Marine Research, Floreat Park, Western Australia

INTRODUCTION

The extent to which humans have altered, perhaps irreversibly, the biological and physical features of the environment has become one of the focal topics of the late 20th century.[1] A pervasive theme that runs throughout this discussion is the lasting impact humans have had through the intentional and accidental introductions of species into new environments.[2] Biological introductions occur when species are transported beyond their natural ranges by the agencies of human activities. Mechanistically, these introduc-

---

1. J. A. Drake, H. A. Mooney, F. diCastri, R. H. Groves, F. J. Kruger, M. Rejmanek and M. Williamson, eds., *Biological Invasions: a global perspective.* Scientific Committee on Problems of the Environment (SCOPE). Report no. 37. Chichester: John Wiley, 1989; J. Lubchenco, A. M. Olson, L. B. Brubaker, S. R. Carpenter, M. M. Holland, S. P. Hubbell, S. A. Levin, J. A. MacMahon, P. A. Matson, J. M. Melillo, H. A. Mooney, C. H. Peterson, H. R. Pulliam, L. A. Real, P. J. Regal and P. G. Risser, "The sustainable biosphere initiative: an ecological research agenda," *Ecology* 72 (1991): 371–412; D. Lodge, "Species Invasions and Deletions: Community Effects and Responses to Climate and Habitat Change," in *Biotic Interactions and Global Change,* eds. P. M. Kareiva, J. G. Kingsolver, R. B. Huey (Sunderland: Sinauer Associates, 1993), pp. 367–87; D. M. Lodge, "Biological invasions: lessons for ecology," *Trends in Ecology and Evolution* 8 (1993): 133–37.

2. J. T. Carlton, "Introduced Invertebrates of San Francisco Bay," in *San Francisco Bay: an urbanized estuary,* ed. T. J. Conomos (San Francisco: California Academy of Sciences, 1979), pp. 427–42; J. T. Carlton, "Man's role in changing the face of the ocean: biological invasions and implications for conservation of nearshore environments," *Conservation Biology* 3 (1989): 265–73; T. J. Case and D. T. Bolger, "The role of introduced species in shaping the distribution and abundance of island reptiles," *Evolutionary Ecology* 5 (1991): 272–90; S. L. Pimm, *The Balance of Nature?* (Chicago: University of Chicago Press, 1991); J. T. Carlton, "Pattern, process, and prediction in marine invasion ecology," *Biological Conservation* 78 (1996): 97–106; J. T. Carlton, "Global Change and Biological Invasions in the Oceans," in *Invasive Species in a Changing World,* eds. H. A. Mooney and R. J. Hobbs (Washington, D.C: Island Press, 2000), pp. 31–53.

tions differ little from the natural range expansions of species into new environments and communities except in spatial and temporal scale.[3] Human-mediated biological introductions can occur over spatial scales greater than 20,000 km (between noncontiguous biotic provinces) and within time scales of days to months. Ecologically, however, these species may present significant alterations to community structure and ecosystem function.[4]

Marine introductions have likely occurred throughout human history but have significantly increased with European expansion over the last 500 years.[5] This has resulted in an exponentially increasing number of species entering ecosystems in which they did not evolve, creating a global homogenisation of the world's biota.[6] The recognition of the degree to which terrestrial and freshwater ecosystems have been invaded by nonindigenous organisms has resulted in several volumes describing and documenting the impacts and case histories of invasions.[7] Despite this work, relatively little

    3. Lodge (n. 1 above); Pimm (n. 2 above).
    4. G. M. Ruiz, P. Fofonoff and A. H. Hines, "Non-indigenous species as stressors in estuarine and marine communities: assessing invasion impacts and interactions," *Limnology and Oceanography* 44 (1999): 950–72; G. M. Ruiz, J. T. Carlton, E. D. Grosholz and A. H. Hines, "Global invasions of marine and estuarine habitats by non-indigenous species: mechanisms, extent and consequences," *American Zoologist* 37 (1997): 621–32; G. M. Ruiz, P. W. Fofonoff, J. T. Carlton, M. J. Wonham and A. H. Hines, "Invasion of coastal marine communities in North America: apparent patterns, processes, and biases," *Annual Review of Ecology and Systematics* 31 (2000): 481–531.
    5. A. W. Crosby, *Ecological Imperialism: the biological expansion of Europe, 900–1900* (Cambridge: Cambridge University Press, 1986), p. 368.
    6. D. Lodge, *Biotic Interactions and Global Change,* (n. 1 above); M. Ribera and C. F. Boudouresque, "Introduced marine plants, with special reference to macro-algae: mechanisms and impact," *Progress in Phycological Research* 11 (1995): 187–68; P. M. Vitousek, C. M. D'Antonio, L. L. Loope and R. Westbrooks, "Biological invasions as global environmental change," *American Scientist* 84 (1996): 468–77; Carlton, *Biological Conservation,* (n. 2 above); A. N. Cohen and J. T. Carlton, "Accelerating invasion rate in a highly invaded estuary," *Science* 279 (1998): 55–58.
    7. Drake et al., *Biological Invasions: a global perspective* (n. 1 above); R. H. Groves and J. J. Burdon, eds., *Ecology of Biological Invasions: An Australian Perspective.* (London: Cambridge University Press, 1986); R. L. Kitching, ed., *The Ecology of Exotic Animals and Plants: Some Australian Case Histories,* (Brisbane: John Wiley & Sons, 1986), pp. 262–70; I. A. W. MacDonald, F. J. Kruger and A. A. Ferrar, eds., *The Ecology and Management of Biological Invasions in South Africa,* (Cape Town: Oxford University Press, 1986); H. A. Mooney and J. A. Drake, eds., *Ecological Studies,* vol. 58, *Ecology of Biological Invasions of North America and Hawaii,* (New York: Springer-Verlag, 1986); W. Joenje, "The SCOPE Programme on the Ecology of Biological Invasions: An Account of the Dutch Contribution," in *The Ecology of Biological Invasions,* eds. W. Joenje, K Bakker and L Vlijm, Proceedings of the Komnklijke Nederlandse Akadamie Wetenschappen C 90 (1987), pp. 3–13; F. R. S. Kornberg and M. H. Williamson, eds., *Quantitative Aspects of the Ecology of Biological Invasions* (London: The Royal Society, 1987); F. diCastri, A. J. Hansen and M. Debussche, eds., *Invasions in Europe and the Mediterranean Basin* (Berlin: Kluwer Academic, 1990); T. F. Nalepa and D. W. Schloesser, eds., *Zebra Mussels: Biology, Impacts, and Controls*

was discussed concerning *marine* bioinvasions. Only in the last 2 decades has considerable attention been directed to bioinvasions in the marine environment.[8]

Recent floral and faunal surveys of estuarine and coastal marine systems indicate large-scale introductions have occurred historically and are continuing to occur (table 1).[9] This increased recognition of introduced and cryptogenic (species whose native origin is not discernable[10]) species provides an indication of the breadth of taxa that have successfully entered new systems. In San Francisco Bay, California, 164 species were identified in estuarine and marine areas that have established reproductive and self-sustaining populations[11] representing 9 animal phyla and 3 algal divisions introduced over a period of at least 133 years. Similarly, Pearl Harbor, Hawaii, has received over 63 estuarine and marine species from 7 animal phyla and 1 algal division in the last 100 years.[12] In Port Phillip Bay, Victoria, Australia, a list of 101 estuarine and marine species from 9 animal phyla and 4 plant divisions

---

(Boca Raton: Lewis, 1993); A. Rosenfield and R. Mann eds., *Dispersal of Living Organisms into Aquatic Ecosystems* (College Park: University of Maryland, 1992).

8. A. N. Cohen and J. T. Carlton, *Nonindigenous Aquatic Species in a United States Estuary: a case study of the biological invasions of the San Francisco Bay and Delta* (Washington, D.C.: United States Fish and Wildlife Service, 1995), p. 246; Cohen and Carlton, *Science* (n. 6 above); S. L. Coles, R. C. DeFelice, L. G. Eldredge and J. T. Carlton. "Biodiversity of Marine Communities in Pearl Harbor, Oahu, Hawaii with Observations on Introduced Exotic Species." *Bishop Museum Technical Report*. Report no. 10. Honolulu, Hawaii: Bishop Museum, 1997, p. 76; H. J. Cranfield, D. P. Gordon, R. C. Willan, B. A. Marshall, C. N. Battershill, M. P. Francis, W. A. Nelson, C. J. Glasby and G. B. Read. "*Adventive Marine Species in New Zealand*," *NIWA Technical Report*. National Institute of Water & Atmospheric Research. Report no. 34. Wellington, New Zealand: NIWA, 1998, ; C. L. Hewitt, M. L. Campbell, R. E. Thresher and R. B. Martin, eds. "*Marine Biological Invasions of Port Phillip Bay, Victoria*," *CRIMP Technical Report*. Center for Research on Introduced Marine Pests. Report no. 2. Hobart: CSIRO Marine Research, 1999, p. 344; Ruiz et al., *American Zoologist* (n. 4 above); Ruiz et al., *Annual Review of Ecology and Systematics* (n. 4 above).

9. Cohen and Carlton (n. 8 above); Cohen and Carlton, *Science* (n. 6 above); Coles et al. (n. 8 above); Cranfield et al. (n. 8 above); K. R. Hayes, "Ecological risk assessment for ballast water introductions: a suggested approach," *ICES Journal of Marine Science* 55 (1998): 201–12; P. Hutchings, J. Van der Velde and S. Keable, "Baseline Study of the Benthic Macrofauna of Twofold Bay. N.S.W., with a Discussion of the Marine Species Introduced into the Bay," Proceedings of the Linnean Society of N.S.W. 110, 1989, p. 339–67; K. Jansson. *Alien Species in the Marine Environment*. Swedish Environmental Protection Agency. Report 4357. Solna, Sweden: Swedish Environmental Protection Agency, 1994, p. 68.

10. J. T. Carlton, "Biological invasions and cryptogenic species," *Ecology* 77 (1996): 1653–55.

11. Cohen and Carlton, *Science,* (n. 6 above); Note that Cohen and Carlton report 212 species from freshwater, estuarine and marine habitats in San Francisco Bay, however, a re-analysis of the data results in 164 estuarine and marine species in order to make the data comparable to other locations.

12. Coles et al. (n. 8 above).

TABLE 1.—RECENT REGIONAL AND LOCAL SURVEYS FOR INTRODUCED MARINE AND BRACKISH WATER SPECIES

Location	No. of Introduced Species	Reference
United States	298[a,e]	Ruiz et al. 2000
Baltic Sea	96[a]	Gollasch and Leppäkoski 1999; Leppäkoski and Olenin 2000
New Zealand	167[a]	Cranfield et al. 1998
United Kingdom	50[b]	Eno et al. 1997
Black Sea	35[a,b]	Zaitsev and Mamaev 1997
Mediterranean Sea	240[a,c,d]	Por 1978; Ruiz et al. 1997
South Africa	58[a,c]	de Moor and Burton 1988
Australia (1990)	62	Pollard and Hutchings 1990[a,b]
Australia (2001)	>215	Hewitt et al. 1999; Hewitt (2002)

[a] Includes marine, brackish, freshwater, and salt marsh species.
[b] Partial evaluation of species.
[c] Includes all species records, not limited to establishment.
[d] Includes Lessepsian migration as well as human mediated introductions.
[e] Limited to continental United States (including Alaska).

TABLE 1 REFERENCES

H. J. Cranfield, D. P. Gordon, R. C. Willan, B. A. Marshall, C. N. Battershill, M. P. Francis, W. A. Nelson, C. J. Glasby and G. B. Read, "Adventive Marine Species in New Zealand," (*National Institute of Water and Atmosphere Technical Report 34*, Wellington, 1998).
S. Gollasch and E. Leppäkoski, eds., *Initial Risk Assessment of Alien Species in Nordic Coastal Waters*, (Copenhagen: Nordic Council of Ministers, 1999).
C. L. Hewitt, "Distribution and diversity of Australian tropical marine bio-invasions," *Pacific Science* 56 (2002): 215–22.
C. L. Hewitt, M. L. Campbell, R. E. Thresher and R. B. Martin, eds., "Marine Biological Invasions of Port Phillip Bay, Victoria," no. 20, *CRIMP Technical Report*[, (Hobart: CSIRO Marine Research, 1999).
E. Leppäkoski and S. Olenin, "Non-native species and rates of spread: lessons from the brackish Baltic Sea," *Biological Invasions* 2 (2000): 151–63.
I. J. de Moor and M. N. Burton, "Atlas of Alien and Translocated Indigenous Aquatic Animals in Southern Africa," no. 144, *South African National Scientific Programmes Report*, (Grahamstown, South Africa: JLB Smith Institute of Ichthyology, 1988).
D. A. Pollard and P. A. Hutchings, "A review of exotic marine organisms introduced to the Australian Region. I. Fishes," *Asian Fisheries Science* 3 (1990a): 205–21.
D. A. Pollard and P. A. Hutchings, "A review of exotic marine organisms introduced to the Australian Region. II. Invertebrates and algae," *Asian Fisheries Science* 3 (1990b): 223–50.
G. M. Ruiz, J. T. Carlton, E. D. Grosholz and A. H. Hines, "Global invasions of marine and estuarine habitats by non-indigenous species: mechanisms, extent and consequences," *American Zoologist* 37 (1997): 621–32.
G. M. Ruiz, P. W. Fofonoff, J. T. Carlton, M. J. Wonham and A. H. Hines, "Invasion of coastal marine communities in North America: apparent patterns, processes, and biases," *Annual Review of Ecology and Systematics* 31 (2000): 481–531.
Y. Zaitsev and V. Mamaev, *Marine Biological Diversity in the Black Sea: a Study of Challenge and Decline*, (New York: United Nations Publications, 1997).

has been compiled from recognised introductions over 138 years.[13] In total, currently recognised marine and estuarine introductions around the world comprise representatives from over 10 phyla and 7 divisions, with taxa representing differing life history characteristics, often with multiple stages within individual taxa.

The mechanisms of marine transfers of organisms include hull fouling, wooden hull boring, mariculture transfers, intentional introductions, dry

13. Hewitt et al. (n. 8 above).

ballast and water ballast, amongst others. Current estimates are that as many as 3,000 species are transported around the world on a daily basis.[14] The continuing transport and introduction of marine species from one region to another has resulted in 100s of thousands species being moved around the world.

Most biosecurity[15] efforts have been terrestrial or freshwater in focus, however, recent research has indicated that marine invasions are of equivalent magnitude to those in terrestrial and freshwater habitats.[16] More importantly, the evidence demonstrates that introductions in marine ecosystems are accelerating[17] (fig. 1). As trade increases,[18] shipping-related introductions appear to be rising.

The challenges to scientists, managers, and policy makers involved with marine invasion science and biosecurity are to identify the current status of invasions in a local and regional context (e.g., the baseline numbers of invaders and the rate of new invasions); establish mechanisms and policies to reduce the rate of new invasions; develop methodologies and response strategies to understand and predict new invasions; and develop control and eradication methodologies, strategies, and policies. Here I provide a short review of the current status of research in these areas and provide insight into the Australian approaches.

CURRENT STATUS OF MARINE INVASIONS

Marine (and estuarine) biological introductions have been detected in all oceans of the world (table 1). However, these studies differ in the temporal, spatial, and taxonomic scales of evaluation, as well as in the methodologies used. Two primary methods exist to identify the current status of invasions: literature and/or specimen collection evaluations; and field surveys, targeting those habitats and areas most linked with overseas vectors of transport. Literature and museum collection evaluations provide the broadest coverage for a region; however, they are inconsistent in scope and effort.

14. J. T. Carlton and J. B. Geller, "Ecological roulette: the global transport of nonindigenous marine organisms," *Science* 261 (1993): 78–82 (n. 12 above).
15. Biosecurity, or biological security, is defined as the activities and strategies concerning protection of native biodiversity including the prevention (quarantine and barrier control efforts) and post-incursion response (eradication and/or control) to invasive species.
16. H. A. Mooney and R. J. Hobbs, eds., *Invasive Species in a Changing World* (Washington, D.C.: Island Press, 2000), p. 457.
17. Cohen and Carlton, *Science* (n. 6 above).
18. B. J. Abrahamsson, "International Shipping: Developments, Prospects, and Policy Issues," *Ocean Yearbook* 8, ed. E. Mann Borgese and N. Ginsburg (Chicago: University of Chicago Press, 1989), pp. 158–75.

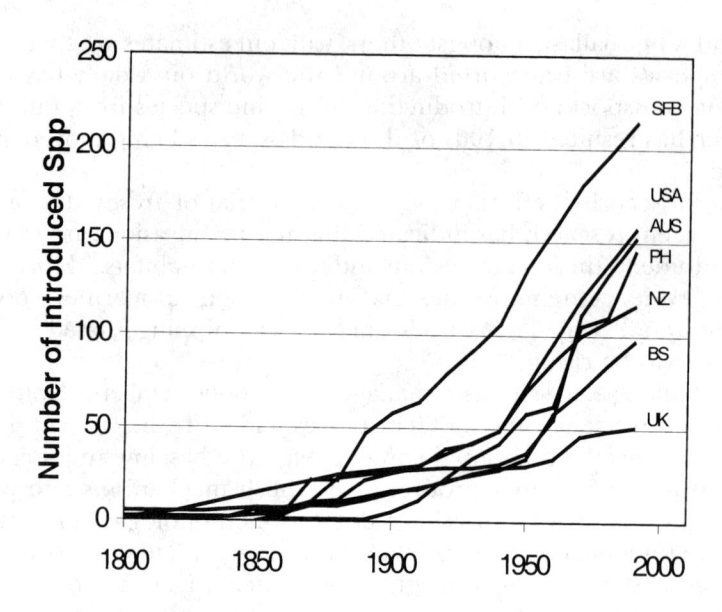

FIG. 1.—Increase in numbers of recognized established introduced species in freshwater (for SFB and BS), estuarine and marine habitats through time (based on data from references in Table 1). Locations are as follows: SFB = San Francisco Bay; USA = United States of America; AUS = Australia; PH = Pearl Harbour; NZ = New Zealand; BS = Black Sea; and UK = United Kingdom.

Patterns derived from these sources alone can result in misleading indications of invasion rates and vector strengths.[19]

The Australian approach has focused on the need for a series of baseline evaluations to determine the current scale and scope of marine invasions in Australian coastal waters. The Australian Ballast Water Management Advisory Council (ABWMAC), the Sub-Committee on Agriculture and Resource Management (SCARM), and the Australia and New Zealand Environment and Conservation Council (ANZECC) State of the Environment (SoE) Reporting Task Force, have all recognised the need for baseline studies of the extent to which introduced species have established in Australian waters. The Joint SCC/SCFA National Task Force on the Prevention and Management of Marine Pest Incursions Report[20] recommends that baseline evalua-

19. Coles et al. (n. 8 above); Hewitt et al. (n. 8 above).
20. *Report by the National Task Force on the Prevention and Management of Marine Pest Incursions.* Joint SCC/SCFA National Task Force on the Prevention and Management of Marine Pest Incursions. Canberra: Commonwealth of Australia, 2000, p. 183.

tions be undertaken to determine the extent of introductions in Australian coastal waters. Australian State of the Environment reporting guidelines have identified introduced species distribution and abundance as one of the key biodiversity threat indicators.[21]

The Australian Ballast Water Management Strategy[22] recognises that "there is no known total solution to the problem at this point in time, but there are measures that can be taken to minimise the risk." A prerequisite for the adoption of a risk management approach to controlling the spread of introduced marine pest species by shipping is a knowledge of the distribution and abundance of nonindigenous species in Australian ports.[23] This information has been lacking for all but a few Australian ports.

In 1995, the Commonwealth Scientific and Industrial Research Organisation Division of Fisheries (now Marine Research) established a Centre for Research on Introduced Marine Pests (CRIMP). CRIMP, in conjunction with the Australian Association of Ports and Marine Authorities (AAPMA) established a National Introduced Species Port Survey Programme. One foundation was the development of a set of standardised survey design protocols and sampling methodologies (the CRIMP Protocols)[24] to be implemented in all Australian ports, regardless of survey organization. The various survey designs and final reports are vetted by the Research Advisory Group of ABWMAC to guarantee consistency.

Under the current International Maritime Organization (IMO) ballast water management guidelines (Resolution A.868(20)),[25] "Port States are encouraged to undertake biological surveys and monitoring in their ports." Similarly, the Convention on Biological Diversity, Subsidiary Body on Scien-

---

21. Australian State of the Environment Committee. *Australia State of the Environment, 2001*. Independent report to the Commonwealth Minister for the Environment and Heritage. Canberra: CSIRO Publishing, p. 130.

22. D. Paterson and K. Colgan, *Invasive Marine Species: An international problem requiring international solutions* (Canberra: AQIS, 1998), p. 30.

23. K. R. Hayes and C. L. Hewitt. "*A Risk Assessment Framework for Ballast Water Introductions*," *CRIMP Technical Report*. Center for Research on Introduced Marine Pests. Report no. 14. Hobart: CSIRO, Division of Marine Research, 1998, p. 75; K. R. Hayes and C. L. Hewitt, "Quantitative Biological Risk Assessment of the Ballast Water Vector: An Australian Approach," in *Marine Bioinvasions, Proceedings of the First National Conference, 24–27 January*, ed. J. Pederson (Boston, Massachusetts: Massachusetts Institute of Technology, Sea Grant College Program, 2000), pp. 370–86.

24. C. L. Hewitt and R. B. Martin. "*Port Surveys for Introduced Marine Species—Background considerations and sampling protocols,*" *CRIMP Technical Report*. Center for Research on Introduced Marine Pests. Report no. 4. Hobart: CSIRO, Division of Fisheries, 1996, p. 40; C. L. Hewitt and R. B. Martin. "*Revised protocols for baseline port surveys for introduced marine species—design considerations, sampling protocols and taxonomic sufficiency,*" *CRIMP Technical Report*. Center for Research on Introduced Marine Pests. Report no. 22. Hobart: CSIRO Marine Research, 2001, p. 46.

25. Accessed 20 August 2002 on the World Wide Web: <http://globallast.imo.org>.

tific, Technical, and Technological Advice notes that baseline evaluations are a fundamental first step in determining the current and future status of introductions.[26]

The International Maritime Organization, in conjunction with the Global Environmental Facility and the United Nations Development Programme have jointly established a Global Ballast Water Management Programme in a concerted effort to provide technical assistance to member countries in the implementation of IMO guidelines in relation to minimising the threat of further invasions by nonindigenous species.[27] Six demonstration sites have been identified in developing countries and economies in transition for the transfer of ballast water management technologies and management practices. One aspect of the Global Ballast Water Management Programme is the implementation of baseline surveys to determine the extent of introduction in the six demonstration sites. The CRIMP Protocols will be tested in the demonstration sites during 2001.[28] Similarly, the New Zealand Ministry of Fisheries Biosecurity Programme has recommended these protocols as guidelines for a National Survey Project to be implemented in 2001–02.[29]

RATES OF NEW INVASIONS

Biological invasions in the marine environment appear to be increasing at an accelerating rate (fig. 1). Cohen and Carlton[30] have demonstrated that in San Francisco Bay a new species has arrived approximately every 14 weeks since 1960 in freshwater, estuarine, and marine habitats. When San Francisco data were reanalysed for only estuarine and marine species, it was estimated that a new species has arrived every 32 weeks since 1960. This is by far the greatest measured rate in world terms, but is likely to be a consistent pattern once full scale and repeated surveys are conducted in other large shipping areas. For example, estimates of invasion rates for estuarine and marine species alone demonstrate that several other regions are comparable to San Francisco bay (fig. 2).[31]

26. UNEP/CBD/SBSTTA/6/7.53.
27. S. Raaymakers, "Port Surveys Underway," *Ballast Water News*, no. 4 (January–March 2001), pp. 3–5.
28. M. L. Campbell, Global Port Survey Coordinator, CPM, pers. comm.; Raaymakers (n. 27 above).
29. C. O'Brien, NZ Ministry of Fisheries, pers comm.; B. Taylor (project leader). *New Zealand under siege: a review of the management of biosecurity risks to the environment*. Office of the Parliamentary Commissioner for the Environment. Te Kaitiaki Taiao Te Whare Paremata. 2001,
30. Cohen and Carlton, *Science* (n. 6 above).
31. R. E. Thresher, C. L. Hewitt and M. L. Campbell. "Synthesis: exotic and cryptogenic species in Port Phillip Bay," *The Introduced Species of Port Phillip Bay*,

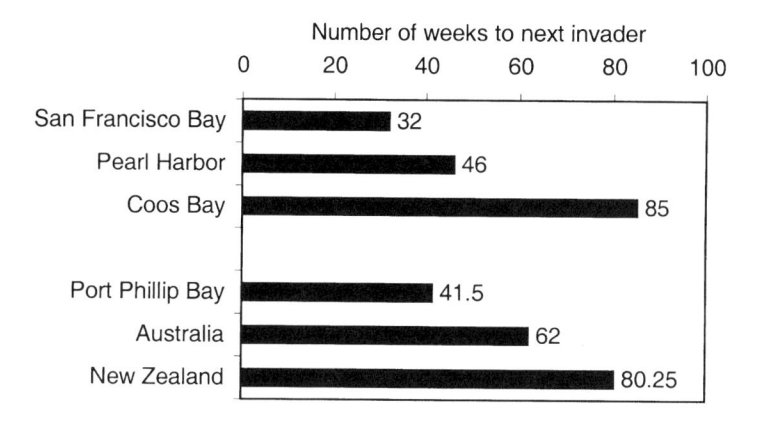

FIG. 2.—Number of weeks to the next invader (since 1960), based on data from references in Table 1.

Location	Number of weeks to next invader
San Francisco Bay	32
Pearl Harbor	46
Coos Bay	85
Port Phillip Bay	41.5
Australia	62
New Zealand	80.25

These estimates are also likely to be biased towards underestimates given our lack of consistent sampling effort through time. For example, Australia has experienced 12 recognised incursions in the marine environment, four of which are new introductions into Australia (table 2) since 1995, far exceeding the calculated expectancy of one invasion every 62 weeks (fig. 2).

The reasons behind these changes in invasion rates are difficult to discern. One suggestion is that as international shipping increases, the supply of potential invaders increases. Clearly, this is one potential mechanism for increase, but others exist. Carlton[32] reviewed the various alterations that could potentially lead to an apparent increase in invasions. These include changes to donor regions, transport vectors, and recipient regions. Changes in donor regions may include new invasions to or environmental alterations (e.g., heated effluent discharged into the port) of donor regions creating a more suitable mix of available species. The opening of new donor regions as trading partners has the ability to create a large number of new invasions by species that have previously not had the opportunity.

Alternately, changes to the transport vector can create invasion opportunities. Historically, wooden hulled vessels were slow moving with poor antifouling technologies. Consequently hull fouling and wooden hull boring species were capable of being transferred. These vessels typically had dry (or

---

Victoria. Centre for Research on Introduced Marine Pests. Technical Report no. 20. ed. C. L. Hewitt et al. Hobart: CSIRO Marine Research, 1999, pp. 283–95.

32. Carlton, *Biological Conservation* (n. 2 above).

TABLE 2.—NUMBER OF RECOGNIZED INVASION EVENTS
IN AUSTRALIA SINCE 1995

	Incursions	
Date	Species	Location
1995	*Asterias amurensis* (seastar)	Port Phillip Bay, Victoria[1]
1996	*Sabella spallanzanii* (fanworm—eradicated)	Devonport, Tasmania[2]
1996	*Undaria pinnatifida* (kelp)	Port Phillip Bay, Victoria[1,3]
1997	*Sabella spallanzanii* (fanworm)	Eden, New South Wales[4]
1998[a]	*Mytilopsis sallei* (mussel—eradicated)	Darwin, Northern Territory[5,6,7]
1998[a]	*Perna viridis* (mussel—eradicated)	Port River, South Australia[8]
1999	*Caulerpa taxifolia* (seaweed)	Sydney, New South Wales[9]
1999	*Caulerpa taxifolia* (seaweed)	Port Hacking, New South Wales[9]
1999	*Caulerpa taxifolia* (seaweed)	Lake Conjola, New South Wales[9]
2000[a]	*Charybdis japonica* (crab—one male found)	Adelaide, South Australia[10]
2000	*Sabella spallanzanii* (fanworm)	Devonport, Tasmania[11]
2001 (1999)[a]	*Hydroides sanctaecrucis* (worm)	Cairns, Queensland[12]

[a]New introductions for Australia.

1. S. Talman, J. S. Bite, S. J. Campbell, M. Holloway, M. McArthur, D. J. Ross and M. Storey, "Impacts of some introduced species found in Port Phillip Bay," in *Marine Biological Invasions of Port Phillip Bay, Victoria*, no. 20 of *CRIMP Technical Report*, ed., C. L. Hewitt et al. (Hobart: CSIRO Marine Research, 1999), p. 261–74
2. R. B. Martin, C. L. Hewitt, S. Rainer, M. L. Campbell, K. M. Moore and N. B. Murfet. "Introduced Species Survey of Devonport, Tasmania," *A CRIMP Report for the Devonport Port Authority*. Hobart, Tasmania: CSIRO Division of Fisheries, 1996.
3. S. J. Campbell and T. R. Burridge, "Occurrence of *Undaria pinnatifida* (Phaeophyta: Laminariales) in Port Phillip Bay, Victoria, Australia," *Marine and Freshwater Research* 49 (1998): 379–81.
4. C. L. Hewitt, P. Gibbs, M. L. Campbell, K. M. Moore and N. B. Murfet. "Introduced Species Survey of Twofold Bay (Eden), New South Wales," *A CRIMP Report for the New South Wales Office of Marine Affairs*. Hobart, Australia: CSIRO Division of Fisheries, 1997. 5. N. Bax, "Eradicating a dreissenid from Australia," *Dreissena!* 10 (1999): 1–5.
6. R. C. Willan, B. C. Russell, N. B. Murfet, K. L. Moore, F. R. McEnnulty, S. K. Horner, C. L. Hewitt, G. M. Dally, M. L. Campbell and S. T. Bourke, "Outbreak of *Mytilopsis sallei* (Recluz, 1849) (Bivalvia: Dreissenidae) in Australia," *Molluscan Research* 20 (2000): 25–30.
7. R. Ferguson, "The Effectiveness of Australia's Response to the Black Striped Mussel Incursion in Darwin, Australia," *A Report to the Marine Pest Incursion Management Workshop*, 27–28 August 1999, (Canberra: Department of Environment and Heritage, 2000).
8. V. Neveraskus, S.A. Primary Industries and Resources, pers. comm.
9. B. Schaffelke N. Murphy and S. Uthicke, "Using genetic techniques to investigate the sources of the invasive alga *Caulerpa taxifolia* in three new locations in Australia," *Marine Pollution Bulletin* 44 (2002): 204–10.
10. J. Gilliland, S.A. Primary Industries and Resources, pers. comm.
11. L. Gray, "Taking on the intruders!," *Southern Fisheries* 8 (Autumn 2001): 6–8.
12. K. R. Hayes and C. Sliwa, "Identifying Australia's next marine pests—a deductive approach," *Marine Pollution Bulletin* (in press).
13. P. Waterman, pers. comm.
14. J. Lewis, DSTO, pers. comm.

semi-dry) ballast consisting of rock, sand, or cobble, which was discharged in the recipient port. When steel hulls replaced wooden hulls (begun in the mid-1800s), the vector for boring organisms was significantly reduced.[33] Almost simultaneously, dry ballast was replaced by water ballast resulting in a change from benthic and meiobenthic organisms in dry ballast to planktonic organisms in water ballast.

Vessels have become faster, resulting in more species surviving transit due to shorter journey duration. Similarly, the reduction in organotin tributyltin (TBT) antifouling paints through IMO Assembly Resolution (A.895(21)) in 1999, and soon by international convention[34] is likely to result in an increase in fouling species settlement and survival if non-TBT antifouling technologies prove to be less effective. Many non-TBT alternatives are being investigated, with several positive results. Additional vectors have also become available including increased mariculture and aquaculture activities and transfers, live seafood and live bait transport, as well as transfers facilitated by scientific study.

Similar to changes in donor regions, changes to recipient regions may cause an increase in invasions. There is a large body of evidence building that suggests introductions can act in synergy with one another to enhance further invasions.[35] This mechanism alone may be sufficient to create an "acceleration" of new invasions.

PREVENTION OF NEW INVASIONS

The prevention of new invasions must begin by reducing the flow of invaders. Despite numerous agreements (e.g., Rio Convention on Biodiversity Article 8h and others[36]), this is proving to be a difficult task. Ballast water continues to be the primary focus of management activities despite increasing

---

33. M. L. Campbell and C. L. Hewitt. "Vectors, shipping and trade," *The Introduced Species of Port Phillip Bay, Victoria*. Centre for Research on Introduced Marine Pests. Technical Report no. 20. ed. C. L. Hewitt et al. Hobart: CSIRO Marine Research, 1999, pp. 45–60.

34. International Convention on the Control of Harmful Antifouling Systems to be adopted at a conference in October 2001.

35. D. Simberloff and B. von Holle, "Positive interactions of nonindigenous species: invasional meltdown?," *Biological Invasions* 1 (1999): 21–32.

36. L. Glowka, "Accountability and Legislation," *Symposium Proceedings of the Best Management Practices for Preventing and Controlling Invasive Alien Species* (Cape Town: the South Africa—United States of America Bi-National Commission, 2000), pp. 68–81.

evidence that hull fouling remains a significant vector of modern invasions.[37] In evaluations of nonindigenous species in San Francisco Bay and Delta,[38] Hawaii,[39] New Zealand,[40] and Australia,[41] both hull fouling and ballast water represent the strongest vectors of nonindigenous species transport.

In 1990, IMO's Marine Environmental Protection Committee (MEPC) formed a Ballast Water Working Group, and in 1997 the IMO adopted Guidelines for the Control and Management of Ships' Ballast Water to Minimize the Transfer of Harmful Aquatic Organisms and Pathogens.[42] These voluntary Guidelines are intended to minimise the risk of introducing harmful aquatic organisms and pathogens from ship's ballast water and associated sediments. Paramount in this discussion has been the need to protect ship safety and develop methods that are practical, environmentally friendly, and cost effective. The 1997 Guidelines have provided a basis for further international development of sound ballast water management practice. The MEPC has continued to actively work on a mandatory international vehicle to regulate and control ballast water.

In the meantime, ballast exchange at sea remains the primary method to mitigate risks. Rigby[43] has reviewed the ballast water management requirements and guidelines in several countries (summarised in table 3). The lead regulatory agency in Australia for the management of international ballast water is the Australian Quarantine and Inspection Service (AQIS). The long dedication to developing a management solution to ballast water-mediated invasions has included the adoption of voluntary ballast water management guidelines in 1991, the Australian Ballast Water Management Strategy of

37. Coles et al. (n. 8 above); Cranfield et al. (n. 8 above); Hewitt et al. (n. 8 above); R. E. Thresher, "Diversity, impacts and options for managing invasive marine species in Australian waters," *Australian Journal of Environmental Management* 6 (September 1999): 137–48; R. E. Thresher, "Key threats from marine bioinvasions: a review of current and future issues," *Marine Bioinvasions, Proceedings of the First National Conference*, 24–27 January, ed. J. Pederson (Boston, Massachusetts: Massachussetts Institute of Technology, Sea Grant College Program, 2000), pp. 24–24.

38. Cohen and Carlton (n. 8 above); Cohen and Carlton (n. 6 above).

39. Coles et al. (n. 8 above).

40. Cranfield et al. (n. 8 above).

41. R. E. Thresher, C. L. Hewitt and M. L. Campbell. "Synthesis: exotic and cryptogenic species in Port Phillip Bay," *The Introduced Species of Port Phillip Bay, Victoria*. Centre for Research on Introduced Marine Pests. Technical Report no. 20. ed. C. L. Hewitt et al. Hobart: CSIRO Marine Research, 1999, pp. 283–95.

42. Annex to Resolution A.868(29), 20th IMO Assembly, 1997, which updates the 1993 IMO Guidelines for Preventing the Introduction of Unwanted Aquatic Organisms and Pathogens from Ships' Ballast Waters and Sediments Discharges (IMO Assembly Res. A.774 (18)).

43. G. Rigby. *Ballast Water Treatment to Minimise the Risk of Introducing Nonindigenous Marine Organisms into Australian Ports*. Agriculture, Fisheries and Forestry—Australia. Report no. 13. Canberra: AFFA, January, 2001.

TABLE 3.—MANDATORY BALLAST WATER REPORTING AND
MANAGEMENT GUIDELINES AND RECOMMENDATIONS

Location	Ballast Water Reporting[a]	Ballast Water Management[b]
Argentina	?	Mandatory (1999)
Australia	Mandatory	Mandatory (2001)
Brazil	Mandatory	?
Canada	Great Lakes—Mandatory (1989)	Great Lakes—Mandatory (1994)
	Vancouver, Nanaimo and Fraser River—Mandatory (1997)	Vancouver, Nanaimo and Fraser River—Mandatory (1997)
	All other ports—Mandatory (2000)	All other ports—Voluntary
Chile	?	Mandatory (1995)
Denmark	None	None
Finland	None	None
Germany	None	None
Greece	None	None
Israel	Mandatory (1994)	Mandatory (1994)
Italy	None	None
Japan	None	None
Hong Kong, China	Recommended compliance with IMO regulations	None
Netherlands	None	None
New Zealand	Mandatory (1998)	Mandatory (1998)
Norway	None	None
Singapore	None	None
Sweden	None	None
United States of America	Great Lakes—Mandatory (1993)	Great Lakes—Mandatory (1994)
	Hudson River—Mandatory (1994)	Hudson River—Mandatory (1994)
	All other ports—Mandatory (1999)	West Coast—Mandatory (2000)
		All other ports—Voluntary (1999)
United Kingdom	Orkney Islands—Mandatory (1998)	None
	All other ports—None	

Source.—G. Rigby. n. 43, above.
[a] Ballast Water reporting requirements refer to filing information about the origin and disposition of ballast water on a vessel. In general this refers to a per tank requirement.
[b] Ballast Water Management encompasses proscribed activities including ballast water exchange at sea through empty-refill or three times volumetric exchange.

1995, and more recently, the revised Ballast Water Management Guidelines in 1999. Australia's Oceans Policy, announced in December 1998,[44] gave explicit support to the implementation of a single national ballast water management regime moving towards mandatory ballast water management in concordance with agreed IMO measures.

The amendment of the Quarantine Act of 1908 (the Act) through the Quarantine Amendment Act 1999[45] allowed the adoption of the new mandatory ballast water management arrangements[46] that were implemented 1 July 2001. The new arrangements incorporate the implementation of a species-specific, risk assessment-based Decision Support System (DSS)[47] that provides an assessment of the likelihood of target species being present on a vessel for each voyage.

Hull fouling issues have yet to be addressed adequately, despite a report by the Australia and New Zealand Environment and Conservation Council (ANZECC)[48] highlighting hull fouling issues. The incursion of the hull fouling associated black striped mussel, *Mytilopsis sallei,* in Darwin, Northern Territory,[49] brought to the fore the issues of "other vectors than ballast water." As a direct result, a National Taskforce was established by the Joint Standing Committee on Conservation (SCC)/Standing Committee on Fisheries and Aquaculture (SCFA) (reviewed below under Eradication and Control Strategies).

In recent years, funding in Australia for marine biosecurity issues has been derived from two primary sources: a research and development levy[50] and the Environment Australia and Natural Heritage Trust's Coast and Clean Seas Programme. The Industry Levy[51] supported the Research and Development Programme of the ABWMAC, which has funded numerous

---

44. *Australia's Oceans Policy.* Canberra: Environment Australia, 1999, Accessed 20 August 2002 on the World Wide Web: <http://www.oceans.gov.au/content_policy_v1/policyv1.pdf>.
45. Accessed 20 August 2002 on the World Wide Web: <http://scaletext.law.gov.au/html/comact/10/6106/top.htm>.
46. Accessed 20 August 2002 on the World Wide Web: <http://www.affa.gov.au/corporate_docs/publications/pdf/quarantine/ballast/bwbroch.pdf>.
47. Developed by CSIRO CRIMP, Hayes and Hewitt (n. 23 above).
48. Australia and New Zealand Environment and Conservation Council, *Working Together to Reduce Impacts from Shipping Operations: ANZECC strategy to protect the marine environment,* vols. 1–3, (Canberra: Australia and New Zealand Environment and Conservation Council, 1996).
49. N. Bax, "Eradicating a dreissenid from Australia," *Dreissena!* 10 (1999): 1–5; R. C. Willan, B. C. Russell, N. B. Murfet, K. L. Moore, F. R. McEnnulty, S. K. Horner, C. L. Hewitt, G. M. Dally, M. L. Campbell and S. T. Bourke, "Outbreak of *Mytilopsis sallei* ([Recluz, 1849] [Bivalvia: Dreissenidae]) in Australia," *Molluscan Research* 20 (2000): 25–30.
50. Under the Ballast Water Research and Development Funding Act 1998 and the Ballast Water Research and Development Funding Levy Collection Act 1998.
51. Maximum of AU$2 million over 3 years.

activities, including the development of the risk assessment framework and models underlying the DSS, research into ballast water sampling and testing, an industry awareness programme, and support for the National Port Survey Programme. The Environment Australia and Natural Heritage Trust's Coast and Clean Seas Programme has funded introduced marine species research such as the development of a Control Toolbox,[52] a National Introduced Marine Pests Information System (NIMPIS), Community Detection Kits, and the development of a Next Pest identification methodology. In addition, the CSIRO Centre for Research on Introduced Marine Pests (CRIMP), established in 1994, has developed a greater understanding of marine pest issues in Australia through baseline evaluations, risk assessment, vector investigation, and undertaking research on potential control methodologies.

New Zealand implemented mandatory ballast water controls for vessels departing Tasmania to protect against invasion by the northern Pacific Seastar (*Asterias amurensis*) in 1996.[53] In 1998, the New Zealand Ministry of Fisheries established mandatory ballast exchange for all vessels under the Biosecurity Act of 1993.[54] The announcement in June 2000 of a funding package of NZ$9.8 million over 5 years for research and management in marine biosecurity combined with NZ$14.1 million over 5 years for biodiversity research will help to create a cohesive marine biosecurity and biodiversity strategy. Already the package is funding port baseline evaluations throughout New Zealand and the identification of both ballast water discharge contingency zones and 'next pests' relevant to a New Zealand context.

UNDERSTANDING AND PREDICTION OF NEW INVASIONS

The ability to identify extant invaders is often difficult, yet managers and policy makers desire an ability to predict new invaders prior to their arrival to protect the marine environment. Several attempts have been made to determine the suite of characteristics that make for an 'ideal' invader.[55] Any

52. N. J. Bax and F. McEnnulty. *Rapid Response Options for Managing Marine Pest Incursions*. Final Report for National Heritage Trust Coast & Clean Seas Project 21249. Hobart: CSIRO Marine Research, 2001. Accessed 20 August, 2002 on the World Wide Web: <http://crimp.marine.csiro.au/nimpis/controls.htm>.
53. New Zealand Biosecurity Act of 1993, Annex 1.
54. Taylor (n. 29 above).
55. U. Ritte and U. N. Safriel, "The Theory of Island Biogeography as a Model for Colonization in More Complex Systems," Proceedings of the 8th Scientific Meeting of the Israel Ecological Society, 1977, pp. 288–94; U. N. Safriel and U. Ritte, "Criteria for the identification of potential colonizers," *Biological Journal of the Linnean Society* 13 (1980): 287–97; P. A. Parsons, *The Evolutionary Biology of Colonizing Species*, (London: Cambridge University Press, 1983); R. C. Willan, "The mussel *Musculista senhousia* in Australia: another aggressive alien highlights the need for quarantine at ports," *Bulletin of Marine Science* 41 (1987): 475–89; Carlton, *Biological Conser-*

attempt to explicitly delineate the life history attributes of invading species (in whatever environment), however, will ultimately fail due to the overbearing weight of exceptions to a rule.[56] Generalities should therefore be sought not in the complete group of successful invaders or even in the identities of those invasions that fail but rather in the subsets of species associated with specific transport mechanisms.[57]

It has long been proposed that biological invasions are associated with disturbed environments, as demonstrated by the apparent increase in susceptibility of estuaries.[58] This in part has been derived from the verbal paradigm erected by Elton[59] that states that species-rich food webs [communities] will be less susceptible to invasion due to increased interaction strengths. This theory has been partially supported by mathematical models,[60] but has received relatively little direct empirical support.[61]

Identification of "next pests" also suffers from the inability to predict the impact of an invader in a new environment. Indeed, many classic biocontrol releases have resulted in unexpected consequences.[62] Even recognised pest species such as the European green crab, *Carcinus maenas*, can have

---

vation, (n. 2 above); P. Kareiva, "Developing a predictive ecology for non-indigenous species and ecological invasions," *Ecology* 77 (1999): 1651–52.

56. M. J. Crawley, "What Makes a Community Invasible?" in *Colonization, Succession, and Stability*, ed. A. J. Gray et al. (Oxford: Blackwell Scientific, 1987), pp. 429–53; M. J. Crawley, "Chance and Timing in Biological Invasions," in *Biological Invasions: A global perspective, SCOPE 37*, ed. J. A. Drake et al. (New York: John Wiley & Sons, 1989), pp. 407–23; D. S. Simberloff, "Which Insect Introductions Succeed and Which Fail?" in *Biological Invasions: A global perspective, SCOPE 37*, ed. J. A. Drake et al. (New York: John Wiley & Sons, 1989); Lodge, *Biotic Interactions and Global Change* (n. 1 above); Lodge, *Trends in Ecology and Evolution*, (n. 1 above).

57. U. N. Safriel and U. Ritte, "Universal Correlates of Colonizing Ability," in *The Ecology of Animal Movement*, eds. I. R. Swingland and P. J. Greenwood (Oxford: Clarendon, 1983), pp. 215–39.

58. Ruiz et al., *American Zoologist* (n. 4 above).

59. C. S. Elton, *The Ecology of Invasions by Animals and Plants* (London, UK: Methuen and Co, Ltd, 1958).

60. J. A. Drake, "The mechanics of community assembly and succession," *Journal of Theoretical Biology* 147 (1990): 213–33; T. Case, "Invasion resistance, species build-up and community collapse in metapopulation models with interspecies competition," *Biological Journal of the Linnean Society* 42 (1991): 239–66.

61. G. R. Robinson, J. F. Quinn and M. L. Stanton. "Invasability of experimental habitat islands in a California winter annual grassland," *Ecology* 76 (1995): 786–94; but see J. J. Stachowicz, R. B. Whitlach and R. W. Osman, "Species diversity and invasion resistance in a marine ecosystem," *Science* 286 (1999): 1577–79.

62. D. Simberloff and P. Stiling, "Risks of species introduced for biological control," *Biological Conservation* 78 (1996): 185–92; H. G. Zimmermann, V. C. Moran and J. H. Hoffmann, "The renowned cactus moth, *Cactoblastis cactorum*: its natural history and threat to native *Opuntia* floras in Mexico and the United States of America" *Diversity and Distributions* 6 (2000): 259–69.

differential impacts in different invaded locations.[63] This inability to predict impacts may be a function of too few process-oriented studies, and too many 'case studies.'[64]

ERADICATION AND CONTROL STRATEGIES

The driving factor in biosecurity is the ultimate cost of failure. The economic costs of invasives (terrestrial, freshwater, and marine) have recently been tallied for the United States resulting in an estimated US$137 billion per year.[65] These costs are largely associated with impacts and damages, however, an approximated US$40 billion is a result of control efforts. Eradication and control efforts in the marine environment are few and vary widely in costs.

The incursion of a sabellid worm into abalone mariculture facilities in California was stemmed through the innovative application of ecological theory, resulting in a relatively inexpensive control effort (estimated cost US$5200).[66] In contrast, the detection and eradication of the black striped mussel, *Mytilopsis sallei,* in Darwin may have cost the Northern Territory government as much as AU$2.4million.[67] In part, the Darwin costs may have been attributed to a lack of incursion response readiness and a 6-month delay before detection. No clear lines of responsibility were delineated between Commonwealth, Territory, and other States. In addition, inadequate legislation required rapid (24 hour) amendment of the Fisheries Act.

As a direct consequence of the black striped mussel incursion, a National Task Force on the Prevention and Management of Marine Pest Incursions was established in August 1999 by the Joint SCC/SCFA. The Taskforce made recommendations on the immediate actions required to establish a national ready-response capability as well as to determine the longer term reforms necessary for a permanent and comprehensive national system for the prevention and management of introduced marine pests in a report submitted in December 1999.[68] This report outlines a National System for

---

63. E. D. Grosholz and G. M. Ruiz, "Predicting the impact of introduced marine species: lessons from the multiple invasions of the European Green Crab *Carcinus maenas,*" *Biological Conservation* 78 (1996): 59–66.

64. Carlton, *Conservation Biology* (n. 2 above).

65. D. Pimentel, L. Lach, R. Zuniga and D. Morrison, "Environmental and economic costs of nonindigenous species in the United States," *BioScience* 50 (January 2000): 53–64.

66. C. S. Culver and A. M. Kuris, "The apparent eradication of a locally established introduced marine pest," *Biological Invasions* 2 (2000): 245–53.

67. Bax, *Dreissena!* (n. 49 above); Willan et al., *Molluscan Research* (n. 49 above).

68. *Report by the National Taskforce on the Prevention and Management of Marine Pest Incursions,* (n. 20, above).

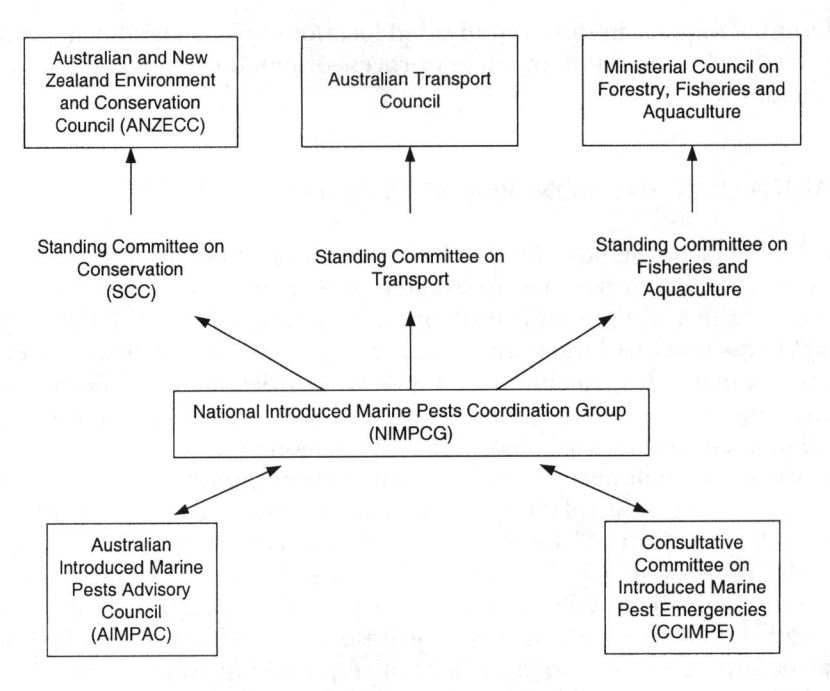

FIG. 3.—Interim arrangements established to develop a National System for the Prevention and Management of Introduced Marine Pests (Report by the National Task Force on the Prevention and Management of Marine Pest Incursions, n. 20 above.)

all transport vectors and phases of marine invasions, and proposes a National Policy Framework as a single document developed and agreed by Commonwealth, State, and Territory governments and in keeping with international policy development.

As a result of the National Task Force Report, interim coordination arrangements have been implemented as outlined in fig. 3. The ABWMAC is transformed into the Australian Introduced Marine Pests Advisory Council (AIMPAC) to reflect a broader focus than ballast water. The National Introduced Marine Pest Coordinating Group (NIMPCG) has been established to oversee policy coordination and development and to secure long-term funding. NIMPCG reports directly to three Ministerial Councils through separate Standing Committees. To fulfill the needs of an immediate incursion response capability, a Consultative Committee on Introduced Marine Pest Emergencies (CCIMPE) has been created. CCIMPE has responded to two pest incursions in its first year—the alga, *Caulerpa taxifolia* in New South Wales and the tubeworm, *Hydroides sanctaecrucis* in Queensland.

CHALLENGES

The challenges that face scientists, managers, and policy makers involved with marine invasions are multiple. We are unlikely to stop all new marine introductions or even eradicate all existing marine pests in the foreseeable future. While a seemingly daunting task lies ahead, a clear pathway is discernable. Williamson,[69] Vermeij,[70] Mooney,[71] and Mooney and Hobbs[72] (among others) have suggested systematic approaches to developing the field of invasion biology. Williamson and Vermeij clearly delineate a perspective driven by scientific research; what biological and ecological information is missing, and what can invasions tell us about the world? In contrast, Mooney and Mooney and Hobbs describe an operational approach (elucidated by the Global Introduced Species Program (GISP)) in which science, policy, and society are drawn together to tackle these issues. This integrated approach is more likely to provide direct outcomes and improve the societal and political will.

In the meantime, we need to progress in the identification of the current status of invasions. A lack of consistency between marine evaluations on a global basis prevents the synthesis necessary for clear understanding. Surveys differ in the degree of spatial, temporal, and taxonomic effort. These efforts will focus our attentions on the greatest threats to the environment and aid in allocating the limited resources available for biosecurity.

We must reduce the rate of new invasions by identifying the relative strengths of vectors and determining the most efficient (and efficacious) treatment methods. Many organisations and private concerns are currently pursuing novel solutions for ballast water treatment, but they require appropriate standards to work towards. Once treatments have been identified, these need to be implemented in an international context and within international and regional frameworks. Lastly, risk analyses and assessments may help us in the process of identifying cost-effective or even necessary treatment activities to mitigate the risks. Australia's Decision Support System (DSS) for ballast water is a prime example. These processes need to be monitored and refined based on the findings of monitoring efforts.

The theoretical base of invasion biology must be refined to develop a more predictive power. Our current understanding of which species invade

---

69. M. Williamson, *Biological Invasions* (London: Chapman & Hall, 1996).

70. G. J. Vermeij, "An agenda for invasion biology," *Biological Conservation* 78 (1996): 3–9.

71. H. A. Mooney, "Global Invasive Species Program (GISP), in *Invasive Species and Biodiversity Management,* eds. O. T. Sanderlund, P. J. Schei and A. Viken (Dordrecht, Netherlands: Kluwer Academic Publisher, 1999), pp. 407–18.

72. H. A. Mooney and R. J. Hobbs, "Global Change and Invasive Species: Where do we go from here?" in *Invasive Species in a Changing World,* eds. H. A. Mooney and R. J. Hobbs (Washington, D.C.: Island Press, 2000).

and which communities are invaded is poor. Empirical research is a fundamental step in this process.

We must develop appropriate local, national, and regional strategies for the early detection, control, and eradication of marine invaders. Early detection significantly increases the ability to eradicate a species once it establishes. Consequently, monitoring activities that incorporate community groups (including industry) and scientific researchers are essential for broad spatial and temporal coverage of coastlines. In support of rapid response, we need to identify appropriate eradication methods for risk species or broad taxonomic groups and pre-evaluate their utility (e.g., identify collateral effects, undertake risk evaluations to determine the cost:benefit ratio). Lastly, policy frameworks and legislative vehicles for achieving protection against marine invaders are necessary. Glowka[73] reviews the international, regional, and local legislation that exists and suggests that the "guiding principles" being developed by the Convention on Biological Diversity Subsidiary Body on Scientific, Technical and Technological Advice may aid in the creation and establishment of clear legislation. In many regions, little or no attention has been paid to marine introductions. While many international vehicles are in place that may aid in the protection of the marine environment against further invasions,[74] the practical activities must occur at the national and regional levels.

73. Glowka (n. 36 above).
74. Discussions in various working groups, for example, CBD (Convention on Biological Diversity) Subsidiary Body on Scientific, Technical and Technological Advice (SBSTTA), IMO MEPC Ballast Water Working Group, ICES (International Council for the Exploration of the Sea) Working Group on Introductions and Transfers of Marine Organisms, and IUCN (International Union for the Conservation of Nature: now the World Conservation Union)/SCOPE, Global Invasive Species Program.

Transportation and Communications

Ballast and Biosecurity: The Legal, Economic and Safety Implications of the Developing International Regime to Prevent the Spread of Harmful Aquatic Organisms and Pathogens in Ships' Ballast Water†

Moira L. McConnell
Dalhousie University Law School, Halifax, Canada

INTRODUCTION

Our current enthralment with the mystique of e-trade, e-commerce, e-mail, and virtual space suggests that the web of global connection, influence, and interdependence is a late 20th century development. But this is not the case. The carriage of goods and people by sea—the shipping industry—is the original activity that began the phenomenon we now describe as globalization. One of the most significant and irremediable impacts of early international commercial shipping was, in fact, a by-product—the process of diversification also known as colonisation/immigration and the introduction of nonindigenous plant and animal species and people. Today numerous international organizations, members of the public, and national governments are again concerned about the problem of unwelcome mass colonisation (otherwise known as "invasions of alien species") that endanger the human and ecological security—biosecurity—of each State. These 21st century colonisers are entering the terrestrial and marine environment of countries in a number of insidious ways.[1] One important unintentional path for marine ecosystem invasions is through ships' ballasting operations. Species can stow away in the water taken on board ships as ballast in one port and discharged

---

† EDITORS' NOTE.—An earlier version of this paper was presented in June 2001 to the University of the Aegean—2nd International Conference 2001 "Safety of Maritime Transport." The author gratefully acknowledges the help provided by the faculty, staff and students of the World Maritime University (Sweden) where the author carried out the research for this paper. She also thanks Capt. Dandu Pughiuc, Chief Technical Advisor to the GEF/UNDP/IMO Global Ballast Water Management Programme for his comments. All opinions and any errors or omissions are the sole responsibility of the author.

1. It can be intentional, for example, importing exotic species or seeds for farming or other activities, or it can be unintentional, for example, seeds in dirt caught in car tires on international travel or attached to tourists' shoes.

214   *Transportation and Communications*

in another, when cargo is picked up. The potential enormity of the problem is revealed when one considers the following:

> Globally, it is estimated that about 10 billion tonnes of ballast water are transferred [between ports] each year. Each ship may carry from several hundred litres to more than 100,000 tons of ballast water, depending on the size and purpose of the vessel.[2]

The public and governmental response to the problem of species transfer is revealed in headlines found in international agency newsletters, media releases, and conferences that proclaim, "Super invaders spreading fast"[3] and "UN Moves on Alien Invaders."[4] These are more than hyperbole; they reflect the contemporary social conceptualisation of the issue. This fear is exacerbated by the fact that, as was the case in the early colonisations, invasive species not only threaten the indigenous culture,[5] they also present the spectre of possible genocide through the spread of infectious diseases, (e.g., cholera) by pathogens also carried in ballast water.[6]

International transport and the movement of goods and people, in particular by ships, is again the unwitting culprit. Activities such as discharging ballast water, putting down an anchor, or unloading cargo—the ordinary activities of shipping and transport operations—are pathologized as "vectors" or carriers of disease and invasive species. The problem is largely the result of increasingly seamless transport systems and larger vessels moving more rapidly between ports on continuous routes. International trade and environment organizations cast the problem in terms of "Trade and Transportation Corridors" that facilitate the movement of invasive species and

---

2. International Maritime Organization, *Focus on IMO: Alien invaders—putting a stop to the ballast water hitchhikers* (London: IMO, 1998), p.1. Accessed 15 August 2002 on the World Wide Web: http://www.imo.org.

3. North American Commission for Environmental Cooperation, "Super invaders spreading fast," *Trio* (Winter 2000-2001).

4. International Maritime Organization, "UN moves on alien invaders," Media Release (10 July 2000) London: IMO, p.1. Accessed 15 August 2002 on the World Wide Web: http://www.imo.org.

5. E. Lazlo, "Human evolution in the third millennium," *Futures* 33 (2001): 649–58 provides an interesting analysis of social evolution through extensive evolution, the objectives of which, according to Lazlo, are "conquest, colonization and consumption." Lazlo advances an alternative possibility based on an expansionary perspective called intensive evolution. Aside from the present paper, the analogy with human colonizing activity is also found in other writing in this field: see for example, C. Shine, N. Williams and L. Gundling, *A Guide to Designing Legal and Institutional Frameworks on Alien Invasive Species* (Gland, Switzerland, Cambridge and Bonn: IUCN, 2000).

6. G. Casale and H. Welsh, "The international transport of pathogens in ships' ballast water," *Journal of Transportation Law, Logistics and Policy* 66 (1997): 79–87.

pose significant risks to biodiversity and biosecurity. Thus shipowners have suddenly discovered that they are also operators of vectors that form part of a transport corridor for invasive species that pose a danger to human and ecological security. Port authorities will find that they are cast as either guardians or gaps in the biosecurity of the State. In both cases the reality of a world of biosecure ports is on the horizon.[7]

The US$10.2 million Global Ballast Water Management Programme (GloBallast)[8] created by GEF/UNDP and IMO in 2000 states the following:

> The introduction of invasive marine species into new environments by ships' ballast water, attached to ships' hulls and via other vectors has been identified as one of the four greatest threats to the world's oceans. The other three are land-based sources of marine pollution, overexploitation of living marine resources and physical alteration/destruction of marine habitat.[9]

What are the implications of this situation for the maritime industry?

This article outlines the various dimensions and characterisations of the problem, the emerging international and national regulatory responses, and current technical and operational developments. It concludes with a discussion of the implications of this issue for the various sectors of the maritime industry.

It is argued throughout that it is imperative that the maritime industry develop a more nuanced appreciation of this "technical shipping issue" in the broader context of the human security and sustainable development agenda. That is, the maritime sector should be viewed as a key contributor, even a leader, rather than a threat to long-term global sustainable development. This is happening to some degree with, for example, the Green Awards programme, the European Commission's ECOPORTS initiative, and recent innovations such as a smokeless diesel engine (the EnviroEngine) for a class of cruise ship[10] and the forward-thinking shift by some operators

---

7. Michael Grey, "More muscle for port health," *Lloyd's List Maritime Asia* (May 2001): 10 comments on this point in the context of the foot-and-mouth disease restrictions.

8. Global Ballast Water Management Programme, *The Problem*, (London: IMO, 2000). Accessed 15 August 2002 at GloBallast Programme Web site on the World Wide Web: http://globallast.imo.org/problem.htm.

9. Ibid.

10. World Infodesk (April 2001). "Vessel with first smokeless diesel engine," *Marine Talk*, 26 (electronic newsletter). Accessed 15 August 2002 on the World Wide Web: http://www.marinetalk.com/infodesk (it is found under the industry News link on this Web page).

to "green" (biocide-free) antifouling paint ahead of the organotin ban.[11] However, shipping industry recalcitrance on some issues has led to public and governmental frustration and to an increased incidence of what is described as unilateralism. This is short sighted on the part of the industry and risks endangering the role of the International Maritime Organization (IMO) as a forum for facilitating harmonized international rule-making and environmental protection. A more sophisticated awareness of the dynamics of the emerging and increasingly integrated international trade and environmental protection regimes is required to enhance the position of the maritime sector as a key contributor to global security and development.

THE PROBLEM: SHIPS' BALLASTING OPERATIONS AND THE TRANSFER OF ORGANISMS AND PATHOGENS

The problem of aquatic organism and pathogen transfer in ships' ballast water is both simple and complex. On a macro system or global governance level it can be seen as a point of intersection or convergence that operates as a catalyst for the increasing integration of international and national regulatory activity based on a more holistic or ecosystemic understanding of the relationship between human activity and the environment. This change in ideas about governance was heralded by documents such as the 1972 Stockholm Declaration,[12] the 1982 United Nations Convention on the Law of the Sea (UNCLOS),[13] and Agenda 21[14] in 1992. It is also evidenced in institutions such as the United Nations Commission for Sustainable Development (CSD), which promotes integrated management at all levels of governance and the new General Assembly process called the United Nations Open-ended Informal Consultative Process on Oceans and the Law of the Sea (UNICPOLOS). The integration of the ocean and land governance systems will have a significant impact on the institutional actors in the regime governing shipping and many other sectors. This is because the emerging ecosystemic understanding does not neatly fit into the existing institutional jurisdiction packages. Historically, shipping has been dealt with as a specialized

---

11. B. Reyes (May 2001). "Owners and yards move to 'green' paints ahead of organisations' ban," *Lloyd's List* 3 (electronic newsletter).

12. Stockholm Declaration of the United Nations Conference on the Human Environment, 16 June 1972, UN Doc. A/CONF.48/14/Rev.1. Accessed 15 August 2002 on the World Wide Web: http://www.unep.org/documents/default.asp?DocumentID=97.

13. UN Doc. A/CONF.62/122. Accessed 15 August 2002 on the World Wide Web: http//www.un.org/depts/los/index.htm.

14. Agenda 21: Programme of Action for Sustainable Development, June 1992, UN Doc. A//CONF.151/26/Rev.1, Vol. 1 (1992). Accessed 15 August 2002 on the World Wide Web: http://www.unep.org/documents/default.asp?DocumentID=52.

or segregated sector—*sui generis*. However, recent State-level actions, labelled unilateralism,[15] may in fact prove to be part of a process of integrating shipping into the broader environmental and industry regulatory regimes (mainstreaming). Arguably, the need for international uniformity in standards to facilitate trade may also be addressed through a combination of international standard setting under multilateral environmental agreements (MEAs) and the World Trade Organisation (WTO) disciplines.

The next Section of this article will describe the problem from a practical, ship-operations perspective.[16] The third Section will examine the impact of marine species and pathogens transfer from the perspectives of human and ecological security, economic/commercial concerns, and safety/technical issues. All three are interrelated, but for purposes of discussion in this article they are separated.

The Activity: Ship's Ballasting Operations and the Transfer of Organisms and Pathogens

In simple terms, ballast and the process of ballast discharge and intake (ballast management) keeps ships balanced or stable and mitigates the stresses that the ocean's movements place on the vessel's superstructure. Ballast is functionally critical to a ship's safety, particularly when it is not fully laden. Ballast in this sense is simply a concept or a function rather than any particular substance. Various materials have been used as ballast through the centu-

---

15. See for example, the U.S. Environmental Protection Agency involvement and concerns about the Annex VI of MARPOL (ship source air emission controls): Intertanko (31 August 2001). U.S. Environmental Protection Agency retreats from MARPOL Annex VI. *Weekly News* no. 35/200 (electronic newsletter); the possibility that the EPA will also become more involved in ballast water management: Sandra Spears (21 September 2001). "Green is the red hot topic," *Lloyd's List*, p. 17; or the Indian government's decision to transfer port development approvals from the Shipping Ministry to the Environment Ministry: author unknown (August 2001). "Port approvals get tougher," *Fairplay Daily News.* 2 (electronic newsletter). In the case of Annex VI, this shift has been construed by the shipping industry as an act of unilateralism and a challenge to the hegemony of the IMO regime.

16. At yet another level the problem serves to highlight the dynamic relationship between technical developments and the rate and direction of regulatory activity, particularly with reference to the question of what really leads or creates change. This question has been explored at length by regime theorists such as Oran Young and others: See for example O. Young, ed., *The Effectiveness of International Environmental Regimes. Causal Connections and Behavioural Mechanisms* (Cambridge, Massachusetts: MIT Press, 1999). Others have examined international rulemaking from the perspective of discourse analysis: see for example an interesting study by K. Bäckstrand, *What Can Nature Withstand? Science, Politics and Discourses in Transboundary Air Pollution Diplomacy* (Lund, Sweden: Monograph, Lund University, Political Studies 115, 2001).

ries, however, since the development of steel hulled vessels in the 19th century, seawater has been used for reasons of economy and efficiency.[17] Modern vessels are equipped with various types of ballast tanks located at strategic points, relative to the cargo or passenger spaces, in the vessel's hull. Depending on the vessel structure, many ballast tanks have extensive internal piping or other formations that facilitate the build up of sludge or sediment in which organisms can thrive.[18] Depending on the voyage conditions, whether it has any cargo on board and the size and function of the ship, that is, bulk carrier, oil tanker, ferry, fish factories, differing quantities of ballast water are taken on to maintain stability. It is a by-product of this core operational process, one that is intrinsically related to the operation of ships as carriers, that is causing the problem.

Because the quantity of ballast required at any one time is directly related to the loading or unloading of cargo and the particular vessel's stability requirements, the discharge or intake of ballast usually occurs either in or en route to and from port areas, or in sheltered waters close to the coastline of a country. The coastal zone is replete with plant and animal organisms in various stages of their life cycles. It is also host to pathogens that may have entered port waters through municipal sewage outlets, discharge from other vessels or other land-based marine pollution sources. These organisms can live for long periods of time in ships' tanks. Estimates suggest they can survive up to 3 months or even longer in the water and sediment taken from coastal waters and pumped into the ballast tanks.

The microscopic size of many organisms and the point in their life cycle also means that the ballast water filters currently in use are of limited utility. The intake of organisms is exacerbated if the ballast is taken in very shallow or turbulent waters close to shore and at night, when many species move to the surface of the ocean. It is believed that at any one time "ballast water may be transporting 3000 species of animals and plants a day around the world."[19] More recent estimates suggest that as many as 4,500 organisms are in transit at any one time.[20] With faster vessels going to more ports of call on each voyage the problems are magnified. While the operational activity

17. *Alien invaders* (n. 2 above).
18. Tanks vary depending on the ships' function. Modern ships have segregated ballast tanks (SBT), that is, tanks devoted only to the ballasting operation. Some older ships still operate with integrated systems but these are now being phased out. Although I have not researched the point, it is possible, as suggested to me by a student at the World Maritime University, Shafiq Islam, that the requirement for dedicated clean or segregated tanks (i.e., no oil or other substances mixing in with the ballast water) may have inadvertently created a more hospitable environment for invasive species.
19. *Alien invaders* (n. 2 above), p. 1.
20. Comments by Dandu Pughiuc, Chief Technical Advisor, GloBallast, during a lecture to the Maritime Administration Students in the Master of Maritime Affairs Degree Programme at the World Maritime University, January 2001, Malmö, Sweden.

causing the problem is reasonably simple to understand, given the key role of ballast in vessel operations, the solution is much less so.

The proposed solutions and associated difficulties will be discussed in more detail later in this article. Suffice at this point to say that the most viable solution developed to date, aside from precautionary procedures to prevent or limit the initial intake of species, is to exchange coastal ballast water for mid-ocean water that does not contain or support the coastal organisms. This preventative method was developed by IMO Member States over the last decade, and is described in the 1997 Resolution A.868 (20), *Guidelines for the control and management of ships' ballast water to minimize the transfer of harmful aquatic organisms and pathogens.*[21] Open sea exchange does not totally eliminate the problem but it can significantly reduce the risk of species transfer. However, mid-ocean or open sea exchange is anathema to most seafarers and is seen as posing unacceptable safety risks to vessels and lives at sea, possibly in contravention of the annexes to the International Convention for the Safety of Life at Sea, 1974 and its Protocol of 1978 (SOLAS).[22]

The ballast water problem has come to international attention, particularly in the last 2 decades, as both ship speed and international trade have grown. This has been combined with the development of awareness of biodiversity maintenance as a core environmental and human security concern. However, the transfer of species in ballast is not a new phenomenon. The problem as it relates to transfer in ballast water was documented as early as 1903 in the North Sea.[23] More benign forms of alien species transfer in ballast have been noted as occurring, for example, in connection with the introduction of soybeans to North Carolina where they are now a significant cash crop.[24] It is believed that soybeans came as ballast in merchant ships from China at some point prior to the American Civil War. Regulatory controls of ballast discharge, not unlike those currently under discussion internationally, also existed well before the 20th century. Cohen and Foster comment on the experience in the United States as follows:

21. International Maritime Organization, Guidelines for the Control and Management of Ships' Ballast Water to Minimize the Transfer of Harmful Aquatic Organisms and Pathogens, 1997, in Resolution A. 868 (20) (London: IMO) (1998).

22. International Convention for the Safety of Life at Sea, 1974 and its Protocol of 1978 (London: IMO). See also: Harmful Aquatic Organisms in Ballast Water. Alternative Ballast Water Treatment Method, submission by Japan, 15 February 2001, MEPC 46/INF.19 (London: IMO).

23. S. Gollasch, *Removal of Barriers to the Effective Implementation of Ballast Water Control and Management Measures in Developing Countries* (London: GEF/UNDP/IMO, 1997).

24. *"Goodness Grows in North Carolina Soybeans,"* in text ed. note, 5/28/2002, North Carolina Department of Agriculture and Consumer Services. Accessed 23 August 2002 on the World Wide Web: http://www.ncagr.com/agscool/commodities/soykid.htm.

220   Transportation and Communications

Ballast dumping came under regulatory control during the 19th century, as harbor masters barred ships from dumping rock, sand, mud and miscellaneous debris carried as ballast into harbors and channels, to prevent shoaling. In many areas, ballast dumping was banned by statute, both to protect channel depths, and in some cases, to prevent the fouling of waters. "Ballast grounds" were set up where ballast could be legally disposed of, and professional "ballast haulers" and guilds of "ballast heavers" serviced the merchant shipping industry. Even on America's wild frontier, laws and regulations prohibited the dumping of ballast into harbors, although . . . ships on the California coast frequently violated them[25]

Although ballast water is the focus of this article, from the point of view of ecological security, commercial efficiency, and effective regulatory design, it is also important to be aware of the fact that organisms are transferred between countries in other ways related to vessel operations. These include attaching to the ship's hull (a process called fouling), sea chest, the anchor, and other parts of a vessel as well as cargo, cargo packaging, and loading equipment. Of these, arguably ballast water operations pose the largest problem. Concerns have been expressed about these other vectors in various fora, but so far there is no specific international regulatory development.[26]

What are the problems affiliated with the transfer of species and why is it proving to be difficult to find a viable solution? As suggested earlier, the problems are multiple and cut across a range of regulatory and sectoral concerns. The next section examines the issue from various perspectives to illuminate the problem.

---

25. A. Cohen and B. Foster "The regulation of biological pollution: Preventing exotic species' invasions from ballast water discharged into California coastal waters," *Golden Gate University Law Review* 30 (Spring 2000): 787. N.B.: Citations in the original text have been omitted.

26. Some States such as Australia and New Zealand also check for hull fouling. Interestingly, an electronic list serve posted a notice in early July 2001 of a proposed "Planning Meeting: Workshop on Ship Fouling and Biological Invasions in Aquatic Ecosystems" (notice on file). The Workshop was proposed by a member of the U.S. Navy, Naval Surface Warfare Centre and a member of the USCG Environmental Standards Division. The proponents note the following:

> "Historically, hull fouling has been the most important means by which shipping has transported non-indigenous species . . . impending limitations on the use of the most effective antifouling paint [organotin based] and on the conduct of hull cleanings, may result in increased fouling of ships and the subsequent transport of non-indigenous species."

The issue has also been raised in the meetings relating to the Convention on Biological Diversity: See for example, SBSTTA/6/7 paras 20–22. Accessed 15 August 2002 on the World Wide Web: http://www.biodiv.org.

Perspectives on the Problem of Aquatic Organism and Pathogen Transfer in Ships' Ballast Water

Biosecurity—Human and Ecological Security
There has always been some natural movement of species through the medium of water, however, the combination of distance, weather, differing water temperatures, salinity, and food sources in the various marine ecosystems of the southern and northern waters has limited the scope and range of natural migration. Human-assisted species transfer does not easily fit the traditional paradigm of human activity resulting in pollution. However, in the last 2 decades and in particular since 1992, environmental law and international environmental institutions have embraced a systemic view of the interaction between human activity and the physical environment. This system or ecosystem is understood to be dynamic and is not easily subject to the more usual point-in-time evaluations of cause, effect, and singular responsibility. The significance of the environment in the maintenance of human health and economic security is now also part of national security agendas. This acceptance of these ideas is evidenced by the extremely high ratification level (184 States as of August 2002) of the 1992 Convention on Biological Diversity (CBD).[27] The CBD focuses on States' obligations to ensure and protect biodiversity in areas and activities under their control. The Secretariat and CBD are situated under the institutional umbrella of the United Nations Environment Programme (UNEP). In March 2001 the CBD Secretariat published the Strategic Plan for the Convention, noting the following:

> . . . the rate of biodiversity loss is increasing at an unprecedented rate. Moreover, this rate of loss is threatening the very existence of life as we know it and, in turn, global security.[28]

This shift from a narrow focus on preventing pollution to a broader approach aimed at supporting and maintaining the existing ecosystem and its chain of interdependence as intrinsically valuable clearly encompasses the question of human intervention in the ecosystem through activities such as transport systems that transfer species. Despite this conceptual shift found in modern multilateral environmental agreements (MEAs), questions of enforcement and compliance are rendered somewhat more difficult by the more traditional polluting substance orientation of many international and

---

27. Convention on Biological Diversity, 1992, UN Doc. UNEP/Bio.Div/N7-INC.5/4UNEP. Accessed 15 August 2002 on the World Wide Web: http://www.biodiv.org/convention.
28. *Strategic Plan for the Convention on Biological Diversity*, 13 March 2001, UNEP/CDB Secretariat, accessed 15 August 2002 on the World Wide Web: http://www.biodiv.org.

domestic regulatory regimes. As discussed earlier, this shift to the concept of biodiversity also has broader implication for international and domestic governance. The divide between land-based marine pollution and environmental protection (UNEP) and ocean activities (IMO et al.) is closed. At a national level, the former is usually dealt with by an environmental ministry, while shipping, for example, is dealt with by maritime transport administrations. This issue will be discussed in more detail in the Section dealing with the regulatory response.

Aside from the human impact of disease carrying microbes or toxic dinoflagellates, the question of whether species migration is a "natural" event and whether an organism is invasive or harmful in an absolute sense is difficult. In many cases, a species may not be invasive or a pest in its home state, where it forms part of the ecosystem (which includes natural predators or other factors that limit its growth). However, it may become a pest in another welcoming host environment where there are no natural limits on its growth. In these cases it may become a predator on indigenous species or it may disrupt and even destroy the food chain or ecosystem to which it has emigrated. This can have a significant impact on indigenous species in the region, and in particular, on fisheries. The case of the American comb jellyfish that destroyed the entire anchovy fishery when it migrated to the Black Sea is infamous. That same species has now migrated, probably in ballast water, east to the Caspian Sea, endangering the seal and other species populations.[29] This is only one case out of many.[30] There are the obvious commercial consequences arising from the destruction of a marine capture fishery. In addition, this issue threatens coastal aquaculture species that are often more vulnerable. On a broader level, this poses a significant risk to the success of States working with international organizations such as the United Nations Food and Agriculture Organization (FAO). Because environmental changes and overexploitation have resulted in a loss of marine capture stocks, the FAO is encouraging aquaculture/mariculture as the best way to meet the escalating world demand for protein and food security.[31] The recent impact of animal diseases such as foot-and-mouth on the consumer market highlight the significance of aquatic species as a food source.

Some aquatic organisms such as algal blooms or toxic dinoflagellates pose a significant danger to human health when they enter the food chain. An IMO case study summary of the Australian experience with "red tide" is a good illustration of the specific and general challenges posed by the spread of some organisms:

29. A. Morgan and D. Harrison, "Invading jellyfish crisis for Caspian seals," Nature Watch, *Sunday Telegraph* (5 November 2000), London.
30. Gollasch (n. 23 above).
31. Food and Agricultural Organization (FAO). *The State of World Fisheries and Aquaculture 2000*. Rome: FAO, 2000. Accessed 15 August 2002 on the World Wide Web: http://fao.org/DOCREP/003/x8002E/x8002E00.htm.

Toxic dinoflagellates are a type of algae known to cause paralytic shellfish poisoning in humans. Evidence suggested that the toxic dinoflagellate *Gymnodinium catenatum* became established in Australian waters after arriving in ballast water—the species was already present in waters of Argentina, Japan, Mexico, Portugal, Spain, Venezuela and in Mediterranean sea ports . . . Dinoflagellates can reproduce simply by splitting in two, allowing multiplication wherever conditions are favourable. *Gymnodinium catenatem* also has a type of reproduction in unfavourable conditions, which can result in a tough encased spore that can survive different conditions by staying dormant in sediment. These spores remain viable for 20–30 years, germinating in the usual swimming form when conditions are suitable, and entering the food cycle of shellfish causing the shellfish to become toxic to humans.[32]

The same study also notes that similar problems resulting from ballast water-introduced dinoflagellates have been experienced in other countries, including China and India.[33]

The fact that many species and pathogens can survive in adverse conditions and remain undetected in a new environment for a long period of time after the transfer, means that both their detection and a useful response is difficult, as is the attribution of specific blame. The problem is compounded in the case of pathogens. A related problem is that most countries have very little scientific knowledge about the range of organisms in their waters to determine whether they have a problem organism in their coastal water that they may be exporting or whether their systems will have a problem with a species that might be imported. This means that the determination of the current level of biodiversity is itself an inexact process.[34]

At the same time it must be understood that the majority of species will not adapt to new environments, particularly if there is a great variation in temperature or conditions in the ecosystems. However, unlike oil and other pollutants, once an invasive organism is introduced it is virtually impossible to remediate the environment. There have been some instances of physical removal or introduction of predators but they are relatively few and may

32. *Alien invaders* (n. 2 above), p. 9.
33. Ibid. See also: J. C. J. M. van den Bergh, et al. (2002). Exotic harmful algae in marine ecosystems: an integrated biological-economic-legal analysis of impacts and policies. *Marine Policy* 26.: 59–74 for a discussion of the growing problem on European coasts.
34. There is now an effort to encourage ports to conduct base line port surveys: See S. Raaymakers, "Port surveys underway," *Ballast Water News* 4 (2001): 3–5; C. L. Hewitt and R. B. Martin, Centre for Research on Introduced Marine Pests, "Revised protocols for baseline port surveys for introduced marine species: survey design, sampling protocols and specimen handling," *Technical Report* 22 (2001) (Hobart, Australia: CSIRO) 15 August 2002 on the World Wide Web: http://crimp.marine.csiro.au/reports/techreport22.html.

pose their own problems.[35] Most responses have focused on containment strategies.[36] Once a new species is introduced, the host ecosystem or environment is changed forever. This explains why regulatory strategies have focused on preventing the introduction of alien species and pathogens: there is no viable cure.

Economic and Commercial Impact

The commercial/economic impact of ballast water management to prevent the transfer of harmful species differs between affected groups. For purposes of this discussion, four points of view, labelled as shipowner, port State, flag State, and coastal State, will be discussed. It is recognized that this is an artificial distinction and one actor will likely have more than one level of concern, however it serves to illuminate the multiplicity of interests involved.

Shipowner/Operator.—From a shipowner/operator perspective, the mere fact that a ship may carry organisms in addition to cargo and ballast water is per se of negligible commercial impact, except perhaps in the case of hull fouling, which can affect vessel speed and impose higher maintenance costs. The regulatory response to the issue may, however, have significant commercial consequences. These consequences can include direct costs of compliance such as altering ballast water tank configurations, purchasing new equipment and/or chemicals, and requiring additional training for on-board personnel and insurance coverage for potentially expanded liability for incidents or noncompliance.[37]

Obtaining additional certificates and classification society surveys can add substantially to the cost of vessel operations. A recent Australian study of the costs of compliance with the various methods of ballast treatment currently available found that, "[c]osts range from 2.5–3.7 Aus cents/m^3, depending on the ship for mid-ocean exchange, to 11.7–32.1 cents/m^3 for filtration, to 10.8–43.7 cents/m^3 for hydrocyclonic treatment. Land-based treatment using dedicated terminals costs up to Aus\$40/m^3."[38] Other comparative cost studies by a number of maritime administrations and research-

---

35. F. McEnnulty, N. Bax, B. Schaffelke and M. Campbell, *A Rapid Response Toolbox: Strategies for the control of ABWMAC listed species and related taxa in Australia* (posted draft August 2000) Centre for Research on Introduced Marine Pests (Hobart, Australia: CSIRO). Accessed 15 August 2002 on the World Wide Web: http://crimp.marine.csiro/au/reports/toolbox.pdf.

36. *The Problem* (n. 8 above).

37. See for example: L. K. Terpstra, " 'There goes the neighbourhood'—the potential private party liability of the international shipping industry for exotic marine species introduced via ballast water in England," *Transnational Lawyer*, vol. 11 (1998): 277–309; Cohen and Foster (n. 25 above).

38. "Exchange of views," *NUMAST Telegraph* (May 2001): 21; "Ballast water treatment 2001," *MER* (May 2001): 14–15.

ers are summarized in a comprehensive review of the species transfer in ballast water problem prepared by Stephan Gollasch[39] as a lead up to the GloBallast[40] programme.

There are more significant indirect costs. Compliance with the operationally based, recommended risk minimisation strategies may add time to a voyage and add physical stresses to the vessel, thereby affecting its safety and stability and possibly also the longevity of the vessel. In an era of increasingly rapid port turnaround times, the spectre of adding inspections, increasing time to ballast discharge and cargo loading because of water testing, and discharging water or ballast sediment into foreseeably inadequate and inefficient reception facilities, together with the need for obtaining administrative fees or even possibly individual ballast transfer permits is frightening. The possibility that a ship may be refused entry or told that its ballast water cannot be discharged, thereby eliminating or significantly reducing its carrying capacity (depending on the type of vessel) will generate nightmares for shippers and shipowners alike.

There are also other more subtle commercial impacts for the shipowner relating to changing regulatory relationships. Because there is no international convention providing uniform procedures for regulating ballast water management, to address this issue, many States are adopting their own legislative responses. To varying degrees they reflect the 1997 IMO *Guidelines*. These national, or even subnational (where legislative jurisdiction is divided e.g., states, provinces, regions, and even port authorities, e.g., Vancouver, Canada) responses are increasingly more diverse in their operational and administrative requirements. The fact that the lead administrative agency will vary between jurisdictions is an additional factor that can cause delay and frustrate shipowners who may end up dealing with an agency that operates with norms and concerns that differ significantly from the culture of the specialized maritime administrations.

From this brief overview it is obvious that the commercial impact of the ballast water issue can be quite significant depending on the regulatory response and whether a simple and reasonably effective and economical "technical fix" can be found. It is also clear that the commercial and economic implications of this issue may go well beyond anything experienced in connection with oil pollution, even including the acrimonious battle over double hulls.

Port State.—It is obvious that a port State is a coastal State. Although UNCLOS created a specific legal status for port State functions, this is not the concern at this point. The reason for distinguishing the two perspectives relates to the possibly divergent commercial impacts and obligations and differing economic and commercial agendas relating to each role. Commer-

39. Gollasch (n. 23 above).
40. *The Problem* (n. 8 above).

cial problems for a State operating in a port State capacity are multiple. Although international attention has been focused primarily on ships, the issue is much more important for ports as they face a period of privatization, increased competitiveness combined with increased regulatory pressure to meet biosecurity, and other public interest requirements.

In general, ports will be concerned about ensuring that despite pressures to act defensively to protect biosecurity and human health, the response adopted is economically viable and commercially attractive. Depending on their trade leverage, port States will be more or less inclined to adopt a stringent regime. In the short term if the response adopted is too stringent and causes delays it may reduce the attractiveness of a particular port, especially if there are nearby competitors with lesser requirements. The lack of reception facilities for sediment and emergency discharge of ballast water, and the problem of increased port State enforcement responsibilities where port inspectors are often stretched beyond capacity and will not have the necessary training is another economic factor for port States. The need for additional documentation for incoming ships, and even more problematically, for documentation relating to safe discharge zones, safe intake areas, or water quality, significantly affects traditional maritime administration authorities. To meet sampling requirements for ships and for ports, labs must satisfy the concerns of quarantine, health, or fisheries officials and the timing requirements of ship operators. This will add to administrative costs. Where a risk assessment approach is developed, personnel will need to be trained to make these evaluations and to deal with the reporting requirements.

At the same time, and in the longer term, failure to respond positively and proactively may mean that a port will become less attractive. Vessels that have problems at subsequent ports because they have no information about their ballast water quality or because they lacked a critical inspection may be reluctant to return. For example, in a recent submission to IMO, Brazil commented:

> The guidelines on preventative measures regulating ballast water uptake at the port of origin are essential. We should not allow the ballasting of inadequate contaminated water, which might cause a serious problem at the port of destination.[41]

The port State will also need to evaluate the question of who should pay for the port-related costs to respond to this problem—the ship-owning industry, the shippers (importers/exporters), or the public through the State.

As noted earlier, the possibility that a vessel may be turned away if it

---

41. *Standards for the Management and Control of Ballast Water*, submitted by Brazil (6 February 2001), MEPC 46/3/14 (London: IMO): Annex, p. 3.

does not have documentation regarding the ballast water that it is carrying, or because of concern about the previous port's water, is clearly a situation that can have a major commercial impact.

It is very clear that ports will have to come to terms with these issues and to decide whether they will be the guardians of the transport corridor (a biosecure port) or a poorly maintained gate and a threat to a State's security.

Flag State.—International law places primary responsibility on the flag State to regulate the ships it has registered. This includes ensuring that ships meet all international standards for ship safety and environmental protection. This is primarily carried out through ship surveys, ensuring crew competency and cooperation with port States that have a monitoring/audit function. Currently, there are no specific flag State standards, other than the obligation found in UNCLOS and the CBD (discussed later) that States prevent the spread of aquatic organisms and pathogens. When an international convention is adopted, flag States will have additional certification and survey obligations to ensure that their vessels comply with international standards. If the State is also a crew supply State it will need to review its training programmes and competency certification, for all on-board personnel as well as maritime administrators. The question of where international financial responsibility for incidents will be placed if the regime is developed is currently an open question.

Coastal State.—As noted earlier, the distinction between a port and coastal State role sometimes is difficult to draw. However, in their coastal State capacity countries are and will be primarily concerned about preventing the import of organisms into their waters, whether caused by open seas discharge that may adversely affect nearby island states, or discharges in their ports or coastal areas. The concerns, as noted earlier, will tend to relate to ensuring that import is prevented at the borders. There has been relatively less emphasis placed on preventing the export of problem organisms. Where an invasive species establishes itself, the costs can be enormous and unexpected. The example of the Black Sea fishery was mentioned earlier. Other examples include the ongoing problem of controlling seaweed and coastal vegetation invasions in the United States and the Mediterranean. The import of the zebra mussel into the Great Lakes has cost billions of dollars since 1989 to simply control the recurrent growth of the mussels.[42] The import of the cholera pathogen, which is believed to have been reintroduced to South America and also the United States in ballast water in the early 1990s, resulted in the death of over 8,000 (South America) and 100 (USA) persons. These costs are incalculable.[43]

42. Unknown author, "Unwanted passengers," *Shipping World and Shipbuilder* (March 2001): 22.
43. Casale and Welsh (n. 6 above).

It is not surprising to find that many States are adopting a hard-line approach. Irrespective of the commercial concerns of shipping and port interests, Australia is developing regulatory responses that maximize risk prevention and precaution by requiring reports, water exchanges outside 12 nautical miles (M), or water samples before permission is given to discharge ballast water.[44] However, the costs of preventing the import and fulfilling a correlative duty not to export invasive species are not small.[45]

One of the primary costs involves carrying out baseline harbor and coastal zone habitat studies to develop the scientific information needed to make the system effective. Setting up information and reporting systems and developing integrated institutional structures to ensure the coordination and intersection of health, quarantine, fisheries, transport, and biodiversity/environmental interests at all levels of government can be an extremely expensive process.

Technical/Safety Concerns
These concerns, like ecological security, are relevant to all the perspectives outlined earlier. Inevitably, technical and safety issues relate to commercial and human security concerns, although industry versus regulator debates have tended to explicitly position biodiversity protection and ship safety/safety of life at sea as countervailing values. This positioning is symptomatic of the larger sustainable development debates and reflects the tension between long-term and shorter term or more immediate interests. In the long term the two issues converge because maintaining human life is central to both biodiversity and safety. The technical and safety concerns are dealt with together in this article because they are intrinsically related: the latter is simply the manifestation of the former.

Disregard of either safety or biodiversity issues is not a politically or morally tenable stance, and a great deal of effort is currently devoted to finding a technical solution to the problem. Given the earlier-noted commercial concerns and complications for both the industry and for governments facing internally competing political, economic, and legal agendas, it is no wonder that it has often been said that the race is on to find the solution. The fact that the current recommended best practices minimize

---

44. The issue is governed by the Australian Quarantine Act 1908, as amended by the Quarantine Amendments Act, No. 137, 2000 and amendments to Quarantine Regulations 2000. See: *Australian Ballast Water Management Requirements,* Australian Quarantine and Inspection Service (AQIS) Department of Fisheries and Forestry. Canberra, Australia: 2001. See also: http://www.affa.gov.au/, accessed 15 August 2002 on the World Wide Web.

45. The question of States' international responsibility to prevent export of alien species or, at a minimum, a duty to warn, has not received much attention. The contingencies associated with invasions make it difficult to ascertain fault and causation.

risk, but do not assure that there is no transfer, adds impetus to the research agenda. It has been estimated that "the current market for an effective solution is in the range of $2 billion."[46]

The search for a technical answer has also generated a great deal of discussion about whether the solution should be based on vessel operational procedures and technical equipment requirements or whether it should focus on water quality standards for ballast water discharge. The choice largely impacts on the point of monitoring, the location of regulatory control, and who will bear the cost. This will be discussed later in the Section on regulatory responses.

The overview section on ballasting operations generally outlined the role of ballast and its relationship to ship stability. The current practices recommended by the IMO-Member States in 1997 (and earlier) in Resolution A.868(20)[47] are prevention by intake precautions, safe uptake and discharge zones, reception facilities, and risk minimisation. As noted earlier, the main risk minimisation procedure developed to date is mid-ocean exchange. This procedure, which is not 100 percent effective, has generated ship safety concerns in the shipping industry.

Appendix 2 of the 1997 IMO *Guidelines*, "Guidance on safety aspects of ballast water exchange at sea," describes the recommended procedures and safety factors for deep or open ocean exchange. They are set out in full below because they illustrate the complexity facing the crew and any shipowner in developing and implementing a ballast water management plan based on open sea exchange. They also illustrate the regulatory challenge in that the multiple factors affecting open sea exchange necessarily mandate a high level of discretion to the ship's master.

. . .
   1.3 In the absence of a more scientifically based means of control, exchange of ballast water in deep ocean areas or open seas currently offers a means of limiting the probability that fresh water or coastal aquatic species will be transferred in ballast water. Two methods of carrying out ballast water exchange at sea have been identified:
      .1 the sequential method, in which ballast tanks are pumped out and refilled with clean water; and/or
      .2 the flow-through method, in which ballast tanks are simultaneously filled and discharged by pumping in clean water.
   2.   Safety precautions
   2.1 Ships engaged in ballast water exchange at sea should be provided with procedures which account for the following, as applicable:

---

46. Pughiuc (n. 20 above).
47. *Guidelines* (n. 21 above).

.1 avoidance of over- and under-pressurization of ballast tanks;
.2 free surface effects on stability and sloshing loads in tanks that may be slack at any one time;
.3 admissible weather conditions;
.4 weather routeing in areas seasonably affected by cyclones, typhoons, hurricanes, or heavy icing conditions;
.5 maintenance of adequate intact stability in accordance with an approved trim and stability booklet;
.6 permissible seagoing strength limits of shear forces and bending moments in accordance with an approved loading manual;
.7 torsional forces, where relevant;
.8 minimum/maximum forward and aft draughts;
.9 wave-induced hull vibration;
.10 documented records of ballasting and/or de-ballasting;
.11 contingency procedures for situations which may affect the ballast water exchange at sea, including deteriorating weather conditions, pump failure, loss of power, etc.;
.12 time to complete the ballast water exchange or an appropriate sequence thereof, taking into account that the ballast water may represent 50% of the total cargo capacity for some ships; and
.13 monitoring and controlling the amount of ballast water.

2.2 If the flow-through method is used, caution should be exercised, since:
.1 air pipes are not designed for continuous ballast water overflow;
.2 current research indicates that pumping of at least three full volumes of the tank capacity could be needed to be effective when filling clean water from the bottom and overflowing from the top; and
.3 certain watertight and weathertight closures (e.g., manholes), which may be opened during ballast exchange, should be resecured.

2.3 Ballast water exchange at sea should be avoided in freezing weather conditions. However, when it is deemed absolutely necessary, particular attention should be paid to the hazards associated with the freezing of overboard discharge arrangements, air pipes, ballast system valves together with their means of control, and the accretion of ice on deck.

2.4 Some ships may need the fitting of a loading instrument to perform calculations of shear forces and bending moments induced by ballast water exchange at sea and to compare with the permissible strength limits.

2.5 An evaluation should be made of the safety margins for stability and strength contained in allowable seagoing conditions specified in the approved trim and stability booklet and the loading manual, relevant to individual types of ships and loading conditions. In this regard particular account should be taken of the following requirements:

.1 stability to be maintained at all times to values not less than those recommended by the Organization [IMO] (or required by the Administration);

.2 longitudinal stress values not to exceed those permitted by the ship's classification society with regard to prevailing sea conditions; and

.3 exchange of ballast in tanks or holds where significant structural loads may be generated by sloshing action in the partially filled tank or hold to be carried out in favourable sea and swell conditions so that the risk of structural damage is minimized.

2.6 The ballast water management plan should include a list of circumstances in which ballast water exchange should not be undertaken. These circumstances may result from critical situations of an exceptional nature, force majeure due to stress of weather, or any other circumstances in which human life or safety of the ship is threatened.

3. Crew training and familiarization

3.1 The ballast water management plan should include the nomination of key shipboard control personnel undertaking ballast water exchange at sea.

3.2 Ships' officers and ratings engaged in ballast water exchange at sea should be trained in and familiarized with the following:

.1 the ship's pumping plan, which should show ballast pumping arrangements, with positions of associated air and sounding pipes, positions of all compartment and tank suctions and pipelines connecting them to ship's ballast pumps and, in the case of use of the flow—through method of ballast water exchange, the openings used for release of water from the top of the tank together with overboard discharge arrangements;

.2 the method of ensuring that sounding pipes are clear, and that air pipes and their non-return devices are in good order;

.3 the different times required to undertake the various ballast water exchange operations;

.4 the methods in use for ballast water exchange at sea if applicable with particular reference to required safety precautions; and

.5 the method of on-board ballast water record keeping, reporting and recording of routine soundings.

It can be seen from this extensive list of safety precautions that the core technical problem in using open sea exchange resides in the fact that stability is a contingent concept resting on a range of variables such as sea and weather conditions, the specific cargo, and the particular vessel design.[48] The emphasis on the overriding interest in ship safety is reinforced in the 1997 *Guidelines*, Rule 11.3 that provides as follows:

> ... Port States should not require any action of the master which imperils the lives of seafarers or the safety of the ship.

Given the safety culture of seafaring, it is not surprising to find that States with a domestic regime implementing the *Guidelines* on a mandatory or voluntary basis are finding that there is a low compliance rate. For example, one report indicates that many shipowners are ignoring the U.S. Coast Guard's ballast water reporting requirements. A study released in October 2000 found that only 6.3 percent of ships entering U.S. waters complied.[49] However, the Canadian experience 5 months after the revised rules came into force in 2000 showed a 50 percent compliance rate in the St. Lawrence Seaway.[50] This is despite the efforts of the Shipping Federation of Canada to encourage compliance, primarily in the form of developing a Code of Best Practices for Ballast Water Management to pre-empt more stringent regulations that are being developed at the subnational levels in both the United States and Canada.[51]

The human factor—that is, seafarer culture (practice, beliefs, and attitudes)—regarding mid-ocean exchange of ballast water needs to be seriously considered as it will have a significant impact on compliance with any

---

48. The individuality of each ship is reflected in, for example, G. A. B. King, *Tanker Practice. The Construction, Operation and Maintenance of Tankers*, 4th ed. (London: The Maritime Press, 1965), p. 92ff that advises of the need for masters to keep detailed notes on the ballasting and response of the vessel particularly in the first year or 18 months of the ship's life to provide a handling guide to future operators.

49. Author unknown. (15 March 2001). "Ballast under the spotlight," *Fairplay*, p. 23.

50. Author unknown. (14 December 2000). "Good water out, bad water in," *Fairplay*, p. 22. A more recent survey on the East Coast suggests this is improving; however, there are still questions that might arise regarding auditing of compliance reporting by ships.

51. The Shipping Federation of Canada, 28 September 2000. Accessed 15 August 2002 on the World Wide Web: http://www.shipfed.ca/library/ballastwater/ballastwaterbestpractices.html.

The Code essentially replicates the IMO *Guidelines*. The Shipping Federation has also been active in discouraging U.S. State level regulations that adopt approaches not in conformity with the federal practice. See: *Submission of The Shipping Federation of Canada to the Senate of Michigan Natural Resources and Environmental Affairs Committee in Respect of Senate Bill No. 955*. This was a proposal by Michigan to require sterilization of all ballast water. The Michigan approach would have dealt with the issue on the basis of permits. Accessed 15 August 2002 on the World Wide Web: http://www.shipfed.ca/ballastwater/ballastwaterpresentation.html.

regulatory system. There are also a number of other safety variables that can affect the viability of the management measure that is adopted. For example, the amount of time involved in an open sea ballast exchange using a sequential method may be lengthy and weather conditions may alter unexpectedly, leaving the ship in a vulnerable position. Other factors also relate to the location of the tanks in the vessel, which may provide a problem for the crew in sampling or monitoring water quality. Modern vessels usually have automatic monitoring systems that can evaluate the water level at any one time. However, a technical problem resides in the fact that, as the water level is adjusted, it results in a variable percentage of water surface that can freely move. The combination of ocean movement, the ships' movement, and the impact of tonnes of water moving from side to side (sloshing) in response means that it is difficult for a ship's master to control and make adjustments to compensate.

Although SOLAS does not directly address this issue, a number of its provisions and the circulars produced by the Maritime Safety Committee of IMO, set additional limits on the feasibility of open sea exchange. SOLAS 1974, Chapter II-1, Regulation 22 requires that the vessel's master be supplied with information permitting him or her to quickly and easily calculate the stability of the vessel under varying conditions of service. Given the multiple variables affecting stability in a ballast exchange while the vessel is enroute, compliance may prove difficult. Certainly provisions in SOLAS will need to form part of the overall assessment of safety and vessel stability and strength.[52]

The proposed flow-through method[53] tries to avoid the ship stability and stress/safety problem posed by the sequential method, but it imposes significant problems of delay, possible over-use of the ballast pumps, and some problems relating to the divisions within ballast tanks. In addition, unless the ballast water piping system is altered it may be ineffective, as the waters will inevitably mix. The requirement of three cycles of change also raises the problem of fuel consumption and related concerns about ship source air emissions.[54]

Neither method addresses the problem of organisms residing in the sediment that will remain in the tanks. Monitoring and compliance requirements that require water sampling and easy access for inspection and cleaning will also be a problem for many existing vessel designs. Lloyd's Register recommends that the IMO Guideline requirements be taken into account by architects when designing new vessels or modifying existing vessels.[55] In-

---

52. Lloyd's Register, "Practical solutions to new ballast-water legislation," *The Naval Architect* (January 2001): 24–26.
53. There is another variation called the dilution method that operates on much the same basis.
54. P. Van Dyck, "Feasible?," *The Motor Ship* (August 2001): 25–31.
55. Lloyd's Register, (n. 52 above).

terestingly, this concern with ballast tanks from an organism transfer point of view is coinciding with an increasing interest in the condition of ballast tanks for other vessel condition assessment reasons. A recent article in *Fairplay Solutions*, "Ballast tanks take centre stage: Is there a problem waiting in the wings?",[56] notes that a post-*Erika* consequence and the increasing number of double-hull vessels have changed priorities. Previously "ballast tanks have, for many owners, been regarded as very much out of sight out of mind in that naval architects focused on ballast tanks only for vessel strengthening and crew rarely visited the tanks."[57] Now, however, the condition of ballast tanks will be scrutinized more closely as indicative of "the real condition of the vessel and the true level of shipcare and maintenance."[58] This issue largely relates to the kinds of coating used in the tanks to prevent corrosion. This is a major cost item. The move to double hulls adds to this problem and also issues regarding increased tank area and problems for removal of sediment cleaning. The requirement for ballast tank ventilation and access will compound the problem and raises concern about the need for inert gas requirements.[59] There are many other ship and equipment design issues that are involved in responding effectively to the problem, including ballast pump capacity, ballast water and seawater suction pipe design, location and configuration of the piping system within the vessel and the ballast tanks and the design of sounding pipes, as well as a number of specific issues depending on the ballast water management or treatment process that is adopted by each vessel.[60] These issues will need to be factored into future ship building requirements and also, possibly, into modifications of existing ships.

Summary

The foregoing has outlined the various perspectives on the problem. From both a ship operator and a port/coastal State perspective, an on-board tech-

---

56. Author unknown, "Paints and coatings. Ballast tanks take centre stage. Is there a problem waiting in the wings?," *Fairplay Solutions* (February 2001): 12–13.
57. Ibid.
58. Author unknown, (n. 56 above).
59. Other issues that may be raised on this topic relate to the use of chemicals, such as those in the chlorine class, to treat ballast water, the interaction of these chemicals with tank coatings, and crew occupational health and safety rules in storing and administering the chemicals. There are also concerns relating to the environmental impact of the treatment process itself.
60. IMO Committee Report and Draft text of an International Convention for the Control and Management of Ships' Ballast Water and Sediments, December 2000, MEPC 46/3 (London: IMO, 2000). See especially Annex 3 dealing with a circular to consider design suggestions for ballast water and sediment management options.

nical treatment solution that provides as close to a 100 percent risk elimination as possible and avoids the less easily calculated costs of risk assessment, monitoring, compliance, and enforcement, is preferable. Regulatory requirements that translate into additional equipment or altered designs or purchase of products may appear excessive. However, from a commercial perspective they are reasonably easy to calculate, can be written off in most tax systems, are predictable, and can be factored into overall operational costs and carriage charges. In contrast, the costs associated with possible port delays or even refusal of permission to discharge, reporting, sampling, and inspections are difficult to account for and can wreak havoc with most commercial relations and carriage agreements. An added complication is that the regulations may vary from port to port. The main issue for most players in the maritime industry will be to ensure that whatever is adopted as the best solution will be recognized at an international level and adopted by all ports. The next section outlines the international regulatory and industry response and the final section then considers the implications of this emerging regulatory regime.

THE EMERGING REGULATORY AND INDUSTRY RESPONSE

Earlier it was pointed out that although there is a great deal of recent interest in this problem, particularly in the last decade, the issue has a much longer history. The problem of aquatic organism transfer has been dealt with in a number of international regulatory instruments since the 1970s. In fact the current debate on State obligations and the call for an international ballast water convention is sometimes misunderstood. International obligations regarding marine and other alien species transfer have existed for some time. The current debate is related to the need to develop a regulatory regime specific to the ballast water aspects of the pre-existing international obligations. This section outlines the current international obligations and the developing IMO ballast water convention and provides a brief overview of national and industry responses.

International Obligations and Responses

There are two main sources of States' international obligation to prevent the spread of alien species transfer. The first and earlier regime, the law of the sea regime with its careful balances and delineation of flag, port, and coastal State responsibilities and control over activities and actors, recognizes a range of oceanic actors, with matters relating to shipping as an ocean activity primarily situated in IMO. The later regime, the biodiversity regime, has grown up within and is more clearly placed in the MEA system of UNEP and affiliated institutions and actors. The regimes are consistent with each other in terms of the convention objectives, however, there are significant differences in their institutional and management frameworks. This article is not specifically focused on the issue of integration of global governance per se, however, it is part of the regulatory context because the tension that results from this difference is played out at the domestic implementation level.

The 1982 United Nations Convention on Law of the Sea
The United Nations Law of the Sea Convention (UNCLOS)[61] was adopted in 1982 after nearly a decade of negotations. It came into force in 1994. As of May, 2002 it is binding on 138 States, with another 19 states having signed, but not yet ratified, the Convention.

UNCLOS was one of the first attempts by the global community to provide a comprehensive regime for managing an international space. It also introduced an holistic framework for addressing environmental rights and responsibilities.[62] Article 196 of the Convention specifically addresses the problem of alien species and state obligations.

Article 196
Use of technologies or introduction of alien or new species

1. States shall *take all measures necessary to prevent, reduce and control pollution of the marine environment resulting* from the use of technologies under their jurisdiction or control, or the intentional or *accidental introduction of species, alien or new, to a particular part of the marine environment, which may cause significant and harmful changes thereto.*
2. This article does not affect the application of this Convention regarding the prevention, reduction and control of pollution of the marine environment. (Emphasis added)

This provision places an obligation on all States to prevent the transfer of species that maybe harmful to another marine environment. One of the difficulties that has arisen in connection with Article 196 relates to the distinction seemingly being drawn in Subsection 2 between marine pollution, defined in Article 1 (4) of UNCLOS.

1. For the purposes of this Convention:
 ... (4) "pollution of the marine environment" means the introduction

61. UNCLOS (n. 13 above).
62. J. Charney, "The marine environment and the 1982 Law of the Sea Convention," *The International Lawyer* 28 (1994): 879–901.

by man, directly or indirectly, of substances or energy into the marine environment, including estuaries, which results or is likely to result in such deleterious effects as harm to living resources and marine life, hazards to human health, hindrance to marine activities, including fishing and other legitimate uses of the sea, impairment of quality for use of sea water and reduction of amenities;

This definition has been adopted in many national laws and may need to be amended to cover aquatic alien species transfer and introduction. The negotiating history of Article 196 indicates that, in the course of developing this text, there were two distinct duties in mind, one of preventing pollution and the second (closer to the more recent biodiversity concept) of maintaining the natural state of the marine environment.[63] Although it did not survive the final negotiations it is also interesting that one version of the text imposed a responsibility to restore affected environments to their pre-alien species transfer state.[64]

Another related question arises as to whether a State can take action to prevent the risk of a transfer of harmful aquatic organisms and pathogens. This question might itself comprise a paper. UNCLOS does not specifically address this question, however, it is clear that a State has a sovereign right to determine the basis of entry into its internal waters (i.e., most ports), subject to the customary practice regarding situations where human lives are in danger.[65] The coastal State can also pass laws governing the Innocent Passage (defined in Art. 19) of foreign ships through its Territorial Sea (12 M) to, inter alia, preserve the environment of the coastal State (Art. 22 (1) (f)); conserve living resources and prevent the infringement of fisheries regulations (Art. 22(1) (d) (e)); and, prevent infringement of sanitary laws and regulations (Art. 22(1)(h)). However, this legislative authority is subject to the important requirement in Article 21(2):

Such laws and regulations shall not apply to the design, construction, manning or equipment of foreign ships unless they are giving effect to generally accepted international rules and standards.

State marine pollution prevention and enforcement obligations and rights are set out in Part XII of the UNCLOS. These rights are very complex and

---

63. M. Nordqvist, ed. in chief, *United Nations Convention on the Law of the Sea 1982. A Commentary,* vol. IV, (Dordrecht: Martinus Nijhoff, 1991), pp. 73–76.
64. R. Platzöder, ed., *Third United Nations Conference on the Law of the Sea: Documents,* vol. X, (New York: Oceana Publications, 1986), p. 453.
65. R. R. Churchill and A. V. Lowe, *The Law of the Sea,* 3rd ed. (Manchester: Juris Publishing, 1999), pp. 62–65 regarding the right to set conditions for access to the port as stated in *Nicaragua,* (1986) ICJ Rep. 14 at 111.

depend on a range of factors[66] including restrictions—safeguards—placed upon the right to inspect and detain ships (e.g., Art. 226). There is a clear duty under Article 194 on States to prevent, control, and reduce marine pollution caused by activities under their control and to prevent damage to other States, including the duty to prevent pollution from vessels by, inter alia, "... preventing intentional and unintentional discharges and regulating the design, construction, equipment, operation and manning of vessels." Articles 194, 211, and 217 are the source of flag State responsibility for the primary regulation of ships. The omission of the word operation from Article 21(2) appears to allow a coastal State to adopt domestic standards in the Territorial Sea with respect to ships' operations without offending the right to Innocent Passage, although any domestic legislation will be subject to the requirement of nondiscrimination (Art. 24; Art. 227). In the absence of an internationally binding standard, this point is relevant to a coastal State's choice regarding the method of ballast water management (equipment based or operational procedures).

The UNCLOS regime, which recognized that "problems of ocean space are closely interrelated and need to be considered as a whole" (Preamble), was based on a careful balancing of rights and claims. It remains a key source of State responsibility for protection of the marine environment. However, since 1982 the evolution of global comprehension of the relationship between human activities and the environment and the concept of sustainble development has taken the next step to an even more holistic or integrated approach based on an ecosystemic view. In a sense this has simply played out the logic articulated in UNCLOS. It means that, aside from questions of interpreting national legislation and coastal State and port State enforcement rights, the later and even more more broadly supported Agenda 21[67] and 1992 CBD have arguably subsumed or at least significantly altered the understanding and implications of the UNCLOS marine pollution provisions.

Agenda 21 and the 1992 Convention on Biological Diversity
Agenda 21[68] was endorsed by the international community in 1992. It is not an international convention. Rather it is a comprehensive global management plan to achieve sustainable development in the 21st century. The document covers almost all sectors of human activity and environmental interaction. Chapter 17 deals with oceans and seas. Section 17.30 provides as follows:

> States, acting individually, bilaterally, regionally or multilaterally and within the framework of IMO and other relevant international organiza-

66. For example, Articles 211, 217, 218, 219 and 220 all require a detailed consideration of the ship's location and standard of proof.
67. (n. 14 above).
68. Ibid.

tions, whether subregional, regional or global, as appropriate, should assess the need for additional measures to address degradation of the marine environment:
a. From shipping, by:
vi. Considering the adoption of appropriate rules on ballast water discharge to prevent the spread of non-indigenous organisms;

The 1992 Convention on Biological Diversity,[69] adopted at the same time as Agenda 21, came into force several years later. Article 8, In-Situ Conservation, requires, inter alia, the following:

Each Contracting Party shall, as far as possible and as appropriate:
h) Prevent the introduction of, control or eradicate those alien species which threaten ecosystems, habitats or species;

These obligations apply not only to biodiversity in the party State's territory but also to the effects on biodiversity elsewhere. Article 4, Jurisdictional Scope, provides the following:

Subject to the rights of other States, and except as otherwise expressly provided in this Convention, the provisions of this Convention apply, in relation to each Contracting Party:
(b) In the case of processes and activities, *regardless of where their effects occur,* carried out under *its jurisdiction or control,* within the area of its national jurisdiction or *beyond the limits of national jurisdiction.* (Emphasis added)

It is clear that most States already have an international obligation to address the problem of alien species transfer to the extent that it occurs within their jurisdiction or because of an activity under their control. This includes the role of flag States and ship-owning and -operating States. It is clearly relevant to the question of State international responsibility to prevent *both* the export and the import of alien species and pathogens in ships' ballast water.

The 1997 IMO Resolution A. 868(20), Guidelines for the control and management of ships' ballast water to minimize the transfer of harmful aquatic organisms and pathogens[70]
In 1973, an International Conference on Marine Pollution organized by IMO passed Resolution 18 *Research into the effect of discharge of ballast water containing bacteria of epidemic diseases.*[71] In the late 1980s and early 1990s a number of States presented research and argued for international rules on

69. (n. 27 above).
70. (n. 21 above).
71. *Alien invaders* (n. 2 above), p. 15.

this issue in IMO's Marine and Environmental Protection Committee (MEPC).[72] In 1991, nonbinding rules entitled Guidelines for Preventing the Introduction of Unwanted Organisms and Pathogens from Ships' Ballast Waters and Sediment Discharges, originally drafted by Canada and modified in a working group, were adopted by the MEPC.[73] These were further developed in light of more experience and adopted in 1993 by the IMO General Assembly.[74] In 1994 a Working Group began to examine the possibility of legally binding regulations that tried to address the safety issues. In 1997 the IMO General Assembly adopted Resolution A.868 (20) that revised the earlier Guidelines. One of the more significant features of the revision was the formal adoption of a risk minimization management approach to the problem, as reflected in the new title Guidelines for the Control and Management of Ships' Ballast Water to Minimize the Transfer of Harmful Aquatic Organisms and Pathogens. IMO is currently focusing on developing the text of a new legally binding convention by 2003 or 2004.

The Guidelines differ from the more usual IMO regulatory strategy that emphasizes flag State responsibility and control. The Guidelines apply to all ships and encourage adoption of uniform rather than unilateral state practices. However, they also state that,

> 11.2 Member States have the right to manage ballast water by national legislation. However, any ballast water discharge restrictions should be notified to the Organization.

The majority of the provisions in the Guidelines either are directed to port/coastal States or simply recommend that ships have a Ballast Water Management Plan (BWMP) and keep a record of ballast water intake and discharge that can be reported to port authorities. Both port administrations and ships are to make use of a standardized Ballast Water Reporting Form. Governments are required to ensure training for ships' crews and masters to ensure proper implementation of the BWMP. The Guidelines also recommend that ships adopt precautionary approaches to try to prevent or reduce the risk of uptake or discharge of harmful organisms. Precautions include avoiding uptake of ballast water at night, in very shallow water, or where a propeller may stir up sediment; removing tank sediment regularly; and practicing either open sea exchange, minimal or no release of ballast water discharge into reception facilities, or use of treatment options. Under the Guidelines, ports are required to provide information to vessels corresponding to the

---

72. Canada and Australia were the earliest countries to pursue this issue as it related to species transfer. In 1988 Canada presented a study report, *The Presence and Implication of Foreign Organisms in Ship Ballast Water Discharged in the Great Lakes.* 4 July 1988, MEPC 26/4, (London: IMO).

73. *Alien invaders* (n. 2 above).

74. IMO Resolution A.774(18).

operational requirements. For example, a port State is required to inform vessels about its ballast water management requirements, reception facilities, alternate discharge zones, and other port contingency requirements. In addition, the port State is required to support ships' measures to avoid the intake of organisms and pathogens by providing information on the following:

> 8.2.2 . . . areas with outbreaks, infestations or known populations of harmful organisms and pathogens; areas with phytoplankton blooms (algal blooms, such as red tides); nearby sewage outfalls; nearby dredging operations; when a tidal stream is known to be the more turbid; and areas where tidal flushing is known to be poor.[75]

The Guidelines also recommend a risk minimization approach that involves consideration by the port State of factors that put a vessel at low risk for species transfer. The two factors mentioned that can reduce the risk of an invasive species establishing in the coastal zone are disparate conditions between the place of ballast water intake and the receiving port and the age of the ballast water. An example of this approach called the Australian Ballast Water Decision Support System (DSS) is a key feature of the Australian regulations that became fully mandatory in July 2001.[76] Det Norske Veritas (DNV) in cooperation with the EU Concerted Action Group on Ballast Water has also developed an on-line risk assessment tool called EMBLA that uses biogeographical data as a predictor of high-risk journeys/ high-risk vessels.[77]

The Guidelines are important because they apportion responsibility for prevention to both ships and the port/coastal State. As will be discussed in the next section, the text of the draft Convention has adopted an approach that reflects the more traditional IMO regulatory strategy with its focus on the flag State management/certification rules, with little or no emphasis on port State export prevention responsibilities.

In addition to its work to develop an international convention, IMO is also host to the joint UNDP/GEF/IMO Global Ballast Water Management (GloBallast)[78] initiative referred to earlier. The Programme is focused on implementation and development concerns and is working with six countries (China, India, Iran, Ukraine, Brazil, and South Africa) to implement the Guidelines and identify best practices and model regulations. It is also aimed at raising greater awareness of the issue and facilitating collaboration

75. (n. 21 above).
76. *New Ballast Water Management Arrangements for International Shipping Visiting Australia*, by Australia (16 February 2001), MEPC 46/3/5, (London: IMO).
77. Det Norske Veritas (DNV), *The EMBLA Methodology*, (Norway: Det Norske, 1999). Accessed 15 August 2002 on the World Wide Web: http://projects.dnv.com/embla/emblaset.html.
78. *The Problem* (n. 8 above).

in technological and scientific research. The combination of these activities will enable rapid implementation of a legally binding instrument when the IMO-Member States adopt it.[79]

The IMO Draft International Convention for the Control and Management of Ships' Ballast Water and Sediments
The IMO MEPC has considered various drafts of a consolidated text of an International Convention for the Control and Management of Ships' Ballast Water and Sediments.[80] This draft text was agreed to in principle at the April 2001 meeting, with some points left open for further negotiation. It provides a good indication of the approach that will be reflected in the final Convention, scheduled for adoption in 2003.

The Preamble to the draft convention refers to the UNCLOS and the Convention on Biological Diversity regimes, public health, and the need for a precautionary approach. Also noted are concerns about unilateral action and the need for globally applicable regulations and guidelines for effective implementation and uniform interpretation. This Preamble firmly connects the ballast water issue and the draft convention to the UNEP/WHO biosecurity/state responsibility agenda and the UN Office for Ocean Affairs (UNCLOS Secretariat) as well as the more traditional IMO concerns about ship safety and uniformity. At a macro-system level, this reflects the increasing integration and, perhaps, even overlapping oceanic interests of the various UN agencies.

The draft convention text reflects the same structure and regulatory strategy as IMO's other ship-source marine pollution prevention instrument MARPOL73/78 dealing with oil, chemicals, harmful substances in packaged forms, sewage, garbage, and air emissions.[81] In fact, much of the text is drawn from MARPOL, Annex 1, Regulation from Prevention of Pollution by Oil, which regulates operational discharges of oil from ships. The draft convention is structured as a short agreement setting out general rights and responsibilities. Interestingly, it affirms in Article 3(3),[82]

79. Convention is now planned for 2003 or early 2004 adoption.
80. *Consolidated text of an International Convention for the Control and Management of Ships' Ballast Water and Sediments,* draft prepared by the USA (19 January 2001), MEPC 46/3/2, (London: IMO); the latest version is *Draft International Convention for the Control of Ships' Ballast Water and Sediments,* MEPC 48/2 (April, 2002). 23 August 2002 on the World Wide Web: http://globallast.imo.org.
81. The International Convention for the Prevention of Pollution from Ships 1973, as modified by the Protocol of 1978 relating thereto, reprinted in *MARPOL 73/78 consolidated edition 2002* (London: IMO, 2001).
82. Although the later revision in November 2001 suggests that, if agreement on the use of zones as a means of resolving differing ideas about the level of protection is reached, this provision may not be needed. IMO, MEPC Harmful Aquatic Organisms in Ballast Water, Report of the Ballast Water Working Group, convened during MEPC46, (30 November 2001), IMO doc. MEPC 47/2, (London: IMO).

Nothing in this Convention shall be interpreted as preventing a Party from taking individually or jointly, more stringent measures with respect to the prevention, reduction or elimination of the transfer of harmful aquatic organisms and pathogens through the control and management of ships' ballast water sediments consistent with international law.

Regulations on specific technical matters such as treatment standards and application of the agreement are set out in an annex. Flag State responsibility is the locus of control and responsibility in the draft convention, which provides for certification and recognition of an International Ballast Water Management Certificate. This requires an initial ship survey, monitoring, and regulatory control by the flag State (as delegated in many cases to a Classification Society), with port States monitoring to ensure ongoing ship compliance with the Certificate requirements. There will be a set of "existing ship or new ship" requirements for tank and other equipment design issues, with a schedule under negotiation for phasing out existing ships. As is the case with the MARPOL 73/78 agreements,[83] it also requires efficient reception facilities for sediment disposal, a vessel Ballast Water Management Plan, and a Ballast Water Record Book that is available for inspection. It also provides for inspection and sampling but recognizes potential commercial consequences by providing compensation for "undue delay."

Also similar to MARPOL's designated "special areas" formula found in, for example, MARPOL Annex 1 (Reg 10),[84] the draft convention text adopts a two-tier approach to standards and operating requirements. Tier 1 will be the generally applicable standards and Tier 2 will allow a State to designate (based on internationally accepted criteria) Ballast Water Discharge Control Areas (BWDCA) in which more stringent requirements may be imposed. The draft convention text of December 2000[85] had criteria for BWDCAs, however, the subsequent (current) draft has removed these as the topic was seen as needing greater discussion. One concern has been the need to ensure consistency between the provisions and UNCLOS. Advice provided to the MEPC Working Group by IMO's legal office suggested that even BWDCAs that extend beyond the EEZ of a State would be consistent with UNCLOS, however, State enforcement of the area beyond the EEZ would be a matter for flag States.[86] It is still unclear at this point whether the designation process will be the same as the process currently used for

83. (n. 81 above).
84. Ibid.
85. *Draft Text of an International Convention for the Control and Management of Ships' Ballast Water and Sediments,* 11 December 2000, MEPC 46/3 (London: IMO).
86. *Advice Concerning Legal Aspects of the Draft International Convention for the Control and Management of Ships' Ballast Water and Sediments,* 16 February 2001, MEPC 46/3/4 (London: IMO).

244   *Transportation and Communications*

other substances or whether it will have a different reporting, designation, and administrative structure. There is also debate about whether some ships can be exempted from the Tier 2 requirements.[87]

Unlike the Guidelines, the draft convention text does not specifically address the ballast water treatment or management strategy that ships must adopt. Instead, it requires that each vessel have a BWMP and focuses on the standard of effectiveness required, irrespective of the method used, although it appears that mid-ocean exchange will be phased out. One of the questions still under discussion is whether open sea exchange should be the baseline for measuring the efficacy of other treatments. In that case the standard exchange may be a theoretical 100 percent for sequential refill in open seas, or 95 percent for the flow-through method. The standard may be based on a particular organism and the vessel type, or on the best available technology. Alternatively, the standard may be biological, based on the receiving environment.[88] The problem is more complicated because methods may differ in their effectiveness for the various organisms and pathogens. For example, Japan prepared a report for the April 2001 MEPC meeting that examined the relative merits of several methods by using the parameters of ship safety, cost, environmental impact, and operational demands.[89]

These are simply a sampling of the many matters still to be resolved before a convention is finalized, however, they serve to illustrate the complexity of the regulatory challenge posed by this issue.

To summarize, if the current draft ends up as the framework for a convention then the following will apply to all States party to it:

- All ships (including offshore exploration and exploitation units) carrying flags of IMO member States or operating under the authority of a Party must comply, unless exempted. Ships that do not use ballast water, do not undertake international voyages (i.e., staying within one State's jurisdiction, or operating in one State's waters and the high seas), warships, navy or other government non-commercial vessels are exempt, although the latter are encouraged to comply.
- Crew members must be trained in the convention's requirements and the appropriate operating techniques. An officer must be designated as responsible for assuring compliance with the BWMP and for reporting to port authorities.
- All vessels must have a BWMP (either in Spanish, English, or French and in the working language of the crew) and a Ballast Water Record

---

87. *"Tier 2" Requirements for Ballast Water Management*, submitted by Norway (16 February 2001), MEPC 46/3/9 (London: IMO).

88. *Standards and Continued Technical Development*, submitted by the United States (14 February 2001), MEPC 46/3/3 (London: IMO).

89. *Comparison of Treatment Techniques of Ballast Water and Sediments*, submitted by Japan (16 February 2001), MEPC 46/3/13 (London: IMO).

Book recording all points set out in Appendix II of Convention. A specific survey for vessels over 400 gross tonnage and appropriate procedures for other vessels is required. These surveys will result in an International Ballast Water Management Certificate that will be recognized by other States.
- Inspections are allowed for sampling or to inspect the Certificate and to check conformity between the vessel and Certificate. However, a ship's movements cannot be delayed to wait for the results of the sample, a stricture that may be problematic for coastal State interests. If there are clear grounds, the port State can warn, detain, dismiss, or exclude the ship from its ports, subject to compensation for undue delay.
- There are some exceptions to the requirements that a vessel follow its BWMP procedures. These cover accidental discharges in the absence of wilful recklessness, the avoidance of other pollution damage, the discharge of water of the same origin as the discharge location (as long as there has been no mixing with other water), and safety concerns. The wording of the safety[90] exception has changed from the Guidelines by adopting the usual MARPOL[91] formula for discharges. The draft convention provides an exception for "the uptake or discharge of ballast water and sediments necessary for the purpose of ensuring the safety of a ship or saving life at sea."[92] As currently drafted, it no longer appears to cover the more usual situation of not discharging water in open seas because of safety concerns. In other words, in most cases discharges or intakes will usually occur in the port or coastal waters because it is not considered safe to discharge outside these waters. In that case, there are no safety issues related to discharge in the sheltered port waters. This narrowing of the defence appears to relate to the emerging view that a technical on-board treatment system, rather than mid-ocean exchange, will become the norm.

At this point it is unclear how the future convention and the Guidelines will relate to each other or whether the Guidelines will become part of the convention regime, perhaps as an appendix. The emerging problems of diverse national practices and the boundaries of the convention's right to adopt more stringent standards consistent with international law are, as yet, unresolved. A further question that also may arise relates to consistency with existing State commitments to limit documentation requirements for port entry under IMO's 1965 Convention on Facilitation of International Maritime Traffic (FAL).[93]

90. Annex Reg. A-3 1
91. MARPOL 73/78 (n. 81 above).
92. (n. 90 above).
93. (1965) I.L.M. 4: 501 as amended to 2001.

It is troubling that, despite some consideration in the MEPC working group for consistency in format between the draft convention and the recently adopted Anti-fouling Convention,[94] the former has not expanded to cover the other ways that ships carry organisms, such as on anchors and other equipment. The problem with the Anti-fouling Convention process is that whilst the decision to ban organotin based anti-fouling paint is laudable and sensible, unless a substitute can be found that is equally effective the risk of alien species transfer will be increased with increased vessel fouling. The increased speed of the ships may also mean that transfer by fouling will be increased. This means that regulators must also be prepared to inspect vessel hulls as well as ballast water to ensure that there are no invasive species. As pointed out earlier, the fact that parts of the vessel other than those associated with ballast are not addressed in the ballast water regime has been noted in meetings related to the CBD, and concern has been expressed about a piecemeal, gap-filling approach to dealing with the transfer of harmful aquatic organisms.[95]

Industry and Domestic Responses

Much of the impetus at IMO to develop an international convention and uniform standards came from growing concerns about national and unilateral State action in this matter. It is not possible within the confines of this article to describe in detail the various national and industry responses.[96] However, it is clear that as awareness grows and countries develop their national biodiversity plans, many states will adopt ballast water management regulations. States and authorities that have taken action for reasons of health or species transfer include Argentina, Brazil, Chile, Israel, Orkney Islands, Canada, United States, California, Michigan, Australia, New Zealand, and more recently Japan. In the next year or two, the six regionally critical demonstration port States (India, China, South Africa, Iran, Ukraine, and Brazil) under the GloBallast[97] Programme will develop State, and, in the future, possibly regional regimes.

Some domestic regimes such as Canada and the U.S. federal regime purport to reflect the IMO Guidelines. Others do not. Even more problem-

94. International Convention on the Control of Harmful Anti-Fouling Systems, IMO Doc. AFS/Conf/26 (18 October 2001), MEPC (London: IMO).
95. Invasive Alien Species, Options for Future Work, SBSTTA VI/8, 20 December 2000, Accessed 23 August 2002 on the World Wide Web: http://www.biodiv.org/doc/meetings/sbstta/sbstta-06/official/sbstla-06-08-en.doc.
96. An online resource with links to ports and port or country regulations is now maintained by INTERTANKO on the World Wide Web: http://www.intertanko.com/tankerfacts/environmental/ballast/ballast reg.htm.
97. *The Problem* (n. 8 above).

atic from a shipping industry perspective, there is often variation in the lead agency for this issue. In some countries transfer of invasive species is treated as a health or food and agriculture/quarantine issue, in others it is a Coast Guard concern, in others it is the responsibility of the fisheries department, the navy, biosecurity, or a maritime administration or, in some cases, the harbour or port authority. Compliance with these differing regimes can become a significant paperwork burden for a ship's master and the crew. The difference in the institutional placement in each country reflects the regulatory dilemma posed by this issue. It is by no means clear that the organism transfer problem should be treated the same way as, for example, operational oil discharge, or even that it should necessarily be a shipping industry rather than coastal State issue.[98]

Despite the safety concerns, many in the shipping industry have understood the importance of this issue and taken action. The Code of Best Practices developed by the Shipping Federation, referred to earlier, is one example. A more well-known response is *The Model Ballast Water Management Plan* developed in 1999 by the International Chamber of Shipping and INTERTANKO.[99] To try to avoid the problem of numerous subnational regulations along the St. Lawrence Seaway and Great Lakes, "[a] new lobbying group, the Ballast Water Management Coalition (BWMC), has been formed by more than 125 companies and organisations on the Lakes to attempt to work out a compromise with U.S. and Canadian legislators."[100]

Classification Societies have also begun to take account of their future role in this regulatory regime. For example, Lloyd's Register has developed a BWMP notational system and last year a Principal Surveyor[101] carried out a thorough study of various treatment methods (in its Advanced Studies Group). Germanischer Lloyd's *1999/2000 IMO Pilot* outlines the ballast wa-

---

98. For example, an NGO has recently filed an action arguing that ballast water discharges should be covered by the U.S. Clean Water Act: R. Nelson, "Shipping now on states of alert," *Lloyd's List* (6 August 2001): 4. See also, the U.S. government response to this claim, "USA Office of Water, Office of Wetlands, Oceans and Watershed, EPA, Aquatic Nuisance Species in Ballast Water Discharges: Issues and Options Draft Report for Public Comment," (September 2001). Accessed 15 August 2002 on the World Wide Web: http://www.epa.gov/owow/invasive_species/ballast_report/report1.html. Others have also examined other domestic regulatory options that fall within the broader environmental regime: See, for example, E. Biber, "Exploring Options for Controlling the Introduction of Non Indigenous Species to the United States," *Virginia Environmental Law Journal* 18 (1999): 375–465.

99. INTERTANKO, International Chamber of Shipping, *Model Ballast Water Management Plan*, 2nd ed., (London, Norway: INTERTANKO, ICS, 1999, 2000). It is now under revision for consistency with the prepared Convention.

100. "Good water out, bad water in," (n. 50 above).

101. L. Karaminas, *An Investigation of Ballast Water Management Methods with Particular Emphasis on the Risks of the Sequential Method* (UK: Lloyd's Register, 2000).

248   Transportation and Communications

ter rules as an issue for new building and ship design.[102] As pointed out earlier, DNV has been working on ballast water management and risk assessment in its EMBLA project. It also provides services to shipowners needing to develop a BWMP, including the calculations needed to conduct open seas exchange as well as other techniques.[103] These are just a few examples among many. The point is that parts of the maritime industry have appreciated the fact that the burgeoning interest in biosecurity and the limited remedial possibilities means that the economic and operational consequences of an invasion are potentially far more significant on shipping than the problem of oil spills or specific substance discharges.

The discussion earlier on commercial issues pointed out that developing a solution that is more effective than open sea exchange, with its limited effectiveness and safety concerns, is driving technical research and development activities. There are a number of alternatives that are being explored and promoted. They can be generally characterized[104]:

- Exchange—replacing the coastal water with deep seawater by using a sequential or flow-through method (currently endorsed by the IMO *Guidelines*);
- Treatment—treating the water on board to eliminate organisms by using mechanical (e.g., filters, cyclonic separator); physical (e.g., thermal heat, ultraviolet, ultra sound); or chemical (e.g., organic biocides, disinfectants) methods, or some combination of these;
- Isolation (e.g., reception facilities).

Other ideas being explored include certified clean ballast water, nonrelease of ballast water, and reliance on scientific assessments of the differences in temperature and salinity of aquatic ecosystems.[105]

In March 2001, the GloBallast[106] Programme hosted an international Ballast Water Treatment R&D Symposium to discuss technical solutions. The Programme has also developed a Ballast Water R&D Directory available online[107] that includes the Symposium papers. A review of the numerous papers presented at the March 2001 Symposium indicates that aside from open sea exchange, most of the research efforts and trials are focusing on either one or a mix of several methods, primarily involving heat, filtration, irradiation,

---

102. 1999/2000 IMO Pilot, Germanischer Lloyd, Germany. Accessed 15 August 2002 on the World Wide Web: http://germanlloyd.org. Web site path: Facts and publications/fleet/shipsafety/IMO Pilot.
103. Det Norske Veritas (n. 77 above).
104. Karaminas (n. 101 above), p. 2.
105. Alien invaders (n. 2 above).
106. *The Problem* (n. 8 above).
107. See http://globallast.imo.org/R&Ddirectory8thed.doc. Accessed 23 August 2002.

chemical or natural biocide treatment of the water, ozonation and deoxygenation. Much of the debate is centred on the standard of effectiveness that is required, that is, 100 percent or less? A few examples of recent or current projects described on this on-line resource include a Japanese study of heat treatment using the main engine cooling system that reported variable effectiveness. Japan has also reported success with a mechanical sterilization system using a pipe inserted in the piping system of ballast tanks.[108] Another project, based in New Zealand and Australia, studied the effectiveness of an on-board heating system on a RoRo vessel and found a complete kill of organisms in 6–10 hours.[109] One research unit has had success using a combination of mechanical separators and UV treatment in cruise liners, and a similar approach is being explored in Canada.[110] A number of researchers have had success with a filtration system operating at 25–50 microns resulting in approximately 96 percent removal of organisms, with less impact on bacteria.[111] Filtration, combined with other methods, has also been explored with some success by researchers in Singapore and the United States.[112] The addition of biocides or chemicals to ballast water is also the focus of a great deal of interest. One of the concerns with this approach relates to the potential environmental impact of water treated with these products. There are also concerns about occupational health and safety of crew working with the substances and the need to ensure that there is no corrosive interaction with the ballast tank or tank coating.[113]

There is still interest in finding a viable operational (exchange) approach that does not require significant vessel design changes but meets the safety and fuel consumption concerns raised by the flow-through or sequential exchange methods. For example, Teekay Shipping (Japan) Ltd., has recently patented a method called the "Natural ballast water treatment method," which makes use of the effect of gravity and pressure (the force of the ships' movement) to push water through the tanks.[114] There are numerous other options being explored including modified vessel and ballast tank design options.

108. MEPC 46/3/13, 16 February 2001; The "special pipe" has been developed under the auspices of the Japan Association of Marine Safety with funding support from the Nippon Foundation. This is also a low cost, minor retro fitting, low maintenance, environmentally sensitive method. However, it has not yet proved as effective against pathogens. Research is still underway (interview 25 July with Capt. Takeaki Kukuchi, General Manager, Marine Pollution Dept., JAMS).
109. Ballast Water Treatment. (August 2002). Accessed 23 August 2002 from the World Wide Web: http://globallast.imo.org/R&Ddirectory8thed.doc.
110. Ballast Water Treatment (n. 109 above).
111. Ibid.
112. Ibid.
113. Karaminas (n. 101 above).
114. The method is described in a submission by Japan to IMO: Alternative Ballast Water Treatment Method (15 February 2001), MEPC 46/INF.19 (London: IMO); See also Van Dyke (n. 54 above), p. 26.

Despite all of these efforts, the search for the most effective and efficient method still continues. The EU has recently funded a 3-year project involving 25 partners that will conduct comprehensive studies and trials to evaluate "On Board Treatment of Ballast Water (Technologies Development and Applications) and Application of Low Sulphur Marine Fuel (MARTOB)." The project is expected to provide recommendations to governments and industry on these two topics taking into account the "limitations of ballast water treatment on-board ships, environmental, economic impacts, risk and safety issues."[115]

The point to take from this brief review is that much of the effort to resolve the apparently intractable conflict between the values of biosecurity and on-board safety is focused on developing a technical solution to resolve the problem and to also deal with the commercial implications of open sea exchange and port/flag State monitoring.

IMPLICATIONS AND CONCLUSIONS

Various perspectives on the problem of organism transfer in ballast water have been described and the international regulatory response has been outlined. This concluding Section will comment on the implications of the proposed international convention for shipowners, flag, port, and coastal States. As noted earlier, it is expected that international standards that will be legally binding will be adopted in 2003. Will it make any difference? The fact that a large number of States already have a binding obligation under UNCLOS and/or the CBD to prevent the import and export of alien species has been noted. International standards are useful, but as the foregoing overview suggests, there are already a number of industry and regulatory initiatives underway. It may well be that a new convention will simply operate as a codification of the general practice.

Whether the international regulatory agenda is now set by the so-called unilateral actions of States with the greatest (economic) leverage in a particular issue is a broader question raised by this issue. However, State practice, which is closer to and more directly responsive to the changing needs of society, has always had a shaping force on the formation of international law. This is how change occurs. It may also be, as suggested at the beginning of this paper, that we are simply seeing an integration of shipping into the more general environmental regime. Certainly, the often excruciatingly slow and expensive process of international negotiations has given rise, in some

---

115. P. Zhou and V. Lagogiannis, "Ballast Water Treatment by Heat—EU Shipboard Trials" (paper presented to the GloBallast R&D Symposium, London: IMO, 26–27 March 2001) Accessed 15 August 2002 on the World Wide Web: http://globallast.imo.org/index.asp?page=Abstracts.htm.

States, to impatience and a loss of faith in the international consensus building process. The increase in large free trade voting blocks will exacerbate this process. These are part of the larger sociopolitical forces shaping the evolving form of governance described as globalisation.

The fact that the draft convention has avoided becoming tangled in the debate over choice of method and is instead focusing on the fact that the BWMP will vary for each vessel also injects room for flexibility and accommodates future solutions. It reflects the view that technology will develop a solution. But, even if the answer in some cases may be simply, for example, inserting a super filter or a special pipe or installing a heat exchange unit, or a combination of methods, are there other implications arising from a new convention?

Shipowner/Operator

Even if a technical fix is found, it is important to understand that the draft convention recognizes the right of each State to inspect and also impose, with notice, more stringent regulations. This means that carriage arrangements must take into account potential delays for inspections or sampling and for sediment discharge. Ballast tanks will need to be a core part of regular maintenance activity. The severe commercial consequences of a vessel being ordered to retain ballast water or refused entry to a port if the ballast water records are not reliable or complete means that the crew must be thoroughly conversant with the BWMP. Even if seafarers have a generic training to carry out these operations, each ship or ship class will have unique operational constraints and treatment procedures. The certification process and possible design modifications required will, of course, add to the operating costs of the vessel and will impose further paperwork and demands on the crew. However, as pointed out earlier, these are calculable and can be factored into operational costs.

The fact that the biodiversity/security agenda is almost universally endorsed means that more States will be proactive on this matter. Shipowners and shippers will need to negotiate with each State and/or port to determine who will bear the direct costs of activities such as sampling or sediment or ballast water reception facilities. Increased scrutiny of transport corridors and the fact that a ship has multiple vectors means that, as much as possible, the shipping industry should encourage regulators at all levels of governance to adopt an integrated approach to dealing with biosecurity issues. Shipowners will also need to encourage more "whole ship inspections" to avoid multiple inspections. In some regions or closely linked ocean areas, encouraging a regional inspection at the first port of call will, as with the other suggestions, support both a biosecurity and commercial agenda. The idea of co-management is also a useful approach to explore. Australia has developed a system of negotiated compliance agreements with companies that are frequent users of its ports.[116]

The port reporting systems under the Guidelines require that the master or shipowner/operators provide advance data about the characteristics of the water and species or pathogens found in the water of each port of water intake.[117] Maritime administrators are not ordinarily informed about the varieties of plants or marine species or the specific location of sewage outfalls near the port.[118] The shipping industry should try to encourage the development of more clearinghouse or research databases to facilitate procedures, such as the Australian DSS for high- or low-risk vessels, which can alleviate some uncertainty and further enable co-management arrangements for vessels that ply lower risk routes.

Flag State

Under the draft convention the flag State is given responsibility for ensuring that its vessels have a BWMP and that it is fully implemented according to the rules of each jurisdiction. The first is reasonably straightforward and the resulting International Ballast Water Management Certificate simply adds to the list of international certificates that the flag State is supposed to provide and verify, usually through a Classification Society or its national surveyors. The second issue, ongoing monitoring and enforcement, presents the flag State with the same problems as the other IMO international certificates. Reasons for noncompliance with, for example, the long-standing oily water discharge standards under Annex 1 of MARPOL 73/78, include a range of factors often relating to the corporate culture of the particular shipowning company. If compliance adds to time/costs, then there may be pressure on the crew to take shortcuts.

The flag State will have to develop its domestic legal and administrative regime to include this responsibility. For example, it will have to amend its registry criteria and ensure that its legal system is able to sanction breaches of international ballast water standards by its ships in other jurisdictions for failure in ballast water management. It will also be important to educate maritime personnel to review documentation, verify surveys, and develop the rules for other surveys or evaluations for vessels less than 400 gross tonnes. Concern about the potentially disastrous impact of a spread of an epidemic or invasive species in a State will mean that informed port States

116. (n. 76 above).
117. It must be noted that the draft convention does not require filing of these forms as it relies on an international certificate system.
118. Although, in some cases, they may be aware because of port client complaints about ship corrosion resulting from contact with municipal runoff and sewage in the water.

will not be inclined to allow vessels with known doubtful practices or equipment to enter the port or discharge water. The level of port State intervention is likely to be much higher than in other pollution matters, resulting in little or no commercial incentive for flag States to fail to regulate properly.

This does not, of course, address the problem of impoverished States where there may be problems in developing *any* flag State capacity and where there are simply insufficient personnel to set up new procedures, inspect vessels or water, or carry out the requisite surveys. Although the formula under the draft convention places responsibility for the vessel BWMP in the hands of the flag State, it is clear that in fact it is the port State regime that will determine the effectiveness of the regulatory strategy. The fact that the port/coastal State will be making a decision to allow the discharge of ballast into its waters creates a very different dynamic from the oil and other discharge issues. It places substantially more de facto enforcement power in the hands of the port State. Irrespective of the convention's emphasis, this may well signal a shift away from the supremacy of flag State jurisdiction.

Port State/Coastal State

These two perspectives are dealt with together in connection with future implications of a new convention because an effective domestic response to the problem requires more integration of ports with broader coastal management and economic development activities of States.

It was pointed out earlier that each country that is party to either the CBD and/or UNCLOS has an obligation to prevent both the import and export of invasive species. The degree of activity and immediate interest in this issue will turn to some degree on each country's trading patterns and partners. If it is primarily an import State, vessels will rarely arrive with much, if any, ballast water and the issue will be that of exporting species and pathogens. The reverse is true if the trade is primarily export (subject to variables depending on the kind of cargo exported). Trading patterns are also important. For example, if one State has a strong interest in this issue and has developed a ballast water management regime, this will influence its trading partners to adopt a similar regime.

This means that each State, and port areas in particular, will need to know, with a high degree of reliability, the species and bacteria in the coastal waters at any one time. This is particularly the case if a risk assessment process is adopted. Ports will need to be able to reassure vessels and other States about the likely range of species or pathogens in the water. Harbour survey or profile studies need to be done, as well as evaluations of the character of the coastal waters—temperature, food chain, salinity, and so on. This means that a State must train people in sampling techniques and develop a well-integrated and rapid sampling, inspection, and evaluation system with adequate staff and a good communication system for reporting and responding to requests for information. This has begun on a pilot project basis with the GloBallast Programme.[119] Unlike the case of oil or other pollutants, a port that does not impose ballast water restrictions is not going to be a desirable port of call because it will mean that the ship will then face a higher level of scrutiny at other ports of call. These ports—Biosecure Ports—may encounter some costs initially, but will ultimately obtain a competitive advantage. A regulatory framework must be developed that ensures effectiveness and expedition. For example, combining inspections to cover quarantine, ballast water, and fouling and possibly even the requisite MARPOL[120] inspections will assist the shipping industry. This will require additional training for inspectors at ports. Centralizing the reporting and decision-making system for easy communication with ships will encourage vessels to comply with any reporting requirements. As noted earlier there is a clear need, and indeed, an obligation in some States party to FAL to minimize the documentation requirements. However, this means that all sectors and governmental departments that are affected must find a means of working together on this topic to ensure that the biosecurity and commercial/economic agendas are served. The development of reception facilities remains, as it is with the MARPOL[121] regime, a good idea in principle, but the cost means that for many countries it is not possible and another sediment disposal method must be found. The cost of all these services will need to be evaluated and the question of who pays will need to be negotiated with the shipper and shipowner community.

Another regulatory dimension exists for States that are members of the World Trade Organisation (WTO) and/or part of economic blocks (North American Free Trade Agreement, European Community) that adhere to the WTO trade liberalisation practices. The recognition of a precautionary approach[122] mandated by MEAs, such as the CBD, and the trade liberalisation antidiscriminatory philosophy of the WTO regime creates some problems, especially if the application of the regulations is not applied to every flag or every voyage.[123] A State may find its ballast water management regula-

119. Raaymakers (n. 34 above).
120. MARPOL 73/78 (n. 81 above).
121. Ibid.
122. Convention on Biological Diversity, COP 5, Decision V/8. *The Interim Guiding Principles for the Prevention, and Mitigation of the Impacts, of Alien Invasive Species,* have as the first principle, the precautionary approach. Accessed 15 August 2002 on the World Wide Web: http://www.biodiv.org/decisions.
123. P. Jenkins, (1999). *Global Policy Changes Needed to Stop Biological Invasions Caused by International Trade.* Presented at the Workshop on the Legal and Institutional Dimensions of Alien Invasive Species Introduction and Control. Held at IUCN, Environmental Law Centre, Bonn, Germany, 10-11 December 1999. Accessed 23 August 2002 on the World Wide Web: http://www.invasives.org/publications.html.

tions challenged under WTO rules such as the Agreement on the Application of Sanitary and Phytosanitary Measures.[124]

This gives a further incentive to the development of internationally binding standards. This will not necessarily resolve the matter in the WTO forum, but the fact that a nation has adopted an international standard when imposing domestic requirements on ballast water treatment or retention will be helpful. The relationship between the WTO and other international social and environmental regimes is a central political issue in the globalization debates.[125]

Conclusions

The global biosecurity imperative stems from interrelated concerns about health, economy, ecology, and national defence. In effect, the agenda is survival. It is already a significant force shaping international regulatory policy. It will have an enormous impact on the priorities of governments, as evidenced by the recent foot-and-mouth epidemic and the increasingly defensive global response, irrespective of trade slow-downs or other commercial consequences. The agenda of international institutions is global biosecurity and integration. However, national and subnational political agendas and public sympathy are increasingly tending to favour protection of domestic interests—an all too human "fortress building" response to the lack of an effective global regulatory structure. The highly emotive issue of alien or harmful aquatic organisms and pathogen transfer through international transport vectors such as ships' ballast water will trigger this action as public awareness grows. Alternatively, from a more optimistic point of view, it may be that we are seeing the integration of ship-sea-land governance at all levels.

Irrespective of the ultimate meaning of these events, it is imperative that the maritime sector develops its awareness and appreciation of the potentially explosive nature of the issue, particularly as it relates to the realignment of international regulatory systems. Rather than being cast as the carrier of dangerous invading species, plagues, and pestilence, the shipping industry must actively work within a number of international biosecurity fora to be viewed as a vital contributor to achieving future ecological and economic security. It must become part of the solution, not part of the problem.

124. D. Wilson and D. Gascoine, "National Risk Management and the SPS Agreement" (paper presented at 1999 Conference, *Globalisation and the Environment—Risk Assessment and the WTO*, Melbourne Business School, Australia, 1999). Accessed 23 August 2002 on the World Wide Web: http://www.dpie.gov.au/content/publications.cfm.
125. N. Bankes, "International environmental law for the new millennium: the challenges ahead," *CCIL Bulletin* (Winter 2000): pp. 13–15.

Environment and Coastal Management

Beyond Fiduciary Duty: Aboriginal Rights and Integrated Coastal Zone Management in New Zealand and Canada*

Matthew Heemskerk
Dalhousie University Law School, Halifax, Canada

INTRODUCTION

In Canada and New Zealand, Aboriginal access to coastal resources is changing. Currently, Canada lacks a national Integrated Coastal Zone Management (ICZM) plan and Aboriginal involvement in the administration of resources tends to be erratic, varying regionally and depending on the particular First Nations group. On the Pacific and Atlantic coasts of Canada, dwindling fishing stocks off both the west and the east coast combined with an increased recognition of Aboriginal rights to these resources have created intense conflict between Aboriginals and non-Aboriginals. In northern coastal regions, federal and provincial governments have recognized First Nations self-government and created extensive land claims agreements.

By comparison in New Zealand a period of dramatic legislative reform has increased Maori involvement in resource management. The Resource Management Act[1] mandates extensive consultation in situations where resource development impacts Maori interests. Efforts to settle Maori claims to fisheries led to the allocation of over 20 percent of the commercial fishery to the Maori people. These dramatic changes have generated discussions and disagreements over the scope of Maori rights in New Zealand.

This article will examine the scope of Aboriginal rights to coastal resources in New Zealand and Canada and the level of indigenous participation in coastal management schemes. One of the primary goals of Integrated Coastal Zone Management is the participation of all stakeholders in resource management. This presents a special set of issues for former colonizing nations where indigenous peoples have been marginalized and isolated from contributing in decision-making processes. At first glance, this seems to be another difficulty preventing the implementation of comprehensive

---

*The author would like to thank Lucia Fanning for her direction with research for this article and Maria Heemskerk for her editing suggestions.
   1. Resource Management Act, Statutes of New Zealand, 1991, No. 69.

© 2003 by The University of Chicago. All rights reserved.
0-226-06620-7/03/0017/0012$01.00 *Ocean Yearbook* 17:256–289

oceans and coastal management schemes. Further inquiry reveals that indigenous rights and ICZM principles are quite complementary. Through a comprehensive resource management scheme, Aboriginal rights can be acknowledged and protected. Through the protection of Aboriginal rights and incorporation of indigenous philosophies resource management programs can be developed in a more holistic manner.

Integrated Coastal Zone Management

International agreements and conferences have pointed to Integrated Coastal Zone Management as an important component of resource management. ICZM seeks to meet the problems in coastal regions of the world through integrated multisectoral resource administration. Cicin-Sain and Knecht describe ICZM as "first and foremost, . . . designed to overcome fragmentation inherent in both the sectoral management approach and the splits in jurisdiction among levels of government at the land-water interface."[2] Proponents characterize ICZM as having five main attributes:

1) It is a process that continues over a considerable time. ICZM is a dynamic programme that usually requires continual updating and amendments. ICZM is not a one-time program.
2) It has a geographic boundary that defines a space that extends from the ocean environment across transitional shore environments to a specified inland extent.
3) There is a management arrangement to establish the policies and process for making allocation decisions.
4) The management arrangement uses one or more strategies to rationalize and systematize for making allocation decisions.
5) The management strategies selected are based on a systems perspective that recognizes the associations between coastal resources and processes. The systems perspective usually requires that a multisectoral approach be used in the design and implementation of the management strategy.[3]

The international community suggests that ICZM is the most appropriate way to meet challenges in coastal areas. The 1992 United Nations

---

2. B. Cicin-Sain and R. Knecht, *Integrated Coastal and Ocean Management: Concepts and Practices* (Washington, D.C.: Island Press, 1998), p. 39.
3. L. Hildebrand and E. Norrena, "Approaches towards effective integrated coastal zone management," *Marine Pollution Bulletin* 25 (1992): 94–95. See also J. Sorenson, "The international proliferation of integrated coastal zone management efforts," *Ocean and Coastal Management* 21 (1993): 47–8.

Conference on the Environment and Development (UNCED) produced *Agenda 21: Programme of Action for Sustainable Development*,[4] a nonbinding document articulating a global commitment to sustainable development. Chapter 17 of the Agenda concerns oceans and coastal areas. Its recommendation includes a call for integrated and sustainable management of the coastal areas.[5] Soon after UNCED, the 1993 World Coast Conference issued the Noordwijk Guidelines for Integrated Coastal Zone Management. This document recognizes the need for coastal states to adopt ICZM programs that are specific to their needs. It recommends a holistic, multisectoral approach to coastal management.[6]

Central to ICZM programs are the concepts of sustainable development and integration. Sustainable development is "development that meets the needs of the present without compromising the ability of future generations to meet their own needs."[7] The concept encompasses principles of equity between generations, among members of society, and among nations and peoples in the international community.[8] The goal of sustainable development is the responsible and equitable use of resources evenly without causing environmental degradation.

For ICZM, integration means ensuring that a holistic multisectoral approach is taken when planning and managing the coastal environment.[9] Cicin-Sain and Knecht describe five different categories of integration:

1) *Intersectoral integration.* Integration among different sectors involves both "horizontal" integration among different coastal and marine sectors . . . and integration between coastal and marine sectors and land-based sectors . . . Intersectoral integration addresses conflicts among government agencies.
2) *Intergovernmental integration,* or integration among levels of government . . .
3) *Spatial integration,* or integration between the land and ocean sides of the coastal zone . . .
4) *Science-management integration,* or integration among the different disciplines important in coastal and ocean management . . .
5) *International integration.*[10]

4. Agenda 21: "Programme of Action for Sustainable Development" (Rio de Janeiro: UNCED, 1992). Accessed 14 April 2000 on the World Wide Web: http://www.unep.org.
    5. Agenda 21, Chapter 17; also Cicin-Sain and Knecht (n. 2 above), p. 86.
    6. World Bank, *Noordwijk Guidelines for Integrated Coastal Zone Management* (Noordwijk: World Coast Conference, 1993).
    7. World Commission on Environment and Development, *Our Common Future* (Oxford: Oxford University Press, 1987), p. 8.
    8. Ibid.; also see Cicin-Sain and Knecht (n. 2 above), p. 83.
    9. See Hildebrand and Norrena (n. 3 above), pp. 94–5.
    10. Cicin-Sain and Knecht (n. 2 above), p. 45.

Integration ensures that management of the marine environment incorporates the participation of all stakeholders. Furthermore, it supports the implementation of principles of sustainable development in management of the coastal zone.

Aboriginal peoples can draw upon ICZM principles to enable access and to reinforce claims to coastal resources. Integration and sustainable development suggest that former colonial nations should begin to bring indigenous peoples into the management of resources through consultation and equitable redistribution of the economic benefits of resource development.

International Law Sources of Aboriginal Rights to Coastal Resources

The rights of indigenous peoples are a developing area of international discussion. Aboriginal participation in the management and development of resources is an important component of this discussion. The UNCED principles contain a provision recognizing the importance of indigenous people in environmental management. Principle 22 of the Rio Declaration states the following:

> Indigenous people and their communities, have a vital role in environmental management and development because of their knowledge and traditional practices. States should recognize and duly support their identity, culture and interests and enable their effective participation in the achievement of sustainable development.[11]

Article 27 of the United Nations Draft Universal Declaration on the Rights of Indigenous Peoples[12] resembles Principle 22 of the Rio Declaration, in protecting indigenous access to coastal resources:

> Indigenous people have the right to own, develop, control and use the lands and territories, including the total environment of the lands, air, waters, coastal seas, sea-ice, flora, fauna and other resources which they have traditionally owned or otherwise occupied or used. This includes the right to full recognition of their laws, traditions, and customs, land tenured systems and institutions for the development and management of resources, and the right to effective measures by States to prevent any interference with, alienation of or encroachment upon these rights[13]

---

11. Rio Declaration on Environment and Development, 13 June 1992, U.N. Doc. A/CONF. 151/5/Rev.1 at Principle 22. Accessed 14 April 2000 on the World Wide Web: http://www.unep.org.
12. U.N. Doc. E/CN.4/Sub.2/1994/2/Add.1(1994).
13. Ibid., art. 27.

The Declaration is a rights-orientated document and recognizes the right of indigenous people to self-determination, as well as the right to control and develop their "distinct political, economic, social and cultural characteristics."[14] Other articles assert the right of indigenous people to participate and have some control over resource development and management, making "just and fair" compensation an appropriate remedy for infringement of these rights.[15] Another article prohibits the storage of hazardous material on traditional Aboriginal territories.[16] Many of the provisions echo case law from United States, New Zealand, Canada and Australia that tend to construe rights broadly. As one commentator notes,

> The draft declarations, in their tortuous way through the channels of the United Nations and the Organization of American States, have encountered virtually no government opposition regarding these issues. With respect to these claims, a consensus has emerged, and has been translated, with whatever imperfections, into widespread, virtually uniform state practice.[17]

However, while international sources defining indigenous rights are informative legally, they are of little weight, acting only as optimistic principles. It is ultimately the constitutional framework of a nation that determines the scope of Aboriginal rights.

MAORI RIGHTS TO MARINE RESOURCES IN NEW ZEALAND

New Zealand Constitutional Law and Aboriginal Rights

Among western nations, New Zealand has an unusual constitutional structure. New Zealand is a unicameral nation with limited regional autonomy through regional and territorial bodies. The Constitution Act, 1986[18] is an ordinary Act of Parliament and can be altered through a simple vote in the legislature. The Act is very short, defining the process and powers of the government of New Zealand. There is no complex division of power issues nor a constitutional bill of rights, and thus, unlike Canada, legislation cannot be challenged on the basis of its constitutionality. This has led to a distinction in New Zealand between what is *legal* and what is *constitutional.* Paul McHugh explains "[p]arliament has the *legal* power to legislate the slaugh-

---

14. Ibid., arts. 3, 4.
15. Ibid., arts. 19, 21 and 22.
16. Ibid., art. 28.
17. S. Wiessner, "Rights and status of indigenous peoples: A global comparative and international legal analysis," *Harvard Human Rights Journal* 12 (1999): 109–10.
18. Statutes of New Zealand, 1986, No. 114.

ter of blue-eyed babies but to do so would be *unconstitutional*."[19] In Canada, there is no such distinction between these concepts.

Maori rights are therefore not constitutionally protected. Rather they are found either in the common law or as through constitutional convention.[20] The basis of any common law or conventional assertion of Aboriginal rights stems from the 1840 Treaty of Waitangi.[21] Maori and non-Maori have challenged the legitimacy of the Treaty as a source of Aboriginal rights.[22] However, legally and politically there is no question of the importance of the document as the starting point for Maori rights in New Zealand.

The Treaty was written in both English and Maori. This has led to interpretation issues. In Article I of the English text, the Maori chiefs cedes "all rights and powers of Sovereignty" to the Queen of England.[23] Article II confirms and maintains Maori rights:

> The Queen of England confirms and guarantees to the Chiefs and Tribes of New Zealand and to the respective families and individuals thereof the full and exclusive and undisturbed possession of their Lands and Estates Forests Fisheries and other properties which they may collectively or individually possess so long as it is their wish and desire to retain the same in their possession.[24]

Article III of the Treaty promises the full right of participation as British citizens. For the British, "the 1840 Treaty of Waitangi guaranteed the Maori control over their ancestral lands and natural resources, and full citizenship and protection under the British Crown, in return for British sovereignty over New Zealand and a British right of first refusal with regard to Maori land sales."[25]

In contrast to the English text, the Maori version of the Treaty suggests a broader understanding of Maori rights. In the Maori text, the Chiefs ceded

---

19. P. McHugh, *The Maori Magna Carta: New Zealand Law and the Treaty of Waitangi* (Toronto: Oxford University Press, 1991), p. 13.
20. Ibid., p. 18–19
21. This document represents negotiations between the Crown and Maori people similar to the Mi'kmaq Treaties of 1760–61, recognized by the Supreme Court of Canada in R v. Marshall 177 D.L.R. (4th) 513.
22. See McHugh (n. 19 above), p. 3–9; F. M. Brookfield, "The New Zealand Constitution: The search for legitimacy," in I. H. Kawharu, *Waitangi: Maori and Pakeha Perspectives of the Treaty of Waitangi* (New York: Oxford University Press, 1989), p. 1.
23. Treaty of Waitangi, Accessed 27 December 2001 on the World Wide Web: http://www.govt.nz.aboutnz/treaty.php3.
24. Ibid.
25. B. Kahn, "The legal framework surrounding Maori claims to water resources in New Zealand: In contrast to the American Indian experience," *Stanford Journal of International Law* 35 (1999): 49 at 59.

kawanatanga to the Queen but maintained tino rangatiratanga under Article II.[26] According to commentators, the difference between the two terms is subtle and "the Maori chiefs' probable comprehension of the term kawanatanga must be placed alongside the view of rangatiratanga."[27] Interpreters explain that the term kawanatanga refers to "governorship" rather than sovereignty because the Maori had no understanding of this concept.[28] Rangatiratanga refers to a degree of self-determination apart from the ability "to make war, exact retribution, consume or enslave their vanquished enemies and generally exercise their power over life and death."[29] The difference between the terms meant the Maori Chiefs "would have believed they were retaining their rangatiratanga intact apart from the licence to kill or inflict material hurt on others, [and] retaining all of their customary rights and duties as trustees for their tribal groups."[30] Thus, the Maori version suggests the Chiefs' maintained considerably more autonomy from the Queen than the English text suggests.

Initially, the Treaty of Waitangi was rejected as a source of Aboriginal rights at common law. In the Privy Council decision, Te Heuhue Tukino v. Aotea District Maori Land Board,[31] Viscount Simon held that "it is well settled that any rights purported to be conferred by such a treaty of cession cannot be enforced in the Courts except in so far as they have been incorporated into municipal law"[32] This trend was followed by the New Zealand Courts. In Wi Parata v. Bishop of Wellington,[33] the court rejected Maori claims holding that Aboriginal rights existed only through the grace of the Crown. Until recently, the courts continued to embrace this traditional viewpoint, ruling that the Treaty was not an independent source of Maori rights or that its provisions had been extinguished.[34] Indeed, it was not until the 1980s that the judiciary began to reconsider this orthodox position.

In Te Weehi v. Regional Fisheries Officer[35] and New Zealand Maori Council v. New Zealand (Attorney General)[36] the court recognized the Treaty as an independent source of common law Maori rights. The Te Weehi

26. McHugh (n. 19 above), p. 3–4.
27. McHugh (n. 19 above), p. 3.
28. Ibid.
29. Kawharu, 'Sovereignty vs. Rangatiratanga' (unpublished paper), cited in McHugh (n. 19 above), p. 4.
30. Kawharu, cited in McHugh (n. 19 above), p. 4.
31. [1941] NZLR 590.
32. [1941] NZLR 590, p. 596–7.
33. (1877) 3 Jur (N.S.) SC 72.
34. For Example, Re Ninety Mile Beach [1963] NZLR 461 where the court held that Maori rights under the treaty to the foreshores of Marlborough Sound had been extinguished; also N.Z. Maori Council v. A-G, [1992] 2 NZLR 576 where the traditional doctrine that the treaty is unenforceable was upheld.
35. [1986] NZLR 682. (hereinafter Te Weehi)
36. [1987] 1 NZLR 641.(hereinafter Maori Council)

case concerned Aboriginal title and the interpretation of s. 88(2) of the Fisheries Act, which stated that regulations would not affect 'Maori Fishing Rights.'[37] In his decision, Williamson J. rejected arguments that the government had extinguished the indigenous fishing rights under the Treaty. Relying on Canadian precedents,[38] he explained that to extinguish an Aboriginal right Parliament must have enacted "specific legislation that clearly and plainly takes away the right."[39]

This line of reasoning was followed a year later in Maori Council. Once again, the case involved the interpretation of a provision guaranteeing Maori rights in a statute. Section 9 of the State Owned Enterprises Act stated that regulations could not violate the principles of the Treaty of Waitangi.[40] The Court of Appeal explained that Government and Maori representatives were bound to the terms of the Treaty and both groups are required to "act towards each other reasonably and with the utmost good faith."[41] They described the Treaty as a "dynamic" contract between the Crown and the Maori not limited to a historical understanding of the terms of the treaty.[42] McHugh explains the importance of the decision:

> The Court of Appeal's judgement stressed the mutuality and reciprocality of the Treaty of Waitangi. On the one hand, the Crown undertook responsibility for the protection of Maori society and interests, not least the mana of the chiefs and tribal property rights. On the other hand, the Maori agreed to the colonization of the country under royal aegis. This mutuality in which each side extracted certain benefits whilst also accepting concomitant responsibility provides the basis for the principles of 'partnership' and 'the reciprocal obligations of the parties to act with reasonableness and the utmost good faith.'[43]

Despite the importance of these two decisions, they have not definitively ended the debate over whether Maori rights exist at common law. Critics point out that in both of these cases a statutory provision existed that pointed to the Maori rights. Thus, the traditional position that the Treaty is not a relevant source of law except as incorporated into domestic law remains intact.[44] This position has been captured by one commentator describing the position of the government in relation to Maori land claims settlements:

37. Statutes of New Zealand, 1983 No. 014.
38. He considered Calder v. Attorney General (1973) 1 S.C.R. 313 and Guerin v. The Queen (1984) 2 S.C.R. 335.
39. Te Weehi (n. 35 above), p. 688.
40. Statutes of New Zealand, 1986, No 69.
41. Maori Council (n. 36 above), p. 667.
42. Maori Council (n. 36 above), p. 703; and see McHugh (n. 19 above), p. 4.
43. McHugh (n. 19 above), p. 5.
44. See M. Wharepouri, "The phenomenon of agreement: A Maori view," *Auckland University Law Review* (1995): 604, at p. 612.

As Minister Graham states, the Treaty of Waitangi "is not subject to [New Zealand's] domestic law" unless statutorily incorporated. In sum, "[t]he doctrine of Parliamentary supremacy enables Parliament to make or unmake whatever law it likes, including any law concerning rights that Maori might claim under the Treaty of Waitangi." At best, the Treaty of Waitangi limits the Crown's prerogative powers, requiring Parliamentary action before the New Zealand government may act inconsistently with the principles or provisions of the Treaty of Waitangi but ensuring the legality of any such overriding legislation.[45]

This characterization may be accurate because even if Maori rights exist at common law they are not protected by a constitutional document. Therefore, any act of Parliament could override Maori treaty rights.

Fortunately, for Maori interests, rather than eliminating references to the Treaty, New Zealand has embraced the principles of the Treaty as a constitutional convention. The Treaty is now referred to in a plethora of legislation and the government has recognized Maori rights and committed to settling Maori claims.[46] Therefore, although a simple act of legislation *could legally* eliminate Maori Treaty rights, such a move would violate New Zealand's constitutional conventions.

Coastal Zone Management in New Zealand

Statutory Framework
The Resource Management Act[47] (RMA) and the Local Government Act[48] establish the statutory framework for coastal policy in New Zealand. These two pieces of legislation complement and support each other. They arose out of a period of reform in New Zealand after the collapse of the economy in the 1980s. Cocklin and Furuseth explain:

> During the late 1980s, New Zealand underwent a period of dramatic economic, social, and administrative restructuring. Among the most fundamental reforms was the establishment of sustainable management as the guiding principle for decisions affecting the allocation and use of natural resources and the maintenance of environmental quality.[49]

45. Kahn (n. 25 above), p. 96.
46. Ngai Tahu Claims Settlement Act, 1998 (the Settlement Act), Statutes of New Zealand, 1988 for a discussion of Maori and Government positions in relation to the act, see Kahn (n. 25 above), p. 133–4.
47. Resource Management Act (n. 1 above).
48. Statutes of New Zealand, Stat. 66 (1974).
49. O. Furuseth and C. Cocklin, "An institutional framework for sustainable resource management: The New Zealand model," *Natural Resources Journal* 35 (1995): 243, at p. 256.

The Local Government Act divided New Zealand into regional and territorial authorities largely based on water catchment areas.[50] Furuseth and Cocklin describe the novelty of the approach:

> While there were some deviations from catchment borders based on localized social factors, called "communities of interest," the overriding definitional criterion was hydrologic. Although the environmental orientation of the regional council boundaries has been criticized for ignoring traditional political and economic considerations, the application of "natural system" boundaries for environmental and natural resource planning has long been advocated, but rarely used on a comprehensive scale.[51]

The Act was thus a conscious effort to ensure the sustainable management of resources. The Chair of the Local Government Commission explained that "the water catchment will be fundamental in dividing New Zealand into identifiable regions to allow for important decision making in respect to the natural resources of land, sea, air and water, and what may be extracted from or added to these resources."[52]

The RMA establishes one of the most extensive resource management regimes in the western world. The first part of the Act contains four sections setting the stage for sustainable resource development and respect for Maori rights. The purpose of the RMA is stated in section 5 as the need to promote sustainable development.[53] Section 6 lists matters of 'national importance' including:

(a) The preservation of the natural character of the coastal environment (including the coastal marine area), wetlands, and lakes and rivers and their margins, and the protection of them from inappropriate subdivision, use, and development . . .
(e) The relationship of Maori and their culture and traditions with their ancestral lands, water, sites, waahi tapu, and other taonga.

Section 7 contains broad principles that must be considered by authorities empowered by the Act such as environmental protection and 'efficient use and development' of natural resources.[54] Section 8 signifies the importance

---

50. Ibid.
51. Ibid.
52. Cited in L. Burton and C. Cocklin, "Water resource management and environmental policy reform in New Zealand: Regionalism, allocation, and indigenous relations, Part I," *Colorado Journal of International Environmental Law and Policy* 7 (1996): 75, at p. 85.
53. Resource Management Act (n. 1 above), s. 5.
54. Resource Management Act (n. 1 above), s. 7.

of the principles of the Treaty of Waitangi.[55] The Ministry of Conservation describes the Act as follows:

> This Act is very different from previous environmental legislation in New Zealand, and the vast majority of environmental legislation throughout the world, because
>
> - It legislates for land, air and water resources under one over-arching Act. Elsewhere, each major resource is normally legislated for separately.
> - It assumes that local communities are the best judges of their own environmental problems and of how to go about dealing with them. Most countries have more central government control.
>
> The Act is "effects based": that is, it is the environmental effects of one's proposed activity that is the deciding factor whether or not a particular activity can go ahead. Overseas legislation often sets hard and fast rules over what activities can and can't be done.[56]

The lead agency for the implementation of the administrative provisions of the RMA is the Ministry of Conservation. The Ministry of the Environment is also given the responsibility to provide advice with respect to environmental issues. The RMA mandates that the Ministry of Conservation coordinate national, regional, and territorial bodies in resource management. The Act further requires the national and local governments to prepare policy statements and plans for resource administration. As Haward explains, the RMA "established a four tier administrative regime; national policy statements, regional policy statements, regional plans and district plans."[57] This structure ensures that all resource development is managed and controlled by a central process while maintaining a high level of local control over decision making.

Under section 57 of the RMA the Minister of Conservation is required to issue a national coastal policy statement in accordance with the broad principles of the RMA.[58] This was released in 1994. It is a very comprehensive document, the scope of which is evident by a perusal of the chapter titles:

55. Resource Management Act (n. 1 above), s. 8.
56. Ministry of the Environment, New Zealand, "Introduction to the Resource Management Act," Accessed 14 April 2000 on the World Wide Web: : http://www.mfe.govt.nz/management/rma/rmaintro.htm.
57. M. Haward, "Recent developments and announcements: Ocean and coastal zone management in New Zealand and Australia: Current initiatives," *Ocean and Coastal Management* 19, no. 3 (1993): 297.
58. S. Davidson, "Current legal developments: New Zealand: New Zealand Coastal Policy Statement," *International Journal of Marine and Coastal Law* 10 (1995): 431.

Chapter 1   National priorities for the preservation of the coastal environment including the protection from inappropriate subdivision use and development;

Chapter 2   The protection of the characteristics of the coastal environment of special value to the *tangata whenua* [Maori] including *waahi tapu tauranga waka, mahunga maataitai* and *taongo raranga;*

Chapter 3   Activities involving the subdivision, use or development of areas in the coastal environment;

Chapter 4   The Crown's interest in land of the Crown in the coastal marine area;

Chapter 5   The matters to be included in any or all regional coastal plans in regard to the preservation of the natural character of the coastal environment including the special circumstances in which the Minister of Conservation will decide resource contents;

Chapter 6   The implementation of New Zealand's international obligations impacting the coastal environment;

Chapter 7   The procedures and methods to be used to review the policies and to monitor their effectiveness;

Schedule 1   The circumstances in which activities that have significant or irreversible adverse effect on the coastal marine area will be made restricted coastal activities.[59]

The policy statement contains clear instructions to the Ministry of Conservation and regional authorities to establish plans in accordance with the statement. Haward explains as follows:

> The statement makes clear and unequivocal reference to the level of direction to and independence of regional and / or district Authorities. Two levels of directions are given in the National Policy Statement. Where a provision involves the term *shall,* the Regional and/or District Plan, and relevant authorities must address this issue to a level and in a manner which is appropriate to that region or district. Where a provision involves the term *should,* Regional or District Plans must address the issue if it is appropriate in that region or district."[60]

This arrangement is similar to the U.S. framework in the Coastal Zone Management Act[61] and "enhances the opportunities for integrated, rather than

---

59. Davidson (n. 58 above).
60. M. Haward, "Institutional design and policy making 'Down Under': Developments in Australia and New Zealand coastal management" *Ocean and Coastal Management* 26, no. 1 (1995): 2–87.
61. Coastal Zone Management Act (CZMA) 302, 16 U.S.C. 1451.

sectoral, management through the mandatory requirement that the hierarchy of coastal policy plans are consistent with the national policy statement."[62]

The RMA assigns responsibility to regional and district authorities to develop local policies and plans for resource management in combination with the national Coastal Policy Statement. Sections 59–62 outline the requirements for a regional policy statement. Section 59 explains that the purpose of regional plans is to implement the provisions of the RMA with concern for the regional issues. Section 61 lists matters to be considered in the development of a policy while section 62 states the possible contents of a policy statement. Sections 63 to 71 detail the requirements, rules and considerations for regional plans. Sections 72 to 77 outline details required and matters to be considered for territorial plans. Regional and Territorial authorities must comply with these provisions and implement policies and plans. In addition, the Ministry of Conservation has published guidelines for regional policy statements, regional plans and district plans in relation to coastal areas. They explain the purpose of the regional policy statement as follows:

> A regional policy statement gives an overview of the natural and physical resource management issues and priorities. It describes the policies and the methods that will be used to manage these resources. Each regional council must always have one policy statement.
> Regional plans deal with specific resource management issues. Regional plans are not compulsory and there may be more than one. Any person may request the preparation of a regional plan for a matter of serious concern, but the council is able to recover certain costs from the applicant for doing so.
> There must be one regional coastal plan for the coastal marine area of each region. It may be included as part of the regional plan.
> Each territorial authority must have one district plan to help them carry out their functions. The plan must not be inconsistent with any national policy statement or the regional policy statement. The district plan may include rules that prohibit, regulate or allow activities.[63]

Like the national policy statement, regional policies and plans provide important guidelines that aid in the implementation of the RMA and the principles of sustainable development and ICZM.[64]

---

62. Ibid., p. 108.
63. Ministry of the Environment, New Zealand, "The Resource Management Act: An overview." Accessed 14 April 2000: http://www.mfe.govt.nz/management/rma/rma11.htm.
64. For example, the Regional Coastal Plan for Wellington Regional Area.

Consent Process
The RMA establishes a formal consent process for any activities that exert an effect on the marine environment. The Ministry of Conservation and national, regional, and territorial bodies are responsible for establishing the consent process. Applicants are advised to follow the guidelines in their own particular region and territory.

Burton and Cocklin explain one such process concerning "a water development project in one of the country's more remote pastoral regions (the Mangakahia River case)."[65] The Mangakahia River is in the Northland Region on the North Island of New Zealand. The proposed project came from farmers concerned with improving irrigation of a dairy pasture. They wanted to divert waters from the river. The example demonstrates the process of obtaining a consent licence for a project affecting resources in New Zealand.

The process began in 1993 when several landowners "operating under the name of the Mangakahia Irrigation Committee (MIC)" submitted requests for consent.[66] The group wanted significant control over the water, and claimed that by acting as a group they would be able to "play a role in monitoring and maintaining the conditions of the resource consents (i.e., water use permits)."[67] The group applied under s. 88 of the RMA to the Northland Regional Council [NRC] to consider their proposal. Burton and Cocklin explain:

> [A]s the consent authority, the NRC was bound under Part VI, s. 104 of the RMA to consider, in making a decision, any actual or potential effects of proceeding with the proposals as well as any relevant rules, policies or objectives of a regional policy statement or plan. Section 104 further required that the consent authority must have regard for Part II of the RMA which includes the statement of purpose and principles, matters of national importance to be considered, other matters (e.g., Kaitiakitanga, intrinsic values, maintenance, and enhancement of the quality of the environment) and the principles of the Treaty of Waitangi.[68]

Once the NRC or another regional group receives an application for consent the process resembles a by-law application process in Canada. A public hearing is set and stakeholders are invited to participate in the decision-

---

65. L. Burton and L. Cocklin, "Water resource management and environmental policy reform in New Zealand: Regionalism, allocation, and indigenous relations, Part II," *Colorado Journal of International Environmental Law and Policy* 7 (1996): 331, at p. 334.
66. Ibid., at p. 337.
67. Burton and Cocklin (n. 65 above), p. 338.
68. Ibid.

making process in a highly consultative nature.[69] The Ministry of Conservation prints literature that outlines and gives examples of sample agendas and priorities for the establishment of these community hearings.[70]

In this particular case, a hearing was set in October 1993; however, several applicants withdrew and the group as a whole asked to have the hearing set back. When a new hearing was set for September 1994 "the MIC commissioned a more detailed assessment of the environmental effects of the irrigation proposals."[71] Burton and Cocklin suggest that this was done to ensure that the project would meet the policy guidelines established by the Northland Regional Council for water diversion programs.

Prior to a hearing, the regional body requests submissions related to the project. In the Mangakahia River proposal,

> [T]he Council received a total of twenty-three responses by interested parties. Seven of these submissions were in support of the proposals, most of which referred to the regional economic benefits that might be realized through the higher levels of dairy production. Among these was from a company that would supply irrigation equipment, the local milk processing company (Northlands Dairy Company), the local chamber of commerce, and agricultural consultants.[72]

Opposition came from the Maori people, who objected on the basis that "there had been inadequate consultation with the *tangata wenua*."[73]

After receiving submissions from the different groups, the body responsible for approving proposals moves to a hearing process.[74] During this time, parties can make oral submissions in an attempt to persuade the members of the merits of their position. The Northlands Regional Council accepted the MIC proposal, though the committee granted the licence on allocation rates lower than those proposed.[75]

Pursuant to sections 120 and 247–308 of the RMA, decisions by regional bodies can be appealed to the Environment Court.[76] The Environment Court "hears appeals and references on decisions made by councils, and considers applications for declarations. It is also responsible for enforcement and prosecution matters."[77] The MIC wanted an increase in the

---

69. New Zealand, "Your Guide to the Resource Management Act (Draft)," (Wellington: Ministry of Conservation, 1999), at p. 19–30.
70. Ibid.
71. Burton and Cocklin (n. 65 above), p. 339.
72. Ibid., p. 339–40.
73. Ibid., p. 340.
74. Draft RMA (n. 69 above).
75. Burton and Cocklin (n. 65 above), p. 341.
76. Formerly, the Planning Tribunal, see Resource Management Amendment Act, Statutes of New Zealand, 1996 no. 160, s. 6.
77. Draft RMA (n. 69 above), p. 21.

amount of water allocated, while the Maori wanted the decision to be overturned.[78] The appeal by the MIC was partially successful, in that water volume amounts were increased. The Maori claims were rejected.[79]

Despite the disappointing result for the Maori, the administrative consent structure created by the RMA offers two advantages not present in the Canadian resource decision-making process. First, it encourages the participation of all stakeholders and creates a forum for the articulation of viewpoints. This is very different than the Canadian process, where Aboriginals and others are usually on the outside of the resource management process and forced to resort to the courts after decisions have been made. Second, the establishment of an Environment Court ensures the existence of resource management expertise in the judiciary. This means that judicial decisions are made in the context of the principles of sustainable development and ICZM as prescribed by the RMA. In Canada, the establishment of a similar court would be beneficial to a comprehensive resource management process.

Maori Involvement in Coastal Zone Management Initiatives in New Zealand

Consultation Process
The RMA contains four provisions that ensure the participation of Maori people in resource management.[80] First, section 6(e) assures that agencies responsible for implementing the Act take into account the relationship of Maori people "and their culture and traditions with their ancestral lands, water, sites..." Second, section 7 states that the powers exercised under the Act be conducted in accordance with the principle of kaitiakitanga (meaning stewardship or guardianship).[81] Third, section 8 requires policy makers to take into account "the principles of the Treaty of Waitangi." Fourth, the first schedule of the RMA mandates the consultation of Maori people in the consent process or when establishing or changing policy statements and plans.[82] In addition to these provisions, the New Zealand National Coastal Policy Statement authorizes Maori participation in resource management.[83]

The most significant of these provisions is the consultation requirement. The duty to consult Maori authorities has been controversial and litigation in this area has been prolific. The Environment Court has identified several factors to consider in determining the level of consultation required.

78. Burton and Cocklin (n. 65 above), p. 343
79. Ibid.
80. List derived from Kahn (n. 25 above), p. 128.
81. Draft RMA (n. 69 above), p. 68.
82. Resource Management Act (n. 1 above), First Schedule s. (I)(3)(1)(d).
83. See Davidson (n. 58 above).

First, the level of consultation varies on a case by case basis.[84] Second, if a plan affects a site that is known to be a place sacred to Maori a higher level of consultation will be required.[85] Similarly, a lower level of consultation is needed when the consent application will have no perceivable impact on the Maori.[86] Third, the duty to consult can be met even if the Maori walk away from the process. In Rural Management v. Banks Peninsula District Council,[87] the court held that when the Maori withdraw from the consultation process without explanation, they cannot later return and argue that the principles of the Treaty have been violated.[88] Finally, the duty to consult does not mean that the parties have to agree with each other. Thus, in cases where clear agreement has not been obtained, the courts have granted resource development permits and licences.[89]

To ensure that proper consultation with the Maori is completed, the Ministry of Conservation has published guidelines suggesting the appropriate level of consultation for a consent application. The Ministry summarizes the case law describing appropriate consultation as follows:

The elements of consultation can be summarized as including, but not limited to, the following:

- Consultation is the statement of a proposal not yet finally decided upon.
- Consultation includes listening to what others have to say and considering responses.
- Sufficient time must be allowed and a genuine effort must be made.
- There must be enough information made available to the party obliged to consult, to enable the consultee to be adequately informed so as to be able to make intelligent and useful responses.
- The party obliged to consult must remain open minded and be ready to change and even start afresh. However, the party consulting is entitled to have a working plan already in mind.
- Consultation is an intermediate situation involving meaningful discussion.
- The party obliged to consult holds meetings, provides relevant information and further information on request, and waits until those being consulted have had a say before making a decision.

---

84. New Zealand, *Case Law on the Tangata Whenua Consultation* (Wellington: Ministry of Conservation, 1999) at 19–30 [hereinafter Draft RMA] at 12 [hereinafter Case Law].
    85. Mason-Riseborough v. Matamata-Piako District Council (1997) A143/97.
    86. Banks v. Waikato Regional Council (1995) A12/95.
    87. [1994] NZRMA 289.
    88. See Summary Case Law (n. 84 above), pp. 12–16.
    89. Mason-Risbourough v. Matamata-Piako District Council (n. 85 above).

Consultation is not

- merely telling or presenting; or
- intended to be a charade; or
- the same as negotiation, although a result of consultation could be an agreement to negotiate.

There are no universal requirements as to the form consultation must take. Any manner of oral or written interchange that allows adequate expression and consideration of views will suffice. Nor is there a universal requirement as to duration required for consultation to be adequate. Consultation could range from one telephone call to years of formal meetings.[90]

These guidelines are informative and useful for parties making a consent application.

The consultation process in New Zealand, although far from perfect, is a step in the right direction. The lack of a formal consultation process with Aboriginals in Canada means that often Aboriginal consultation occurs after the decision-making process through the courts. Furthermore, the establishment of a consultation process for consent applications means that corporations and individuals involved in resource development must consult with the Maori. Essentially, this framework has the effect of establishing procedures that can meet the fiduciary obligations of the Crown and the integration principles for ICZM.

Treaty of Waitangi (Fisheries Claims) Settlement Act[91]
The management of fisheries is outside of the auspices of the RMA. Recently, the government introduced the Treaty of Waitangi (Fisheries Claims) Settlement Act, which seeks to formalize Maori access to the fisheries of New Zealand. Davidson summarizes the purposes of the Act as follows:

(a) To give effect to the settlement of claims relating to Maori fishing rights; and
(b) To make better provision for Maori noncommercial traditional customary fishing rights and interest; and
(c) To make better provision for Maori participation in the management and conservation of New Zealand's fisheries.[92]

In financial terms, the Act mandates the transfer of 150 million New Zealand dollars to the Maori[93] as compensation, and transfers over 20 percent of the

90. *Case Law* (n. 84 above), pp. 11–12.
91. Statutes of New Zealand 1992, No. 121.
92. S. Davidson "Current legal developments: New Zealand: Maori fishing rights" *International Journal of Marine and Coastal Law* 9 (1994): 408.
93. Settlement Act (n. 91 above), s. 7.

New Zealand commercial fishery to the Maori. In exchange, Maori court actions in relation to the customary fishery are ended and the rights contained in the Treaty of Waitangi are extinguished.[94] Both the cash settlement and access to the fishery come from a government purchase of half of the quota allocations from Sealord, New Zealand's largest commercial fishing company. The Act also did not limit any further acquisition of the fishing quota by Maori fishery and "guaranteed the Maori twenty percent of any new quota issued."[95] Not surprisingly, the Act was met with controversy. On the one hand, the Maori were concerned about the implications of the extinguishment of Treaty rights. The Sealord quota purchase was below Maori negotiators' expectations. Price explains:

> [T]his Sealord deal fell short of the [Maori] negotiators' mandate to achieve the Maori goal of 50 percent ownership of commercial fishing quota, because, in the Sealord deal, the total would be half of Sealords 26 percent quota (i.e., 13 percent) plus the 10 percent already in hand from the 1989 interim fisheries settlement, in other words about 23 percent of the total commercial fishing quota in New Zealand.[96]

On the other hand, commercial fishers and industrial leaders questioned the Act as intruding on their business interests.[97]

Davidson identifies one major problem with the Act. He complains that, under section 10, the Crown through the Ministry of Agriculture is empowered to make regulations and provide for the sustainable management for the traditional Maori fishery. While the section acknowledges Treaty obligations, it states that the provisions of the Treaty have no legal effect "except to the extent that such rights or interests are provided in regulations under section 89 of the Fisheries Act."[98] Davidson comments that "the Act therefore creates the curious situation that the Crown is under obligations created by the Treaty of Waitangi, but the Maori have no enforceable legal rights unless such are implemented by the adoption of delegated legislation."[99]

---

94. Davidson (n. 92 above), pp. 410–11; Settlement Act, (n. 91 above), ss. 7, 9 and 11.
95. Kahn (n. 25 above), p. 131.
96. R. T. Price, "Assessing Modern Treaty Settlements: New Zealand's 1992 Treaty of Waitangi (Fisheries Claims) Settlement and Its Aftermath," cited in Kahn (n. 25 above), p. 131.
97. Davidson (n. 92 above), p. 408.
98. Davidson (n. 92 above), p. 411.
99. Davidson (n. 92 above), p. 411.

ABORIGINAL RIGHTS TO MARINE RESOURCES IN CANADA

Constitutional Law and Aboriginal Rights to Marine Resources

Analysis of Aboriginal Rights, Treaty Rights and Aboriginal Title
Prior to the adoption of the Constitution Act, 1982,[100] Aboriginal rights were protected only in the common law. This meant that Aboriginals in Canada faced similar problems as the Maori when attempting to assert treaty rights. Since the introduction the Constitution Act, 1982, Canadian Aboriginals enjoy Constitutional protection of their rights. Section 35 of the Constitution Act reads as follows:

> 35(1) The existing Aboriginal and treaty rights of the Aboriginal people are hereby recognized and affirmed.
> (2) In this act, Aboriginal peoples of Canada include the Indian, Inuit, and Metis peoples of Canada.
> (3) For greater certainty, in subsection (1) "treaty rights" includes rights that now exist by way of land claims agreements or may be so acquired.[101]

Combined with section 52(1), section 35 (1) ensures that legislation that infringes Aboriginal rights is "to the extent of the inconsistency, of no force or effect."[102] These two sections mean that any legislation enacted by the provincial or federal governments are subject to existing treaty and Aboriginal rights.

In several important decisions, the Supreme Court of Canada has defined the scope of section 35(1). One of the leading decisions is the 1990 Sparrow[103] decision. The case involved a member of the Musqueam Indian Band in British Columbia who was charged with fishing with a net which is contrary to restrictions imposed by the Fisheries Act. The Court held that the Crown has a responsibility to act in a fiduciary capacity with respect to Aboriginal peoples. Relying on Guerin,[104] the Court ruled:

> [T]he government has the responsibility to act in a fiduciary capacity with respect to Aboriginal peoples. The relationship between the gov-

---

100. Constitution Act, 1982, R.S.C. 1985, App. II [hereinafter Constitution Act, 1982].
101. Constitution Act, 1982 (n. 100 above), s. 35(1)
102. Constitution Act, 1982 (n. 100 above), s. 52(1).
103. [1990] 1 S.C.R. 1075, 70 D.L.R. (4th) 385. (hereinafter Sparrow–[cited to D.L.R])
104. Guerin (n. 38 above).

ernment and Aboriginals is trust-like, rather than adversarial, and contemporary recognition and affirmation of Aboriginal rights must be defined in light of this historic relationship.[105]

This decision established a four-part legal test to be used by the courts in assessing whether or not a person has a protected Aboriginal right. Lamer J. summarized the test succinctly in R. v. Van der Peet:[106]

> First, a court must determine whether an applicant has demonstrated that he or she was acting pursuant to an Aboriginal [or treaty] right. Second a court must determine whether that right has been extinguished. Third a court must determine whether that right has been infringed. Finally a court must determine whether the infringement is justified.[107]

To meet the first part of the section 35(1) test, a claimant must demonstrate, through extrinsic evidence, the existence of an Aboriginal right or a treaty right. Treaty rights are found in documents signed by the Crown and Aboriginal people, or even through documents that demonstrate a British agreement with Aboriginal people.[108] The courts have held that treaties are special agreements between natives and the Crown, and as the honour of the Crown is at stake, ambiguities should be resolved in favour of the Aboriginals.[109]

The existence of an Aboriginal right is more complex. In Van der Peet, Lamer J. explained that to be an Aboriginal right, an activity "must be an element of practice, custom or tradition *integral to the distinctive culture* of the Aboriginal group claiming the right (emphasis added)."[110] Lamer J. explained further that a claimant must establish that the activity was of central significance to the Aboriginal society prior to European contact.[111]

At the next stage of the analysis, the onus shifts to the government to show that the right has been extinguished. To prove this, the government must demonstrate that the extinguishing act occurred prior to the adoption of the Constitution Act, 1982. Courts have tended to construe Aboriginal rights broadly and require the "clear and plain" legislative intent to extin-

---

105. Sparrow (n. 103 above), p. 408.
106. [1996] 2 S.C.R. 507. (hereinafter Van der Peet)
107. Ibid. at para. 2.
108. See The Queen v. Simon, [1985] 1 S.C.R. 387; R. v. Sioui, [1980] 1 S.C.R. 1025. Although the cases were decided outside of the 35(1) context they outline factors courts will consider when determining whether a document is a treaty. See R. v. Marshall 177 D.L.R. (4th) 513 for treaty interpretation in the 35(1) context.
109. Simon (n. 108 above), p. 402; Sioui (n. 108 above), p. 1035.
110. Van der Peet (n. 106 above), para. 46.
111. Ibid., para. 60–67.

guish an Aboriginal right.[112] For example, in Sparrow, the existence of long-standing fishing regulations was found to be an insufficient extinguishing act of an Aboriginal right.

If the court finds that the right exists and it has not been extinguished, the claimant must then demonstrate that the right has been infringed. To determine if there has been infringement of a right, the court will consider several factors, including whether there has been undue hardship to Aboriginal people as a result of government legislation or regulations or other unreasonable government action.[113]

In the final step of the analysis, the court makes an assessment of whether the infringement of the right is justified. In Sparrow, Dickson J. and Laforest J. explain:

> The justification analysis would proceed as follows. First, is there a valid legislative objective? Here the court would inquire into whether the objective of Parliament in authorizing the department to enact regulations regarding fisheries is valid. The objective of the department in setting out the particular regulations would also be scrutinized. An objective aimed at preserving s. 35(1) rights by conserving and managing a natural resource, for example, would be valid. Also valid would be objectives purporting to prevent the exercising of s. 35(1) rights that would cause harm to the general populace or to Aboriginal peoples themselves, or other objectives found to be compelling and substantial.[114]

In R. v. Delgamuukw,[115] while considering Aboriginal title, Lamer J. seems to greatly increase the possible legislative objectives that would justify infringing Aboriginal rights:

> In my opinion, the development of agriculture, forestry, mining and hydroelectric power, the general economic development of the interior of British Columbia, protection of the environment or endangered species, the building of infrastructures and the settlement of foreign populations to support those aims, are the kind of objectives that are consistent with this purpose and, in principle, can justify the infringement of Aboriginal title.[116]

He explains that when there has been infringement because of a legitimate government objective, fair compensation is required.[117] Whether this will be

112. Sparrow (n. 103 above), p. 401.
113. See McLachlin in dissent, Van der Peet (n. 106 above), paras. 297–300.
114. Sparrow (n. 103 above), at 412.
115. [1997] 3 S.C.R. 1010. (hereinafter Delgamuukw)
116. Delgamuukw (n. 115 above), para. 165.
117. Ibid., para. 169.

permitted as justification for infringement of an Aboriginal right has not yet been judicially tested.

The Delgamuukw decision is of primary importance with respect to Aboriginal access to resources in Canada. The case involved the Nisga'a people of British Columbia and their attempt to assert Aboriginal title over land in Northern British Columbia. Although the case was sent back on a technicality, the Court clarified the definition and scope of Aboriginal title. The Court explained that Aboriginal title is *sui generis* [unique], and cannot be transferred or sold to anyone except the Crown. It is a burden on the Crown's title to the land and "arises from [Aboriginal] possession before the assertion of British Sovereignty."[118] Lamer J. held that the concept of Aboriginal title can be summarized by two principles:

> [F]irst Aboriginal title encompasses the right to exclusive use and occupation of the land held pursuant to that title for a variety of purposes, which need not be aspects of those Aboriginal practices, customs and traditions which are integral to distinctive Aboriginal cultures; and second that those protected uses must not be irreconcilable with the nature of the group's attachment to the land.[119]

The Supreme Court of Canada explained that because Aboriginal title is a broad entitlement to land, an infringement of title must be accompanied by fair compensation.[120]

In a sense, Aboriginal rights in Canada appear to have more protection than those in New Zealand, because they are constitutionally enshrined. However, despite these liberal decisions, the Supreme Court of Canada has not articulated the full scope of Aboriginal rights. These decisions leave plenty of room for interpretation in individual cases and suggest that governments negotiate with natives to determine the scope of the rights. As well, many of these cases are criminal or quasi-criminal in nature, challenging existing regulatory schemes or other legislation. This is a slow way to develop Aboriginal policy. Given the Court's recent decisions, it would be more effective for the federal government to develop a comprehensive policy addressing Aboriginal rights or title.[121] This could be achieved either through

---

118. Ibid., para. 112–3.
119. Ibid., para. 117.
120. Ibid., para. 169.
121. In a situation where the development of a policy has not occurred, problems are created for decision makers in relation to attempting to determine when an Aboriginal right or title is involved: see H. M. Braker and R. Freedman, "Consultation with First Nations Prior to Major Natural Resources Development and Other Projects," (Environmental Law and First Nations Conference, Pacific Business and Law Institute, Vancouver, 18 November 1999) (unpublished) at 3.12. (hereinafter Braker and Freedman)

legislation or through adaptation of clear policy guidelines in reference to Aboriginal rights.

On a positive note for the development of Aboriginal law, the decisions make it clear that any government program or legislation that infringes Aboriginal rights will be found to be unconstitutional to the extent that it offends the right. Thus, in the development of an ICZM program, Aboriginal rights must be taken into consideration. This means that the government must consult with Aboriginal peoples in the development of a program.

The Duty to Consult
Leading Supreme Court decisions regarding Aboriginal rights and Aboriginal title have emphasized the importance of consultation with Aboriginal peoples whenever governmental decisions affect their rights. Developing case law suggests that there may be a duty to consult in determining whether an infringement is justified. The emerging duty to consult is based upon the fiduciary duty of the Crown protected by section 35 of the Constitution and in common law. The Supreme Court of Canada has upheld the duty to consult in the case of both Aboriginal rights and Aboriginal title.[122] The Sparrow decision concerning Aboriginal rights explains that courts will consider consultation when determining if an infringement is justified:

> Within the analysis of justification, there are further questions to be addressed, depending on the circumstances of the inquiry. These include the questions of whether there has been as little infringement as possible in order to effect the desired result; whether, in a situation of expropriation, fair compensation is available, and whether the Aboriginal group in question has been consulted with respect to the conservation measures being implemented. The Aboriginal peoples, with their history of conservation-consciousness and interdependence with natural resources, would surely be expected, at the least, to be informed regarding the determination of an appropriate scheme for the regulation of the fisheries.[123]

In Delgamuukw, the court explains that to justify an infringement of Aboriginal title there must be consultation with Aboriginal people. Lamer J. writes:

> This aspect of Aboriginal title suggests that the fiduciary relationship between the Crown and Aboriginal people may be satisfied by involvement of Aboriginal peoples in decisions taken with respect to their lands. There is always a *duty of consultation*. (Emphasis added)[124]

122. Braker and Freedman (n. 121 above), at 3.1.
123. Sparrow (n. 103 above), at 416–17.
124. Delgamuukw (n. 115 above), para. 168.

These decisions serve as a warning to the Crown and third parties relying on Crown decisions that a failure to consult with First Nations may lead the court to conclude that an infringement of an Aboriginal right or Aboriginal title is not justified. Decisions related to resource development or allocation may be suspended until negotiations are completed. As well, government actors will be forced to compensate Aboriginals for the infringement if the infringement is not justified.[125] Therefore, the government must be careful to consult with First Nations in the development of ICZM programs.

Lamer J.'s dicta in Delgamuukw was a conscious intention to provide the means for some reconciliation between Aboriginal peoples and the Crown. It was the hope of the Court that these clear rights would encourage negotiation.[126] For many, the decision appeared

> [T]o establish new constitutional benchmarks in the relationship between the Crown and First Nations. It held that Aboriginal title is protected as a matter of constitutional right, and affirmed that the Crown is under a duty to consult with a First Nation's Aboriginal title. The Court's affirmation of the Crown's duty to consult is especially significant in light of repeated judicial calls for First Nations and the Crown not to take the institutional competence of the judiciary by excessive litigation of disputes and instead reach negotiated settlements.[127]

Unfortunately, the requirement of a duty to consult in Delgamuukw has increased litigation by Aboriginals surrounding the duty to consult. The main problem is that the scope of the duty to consult is unclear. In Delgamuukw, the Court explained that the level of consultation required varies in different circumstances:

> In occasional cases, when the breach is less serious or relatively minor, it will be no more than a duty to discuss important decisions that will be taken with respect to lands held pursuant to Aboriginal title. Of course, even in these rare cases when the minimum acceptable standard is consultation, this consultation must be in good faith, and with the intention of substantially addressing the concerns of the Aboriginal peoples whose lands are at issue. In most cases, it will be significantly deeper than mere consultation. Some cases may even require the full

---

125. For commentary on the implications of Delgamuukw, see N. Bankes, "Delgamuukw: Division of Powers and Resources" 32 UBCLR 312 (1998).
126. Delgamuukw (n. 115 above), at para 187.
127. S. Lawrence and P. Macklem, "From consultation to reconciliation: Aboriginal rights and The Crown's duty to consult," *Canadian Bar Review* 252 (2000): 254.

consent of an Aboriginal nation, particularly when provinces enact hunting and fishing regulations in relation to Aboriginal lands.[128]

Case law has not developed an acceptable test to determine if meaningful consultation has been achieved. At a minimum, it is clear from Delgamuukw that the duty is engaged whenever the Crown makes a decision that affects Aboriginal rights. The scope of that duty seems to range from mere discussions to a requirement of First Nations' consent for infringement.[129] Actual consent seems to be necessary when there is a legislative framework in place and the level of infringement is very high.[130] Significantly, for ICZM proponents Delgamuukw suggests that full consent of First Nations may also be required for hunting and fishing regulations affecting Aboriginal lands. Braker and Freedman write that "consultation arises where the Crown is implementing conservation measures . . . and it is also required where any Crown measure, such as permits and application approvals, may infringe Aboriginal rights or title."[131]

Other decisions seem to ignore the use of a sliding scale approach introduced in Delgamuukw. As Lawrence and Macklem explain, some courts only look to an acceptable minimal standard of consultation and do not even consider if full consent is required.[132] Rather, they argue that some judges seem to rely on a three-part test that was first suggested in the British Columbia Supreme Court decision of Halfway River v. British Columbia (Ministry of Forests).[133] This test requires that, in consultation, the Crown must

> (a) provide a First nation that may be affected by government legislation or a decision with "full information" on the proposed legislation or decisions (b) fully inform itself of the practices and views of the First Nation; and (c) undertake meaningful and reasonable consultation with the first nation.[134]

The first requirement suggests that the Crown should disclose all information necessary for the First Nations group to develop an informed position.[135] In the appeal of Halfway River, the British Columbia Court of Appeal

---

128. Delgamuukw (n. 115 above), at para. 168.
129. Braker and Freedman list 15 principles the courts rely upon when examining the topic of whether or not the duty to consult has been met, (n. 120 above), at 3.2–3.4.
130. Cree School Board v. Canada (Attorney General) [1998] 3 C.N.L.R. 24 (QueSC). (hereinafter Cree School Board)
131. Braker and Freedman (n. 121 above), at 3.3.
132. For a full list of cases ignoring the possibility of consultation see Lawrence and Macklem (n. 127 above), at 263.
133. [1997] 4 C.N.L.R. 45 (B.C.S.C.). (hereinafter Halfway River S.C.)
134. As adapted by Lawrence and Macklem (n. 127 above), at p. 264.
135. Ibid.

quashed the approval of a forestry permit based upon the Crown's failure to consult with First Nations before granting such permits.[136] Finch J. explained:

> The Crown's duty to consult imposes on it a positive obligation to reasonably ensure that Aboriginal peoples are provided with all necessary information in a timely way so that they have an opportunity to express their interests and concerns, and to ensure that their representations are seriously considered and, wherever possible, demonstrably integrated into the proposed plan of action.[137]

In Chelsatta Carrier Nation v. British Columbia (Environmental Assessment Act, Project Assessment Director),[138] the Court overturned the director's decision for failing to provide the Chelsatta Nation with maps regarding potential harm to wildlife in the area.[139]

The second part of the test requires that the Crown become aware of the nature of an Aboriginal right. In R. v. Bones the court chastised a DFO official for being unaware of native fishery practices.[140] As well, the Crown may be required to consult regardless of whether an Aboriginal right has been asserted. In the Halfway River British Columbia Supreme Court decision, Dorgan writes the following:

> The MOF submits that the duty to consult does not arise until the Aboriginal group has established a prima facie infringement, citing Sparrow, where consultation is not considered until the second stage of the infringement test. In my view, this approach is inconsistent with the cases referred to and is inappropriate given the relationship between the Crown and Native people.
> Based on the Jack, Noel and Delgamuukw cases, the Crown has an obligation to undertake reasonable consultation with a First Nation which may be affected by its decision. In order for the Crown to consult reasonably, it must fully inform itself of the practices and of the views of the Nation affected. In so doing, it must ensure that the group affected is provided with full information with respect to the proposed legislation or decision and its potential impact on Aboriginal rights.[141]

---

136. Halfway River S.C. (n. 133 above).
137. [1999] B.C.J. No. 1880 (QL)(C.A.), at para. 191. (hereinafter Halfway River CA) at para. 160.
138. [1998] B.C.J. No. 178(QL) (S.C.).
139. Ibid., at para. 58.
140. [1990] 4 C.N.L.R. 37 (B.C.Prov.Ct.) and see Lawrence and Macklem (n. 127 above), at 265.
141. Halfway River S.C. (n. 133 above), at para. 131–2.

The duty seems to require consultation with First Nations when there have been some negotiations, or where Aboriginal rights have been acknowledged, or where there is some legislative requirement to consult.[142]

The final requirement regarding meaningful and reasonable consultation appears to be as ambiguous as the Delgamuukw sliding scale approach. In R. v. Noel,[143] the Court explained that holding a meeting does not itself amount to meaningful consultation, and that something more is required:

> The one meeting that followed was to explain the proposed regulation and the government was made aware of the objections of the representatives of the Aboriginal people to be affected. Other alternatives were suggested which the government representatives either ignored or did not consider seriously. In my view this does not amount to meaningful consultation or a deliberation of people on a subject.[144]

The Court elaborated on the scope of meaningful consultation in the Halfway River Supreme Court decision. It was explained that simply writing letters after a decision has been made does not constitute meaningful consultation. Rather, First Nations should be involved in the decision-making process itself.[145] Courts have explained that meaningful consultation must occur early in the development of a decision affecting First Nations' rights. In R. v. Sampson, regarding a fisheries violation, the Court wrote,

> In our respectful view, the requirement of consultation, as set out in Sparrow, is not fulfilled by the DFO merely waiting for a band to raise the question of its Indian food fish requirements, discussing those requirements, and attempting to fulfil those requirements. Consultation embraces more than the foregoing. It includes being informed of the conservation measures being implemented.[146]

Regardless of the approach the courts take, Crown agents must be careful to ensure that consultation with First Nations is effected whenever a decision may affect an Aboriginal right or Aboriginal title. Otherwise, the courts may determine that a decision or a regulation is invalid. This could be a serious inconvenience to third parties relying on these decisions (for example, a forestry company awaiting a permit to log a parcel of Crown land). Braker and Freedman explain "a third party . . . is effectively caught by the consulta-

---

142. See Cree School Board (n. 130 above) and Nunavut Tunngavik Inc. v. Canada (Minister of Fisheries and Oceans) [1998] F.C.J. No. 1026 (C.A.).

143. [1995] 4 C.N.L.R. 78 (Y.T.T.C.).

144. Ibid., at 94–5.

145. Halfway River S.C. (n. 133 above), at 140–5 affirmed Halfway River CA (n. 137 above).

146. R. v. Sampson, 131 D.L.R. (4th) 192 at 220.

tion requirements in an indirect way, as the approval for a project is subject to the Crown's duty to consult."[147] Furthermore, in relation to Aboriginal title, the government may be required to pay compensation for its failure to consult Aboriginal peoples. Thus, there is a strong incentive to ensure that meaningful consultation takes place.

The courts have articulated some limitations to the duty to consult. In Gitanyow First Nation v. Canada,[148] the Court allowed an application for a declaration that Crown had a legal (not just a moral) duty to negotiate in good faith. The court made it clear that the Crown has a duty to negotiate in good faith once treaty negotiations begin, but does not have a duty to conclude a treaty.[149] In Kelly Lake Cree First Nation v. Canada (Ministry of Energy and Mines),[150] the Court stated that consultation is a two-way street, and that First Nations have a reciprocal duty to negotiate with the Crown. First Nations cannot refuse to participate and then argue that there was no adequate consultation. This was echoed in Ryan v. British Columbia,[151] where Macdonald J. rejected Aboriginal claims on the basis that the Gitskan First Nations could not argue that the Crown had failed to consult where the Gitskan had refused to speak with the Ministry of Forests. He writes,

> Instead while refusing to engage in any discourse as to the nature of the rights they claim in this area, they insist that nothing should occur with out their consent . . . Consultation did not work here because the Gitksan did not want it to work."[152]

These decisions make it clear that First Nations must participate in meaningful, good faith attempts at consultation if they wish to rely on legal remedies to enforce their rights.

The difference between the consultative models in Canada and New Zealand is significant. Under the RMA, consultation is built into the decision making process of resource management and is overseen by a specialized court. In Canada, the courts have been left to articulate the scope of consultation and there is no formal process in place. In the Delgamuukw decision the Supreme Court of Canada indicated its disapproval of First Nations and government resolving their disputes through the courts, and recommended that such disputes be resolved through negotiation.[153] However, in reality,

147. Braker and Freedman (n. 121 above), at 3.5.
148. Gitanyow First Nation v. Canada [1998] B.C.J. No. 2732, Vancouver Registry No. C981165 (November 24, 1998), (B.C.S.C.).
149. Ibid.
150. Kelly Lake Cree First Nation v. Canada (Ministry of Energy and Mines), (1998) B.C.J. No. 2471 (Oct. 23, 1998) (B.S.S.C.).
151. Ryan et al. v. Fort St. James Forest District (District Manager) [1994] B.C.J. No. 2642.
152. Ibid. at para. 21–24.
153. Delgamuukw (n. 115 above), at para. 195–6.

the decision has resulted in an increase in the amount of litigation attempting to define the level of consultation required when Aboriginal rights or Aboriginal title are affected. Adopting a regime similar to New Zealand would help to enable effective negotiation.

Coastal Zone Management in Canada

Historically, lack of political will and complex jurisdictional issues have stalled the implementation of a comprehensive ICZM program in Canada. However, as Huggett writes "it would be wrong to conclude the CZM [ICZM] does not exist in Canada, nor that the federal government is involved."[154] Programs such as the Atlantic Coastal Action Program [ACAP] are evidence of community-based ICZM programs. ACAP is a coastal management program initiated by Environment Canada in 13 Atlantic Region coastal communities through its Green Plan initiative.[155] It began in 1991 "in response to both an increasing concern by the public about the environmental quality of the Atlantic coastal zone and their [the communities] growing demand to be involved in decisions concerning their future."[156] The program's main goal is to allow communities to take responsibility for the coastal area and "assume a leadership role for the planning and management of regional coastal ecosystems throughout the Atlantic region."[157] The intended outcome of the program is the creation of a Comprehensive Environmental Management Plan (CEMP) to manage the local ecosystem. This is achieved through consensus and with the participation of local stakeholders according to the principles of ICZM. The program continues and has had some success in achieving its goals.[158]

The introduction of the Oceans Act[159] has increased the possibility of implementing a national ICZM program in Canada. The Act came into force in January 1997 and established the Department of Fisheries and Oceans (DFO) as the lead agency of ocean management. The Preamble of the Act contains references to the principles of sustainable development and Canada's commitment to ICZM.[160] It is divided in three parts: Part One asserts Canada's jurisdiction over oceans; Part Two concerns the development of

---

154. D. Huggett, "The role of federal government intervention in coastal zone planning and management," *Ocean and Coastal Management* 39 (1998): 46.
155. J. Ellsworth et al., "Canada's Atlantic Coastal Action Program: A community-based approach to collective governance" *Ocean and Coastal Management* 36 (1997): 1–3, p. 121.
156. Ibid. at p. 126.
157. Ibid.
158. Ibid.
159. Oceans Act, S.C. 1996, c. 31.
160. Ibid., at Preamble.

an Oceans Management Strategy; and Part Three consolidates federal responsibility over the oceans.

Part One of the Oceans Act, Canada's Maritime Zones, articulates Canada's jurisdiction over the oceans. The Act asserts Canada's jurisdiction over adjacent oceans as follows: the Territorial Sea extends 12 nautical miles (M) from the low water line,[161] the Contiguous Zone stretches 24 M from the low water line,[162] the Exclusive Economic Zone reaches 200 M from the baseline,[163] and the Continental Shelf jurisdiction extends as "determined in the manner under international law that results in the maximum extent of the continental shelf of Canada."[164]

Part Two of the Act, the Oceans Management Strategy (OMS), establishes the Minister of Fisheries and Oceans as responsible for developing a comprehensive OMS. Section 29 mandates that the Minster, in consultation with other federal, provincial, territorial, affected Aboriginal bodies, coastal communities, establish a national strategy:

> . . . for the management of estuarine, coastal and marine ecosystems in waters that form part of Canada or in which Canada has sovereign rights under international law.[165]

Section 30 of the Act sets out the principles for a national strategy:

(a) sustainable development, that is, development that meets the needs of the present without compromising the ability of future generations to meet their own needs;
(b) the integrated management of activities in estuaries, coastal waters and marine waters that form part of Canada or in which Canada has sovereign rights under international law; and
(c) the precautionary approach, that is, erring on the side of caution.

Despite these broad principles, the Oceans Act does not establish a clear OMS, ICZM program or a comprehensive resource management program committed to sustainable development. Rather, it is better described as a preliminary document establishing guidelines for the development of these programs. The use of the directive language "shall" in Section 29 does not mean the Minister will be legally obliged to implement a national strategy. Chao explains:

> First, the mandatory word, shall, is used only with respect to the leading and facilitating of the development and implementation of the OMS.

161. Ibid., at s. 4(a).
162. Ibid., at s. 10.
163. Ibid., at s. 13.
164. Ibid., at s. 174.
165. Ibid., at s. 29.

Secondly the provision only provides that the Minister is to "collaborate" with these other government, bodies, individuals, rather than requiring the Minister to integrate or to reach a consensus with these parties.[166]

Similarly, Sections 31 and 32, regarding the implementation of an ICZM program, are extremely vague, containing a mixture of permissive and directive commitments. There is no clear statement to direct the development of a comprehensive management strategy. This has led some to criticize the Act as failing to take the principles of ICZM seriously.[167]

Aboriginal Involvement in Coastal Zone Management Initiatives in Canada

The extent of Aboriginal involvement in coastal zone management and resource management has been extremely limited with several notable exceptions in the northern regions of Canada (due in a large part to different demographics and the remoteness of settlement areas).[168] The provisions of the Oceans Act suggest the need to consult Aboriginal people in the establishment of a national OMS and ICZM program. This is reiterated by the recent decisions compelling the Crown to ensure First Nations are consulted whenever an Aboriginal right or treaty may be affected be a decision. However, the Act establishes no clear guidelines for First Nations involvement in resource management. The lack of a clear national policy on Aboriginal involvement is problematic and needs to be overcome. The intense reaction to the implications of the Marshall and the Delgamuukw decisions underscores the need to develop these policies.

In Northern Canada, the creation of Nunavut, the Inuvialuit Final Agreement,[169] and the Nisga'a Agreement demonstrates a shift in allowing

---

166. G. Chao, "The Emergence of Integrated Coastal and Ocean Management in Canada's Ocean Act" (August 1999) (thesis, archived at Sir James Dunn Library at Dalhousie Law School) at p. 64–65.

167. See A. Chircop, et al., "Legislating for integrated marine management: Canada's proposed Oceans Act of 1996," *Canadian Yearbook of International Law* 33 (1995): 305.

168. For example, the Inuit Circumpolar Conference, which set international guidelines and created opportunities for future cooperation of governments and peoples in the Arctic. For commentary, see C. Reimer, "Moving toward co-operation: Inuit circumpolar policies and the Arctic environmental protection strategy." Accessed 27 December 2001 on the World Wide Web: http://www.carc.org/pubs/v21no4/moving.htm.

169. For a detailed analysis, see M. E. Turpel, "Aboriginal Peoples and Marine Resources: Understanding Rights, Directions for Management," in *Canadian Ocean Law and Policy*, ed. D. VanderZwaag (Toronto: Butterworths, 1992), 393–429, at pp. 412–20.

First Nations access to and control of coastal resources. These agreements may help to shape the development of similar agreements further south. The Nunavut agreement establishes a third territory in Canada.[170] The territory arose out of a settlement agreement between the Government of Canada and the Inuit people of the region. In exchange for extinguishments of land claims, Canada agreed (among other things) to compensate the Inuit through a cash settlement, a guarantee of royalties from oil, gas, and resource development on Crown land, and the creation a new territory.[171]

The Inuvialuit Final Agreement concerns a land claims settlement with the Inuvialuit people in Northwestern Arctic encompassing both the Yukon and the Northwest Territories. The agreement extinguishes Inuit claims to Aboriginal title to land in exchange for compensation and a guarantee of lands set aside "for the right of the Inuvialuit to hunt, fish, trap, and carry on commercial activity with the Inuvialuit Settlement area."[172] The agreement gives the Aboriginal people priority for a sustenance fishery and grants access to a commercial fishery. It also indicates that the Inuvialuit Game Council will receive information and advice from Hunter and Trapper Committees. In turn, the Game Council will advise and assist the Wildlife Management Advisory Council, a body established by the agreement, comprised of Inuvialuit and government representatives.[173] Although not specifically an ICZM program, the agreement allows Aboriginal input in the decision-making process of resource management.

The Nisga'a Agreement[174] is similar to the Inuvialuit Final Agreement in that it cedes full title to a territory of land and grants compensation to the Nisga'a for past infringement of Aboriginal rights. The agreement delegates some form of self-government, but provincial and federal legislation and regulations still apply to the area.[175] The Nisga'a Agreement does not establish an ICZM scheme. Rather, the agreement is designed to fit in with existing provincial and federal management processes. Thus the agreement is divided on the basis of traditional sectoral resource management with individual chapters devoted to 'forestry and other resources'; 'wildlife and migratory'; 'fisheries'; and 'environmental assessment and protection.' However, the agreement does provide a high degree of participation in resource

---

170. Nunavut Act, S.C., c. 28 (1993).
171. Ibid., s. 23. Also see "Land Claim Overview" Accessed 27 December 2001 on the World Wide Web: http://npc.nunavut.ca/eng/nunavut/ for a tertiary explanation of the agreement.
172. Preamble to the Western Arctic (Inuvialuit) Claims Settlement Act cited in Turpel (n. 169 above), at p. 413.
173. For a full explanation of the agreement, see Turpel (n. 169 above).
174. Nisga'a Treaty: Final Agreement Act, S.B.C. 1999, c. 2. See also Nisga'a Treaty: Final Agreement Act; S.C. 2000, c. 7 (2000). Accessed 27 December 2001: www.aaf.gov.bc.ca/aaf/treaty/nisgaa/nisgaa.htm.
175. Ibid., Chapter 2 s. 9, 13, but also see s. 14.

management decisions. For example, a Joint Fisheries Management Committee[176] is given vast responsibility to participate in the management of the coastal fishery in Northern British Columbia.[177]

CONCLUSION

Canada and New Zealand approach resource management in very different ways. New Zealand has implemented a comprehensive national regime that readily incorporates the perspectives of the Maori people and allows the implementation of ICZM. The Canadian experience is more fragmented. Complex jurisdictional issues and a lack of political will have led to the creation of a weak ocean management regime in the Oceans Act. Furthermore, Aboriginal peoples are often left out of the resource management decision-making process. This must change. Canada should learn from the example of New Zealand and implement a similar administrative framework. Despite the differences between the two countries, there is no reason why a similar structure, centered on resource management legislation, could not work in Canada. With political will, even difficult division of power issues could be overcome. Canada should develop a coastal management regime that

- Requires the implementation of national and regional policy statements;
- Develops comprehensive resource management plans at both the national, provincial and regional levels through incentive style legislation;
- Oversees environmental management based on an 'effects' basis;
- Takes into account Aboriginal rights to resource management and incorporates Aboriginal peoples as partners in the administration of an ICZM program; and
- Creates a national specialized court empowered to resolve disputes arising from the regulatory framework.

Adopting these measures would ensure that the principles of ICZM are attainable, and would protect First Nations' access to coastal resources.

176. Nisga'a Treaty (n. 174 above).
177. Nisga'a Treaty (n. 174 above), at Sced., Chapter 8.

Environment and Coastal Management

Integrating Risk Assessment and Management and Disaster Mitigation in Tourism Development Planning in Jamaica*

Bevon Morrison
Call Associates Consultancy Limited, Kingston, Jamaica

INTRODUCTION

Natural hazards have posed significant risks to Jamaica and other Caribbean islands. During the period 1983–89, major natural disasters in the Caribbean have resulted in loss of life, disrupted the lives of 1,845,000 people, and caused more than US$2,000 million in property damage.[1] The distribution of geological and hydrologic hazards (earthquakes, volcanic eruptions, floods, and landslides) in the Caribbean are as a result of its characteristic "on-shore and off-shore geologic-tectonic-geophysical framework, which is common to many of the island nations by virtue of their geographic location within the plate boundary zones of the Caribbean Plate."[2] Global warming and sea-level rise, El Niño and La Niña are changing weather patterns in the world and increasing the risk of natural hazards in the region.

Jamaica and the other islands of the Caribbean are also vulnerable to natural disasters because they can be affected over their entire area, and major infrastructure and economic activities may be crippled by a single event. Disasters often create irreversible damage to the natural resource base, and scarce resources earmarked for development projects have to be diverted to relief and reconstruction, setting back economic growth. Landslides and accelerated erosion caused by tropical storms and hurricanes reduce agricultural productivity and destroy marine resources. Tourism facili-

---

*This article draws upon a report prepared for the International Ocean Institute. The research reflected in this article would not be possible without the assistance of those persons interviewed as part of this study. They agreed to meet with us to share their insights and experiences. We are grateful for their time, suggestions and frank opinions. Special thanks must also be given to Mr. Sean Henry and Mr. Agostino Pinnock for their research assistance during the original project.

1. R. Ahmad, ed., *Natural Hazards in the Caribbean, Special Issue No. 12*. The Journal of the Geological Society of Jamaica, (Mona, Jamaica: Geological Society of Jamaica, 1992). 108 pp.
2. Ibid.

ties are normally sited in hazard-prone, low-lying areas on the coast, which makes them vulnerable to hurricanes, storm surges and wave action.

Two hurricanes hit Jamaica in the 1980s. In 1980, Hurricane Allen passed approximately 30 miles north of the island causing coastal damage through storm surge and wave action. Many resorts suffered significant property damage, some of which could have been avoided if risk mitigation measures were in place. In 1988, Hurricane Gilbert wreaked havoc, causing damage to infrastructure and halting various economic activities.

The Planning Institute of Jamaica (PIOJ) estimated the direct impact of Hurricane Gilbert on Jamaica. The losses were estimated at approximately US$956 million: nearly 50 percent from losses from agriculture, tourism and industry; 30 percent from housing, health, and education infrastructure and 20 percent from economic infrastructure. Economic projections for 1988 were adjusted dramatically, based on expected losses in export earnings of US$130 million and lost tourism earnings of more than US$100 million. Instead of a growth in Gross Domestic Product (GDP) of 5 percent, a decline of 2 percent was projected. Other changes induced by the disaster were expected increases in inflation (30 percent), government public expenditures (US$200 million), and public sector deficit (from 2.8 to 10.6 percent of GDP).

The underwriting losses suffered by reinsurers in the aftermath of the recent hurricanes in the region have reduced the availability and increased the cost of reinsurance. This in turn has led to a defensive reaction on the part of insurance companies and agents in the region, who are advising their clients that hurricane coverage will not be as readily available as before, and that, when available, costs will be higher, and property owners will have to bear a larger share of the risk.

As a consequence, property insurance, the traditional mechanism for reducing economic risk from catastrophic events, may no longer be as available or affordable as in the past. This development is forcing property owners and developers to seriously look at other mechanisms to minimize the consequences of natural disasters. The time has come to practice disaster loss reduction in a systematic way, as an integral part of ongoing development planning and investment. Therefore, risk analysis must become a critical part of the development process.

There are other risks that the tourism industry face, however, particularly in developing countries—social risks such as crime, drug peddling, visitor harassment and squatting—which affect the profitability and viability of the industry. These risks are not insurable and can affect the long-term sustainability of the tourism industry. Social risks, such as squatting, can result in deforestation, soil erosion, pollution of coastal water resources, and resultant damage to coral reefs, mangroves, and other marine resources. Damage to these natural protective systems increases the risks of landslides, flooding, wave action and storm surges. Therefore, it is important that risk analysis and management techniques be applied to these social risks. Figure

292 Risk Assessment and Disaster Mitigation

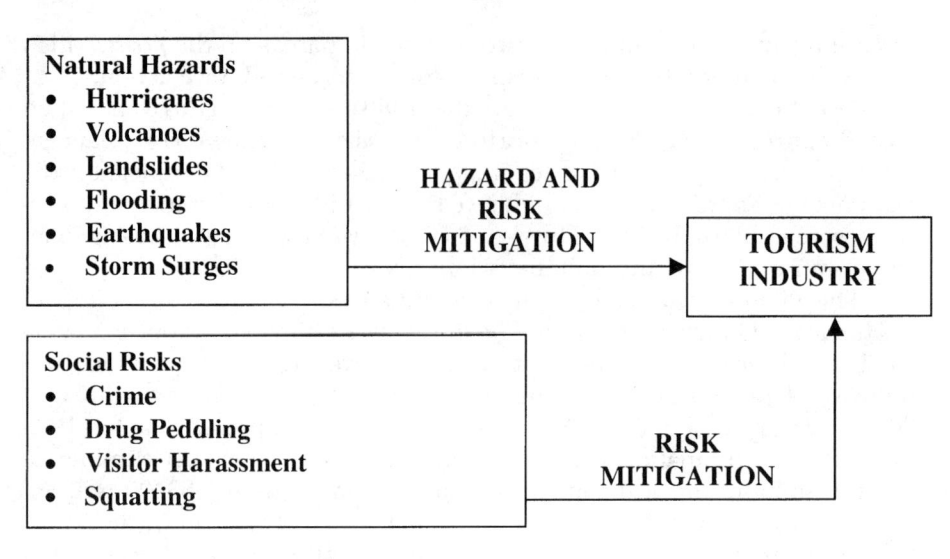

```
┌─────────────────────────┐
│ Natural Hazards         │
│  • Hurricanes           │
│  • Volcanoes            │
│  • Landslides           │    HAZARD AND
│  • Flooding             │       RISK           ┌──────────┐
│  • Earthquakes          │    MITIGATION        │ TOURISM  │
│  • Storm Surges         │ ───────────────────▶ │ INDUSTRY │
└─────────────────────────┘                      └──────────┘
┌─────────────────────────┐                           ▲
│ Social Risks            │                           │
│  • Crime                │         RISK              │
│  • Drug Peddling        │      MITIGATION           │
│  • Visitor Harassment   │ ──────────────────────────┘
│  • Squatting            │
└─────────────────────────┘
```

FIG. 1.—Natural hazards and social risks facing the tourism industry

1 summarizes the natural hazards and social risks facing the tourism industry. A summary of these issues is outlined in the section on findings.

Importance of Tourism in the Jamaican Economy

In Jamaica, export growth over the last decade has been driven by the tourist industry, and it is now the dominant export service sector. It comprises the following subsectors:

- Accommodations (hotels and non-hotels);
- Food and Beverage (restaurants, bars);
- Attractions (e.g., Dunns River Falls);
- Entertainment (music festivals, sports, and recreation);
- Agriculture;
- Transportation (taxis, car rentals, and inland air travel); and
- Shopping (in-bond facilities, clothing, craft items, and souvenirs).

The tourism sector is responsible for approximately 51 percent of Jamaica's foreign exchange earnings. In 2000, its contribution to GDP was approximately 9.2 percent.[3] The daily expenditure for the main categories of visitors

3. Planning Institute of Jamaica, *Economic and Social Survey* (Kingston, Jamaica: 2000).

TABLE 1.—AVERAGE VISITOR EXPENDITURE PER DAY, 1999–2000

Average Expenditure Per Day, US$	1999	2000	Percent Change
Stopover Visitor	99.05	98.70	−0.4
Cruise Ship Visitor	79.95	80.62	0.8

SOURCE.—Jamaica Tourist Board, *Annual Travel Statistics, 2000.* (Jamaica, 2000).

can be seen in Table 1. The figures for the stopover visitor show a slight decrease in expenditure.

In 2000, Jamaica earned approximately US$1.3 billion from the tourism sector, representing an increase of approximately 4 percent over the US$1.2 billion earned in 1999. Most of the tourists were from the United States (71.3 percent), Europe (15 percent) and Canada (8.1 percent).[4] The main resort areas in Jamaica are Montego Bay, Ocho Rios, Negril, and Port Antonio.

The tourism industry in Jamaica accounts for 13 percent of the economy's value added products,[5] employs approximately 62,000 full-time workers in the low season (May to November) and 97,000 in the high season (December to April). Estimates of both direct and in-direct employment in the tourism sector are approximately 217,000 equivalent full-time jobs, or approximately 23 percent of total employment.

A review of the Jamaica Tourist Board's annual tourism statistics indicates that during 2000, visitor arrivals to Jamaica numbered 1,322,690, representing an increase of 6.0 percent compared with 1999's visitor arrivals of 1,248,397. Both major categories of visitors posted record performances during the year. Cruise arrivals showed an 18.7 percent increase over 1999's visitor arrival of 764,341 to 907,611 in 2000. The foreign national stopover visitor category increased by 6.3 percent from 1,147,135 in 1999 to 1,219,311 in 2000. Visitors from the United States amounted to 942,561 in 2000 (up 8.3 percent) compared with 870,019 in 1999. There was a decline in most of the major markets. There were declines in visitors from Europe, Latin America and Japan.

Jamaica has proved to be a popular stopover on itineraries of several cruise companies. The island has deep-water ports in Kingston, Montego Bay, Ocho Rios, and Port Antonio. Montego Bay handles approximately 25 percent of cruise arrivals and has five berths plus a modern terminal building. Ocho Rios is favoured for day stops and has three deep-water piers, with

4. Jamaica Tourism Board, *Annual Travel Statistics.* (2000).
5. World Bank, *Jamaica Public Expenditure Review* (Washington DC: The World Bank, 1996).

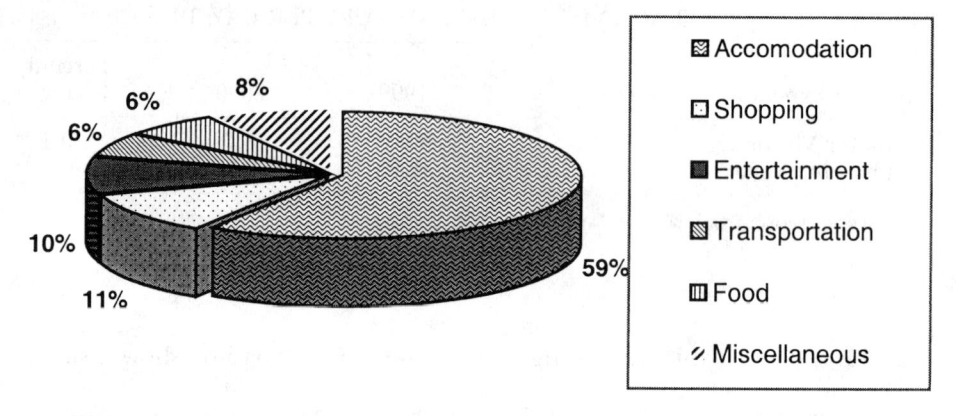

FIG. 2.—Distribution of stopover visitor expenditures in 2000

capacity to handle smaller ships. Kingston, with the seventh largest harbour in the world, is hoping to re-establish itself as a port of call.

Global receipts from visitors make tourism one of the largest sectors in the Jamaican economy. The direct value added (GDP) is approximately one-third of tourism expenditure.[6] Visitor expenditures by stopover visitors were:

(a) Accommodations;
(b) Food and Beverages;
(c) Transportation;
(d) Shopping; and
(e) Miscellaneous items.

In 2000, their average expenditure expressed as a percentage can be seen in Figure 2.

The Attractions subsector is an important subsector both in terms of investments and as a foreign exchange earner. In a recent survey conducted on 31 attractions in Jamaica,[7] the value of these 31 attractions was in excess of US$18.2 million. Projected earnings from these 31 attractions was expected to be US$4.54 million for the fiscal year 1998–99.[8]

6. Organization of American States, *Economic Analysis of Tourism in Jamaica*, Technical Report of the OAS National Programme of Technical Cooperation with the Jamaica Tourist Board and the Ministry of Industry, Tourism and Commerce. (Washington DC: Department of Regional Development and Environment Executive Secretariat for Economic and Social Affairs, General Secretariat, Organization of American States, 1994).
7. L. Dunn, *Tourism Attractions: A Critical Analysis of this Sub-Sector in Jamaica* (Mona: Canoe Press, University of the West Indies, 1997).
8. Ibid.

The World Tourism Organization (WTO) indicates that there is a shift in the tastes of tourists from sun, sand, and sea tourism to special interest areas such as eco-tourism, health tourism, and other types of tourism that allow them to interface more closely with the culture and people at the destination. This type of tourism necessitates increased interaction with local communities.

Purpose of the Research Project

The purpose of this article is to assess how risk analysis and management may be incorporated in tourism development planning, particularly as it relates to natural hazard and social risks.

Definitions

For the purposes of this article the following terms are defined: *Risk* is the potential damage that could arise as a result of the occurrence of a hazard, with a given degree of uncertainty. There is also a need to distinguish between risk and hazard. In general terms, *hazard* may be defined as source of danger. In simple terms therefore, a hazard is a source of danger, while risk involves the likelihood of a hazard developing into some adverse occurrence that may cause loss, injury, or some other form of damage. Risk may also be defined as:

$$\text{Risk} = \text{Hazard} \times \text{Probability of Occurrence}$$

It should be noted that the consequences of risk may be contained if safeguards are put in place. However, hazards cannot be reduced to zero unless the hazard itself is removed. Further, it is necessary to define the terms risk analysis and risk assessment. Specifically, *risk analysis* allows for an evaluation of what hazard can occur, the likelihood of its occurrence (probability of occurrence) and the consequences of its occurrence. *Risk assessment* takes this one step further to address the importance of these consequences, if they do occur. Based on these definitions, it can be seen that risk analysis may be carried out in an objective manner, while risk assessment is much more subjective and should include public policy makers. Finally, it is necessary to include the term *risk management*. This stage of the process includes all of the actions required to quantify, mitigate, and control risk.

In summary, the risk analysis and assessment phases require the interaction of scientific analysts with public policy makers. The scientific analyst provides the required information and puts the risk analysis in context. It is imperative that the policy makers understand fully the nature of the risks and the cost implications of alternative remedial courses of action, in order to represent these properly to the population at large.

One other term that must be introduced is that of *vulnerability*. This is defined, from a technical standpoint, as the proportion (as a percent or as an index from 0 to 1) of what could be damaged (human life, property, etc.) in a given place in the case of the occurrence of a given natural phenomenon.

Typically, several criteria can be used to determine levels of vulnerability. These tend to include population density and annual growth rates, Human Development Indicators set against long-term urban growth rates, and real adjusted GDP per inhabitant set against illiteracy percentages, or set against child mortality rates. Of these criteria, population density and growth rates provide a "first-cut" assessment of a country's vulnerability, on the basis that countries with higher population densities are more vulnerable. Another major vulnerability criterion is poverty, which is useful in characterizing the sectors of a society that are most vulnerable to disasters. When vulnerability data are compared with natural hazard information, then it becomes possible to define potential risk levels.

Another risk that this article evaluates are *social risks*, primarily as they apply to the tourism industry. These risks are man-made and are normally a result of

   (a) poverty,
   (b) inadequate access to housing, education and medical care, and
   (c) lack of economic opportunities.

They include crime, visitor harassment, squatting, etc. Social risks directly and indirectly affect the industry and may be increased during episodes of natural disasters.

APPROACH AND METHODOLOGY

The researcher and author conducted a literature review of the natural hazard and mitigation studies undertaken by the Caribbean Disaster Management Programme sponsored by the Organization of American States, studies conducted for the Master Plan for Sustainable Tourism Development being conducted by the Office of the Prime Minister (Tourism Division), and other relevant information. The article relies on a wide range of studies.[9]

9. Disaster Risk Reduction as a Development Strategy by Jan C. Vermeiren (Organization of American States [OAS]/Caribbean Disaster Mitigation Program); Master Plan for Sustainable Tourism Development—Planning Issues. Prepared for the Office of the Prime Minister, Tourism Division by Pauline McHardy; Insurance, Reinsurance and Catastrophe Protection in the Caribbean, prepared by the World

Interviews and discussions were held with the main stakeholders involved in risk analysis and management and tourism development planning in Jamaica including the Office of the Prime Minister (Tourism Division), Tourism Product Development Company (TPDCO), Jamaica Hotel and Tourist Association (JHTA), Jamaica Tourist Board (JTB), Jamaica Promotions Ltd. (JAMPRO), Sandals Resorts International, Office of Disaster Preparedness and Emergency Management (ODPEM), Natural Resources Conservation Authority (NRCA), Jamaica Association of General Insurance Companies (JAGIC) and Town Planning Department, Negril Environmental Protection Trust (NEPT), and Port Royal Development Company (PRDC). The study was originally conducted over a four-week period during January 2000. The paper was revised in September 2001.

INTEGRATED COASTAL ZONE MANAGEMENT IN JAMAICA

Jamaica is located in the northwestern Caribbean and is the third largest country in the region. Jamaica's population is 2.5 million and approximately 65 percent of the population live within 5 km of the coast.[10] Jamaica's coastline is 795 kilometers (494 miles) long and is highly irregular with diverse ecosystems, including bays, beaches, rocky shores, estuaries, wetlands, cays, seagrass beds, and coral reefs. Most of Jamaica's tourism facilities are located along the coast. Coastal and marine ecosystems protect land-based communities from natural disasters and act as stabilizers of global systems, especially our climate.

Expanding urbanization along the coast and on steep slopes increases pollution of coastal waters and destroys natural protective systems such as forests, coral reef, and mangroves. This increases Jamaica's coastal vulnerability to natural disasters. Direct damage from Hurricane Gilbert, which hit Jamaica in 1988, stood at US$956 million. Roughly 50 percent of beaches were seriously eroded, 60 percent of mangrove trees were lost, 50 percent

---

Bank and the OAS/Caribbean Disaster Mitigation Program; Master Plan For the Sustainable Tourism Development—Social and Economic Integration of Local Communities with the Tourism Industry. Prepared for the Office of the Prime Minister (Tourism Division) by Call Associates Consultancy Ltd.; The Role of Science and Technology in Poverty Eradication. Prepared by Bevon V. Morrison for the Office of the Prime Minister and the Organization of American States; Growth in Squatter and Informal Areas in Montego Bay, Jamaica, 1958 to 1978 by Alan Eyre; Report on Visitor Harassment and Attitudes to Tourism and Tourists in Negril by Dunn and Dunn (1994). Submitted to the Tourism Action Plan, Jamaica Tourist Board and the Negril Resort Board; Natural Hazards in the Caribbean (1992), The Journal of the Geological Society of Jamaica, edited by Rafi Ahmad.

10. Natural Resources Conservation Authority. *The State of the Environment,* 1997 Report. Kingston, Jamaica.

of the oyster resources were unsalvageable, marine water quality deteriorated, and landslides were widespread.[11]

Jamaica does not have an integrated coastal zone management plan. However, it has taken the following initiatives:

1. established a multi-sectoral Commission on Ocean and Coastal Zone Management (including the tourism sector);
2. developed a Manual for Integrated Coastal Planning and Management in Jamaica;
3. developed several coastal protection policies; and
4. pursued an Integrated Coastal Zone Management Approach.

The policies developed to date include the National Policy for the Conservation of Seagrass, Mangrove and Coastal Wetlands Protection Draft Policy/Regulations, the Beach Policy, National Policy on Mariculture and Policy for the Protection of Mangroves, Coral Reef Protection and Preservation Policy and Regulations (Draft).

In these policies, the government of Jamaica recognizes that construction in coastal high hazard areas, including coastal wetlands, increases the risk of property damage and personal injury. It is recommended that hazard mitigation techniques be incorporated in the site plan and structural design of developments approved for wetlands and/or adjacent areas. However, there are few initiatives to implement disaster management and mitigation measures in integrated coastal zone management planning.

The Historical Development of Tourism in Jamaica

An outline of the historical development of tourism is necessary to understand the pattern of development in the industry and the basis for its vulnerability to natural hazards and social risks. It is argued that tourism in Jamaica had its genesis in the 17th century. At that time, families who did not live in Jamaica owned and maintained large residences on estates, which they periodically visited with friends. The first legislation to promote tourism in Jamaica was enacted as early as 1890. Fiscal incentives, which spoke to the issue of low interest rates, and duty free imports on building materials and furnishings, were granted to encourage the construction of approved hotels to accommodate visitors attending the Jamaica International Exhibition of 1891.[12] As a result of this first Jamaica Hotel Law, four new hotels were con-

11. United Nations Environment Programme, *Assessment of the Economic Impact of Hurricane Gilbert on Coastal and Marine Resources in Jamaica*. Regional Seas Reports and Studies No. 110. (Kingston: UNEP, 1989).
12. Town Planning Department, *National Atlas of Jamaica*. (Kingston, 1989).

structed. In the years preceding the outbreak of war in 1914, visitor arrivals were quite low, scarcely reaching 10,000 per year.[13]

The hotel industry began to flourish significantly after World War I and this resulted in a diversification of the Jamaican economy. In 1922, the Jamaica Tourist Trade Development Board (JTTDB) was established and was responsible for the promotion and development of tourism. The JTTDB was later replaced in 1954 by the Jamaica Tourist Board (JTB), which was reconstituted in 1963 with its members being directly appointed by the government of Jamaica and its scope of operations widened. Currently, the JTB is involved in all aspects of inspecting and licensing of the tourism product and its marketing. While the JTB has become the agency responsible for the promotion and marketing of tourism, its role in economic and physical development planning has not been as important.

The Hotels Incentive Act of 1968[14] and Resort Cottages Incentive Act of 1971[15] mirrored the first legislation of 1890, with an expansion to the list of capital items for the construction and furnishing of the properties. This further encouraged the development of the industry. However, no concurrent structured development and social plans were done for the fledgling industry.

By the 1970s, total visitor arrivals (including stopovers, visitors and cruise ship passengers) had reached nearly 500,000 per year. With the 1970s depression in the tourism industry, the government of Jamaica saw it prudent to establish the National Hotel and Property Ltd., a subsidiary of the Urban Development Corporation (UDC) in 1973. Through this organization, the government constructed and bought hotels, which not only illustrated a confidence in the industry but also encouraged local investment in the sector.

Although never adopted, the National Physical Development Plans for 1970–1990 and 1978–1998 outlined tourism development plans for the island. Both plans recommended the following goals:

- To carry out the development of tourism in a manner that would create minimum disturbance to the natural environment;
- To link tourism development with the creation of new jobs;
- To use tourism development as a tool to supplement and augment existing towns and villages. However, the mechanisms to achieve this were never detailed and any recommendations made were that these mechanisms be incorporated in regional and urban planning;
- To stage and coordinate the planning and development of tourist

---

13. D. Buisseret, *Historic Jamaica—From the Air* (Kingston: Ian Randle Publishers, 1996).

14. The Hotels (Incentives) Act of 1968, Vol. X, Revised Laws of Jamaica.

15. Resort Cottages (Incentives) Act of 1971, Vol. XXIV, Revised Laws of Jamaica.

accommodation and facilities with the improvement of ground transport, the provision of water supplies, and all utilities and services required by the support population;
- To transform resort centres into more interactive communities and thus de-emphasise the demarcating line between tourist facilities and local community facilities; and
- To aid in reducing the level of economic evaporation from tourist expenditure so as to ensure that a higher percentage of the tourist dollar remains in the country and in particular within the area where it is spent.

Whereas the tourism industry recognized the importance of disaster management, there were no strategic initiatives by government or the private sector to incorporate hazard and risk mitigation measures throughout the industry.

There were few strategic initiatives to deal with social risks. It is obvious that if the mechanisms had been developed to ensure that tourism development was used as a tool to augment existing towns and villages, there would have been greater social and economic integration of local communities with the tourism industry, resulting in lowered social risks. However, there was little attempt to transform resort centres into more interactive communities. Therefore, tourism enclaves developed that increased the demarcation between tourist facilities and local communities.

In 1980, the first Ministry of Tourism was established. In the late 1980s, the hotels that the government acquired in the 1970s were divested to the private sector. In 1988, the Tourism Action Plan Ltd. (TAP) was established as a private limited company formed through a joint venture arrangement between the government of Jamaica and the Jamaica Hotel and Tourist Association. The TAP was charged with the responsibility of assisting with the development of tourism in Jamaica. The organization's coordinating and facilitating role stimulated joint action between public and private sector interests.

In 1994, TAP was reorganized as TPDCO, a public sector company with responsibility for the development, promotion and maintenance of standards and guidelines for the industry; inspection and licensing of all tourism entities; human resource development; and special projects such as beautification and infrastructural works. In the same year, the Resort Boards were revived in each resort area to oversee tourism activities and to manage certain projects. It provided a mechanism through which the community could voice its opinion on product development and marketing. However, it did not facilitate broad-based participation of community members in tourism planning and development.

Today, the tourism trade in Jamaica is a year-round activity and most of the tourists are from the United States, Canada, and Europe. The visitors to Jamaica are demanding harassment-free vacations in a country that focuses on sustainable tourism development through social integration with local communities. Therefore, a sharp contrast exists between a product that was offered for three decades, and what is demanded at present. Jamaica has to constantly examine its development strategies in order to ensure that the balance exists between developing local communities and the demands made by international tourists.

Sustainable Tourism Development: The Evolving Pattern of Tourism in Island Communities

Tourism is based on a clean, healthy environment and stable socio-economic conditions. There is a growing recognition among tourism practitioners that tourism development and practices need to be redefined to ensure the industry's survival. The WTO defines sustainable tourism

> ... as tourism that meets the needs of present tourists and host regions while protecting and enhancing opportunities for the future. It should encourage the management of all resources in such a way that economic, social and aesthetic needs are fulfilled while maintaining cultural integrity, essential ecological processes, biological diversity and life support systems.[16]

As part of the criteria for sustainable tourism, sustainable tourism products are operated in harmony with the environment, local communities and cultures, so that these become the permanent beneficiaries and not the victims of tourism development. The WTO's guidelines for sustainable tourism development are:

   a. travel and tourism should assist people in leading healthy and productive lives in harmony with nature;
   b. tourism development should be appropriate to place, reflecting its scale and character;
   c. environmental protection (including natural hazard planning and mitigation) should constitute an integral part of the tourism development process;
   d. sustainable tourism development should ensure a fair distribution of benefits and costs;
   e. tourism development issues should be handled with the participation of all stakeholders, with decisions being taken at the local level;

---

16. World Tourism Organisation, *Changes in Leisure Time: The Impact on Tourism* (Madrid, 1998).

f. travel and tourism should use its capacity to create employment for women and local people to the fullest extent; and
g. tourism development should recognize and support the identity, culture and interests of local people.

While the concept may be well defined, it is harder to apply in practice. Sustainable tourism development involves making hard choices based on complex social, economic, and environmental trade-offs.[17] Despite these challenges, the government of Jamaica has indicated that it is embarking on a sustainable tourism development path.

Other patterns that have been emerging in small island states include nature tourism, cultural and historical attractions and eco-tourism. For example, in Jamaica, the Blue and John Crow Mountain National Park, the Montego Bay Marine Park, and Port Royal Heritage Trust can be utilized as historical and eco-tourism sites.

Social and Economic Context

In order to understand the context in which the modern tourism industry now operates, it is important to understand the societal context within which it has developed. One might argue that the social and economic issues facing the tourism industry such as squatting, illegal vending, access to beaches, visitor harassment are due to a complex set of interacting factors that are operating in the country. These factors are based on:

(a) the socio-cultural development of the modern Jamaican society;
(b) economic problems facing the country as a result of the structural adjustment programme (SAP) and the concomitant economic policies it has undergone over the last two decades; and
(c) an approach taken by successive governments to develop the industry as a separate sector rather than one which is fully integrated into the economy.

Pre-emancipation Jamaica was primarily an agrarian society in which slave labour was the predominant means by which plantation owners produced sugarcane on large acreages of land. These plantation owners were white and were mainly from England. During this period, "slaves generally received all their income in kind."[18] They were also given a small plot of

17. Halcrow Ltd., *South Coast Sustainable Development Study*, Technical Report 8–Tourism and South Coast Sustainable Development Study, 1998.
18. G. Eisner, *Jamaica, 1830–1930, A Study in Economic Growth* (Westport: Greenwood Press, 1961).

land—provision lands—to produce staple foods. In 1834, slavery was abolished.

Ken Post, in his analysis[19] of the modern Jamaican society argues, that when the four-year apprenticeship, which was the last phase of slavery, ended 1st August 1838, the old Jamaica had been struck a deadly blow. There was some reluctance on the part of the ex-slaves to continue to work for their former masters. Most of the ex-slaves cherished the thought of the provision grounds that they had tilled for themselves while in servitude and now sought land to become farmers in their own right. After slavery, however, the peasants were essentially a landless, disenfranchised proletariat. Most of them had to resort to squatting on lands in order to produce crops and to provide housing for themselves.

The ex-slaves sought to turn the Jamaican economy into one dominated by independent peasants. Small cane growers came to play a significant part in the sugar production, though until the 20th century they met local demands rather than those for exports.

During the post-emancipation period, the class/caste system that had developed and had been maintained during slavery was replaced by a different class structure. As early as the 1850s, differentiation was appearing among the new peasants based on those who specialized in export crops and sugar for local consumption, and those who grew food crops purely for local use. Post argues that an observer noted in 1860 that "A sort of better class is . . . gradually emerging from the masses." Whereas the planters suffered after emancipation, the urban-based merchants tended to prosper because of an increased demand for commodities, and soon some began to put their capital into land, including some of the Jewish merchants and also a number of the free brown elements (mulattos). After 1865 these trends continued. So too, it must be emphasized, did the development of what we must begin to discuss as poor, middle and rich peasants.[20]

Post argues that the hybrid relations of production in the countryside, whereby poor peasants were poised between working their own plots and selling their labour to the big estates or government bodies, coupled with the deliberate policy of rotating work to new employees over short periods of time by the government and the capitalists, led to the consolidation of a casual labour market. This trend continues today.

Between 1880 and 1920, more than 146,000 Jamaicans—practically all from rural Jamaica—moved overseas in search of employment.[21] Emigration was their response to poverty. The large rural–urban migration which re-

---

19. K. Post, *Arise Ye Starlings–The Jamaican Labour Rebellion of 1938 and its Aftermath* (London: Martinus Nijhoff, 1978).
20. *Ibid.*
21. In 1881, the total population in Jamaica was 580,000 and in 1920 it was 858,000. Eisner (n. 18 above).

sulted in 69,500 persons moving to Kingston and St. Andrew and other urban towns during this period, meant that large numbers of the urban unemployed could not fall back on their own land.[22] Therefore, they were compelled to engage in "hustlings" such as collecting and selling fruit, peddling sweets and cigarettes, shoe-cleaning, chopping firewood and so on. Vending became an important supplement to their incomes. Of course, vending was something that was rooted into the tradition of the Jamaican people, because as slaves they were allowed to farm their plots and sell their produce at Sunday Market. In fact, the slaves from West Africa (mainly women) had a long history in trading goods. These women were aggressive salespersons, a trait that still persists today. Therefore, some of the "aggressive" behaviour that is considered "harassment" in the tourism industry (such as pressure to buy things that the tourists do not wish) should rather be seen as traditional means of selling goods which have been further exacerbated by the poor economic conditions in Jamaica.

In the late 19th century, banana production became important to Jamaica due in part to the efforts of Captain D. Baker and Captain Busch, who settled in Port Antonio. They proved that bananas could be shipped to the USA without spoilage. The development of the banana industry in the early 19th century augmented the development of the tourism industry in Jamaica.

The rapid economic growth that took place in the 1950s and early 1960s led to significant transformations in the island's economy and social structure. Ambursley[23] argues that in 1948 traditional exports—sugar, bananas and other agricultural exports—accounted for 96 percent of visible export receipts. Twenty years later, their share in the value of visible exports had shrunk to 37 percent, and as a proportion of total export value, it was only 22 percent. This was due to the growth of three main exports: bauxite/alumina, tourism and manufactured goods. In 1968, tourism accounted for 22 percent of the value of export receipts. During this period, foreign trade increased eightfold, nominal GDP grew sevenfold and per capita national income also grew by 700 percent.

The economic importance of the new growth sectors did not result in a significant reduction in unemployment, and the Jamaican economy failed to achieve self-sustained economic growth. During the late 1970s to mid-1990s, Jamaica embarked on Structural Adjustment Programmes (SAPs). This resulted in successive devaluations, reduced capital expenditure on economic, and social infrastructure, and reduced public sector expenditure.[24]

22. P. Sherlock and H. Bennett, *The Story of the Jamaican People* (Kingston: Ian Randle Publishers, 1998); F. Ambursley and R. Cohen, *Crisis in the Caribbean* (Kingston: Heinemann, 1983).

23. Ambursley and Cohen (n. 22 above).

24. K. Levitt, *The Origins and Consequences of Jamaica's Debt Crisis 1970–1990*, (Kingston: Consortium Graduate School of the Social Sciences, 1991).

During this period, there was growth in the informal sector. Unemployment remains a seemingly intractable problem for the Jamaican economy. Palmer[25] attributes this to the fact that "[t]he Caribbean has been unable to achieve the kind of economic development that would widen the range of job opportunities and allow its economy to absorb the incremental growth of its labour force." The Jamaican scenario only served to lend more credence to this belief as the process of structural adjustment only exacerbated this preexisting problem.

Growth in the Informal Sector

The Informal Sector in Jamaica is defined as small businesses employing less than 10 workers each.[26] The seasonality of demand within the tourism and agricultural sectors saw the emergence of large pockets of skilled and unskilled workers being unemployed during off-seasons. The effect of Jamaica's SAPs, occupational changes and displacement of residents as a result of drastic changes in economic activities in certain communities (e.g., the conversion of Ocho Rios from a fishing village to a resort area) has accelerated the growth of the informal sector. With reference to structural adjustment, Witter and Anderson[27] posit that:

> The selective pressures and incentives which the structural adjustment programme imposed on different sectors of the economy had their counterpart in a tremendous reorganization in modes of livelihood among the Jamaican people.

Similarly, the World Bank also views the adverse economic conditions during the past two decades, as well as the reduction in public sector employment which often follows such structural changes as the two central phenomena behind the increased activity in the micro-enterprises sector.[28] The businesses may be categorised in two ways:

1. those operated by the poor for subsistence purposes; and
2. some establishments that are relatively stable and financially viable family businesses, which have the potential to move up to the small-scale end of the formal sector.

---

25. R. Palmer, *In Search of a Better Life* (New York: Praeger, 1990).
26. World Bank (n. 5 above).
27. P. Anderson, *Consequences of Structural Adjustment: A Review of the Jamaican Experience* (Mona: The Consortium Graduate School of Social Sciences. Canoe Press, 1994).
28. World Bank (n. 5 above).

Most entrepreneurs in this sector exhibit a high level of commitment to their businesses. Women operate at the smaller end of the sector compared with males, and are concentrated within a narrower range of activities, such as the sale of merchandise. The expansion of the informal sector is constrained by access to credit from financial institutions, the high lending rates of domestic financial institutions, and inadequate market access.[29] Players in the informal sector attempt to solve the problems related to credit access by borrowing from friends and family or "playing partner."

Available estimates indicate that the growth rates of the informal sector's employment and income during the past fifteen years were more than 5 percent and 2 percent per year respectively. In contrast, the formal sector's employment and GDP growth rates during the same period were about 2.2 percent and 1.6 percent per year respectively.[30] The sector is dominated by sole proprietors engaged in import and/or production and sale of clothes and garments, food, furniture, plastic household items (cups, plates, buckets), toiletry, perfumes, jewellery, plastic footwear, and small electrical parts. To a lesser extent, activities in the transport and services (e.g., repair or domestic services) sectors are also a source of employment and income.

Current Macro-economic Trends

The country has been making steady progress toward transforming its economy into a more market-oriented, export-led system with private investment as the engine of growth. Despite the macro-economic progress that Jamaica has experienced, real growth in GDP has been slow, averaging only 1 percent during the 1991–1994 period and negative growth of −1.7 percent in 1996.[31] The two main reasons for this low growth are the high interest rate regime and inflation during this period. These factors increased economic uncertainty and diverted private resources towards short-term financial instruments with higher returns. Formal unemployment currently stands at approximately 16 percent and one-third of the population lives in poverty.[32]

There has been a major shift from manufacturing to services both in terms of GDP and employment. In 1996, the contribution to GDP from the service sector was 76.7 percent and its contribution to employment was approximately 57.7 percent. Central to the expansion of the service sector is the growth in the tourism sector.

The National Industrial Policy of Jamaica targets tourism as a strategic

29. Ibid.
30. Ibid.
31. Planning Institute of Jamaica, *Economic and Social Survey*. (Kingston, Jamaica, 1996).
32. Ibid.

sector for growth and development in Jamaica. This is due not only because of its contribution to foreign exchange earnings but because of its direct linkages with the Agricultural, Manufacturing, Music and Entertainment sectors of the economy, and hence its potential to assist macro-economic growth. The gross foreign exchange earnings from tourism during 1997 were estimated at US$11,140 million, an increase of 4.4 percent over 1996.[33]

The centrality of the tourism industry to national development is reflected in the large number of persons who make a living from the sector: hoteliers, tourists, government, taxi drivers, shop-owners, craft vendors, restaurant owners and tour guides. However, the lure of employment opportunities without adequate infrastructure to accommodate persons migrating to the main resort areas may cause several social problems in the communities around the resort towns.

The Risk of Natural Disasters: Direct and Indirect

By their location and physiographic nature, the Caribbean Basin nations are subject to strong atmospheric, hydrologic and geologic extremes. Meteorologic hazards such as tropical storms and hurricanes may pose the most frequent threats, yet earthquakes and volcanic eruptions were responsible for the greatest loss of life during the modern history of the Caribbean.[34] Heavy rains carried by tropical storms and hurricanes, with storm surges compounding the situation in coastal areas, generally trigger flooding. Drought occurs when normal rainfall patterns are disrupted over extended periods and the combination of steep topography and unstable soils found in much of the countries in the region can lead to severe erosion and frequent landslides.

The number of people affected by disasters and their losses are on the increase worldwide. In the period 1910–1930, North Atlantic hurricanes averaged 3.5 per year, and increased to an average of 6.0 per year between 1944–1980. The Office of Foreign Disaster Assistance of the United States Agency for International Development (USAID) has been collecting yearly figures of these damages. According to their figures, an estimated 574 million people have been affected during the six years from 1980–1986, principally by drought, floods, and tropical cyclones, compared with 277 million people affected by disasters during the sixties.[35] In these six years, total losses

---

33. Planning Institute of Jamaica. *Economic and Social Survey Jamaica 1997.* (Kingston, Jamaica, 1997).
34. J. Tomblin, "Earthquakes, Volcanoes and Hurricanes: A Review of Natural Hazards and Vulnerability in the West Indies," *Ambio* 10, 6 (1981): 340–345.
35. Office of Foreign Disaster Assistance (OFDA), *Disaster History: Significant Data on Major Disasters Worldwide, 1900–Present* (Washington, D.C.: U.S. Agency for International Development, 1988).

of US$75.4 billion were reported, excluding losses in the USA and USSR. Natural disasters directly threaten a country's development strategy and socio-economic performance by destroying infrastructure and productive capacity, interrupting the production process, and creating irreversible changes in the natural resource base.

The United Nations Disaster Relief Organization distinguishes between direct, indirect and secondary effects of a disaster in evaluating its impact on a country.[36] This classification has been adapted by the Economic Commission for Latin America and the Caribbean (ECLAC) in its assessment of sectoral damages and impact on overall economic performance and living conditions with the following definitions:[37]

- The direct effects on the property of state, business enterprises and population affected by the disaster: these include damage to social and economic infrastructure and losses of capital stock and inventories;
- The indirect effects which result from the decline in production and in the provision of services: these include loss of revenue due to the disruption of production and services, and increased costs of goods and services;
- The secondary effects which may appear some time after the disaster: decreases in economic growth and development, increased inflation, balance of payment problems, increases in fiscal expenditures and deficit, decreases in monetary reserves, etc.; and
- Disasters may weaken a system leaving it more vulnerable to disease and future disasters.

In focusing on the economic and social aspects of the impact of natural disasters, the ECLAC definition fails to take into account the effect of disasters on the resource base and the corresponding impact on development, or in other words, the linkage between environment and economic development.

Insurance Industry Practice—Natural Hazards and Social Risks

Accelerated rise in sea level is one of the more certain responses to global warming and presents a major challenge to coastal resort areas. Rising sea level causes erosion, submergence, salination and a greater risk of impacts

---

36. United Nations Disaster Relief Organization (UNDRO), *Disaster Prevention and Mitigation*, Vol. 7. (New York, N.Y.: United Nations, 1979).

37. R. Jovel, *Economic and Social Consequences of Natural Disasters in Latin America and the Caribbean* (Santiago, Chile: United Nations Economic Commission for Latin America and the Caribbean (UN-ECLAC), 1989).

from flooding and storms.[38] The relative or actual sea-level rise has often been much higher for many coastal areas because of the additional problem of land subsidence. Thus, without any change in present trends, significant problems exist in coastal zones. While the magnitude of future sea-level change is uncertain, the consensus is that the sea will rise in response to global warming.

In recent years, the Caribbean area has had its share of natural disasters and is at risk from the hazards associated with sea-level rise. As a result of the many hurricanes, volcanic eruptions, and earthquakes that have plagued the region, the Caribbean's insurance markets have been through one of their most difficult periods. Under its umbrella organisation, the Jamaican Association of General Insurance Companies (JAGIC), the industry has recommended two basic responses to natural hazard catastrophes:

1. Hazard mitigation and vulnerability measures adopted prior to a hazard event to optimize protection from damage: Many insurers and reinsurers require that beach-front properties take the necessary steps to improve the infrastructure's ability to withstand certain events—this helps to lower premiums, although this is not always the case.
2. Economic mechanisms aimed at prefinancing the repair of the damage caused by disasters.

The former is more efficient than the latter, which does not prevent or minimize the impact of the damage. The use of mitigation as a primary strategy for insurers is gaining momentum in Jamaica, as expressed by some members in the field.

For the most part, insurance companies have traditionally assessed the potential impact of catastrophe losses in collaboration with their reinsurers. These approaches have included tabulating insured risks on catastrophe maps that display perceived hurricane tracks and seismic zones. The Caribbean is often seen as a 'hard market' as it relates to reinsuring property against natural hazards, especially hurricanes. A good insurance programme should adequately finance the replacement of assets whether it is the damaged property, or loss of operational time. Loss of profits or increased cost of working insurances will reimburse revenue losses incurred as a result of reduced business activity within a predefined period of time following the loss. This "period of indemnity," combined with the sum insured, are backbones of consequential loss protection. However, while the insurer can bring experience in getting businesses back on their feet, it can only mitigate the destruction that has already occurred. These indemnity periods under increased cost of working or loss of profits covers are as crucial as the sum insured.

38. National Research Council, *Responding to Changes in Sea Level: Engineering Implications* (Washington, D.C.: National Academy Press, 1987).

Insurers and reinsurers assess total exposures, split by distinctive disaster type. A Probable Maximum Loss (PML) percentage factor is applied to reflect assumptions as to the vulnerability of distinctive areas, construction types and structure occupancies. The PML is used to determine the premium amount to be applied to the client. Consultation with the sector has indicated that the industry is moving towards individual evaluation and as part of its evaluation process takes strongly into account mitigation mechanisms implemented by the client.

It is also germane to point out that throughout the Caribbean, natural hazard insurance is perceived to be unaffordable by many small and midsize businesses. In most cases, it is found that underinsurance is widespread. Although competition in the market is tight, the majority of insurers have high and almost identical premiums. Therefore, businesses are forced to pay a high premium or not insure at all.

Institutions Involved in Disaster Management in Jamaica

There are several institutions involved in disaster management in Jamaica. The Office of Disaster Management and Emergency Management (ODPEM) is the agency with responsibility for disaster management in Jamaica. However, there are several other institutions that assist ODPEM, such as the Ministry of Transport and Works, the National Environment and Planning Agency (NEPA), the Environmental Control Division of the Ministry of Health, the Water Resources Authority, and the Mines and Geology Division of the Ministry of Mining and Energy.

The local ODPEM has been one of the leading proponents of the incorporation of vulnerability reduction measures in development planning at all levels. In 1982, the ODPEM developed a Natural Hazards Management Plan for the country. This plan, which was completed in 1984, recommended a combination of structural and nonstructural mitigation measures to reduce the island's vulnerability to disasters and also recommended the development of a comprehensive hazard-mapping project. Recommendations from this plan were incorporated in the Kingston Metropolitan Regional Plan, which was completed in 1986, and also led to the Flood Plain Mapping Project. This system regulates and guides residential development, location of roads and critical lifeline facilities.

The first automated real-time flood warning system for Jamaica was installed in the Rio Cobre basin as a part of the flood loss reduction programme sponsored jointly by the United Nations Development Programme and the Government of Jamaica and executed by the ODPEM and the World Meteorological Organisation.

The system collects rainfall and stream flow data at remote sites in the upper watershed regions of the Rio Cobre. The data is transferred by VHF signal to a base station where it is received and analyzed by computers. An alarm is activated if the predetermined limits of precipitation and water rise are exceeded. This allows for the efficient evacuation of communities facing the threat of flooding and also allows for the utilization of flood impact reduction measures.

Another output from this programme was the development of flood plain maps for the Rio Cobre. The need for improved understanding of storm hazard and risk in the Caribbean has been driven home by the recent upsurge in insurance costs following the underwriting losses suffered by insurance and reinsurance companies in the aftermath of Hurricanes Gilbert, Hugo, and Andrew. Accurate hazard maps will allow the insurance industry to improve risk management and underwriting practices.

The ODPEM is also an important actor in disseminating the information generated in the storm hazard mapping exercise of the OAS/CDMP. There is evidence that both public and private sector agencies/companies, including private engineering companies, are using the information. The ODPEM regularly receives requests for information generated by the project, from both the public and the private sectors. Additionally, some insurance companies are contracting services to help determine their exposure to hazards, including their probable maximum loss from extreme events.

A substantial part of the OAS/CDMP project's resources are directed at training and technology transfer. Staffs of the collaborating agencies are trained in the use and applications of the storm hazard assessment model. Workshops are held for planners and representatives from the insurance and development financing sectors in the use of hazard mapping. The investment in training is aimed at ensuring that the activity can be sustained beyond the project's timeframe.

Another initiative taken by the ODPEM is the National Zonal Programme. Developed in the mid-1990s, the programme focuses on improving communities' preparedness and response to natural disasters. There are approximately 40 zonal committees island-wide, each made up of a Chairperson, Vice Chairperson, a Secretary and 3 Subcommittees. Subcommittees are focused on welfare and relief; public education and fund raising; and emergency operation and communications respectively. Zonal committees report to the parish disaster coordinator in each parish council, who in turn liaises with the four regional coordinators at the ODPEM offices. These regional coordinators are responsible for the four larger zones in the island (Northern, Southern, Western and Eastern).

The ODPEM offers training to these zonal committees in the areas of:

- shelter management;
- emergency operations center management; and
- disaster management training.

Initially, the Adventist Disaster Relief Agency was used to spearhead the development of these zones because of their wide community network and their availability of persons who could facilitate the process. Now, the ODPEM targets different community groups to try and get them to introduce disaster management into their other activities. The aim is that eventually a mitigation component of the programme will be implemented, where communities are taught how to reduce the risk of hazards by working in harmony with their natural surroundings.

DEALING WITH NATURAL HAZARDS: INITIATIVES TAKEN BY JAMAICA

Tourism facilities that are sited in the coastal zone may be vulnerable to disasters for the following reasons:

- vulnerability inherent in its location;
- vulnerability inherent in its structural characteristics; and
- vulnerability introduced by the loss of natural protective qualities of the landscape in which the development is situated.

The vulnerability of the tourism industry is not confined to its own capital stock, as was demonstrated by the Jamaican experience. Damage to roads, utilities, airports, harbors, and shopping centers also affected the industry after Hurricane Gilbert. The recent experience of the Eastern Caribbean after Hurricane Lenny (13–23 November 1999) would suggest that the failure to implement hazard mitigation measures or ensuring that there is proper siting of hotels could result in significant damage to hotels and attractions. For example, the Cap Juluca Hotel, an upscale property in Anguilla, was severely damaged by Hurricane Lenny because of the improper siting of the hotel.

Vulnerability Inherent in Location

Zoning guidelines, accurate and up-to-date natural hazard information and mapping of the resources in the coastal zone are required to minimize the vulnerability of tourism facilities in the coastal zone. These factors should also be considered in the tourism development planning process. Jamaica's initiatives in this regard are outlined below.

Risk Information
ODPEM has developed hazard maps for the island as it relates to hurricanes, earthquakes, storm surges, riverine flooding and landslides. The Earthquake

Unit, University of the West Indies, Mona (UWI, Mona) has done considerable work on the earthquake risk on the island. Therefore, there is information on earthquake frequency and the vulnerability of certain regions of the island to earthquakes. However, there is very little localized information, which can be used to guide developments in specific areas of the island. For example, studies have been conducted on the vulnerability of Montego Bay to storm surges, but there is very little information for Negril, Ocho Rios, and Port Antonio, which are main resort areas in Jamaica.

The recent work on the vulnerability of Montego Bay to storm surges conducted by the OAS/CDMP indicates that based on the mapping of the 1 in 25 year return period surge, approximately 34 percent (480 hectares) will be impacted by storm surge. The areas that will be most seriously affected include land areas designated as resort areas, conservation and transport (airport). This is clearly the kind of information that is required to guide development planners as to where developments should be sited.

Multi-hazard mapping and analysis are also needed to determine the overall risk of natural hazards to tourism facilities and related infrastructure such as roads, airports and communication systems. For example, the coast of Montego Bay is vulnerable to storm surges, whereas the hills are more susceptible to landslide as was demonstrated by the 1957 earthquake.

Coastal Zone Mapping
The National Environment and Planning Agency (NEPA) has been involved in mapping coastal resources and has developed a coastal atlas in Jamaica. The research has focused on demarcating coral reef areas, mangroves and seagrasses. However, the information is not very detailed, presenting an aerial view of these resources in relation to the island. In its present state, the coastal atlas cannot indicate the physical state of the natural resources. In order to determine levels of deterioration (information that would help with identifying areas at risk), detailed mapping is needed. Mapping of the coastal zone should also include critical infrastructure and settlements sited along the coast.

Zoning Guidelines and Regulations
Zoning regulations and guidelines and building practices are important in minimizing the effect of natural disasters in the tourism industry. The Town and Country Planning Act of 1957[39] and the Local Improvements Act of 1944[40] are the two main statutes that control the development and subdivision of land. The former provides not only statutory requirements but also guidelines for the preparation of Development Orders.

---

39. Town and Country Planning Act of 1957, Vol. XXVI, Revised Laws of Jamaica.
40. Local Improvements Act of 1914, Vol. XVII, Revised Laws of Jamaica.

McHardy[41] indicates that the Development Order is a zoning ordinance and, like all zoning ordinances, consists of two parts, a zoning map and a text of regulations. The map divides the area into functional zones; the text states how land may be used within each class of zone.

Zoning influences the way in which private land may be used. Zoning is therefore one of the most important tools in the private land use decision process, as it allows a public body to influence the use of land by a private owner by subjecting his use to reasonable constraints. The development order is therefore a powerful tool that is used in guiding land use activities and thus protecting the environment in terms of the type and level of development.

Under section 3(1) of the Town and Country Planning Act, the Minister shall appoint a person or persons to be the Town and Country Planning Authority. Under section 5(1) this Authority

> . . . may after consultation with the local authority concerned prepare so many or such provisional development orders as the Authority may consider necessary in relation to any land, in any urban or rural area, whether there are or are not buildings thereon, with the general objective of controlling the development of the land comprised in the area to which the respective order applies, and with a view to securing proper sanitary conditions and conveniences and the coordination of roads and public services, protecting and extending the amenities, and conserving and developing the resources, of such area.

Development Orders are prepared by the Town and Country Planning Authority in consultation with the parish council for the area. It should be noted that the Town Planning Department, although not mentioned in the Town and Country Planning Act, operates as a technical department in support of the Authority and prepares Development Orders for approval by the Authority.

A prerequisite for planning permission under the Town and Country Planning Act is that the site of the development should be located within a Development Order area. The permission granted by the Development Order may be granted either conditionally or subject to such conditions or limitations as may be specified in such order. In areas where there are no Development Orders, permission for the erection of buildings falls under the Building Law (Local Building Regulations), which is the responsibility of the parish councils. In addition, the entire island is not covered by Development Orders and therefore large portions of the island do not fall under the regulations of the Town and Country Planning Act of 1957.

41. P. McHardy, *Master Plan for Sustainable Tourism Development in Jamaica. Planning Issues*. Prepared for the Office of the Prime Minister, Tourism Division, 1998.

McHardy[42] also argues that until very recently, Development Orders were very general documents with major land uses not divided into subclasses. In addition, many of the Development Orders were prepared so long ago they bear no relevance to the situation on the ground. Therefore, as a tool for guiding growth they are not very effective. Recently, the Town Planning Department has been revising these documents using the urban plan as a guide in their preparation. The new Development Order for Ocho Rios/St. Ann's Bay has been prepared in this manner and a new Negril–Green Island Development Order will also be prepared in this manner. New construction sites should be evaluated for susceptibility to hazards and setback distance from the shoreline should be enforced.

Vulnerability Inherent in Structural Characteristics

Tourism facilities may be vulnerable to natural hazards based on the structural characteristics of the buildings. In 1989, at the request of the government of Jamaica, the OAS conducted an assessment of the vulnerability of the tourism sector to natural hazards and recommended mitigation actions.

The assessment disclosed that much of the damage to tourism facilities, and other buildings, was due to lack of attention to detail in construction and maintenance, particularly in roof construction. For example, roof sheeting was poorly interlocked, tie-downs of roof structures were inadequate, nail heads were rusted off, and termites in the roof caused a reduction in timber and metal strength by corrosion. Much glass was needlessly blown out because of faulty installation and poor design criteria, but also because windows were not protected from flying debris. Drains clogged with debris caused excessive surface runoff, resulting in erosion and scouring around buildings. Local water shortages developed because the lack of back-up generators prevented pumping. Therefore, faulty building practices and maintenance deficiencies were a major contributor to the damage. It was calculated that proper attention to these matters would have increased the cost of construction by less than one percent.

There is the legal framework to ensure that buildings withstand the effects of hurricanes and earthquakes (but not storm surge and wave action). The Parish Councils Building Act[43] gives the power to the Parish Council to make bylaws for the erection, alteration and repair of buildings within the limits of any town or any rural area defined by the Parish Council. The Building Bylaws of the Parish Councils and the Building Act of the Kingston and St. Andrew Corporation (KSAC)[44] are used to guide building activity in Ja-

42. *Ibid.*
43. Parish Councils Building Act of 1908, Vol. XIX, Revised Laws of Jamaica.
44. Kingston and St. Andrew Building Act of 1883, Vol. XIV, Revised Laws of Jamaica.

maica. The Building Bylaws of the parishes were enacted in 1949, while the KSAC Building Act was enacted in 1908 and revised periodically. The Laws and Regulations (made under the laws) prescribe detailed construction procedures for buildings in accordance with the technologies at the time.

McHardy[45] indicates that the engineers and architects using the Regulations found conflicts between the Regulations, current building practices and use of available materials. The government of Jamaica, recognizing the need for a comprehensive Building Code, developed such a document, which was issued as a policy statement pending the enactment of the necessary legislation which would make use of the Building Code for the design of all buildings in Jamaica. The Building Code, although not yet enacted into law, is being used by architects and engineers and their experience in the use of the code is reported to the Bureau of Standards, which regularly updates the document through a Standing Committee. One of the drawbacks of the Building Code is that it does not assign higher standards to buildings of high importance (such as hospitals, churches, and tourism facilities).

For areas not covered by development orders, the regulation of building activity is governed by the Building Regulations, as mentioned earlier. However, this situation often results in charges for violations under the Regulations being null and void, as there is often confusion as to which areas fall outside of the development order area and violations are often mistakenly dealt with under the Town and Country Planning Act.

Most provisions in the Building Regulations relate to the technology that was available in 1949 when the regulations were produced. There is a tendency, however, to ignore the fact that outdated provisions and building structures are inspected only for general compliance with accepted engineering principles. In addition, most Parish Councils (who do not always have the requisite technical staff) rely on the competence of the architects and engineers who designed the buildings and therefore, do not carry out detailed checks.

Vulnerability Introduced by the Loss of Natural Protective Qualities of the Landscape in which the Development is Situated

Expanding urbanization of reclaimed land in the narrow coastal fringe and on steep slopes increases the risks from natural disasters. Increased soil erosion as a result of improper development can for example raise the levels of streambeds, thereby contributing to flooding. In extreme rainstorms, sediments can fill up the stream channel causing sudden shifts in the course

---

45. McHardy (n. 41 above).

of the streams. When attractions and other tourism infrastructure are located in these areas, they are at risk from flooding.

The removal of mangroves reduces the protective coastal barriers and exposes many coastal developments to storm surges and increased long-term wave erosion. Valuable beach material is lost and this in turn affects the tourism product. The loss of beach material does not only affect the product being offered, but also poses a threat to developments found on the coast. For example, in Negril, there is significant coastal erosion due to increased wave action as a result of the destruction of coral reefs that normally break the waves. Goreau[46] estimates that approximately 90 percent of the coral reefs on the Negril coast has been destroyed as a result of hurricanes.

The OAS study assessed the impact of Hurricane Gilbert on the tourism sector. The study recommended the protection of beach vegetation, sand dunes, mangroves, and coral reefs, all of which help to protect the land from wave and wind action, and the quality of sewage outfall should be maintained to protect live coral formations.[47]

There is a need for an Integrated Coastal Zone Management approach in order to properly manage the zoning of tourism facilities and to reduce the land-based sources of pollution in the coastal zone. These initiatives will in the long-run reduce the natural hazard vulnerability of the island. Multi-hazard mapping and analysis are also needed to determine the overall risk of natural hazards to tourism facilities and related infrastructure such as roads, airports and communication systems. In determining the vulnerability of the tourism sector to various natural hazards, it is important that the related infrastructure sectors and the communities that surround the resort areas are evaluated.

DEALING WITH SOCIAL RISKS

Tourism is Jamaica's main source of foreign exchange, a reality that implies that every Jamaican has a vested interest in ensuring the continued prosperity of the industry. This article indicates that there are several social and environmental risks that have clear implications for the industry and that may threaten the long-term viability of the product. Tourists (stopover tourists and cruise ship arrivals) and locals are both exposed to these social risks.

Based on the findings, we may conclude that the main environmental and social risks that could affect the tourism industry are environmental

46. T. Goreau, *Negril Environmental Protection Area Water Quality Monitoring Program—Draft Proposal*. Produced for Negril Environmental Protection Trust/Negril Coral Reef Protection Society, 1995.

47. Organization of American States (OAS), *Insurance, Reinsurance and Catastrophe Protection in the Caribbean* (Washington DC: USAID/OAS Caribbean Disaster Mitigation Project, 1996).

degradation, visitor harassment, crime, squatting, and drug peddling. These problems have been exacerbated by the fact that the employment and economic opportunities in the tourism sector have attracted persons from other parts of the island to the main resort towns. However, most of these problems are manifestations of larger structural social and economic relations. Therefore, these underlying social and economic concerns must be dealt with in order to ensure that there is an appropriate social and economic climate for tourism.[48]

It must be stressed that the tourist industry does not operate in a social vacuum. Many of the social risks that are faced by the tourism sector are as a result of a combination of many phenomena. These include poverty; inadequate access to housing, education, and medical care; and lack of economic opportunities.

The following section outlines the main social risks and the initiatives taken to deal with these risks.

Squatting

The tourism industry has experienced sustained growth during the last twenty years. As a consequence of this, and the perception that the industry is one with significant amounts of economic opportunities, people from other parts of the island move to resort towns hoping to improve their standard of living. Some of those migrating to the tourist centers can earn a livelihood, either through direct trade with the tourists or offering them services. Other migrants seek steady employment with the formal tourism sector such as hotels, restaurants and major attractions.[49]

As more people flocked to these urban centers to gain employment, little planning for housing to accommodate workers in the tourist industry was done. Consequently, several squatter settlements developed on the outskirts of resort areas including the squatter settlements at Roaring River, Belmont, Bogue, and Mansfield in Ocho Rios; Flankers, Norwood, and Rosemount in Greater Montego Bay; and Whitehall in Negril.[50] Squatters tend to construct housing with temporary material such as wood, zinc and cardboard that have no guarantee of secure tenure and because these materials are less expensive than the permanent alternatives.

48. Call Associates Consultancy Ltd., *Master Plan for the Sustainable Tourism Development—Social and Economic Integration of Local Communities with the Tourism Industry*, (Kingston: Prepared for Office of the Prime Minister, Tourism Division, 1998).
49. *Ibid.*
50. N. Isaacs, "History of Site and Services and Squatter Upgrading in Jamaica," *Housing and Finance*, (Kingston: Building Societies Association of Jamaica Ltd., 1989).

These unplanned settlements have no physical or social infrastructure such as: potable water supply, electricity, roads, garbage collection, and sewage collection and disposal. This situation contributes significantly to environmental degradation.

There is also pollution of coastal waters and beaches due to inadequate disposal of human waste and garbage.[51] In some of the squatter settlements, limestone is the dominant form of rock formation. The squatter communities are not sewered and pit latrines and other sewage disposal solutions are used to dispose of human wastes. Limestone is very porous, however, and as a consequence, sewage effluent may seep into the ground water.

Additionally, squatter settlements adjacent to beaches, for example, fishing communities in White River, Negril, tend to be at risk from natural hazards such as flooding and storm surges. The damage to coastal resources as a result of squatting may result in the destruction of natural protective systems and further increase the vulnerability of certain sites to natural hazards.

The lack of social and physical infrastructure for squatter settlements is a threat to environmental sustainability in tourist areas. As the population density of the tourist areas increases, the carrying capacity of the areas, for example supplying water resources or accommodating housing, is threatened and this in turn has baneful consequences for the quality of life of the communities. Moreover, the economic base of the area—the tourism product—is endangered.

Taking into account carrying capacity is a key consideration for aspirations for achieving sustainable development of the resort areas. Weeks[52] advances that carrying capacity can be construed as the ability of the environment to sustain a given population at a particular standard of living. The state of the physical environment is even more important in the context of tourism.

An important opportunity to facilitate the development of the area in a sustainable manner while still providing employment was missed. Operation Pride is a step in the right direction but not withstanding protracted economic constraints, the efforts fall far short of the needs of the people. The Ministry of Water and Housing is currently executing a number of projects to improve housing availability in resort areas such as Ocho Rios, Negril and Montego Bay. This includes the following projects: Beecher Town, Shaw Park Estates (1 and 2), Belle Air (1 and 2), Mount Edgecombe, Mamee Bay, Alexandria, Seville Heights, and Seville Views, which are all located in Ocho Rios and its environs. These projects in Ocho Rios and its environs will provide more than 6,000 housing solutions.

51. Call Associates (n. 48 above).
52. J. Weeks, *Population: An Introduction to Concepts and Issues* (Belmont, California: Wadsworth Publishing Company, 1996).

Visitor Security and Harassment

Visitor security and harassment may affect the long-term sustainability of tourism in Jamaica.

Crime and Security

Although violence is clearly not confined to either urban or economically challenged areas, violent crime tends to be geographically concentrated in economically depressed built-up areas. The effects of crime extend beyond these areas and are felt at all levels of the society and the economy.

The impact of crime at the macroeconomic level is apparent, although difficult to quantify. For example, the tourism industry is affected by press reports on crime. Although the statistical relationship between crime and tourist arrivals is inconclusive, the tourism industry is particularly sensitive not only to changes in Jamaican and international visitor confidence, but to the image projected of the island to potential vacationers. Increasing levels of crime hurt the tourism industry as people choose other less risky destinations.

It is also worth noting that crime and security issues can have an impact as relatively independent variables. A cogent example of this is recent disturbances that captured the attention of the international media in early July 2001. There is certainly no information to suggest that the disturbances were aimed at tourists. Indeed, the geographical locality in which the violence occurred all but precluded this possibility. Nevertheless, the occurrences amplified by the presence of international media did untold damage to the industry. In the tourism industry, the country's image is a part of the product.

Crimes Committed against Tourists

A review of data over a ten-year period which were obtained from the Community Relations Branch, Tourism Liaison Office of the Jamaica Constabulary Force (JCF) indicates that there was a gradual increase in crimes against tourists between 1989 and 1992 (Figure 3).

In 1989, the number of crimes committed against tourists started to increase and the trend continued until 1992. However, since 1992 there has been a 78 percent reduction in crimes against tourists in resort areas. Major crimes committed against tourists include robberies and larceny. In 1992, the highest levels of robberies committed against tourists were reported, with 1989 being the lowest so far for the period.

The resort towns of Montego Bay, Negril, and Portland have seen a decline in crimes against tourists. In summary, one may conclude that:

- There has been a 78 percent decrease in crimes against tourists between 1992–2000. Notwithstanding this, Jamaica is still perceived as

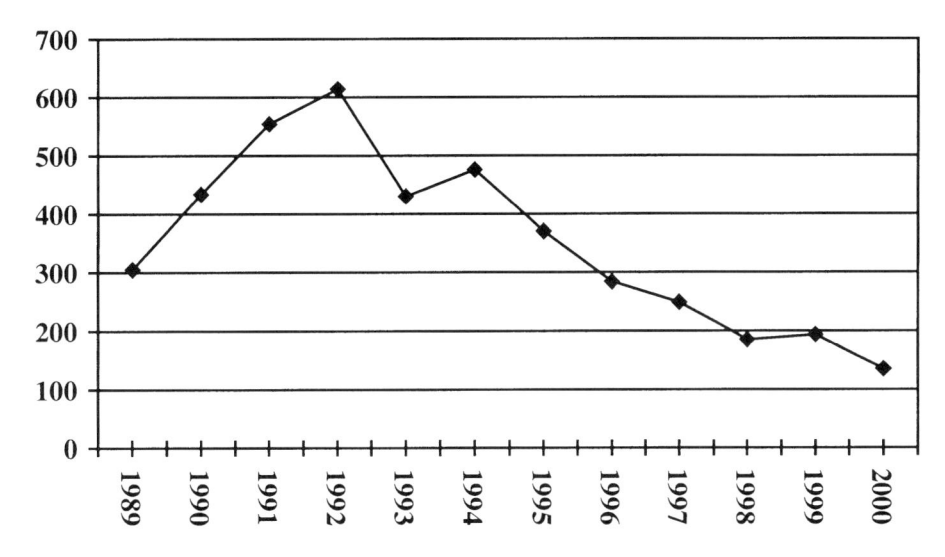

FIG. 3.—Crimes committed against tourists, 1989–2000

a tourism destination with significant security problems. The level of crime in the resort areas is higher than in the non-resort areas.[53]

- In general, the existing laws are suitable in defining and dispensing justice for major crimes such as murder, larceny, robbery, assault, supply and use of illegal drugs in the resort areas. However, laws related to petty offences under the Towns and Communities Act[54] and Tourist Board Act,[55] which may be defined as harassment are not adequate in defining and dispensing justice for these "specific crimes against tourists."[56]

Visitor Harassment

While a definition and the contours of crime are defined by law and are fairly clear-cut, the situation for harassment is not the same. Harassment is a problematic phenomenon to define and thus it is very subjective. It has different meanings to different interests. However, there are some factors that can be generally viewed as constituting harassment. Stone[57] defines harassment as:

   53. Call Associates (n. 48 above).
   54. Towns and Communities Act of 1843, Vol. XXVI, Revised Laws of Jamaica.
   55. Tourist Board Act of 1955, Vol. XXVI, Revised Laws of Jamaica.
   56. Call Associates (n. 48 above).
   57. C. Stone, *Report on Tourist Harassment* (Kingston, Jamaica, 1989.) (Copy on file with the author.)

TABLE 2.—PROPORTION OF TOURISTS EXPERIENCING HARASSMENT
IN 1989

Resort Areas	Levels of Harassment (%)	Visitors Willing to Return to Jamaica (%)
Ocho Rios	45	80
Montego Bay	51	82
Negril	32	96
Port Antonio	29	95

SOURCE—C. Stone, *Report on Tourist Harassment* (Kingston, Jamaica, 1989).

- Being followed by persons who wanted to sell some service or item;
- Being shoved into taxis they did not wish to use, or into vending areas;
- Suffering abuse, insults or threats from persons who refused to take no for an answer;
- Being approached to buy drugs, sexual favours, or for sex.

Dunn and Dunn[58] and Dixon[59] in their studies of Negril and Montego Bay, respectively, also used this definition of harassment. They argued that tourism harassment includes a wide range of activities and behaviour and large numbers of perpetrators. However, none of the harassers would classify the tourist as their enemy or opponent. Dunn and Dunn found that most of the offenders were from outside the Negril community, drawn mainly from the unemployed youth of neighbouring communities and parishes, and that an increasing number of traders, prostitutes and touts were coming in from Kingston and other major urban centres. Call Associates Consultancy Ltd.,[60] found similar results to Dunn and Dunn and Dixon in their studies of visitor harassment in the main resort towns in Jamaica.

Stone[61] observed that Montego Bay and Ocho Rios were "experiencing a higher level of tourist harassment than either Negril or Port Antonio." Table 2 outlines the proportion of tourists who reported that they experienced harassment in the major resort areas in 1989.

Stone also found that the two dominant categories of harassment were:

58. L. Dunn and H. Dunn, *Report on Visitor Harassment and Attitudes to Tourism and Tourists in Negril.* Submitted to the Tourism Action Plan, the Jamaica Tourist Board and the Negril Resort Board, 1994.
59. A. Dixon, *Informal Sub-sector Activity in the Tourism Industry in Montego Bay,* Submitted to the Tourism Action Plan, January 1995. Kingston, 1995.
60. Call Associates (n. 48 above).
61. Stone (n. 57 above).

- Pressure to buy things they do not wish (35 percent); and
- Attempts to sell them drugs (29 percent).

The Jamaica Tourist Board's 1999 Visitor Satisfaction Study and our own findings indicate that these two categories were the main forms of harassment experienced by visitors to the island. The 1999 Visitor Satisfaction Survey conducted by the Jamaica Tourist Board indicates that 43 percent of the visitors to Jamaica reported that they were harassed. When the category of 'visitors' was disaggregated into different kinds of accommodation, some seemed more prone to harassment than others. It is intriguing that of all the types of accommodation that exist, all-inclusive hotels had the lowest level of harassment at 41 percent. This is the type of accommodation that would likely limit the exposure of the guest to potential harassment. The implication of the findings is that in all the other cases where exposure to the public is great, harassment is also that much higher. Visitors staying in villas and apartments had the highest levels of harassment at 51 percent and 58 percent, respectively.

Stone argued that the "more harassment intensifies the more tourists will adopt coping devices which will reduce the flow of spending outside of the hotels where they are guests." Paradoxically this only seems to increase harassment as it makes those who engage in harassment feel it necessary to increase their activities. This seems to hold well with the current trends in the industry, but as our analysis indicates, there is a perception that enclave tourism was a contributing factor in the reduction of spending outside the hotels.

Many persons felt that all-inclusive hotels wanted to have an overwhelming proportion of the revenue from the industry. The conflict that this entails reverberates considerably throughout the spectrum and hierarchy of key players across the industry, especially in a context of national economic difficulties and in the low season of the tourist year.[62] This scenario seems to be verified by the Visitor Satisfaction Study, as the only locality in which there was a substantial decrease in the incidence of visitor harassment was the airport, an area in which the authorities probably have the most control. Moreover, between 1997 and 1999 there was actually a precipitous 17 percent increase in the number of visitors who were harassed on our beaches.

Of additional and long-term significance are the responses from tourists on whether they would have an interest in returning to Jamaica or would recommend Jamaica to a friend as a vacation destination. Stone's findings in Table 2 illustrate that in 1989, 80–95 percent of the visitors were willing to return to Jamaica. A decade later the 1999 Visitor Satisfaction Study finds only 26.7 percent of those who were harassed would be extremely likely to return to the island. In the same vein, the study finds that of the visitors

62. Call Associates (n. 48 above).

who suffered harassment, less than 50 percent would definitely recommend Jamaica to a friend. In contrast, of those tourists who were not harassed 71 percent of this group stated that they would definitely recommend Jamaica to a friend. These findings have significant implications for the growth of the industry, as new and repeat business is harmed by harassment.

Tourism Product Development Co. Ltd.,[63] carried out a study that revealed five main "traits" of harassment. These were:

1. airport;
2. transport;
3. hotels;
4. watersports; and
5. shops, street and attractions.

At the airport, baggage handlers, peddlers, and others attempt to establish personal and business contacts in the arrival areas. Within the transportation system, legal and illegal taxis and car rental operators sometimes aggressively promote and overcharge for services while they vie for passengers continually. The study also noted that tour company operators working for hotels, confront guests on arrival to pressure them into paying for non-refundable packaged tours. In the case of water sports, jet ski operators have received strong criticism from members of the public and from visitors, whose complaints deal with issues of safety, noise pollution, environmental impact and harassment. Operators harass visitors to rent rides and to purchase drugs and other goods.

Harassment also occurs in the shops, street and at attractions. Roving vendors and tours are widely dispersed and highly mobile throughout the resort areas. Some of these persons offer only a few products but use this as a cover to offer illegal items and services such as drugs, sex, unsanctioned tour guide services, and water sports facilities. Harassers move quickly to locations where tourists congregate and they relocate just as quickly to avoid police and resort patrol officers. Roving vendors are able to offer products at a cheaper rate than stall or store operators, and their aggressive and sometimes abrasive approach tarnishes the reputation of official traders. The attitudes and alleged unfair competition of illegal vendors have given rise to tensions between them and fixed location operators. Harassers cut across all classes, groups and sectors of the industry and are therefore not confined to the informal vendors. However, the street trader is the most visible and obvious source of harassment.

63. Tourism Product Development Co. Ltd., *Solving the Harassment Problem*, Presented at a workshop by the Office of the Prime Minister—Tourism Division, Tourism Product Development Company Ltd. and the Jamaica Tourist Board. Renaissance Jamaica Grande. (1997). *Not published.*

Dunn and Dunn[64] concluded, "visitor harassment is a problem related to several social, economic, political, cultural and psychological factors." Our findings indicate that visitor harassment stems from the lack of adequate economic and employment alternatives that offer good potential earnings.

A profile of a 'tourist harasser' based on our own research and the Dunn and Dunn study is as follows:

> The offenders are primarily young males, although women are prime offenders for soliciting and hair braiding. Men are rarely arrested for soliciting although the 'rent a dread' phenomena occur in all areas under review.

Harassers that were interviewed as part of this study see their activities as gainful employment and an alternative to working for someone for minimum wage. Dixon[65] noted in her study that:

> The majority of male harassers agreed that there was a level of engagement that involved forcing some visitors into options which they probably would not otherwise contemplate. However, they argued that they receive some requests from tourists to find locations of interest, take photographs, . . . take them to swim and to various heritage sights . . . It was generally agreed that illegal activities opened up doors for the abuse of visitors. The expressed view is that their relationship with the sector should be formalized.

The studies conclude that these problems could be minimized if there were community policing of the resort towns in addition to the resort patrols. However, community policing would be effective if the community perceives that there are benefits from maintaining a sustainable product.

Drug Peddling

At present one of the most insidious threats to the security of small states is the trafficking of illicit narcotics. Jamaica is ostensibly a major transshipment port for the exportation of illegal drugs, especially cocaine. This poses a threat to the political as well as economic stability of the country affecting many industries including tourism. Discussions with Senior Superintendent Spence (Narcotics Division, Jamaica Constabulary Force) indicate

---

64. Dunn and Dunn (n. 58 above).
65. Dixon (n. 59 above).

that this has resulted from poor security at the ports and also because of the many trade destinations that our ports service.

Additionally, there are also many local farmers that routinely plant marijuana among crops of bananas, cocoa and yams. These small-scale operations range from supplying the ganja farmers to other community members and eventually the tourist industry. Griffith[66] suggests that a seminal problem for the region is its dual-proximity to the US, which is the largest single market for narcotics in the world, as well as South America, the world's largest producer of illegal drugs. Griffith points out that in 1995 General Barry McCaffrey, then U.S. drug "czar," posited that 52 percent of the approximately 575 metric tons of cocaine available worldwide in 1994 were consumed in the United States, and six Latin American countries produce 80 percent of the world's cocaine.

Within this context it is not surprising that the JTB's 1999 Visitor Opinion Survey indicates that of the people who were harassed, 51 percent stated that the form of harassment took the pressure to purchase drugs. This was secondary only to the pressure to buy other items. Official statistics of the Narcotics Division[67] of the JCF indicates the amount of marijuana production in hectares in Jamaica decreased by approximately 42 percent between 1999–2000. The data collected for the resort towns in Jamaica indicate that illegal drug activities continue to occur.

Where the growth of marijuana is concerned, in many cases hill slopes are denuded of natural vegetation to produce this illegal product, which has resulted in landslides in some rural areas. Where illegally cultivated fields are identified, burning of the crop on site or at another location further exacerbates the problem of landslides and soil erosion. Moreover, further information provided by the Narcotics Division indicates that in 2000 there was a considerable increase (162 percent) in the growth of marijuana over the previous year.

Not only are there environmental problems associated with illegal and unregulated cultivation, but also the societal effects are potentially devastating.

Initiatives Taken to Reduce Social Risks Associated with Tourism in Jamaica

The aim of the Tourism Master Planning Process in Jamaica is to develop a sustainable tourism product in which the benefits to local communities and the economy are enhanced and social risks are minimized. In the past,

---

66. I. Griffith, *Drugs and Security in the Caribbean: Sovereignty Under Siege* (University Park: Pennsylvania State University Press, 1997).
67. Provided by Senior Superintendent Carl Williams.

stakeholders have taken several initiatives to reduce the social risks associated with the tourism industry, which have had varying degrees of success in dealing with issues such as visitor security and harassment, drug trafficking and squatting.

The government has made several interventions to improve the infrastructural base of the various tourist towns. They have initiated the Northern Jamaica Development Project (NJDP) that includes the Montego Bay and Ocho Rios Sewerage Improvement Project, Lucea, Negril and Great River/Montego Bay Water Supply, Northern Coastal Highway Improvement, Montego Bay Drainage and Flood Control and the Ocho Rios Port Expansion. These projects are at various stages of completion.

During the last four decades, successive governments have adopted and implemented several permutations and combinations of physical and financial solutions in providing housing to low income groups. Of significance in targeting squatters was the Sites and Services Project of the 1970s. In 1995, the Operation Pride Programme was initiated. Between 1995–97, US$300 million had been spent on planning and site improvements for low-income housing, which seeks to upgrade squatter settlements by providing land at an affordable cost to those who need it.

The Resort Security Programme (RSP) has been designed to combat crime against tourists in the resort areas. The RSP's objective is to create an environment in which residents and visitors can be secure. As part of its mandate, the Resort Patrol Service enforces anti-litter statutes and monitors all water sports operations. There are regular patrols along 'problem strips' in all resort towns to reduce harassment. It is the view of the consulting team that the introduction of the Resort Patrol Service is a short-term solution to the visitor security and harassment problem. In the 1970s, the then government initiated an anti-crime programme that included "a stronger security presence in the resort areas and increased home guard service in all communities."[68] However, this programme only proved to be a short-term solution to a long-term problem. In the long-run, programmes will need to be developed to improve the socio-economic conditions of the community.

Anti-Harassment Programme
The government of Jamaica has developed an Anti-Harassment Programme that aims to reduce harassment in resort areas through enforcement, public education, and social upliftment.[69] Currently, the programme focuses on enforcement, and will be expanded to include public education and social upliftment. A pilot project has been started in Ocho Rios to enforce the

68. M. Manley, *Importance of Tourism to Jamaica, in Caribbean Tourism Policies and Impacts* (Kingston: Caribbean Tourism Research and Development Centre, 1977).
69. Based on discussions with Mr. Carl Miller—Director, Anti-Harassment Programme.

laws related to visitor security and harassment. Enforcement is effected through the Port Security Corps. Based on discussions with the resort patrols and community leaders, one may conclude that Phase 1 of the Anti-Harassment Programme has been effective in reducing the levels of harassment in the resort areas. There has been a reduction in visitor harassment on the beaches and on the streets in Ocho Rios.

The programme aims, through public education, to empower persons in the community, foster a feeling of responsibility towards the industry and develop a feeling of being a part of the industry by outlining linkages and benefits to the community. It is the view of the consulting team that the public education and social upliftment components of the programme are appropriate long-term solutions to the problem of visitor security and harassment.

Crime, violence, and harassment in the various resort towns are partially the result of high unemployment levels, within and external to 'tourist' towns. Through the activities of the non-governmental organisations (NGOs) and government programmes such as Learning for Earning Activity Programme (LEAP), Team Jamaica, and the National Youth Service Programme (NYS), employment and training which support the tourism sector are being done in towns surrounding the resort areas. Training and seminars conducted by the Jamaica Tourist Board (JTB), Tourism Product Development Company (TPDCO), and JAMPRO have assisted in improving the property standards and quality of services delivered in the industry.

Private Sector Initiatives
Several private sector initiatives have been taken to improve social and economic integration of local communities with the tourism industry in Jamaica. These initiatives have had some success and have resulted in reduced crime, less harassment, and more effective linkages between local communities and the tourism industry. Three case studies are presented below that outline past and current initiatives to improve social and economic integration of local communities with the tourism industry.

Case 1: Countrystyle Holdings working with communities on the South Coast

Countrystyle Foundation Limited is one of the most established operators in the field of community tourism. Community tourism may be described as tourism activities that actively involve the community in presenting Jamaica to the visitor. With community tourism, Jamaica becomes more than sand, sea and sun. It is tourism that gives the tourist an insight into the lives of members of the community, while maintaining the cultural balance in the community.[70]

There are approximately 25–30 villages in the south coast areas in the Countrystyle Village programme. Countrystyle Holding, as part of its community tourism strategy organises tours and activities in the local communities. Guests are taken to the bars, the villages, the churches, historical sites, national parks and other interesting venues in the area. Guests are encouraged to get involved in community projects, as is the case with various communities of the central and south coast areas of the island. In addition, there are private homes which are registered as part of the Countrystyle programme and which provide accommodation for visitors.[71] These homes add up to between 100–150 rooms (approximately 5 in each village).

Countrystyle is currently developing the following projects and programmes:

(a) Projects to reduce the level of environmental degradation in the selected communities. This will involve the physical improvement in attractions, implementation of sanitation projects and the development of recycling programmes.
(b) Projects to increase the income of rural communities. This will involve an increase in crops grown using organic methods and sold to specialized markets, and other projects to increase incomes in target communities.
(c) A programme to develop an alternative tourism product that fosters cultural exchange and encourages community development. This will involve visitors staying in communities and taking part in day-to-day activities of the community. In addition, visitor information centres and rest stops would be established.

Discussions with community members involved in the Countrystyle Programme indicate that they welcome this programme involving their communities.

Case 2: All-Inclusive Hotels: Sandals

The All-Inclusive Hotels have taken several initiatives to foster social and economic integration of local communities with the tourism industry. These initiatives taken by Sandals, one of the largest groups of all-inclusive hotels in Jamaica, are outlined below.

70. Call Associates (n. 48 above).
71. Halcrow Ltd., (n. 17 above).

Sandals

The Sandals Group has implemented several initiatives to facilitate social and economic integration of local communities with the tourism industry. This includes the provision of scholarships and other training for employees (US$12.3 million) and assistance to small farmers under the Sandals/Rural Agricultural Development Agency/Small Farmers Project.[72]

Sandals, in association with the Rural Agricultural Development Agency (RADA), have been assisting local farmers with the marketing and export of their fresh produce under the RADA/Sandals/ Small Farmers Project. Sandals is working with farmers in St. Ann and St. James. This project was developed as a result of concerns raised about the quality of produce received from farmers at a Continuing Educational Programme for Agricultural Training/ University of the West Indies (CEPAT/UWI) training programme in November 1995, and the limited production of some "exotic" vegetables. This programme has now been extended to St. Lucia. Last year Sandals purchased agricultural produce valued at J$400 million from local farmers.

Under this program, Sandals/RADA have provided training, fertilizers, insecticide, and seeds and other inputs to facilitate small farmers in the production of certain crops demanded by the resort year-round. Participating farmers benefit by earning additional income and acquiring new skills and techniques in the production of new crops such as hybrid Duke tomato, snow peas, sweet corn, yellow and red sweet pepper, zucchini, yellow squash, and red cabbage. There has been a favourable response to the programme by the small farmers.[73]

The Project consists of two phases:

Phase 1—Initial Financial and Technical Assistance
In Phase 1 of the program, the following activities will be conducted:

a. The purchase of seeds for exotic crops by Sandals;
b. Provision of insurance coverage for the capital cost of over $100,000 by the Inter-American Institute for Cooperation in Agriculture (IICA);
c. The forming of farmers into a marketing entity to facilitate the sale of fresh produce.

Phase 2—Production of Exotic Vegetables
In Phase 2 of the program, the following activities will be conducted:

a. Distribution of seeds;
b. Production and marketing of crops.

72. Call Associates (n. 48 above).
73. Call Associates (n. 48 above).

Under this project RADA assists by making production and marketing links with the farmers and exporters through its marketing and extension services. The produce is sold to Sandals while excess volume is marketed to supermarkets and other hotels.

The group also undertakes other social programmes and services. For example, every property has adopted a nearby school. In Flankers, plans are in place to raise the literacy level of the school population by hiring a dyslexia specialist. Significant contributions are also made to the areas of health and sport. The Sandals chain also contributes to approximately two hundred community projects in all the islands in which it operates. These projects deal with a plethora of social issues such as dispute resolution in the form of the Social Conflict and Legal Reform Project under the auspices of the Canadian International Development Agency (CIDA).

While these activities constitute a step in the right direction, these initiatives need to be broadened and deepened. A particular weakness of the approach is that there does not seem to be any one organization overseeing the various activities and monitoring the effectiveness of said initiatives. Moreover, there is no degree of uniformity of the programmes, as not all the programmes are offered in the various communities. There is an effort to raise the importance of tourism in these areas. However, it has to be taken into account that a significant number of the people attempting to derive benefit from tourism come from outside those communities in close proximity to the tourist areas. A formal structure to execute these objectives could be the vehicle through which there can be stimulation of micro-enterprises, strengthening of the linkages between the tourism industry and other economic sectors at the community level, development of community centres and the expansion of the cultural and sporting activities in the community.

Case 3: Port Royal Developments Ltd. Working with the Town of Port Royal

The Port Royal Development Company Ltd. (PRDCL) plans to transform the sleepy fishing village of Port Royal to a world-class heritage and environmentally friendly tourist attraction. Port Royal's development will capitalize on the global trends that have been emerging in the travel and tourism industry. Tourists now desire destinations that provide educational, historical and cultural interaction. Therefore, the architects of the Port Royal development project will seek to attract cruise ship passengers and other tourists to this cultural theme attraction in the Caribbean.

Heritage tourism is one of the fastest growing subsectors of the tourism industry. The value of heritage and nature tourism in the developing world is estimated to be in the region of US$5–10 billion, with a potential growth rate of approximately 10 percent annually.[74] Jamaica has a growing propensity to historical tourism. Historical sites on the island's south coast will receive funds from the IDB aimed at restoring such sites.

The Port Royal project envisages an 80-acre theme park built around the rich historical background of the town. Once known as the wickedest city on earth, Port Royal was one of the largest towns in the English colonies during the late 17th century. A haven for pirates and privateers (known as buccaneers), Port Royal's geographic location made it an excellent base from which to prey upon heavily laden treasure fleets departing the Spanish Main. On the morning of June 17, 1692, a massive earthquake rocked the sandy peninsular that encloses much of Kingston Harbour, and a significant portion of the town, which was built on the point, slipped beneath the sea and the rest was consumed by fire.

Phase One of the project was anticipated to come on stream by mid-2000, but did not start until 2001. The project will be gradually developed and upon completion will boast entertainment, shopping facilities, museums, art and craft, dining facilities and tours in and around Port Royal. In later phases, the tours will encompass the wider Kingston and Spanish Town areas.

The PRDCL recognizes that if the project is to be attractive and sustainable, there needs to be active involvement of the community during the planning and development stages, and key infrastructure development has to be done. There has been extensive involvement of the citizens of Port Royal in the project. The residents are an integral part of the development by providing a readily available pool of labour of 1,300 persons for the project. The PRDCL has created a Community Learning Centre to improve the skills of the community in an informal way. Various sections of the community are participating in this programme and are acquiring new skills. The residents will not be relocated from the Port Royal area. To ascertain the population size, a house-to-house survey and questionnaire was developed and administered with the assistance of the Port Royal Citizens' Association (PRCA).

Infrastructure such as roads, utilities, schools, disposal systems, security, health and protected services have been designed into the development plan. Dialogue has been ongoing with the relevant Government Ministries, National Water Commission (NWC), Cable and Wireless Jamaica Ltd., Jamaica Public Service Company (JPSCO), Metropolitan Parks and Markets (MPM), Port Royal communities, cruise lines and other private sector organisations. The National Housing Trust is planning new housing schemes for this area. A Type-Three Health Centre capable of handling emergencies is also proposed.

74. L. D. Dunn, *Tourism Attractions—A Dynamic Sub-Sector.* Jamaica. (1998). *Not published.*

A community-based non-governmental organisation called the Port Royal Environmental Management Trust (PREMT) was launched with a mandate to protect the area's natural habitats for a variety of wild bird life, crocodiles, fishes, among other species and the sunken cultural treasures and underwater and surface resources. The PREMT has applied to the National Resources and Conservation Authority (NRCA) to have responsibility for the management of the proposed protected areas of Palisadoes, Port Royal and the offshore cays. The PREMT is also responsible for achieving a financial framework to support the organization's efforts in preserving the environment. The community manages PREMT with support from PRDCL. The PREMT has not been granted the right by the NRCA to manage the aforementioned areas.

The PRDCL sought to:

1. develop a shared vision among main stakeholders;
2. provide infrastructural services for the community such as improving roads, providing water and sanitation services, etc.; and
3. improve skills of the community by establishing a community learning centre.

These skills can be used in the tourism activities earmarked for Port Royal or in other economic activities. The PRDCL takes the view that the plans will not be successful if a concerted effort is not made to bring tourism to the people.

The Port Royal Development has attracted investors who are willing to fund the total project, including the community development and environmental aspects of the project because it is recognized that these components are essential for the long-term sustainability of the project. This does not however mean that funding is not a problematic issue. Funding efforts, both locally and internationally have been hampered by the upsurge of violence in early July in Jamaica, as well as by the terrorist attacks of September 11, 2001, in the United States. Furthermore, most of the funding, in the form of equity, has come from government as the private sector has only contributed 30 percent.

NGO Initiatives

Uplifting Adolescents Project (UAP)
In order to address some of the social risks in the island, a number of non-governmental organisations are focused on empowering youths in order to stop the cycle of unemployment, harassment, crime and squatting. The Uplifting Adolescents Project (UAP), for example, is a joint venture of the governments of Jamaica and the United States to improve literary, social and

vocational skills of at-risk youths, offering them the opportunity to become responsible adults and hence, reduce the risks they may pose to the tourism industry in the future. The UAP, financed by USAID and assisted by 15 NGOs, is delivering services in 4 technical areas to youths aged 10 to 18 years. These areas include:

1. Literacy and Remedial Action;
2. Personal and Family Development;
3. Reproductive Health; and
4. Technical and Vocational Training.

Other similar institutions include the Western Society for the Upliftment of Children (Montego Bay), the St. Patrick's Foundation that works in inner cities in Kingston and focuses on helping youths transcend the constraints of their socio-economic environment to become responsible citizens.

These institutions are focused primarily on the younger generation because many of the country's youths are dropping out of school and are becoming "vagrant citizens" that pose a threat to citizens and visitors alike. In attempting to empower these youths through training and skills development, it is hoped that gainful employment can be found and this would ultimately remove the risk from the streets.

Summary

It is obvious that if the mechanisms had been developed to ensure that tourism development is used as a tool to augment existing towns and villages, there would have been greater social and economic integration of local communities with the tourism industry. However, there was little attempt to transform resort centres into more interactive communities. Therefore, tourism enclaves developed, increasing the demarcation between tourist facilities and local communities.

Successive governments did not achieve the objective of reducing the level of economic evaporation from tourist expenditure so as to ensure that a higher percentage of the tourist dollar remains in the country and in particular within the area where it is spent. Dixon[75] argued that there is very little re-investment of the income from tourism into Montego Bay. This could be easily argued for Negril and Ocho Rios.

It is obvious that the initiatives taken to integrate local communities with the tourism industry need to be broadened and deepened. Communities island-wide have articulated the changes that are required to achieve this goal at Community Consultations Meetings which were organised as a

75. Dixon (n. 59 above).

part of the Tourism Master Planning Process.[76] These meetings were held in various communities in St. Thomas, St. Mary, Trelawny, St. Ann, and Hanover. These meetings facilitated discussions with local communities on issues and options related to the tourism planning and development, and the strategies required to improve community participation in tourism activities. Community consultations were also organised under the South Coast Development Planning Process to assess the views of communities on the type of tourism product they envisage on the South Coast.[77] The communities indicated that they would like a more diversified product in which communities could participate more fully—community tourism—in which they could expose their culture, history, and natural environment.

POLICY MECHANISMS REQUIRED TO INTEGRATE RISK MANAGEMENT AND DISASTER MITIGATION IN TOURISM DEVELOPMENT PLANNING

The findings outlined above indicate that urgent steps need to be taken to ensure that risk analysis and management and disaster mitigation are integrated into tourism development planning. Natural hazards can result in severe damage to tourism facilities and retard economic growth if risk mitigation measures are not implemented. Social risks such as squatting can result in deforestation, soil erosion, and pollution of coastal water resources and resultant damage to coral reefs, mangroves, and other marine resources. Damage to these natural protective systems increases the risks of natural hazards such as landslides, flooding, wave action, and storm surges. Therefore, it is important that risk analysis and management techniques be applied to both natural and social risks.

Risk management techniques need to be applied to natural, environmental and social risks. Disregarding natural and social risks may be due to the following reasons:

1. The decision-maker is unaware of the existence of risk, or has prejudged it as insignificant;
2. The decision-maker recognizes the risk, but accepts it as inevitable, since he or she does not know what to do about it;
3. The decision-maker is aware of the risk, has some knowledge about response options, but cannot determine which response is worth implementing; and
4. Inadequate strategies for improving risk perception and response.

---

76. J. Wilson, Report on Community Consultations Meetings Organized as part of the Master Plan for Sustainable Tourism. Kingston, 1997.
77. Halcrow Ltd. (n. 17 above).

Hence, risk *perception* is critical to the process of managing risk to reduce the impacts in the tourism industry. Understanding how a community perceives a particular risk allows decision-makers to determine the level of communication needed in preparing communities against future hazards. It is also the first step in determining public response to natural and social hazards and developing a comprehensive risk management policy.

Policy mechanisms are required to ensure integration of risk analysis and management in the tourism industry. The following sections outline some of the mechanisms that can be taken in facilitating this integration.

Mechanisms Required to Ensure Integration of Risk Analysis with the Tourism Industry (Natural Hazards)

Building Consensus on Risk Management in the Tourism Industry

Vermeiren[78] proposed that any strategy for risk management should be sector specific, and should be carried out with maximum involvement of the different interest groups of the sector, and in particular those that are directly affected by disasters. This approach is effective in raising the hazard awareness of decision-makers in the sector, and in facilitating implementation of recommended loss-reduction measures. It would consist of the following four steps:

1. Identify interest groups in the tourism sector and related sectors (e.g., infrastructural sector) directly affected by disasters;
2. With direct involvement of those interest groups, identify possible mitigation options;
3. Prepare a systematic economic analysis of the loss-reduction alternatives for the sector. (Realistic estimates will be needed for the costs of mitigation measures and of the losses avoided to the sector and the economy as a whole by implementing these i.e., a cost-benefit analysis of this approach); and
4. Adopt an effective institutional framework in which public, private and community interests can agree to set priorities among loss-reduction measures and cooperate in their implementation.

Development of a Comprehensive Risk Management Policy and Plan for the Tourism Industry

It has been established that natural disasters retard economic growth and result in social dislocation. Despite the frequency with which natural disas-

---

78. J. Vermeiren, *Natural Disasters: Linking Economics and the Environment with a Vengeance,* Report prepared under the OAS/Caribbean Disaster Mitigation Program. Washington, DC, 1989.

ters occur, there is little natural hazard vulnerability analysis in the development planning process. The ODPEM reviews some of the development applications at the request of the Town Planning Department in Jamaica. However, it is not a requirement by law. Where disaster management and risk assessment processes exist, they are often myopic and reactive.

Decision makers give risk management a low priority rating and a reluctance to adopt mitigation policies that include:

- land use regulations;
- building and safety codes; and
- disaster insurance programmes.

However, a comprehensive risk management policy and plan need to be developed. Risk management and disaster mitigation call for long-term planning and involve the following steps:

- Hazard identification and assessment;
- Collection of Risk Information (including detailed mapping of natural and social resources in the coastal zone, use of models to forecast changes in climate and impacts of various hazards)
- Identifying hazard management goals and objectives;
- Identifying potential strategies and developing a hazard management plan;
- Implementation; and
- Review.

Careful identification and assessment of hazards that may affect communities and the tourism industry should be the basis of any risk management policy and plan which is developed for the main resort areas. The Risk Management Policy and Plan should be developed within the context of an integrated coastal zone management process. Out of this identification and preliminary assessment, a clearer direction of hazard management goals can be constructed. All of the potentially affected groups should be consulted and involved through an interactive planning process. Pursuing potential strategies identified will be determined by the constraints of the community and industry resources and the objectives of the risk management programme. Any negative consequences of the selected strategy should be addressed in the risk management plan, and this plan should undergo continuous review and evaluation.

In terms of the tourism industry, integrating these varying steps are necessary to reduce the social dislocation associated with natural hazards. Pre-impact planning can protect small and large-scale operators alike and reduce the "downtime" experienced in the industry. Policy addressing development needs to consider future impacts of disasters and should make preparation both in legislature and infrastructure to minimize these impacts. It is also imperative that future tourism management strategies be designed to accommodate flexibility in decision making. For this to be successful, local communities and business enterprises have to be included in all elements of disaster management.

It should be borne in mind that the ultimate goal of any integrated risk management programme is to anticipate rather than suffer the consequences of natural disasters; to rapidly reduce the risk or mitigate the consequences and to do so with minimum burden on social and economic resources.

Enforcement of Building Codes and Zoning Regulations
To reduce vulnerability in the tourism sector, new structures must be located in areas safe from hazards or they should be able to resist their impacts. This requires changes in public and private approaches to location, design, construction, and maintenance of structures. Techniques for eliminating or reducing property losses due to hurricanes and earthquakes are well known. More education and training are needed to transfer these techniques to designers and builders in both formal and on-the-job settings.

In Jamaica, the Building Code should be modified to ensure that it assigns higher standards to buildings of high importance (hospitals, churches, and tourism facilities). An Integrated Coastal Zone Management approach should be used to properly manage and reduce the land-based sources of pollution in the coastal zone. Zoning regulations should be enforced and setback regulations must be based on risk information and not on the characteristics of the shoreline such as foreshore slope.

Engage Communities in Natural Hazard Mitigation
The tourism industry needs to foster community participation as part of its hazard reduction objectives. Ideally, there should be community-based hazard assessment. This assessment should build on the community-based disaster preparedness committees of the ODPEM (which are currently focused on emergency response) to develop hazard mitigation. At the national level, the country must incorporate hazard mitigation into all its activities, from the traditional realms of emergency management and development control through economic planning, education, tourism, and infrastructure development.

Establishment of Fund Facility and Incentives for Natural Hazard Mitigation in the Small Hotel Sector
As a result of the economic challenges facing Jamaica, the small hotel sector is experiencing tremendous difficulties with maintaining viability and maintaining the capital stock. Therefore, the small hotel sector and the attractions subsector may not be able to implement natural hazard mitigation mea-

sures unless it has access to relatively low cost funds and/or duty free concessions to obtain the equipment necessary to implement structural mitigation measures. Therefore, a fund facility should be developed to provide funding for the implementation of risk mitigation measures in these facilities. The insurance sector could also offer lower insurance premiums to clients who implement mitigation measures as an incentive.

The Social Risk Management Mechanism

Most of the risks are due in part to the lack of social and economic integration of local communities with the tourism industry and the lack of economic opportunities in the country. In order to ensure the long-term sustainability of the tourism industry, further analysis of the social risks and the mechanisms that are required to ameliorate these risks in the tourism sector must be conducted. However, our initial assessment indicates that the initiatives that are required to mitigate these social risks reside in eradicating poverty and creating economic opportunities for the poor.

The following initiatives should be taken to reduce the social risks affecting Jamaica's tourism industry.

Assess the Vulnerability of the Resort Areas Based on Social Risks
In order to build consensus in the tourism industry to address social risks, vulnerability assessments of the various resort areas in Jamaica should be conducted. With direct involvement of the main stakeholders in the industry, identify possible mitigation options. A systematic economic analysis of the loss-reduction alternatives for the sector should be prepared. Realistic estimates will be needed of the costs of mitigation measures and of the losses avoided to the sector and the economy as a whole by implementing these i.e., a Cost-Benefit Analysis of this approach. Finally, the stakeholders should adopt an effective institutional framework in which public, private, and community interests can agree to set priorities to deal with these social risks and cooperate in their implementation.

Interministerial Committee to Facilitate Social and Economic Linkages with the Tourism Industry
The economic sectors that could be linked with the tourism industry include music, entertainment, agriculture and information technology. Some of the challenges that the industry faces are poor infrastructure, visitor security and harassment and environmental degradation. These issues fall under different ministries and therefore, to facilitate and strengthen the economic linkages with the tourism industry there is need for an interministerial committee that reports to Cabinet to ensure coordination of these sectors. For example, small farmers could be organized into a cooperative. This coopera-

tive could buy seeds and other inputs in bulk and provide training and technology to improve their efficiency and sell these products to the tourism industry at a reasonable price. Under this scheme, the science and technology agencies should develop or adopt technologies to improve agricultural production efficiency, thereby reducing the costs of these products.

Creation of a Fund Facility for Social Hazard Mitigation
The social risks associated with the tourism industry are due in part to poverty and the lack of economic opportunities available to the poor. Projects that encourage the economic independence of the poor, micro-enterprises and the small business sector could be financed by a fund created by the tourism sector. Charging an additional 50 US cents in room rates could finance social and environmental projects in the resort areas. Alternatively, members of the tourist trade could establish a non-profit organization whose mandate is to provide economic opportunities to the poor, and to develop linkages between the tourism industry and other economic sectors in Jamaica.

More specifically, the development of a fund facility for social risk mitigation could be explored. Discussions have been held with members of the tourism industry regarding the establishment of a fund facility for social risk mitigation. The Jamaica Hotel and Tourist Association and several tourism agencies have expressed their support for the fund facility. However, the modalities of financing and implementing the fund need to be agreed on among the parties.

For the purposes of this discussion, social funds are defined as agencies that finance small projects in several sectors targeted to benefit a country's poor and vulnerable groups based on a participatory manner of demand generated by local groups and screened against a set of eligibility criteria.[79] The proposed fund facility for social risk mitigation will operate as an agency that appraises, finances and supervises implementation of social investments identified and executed by a wide range of actors. It will also serve a catalytic role in the development and facilitation of projects by community members to improve linkages between players in the tourism sector and the other economic sectors. One of the main objectives of the fund facility will be to facilitate long-term poverty eradication and social capital creation in marginal areas and populations primarily in the resort areas.

Beneficiary assessments among the target groups should be used to identify the priorities and needs of the very poor in the community, and ensuring participation and ownership among the community members. Promoting participation in determining the objectives and functioning of the fund is a

---

79. S. L. Jorgensen and J. Van Domelen, "Helping the Poor Manage Risk Better: The Role of Social Funds," in *Shielding the Poor: Social Protection in the Developing World,* ed. N. Lustig (Washington, DC: The Brookings Institution and Inter-American Development Bank, 2000), chap. 5.

vital concomitant to success of the effort. Moreover, participation builds the ability of beneficiaries to assume responsibilities throughout the life of the project. Ultimately, this will enable community ownership of the project.

A combination of research and evaluation tools, including in-depth interviews, participatory learning and action (PLA) techniques, focus group discussions and other participatory techniques should be employed by the fund facility. The key method that will be used at a community level will be the participatory learning and action tools combined with in-depth interviews with key informants. Some of the PLA tools that may be used depending on the community and the issues are: listings, ranking, scoring, matrices, daily round, calendar, trend analysis, causal flow, impact analysis, community action plan, pie chart, drama/animation, community cartoons, and Roti diagram.

The development of the fund facility would involve the implementation of an industry-wide approach within the context of poverty eradication through community empowerment. Members of the tourist trade could establish a not-for-profit organization whose mandate is to provide economic opportunities to the poor, and to develop linkages between the tourism industry and other economic sectors in Jamaica. For example, facilitating further linkages between the tourism and agriculture sectors, where there could be more cooperation between hoteliers and small farmers.

This not-for-profit organization could be financed from membership fees, with contributions varying based on the member company's annual turnover. It is envisaged that the small hotel sector will provide in-kind contribution instead of cash since the sector is currently facing severe economic challenges (particularly after the September 11, 2001, disaster and the upsurge of violence in a section of Kingston, Jamaica, in July 2001).

This not-for-profit organization, whose membership would comprise an industry coalition of hoteliers, the attractions subsector, the cruise shipping sector and other segments of the tourist sector, would be created to mitigate the social risks in the tourism sector. An executive director would head the organization supported by administrative and technical staff.

The membership of the proposed NGO that will administer the fund facility should be focused on the following activities:

1. Identifying social risks specific to the industry;
2. Establishing a non-profit organization whose mission will be to educate, increase public awareness of the impacts of the social risks affecting the tourism sector, and empowering community groups to reduce and mitigate social risks in their communities;
3. Based on the findings of the coalition, suggest mechanisms for improving responses to social risks by the industry and develop effective partnership with community groups and government to reduce the risks.

It is suggested that the coalition be formed from key industrial players who are willing to fund the programme. It is recognized that any steps to be taken will require the joint effort of the government, NGOs, community groups and the tourism industry. Initially focusing on identifying the main causes of social risks unique to the industry, the coalition can create a roadmap for dealing with these risks.

A similar model has been used in the food and beverage industry in Jamaica to promote litter reduction and recycling in Jamaica and has had reasonable success. In 2000, 24 million PET plastic bottles were recycled in Jamaica. There was a 90 percent increase in the number of collection centers for plastic containers and significant improvement in the recycling infrastructure of the island under the recycling programme. The budget of Recycle For Life—the not-for-profit organization established by the soft drink industry to promote litter reduction and recycling in Jamaica is approximately J$30 million (US$750,000) is financed mainly by the soft drink industry (70 percent), contributions by other private sector companies and sale of PET plastic regrind overseas. The Recycle For Life Programme has been a tool in poverty eradication, community development, and solid waste management.

This model has been developed in collaboration with community groups, NGOs, and the government. Participating companies are recognized for their involvement in the program. Currently, Pepsi Cola, Coca-Cola, Jamaica Beverages Ltd, and Jamaica Drink Ltd, are the major sponsors of the program. Other sponsors include Shell Company (W.I.), McDonalds, Burger King, WISYNCO Trading, West Indies Synthetics Ltd., Seprod Jamaica Ltd, and Grace Kennedy. Peak Bottling Ltd., Cool Runnings, and Dunlop Corbin Communications are involved in the program. Other private sector companies have expressed an interest in the program.

CONCLUSIONS

The Caribbean region has experienced a dramatic upsurge in the number of hurricanes and tropical storms passing through the region. These events have focused attention on the destabilizing effects of natural hazards. While storms and related flooding and landslides are the most frequently experienced hazards, earthquakes and volcanoes also pose significant risks to the region.

As a consequence, property insurance, the traditional mechanism for reducing economic risk from catastrophic events, may no longer be as available or affordable as in the past. This development is forcing property owners and developers to seriously look at other mechanisms to minimize the consequences of natural disasters. It is now being recognized that integrating disaster risk assessment in the tourism sector is vital to the protection of the tourism product. Therefore, the tourism sector should develop a "practical and effective loss reduction strategy and programme in response to the risks posed by natural disasters to the industry." Similarly, the identification of social risks and their possible mitigation are useful in protecting the tourism product.

In order to integrate risk assessment and management and disaster mitigation in the tourism development process, a public/private sector partnership is required. This public/private partnership should include the main stakeholders in the tourism industry and other related sectors, the insurance industry and community-based organizations. Under this arrangement, government would provide the regulatory framework that is required to encourage nonstructural mitigation measures (e.g., land use regulations, fiscal incentives, natural resources management policy, and structural measures such as building codes). The private sector should finance schemes to implement mitigation measures for natural and social risks.

Initiatives Required to Reduce Risks from Natural Hazards

In order to reduce the risks associated with natural hazards in the tourism industry, certain policy initiatives need to be taken by the industry. These policy initiatives include:

a) Building consensus on risk management in the tourism industry among the main stakeholders;
b) Developing a comprehensive risk management policy and plan for the tourism industry;
c) Enforcing building codes and zoning regulations;
d) Engaging communities in natural hazard mitigation; and
e) Establishing a fund facility and incentives for natural hazard mitigation.

Specific interventions are required to ensure a multistakeholder approach involving community groups, NGOs, private sector and government. There is a need for proper enforcement of the zoning regulations and building code in Jamaica. Stricter standards should be assigned to buildings of higher importance (hospitals, churches, and tourism facilities). Therefore, the enforcement capabilities of the Parish Councils need to be improved.

There is need for an Integrated Coastal Zone Management approach in order to properly manage the zoning of tourism facilities and to reduce the land-based sources of pollution in the coastal zone. These initiatives will in the long-run reduce the natural hazard vulnerability of the island. Multi-hazard mapping and analysis are also needed to determine the overall risk of natural hazards to tourism facilities and related infrastructure such as roads, airports and communication systems. In determining the vulnerability of the tourism sector to various natural hazards, it is important that the related infrastructure sectors and the communities that surround the resort areas are evaluated.

Initiatives Required to Reduce Social Risks in the Tourism Industry

In order to reduce social risks in the tourism industry, mechanisms need to be put in place to facilitate social and economic integration of local communities with the tourism industry. Call Associates[80] argues that based on their analysis of the main initiatives taken to facilitate social and economic integration of local communities with the tourism industry, one may conclude that the positive impacts of tourism can be extensive if community involvement is paramount in the planning and implementation process.

In the parts of the country where there is a deliberate attempt to facilitate social and economic integration of local communities with the Jamaican tourism industry, the following results may be seen:

(a) a more sustainable product has emerged;
(b) virtually no crime or harassment; and
(c) community members have developed additional skills which allow them to gain better access to the benefits of the tourism industry and other economic activities in the community.

Turner and Ash[81] and De Kadt[82] found similar results in their study of the impacts of tourism on communities in Canada, the United Kingdom, and the Caribbean when the community is involved.

It is only in recent times that there have been concerted attempts to involve communities in tourism development and management in Jamaica. The views articulated by various communities are that they have been marginalized in the tourism development process, and they wish to fully participate in the process. They have indicated that community tourism and ecotourism are alternative tourism products in which they can be involved based on the community consultations organised by the Office of the Prime Minister (Tourism Division).

The social effect of tourism development can be represented by the

80. Call Associates (n. 48 above).
81. L. Turner and J. Ash, *The Golden Hordes: International Tourism and the Pleasure Periphery* (London: Constable, 1995).
82. E. De Kadt, "Social Planning for Tourism in the Development Countries," *Annals of Tourism Research* 6, 1 (1979): 36–48.

polarities of a social interaction continuum. At one extreme, tourism-induced social change can lead to development and socio-economic benefits, while at the other end tourism can lead to dependency and social discrepancies. These two extremes are evident in the island. Within the continuum, island destinations often find themselves dependent "on two major factors: the historic events that frame its current touristic conditions and the extent to which its political economy is dominated by exogenous forces."[83]

The social and economic issues facing the tourism industry are due to a complex set of issues. These issues are based on the socio-cultural development of the modern Jamaican society, economic problems facing the country as a result of the SAP (and by extension the economic policies) it has undergone over the last two decades, and the development of the tourism industry as a separate sector which is not integrated into the economy.

Under the SAP taken by the government of Jamaica, the economic policies that were pursued included devaluation, reduced spending on physical and social infrastructure in the communities and the growth of the informal sector. These social discrepancies have been exacerbated because there has been no deliberate and targeted intervention to facilitate social and economic integration of local communities with the tourism industry by successive governments in Jamaica. Therefore, it has occurred on an ad hoc basis. In some resort towns (e.g., Ocho Rios), it would appear as if the communities have reached the final stages of Doxey's Irridex, as highlighted in Figure 4.

Doxey's Irridex[84] and Milligan's modification of the index as illustrated by Ryan,[85] documents the process of tourism development over time and the effects of increased tourist pressure on an area. These studies highlighted the relationship between tourism, the community and the external environment. Unplanned tourism with little involvement of the local community results in hostility towards the visitor, a rise in crime and violence and desperate attempts to earn a living which manifest itself as harassment. The dislike of tourists because of culturally unacceptable behaviour, congestion, noise, crowding, and squatting in communities, is manifested in exploitation and abuse of tourists. Excessively inflated prices, outright aggression and physical abuse of tourists are outward demonstrations of a community's

83. P. Burns and A. Holden, *Tourism: A New Perspective* (London: Prentice Hall, 1995).

84. G. Doxey, "A Causation Theory of Visitor-Resident Irritants: Methodology and Research Inferences, *Proceedings of the Travel Research Association*, 6th Annual Conference, San Diego, California, 1975.

85. C. Ryan, *Recreational Tourism: A Social Science Perspective* (London: Routledge, 1991).

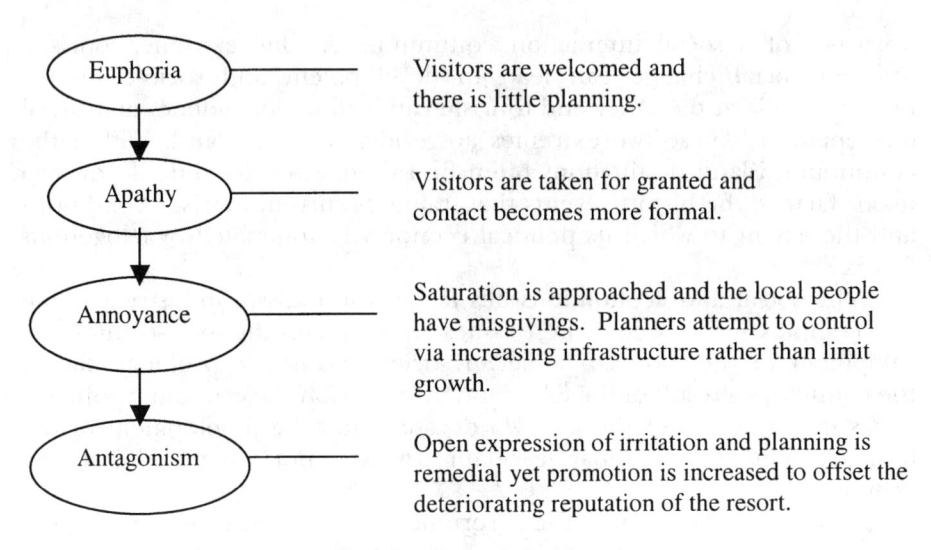

```
Euphoria ──────── Visitors are welcomed and
                  there is little planning.
   ↓
 Apathy  ──────── Visitors are taken for granted and
                  contact becomes more formal.
   ↓
                  Saturation is approached and the local people
Annoyance ─────── have misgivings. Planners attempt to control
                  via increasing infrastructure rather than limit
                  growth.
   ↓
                  Open expression of irritation and planning is
Antagonism ────── remedial yet promotion is increased to offset the
                  deteriorating reputation of the resort.
```

FIG. 4.—Doxey's irridex

intolerance to tourism. In other words, both the economic and community capacity for tourism is surpassed.[86]

It is obvious that if the mechanisms had been developed to ensure that tourism development is used as a tool to augment existing towns and villages, there would have been greater social and economic integration of local communities with the tourism industry. However, there was little attempt to transform resort centres into more interactive communities. Therefore, tourism enclaves developed, increasing the demarcation between tourist facilities and local communities.

Successive governments did not achieve the objective of reducing the level of economic evaporation from tourist expenditure so as to ensure that a higher percentage of the tourist dollar remains in the country and in particular within the area where it is spent. Dixon[87] argued that there is very little re-investment of the income from tourism into Montego Bay. This could be easily argued for Negril and Ocho Rios.

Call Associates[88] indicated that the following characteristics are observed in all the case studies of successful integration of local communities with the tourism industry:

86. P. W. William and A. Gill, *Carrying Capacity Management in Tourism Setting: A Tourism Growth Management Process* (Vancouver, BC: Centre for Tourism Policy and Research, Simon Fraser University, 1991).
87. Dixon (n. 59 above).
88. Call Associates (n. 48 above).

a. Shared vision of development among the main stakeholders (government, private sector and communities);
b. Tourism industry, in partnership with the private sector and the community, provides social amenities such as clinics and basic schools in the community;
c. Broadening of the economic base of the area in which the tourism activity is being conducted; and
d. Strong linkages between the tourism industry and other economic activities in the community (e.g., agriculture, craft industry, etc.).

The initiatives taken to integrate local communities with the tourism industry need to be broadened and deepened in Jamaica.

Environment and Coastal Management

The Prince Edward Islands: Southern Ocean Oasis[†]

Evgeny A. Pakhomov
Department of Zoology, University of Fort Hare, South Africa

Steven L. Chown
Department of Zoology, University of Stellenbosch, South Africa

INTRODUCTION

In the austral summer of 1947–1948, South Africa annexed the sub-Antarctic islands, Marion and Prince Edward, mostly for strategic reasons that had been the topic of considerable discussion between General J.C. Smuts and the U.K. government. With the stroke of a pen, some formidable logistics, and a great deal of initial secrecy, Operation Snoektown effectively added a whole series of ecosystems to South Africa's already rich natural heritage. South African scientists and their colleagues abroad immediately recognized the opportunities for research into the functioning of Southern Ocean marine and terrestrial systems, and scientific work followed soon after the establishment of the weather station on the east coast of Marion Island.

Since that time, an immense fund of knowledge concerning the islands and the surrounding oceanic ecosystems has accrued. However, as with all human endeavours, this knowledge has come at some cost to the natural heritage of the islands, mostly as a consequence of the introduction of several destructive alien species. Although sealers interested in both oil and fur contributed several exotic species early on, including house mice and the grass, *Poa annua*, the list of introduced species has grown substantially since the annexation of the islands. Many of these species have also spread widely around Marion, and in some cases Prince Edward Island. These species, and

---

[†] EDITORS' NOTE.—The authors wish to note that this synthesis is based on the long-term research efforts of many South African and international researchers, too numerous to name, who have worked and are working on and around the Prince Edward Islands and have made considerable contributions. The South African Department of Environmental Affairs and Tourism (DEA&T) and several South African universities provided logistic and financial support for research undertaken on the islands. The authors gratefully acknowledge many fruitful years of cooperation with the Directorate Antarctica and Islands of the DEA&T. The authors wish to especially thank William Froneman, who inspired the title of this synthesis.

© 2003 by The University of Chicago. All rights reserved.
0-226-06620-7/03/0017-001301.00 *Ocean Yearbook* 17:348–379

the likelihood of other biological invasions, especially given the rapid climate change (warming and drying) at the islands, continue to constitute significant conservation threats. In consequence, over the last decade there have been increasing efforts to better manage both research and conservation at the islands.

Here we provide an overview of the outcome of the research that has been undertaken since the Prince Edward Islands were annexed, highlighting some of the more important ecological lessons that have emerged from this work. We also explore conservation threats to the islands, associated especially with climate change and invasive species, and how research at the islands has to be managed to minimize these threats while maximising the scientific returns. We pay specific attention to the Management Plan that has been adopted for the conservation management of these IUCN Category I Special Nature Reserves, and the kinds of research that this plan calls for. In addition, we draw attention to research information that has also been essential for ongoing management of the islands. We conclude by pointing out that while the Prince Edward Islands present significant opportunities for understanding marine-terrestrial interactions, the effects of global climate change on these interactions, and on the outcomes of biological invasions, we emphasize the fact that this work will have to be carefully managed in the context of the conservation of the marine and terrestrial systems.

ENVIRONMENTAL AND HISTORICAL BACKGROUND

The Prince Edward Archipelago, including Marion and Prince Edward Islands, is situated halfway between the continents of Africa and Antarctica, approximately 2000 km southeast of South Africa (fig. 1). The nearest landfall to the archipelago, the Crozet Islands, is ca. 950 km to the east. Marion Island is the larger island at approximately 250 km^2 in extent, while Prince Edward Island, which is 22 km northeast, is only about 45 km^2 in area. Rising steeply from depths of >3000 m, the islands are separated by a shallow saddle, which varies between 45 and 260 m in depth.[1] The Prince Edward Islands (PEI) are relatively young, ca. 450,000 years old,[2] thin soiled volcanic islands with many terrestrial communities deeply subsidized by the

---

1. C. Hännel and S. Chown, *An Introductory Guide to the Marion and Prince Edward Island Special Nature Reserves 50 Years After Annexation*, (Pretoria, South Africa: Department of Environmental Affairs and Tourism, Directorate Antarctica and Islands, 1998), pp. 1–80.
2. I. McDougall, W. J. Verwoerd and L. Chevallier, "K-Ar geochronology of Marion Island, Southern Ocean," *Geological Magazine* 138 (1978): 1–17.

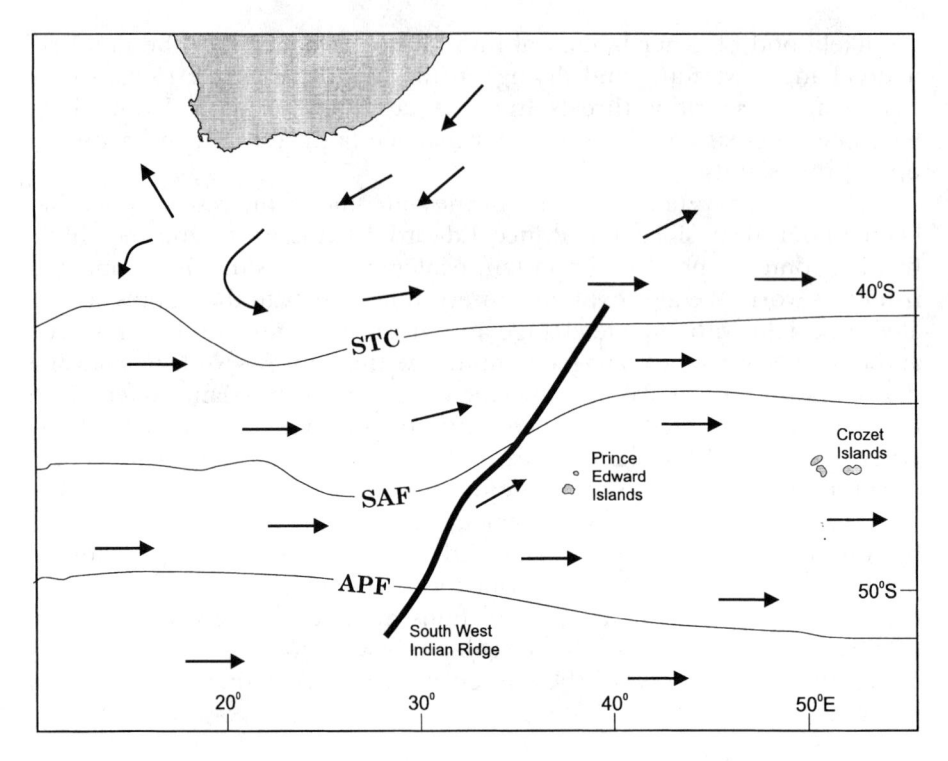

FIG. 1.—The positions of the Prince Edward and Crozet islands in relation to African continent, major frontal systems (average position) of the Southern Ocean and the South West Indian Ridge. STC: subtropical convergence, SAF: sub-Antarctic front, APF: Antarctic polar front.

allochthonous[3] input of energy and nutrients from the surrounding seas through the medium of land-based marine predators.

Situated in the roaring forties, climatic conditions on the islands are characterized by low temperatures with little seasonal change (<5°C), hyperoceanic humidity (annual average 83 percent) and high, but declining precipitation (rainfall ca. 2000 mm per annum).[4] Westerly winds predominate on the islands and winds of gale force may prevail on 150 days per year.[5] Frequent cloudy conditions limit incoming radiation, and, on average, only

3. Allochthonous input—input of productivity from other systems, which usually increases resource availability in a local system. This is an opposite of an autochthonous input, which is a result of a local productivity enhancement.
4. E. M. van Zinderen Bakker Sr., "Origin and general ecology of the Marion Island ecosystem," South African *Journal of Antarctic Research* 8 (1978): 13–21.
5. Ibid.

ca. 210 calories/cm^2/day reach the island's surface, thus limiting plant growth.[6]

Like other sub-Antarctic islands, the PEI lie directly in the path of the easterly flowing Antarctic Circumpolar Current (ACC).[7] Two major frontal systems, namely the Sub-Antarctic Front (SAF) to the north and the Antarctic Polar Front (APF) to the south, are found in the vicinity of the PEI (fig. 1). As a consequence, the climate of the PEI is influenced by oceanic and atmospheric circulation of both Antarctic and sub-Antarctic origin. The water flow conditions in the vicinity of the PEI appear to be dominated by this through-flow regime.[8] At times, however, water can be trapped between the islands for prolonged periods as a consequence of eddy formation.[9] The dynamics of this process in the inter-island shelf region are not yet comprehensively understood. The position of frontal systems, particularly the SAF, in the vicinity of the PEI has been hypothesized to play an important role in forming the macro- and mesoscale water circulation around the islands.[10]

After the discovery of PEI in 1663, the first observations on the marine environment in the vicinity of the islands were undertaken during the sealing period in April 1840, when the HMS *Erebus* of Captain James Clark Ross's expedition conducted soundings and some dredges at the PEI. Thirty-three years later, the British scientific vessel HMS *Challenger* visited the islands on Christmas Day 1873. Again, extensive soundings as well as dredging and trawling took place. The chart that was compiled after a few days' visit remained the only reliable graphic source of information for 74 years until the time of the Islands' annexation by South Africa in December 1947/January 1948. Since 1873 and after annexation until the mid-1970s, little offshore biological work was undertaken in the vicinity of the PEI. In 1976, French and South African scientists on board the research vessel MS *Marion-Dufresne* studied benthos, oceanography, primary productivity, plankton, fish, and seabirds of the PEI region.[11] In 1978 the South African Scientific

---

6. V. R. Smith, "A quantitative description of energy flow and nutrient cycling in the Marion Island terrestrial ecosystem," *Polar Record* 18 (1977): 361–70.

7. J. R. E. Lutjeharms, "Location of frontal systems between Africa and Antarctica: some preliminary results," *Deep-Sea Research* 32 (1985): 1499–1509; I. J. Ansorge and J. R. E. Lutjeharms, "Twenty-five years of physical oceanographic research at the Prince Edward Islands," *South African Journal of Science* 96 (2000): 557–65.

8. van Zinderen Bakker Sr. (n. 4 above).

9. Ibid.

10. E. A. Pakhomov and P. W. Froneman, "The Prince Edward Islands pelagic ecosystem, south Indian Ocean: a review of achievements, 1976–1990," *Journal of Marine Systems* 18 (1999): 355–67; van Zinderen Bakker Sr. (n. 4 above).

11. P. G. H. Frost, J. R. Grindley and T. H. Wooldridge, "Report on South African participation in cruise MD08 of MS *Marion Dufresne*, March–April 1976," *South African Journal of Antarctic Research* 6 (1976): 28–29; J. R. Grindley, "Marine ecosystems of Marion Island," *South African Journal of Antarctic Research* 8 (1978): 38–42.

Committee for Antarctic Research (SASCAR) recognized, based on the results of this expedition, the importance of the offshore research in the vicinity of the PEI to enhance the understanding of the functioning of the Islands' terrestrial ecosystem.[12]

Six exploratory surveys in the vicinity of the PEI took place between 1980 and 1985. These surveys provided much information on pelagic ecosystem dynamics and highlighted an urgent need for further interdisciplinary studies of the offshore environment to clarify trophic links between the marine and terrestrial systems. The presence of millions of birds and seals on the Islands raised a crucially important question as to the origin of their food. To address this question, the Marion Offshore Ecological Study (MOES, 1987–1990) was initiated.[13] The main aims of the MOES were a) to establish the existence of an "island mass effect,"[14] to examine its dynamics and clarify the physical mechanisms for its existence; b) to study factors responsible for enhanced primary production within the inter-island area and the ways in which primary production is partitioned between the pelagic and benthic food webs; and c) to examine the relationships between these trophic subsystems and the land-based predators.[15] The investigations were largely carried out during two dedicated cruises of the *SA Agulhas*, MOES-I (1987) and MOES-II (1989).

The follow-up study, the Marion Island Oceanographic Survey (MIOS), was initiated in 1996 as a major part of the 5-year programme called "The Prince Edward Island's life support system and variability of living resources in the Southern Ocean." The main aim of the programme was to provide scientifically based information to the Prince Edward Islands Management Committee (the body responsible for the conservation management of the Islands), with a major emphasis on the variability in food supply to the biological communities of the PEI in response to the interaction between the Antarctic Circumpolar Current (ACC) and the Islands.[16] In total, five cruises

---

12. SASCAR, "South African southern ocean research programme," *South African National Scientific Programme Report* 134 (1987): 1–58.

13. J. R. E. Lutjeharms, "A history of recent South African marine research in the Southern Ocean," *South African Journal of Antarctic Research* 21 (1991): 159–64.

14. The "island mass effect" is enhancement of biological productivity and biomass around the island systems. This may include increases in chlorophyll and plankton concentrations in surrounding waters, retention of larval and juvenile stages of fish, recirculation and increases in larval abundance, eddies trapped and created on the lee side of the island systems.

15. R. Perissinotto, B. P. Boden and C. M. Duncombe Rae. *Characterization of the Physical and Chemical Environment of the Prince Edward Island Seas. The Origins and Distribution of Potential Energy Production in the Prince Edward Island Seas.* South African National Antarctic Research Programme (SANARP). Final Project Report, FRD-CSIR. Pretoria, South Africa, 1990, p. 30.

16. E. A. Pakhomov, P. W. Froneman and C. D. McQuaid. *The Prince Edward Island's Life-Support System and the Variability of Living Resources in the Southern Ocean.*

(MIOS I to V, annually) were carried out to the vicinity of the PEI between 1996 and 2001.

In April 2001, a 3-year Marion Offshore Variability Ecosystem Study (MOVES) was initiated with the objective of assessing variability in the PEI upstream pelagic ecosystem resulting from interactions between the ACC and the South-West Indian Ridge. This programme aims to improve the formulation of a management strategy to cope with environmental changes and to develop conservation policies for sensitive biological populations on the Islands by studying mesoscale plankton dynamics (accumulation and advection) upstream of the PEI.

The PEI accommodate up to 5 million top predators seasonally, including flying seabirds, penguins, and seals.[17] These islands support commercially exploited demersal fish populations and attract migrating whales to rich feeding grounds.[18] Birds and seals come to the sub-Antarctic islands, including the PEI, for two major reasons: either to moult or to reproduce. As both of these activities are energetically expensive, identifying the "life support system" of these populations and the interactions between the marine and terrestrial systems has long been a goal of research at the Islands. Hence, our aims here are to provide a synthesis of the available information on the "life support system," to briefly examine the effects of marine allochthonous energy inputs on the terrestrial ecosystem of the islands, and to provide an overview of the way in which this research information has been used in the development of management policies for the PEI.

MINERAL CYCLING AND MARINE INPUT TO THE ISLAND'S ECOSYSTEMS

Small oceanic islands, including the PEI, depend partly on their own resources and atmosphere for mineral requirements, but mainly on the surrounding ocean.[19] Mineral cycling in the PEI includes six principal processes:

South African National Antarctic Research Programme (SANARP). Final Project Report. Pretoria, South Africa: 2001, pp. 1–53.

17. A. J. Williams, W. R. Siegfried, A. I. Burger and A. Berruti, "The Prince Edward Islands: a sanctuary for seabirds in the Southern Ocean," *Biological Conservation* 15 (1979): 59–71; P. R. Condy, "Annual food consumption and seasonal fluctuations in biomass of seals at Marion Island," *Mammalia* 45 (1981): 21–30; C. Guinet, Y. Cherel, V. Ridoux and P. Jouventin, "Consumption of marine resources by seabirds and seals in Crozet and Kerguelen waters: changes in relation to consumer biomass 1962–85," *Antarctic Science* 8 (1996): 23–30.

18. CCAMLR, *Report of the Fifteenth Meeting of the Scientific Committee,* (Tasmania, Australia: CCAMLR, 1996), p. 119.

19. van Zinderen Bakker Sr. (n. 4 above).

1. Rock decomposition: The basalts, pyroclasts, and tuffs of the island degrade as a consequence of high humidity and regular frost, contributing to the soil and water chemistry. The contribution is limited due to low rates of decomposition.[20]

2. Evaporation of ocean water and salt spray: It is not only the coastal regions that are affected by this process, because salt is deposited by rain all over the islands. As a consequence, the ionic dominance and chemical composition of the water on the islands are closely related to that of the surrounding ocean.[21]

3. N_2-fixation in mires: In the extensive mires, the gelatinous mass of *Cyanophyta*, bacteria, and lichens are major groups responsible for the fixation of atmospheric nitrogen.[22] This process is a significant contributor to the overall nitrogen budget, especially during brief periods of sunshine when the mires act as energy traps.[23] However, the overall nutrient requirement on Marion Island is high owing to considerable primary production (400–2000 $gC/m^2/year$), mostly because of a long growing season despite the reasonably adverse environment. This high primary production on the island anticipates a large requirement for nutrients. In coastal areas, pelagic vertebrate colonies act as a source of nutrients, as do burrowing seabirds inland (see later). In other areas plants are dependent on nutrient release from either live or decaying material. However, there are no macroherbivores on the island, and insects play only a small role as herbivores. Nonetheless, the islands support large numbers of soil arthropods. Thus, these organisms, together with the soil micro-organisms, are responsible for the bulk of nutrient release.[24] Consequently, most of the energy and nutrients incorporated in primary production in areas that are not directly influenced by seabirds and seals are made available through detritus, rather than by a grazing cycle.[25] Because most of the plant communities occur on soils with low

20. E. J. D. Kable, A. J. Erlank and R. D. Cherry, "Geochemical Features of Lava," in *Marion and Prince Edward Islands: Report on the South African Biological and Geological Expeditions* 1965–1966, eds. E. M. van Zinderen Bakker, J. M. Winterbottom and R. A. Dyer (Balkema, Cape Town: 1971), pp. 78–88.

21. J. U. Grobbelaar, "The lentic and lotic freshwater types of Marion Island (sub-Antarctic): a limnological study," *Verhandlungen der Internationalen Vereinigung Theoretische und Angewandte Limnologie* 19 (1975): 1442–49.

22. R. L. Croome, "Nitrogen fixation in the algal mats on Marion Island," *South African Journal of Antarctic Research* 3 (1973): 64–7; H. J. Lindeboom, "Chemical and Microbiological Aspects of the Nitrogen Cycle on Marion Island (sub-Antarctic),"(Ph.D. Thesis, Gruningen University, 1979).

23. V. R. Smith, "Animal-plant-soil nutrient relationships on Marion Island (sub-Antarctic)," *Oecologia* 32 (1978): 239–58; V.R. Smith, "The environment and biota of Marion Island," *South African Journal of Science* 83 (1987): 211–30.

24. V. R. Smith and M. Steenkamp, "Soil macrofauna and nitrogen on a sub-Antarctic island," *Oecologia* 92 (1992): 201–06.

25. V. R. Smith and M. Steenkamp, "Classification of the terrestrial habitats on Marion Island based on vegetation and soil chemistry," *Journal of Vegetation Science* 12 (2001): 181–98.

levels of nutrients, the nutrient mineralization from organic reserves appears to be the main bottleneck in nutrient cycling and primary production.[26] The absence of herbivores and the reasonably slow rates of decomposition also mean that plant material is transformed into thick peat deposits under favourable conditions of high humidity and low pH, at rates of 0.2–0.9 mm per annum.[27]

4. Freshwater primary production: Pelagial primary production in numerous small freshwater bodies on the island is very low, ca. <100 mgC/m^2/day but may reach 6000 mgC/m^2/day in wallows that are highly nutrient enriched.[28] The primary production of the benthic algal felts and macrophytes may be, however, as high as 42 and 200 times, respectively, that of phytoplankton.[29] Overall, there is little primary production in the unfertilised waters, and even in the biotically fertilized water bodies, there are no fish and the food chain ends with zooplankton.[30]

5. Allochthonous input: Numerous birds and seals transport to the islands large quantities of organic and inorganic matter obtained from kelp, benthic and pelagic communities found over the island shelf, and from pelagic communities far afield. It is recognized that the main source of nitrogen and phosphorus to the island's terrestrial ecosystems is via remains, dung, urine, and guano of sea-going animals. Annually, birds (no data available for seals) deposit on the islands ca. 4318 tonnes dry weight (~ 6.3 × 10^{10} kJ) of guano, carcasses, feathers, and eggshells. In terms of energy (kJ), guano accounts for 83 percent of the total amount, followed by feathers (14 percent), while the contribution of carcasses/eggshells is only 3 percent. This contribution is equivalent to 565 tonnes of nitrogen, 96 tonnes of phosphorous, and 194 tonnes of calcium (table 1). Because the island's soils are deficient in these elements, the plant growth is enhanced in regions of guano deposition.[31] Virtually all deposition (ca. 96 percent of the total), however, occurs on the coastal plain, particularly along the eastern and northern coasts, and deposits are small (<4 percent) in vegetated areas. On average, ca. 0.4 tonnes dry weight/hectare/year of avian-derived matter are deposited on Marion Island's coastal lowland, assuming an even spread of birds over its 100 km^2 area. The avian-derived nitrogen alone roughly amounts to ~12 percent of the total nitrogen content in the above- and below-ground

---

26. V. R. Smith and M. Steenkamp, "Climatic change and its ecological implications at a subantarctic island," *Oecologia* 85 (1990): 14–24.

27. van Zinderen Bakker Sr. (n. 4 above).

28. J. U. Grobbelaar, "The limnology of Marion Island: southern Indian Ocean," *South African Journal of Antarctic Research* 8 (1978): 113–18.

29. J. U. Grobbelaar, "Primary production in freshwater bodies of the sub-Antarctic island Marion," *South African Journal of Antarctic Research* 4 (1974): 40–45.

30. Grobbelaar (n. 28 above).

31. N. J. M. Gremmen, "The Vegetation of the Subantarctic Islands Marion and Prince Edward," (The Hague: Dr. W. Junk, 1981).

TABLE 1.—ANNUAL PRODUCTION (kg/year) OF NITROGEN, PHOSPHORUS AND CALCIUM BY BIRDS AT MARION ISLAND

	Carcasses[a]	Guano[b]	Feathers[c]	Eggshells[d]
Nitrogen	9758	511880	41879	1806
Phosphorus	779	94990	151	194
Calcium	2441	183190	517	8274

SOURCE.—a) A. J. Williams, A. E. Burger and A. Berruti, "Mineral and energy contributions of carcasses of selected species of seabirds to the Marion Island terrestrial ecosystem," *South African Journal of Antarctic Research* 8 (1978): 53–59; b) A. E. Burger, H. J. Lindeboom and A. J. Williams, "The mineral and energy contribution of guano of selected species of birds to the Marion Island terrestrial ecosystem," *South African Journal of Antarctic Research* 8 (1978): 59–70; c) A. J. Williams and A. Berruti, "Mineral and energy contributions of feathers moulted by penguins, gulls and cormorants to the Marion Island terrestrial ecosystem," *South African Journal of Antarctic Research* 8 (1978): 71–74; d) W. R. Siegfried, A. J. Williams, A. E. Burger and A. Berruti, "Mineral and energy contributions of eggs of selected species of seabirds to the Marion Island terrestrial ecosystem," *South African Journal of Antarctic Research* 8 (1978): 75–87.

plant matter of the island's lowland terrestrial vegetation.[32] However, as a consequence of the high rainfall, there are no large accumulations of guano and feathers on the islands. The nutrient deposition and leaching are highly seasonal and affect significantly the nutrient status of the oceanic waters adjacent to the islands, providing a feedback mechanism for recycling nutrients over the shelf region. Seabirds may also cause the erosion of bare rocks and soil on the islands.[33]

6. Drainage to ocean: Most of the nutrients made available by the earlier-mentioned processes are lost to the ocean by leaching and drainage as a consequence of the substantial rainfall.[34] Thus, the distribution of nutrients in the vicinity of the PEI is complex and generally coupled with physical parameters.[35]

LIFE-SUPPORT SYSTEM OF THE PEI

The islands' vertebrate fauna is characterized by 29 bird species (of which four are penguins), three seal species, and since the eradication of feral cats in 1991, the only alien mammal species is the house mouse *Mus musculus*, which is present only on Marion Island.[36] Based on density data and average

32. V. R. Smith, "The effect of burrowing species of Procellariidae on the nutrient status of inland tussock grasslands on Marion Island," *South African Journal of Botany* 42 (1976): 265–72.
33. W. R. Siegfried, "Ornithological research at the Prince Edward Islands: a review of progress," *South African Journal of Antarctic Research* 8 (1978): 30–34.
34. van Zinderen Bakker Sr. (n. 4 above).
35. Pakhomov and Froneman (n. 10 above).
36. *Prince Edward Islands Management Plan* (Pretoria, South Africa: Department of Environmental Affairs and Tourism, 1996), pp. 1–64.

individual body weight measurements,[37] the overall biomass of top predator populations has been estimated to be ca. 10,300 tonnes of wet weight. From this amount, truly offshore feeders (animals foraging beyond the shelf region of the islands, e.g., seals, some penguins, and procellariiformes) account for 86 percent of total biomass. The inshore feeders (forage mainly over the island shelf, mostly penguins and small flying birds) contributed only ~14 percent to total biomass, while species obtaining their food from the islands, for example, mice and some flying birds, was negligible (< 1 percent). Although the rates of food consumption by each of three groups may change the estimated values, these proportions highlight the importance of the oceanic environment, not only in providing food for top predators, but also in supplying nutrients to the islands' terrestrial ecosystems.

The uniqueness of the PEI ecosystem is a result of a close marine-terrestrial interaction known as the "life-support system," which provides the food for the entire community of numerous seabirds and mammals on the islands. The life-support mechanism is thought to function as a long-term alternation between two components, an inshore (autochthonous) component, which is of crucial importance for only a few land-based top predators, and an offshore (allochthonous) component, which supports most of the top predators on the PEI.[38]

Inshore Component of the Life-Support System

In several studies conducted during the spring and summer seasons between November and May, the regular occurrence of algal blooms in combination with trapped eddies over the PEI inter-island shelf has been observed,[39] suggesting that the archipelago generates an "island-mass effect."[40] During these surveys, photosynthetic pigment (chlorophyll-*a*) concentrations often exceeded 1.5 mg/m^3, reaching at times 3.0 mg/m^3, and algal concentrations, chiefly the chain-forming diatom *Chaetoceros radiacans*, were as high

---

37. Ibid.
38. Pakhomov and Froneman (n. 10 above).
39. R. Perissinotto and C. M. Duncombe Rae, "Occurrence of anticyclonic eddies on the Prince Edward Plateau (Southern Ocean): effects on phytoplankton biomass and production," *Deep-Sea Research* 37 (1990): 777–93.
40. B. R. Allanson, B. Boden, L. Parker and C. Duncombe Rae, "A contribution to the oceanology of the Prince Edward Islands," in *Antarctic Nutrient Cycles and Food Webs*, eds. W. R. Siegfried, P. R. Condy and R. M. Laws (Berlin Heidelberg: Springer-Verlag, 1985), pp. 38–45; S. Z. El-Sayed, P. Benon, P. David, J. R. Grindley and J.- F. Murail, "Some aspects of the biology of the water column studies during the "Marion-Dufresne" cruise 08," *Comite National Francais des Recherches Antarctiques* 44 (1979): 127–34.

as 10^6–10^9 cells/litre.[41] Regression analysis showed that water-column stability and mixed-layer depth accounted for most of the variance associated with integrated productivity and photosynthetic capacity of the local algal community.[42] Mesoscale distribution of nutrients, particularly ammonia and urea, also showed a strong correlation with algal biomass and significantly contributed to its variability.[43] It was suggested that both water-column stratification and high nutrient concentrations entrained by the water circulation represented the most important factors initiating and controlling algal blooms in the inter-island region of the PEI (fig. 2).[44] The island-mass effect of the PEI appears to be related to interactions of the free-stream ACC with the local topography and to nutrient inputs via run-off from the islands themselves.[45]

Zooplankton grazing during bloom conditions is responsible for removal of only a small (<20 percent) proportion of primary production.[46] Unexploited organic matter may be horizontally dispersed but mainly sinks into deeper water (fig. 2),[47] providing a plentiful food source for the benthic community. As a result, the benthic community of the PEI is diverse (~550 species), rich in biomass (up to 6000 g wet mass/m^2), and dominated by animals (bryozoans, cnidarians and sponges) that feed upon suspended organic particles and small plankton.[48] Besides consuming sedimented algal cells and spores after blooms over the island shelf, benthic animals may directly consume algal cells on the shallow parts of the shelf.[49] Furthermore,

---

41. Pakhomov and Froneman (n. 10 above); B. P. Boden, "Observations of the island mass effect in the Prince Edward Archipelago," *Polar Biology* 9 (1988): 61–68.

42. R. Perissinotto, C. M. Duncombe Rae, B. P. Boden and B. P. Allanson, "Vertical stability as a controlling factor of the marine phytoplankton production at the Prince Edward Archipelago (Southern Ocean)," *Marine Ecology Progress Series* 60 (1990): 205–9.

43. R. Perissinotto, J. R. E. Lutjeharms and R. C. van Ballegooyen, "Biological-physical interactions and pelagic productivity at the Prince Edward Islands, Southern Ocean," *Journal of Marine Systems* 24 (2000): 327–41.

44. Perissinotto and Duncombe Rae (n. 39 above).

45. Allanson et al. (n. 40 above).

46. R. Perissinotto, "Mesozooplankton size-selectivity and grazing impact on the phytoplankton community of the Prince Edward Archipelago (Southern Ocean)," *Marine Ecology Progress Series* 79 (1992): 243–58.

47. R. Perissinotto, B. R. Allanson and B. P. Boden, "Trophic relations within the island seas of the Prince Edward Archipelago, Southern Ocean," in *Trophic Relationships in the Marine Environment*, eds. M. Barnes and R. N. Gibson (Aberdeen: Aberdeen University Press, 1990), pp. 296–314.

48. L. D. Parker, "A Contribution to the Oceanology of the Prince Edward Islands," (MSc thesis, Rhodes University, 1984), p. 97; G. M. Branch, C. G. Attwood, D. Gianakouras and M. L. Branch, "Patterns in the benthic communities on the shelf of the subAntarctic Prince Edward Islands," *Polar Biology* 13 (1993): 23–34.

49. Perissinotto et al. (n. 47 above).

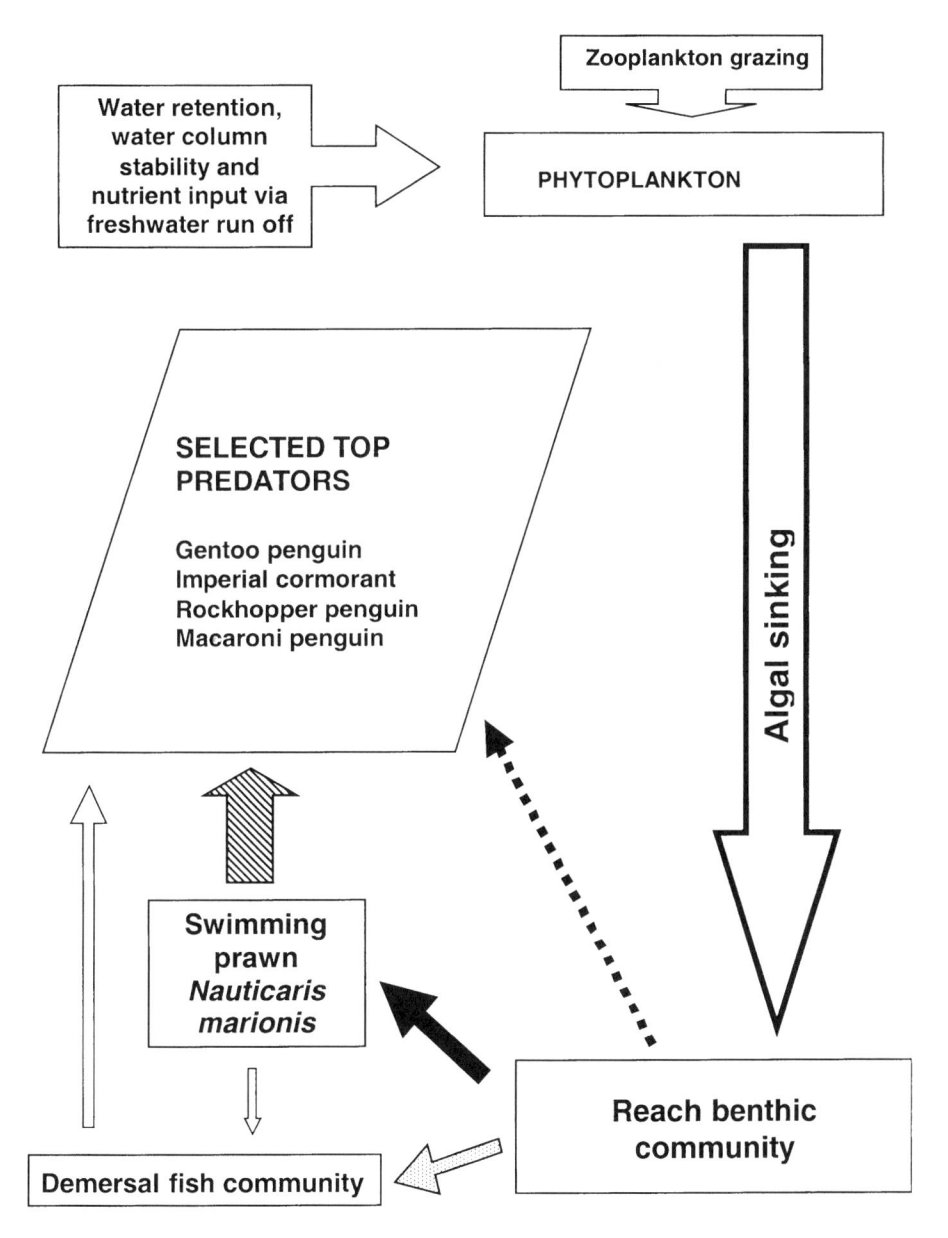

FIG. 2.—Diagram illustrating the autochthonous component of the life-support system of the Prince Edward Islands. See explanation in the text.

late larval stages of the benthic shrimp *Nauticaris marionis*, which are abundant over the island's shelf, undertake marked diel vertical migrations and are also able to utilize algae directly.[50] Bottom-dwelling fish and the shrimp *N. marionis*, both consuming mainly benthic animals, provide an important link between benthic production and selected top predators on the islands (fig. 2).[51]

Offshore Component of the Life-Support System

It is obvious that owing to the higher number of offshore-feeding predators on the islands, the offshore/allochthonous component appears to be the more significant one for sustaining high populations of top predators on the PEI. Two major food supply mechanisms have been proposed for the offshore component of the island's "life-support system."

The Replenishing and Pulsing Mechanism
During several surveys conducted in the vicinity of the PEI (e.g., April/May 1989, 1997–2000), through-flow regimes dominated between the islands,[52] resulting in no water trapping and low (<0.5 mg/m^3) chlorophyll-*a* concentrations within the inter-island region. During such situations, the advection of zooplankton from the upstream region of the islands should be able to supply much of the food necessary for the survival of the land-based offshore feeding predators (fig. 3a).[53] Preliminary calculations showed that ca. 310 tonnes of myctophid (pelagic) fish and ca. 3200 tonnes of zooplankton could be replenished between the PEI every 24 hours through advection. Just nocturnal advection of allochthonous zooplankton into the islands may exceed the local algal production by twofold.[54] The advected plankton is

50. R. Perissinotto and C. D. McQuaid, "Role of the sub-antarctic shrimp *Nauticaris marionis* in coupling benthic and pelagic food-webs," *Marine Ecology Progress Series* 64 (1990): 81–87.

51. Ibid; W. O. Blankley, "Feeding ecology of three inshore fish species at Marion Island (Southern Ocean)," *South African Journal of Zoology* 17 (1982): 164–70; E. A. Pakhomov, P. W. Froneman, P. J. Kuun and M. Balarin, "Feeding dynamics and respiration of the bottom-dwelling caridean shrimp *Nauticaris marionis* Bate, 1888 (Crustacea: Decapoda) in the vicinity of Marion Island (Southern Ocean)," *Polar Biology* 21 (1999): 112–21.

52. Perissinotto et al. (n. 43 above); B. P. V. Hunt, "Mesozooplankton community structure in the vicinity of the Prince Edward Islands (Southern Ocean) 37°50′E, 46°45′S" (MSc thesis, Rhodes University, 2000).

53. R. Perissinotto and C. D. McQuaid, "Land-based predator impact on vertically migrating zooplankton and micronekton advected to a Southern Ocean archipelago," *Marine Ecology Progress Series* 80 (1992): 15–27.

54. R. Perissinotto, "The structure and diurnal variations of the zooplankton of the Prince Edward Islands: Implications for the biomass build-up of higher trophic levels," *Polar Biology* 9 (1989): 505–10.

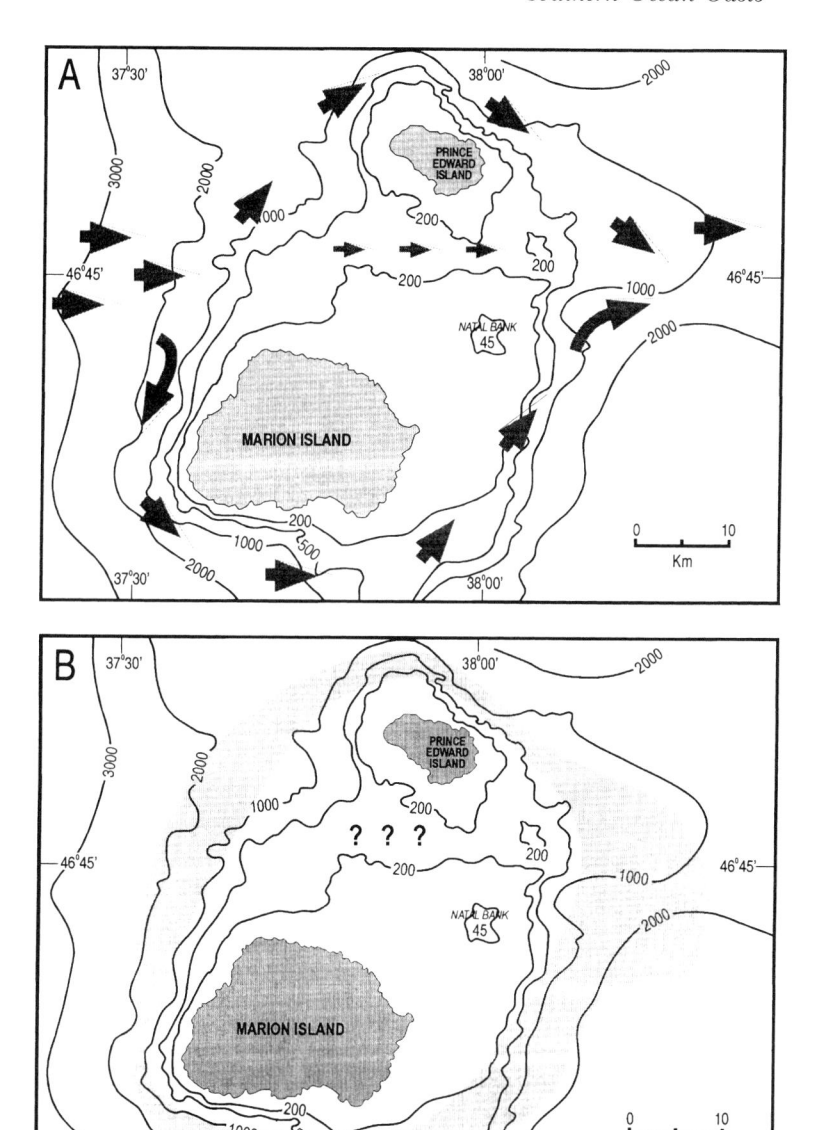

FIG. 3.—Diagram illustrating allochthonous component of the life-support system of the Prince Edward Islands. A: Generalized current system around the Prince Edward Islands. Size of arrows shows major flow pattern of water indicating that most of the water moves around the island shelf and only limited portion flows between the islands. B: Diagram showing elevated concentrations of myctophid fish (shaded area) around the islands. Question marks indicate that very limited amounts of myctophid fish may be washed over the inter-island trench region. For further explanation see the text.

likely to be trapped by the island's shallow topography, depleted between the islands during the day by visual predators, and replenished by advection from the upstream during the following night. Predators on the archipelago, excluding bottom-dwelling fish, may consume daily as much as 900 tonnes of crustaceans and up to 1770 tonnes of myctophid fish.[55] Thus, advection supplies all zooplankton needs of top predators but it is still inadequate to support myctophid-feeding predators. These predators would, therefore, have to forage in the deeper offshore waters, which is confirmed by foraging behaviour studies on king penguins.[56] In fact, myctophid fish generally avoid advection to the inter-island region and are largely diverted around the islands, forming a belt of elevated concentrations in close proximity to the island shelf, particularly in the lee of the islands (fig. 3b).[57] These areas thus represent important foraging grounds for the land-based predators. This is indirectly supported by the distributional pattern of all the major penguin colonies, for example, king, rockhopper, and macaroni penguins, which are concentrated on the eastward side of the PEI[58] and foraging trip patterns of some albatrosses from the Crozet Islands.[59]

It was previously mentioned that relatively higher acoustic signals (planktonic concentrations) were observed within the deepest channel between the islands.[60] To date, however, little data are available on the potential through-flow capacity of the inter-island saddle. Recently, it was demonstrated that even during a through-flow regime, pronounced water pulses might occur across the inter-island region of the PEI.[61] The location of the SAF to the north of the island plateau appeared to determine water dynamics (trapping or advection) on different time scales in the trench between

55. Perissinotto and McQuaid (n. 53 above).
56. N. J. Adams, "Foraging range of King penguins (*Aptenodytes patagonicus*) during summer at Marion Island," *Journal of Zoology,* London 212 (1987): 475–82.
57. E. A. Pakhomov and P. W. Froneman, "Macroplankton/micronekton dynamics in the vicinity of the Prince Edward Islands (Southern Ocean)," *Marine Biology* 134 (1999): 501–15.
58. A. J. Williams, "Geology and the distribution of macaroni penguin colonies at Marion Island," *Antarctic Record* 19 (1978): 279–87; N. J. Adams and M.-P. Wilson, "Foraging parameters of gentoo penguins *Pygoscelis papua* at Marion Island," *Polar Biology* 7 (1987): 51–56; B. P. Watkins, "Population sizes of king, rockhopper and macaroni penguins and wandering albatrosses at the Prince Edward Islands and Gough Island, 1951–1986," *South African Journal of Antarctic Research* 17 (1987): 155–62.
59. H. Weimerskirch, "Unpublished Invited Contribution at the 7th SCAR Biological Symposium Held in Christchurch, New Zealand, 1998."
60. D. G. M. Miller, "Results of a combined hydroacoustic and midwater trawling survey of the Prince Edward Islands group," *South African Journal of Antarctic Research* 12 (1982): 3–10.
61. E. A. Pakhomov, I. J. Ansorge and P. W. Froneman, "Variability in the inter-island environment of the Prince Edward Islands (Southern Ocean)," *Polar Biology* 23 (2000): 593–603.

the islands (see later). The biological consequences of water pulses observed within the inter-island trench with regard to the food supply to top predators were, however, found to be minimal.[62] The frequency of water pulses and their importance as potential carriers of plankton and fish stocks over the inter-island saddle as well as the effect of the SAF on the inter-island environment should be further investigated, particularly including a temporal aspect of these events.

Frontal Movements and Mesoscale Anomalies/Eddies Mechanism
Recent investigations have shown that frontal systems, found in the vicinity of the PEI, are characterized by enhanced plankton and pelagic fish standing stocks.[63] The position of these fronts demonstrates a high degree of latitudinal variability.[64] Although many of the land-based top predators are able to travel long distances to forage within the nearby and distant frontal systems,[65] it is hypothesized that elevated plankton and fish biomass associated with the SAF and APF may be directly transported to the island ecosystem via physical mechanisms during periods when fronts are found in close proximity to the islands.[66] Trapped in the leeward region of the island's plateau, these stocks would become more accessible to both inshore and offshore foraging top predators.

Most recent studies showed that many flying birds and seals have a strong association with mesoscale oceanographic anomalies/eddies during their foraging trips.[67] In the upstream region of the PEI, such anomalies are

62. Ibid
63. E. A. Pakhomov and P. W. Froneman, "Composition and spatial variability of macroplankton and micronekton within the Antarctic Polar Frontal Zone of the Indian Ocean during austral autumn 1997," *Polar Biology* 23 (2000): 410–19; M. Barange, E. A. Pakhomov, R. Perissinotto, P. W. Froneman, H. M. Verheye, J. Taunton-Clark and M. I. Lucas, "Pelagic community structure of the Subtropical Convergence region south of Africa and in the mid-Atlantic Ocean," *Deep-Sea Research* I 45 (1998): 1663–87; E. A. Pakhomov, R. Perissinotto, C. D. McQuaid and P. W. Froneman, "Zooplankton structure and grazing in the Atlantic sector of the Southern Ocean in late austral summer 1993. Part 1. Ecological zonation," *Deep-Sea Research* I 47 (2000): 1663–86.
64. Lutjeharms (n. 7 above); Ansorge and Lutjeharms (n. 7 above).
65. C. A. Bost, J. Y. Georges, C. Guinet, Y. Cherel, K. Pütz, J. B. Charrassin, Y. Handrich, T. Zorn, J. Lage and Y. Le Maho, "Foraging habitat and food intake of satellite-tracked king penguins during the austral summer at Crozet Archipelago," *Marine Ecology Progress Series* 150 (1997): 21–33; F. C. Jonker and M. N. Bester, "Seasonal movements and foraging areas of adult southern female elephant seals, *Mirounga leonine*, from Marion Island," *Antarctic Science* 10 (1998): 21–30.
66. Pakhomov and Froneman (n. 10 above).
67. P. G. Rodhouse, P. A. Prince, P. N. Trathan, E. M. C. Hatfield, J. L. Watkins, D. G. Bone, E. J. Murphy and M. G. White, "Cephalopods and mesoscale oceanography at the Antarctic Polar Front: satellite tracked predators locate pelagic trophic interactions," *Marine Ecology Progress Series* 136 (1996): 37–50; N. T. W. Klages and M. N. Bester, "Fish prey of fur seals *Arctocephalus* spp. at subantarctic Marion Island," *Marine Biology* 131 (1998): 559–66; D. C. Nel, J. R. E. Lutjeharms, E. A. Pak-

created as the ACC crosses the South-West Indian Ridge (fig. 1). They drift closer to the PEI and become significant foraging grounds for land-based top predators thus contributing to the offshore component of the PEI life-support system.[68] Determining the relative importance of these two major mechanisms depends on information on trophic links and origins of energy and nutrients for consumer species.

Marine Food Web Trophic Structure in the Vicinity of the PEI

In a single study on lipid content and fatty acid composition of zooplankton, the herbivorous feeding mode dominated for all sub-Antarctic species, suggesting that a "simple" and short food chain (primary production → zooplankton → top predators) links marine and terrestrial systems of the islands.[69] Recently, the origins (autochthonous or allochthonous) and pathways of organic matter in various marine communities in the vicinity of the PEI were studied in detail using stable-isotope analysis.[70] Four major assemblages, comprising zooplankton, kelp-associated species, and inter-island and nearshore benthos were considered. Both pelagic (zooplankton) and benthic inter-island communities ultimately derived their diets from pelagic algae. However, zooplankton fed primarily on allochthonous (open ocean production) algae, while inter-island benthos seemed to rely mostly on autochthonous (inter-island blooms) algae.[71] In contrast, kelp-associated animals derived a high proportion (>50 percent) of their diet from kelp. The nearshore benthic community had an intermediate position between kelp-associated and inter-island communities. These findings for the first time have shown that autochthonous sources of organic matter (e.g., kelp-derived and inter-island blooms) are important components of the diets of all but the zooplankton community at the PEI. Depending on the length of the foraging trips, some land-based top predators may consume organic matter that ultimately originates at least partly from kelp-derived matter.[72] Stable-isotope analysis of the top predators (currently in progress) would, there-

---

homov, I. J. Ansorge, P. G. Ryan and N. T. W. Klages, "Exploitation of mesoscale oceanographic features by grey-headed albatrosses *Thalassarche chrysostoma* in the southern Indian Ocean," *Marine Ecology Progress Series* 217 (2001): 15–26.
    68. Nel et al. (n. 67 above).
    69. C. G. Attwood and K. D. Hearshaw, "Lipid content and composition of sub-Antarctic euphausiids and copepods from the Prince Edward Islands," *South African Journal of Antarctic Research* 22 (1992): 3–13.
    70. S. Kaehler, E. A. Pakhomov and C. D. McQuaid, "Trophic structure of the marine food web at the Prince Edward Islands (Southern Ocean) as determined by $\delta^{13}C$ and $\delta^{15}N$ analysis," *Marine Ecology Progress Series* 208 (2000): 13–20.
    71. Ibid
    72. Ibid.

fore, shed more light on the importance of the coupling of the land-based and nearshore marine systems.

Oceanographic Forcing of the Life-Support System

The importance of the two major components of the life-support system that supply energy and nutrients to the top predators at the islands is a function of oceanographic forcing.[73] Results of previous studies in the region of the PEI indicated that the geographical position of the SAF in the proximity of the island group plays an important role in forming the macro- and meso-scale oceanographic conditions.[74] When the SAF lies far north of the islands, the interaction between the ACC and the archipelago results in water retention over the inter-island region and algal bloom development.[75] In contrast, when the SAF is found in close proximity of the islands, advection forces prevail and the through-flow system is established between the islands with a pronounced meandering wake observed in the downstream region of the islands.[76] These scenarios have important implications for both the inter-island and surrounding realm dynamics because it is likely that they determine the predominance of either inshore or offshore components of the life-support system of the PEI. The alteration between both components likely provides the basis for the either overwhelming stability or long-term changes of the top predator populations on the PEI.[77]

TEMPORAL VARIABILITY IN THE VICINITY OF THE PEI

Results of daily sampling during April 2000 near Marion Island, on the lee side of Marion Island, and between the islands, showed substantial variability of the inter-island physical and biological environment on the scale of days.[78] General consistency between seawater temperature and algal and zooplankton densities indicated that a through-flow scenario was observed. However, even during the through-flow regime between the islands some degree of

---

73. Lutjeharms (n. 7 above); Ansorge and Lutjeharms (n. 7 above); Hunt (n. 52 above).
74. Pakhomov et al. (n. 61 above).
75. Perissinotto and Duncombe Rae (n. 39 above).
76. Perissinotto et al. (n. 43 above).
77. E. A. Pakhomov, P. W. Froneman, I. J. Ansorge and J. R. E. Lutjeharms, "Temporal variability in the physico-biological environment of the Prince Edward Islands (Southern Ocean)," *Journal of Marine Systems* 26 (2001): 75–95.
78. E. A. Pakhomov, B. P. V. Hunt and L. J. Gurney, "The fifth cruise of the Marion Island Oceanographic Survey (MIOS-V), April to May 2000," *South African Journal of Science* 96 (2000): 541–43.

water retention over the island shelf occurs.[79] To date, however, the mechanisms responsible for short-term variation in the shelf environment are poorly understood.[80]

Inter-annual variability of the PEI ecosystem was investigated during six surveys conducted in late austral summer (April/May) from 1996 to 1999.[81] The positions of the SAF and APF appeared to have a significant impact on plankton biomass in the vicinity of the PEI, possibly through the alternation of local oceanographic flow dynamics. Overall, a close coupling between environmental parameters (seawater temperature, salinity, chlorophyll-*a*), location of the SAF, and zooplankton dynamics was observed. Similarly with short-term variability (pulsing dynamics), it was evident that a shift in the position of the SAF accounted for the most variability in zooplankton in the vicinity of the PEI.[82]

Long-term variability in the vicinity of the PEI has been investigated by using historical planktonic data collected since 1976. A combined data set (1976–2000) shows a trend of increased sea surface temperature with time,[83] corresponding well with the long-term trend of warming in air and sea surface temperature (see later) observed by Smith and Steenkamp.[84] This increased temperature may reflect a possible long-term southward shift in the mean position of the SAF, as suggested by a data set of more than thirty crossings of the SAF upstream of the PEI collected between 1959 and 2000 (fig. 4d). Such a shift would not only increase seawater temperature in the vicinity of the islands, but also unavoidably alter the occurrence of alternating flow modes by favouring a through-flow mode and preventing water retention and the development of algal blooms over the island shelf. The subsequent decrease in average chlorophyll-*a* concentrations and changes in zooplankton composition (fig. 5) observed in the vicinity of the islands[85] since the first oceanic survey in 1976 supports the suggestion of a southward movement of the SAF. The long-term southward shift in the location of the SAF should have a visible effect on marine-terrestrial interactions of the islands.

CLIMATE CHANGE IMPLICATIONS FOR MARINE-TERRESTRIAL INTERACTIONS

Global climate change, accelerated by anthropogenic emissions of greenhouse gases is expected to have its greatest effect in polar ecosystems. In

   79. Pakhomov et al. (n. 61 above).
   80. Pakhomov et al. (n. 78 above).
   81. Hunt (n. 52 above).
   82. Ibid.
   83. Ibid.
   84. Smith and Steenkamp (n. 26 above).
   85. Hunt (n. 52 above).

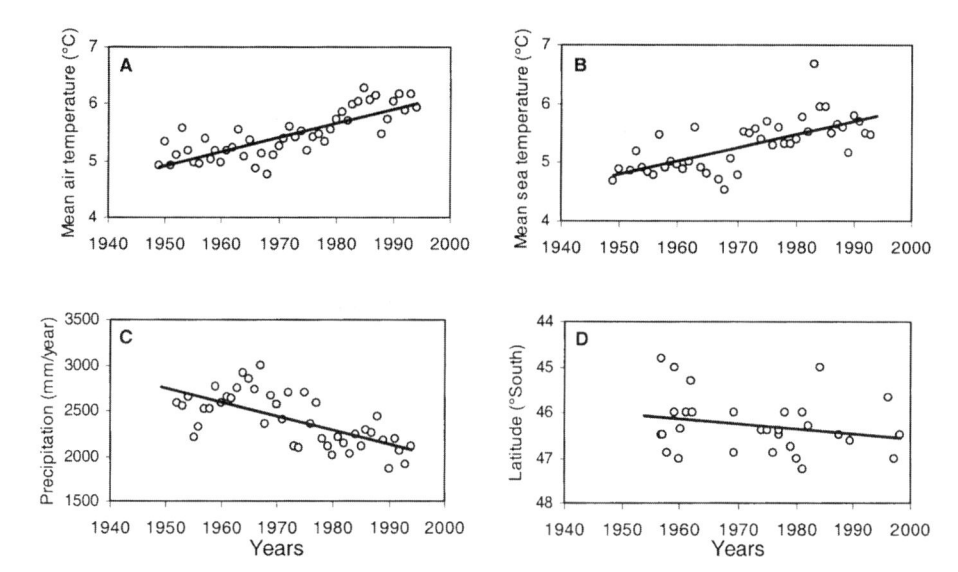

FIG. 4.—Mean surface temperatures of air (A, T°C) and seawater (B, T°C), annual precipitation (C, mm/year) at Marion Island, and the latitudinal position (°South) of the sub-Antarctic front (SAF) (D) in the upstream region (between 28 and 37°E) of the Prince Edward Islands between 1949 and 1998. Sources: Data bank of the S.A. Weather Bureau, Pretoria, South Africa; J. R. E. Lutjeharms and H. R. Valentine, "Southern Ocean thermal fronts south of Africa," *Deep-Sea Research* 31 (1984): 1461–75; J. R. E. Lutjeharms and H. R. Valentine, "Oceanic thermal fronts in the Southern Ocean: A statistical analysis of historical data south of Africa," *CSIR Research Report* 558 (1983): 1–187; E. A. Pakhomov, I. J. Ansorge and P. W. Froneman, "Variability in the inter-island environment of the Prince Edward Islands (Southern Ocean)," *Polar Biology* 23 (2000): 593–603.

view of this, the Southern Ocean has been identified as a critical area for the investigation of the effects of global climate change. The sub-Antarctic islands, with their relatively simple terrestrial ecosystems, are ecologically sensitive and consequently offer ideal sites to study responses to climate change.[86] All the sub-Antarctic islands of the Southern Ocean accommodate large populations of top predators, which are sustained mainly by allochthonous food resources and/or forage mainly within the major Southern Ocean frontal systems located in close proximity to the islands. The simple

86. V. R. Smith, "Climate change and its ecological consequences at Marion and Prince Edward Islands," *South African Journal of Antarctic Research* 21 (1991): 223–24; D. Bergstrom and S. L. Chown, "Life at the front: history, ecology and change on southern ocean islands," *Trends in Ecology and Evolution* 14 (1999): 472–77.

Environment and Coastal Management

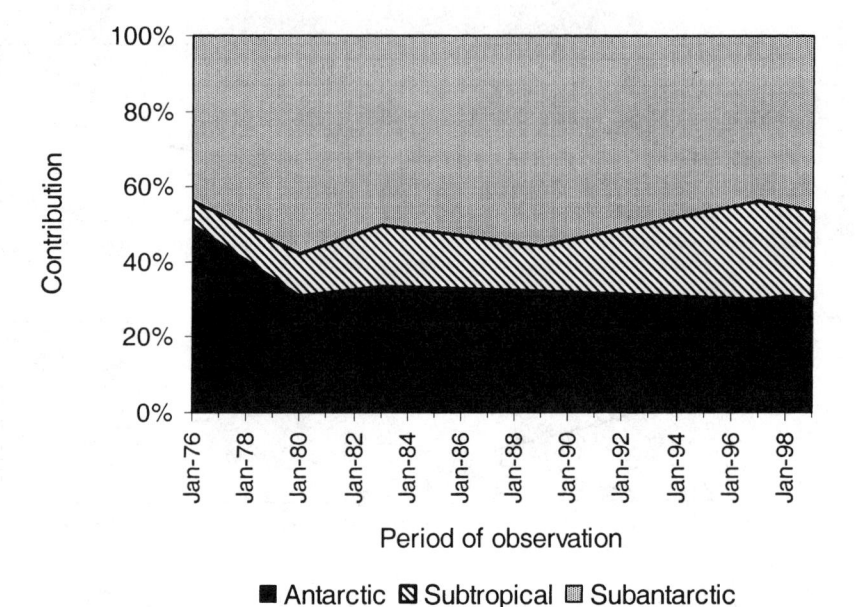

FIG. 5.—Contribution of Antarctic, subtropical, and sub-Antarctic species to total zooplankton encountered around the Prince Edward Islands. Over the past 2 decades the contribution of Antarctic zooplankton species to the total decreased by approximately 20 percent, whereas the number of subtropical species found in the area has increased from 6 to 26 percent. This clearly suggests that warmer-water zooplankton are intruding into the vicinity of the PEI more frequently most likely due to a southward shift in the position of the SAF. Sources: J. R. Grindley and S. B. Lane, "Zooplankton around Marion and Prince Edward Islands," *Comite National Francais des Recherches Antarctiques* 44 (1979): 111–125; D. G. M. Miller, "Results of a combined hydroacoustic and midwater trawling survey of the Prince Edward Island group," *South African Journal of Antarctic Research* 12 (1982): 3–10; B. P. Boden and L. D. Parker, "The plankton of the Prince Edward Islands," *Polar Biology* 5 (1986): 81–93; I. J. Ansorge, P. W. Froneman, E. A. Pakhomov, J. R. E. Lutjeharms, R. Perissinotto and R. C. van Ballegooyen, "Physical-biological coupling in the waters surrounding the Prince Edward Islands (Southern Ocean)," *Polar Biology* 21 (1999): 135–145; B. P. V. Hunt, "Mesozooplankton community structure in the vicinity of the Prince Edward Islands (Southern Ocean) 37°50′E, 46°45′S" (MS thesis, Rhodes University, 2000).

food chains comprising top predators and zooplankton are also expected to be sensitive to climate change.

Environmental observations at the Marion Island Weather Station indicate that the mean annual surface air temperature has increased by >1°C (approximately 0.025°C/year) over the period 1949 to 1998 (fig. 4a). It was strongly coupled with corresponding changes in seawater surface temperature and closely associated with a significant decrease in annual precipitation (fig. 4b, c). Elevated primary production rates on land, an overall decrease in rates of nutrient cycling, and an increase in the introduced house mouse *Mus musculus* population have all been linked to these climatic trends.[87]

Since 1970, populations of seabirds at Marion Island have undergone disparate trends[88] (fig. 6). Populations of offshore feeding seabirds with a foraging range of >300 km, such as northern giant petrel, king penguin, and grey-headed and wandering albatrosses, have generally increased over the last 2 decades. In contrast, populations of inshore feeding species, for example the imperial cormorant, and rockhopper and gentoo penguin, have decreased, especially in the most recent years.[89] The population of the macaroni penguin, which has an intermediate foraging range, was relatively stable (fig. 6). The populations of Antarctic and sub-Antarctic fur seals (both offshore feeders) at Marion Island increased, on average, by 11 and 15 percent per annum, respectively.[90] These increases are attributed to rapid growth following recolonisation. On the other hand, there has been an enigmatic, steady decline (on average 3.4 percent per annum, but now stabilizing) of the numbers of southern elephant seals at Marion Island.[91] Similar trends in populations of seabirds and seals have been reported from other sub-Antarctic islands.[92] Because all sub-Antarctic islands lie within a narrow latitudinal band, similar mechanisms may be responsible for changes in their populations of top predators.

Smith[93] postulated that the climatic trends and the subsequent responses within the terrestrial ecosystem on the Prince Edward Islands were linked to a shift in atmospheric and oceanic circulation patterns. A change

---

87. Smith (n. 86 above); Smith and Steenkamp (n. 26 above); S. L. Chown and V. R. Smith, "Climate change and the short-term impact of feral house mice at the sub-Antarctic Prince Edward Islands," *Oecologia* 96 (1993): 508–16.

88. R. J. M. Crawford, J. Hurford, M. Greyling and D. C. Nel, "Population size and trends of some seabirds at Marion Island," *New Zealand Natural Sciences* 23 Supplement (1998): 42.

89. Ibid.

90. G. J. G. Hofmeyr and M. N. Bester, "Changes in population sizes and distribution of fur seals at Marion Island," *Polar Biology* 17 (1997): 150–58.

91. P. A. Pistorius, M. N. Bester and S. P. Kirkman, "Dynamic age-distributions in a declining population of southern elephant seals," *Antarctic Science* 11 (1999): 445–50.

92. P. Jouventin and H. Weimerskirch, "Changes in the Population Size and Demography of Southern Seabirds: Management Implications," in *Bird Population Studies: Their Relevance to Conservation and Management,* eds. C. Perrins, J. D. Lebreton and G. Hirons (Oxford: Blackwell Science Publication, 1991), pp. 297–314; D. M. Cunningham and P. J. Moors, "The decline of rockhopper penguins *Eudyptes chrysocome* at Campbell Island, Southern Ocean and the influence of rising sea temperatures," *Emu* 94 (1994): 27–36; E. J. Woehler and J. P. Croxall, "The status and trends of Antarctic and sub-Antarctic seabirds," *Marine Ornithology* 25 (1997): 43–66.

93. Smith (n. 86 above).

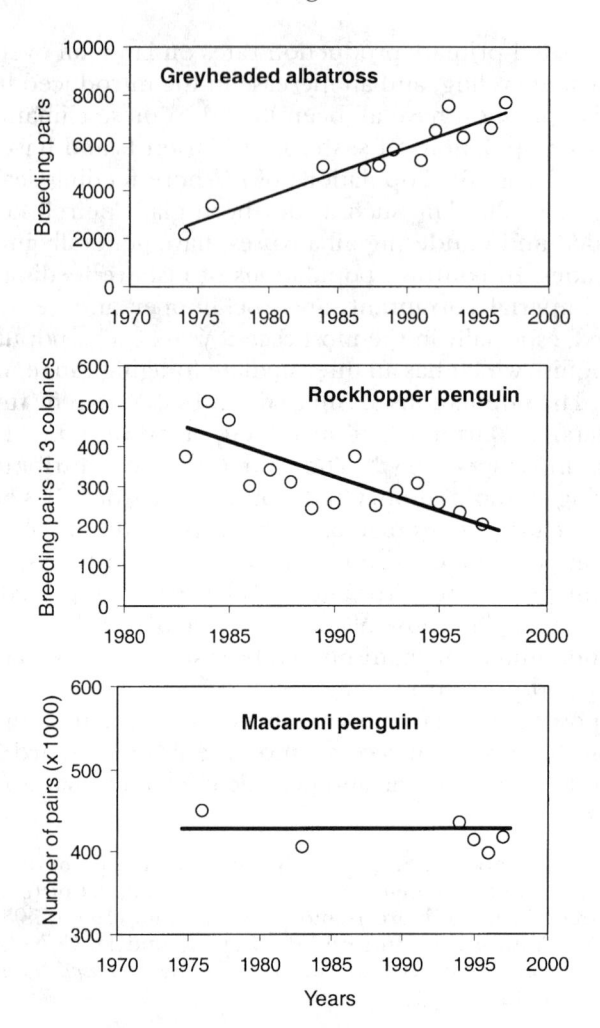

FIG. 6.—Trends in population of selected seabirds at Marion Island. Greyheaded albatross *Diomedea chrysostoma* (example of offshore feeding species), rockhopper penguin *Eudyptes chrysocome filholi* (example of inshore feeding species) and macaroni penguin *Eudyptes chrysolophus* (example of species with an intermediate foraging range). Sources: A. J. Williams, W. R. Siegfried, A. I. Burger and A. Berruti, "The Prince Edward Islands: a sanctuary for seabirds in the Southern Ocean," *Biological Conservation* 15 (1979): 59–71; J. Cooper and C. R. Brown, "Ornithological research at the sub-Antarctic Prince Edward Islands: a review of achievements," *South African Journal of Antarctic Research* 20 (1990): 40–57; A. J. Williams, A. E. Burger, A. Berruti and W. R. Siegfried, "Ornithological research on Marion Island, 1974–75," *South African Journal of Antarctic Research* 5 (1975): 48–50; J. Cooper, A. C. Wolfaardt and R. J. M. Crawford, "Trends in population size and breeding success at colonies of macaroni and rockhopper penguins, Marion Island, 1979/80–1995/96," *CCAMLR Science* 4 (1997): 89–103; B. P. Watkins, "Population sizes of king, rockhopper, and macaroni penguins and wandering albatrosses at the Prince Edward Islands and Gough Island, 1951–1986," *South African Journal of Antarctic Research* 17 (1987): 155–62; Percy FitzPatrick Institute of African Ornithology (courtesy of R. J. M. Crawford).

in atmospheric circulation is evident from increased incidence of exotic pollen at the islands. Prior to 1965, only 0.02 percent of total pollen at Marion Island was exotic, whereas by 1981 this had increased to 1.3 percent.[94] The proportion of exotic pollen grains has increased during the last decade, and may contribute up to 40 percent of the pollen content of some surface samples (V. R. Smith, personal communication April, 2000). Another example is the increased frequency of occurrence of the cosmopolitan painted lady butterfly *Vanessa cardui* since the mid-1970s. Although this species has been observed sporadically since the 1950s, it does not breed on the PEI and thus has to rely on airflow patterns at and around the islands for colonization.[95]

It is expected that latitudinal shifts in positions of oceanic fronts may be linked to atmospheric changes and will become more dramatic and frequent in response to global climate change.[96] Historical data on crossings of the SAF in the upstream region of the PEI indicate a gradual southward migration in the geographic position of this front (fig. 4d). The changes in plankton composition supporting this were discussed earlier (long-term variability). Thus by warming the sub-Antarctic seas,[97] climate change may have already dramatically altered marine ecosystem functioning through changes in pelagic and benthic community structure.[98] A southward shift in the geographic location of the SAF also appears to have favoured the through-flow mode, which advects allochthonous zooplankton and micronekton to the islands, thus benefiting the offshore foragers on the PEI. As a consequence, their populations have shown an increasing trend (fig. 6). Incidence of the trapped-eddy mode has declined and should continue to decrease, leading to less food for inshore foragers and explaining a subsequent decline in their populations. Populations of foragers whose food is presumably provided by both situations have been stable (fig. 6). These findings indicate that southward migration of the SAF in the upstream proximity of the PEI, brought about by global climate change, has promoted long-term changes in land-based top predator populations at the islands through food-chain modifications. Of course, the situation is likely to be more complicated than this polarized characterization of the PEI life-support system and should be further investigated. Nonetheless, the present evidence shows that the sub-Antarctic, and perhaps the entire Southern Ocean,[99] has experienced substantial change in the past half a century and this change can be expected to continue, with important consequences for the associated ecosystem.

94. Smith and Steenkamp (n. 26 above).
95. Ibid.
96. SCOR, *Oceans, Carbon and Climate Change* (Halifax, Canada: Scientific Committee on Oceanic Research, 1990), p. 1–13.
97. S. Levitus, J. I. Antonov, T. P. Boyer and C. Stephens, "Warming of the World Ocean," *Science* 287 (2000): 2225–29.
98. SCOR (n. 96 above).
99. Levitus et al. (n. 97 above); W. K. de la Mare, "Abrupt mid-twentieth-century decline in Antarctic sea-ice extent from whaling records," *Nature* 389

CONSERVATION ISSUES

Since their annexation, both Marion and Prince Edward Islands have been declared nature reserves in the widest sense. Presently, the legal protection of the islands is regulated by several acts. These include Acts No. 43 of 1948 (Prince Edward Islands Act) and No. 46 of 1973 (The Sea Birds and Seals Protection Act). The latter act provides for the protection and control of the capture and killing of most species of seabirds and seals occurring on the islands. Furthermore, the Sea Fishery Act 1988 (Act 12 of 1988) and the Maritime Zones Act 1994 (Act 15 of 1994) provides for the control and conservation of sea fisheries in the island's territorial waters below the low-water mark (12 nautical miles) and fishing zone (200 nautical miles). The land areas of Marion and Prince Edward Islands above the low-water mark have been declared Special Nature Reserves under the Environment Conservation Act, 1989 (Act 73 of 1989). Finally, the PEI fall within the area of application of the international convention on the Conservation of Antarctic Marine Living Resources (CCAMLR), Canberra, 1980, of which South Africa is an original signatory. The South African Department of Environmental Affairs and Tourism is responsible for the management of the archipelago, sponsoring expeditions and ensuring that all visitors to the islands comply with conservation requirements and measures to prevent interference with animal and plant life and the introduction of alien species.

Notwithstanding this regulatory framework, there are three major threats to the conservation of the biota and ecosystems of the Prince Edward Islands. As is the case for many other islands (and continental areas), alien, invasive species clearly constitute one of the most significant threats to the terrestrial systems.[100] Both invasive plants and animals have been shown to have significant impacts on the terrestrial system by way of direct and indirect impacts on the local biota. For example, the introduced grass, *Agrostis stolonifera*, is currently expanding its range and has been shown to substantially reduce the diversity of other plant species in areas where it comes to predominate.[101] House mice are also having major effects directly on the invertebrate biota by reducing the body size and densities of several species. These mice are likely to have large effects on ecosystem functioning because

---

(1997): 57–60; V. Loeb, V. Siegel, O. Holm-Hansen, R. Hewitt, W. Fraser, W. Trivelpiece and S. Trivelpiece, "Effects of sea-ice extent and krill or salp dominance on the Antarctic food web," *Nature* 387 (1997): 897–900; C. Barraud and H. Weimerskirch, "Emperor penguins and climate change," *Nature* 411 (2001): 183–86.

100. S. L. Chown and K. J. Gaston, "Island-hopping invaders hitch a ride with tourists in the Southern Ocean," *Nature* 408 (2000): 637.

101. N. J. M. Gremmen, S. L. Chown and D. J. Marshall, "Impact of the introduced grass *Agrostis stolonifera* on vegetation and soil fauna communities at Marion Island, sub-Antarctic," *Biological Conservation* 85 (1998): 223–31.

many of the invertebrate species play key roles in nutrient recycling.[102] The small introduced midge, *Limnophyes minimus*, could potentially also affect ecosystem functioning,[103] illustrating that it is not only large vertebrate species or plants that should be considered potentially damaging species.[104] The problems that arise as a consequence of the introduction of alien species, and the costs of mitigating the problem are nicely illustrated by the introduction and subsequent eradication of feral cats on Marion Island.[105] The feral domestic cat *Felis catus* population originated from five pets introduced in 1949 at the scientific station to address problems caused by house mice. The first feral cat was seen in 1951. Following this sighting, the population increased at an annual rate of 23 percent to an estimated 2140 by 1975, of which most individuals were feeding predominantly on medium-sized petrels. In 1977, when the population was estimated to be 3400 cats, a viral disease, *feline panleucopaenia*, was introduced as a primary control measure. Following introduction of cat flu, the population decreased substantially, and in 1986 a multi-million rand, extensive programme of hunting, trapping, and poisoning ensured the successful eradication of feral cats by 1991.[106]

Recent work based on an assessment of many Southern Ocean islands has also demonstrated a positive relationship between the numbers of humans occupying an island, the mean annual temperature of an island, and the number of alien species it houses.[107] Essentially, warmer islands and those that have larger numbers of human visitors tend to have more alien species.[108] These findings do not bode well for conservation of the islands because of increasing numbers of tourists in the region[109] and the continuing increase in temperature associated with global warming.[110] In consequence,

---

102. Smith and Steenkamp (n. 26 above); Chown and Smith (n. 87 above).

103. C. Hänel and S. L. Chown, "The impact of a small, alien macro-invertebrate on a sub-Antarctic terrestrial ecosystem: *Limnophyes minimus* Meigen (Diptera, Chironomidae) at Marion Island," *Polar Biology* 20 (1998): 99–106.

104. Bergstrom and Chown (n. 86 above).

105. J. P. Bloomer and M. N. Bester, "Control of feral cats on sub-Antarctic Marion Island, Indian Ocean," *Biological Conservation* 60 (1992): 211–19.

106. M. N. Bester, J. P. Bloomer, P. A. Bartlett, D. D. Muller, M. van Rooyen and H. Büchner, "Final eradication of feral cats from sub-Antarctic Marion Island, southern Indian Ocean," *South African Journal of Wildlife Research* 30 (2000): 53–57.

107. S. L. Chown, N. J. M. Gremmen and K. J. Gaston, "Ecological biogeography of southern ocean islands: Species-area relationships, human impacts, and conservation," *American Naturalist* 152 (1998): 562–75.

108. A. G. A. Gabriel, S. L. Chown, J. Barendse, D. J. Marshall, R. D. Mercer, P. J. A. Pugh and V. R. Smith, "Biological invasions on Southern Ocean islands: the Collembola of Marion Island as a test of generalities," *Ecography* (2001), in press.

109. Chown and Gaston (n. 100 above).

110. Bergstrom and Chown (n. 86 above).

it has been suggested that both tourist and scientific visits to some islands should be prohibited.[111]

Climate change is the second major threat to both terrestrial and marine systems at the islands. From a terrestrial perspective, the direct effects of the warming and drying trend on the PEI on indigenous species have not been carefully examined. However, the changing climate will exacerbate the effects of alien species on the indigenous biota. Not only will warmer climates allow alien species to establish more easily, but they will also mean that species already present on the island will expand their ranges, or their direct/indirect effects on local species and systems will be exacerbated. Changes in the status of several alien plant species are already being noted,[112] and it appears that alien insect species will respond more favourably to warming than will indigenous ones, owing to preferences for warmer conditions and more rapid life cycles in the former species.[113] There is also evidence that the effects of alien house mice on terrestrial ecosystem functioning will be considerably enhanced under a scenario of continued warming.[114] There already appears to be a negative indirect effect of mice on the endemic subspecies of the lesser sheathbill.[115]

As a consequence of the interactions between alien species and the changing climate, and the lower numbers of alien species on Prince Edward Island, this island makes a suitable strict entry reserve that will ensure long-term survival of the biota, compared with nearby Marion Island that is more heavily influenced by its long-standing human occupation. While the islands are certainly managed this way,[116] the utility of Prince Edward Island as a reserve will depend largely on the extent to which activities at both islands are regulated (given the likelihood of natural dispersal of plant and other invasives across the 22 km ocean barrier separating the islands)[117] and the extent to which populations of indigenous species can be considered genetically equivalent. Currently work is underway to assess these issues.

In the case of marine systems, we have already indicated how changes to the positions of the major fronts will change the life support system of the islands, thus differentially affecting the seabird and seal populations that use the islands as breeding and moulting platforms. However, marine re-

111. Ibid.
112. N. J. M. Gremmen and V. R. Smith, "New records of alien vascular plants from Marion and Prince Edward Islands, Sub-Antarctic," *Polar Biology* 21 (1999): 401–09.
113. J. Barendse and S. L. Chown, "Abundance and seasonality of mid-altitude fellfield arthropods from Marion Island," *Polar Biology* 24 (2000): 73–82.
114. Chown and Smith (n. 87 above).
115. O. Huyser, P. G. Ryan and J. Cooper, "Changes in population size, habitat use and breeding biology of lesser sheathbills *Chionis minor* at Marion Island: impacts of cats, mice and climate change?," *Biological Conservation* 92 (2000): 299–310.
116. Prince Edward Islands Management Plan (n. 36 above).
117. Gremmen and Smith (n. 112 above).

sources also face another major threat. A small shelf area surrounds the PEI. There is virtually no information on any fish, squid, or crustaceans that may occur in economically exploitable quantities around the islands. Nevertheless, a legal, exploratory fishery on the Patagonian toothfish *Dissostichus eleginoides* commenced in 1996. However, unknown numbers of fishing boats have been catching the Patagonian toothfish in the area illegally prior to 1996, resulting in the near-collapse of the highly lucrative toothfish fishery. Since the beginning of exploitation, the total catch of toothfish by licensed fishing boats rose to 1332 tonnes per year in 2000. The signals of overfishing are noticeable, as average body size of toothfish has decreased dramatically (B. Watkins, personal communication March 2001). The boats exploiting the toothfish use longlines with thousands of hooks. In consequence, many birds, but particularly small albatrosses and some petrels, have been caught and killed during fishing operations, leading to the expression of considerable conservation concern. Therefore, the vessels, which are licensed by the South African government to fish in the area, are required to take every precaution to reduce the incidental mortality of seabirds.[118] Such measures include limitation of line setting to the hours of darkness, adequate weighting of lines to maximize sinking rates, avoidance of dumping of offal during setting, minimization of deck lighting, and the use of bird deterrents during setting. Marine observers on licensed vessels in the PEI toothfish fishery have shown that these measures are successful and reported an 87 percent reduction in seabird by-catches between 1997 and 1999.[119]

PEI MANAGEMENT PLAN: OBJECTIVES AND POLICIES

In November 1995 the PEI were proclaimed Special Nature Reserves under the Environment Conservation Act (Act 73 of 1989), which requires the development of a management plan for such reserves. The continuity of both the terrestrial and marine research programmes on the islands insured that a comprehensive, long-term data set (spanning some 50 years) was available to be consulted for the drafting of the management plan. This meant that the management plan was grounded on a thorough understanding of ecosystem functioning and variability at the islands, and also that an invaluable baseline for long-term assessment of management actions was and continues to be available.

This Prince Edward Islands Management Plan was published in 1996.[120] Based on the available information, the plan provides a set of management

---

118. Prince Edward Islands Management Plan (n. 36 above).
119. C. de Villiers, "South Africa hosts talks to end seabird carnage," *SANCOR Newsletter* 166 (2001): 9–10.
120. Prince Edward Islands Management Plan (n. 36 above).

objectives and a framework for decision making (legal protection, administration, permits, support services, management zones, waste and sewage disposal, visits to the islands, research, historical conservation, protection and management of flora and fauna, control and prevention of alien species, international cooperation) with the overall goal of conserving the Prince Edward Islands' biological diversity, ecological uniqueness, and integrity. Consequently, the main management objectives set out in the plan are the following:

1. To maintain biological diversity, including genetic diversity, species diversity, and the diversity of ecological processes;
2. To maintain geological and scenic objects;
3. To minimize interference with natural processes and the destruction or degradation of natural features resulting from human interference;
4. To ensure that obligations to, and the provisions of, the Convention on the Conservation of Antarctic Marine Living Resources are met;
5. To protect historic features and objects from human interference;
6. To encourage activities aimed at restoring and rehabilitating damage due to local human activities;
7. To encourage research applicable to objectives (1) through (6) above;
8. To seek cooperation with all parties interested in the conservation of the Southern Ocean and its islands;
9. To create an awareness of the value and fragility of the Islands' ecosystems;
10. To allow scientific research not in conflict with objectives (1) to (9).

Management policies are generally directed at the achievement of these objectives, take many forms, and are also subject to change. Perhaps one of the most pivotal of the management policies is that concerning zonation of activities at the islands. Essentially, four management zones have been demarcated and access to these zones is regulated via a permitting system in accordance with the conservation status of each of the zones.[121] Zones 1 and 2 are restricted to Marion Island. Zone 1 (Service Zone) includes the Marion Island scientific station and a small area surrounding it. The zone allows for the construction, alteration, and removal of buildings, facilities, and scientific equipment, and houses the support and scientific staff. Zone 2 (Natural or Buffer Zone) forms a triangle from the northern shore of Ships Cove to the peak of Junior's Kop and to the Fault at Trypot Beach. Limited free walking in the area is permitted provided that a permit is held. The areas around each of the existing field huts are zoned as Natural Zones.

121. Ibid.

Beyond Natural Zone is Zone 3 (Wilderness Zone), which provides a high degree of protection to the environment. This zone is open for research purposes, which have to be approved by the regulatory authority, the Prince Edward Islands Management Committee (PEIMC), and is generally closed to members of the relief and management teams. Zone 4 (Protected Zone) areas are special entry areas afforded maximum protection. Prince Edward Island is demarcated as Zone 4 and all entry is prohibited except for visitors that hold a Special Entry Permit issued by the Director-General of the Department of Environmental Affairs and Tourism (DG DEA&T). Visits to Prince Edward Island are strictly regulated and limited to not more than one visit per annum by up to six persons for a maximum of 4 days per visit. On Marion Island, Zone 4 includes perimeters extending 100 m around gentoo and southern giant petrel colonies, and 200 m around wandering albatross demographic study colonies and the grey-headed albatross colony at Grey-headed Albatross Ridge. Finally, all historical sites fall within Zone 4.

Access to the islands is controlled subject to the stated management aims and objectives, but also bearing in mind the stated vulnerability of the biota to human-mediated disturbance. Entry, research, and collection permits issued by the DG DEA&T regulate activities on the islands. However, in terms of the Environment Conservation Act, visits other than those undertaken for research and management purposes are not prohibited. At present, no tourism, either ship- or land-based, is envisaged for the islands, although the situation may change in the future.[122] Given this possibility, and especially because of an increase in tourism to the overall region,[123] the PEIMC recently commissioned an environmental impact assessment of tourism to Marion Island. The assessment was completed in 2000, and it provided a set of recommendations for the management of such activities to ensure that the environment would be least compromised. These recommendations are currently under review by the DG DEA&T and it seems likely that for the immediate future tourism will not be permitted.

Nonetheless, the research station will be maintained and consequently the activities of scientists and the associated support staff will continue. Because alien plants, animals, and other organisms pose a threat to indigenous fauna and flora on the islands, no plants or animals (including pets) can be taken onto the islands. Likewise, no organic material, ornamental plants, or fresh fruit and vegetables are allowed ashore on the islands. It is also expected that all relief and overwintering personnel take precautions to prevent the transfer of invasive organisms to the islands in the treads of helicopter tires, with building material, on field equipment, or on personal belongings.

122. R. Heydenrych and S. Jackson. *Environmental Impact Assessment of Tourism to Marion Island*. South African Department of Environmental Affairs and Tourism. Pretoria: Department of Environmental Affairs and Tourism, 2000.

123. Chown and Gaston (n. 100 above).

Although the risks of transferring propagules to the islands are one of the major drawbacks of scientific work done there, it is widely held that scientific knowledge of the PEI ecosystems is essential for their effective management. Nonetheless, because the islands exhibit tremendous natural heritage value, the research policy serves to ensure that scientific research is conducted to assure protection of the natural ecosystems and mitigate against lasting changes in indigenous wildlife populations or community relationships; to avoid conflict with essential management operations; to prohibit the collection of specimens except as part of scientific research or for management purposes; and to arrange, facilitate, and support scientific research for better management of the islands.

The Marine Zone policy does not form part of the PEI Management Plan (given that the Special Nature Reserves do not extend beyond the low-water mark) and has as its main goals the protection of all fish species and aquatic plants around the islands below the low-water mark to the outer boundary of the Fishing Zone (i.e., 200 nautical miles). The DG DEA&T is responsible for issuing and revoking permits for this zone and for setting out the conditions with respect to the protection of fish and collection of aquatic plants.

As a consequence of its International Cooperation policy, the Department of Environmental Affairs and Tourism is encouraged to participate in all relevant international fora and meetings concerning the sub-Antarctic islands, including those that have been created to foster scientific cooperation and the exchange of scientific data and information, with the aim of improving the overall understanding of sub-Antarctic and Southern Ocean ecosystems.

The Management Plan also makes provision for the monitoring and revision of the Management Plan and its policies and stipulates that it will be revised every 5 years to ensure that its overall goals are being met.[124] As part of this process the South African Department of Environmental Affairs and Tourism has also reached the conclusion that the islands, but particularly Prince Edward Island, are amongst the least impacted, yet most species-rich islands in the region.[125] Consequently, a decision has been taken to nominate the islands for inclusion on the World Heritage Site list under UNESCO's World Heritage Convention.

PROSPECTS FOR COOPERATION

Like many remote oceanic islands, the PEI provide considerable opportunities for studying ecological and evolutionary processes, for monitoring eco-

124. Prince Edward Islands Management Plan (n. 36 above).
125. S. L. Chown, A. S. Rodrigues, N. J. M. Gremmen and K. J. Gaston, "World Heritage status and the conservation of Southern Ocean islands," *Conservation Biology* 15 (2001): 550–57.

logical changes, and for conserving a unique component of the planet's biological diversity. The similarity of the biotas, ecological processes, and long-term trends in response to climate change on all of the sub-Antarctic islands provides the natural basis for collaboration at least among those countries responsible for the sub-Antarctic territories, including South Africa, France, Australia, and New Zealand. Such collaboration may prove to be especially fruitful because biologists working in the region have come to the realization that studies of the biotas and systems on the islands are more broadly applicable. Indeed, the islands can be thought of as microcosms of the larger world that offer considerable opportunities for research in ecology and conservation biology.[126] Thus, it is unfortunate that interest in biological research in the sub-Antarctic territories appears to have declined over the last decade.[127] Nonetheless, this situation is changing, as is evidenced by the Scientific Committee on Antarctic Research's (SCAR) Regional Sensitivity to Climate Change Programme, which is concerned especially with work on the Southern Ocean Islands, and by the proposal of French scientists to hold a Workshop on Oceanography and Biology of the sub-Antarctic regions. It is our view, therefore, that scientific interest in these islands will increase over the medium term, and will increasingly involve multinational collaborative ventures. While the benefits of such an approach will be considerable from a scientific perspective, the conservation risks are likely to be substantial. Hence, the future of both research on and conservation of these islands depends on ongoing compromises between research and conservation. In a world where use and conservation seem increasingly incompatible it would be a fitting demonstration of our resolve to conserve these systems. A substantial body of sound scientific work should emerge from these islands with a minimum of environmental disturbance.

---

126. S. L. Chown, K. J. Gaston and N. J. M. Gremmen, "Including the Antarctic: Insights for ecologists everywhere," in *Antarctic Ecosystems: Models for Wider Ecological Understanding*, eds. W. Davison, C. Howard-Williams and P. A. Broady (Christchurch: New Zealand Natural Sciences, 2000), pp. 1–15.

127. S. L. Chown, W. Block, P. Vernon and P. Greenslade, "Priorities for terrestrial Antarctic research," *Polar Record* 33 (1997): 187–88.

Environment and Coastal Management

Integrated Coastal Management in Decentralizing Developing Countries: The General Paradigm, the U.S. Model, and the Indonesian Example*

Jason M. Patlis
Environmental Law and Law Development Associates, Washington, DC

Maurice Knight
Coastal Resources Center, University of Rhode Island, Narragansett

Jeff Benoit
J.R. Benoit Consulting, Arlington, Virginia

BACKGROUND

Many countries around the world are undertaking measures to decentralize systems of governance from national, authoritarian regimes, to subnational, democratic regimes. Economic conditions, resource allocations, demographic pressures, and ethnic strife are just some of the reasons that are leading central governments to cede more power to subnational power centers. This is particularly true in management of natural resources, including coastal resources. A general failure to manage natural resources in a sustainable manner, compounded by poor redistribution of revenues derived from these resources, is one of the more significant factors driving decentralization efforts in many countries.

Developing countries seek various benefits through decentralization, such as equitable allocation of resources, more efficient delivery of services, and a more responsive form of government. While the manner in which decentralization takes place can vary widely from country to country, there is a general paradigm that exists. Essentially, developing countries generally start with strong central governments that expand the economy and improve infrastructure and then progress through a series of steps that usually range from deconcentration of governance to full devolution of authority.

The results of these efforts can vary widely, depending on the political,

---

*Research for this article was conducted in Indonesia with funding through a Fulbright Senior Scholarship to Jason M. Patlis.

© 2003 by The University of Chicago. All rights reserved.
0-226-06620-7/03/0017-0014$01.00 *Ocean Yearbook* 17:380–416

social, economic, and legal conditions facing the country. Often, central governments decentralize more quickly than subnational governments can adequately assume new authorities, and more quickly than the central government will reallocate financial resources for such authority. This often amounts to a situation in which natural resources under stress fall under greater stress at the hands of local governing bodies that have little institutional capacity and little financial capability to manage them.

The challenge, then, is to devise a stable system of governance that can mitigate the stresses on natural resources at the same time it can accommodate the realistic limitations of the existing governing institutions. This is a tall order, given that even the most basic resource protection often requires efforts greater than local institutions can provide. Ideally, such a system would provide enabling conditions for the future: it would not only mitigate resource depletion, but promote resource conservation; it would not only accommodate realistic limitations, but develop institutional capabilities.

The system of governance for coastal resources in the United States—centered on the Coastal Zone Management Act (CZMA) of 1972[1]—provides a model that could effectively be used for natural resource management, particularly coastal resource management, in decentralizing developing countries. While the CZMA has long been heralded as the definitive model for integrated decision making, it also serves as a superlative model for a voluntary, incentive-based decision-making program that fits well with the conditions of decentralizing developing countries.

Despite the obvious differences in economic positions, many of the conditions that existed in the United States in 1972 with respect to coastal resource management are the same conditions that currently exist in many decentralizing developing countries with respect to natural resource management. Specifically, coastal resources were under great pressure due to increasing coastal populations and sector-based management of activities affecting the coast. There were no coordinated coastal resource management policies either at federal, state, or local levels of government, and little infrastructure or know-how at state and local levels to undertake a coordinated coastal resource management program. At the same time, a strong awareness was emerging among the scientific community and the public at large that change was necessary.

The CZMA established a voluntary, incentive-based program that does not require the U.S. states to comply with its prescriptions but provides that if a state were to comply with them, the federal government would provide various benefits. The prescriptions are broad, and almost entirely procedural in nature, with the premise that a sound process will lead to sound management. The federal government assists states with development of

1. Coastal Zone Management Act (CZMA) 302, 16 U.S.C. 1451.

their programs, certifies them if they comply with the federal requirements, provides benefits, and monitors and reviews implementation to ensure continuing compliance. The state and local governments take the lead in developing and implementing their own programs as they see fit, within the confines of the federal prescriptions.

An inherent part of decentralization is movement away from authoritarian central government mandates. Consequently, voluntary guidelines or criteria are more palatable. In addition, enforcement at the central government level is generally poor, so that a mandatory rule would largely remain unenforced. Local capacity to manage resources is often poor, so mandatory rules often go unimplemented even if they can be enforced. Practically speaking, a voluntary program with appropriate incentives could be more efficacious than a mandatory program with little compliance and little enforcement.

Indonesia offers an illuminating application of this model. After practically 50 years of authoritarian rule, the fall of President Suharto precipitated a rapid (sometimes violent) process of decentralization in which the power center shifted from Jakarta to the approximately 350 district-level governments in the country. As part of this process, the great bulk of financial revenues derived from natural resources were redirected to the districts of origin, rather than Jakarta. However, these districts have little capacity to begin writing laws for managing natural resources.

Given the urgent need for conservation of a rapidly depleting resource base, particularly in timber and fisheries, the situation offers many difficult challenges. There is little hope that another mandatory program coming from Jakarta, such as one for coastal resource management, will be accepted or followed. At the same time, there is little hope that the 265 districts with a coastline will be able to develop adequate coastal management programs any time in the near future. Consequently, the best alternative is to work through the central government and the provinces, in a tiered, cooperative, coordinated manner, in an effort to establish at least some basic prescriptions at the district level. A voluntary, incentive-based program would be the most appropriate means to achieve this goal.

This article discusses a new view of the CZMA as a voluntary, incentive-based law that can apply to developing countries currently decentralizing natural resource management. The next section discusses some general patterns related to decentralization in developing countries. The third section discusses in more detail the U.S. CZMA and sets the stage for a discussion of how it could be used as a model for decentralizing developing countries. The fourth section demonstrates how this program can be effective in decentralizing developing countries today. The final section discusses conditions in Indonesia and how the ideas presented in this article could be applied there.

THE GENERAL PARADIGM

Numerous developing countries around the world have recently undertaken serious efforts to decentralize their systems of governance. A recent World Bank study indicates that "out of 75 developing and transitional countries with populations greater than 5 million, all but 12 claim to have embarked on some form of transfer of power to local government."[2] As Lowry describes, within these 75 countries, definitions, contexts, and reasons for decentralization all vary widely.[3] There are, however, certain common trends and conditions that can be teased out of the variety of experiences seen among decentralizing developing countries.

In general, developing countries begin with strong centralized governments. A strong central State apparatus is virtually required for progress as investment in public goods and services with low or negative return ratios such as electricity, health, education, and transportation infrastructures would not be made by the private sector. This is often true either because the private sector is not of a sufficient size to make the necessary investments, or the private sector will forgo investment in public goods and services in lieu of more profitable investments. In this way, a strong central State apparatus builds and maintains the necessary public infrastructure for an emerging private sector. It is such an 'autocratic-paternalistic system of rulership' that has fueled the economic growth in Asia in recent years.[4]

Reasons for decentralization are generally attributed to an effort to improve public service delivery in legal, social, and economic arenas, first by improving 'allocative efficiency'—tailoring services to regional needs—and second by improving 'productive efficiency'—addressing accountability, bureaucracy and costs.[5] Whether and how a country chooses a certain path to decentralization are dictated by political circumstances.

---

2. A. Agrawel and J. C. Ribot, *Accountability in Decentralization: A Framework with South Asian and West African Cases* (Washington, DC: World Resources Institute, 2000), p. 23.

3. K. Lowry, "Decentralized Coastal Management" (unpublished report prepared for Coastal Resources Management Project, USAID, University of Hawaii, 2001).

4. A. Cheung, "The Paradox of Administrative Reform in Asia: Comments on the BICA Report 2001," in *Public Sector Challenges and Government Reforms in South East Asia*, pp. 15–25. (No editor listed. Conference Proceedings, Building Institutional Capacity in Asia (BICA), 12 March 2001, Jakarta).

5. World Bank, 2001. "Decentralization and governance: does decentralization improve public service delivery?," *Poverty Reduction and Economic Management (PREM) Network*, 55. Accessed Sept. 2001 on the World Wide Web: http://www1.worldbank.org/publicsector/legal/PREMnote55.pdf (Washington, DC); See generally O. Azfar, S. Kahkonen, A. Lanyi, P. Meagher and D. Rutherford, *Decentralization, Governance and Public Services, The Impact of Institutional Arrangements: A Review of the Literature* (Rockville, Md.: IRIS Center, University of Maryland, 1999).

Patterns of Decentralization

For those countries that engage in decentralization, three patterns[6] generally emerge that may follow each other or happen independently and remain static. An early expression of decentralization is deconcentration, which essentially extends the central government apparatus into the regions. Deconcentration usually involves the development of offices of national agencies in regional areas outside the direct control of the central government. While this is a form of decentralization, these offices are effectively extensions of the central government under direct line-authority of the same central government ministries and are maintained on the central government payroll. In this case, real decentralized authority is absent.

As economic development, governing capacity, and resource utilization expands in the regions, regional governments develop identities and limited capacities separate from the central government, especially related to allocation and conversion of locally based natural resources. At this point, the second form of decentralization—delegation of authority—is likely to occur. Regional governments are given more autonomy for decision making through delegation of specific authorities to regional governing offices that are either semi-autonomous units of the national government, or somewhat wholly autonomous local government units. Usually these offices are still in some way accountable to the central government (often through a system of officials appointed by the national government and controlled budgets).

At this point, the stage is set for the emergence of full decentralization or devolution. Devolution involves the full transfer of administrative and decision-making authorities to regional and local government bodies based on specific administrative and geographic boundaries.[7] Planning and decision making are fully assumed by local authorities with concurrent transfer of authority over fiscal controls.

Concurrent with each of these types of decentralization, central governments can be described as having three distinct positions relative to sharing regional and local governance power. Central governments can and often do take resistant positions, characterized by continued central control of all decision-making authorities and fiscal planning. As with deconcentration, some management authority is shifted to the regional offices but virtually all decision making is retained by the central government. Often, resistant governments are characterized by strong military control of the central government position. The second description—a capitulating government—is often associated with delegation of central government authority. Capitulating governments are responding to both the increasing demand for higher

---

6. D. Rondinelli and S. Cheema, eds., *Decentralization and Development: Policy Implementation in Developing Countries* (London: Sage Publications, 1983).

7. Agrawal and Ribot (n. 2 above).

decision-making authority in the regions and the increasing difficulty of managing the complexity of governance in the regions as development proceeds. The third description—the progressive position—is characterized by active engagement of the decentralization process and is most associated with full devolution of authority to regional and local governments, including fiscal controls. Progressive governments are also characterized by a higher degree of civil control and governing capacity.

Conditions Relating to Decentralization

Decentralization can be a methodical process under a progressive government, or it can be a fast, chaotic, often violent, process under a recalcitrant government. The pace and form of movement from centralized to decentralized governance largely depends on the conditions facing individual countries. Conditions fall into two broad categories: precipitating conditions, or conditions that bring about the need for decentralization, or some alternative form of governance; and resulting conditions, or those conditions that arise from decentralization efforts and the introduction of a new form of governance. These conditions generally overlap, as the conditions that both spark decentralization and stem from decentralization can last for many years and coincide with each other. For this reason, both sets of conditions need to be addressed, even though the practical distinction among them can be difficult to discern.

Nevertheless, it remains useful to discern between cause and effect. Precipitating conditions will include a range of environmental, institutional, political, and socioeconomic factors, such as stress on the natural resource base, inequitable funding/resource allocation among regions, institutional and governance problems at the national level, such as accountability, transparency, bureaucracy, et cetera, and growing community awareness at the local level.[8] Resulting conditions will generally relate to legal, financial, and institutional factors, such as a transfer of authority and greater financial allocation to local institutions, together with governing capacity or experience, institutional structure, and the extent and clarity of a legal system at the local level.

The question, then, is what conditions are conducive to effective decentralization. Dillinger and Fay[9] writing on behalf of the IMF, state that decentralization must be based on a set of 'coherent, explicit and stable set of

---

8. H. I. Vista-Baylon, ''Decentralization: What, When and How,'' in *To Serve and To Preserve: Improving Public Administration in a Competitive World*, eds. S. Schiavo-Campo and P. S. A. Sundaram, (Manila, Philippines: Asian Development Bank, pp. 156–186.

9. W. Dillinger and M. Fay, ''From Centralized to Decentralized Governance,'' in *Finance and Development*, IMF, (1999) vol. 36, no. 4, p. 1.

rules' that must encompass three aspects of decentralized governance: 'the division of national political power between national and sub-national interests; the functions and resources assigned to sub-national governments; and the electoral rules and other political institutions that bind local politicians to their constituents.' Regional governments should maintain a 'principal-agent' relationship with the central government, in which they act as both agents of the central government and as principals in the delivery of local services. As a universal matter, finance must follow function, so that as regional governments get more authority, they receive additional funding to carry out that authority. Lastly, in binding local officials to their constituents, it comes down to one word: accountability. Azfar[10] looks at the need for transparency, citizen participation, and capacity, in terms of human and physical capital.

The U.S. Agency for International Development has a lengthy set of indicators to measure whether a country has achieved the conditions for effective decentralization in three broad areas: administrative, financial, and political dimensions. For example, in administrative functions, it looks at the number of responsibilities reserved for local government, number of local government actions overturned by central governments, numbers of laws passed by local governments, and percentage of staff hired by local governments; in financial functions, it looks at local authority to collect taxes and fees, local sources of revenues, percentage of locally generated revenue, percentage of local budget mandated by central government; in political functions, it looks at constitutional and legal reforms transferring power, legal status of local governments, level of laws supporting free speech and association, and percentage of citizens who vote.[11]

There are four conditions that are unanimously considered the pillars of good governance: accountability, transparency, predictability, and participation.[12] Effective decentralization generally requires an effective central government, at least in terms of supervision of process. These conditions generally require adequate 'capacity' among the governing bodies, which Schiavo-Campo and Sundaram[13] define to include institutional development, organizational development, human resource development, and informatics development. In the context of decentralization, capacity is stronger initially at the central level, so that it is incumbent on the central level to lead the effort to develop capacity at the regional level.

10. Azfar et al. (n. 4 above).
11. United States Agency for International Development, *Decentralization and Democratic Local Governance Programming Handbook* (Washington, DC: U.S. AID, 1999).
12. S. Schiavo-Campo and P. S. A. Sundaram, eds., *To Serve and Preserve: Improving Public Administration in a Competitive World* (Manila, Philippines: Asian Development Bank, 2001), p. 10.
13. Ibid., p. 15.

While there is a consensus on what characteristics constitute effective decentralization, there appears to be little actual success in achieving these characteristics. It is not easy to generalize among countries decentralizing their governance system as to why success has been so difficult to achieve, it is no stretch to say that almost all governments—progressive, capitulating, and recalcitrant—have enacted decentralization policies more rapidly than can be effectively implemented. Generally, the central government faces sufficient pressure under precipitating conditions to force it to decentralize various managerial and financial authorities before adequate competence is attained at the regional level. It is much easier to cast off authority than it is to assume it.

The result is often a lack of focused development of regional and local governance structures. Problems exist at the institutional level, such as poor decision-making models, and spill into the personal level, with lack of technical knowledge and training. Overall, decentralization often fails to improve the ability to discover and achieve the aspirations of civil society through local governance structures any more than it succeeded at the central level.

With respect to natural resources, these conditions often combine to create a situation in which natural resources under stress often fall under greater stress at the hands of local governing bodies that have little institutional capacity and little financial capability to adequately manage them. Often the negative conditions that precipitate decentralization exacerbate the negative consequences of decentralization.

This is particularly true with coastal resources. In most developing countries, marine areas and territorial seas have been managed entirely by the central government. This included authority over activities ranging from oil and gas drilling to local fishing practices. Regional governments under centralized regimes generally have had no authority over marine areas, including inshore and nearshore waters. These areas operated essentially in a regulatory void, as there was little management at the local level, and little enforcement from the central level. With decentralization in many countries, however, regional governments are suddenly given authority over inshore and nearshore seas. Even the most basic data, such as measurements of boundaries, are often not known, giving rise to chaotic and conflicting management. Worse, some local governments might begin imposing restrictions within their new marine jurisdictions that conflict with either neighboring or national laws. Such is the case in Indonesia, as discussed later.

In the Philippines, for example, the central government rewrote the Constitution in the aftermath of the 'People's Power' (EDSA) revolution of 1986. The Constitution delegated significant authorities and revenues to regional governments. The national legislature expanded its decentralization push with enactment of the Local Government Code of 1991, which specifically devolved authority for health, agriculture, public works, and social welfare to regional governments. The central government in the Philippines has thus moved quickly and thoroughly towards decentralization.[14]

The rapid decentralization put a great burden on regional governments in the Philippines, however, to measure up to their new authorities. Many regional governments struggled with their new roles, while not many lived up to them. Particularly in the case of coastal resources, their management suffered. The 1991 law gave municipalities authority to manage marine areas up to 15 km from the coastline. In the 1998 Fisheries Code, the central government required that every village set aside at least 15 percent of its marine area for conservation. The implementation of this combination of authority and requirements was zealous if not a bit chaotic. On the one hand, the number of community-based marine sanctuaries has skyrocketed in recent years, from approximately 20 before passage of the 1991 Local Government Code, to approximately 210 in 1998.[15] On the other hand, one study estimated that only 19 percent of the marine sanctuaries established in the Philippines have been successful.[16]

In summary, the general paradigm that one can draw from decentralization efforts in developing countries is that (1) there are weaknesses in the central government that create pressure towards decentralization, such as lack of transparency, lack of participation, lack of accountability; (2) there are weaknesses in the social and economic conditions of the country that create pressure towards decentralization, such as ethnic tension, inequitable income distribution, and inequitable resource allocation; (3) a relatively strong central government shift, in one form or another, power and decision-making authority to relatively weak regional governments; and (4) regional governments assume this authority with little experience, capability, or funding.

Furthermore, while there is a solid consensus among scholars and government officials as to what constitutes effective decentralization, there is little evidence that those conditions are actually achieved on a widespread

14. L. Carino, "Governance in Local Communities: Towards Development and Democracy," in P. D. Tapales, J. C. Cuaresma, and W. L. Cabo, *Local Government in the Philippines: A Book of Readings, vol. 1: Local Government Administration,* Center for Local and Regional Governance and National College of Public Administration and Governance, University of Philippines, (Manila, Philippines: Kadena Press, 1998), pp. 67–83.

15. B. Crawford, M. Balgos and C. Pagdilao. *Community-Based Marine Sanctuaries in the Philippines: A Report on Focus Group Discussions.* Coastal Management. Report no. 2224. PCAMRD Book Series no. 30. Los Banos, Philippines: Coastal Resources Center, University of Rhode Island and Philippine Council for Aquatic and Marine Research and Development, 2000.

16. R. S. Pomeroy and M. B. Carlos, 1997. "Community-based coastal resource management in the Philippines: a review and evaluation of programs and projects, 1984–1994," *Marine Policy,* vol. 21, no. 5, pp. 445–64.

basis. Azfar[17] states that there are examples where regional disparities increase as a result of decentralization; public services may suffer and deteriorate as a result of decentralization; and corruption may increase at the local level to beyond what previously existed at the central level. All of this is to say that the effectiveness of decentralization efforts 'depends on the design of decentralization and institutional arrangements that govern its implementation.'[18] It is with the need for an appropriate design and institutional arrangement for decentralization that we turn to the U.S. CZMA.

THE U.S. MODEL

Prior to examining the CZMA itself, we should consider the setting for the law. In terms of resource and demographic profiles, approximately 60 percent of the U.S. population lives along the 19,800 km (12,300 mile) coasts of the Pacific and Atlantic Oceans, the Gulf of Mexico and the Great Lakes. From 1945 to 1970, the coastal areas saw tremendous expansion of residential, commercial, and industrial use, increasing pressures on coastal resources, and increasing conflicts for those resources.

Beginning with the release of a National Academy of Science's Committee on Oceanography (NASCO) report in 1959, *Oceanography 1960–1970*, the federal government slowly began increasing awareness of the need for more science dedicated to the oceans. Ten years later, the *Our Nation and the Sea* report of the special "Stratton Commission" was the first study that explicitly acknowledged the threats to the U.S. coastal areas. That report recommended the establishment of a National Coastal Management Program. Congress responded to both the Stratton report and the growing political pressure to preserve the unique quality of the coasts, by passing the CZMA in 1972.

The U.S. CZMA was only one in a spate of legislation enacted by Congress and signed into law during a relatively short period between 1972 and 1976. These included the Clean Water Act,[19] the Marine Mammal Protection Act,[20] the Endangered Species Act,[21] the Fisheries Conservation and Management Act,[22] and the Outer Continental Shelf Lands Act Amendments.[23] All

17. Azfar et al. (n. 5 above).
18. Ibid.
19. Clean Water Act, 33 U.S.C. s. 1294 (1977).
20. Marine Mammal Protection Act, 16 U.S.C. s. 1361 (1972).
21. Endangered Species Act, 16 U.S.C. s. 1531 (1973).
22. Magnuson-Stevens Fishery Conservation and Management Act, 16 U.S.C. s. 1801 (1976).
23. Outer Continental Shelf Lands Act Amendments, 43 U.S.C. s. 1331 (1953); Outer Continental Shelf Lands Act 1978 Amendments, 43 U.S.C. s. 1801 (1978), Public Law 95-372; Outer Continental Shelf Lands Act Amendments, (Public Law 99-272). This Public Law enacted no currently effective sections.

of these laws were based on well-defined regulatory frameworks focusing on a single issue or sector. The CZMA, however, was fundamentally different in two respects: (1) it did not focus on one particular issue or sector, but rather it focused on all the issues affecting one place—the coast; and (2) it was nonregulatory in nature.

In enacting the CZMA, Congress found that (1) the coastal zone has tremendous natural, commercial, recreational, ecological, and esthetic resources that benefit the entire nation; (2) those resources are being threatened by increasing and competing demands for them; (3) the states have authority to develop programs for the use of these resources; and (4) there is a national interest in effective management of these resources.[24] These findings lay the foundation for a collaborative federal-state approach to protecting coastal resources.

One of the greatest strengths of the CZMA is its recognition of state management authority. Unlike almost all other environmental laws enacted by the federal government in the United States, this law does not rely heavily on the notion of federal supremacy in mandating certain acts, standards, or prohibitions with which the states need to comply. To the contrary, the CZMA imposes no mandatory substantive requirement upon a state with respect to coastal zone management. It allows the states to manage their coastal resources as they deem appropriate. Not only does the law recognize state management authority, but it defers to it—that is, under certain conditions.

The CZMA is the foundation of federal and state policy for managing coastal areas, and is implemented by the National Oceanic and Atmospheric Administration (NOAA), within the U.S. Department of Commerce. The law operates on three basic principles: voluntary state participation, financial and legal incentives for encouraging state participation, and a balance of environmental and economic interests in devising state programs. By allowing states to participate on a voluntary basis, Congress recognized that many conditions—political realities, economic and social structures, resource abundance and resource consumption—varied from state to state. As a result, the CZMA set a framework of broad national goals and left it up to coastal states to individually determine how to assume responsibility for meeting those goals within certain basic requirements. If any state satisfied those basic requirements, the federal government would certify the state program, which in turn entitled the state to certain federally provided incentives.

CZMA Requirements for State Programs

The requirements of the CZMA can be characterized in three ways: they are broad and general, refraining from specific mandates; they are descriptive

24. Coastal Zone Management Act (n. 1 above)

in nature, refraining from normative requirements; and they are procedural in nature, refraining from substantive outcomes.

The breadth and generality of the requirements gives the states tremendous latitude in how to implement the law. For example, the 'coastal zone' is defined as the "coastal waters . . . and adjacent shorelands . . . strongly influenced by each other and in proximity to the shorelines . . . ."[25] There is no more specificity than that; states are free to work with any specific application of this general definition that they believe is appropriate. As another example, 'coastal resource of national significance' is defined as any area that is "determined by a coastal state to be of substantial biological or natural storm protective value."[26] Thus, not only can each state determine what constitutes coastal resources within its jurisdiction, each state can determine for itself which of its own resources constitute those of national significance. As yet another example, where the CZMA does contain substantive requirements, it often allows the states significant leeway in devising alternatives to comply with them. The law requires each state to maintain control over the land and water uses of the coastal zone in its jurisdiction; however, it allows the state any combination of measures to exercise this control, such as state establishment of criteria and standards for local implementation, direct state regulation, or state administrative review for consistency.[27] At the same time the CZMA recognizes state discretion over coastal resource management, it also balances this discretion with federal prescriptions for management.

The second characteristic of the CZMA requirements is that they are descriptive rather than normative. For example, the state management plan must identify the boundaries of the coastal zone, define permissible land and water uses within the zone, inventory areas of particular concerns, identify means of state control over land and water uses, and describe the legal and organizational structure to manage the program.[28] There is nothing in the CZMA that requires the states to limit the land and water uses allowed, to control land and water uses in any particular manner, or to adopt any particular organizational structure.

The third characteristic of the CZMA requirements is that they are procedural in nature, rather than substantive in scope. Indeed, the entire structure of the CZMA seeks to influence the process rather than the substance or results of state coastal management planning efforts. This is done through certification requirements. For example, the state management program must be adopted "with opportunity of full participation by [all]

---

25. CZMA 304(1), 16 U.S.C. 1453(1).
26. CZMA 304(2), 16 U.S.C. 1453(2).
27. CZMA 306(d)(11), 16 U.S.C. 1455(d)(11).
28. CZMA 306(d)(2), 16 U.S.C. 1455(d)(2).

interested parties and individuals,"[29] and with public hearings during program development.[30] The program must also contain a planning process for managing beaches, developing energy facilities, and addressing erosion.[31] The state must also have coordinated its program with other local and regional plans and established a mechanism for consultation and coordination.[32]

Christie and Hildreth[33] describe the CZMA requirements in a slightly different manner, but with essentially the same observations. They place the requirements in one of three categories: (1) informational and definitional; (2) institutional and organizational; and (3) planning. Informational and definitional requirements include setting boundaries for areas covered under the coastal management plan, defining water and land uses allowable within these boundaries, and biophysical characterization of the coastal area leading to identification of areas of special concern. Institutional and organizational requirements include identifying the organizational and structural means and legal authorities through which the coastal management program will be implemented, including showing funding sufficient for program implementation. This includes identifying the intergovernmental coordination mechanisms and mechanisms for meaningful public participation in decisions concerning coastal use and protection. The requirements for planning include developing a system for prioritizing coastal dependent uses, establishing conservation and preservation areas, dealing with coastal hazards, planning for recreation, public access and historical areas, and finally, a process for dealing with issues of national concern.

Oversight by NOAA is not overly intrusive. Federal approval, or certification, of the state management program is a one-time approval. Federal review of the state management program is to consider "the extent that states have implemented and enforced the program approved by [NOAA], addressed coastal management needs identified in [the CZMA], and adhered to the terms of any grant, loan or cooperative agreement funded under [the CZMA]."[34] Each state is free to amend or modify its program as it desires. The only condition is that the state must notify NOAA and submit the changes for NOAA's approval. NOAA has a short period—30 days, with the opportunity for a 120-day extension—in which to review the changes, and if it fails to respond within that period, the change is deemed automatically approved.[35]

29. CZMA 306(d)(1), 16 U.S.C. 1455(d)(1).
30. CZMA 306(d)(4), 16 U.S.C. 1455(d)(4).
31. CZMA 306(d)(2), 16 U.S.C. 1455(d)(2).
32. CZMA 306(d)(3), 16 U.S.C. 1455(d)(3).
33. D. R. Christie and R. G. Hildreth, *Coastal and Ocean Management Law* (St. Paul, Minn: West Group, 1999), p. 63.
34. CZMA 312(a), 16 U.S.C. 1458(a).
35. CZMA 306(e), 16 U.S.C. 1455(e).

CZMA Incentives for State Participation

The CZMA includes incentives to encourage states to develop and implement coastal management programs consistent with the law. First are program development grants, which go to states that do not have certified programs, but that desire to develop one. Development grants are matched at 4 to 1, federal to state funds. Second, administration grants are for administering or implementing certified state programs. Matching grants are generally 1 to 1. Evident from different matching ratios of the two types of grants, Congress placed much greater emphasis on developing new programs rather than on administering existing programs.

There are two other types of grants that go beyond basic development and administration of the state management program. One is a resource management improvement grant, which is available to states with certified programs, for special needs and activities to improve degraded habitats, facilities, and beaches, such as redevelopment of deteriorated ports and waterfronts, increasing access to public beaches, and restoration of special management areas.[36] Another is a coastal zone enhancement grant, which is available to support changes to previously approved state management programs. The changes must enhance coastal zone management and protection, such as restoring and creating coastal wetlands, reducing development in hazard-prone areas, reducing marine debris, addressing secondary impacts of coastal growth and development, et cetera.[37]

Perhaps the most important incentive is the requirement of 'federal consistency.' In sum, federal activities within or outside the coastal zone of any state, that affect the land or water use or natural resources of the coastal zone, shall be carried out in a manner that is consistent, to the maximum extent practicable, with the enforceable policies of an approved state management program.[38] Even applicants for federal permits need to comply with this requirement.[39] This federal consistency gave unprecedented power to U.S. state governments, essentially reversing the general notion in the United States of federal supremacy over state law, at least with respect to federal government actions affecting coastal resources. It has proven to be a very effective incentive for state participation.

Finally, the CZMA program made technical assistance available to states developing programs and to those with already approved coastal management programs.[40] These technical assistance services continue to provide a wide array of technical and planning expertise to states and serve as the

---

36. CZMA 306a(b), 16 U.S.C. 1455a(b).
37. CZMA 309(a), 16 U.S.C. 1456b(a).
38. CZMA 307(c)(1)(A), 16 U.S.C. 1456(c)(1)(A).
39. CZMA 307(c)(3)(A), 16 U.S.C. 1456(c)(3)(A).
40. CZMA 310(a), 16 U.S.C. 1456c(a).

basis for developing close working relationships between state and federal agencies. These relationships prove particularly useful during the periodic evaluations conducted by NOAA.

A Measure of Effectiveness

The CZMA is widely regarded as a success in raising awareness and instituting programs to protect coastal resources. Almost 30 years after the enactment of the CZMA of 1972, states have developed a broad range of coastal management programs reflecting the different conditions of each state. These programs generally fit within one of the three approaches recognized by the CZMA[41]: (1) state establishment of criteria and standards for implementation of coastal programs at the local level; (2) direct implementation by state land and water use planning and regulations; and (3) direct state review of all development plans, projects, or proposed land and water uses.

Furthermore, the CZMA has come to anchor all coastal management activities at all levels of government.[42] The federal government interprets the statute though rules and guidelines, oversees state implementation, and awards grants to states; the state governments are the "action arm of the coastal management system," while local governments "are often primary implementers of state coastal policies and programs."[43] There is thus solid evidence that the CZMA—as a voluntary, incentive-based program—has been effective in shaping individual state coastal resource management policies and programs in the United States, consistent with a national interest in how those coastal resources are managed.

There are three caveats, however. The first is that integrated coastal zone management was, in many respects, very slow to catch on in the United States. For example, it took more than 25 years for all 35 coastal states to develop management programs; many states were either not interested in the first place, or never committed the resources necessary to move quickly towards completion of their programs. This lack of urgency was due to the voluntary nature of state participation. The second caveat is that there are still no systematic, decisive data on how effective the CZMA has been in delivering the desired outcomes in coastal management. Because the emphasis in the CZMA is almost entirely on process, there is no sense that the substantive outcomes for improved coastal management have been achieved on a widespread basis attributable to the CZMA.[44] The third is that a volun-

41. Christie and Hildreth (n. 33 above), p. 68.
42. M. J. Hershman, J. W. Good, T. Bernd-Cohen, R. F. Goodwin, V. Lee, P. Pogue, "The Effectiveness of Coastal Zone Management in the United States," *Coastal Management* 27 (1999): 113–38.
43. Ibid., p. 115.
44. Hershman et al. (n. 42 above).

tary regime cannot work effectively by itself. One reason that the CZMA works well as a voluntary program in the United States is because it does not carry the burden of larger, broader environmental issues, which are regulated by a panoply of other mandatory laws—the Clean Air Act,[45] the Clean Water Act,[46] the Magnuson Fisheries Conservation and Management Act,[47] the Resource Conservation and Recovery Act,[48] the Comprehensive Environmental Response, Cleanup and Liability Act (CERCLA),[49] and so on.

So it is the very attributes of the CZMA that make it unique and successful—that it is a voluntary, incentive-based program that focuses on procedural issues such as coordination and integration—are the same attributes that have ironically limited its success in accomplishing its substantive goals. The question then arises as to how applicable, and how useful, is the CZMA as a model for not only coastal resource, but natural resource, management in other countries that are both developing and decentralizing.

APPLICATION OF THE U.S. MODEL TO THE GENERAL PARADIGM

Comparison of Conditions

Before discussing why the CZMA serves as a model for coastal and natural resource management in decentralizing developing countries, one may find it worth considering the similarities between the United States at the time the CZMA was enacted, and decentralizing developing countries now. The similarities exist in three basic arenas: political governance systems; relative institutional structures; and social, economic, and environmental conditions.

Politically, the U.S. system is based upon the principle of federalism, in which sovereign subnational states coexist within a federal republic. While states maintain authority over activities within their borders, these activities can be limited to some extent by the federal government on the basis of the national interest and in accordance with the Constitution. The federal government occasionally tests the boundaries of its authority over states, such as the early 1970s, when it enacted the host of environmental laws. The only law that met significant resistance concerned a statute on land-use planning; states vehemently resisted any effort by the federal government

---

45. Clean Air Act, 42 U.S.C. s. 7401 (1955).
46. Clean Water Act (n. 19 above).
47. Magnuson-Stevens Fishery Conservation and Management Act (n. 22 above).
48. Resource Conservation and Recovery Act, (Public Law 94-580). This Public Law enacted no effective sections.
49. Comprehensive Environmental Response, Compensation and Liability Act, 42 U.S.C. s. 9601 (1980).

to engage in land-use planning, which was considered an inherently local activity. The law most closely related to this effort was the CZMA, a law that prescribed certain standards on coastal land use in exchange for certification. To be sure, it was a mere shadow of the larger land-use planning efforts, but it succeeded where the others failed. And it succeeded precisely because it was only a shadow of the original bill—a largely procedural, voluntary program that looked to guide land-use planning rather than mandate it. As noted earlier, Section 302(i) of the CZMA fully recognized states' rights over its land and marine areas, stating that the "key to more effective protection and use of the . . . coastal zone is to encourage the states to exercise their full authority over the lands and waters of the coastal zone . . . in developing land and water use programs . . ."

This political dynamic—in which subnational governments are seeking to gain or retain significant authority from a central government—is essentially what is taking place in countries that have recently decentralized or are decentralizing. The shift in authority provides a more equitable balance of power between the central government and regional governments, which is one of the hallmarks of the U.S. system of government. Furthermore, once a country has already begun the process of decentralization, it is politically difficult for the central government to impose new mandates and subsume new authorities. Indeed, there are some scholars who go as far as to say it is immoral and unjust, stating that subsidiarity—a philosophical underpinning of decentralization—has its roots in socioreligious dogma.[50] Thus, in pursuing governance reform through decentralization, the central government must enact laws that recognize regional authority and build regional capacity. The CZMA is such an example.

Institutionally, during the 1960s, the U.S. federal government had begun to place a concerted effort into improving its scientific knowledge base and expand its research programs. Commissions were convened, new agencies were established, and research projects funded and carried out. The states, however, lagged behind the federal government, particularly in the area of coastal issues. Consequently, any program to address coastal issues needed to begin slowly, giving states time to improve their knowledge base, their institutional capacity, and their legal systems. In addition, given the relative institutional strength of the federal government vis-à-vis the states, such a program also needed to give the states assistance—technical and financial—in making these improvements.

This institutional positioning also exists in decentralizing developing

50. W. Meyer, "Promoting Decentralization and Strengthening Local Government: The Perspective of the Konrad Adenauer Foundation," in *East and Southeast Asia Network for Better Local Governments: New Public Management: Local Political and Administrative Reforms*, (Pasay City, Philippines: Konrad Adenauer Stiftung, Local Government Development Foundation (LOGODEF), 2000), pp. 45–59. No editor named.

countries: the central government is relatively much stronger than regional governments in terms of legal systems, information, knowledge and data gathering, staffing, and finances. A law addressing coastal and natural resource management must take advantage of these relative institutional positions. The CZMA took such an approach, giving the states no time limit to decide whether to undertake a coastal management program, a long window in which to develop it if a state so decided, and significant assistance in a variety of forms to develop its program once a state started.

Among the social, economic, and environmental conditions facing the United States in the late 1960s and early 1970s were population growth and a rapidly expanding economy, particularly along coastal areas. In addition, there was a growing awareness among both the scientific community and the general public as to the environmental problems stemming from this economic and demographic expansion.

These conditions also exist in many decentralizing developing countries. As these developing countries improve their economic base and industrial infrastructure, effects are being felt along the coasts more than any place else.[51] With assistance from multinational projects and international organizations, many of these countries have benefited from scientific studies, socioeconomic analyses, and legal assessments to determine the status of coastal resources in the countries and to develop policies to manage coastal resources. Political awareness and scientific understanding is growing among these populations. In this broad sense, the dynamic is the same in these countries as it was in the United States in the 1960s and 1970s.

The CZMA in a New Light

These comparisons highlight the similarities between the United States and decentralizing developing countries (despite the obvious differences in relative economic and political strength) and underscore the appropriateness of using the CZMA as a model in these countries. To promote the U.S. CZMA as a model for integrated coastal management is nothing new. Since its enactment it has been cited countless times as the original model for a national, integrated program for managing and conserving coastal resources.[52] Numerous developing countries around the world have recently sought to integrate decision making relating to coastal and marine resources. Recent entries include Sri Lanka, the Philippines, South Africa, Tanzania, Ecuador, and Indonesia, among others.

51. D. Hinrichsen, *Coastal Waters of the World: Trends, Threats and Strategies* (Washington, D.C.: Island Press, 1998).
52. B. Cicin-Sain and R. W. Knecht, *Integrated Coastal and Ocean Management: Concepts and Practices* (Washington, DC: Island Press, 1998), pp. 317–19.

What is new, however, is that the U.S. CZMA is a particularly appropriate model for a certain group of developing countries—those that are in the throes of decentralization. It is an appropriate model not so much for a particular approach to integrated (vis-à-vis sectoral) coastal management, but for a particular approach to voluntary (vis-à-vis mandatory) coastal management. To say it another way, the importance of the CZMA as a model is not so much in the decision-making processes it establishes for integrating coastal management decisions, but rather in the incentive-based approach it creates for encouraging such decisions. In this view, the CZMA serves as a model not just for coastal resource management, but for all natural resource management. The model becomes one of a voluntary, incentive-based natural resource management program rather than one of an integrated coastal management program.

As discussed earlier, for a variety of reasons the central government in decentralizing, developing countries does not have the institutional capacity, political wherewithal, legal structure, or social inclination to impose a mandatory regime for coastal or natural resource management upon the regional levels. Furthermore, the regional levels generally do not have the institutional capacity or legal structure to undertake such a regime if it were to be imposed upon them. A voluntary program, by its very nature, can be implemented much more readily and effectively in such countries than can a mandatory program. Of course, as seen with the CZMA discussed earlier, a voluntary program may sacrifice a great deal in terms of the extent and nature of the goals it seeks to achieve. The question is whether the benefits outweigh the disadvantages.

Before attempting to define the benefits of a voluntary, incentive-based law, it is perhaps useful to take a short detour into the realm of critical legal theory and to examine the question of why a voluntary law would work in the first place. There are two general theories on the nature of law—either law is descriptive, in that it reflects the behavioral patterns of society; or law is prescriptive, in that it serves to establish behavioral norms of society.[53] There are scholars who believe that behavioral patterns exist independent of a rule of law—people act the way they do for reasons other than the existence of a law.[54] For example, people wear seat belts not because it is required, but because it can save their lives. However, a rule of law constitutes one reason why people may change their behavioral patterns. For example, people apply for drivers' licenses and pay registration fees for their cars not because it will save their lives, but because there is a law requiring them to do so. Law, therefore, is one tool to shape behavioral patterns, even if it is not the only tool. As such a tool, the law can act in one of two ways: it

---

53. H. L. A. Hart. *The Concept of Law* (Oxford: Oxford University Press, 1961).
54. A. Seidman, R. B. Seidman and N. Abeyesekere, *Legislative Drafting for Democratic Social Change: A Manual for Drafters* (London: Kluwer Law International, 2001).

can either shape a direction for certain behavioral patterns, or it can simply mandate those behavioral patterns. A voluntary law acts as a tool in the former capacity—nudging behavioral patterns in a particular direction without necessarily forcing those patterns.

Most significantly, a voluntary regime eliminates much of the enforcement pressure that a mandatory law faces to ensure implementation. Mandatory regimes, by nature, are impositions of the will of the governing body upon a segment of society.[55] They may or may not reflect the will of that segment of society; they may or may not reflect the capabilities of that segment of society. Thus, enforcement of such regimes is critical to their success. This generally requires a significant amount of investment in resources—personnel, funding, and training—that the governing body may or may not have. If the law reflects the will of society, or if the governing body can maintain a strong enforcement presence, then the law will succeed. However, if the law does not reflect the will of society, or if the law does not reflect the capability of society, and there is little enforcement presence, the law will fail. In almost all developing countries, it is this latter scenario that prevails. Consequently, mandatory regimes that appear strong on paper all too often are weak and ignored in actuality.

Many scholars will often profess that a developing country has good laws that are not implemented. This is an oxymoron. A good law, by its definition, is one that is implemented by the governing body and is followed by the society. Seidman[56] states that "[a] law always addresses two sets of persons: the law's primary addressees whose behaviors the law-makers principally propose to change, and those who work for the agency responsible for implementing the law." Consequently, the implementing agency must be capable of overseeing, implementing, and enforcing the law. A mandatory rule is generally prescriptive in nature, but it requires a reasonably strong state apparatus to implement and enforce it. On the other hand, a descriptive rule would merely reflect the existing behavioral patterns of society, and would not change behavior; therefore it would not require a strong state apparatus. In decentralizing, developing countries, the optimal rule is one that is a combination of both types of rules: something that is based on descriptive behavioral patterns, but moves toward desired prescriptions. Such a rule avoids the unrealistic expectations that so often accompany a prescriptive, mandatory law that has little hope of being enforced because of weaknesses within the governing body and implementing agency.

The U.S. CZMA is such a combination of a descriptive and prescriptive rule. It allows subnational governments to manage coastal resources under their own authorities, but attempts to shape the way in which these governments use those authorities. Its recognition of regional government author-

55. Hart (n. 53 above).
56. Seidman et al. (n. 54 above), p. 16.

ity is exactly the recognition that many central governments are now making in decentralizing developing countries. At the same time, the CZMA is based upon a package of incentives and assistance to develop capacity among regional governments. In the United States, capacity development assistance was particularly important as states initially had very little capacity for environmental management in the complex coastal zone. This as well is true of decentralizing, developing countries. The nature of the CZMA program capitalizes on a strong institutional base at the national level without compromising subnational authorities and rights to govern autonomously.

As discussed earlier, effective decentralization requires an effective central government. The CZMA model provides an appropriate role for central government: develop guidance to the regional governments and communities that now have authority to manage their coastal resources, but as of yet do not have the ability to do it themselves. Particularly in natural and coastal resource management, such guidance is critical. To be sure, the limitations discussed earlier regarding the CZMA as a voluntary, incentive-based program apply in a decentralizing, developing country as well: additional mandatory rules would likely be required to supplement the program; the behavioral patterns of society may not change as quickly as they otherwise could; and as much as such a program is process oriented, specific outcomes may be uncertain. Taking these limitations into account, one of the most appropriate and effective means of natural resource management in countries that are developing and decentralizing will be a voluntary, incentive-based program. Indonesia provides an excellent example.

THE INDONESIAN EXAMPLE

Even though Indonesia is the largest archipelago state in the world, with the second longest coastline behind Canada, integrated coastal management (ICM) has only recently become a subject receiving any significant attention from the central government.[57] The government first addressed it in its fourth Five-Year Development Plan (Repelita) in 1984, but it was not until 1994, in Repelita VI, that the national government considered the marine sector independent from other institutional and economic sectors.[58] Since then, great strides have been made in promoting marine and coastal management issues—food security and fish production, hazards mitigation and control, land-based pollution and environmental protection of marine areas—within larger planning efforts. The strides have been assisted by out-

57. R. Dahuri and I. M. Dutton, "Integrated coastal and marine management enters a new era in Indonesia," *Integrated Coastal Zone Management* 1 (2000): 11–16.
58. Badan Perencanaan Pembangunan Nasional (BAPPENAS), Repelita (Rencana Pembangunan Lima Tahun) VI, 1994/95–1999.

side organizations, but received a tremendous boost from the central government itself with the creation of a new department of marine affairs and fisheries in 1999.[59] With this new ministry, there is now an opportunity for the development of a strong national program for integrated coastal management.[60]

At the same time that these efforts are getting underway in the central government, the reformasi movement has triggered a tremendous push to decentralization. Since independence in the 1945, and particularly since the New Order in 1965, Indonesia has operated under a centralized governance structure, with virtually all mandates emanating from the central government in Jakarta.[61] This regulatory structure is implemented at the regional level through Perdas issued by the provincial level (enactments by the Governor and DPRD I), and regency level (Regent, [or Bupati] and DPRD II).[62] With the rise of democracy in Indonesia since the fall of President Suharto, there has been a growing demand for transparency, honesty, and especially autonomy from the central government. The central government has responded with a series of laws shifting both the political power and the financial control within the country from the central government to individual regencies, and enacting legislation regarding corruption, collusion, and nepotism (KKN). The consequence is nothing less than a revolution in governance.

Overview of Decentralization

With the enactment of two laws in particular—Act No. 22 on decentralization and Act No. 25 on revenue sharing in 1999—regional autonomy has become a fast reality in Indonesia. These two laws—and several regulations

---

59. Departemen Kelautan & Perikanan (DKP) 2001. Annual Report 1999–2000.

60. Minister S. Kusumaatmadja, National Marine Exploration and Fisheries Policy, Statement by the Minister, 14 February 2000.

61. C. MacAndrews, *Central Government and Local Development in Indonesia*, (Oxford: Oxford University Press, 1986).

62. A note on terminology: this article uses the terms 'regency' and 'district' interchangeably, translating into 'kabupaten.' Use of these terms also includes 'cities' ('kota'), which, under Indonesian law, have the same jurisdictional authority as kabupaten. 'Regional government' refers to both the regency and provincial levels. This article also uses the term 'act' to describe an 'undang-undang,' a law that is enacted by the DPR and signed by the President of the Republic of Indonesia. 'Undang-undang' is often translated as 'law,' but as noted by Pak Koesnadi Hardjasoemantri, the term 'law' is a general reference to governing rules and regulations, rather than the particular type of rule constituting an 'undang-undang'; O. Podger, *Government in Regions: An Overview of Government Organization within Provinces and Regions of Indonesia* (Surubaya: publisher unknown, 1994).

with greater detail—create the legal and financial framework for governance primarily by the regency, with assistance from both provincial and central levels of government.[63] In sum, the role of the central government has shifted from one of heavy-handed regulation and detailed management to one of guidance and policy direction. The role of the 332 regencies has shifted from one of administration of central policy to primary management over their jurisdictions.

Act No. 22/1999 (Article 4) sets the general tone, that the law is intended to arrange and organize local societies, through their own decisions, based on their own aspirations. Article 7(1) provides that this authority covers every governance field except foreign affairs, defense and security, justice, finance, and religion. The central government retains authority to develop policy regarding a host of subjects, including natural resource use and conservation. The primary regulation implementing the Act defines 'policies'—in addition to guidelines, criteria, standards, and supervision—with language that conveys that subsequent, more specific action is required (Reg. No. 25/2000). Thus the role of the central government is primarily one of indirect action rather than direct regulation and control, with specific action to follow at the regional level.

The change in governance (and the difference between policy and management) is underscored in the treatment of coastal waters. Act No. 22/1999 (Article 3) establishes, for the first time, a local marine area under the jurisdiction of the province, up to 12 nautical miles (M) the island baseline, in which the province is given authority over exploration, exploitation, conservation, and management of the sea. Pursuant to Article 10(3), the regency may establish jurisdiction over one-third of the provincial waters, seaward from the island baseline.[64] With respect to the maritime areas within the jurisdiction of the central government, specifically within the Exclusive Economic Zone (EEZ) beyond the 12-M mark, the central government maintains direct responsibility for activities. According to Regulation No. 25/2000, (Art. 2(3)(2)(a)), it can determine conduct on exploration, conservation, processing, and exploitation of natural resources in those waters. The difference between the role of the central government generally and its role within its own jurisdictional territory is illustrated by the language in Regulation No. 25 regarding natural resource conservation: generally, the central government is to "determine guidelines on manage-

---

63. J. Alm and R. Bahl, *Decentralization in Indonesia: Prospects and Problems* (Jakarta, unpublished report prepared for USAID, 1999).
64. There are two notable exceptions to these new regional authorities. First, the seabed underneath the sea territory is not explicitly included in the maritime area, so that authority for management of the seabed remains under central government control. This includes rights to conduct activities on the seabed, such as oil, gas and mineral extraction. Second, traditional fishing rights are not restricted by the regional territorial sea delimitation.

ment and protection on natural resources" (Art. 2(4)(g)); but within its own jurisdiction, the central government is to "manage and to implement protection of natural resources in maritime areas beyond twelve miles" (Art. 2(4)(h)).

While management authority has thus been shifted to the regency, it is not absolute. Apart from the five areas of governance explicitly withheld for the central government, the central government can withhold other areas of governance through regulation.[65] Regencies must still abide by central government laws that the central government can still enforce. In addition, the provinces have certain managerial authority, although it is still largely undefined and vague. Specifically, according to Article 9 of Act No. 22/1999, the province maintains authority in three circumstances: (1) authority over intersectional regency government affairs, such as matters that affect two or more regencies; (2) in lieu of the regency for matters not yet, or not able to be, handled by the regency; and (3) administrative authority delegated from central government.

As with the jurisdiction of the central government within the EEZ beyond 12 M miles, Regulation No. 25/2000 (Article 3(5)) gives the province clear autonomous authority within the territorial waters between 4 M miles and 12 M miles. Specifically, the province is to organize and manage maritime waters, and to explore, exploit, conserve, and manage maritime resources limited to the maritime areas under provincial authorization, including the supervision of fishery resources and licensing of permits for cultivating and catching fish, and management non-oil mineral and energy resources.

If Act No. 22/1999 is the vehicle for decentralization, then Act No. 25/1999 is the engine. It provides for an almost complete shift of budgetary management from the central government to the regional government, particularly with revenues derived from natural resource consumption. The central government used to collect 100 percent natural resource revenues and distribute it regardless of the place of derivation. Under the new formulas, the central government gets 20 percent of natural resource revenues, specifically forestry, fishing and mining, while the regional governments with jurisdiction over the resources get 80 percent (Art. 6(5)). From oil production, the central government gets 85 percent and the regional government gets 15 percent, and from natural gas production, the central government gets 70 percent and the regional government gets 30 percent (Art. 6(6)). Regulation 104 (Articles 9 and 10) elaborates upon those allocations, providing that of the 80 percent revenues that go to regional governments, 16 percent go to the provincial governments and 64 percent goes to the regencies according to various distributions, with the bulk going to the particular regency in which the activity is taking place.

65. Act No. 22/1999, Art. 7(2).

Fisheries revenues[66] are handled differently. They are to be distributed to the regencies, but "in equal sums to regencies throughout Indonesia."[67] This is a fundamental difference compared with regional revenues from other natural resource uses, which are distributed primarily to the regency of origin. This difference highlights the fact that fisheries are treated as true commonly owned, national resources to be shared by all. The result of this difference is that an individual regency will receive significantly less revenue from fishing activities within its own jurisdiction than other natural resource activities. This provision removes much of the pecuniary interest—and the immediate incentive—for regency governments to sell off fishing rights, as they are already doing with concessions in the forestry sector.

While these four laws—Acts No. 22/1999 and 25/1999, and Regulations No. 25/2000 and 104/2000—form the central pillars of decentralization, there are almost 1000 other regulations, decrees, and guidelines that will need to be modified to be consistent with them.[68] Numerous questions remain as to the extent of central and provincial authority, and exactly how the authority is to be exercised in light of the emphasis on regency authority. There is an effort by the central and provincial governments to revise the newly established system to restore some authority to themselves. For example, DPR recently commissioned a study to revise Act No. 22, which recommended that regional jurisdiction over territorial seas within 12 M miles of the island baseline be revoked, with jurisdiction of those waters being returned to the central government.[69]

Other questions exist. Act 25/1999 provides that the regencies will receive most of the public revenues. However, as the bulk of income is derived from natural resource use, the revenue distribution will vary enormously from region to region.[70] More importantly, the bulk of the income is to be used for administrative expenditures, such as operating new bureaucracies in the regions, and to support the transfer of thousands of civil servants from the central government to the regional governments.[71] For example, in two

---

66. Section (1) of Article 11, Regulation No. 104/2000, defines these revenues to include levies on fishery exploitation and levies on fishery production.

67. Reg. 104/2000, Art. 11(2).

68. GTZ (Deutsche Gesellschaft fur Technische Zusammenarbeit), "Project support for decentralization measures (SFDM)," Decentralization News Issues, no. 1–8 (2001). Accessed Aug. 2001 on the World Wide Web: http://www.gtzsfdm.or.id.

69. B. S. H. Hoessein et al., Emikiran Filosofis, Yuridis Dan Sosiologis Revisi UU No. 22 Tahun 1999 Dan UU No. 25 Tahun 1999, Sekretariat Jenderal Dewan Perwalikilan Rakyat, Republik Indonesia and United Nations Development Programme, 2001.

70. T. H. Brown, *Economic Crisis, Fiscal Decentralization and Autonomy: Prospects for Natural Resource Management,* (Jakarta: Natural Resources Management Project, August 1999); U.S. Embassy, *Economic Report: Where the (Natural Resource) Wealth Is,* (Jakarta: 18 May 1999).

71. GTZ (n. 68 above).

kabupatens in central Java, it is estimated that upwards of 86 percent of the new funding will go to pay civil service salaries.[72] Thus, very little new revenue will go to development projects and resource conservation.

Fitting into the Paradigm

In terms of pressure on resources, failure of legal regimes at the national level, lack of capacity at local levels, and a growing awareness of these issues, Indonesia fits squarely in the general paradigm described earlier, and therefore presents an excellent case study for how a voluntary, incentive-based coastal resource management program can work. On the one hand, there is a decent infrastructure for marine and coastal resource management at the central level. However, with decentralization, management authority has shifted to the regional level. Given the speed with which this shift has occurred, there is the risk of 'a vacuum of authority' in which there is little adequate management of the resource.[73]

Natural resource conservation has suffered in Indonesia in part to the deficiencies in the legal structures at the national level. There are generally three reasons for the profound number of conflicts, gaps, and overlaps in Indonesian law. First, the laws themselves are so vague and broad that conflicts often arise even within a single act. One act may offer two or more broad goals or principles that, when applied in specific circumstances, may conflict. For example, in Act No. 9/1985 relating to Fisheries, Article 7(1) prohibits damage to the marine habitat, yet the Act also allows bottom trawl fishing and other gear types that, depending on the situation, can be very destructive to surrounding habitat.

Second, the rules of statutory construction for resolving differences among laws are vague and broad. As in most countries, Indonesia recognizes the premise that laws enacted later in time take priority over laws enacted earlier in time, and laws that are more specific take priority over more general laws. These rules of legal interpretation are not codified, however, so there is no consistent application by the judiciary.[74] Furthermore, the rule of interpretation that is codified in a typical act is extremely weak: each act typically states that previous laws remain valid unless specifically in conflict with the new act. Rather than explicitly replacing one law for another, the

---

72. R. MacClellan, Chief Science Officer, U.S. Embassy, Jakarta, personal communication, 22 March 2001.

73. BICA (Building Institutional Capacity in Asia), "Public Sector Challenges and Government Reforms in South East Asia"ed. University of Sydney. Conference Proceedings, 12 March 2001, The University of Sydney, Research Institute for Asia and the Pacific, Jakarta, p. 42.

74. I. P. Diantha, Fakultas Hukum Unud, personal communication, 12 July 2001.

Act offers only an implicit replacement. Such an implied repeal is often very difficult to interpret.

Third, where conflicts do arise, they are generally not resolved through the judiciary.[75] Rather, they historically have been resolved with the issuance of a Presidential Decree or Ministerial Decree. This approach—where the executive branch of government resolves disputes among laws enacted by the legislature—makes a highly politicized legal system with little certainty, as opposed to an approach in which the judiciary resolves disputes and adheres to its own precedents.[76]

These conflicts are exacerbated in coastal management issues, because coastal management involves a particular biogeographic space in which many sectors operate, rather than involving a particular sector.[77] The existing legal regime governing coastal resources in Indonesia is, in a word, sectoral, meaning that coastal resources are not managed as a whole, but as individual elements. There are approximately 20 acts that relate to coastal resource management.[78] As one example of a conflict between marine and forestry sectors, Act No. 41/1999 relating to Forestry allows for harvest of coastal mangrove forests; however, such harvest conflicts with the prohibitions against damaging habitat of fishery resources, contained in Article 7(1) of Act No. 9/1985 relating to Fisheries. As another example between the fisheries and natural resources, Act No. 9/1985 has an extremely broad definition of the term "fish" that can be harvested under that law, including sea turtles, marine mammals such as whales and manatees, the sea cucumber, and corals; however, Act 5/1990 relating to Conservation of Natural Resources protects fish and wildlife that are threatened with extinction.

Conflicts are also exacerbated with respect to enforcement. Different acts have different sanctions and liability for similar offenses. Sanctions, such as criminal versus civil penalties, vary widely. Different acts also have different standards of liability, such as negligence, intentional or strict, for almost identical violations. This complicates enforcement and prosecution efforts. There are countless other examples, especially in looking at regulations and decrees. There is a profound need to develop a new umbrella law that serves to coordinate existing laws and create new mechanisms to resolve legal discrepancies. This is the primary reason why a new national program is necessary.

75. BICA (n. 73 above), p. 43.
76. L. Heydir, 1986. Skripsi, Peran Sekretariat Negara Dalam Proses Pembentukan Peraturan Perundangan, Departemen Pendidikan dan Kebudayaan, Universitas Gadja Mada, Fakultas Hukum.
77. T. H. Purwaka, "Policy on marine and coastal resource development planning," Occasional Paper Series 8, (Bandung: Center for Archipelago, Law and Development Studies, 1995); S. Putra, Head of Sub-Directorate of Integrated Coastal Zone Management, Direktorat Jenderal Pesisir dan Pulau-Pulau Kecil, DKP, personal communication, March 2001.
78. Putra (n. 77 above).

Another condition is the growing awareness of coastal resource issues, and attempts to address them. For example, as mentioned earlier, in 1994, for the first time, the national Five-Year Development Plan identified marine sector as an independent sector. In 1999, a new Ministry of Marine Affairs and Fisheries was established, taking portfolios from the Ministry of Forests and Ministry of Environment. (It was originally named Ministry of Marine Exploration and Fisheries, but the name was changed shortly thereafter to reflect a larger mandate). A new Maritime Council was established as an interagency executive body to supervise maritime issues. In addition to these activities by the central government, international nongovernment organizations, donor agencies, and local groups have all engaged in numerous projects amounting to billions of dollars in coastal resource conservation and management in the last 10 years.[79]

These projects have raised awareness, developed capacity and skills for resource management, and established conservation areas. Since the mid-1990s, there has been a growing realization that greater autonomy and community-based governance was likely to be more effective in protecting the environment.[80] Since then numerous projects have been carried out in Indonesia that support community-based management of natural resources, with good success.[81] This guidance would rely on the community-based models that already exist, and shape new models for the future.[82] There is a desire among the central government and other groups to establish a national mechanism to replicate such projects.[83]

At the same time, governing capacity at the regional level is minimal,

---

79. F. Sofa, *Program Pengalolaan Pesisir dan Kelautan di Indonesia: Sebuah Tinjauan, Proyek Pesisir Technical Report. Proyek Pesisir* (Jakarta: Coastal Resoures Management Program, 2000).

80. CIDE (Center for International Development and Environment), World Resources Institute, and the Foundation for Sustainable Development, *Strengthening Local Community-NGO-Private Sector-Government Partnerships for the Environmental Challenges Ahead: Indonesia Environmental Threats Assessment and Strategic Action Plan* (Jakarta: US Agency for International Development, April 1995); A. T. White, L. Z. Hale, Y. Renard and L. Cortesi, *Collaborative and Community-based Management of Coral Reefs* (West Hartford, Conn.: Kumarian Press, 1994).

81. I. M. Dutton, K. Shurcliffe, D. Neville, R. Merrill, M. Erdmann and M. Knight, 2001. "Engaging Communities as Partners in Conservation and Development," *Van Zorge Report on Indonesia,* Vol. III, No. 8, p. 4.

82. B. R. Crawford, I. A. Dutton, C. Rotinsulu and L. Z. Hale, "Community-based Coastal Resources Management in Indonesia: Examples and Initial Lessons from North Sulawesi," in *Proceedings of International Tropical Marine Ecosystems Management Symposium,* eds. I. Dight, R. Kenchington and J. Baldwin, (Townsville, Australia: International Coral Reef Initiative, 1998), pp. 299–308.

83. B. R. Crawford and J. Tulungen, "Scaling-up Models of Community-based Marine Sanctuaries into a Community-Based Coastal Management Program as a Means of Promoting Marine Conservation in Indonesia," Working Paper, (Narragansett, RI: Center for Marine Conservation, University of Rhode Island, 1999).

so there is currently little alternative for adequate management of natural resources. Given a small handful of community-based conservation projects and regencies that have proven successful, there is a sense that the most efficient means to develop capacity is to begin with these projects and copy their formula. Application of the model discussed earlier perhaps would be a better strategy: it would rely on the information acquired through individual projects, but it would feed more information, more direction, to more regencies than could be done through any other means.

Application of the Model: Developing a Voluntary, Incentive-Based Coastal Resource Program in Indonesia

An integrated, decentralized coastal resource management program like the CZMA can fit comfortably into Indonesia's new governance structure. The general framework entails promulgation of national guidelines and standards to be implemented at the regional level, which is exactly the vision behind Act No. 22/1999 and Act No. 25/1999.[84] This section addresses four overarching questions: (1) how would the central government implement the program; (2) how would the regional government implement the program; (3) what incentives exist to implement the program; and (4) how would the program be funded?

The new role of the central government under Act No. 22 and its implementing regulations is to develop guidelines and policies rather than directly control and manage activities.[85] Can it require adherence to these guidelines and policies if management authority rests with the regencies? As much as the central government can establish policies and guidance under Article 7(2) of Act No. 22/1999, and can enforce laws and regulations under Article 7 of Regulation No. 25/2000, it is not clear exactly what authority the central government maintains to enforce its policies and guidance. Even if it has authority to require such adherence, can it, as a practical matter, enforce such adherence? Guidelines and policies are included within the framework of Article 7 of Regulation No. 25/2000, so the answer to the first question is yes. However, the answer to the second question is likely no. First, with implementation of policy, budgetary, and financial matters now being exercised at the regional level, policy emanating from the national level may have little meaning or respect in the field. In addition, any national policy will need to be broad and general enough to cover regional differ-

84. J. M. Patlis, R. Dahuri, M. Knight, J. Tulungen, "Integrated coastal management in a decentralized Indonesia: how it can work," Jurnal Pesisir & Kelautan, *Indonesian Journal of Coastal and Marine Resources*, vol. 4, no. 1 (2001), Bogor, Indonesia, pp. 24–38.
85. G. F. Bell, "The new Indonesian laws relating to regional autonomy: good intentions, confusing laws," *Asian-Pacific Law and Policy Journal*, (2001), vol. 2, p. 1.

ences, creating lots of room for differing interpretations of the policy, thereby making enforcement extremely difficult.

A voluntary program would avoid the obvious questions about the extent of central government authority in enacting and enforcing a mandatory program. First, even though a mandatory program may seem to be the stronger alternative, if implementation is not likely to follow at the local level and enforcement is not likely to come from the national level, then a voluntary program would be more efficacious. Second, a voluntary program would be acceptable to the community implementing it—by its nature as a voluntary program—so it would be better implemented and better enforced by the community itself. This is the case with community-based projects in Northern Sulawesi in which villages have adopted ordinances drafted under voluntary programs.[86]

A voluntary program would also allow regional and central governments to effectively transcend the confines of Act No. 22 and Regulation No. 25, because those laws recognize such mutually agreeable arrangements. Specifically, Article 3(d) and Article 4 of Regulation No. 25 provide the flexibility for such arrangements. Article 3(d) provides the general authority for delegation agreements. Article 4(a) states more specifically that regencies can delegate any portion of their authority to the province; under Section (i), the provinces can delegate any portion of their authority to the central government; and under Section (j), the central government or province can redelegate these authorities. Thus, a voluntary arrangement would allow the various levels of government to delegate different responsibilities and activities among each other based on their respective strengths and weaknesses (see fig. 1).

The question then becomes how to encourage voluntary implementation. The answer lies in the central government's ability to craft a package of incentives that would entice provincial and district level governments to adopt and implement an ICM program. This package would include financial and technical assistance, in the form of grants and loans, advice and guidance, and training and outreach, which is consistent with the role of the central government as envisioned in Act No. 22 and its regulations.

The central government could offer additional incentives: for example, the central government could agree that its own activities must comply with the provisions of any regency ICM program as with the CZMA. This compliance is not required under Act No. 22, particularly for areas of governance enumerated in Article 7. However, as incentive for regencies to adopt ICM programs, the central government can commit to this approach. For example, if a regency were to develop a certified ICM program consistent with the requirements of the central government law, then activities by the central government will only go forward with prior approval granted by the regency.

86. Crawford et al. (n. 82 above).

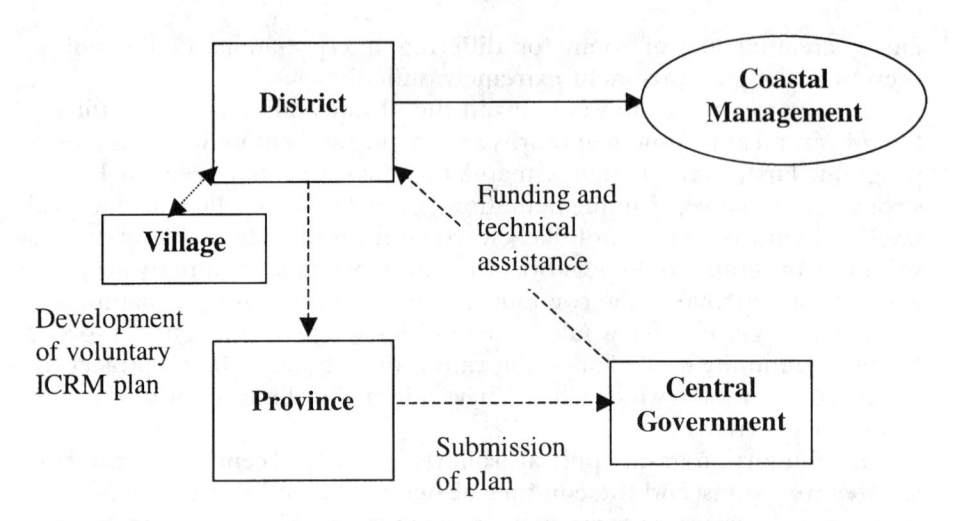

FIG. 1.—Overview of a voluntary integrated coastal management program. The kabupaten has authority to manage coastal resources directly, or it will have the option to work with desas and the province to develop a plan for submission to the central government. If the plan is approved, the central government will provide funding and technical assistance for coastal management.

Such an arrangement also furthers the spirit of decentralization, providing even greater deference to local governments than required under Act No. 22/1999.

Such benefits and incentives should not be given to regional governments without any obligations, however. There must be some standards and criteria that they must follow to ensure that they develop and implement an ICM program that deserves those benefits. Again, this is where the central government comes in. Article 2(3)(2)(d) of Regulation 25/2000 specifically recognizes that the central government has authority to set standards for management of the coasts, including three types of minimal requirements with which the local governments need to comply to receive any benefits.

First are general environmental and public health requirements on activities affecting coastal resources and populations—wastewater treatment and discharge requirements, solid waste disposal, pesticide and herbicide use, extraction of renewable coastal resources, such as with fishing quotas and mangrove harvest yields, et cetera. Second, it should include basic substantive, spatially related requirements specifically for coastal development including mandating priorities for coastal-dependent uses, issuing standards for spatial planning, and identification of special management areas for environmental protection or hazards control. Third, it should impose procedural requirements to ensure coordination and transparency, such as inter-agency review, permit review processes, public participation, stakeholder involvement, village involvement, dispute resolution, et cetera. The central government would help local governments develop ICM plans that met these requirements, formally approve the plans that satisfied them, and provide the incentives and benefits to any such approved plan. Lastly, the central government must monitor, review and enforce such plans to ensure they are faithfully carried out consistent with the national interests and objectives.

As an example, regional governments should be allowed to define the 'coastal area' in a way that suits their particular needs. The geographic and ecological nature of coastal areas, resources, and uses varies tremendously from region to region. It is possible to define the coastal area in a number of different ways based on these variations ranging from narrow political or otherwise arbitrary boundaries, to broad ecosystem-based boundaries covering large inland areas.[87] At the same time, the central government should provide minimal standards and guidelines to regions in defining the coastal area. For example, a minimal standard might require all regional definitions to be based on ecological criteria or allow regional governments to define certain political boundaries such as local waters of a certain mileage. Minimal guidelines also might include the methodologies for determining coastal areas, such as use of geographical information systems or certain scales of maps.

The next question is how the central government would establish and implement such a program. The key to ICM is the development of a procedural mechanism for coordinating management and budgetary decisions. The most obvious mechanism is the establishment of an inter-agency council with adequate authority over, or delegated from, the sectoral agencies. Consequently, the central government must accomplish two tasks: it must establish a coordinating structure for itself, and it must establish the parameters for establishing coordinating structures at the regional level. This latter task requires a balance between specific guidance and great flexibility to take into account specific circumstances among various regions. Central government coordination has already begun with the recent establishment of the Ministry of Marine Affairs and Fisheries (Departemen Kelautan dan Perikanan [DKP]). This agency has taken the lead in development of a new national ICM Act currently under preparation.[88] The new Act will be submitted to the national parliament for approval, allowing for interagency coordination and national planning and implementation of coastal programs.

The big winners under Act No. 22 and its regulations are currently the

---

87. S. B. Suominen and C. Cullinan, *Legal and Institutional Aspects of Integrated Coastal Area Management in National Legislation* (Rome: Food and Agriculture Organization of the United Nations, 1994).

88. Departemen Kelautan & Perikanan (DKP). Direktorat Jenderal Pesisir dan Pulau-Pulau Kecil, Naskah Akademik Pengelolaan Wilayah Pesisir, Jakarta, Indonesia (2001).

regencies. Except for the few areas of governance withheld under Act No. 22, they essentially have authority for all decision making within their jurisdiction, unless otherwise stipulated by central government regulation, or in certain circumstances in which provinces have been given authority. With respect to coastal management, regencies are where the bulk of activities, if not authority, should be housed, with the caveat that regencies work in conjunction with the lower levels of community government within their jurisdiction. The continuing question is capacity, which in the regions is still strongest at the provincial level. However, a national program can provide direction and assistance directly to the district level.

Compared with central and provincial levels, regencies are best positioned to develop ICM programs tailored to resource conservation and use within their jurisdictions. Regencies are effectively close enough to the resources and yet large enough to coordinate among neighboring villages. Under a voluntary program as described here, regency governments would first decide whether to initiate an ICM program sponsored by the central government and then be responsible for developing ICM plans for activities within its jurisdiction. Development of this plan would be done in cooperation with the provincial and central governments, as well as local stakeholders. It is incumbent that any ICM program provides for meaningful participation, regardless of the level of government. In the case of regional programs, this means down to the most local level. This will not be easy given that the relationship between regencies and villages varies widely across Indonesia. However, through subregency (kecamatan) offices, regencies have strong connections with village and subvillage governing bodies.

Within the framework established by the central government, the regency would develop the necessary procedural mechanisms for coordination and collaboration and would ensure that the necessary substantive requirements are satisfied. Beyond satisfying these minimal requirements, the regency would have flexibility to structure its ICM plan however appropriate to match local aspirations and to use whatever mechanisms suitable to satisfy the broader goals and objectives of the national ICM program. Upon approval of the plan by the central government, the regency would then implement the plan with the assistance provided by the central government.

While the regency is the most logical level for management of coastal resources, it might not be the most logical level for direct coordination with the central government. Of 332 regencies, 245 have a coastline. While not all regencies would be expected to take part in a voluntary ICM program, it is to be hoped that most would. In any event, the number can potentially be huge, creating a tremendous logistical challenge for the central government in assisting, approving, monitoring, and enforcing each local ICM plan. This dilemma can be solved through delegation of coordination authority to provinces for regencies within their jurisdictions.

The provinces are currently the wild card in the new decentralized regime in Indonesia. On the one hand, they appear as the big losers in the new power structure, with authority and funding—which has gone directly from central to regency levels—almost completely bypassing them. Under Act No. 22, the provinces have been largely cut out of any meaningful role of governance. Even were they to have one, under Act No. 25/1999, they have little financial means to carry it out. On the other hand, the provinces are not to be completely dismissed just yet. While Article 9 of Act No. 22/1999 limits their authority to three situations, these situations are presently very vague but potentially very broad. It is likely that the role of the provinces will be decided on a case-by-case basis, where strong governors may very well take advantage of the law's ambiguity and try to grab significant amounts of authority, while weaker governors will not be able to resist the general push towards district-level management. In addition, Act 22/1999 is currently under revision by the Ministry of Home Affairs. Indications are that considerable authority will be moved back to the central and provincial governments, although the extent of this is not yet clear.

Regardless of the authority that the province can attain for itself, enjoying that authority may prove difficult because it currently has virtually no funding. And that situation is unlikely to change. The struggle for revenues, particularly revenues derived from natural resource consumption, is largely going to play out between the central government and the regencies. The provinces will likely be relegated to the sidelines. In any case, the role of the provinces will continue, almost as a matter of default, to take on a tone of guidance and policy rather than actual management.[89] On cross-boundary issues they may have a stronger hand in shaping policies, coordinating activities, and settling disputes, but it is doubtful it will amount to more than that.

Such a role for provinces would be consistent with the ICM program envisioned. Indeed, provinces would be absolutely integral to effective ICM programs in three ways, each perfectly valid under Act No. 22. First, the province could prepare guidelines and standards to elaborate upon the central government guidelines. Given the breadth and generality of guidelines and standards that are coming from the central government, more specific guidelines and standards are needed. The differences among provinces that must be addressed in ICM are enormous, with some provinces having lots of available information and great capacity while others have none; some provinces being resource rich while others are resource poor; and some being industrial and urban while others are completely rural. These differences can be more adequately addressed at the provincial level than at the central level, with guidelines tailored to address these differences.

Second, the province could review regency plans and package them to

---

89. L. M. Kaimudin, "Decentralization of Coastal and Marine Reosurce Management," (paper presented at the Second Conference on the Management of Indonesian Coastal and Marine Resources, Makassar, 15–17 May 2000).

facilitate central government approval. It may even be possible to have the central government certify provincial programs and then delegate review and approval authority to provinces for regency plans. In this way, the entire burden on central government resources is avoided, with resources focused on the far fewer number of provinces. Even if provincial governments do not have formal control over regency/city ICM planning, they could play an important role in facilitating and coordinating review of regency/city plans by the central government. It could also make recommendations both to the local and central governments as to improvements to the plans. Third, the province could serve as the liaison for technical assistance to help implement local programs.

In terms of funding a new ICM program, there are several possibilities. The most straightforward possibility is that the central government, most likely DKP, dedicates a portion of its budget for grants for ICM program development and implementation. In addition to grants, the central government can use its own funds to establish a revolving loan fund for projects. However, given the lack of funding at the central government level as a result of Act No. 25/1999, this is likely to be a relatively small program and may not provide adequate incentive to regional governments. As an illustration, the 2000 budget for DKP is 498 billion rupiah.[90] Of this, Rp 70 billion funds the Directorate General for Coast and Small Island Affairs (Direktorat Jenderal Pesisir dan Pulau-Pulau Kecil), which is responsible for implementing ICM in DKP. Of this, Rp 13 billion are used for grants to the provinces for coastal resource management and conservation. Funds are distributed based on proposals submitted to DKP from the provinces.[91]

A second possibility is that the central government can use special allocation funds (dana alokasi khusus-DAK) to support an ICM program. These monies, pursuant to Article 8 of Act No. 25/2000, are not required to be doled out to regional governments, but are available for specific and special needs. According to Article 8(2)(b), this includes national priorities that certainly can be enunciated to include ICM. The central government could make distributions of DAK funds to regional governments that have developed ICM plans approved by the central government. While it might be worth exploring whether any funds can be made available from the general allocation funds under the general public distributions (dana alokasi umum), it does not appear that the central government has any discretion to change these regional allocations, or to attach any conditions to these general formula distributions under Article 7 of Act No. 25/2000.

90. At the time of writing the currency exchange rate was approximately 10,000 Indonesian rupiah to US$ 1.
91. M. E. Rudianto, Head, Sub-Directorate for Rehabilitation and Coastal Resources Utilization, Direktorat Jenderal Pesisir dan Pulau-Pulau Kecil, DKP, personal communication, March 2001.

A third, more visionary, possibility would require a new Act and amendments to Act No. 25/1999. It would also cure the most profound shortcoming in the new financial decentralization scheme. This shortcoming concerns the current freedom of the regional governments to use natural resource revenues for any purpose whatsoever. These revenues can be used for administration, development, physical infrastructure, social infrastructure, et cetera. The freedom, of course, is desirable, but what is missing is a requirement that some of those revenues be reinvested in the management and conservation of natural resources—the very resources responsible for generating these revenues in the first place. A shortsighted regional government will extract natural resources to the point that they are depleted or overexploited, thus destroying its future revenue stream. Consequently, the central government should amend the fiscal decentralization regime to impose a requirement that regional governments use some percentage of their revenues generated from natural resources for natural resource conservation. Under Act No. 25/1999, regional governments have several sources of new funding including original revenue receipts, general allocation funds, and loans. It is only a portion of the general allocation fund—that portion which, according to Article 6(1)(a), is derived from natural resources—that would be subject to this new requirement. Consequently, the restrictions would not be too onerous, with complete regional autonomy still available for other revenue sources.

Under this hypothetical scenario, ICM programs could be funded through the revenues derived from natural resource use, specifically revenues derived from fisheries. These funds are divided equally across all regional governments. However, under a new law, the central government could hold them in escrow for individual regional governments until those governments engaged in an approved ICM program. This may be politically infeasible at this point, but, given the constant shifts taking place in implementation of decentralization, it should be entertained.

CONCLUSION

As many developing countries undertake efforts to decentralize their systems of governance, they often create new problems as well as address old ones. One of the largest problems is in delegating authority for management and budget to levels of government that have little experience or capacity to handle such authority. Such countries must be cognizant of the risks of decentralization as well as the supposed benefits. Often in the case of natural resources, the risks of decentralization often far outweigh the benefits, because significant degradation of habitat and resources can occur in a relatively short amount of time. Numerous studies discuss issues relating to legal reform and governance policy, attempting to identify conditions for good governance, such as participation, transparency, accountability, et cetera, and attempting to identify methods to achieve these conditions.

The method proposed here is to develop a voluntary, incentive-based program to encourage different aspects of good governance. This would accommodate the institutional weaknesses of the central government in its inability to enforce authoritative laws on the books. At the same time, it would accommodate the institutional weakness of regencies. It would not place onerous, unrealistic burdens on communities, but rather would provide local discretion on whether and how to undertake such a program. Lastly, it would accommodate the relative strength of the central government vis-à-vis regional governments. Central government entities could supply expertise and experience to regional governments in management of the coastal areas. This voluntary, incentive-based program is centered on guidance and criteria emanating from the central government to help guide regional governments in their own management actions and decisions based on local aspirations. In addition, such a program entails a transfer of funds and assistance in the form of grants from central to regional governments, supporting the notion that finance must follow function in decentralization.

The U.S. CZMA provides an excellent example of how a tested program can work in decentralizing, developing countries such as Indonesia. While there are some inherent limitations in the application of a voluntary, incentive-based program, for the most part this model is one of the most promising to guide natural resource management in such countries.

Environment and Coastal Management

Communicating the Harmful Impact of TBT: What can Scientists Contribute to EU Environmental Policy Planning in a Global Context?*

Cato C. ten Hallers-Tjabbes
CaTO Marine Ecosystems, The Netherlands

Jan P. Boon
Royal Netherlands Institute for Sea Research (NIOZ), The Netherlands

José Luis Gómez Ariza
Department of Chemistry and Material Sciences, Escuela Politécnica Superior, Huelva, Spain

John F. Kemp
University of Nottingham, United Kingdom

INTRODUCTION

The project "Action to demonstrate the harmful impact of TBT, Effective Communication Strategies between Policy makers and Scientists in support of Policy Development (HIC-TBT)," offers a case study whereby independent scientists engaged in direct communication with decision makers' processes and the general public.

*The funding of the EU Life Programme was crucial in designing and performing this cooperative project. The ongoing participation of the policy network of this project has been of great value. We thank the members for their continuing interest and willingness to participate in formal and informal discussions throughout the project, in particular Mr L. Burgel, Mr H. Foeken and Ms M. van Binsbergen (Directorate General for Transport, NL Ministry for Transport, Public Works and Water Management), Mr S. Josephus Jitta (Directorate General for the Environment, NL Ministry for Public Housing, Physical Planning and the Environment), Professor Dr E. Hey (International Natural Resources Law, Erasmus University, The Netherlands), Mr H. Van Hoorn (NL Ministry for Transport, Public Works and Water Management), Ms S. Lintu (International Maritime Organization), and Mr M. Leonard-Williams (Directorate General for Transport, European Commission). We are grateful to the many representatives from policy and from interested organisations for their preparedness to present their views at the project public meetings and at special seminars and their willingness to join in the public discussions. This is NIOZ Publication No. 3676.

© 2003 by The University of Chicago. All rights reserved.
0-226-06620-7/03/0017-0015$01.00 *Ocean Yearbook* 17:417–448

A strategy for the mutual exchange of information between scientists and policy makers was developed alongside a scientific programme to investigate some responses of the marine environment to the harmful input of antifouling paints. Scientific work concentrated on the most sensitive bioindicator, the sexual development of predatory snails. The strategy was based on the understanding that scientists can assist in the interpretation and judgement of the relevance of their findings to policy processes. It followed a path from the first scientific signals of environmental harm in the target environment to a global decision process. The case study comprised the impact of tributyltin (TBT)—a highly toxic antifouling paint for marine ships—from offshore shipping on the marine environment, and the policy target was the decision process, leading toward a worldwide TBT ban, at the International Maritime Organization (IMO) and its Marine Environment Protection Committee (MEPC).

In demonstrating the adverse impact of TBT from antifouling paints used by shipping in the seas off western Europe and in the Mediterranean Sea, the scientists showed that direct and flexible communication based on mutual understanding between scientists and policy makers could stimulate the development, implementation and enforcement of environmental policies. This resulted in the project being brought to the attention of IMO-MEPC in two joint submissions by the participating States. The results were also disseminated through scientific channels and were passed on in direct communication between scientists and policy makers and through several channels to a wider public. The project thus made both science and policy making processes more transparent and participatory, which contributed to attaining general support for implementation.

The now widely adopted precautionary principle motivated the scientists to explore its potential for communicating underlying scientific principles for recognition of environmental threats, which helped to raise environmental awareness while scientists endeavored to understand the processes that are relevant in decision making. The project drew attention from policy makers and scientists from parts of the world other than Europe, in particular from Mexico and Brazil. General lessons from the project and the strategy employed are being discussed for possible application to other environmental problems, with specific attention being given to related marine environmental problems, such as TBT in harbour dredge spoil and the discharge of ballast water from ships resulting in the introduction of alien organisms to regional marine ecosystems.

Ten years after the first signs in September 1991 that subtidal snails in continental shelf seas off the coast of the North Sea suffered from exposure to TBT, the IMO adopted an International Convention on the Control of Harmful Anti-fouling Systems on Ships (5 October 2001).[1] The Convention

---

1. International Convention on the Control of Harmful Anti-fouling Systems on Ships. Adopted 5 October 2001. IMO Documents on the International Confer-

will prohibit the use of harmful organotins in antifouling paints used on ships and will establish a mechanism to prevent the potential future use of other harmful substances in antifouling systems.

Natural sciences often play a major role in political decision-making processes, which, in turn, generates a societal context for scientists to offer support and justification, if not a rationale, for policy development. The role of science in international policies is well recognised. The current international agenda would have been devoid of many of the present environmental issues, if not for the input from science that made them recognised as problems.[2] Although conclusions from scientific research have influenced decision making, they have seldom been decisive for the outcomes.[3] Rational management of the environment needs an interpretation of findings and insights from relevant research so as to attain a practical basis for policy decisions. No matter how firm a knowledge base is, policy cannot simply be derived from knowledge. "Research findings rarely, if ever speak for themselves; no documentation of damage to nature by itself prescribes the optimal cure. Only when interpreted in a particular context and related to some particular concerns, interests and values, can knowledge be used by decision-makers."[4] Many pleas have been made for positive feedback between scientists and

---

ence on the Control of Harmful Anti-fouling Systems on Ships, Agenda Item 8: "Adoption of the Final Act of the Conference and any Instruments, Recommendations and Resolutions resulting from the Work of the Conference. International Convention on the Control of Harmful Anti-fouling Systems on Ships, 2001. Text adopted by the Conference." IMO, AFS/CONF/25, 8 October 2001; IMO, AFS/CONF/26, 18 October 2001. Entry into force: 12 months after 25 States, representing 25 percent of the world's merchant shipping tonnage have ratified it. The harmful antifouling systems to be prohibited under the regime of the Convention are defined in Annex 1 to the Convention. At present only organotins used as biocide in antifouling paints are listed; the Annex will be updated as and when necessary. Annex 1 states that by an effective date of 1 January 2003, all ships shall not apply or re-apply organotin compounds which act as biocides in antifouling systems. By 1 January 2008 (effective date), ships either:
(a) shall not bear such compounds on their hulls or external parts or surfaces; or
(b) shall bear a coating that forms a barrier to such compounds leaching from the underlying non-compliant antifouling systems.
The Diplomatic Conference adopted a Resolution on Early and Effective Application of the Convention, which invites Member States of the Organization to do their utmost to prepare for implementing the Convention and also urges the relevant industries to refrain from marketing, sale and application of the substances controlled by the Convention (Resolution 1 to the Convention).

2. A. Underdal, "Science and Politics: the anatomy of an uneasy partnership," in *Science and politics in international environmental regimes: Between integrity and involvement*, (eds.), S. Andresen, T. Skodvin, A. Underdal and J. Wettestad, (Manchester: Manchester University Press, 2000), p. 3.

3. J. Wettestad, "Dealing with land-based marine pollution in the North-East Atlantic: The Paris Convention (PARCON) and the North Sea Conferences," in Andresen *et al.* (n. 2 above).

4. Underdal (n. 2 above), p. 80.

policy makers, for both coastal and offshore marine areas;[5] a lack of coordination between research and policy had been recognised for many years.[6]

Scientists' understanding of the natural system can complement the data with information that comments on the value of the data for the specific policy process and so acknowledges both the complexity of natural systems and processes and the context of any given decision-making process. An analogous approach can be encountered in disease processes where medical symptoms can only be properly interpreted by passing on information about the patients' world to the treating physician.

The independent spheres of decision making and scientific research render the relationship between them complex, precarious, and often tenuous. Traditionally, communication between scientists and policy makers occurs in institutional structures, with the data presented in a scientific format and then passed on to a world more oriented toward decisions. The often-voiced perception: "Let the scientists do their science, pass it on and let the policy makers decide how to use the results," leaves the area of interpretation of scientific results in a policy context untouched.

Lack of understanding between policy and science may result in stereotyped perceptions, such as "Policy makers ask for absolute answers and they want them yesterday" or "Scientists forever discuss the results; they are never certain." Such perceptions obviously reflect the different contextual frameworks and the specific requirements for proper performance in either science or policy making. Transparency of the relevant science, policy, and the underlying processes will benefit from bridging such existing gaps in perceptions. For a specific environmental issue, such transparency could likewise encourage participation in the process by other relevant actors, such as the users of products and technologies, the associated industries, the public at the receiving end of the environmental effects, and the scientists involved in relevant research. The policy process could then benefit from input from such participants that, in turn, would contribute further to transparency of the process.

An early and agenda-setting example of scientists taking an initiative to bring their scientific concerns forward to the decision-making community in the field of nuclear arms control, materialised in the Pugwash Conferences, where nuclear scientists from Western and Eastern Europe convened to discuss disarmament and gave an intellectual impetus to arms control.[7] The

---

5. A. Vallega, "Ocean governance in post-modern society – a geographical perspective," *Marine Policy* 25 (2001): 399–414.

6. C. C. Ten Hallers-Tjabbes and G. Peet, "Research interests in the North Sea," *Marine Policy* 6(1) (1982), p. 63–66.

7. J. Rothblatt, *History of the Pugwash Conferences* (London: Taylor & Francis, 1962), in E. Adler "The emergence of cooperation: national epistemic communities and the international evolution of the idea of nuclear arms control," *International Organization* 46 (1992): 101–146.

initiating scientists' concerns followed on the development and subsequent deployment of nuclear arms at the end of the Second World War. Nuclear scientists in the U.S. took similar initiatives.

In the present study, we thus follow on a tradition where scientists have had a pivotal input into the current international agenda by being the first to identify a potential environmental problem and bring it to the attention of decision makers.

In particular, communication between science and policy making has been developed in the framework of the precautionary approach, which allows the employment of scientific data that are indicative of environmental degradation, without requiring exclusive full proof.[8]

PRECAUTION

The precautionary principle, as applied to environmental protection, is set out as one of the objectives of the United Nations Conference on Environment and Development (UNCED) and has since been incorporated in an IMO resolution. Agenda 21, section 17.22 of UNCED provides that States, in accordance with the provisions on the protection and preservation of the marine environment of the United Nations Convention on the Law of the Sea, commit themselves to "apply preventative, precautionary and anticipatory approaches so as to avoid degradation of the marine environment as well as to reduce the risk of long-term or irreversible effects upon it."

Paramount to precaution is a clear understanding of the nature of scientific evidence, and of its inherent limitations,[9] since the scientific state of the art is never to be absolutely certain. Scientists' uneasy relationship with (level of) certainty is also a recognised factor in political decision-making processes, where there is a need to take account of "uncertainties."[10]

It is our experience that scientists can assist in addressing "uncertainty" about the interpretation of their scientific findings by being open about the content and the background of their message and by giving insight into the process of acquiring scientific "evidence." Commenting on scientific uncertainties can explain how more is always happening in the environment than merely the strict subject of the specific research. Understanding of

---

8. D. Freestone and E. Hey, (eds.), *The Precautionary Principle in International Law, The Challenge of Implementation* (The Netherlands: Kluwer Law International, 1996), 274 pp.; Underdal (n. 2 above); T. Skodvin and A. Underdal, "Exploring the dynamics of the science-politics interaction," in Andresen *et al.* (n. 2 above).

9. D. Santillo, R. L. Stringer, P. A. Johnston and J. Tickner, "The Precautionary Principle: Protecting against failures of scientific method and risk assessment," *Marine Pollution Bulletin* 36 (1998): 939–950.

10. B. Wynne and S. Mayer, "How science fails the environment," *New Scientist* (5 June 1993), p. 33–35.

causes for uncertainty as reflected in decision-making processes can illuminate the complex context of arriving at policy decisions.

Within Europe, communication between science and policy is on the current agenda. The European Commission has identified the need for a new relationship between science and governance in European environmental policies, including greater openness. The Commission therefore intends to meet the requirements by establishing guidelines for applying the precautionary principle and for the development, integration, and management of European research.[11]

In European marine environmental policies, a well-known example of communication between scientists and decision makers that was initiated by governmental processes is the role of science in the North Sea Conferences where, in 1984, a place was granted for nongovernmental participants in the policy process. This initiative is an example of the changing perception of environmental resources and of the role that science can play. The precautionary principle was applied and preliminary decisions were revisited and revised as a result of changed scientific insights.[12] The process has now shifted from measuring chemical and physical parameters in the wake of contaminant inputs to considering the state of the ecosystem and the assessment of biological parameters. The development of the Quality Status Reports (North Sea 1987, 1993 and Northeast Atlantic, Region II, the North Sea, 2000)[13] reflects this shift in priorities and illustrates the improved effectiveness of the process and the increased acceptance of scientific evidence.[14] The input of scientists in this process was initiated by government actors, who also introduced a then unique openness to nongovernmental bodies as participants in the process.

Other examples of effective feedback loops in a decision process in which scientists are part can be identified within Europe. A successful cooperative structure and process for agreement on the acceptability of environmental impact—and feedback monitoring—was established during the construction phase of the Sund bridge in Denmark. The monitoring programme that ran alongside the construction activities was empowered to start a wide-ranging survey as soon as a particular signal of environmental damage was encountered and, if the survey proved that the activities were

---

11. S. Funtowicz, I. Shepherd, D. Wilkinson and J. Ravetz, "Science and governance in the European Union: a contribution to the debate," *Science and Public Policy* 27 (2000): 327–336.

12. E. Hey, "The international regime for the protection of the North Sea: from functional approaches to a more integrated approach," *The International Journal of Marine and Coastal Law* 17 (2002), (forthcoming).

13. OSPAR Commission for the Protection of the Marine Environment of the North-East Atlantic. *Quality Status Report 2000, Region II Greater North Sea*, (London: OSPAR Commission, 2000).

14. Wettestad (n. 3 above).

responsible for the environmental degradation, to stop the activities.[15] A Control Panel was responsible for the feedback-monitoring concept, and involved not only the authorities, but also the companies, the scientists and the environmental community. The process was initiated by decision makers.

The role of a well-informed public has been exemplified in the case of the Economic Commission for Europe Convention on Long-Range Transboundary Air Pollution. The process of the formal reporting bodies contributed somewhat to acceptance of scientific evidence, but the true impetus in 1983 to consider sulphur reduction as an option in the interest of the decision process, was predominantly driven by public (and media) attention.[16] The HIC-TBT study took this experience to heart and spent considerable effort to inform the public and the press.

THE CASE STUDY

TBT: A Major Marine Environmental Problem

At present, TBT is the predominant biocide in marine antifouling paints to prevent organisms from attaching to ships' hulls. TBT leaches into the marine environment, accumulates in sediment and biota, and causes serious adverse effects in nontarget organisms, even at very low concentrations.[17] TBT has been recognised as one of the most toxic substances ever introduced in the marine environment.[18] The most sensitive indicator of TBT contamination as yet known in the marine environment is the development of imposex (masculinisation of females) in marine snails, at extremely low concentrations of TBT (1 ng/l), which results in reproductive failure and declining snail populations.

As early as the 1970s, the major toxic action of TBT to nontarget organisms had been recognised when organisms in marine harbours were found to die, irrespective of very low, and eventually at nondetectable levels of TBT.[19] When evidence was found that TBT from local yachting was detri-

---

15. J. S. Gray, "Integrating precautionary scientific methods into decision-making," in Freestone and Hey (n. 8 above), p. 133–146.
16. Wettestad, "From common cuts to critical loads: the ECE Convention on long-range transboundary air pollution," in Andresen et al. (n. 2 above).
17. K. Fent, "Ecotoxicology of organotin compounds," *Critical Reviews in Toxicology* 26(1) (1996), pp. 1–117.
18. J. Ward, "Antifouling paints threaten fisheries resources," *Naga: The ICLARM Quarterly*, (Manila, Philippines: International Center for Living Aquatic Resources Management, 1988), p. 15; L. D. Mee, "Scientific methods and Precautionary Principle," in Freestone and Hey (n. 8 above), pp. 109–131.
19. D. Dundee, Personal communication, University of New Orleans (1980).

mental to oyster cultures in the Bay of Arcachon,[20] France was the first State to act by banning TBT from ships not longer than 25 meters. Following subsequent findings of imposex and decline of intertidal dogwhelk populations along the coast, similar bans were imposed in other States (UK, 1987; USA, Canada, 1988; EU, 1989; as well as several others). The International Maritime Organization followed suit in 1990, when the Marine Environment Protection Committee recommended banning TBT from vessels not longer than 25 m and set a maximum leaching rate of TBT for larger vessels (Resolution MEPC 46(30)),[21] on the assumption that TBT concentrations in the open sea were too low to cause an environmental impact. During the following years, monitoring for the impact of TBT focused on coastal regions but, while partial recovery of snails was reported in some areas, more often the snail populations remained affected.[22] It has been found that TBT accumulates in sediments, where it can remain for decades, in particular in the frequently occurring anaerobic conditions. As a result, TBT is eligible to be identified as one of those "stored potential pollutants that do not necessarily display biological effects within the present environmental conditions."[23] A continuing decline of snail populations has been reported in areas such as the North Sea since the early 1970s when TBT was first used on marine craft.[24]

The toxic action of TBT is, as yet, unpredictable, and in this sense the compound exemplifies the limitations of predicting environmental impact on the basis of generally adopted standard methods. Such methods can be used for making predictions, on the basis of physico-chemical properties, environmental toxicity and the accumulation behaviour of man-introduced chemicals, or "xenobiotics" into the environment. However, the Quantitative Structural Activity Relationships (QSARs) could not identify the toxicity of tributyltin.[25] The compound TBT, with androgenic physical and endocrine disruption as the most striking effects,[26] is also neuro-

20. C. Alzieu, M. Héral and J.-P. Dreno, "Les peintures marines antisalissures et leur impact sur l'ostréiculture." *Equinoxe* 24 (1989), p. 22–31; C. Alzieu. "Environmental problems caused by TBT in France: Assessment, regulations, prospects," *Marine Environmental Research* 32 (1991): 7–17.
21. Resolution MEPC 46(30), "Resolution on the Control of Harmful Antifouling," (London: IMO, 1990).
22. M. M. Santos, M. N. Vieira and A. M. Santos, "Imposex in the dogwhelk, *Nucella lapillus* (L.), along the Portuguese coast," *Marine Pollution Bulletin* 40 (2000): 643–646.
23. Mee (n. 18 above).
24. C. C. Ten Hallers-Tjabbes, J. M. Everaarts, B. P. Mensink and J. P. Boon. "The decline of the North Sea whelk (*Buccinum undatum* L.) between 1970 and 1990: a natural or a human-induced event?," *Marine Ecology PSZN I*, 17 (1996): 333–343.
25. Gray (n. 15 above).
26. N. Spooner, P. E. Gibbs, G. W. Bryan and L. Goad, "The effect of tributyltin upon steroid titres in the female dog-whelk, *Nucella lapillus,* and the development

toxic and immunotoxic and is known to affect several other biological processes.

TBT and the Present Study

The study reported here explored the benefits of an approach based on communication between scientists and policy makers as had been employed in earlier studies on the specific environmental problem of TBT as a biocide in antifouling paints.[27] In the earlier studies, direct communication between scientists and policy makers had been remarkably effective in stimulating decision-making processes.

The first ever investigations of the impact of TBT in offshore waters in the early 1990s, showed the effect of TBT (imposex) in the subtidal snail species, *Buccinum undatum*, while the imposex incidence was clearly correlated with the number of passing ships.[28] The cause-effect relationship between imposex in juvenile *B. undatum* and TBT was shown in experimental studies.[29] Following the North Sea studies, a similar relationship between the presence of imposex in offshore seas and shipping was shown in snail species in Southeast Asia.[30] Communication of the above-mentioned findings of imposex in offshore marine waters provided a clear message to those involved in the relevant policy fields as well as to a wider public.

The developments in decision making in response to the findings of the impact of TBT in seas off the coast previous to the present study had illustrated the benefits of a clear scientific message from scientists and the

---

of imposex," *Marine Environmental Research* 32 (1991), p. 37–49; E. Oberdörster and P. McClellan-Green, "The neuropeptide APGWamide induces imposex in the mud snail, *Ilyanassa obsoleta*," *Peptides* 21 (2000): 1323–1330.

27. C. C. Ten Hallers-Tjabbes, "Tributyltin and policies for antifouling," *Environmental Technology* 18 (1997), p. 1265–1268.

28. C. C. Ten Hallers-Tjabbes, J. F. Kemp and J. P. Boon, "Imposex in whelks (*Buccinum undatum*) from the open North Sea: relation to shipping traffic intensities," *Marine Pollution Bulletin* 28 (1994), p. 311–313.

29. B. P. Mensink, J. M. Everaarts, H. Kralt, C. C. Ten Hallers-Tjabbes and J. P. Boon, "TBT exposure in early life stages induces the development of male sexual characteristics in the common whelk, *Buccinum undatum*," *Marine Environmental Research* 42 (1996), p. 151–154; B. P. Mensink, "Imposex in the common whelk, *Buccinum undatum*," (Ph.D. Thesis, Wageningen University, The Netherlands, 1999), 125 pp.; B. P. Mensink, H. Kralt, A. D. Vethaak, C. C. Ten Hallers-Tjabbes, J. H. Koeman, B. Van Hattum and J. P. Boon, "Imposex induction in laboratory reared juvenile *Buccinum undatum* by Tributyltin (TBT)," *Environmental Toxicology and Pharmacology* 11 (2002): 49–65.

30. C. Swennen, N. Ruttanadakul, S. Ardseungnern, H. R. Singh, B. P. Mensink, and C. C. ten Hallers-Tjabbes, "Imposex in sublittoral and littoral gastropods from the Gulf of Thailand and Strait of Malacca in relation to shipping," *Environmental Technology* 18 (1997): 1245–1254.

need for them to present a scientific rationale in a field where concerns on TBT had been long-present,[31] although not recognised, in global policy processes.[32] The findings of the Netherlands Institute for Sea Research (NIOZ) in the North Sea and in Southeast Asia, influenced policies concerning the use of TBT.[33] With the support of scientists and policy makers from Southeast Asia, the North Sea States took the scientific concerns to IMO-MEPC (Marine Environment Protection Committee of the International Maritime Organization, the international body responsible for shipping), in 1996.[34] This was sufficient to arouse wider support for banning TBT (and other organotins) as antifouling biocides from all ships.[35] The process to phase out TBT worldwide within 10 years, is now well on its way, with the full support of the European Union (EU).[36] The combined science-shipping-policy processes have stimulated massive efforts to develop effective alternatives of less harmful antifouling technologies by the paint and related industries, together with an urgent commitment to apply such alternatives by shipping companies.[37]

The present study was started in 1998 when the global decision-making process—aimed at phasing out TBT and other organotins from ships' antifouling—had just been initiated,[38] and aimed to build support for implementing the upcoming ban within the EU. The scientists initiating this project[39] actively sought to communicate their research findings and the concern those findings raised to marine decision makers and to the public. It was understood that the development, implementation, and enforcement of environmental policies would benefit when scientists and policy makers were willing to communicate directly and in a flexible way, while being aware of the differences in the decision-making process versus the scientific worlds.

31. Ward (n. 18 above)
32. Resolution MEPC 46(30); MEPC 30, Report of Meeting (London: IMO, 1990); MEPC 35, Report of 35th Meeting Marine Environment Protection Committee (MEPC). (London: IMO, 1994).
33. Swennen et al. (n. 30 above).
34. MEPC 38, 1996. Report of Meeting, (London: IMO, 1996); MEPC 38/INF.16: "Harmful effects of the use of antifouling paints for ships. Impact of antifouling paints in South-East Asian seas," Submitted by the Netherlands (London: IMO, 1996).
35. MEPC 42, Report of Meeting, (London: IMO, 1998).
36. MEPC 46, Report of Meeting, (London: IMO, 2001); Ten Hallers-Tjabbes (n. 28 above).
37. HIC-TBT, Report of Malmö Seminar, World Maritime University (2000).
38. MEPC 42, Report of Meeting, (London: IMO, 1998).
39. The project consortium partners Netherlands Institute for Sea Research (The Netherlands), CaTO Marine Ecosystems (The Netherlands), Institute for Environmental Studies, Free University (The Netherlands), Department of Chemistry and Material Sciences, University of Huelva (Spain), Department of the Environment, ENEA-CASACCIA (Italy), with additional assistance by Professor J. F. Kemp, University of Nottingham (U.K.) and Mr M. Santos, University of Porto (Portugal).

Rationale for this Study

The scientists involved acted on the underlying belief that they had a clear social responsibility to explain their work to the world outside the scientific arena. They recognised that, although the global decision-making process to ban organotins from antifouling systems was now well under way, in several regions the problems of TBT were not appreciated. Lack of such recognition might hamper a proper implementation of the ban and could, for instance in an area such as the Mediterranean Basin, lead to the transfer of an environmental problem to areas in other parts of the basin that were less familiar with TBT's environmental risk. The project built on earlier experience, where communication between scientists and policy makers had helped the decision-making processes along.

Traditionally in teaching and supervising we try to make students aware of the role of science in society and, in particular, to promote an understanding of the ecosystems below the sea surface. For humans the sea is a more alien environment than the land.[40] While onshore environmental concerns are often quite visible and may even surface in "people's backyards," in the seas off the coast, nobody is that close to the problem. In coastal areas, the intertidal zone becomes frequently visible, and coastal communities are often supported by direct coastal zone related activities.[41] Hence the intertidal zone can act as a people's backyard. The subtidal seabed differs from a terrestrial surface in being hidden from the eye and often having a turbid and dark environment. Similarly, environmental problems in the sea off the coast do not often surface and may go unnoticed for lengthy periods of time. TBT itself is a good example to illustrate such differences between the strictly coastal marine environment and the subtidal part of even shallow seas, such as the continental shelf seas studied in this project. Once it had been recognised in European coastal areas, imposex, resulting from exposure to TBT, was found in intertidal snails near yachting and marine harbour areas in other regions such as Indonesia and Japan.[42] The subtidal sea off the shore, where the majority of the much larger ships of the merchant fleet sail, was left alone both by decision makers and scientists alike.[43]

---

40. A. C. Hardy, *The Open Sea – its Natural History.* Part I, The World of Plankton, (London: Collins, 1956), p. 1–2.

41. R. Johnson, K. Bell, and D. Huppert, "Public Perceptions, Attitudes, and Values: Coastal Resident Survey," Chapter 13 in *Pacific Northwest Coastal Ecosystem Regional Study (PNCERS), Annual Report* (2000), p. 125–142.

42. T. Horiguchi, H. Shiraishi, M. Shimizu and M. Morita, "Effects of triphenyltin chloride and five other organotin compounds on the development of imposex in the rockshell, *Thais clavigera*," *Environmental Pollution* 95 (1997), p. 85–91; D. V. Ellis and L. A. Pattisina, "Widespread neogastropod imposex: a biological indicator of global TBT contamination?" *Marine Pollution Bulletin* 21 (1990): 248–253.

43. MEPC 35 (1994) (n. 32 above).

The scientists participating in the HIC-TBT project had a good understanding of the environmental problems related to TBT from their own scientific work. They were familiar with the present state of affairs in research on TBT and its impact on the marine environment. The HIC-TBT study aimed to create channels for scientists who recognised an environmental problem in their scientific work, to pass on their concerns quickly, and in a clear manner to stimulate participation in, and support for, effective implementation of the global ban on TBT and other organotins as antifouling agents on ships.

As to regulation of antifouling paints, the EU follows the International Maritime Organization, but implementation of a ban will benefit from regional and national recognition of the problems of TBT.[44] In the late 1990s, support for stricter regulation of TBT was present in northern Europe, and legislation in EU States had been put in place to ban TBT from use on small craft of less than 25 metres in length, following the EU Directive in 1989.[45] In southern Europe the problem of TBT was less obviously present on the environmental agenda. If support for stricter regulating of TBT is lagging behind in European regions, its effects will continue to build up with demonstrable adverse impacts on the marine environment and unknown impact on humans who use the marine system as a resource for food. In southwestern Spain previous studies revealed the presence of significant levels of TBT in water, sediments and biological tissues from near-harbour areas.[46] However, regulation of TBT was limited, and no studies were conducted in the continental shelf area. Elsewhere several studies indicated continuing (illegal) use of TBT on small vessels, thus underpinning the need of support for effective implementation of international regulations from national/regional policy makers and the public. Ineffective implementation of the ban on TBT for small craft and lack of knowledge of the presence and impact of TBT still contributes to illegal use of TBT, as has been reported in Australia.[47]

A global ban can only be effective when regionally implemented. Implementation needs support from policy makers, users, and the public in the region. Debate of the issues is necessary, and there is a need for the participants in the debate to be properly informed.

The Strategy

The Communication Strategy programme of the study group was intended to optimise information exchange between science and policy in different

44. Ten Hallers-Tjabbes *et al.*, (n. 28 above).
45. EU Directive 89/677/CEE (1989).
46. J. L. Gómez Ariza, E. Morales, I. Giráldez, R. Beltrán and M. A. F. Recamales, "General appraisal of organotin in the Southwest Spain coastal environment," *Química Analítica* 16 (1997): 451–455.
47. B. Kettle, Paper presented at LC/SG 23, (Townsville, Australia, May 2000), Report of Meeting LC/SG 23 (London: IMO, 2000).

entities and to contribute to rendering the scientific and policy-making processes more transparent and participatory. The HIC-TBT scientists investigated the impact of TBT in the open sea in relation to shipping, sought cooperation with regional and national policy makers and passed their findings on to users, to the public and to the press. Throughout the project, communication played a pivotal role, which was also reflected in the character of meetings and other activities. Mutual understanding between scientists and policy makers of each other's positions, perceptions and "Maps of the World,"[48] was a guiding principle.

For optimising communication with the world outside the scientific community, guidance was sought from experts in the relevant decision-making processes. In the course of the study key factors in the approach were identified that could be applied in other environmental processes where precaution is a part.

As in the late 1990s, knowledge about the extent of ecological damage by TBT was limited in southern Europe and, recognising the need for proper implementation in the EU of the phasing out of TBT, cooperation was sought with scientists in southern EU countries. We therefore set out to investigate the impact of TBT in the seas off southwestern Spain, Portugal, and Italy as well as in the North Sea, in relation to shipping densities and to employ the results as a basis for communication between the scientists involved, policy makers, and the public in the regions. The project offered an opportunity to pass on expertise on studying the impact of TBT in relation to commercial shipping as well as transferring strategies for communication between scientists and policy makers, a wider public and other interested bodies.

The scientists, aiming at transparency in passing on their message to the world outside science, needed to step away from the mode of expressing scientific results in the format traditionally used in communication among scientists. For understanding the background of such a shift in communication channels and methods, one has to realise that, at least in the natural sciences, credit amongst the scientific community is strictly evaluated on the basis of publications in the (refereed) scientific press and on citations in such publications by fellow scientists. Traditionally, a strong body of opinion among the scientific community favours the belief that scientists should

---

48. Map of the World: the internal representation of human beings as based on their reflection and interpretation of all sensory information as received from the external world, as well as internal sensory (proprioceptive) signals. The interpretation reflects people's backgrounds, such as the community in which they operate and its rules, structure and processes, the values of the individual human and of relevant peer groups, as well as character traits and structures. See A. Korzybski, *Science and Sanity: An Introduction to Non-Aristotelian Systems and General Semantics*, (Lakeview, Conn., USA: The International Non-Aristotelian Library Publishing Co., 1958).

strive for maximising such scientific output and that diverging onto the path of the popular press is draining the scientific potential. And although a different opinion about relevant scientific communication is often voiced in the world outside the scientific community, and is recognised by many scientists as well, the pressure to steer away from communication outside the strictly scientific press remains strong in the perception of a majority of natural scientists, at least in Europe. Time and again funding and peer review channels demonstrate the value of such attitudes by the scientific community. In this study public participation was sought while recognising the two facets of public involvement in policy processes, creating support for regulatory measures, and stepping up pressure on decision makers to take measures. Societal participation can be pivotal in creating support for the implementation of political decisions, and thus limiting the need for enforcement, in the sense of fines or measures involving coercion. Societal participation can stimulate the concerned public to make their views clear to policy makers and in this way augment the pressure on policy makers to take precautionary decisions.

By communicating with the world of decision makers and users, the case study explored the potential that a precautionary approach offers to scientists to contribute to policy planning and to pass on their scientifically based message on the environment in which they work. If precaution is to be implemented, then open and effective communication between science and policy is a priority and initiatives taken by scientists to bring environmental problems to the attention of the world outside in a transparent manner, should be encouraged. The participation of independent scientists should also be encouraged. This would guarantee the freedom for scientists to pass on their concerns, whether the concerns are indicatory of a need for refinement in policy planning or whether they generate new information that deviates from the present perception of what constitutes a specific environmental problem. This may point to a different direction in political choices such as to identify the information needed to evaluate the environment for a proper decision-making process.

The case study was inspired by the now widely adopted precautionary principle, which sets a framework for scientists to communicate underlying explanations for potential environmental threats. This is an alternative to absolute statistical proof, which can seldom be given when evaluating the state of the environment. By such an approach, scientists can help in raising environmental awareness and so generate support for the development of environmental policy planning.[49] Once the impact of TBT has been identi-

49. E. Hey, "The Precautionary Concept in Environmental Policy and Law: Institutionalizing Caution," *Georgetown International Environmental Law Review* 6 (1992), p. 257–318; D. Freestone and E. Hey, "Origins and development of the Precautionary Principle," in Freestone and Hey (n. 8 above), pp. 3–15; Gray (n. 15 above); Mee (n. 18 above); Santillo et al., (n. 9 above).

fied in an area and policy decisions have been implemented in response, scientific findings can also serve as a basis for assessing the effectiveness of the environmental policy measures. This can be done by monitoring the development in the marine environmental conditions, where the impact of the specific activity was detected in the first place.

Meeting the Challenges of the Approach

Challenges of the project were developing and employing the new skills required from the participating scientists from Spain and Italy. They had to acquire new scientific skills for assessing and interpreting the biological impact of TBT, in addition to their chemical expertise in analysing burdens of TBT. More challenging was for them to embark on active communication with policy makers, users, and the public, on their own initiative. Thanks to the support we provided for developing this specific skill, some initial confusion was soon replaced by confidence, in particular by the Spanish partner, who developed new and original initiatives early on. The interested participation of international policy makers and political-legal experts from the network-steering group throughout the project was an invaluable stimulus to the communication process.

MATERIAL AND METHODS

Communication

Scientific partners demonstrated the impact of TBT in the continental shelf and continental margin seas off Spain, Portugal, Italy, and in the North Sea in relation to shipping density, shipping routes, and other input sources of TBT.

The scientists learned the specific requirements for policymaking and developed scientific strategies and recommendations that made sense within the policy framework. These recommendations were communicated as appropriate.

The assessment and communication processes were aimed at arousing recognition of the problems associated with TBT antifouling in Spain, Italy, and Portugal and at drawing attention to the continuing problems of TBT in the North Sea in support of environmental policy planning. The investigations were also intended to develop a basis for assessment of the effectiveness of global and regional measures to reduce the impact of TBT-based antifouling in relation to the upcoming ban on the use of organotin in ships' antifouling paints.

The scientists and network steering group members met on a regular basis. At the inception of the project they evaluated the initial contacts in respect of the communication strategy, and demonstration objectives and strategies were reconsidered. Throughout the project, feedback from policy makers was incorporated in the fine-tuning of the biological and chemical scientific programme so as to optimise communication. The adequacy of the communication strategy was evaluated halfway through the project, readjusted, and once again optimised. In a final stage the process and the strategies employed were evaluated for their potential to learn from the process of this case study for future communication between policy makers and scientists, by answering the questions: "What went well?" "What could be improved?" "What have we learned?"

Specific tools were employed during the project to communicate findings and strategies to relevant target groups (such as specific user groups of TBT, local authorities, interested public and the press). Apart from informal communications to policy makers and users, several other tools were employed for communication. The public was approached through press releases and a press conference, by project-related general and regional worldwide web sites, and through presenting the project results to wider audiences. In the southern European partner countries, open public meetings, with participation of regional and national policy makers were organised (Palermo in 1999 and Huelva in 2000). The press releases disseminated in association with each meeting were met with considerable interest. A press conference in Spain, prior to the Huelva meeting, resulted in wide radio and TV coverage of the meeting itself. Additional to the scheduled project meetings, two seminars (in Southampton and Malmö) were organised to involve audiences in northwest Europe. Papers were presented at the seminars by project members, an IMO-related policy maker and by representatives from the shipping community and from the (alternative) paint industry. All partners made a regional project web site, cross-linked to the general project web site.[50] A Netherlands booklet on TBT was translated into Spanish and Italian. In Spain, a questionnaire stimulated awareness about the use of TBT, associated environmental problems and existing regulations. At a final stage, the project results and their future potential were communicated to a wider audience at the EU.[51]

Scientific Demonstration

As a measure for the input of TBT, we used three levels of shipping densities, expressed as the number of ships of 100 gross tons (gt) or greater per day passing within a given distance of the sampling site. The three density levels

50. Available online: <http://www.nioz.nl/projects/tbt>.
51. EU, Green Week Conference, (Brussels: 24–28 April 2001).

were: High (more than 10 ships per day); Intermediate (between 5 and 10 ships); and Low (less than 5 ships). The distance of passing was calculated to take specific local conditions on board. In the North Sea and the Bay of Cadiz the distance was within 15 nautical miles (M) of a sampling station and within 10 M of a sampling station in the Tyrrhenian Sea, where tidal streams have a much smaller effect on dispersing water laterally.

Target organisms (gastropods or "snails") and sediments were collected from the North Sea, the Mediterranean Sea off the coast of northern Sicily, and the Atlantic Ocean off the western Iberian Peninsula. Additional (coastal) sources of TBT in the areas of investigation were inventoried.

Imposex and levels of TBT were recorded in snails and in sediments, and the sediments were characterised according to grain size and organic carbon content. In order to isolate the impact of TBT from other influences in the North Sea, the presence or absence of damage from benthic (bottom) fisheries was also examined. Snails were investigated on board ship or on land, while samples of snails and sediment were analysed for organotin on land. As a parameter for past presence, the ratio of live snails and empty shells per trawl was noted at sampling sites in the North Sea and off Iberia. Specific sampling logistics and snail species differed per region.

The gastropod species common whelk (*Buccinum undatum*) and red whelk (*Neptunea antiqua*) were sampled at 33 locations throughout the North Sea, during an extended research cruise in 1999 by the Research Vessel *Pelagia* (NIOZ).

Ten snail species (including *Bolinus brandaris* and *Hexaplex trunculus*) and sediments, were sampled at 11 locations off southwestern Spain and Portugal during two research cruises in May/June 1999 (R.V. *Patrimonio* (U. Huelva) and January 2000 (R.V. *Pelagia*). Off Sicily *Hexaplex trunculus* and *Bolinus brandaris* and sediments were sampled at five locations mainly by scuba diving during sampling trips in 1999 and 2000.

RESULTS

The Marine Scientific Programme

Imposex was present in all areas investigated, while TBT levels in snails were elevated in areas with higher shipping densities, compared with areas with lower shipping density. Off western Iberia and in the North Sea, the rate of imposex and the levels of TBT correlated well with shipping density. In Sicily the omnipresent imposex could not be directly related to shipping density alone, as several additional input sources were present in the sampling areas.

Imposex incidence and organotin levels in snails and sediment reflect an increased impact of TBT in relation to shipping in the North Sea and off western Spain and Portugal, imposex being higher in areas with higher shipping density. Differences in sensitivity responses between different snail species were detected in both the North Sea and the western Iberian Atlantic Ocean. In the revisited North Sea, a relationship with local hydrology was found to be crucial for interpreting the dose-effect relationship in the open sea. The northern North Sea is often stratified in the summer, while the Skagerrak is stratified year-round, with the thermo- or halocline—boundary layers separating water masses of different temperature or salinity—forming a barrier to a direct vertical transport of surface input to the local seabed, the habitat of the investigated gastropods. Adequate interpretation in terms of relationships between surface input and local effects, proved only valid in areas with a year-round mixed water column. This mechanism has a major bearing on interpreting data, both for new findings of impact from surface-bound input sources, and for monitoring efficacy of policy measures taken to reduce input sources, such as TBT, when the International Convention on the Control of Harmful Anti-fouling Systems on Ships comes into effect.

In the North Sea high shipping density was significantly correlated with imposex development and with body burdens of organotin compounds in *B. undatum*, in areas with a year-round mixed water column.[52] In areas where both the common whelk and the red whelk, *N. antiqua*, co-occurred, the latter species systematically showed the higher level of imposex, indicating a higher sensitivity to TBT in this species. In sediments and organisms from stratified locations, with low or intermediate shipping density, butyltin and phenyltin compounds were below or close to detection limits. The highest concentrations were found in mixed waters with high shipping intensity, in the southern North Sea and the German Bight.

Off southwestern Spain and Portugal, imposex was found in five snail species from different locations, with a 100 percent imposex incidence in two locations closer to the coast. Differences in sensitivity between species were observed, which interfered with comparisons between locations with different snail species. Organotin concentrations in sediments ranged from 21 to 185 ng Sn/g and were positively correlated to shipping, the higher values being found in the near vicinity of shipping routes. The concentrations of organotins in snail tissue ranged from <5 to 196 ng Sn/g.[53]

In the coastal seas along northern Sicily, imposex was omnipresent in snails at all locations except one, despite low organotin concentrations being

---

52. C. C. Ten Hallers-Tjabbes, J.-W. Wegener, A. G. M. Van Hattum, J. F. Kemp, E. Ten Hallers, T. J. Reitsema and J. P. Boon, "Imposex and organotin concentrations in *Buccinum undatum* and *Neptunea antiqua* from the North Sea: relationship to shipping density and hydrographical conditions," *Marine Environmental Research* (2002), (in press).

53. J. L. Gómez-Ariza, M. M. Santos, N. Vieira, E. Morales, I. Giráldez, D. Sánchez-Rodas, A. Velasco, J. P. Boon and C. C. Ten Hallers-Tjabbes, "The Impact of organotins in open Atlantic sea along the Iberian coast in relation to shipping," (2002), (in press).

found in sediment and snails. A correlation of imposex and organotin levels with shipping densities could not be isolated from the influence of other input sources of TBT.[54] Additional input sources of TBT are expected to exist in the near harbour and most other coastal areas.

The assessment and communication of these findings led to increased recognition of the problems associated with TBT antifouling in Spain, Italy and Portugal and the continuing problems of TBT in the North Sea. The wide geographical coverage of the investigations can serve as a basis for future monitoring of the effectiveness of environmental policy measures.

The Communication Programme

When the findings on the impact of TBT were presented, regional policy makers and the public participated in discussions in project consortia and in public communication meetings. In addition, the mutual communication between network and project members was highly productive throughout the project. In particular the joint meetings between network and project members, where some time was also reserved for informal exchange of ideas, enhanced our understanding of the specifics of the policy processes encountered.

After the stage was set in The Netherlands with full participation of network steering group members, the wider audience meetings in Italy and Spain (including Portugal), had regional, national, and international policy makers participating as speakers and in discussion panels. Interested organisations, users, authorities, and NGOs also joined in the discussions.

In the southern European countries, press reports appeared in the major daily newspapers and radio and TV channels. For instance, in Spain: *El Pais, Sociedad, La Prensa, Huelva Information,* National TV, TV Andalucia; in Italy: *Scienza e Societa, Ecologia Territorio, il Jornale della Natura;* and in Portugal: *Correio da Manha, Jornal Noticias.* Internationally, a more specialised maritime press responded to the activities, such as: *Naval Architect, Motor Ship, Lloyds List, Greenpeace Journal* (NL), and, the more general, BBC World Service. The presence and full participation of a Mexican delegation at the meeting in Spain indicated a wider interest in the project results.

The two additional seminars in the UK (Southampton Institute, November 1999) and in Sweden (World Maritime University, Malmö, September 2000), were well attended. The decision to organise the stand-alone seminars, as a contribution to informing the debate concerning the control of TBT in ship antifouling paints, only emerged during the project and was

---

54. S. Chiavarini, P. Massanisso, P. Nicolai, C. Nobili and R. Morabito, "Butyltins concentration levels and imposex occurrence in snails from the Sicilian coasts (Italy)," (2002), (in press).

facilitated by the generosity of the co-organising institutes and speakers. The seminars were highly successful in achieving their aim of presenting the viewpoints of the main players in the TBT debate and encouraging informed discussion of the issues in the context of the precautionary approach. The seminar attendees represented the relevant interest groups, from the maritime and shipping community, the shipping and paint industry, maritime and harbour authorities, environmental NGOs, policy makers, scientists, and the press.

The project reports and press messages are made accessible at the general and regional project web sites,[55] and the translated brochure on TBT[56] was added to the regional web sites.

Enquiries about the use of TBT in the partner countries revealed limited knowledge and lack of understanding of environmental problems and regulations. A questionnaire developed in response to the findings of initial local inquiries among fishermen, other users, and authorities was circulated in Spain. A first mailing to 200 target addresses met with an extremely high return of 40 percent, but only 20 percent of the respondents indicated that they were aware of TBT, associated environmental problems and existing regulations. Apart from this knowledgeable minority, most users indicated that they were not aware of the composition of the antifouling paints they used. The outcome of the questionnaire was consistent with the earlier reported lack of understanding of environmental problems associated with TBT, in Spain as well as in Italy.

When the findings were promulgated beyond the scientific community, there was an increased recognition of the problems caused by TBT. Regional policy makers acted accordingly and direct links between the scientists and policy makers were established. The study thus enhanced transparency and increased participation in policy processes. The wider recognition of the problems associated with TBT resulted in joint policy contributions of the research findings to the IMO-MEPC, by The Netherlands, Spain, Portugal, and Italy (1999: MEPC 44/INF.11;[57] 2000: MEPC 46/INF.2[58]).

During the study the interest of entities outside the participating EU countries, apart from that shown by the Mexican delegation, was expressed

---

55. (n. 51 above).
56. RIKZ, *An issue of substance, TBT in marine antifouling paints*, (The Netherlands: Ministry of Transport, Public Works and Water Management, 1998).
57. MEPC 44/INF.11, "Harmful effects of the use of antifouling paints for ships. Information on TBT levels and the occurrence of imposex in certain marine species in the North Sea, the Mediterranean and the coastal waters of Portugal," submitted by Italy, the Netherlands, Portugal and Spain, (London: IMO, 1999).
58. MEPC 46/INF.2, "Harmful effects of the use of antifouling paints for ships. Information on TBT levels and the occurrence of imposex in certain marine species in the North Sea, the Mediterranean and the coastal waters of Portugal," submitted by Italy, the Netherlands, Portugal and Spain, (London: IMO, 2000).

by invitations to speak at external events (Brazil, 2000;[59] WWF, 2000[60]) and to present information at London Convention Meetings (LC/SG 23 and 24).[61] The HIC-TBT concept and strategies were presented in a wider perspective at the Southampton and Malmö seminars, and at the EU Green Week conference. The approach was discussed in the context of experience with societal participation in other environmental issues and related policy processes, with specific attention to what causes the balance in the decision processes to tip. For several environmental issues the perceived technical and economic benefits and the societal appreciation of environmental consequences and constraints were analysed according to factors that contributed to making the pendulum swing to an environmentally based policy path. The experiences acquired in the HIC-TBT study demonstrated that public participation can play a major role in such decision processes. Figure 1 illustrates the role of emerging public awareness of an environmental problem and the consequent motivation to develop environmentally more friendly alternatives by industries and markets thus tipping the policy balance in favour of more sustainable options.

Two directly related environmental policy processes that can benefit from the experience of this study were identified. The first issue concerns the policies needed to address the problems associated with the presence of TBT in harbour-dredged material, which is dumped in large quantities in the world's coastal seas. The presence of TBT in this matrix is at present not screened to assess its suitability for dumping at sea, although the impact of TBT in harbour-dredge spoil has been demonstrated.[62] A second issue is the policy process to address discharges of ballast water from ships' tanks and the associated risk of introducing alien species and compounds to other marine areas, which is at present an issue on the agenda at IMO. Other environmental decision processes related to shipping that were analysed for factors that could influence the swing of the environmental policy balance toward a more environmentally progressive process were leakage of CFCs from refrigerated container vessels, sulphur exhaust of bunker oils, and PAHs from incomplete combustion.

59. TBT Seminars, Rio de Janeiro and Brasilia (Brazil: Ministry of the Environment, 28–30 August 2000).

60. WWF-initiative Global 2000 – TBT: Conference "A threat to the oceans shown in three ecoregions – environmentally sound ship-paints as alternatives," (Hanover EXPO, 19 September 2000).

61. LC/SG 23, Report of Meeting (2000), (London: IMO, 2000); LC/SG 24, Report of Meeting (2001), (London: IMO, 2001).

62. C. C. Ten Hallers-Tjabbes and B. van Hattum, "Imposex and TBT in *Buccinum undatum* and in sediments in the North Sea," Report re STOF*CHEMIE, RIKZ-RWS, 1996). Also LC/SG 19/INF.18, Submitted by the Netherlands, May 1996; M. M. Santos, M. N. Vieira, A. M. Santos, J. L. Gomez-Ariza, I. Giraldez and C. C. Ten Hallers-Tjabbes, "Impact of Butyltins released by discharge of dredged spoil material," (Submitted, 2002).

The Environmental Policy Balance

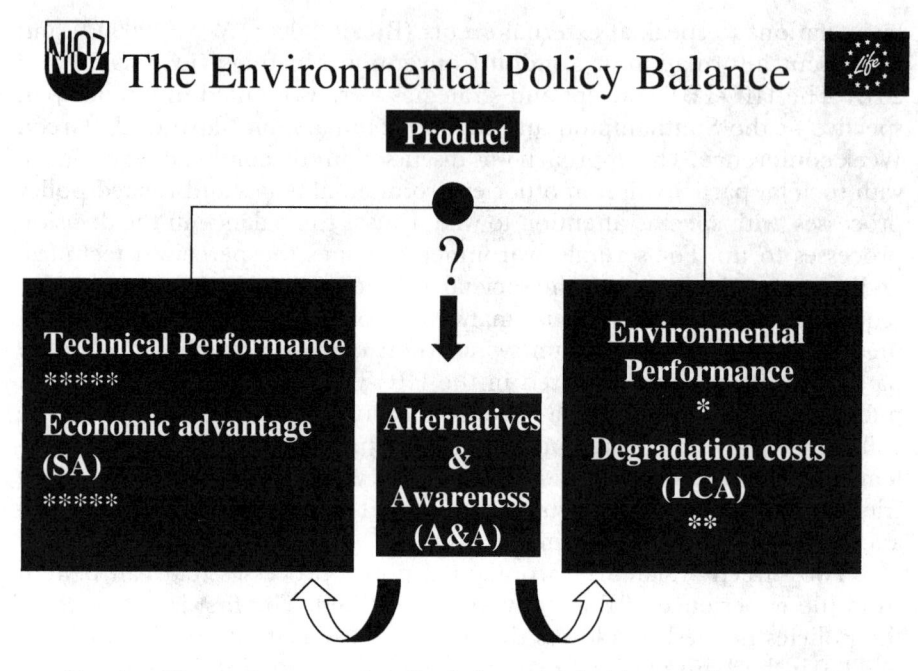

FIG. 1.—The environmental policy balance in decision-making processes.

DISCUSSION, CONCLUSIONS AND POTENTIAL OF THE CASE STUDY FOR A WIDER CONTEXT

The study has resulted in regional recognition of the environmental problems associated with TBT from ships' antifouling systems. This gave momentum to the regional decision-making process, as is reflected by the active participation of the three southern European countries in the international policy process and by the advisory groups on antifouling and the problems of TBT, as established by the governments of Spain, Portugal, and Italy, where project scientists participate in a senior advisory role. Moreover, the study stimulated processes beyond the strict scope of the project, such as the joint submissions to MEPC, a first example of effective co-operation between southern and northern EU States in this context, and the interest of Latin American States and involvement in the London Convention 1972 framework.

The project has shown the importance of early recognition of an adverse environmental impact. It also demonstrated how understanding the relevant policy-making process helps effective transfer of knowledge from scientists to policy makers and how direct communication between science and policy can play a role in development of policies. Last, but not least, it set an example of how monitoring of the marine environment can play a role in the dynamics of development, implementation and enforcement of EU policies.

Precaution is sufficiently understood to enable scientists to pass on their message when their research arouses serious environmental concerns, and the project has demonstrated the potential of a precautionary approach for scientists who are willing to pass on such concerns.[63] Scientists can play an active role in the dissemination of their concerns through policy channels in a communication strategy that facilitates a two-way exchange of information and ideas. A network, in which policy makers contribute their experience and understanding along with the scientists' message, enhances the process of communication and creates confidence in scientists. The members of the project policy network (international, EU, and national) were happy to participate, to stimulate developments, to facilitate new contacts, and to discuss the project process in the light of relevant policies. The success of the project appears to be related to the scale of the study that was limited by taking up only one environmental issue and one specific target decision process. One advantage of such a scale was the associated group size, which allowed for social interaction and something that may be called "care," in the sense that one was interested whether messages were perceived and understood by others, in the way that was intended. It is a natural aspect of human (and other animal) behaviour that displayed behaviour intended as communicating signals can be adjusted when the perceived feedback to a message or a signal is not consistent with the intent of the signal.[64]

As the HIC-TBT project progressed, the differences in spheres and positions between the scientific world and the world of policy makers emerged in all meetings where the policy network and the project scientists sat together. As was observed by a network member: "The HIC-TBT project clearly illustrates the different perceptual frames of scientists and policy makers. Where the former most often refer to a regime of uncertainty, findings being true only as far as they cannot be falsified, policy makers, often steered by their political masters, seek to express themselves in black and white perspectives and require certainty and absolute facts."[65] This was followed by a further observation that the project had contributed to activating policy makers in the relevant countries, establishing communication chan-

---

63. C. C. Ten Hallers-Tjabbes, "Science communication and Precautionary Policy: A marine case study," in J. Tickner (ed.), *Precaution, Science and Preventive Public Policy* (Washington, DC: Island Press, 2002), (in press).

64. Korzybski (n. 48 above); R. A. Hinde, *Biological basis of human social behaviour* (New York: McGraw-Hill, 1974).

65. HIC-TBT Project, Report of 2nd Progress Meeting, available at: <www.nioz.nl/projects/tbt> (2000).

nels that made science and policy transparent, and initiating participatory structures between scientists and policy makers that appeared to have sufficient potential to be sustainable. Once more this illustrates the importance of mutual responsiveness and involvement, in particular when autonomy and integrity are part of the process. "In the science-politics interface, relationships gain if science enjoys great autonomy and is only to a moderate extent engaged in the policy process. Responsiveness should best focus on the concerns and questions that decision-makers are wrestling with, and not aim at accommodating preferences in view of the substance of the answers."[66]

The mutual confidence between scientists and policy makers, as created by the project, was reflected by initiatives in communication by the participating scientists, and by support for such initiatives from policy makers. Such mutual confidence can be a major asset in stimulating scientists to pass on early research findings that arouse concern as to environmental consequences, rather than holding such information back until full proof or statistical "certainty" (a contradiction in terms) has been established. The present openness is in accordance with trends such as have been identified in the North Sea Conferences,[67] where consecutive Quality Status Reports reflect the progress in scientific input into the decision process.[68]

Employing press releases, conferences, and meetings for a wider audience in order to pass on clear information on science, policy, and policy processes has stimulated public, policy, and societal awareness of the presence of TBT on ships and associated problems.

The questionnaire, by identifying a public that lacks information, has revealed the need to inform the nautical community about TBT and associated problems and has also stimulated participation in environmental efforts. Public attendance at meetings and the response to the Spanish questionnaire reflected the interest of relevant bodies. These included nautical communities (such as shipping, fisheries, and ports), industries (such as marine paints, shipbuilding, and repair), and environmental NGOs, as well as scientists, local and regional authorities, the press and interested representatives of the public.

Future seminars of a similar structure to the Southampton and Malmö Seminars organised in this study could have a role as further contributions to the TBT debate, or for informing and encouraging debate on other issues of environmental concern in the context of a precautionary approach. Experience with the two extra seminars suggests the need for specific funding, which was not foreseen within the project budget. Organisational activities for the seminars had to be done at low budget conditions, drawing on volun-

66. Underdal (n. 2 above), p. 11.
67. Wettestad (n. 3 above).
68. Hey (n. 12 above).

tary contributions, especially from the host institutions. Such voluntary contributions should not be taken for granted in planning future seminars.

The approach of the project was original and had not been explored before in such an EU framework. Initiating and maintaining flexible communication on a specific actual environmental issue between the independent scientists working on the issue and relevant policy makers has great potential for developing effective action and, as such, could be complementary to institutional communication between science and policy.

The present study deviated from processes such as the North Sea Quality Status Reporting.[69] First, the HIC-TBT project focused on the environmental impact of one specific input source, although protecting the marine environment from such impact was within the competence of two separate decision-processes (IMO-MEPC and LC 72). The current target decision-making process is for regulating maritime shipping and not for the dumping of matter where the compound has accumulated, such as harbour dredged material. Second, the study itself was initiated by a small group of scientists who took the initiative, on a voluntary basis, to engage decision makers, users, and the public in the process. The feedback loop in this process was initiated by the scientists and has since been maintained by scientists and decision makers alike.

The scientists of this study were independent and not associated with governmental departments responsible for policy planning for the marine environment. They belong to a pool of scientists that may be asked for advice on the initiative of decision making institutes or bodies, yet are not particularly encouraged to take advisory initiatives themselves. Initiatives taken by independent scientists under their own steam, to bring an environmental concern to the attention of policy makers, sometimes tend to be considered as just a means to raise grant money. Notwithstanding the benefits of institutionalised structures to advise government policies, the practice of channelling scientific advice through such structures also allows decision makers to select the environmental issues considered relevant to be investigated. This process leaves only limited opportunities for scientists to draw attention to environmental threats which are unforeseen by policy making processes.[70]

The direct and often inter- or intra-institutional links between government-related advisory structures and the policy making process itself seldom allow for the underlying science and policy processes to surface to the public eye. The HIC-TBT project demonstrates the potential of independent scientists to unearth phenomena and ideas that cannot logically be inferred from

---

69. North Sea Task Force, *North Sea Quality Status Report*, (London: Oslo and Paris Commissions, 1993), 132 pp.; OSPAR (2000) (n. 13 above).

70. A. Underdal and J. Wettestad, "Science and politics in international environmental regimes: Between integrity and involvement," in Andresen *et al.* (n. 2 above).

former research, nor be deduced through the channels of policy-driven research structures. By questioning the international consensus that TBT posed no problem in seas beyond the coastal area, the extent of the problem and the contribution of offshore shipping was discovered. The recognition of the role of hydrographical conditions, and more specifically stratification, on the dispersal of surface contaminant input, was only possible by investigating hydrographics along with the study of the target impact and compound, and by not focusing on burdens and the impact of TBT alone.

The rather strict time-frame inherent to policy making, leaves little time for a higher flexibility in the mode of planning, such as science often requires, and hence sets limits to the freedom of investigation and interpretation of scientific results. This is certainly true for marine environmental issues, where the field conditions are not only unpredictable, but also can often not be captured in predetermined methodological frameworks. A plea has been made for fora where interpretation of environmental knowledge is done jointly by scientists and policy makers.[71] The added value of joint platforms where scientists, policy makers and interest bodies discuss marine environmental issues in relation to policy processes for the North Sea has also been illustrated.[72]

Scientific input in decision-making processes should be based on confidence in scientists as a reliable knowledge base, with the understanding that decision makers would only utilise such inputs if they could recognise them as relevant to their concerns. It is there where mutual involvement is required, as the standard reporting of research results rarely offers ready-to-use knowledge, but needs translation and adaptation to be utilised by decision-makers.[73]

The approach of the HIC-TBT project was very different from another widely explored participatory process, Integrated Coastal (Zone) Management (ICM) with stakeholder participation. The ICM constitutes a spatially defined management process, where stakeholders are groups that have a special interest in the use of the coastal resources.[74] The ICM-related mechanisms in support of the decision-making processes have had a longer tradition in North America than in Europe and in other parts of the world.

71. T. Skodvin and A. Underdal, "Exploring the dynamics of the science-politics interaction," in Andresen *et al.* (n. 2 above).

72. A. Bijlsma and C. C. Ten Hallers, "Executive summary to the proceedings," in *3rd North Sea Seminar: Distress Signals. Signals from the Environment in Policy and Decision-Making*, Ten Hallers-Tjabbes and Bijlsma, (eds.), (Amsterdam: Werkgroep Noordzee Foundation, 1989), pp. 9–13.

73. Underdal (n. 2 above).

74. L. F. Scura, T.-E. Chua, M. D. Pido and J. N. Paw, "Lessons for integrated coastal zone management: the ASEAN experience," in *Integrative Framework and Methods for Area Management*, T.-E. Chua and L. F. Scura, (eds.), ICLARM Conference Proceedings No. 37 (Manila: ICLARM, 1992), pp. 1–70; B. Cicin-Sain and R. W. Knecht, *Integrated Coastal and Ocean Management: Concepts and Practices* (Washington, DC: Island Press, 1998).

The concept of ICM with stakeholder participation originally derives from management theory. The concept is based on inclusion or representation of all relevant parties in a corporate decision-process, as groups other than users have a legitimate right to be consulted before decisions are made.[75] The concept, employed for coastal zone management and some other marine contexts, such as fisheries management,[76] strives for representation of interests in the course of a decision-making process. The approach mainly deals with groups of many members that need to be represented.

ICM encompasses a systematic, cyclic process, and has an essentially different structure than that of the HIC-TBT project. The HIC-TBT study was not designed on the initiative of governmental departments nor by decision-making entities and, where a common framework is paramount to ICM, the HIC-TBT project started with an unstructured basis aimed at facilitating flexibility. This flexible approach allowed for action to fulfil the objectives by recognising, appreciating, and utilising emerging options. The two joint contributions to IMO-MEPC and the organisation of nonscheduled seminars are both examples of the benefits of such an approach, as is the Spanish questionnaire. And, whereas ICM projects aim at consensus building and a shared vision of stakeholders,[77] such was not the aim of HIC-TBT. Instead, the HIC-TBT project focused on raising awareness in groups that might have been identified as stakeholders, if that concept had been chosen as the approach to be followed. The initiating scientists were not seeking to have their interests represented; they acted on the understanding that our small epistemic group had a role to play in developing a regional rationale to support a global decision-making process. The case study was intended to set the stage and to explore a strategy that had the potential to serve as an example of the role of scientists in an environmental decision-making process, given a clear-cut scientific understanding of the environmental problem and of the relevant (global) decision processes. As such, the activities and concept of the HIC-TBT project have the potential to be applied to ICM processes and projects.

A difference of a more physical nature between ICM and HIC-TBT is the use of the term "coastal" versus "offshore." Coastal areas in a regulatory context are commonly understood as the coastal part of the sea that is governed by a coastal State, often coinciding with the Exclusive Economic Zone

---

75. R. K. Mitchell, B. R. Agle and D. J. Wood, "Toward a theory of stakeholder identification and salience: defining the principle of who and what really counts," *The Academy of Management Review* 22 (1997), p. 853–886; K. H. Mikalsen and K. Jentoft, "From user groups to stakeholders? The public interest in fisheries management," *Marine Policy* 25 (2001): 281–292.

76. Mikalsen and Jentoft (n. 75 above).

77. T.-E. Chua, "An analysis of the application of integrated coastal management – Linking local and global environmental concerns." Proceedings, *Oceans and Coasts Rio+ 10* (Paris: UNESCO, December 2001).

(200 M). Offshore is then understood as the seas beyond this zone. In HIC-TBT and in related contexts that deal with maritime regulation of TBT-based antifouling, whether research or decision making, the major break point is between the regulation for ships not larger than 25 m, referring to environmental effects in coastal, intertidal organisms, versus the seas off the shore where the majority of the world's fleet operates, the realm of the offshore, subtidal species. Spatially, in the North Sea, an EEZ of any State would grossly interfere with that of several other States, if the 200-M zone were to be maintained. We would rather, for this study, in particular for the North Sea, distinguish between the estuarine and near-shore ("internal") waters versus external waters, if we must comply with terms as they are understood in a regulatory framework.

Having said that, ICM and the present study share common grounds. Both are interactive, adaptive, participatory, and consensus-building processes, in which research is an essential part. ICM, as successfully applied in East Asia, and HIC-TBT, both aim at support for implementation of national legislation and international instruments. The target of HIC-TBT, the environmental decision process in IMO, is also one of the multilateral environmental instruments addressed by the Environmental Strategy developed for the East Asian ICM-projects.[78] Both the East Asian ICM projects and HIC-TBT have recognised the need to strengthen linkages between local and global environmental concerns and to act upon it. A common denominator for both the success of the East Asian ICM projects and HIC-TBT may also be the confined scale but, whereas the ICM projects are spatially confined, the HIC-TBT project was confined in subject matter and in the decision process.

The HIC-TBT Scheme as a General Scheme for Application to other Environmental Issues

Since this has been a case study, we are seeking to employ the knowledge gained for other environmental issues beyond the scope of the case study and for geographical areas beyond the project region. While benefits for geographical areas beyond the EU have been demonstrated within the project, in the following sections we try to analyse the strategy and logistics of this study with a view to generalisation. Two related environmental issues, where such a scheme may have potential, have already been mentioned.

The scheme as adopted in the HIC-TBT project is described below and examined for its general value for serving as a policy scheme for other environmental issues. Essential to our approach is recognition of the potential of the 'Precautionary Approach' for scientists to take up their societal role

---

78. Available online: ⟨http://www.pemsea.org⟩.

and to inform policy makers and the public of scientific findings that they believe raise environmental concerns.

The Project Strategy and Process

1. Set the goal.
2. Identify who is interested.
3. Identify funding sources.
4. Involve policy makers from the beginning.
5. Set the scientific frame and process requirements.
   Collect and collate existing research on the environmental problem under consideration. Note where there are gaps in the knowledge needed to assess the extent of the problem and to point the way to solutions.
6. Develop communication channels.
   Seek and employ links with policy makers from the relevant field, who are interested in exploring the merits of mutual communication. Identify channels that are appropriate for communication of the specific issue and identify potentially interested audiences and representatives of the press. Employ existing channels and develop additional ones if necessary. Be transparent in communicating the scientific concerns and do so with an understanding of the requirements of policies in the relevant field.
7. Conduct original research.
   Conduct original scientific research, where needed to supplement existing research, to provide a firm basis for identifying and quantifying the environmental problem in scientific terms. Also, where needed, conduct research to inform the debate for formulating practical solutions to the identified problem.
8. Organise open meetings.
   Convene meetings amongst all interested parties including, but not limited to, scientists, commercial operators, manufacturers of substances or products that contribute to the identified problem, manufacturers of safer alternatives, environmental agencies, people affected by the problem, and policy makers. Convene meetings amongst scientists, experts and policy-makers to identify bodies where effective legislation might be introduced. Then draft submissions to the identified bodies providing information on which future regulations may be based, bearing in mind the precautionary principle.
9. Allow for developing extra seminars.
   Ensure that the research effort, the meetings and the dissemination of information take place in representative parts of the European

community or other relevant geographical areas, but co-ordinated so that results contribute to an effective overall programme.
10. Develop press contacts, disseminate press messages and public information.
Disseminate information to all forms of media whenever significant findings, meetings or other activities are taking place so that the general public is made aware of the environmental problems and the implications of possible solutions. Indicate contact points where concerns can be voiced or where questions can be addressed.
11. Reconsider the goal.
At a specific moment in time (deliberately chosen or when needed) evaluate the state of the project, the activities and the outcomes, and compare with the objectives in terms of what has been achieved so far, as well as whether the objective, as set at the outset, still stands. Adapt the programme of work, taking account of what has been learnt, and, if need be, fine-tune or adapt the objectives. Keep an open eye for emerging other doors, barriers or side tracks and act accordingly by considering them in terms of the continuing adequacy of the objectives and logistics as set at the outset.
12. Disseminate scientific and policy papers.
13. Explore generalisation of the concept, the strategy and the process.

Two Related Environmental Issues Where the Scheme Might Serve

Two environmental issues bear close relationships to the problems of TBT in ships's antifouling. One is TBT in harbour dredge spoil, referring to the same environmental concern, and a similar impact, although policies are set in another global forum (LC 1972). The other is ballast water discharged from ships, which raises other environmental concerns (introduction of alien species, alien compounds, and other effects), but policies are being addressed in the same global forum (IMO-MEPC).

The TBT load of harbour-dredge spoil, as a legacy in harbours, is closely associated with TBT from ships. When TBT is no longer present on ships' hulls, a depository of TBT will remain present in harbour sediments that are at present lacking a regime to prevent their introduction into the marine environment as dredge spoil. The London Convention 1972 and the 1996 Protocol, the global forum for dumping at sea, does not at present provide a regulatory regime to deal with materials which are not acceptable for dumping at sea.[79] The link between TBT in relation to shipping and in relation to harbour dredged material was discussed in the 23rd Consultative

---

79. LC 22/20, Report of 22nd Consultative Meeting London Convention 1972 (LC 72), (London: IMO, 2000).

Meeting of LC 72.[80] TBT in such a matrix is recognised as an emerging issue that needs recognition,[81] and the approach employed in this case study could be used to explore the problem and to stimulate preventive policies. The process in LC 72 would gain from increased transparency, in which the concept of the HIC-TBT project and the experience and capacities of initiators and participants in HIC-TBT may assist.

Environmental policy-making processes for preventing marine environmental damage from discharges of ballast water from ships follow a decision route within IMO that bears similarities with that of antifouling and TBT. As the latter issue will be on the agenda of IMO to resolve for the coming years, a similar concept and similar strategies as employed in HIC-TBT could be explored to create transparency and to encourage participation in the development and implementation of policies.

CONCLUSIONS AND FUTURE PERSPECTIVES

The case study illustrates the benefits of maintaining flexible communication on a specific environmental issue between the relevant scientists and policy makers as complementary to institutional communication between science and policy. Establishing direct links between regional/national scientists and policy makers in advisory channels sets an example for communication on other environmental issues and in other areas of the world—most notably in Latin America by employing the project languages.

Dissemination to the public has been more effective in Spain than in any other of the partner countries, which all did well; dissemination to the international specialised press and to all national MEPC related policy makers and to the scientific community has also been effective. Policy and public recognition of the relevance of the issue in all three southern EU States has leapt forward. The awareness of the problem extended to recognising possible solutions, such as the employment of less harmful antifouling techniques without TBT, and addressing the concerns of specific sectors that are particularly vulnerable to the impact of TBT (fisheries and aquaculture).

Within the IMO, TBT is recognised as a case study for implementing the precautionary approach.[82] The two joint documents by Spain, Italy, Portugal, and The Netherlands, submitted to IMO – MEPC, were also relevant to the EU legislative framework since the EU follows the IMO. The project

80. LC 23/16, Report of 23rd Consultative Meeting London Convention 1972 (LC 72), (London: IMO, 2001).
81. LC/SG 23, 2000, Report of 23rd Meeting Scientific Group LC 72, (London: IMO, 2000). LC/SG 24, 2001, Report of 24th Meeting, (London: IMO, 2001).
82. MEPC 41/10, 1997, "Harmful effects of the use of antifouling paints for ships. Final report of the correspondence group," Submitted by The Netherlands acting as lead country of the correspondence group. (London: IMO, 1997).

could stimulate Europe to develop long-term and interim policy objectives for antifouling, which, by being beneficial to the marine environment, could offer additional tools for EU policy practice.[83]

As the concept has proven to be feasible and effective, it suggests a general strategy and framework, which could be applied to other environmental issues in support of environmental decision-making processes within the framework of a precautionary approach. The project model: identify an environmental problem and invest in communication to policy and public as outlined in the step-by-step process, has the potential to be developed as a model for the role of science in environmental planning for the EU and other policy bodies. A future goal of the project is to carry the concept beyond the EU, and preliminary contacts outside the region indicate that this would be a valuable and welcome initiative.

---

83. Underdal and Wettestad (n. 70 above).

Environment and Coastal Management

Empowering Local Communities: Case Study of Votua, Ba, Fiji

Joeli Veitayaki, Bill Aalbersberg, and Alifereti Tawake
University of the South Pacific, Suva, Fiji

INTRODUCTION

A new consciousness is developing to involve people in local communities in the management of their marine resources using the participatory approach throughout the Pacific Islands. This emphasis on community-based marine resource management is developing as a result of the collaboration of government institutions, a network of nongovernmental organizations (NGOs) and local communities. The partnerships target the involvement of local communities in the management of their marine resources. These local communities own the majority of the resources in the coastal areas but are facing escalating pressure to exploit them. The initiative, which is referred to as the Locally Managed Marine Areas (LMMA) network, is based on the empowerment of local communities so that they can make rational decisions on how the marine resources under their care are used and protected for their use in the future. The communities are adopting the precautionary approach and using customary resource management arrangements that do not rely on the nonexistent scientific knowledge.

The partnerships are working well and need to be publicized widely. Local communities are the most knowledgeable about the changes taking place within their realm and how best to address resource management issues in their areas. Local people adjudicate on issues such as the use of their resources and their management. However, in recent times, local communities are facing escalating pressure to utilize their resources to enhance their economic activities as well as to improve their living standards. Consequently, people are lured by commercial cutthroats and advisers whose main aim is to acquire quick profit from the utilization of the resources without any consideration of the associated environmental costs. Such liaisons have led to the widespread environmental degradation that is evident today throughout the Pacific Islands.

In an attempt to promote sustainable development in local communities, the network of NGOs working collaboratively with specific government departments has involved local communities in the new initiative. The new initiative is rooted in a process that empowers people and involves them in deciding and implementing their resource management decisions. The approach acknowledges the existence of complex economic, social, and cultural forces that affect people and adopts an appropriate and realistic method to address the main resource management issues concerning the different local communities.

Experience in dealing with local communities has been widely discussed in the development literature.[1] It has been highlighted that development activities need to achieve their targets if they are to be successful. In addition, there have been queries about the effectiveness of development assistance in developing countries. In the Pacific Islands, critics and development scholars have referred to the Pacific Paradox, which is the unfavorable economic growth rates experienced in the region in spite of high investment ratios, and how the countries have become aid dependent.

This new approach attempts to achieve more success by involving local communities in managing their marine resources. To achieve this goal, the people of the LMMA network have formulated a new way of involving people to ensure that their support is maintained. This is a major challenge. The commitment of the local communities is obtained through extensive consultation using the Participatory Learning and Action (PLA) methods, which consist of facilitating the involvement of people in designing and implementing resource management activities that suit their particular situations, have enabled the articulation of locally determined and therefore appropriate resource management practices which are more likely to work.

The results to date are pleasing. This case study illustrates the empowerment process and what has been done to ensure that the people are genuinely involved in managing their marine resources. The aim of this article is to describe the empowerment process and to explain the application of the process in a specific place. Although the case study may not qualify as proof, the fact that people are acting without waiting for government to

---

1. R. Chambers, Rural Development: *Putting the Last First.* (London: Longman, 1983); R. Chambers, *Whose Reality Counts: Putting the First Last.* (London: Intermediate Technology Publications, 1997); D. Porter, B. Allen and G. Thompson, *Development in Practice: Paved with Good Intentions.* (London: Routledge, 1991); A. Carr, 1994. *Grassroots and Green Tape: Community-based Environmental Management in Australia,* Ph.D. Thesis, Australian National University, Canberra; R. Beilin, "Participatory Rural Appraisal: an inclusive methodology for community and environment planning," in H. Wallace (ed.), *Developing Alternatives: Community Development Strategies and Environmental Issues in the Pacific.* (Melbourne: Victoria University of Technology, 1996), p. 67–83; P. Blaikie, 1996. "New knowledge and rural development: a review of views and practicalities," Paper for the 28th International Geographical Congress, Hague (*not published*); G. H. Axinn and N. W. Axinn, *Collaboration in International Rural Development: A Practitioner's Handbook.* (New Delhi: Sage Publications, 1997).

introduce management regulations is an indication that desired change is taking place. The commitment people show for the management of their resources is indicative of their desire to make a difference, while the role of the Institute of Applied Science (IAS) team from the University of the South Pacific demonstrates the use of the top-down and bottom-up approaches.

THE METHOD

Empowering local communities and involving them in the management of their marine resources has been widely acknowledged.[2] However, until recently, very little has been achieved in involving people in resource management arrangements that affect them. Consequently, all the attempts to incorporate traditional knowledge in contemporary management practices or in involving people in designing and implementing marine resource management systems have only been partially successful.[3]

Over the years, community groups, NGOs, and development agencies have used different strategies to empower and involve people and incorpo-

---

2. E. Hviding, "Customary Marine Tenure and Fisheries Management: some challenges, prospects and experiences," in G. R. South, D. Goulet, S. Tuqiri and M. Church (eds.), Traditional Marine Tenure and Sustainable Management of Marine Resources in Asia and the Pacific. (Suva: International Ocean Institute–South Pacific, 1994), 88–100, at p. 91; R. Johannes, "Design of Tropical Nearshore Fisheries Extension Work Beyond the 1990s," in G. R. South, D. Goulet, S. Tuqiri and M. Church (eds.), *Traditional Marine Tenure and Sustainable Management of Marine Resources in Asia and the Pacific.* (Suva: International Ocean Institute–South Pacific, 1994), 162–174, at p. 162; R. S. Pomeroy, "Traditional Base for Fisheries Development: revitalising traditional community and resource management systems in Southeast Asia," in G. R. South, D. Goulet, S. Tuqiri and M. Church (eds.), *Traditional Marine Tenure and Sustainable Management of Marine Resources in Asia and the Pacific.* (Suva: International Ocean Institute–South Pacific, 1994), 226–252, at p. 226; K. Ruddle, "Traditional Marine Tenure in the 90s" in G. R. South, D. Goulet, S. Tuqiri and M. Church (eds.), *Traditional Marine Tenure and Sustainable Management of Marine Resources in Asia and the Pacific.* (Suva: International Ocean Institute–South Pacific, 1994), 6–45, at p. 6; J. Veitayaki, "Introduction", in G. R. South, D. Goulet, S. Tuqiri and M. Church (eds.), *Traditional Marine Tenure and Sustainable Management of Marine Resources in Asia and the Pacific.* (Suva: International Ocean Institute–South Pacific, 1994), 1–5, at p. 1; J. Veitayaki, *Fisheries Development in Fiji: The Quest for Sustainability.* (Suva: Institute of Pacific Studies and the Ocean Resources Management Programme, 1995), p. vii; J. Veitayaki and G. R. South, "Coastal Fisheries in the Tropical South Pacific: a question of sustainability," in B. Japar Sidik, F. M. Yussof, M. S. Mohd Zaki and T. Petr (eds.), *Fisheries and the Environment Beyond 2000.* (Serdang, Malaysia: Universiti Putra Malaysia, 1997).

3. J. Veitayaki, "Cooperation among Small Island Developing States in the Area of Marine Biodiversity: A Case Study of SPREP's South Pacific Biodiversity Conservation Programme," Submitted to Ocean and Coastal Management (in press).

rate wherever possible traditional knowledge in contemporary resource management systems. People went from using Rapid Rural Appraisals (RRA) to Participatory Rural Appraisals (PRA) as the search continued for "an approach and methods for learning about rural life and conditions from, with and by rural people."[4] The RRA method was inadequate because it did not allow for the gathering of accurate information from rural people because they were based on quickly organized short visits. The RRA method, therefore, did not allow for the involvement of rural communities in decision-making matters that concern them.

Compared with RRA, PRA is much more community oriented but is still largely externally driven and specially suited to rural communities that are targeted in these initiatives. Experiences have shown that although participation was emphasized, PRA methods have not allowed the local communities to be involved in the decision-making process.[5] Initiatives were largely formulated externally and were taken to the people for implementation. Innovative additions to the RRA methods from PRA include facilitating self-critical awareness, and responsibility and sharing.[6] People are encouraged to facilitate learning, analysis, presentation, and learning. In addition, external partners are continually examining their behavior and performance and trying to do better—welcoming error as an opportunity to learn and do better. Sharing of information and ideas is encouraged at all levels to ensure that outsiders' attitudes, behavior and action are all focussed on empowering local communities. Meanwhile, local communities are encouraged to strive for self-determined development activities.

The PLA method used in this case study is closely related to PRA in its emphasis and aims to facilitate the involvement of people in designing and implementing resource management activities that suit their purpose and situation. With PLA, it is recognized that the PRA is more than just a method for use in rural areas. In addition, there is emphasis on monitoring and adaptive management. The assumption is that community members have the best understanding of their resources, cultures, and dynamic community life and should be in a better position to know what resource management activities will and will not work with them. The PLA is action oriented and emphasizes learning by doing. Group discussions and activities are the bases of the learning process, while the management decisions made are imple-

4. R. Chambers 1992. "Rural Appraisal: rapid, relaxed and participatory," International Development Studies (IDS) Discussion Paper 311, p. 5.
5. J. Carew-Reid, 1989. "Environment, Aid and Regionalism in the South Pacific," Pacific Research Monograph No. 22, National Centre for Development Studies, Australian National University, Canberra; P. Schoeffel, 1996. "Sociocultural Issues and Economic Development in the Pacific Islands," Pacific Studies Series, Asian Development Bank, Manila.
6. Chambers (n. 4 above), p. 15.

mented through the involvement of the local communities and their external partners.

The PLA methods have been used in the Pacific Islands and Asia to involve local communities in managing their marine resources.[7] Parks and Salafsky[8] described the use of the approach to assist people to formulate ways of addressing their marine resource management problems. The experiences provide useful lessons that must be considered by those who are dealing with similar situations. The use of the PLA method is appropriate in training individuals from the communities who can continue to formulate, implement, and follow up on resource management arrangements that meet their needs.[9] This is a challenge because the involvement of people in community-based resource management requires a great deal of consultative work, goodwill, trust, convincing, and commitment. The process can not be shortened and requires patience and understanding.

The PLA methods emphasize the active participation of local people to strengthen their capacity to learn and act.[10] The PLA methods emphasize implementation and the importance of monitoring to the PRA principles. The PLA consists of three phases: before the workshop, the PLA workshop, and after the workshop. There are five steps: two steps each under phases 2 and 3 and one in phase 1. The steps emphasize the important parts of the process that need to be carefully addressed. Step 1 under phase 1 is on background research and coordination, which is needed to ensure that the preparation is relevant and appropriate. For this reason, the IAS team was purposefully made up of people who could converse in the vernacular and therefore better appreciate the position of the people they are working with. In addition, the use of the vernacular will encourage the people to raise issues that matter to them. Steps 2 and 3 under phase 2 are on training the PLA practitioners and PLA fieldwork, while steps 4 and 5 under phase 3 are on analysis and follow-up, respectively.

The training is essential to promote the use of the method and ensure that the training is continuous. The PLA fieldwork emphasizes that the workshop and their outcomes should be set in a context. In many instances, re-

---

7. WWF (World Wildlife Fund for Nature) South Pacific Program, Community Resource Conservation and Development: a toolkit for community-based conservation and sustainable development in the Pacific, (Suva: WWF South Pacific Program, 1996).

8. J. Parks and N. Salafsky (eds.), "Fish for the Future? A collaborative test of locally-managed marine areas as a biodiversity conservation and fisheries management tool in the Indo-Pacific region," Report on the Initiation of a Learning Portfolio. (Washington DC: World Resources Institute, 2001).

9. ECOWOMAN, 2000. Participatory Learning and Action: A Trainers Guide for the South Pacific. ECOWOMAN and SPACHEE, Suva. p. 1.

10. Ibid.

454 Environment and Coastal Management

source management arrangements are finalized externally and are taken to the communities to adopt and observe them. This is difficult to achieve let alone enforce. As with national laws, people are often noncommittal and there are needs for enforcement measures to support these initiatives. The analysis in step 4 thus is important to illustrate the significance of information and data gathered from the people who need to be convinced that the decision made is directly related to the discussion in the workshop. Furthermore, the people make the decisions and therefore will be committed to their implementation. Lastly, step 5 on follow-up is critical because the people need to be convinced that the method is responsive to their needs and pleas and that the changes that are agreed to and implemented are to produce the desired results. The workshop participants are required to promote the messages from the workshops within their communities and need to be supported by their external partners through follow-up activities.

Before the Workshop

The organization of the marine resource management workshop in Votua, Fiji, marked the cultivation and nurturing of an initiative that had originated within the community. One influential elder in the village had read in the media about the Community-based Resource Management undertaken by the IAS in Verata tikina (a local district) of Fiji and sought the assistance of the IAS. This process is significant because of the PLA emphasis on local initiatives and self-determination. This show of commitment is important because local communities need to be convinced of the importance of an initiative if they are to be committed to it. Members of the LMMA network believe that this is a prerequisite for success as the villagers feel that they own the initiative and are committed to it. In this case, the elder, who was the head of the village Fisheries Committee, which was responsible for the management of the villagers' fishing activities, was convinced that the IAS was the organization to assist them in their quest to better manage their fisheries resources. The IAS had been involved in the Biodiversity Conservation Network (BCN) project for the formulation of a community-based resource management plan in Verata.[11] The management activities in Verata included the establishment of *tabu* (total no-take or species-specific) areas and the training of villagers to do resource monitoring to provide evidence of the regeneration of the resources in the managed areas.[12]

11. Biodiversity Conservation Network (BCN), "Summary Report on the 1997 Verata Tikina Marine Resource Monitoring Workshop," Ucunivanua. 22 April–2 May 1997.
12. A. Tawake, J. Parks, P. Radikedike, W. Aalbersberg, V. Vuki and N. Salasfsky, "Harvesting Clams and Data: involving local communities in implementing and monitoring a marine protected area. A case study from Fiji," Submitted to Conservation Biology in Practice (in press); A. Tawake and W. Aalbersberg, "Community-

For the Votua project, the IAS secured a grant of F$10,000 from the United Nations Development Programme's (UNDP) Small Grants Scheme to consult the villagers and formulate a marine resource management plan.[13] As part of the preparation for the workshop (phase 1 and step 1), consultative meetings were organized as the IAS team members traveled to Votua to get acquainted with the villagers by attending village meetings. Local government officials in Suva and the Ba area including the Assistant *Roko* and fisheries extension officers were invited as part of the integrated IAS team. Community representatives were also brought to meetings organized in other parts of the country where community-based resource management initiatives were discussed. For example, IAS invited three people from Votua, including the elder, to a training for trainers' workshop at the University of the South Pacific where the people of Verata discussed the results of their marine resource management activities.[14]

A larger grant was secured from the Packard Foundation to support the work in Votua and extend the activity to four or five other districts and villages in Fiji. The funding arrangement was notable in its flexibility, which allowed close consultation that ensured that the project and its activities are properly introduced to the people. The process allowed the people to determine how they want to address the problem and more importantly to agree as a group on their course of action. This collective agreement and consensus distinguish the "Pacific Way" of doing things, which is unanimously adhered to throughout the region. Such decision-making processes are significant because everyone is part of the decision making, which in turn ensures support for the decision and identification of the course of action. Unfortunately, this consensus has not been properly attained with many of the hurriedly put together development assistance activities that involve local communities.

The preliminary consultation with villagers in Votua was delicate and sensitive as the IAS team tried to involve all the villagers in the decision making. This is a challenge that is faced in all areas where outsiders are working with local communities. In Votua, it was evident from the beginning that there were different groupings in the village. The village was divided into three main *mataqali* (social units consisting of closely related extended family groups), each of which was led by a head of *mataqali*. The consultative meetings were tricky because the facilitators were mindful that their initiative should not be divisive in the community. It was intended that the facilitators would win the confidence and trust of all the people to allow them to

---

based Refugia Management in Fiji," Paper presented at the Pacific Science Inter-Congress in Guam, June 1–6, 2001.

13. E. Radrodro, "UNDP to help Ba Province Marine Project," The Fiji Sun, 21 March 2000, 5.

14. Tawake et al. (n. 12 above); Tawake and Aalbersberg (n. 12 above).

openly discuss their concerns. The facilitators were also aware of the dangers of putting forward certain individuals in the village, including the elder, who may have ulterior motives to highjack the initiative.

The consultative visits over an eight-month period were important to ensure that the formulation of a community-based resource management plan involved the majority of the villagers. The Ba Provincial office advised the IAS team about the situation in the village and how they should proceed. The consultative meetings were opened for everyone but the first one was attended by only about 15 men. The second and third meetings were better attended, but still there was a significant number absent and there were few women. It was noted that different people were attending different meetings and that there were social divisions and groupings within the village.

Throughout the consultative meetings, it was emphasized that the decision on whether the workshop took place rests entirely with the villagers. It was also emphasized that the community effort could only be effective in pursuing its resource-management plans if the villagers as a whole worked together. It was evident that the influential elder was not representing the whole community. The elder's support was in the course of the work diminished to the point of being withdrawn, but in its place the rest of the community became more unified and active.

Concerns raised during the consultation meetings related to the timing of the workshop and the toilet facilities in the village, which were considered unsuitable for visitors. Questions were also raised about the catering of participants and their composition. The timing of the workshop was related to the fact that the chosen time fell in the middle of the cane-harvesting season, which would affect the attendance. In addition, there was concern that all three *mataqali* in the village and those from the nearby villages of Nawaqarua and Natutu, with whom the people of Votua share the use of the fishing grounds, should be involved in the workshop.

The toilet facilities were an issue because of the request made during the first meeting by the elder that the IAS team fund a toilet block for the village. This request embarrassed the rest of the villagers who argued that the villagers should be responsible for their toilet and that the IAS team should be invited only to conduct the training for trainers' workshop as proposed. Similarly, it was decided in the last consultative meeting that the villagers would cater for the facilitators as well as the participants at the meeting. Each of the *mataqali* was to host the meeting participants for a day.

The consultative meetings were flexibly organized between the IAS team, the Provincial office and the villagers. On a number of occasions, the scheduled meetings were deferred due to a death or other social events in the village or some prior commitment at the university. All activities at this stage were geared toward gaining acceptance of the IAS team by the villagers in order to allow for meaningful collaboration.

Participatory Learning and Action Workshop

Between August 29 and September 1, 2000, 25 people from the villages of Votua, Natutu, and Nawaqarua participated in a four-day workshop that aimed at formulating a community-based marine resource management plan for the villages. Despite their huge customary fishing ground, which extends to the Yasawa Islands covering approximately 1,347 square kilometers of ocean area, the people of Votua were convinced of the impoverished nature and overexploited state of their fisheries resources. After years of using their fishing grounds and granting many commercial licenses, the villagers were aware of the impact of their indiscriminant use of dynamite, extensive employment of gillnets, and increasing contamination from other people's upland economic activities that included logging, sugarcane farming and milling, and Ba town.

The participants at the workshop were from different ages and positions and represented their peer groups in the discussion. About 50 percent of the participants were women. There were equal numbers of elders and youths. These participants would promote the matters agreed to at the workshop to their colleagues in the villages. The resource people consisted of scientists and social scientists from IAS, government officials from the Ministries of Fijian Affairs and Agriculture, Fisheries and Forestry and colleagues from NGOs. The resource people were conversant in the Fijian language and knew about the people and their situations.

The workshop using the PLA methods commenced after the Fijian cultural formalities of making special announcements and exchanging welcome between the visitors and their hosts were completed. The workshop activities began with the discussion of the objectives and the rationale of the workshop. It was emphasized that the community-based management activities would be as the villagers wanted.

Sessions on the first day focussed on the definition of a marine resource management plan, the use of resource mapping, marine resources transects, and the historical profile. Issues that were discussed included marine ecology, marine resource impact flow, and relationship diagrams. The discussions also covered the community marine resource survey, the vision map and a monitoring plan. The Votua community action plan listed the activities that the participants agreed would enable them to address the threats to their fisheries resources. These discussions allowed the people to conceptualize the changing situation with their fisheries resources and what was required to ensure their sustainable use.

On day two, the participants took a field trip to the shoreline and reef. The exercise was important to ensure that the people understood the resources they were dealing with and allowed the IAS team a chance to see the nature and condition of the fishing ground. The rest of the day was spent doing the marine resource analysis, time line of abundance and diversity of some of the resources and marine ecology, emphasizing the impacts of broken food pyramids.

Day three was spent on discussing the critical resource management issues and their likely solutions. It was emphasized that the management activities include measures that people felt could address the threats that were causing the problems. Threats such as the use of destructive fishing methods, the increasing number of fishers, and waste management were discussed. In the afternoon, a proposed monitoring plan was introduced.

The discussion on monitoring continued on day four. Monitoring, it was explained, was required to demonstrate the recovery of the resources. In addition, the monitoring would allow the collection and analysis of scientific data. Time was also spent discussing the benefits of locally managed marine areas and the impacts of management activities. Locally managed marine areas or marine protected areas are more suitable in Fijian communities than the marine parks because people rely on their marine resources for livelihood and subsistence. Marine parks are permanently no-take zones, which is difficult to observe in subsistence societies like Votua. In addition, the concept of user pays would be difficult to enforce in such societies. On the other hand, the concept of marine protected areas is suited to the tradition of periodically prohibiting the use of a resource whenever some form of management is needed.[15]

Although the IAS team formulated the workshop structure and chose the topics for discussion and the tools to be used, the details on the resources, their usage and the management activities were obtained from the people. The participants were divided and worked in groups that related to their role in the community. The groups of five to six members, which remained the same from day one to four, were based on the main fishing habitats the participants used: mangrove and river, lagoon, reef and deep sea. In addition, there was also a group made of up of village elders who were too old to do any fishing but who were the important decision makers. Under each of the topics discussed, the participants related issues that influenced their activities in their chosen habitats. The contributions from each of the groups were then presented and discussed in the plenary sessions where the rest of the groups commented on the proposals from the presenting groups. The result of the workshop was a wealth of information on all the major fisheries in Votua.

The village group discussed the role of decision makers in the village in determining how the fisheries resources are used and managed. The group also explored the different ways in which effective management could be achieved. The mangrove and river group discussed the fresh water mussel

---

15. J. Veitayaki, "Traditional Resource Management Practices in the Pacific Islands: an agenda for change," Ocean and Coastal Management, 37, 1: 123–136.

(*Batissa violacea; kai ni waidranu, tave*) and crab fisheries within their areas and made suggestions relating to their management. Likewise, the lagoon, reefs, and the deep-sea groups explored the different marine resources in their areas and the best ways of addressing the issues related to their management. It was interesting to find the different products of the different sections of the fisheries and the different users. Furthermore, there were distinct levels of specialization. For example, most of the older women specialized in freshwater and crab fishing while the younger men were involved in net fishing and the older men used handlines.

The main threats that the participants mentioned included overfishing, unauthorized fishing, use of destructive fishing methods, waste disposal, and the pollution of riverside areas. The threats included those that were caused by the villagers and those caused by people with whom the villagers shared their fishing grounds. The participants were aware of their increasing population and acknowledged their use of destructive fishing methods such as dynamite, gillnets and fish poison, but were unfamiliar with the impacts.

A Resource Management Plan for Votua was formulated at the end of the workshop. The proposed management activities that the participants agreed to was fascinating because they included action that could be taken by the community to reduce the impact of the threats that were identified to ensure the health of the fisheries and their habitats. This part of the workshop demonstrated that the people knew what they wanted and what they are capable of doing without government input. It was also decided that the workshop proposals be reported to the three villages in the district and if approved by these bodies to be adopted as the villagers' plan for the management of their marine resources.

After the Workshop

The Resource Management Plan proposed by the workshop was endorsed by the village and *tikina* and is now incorporated into what the villagers perceived as the main solution to the identified threats to their fisheries resources. The people agreed that about 5 percent of their total fishing areas be declared protected (*tabu*) or prohibited. This area, which amounts to 65.8 square kilometers, includes 17.25 square kilometers of reefs, 38.45 square kilometers of lagoon areas and 10.13 square kilometers of mangrove forests. The protected area extends from the mangrove forests through the lagoon and on to the reef area and functions as a sanctuary to stimulate the recovery of the resources in surrounding areas. The area is to be monitored by the people as well as the IAS to illustrate the effects of no-take on the recovery of the fish stocks.

To complement the recovery effort, the villagers also agreed to reduce the number of licenses offered for their fishing grounds. In 2001, only 35 licenses were approved for people from outside the community, compared with the 70 Inside Demarcated Area (IDA) licenses that were offered in 2000. The villagers were aware of the reduced income but are convinced that the state of the fishing grounds can only improve if fishing is reduced. The villagers also decided to offset the loss by raising the price of each license, but the increase is not based on any rational economic basis. In addition, the use of gillnets was to be allowed in five cases. The rest of the licensees are to use only lines. The villagers also banned the use of small gillnets that are considered destructive because they catch undersized fish.

The villagers also decided to exert more control on their fishing areas by organizing more surveillance and patrol. Six months after the formulation of the management plan, the villagers had apprehended four people who were illegally fishing within Votua waters. In one instance, a compressor was also confiscated.

The villagers have also agreed to use better waste disposal methods and asked the IAS to do water monitoring studies on the Ba river because of upstream land use activities.

The villagers have now undertaken resource management activities over the last 24 months and have completed their first monitoring training. The villagers are claiming to have seen positive signs relating to the recovery of their fisheries. This was proven to be the case when the biological survey was conducted in February 2002.[16] In addition, the people are continuing to seek advice from the IAS on how they use their fisheries resources. In July 2001, the people invited the IAS team to advise them on a proposal they were considering for the community to be involved in the aquarium trade. The experience in Votua has shown how the people in the communities can contribute to the effective management of their marine resources and demonstrate what local communities can do with assistance from outside partners. The recovery of the fisheries may be still a long way away, but the initiative is indicative of what the people can do to better use their marine resources.

LESSONS LEARNED

The participant turnout at the workshop was very encouraging given the different groups that were represented. The people who attended represented a good cross-section of the community in Votua. This increased the probability that the management decisions the group made could quickly be implemented, as people from all groups and backgrounds were involved in the decision, and thus would be supportive of their implementation. The

---

16. Results of this study may be found at the following Web site: www.lmma-network.org.

involvement of the elder men and women was as welcome as that of the men and women who were doing the bulk of the fishing because of their role in decision making in the community. The workshop discussions allowed for some meaningful dialogue and the sharing of knowledge and information that affected the people.

The villagers have relayed their marine resource management decisions to the other levels of the local and provincial governments where these decisions have been supported and upheld by those that are affected by these decisions, including those who are excluded from fishing.

The awareness and empowerment witnessed as result of the workshop was revealing. It is true that the people who used and owned the resources are the best to determine what is wrong with the fishing ground and the associated strategies to address these. At the workshop, it was not difficult for the villagers to identify the main threats and propose remedies for addressing the problems. For example, the participants overwhelmingly supported the need to reduce the number of licenses offered to outsiders. The 70 IDA licenses provided an average annual income of $13,000 to the village. The people have reduced the licenses by about 50 percent to allow the resources to regenerate.

There were still some misconceptions to be corrected, however. For example, there was the belief amongst some of the people that the resources would regenerate as they had always done in the past. Fortunately, the villagers still remembered their recent experiences with bêche-de-mer and how the resource quickly dissipated due to intensive use. This illustration is fresh in people's memory because they remembered how quickly the bêche-de-mer disappeared at the height of the trade in their area and the long time it has taken since then for them to find any sizeable animal within their reefs. The villagers remember the influx of traders and how quickly they disappeared after the bêche-de-mer were fished out.

The villagers blame others for the deteriorating state of their fisheries. In doing so the people overlooked their own role in terms of the higher population, their wastes and their use of destructive fishing methods. The village people are fishing daily and are the ones offering the licenses. People also use dynamite, which they believe is less destructive than the use of gill nets. According to the villagers, the continuous use of dynamite in the past was not as destructive as the continuous use of monofilament gillnets.

The villagers are fascinated by the interrelations within the ecosystems. The interchange between the mangroves and the reefs, and the interdependence of organisms that are part of these ecosystems were unknown to villagers. This was the reason why the people were happy at the end of the workshop to declare a protected area that covers the mangrove, seagrass beds, and adjacent reefs. People realized that the protection of their fisheries resources required that these intricate interrelationships be understood. Votua is located at the mouth of the heavily silted and polluted Ba River. The forest, milling and farming activities in the upland areas of the Ba River, as well as the sources of pollution from the sugar cane farms, the Rarawai Sugar Mill and Ba Town need to be addressed to complement the attempt in Votua to better manage their fisheries resources.

It is imperative that the people appreciate the vulnerable nature of their environmental resources and address the problems associated with their management. It is rewarding to see how easily people can be mobilized when they are properly empowered. Furthermore, the people have embraced the value of self-determination. For instance, after all the controversy surrounding the toilets, the people in the end used the subsistence allowance that the IAS team gave them for looking after the workshop participants to build a toilet block next to the village hall.

However, people must not just blame outsiders for the destruction of their resources. The depletion of bêche-de-mer is a strong reminder of the need to adapt management measures in the village. The villagers need to face the truth regarding their role in the degradation of their fisheries resources. Moreover, it seemed that a better system is required in the offer of consent for license. At the moment, different people in Votua offer fishing licenses with no set guidelines on the total allowable catch from the area. Furthermore, the resources of the village's Fisheries Committee that looks after the fishing license fees need to be better managed. The bottom line is that the fishing grounds belong to the people who should be aware of their vulnerability and must support the attempts to manage their resources. However, the management activities must be executed in a fair, transparent, and firm manner.

CONCLUSION

The workshop in Votua was successful in empowering the people to manage their marine resources. The discussions were open and free flowing and the decisions were unanimous. It was based on the philosophy of providing people with the alternatives and encouragement to enable them to decide on how they want to manage their fisheries resources. Judging from the comments passed in the workshop and how we were treated, catered for and how things have progressed since the workshop, the rationale was well received.

The villagers have decided that the IAS team works with them in their management of their marine resources. The IAS team has conducted the training in resource monitoring skills and has conducted the biological survey. The IAS team will return for a workshop on social and economic monitoring. In addition, the IAS team is exploring the possibilities of alternative income. After close to 24 months of observing their community-based resource management area, the people are testifying to the recovery of the fish stocks within the no-take zone and surrounding areas. In line with the

PLA philosophy of involving and empowering local communities, the decisions on what the people must do to manage their activities and their fisheries must remain with them. Outsiders like the IAS team and the donors must only facilitate the process.

Major challenges remain, however. Although the people have agreed to the implementation of the management plan, the challenges would be to control the temptation people in and around the area face in observing the conditions within the no-take area. The temptation is expected to increase in time as the resources within are expected to be more abundant than the adjacent areas. For this reason, the protected area needs to be clearly marked and widely publicized. Everyone, including the people from outside Votua, must respect the no-take area. In addition, the people must have the means to enforce and patrol their protected area.

In time, people will need to decide on honorary fish wardens to administer their protected areas. These wardens must have the equipment and resources to effectively conduct their work. These requirements however must not burden the community, which may find that the declaration of the protected area is the easiest part of management.

The workshop in Votua illustrated how the people in communities discuss the information, analyze and use it to determine the course of action they take. The PLA method worked well based on the decisions that were taken for the management of the resources and the relations that exist now between IAS and the people of Votua.

In an attempt to mainstream the lessons learnt from the initiative in Votua, the IAS has joined a learning portfolio on locally managed marine areas. Through this network, the IAS shares the lessons from its site such as Votua with other practitioners and donors. The satisfaction of doing this type of work is tremendous and is associated with the knowledge that the people are addressing their resource use problems using the methods they have chosen. The local communities are quick to act because they are familiar with what they are required to do and the reasons. The people are acting responsibly without waiting for a government directive because they would be the main beneficiaries. The experience is enriching to all.

Security and Military Activities

Technology Cooperation and Transfer, Piracy and Armed Robbery at Sea: A Discussion Paper in Two Parts for UNICPOLOS II[†]

Elisabeth Mann Borgese and François N. Bailet
International Ocean Institute, Halifax, Canada

INTRODUCTION

The UNICPOLOS I Report was considered by the General Assembly on 26 October 2000. Undoubtedly, it gave a new direction to the discussion of the Secretary-General's Report and to the extremely comprehensive resolution on oceans and the law of the sea adopted by the Fifty-fifth Session of the General Assembly.[1] Resolution 55/7 reaffirms the importance of the annual consideration and review of developments relating to ocean affairs and the law of the sea by the General Assembly as the global institution having the competence to undertake such a review, and took note of the outcome of the first meeting of the United Nations Open-ended Informal Consultative Process on Oceans and the Law of the Sea (UNICPOLOS).[2] Among its numerous recommendations, the Resolution stresses the need to consider as a matter of priority the issues of marine science and technology and to focus on how best to implement the many obligations of States and competent international organizations under Parts XIII and XIV of the 1982 United Nations Convention on the Law of the Sea (UNCLOS).[3] The Resolution also urges all States, in particular coastal States, in affected regions to take all necessary and appropriate measures to prevent and combat incidents of pi-

---

[†] EDITORS' NOTE.—This article is an edited version of the paper "UNICPOLOS II: A Discussion Paper Compiled by the International Ocean Institute," presented by Elisabeth Mann Borgese at the second meeting of UNICPOLOS in New York from 7 to 11 May 2001.

1. Published as Draft Resolution A/55/L.10. Adopted by the General Assembly on 30 October 2000 and issued in final version as A/RES/55/7.

2. Report on the Work of the United Nations Open-ended Informal Consultative Process on Oceans and the Law of the Sea at its First Meeting (A/55/274). Letter dated July 28, 2000 from the Co-Chairpersons of the Consultative Process addressed to the President of the General Assembly. The first meeting of UNICPOLOS was held on 30 May–2 June 2000.

3. Resolution A/55/L.10, para 32.

racy and armed robbery at sea, including through regional cooperation, and to investigate or cooperate in the investigation of such incidents wherever they occur and bring the alleged perpetrators to justice in accordance with international law.[4]

It reaffirms its decision to undertake an annual review and evaluation of the implementation of UNCLOS and other developments relating to ocean affairs and the law of the sea, taking into account the establishment of UNICPOLOS[5] to facilitate this review. It requests the Secretary-General to convene the second meeting of UNICPOLOS in New York from 7 to 11 May 2001[6] and recommends that, in its deliberations on the report of the Secretary-General, UNICPOLOS should organize its discussions around two specific issues, that is, marine science and the development and transfer of marine technology as mutually agreed, including capacity building in this regard, and coordination and cooperation in combating piracy and armed robbery at sea.

The selection of these two issues is in accordance with recommendations J and K of the "Output" of UNICPOLOS I. Both of these issues are extremely timely. Both will have to be discussed "with an emphasis on identifying areas where coordination and cooperation at the intergovernmental and interagency levels should be enhanced."[7]

Capacity building and technology cooperation and transfer are now splintered in the marine sciences. UNCLOS had envisaged a system of technology cooperation that reached from capacity building at the national level, providing the essential basis for international cooperation, through the regional level, where Articles 276 and 277 mandated the establishment of regional centres, to the global level, where the specialized agencies—especially the Intergovernmental Oceanographic Commission of UNESCO (IOC), the United Nations Industrial Development Organization (UNIDO), the Food and Agriculture Organization (FAO), the United Nations Environment Programme (UNEP), and the International Maritime Organization (IMO)—should have made their contributions. But coordination and integration of efforts have left much to be desired. The United Nations Conference on Environment and Development (UNCED) conventions, agreements, programmes, and action plans, each limiting capacity building and technology cooperation to sectoral fields, have further complicated the picture. Within the limits of the strictly sectoralized structure of the UN institutions and secretariats, it is indeed immensely difficult to initiate a really integrated approach. A breakthrough is needed. The time has come to take up the

4. Ibid., para 33.
5. UNICPOLOS was established by the General Assembly on 24 November 1999 in its Resolution 54/33 to facilitate the annual review by the Assembly of developments in ocean affairs.
6. Resolution A/55/L.10, para 40.
7. Resolution 54/33.

challenge. New approaches are now possible within the framework of the "revitalization of the Regional Seas Programme," using the implementation of the Global Programme of Action for the Protection of the Marine Environment from Land-based Activities (GPA) as a trigger mechanism. *One* system, regionally decentralized, for capacity building in the sciences and technology cooperation and transfer should and could now be designed *to serve the needs of all conventions, agreements, codes, programmes, and action plans.*

The International Ocean Institute (IOI) has done policy research on such a system for many years. UNICPOLOS II has a unique opportunity to make a real breakthrough and get it followed up by the Fifty-sixth Session of the General Assembly.

Similar arguments could be made with regard to the second topic. The suppression of piracy and armed robbery at sea has become an urgent matter of vital importance for the implementation and enforcement of the whole UNCLOS/UNCED process. The IMO has been designated as the lead agency and has contributed, and continues to contribute, most valuable studies on the subject,[8] thus providing a basis for action. But action is needed—new forms of regional cooperation between navies and coast guards—if the problem is really to be solved. If crime has been "globalized," so too must be the suppression of crime. Experience has already demonstrated amply that individual States, especially if they are small and poor and responsible for very large pieces of ocean space, are not able to cope with the problem. In the Mediterranean, the Euro-Mediterranean Process has had some discussions on the very forward-looking but highly controversial proposal for the establishment of a Mediterranean Regional Coast Guard. This is a concept that deserves to be studied in this context, and perhaps the IMO could be requested to prepare a Protocol for its implementation.

One should look at it, however, from the point of view of integration. Obviously, piracy and armed robbery are not the only crimes at sea, and in many cases they are linked directly to other crimes at sea, such as drug trafficking. If a cooperative instrument for implementation and enforcement were to be created at the regional level, it could do far more than suppress piracy and armed robbery. It could enhance the whole process of implementation and enforcement, including the suppression of other internationalized crime such as drug trafficking and the illegal transport of persons or hazardous materials, and the provision of humanitarian assistance and intervention.

Again, all these sectors are covered by separate convention regimes, but at the operational level of implementation and enforcement, they should

---

8. For example, IMO-Maritime Safety Committee, Piracy and Armed Robbery Against Ships: Guidance to Shipowners and Ship Operators, Shipmasters and Crews on Preventing and Suppressing Acts of Piracy and Armed Robbery Against Ships (MSC/Circ.623/Rev.1) (London: IMO, 16 June 1999).

be integrated into one system of multipurpose naval cooperation. Such an approach, "with an emphasis on identifying areas where coordination and cooperation at the intergovernmental and interagency levels should be enhanced," would also mean a continuation and development of the subjects covered by UNICPOLOS I: for both the suppression of illegal, unregulated, and unreported (IUU) fishing and the mitigation of the economic and social consequences of pollution through improved surveillance and enforcement would greatly benefit. Here again, UNICPOLOS II has a unique opportunity to be innovative and initiate a breakthrough.

The two IOI studies following this introduction are intended as a contribution to the discussions. One of the IOI's basic guidelines is to think ahead of governments—not so far as to be utopian, but far enough to be stimulating. The models we are proposing are merely illustrative, but they should contribute to directing thoughts to possible futures, implicit in numerous official statements and recognitions, but not yet spelled out. The useful function of models of this sort is to make issues concrete and serve as a basis for discussion.

PART I. TECHNOLOGY COOPERATION AND TRANSFER: THE NEW INTERNATIONAL TECHNOLOGICAL ORDER

Background

The United Nations Convention on the Law of the Sea, the "Constitution for the Oceans," has fundamental implications for the future. Based on the recognition of the unity of ocean space and the systemic interaction of all its subsystems as well as on the revolutionary concept of the Common Heritage of Mankind, it implies and indicates the direction toward a new sociopolitical, community-based world order, a new economic system, transcending both centralized and market-based systems, and a new international technological order, founded on a new science paradigm and high technology,[9] qualitatively different from the old technologies, on which the

9. Marine technological developments are intimately connected with other high-tech endeavours especially in the fields of microelectronics, robotics, genetic engineering and biotechnology, space and lighter-than-air technologies, new materials technology, and superconductivity. There is a need to monitor developments in these fields for their complementary and competitive impacts. The complementary impacts would be the developments they could generate in marine technology (e.g., space technology and microelectronics on environmental pollution monitoring, biotechnology on leaching of minerals and on mariculture, and superconductivity on submarine cables for the transmission of electricity). At the same time, the impact of these developments on terrestrial industries has also to be taken into account. For a resource to be exploited, not only does it have to be accessible with the technology available to exploit it, but exploitation also has to be economic. In the case of ocean resources, it would mean extracting resources at an economic price competitive with

previous international technological order was founded. The well-known differences were recapitulated in our previous contribution to UNICPOLOS I.[10] In previous phases, technology was resource- and capital-intensive; it was massive; it could be sold, bought, and transferred in a one-time transaction, and neither the producer nor the consumer was concerned with what it might do to the environment; it could be serviced by a mass of fairly untrained workers in a centralized system of production. Today's technology is knowledge- and information-based. It cannot really be "bought" and "transferred"; it has to be "learned." It is far less resource intensive, because, the more effective it gets, the smaller it grows ("miniaturization"). It calls for an ongoing relationship between the producer and the consumer. The consumer gets involved in the product design and has to learn, often through extensive training, the use of the technology, its maintenance, repair, and upgrading. Producer and consumer in a way are linked in an ongoing joint undertaking for the duration of the product. The longer this duration, the higher will be the utilization value of the product, which becomes more of a "process" than a "product" in the old sense. The only effective method of technology transfer in this new phase of the industrial revolution is the joint venture in research and development. Such joint ventures became an important component of the new industrial system within and among industrialized States in the 1980s. Research and development (R&D) in high technology is extremely costly. And not only is it costly, it is also extremely risky, especially in the early phases. The rate of failure has been estimated variously at between 7 to 1 and 20 to 1, or even 40 to 1. The farther back in the process of invention we go, the more overwhelming the rate of failure.[11]

This, among other things, triggered the trend toward the formation of R&D consortia, to share and reduce the cost and spread the risk inherent in R&D in high technology during the early phase—the "precompetitive" phase, as it is termed. At the same time, we saw an increasing involvement of the banks first and then of governments in the financing of this R&D. Even in the United States, the staunchest defender of private enterprise, over 50 percent of R&D today is paid for by the federal government. Without government participation, none of the modern high technologies would have passed the first stage of research and development. In this sense, the boundaries between public sector and private sector are getting blurred.

For most countries, however, even this public-private co-investment was

---

resources from terrestrial sources. So the increasing efficiency of terrestrial mining due to technological advance would have a competitive impact on ocean technology. See K. Saigal, *Feasibility Study* (Malta: International Ocean Institute, 1988).

10. E. Mann Borgese, *The UNCLOS/UNCED Process: A Comparative Study of Eight Documents* (Halifax: International Ocean Institute, 2000), Part 3, p. 65.

11. M. B. Spangler, *New Technology and Marine Resource Development* (New York: Prager, 1970).

not strong enough to make technology development competitive on an international scale. Thus, what we saw emerging was a new form of international public-private cooperation, exemplified by systems like the European programme, EUREKA, and half a dozen others. The growth in international, interfirm technical cooperation agreements represented one of the most important novel developments of the first half of the 1980s.[12] The sharing of risk and cost served in many cases to encourage firms to spend more on R&D than they would have done otherwise. The synergy between private and public investments at the regional, European level generated billions of dollars of investments in R&D, and this is a point not to lose sight of.

This development remained, however, restricted to the industrialized States. Less than 3 percent of all funds invested in science and technology throughout the world was allocated to projects being executed in developing countries. More than 90 percent of scientists and experts lived in industrialized countries and 93 percent of patents were taken out in these countries.

The picture changed in the 1990s. Joint ventures in research and development have been overtaken by the sweeping merger movement and globalization—with benefits, clearly, going to the stronger partners. The rich are getting richer, the poor, poorer.

At the same time, however, the proportion between scientists in developed and developing countries is rapidly changing. The number and the quality of scientists trained in China and India or Brazil, Mexico, and Cuba is awesome. And, although most of the smaller and poorer countries are still left out of this development, it is safe to predict that, in another decade or two, the proportion between scientists in developed and developing countries will be reversed and the majority of experts and scientists will be from developing countries. This will undoubtedly affect the philosophy of science, which presently has undergone some fundamental changes in the West, as well as the direction of research and development.

Technology today is knowledge- and information-based. It depends on the development of human resources. Human resources are what developing countries have, and the time has come for a New International Technological Order. The legal framework for this has already been created, starting with UNCLOS. The challenge now is to develop the institutional framework for the implementation of this legal framework.

The New International Technological Order Emerging from the United Nations Convention on the Law of the Sea

The framers of UNCLOS were fully aware of the importance of science and technology development and technology transfer. It may be enough to re-

---

12. M. Sharp and C. Shearman, *European Technological Collaboration* (London: Royal Institute of International Affairs, 1987).

member that almost one-third (about 100) of the 320 articles of this Convention touch in one way or another on science and technology. Technology development and transfer are provided for in three ways—at the national, regional, and global levels.

The National Level

First, the Convention imposes on the "competent international organizations" (FAO, IOC, IMO, UNEP, and others) the duty of assisting developing countries to acquire the technologies they need to benefit from, and comply with, the provisions of the Convention. Thus Article 202, in Part XII of the Convention that covers the protection of the marine environment, provides that each State, directly or through competent international organizations, shall promote the training of scientific and technical personnel in developing countries, supply them with the necessary equipment and facilities, and enhance their capacity to manufacture such equipment. Article 271 provides for international cooperation for the development and transfer of technology through existing programmes or new programmes to be established, bilaterally, or through the competent international organizations. There are half a dozen other articles calling for these organizations to assist developing States.

The efficacy of these articles is, of course, limited by the budgets and organizational capacities of these organizations, which, in fact, are quite inadequate. While they have done their best, they have not really been able to narrow the technology gap. The important point, however, is that the Convention fully reflects the awareness that technology development must start at home. Developing countries themselves must lay the foundation: they must build, as a matter of priority, the scientific and technological infrastructure on which international cooperation can be based. It is a matter of goal setting and political will. *Pacem in Maribus XVI,* in Canada (1988), took up a recommendation made by the Third World Academy of Science:

> Every developing country should earmark a certain percentage of its educational budget for the advancement of science and technology, including marine technology. The Third World Academy of Science recommends that 4 percent of the educational budget should thus be earmarked for fundamental science, another 4 percent for applied research, and 10 percent to research and development.[13]

> The building of national infrastructure might include a policy-making and implementing agency; the building, under the auspices of such an

---

13. See "Recommendations" in *Ocean Technology, Development, Training and Transfer: Proceedings of Pacem in Maribus XVI August 1988,* eds. J. Vandermuelen and S. Walker (Oxford: Pergamon Press, 1991).

agency, of a strong industrial information system; the establishment of engineering design and consultancy organizations; and the establishment of R&D laboratories to provide specialised advanced training, do applied research, and assist the policy-making agency and industrial enterprises in identifying, selecting, and negotiating with foreign technology suppliers.[14]

These recommendations were based on the Indian experience, which, as is well known, has been one of the most successful in the development of marine technology in, for example, seabed mining, offshore oil production, Antarctic research, and remote-sensing applications.

In the 1970s, when the Convention was framed and the basis for the new international technological order was being laid, awareness of the importance of local, indigenous traditional technological experience was far less understood than it is today. Instances where ancient wisdom and experience is often ahead of modern science or, in many cases, reconfirmed by modern science include the utilization—medical, culinary, and otherwise—of genetic resources, the sustainable use of living resources, vermiculture, and water management, including irrigation. This source of knowledge most certainly should be fully utilised in national policy making. This can best be achieved, as is generally recognized today, through community-based co-management systems, an essential component of integrated coastal management, comprising technology management and development. The inclusion of local communities among the "major groups" participating in regional and global fora of decision-making today should encourage the development of socially and environmentally sustainable technologies, "eco-technologies" that recombine ancient traditions with modern science-based technology, thus enhancing efficiencies while facilitating social acceptability. It is on the basis of such national efforts that *regional cooperation*, both South-South and North-South, can become effective.

The Regional Level
Regional cooperation is provided for in Articles 276 and 277 of UNCLOS. These articles prescribe the establishment of regional centres for the advancement of marine sciences and technologies, in accordance with yet another resolution adopted by the Third United Nations Conference on the Law of the Sea (UNCLOS III), the Resolution on Development of National Marine Science, Technology, and Ocean Service Infrastructures.[15] This Reso-

14. Ibid.
15. United Nations Convention on the Law of the Sea, with Index and Final Act of the Third United Nations Conference on the Law of the Sea (A/CONF.62/122) (New York: United Nations, 1983), Annex VI to the Final Act of the Third United Nations Conference on the Law of the Sea. Accessed 10 March 2001 on the World Wide Web: <http://www.un.org/depts/los/>.

lution is important in that it expresses awareness of the rapid advances being made in the field of marine science and technology and the need for the developing countries to share in these achievements if the goals of the new ocean regime are to be met. And it warns that, unless urgent measures are taken, the marine scientific and technological gap between developed and developing countries will widen further and thus endanger the very foundations of the new regime. Regional centres, which foster South-South as well as North-South cooperation and are based on cost-sharing and economies of scale, can play a crucial role in narrowing this gap.

The scope of the activities of these centres is described in some detail in the two earlier-mentioned articles. They include the acquisition and processing of marine scientific and technological data and information. The range of technologies considered covers marine biology, (including the management of living resources), oceanography, hydrography, engineering, geological exploration of the seabed, mining and desalination technologies, as well as technologies related to the protection and preservation of the marine environment and the prevention, reduction, and control of pollution.

The Convention does not identify the regions within which such centres are to be established. Nor does it give a time schedule for their establishment or any indication as to how they should be financed. Quite a number of international—regional or global—institutions have been established over the past quarter of a century,[16] and all of them have been performing extremely useful functions. They have provided training to scientists and engineers from developing countries, facilitated networking among scientific and technological institutions, conducted pilot projects, and enhanced international cooperation, both South-South and North-South. But, apart from their financial limitations and the limitations entailed by their sectoral mandates, there is no linkage whatsoever between these institutions and Articles 276 and 277, which have not been implemented as yet. The only attempt to implement these articles, and thus to build an essential component of

16. For example, the International Centre for Genetic Engineering and Biotechnology, with headquarters in Trieste and Delhi, and the International Centre for Science and High Technology (ICS) in Trieste, both established by UNIDO. At the global level, a Multilateral Ozone Fund has been established to cover all "incremental costs" arising from the introduction of environmentally safe technologies. Similar arrangements have been made for the implementation of technology transfer under the Biological Diversity and Climate Change Conventions. At the regional level of the Mediterranean, one should mention the UNDP-initiated Centre for Environment and Development for the Arab Region and Europe (CEDARE) which was established to enhance European-Arab cooperation in sustainable development and which has a technology cooperation component. In Greece, a mechanism was established to facilitate technology cooperation in the private sector. In Spain, a centre for the advancement of environmentally sustainable technology was established and funded by the Government of Spain.

the new international technological order postulated by the Convention, was made by the International Ocean Institute a quarter of a century ago. It may be worthwhile to recall this effort. It was premature at its time, but its time may have come and it could easily be revised, updated, and adjusted to today's and tomorrow's technological, economic, and sociopolitical trends and needs.

The IOI launched its proposal in February 1987, at an international seminar in Malta, organized by the IOI in cooperation with the Foundation for International Studies and the Malta Oceanographic Commission. While aiming at the establishment of centres in each one of the Regional Seas, the proposal focused on a pilot project in the Mediterranean, considered the most advanced of the Regional Seas Programmes. The purpose of establishing a regional centre in the Mediterranean was

- to promote regional cooperation in the peaceful uses of the Mediterranean Sea;
- to promote the implementation of Articles 276 and 277 of UNCLOS that prescribe the establishment of regional centres for marine scientific and technological research to stimulate and advance the conduct of marine scientific research by developing States and foster the transfer of marine technology; and
- to encourage new forms of scientific-industrial cooperation between the developed and the developing countries in the Mediterranean region.

The proposal then gave a brief overview of marine technology cooperation in Europe, drawing attention in particular to the achievements of EUROMAR, the marine technology sector of EUREKA, which had succeeded in mobilizing billions of dollars of public-private investments within an international European framework. The problem was that this promising development was restricted to cooperation among European countries, all of them developed. The proposal suggested that, through the Mediterranean Centre for Research and Development in Marine Industrial Technology, this system could be broadened so as to include the developing as well as the then socialist countries on the southern and eastern shores of the Mediterranean.

The structure of the proposed centre was very simple. It was suggested that each Member State appoint its own national coordinator, to whom industrial enterprises would submit their project proposals. The national coordinators, the technical/advisory body of the Centre, would meet periodically to discuss these project proposals and recommend a selection to the Conference of Ministers. The Conference of Ministers would be the decision-making organ and make the final selection of projects. The Ministers would also appoint the Director General of the Centre and determine the Centre's general policy. Projects adopted by the Conference of Ministers would be financed equally by the industrial enterprises that made the proposal and the Governments of participating States, the latter with the assistance of regional organizations (such as the Trust Fund for the Mediterranean Action Plan) or regional divisions of international organizations (such as UNEP, UNIDO, regional economic commissions and banks). It was these organizations that should have facilitated the participation of the developing countries of the southern Mediterranean in the projects.

The centre itself, when fully developed, would have accommodated 300 professionals and support staff, most of whom would have been seconded by their firms or governments. It would have been furnished with some equipment such as a deep, free surface diving tank for the development and testing of underwater equipment, a pressurized diving tank for deep water system development, a large cavitation tunnel for propeller research, and instrument calibration and ship construction facilities.

It was proposed that the Centre specialise, inter alia, in the following sectors of R&D:

- aquaculture technologies
- desalination technologies;
- alternative energy technologies; and
- pollution prevention, controlling, and abatement technologies.

In conclusion, the proposal recommended, as a first step, the preparation of a feasibility study, for which it gave the terms of reference. The proposal was endorsed by the Government of Malta and the IOI proceeded with a detailed feasibility study, under the direction of Krishan Saigal, then a senior consultant to the UN Division for Ocean Affairs and the Law of the Sea. Saigal visited a number of Mediterranean countries and talked to scientists, industrialists, and government officials. The IOI circulated a questionnaire among hundreds of Mediterranean scientific and technical institutions. This work was supported both by UNEP and UNIDO.

The initial cost of establishing the Centre was estimated as very modest. A figure somewhere between US$100,000 and $500,000 was proposed. When fully developed, the cost was projected to be about US$5.5 million a year.

Upon completion of the feasibility study, UNIDO organized a Mediterranean workshop at the technical level for its discussion. The proposal and the study were well received. After the workshop, UNIDO issued a report, fully endorsing the project. Several Mediterranean countries, besides Malta, offered to host the Centre. This, however, was the end of the success story. Invisible hands began to pull invisible brakes. Most European States considered the proposal "premature."

That was almost a quarter of a century ago. Today, the proposal is no longer premature. It is overdue. The changes that have taken place and make the establishment of such centres, systems, or mechanisms dramatically more urgent, and more possible, are the following:

> The need for technology cooperation and transfer has become more pressing. Today, it is not only the developing countries who are making the requests, it is also the industrialized countries who realize that, without the full cooperation of the majority of the world's people, that is, the developing countries, it would be impossible to implement the conventions, agreements, codes, programmes, and action plans adopted—especially since Rio (1992)—to conserve the environment and biodiversity, react to climate change, and, quite simply, survive on this planet. If, however, developing, small, disadvantaged, poor States are to cooperate, they need the technologies required for implementation. Most industrialized States, therefore, are far more ready to contribute what they far more stingily tried to withhold in the 1970s, when UNCLOS was drafted. Funding options have increased, especially through the establishment of the Global Environment Facility (GEF).

> New and additional funds, however, are still required. Integrating the private sector into the system, by generating the kind of synergism between private and public funding at the regional level as suggested by the IOI, would be one response to this challenge.

> During the past 20 years, also, the Regional Seas Programme has been developed and consolidated. The implementation of the Global Programme of Action for the Protection of the Marine Environment from Land-based Activities (GPA) has acted as a new stimulus to regional development and the revitalization of the Regional Seas Programme. The strong technological input required could also be utilized to satisfy the needs of the other conventions and programmes as well for, as the IOI comparative study for UNICPOLOS I tried to show, the science and the technology needs in each case are the same.

> The Mediterranean Regional Seas Programme, with its revised Barcelona Convention and Action Plan, is still the most advanced of the regional systems and perhaps the most suitable venue for a pilot project. The establishment of the Mediterranean Commission on Sustainable Development, to which the new mechanism or process could conveniently be hitched, enhances the suitability of the Mediterranean for a pilot project.

The need for new mechanisms for technology cooperation and transfer is universal. The Discussion Model for a Regional Protocol we are proposing (see Appendix A) is therefore not limited to any one region, but is flexible enough to be adaptable to any regional sea. Let it be stressed again that the model is purely illustrative, an attempt to address issues that have to be faced as concretely and concisely as possible.

The Global Level

The optimum level for technology cooperation, development, and transfer is probably the regional level. States bordering regional seas and contracting parties to regional seas conventions share the same environmental and security challenges and in many cases—as, for example, in the Mediterranean or the Caribbean, the South Pacific Islands, or Southeast Asia—they have evolved common cultural traits. The regional level thus offers a commonality of interests as well as an economy of scale, facilitating agreement on technological priorities, creativity in eco-technology development, and social acceptance of technologies.

It is therefore not surprising that UNCLOS does not offer much institutional innovation with regard to technology cooperation at the global level. The existing competent international organizations are exhorted to do their share. There is, however, one institution that might make an important contribution to joint technology development and transfer in a sector which, according to some scientists, may well become the most important among the sectors of oceanographic sciences in this new century—the exploration of the deep seafloor and its subsoil. The institution responsible for this sector is the International Seabed Authority (ISA). This potential, however, has been generally neglected, within the Authority as well as outside of it.

The Convention has lavished great detail on the Authority's mandate with regard to science and technology. Article 143 establishes that marine scientific research in the Area shall be carried out exclusively for peaceful purposes and the benefit of mankind as a whole, making it part of the Common Heritage of Mankind. The Authority itself is authorized to carry out marine scientific research in the Area, coordinate the research carried out by States, and disseminate its results. Article 144 empowers the Authority to acquire technology and scientific knowledge and promote and encourage its transfer to developing States.

Detailed rules for the transfer of technology were contained in Article 5 of Annex III on "basic conditions of prospecting, exploration, and exploitation." These provisions have been widely criticized, both by the industrialized and the developing countries. For the industrialized countries and their companies, the provisions appeared to be too stringent, even though it was generally recognized that the language was far more stringent than the substance and there were sufficient loopholes to make it very difficult for any court to enforce them. Developing countries, thus, were dissatisfied because the provisions were not stringent enough and, even in the best of cases, would take so much time to be complied with as to make them useless.

Good or bad, or both, the provisions for technology transfer to the Enterprise and to developing countries were simply abolished by the implementation agreement of 1994,[17] which overrides the provisions of Part XI and the pertinent Annexes of the Convention. While the legality of this "Agreement" has been questioned, the prospects for seabed mining have changed so drastically since the 1970s when Part XI was drafted that this Part, with or without the Implementation Agreement, is to some extent obsolete. The mandate and functions of the International Seabed Authority need to be reconsidered, through an "evolutionary approach" as suggested by the Implementation Agreement.

For various reasons, the future of commercial seabed mining seems to be rather indeterminate. Manganese nodule mining, the original raison d'être of the Authority, has been postponed indefinitely; the mining of seafloor massive sulphides (SMS) appeared to be closer at hand, but even this is still in an initial phase of research and development. Other, most exciting discoveries have been and continue to be made at a fast pace—genetic resources, for example, especially microorganisms such as bacteria, and methane hydrates. New services, such as those provided by the fibre-optic cables crossing the international seabed, have been developed. All in all, the scientific, environmental, and economic importance of the deep seabed and its subsoil is far more important today, and will become even more so during this new century, than the framers of the Convention could imagine in the 1970s.[18] It should be obvious that if the seabed is more important, so too should be the International Seabed Authority. To rise to this challenge, however, the International Seabed Authority has to adjust its functions to the changed and changing scientific and economic perspectives.

Activities today—all unregulated, and outside of, with no reference to, the Authority—are focused on exploration, research and development, and the development of human resources. The challenge is to bring these activities into the scope of the Authority and organize them in a manner that is beneficial to all stakeholders—States, other Convention regimes (e.g., Biological Diversity and Climate Change), and major groups. This could be initiated even within the present mandate of the Authority. Articles 143 and 144, which remain valid, would empower the Authority to proceed along these lines.

Over the last 20 years, the International Ocean Institute has done quite a bit of work to develop a methodology. In cooperation with the Delegation of Austria, a proposal was elaborated and presented to the Preparatory Commission for the International Seabed Authority and the International Tribu-

17. Agreement Relating to the Implementation of Part XI of the United Nations Convention on the Law of the Sea of 10 December 1982 (Resolution 48/263) (1994), Annex.
18. See *The International Sea-bed Authority: New Tasks*. Proceedings of the Leadership Seminar, Jamaica, 14–15 August 1998 (Halifax: International Ocean Institute, 1998).

nal for the Law of the Sea ("the Prepcom") in 1984 and 1985, for the establishment of a Joint Enterprise for Exploration, Research and Development, JEFERAD, for the exploration of a first mine site ("reserved area") for the Authority's Enterprise. This proposal is part of the Official Record of the Prepcom.[19] A similar proposal was developed by the IOI in cooperation with the Delegation of Colombia, in a series of papers entitled *The International Enterprise, 1987–1988.*[20] The purpose of this JEFERAD or International Enterprise would have been the same as that of the Regional Centres for Research and Development in Marine Industrial Technology, that is, cost sharing and reduction, spreading of risk, creating new and additional funding through generating synergies between public and private investments, and enhancing the participation of developing countries. Also, the structure and funding of projects would be analogous to that of the proposed regional centres. In other words, this kind of Enterprise would be a sort of global EUREKA for the advancement of deep-sea exploration, technology development, and development of human resources. The body responsible for the final selection of projects would be the Council of the Authority. Our studies at that time concluded that joint exploration of the mine site would cost about 30 percent less than it would if each Pioneer Investor carried it out independently, in accordance with the obligations as defined by the Convention and Resolution II. Similarly, joint action in R&D and training was projected to produce savings of 67 percent and 33–50 percent respectively over the cost of separate, independent programmes.

Today, one should add to the purposes of the JEFERAD the harmonisation and integration of various Convention regimes. The results of this R&D should not be restricted to the International Seabed Authority. They would serve the needs of the 1992 Biological Diversity[21] and 1992 Climate Change Convention[22] regimes as well. The scope of research, therefore, would be fairly wide.

In 1998, the IOI thus revised the proposal[23] to include the Biological Diversity Convention regime, which now shares the responsibility for the conservation of deep seabed biota with the Authority, as well as the Climate Change Convention regime, for which research on the deep-sea hydrates is of fundamental importance because their destabilisation has important effects on climate change. These regimes now should be co-involved with the

19. LOS/PCN/SCN.2/L2 Add.1 and L.2 Add.2.
20. LOS/PCN/SCN.2/WP14, Add.1 and Add.2.
21. Convention on Biological Diversity, accessed 14 February 2001 on the World Wide Web: http://www.biodiv.org/.
22. United Nations Framework Convention on Climate Change, accessed 12 March 2001 on the World Wide Web: http://www.unfccc.org/.
23. International Ocean Institute, *Mediterranean Centre for Research and Development in Marine Industrial Technology: A Proposal* (Malta: Foundation for International Studies, 1988).

study and long-term monitoring of the living resources, including also the microfauna (genetic resources) and the hydrates in the Area. This revised proposal was discussed informally with both the Authority and the Global Environment Facility. The project would certainly qualify for co-funding by the GEF under its responsibility for biodiversity in international waters and climate change.

In August 1999, the International Seabed Authority organized a Workshop on Proposed Technologies for Deep-Seabed Mining of Polymetallic Nodules. This work was resumed in a second workshop on Mineral Resources of the International Seabed Area in June 2000. The Authority released a beautifully illustrated summary report of the results of both workshops in September 2000. It was gratifying to read in this Report the following:

> Considerable duplication of technical research in exploration, mining development and potential environmental impacts led to a proposal to form a cooperative venture to mine one of the better prospective areas as an environmental demonstration. In this way, environmental effects and mitigating measures could be established under actual working conditions. *New technology would be developed jointly by the participants,* and the costs of operations would be offset by the value of the metals produced, with the resulting net benefits, or costs shared by all. The proposal served as a basis for discussion among the participants and the Secretariat [emphasis added].

The report also presents a useful list of recommended cooperative research and concludes with the following information:

> Currently, the ISA's Secretariat is working with existing explorers who have exploration contracts and with various government and research organizations to facilitate the initiation of these research projects.

> Thus it appears that a new and productive phase of work is opening for the Authority, an important building block in the emerging local-national-regional-global system of technology cooperation and transfer is falling into place, and the concept of joint technology development is coming into its own under the aegis of the institution which today is the custodian of the concept of the Common Heritage of Mankind.

Conclusion

Thus the New International Technological Order, just like the New International Economic Order or the New International Social and Political Order, must be built on three interacting levels (national, regional, and global), the basis of self-reliance, and South-South and North-South cooperation. If any one of these three carrying pillars is missing, the building collapses.

In a broader historical context, we have tried to show that the previous phase of the industrial revolution was based on technologies that were resource- and capital-intensive: technologies that were embodied, so to speak, in pieces of hardware that could be handled by relatively unskilled labour. These pieces of technology could be traded, but the terms of trade were unfavourable to the developing countries that were, and remained, unable to produce them themselves. Under those circumstances, it was extremely difficult for the developing countries to "catch up" with that phase of the industrial revolution. All this took place in the context of an economic system that basically was a war system, both historically and ideologically: historically, because it developed in the age of Western expansionism, imperialism, and colonialism; ideologically, because it was based on competition and conflict rather than on cooperation and the optimization of common interests.

The new phase of the industrial revolution is based on technology that is far less static, that cannot be objectivised, that is a process, that is knowledge, know-how, service, based on human resources in whom capital must now be invested.

Now human resources are what developing countries have. And they can be developed just as fast and effectively as can human resources anywhere. Their development is primarily a matter of goal orientation and political will. The notion that technology can be bought and imported is obsolete and already discredited. Technology must be developed and this development must be based on the development of human resources. This can be done through the proper linkage of the three levels—national, regional, and global—within the UNCLOS/UNCED process.

The previous phase of the industrial revolution produced technology that was dehumanizing, subordinating the human being to the machine on a moving belt. Lenin taught that the factory was the model for the totalitarian State. That technology, dehumanizing, was also destructive of nature. The smokestack was the symbol of progress.

The new technology, the technology of the Service Economy, which is neither resource intensive nor based on the exploitation of cheap labour, need not be destructive of the environment. Based on the development of human resources, it restores to humanity its rightful central place.

Of course, all technology is dual-purpose technology, for constructive or destructive purposes depending on how we want to use it. In the context of an evolving new economic order based on the concept of the Common Heritage of Mankind, which would be an Economy of Peace, this technology might, potentially, become a source for the rise of a new humanism: a humanism, however, that does not set mankind above nature, but knows that it is part of it and depends on a harmonious relationship between the part and the whole.

PART II. PIRACY AND ARMED ROBBERY AT SEA: THE NEW INTERNATIONAL MARITIME SECURITY ORDER

> Piracy is a scourge that disrupts trade, threatens or destroys property and lives and adds to the perils of the sea. It must be brought under control . . . All unnecessary barriers should be removed to smoothen the field.[24]

Background

Piracy and armed robbery at sea are not new phenomena. However, they are increasingly common and violent. A quick overview of current statistics indicates that piracy has risen 57 percent between 1999 and 2000, and acts of violence resulting in injury and death have risen, in the same time period, from 24 to 99 and from 3 to 72, respectively.[25] Incidents of piracy are also becoming more costly in economic terms, as pirates are increasingly efficient at targeting high value cargo and sometimes vanish with entire vessels.

Furthermore, acts of piracy and armed robbery at sea often place the natural environment at risk, as it is commonplace for pirates to leave behind a bound and gagged captain and crew as the vessel continues navigating unattended. One can easily imagine the potential environmental implications of such situations, particularly when they occur in straits and passages or near coastlines.

---

24. A. Chan, *The Dangers of Piracy and Ways to Combat It.* Keynote address to the 1999 Piracy Seminar, Singapore, 22 October 1999, accessed 9 February 2001 on the World Wide Web: http://www.sils.org/seminar/1999-piracy-01.htm. (See also J. Abhyankar, *An Overview of Piracy Problems—A Global Update.* Address to the 1999 Piracy Seminar, Singapore, 22 October 1999, accessed 8 February 2001 on the World Wide Web: http://www.sils.org/seminar/1999-piracy-02.htm. The example of the Strait of Malacca is particularly alarming as there has been a recent dramatic increase in incidents of armed robbery against ships transiting this narrow and shallow waterway which some 600 vessels utilize daily. For a description of the difficulty of navigating this Strait, particularly while under attack, see MaritimeSecurity.com and Special Ops Associates, *Report on Worldwide Maritime Piracy: June 1999* (Electronic version 2.0), accessed 19 December 2000 on the World Wide Web: http://www.maritimesecurity.com/maritime_piracy.htm, p. 9.

25. International Chamber of Commerce—Commercial Crime Services, *Piracy Attacks Rise to Alarming New Levels, ICC Report Reveals,* accessed 1 February 2001 on the World Wide Web: http://www.iccwbo.org/ccs/news_archives/2001/piracy_report.asp.

It is also critical to note that organized crime groups are increasingly becoming involved in acts of piracy and armed robbery at sea. This trend reveals a new level of globalization that characterizes the piracy "industry," as groups increasingly cooperate with each other on all geographic scales and combine various aspects of their activities, including drug trafficking and the smuggling of humans and small arms, to rationalize their modus operandi as well as penetrate and develop new illicit markets.

One might then ask, What is piracy and why is it not being countered effectively? Answers to these questions are difficult and they are, as is often the case with difficult questions of ocean governance, politically charged.

Academics, politicians, and diplomats have debated the definition of piracy for many years. These debates, centred mostly on the geographical scope of the definition of piracy, are rich in ideas and arguments, all of which contribute to a greater understanding of the issue. We would suggest that any eventual elaboration on the current codified definition should not be based on the notion of territoriality, but should be universally applicable to all maritime areas beyond the territorial sea of any nation. This approach allows the circumvention of many problems arising from the definition of piracy, as codified within Article 101 (Part VII: High Seas) of UNCLOS.

Piracy consists of any of the following acts:

(a) any illegal acts of violence or detention, or any act of depredation, committed for private ends by the crew or the passengers of a private ship or a private aircraft, and directed:
  (i) on the *high seas*, against another ship or aircraft, or against persons or property on board such ship or aircraft;
  (ii) against a ship, aircraft, persons or property in a place outside the jurisdiction of any State;
(b) any act of voluntary participation in the operation of a ship or of an aircraft with knowledge of facts making it a pirate ship or aircraft;
(c) any act of inciting or of intentionally facilitating an act described in subparagraph (a) or (b) [emphasis added].[26]

Article 101 seems to clearly limit the definition of piracy to acts on the high seas. This interpretation is further reinforced by Article 86 of the Convention, the first Article of Part VII: The High Seas, which limits the scope of application of the High Seas provisions to maritime zones that are not included within the EEZ, territorial sea, internal waters, or archipelagic waters of any nation.

However, some argue that the provisions within Article 58 (Part V: Exclusive Economic Zone) dampens the purist's interpretation of Article 101, by stipulating that "Articles 88 to 115 and other pertinent rules of international law apply to the exclusive economic zone in so far as they are not incompatible

26. United Nations Convention on the Law of the Sea (n. 15 above), Art. 101.

with this Part."[27] Is the suppression of acts of armed robbery against vessels in the EEZ incompatible with Part V of the Convention? What of the archipelagic waters, sometimes much vaster and more vulnerable than EEZs?

For reporting purposes, the International Maritime Bureau (IMB) of the International Chamber of Commerce's definition of piracy does not limit itself geographically: "Piracy is an act of boarding of any vessel with the intent to commit theft or any other crime and with the intent or capability to use force in the furtherance of that act."[28]

Although not recognized by the international community, this definition of piracy is progressively finding its way into the contemporary parlance. The addition of the concept of "armed robbery," as a piratic act occurring within the jurisdiction of states, further reinforces the universal character of piracy. One now speaks of "piracy and armed robbery at sea," with the latter part of the term being defined as: "any unlawful act of violence or detention or any act of depredation, or threat thereof, other than an act of 'piracy,' directed against a ship or against persons or property on board such a ship, within a State's jurisdiction over such offences."[29]

One hopes this definition will enable discussions to steer clear of the previously mentioned debate and permit the international community to approach the issue with a more functional mind-set; the inclusion of piracy *and* armed robbery at sea as an agenda item for UNICPOLOS II is certainly promising.

Piracy and armed robbery are indeed global phenomena. Through the compilation of national and regional reports on such acts,[30] the geographical distribution of the phenomena has been pieced together. Occurrences are concentrated in West Africa, the Strait of Malacca, the South China Sea, and South America,[31] as well as "hot spots" (in 2000, Indonesia accounted for one-quarter of all incidents; the Strait of Malacca showed the fastest growth in incidents; and Bangladesh the second highest growth rate).[32]

Up to now, efforts to address the problem have been largely undertaken at the national level, with little incentive for the private sector to become involved. The international community needs to be reminded of the golden rule of the oceans—the unity of ocean space. "The problems of ocean space

---

27. Ibid., Art. 58, para. 2.
28. International Chamber of Commerce-International Maritime Bureau. Accessed 4 January 2001 on the World Wide Web: http://www.iccwbo.org/ccs/imb_piracy/weekly/_piracy_report.asp.
29. International Maritime Organization-Maritime Safety Committee, Draft Code of Practice for the Investigation of the Crimes of Piracy and Armed Robbery Against Ships (MSC./Circ.984) (London: IMO, 20 December 2000), Annex 1.
30. Principally under the auspices of the IMO and the ICC-IMB.
31. IMO, "Piracy and armed robbery at sea," *Focus on IMO*. (London: IMO, January 2000).
32. ICC-IMB, *2000 Annual Report: Piracy and Armed Robbery Against Ships.*

are closely interrelated and need to be considered as a whole."[33] Furthermore, as this same community considers sustainability and begins to implement the UNCED Process, notions of security must not be forgotten. Perhaps the Rio Declaration (1992) most clearly expresses this in its Principle 25: "Peace, development, and environmental protection are interdependent and indivisible." As in the case of the New International Technological Order, the challenge now remains to develop the institutional framework for the implementation of the New International Maritime Security Order.

The New International Maritime Security Order

Although this study focuses uniquely on the issue of piracy and armed robbery at sea, one must not forget that instances of illicit uses of the seas are increasingly interrelated. Thus the elaboration of interdiction policies and regimes must provide common solutions applicable to the whole. The illicit uses of the seas represents a clear threat to all levels of society, governments, and the economy.

The National Level
Contemporary incidents of piracy or armed robbery at sea occur mostly in areas under national jurisdiction. The IMO estimates that some 86.5 percent of these incidents occur within the territorial waters of states.[34] If one examines the latest reported cases of armed robbery and piracy, two factors become apparent:

- The majority of the reported cases in 2000 are within the waters of developing nations.
- A large proportion of these attacks occurred in straits.

Furthermore, a cursory examination of the geopolitical profiles of the nations that are most targeted by piratic activities yields the following observations:

- The majority of these nations are currently engaged in either internal or external low-to-medium intensity conflicts that involve maritime jurisdiction issues.
- The majority of these nations have large maritime areas under their jurisdictions, within which they are not able to ensure maritime security.

33. United Nations Convention on the Law of the Sea, Preamble (n. 15 above).
34. IMO, "Piracy and Armed Robbery at Sea," (n. 31 above), p. 4.

These factors clearly underline two major considerations of critical importance to understanding and fighting acts of piracy and armed robbery, namely that these acts occur within the waters of nations that do not possess national capabilities to counter them, and as a consequence, the international community must mobilize itself to aid the development of national capacities to effectively interdict these acts. Even when regional approaches to combating piracy and armed robbery at sea are considered, the deliberations must keep in mind that the collective is only as strong as the abilities of the weakest nation, a nation that is generally unwilling to cede interdiction rights to others within its territorial seas.[35]

Nations possess many tools of enforcement that they can utilize in various combinations to counter illicit activities within their areas of jurisdiction.[36] Often these tools are not fully recognized or utilized by nations and the costly and ineffective traditional approach of executive enforcement measures is employed exclusively. This approach must be modified. The various government agencies that possess maritime mandates, even if they are not enforcement mandates, must work together to establish what interdiction activities are being undertaken by all agencies and how these are carried out. Furthermore, the nongovernmental sector and the private sector must also be involved, for these sectors are often the most directly concerned, possess many resources, and have intimate knowledge of the issues. Their participation also plays a major role in enhancing compliance and is imperative if one is to speak of true governance.[37]

Government departments operating in the maritime areas should seek to rationalize their operations as much as possible so as to eliminate duplication of effort and multiplication of costs. An example of this approach could take the form of air force training flights being routed over areas of known cases of piracy and armed robbery at sea, thus providing a presence that would serve to deter illicit activities. Another example of particular relevance

---

35. A notable exception could have been the United States' Ship Rider Accords that were negotiated between the U.S. and many of its regional neighbours in an attempt to counter regional maritime drug trafficking. However, these accords have been increasingly criticized and labelled as U.S. "extraterritorialism," resulting in wholesale loss of maritime sovereignty for the nations of the Caribbean Region. It is expected that these accords will soon be phased out by a truly regional approach in the form of the Inter-American Drug Abuse Control Commission of the Organization of American States' (OAS-CICAD) Draft Agreement Concerning Co-operation in Suppressing Illicit Maritime Trafficking in Narcotic Drugs and Psychotropic Substances in the Caribbean Area (CICAD/doc.1076/00 rev.1) (28 September 2000), hereafter referred to as the Regional Drug Interdiction Protocol.

36. F. Bailet, "Achieving Compliance in the Maritime Sectors." Paper presented to the Maritime Institute of Malaysia (MIMA), October 2000.

37. E. Mann Borgese and F. Bailet, *Ocean Governance: Legal, Institutional and Implementation Considerations* (Tokyo: Ship and Ocean Foundation, to be published November 2001), in press.

to many nations is that of interdepartmental memoranda of understanding (MOUs) that make provisions for officers of one department (e.g., municipal police) to be embarked on platforms of another department (e.g., coast guard patrol boats) to direct a maritime interdiction operation that otherwise would not be within the mandate of the embarking agency.[38] This type of practice allows for the utilization of resources across agencies, thus rationalizing the high cost associated with the purchase, upkeep, and operation of these platforms and virtually eliminating interdepartmental jurisdictional impediments and disputes.

The nongovernmental sector must also play a role in the suppression of acts of piracy and armed robbery at sea. In the Asian context, the Nippon Foundation serves as an example of positive NGO involvement. This Foundation has undertaken several important projects that centre on three major themes:

- Providing information on piracy and raising public awareness about the problem;
- Encouraging shipping companies and others concerned to share their information and knowledge; and
- Developing and producing inexpensive and maintenance-free warning systems against trespassers (pirates).[39]

Clearly, these activities will not only contribute to a better understanding of the phenomena through an open dialogue between all concerned, they will also provide concrete tools and information that can be applied directly. Furthermore, this type of NGO activity must interface with governance structures as it will then enhance comprehension of the issues, propose creative solutions, and ensure that stakeholders are involved fully in the process.

The private sector is also increasingly becoming involved in supporting initiatives aimed at suppressing illicit acts at sea. The private sector is the primary victim of these acts—acts that result in loss of life and cargoes, rising insurance premiums, increased operational costs, et cetera,—and bears the primary responsibility for the implementation of prevention and response measures.

IMO's Maritime Safety Committee's (MSC) Guidance to Shipowners

---

38. In the case of Canada, various government departments that possess maritime interdiction mandates have negotiated these types of MOUs in order to facilitate rapid and flexible interdiction operations which utilize, for the most part, either Coast Guard or Naval platforms.

39. H. Terashima, "The Role of NGOs in Dealing with Piracy at Sea." Address to the 1999 Piracy Seminar, Singapore, 22 October 1999. Accessed 8 February 2001 on the World Wide Web: http://www.sils.org/seminar/1999-piracy-09.htm.

and Ship Operators, Shipmasters and Crews on Preventing and Suppressing Acts of Piracy and Armed Robbery Against Ships,[40] which outlines steps to reduce risks of attack and measures to be taken in response to attacks, should be considered a blueprint for implementation by the private sector. Further elaboration of these types of instruments should be carried out from time to time in consultation with the private sector, for no one knows the realities of the issues better than this sector.

Furthermore, the reporting of attacks—whether the attacks are successful or not—is an imperative for the private sector. All too often, this reporting does not occur as the financial cost of piracy to corporations may be small compared to the cost of reporting. Indeed, the theft of the crew's personal objects and the few thousand dollars of electronic equipment lost to piracy pales in comparison to the potential costs of reporting such incidents.[41] Local officials, often not motivated to expedite their investigations, can cause delays that have costly consequences such as higher berthing fees (up to US$10,000/day), shipping delay penalties, and even loss of contracts, all of which may lead to the severe reprimand or dismissal of the shipmaster. In addition, the policies of insurance companies must also be factored into the corporate cost equation. Shipping companies often will not report incidents for fear of being exposed to an increase in premiums and the crews may demand high risk premiums through union actions. To minimize these types of disincentives, States must develop national codes of practice and guidelines for the reporting of incidents and their subsequent investigation and ensure that vessels incur no undue delays.

The IMO-MSC's Draft Code of Practice for the Investigation of the Crimes of Piracy and Armed Robbery Against Ships, if approved by the member states at the IMO's Twenty-second General Assembly, will provide the States with "an aide-mémoire to facilitate the investigation of the crimes of piracy and armed robbery against ships."[42] In addition to providing these States with a harmonized code of practice for investigations, the Draft Investigation Code will also serve to codify the broader definition of piratical acts to include those of armed robbery at sea.[43] Furthermore, nations will be

---

40. IMO-MSC, Piracy and Armed Robbery Against Ships, (n. 29 above).

41. It has been estimated that the average economic cost of an incident is about U.S.$5,000. C. Vallar, *The Cost of Piracy—Modern Piracy, Part 3* (1 October 2000). Accessed 8 February 2001 on the World Wide Web: http://www.suite101.com/article.cfm/6236/44265.

42. IMO-MSC. Draft Code of Practice for the Investigation of the Crimes of Piracy and Armed Robbery Against Ships, Article 1 (hereafter referred to as the Draft Investigation Code). The Draft Investigation Code was approved at the IMO-MSC's Seventy-third Session (27 November–6 December 2000) and will be considered at the IMO's Twenty-second General Assembly (19–30 November 2001).

43. Ibid., background.

encouraged strongly to ratify UNCLOS and other relevant international instruments and enact domestic legislation that establishes jurisdiction over these offences and provides for effective interdiction and prosecution mechanisms. The Draft Investigation Code also outlines various aspects to be considered by nations when training their investigators and enumerates a comprehensive investigation strategy which, if followed, will aid in eliminating unnecessary delays and enhancing the transparency of the process.

It is also critical to realize that although the financial costs of piracy and armed robbery at sea may not be very high, these costs are on the rise as organized criminal actors are beginning to take note of the potentially lucrative opportunities. Moreover, the human cost incurred during these illicit acts is also on the rise. Such statistics must be unacceptable to the international community. One also does not need to look beyond the crowded straits through which the world's main sea-lines of communication flow to realize the enormity of the danger that piracy and armed robbery present to the environment.

All of these factors and their true cost must be understood clearly by the private sector and governments, who must reiterate continually to the vessels under their flags the importance of accurate and timely reporting of all acts, successful or not, carried out against them. For without accurate information, it is impossible for governments or the private sector to understand properly the phenomena, let alone begin to mount an effective response.

It seems that the private sector has begun to recognize this imperative, as it has mobilized its own resources to establish the IMB's Piracy Reporting Centre. This Centre receives reports of piratical activities, collates and analyses these data, and issues regular reports on piracy and armed robbery at sea via the Internet,[44] the safetyNET service of Inmarsat-C, as well as in printed format. The organization also provides post-incident support and information guides on prevention measures, reporting procedures, and post-incident management.[45]

The IMB's clear commitment to mobilizing private sector resources and cooperating with governments in an attempt to fight piratical acts is a positive sign that stakeholders are beginning to combine forces in an effective manner. Once again, the Nippon Foundation provides an excellent example of how an NGO that enjoys close working relationships with the private sec-

---

44. The URL for the weekly piracy report is http://www.iccwbo.org/ccs/imb_piracy/weekly_piracy_report.asp.
45. See http://www.iccwbo.org/ccs/menu_imb_bureau.asp; See IMO, "Piracy and Armed Robbery Against Ships," (n. 29 above, p. 7).

tor can contribute to this process. The Nippon Foundation has undertaken a piracy and armed robbery information collection and dissemination programme in partnership with the Japanese shipping industry. This programme supplements the IMB's activities through the dissemination of timely and useful information through an Internet-based database that is publicly accessible.[46]

The use of information technologies is not limited only to information gathering and dissemination. The electronic revolution has also brought low cost vessel tracking systems as well as automated early warning systems, all of which are being adapted by the private sector for anti-piracy and robbery applications. Such technologies should be made widely available to the entire realm of oceanic industries and the international community must make certain that LDCs are provided with opportunities to equip themselves appropriately.

Through this cursory examination of some potential responses a nation may mobilize to prevent and respond to acts of piracy and armed robbery at sea, it becomes evident that the enforcement matrix is complex but rich in opportunities. Nations must be flexible in their approaches. Policy, as well as action plans, must be developed in such a manner as to include all these opportunities at all levels of governance.

However, nations must not only possess the capacities to effectively interdict illicit acts within their territorial waters and beyond, they must also possess a credible judicial system that can convict such acts. Laws relating to the investigation and prosecution of incidents of piracy and armed robbery at sea must be developed and incorporated into criminal or penal codes, as the case may be. The implementation of the IMO's Draft Investigation Code by all nations will undoubtedly aid this process. However, much more work needs to be done in this area, particularly when it comes to provisions for mutual legal assistance, the exchange of information and evidence, and the extradition of suspects. Without these laws in place, nations, in effect, remain powerless, as they are unable to prosecute these illicit acts effectively, not to mention implement their treaty obligations under international law.

These examples of government, nongovernmental, and private sector involvement in the suppression of piracy and armed robbery at sea must all be incorporated into a horizontally and vertically integrated governance structure so that the collective knowledge and actions of all stakeholders are coordinated into a unified action.

46. The English version of the Nippon Foundation Piracy Database is available and was accessed 2 January 2001 on the World Wide Web: http://db01.nippon-foundation.or.jp/cgi-bin/zaidan/index_e.cgi.

The Regional Level

> All States shall cooperate to the fullest extent possible in the repression of piracy on the high seas or in any other place outside the jurisdiction of any state.[47]

Multilateral cooperation for the suppression of all illicit acts (be they maritime or terrestrial) is an imperative: not only by codified customary international law, but also—and perhaps most importantly—because it is the only manner in which the international community will be able to confront these acts successfully.[48] However, as was made evident in the previous section, such cooperation should not limit itself to areas within national jurisdiction. Entire regions or subregions should be considered and all waters concerned must be incorporated into multilateral approaches, regardless of their juridical nature.

The modus operandi of pirates, as well as that of maritime narco-traffickers and smugglers of persons, is often characterized by the use of the latest information technologies and highly adaptive and flexible command and control structures. These characteristics render any kind of response very difficult and force national, regional, and international initiatives to be flexible and proactive if they are to be effective. Furthermore, these illicit acts are multijurisdictional in nature, as one part of the criminal organization may be located in one nation but coordinating activities that occur in the jurisdiction of several other nations. To render matters even more complex, these organizations actively exploit the weaknesses of certain nations' judicial systems, the stressed political relations within and between nations, the limitations imposed on interdiction operations due to maritime boundaries, and the lack of cooperation between victims.

Appropriate existing regional organizations should be utilized wherever possible to ensure the coordination and harmonization of policies and laws.

---

47. United Nations Convention on the Law of the Sea, (n. 15 above), Article 100.

48. It is important to note that the international community has renewed its will to cooperate in the fight against transnational crime through the recently negotiated United Nations Convention Against Transnational Organized Crime and its associated protocols: the Protocol to Prevent, Suppress and Punish Trafficking in Persons, Especially Women and Children and the Protocol Against the Smuggling of Migrants by Land, Sea and Air (Palermo, December 2000). Although this Convention remains sectoral in its approach, the logical next step for the international community would be to expand provisions of this instrument through the adoption of additional protocols and to harmonize it with other related conventions such as the UN Single Convention on Narcotic Drugs (1961, as amended by the 1972 Protocol Amending the Single Convention on Narcotic Drugs), the UN Convention on Psychotropic Substances (1971), and the UN Convention Against the Illicit Traffic in Narcotic Drugs and Psychotropic Substances (1988).

Furthermore, these fora should not limit their activities to the political, but should also serve as funding agencies which, through voluntary contributions by the regional states and the private sector as well as the implementation of novel funding approaches,[49] could raise the necessary funds to develop and maintain regional capacities. These capacities should be multilateral and centralized and must seek to integrate the implementation of various convention regimes at the regional level, while remaining flexible and adaptive in implementing a proactive policy.

Effective coordination will be needed, especially at the initial stage, to ensure that interdiction operations and judicial proceedings are not impeded. In this respect, MSC's Draft Regional Agreement on Co-operation in Preventing and Suppressing Acts of Piracy and Armed Robbery Against Ships[50] should be implemented in all regions. It is essential, however, to keep in mind that cooperative frameworks should not be restricted to dealing with piracy and armed robbery at sea, as there are other illicit activities occurring in the seas, such as maritime narco-trafficking and the smuggling of persons, which need to be dealt with through the same types of cooperative arrangements.

Member States of the Organization of American States (OAS) are currently in the process of negotiating the Regional Drug Interdiction Protocol, which contains provisions identical to the IMO's Regional Piracy Interdiction Protocol. As both types of illicit activities are intimately linked and increasingly occurring in tandem, why are regions not establishing cooperative frameworks that include mandates for the simultaneous interdiction of both activities? The interdiction mandates stipulated by these regional cooperative frameworks should also remain flexible and adaptive so as to incorporate the interdiction of other illicit activities, such as the smuggling of humans, which may be encountered during an operation.

One can easily imagine the logical, and necessary, next step: one mandate and one interdiction force (a regional coast guard, for example) enforcing at the regional level the provisions of all conventions and agreements applicable to the regional sea, suppressing all crimes at sea, and enhancing regional maritime security. Such an approach would virtually eliminate the judicial, territorial, and operational barriers that currently plague response initiatives in all nations and regions. Mr. Alan Chan could not be clearer on this point during his keynote address to the 1999 Piracy Seminar held in Singapore: "Border crossing for a common purpose without exploiting the sea resources should be allowed. The movement of legiti-

---

49. See Mann Borgese and Bailet (n. 37 above).
50. IMO-MSC, Piracy and Armed Robbery Against Ships: Recommendations to Governments for Preventing and Suppressing Piracy and Armed Robbery Against Ships (MSC/Circ.622/Rev.1) (London: IMO, 16 June 1999), Annex, Appendix 5 (hereafter referred to as the Regional Piracy Interdiction Protocol).

mate patrol boats should be further liberalized if joint or harmonized patrol is to perform its duty more efficiently."[51]

Such suggestions have gained considerable support from both academics and diplomats over the past decade as it becomes clear that nations, particularly LDCs and SIDS, are not capable of mobilizing the appropriate level of national resources to mount independent maritime enforcement operations and the very nature of these illicit operations is "going global."

The following, solely illustrative, Draft Protocol for the Creation and Operation of a Regional Coast Guard (see Appendix B) is provided for discussion only, to advance the consideration of ways and means to rationalize limited regional resources and mount an effective, unified response to illicit maritime activities such as piracy and armed robbery.

The Global Level

The dialectic of globalization, between the benefits of increased flows of goods and services and the cost of increasing inequality resulting in the rise of illicit activities of globalized crime syndicates, is only now being recognized by the international community. This dialectic is further complicated by the information revolution, whose own advantages and drawbacks are also clearly emerging.

Although nature is beginning to teach us that the interconnectedness of natural systems must be a central consideration in the elaboration of our governance structures, we seem to hesitate in translating this recognition into strategies to be implemented by these structures. This is particularly evident in the context of piracy and armed robbery at sea and, indeed, throughout the broader matter of illicit maritime activities.

Perhaps this has its roots in the separation of environment and development concerns, on the one hand, and security concerns, on the other, within the United Nations system. The specialization of the United Nations organs, programmes, and funds, including the competent international organizations, certainly contributes to the problem. There is all too often little coordination between these bodies, and the little there is must struggle to achieve progress within the system.

It is only at the level of the General Assembly that this sectoralism can be transcended. The General Assembly is now greatly assisted in this task by UNICPOLOS but, even here, sectoralism among various interest groups is still to be reckoned with.

Throughout this article, it has been underlined that the IMO is taking the lead at the global level in formulating and coordinating the fight against piracy and armed robbery at sea. There is little more that needs to be elaborated upon here, except for fully endorsing the work of the MSC and suggesting that this body should further strengthen close working relations with

---

51. Chan (n. 24 above).

other bodies with similar goals, within the UN system and beyond. Of particular and immediate relevance, cooperation with two UN programmes—the United Nations International Drug Control Programme (UNDCP) and the Global Programme Against Trafficking in Human Beings—should be undertaken. Both of these UN programmes have maritime interdiction components that are to be implemented regionally and nationally by the same bodies. One would also envisage close cooperation and coordination between these UN programmes and organizations such as the Organization of American States' Inter-American Drug Abuse Control Commission (OAS-CICAD), which is, itself, facilitating the negotiation and implementation of a regional maritime interdiction accord.

The work of these intergovernmental organizations should be complemented by the contribution of the global nongovernmental and private sectors. The participation of these two stakeholder groups is becoming increasingly important within the UN system, including UNICPOLOS.

The Secretary-General's Agenda for Peace[52] is another example of the enhancement of law and order, and peace and security at the global level. The document stresses throughout the importance of regional organization and cooperation in ensuring political, economic, and environmental security. The regions referred to are land-centred regions (EU, OAU, etc.), not regional seas, but as we have shown elsewhere,[53] regional seas organizations would qualify equally for inclusion in the global security network proposed by the Agenda. They would, in fact, be a vital component of this network. Under the new Law of the Sea, they could make decisively important contributions to peace and security, including the suppression of crime at sea.

Thus, if regional arrangements along the lines suggested by this study were eventually to be made for the suppression of crimes at sea, proper linkages should be established with the global system under the Security Council and the Secretary-General.

In concluding this section, I feel it may be opportune to recall a great international jurist of the early 20th century who, as early as 1917, suggested the establishment of naval forces by the nascent League of Nations to ensure peace and security in the world. Walther Schuecking, Professor of Law at the University of Marburg and, later, Judge at the World Court in The Hague, noted that international conventions had already been adopted before World War I for the collective implementation of certain tasks, such as the suppression of piracy and slave trade, the control of the North Sea fisheries, and the protection of seafloor cables. He argued that what was then needed was a global agreement on the establishment of a sea police force

---

52. B. Boutros-Ghali, *An Agenda for Peace: Preventive Diplomacy, Peacemaking and Peacekeeping: Report of the Secretary-General* (New York: United Nations, 1992).

53. E. Mann Borgese, *Ocean Governance and the United Nations,* 2nd ed., (Halifax: Centre for Foreign Policy Studies, Dalhousie University, 1996).

to suppress all crimes at sea, in all seas and oceans. This police force should be composed of contingents contributed by individual States, placed under an international command structure. Furthermore, he suggested that straits used for international navigation should be completely internationalized and protected by an adequate presence of the international sea police force. In support of his proposals, he cited, among others, U.S. President Woodrow Wilson, who wanted the world ocean placed under the control of the League of Nations.[54] Evidently, Walther Schuecking was ahead of his time. With the international community's new emphasis on implementation and enforcement, perhaps his time is coming now.

CONCLUSION

The two issues before UNICPOLOS II—(i) marine science and the development and transfer of marine technology as mutually agreed, including capacity building in this regard; and (ii) coordination and cooperation in combating piracy and armed robbery at sea—call for the same logical course of development.

In each case, the regional level appears to be the optimum level for enhancing implementation. In each case, we have a multiplicity of sectoral conventions, each of which contains provisions for the same tools of implementation. In the case of marine sciences and technology development and transfer, each convention, agreement, code, programme, and action plan calls for regional cooperation *in its own sector;* and yet, the sciences to be enhanced and the technologies to be developed and transferred are basically the same for all. Hence, at the regional level of implementation, it becomes logical to establish *one* system—rather than eight or nine—to serve the requirements of *all* the conventions, agreements, codes, programmes, and action plans. Equally, in the case of piracy and armed robbery at sea, there are numerous conventions calling for regional cooperation and joint action for the suppression of crimes at sea. If a mechanism could be created for such joint action in the context of just *one* of these sectoral conventions, it would be logical and cost effective if this mechanism would serve the requirements of *all,* because the mechanism would be the same for each convention.

In each case, there may be a "trigger" convention or agreement to set off the process. The trigger for technology development and transfer may be the GPA, for, clearly, effective implementation of the GPA requires technology development and transfer to developing countries, and the effective implementation of the GPA is under discussion right now. The trigger for

---

54. W. Schuecking, *Der Bund der Voelker. Studien und Vortraege zum organisatorischen Pazifismus* (Leipzig: Der Neue Geist-Verlag, 1918), pp. 149–53.

piracy and armed robbery at sea may be piracy or the suppression of drug traffic, but once a mechanism is created to deal with the suppression of *one* type of crime at sea, it will be logical and economical to use it for the suppression of *all* crimes at sea.

Clearly, the establishment of such common services within regional seas would greatly contribute to the revitalization of the regional seas programme and the implementation of Agenda 21 and all other UNCED and post-UNCED conventions, agreements, codes of conduct, programmes, and action plans. It would be a major contribution towards harmonizing the various convention regimes and integrating their implementation, all of which is in the mandate of UNICPOLOS. It would perhaps be most practical if pilot studies/experiments could be undertaken in selected regional seas, chosen with the greatest chance for success: a pilot programme might be undertaken in the Mediterranean, for example, to create a system of joint research and development in marine industrial technology; and another pilot programme to study integrated maritime enforcement might be undertaken in the South Pacific, where some measures are already in place for joint surveillance and enforcement. Were UNICPOLOS to address such proposals to the World Summit on Sustainable Development in 2002, this would be a concrete and quite important contribution to the building of the new institutional order needed for the implementation of the jurisprudential inheritance of the past half-century.

APPENDIX A: DISCUSSION MODEL FOR A REGIONAL PROTOCOL

PREAMBLE

Convinced that peace, development, and environmental protection are interdependent and indivisible;[55] that peace and human security must be founded on sustainable development; that sustainable development must be founded on peace and human security; and that both human security and sustainable development must be based on equity,

Recognizing that research and development are the basis for technological innovation which is the engine of economic development, and that new forms of cooperation between governments, academia, and industry at the international level are needed to carry out research and development in high technologies in a manner that should reduce costs and spread risks,

Aware that the 1982 United Nations Convention on the Law of the Sea con-

---

55. United Nations Conference on Environment and Development, Rio Declaration on Environment and Development (Rio de Janeiro, Brazil, 3–14 June 1992), Principle 25.

tains two articles, Art. 276 and 277, which promote the establishment of Regional Centres to be established through States, in cooperation with the competent international organizations and national marine scientific and technological research organizations, and that such Centres would greatly facilitate, inter alia, the pursuit of these aims,

Bearing in mind that the conventions, agreements, codes of conduct, programmes, and action plans adopted at the Earth Summit at Rio (1992) and in its wake call for the enhancement of the sciences required to provide a better understanding of complex processes and phenomena involved in pollution of the seas and oceans, the improvement of technological cooperation and exchange of know-how among member States and their scientific and industrial institutions, the exploration of potential applications of renewable resources of energy, the design of improved methods of disposing of solid and liquid waste, and the implementation of long-term pollution monitoring, all of which require research and development in highly sophisticated technologies,

The Contracting Parties to the _____ Regional Seas Convention have decided as follows:

PART I
USE OF TERMS

Article 1: Use of Terms

For the purposes of this Protocol:

(a) "Convention Regime" means the total of rules, regulations, or recommendations contained in any of the conventions, agreements, codes, programmes, or action plans concerning ocean space and its resources and uses.
(b) "Major Groups" means the groups mentioned, inter alia, in the Mediterranean Plan of Action as participants in the Mediterranean Commission on Sustainable Development, i.e., local communities, socio-economic actors, and nongovernmental organizations.
(c) "Network" means the total of entities participating in the activities covered by this Protocol, i.e., States, scientific and technological institutions, industrial enterprises, and other Major Groups.
(d) "Ocean Governance" means the way in which ocean affairs are governed, not only by governments, but also by local communities, industries, and other stakeholders. It includes national and international law, public and private law, as well as custom, tradition, and culture and the institutions and processes created by them.

(e) "Regime." See "Convention Regime" above.
(f) "System" means the structures determining the interrelationship between and functions of the components of the network.
(g) "The UNCED Process" means the development that began with the establishment of the United Nations Commission on Environment and Development and comprises the publication of the Brundtland Report, the convening of the United Nations Conference on Environment and Development in Rio (1992), the series of global or regional conferences on the subject that have taken place between Rio and Rio +10, as well as all the legal and soft-law instruments adopted by these Conferences.

PART II
ESTABLISHMENT AND PURPOSE

Article 2: Establishment

1. There is hereby established a Regional Centre for Research and Development in Marine Industrial Technology (hereinafter referred to as "the Centre").
2. The Centre shall coordinate a network consisting of the following components:
   (a) all Contracting Parties of the _____ Regional Seas Convention (Members)
   (b) all regional scientific and technological centres and institutions as well as international scientific and technological institutions operating in the region (Associate Members)
   (c) Major Groups, including socio-economic actors, local communities, and nongovernmental organizations (Associate Members)

Article 3: Purpose

The purpose of the Centre shall be

1. to enhance technology development and transfer through South-South and North-South cooperation;
2. to create synergies between public and private funding at the international (regional) level, and thus provide new and additional funding for capacity-building, training, and education in the marine sciences and technologies, especially in developing countries; and
3. to contribute to the effective implementation of the Law of the Sea Convention as well as the conventions, agreements, codes of conduct, pro-

grammes, and action plans of the UNCED process by responding to the technological requirements of all of them.

PART III
FUNCTIONS

Article 4: Functions

The functions of the Centre shall be

1. to receive and provide information on environmentally and socially sustainable marine technologies, generating a data bank to enhance this function, in cooperation with the network and the competent international organizations, especially UNIDO;
2. to promote the publication and dissemination to any interested States of the results of that research and information relating to its objectives and methods and, to the extent practicable, to facilitate the participation of scientists from those States in such research;[56]
3. to make periodic studies of the technological requirements of all the convention regimes and programmes in the region;
4. to draw up periodically a priority list of technologies required in the region, in accordance with the process indicated in Part IV of this Protocol;
5. to select R&D projects within the scope of the priority list and organize their implementation and funding;
6. to ensure the full participation of developing countries in such projects;
7. to organize and facilitate training in various aspects of marine industrial technology, using, wherever possible, existing training institutions in the countries of the region or including technicians from developing countries in the R&D projects organized by the Centre; and
8. to provide consultancy and advisory services in order to strengthen the capabilities of national institutions and public and private enterprises.[57]

Article 5: Scope of R&D

The first list of requirements to be compiled in accordance with Article 4, para. 4 above should be based on the original documents establishing the regimes. The technologies required for the effective implementation of these regimes would include, inter alia:

56. FAO, *Code of Conduct for Responsible Fisheries* (Rome, Italy: 1995).
57. K. Saigal, *Research and Development in Marine Industrial Technology: A Feasibility Study* (Unpublished, 20 April 1989).

1. cleaner production techniques, recycling, waste audits and minimization, construction and/or improvement of sewage treatment facilities, and quality management criteria for the proper handling of hazardous substances;[58]
2. waste recovery, recycling, including effluent re-use; sewage treatment technologies; beneficial use of sewage; substitutes for persistent organic pollutants (POPs); water saving technologies; environmentally friendly substitutes for materials, products, processes, and activities that have adverse environmental effects; compacting toilets; innovative technologies to improve septic tanks; containment technologies to prevent leakages from construction sites, aquaculture ponds, etc.; geotextiles; bio-remediation; small-scale bio-gas; development of more degradable packaging material; and planning "industrial symbiosis."[59]
3. computers, software, water management including desalination and rainwater harvesting,[60] pollution control technologies, coastal engineering, and infrastructure research;[61]
4. energy efficient technologies[62] and new and renewable sources of energy, including wind, solar, geothermal, hydroelectric, OTEC, wave, and biomass energy;[63]
5. selective fishing gear, improved techniques for dealing with risk and uncertainty, and technologies for monitoring, control, surveillance, and enforcement, including satellite transmitter systems;[64]
6. navigational aids;[65]

---

58. United Nations Conference on Environment and Development. Agenda 21: Programme of Action for Sustainable Development (Rio de Janeiro, Brazil, 3–14 June 1992), Paragraph 17.21.

59. Global Programme of Action for the Protection of the Marine Environment from Land-based Activities (UNEP(OCA)/LBA/IG.2/7) and the Washington Declaration on Protection of the Marine Environment from Land-based Activities (Annex II of the Washington Conference Report of the Meeting, UNEP (OCA)/LBA/IG/2/6). Adopted at the High-level Segment of the Intergovernmental Conference to Adopt a GPA on 1 November 1995.

60. Programme of Action for the Sustainable Development of Small Island Developing States. Adopted at the Global Conference on the Sustainable Development of Small Island Developing States, Barbados, 25 April–6 May 1994.

61. United Nations Convention on Biological Diversity. Adopted at the UN Conference on Environment and Development, Rio de Janeiro, Brazil on 5 June 1992.

62. United Nations Framework Convention on Climate Change. Adopted in New York, U.S.A., on 9 May 1992.

63. Programme of Action for the Sustainable Development of Small Island Developing States, (n. 60 above).

64. FAO, *Code of Conduct for Responsible Fisheries*, (n. 56 above), accessed 23 March 2001 on the World Wide Web: http://www.fao.org/.

65. United Nations Convention on the Law of the Sea, (n. 15 above), Article 43.

PART IV
STRUCTURE

Article 6: The System

The system shall have three principal components:

1. National Coordinators and Representatives of regional and international competent organizations and Major Groups
2. The Ministers of Science and Technology
3. The Coordinating Centre

Article 7: The National Coordinators

1. Each Contracting Party shall designate a National Coordinator. The National Coordinator shall be located and serviced by the Focal Point established by each Contracting Party for the implementation of the GPA.
2. Each National Coordinator shall solicit projects from both the public and private sectors. To be eligible, projects must
   (a) fall into one of the categories of priority technologies agreed upon by the Contracting Parties, and
   (b) have partners in at least two countries, including at least one developing country.
3. National Coordinators and representatives of regional and international competent organizations shall meet twice a year to make a first selection among proposed projects. To enhance the transparency of the process, Major Groups as referred to in Article 2, may participate in these meetings as Observers.

Article 8: The Ministers of Science and Technology

1. The Ministers of Science and Technology of the Contracting Parties, or other Ministers responsible for Science and Technology, shall meet once a year to make the final project selection, approve the resulting Joint Ventures in Research and Development, and assure the necessary funding. The Meeting shall also appoint the Executive Director of the Coordinating Centre, when such appointment is due.
2. Where Regional Commissions for Sustainable Development exist, these meetings should constitute the High-Level Segment of the Commission, thus ensuring the proper linkage between joint technology development and the goals of sustainability and conservation aspired to by the various UNCED conventions, codes, programmes, and action plans.

3. Where Regional Commissions for Sustainable Development have not yet been established, the Meeting of Ministers should be attached to any other suitable regional institution or mechanism as each Contracting Party may decide.
4. The projects selected by the Ministers shall be funded half by the industries that initiated the proposal and half by their Governments and international or bilateral funding agencies. This should create the desired synergism between private and public investments at the regional level. The participation of developing countries shall be largely, but not necessarily wholly, financed by funding agencies and regional banks, etc.

Article 9: The Coordinating Centre

1. The Coordinating Centre shall be established as a Regional Action Centre (RAC) within the Regional Seas Programme. Its Executive Director shall be appointed by the Meeting of Ministers.
2. The Coordinating Centre should consist of a Core Module and Additional Modules to be established according to needs and available funding.
3. The Core Module shall service the Meetings of National Coordinators and the Meetings of Ministers.
4. The Executive Director shall be the Chief Executive Officer of the Centre. He shall represent the Centre vis-à-vis third parties, prepare the annual budget, draft the annual report, sign contracts, and fulfil any other duties of a CEO within the mandate of the Centre.
5. As soon as possible, there should be an additional module for the organization of training programmes. Training programmes should cover the sciences and technologies involved in the network's projects, and trainees should be directly involved in these projects as much as possible. Training programmes should be of an interdisciplinary nature, cover management and project planning, and give an introduction to regional cooperation and development and the emerging elements of ocean governance as these themes provide the broader framework within which the new international technological order is to evolve. The training module should cooperate with existing training programmes and institutions as well as distance-learning facilities.
6. There should be a legal module to assist in the drawing up of joint venture agreements, the sharing of intellectual property, and other legal questions arising from the projects.
7. There should be a module for data handling, information, and cooperation with technology cooperation systems in other Regional Seas Programmes.

PART V
SETTLEMENT OF DISPUTES

Article 10: Mandatory Peaceful Settlement of Disputes

1. Any dispute arising from the interpretation or implementation of this Protocol is subject to mandatory peaceful settlement in accordance with Part XV of the United Nations Convention on the Law of the Sea and the relevant Annexes.
2. Attention is drawn to the options for Arbitration or Special Arbitration which may be particularly applicable in a regional context.

PART VI
FINAL CLAUSES

1. This Protocol shall be open for signature in ____ from ____ to ____, by any Contracting Party to the Convention. It shall also be open, within the same dates, for signature by the Associate Members mentioned in Article 2.
2. This Protocol shall be subject to ratification, acceptance, or approval. Instruments of ratification, acceptance, or approval shall be deposited with the Government of ____, which will assume the function of Depositary.
3. As from ____, this Protocol shall be open for accession by the Entities referred to in Paragraph 1 above.
4. This Protocol shall enter into force on the thirtieth day following the date of deposit of at least six instruments of ratification, acceptance, or approval of, or accession by Contracting Parties referred to in Paragraph 1 of this Article.

IN WITNESS WHEREOF the undersigned, being duly authorized, have signed this Protocol.

APPENDIX B: DRAFT PROTOCOL FOR THE CREATION AND OPERATION OF A REGIONAL COAST GUARD

PREAMBLE

The Contracting Parties to the present Protocol,

Being Parties to the Conventions _____ and _____,

Taking into account the Protocols related to the Convention _____,

Bearing in mind the relevant provisions of the United Nations Convention on the Law of the Sea, done at Montego Bay on December 10, 1982, as well as all other legal instruments relevant to the suppression of crime at sea,

Recalling the relevant provisions of the United Nations Single Convention on Narcotic Drugs (1961), the United Nations Convention on Psychotropic Substances (1971), and the United Nations Convention Against the Illicit Traffic in Narcotic Drugs and Psychotropic Substances (1988),

Noting the relevant provisions of the United Nations Convention Against Transnational Organized Crime which was adopted by the General Assembly at its Millennium meeting in November 2000,

Noting the work of the Inter-American Drug Abuse Control Commission (CICAD) of the Organization of American States in elaborating a Draft Agreement Concerning Co-operation in Suppressing Illicit Maritime Trafficking in Narcotic Drugs and Psychotropic Substances in the Caribbean Area, as well as the work of the International Maritime Organization in the development of a Draft Regional Agreement on Co-operation in Preventing and Suppressing Acts of Piracy and Armed Robbery Against Ships,

Emphasizing the universal and unified character of the United Nations Convention on the Law of the Sea and its fundamental importance for the maintenance and strengthening of international peace and security as well as for the sustainable use and development of the seas and oceans of their resources,

Noting that Principle 25 of the Rio Declaration (1992) upholds that "peace, development, and environmental protection are interdependent and indivisible," that peace and human security must be founded on sustainable development, that sustainable development must be founded on peace and human security, and that both human security and sustainable development must be based on equity,

Recognizing that the illegal activities in the marine areas are on the rise and their consequences represent a serious danger to the environment and the sustainable development of the Region,

Desirous of protecting and preserving the seas and peoples of the Region from deleterious socio-economic and environmental effects,

504   Security and Military Activities

Aware of the differences in levels of development among the coastal States, and taking account of the economic and social imperatives of the developing countries,

Acknowledging that the main emphasis during the coming decades will be on consolidation, implementation, and enforcement of the vast juridical legacy of the past decades,

Have agreed as follows:

SECTION I
GENERAL PROVISIONS

Article 1: Use of Terms

For the purposes of this Protocol:

(a) "Convention" means the _____.
(b) "Protocol" means the _____.
(c) "Contracting Parties" means the contracting parties to the _____ Regional Seas Convention.
(d) "The Action Plan" _____.
(e) "Organization" _____.
(f) "Assets" means personnel, platforms, and other equipment as well as financial requirements for the effective implementation of the Protocols or any other mandate of the Regional Coast Guard pursuant to Section II of the Protocol.
(g) "The Protocol Area" means the area to which this Protocol applies, as defined in Article 2 of the Protocol.
(h) "Surveillance" means the detection and notification of any state of condition, activity, or event of interest within an area of interest over a period of time.
(i) "Monitoring" means the systematic observation of a specific condition, activity, or event of interest.
(j) "Control" means the executive enforcement act which renders effective the rule of law.

Article 2: Geographical Coverage

1. The area to which this Protocol applies shall be:
    (a) The maritime waters of the Region proper, bound by _____, and including the continental shelf and the seabed and its subsoil;

(b) Waters, including the seabed and its subsoil, on the landward side of the baselines from which the breadth of the territorial sea is measured and extending, in the case of watercourses, up to the freshwater limit;
2. Any Party may also include in the Protocol area wetlands or coastal areas of its territory.
3. Nothing in this Protocol, nor any act adopted on the basis of this Protocol, shall prejudice the rights and obligations of any Party.

Article 3: General Undertakings

1. The Parties shall enhance measures to prevent, abate, and combat illegal activities and environmental degradation in the Protocol Area by establishing a Regional Coast Guard (RCG).
2. The Parties shall mandate the RCG to fulfil, within the Protocol Area, surveillance, monitoring, and control functions as required for the effective implementation of its mandate as defined in Section III of the Protocol.
3. The Parties shall contribute to the operational, personnel, and financial assets of the RCG in accordance with Article 10 of the Protocol.
4. The Parties shall, as soon as possible, endeavour to harmonize their laws and regulations with the international rules, standards, and recommended practices and procedures.

SECTION II
THE MANDATE OF THE REGIONAL COAST GUARD

Article 4: Primary Mandate

1. The activities of the RCG shall include, inter alia, the surveillance, monitoring, and control functions required to implement effectively its mandate as outlined in the Mandate Annex.
2. The implementation of all regional multilateral treaties and agreements applicable to sections of the Protocol Area shall also be enforced by the RCG. The Contracting Parties may agree, by way of a diplomatic conference, to enforce such legislation in the entire Protocol Area.
3. The United Nations Convention on the Law of the Sea, the sea-related parts of the UNCED conventions, agreements, programmes, and action plans as well as all international laws and standards as adopted by the Parties and approved by the Meeting of Contracting Parties shall be enforced by the RCG within the Protocol Area.
4. Furthermore, the RCG shall
   (a) develop and implement, in conjunction with the competent national

authorities of the Contracting Parties, contingency planning as deemed necessary by the Meeting of Contracting Parties. This may be achieved through the establishment of a standing Special Commission on Contingency Planning;
  (b) Pay particular attention to the requirements of Marine Protected Areas, as outlined in the provisions of the _____;
  (c) serve as the Regional focal point for the gathering and dissemination of information relating to, *inter alia*, the safety of life at sea, marine emergencies, and illegal activities; and
  (d) serve as an implementation agency for regional capacity-building in the fields of, *inter alia*, general maritime surveillance, monitoring, and control, contingency planning, response to marine emergencies, humanitarian assistance, and interdiction of illegal activity—all in conformity with generally accepted Regional standards, practices, and procedures.
5. Pursuant to Article 11 of the Protocol, the mandate of the RCG may be altered, expanded, or otherwise modified by a three-quarters majority vote of the Meeting of Contracting Parties. Any such modifications will take the form of an amendment to the Mandate Annex of the Protocol.

SECTION III
ORGANIZATIONAL STRUCTURE

Article 5: The Meeting of the Contracting Parties

1. The Regular Meetings of the Contracting Parties shall determine the policies and activities of the RCG.
2. Extraordinary meetings may be called at any other time deemed necessary, upon the request of the Tasking Commission or at least three Contracting Parties. Parties should include Secretaries of their Navies and/or Coast Guards in the delegations to these sessions.
3. Furthermore, the meetings of the Contracting Parties shall, *inter alia*
  (a) consider the Tasking Commission's Strategic Plan and Yearly Requirement Report, and determine whether these reports satisfy the current mandate of the RCG;
  (b) negotiate, amongst the Parties, the appropriate level and type of contributions to be made for the prescribed period pursuant to Article 10 of the Protocol;
  (c) receive and evaluate any proposed contributions put forth by non-Party entities and forward to the Tasking Commission such proposals for evaluation and eventual inclusion in the Yearly Requirement Report;

(d) consider the Advisory Resolutions put forth by the Advisory Council pursuant to Articles 8(2) and 8(3) of the Protocol;
(e) keep under review the implementation of this Protocol, consider the efficacy of the measures adopted, and the advisability of adopting any other measures, in particular in the form of annexes and appendices;
(f) establish criteria and formulate international rules, standards, and recommended practices and procedures in whatever form the Contracting Parties may agree;
(g) facilitate the implementation of the policies and the achievement of the objectives outlined in Article 3 of the Protocol, in particular the harmonization of national legislations within the Region as well as generally accepted international law;
(h) revise and amend the Protocol and its Annexes pursuant to Article 11 of the Protocol;
(i) approve the programme budget;
(j) elect representatives to the Regional Activity Centre; and
(k) discharge such other functions as may be appropriate for the application of this Protocol to Special Commissions pursuant to Article 9 of the Protocol.
4. In between sessions, the Bureau shall be responsible for carrying out policies and activities.

Article 6: The Tasking Commission

1. The Contracting Parties shall appoint a Tasking Commission of eight members, chosen on the basis of their personal capacity and experience in broad maritime security duties, and observing the principle of equitable geographic distribution.
2. The Chairman of the Tasking Commission shall be the Staff Commander of the Regional Coast Guard.
3. The mandate of the Tasking Commission shall be:
   (a) The coordination and execution of all activities of the RCG in accordance with the requirements set forth in Section III of the Protocol.
   (b) The preparation and presentation to the Bureau on a yearly basis of the Yearly Requirement Report, which will outline the corresponding tasking requirements of the RCG.
   (c) The preparation and presentation to the Bureau on a quarterly basis of the Strategic Plan, which will outline and justify current mission priorities for the Protocol Area.
   (d) The consideration of proposals by non-Party entities to contribute to the RCG.

(e) The discharge of such other functions as may be appropriate for the application of this Protocol to Special Commissions pursuant to Article 9 of the Protocol.
4. Ordinary meetings of the Tasking Commission shall be monthly and extraordinary meetings may be called by the request of three members.
5. The Tasking Commission shall be serviced by a Regional Activity Centre (RAC) to be established in _____.

Article 7: The Regional Activity Centre

1. The Regional Activity Centre shall be composed of representatives of the Contracting Parties elected by the Meetings of the Contracting Parties. In electing the members of the RAC, the Meetings of the Contracting Parties shall observe the principle of equitable geographical distribution.
2. The principal function of the RAC shall be to service the Meeting of the Contracting Parties and the Tasking Commission pursuant to Articles 5(4) and 6(5) of the Protocol.
3. The terms and conditions upon which the RAC shall operate shall be set in the Rules of Procedure adopted by the meetings of the Contracting Parties and annexed to the Protocol.

Article 8: The Advisory Council

1. The Advisory Council shall be composed of:
    (a) The Regional Commission on Sustainable Development.[66]
    (b) Non-State entities that hold a direct interest in the implementation of the RCG's mandate and which contribute to the assets of the RCG pursuant to Article 10(4) of the Protocol.
2. The Advisory Council may adopt, by simple majority, Advisory Resolutions on all matters relevant to the policies and activities of the RCG.
3. The Advisory Resolutions will be considered by the Meeting of Contracting Parties pursuant to Article 5(3)(d).
4. The sessions of the Advisory Council shall follow immediately the regular sessions of the Regional Commission on Sustainable Development.

66. In regions where such a commission does not exist, its establishment should be undertaken and modeled on the highly successful and innovative Mediterranean Commission on Sustainable Development.

Article 9: Special Commissions

1. Special Commissions may be created by the Meeting of Contracting Parties or the Tasking Commission to consider any issues which may require special attention.
2. The Commissions can be of two types:
   (a) Special Standing Commissions: These Commissions can be proposed by the Meeting of Contracting Parties and/or the Tasking Commission and must be approved by a two-thirds majority vote by the Meeting of Contracting Parties. The mandates of these Commissions shall be annexed to the Protocol.
   (b) Ordinary Special Commissions: These ad hoc Commissions can be proposed by the Meeting of Contracting Parties and/or the Tasking Commission and must be approved by a simple majority vote taken only by the proposing body. The mandates of these Commissions shall be in the form of a circular to be distributed to all the Parties and Advisory Council members.
3. The Special Commissions will report directly to the proposing body.

Article 10: Contribution to RCG Requirements

1. The Yearly Requirement Report, drafted by the Tasking Commission pursuant to Article 6(3)(b) of the Protocol and approved by the Meeting of Contracting Parties pursuant to Article 5(3)(a) of the Protocol, shall enumerate the anticipated RCG requirements for the upcoming year.
2. The Contracting Parties will be invited to contribute assets in order to meet the requirements outlined by the Yearly Requirement Report.
3. Minimum yearly contributions by Parties shall be negotiated each year, in consultation with the Tasking Commission, by Meetings of the Contracting Parties.
4. Non-State entities may contribute to the RCG requirements by making available assets, even on an ad hoc basis, but must be willing to relinquish at least operational command if not control. In so doing, representatives of these entities are entitled to become members to the Advisory Council.

Article 11: Amendments to the Protocol and Annexes

1. Any Contracting Party may propose amendments to the Protocol. Amendments shall be adopted by a diplomatic conference which shall be convened by the Bureau at the request of two-thirds of the Contracting Parties.

2. Any Contracting Party may propose amendments to any Annex. Such amendments shall be adopted by a diplomatic conference which shall be convened by the Bureau at the request of two-thirds of the Contracting Parties.
3. Amendments to this Protocol or any Annex shall be adopted by a three-quarters majority vote of the Contracting Parties and shall be submitted by the Depositary for acceptance by all Contracting Parties.
4. Acceptance of amendments shall be notified to the Depositary in writing. Amendments adopted in accordance with Paragraph 3 of this Article shall enter into force between Contracting Parties having accepted such amendments on the thirtieth day following the receipt by the Depositary of notification of their acceptance by at least three-quarters of the Contracting Parties.

Article 12: Standards

It is understood that all management and operational standards utilized by the RCG shall be those most widely accepted and utilized.

Article 13: Mutual Information

The Contracting Parties shall inform one another directly or through the Bureau of measures taken, results achieved, and—if the case arises—difficulties encountered in the application of this Protocol. Procedures for the collection and submission of such information shall be determined by the Meetings of the Contracting Parties.

Article 14: Peaceful Settlement of Disputes

1. Any dispute arising from the interpretation or implementation of this Protocol is subject to mandatory peaceful settlement in accordance with Part XV of the United Nations Convention on the Law of the Sea and the relevant Annexes.
2. Attention is drawn to the options for Arbitration or Special Arbitration which may be particularly applicable in a regional context.

Article 15: Final Clauses

1. This Protocol shall be open for signature in _____ from _____ to _____, by any State invited to the Conference on the Creation and

Operation of a Regional Coast Guard. It shall also be open, within the same dates, for signature by the relevant Regional Intergovernmental Organizations, of which at least one member is a coastal State of the Protocol Area and exercises competence in fields covered by Article 3 of this Protocol.
2. This Protocol shall be subject to ratification, acceptance, or approval. Instruments of ratification, acceptance, or approval shall be deposited with the Government of _____, which will assume the functions of Depositary.
3. As from _____, this Protocol shall be open for accession by the States and Intergovernmental Organizations referred to in Paragraph 1 above.
4. This Protocol shall enter into force on the thirtieth day following the date of deposit of at least six instruments of ratification, acceptance or approval of, or accession to, the Protocol by the Parties referred to in Paragraph 1 of this Article.

IN WITNESS WHEREOF the undersigned, being duly authorized, have signed this Protocol.

Security and Military Activities

Abortions on the High Seas: Can the Coastal State invoke its Criminal Jurisdiction to Stop Them?†

Adam Newman
Dalhousie University Law School, Halifax, Canada

INTRODUCTION

In June 2001, the Netherlands-registered ship *Sea Change* docked in the port of Dublin, attracting considerable media attention. At the time, abortion was legal in the Netherlands,[1] but illegal in Ireland.[2] The crew members of *Sea Change* openly announced their intent to circumvent Irish criminal law. They planned to take pregnant Irish women on board, travel into international waters, and there, in international waters, provide abortions. Ultimately, the plan was unsuccessful because the crew had failed to acquire the abortion clinic license required by Dutch law, so if crew members had performed abortions on board the ship, they could have been subject to prosecution upon their return to the Netherlands.[3] *Sea Change* was spon-

---

†EDITORS' NOTE.—This article was the winning entry in the 2001 *Ocean Yearbook* Student Paper Competition.
  1. The criminal prohibition against abortion in the Netherlands was lifted by the Law on the Termination of Pregnancy in 1981. The Law came into force in 1984 as a result of the Decree on the Termination of Pregnancy. English-language versions of both the Law and the Decree can be found in the volumes 8 and 11 of the *Annual Review of Population Law* (New York: United Nations Fund for Population Activities, 1981) and (New York: United Nations Fund for Population Activities, 1984).
  2. Abortions have been prohibited in Ireland since 1861, when the Offences Against the Person Act made self-induced and aided abortions illegal for both the mother and any person who assists the mother. For a fuller discussion of Irish abortion law, see: D. A. MacLean, "Can the EC kill the Irish unborn?: an investigation of the European Community's ability to impinge on the moral sovereignty of member states," *Hofstra Law Review* 28, no. 2 (1999): 527 at 552–61. See, further: Susan Bouclin, "Abortion in post-X Ireland," *Windsor Rev. of Legal and Social Issues* 13, (2002): 133.
  3. According to news reports, the crew could have faced up to four-and-one-half year's imprisonment if convicted. (12 June 2001). Dutch say abortion boat flouts law. Accessed 12 June 2001 at CNN.com on the World Wide Web: http://www.cnn.com/2001/WORLD/europe/06/12/netherlands.abortion.index.html. (However, just over one year later, on 1 July 2002, Women on Waves issued a press

sored by the Women on Waves Foundation ("Women on Waves"), a charitable foundation based in the Netherlands, which aims to provide offshore abortions to women who live in coastal countries where abortion is prohibited.[4]

The plan proposed by Women on Waves raises issues about the limits of a coastal State's criminal jurisdiction. A traditional view is that a State's criminal jurisdiction ends where its territory ends.[5] According to Article 3 of the United Nations Convention on the Law of the Sea (UNCLOS),[6] a coastal State's territory ends 12 nautical miles (M) from its coastline, at the limits of the territorial sea. Thus, it appears at first glance that activities occurring beyond the 12-M limit of the territorial sea cannot be subject to the criminal jurisdiction of the coastal State. This is the position taken by Women on Waves:

> By performing abortion services outside territorial waters, women and abortion providers cannot be charged or prosecuted when returning to port because applicability of national penal legislation, and thus also of abortion law, extends only to territorial waters. Outside that 12-mile radius it is Dutch law that applies on board a Dutch ship and abortion

---

release stating that the Dutch Minister of Health, Els Boist, had stated that, since in the Netherlands no license is needed to provide pregnancy terminations within the first 45 days of pregnancy, Women on Waves would not be violating Dutch law by providing the abortion pill in such situations on board a Dutch ship.)

4. The Foundation was established in May 1999 by Dr. Rebecca Gomperts, a Dutch gynecologist. Providing abortions is only part of the Foundation's mission. The Foundation's mission, in its entirety, as stated on its Web site: http://www.womenonwaves.org, is:

- Empower women to make conscious, well-informed decisions about family planning.
- Prevent unwanted pregnancy.
- Ensure safe and legal abortion.
- Reduce unnecessary physical or psychological suffering and deaths from illegal abortions.
- Catalyze support for liberalization of abortion laws worldwide.

5. R. v. Keyn (1876) 2 Ex. Div. 63 (CCR). See also s. 6(2) of Canada's Criminal Code, R.S.C., c. C-46 (1985), which states that: "Subject to this Act or any other Act of Parliament, no person shall be convicted or discharged under section 730 of an offence committed outside Canada." However, this is not the only traditional view, as Akehurst notes that the principles of passive personality and universality also have a long history. M. Akehurst, "Jurisdiction in international law," *British Year Book of International Law* 46 (1972–73): 145 at 163–64.

6. Montego Bay, 10 December 1982, in force 16 November 1994. 137 ratifications as of 24 September 2001. Cmnd. 8941; *International Legal Materials* 21 (1982): 1245. Accessed 24 September 2001 on the World Wide Web: http://www.un.org/depts/los/.

is entirely legal in the Netherlands. Thus it is possible to sail to international waters, perform abortions and return all in the same day.[7]

Can this really be so? Could it really be so easy for Women on Waves to circumvent national criminal laws? This article attempts to provide an answer.

The discussion addresses the issue both from the perspective of UNCLOS and from the perspective of general principles of international law. The Netherlands ratified UNCLOS in 1996,[8] so in any dispute involving the Netherlands and another country that has ratified UNCLOS, the provisions of UNCLOS will govern. General principles of international law are relevant to the determination of this question for two reasons. Firstly, many countries that prohibit or limit abortion have not ratified UNCLOS,[9] so their jurisdictional claims will be governed by general principles of international law. Secondly, to a certain extent, even countries that have ratified UNCLOS will be governed by general principles of international law.[10] Apart from UNCLOS, the discussion focuses on domestic common law rules and may not be applicable to other legal systems.

This article is divided into seven parts. The first part provides some background information about the issue of abortion. Part two discusses the principle of territoriality, which informs the legal regime established by UNCLOS. The third part of this article focuses on UNCLOS. Part four discusses other principles under which jurisdiction may be claimed, namely the principles of active personality, passive personality, the protective principle, and the universality principle. Care is taken throughout to consider potential conflicts between a coastal State's claims of jurisdiction and the

---

7. Women on Waves, "Activities," accessed January 22, 2001 on the World Wide Web: http://www.womenonwaves.org.

8. The Netherlands ratified UNCLOS on 28 June 1996.

9. The United States is an example of one country which has not ratified UNCLOS and which limits abortion. Although the criminal prohibition against abortion was struck down by the U.S. Supreme Court in Roe v. Wade, 410 U.S. 113 (1973), there is no absolute legal right to receive an abortion in that country. Rita Simon notes that: "(a)bortion restrictions still vary by state; thirty-five (states) have laws that prevent a minor from obtaining an abortion without parental consent or notice." R. Simon, *Abortion: Statutes, Policies and Public Attitudes the World Over* (Westport, Ct.: Praeger, 1998), p. 35.

10. For example, the Preamble to UNCLOS states, in part, that ". . . matters not regulated by this Convention continue to be governed by the rules and principles of international law." Article 2(3) states that: "The sovereignty over the territorial sea is exercised subject to this Convention and to other rules of international law." Article 58(2) states that "pertinent rules of international law" apply to the EEZ in so far as they are not incompatible with Part V of UNCLOS. Article 221(1) allows states to "take and enforce measures beyond the territorial sea" in certain circumstances, when such measures are "pursuant to international law, both customary and conventional."

jurisdictional claim of the flag State. The fifth part of this article proposes a methodology for resolving competing claims, and refers to potential obstacles to jurisdiction in Ireland and the United States (U.S.). Part six discusses the ability of the coastal State to enforce its jurisdiction pursuant to general principles of international law. The main conclusions of this article are summarized in part seven.

BACKGROUND

On the Foundation's Web site, Women on Waves argues that "safe" abortions by qualified medical practitioners are needed because

- The legal status of abortion makes little difference to overall levels of abortion incidence.
- Where illegal most abortions are done with unsafe methods by persons lacking the necessary skills, causing complications in approximately 40 percent of the cases and more than 70,000 deaths each year . . . Where abortion is legal, safe, and available, complication rates are less than 1 percent.
- 20 million of the 53 million abortions performed annually are unsafe.[11]

The argument is essentially a humanitarian one that can be summarized as follows—each year 7.8 million women suffer complications and over 70,000 women die as a direct result of unsafe abortions.[12] This suffering can only be prevented by providing women with safe abortions performed by qualified medical practitioners. The suffering cannot be avoided by criminalizing

---

11. Women on Waves, "Facts," accessed on 15 January 2001 on the World Wide Web: http://www.womenonwaves.org.
12. Using statistics available on the website of Women on Waves, I have arrived at the figure of 7.8 million annual cases of preventable complications from abortions as follows: (1) Each year, 53 million abortions are performed (this figure includes both safe and unsafe abortions); (2) For safe abortions, the rate of complication is approximately 1 percent. Thus, if all 53 million abortions performed each year were safe, there would be ca. 530,000 complications per year. (3) However, only 33 million safe abortions are performed each year, resulting in 330,000 complications. (4) Twenty million unsafe abortions are performed each year, and 40 percent (or 8 million) of these abortions result in complications; (5) Thus, each year there are 330,000 complications from safe abortions, and 8 million complications from unsafe abortions, for a total of 8,330,000 complications from all abortions (safe and unsafe); (6) When one recalls that if the same number of abortions (53 million) were performed each year, but all abortions were safe, there would only be 530,000 complications, one concludes that 7.8 million complications (out of the total of 8.33 million) are a direct result of unsafe abortions.

abortion, because history proves that the legal status of abortion makes little difference to overall levels of abortion incidence.[13]

The stated goals of Women on Waves are to provide 5,000 abortions per year, and to encourage public discussion to bring about the repeal of criminal legislation related to abortion.

A recent United Nations survey of 189 countries found that unlike the Netherlands, 78 percent of countries surveyed do not allow abortion upon request.[14] Furthermore, even in the relatively few countries that allow abortion upon request, a woman's right to choose is not absolute.[15] Therefore, the services that Women on Waves propose to offer may interest women in most countries in the world. This raises the question of whether such services could be legally provided.

TERRITORIALITY

A coastal State could seek to assert its jurisdiction over the vessel *Sea Change* by invoking the principle of territoriality. The territorial principle is widely recognized as a legitimate source of criminal jurisdiction. In a landmark study of jurisdiction in international law published in 1972, Michael Akehurst commented that:

> (T)he territorial principle is the most frequently invoked ground for criminal jurisdiction; even in continental countries, which also rely on the nationality principle to a far greater extent than common law countries, prosecutions based on the territorial principle far outnumber prosecutions based on the nationality principle.[16]

---

13. Abortion was legalized in the Netherlands in 1981. Four years prior to legalization (in 1977), there were 5.5 abortions per 1,000 women aged 15–44 performed in that country. Four years after legalization (in 1985), the figure had actually declined slightly, to 5.1. United Nations, *Abortion Policies: a Global Review.* Dept. for Economic and Social Information and Policy Analysis. Vol. 2. New York: United Nations, 1993, p. 168.

14. However, 92 percent of the countries surveyed allow abortion in order to protect the life of the mother. United Nations, *Abortion Policies: a Global Review.* Dept. for Economic and Social Information and Policy Analysis. Vols. 1–3. New York: United Nations, 1992, 1993, 1995, pp. 1–158, 1–243, 1–326. The study was published in 3 volumes. Volume 1 was published in 1992; vol. 2 in 1993, and vol. 3 in 1995.

15. For example, as mentioned in n. 9 above, 35 states in the United States currently have laws that prevent a minor from obtaining an abortion without parental consent or notice.

16. M. Akehurst (n. 5 above), at 152. Section 6(2) of Canada's Criminal Code (n. 5 above) provides that, unless otherwise indicated, no Canadian shall be charged with an offence committed outside of Canada. Sharon Williams and J. G. Castel comment that, "(g)enerally speaking, Canada adheres to a territorial theory of criminal jurisdiction by prescribing rules attaching legal consequences to both conduct

The rationale underlying the territorial principle of criminal jurisdiction was expressed by La Forest J. of the Supreme Court of Canada:

> The primary basis of criminal jurisdiction is territorial. The reasons for this are obvious. States ordinarily have little interest in prohibiting activities that occur abroad and they are, as well, hesitant to incur the displeasure of other States by indiscriminate attempts to control activities that take place wholly within the boundaries of those other countries.[17]

There are at least three possible ways in which the territorial principle might be applied by a coastal State to assert jurisdiction over the vessel *Sea Change*. The first two methods are based on the fact that the criminal law of many States applies to all ships, including foreign ships, within the territorial sea.[18] These methods relate to the activity of ferrying women back and forth between territorial waters and international waters. First, if *Sea Change* actually entered territorial waters to ferry women back and forth, the coastal State might attempt to assert jurisdiction over voyages that begin and end in its territorial waters. Second, if *Sea Change* remained in international waters but used boats to communicate with the shore, the common law doctrine of constructive presence might apply so that *Sea Change* would be deemed to be within the territorial waters of the coastal State.

The third method that might be employed by the coastal State has no relation to any ferrying of women back and forth. Under this third method,

---

occurring in Canada and conduct occurring outside of Canada but having effects within her territory." S. Williams and J. G. Castel, *Canadian Criminal Law: International and Transnational Aspects* (Toronto: Butterworths, 1981), p. 10.

17. R. v. Libman, 2 S.C.R. (1985) 178 at p. 183–84.

18. The principle that the criminal jurisdiction of the coastal state applies to foreign ships within its territorial waters has been part of English law since 1878, when the Territorial Waters Jurisdiction Act, 41 & 42 Vict., c. 73, was passed. Commenting on this statute, O'Connell notes that "(t)he objective of the Act was to invest English courts with jurisdiction over indictable offences committed aboard foreign ships in the territorial sea by aliens." D. P. O'Connell, *The International Law of the Sea* (Oxford: Oxford University Press, 1982), p. 936. A similar stance was adopted by the United States Supreme Court in the case of Cunard Steamship Co. v. Mellon, 262 U.S. 100 (1923). The legal situation in many other countries is similar to that in the U.K. and the U.S. O'Connell notes that "(t)he criminal codes of several countries specify that they apply to acts committed aboard *any ships* in the territorial sea" and cites a long list of such countries: Australia, Brazil, Canada, Ceylon, Chile, Costa Rica, Cuba, Cyprus, Ecuador, Guatemala, Ireland, Israel, Jamaica, Lebanon, Netherlands, New Zealand, Nicaragua and Thailand. In addition, he notes that Malta and Somalia have extended their criminal laws to fixed distances, and Sudan, Tanzania, Yemen and Yugoslavia have extended their criminal laws to the contiguous zone. O'Connell, p. 952 (emphasis added). Of the 26 countries listed by O'Connell, only 5 allow abortion upon request (U.S., Canada, Cuba, Netherlands, and Yugoslavia). All the other countries impose some restrictions on abortion. See Simon (n. 9 above), p. 50.

the coastal State might attempt to assert jurisdiction over activities committed by the crew of *Sea Change* in international waters because these activities have *effects* within the territory of the coastal State. This is known as the modified objective territorial principle, which first rose to prominence in the Lotus case.[19] I will examine each of these three methods in turn, focusing throughout on how potential conflicts between these methods and flag State jurisdiction might be resolved.

Voyages that Begin and End in the Territory of the Coastal State

The coastal State might attempt to assert jurisdiction by creating an offence of providing and/or receiving abortions on any voyage that begins and ends within the territory of the State. To determine whether or not such legislation would be consistent with general principles of international law, one may find it useful to consider U.S. legislation aimed at prohibiting offshore gambling. The U.S. State of South Carolina recently proposed legislation that prohibits gambling on a vessel in a voyage that begins and ends within the State, even if such gambling occurs outside the territorial waters of the State, and even if the vessel on which the gambling occurs is flying a foreign flag. Section 1 of the Gambling Cruise Prohibition Act provides that:

> . . . It is the intent of the General Assembly in enacting this act to reenforce long-standing prohibitions on gambling by reiterating that the gambling offenses provided under the Constitution and laws of this State extend to any United States or *foreign documented vessel in this state where voyages begin and end in the waters of this State* . . .[20] (emphasis added)

Significantly, this legislation applies to foreign ships. The wording of this section suggests that its contemplated enactment is pursuant to the territorial principle, because the section applies to voyages that begin and end within the territorial waters of the State. However, Akehurst suggests that this type of legislation should not be justified by the territorial principle:

> . . . (S)everal states in the United States make it a criminal offence to leave the state with the intention of committing a crime outside the state. Donnedieu de Vabres defends these laws as necessary to prevent fraude à la loi—if an act . . . is illegal inside the State but legal in a neighbouring State, the State is entitled to prevent people evading its

---

19. Permanent Court of International Justice, Ser. A., no. 10, p. 23 (1927).
20. Gambling Cruise Prohibition Act, Bill 3009, South Carolina (Introduced 9 January 2001). Accessed 15 January 2001 on the World Wide Web: http://www.lpitr.state.sc.us/bills/3009.htm.

law by going to the neighbouring State and committing the act there. That may be so, but the State surely has a legitimate interest in preventing such acts only by its own citizens or residents, not by people who have merely passed briefly through its territory. It would be better, therefore, if the law asserted jurisdiction on the basis of residence and prohibited residents from doing certain acts abroad, instead of punishing some difficult-to-prove act (formation of intention) occurring inside the State's territory.[21]

If one accepts Akehurst's comments, such legislation should not be justified by the territorial principle. However, the coastal State may still invoke the active personality principle[22] to justify criminal legislation that would assert jurisdiction over its residents.

As for the foreign crew of *Sea Change*, Akehurst's comment implies that foreigners should not be held liable under legislation of the type proposed by South Carolina. His reason for concluding so is that the formation of the requisite mens rea (guilty mind) within the territory of the State would be difficult to prove. With respect, I believe that in this factual situation, the formation of intention on the part of the crew of the vessel *Sea Change* would not be difficult to prove. Women on Waves is a well-organized group, with a foundation set up under Dutch law and a Web site. To reach the shores of a country where abortion is illegal, such as, for example, a country in Africa or Latin America, the crew of *Sea Change* would have to travel a great distance. In my view, the group's high degree of organization and the crew's extensive travel suggest that the formation of intention on the part of the foreign crew of *Sea Change* would be easy to prove. Although the crew's intention to provide abortions might be formed long before *Sea Change,* or its boats, entered the territorial waters of the coastal state, the crew's intention would continue as it went about its activities. In my opinion, no useful purpose would be served by distinguishing between the territory where criminal intention is formed, and the territory where criminal intention is continued. Accordingly, I believe that the crew of *Sea Change* could be found guilty under legislation similar to the legislation proposed by South Carolina.

Constructive Presence

To avoid this result, the crew of *Sea Change* might anchor the vessel in international waters, and use boats to transport women back and forth between the coastal State and international waters. By doing so, the crew of *Sea Change*

---

21. See Akehurst (n. 5 above), p. 157.
22. For a discussion of the active personality principle, see infra n. 83 and accompanying text.

would hope to avoid the territorial jurisdiction of the coastal State, because the vessel *Sea Change* would never enter territorial waters. However, in this case the doctrine of "constructive presence" could apply. According to the doctrine, even if *Sea Change* remained in international waters, the law could presume that *Sea Change* was in territorial waters.

The doctrine of constructive presence was first expressed in an 1877 case off the coast of Australia. Lowe comments as follows:

> In 1877 the American ship *Catalpa* sent its boats to the shores of Australia to assist in the escape of convicts. In their opinion to the Colonial Office on the incident the Law Officers advised that no sound distinction could be drawn between the Catalpa and its boats, and that, the ship having therefore violated British territory, it was liable to be pursued and stopped on the high seas.[23]

The doctrine was invoked again in the 1888 Araunah case involving illegal fishing.[24] Churchill and Lowe trace the subsequent development of the doctrine as follows:

> According to later refinements of the doctrine, illustrated, for example, by some of the diplomatic exchanges concerning the seizure of the *Henry L. Marshall* for violation of the American liquor laws in 1922 and by the Italian case of the *Sito* in 1957, it had no application where a foreign ship communicated with the shore not by means of its own boats but by means of boats sent out from the shore. In such cases the exclusive jurisdiction of the flag State over its ships on the high seas was said to remain intact.[25]

As a result of these developments, two competing interpretations of the doctrine of constructive presence emerged. Some argued that the doctrine should be construed narrowly, as "simple constructive presence." According to this interpretation, the doctrine applies only when a ship uses *its own boats* to communicate with the shore. Others argued that the doctrine should be interpreted broadly, as "extensive constructive presence," and that it should apply when a ship uses *any boats* to communicate with the shore.

Both interpretations were reflected in Article 23 of the 1958 Geneva

---

23. A. V. Lowe, "The development of the concept of the contiguous zone," *British Year Book of International Law* 52 (1981): 109 at 117.
24. R. R. Churchill and A. V. Lowe, *The Law of the Sea*, 3d ed. (Manchester: Manchester Univ. Press, 1999), p. 133.
25. Ibid.

Convention on the High Seas.[26] Article 23(1) supports the interpretation of the doctrine as "simple constructive presence," stating that the doctrine applies when a ship uses *its own boats*. The Article provides, in part, that

> The hot pursuit of a foreign ship may be undertaken when the competent authorities of the coastal State have good reason to believe that the ship has violated the laws and regulations of that State. Such pursuit must be commenced when *the foreign ship or one of its boats* is within the internal waters or the territorial sea or the contiguous zone of the pursuing State . . . .[27]

On the other hand, Article 23(3) supports the interpretation of the doctrine as "extensive constructive presence," stating that the doctrine should apply when the ship uses *any boats* to communicate with the shore. The Article provides, in part, that

> Hot pursuit is not deemed to have begun unless the pursuing ship has satisfied itself by such practicable means as may be available that *the ship pursued or one of its boats or other craft working as a team and using the ship pursued as a mother ship* are within the limits of the territorial sea, or as the case may be within the contiguous zone . . . .[28]

The reference to "other craft" in Article 23(3) suggests that the contemplated boats need not belong to the ship in question. However, Article 23(1) contains no such reference. The Geneva Convention on the High Seas is thus somewhat ambiguous, and does not resolve the issue of the proper scope of the doctrine. However, it is well accepted that, at a minimum, the doctrine of constructive presence applies when a ship uses its own boats. Churchill and Lowe describe this as a rule of customary international law.[29]

26. In force 30 September 1962. Sixty-two ratifications. 450 *United Nations Treaty Series* 11; *NDI*, p. 257 (29 April 1958). Accessed 21 August 2002 on the World Wide Web: http://www.un.org/depts/los/.
27. Emphasis added.
28. Emphasis added.
29. "There is . . . in customary international law, a right to arrest foreign ships which use *their boats* to commit offences within the territorial sea (and, perhaps, now the EEZ) while themselves remaining on the high seas. This is the doctrine of constructive presence, implicitly recognised in the provisions of the High Seas and Law of the Sea Conventions relating to hot pursuit. . . ." See Churchill and Lowe (n. 24 above), p. 215, emphasis added. However, in one Canadian case involving alleged illegal fishing, the court, while it recognized the validity of the doctrine of simple constructive presence in international law, refused to apply this doctrine in the absence of explicit statutory wording. In R. v. Dos Santos, 96 Nfld. & P.E.I.R. 13 (1992), a Portuguese national operating a Panamanian fishing vessel was charged with unlawfully entering Canadian fisheries waters contrary to s. 7(1)(a)(i) of the Coastal Fisheries Protection Act. Although the accused was convicted on other

The contentious issue is not whether the doctrine of constructive presence exists, but whether it should be interpreted broadly, as "extensive constructive presence." Although Article 23(3) appears to support such an interpretation, its drafting history makes clear that there was a lack of consensus on this issue. William Gilmore comments:

> At the Conference the draft article on hot pursuit was examined, on a section-by-section basis, by the Second Committee at its 28th meeting held on 9 April 1958. At that time, Mexico proposed an amendment which called for the insertion of the words 'or other craft working as a team and using the ship pursued as a mother ship' after 'one of its boats' . . . A roll-call vote was held on this amendment on 11 April and it was adopted by 35 votes on 13, with 16 abstentions. The proposal, which was supported by Canada, was opposed by a number of major Western maritime powers including the UK and the USA.[30]

As a consequence of this drafting history, Gilmore concludes that:

> Although the question is not entirely free from doubt, the balance of the authorities support the proposition that in 1958, customary law embraced only the notion of 'simple' constructive presence, and that the more extensive form in which it is found in the convention represented a progressive development of the law.[31]

Further evidence that Article 23(3) may not have been declaratory of customary international law at the time it was enacted is the fact that unlike Article 23(1), Article 23(3) is not operative to establish the rule of hot pursuit. O'Connell comments:

---

grounds, he was held not to have "entered" Canadian fishing waters, despite the fact that he sent his boats into Canadian fishing waters in order to fix his nets. Barry J. held that: "On the matter of whether the accused can be found guilty of entering Canadian fisheries waters, where he stayed outside and merely sent *his small boats* in, I believe that such a 'constructive entry' could be have been legislated by Parliament without international criticism . . . However, I believe that in a matter such as the present, which involves the possible imposition of large fines, if Parliament intends to impose liability for a 'constructive entry,' this intention should be expressly set out in the legislation. An example of where such liability has been imposed in the Criminal Code is in section 350, which says that a person 'enters' a dwelling house 'as soon as any part of his body or any part of an instrument that he uses is within any thing that is being entered.' But there is no such language in the Coastal Fisheries Protection Act providing for 'constructive entry' and, accordingly, I hold that the accused is not guilty of 'entering' Canadian fisheries waters." (at paras. 66–67, emphasis added.)

30. W. C. Gilmore, "Hot pursuit and constructive presence in Canadian law enforcement: a case note," *Marine Policy* 12 (1988): 105 at 110.

31. Ibid.

Paragraph 3 [of Article 23] says that hot pursuit is not deemed to have begun unless the pursuing ship has satisfied itself that the ship pursued or one of its boats, or other craft working as a team and using the ship pursued as a mother ship, are within the zone. *This paragraph is not, like paragraph 1, operative to establish the rule, but circumstantial to its application; and it makes pursuit conditional on team work and use of the vessel as a mother ship, which are not conditions usual in transhipment.*[32]

Whatever the status of the doctrine of extensive constructive presence may have been in 1958, the doctrine was applied by the Nova Scotia Court of Appeal in 1986. In R. v. Sunila and Soleyman,[33] the accused were foreigners charged with criminal and narcotics offences after they were found on a Honduran ship (*Ernestina*), which had delivered a cargo of hashish to a Canadian boat (*Lady Sharell*) for importation into Nova Scotia. The accused argued that their arrest was illegal, because the Honduran ship was in international waters at the time that it was pursued and arrested by Canadian authorities.[34] The Nova Scotia Court of Appeal rejected their argument. Hart J.A. held that

> I am satisfied that the "Ernestina" was a *mother ship* of the "Lady Sharell" and that the "Lady Sharell" was within Canadian waters when the pursuit of the "Ernestina" took place by H.M.C.S. "Iroquois".[35]

Subsequent drug trafficking cases in New Brunswick have similarly applied the doctrine of extensive constructive presence.[36] In R. v. Rumbaut,[37] the accused was a Spanish national standing trial for conspiracy to import cocaine into Canada. A Cypriot vessel (*The Pacifico*) was met just outside Canada's territorial waters by a Canadian vessel (*Lady Teri-Anne*), and both vessels stopped in the water for about 25 minutes. *Lady Teri-Anne* returned to Shelburne, where it was boarded and its fish hold was found to be full of bales

---

32. O'Connell (n. 18 above), p. 1093. Emphasis added.
33. 71 N.S.R. (2d) 300 (1986).
34. Although the Honduran ship was in international waters at the time it was pursued and its crew arrested, the ship had previously been located for a few hours within the territorial sea of Canada, when its crew transferred 13.4 tons of Cannabis Resin to the Canadian boat. The court stated (at para. 8) that: "In total, the 'Ernestina' spent only about five hours within the limits of the territorial sea of Canada." However, the fact that the *Ernestina* had been located for some time in Canadian territorial waters was not relevant to the court's decision, as the issue was whether or not the *Ernestina* was in Canadian waters *at the time that it was first pursued*.
35. At para. 32, emphasis added.
36. R. v. Kirchhoff, 172 N.B.R. (2d) 257 (1995); R. v. Rumbaut, 202 N.B.R. (2d) 87(1998).
37. 202 N.B.R. (2d) 87 (1998).

that contained cocaine. The Canadian authorities then pursued *The Pacifico*, boarded it, and seized evidence. The accused argued that the evidence seized from *The Pacifico* was inadmissible, because the ship was in international waters when it was first pursued. Deschenes J. rejected this argument, and held that

> ... Article 23 of the Geneva Convention and Article 111 of the Montego Bay Convention (UNCLOS) as they relate to the issue of *extensive constructive presence* are declaratory of existing customary international law . . .[38]

In holding so, Deschenes J. referred to the English case of R. v. Mills,[39] which in turn referred to other cases upholding the doctrine of extensive constructive presence, including a 1977 Italian case.[40]

Following this line of cases, an argument could be advanced that even if *Sea Change* remained in international waters, it might be constructively presumed to be within the territorial waters of the coastal State if it acted as a "mother ship" or worked "as a team" with *any vessel* within the territorial waters of the coastal State. The success of this argument may depend on whether abortions are considered analogous to drug trafficking, a situation which seems unlikely at international law.[41] Nevertheless, it seems clear that if *Sea Change* used *its own boats* to transport women back and forth, *Sea Change* could be constructively presumed to be within the territorial waters of the coastal State. In this latter case, *Sea Change* would be subject to the criminal jurisdiction of the coastal State.

---

38. Ibid at para. 32, emphasis added.

39. An unreported decision of the English Circuit Court (10 March 1995), cited by Deschenes J. in *Rumbaut*. See (n. 37 above), at para. 15 and following.

40. J. Deschenes quoted the following passage from R. v. Mills: "After the ratification of the Geneva Convention in 1962, the Doctrine was considered in the Italian Courts in The Pulos 1977 International Year Book of International Law at page 587. In that case there had been a transshipment in international waters of cigarettes to a daughter ship which had come from Italian territorial waters. The daughter ship was pursued to and arrested in territorial waters. It was held that the right of hot pursuit which began immediately on the arrest of the daughter ship was extendible to [the]mother ship."

41. While there is widespread opposition to drug trafficking in international law, opposition to abortions is not nearly as widespread. There have been several international treaties aimed at stopping drug trafficking. A recent example of such a treaty is the United Nations Convention against Illicit Traffic in Narcotic Drugs and Psychotropic Substances, concluded at Vienna on 20 December 1988 and entered into force 11 November 1990, accessed 15 April 2001 on the World Wide Web: http://www.incb.org/e/conv/. In contrast, there have been no such treaties relating to abortion.

Effects within the Territory—The Modified Objective
Territorial Principle

Quite apart from any question relating to ferrying women back and forth, the question arises as to whether the coastal State could claim jurisdiction over *Sea Change* on the basis that, although the ship is located beyond its territorial waters, it is engaging in activities which are producing harmful effects within the territory of the coastal State.

The traditional view is that the principle of territoriality cannot be invoked to assert jurisdiction over alleged crimes that occur beyond territorial boundaries. As O'Connell notes, the converse of the assertion that a State may exercise its criminal jurisdiction within its territorial sea is the assertion that a State may not exercise its criminal jurisdiction beyond the limits of its territorial sea:

> International legal theory respecting jurisdiction has passed through several phases . . . In the first stage, jurisdiction was conceived of in strictly territorial terms: a State had no jurisdiction beyond its territory except over its own ships. As Marshall C.J. said in the leading case on the subject [The Schooner Exchange v. M'Faddon 7 Cranch 116 (1812) at 135], there are two concomitant propositions in the law: 'the jurisdiction of a nation within its own territory is necessarily exclusive and absolute', and 'this full and absolute territorial jurisdiction' is 'incapable of conferring extraterritorial power'. These two propositions were two sides of the same coin, and they supposed an absolute barrier to the exhibition of law at the national frontier—except, that is, in the case of national ships.[42]

Along these lines a recent case in Nova Scotia, Canada concluded that seven accused could not be extradited from Canada to Romania, because the alleged crimes occurred on a Taiwanese-registered vessel on the high seas, but the extradition agreement between Canada and Romania applied only to acts allegedly committed within the territory of either country. In addition, Canada's Extradition Act[43] was held to apply only to acts allegedly committed on Canadian territory. MacDonald J. commented:

> Professor Stephen Toope, a leading authority on international law testified on behalf of the intervenor and gave expert evidence on meaning of "territorial jurisdiction" in the context of both the treaty and the Extradition Act. He concluded that the treaty's reference to "committed within the territory" means "committed within [the geographical boundaries of] the territory. His conclusions were based on established principles

---
42. See O'Connell (n. 18 above), p. 733.
43. R.S.C., c. E-2, 1985.

of international law. He placed particular emphasis on the meaning of "territory" at the time the treaty was first enacted. In other words *in the late nineteenth century the term territorial jurisdiction involved only the concept of geography.* I fully accept Professor Toope's conclusions in this regard.[44]

The law, however, is not static, as is hinted at by the fact that MacDonald J.'s discussion focuses on the meaning of territorial jurisdiction "in the late nineteenth century," The question arises as to the current meaning of territorial jurisdiction. The issue arises in a number of cases that involve transboundary or transnational criminal offences—offences that might arguably be held to occur in more than one jurisdiction. One such case was decided by the Supreme Court of Canada in 1985. In R. v. Libman,[45] the accused ran a telephone operation in Toronto. His employees engaged in telephone sales to U.S. residents. They sold fraudulent shares in nonexistent companies that were supposedly mining gold in Costa Rica. Thus, in that case, the crime of fraud might arguably have occurred in three different countries. After reviewing several precedents from the United Kingdom and Canada, La Forest J. commented that:

> While . . . there were occasional strong expressions of the territorial doctrine, particularly in earlier times, the fact is that the courts never applied the doctrine rigidly. To have done so, as Cockburn C.J. noted in Keyn, . . . would have meant that a state could not apply its laws to offences whose elements occurred in several countries. This would have provided an easy escape for international criminals.[46]

La Forest J. also noted that

> The cases reveal several possibilities . . . One is to assume that jurisdiction lies in the country where the act is planned or initiated. Other possibilities include the place where the impact of an offence is felt, where it is completed, or again where the gravamen, or essential element of the offence took place. It is also possible to maintain that any country where a substantial or any part of the chain of events constituting an offence takes place may take jurisdiction.[47]

---

44. Romania (State) v. Cheng, 158 N.S.R. (2d) 13 at para. 123 (1997), emphasis added. Affirmed: 162 N.S.R. (2d) 395 (C.A.) (1997).
45. 2 S.C.R. 178 (1985).
46. Ibid., p. 208–09.
47. Ibid., p. 185–86. After reviewing the English precedents, La Forest J. held (at 213) that Canada may invoke the territorial principle to assert jurisdiction whenever there is a "real and substantial link" between the alleged offence and Canadian territory. In this case, he held that the test was met because the scheme was devised in Canada, and the whole operation that made it function, including the directing minds and the telephone solicitation, was located in Canada.

The idea that that country "where the act is planned or initiated" may claim jurisdiction has already been discussed, in relation to water taxi service and legislation that aims to prohibit activities on voyages which begin and end within the State.

The idea that the country "where the impact of an offence is felt" may claim jurisdiction is commonly referred to as the "objective territorial principle." This principle was the basis of decision in the Lotus case, in which Turkey successfully asserted jurisdiction over an act of criminal negligence committed by a French officer on board a French ship on the high seas. The French officer's criminal negligence led to a collision with a Turkish ship, which subsequently sank, leading to the death of several Turkish crew members and passengers. According to Ian Brownlie

> The basis of the majority view on the Court . . . was the principle of objective territorial jurisdiction. This principle was familiar but to apply it the Court had to assimilate the Turkish vessel to Turkish national territory. On this view the collision had affected Turkish territory.[48]

Significantly, the jurisdiction of the coastal State (Turkey) was held to apply to acts by foreigners (French nationals) on board a foreign (French) ship on the high seas. This suggests that the objective territorial principle could be used by a coastal State to justify the exercise of criminal jurisdiction over acts by foreigners on the Netherlands-registered *Sea Change*. However, there are two difficulties with this analysis. First, in the Lotus case, a Turkish vessel was damaged. Absent any similar damage to a vessel registered in one of the coastal States, the ruling in the Lotus case might not apply. Second, the Lotus case was decided by the slimmest of margins,[49] and the doctrine set forth in the majority decision has been negatived by Article 97 of UNCLOS.[50]

The objective territorial principle was recently expressed in a Supreme Court of Florida decision, upheld by the United States Supreme Court. In State v. Stepansky,[51] the court held that Florida had criminal jurisdiction to prosecute a U.S. citizen for burglary and attempted sexual battery allegedly

---

48. I. Brownlie, *Principles of Public International Law*, 5th ed. (Oxford: Oxford Univ. Press, 1998), p. 305.
49. Brownlie notes that the case was decided by the casting vote of the President of the Permanent Court of International Justice, as the votes were equally divided, six on either side: (Ibid., p. 304–05)
50. Article 97(3) provides that in the event of a collision on the high seas, "(n)o arrest or detention of the ship, even as a measure of investigation, shall be ordered by any authorities other than those of the flag State." In the event of a high seas collision, the Article prevents a coastal state from acting against a ship flying a foreign flag.
51. 761 So. 2d 1027. Decided: 20 April 2000. Rehearing denied 12 June 2000. Released for publication 12 June 2000. Certiorari denied 30 October 2000 (reported at 2000 U.S. LEXIS 7081).

committed against another U.S. citizen on a Liberian-registered cruise ship on the high seas.[52] The court stated that

> Subject to constitutional limitations, *a State may exercise jurisdiction on the basis of territoriality, including effects within the territory,* and, in some respects at least, on the basis of citizenship, residence, or domicile in the State.[53]

Following this type of reasoning, the offended coastal State could invoke the territorial principle to justify criminal legislation aimed at preventing harmful "effects" within the territory of the State. For example, a State could enact legislation making it a crime to "kill" the fetus, who is a "person" ordinarily residing within the territory of the State. Alternatively, legislation could prohibit "killing" the fetus because it causes grief to a "father" living within the territory of the State.[54] In either case, such legislation could conflict with the flag State jurisdiction of the Netherlands, because the law of the Netherlands allows abortions. Therefore, it is crucial to consider how a conflict between the jurisdiction of the coastal State and the flag State might be resolved. Specifically, the question arises as to whether the Netherlands would be entitled to prevent the coastal State from prosecuting in relation to abortions performed on a Netherlands-registered vessel in international waters. Unfortunately, Stepansky provides little guidance on this issue. In Stepansky, the court held that because the flag State (Liberia) had not attempted to block Florida's prosecution of the offence, the issue of whether a flag State could prevent a coastal State from exercising criminal jurisdiction over activities occurring on the flag State's vessel in international waters was not properly before the court.[55]

---

52. The cruise ship departed from and returned to Florida. Although both the accused and the complainant were U.S. citizens, neither of them were Florida residents. At the time of the alleged crime, the cruise ship was approximately 100 nautical miles from the Atlantic coastline of Florida.

53. See n. 51 above, p. 1032, emphasis in original.

54. However, this latter type of legislation might be invalid, as Akehurst suggests that the territorial principle of criminal jurisdiction should only be used to prohibit an extraterritorial activity that has effects within the territory if the "primary effects" of that activity are felt within the territory of a coastal state. (n. 5 above, p. 154). In the case of offshore abortions, it might be argued that the primary effects of the abortion would occur on board the ship where the abortion takes place.

55. "Stepansky asserts this prosecution is prevented by the flag-state rule set forth in the Geneva Convention on the High Seas . . . However, as Stepansky conceded during oral argument, criminal defendants lack standing to raise a violation of an international treaty that is not self-executing . . . Article 6 of the Geneva Convention on the High Seas is not a self-executing treaty and does not operate to limit the jurisdiction traditionally asserted by the United States over foreign vessels on the high seas . . . Therefore, the question of whether section 910.006(3)(d) is in violation of this treaty is not properly before this Court." See n. 51 above, p. 1032.

Conclusion re: Territoriality

Although the modified objective territorial principle might not be applicable in this case, the coastal State is likely to succeed if it focuses its legislation on the ferrying back and forth of women from territorial waters to international waters. The coastal State could pass legislation making it a crime to receive or provide an abortion on a voyage beginning and ending within the territorial waters of the State. While there is a chance that the territoriality principle might not justify applying this legislation to nationals of the coastal State, the active personality principle[56] could be invoked to justify such legislation.

The coastal State could also rely on the doctrine of constructive presence to assert territorial jurisdiction over the vessel *Sea Change*. To counter the coastal State's attempt to assert jurisdiction, the crew of *Sea Change* could remain in international waters and use boats other than its own to communicate with the shore. The crew would argue that while "simple constructive presence" is a well-established doctrine at international law, "extensive constructive presence" is not. It would note that most of the cases in which the doctrine of extensive constructive presence have been raised involve drug trafficking, and argue that drug trafficking is not analogous to abortion, because the number of nations that condemn abortion is much less than the number of nations that condemn drug trafficking.

THE UNITED NATIONS CONVENTION ON THE LAW OF THE SEA (UNCLOS)

The provisions of UNCLOS may help resolve disputes about criminal jurisdiction for two reasons. First, the Netherlands has ratified UNCLOS, and several coastal States that have ratified UNCLOS prohibit abortion,[57] so the terms of the treaty will be binding in any disputes between those States and

---

56. For a discussion of the active personality principle, see infra n. 83 and accompanying text.

57. For example, Malta ratified UNCLOS on 20 May 1993 and Chile ratified UNCLOS on 25 August 1997. Both these countries prohibit abortion on any grounds (even if the woman's life is at stake) and impose penalties on abortion providers. See Simon (n. 9 above), p. 49. According to the Chilean Penal Code, anyone who performs an abortion *with the woman's consent* is subject to 3 years in prison. See United Nations, *Abortion Policies: a Global Review*. Dept. for Economic and Social Information and Policy Analysis. Vol. 1. New York: United Nations, 1993, p. 83. Similarly, a person who performs an abortion in Malta may be subject to a period of imprisonment ranging from 18 months to 3 years. See United Nations, *Abortion Policies,* Vol. 2, (n. 13 above), p. 133.

530   Security and Military Activities

the Netherlands. Second, UNCLOS may also be relevant to countries that have not ratified the treaty,[58] because it could be argued that UNCLOS is indicative of customary international law.

Within the Territorial Sea

Assuming that Women on Waves did not perform abortions within the territorial sea,[59] but merely passed through the territorial sea while transporting women to and from the abortion ship, the question arises as to whether or not that would be a violation of the right of innocent passage. The question arises whether Women on Waves transports women by boat or by helicopter, because Article 2(2) states that the sovereignty of a coastal State over its territorial sea "extends to the airspace over the territorial sea."

Article 21 allows the coastal State to adopt laws and regulations relating to innocent passage through the territorial sea, and states that foreign ships passing through the territorial sea shall comply with these laws. If these laws are breached, Article 111 allows the coastal State to exercise the right of hot pursuit. However, the contemplated fact situation is unusual, in that the only activities to which the anti-abortionist coastal State is likely to object would occur beyond the limits of the territorial sea. Put another way, it is unlikely that a coastal State would object to its nationals going for a mere pleasure cruise to international waters and back, provided that during the pleasure cruise the nationals did not receive abortions or information related to procuring abortions. This suggests that from the point of view of the coastal State, the objectionable activity is not the passage through territorial waters, but an activity occurring outside of territorial waters.

Some writers suggest that if passage in and of itself is not objectionable, activities that occur beyond the limits of the territorial sea cannot render passage noninnocent. W. Burke comments that ". . . the innocence, or lack thereof, of passage must be determined only by specific acts occurring dur-

58. For example, the United States of America.
59. If Women on Waves performed abortions within the territorial sea of countries that are parties to UNCLOS, the abortion providers could be subject to the criminal jurisdiction of the coastal state, pursuant to Article 27 of UNCLOS. Article 27 allows the coastal state to assert criminal jurisdiction over a foreign ship in the territorial sea if, inter alia, "the consequences of the crime extend to the coastal State" (Article 27[a]), or "the crime is of a kind to disturb the peace of the country or the good order of the territorial sea" (Article 27[b]). If the coastal state considers the fetus to be a national, it seems likely that the consequences of "killing" the fetus would extend to the coastal state. Also, given that abortion is such a divisive issue, and that death threats and attempted killings of doctors who provide abortions occur from time to time, it is reasonable to suppose that if Women on Waves carried out its plan, its activities might "disturb the peace of the country or the good order of the territorial sea."

ing the passage *in* the territorial sea itself."[60] Commenting on this statement, Francis Ngantcha writes that "... it could be claimed that the coastal State's subjective appraisal of passage as noninnocent is limited, restricted objectively by the requirement that the violation occur while the foreign ship is *in* passage."[61] Since the activities of the vessel *Sea Change* within the territorial sea would be limited to ferrying women back and forth, the commentary quoted above suggests that this passage would be innocent. If this analysis is correct, the coastal State may not interfere with the right of *Sea Change* to innocently pass through territorial waters, as guaranteed by Article 24(1)(a).

Beyond the Territorial Sea

Article 27(1) states that a coastal State should not ordinarily exercise criminal jurisdiction on a foreign ship in the territorial sea.[62] Article 27(5) limits the jurisdiction of the coastal State in relation to any crime committed before the ship has entered the territorial sea.:

> *Except as provided* in Part XII [protection and preservation of the marine environment] or with respect to violations of laws and regulations adopted in accordance with Part V [exclusive economic zone], *the coastal State may not take any steps on board a foreign ship passing through the territorial sea to arrest any person or to conduct any investigation in connection with any crime committed before the ship entered the territorial sea,* if the ship, proceeding from a foreign port, is only passing through the territorial sea without entering internal waters.[63]

---

60. W. Burke, *Contemporary Law of the Sea: Transportation: Communication and Flight.* (Kingston, R. I.: Law of the Sea Institute, Univ. of Rhode Island, 1975) [Occasional Paper (Law of the Sea Institute), No. 28] Cited in F. Ngantcha, *The Right of Innocent Passage and the Evolution of the International Law of the Sea* (London: Pinter Publishers, 1990), p. 51, emphasis in original.

61. F. Ngantcha, *The Right of Innocent Passage and the Evolution of the International Law of the Sea* (London: Pinter Publishers, 1990), p. 51, emphasis in original.

62. Art. 27(1) states, "The criminal jurisdiction of the coastal state should not be exercised on board a foreign ship passing through the territorial sea to arrest any person or to conduct any investigation in connection with any crime committed on board the ship during its passage, save only in the following cases: (a) if the consequences of the crime extend to the coastal state; (b) if the crime is of a kind to disturb the peace of the country or the good order of the territorial sea; (c) if the assistance of the local authorities has been requested by the master of the ship or by a diplomatic agent or consular officer of the flag State; or (d) if such measures are necessary for the suppression of illicit traffic in narcotic drugs or psychotropic substances."

63. Emphases added. Similar provisions are found in the Geneva Convention on the Territorial Sea and Contiguous Zone, and the Montego Bay Convention. According to O'Connell: "(Article 19[5] of the Geneva Convention on the Territorial Sea and Contiguous Zone) expressly states that the coastal state may not take

The provision can be interpreted as a general ban on extraterritorial criminal jurisdiction, subject to two exceptions: those listed in Part XII and those listed in Part V. Part XII is not relevant to this discussion, but Part V deserves consideration.

Part V—Exclusive Economic Zone [and High Seas]
Part V of UNCLOS sets out the legal regime for the exclusive economic zone (EEZ). The EEZ begins where the territorial sea ends,[64] and ends 200 miles from the coastline of the coastal State.[65] The EEZ also overlaps with the high seas because Article 58(2) states that Articles 88–115 apply to the EEZ,[66] and those same Articles set out most of the legal regime governing the high seas.[67]

Article 92 states that when a ship is sailing under the flag of a particular State on the high seas, the starting presumption is that the ship is subject to the exclusive jurisdiction of the flag State. In this case, the starting presumption would be that *Sea Change*, which is registered in the Netherlands, would be subject to the exclusive jurisdiction of the Netherlands while on the high seas. However, this presumption may be rebutted in exceptional circumstances. Although none of the exceptional circumstances listed in UNCLOS exactly match the facts of this case, analogies can be drawn.

Article 109—Unauthorized Broadcasting
Article 109 gives a coastal State jurisdiction over unauthorized broadcasting from a foreign ship on the high seas. Article 109(2) defines unauthorized broadcasting as "the transmission of sound radio or television broadcasts from a ship or installation on the high seas intended for reception by the general public contrary to international regulations, but excluding the transmission of distress calls."

---

any steps on board a foreign ship which is only proceeding through the territorial sea without entering internal waters, in respect of any crime committed before the ship entered the territorial sea. (Article 27[5] of) (t)he Montego Bay Convention, which otherwise adopts the text of Article 19, makes this subject to its provisions relating to violations of the coastal State's laws respecting the marine environment and the EEZ." See O'Connell (n. 18 above), p. 957.

64. Article 55 states that the EEZ is an area "beyond and *adjacent to the territorial sea.*" (emphasis added)

65. Article 58.

66. As well as "pertinent rules of international law," but only insofar as they are not incompatible with Part V of UNCLOS, which deals mainly with conservation of living resources.

67. The legal regime for the high seas is set out in Articles 86–120. Therefore, the difference between the high seas and the EEZ is that the following apply in the high seas but not in the EEZ: Articles 86–87 and Articles 116–20. Since none of these Articles (86–87 and 116–20) are particularly relevant in this case, I will treat the EEZ and the high seas as functionally equivalent.

Article 109(3)(e) provides that:
Any person engaged in unauthorized broadcasting may be prosecuted before the court of:
. . .
(e) any State where authorized radio communication is suffering interference.

Article 109(3)(e) thus describes an exceptional circumstance that rebuts the presumption of exclusive flag State jurisdiction on the high seas expressed in Article 92.

To better understand Article 109, one may find it useful to consider its legislative history. O'Connell notes that

> At the Third Law of the Sea Conference, the countries of the European Economic Community submitted proposals on the subject of transmission from the high seas, which was introduced without opposition by the United Kingdom, and instantly supported by twenty-eight delegations.[68]

Since the Official Records of the Third United Nations Conference on the Law of the Sea provide little background information about Article 109, to get a fuller appreciation of its history one must examine the international law upon which it is based. This analytical approach is warranted in view of the fact that Article 58(2) states that "pertinent rules of international law" apply to the EEZ.

Article 109 represents the codification of international law relating to unauthorized broadcasting, such as The Radio Regulations, 1959 and the European Agreement for the Prevention of Broadcasts Transmitted from Stations Outside National Territories (European Agreement),[69] which was opened for signature in 1965. Article 1 of the European Agreement states that

> This Agreement is concerned with broadcasting stations which are installed or maintained on board ships, aircraft, or any other floating or airborne objects and which, outside national territories, transmit broadcasts intended for reception or capable of being received, wholly or in part, within the territory of any Contracting Party, or *which cause harmful interference* to any radio-communication service operating under the authority of a Contracting Party in accordance with the Radio Regulations.[70]

68. See O'Connell (n. 18 above), p. 819.
69. Strasbourg, 22 January 1965. In force 19 October 1967. Eighteen ratifications (as of 28 October 1998) 634 United Nations Treaty Series 239; *NDI*, p. 211, 212 and 270.
70. Emphasis added.

Although the European Agreement refers only to a desire to avoid "harmful interference" with radio frequencies, other reasons may have motivated European countries to attempt to stop unauthorized broadcasting. While N. March Hunnings emphasizes the fact that "pirate" broadcasting stations threatened the broadcasting monopoly of local stations, escaped copyright laws and avoided paying income taxes and other taxes,[71] other writers emphasize other problems. H.F. Van Panhuys comments that

> The appearance of these so-called "pirate" stations raises several important problems of domestic and international law ... [T]he transmissions are intended to be heard and produce an effect within the territory of the coastal States without being subject to the control of any State. Their contents might endanger the security, the *public order*, the mental health or the *good morals* of the coastal States.[72]

Similarly, Delbert D. Smith comments that "[c]ontinuous 'pop' music may be considered socially or *morally objectionable* by government officials."[73] Concerns about morality are also expressed by Horace B. Robertson, Jr., who comments as follows:

> ... [T]he pirate stations presented what might be called a sociological problem. Continuous rock music was considered by the British power structure as somehow *morally decadent* despite its substantial popularity.[74]

Given that several commentators note that a concern about morality may have motivated European countries to adopt the European Agreement, which was the precursor of Article 109 of UNCLOS, it is arguably appropriate for Article 109 to be interpreted in such a way as to give the coastal State jurisdiction to prohibit unauthorized broadcasting from the high seas

---

71. N. M. Hunnings, "Pirate broadcasting in European waters," *International Comparative Law Quarterly* 14 (1965): 410 at 413.

72. H. F. Van Panhuys and M. J. Van Emde Boas, "Legal aspects of pirate broadcasting," *American Journal of International Law* 60 (1966): 303–11, emphases added. Van Panhuys also mentioned concerns that unauthorized broadcasting could fill the airwaves, preventing the transmission of radio broadcasts that are required for the navigational safety of ships and airplanes. Thus, there was a concern that unauthorized radio broadcasts could threaten public safety.

73. D. D. Smith, "Pirate broadcasting," *Southern California Law Review* 41 (1968): 769 at 772, emphasis added.

74. H. B. Robertson, Jr., "The suppression of pirate radio broadcasting: a test case of the international system for control of activities outside national territory," *Law and Contemporary Problems* 45 (Winter 1982): 71 at 76, emphasis added. As to why this might be a threat to morality, Robertson commented that: "It (unauthorized broadcasting) epitomized the social turmoil of the 1960s, an era marked by long hair, eccentric clothes and free life style."

on the sole grounds that it considers the effects of such broadcasting to be immoral.

Furthermore, adopting a broad, purposive approach to the interpretation of Article 109, the coastal State could argue that the application of Article 109 is not limited to regulating the morality of mere radio broadcasts. Instead, the coastal State could argue that Article 109 establishes in UNCLOS a broad principle that allows coastal States to assert jurisdiction with respect to activities on the high seas that are intended to produce effects within the territory of the coastal State that the coastal State considers to be morally objectionable. In making this argument, the coastal State could refer to the "modified objective territorial principle" discussed in the previous part of this article. Admittedly this argument is a bit of a stretch, because a literal interpretation of Article 109 suggests that the Article is merely intended to address the specific problem of unauthorized broadcasting.

If these arguments were to succeed, the coastal State would be entitled to exercise the following remedies.[75] First, if the coastal State reasonably suspected that *Sea Change* was engaging in activity calculated to have effects within the territory of the coastal State that the coastal State considered to be immoral, the coastal State could board the vessel.[76] If *Sea Change* resisted the attempted boarding, the coastal State could use force "as a measure of last resort."[77] Second, if boarding did not discharge the suspicion of the boarding party, the boarding party could proceed to search the vessel,[78] arrest her, and seize her abortion-related equipment.[79] The coastal State could then prosecute those involved in the unlawful activity—no matter what their nationality.[80]

If *Sea Change* attempted to flee, the coastal State could exercise the right of hot pursuit. Pursuant to Article 111, hot pursuit could begin when *Sea Change* was in international waters, provided that *Sea Change* was constructively presumed to be in the territorial waters of the coastal State. Like Article 23 of the Geneva Convention on the High Seas, Article 111 of UNCLOS suggests two different interpretations of the doctrine of constructive presence. While Article 111(1) suggests that the doctrine should be construed

---

75. The outline of the remedies which follows is adapted from: R. C. F. Reuland, "Interference with non-national ships on the high seas: peacetime exceptions to the exclusivity of flag-state jurisdiction," *Vanderbilt Journal of Transnational Law* 22(5) (1989): 1161 at 1226–27.

76. Article 110(1)(c).

77. O'Connell writes that: ". . . (W)hile force may be employed in arrest of foreign ships which resist boarding, this is a measure of last resort." See O'Connell (n. 18 above), p. 1073.

78. Article 110(2).

79. Article 109(4) allows a coastal state to seize the "broadcasting apparatus" of the foreign vessel. By analogy, Article 109(4) might allow the coastal state to seize abortion-related equipment.

80. Article 109(3)(e).

narrowly, as "simple constructive presence," Article 111(4) suggests that it should be interpreted broadly, as "extensive constructive presence." If *Sea Change* remained in international waters but used boats to communicate with the shore, the coastal State could invoke one or both of these interpretations to justify the exercise of hot pursuit.

Conclusion

This discussion of UNCLOS leads to the conclusion that the activities contemplated by Women on Waves do not violate the right of innocent passage that is guaranteed by Article 17, and therefore, pursuant to Article 24(1)(a), the coastal State must not interfere with Women on Waves' right to innocently pass through territorial waters. If the coastal State were to interfere in this way, the Netherlands would have a grievance against the coastal State, and the coastal State would be required to exchange views with the Netherlands, pursuant to Article 283.

Beyond the territorial sea, a broad interpretation of Article 109, though somewhat unlikely, could subject *Sea Change* to the jurisdiction of the coastal State if *Sea Change* were engaging in activities that were having effects within the territory of the coastal State that the coastal State considered to be morally offensive. If so, the coastal State would be allowed to board *Sea Change*, search her, arrest the vessel or the crew, and seize any abortion-related equipment. The coastal State could then prosecute those involved in the unlawful activity—no matter what their nationality. If *Sea Change* attempted to flee, the coastal State could invoke the doctrine of constructive presence to justify the exercise of hot pursuit.

OTHER INTERNATIONAL LEGAL PRINCIPLES

Although the territorial principle set out earlier is well established in the common law systems, Akehurst notes that other principles of criminal jurisdiction, such as the universality principle, and the principle of passive personality, also have a long and venerable history, and continue to be applied today:

> The territorial principle is very deep-rooted in English-speaking countries . . . Some writers from English-speaking countries imagine that the criminal law of all countries was originally based on the territorial principle, and that other bases of jurisdiction are recent (and usually questionable) innovations. But in many continental countries the universality principle is as ancient as the territorial principle in England. It existed in medieval Italy, sixteenth-century Brittany, seventeenth-and-

eighteenth-century France until 1782, and seventeenth-and-eighteenth-century Germany. It was supported by Grotius, Vattel, Paul Voet, Huber and Bynkershoek, not to mention lesser-known writers in sixteenth-and-seventeenth century Belgium. Countries claiming jurisdiction under the universality and/or passive personality principles in modern times include Argentina, Austria, Belgium, Bulgaria, Colombia, Czechoslovakia, Finland, Germany, Greece, Guatemala, Hungary, Italy, Japan, Mexico, Monaco, Peru, Romania, San Marino, South Korea, Switzerland, Turkey, Uruguay, Venezuela and Yugoslavia.[81]

The 1935 Harvard Draft Convention on Jurisdiction with Respect to Crime[82] identified five sources of criminal jurisdiction. Other than the territorial principle, these principles are active personality, passive personality, the protective principle, and the universality principle. I will discuss the application of each of these principles in turn, with particular attention to how any potential conflicts with a claim of flag State jurisdiction might be resolved.

Active Personality

The active personality principle of criminal jurisdiction allows a State to assert jurisdiction over acts committed by its nationals outside of national territory. This principle is well established in international law.[83] The principle has recently been invoked by many nations in an attempt to stop child sex tourism.[84] Writing in 1998, Vitit Muntarbhorn notes:

---

81. See Akehurst (n. 5 above), p. 163–164.
82. E. D. Dickinson, Reporter, *American Journal of International Law* 29 (Supp.) (1935): 439.
83. Akehurst notes that "a state has an unlimited right to base jurisdiction on the nationality of the accused," and gives several examples of instances in which the active personality principle was invoked. See Akehurst (n. 5 above), p. 156.
84. Many countries make it a criminal offence to have sex with a child, but some do not. In an effort to avoid the criminal jurisdiction of their home countries, some people have engaged in "child sex tourism" by visiting other countries in order to have sex with children. This problem was addressed by representatives of 122 countries at the World Congress Against Commercial Sexual Exploitation of Children, which was held in Stockholm in August 1996. At the conclusion of the conference, an Agenda for Action against Commercial Sexual Exploitation of Children was drafted. Paragraph 4(d) of that document urges countries to: ". . . develop or strengthen and implement laws to criminalise the acts of the nationals of the countries of origin when committed against children in the countries of destination ('extra-territorial criminal laws'). . . ." Accessed 15 April 2001 on the World Wide Web: http://www.ecpat.org.

538 Security and Military Activities

> ... almost 20 countries have criminal laws, which can be used to apprehend their nationals, residents or others passing through their territory, for child sexual exploitation committed in other countries.[85]

These laws are sometimes applied to prohibit acts that are legal in the territory in which they are committed. For example, Tim MacIntosh comments on the Australian child sex tourism legislation:

> Significantly, unlike the operation of other world state's extra-territorial laws, the Act does not have a double-criminality requirement, thus obviating the need to establish that the offences created were offences in both the lex fori and the prosecuting state.[86]

MacIntosh also notes that the Australian legislation has been enforced as follows:

> To date there have been approximately 10 prosecutions, involving all stages of the criminal process. These prosecutions, in different States of the Commonwealth, have included pre-trial committal hearings, sentence proceedings in District Court of County Court jurisdictions . . . and one conviction following a jury trial.[87]

Like Australia, Canada is one of many countries that have passed criminal legislation prohibiting child sex tourism, based on the active personality principle.[88] When this legislation was debated in the House of Commons, one Member of Parliament commented as follows:

> People who commit infractions outside the country could be prosecuted here, and this is precisely an essential element of this bill. Canadians, not only Canadian citizens, but also residents, refugees and asylum seekers, who commit crimes in another country could be prosecuted

---

85. V. Muntarbhorn, *Extraterritorial Criminal Laws Against Child Sexual Exploitation* (Geneva: UNICEF, 1998), p. 19. Muntarbhorn lists these countries as: Australia, Austria, Belgium, Denmark, Finland, France, Germany, Ireland, Italy, Japan, Netherlands, New Zealand, Norway, Spain, Sweden, Switzerland, Thailand, the U.K., and the U.S. Of these countries, Ireland prohibits abortion, and some of the other countries impose restrictions upon abortion.

86. T. MacIntosh, "Exploring the boundaries: the impact of the child sex tourism legislation," *Australian Law Journal* 74(9) (2000): 613 at 616.

87. Ibid., p. 617.

88. An Act to Amend the Criminal Code (Child Prostitution, Child Sex Tourism, Criminal Harassment and Female Genital Mutilation). S.C., c. 16 (1997). Section 1 of this Act states that it only applies to Canadian citizens and permanent residents, and that prosecution can only be initiated if the country where the offence occurred asks the Minister of Justice of Canada to prosecute.

in Canada. I believe this is the way to get rid of this reprehensible practice. If the principle of extraterritoriality is not applied, the bill's effect will be too limited and it will not have much significance.[89]

Given that the active personality principle is well established in international law and that prosecutions pursuant to this principle are sometimes successful even in the absence of a double-criminality requirement, the question arises as to how a potential conflict between the active personality principle and the flag State principle might be resolved. Two cases shed light on this issue.

In United States v. Black,[90] the accused were American citizens charged with operating a non-American vessel on a cruise from New York harbor into international waters and back to New York. Once in international waters, a group known as "The Sons of Italy" conducted gaming activities in an area set aside by the ship's master. The accused were charged under a federal statute applicable to "any citizen or resident of the United States, or any other person who is on an American vessel or is otherwise under or within the jurisdiction of the United States, directly or indirectly."[91] The accused argued that even though they were American citizens, they were not subject to U.S. jurisdiction, because the alleged acts occurred on a foreign ship in international waters. The District Court for the Southern District of New York did not accept this argument, and upheld the indictment.

The court's reasoning might be applied in this case to allow the anti-abortionist coastal State to assert jurisdiction over its nationals on board the foreign-registered *Sea Change*. However, this case might not apply to the current factual situation, for two reasons. First, the accused in United States v. Black were U.S. *citizens,* so the case arguably might not apply to *residents* of the coastal State. Secondly, the accused in United States v. Black operated the gambling ship, so the case arguably might not apply to women from the coastal State who are mere passengers on board the vessel *Sea Change*.

Faced with a similar potential conflict between the active personality principle and the flag State principle, a 19th century British court avoided deciding the issue of which principle should prevail. Instead, the court resolved the case before it by resorting to the modified objective territorial principle. Geoffrey Marston comments:

> In 1856, one Conolly, a *British subject,* struck an American seaman who was a fellow crew member of the *American ship Driver* when the ship was an unspecified distance from the southern coast of Ireland. The injured man died on shore at Liverpool . . .

---

89. Mr. Nunez, *Commons Debates,* p. 3580, (10 June 1996).
90. 291 F. Supp. 262, S.D.N.Y. (1968).
91. 18 U.S.C., sec. 1802(a).

540   Security and Military Activities

The accused was duly tried at Liverpool Spring Assizes and convicted, apparently on the basis of section 8 of the statute 9 Geo. 4, c. 31. This section provided that if any person was feloniously hurt on the sea and should die in England then *the crime could be dealt with in the place where the death took place*.[92]

This discussion suggests that although the active personality principle is gaining increasing prominence in international law, its application on board foreign ships may be quite limited.

Passive Personality

According to the passive personality principle, a State may assert jurisdiction over foreigners who harm its nationals. The State thus asserts jurisdiction because the victims of the crimes are its nationals. Pursuant to this principle, the coastal State could create a crime of "killing" its fetuses, or a crime of causing grief and suffering to "fathers" of fetuses who are "killed." However, Brownlie comments that the passive personality principle is "the least justifiable . . . of the various bases of jurisdiction,"[93] so I will not consider its potential application in this case.

Protective Principle

Pursuant to the protective principle, a State may assert jurisdiction over non-nationals where their acts jeopardize the security interests of the State. An example of this type of jurisdiction is the jurisdiction claimed by the U.S. over non-nationals who assassinate, kidnap or assault members of Congress, Cabinet members, Supreme Court Justices, heads of executive departments or their deputies.[94] In this case, the protective principle does not seem to be very relevant.

        92. G. Marston, "Crimes on foreign merchant ships at sea," *Law Quarterly Review* 88 (1972): 357 at 363, emphases added.
        93. See Brownlie (n. 48 above), p. 306.
        94. 18 U.S.C., c.18, s. 351(i). Added Jan. 2, 1971, P.L. 97-285, ss. 1, 2(a), 96 Stat. 1219. Extraterritorial aspect confirmed: United States v. Layton (1998, CA9 Cal) 855 F2d 1388, 26 Fed Rules Evid Serv 988, cert den (1989) 489 US 1046, 103 L Ed 2d 244, 109 S Ct 1178. For discussion of this legislation, see: Petersen, "The extraterritorial effect of federal criminal statutes: offences directed at Members of Congress," *Hastings Int'l & Comp. L. Rev.* 6 (1983): 773. See, further: I. Cameron, *The Protective Principle of International Criminal Jurisdiction* (Aldershot, England: Dartmouth, 1994) at p. 252.

Universality Principle

The universality principle allows a State to assert jurisdiction over international crimes or crimes against the global community (i.e.,—war crimes, crimes against humanity, genocide), regardless of the nationality of the wrongdoer.[95] In this case, the coastal State might argue that the intended provision of 5,000 abortions per year by Women on Waves would amount to a crime against humanity because it is a large-scale systematic plan of "murder" organized by a group.

Recently, criminal legislation justified by the universality principle has been passed by Belgium and Canada. A recent decision of the International Court of Justice considered Belgian legislation that empowers Belgian criminal courts to hear cases of alleged human rights violations committed by foreigners against foreigners in foreign territory.[96] Similar legislation has recently been passed in Canada, allowing alleged perpetrators of crimes against humanity to be prosecuted as soon as they enter Canada, even if they are foreigners and the alleged crimes were committed against foreigners in foreign territory.[97]

---

95. For a review of this topic, see: S. Macedo, ed. *The Princeton Principles on Universal Jurisdiction*. (Princeton, N. J.: Program in Law and Public Affairs, 2001). Accessed August 21, 2002 on the World Wide Web: http://www.princeton.edu/lapa/univ_jur.pdf.

96. In *Arrest Warrant of 11 April 2000* (*Democratic Republic of the Congo v. Belgium*), the Democratic Republic of the Congo contested the validity of an arrest warrant issued pursuant to the Belgium Law of 16 June 1993 (as amended by the Law of 10 February 1999). The case was decided by the ICJ on 14 February 2002. In a dissenting opinion, Judge Oda commented, at 12:

> "It is one of the fundamental principles of international law that a State cannot exercise its jurisdiction outside its territory. However, the past few decades have seen a gradual widening in the scope of the jurisdiction to prescribe law. From the base established by the Permanent Court's decision in 1927 in the Lotus case, the scope of extraterritorial criminal jurisdiction has been expanded over the past few decades to cover the crimes of piracy, hijacking, etc. Universal jurisdiction is increasingly recognized in cases of terrorism and genocide. Belgium is known for taking the lead in this field and its 1993 Law (which would make Mr. Yerodia liable to punishment for any crimes against humanitarian law he committed outside of Belgium) may well be at the forefront of a trend. There is some national case law and some treaty-made law evidencing such a trend.
>
> Legal scholars the world over have written prolifically on this issue. Some of the opinions appended to this Judgement also give guidance in this respect. I believe, however, that the Court has shown wisdom in refraining from taking a definitive stance in this respect as the law is not sufficiently developed and, in fact, the Court is not requested in the present case to take a decision on this point.

Accessed 21 August 2002 on the World Wide Web: http://www.icj-cji.org/icjwww/idochet/iCOBE/icobejudgment/icobe_ijudgment_20020214_oda.PDF.

97. Crimes against Humanity and War Crimes Act, S.C. 2000, c. 24.

The legislation passed by Belgium and Canada provides a model that a coastal State could follow in passing legislation prohibiting the activities proposed by Women on Waves, based on the premise that such activities constitute a crime against humanity of murder. Crimes against humanity are defined in Article 7 of the Rome Statute of the International Criminal Court as follows:

> For the purpose of this Statute, "crime against humanity" means any of the following acts when committed as part of a widespread or systematic attack directed against any civilian population, with knowledge of the attack; (a) murder;[98]

In a document called *Elements of Crimes,* released in November 2000, the Preparatory Commission for the International Criminal Court outlined the elements of the crime against humanity of murder as follows:

1. The perpetrator killed one or more persons.
2. The conduct was committed as part of a widespread or systematic attack directed against a civilian population.
3. The perpetrator knew the conduct was part of or intended the conduct to be part of a widespread or systematic attack against a civilian population.[99]

In applying these elements to the current factual situation, it is clear that the first element would be extremely controversial, as it would be satisfied only if fetuses were held to be "persons" at international law. This condition is unlikely to be met. Abortion is a very divisive issue, and there appears to be no international consensus about whether or not a fetus is a legal person.[100] As a result, the coastal State's argument that Women on

---

98. UN Doc. A/CONF.183/9*, accessed 15 April 2001 on the World Wide Web: http://www.un.org/law/icc/statute/romefra.htm.

99. Finalized Draft Text of the Elements of Crimes. Adopted by the Preparatory Commission for the International Criminal Court at its 23rd meeting on 30 June 2000. Document*: PCNICC/2000/1/Add.2. Accessed 15 April 2001 on the World Wide Web: http://www.un.org/law/icc/statute/elements/english/1_add2.pdf#pagemode=bookmarks.

100. For a discussion of this topic, see: D. Shelton, "International Law on Protection of the Fetus," in S. J. Frankowski and G. F. Cole, eds., *Abortion and Protection of the Human Fetus: Legal Problems in a Cross-Cultural Perspective,* (Dordrecht, Netherlands: Martinus Nijhoff Publishers, 1987), p. 1. International law on this topic is far from clear. The current state of the law in the U.S. exemplifies this confusion. On the one hand, abortions have been permitted since Roe v. Wade, suggesting that the fetus is not a legal person. On the other hand, recent decisions of some state courts have held that a fetus may be protected by criminal prohibitions against homicide, suggesting that the fetus is a legal person. For example, in Hughes v. State of Oklahoma, 868 P.2d 730 (1994), the accused was charged with manslaughter arising

Waves is engaging in a crime against humanity of murder is likely to fail.

Conclusion

As regards the active personality principle, Women on Waves could argue on behalf of its clientele that most countries require double criminality,[101] so if the active personality principle were a principle of customary international law, it would require that double criminality be proved. Obviously, double criminality would be absent in this case, because the Netherlands does not prohibit abortion.

If the coastal State attempted to invoke the universality principle by alleging that a crime against humanity had been committed, such an argument would likely fail because it is not clear that a fetus is a person according to general principles of international law.

OBSTACLES TO THE JURISDICTION OF THE COASTAL STATE

If a coastal State attempted to assert jurisdiction by invoking either the territoriality principle or the active personality principle, such a principle could conflict with the jurisdiction of the Netherlands. As a hypothetical example, the Netherlands could pass legislation making it a criminal offence to interfere with one of its nationals who is performing an abortion on a Netherlands-registered ship. Under this hypothetical example, the Netherlands would claim jurisdiction based on two principles: the flag State principle and the principle of passive personality (because the national of the

---

out of a drunk driving accident which resulted in the driver's 8-month-old fetus being stillborn. The accused argued that she had not committed homicide, since the statute under which she was charged defined homicide as "the killing of one *human being* by another" (emphasis added). She argued that a fetus was not a "human being" within the meaning of the criminal statute. The Oklahoma Court of Criminal Appeals rejected her argument, and stated (at 734) that: ". . . (a) viable human fetus is nothing less than human life . . . Thus the term 'human being' in Section 691—according to its plain and ordinary meaning—includes a viable human fetus."

101. Double-criminality is a concept requiring that an act be criminal in both jurisdictions where two jurisdictions may apply. Women on Waves would argue that the lack of a double-criminality requirement in Australia is an exceptional circumstance. The comment by MacIntosh quoted earlier in this article suggests that this would be a valid argument, since he writes that Australia's legislation is "unlike . . . other world states' extra-territorial laws." In Canada, double criminality is a prerequisite for the prosecution of extra-territorial offences (R. v. Finta, 1 S.C.R. 701 [1994]).

Netherlands who would be providing the abortion would be the victim of the crime of interference). At the same time, the coastal State could assert jurisdiction over the abortion provider based on the territorial principle (for example, the doctrine of constructive presence) and the passive personality principle (because the fetus being "killed" would be the victim of the crime of abortion). This leads to a question of how these conflicting jurisdictional claims could be resolved.

According to Akehurst, there is no solution to this type of problem:

> Since several States can exercise concurrent jurisdiction in many cases, it can happen that an individual is forbidden by one law to do an act which is permitted by another law. In extreme cases the actual content of the law of one of the States concerned may be contrary to international law, or it may be possible to show that the law in question represents an abuse of legislative power and thus a breach of international law. But, apart from such extreme cases, there is probably no way of resolving the conflict, because 'international law . . . does not provide for choosing among competing bases of jurisdiction to prescribe rules of conduct.'[102]

However, according to the Supreme Court of Canada, in the case of "clear conflict" between juridictional claims, international law dictates that the jurisdictional claim with "the most significant connection to the events in question" should prevail.[103]

In determining which State has the most significant connection to the events in question, it may be helpful to consider the five bases of jurisdiction outlined in the Harvard Draft Convention on Jurisdiction with Respect to Crime. After listing these five principles, Paul Arnell states that

> The framework evinced above . . . is such to be useful . . . in forming the basis of a methodology to govern competing claims . . . [However, it] should be noted that an exception to this methodology is perhaps necessary where the State with the closest connection is either complicit in the commission of the crimes or does not desire to prosecute them.[104]

In the contemplated hypothetical situation, *both* States (the Netherlands and the anti-abortionist coastal State) would be complicit in crimes committed against the other, so there seems to be little justification for treating this as an exceptional circumstance. Accordingly, applying Arnell's methodology, one

---

102. See n. 5 above, p. 167–68.
103. *R. v. Cook*, [1998] 2 S.C.R. 597, at para. 137, per Bastarache J.
104. P. Arnell, "Criminal jurisdiction in international law," *Juridical Review* 3 (2000): 179 at 189.

must resolve the competing claims by examining all five principles of jurisdiction, and weighing the claim of one State against the claim of the other. Brownlie's comments imply that he would also support this type of approach:

> The various principles held to justify jurisdiction over aliens are commonly listed as independent and cumulative . . . However, it must be remembered that the 'principles' are in substance generalizations of a mass of national provisions . . . *It may be that each individual principle is only evidence of the reasonableness of jurisdiction.* The various principles often interweave in practice . . . These features have led some jurists, with considerable justification, to formulate a broad principle resting on some genuine or effective link between the crime and the state of the forum.[105]

Assuming that the coastal State succeeded in establishing jurisdiction, Women on Waves could still raise other objections, specific to certain countries. If Ireland were the objecting coastal State, Women on Waves could argue that the Irish prohibition against abortion conflicts with Ireland's obligations under the European Community Treaty.[106] If the objecting coastal State were one of the coastal States of the United States, Women on Waves could argue that U.S. constitutional law prohibits States from enacting extraterritorial prohibitions on abortion.[107]

ENFORCEMENT JURISDICTION AT INTERNATIONAL LAW

Even if the coastal State were able to assert jurisdiction at international law, it might still be unable to enforce it. Brownlie comments:

---

105. See Brownlie (n. 48 above), p. 309. Brownlie's comments are consistent with Canadian jurisprudence, which requires a "real and substantial link" between the alleged offence and Canadian territory before the territorial principle can be invoked to assert jurisdiction (See n. 47 above).

106. The European Community Treaty, which has been ratified by both Ireland and the Netherlands, states that nationals of member states have the right to provide services to nationals of other member states. The right to provide services has been held to include the right to receive services, including the right to receive medical treatment. Assuming that abortions are "medical treatment," the right guaranteed by the Treaty could be held to conflict with Irish law, in which case the Irish prohibition on abortion would have to be struck down. For a recent discussion of this topic, see: D. A. MacLean, "Can the EC kill the Irish unborn? An investigation of the European Community's ability to impinge on the moral sovereignty of member states," *Hofstra Law Review* 28(2) (1999): 527.

107. See: C. S. Bradford, "What happens if Roe is overruled? Extraterritorial regulation of abortion by the states," *Arizona Law Review* 35(1) (1993): 87. In this article, Bradford describes anti-abortion legislation in the U.S. before Roe v. Wade, and he notes that several states sought to apply their anti-abortion legislation extraterritorially, but this caused some constitutional problems.

The governing principle is that a state cannot take measures on the territory of another state by way of enforcement of national laws without the consent of the latter.[108]

Following this principle, if the Netherlands-registered vessel were held to be the "territory" of the Netherlands, the starting presumption would be that general principles of international law would not allow the coastal State to enforce its law on board the vessel *Sea Change*. On this point, O'Connell comments that historically ". . . a ship was thought of as a legal enclave in its own right, a piece of peripatetic national territory."[109] However, he notes that the "floating island" theory was "emphatically repudiated" by the Privy Council in Chung Chi Cheung v. R.[110] Therefore one may conclude that, pursuant to international law, if the coastal State were able to establish jurisdiction over activities occurring on board *Sea Change*, the coastal State would be able to enforce its laws on board the foreign ship.

CONCLUSION

This article has examined the arguments that a coastal State might advance to assert jurisdiction over criminal acts carried out on board foreign flag vessels outside its territorial waters. In this article, the specific situation of abortions carried out on board a ship, operating under the flag of a country that does not criminalize the medical act of abortion has been considered. The coastal State could employ various methods in an attempt to assert jurisdiction over its nationals as well as over foreigners, both within the territory of the coastal State, and beyond. The most difficult questions concern the application of the coastal State's jurisdiction to foreigners on a foreign ship in international waters. However, even here the coastal State might succeed in asserting jurisdiction.

From the point of view of the coastal State, perhaps the most compelling argument for jurisdiction is the doctrine of constructive presence. As we have seen, this doctrine operates both as part of UNCLOS and as a general principle of customary international law. Therefore, this doctrine may be invoked by any State to overrule the starting presumption of the exclusivity of flag State jurisdiction on the high seas. The best that the foreign nationals could do to avoid this doctrine would be to use boats other than their own boats, and argue that the doctrine should be interpreted narrowly, as "simple constructive presence" only.

Anti-abortionist coastal States that have ratified UNCLOS may put for-

---

108. See Brownlie (n. 48 above), p. 310.
109. See O'Connell (n. 18 above), p. 735.
110. AC 160 (1939).

ward other arguments under that treaty. Although arguments under Article 19 (innocent passage) are likely to fail, arguments related to Article 109 (unauthorized broadcasting) may stand a better chance (albeit a slim chance) of success. If a breach of Article 109 were established, the coastal State could exercise a right of visit pursuant to Article 110, and/or a right of hot pursuit pursuant to Article 111.

Some States that place greater restrictions on abortion than does the Netherlands have not ratified UNCLOS. The United States is an example of one such State. To assert jurisdiction, these States would have to advance arguments pursuant to general principles of international law. The most compelling of these arguments relate to the principle of territoriality, including the doctrine of constructive presence already mentioned. However, the active personality principle could also arguably be invoked to allow the coastal State to assert jurisdiction over its nationals. The success of this argument would depend on whether the active personality principle could be invoked in the absence of double criminality. The experience of Australia suggests that this may be so. If a coastal State were to succeed in advancing these arguments, it would likely also be able to enforce its laws on board the foreign ship.

Questions concerning the limits of the jurisdiction of coastal States to respond to actions by foreign flag vessels are very much alive as new challenges arise that engage the interests of coastal States in the zones just beyond the territorial sea. In the context of abortion services offered by foreigners to nationals outside the waters of a State that criminalizes abortion, it appears likely that traditional criminal jurisdiction could be exercised to address this issue: *Sea Change* might not bring about a sea change.

Security and Military Activities

Military Activities in the Exclusive Economic Zone: The Case of Aerial Surveillance†

Ivan Shearer
*Challis Professor of International Law, University of Sydney, Australia
Stockton Professor of International Law, United States Naval War College, Newport, Rhode Island*

INTRODUCTION

The collision between an unarmed United States military surveillance aircraft and a Chinese fighter jet over an area of seas some 50 nautical miles (M) southeast of Hainan Island[1] has drawn attention to the issue of the true juridical nature of the exclusive economic zone in international law. May a coastal state object to surveillance activity, or "spy flights," by foreign States over its EEZ?

On 1 April 2001, a surveillance aircraft of the type EP-3E Aries II operated by the United States Navy was flying on a surveillance mission from its base in Japan. It collided with an F8 fighter jet of the People's Liberation Army Navy (PLAN) over the South China Sea. The two aircraft touched each other in circumstances yet to be clarified, resulting in the loss of the Chinese aircraft and the death of its pilot, and the forced landing of the United States aircraft at the Lingshui military airfield on the island of Hainan. There had been a history of encounters between United States and Chinese military aircraft in the area, including alleged harassment of United States aircraft by Chinese aircraft, but until the incident of 1 April, no collisions. The EP-3E and its crew of 24 members were detained by the Chinese authorities for some days before being released. China lodged vigorous protests against the behavior of the United States aircraft and against the practice of con-

---

†EDITORS' NOTE.—This article is partly based on a presentation "The Law of the Sea and Military Operations" given by the author at a seminar conducted by the Marine and Environmental Law Programme of Dalhousie University in Halifax, Canada on 14 February 2001. The opinions expressed in this article are the author's own and do not express the views of the Naval War College or the United States government.

1. F. L. Kirgis, "United States reconnaissance aircraft collision with Chinese Jet," *American Society of International Law Insight* (April 2001). Accessed 19 August 2002 on the World Wide Web: http://www.asil.org/insights/htm.

© 2003 by The University of Chicago. All rights reserved.
0-226-06620-7/03/0017-0019$01.00 *Ocean Yearbook* 17:548–562

ducting surveillance missions in waters adjacent to its territorial sea. An investigation of the exact circumstances of the collision was opened by representatives of the two governments under the auspices of the Military Maritime Consultative Agreement, established as a confidence building measure in 1998.[2]

According to Chinese sources, the collision occurred in the airspace above the South China Sea at a point 104 km southeast of Hainan Island.[3] The waters at this point are part of the open sea and lie between Hainan and the Xisha (Paracel) Islands. They fall within China's declared Exclusive Economic Zone. The dispute between China and Vietnam over the ownership of the Paracel Islands does not appear to play any role in the affair, although the incident occurred at a point closer to the nearest of the Paracel Islands than to Hainan. Nor do Chinese claims to the disputed islands in the Spratly Group come into account, because these lie well to the south. Chinese sources recognize the area in question as belonging to its exclusive economic zone and argue that the international law right of free overflight of the EEZ does not include the right of surveillance or other activities that threaten the security of the coastal state.[4]

MARITIME ZONES CLAIMED BY CHINA

The People's Republic of China (PRC) claims a territorial sea of 12 M, a contiguous zone of 12 M beyond its territorial sea, and an exclusive economic zone of 200 M, each zone measured from the same declared baselines. A territorial sea of 12 M was first proclaimed by China on 4 September 1958.[5] In the same Declaration straight baselines were proclaimed linking the outermost islands with the mainland from which the territorial sea was measured. The Gulf of Bohai and Hainan Strait were claimed as internal waters. The baselines declared in 1958 without geographical coordinates,

---

2. Agreement between the Department of Defense of the United States of America and the Ministry of National Defense of the People's Republic of China on Establishing a Consultation Mechanism to Strengthen Military Maritime Safety, signed in Beijing, 19 January 1998. The author is indebted to Captain George Galdorisi, U.S. Navy (Ret) for this information. The history and work of the consultation mechanism is the subject of an outside publication by Captain Galdorisi and Lieutenant Commander George Capen, U.S. Navy, "Military Contact is Linchpin in Sino-U.S. Relations," 2001, US Naval Institute, Proceedings, 127:70.

3. (2001, 4 April). Chinese fighter bumped by U.S. military scout. Accessed 19 August 2002 on the World Wide Web: http://www.china-embassy.org/eng/9514.html.

4. Ibid.

5. United States Department of State, "National claims to maritime jurisdictions," *Limits in the Seas* 36 (4th revision, 1981).

and from which only hypothetical baselines could be drawn on a map,[6] were supplanted by the Declaration of 15 May 1996, which clearly indicated the coastal baselines drawn continuously from the Shandong Peninsula at the southern entrance point of Bohai Gulf in the north to the western coast of Hainan Island in the Gulf of Tonkin in the south.[7] Straight baselines were also declared and specified in the same Declaration around the Xisha (Paracel) Islands. No straight baselines have been promulgated by the PRC for Taiwan or Penghu, thus avoiding a direct confrontation with Taiwan, but the Taiwanese-held islands of Quemoy and Matsu close to the mainland are enclosed behind the PRC baselines.

The territorial sea and contiguous zone of the PRC are governed by the Law on the Territorial Sea and Contiguous Zone of 25 February 1992.[8] That Law follows the United Nations Convention on the Law of the Sea, 1982 (LOSC), except for the requirement of prior permission to enter the territorial sea by "foreign ships for military purposes."[9]

The PRC made a declaration claiming an EEZ and a continental shelf accompanying the deposit of its instrument of ratification of the LOSC on 7 June 1996. This declaration stated that "In accordance with the provisions of the United Nations Convention on the Law of the Sea, the People's Republic of China shall enjoy sovereign rights and jurisdiction over an exclusive economic zone of 200 nautical miles and continental shelf."[10] The declaration was followed in 1998 by the enactment of the Law of the PRC on the Exclusive Economic Zone and the Continental Shelf, 26 June 1998.[11]

On the assumption that there is only one China, the extension of the PRC exclusive economic zone from the mainland baselines into the Taiwan Strait would cover the entire Strait. Taiwan (the Republic of China) itself proclaimed an exclusive economic zone in 1998, which may be interpreted as purporting to apply to the whole of China also and not merely to the island of Taiwan.[12] There are thus two rival Chinese declarations, the latter

6. United States Department of State, "Straight baselines: The People's Republic of China," *Limits in the Seas* 43 (1972).

7. Ibid., No. 117 (1996).

8. Ibid.; L. Wang and P. H. Pearse, "The new legal regime for China's territorial sea," *Ocean Development and International Law* 25 (1994): 431.

9. Taiwan (the Republic of China) enacted its own Law on the Territorial Sea and Contiguous Zone of the Republic of China on 21 January 1998: *Chinese Yearbook of International Law* 16 (1997–98): 124.

10. United Nations, *Status of Multilateral Treaties Deposited with the Secretary-General.* Accessed 19 August, 2002 on the World Wide Web: http://untreaty.un.org/English/treaty.asp.

11. United Nations, Division for Oceans and the Law of the Sea, *Law of the Sea Bulletin* no. 38 (1998): 28–31.

12. Law on the Exclusive Economic Zone and the Continental Shelf of the Republic of China, promulgated on 21 January 1998, *Chinese Yearbook of International Law* 16 (1997–98): 129.

of which is rather theoretical in that the government of Taiwan does not control the rest of China and reserves the proclamation of baselines to future action by the Executive Yuan.[13] On the assumption of respect or caution for the de facto control of Taiwan in the Taiwan Strait, third states would assume a broadly equidistant line between Taiwan and the PRC dividing their overlapping EEZs. The state of tension between the PRC and Taiwan, however, makes the matter somewhat uncertain and even hazardous, and one not to be considered as an ordinary issue of delimitation.

NAVIGATION IN THE TERRITORIAL SEA OF CHINA

Although there is no suggestion that the EP-3 aircraft strayed over areas of the territorial sea of China, it is necessary to examine first the conditions attaching to the right of innocent passage in the territorial sea to contrast these with the freedoms of navigation applicable in the exclusive economic zone. It is also necessary to study the attitude of China towards navigation by foreign warships in its territorial sea for the light that it might shed on China's attitude towards navigation by warships and military aircraft in areas adjacent to its territorial sea and over its exclusive economic zone.

There is no need to consider the narrower issue of overflight of the territorial sea, since the LOSC, the earlier Geneva Convention on the Territorial Sea and Contiguous Zone, 1958, and customary international law, exclude by clear implication any right of overflight of the territorial sea by aircraft, whether civil or military, without permission.[14] The only exceptions are for specially authorized flights and for regularly scheduled civil aircraft operating on routes approved by the territorial state under bilateral agreements or the multilateral International Air Services Transit Agreement, 1944.[15] Thus, international law recognizes no counterpart "innocent overflight" to the right of innocent passage of foreign ships in the territorial sea.

The conditions attaching to the right of innocent passage by ships in the territorial sea are set out in the LOSC, Part II, Section 3. These conditions are regarded as rapidly acquiring a parallel status as customary international law and as thus binding on non-parties to the LOSC, such as the

---

13. Y.-H. Song and Z. Keyuan, "Maritime legislation of Mainland China and Taiwan," *Ocean Development and International Law* 31 (2000): 303–45. On the debate within Taiwan concerning the one China aspect of the legislation, see Song and Keyuan, pp. 311–12.
14. R. R. Churchill and A. V. Lowe, *The Law of the Sea* 3rd ed. (Manchester: Juris Publishing, 1999), pp. 75–76. For the proposal at the Hague Codification Conference, 1930, that overflight should be permitted, see D. P. O'Connell, *The International Law of the Sea*, ed. I. A. Shearer (Oxford: Clarendon Press, 1982), 1: 304.
15. *United Nations Treaty Series* 389, vol. 84.

United States.[16] Article 19 of the Convention provides that "passage is innocent so long as it is not prejudicial to the peace, good order or security of the coastal State." It then proceeds to list a number of activities that are to be considered as so prejudicial. These include the following:

> "a. any threat or use of force against the sovereignty, territorial integrity or political independence of the coastal State, or in any other manner in violation of the principles of international law embodied in the Charter of the United Nations;
> c. any act aimed at collecting information to the prejudice of the defence or security of the coastal State;
> f. the launching, landing, or taking on board of any military device;
> j. the carrying out of research or survey activities;
> l. any other activity not having a direct bearing on passage."

Some States, including China, consider that warships are threatening per se and prejudicial to coastal State security if they enter the territorial sea without prior permission or unannounced. They argue that the provisions of the LOSC, Part II, Section 3 on innocent passage do not give a right of passage to warships, and that coastal States have the right to require that the entry of warships be subject to their consent. A list of States maintaining that position contained in the *Commander's Handbook on the Law of Naval Operations*[17] shows that some 27 States require prior permission[18] as a condition of innocent passage through their territorial waters of foreign warships, and some 12 others require prior notification.[19] The United States and other major naval powers contest these positions and argue that the LOSC and customary international law accord to warships the right of innocent passage on an unimpeded and unannounced basis.[20] They point to the clear meaning of the title to Subsection A of Section 3—"Rules Applicable to All Ships"—and to the travaux preparatoires of the 1958 and 1982 Conven-

---

16. Churchill and Lowe (n. 14 above), p. 87.
17. A. R. Thomas and J. C. Duncan, eds., *Annotated Supplement to the Commander's Handbook on the Law of Naval Operations,* International Law Studies no. 73 (Newport: Naval War College, 1999), Table A2-1, pp. 202–3.
18. In addition to China, this group includes Algeria, Bangladesh, Brazil, Cambodia, Denmark, Iran, Myanmar, Pakistan, the Philippines, Poland, Sri Lanka, and Vietnam.
19. This group includes Croatia, Egypt, India, Indonesia, Korea (ROK), and Yugoslavia.
20. *Commander's Handbook on the Law of Naval Operations,* para 2.3.2.4. The Soviet Union dropped its objection to a right of innocent passage by warships in the Jackson Hole Agreement of 1989: *Joint Interpretation of the Rules of International Law Governing Innocent Passage,* attached to the Joint Statement by the United States of America and the Union of the Soviet Socialist Republics, 23 September 1989, *International Legal Materials* 84 (1990), p. 239.

tions, which show that efforts to include in the texts a requirement of advance permission or notification for passage by warships were defeated.[21]

The entry of warships into the territorial sea of China is subject to express permission. This restriction, which has been imposed since at least 1958, was repeated in Article 6 of the Law on the Territorial Sea and Contiguous Zone of 1992. The laws of the Republic of China (Taiwan) also contain a restriction on the right of entry of foreign warships into its territorial sea, but the earlier requirement of permission was modified to a requirement of prior notice in 1998.[22]

NAVIGATION IN AND OVER THE CONTIGUOUS ZONE OF CHINA

Again, although there is no suggestion that the EP-3 was overflying the contiguous zone of China at the time of the incident on 1 April 2001, it is necessary to comment on the nature of the contiguous zone in international law in relation to navigational freedoms, because the practice of China is evidently based on a concern for security that extends beyond the territorial sea to the EEZ.

The contiguous zone is an area of waters adjacent to the territorial sea extending not further than 24 M from the baselines from which the territorial sea is measured (i.e., not further than 12 M from the outer limit of the territorial sea). Article 33 of the LOSC, 1982, provides that in that zone the coastal State may exercise the control necessary to:

  a. prevent infringement of its customs, fiscal, immigration or sanitary laws and regulations within its territory or territorial sea;
  b. punish infringement of the above laws and regulations committed within its territory or territorial sea.

It is important to note that the four categories of laws specified in Article 33 are exclusive. Efforts to include "security" as an additional category were defeated at the Third United Nations Conference on the Law of the Sea

---

21. Center for Oceans Law and Policy, University of Virginia School of Law, *United Nations Convention on the Law of the Sea, 1982—A Commentary*, II, vol. eds. S. N. Nandan and S. Rosenne (Dordrecht: Nijhoff, 1993), p. 155 (17.7); B. H. Oxman, "The regime of warships under the United Nations Convention on the Law of the Sea," *Virginia Journal of International Law* 24 (1984): 809, 854; F. D. Froman, "Uncharted waters: Non-innocent passage of warships in the territorial sea," *San Diego Law Review* 21 (1984): 625, 659.
22. Law of 1998, Article 7. Song and Keyuan (n. 13 above), p. 314.

(UNCLOS III).[23] It is also important to note that those laws do not, as such, apply in the contiguous zone but that the rights given to coastal States in their contiguous zones give them an extended space in which to (a) prevent in-bound vessels from committing offences once they reach the territorial sea, or (b) punish outbound vessels for offences they may have committed while in the coastal state or in the territorial sea.

The contiguous zone is included within the exclusive economic zone, and consequently the high seas freedoms reserved in the exclusive economic zone apply in the contiguous zone subject to the special contiguous zone and EEZ rights specified in the Convention.

China declared a contiguous zone by its Law on the Territorial Sea and Contiguous Zone of 25 February 1992.[24] Article 13 of the Law provides the following:

> The People's Republic of China has the right to exercise control in the contiguous zone to prevent and impose penalties for activities infringing the laws or regulations concerning security, the customs, sanitation, or entry and exit control within its land territory, internal waters or territorial sea.

It will be seen that "security" has been added to the categories of laws that may be enforced in the contiguous zone by China. This is clearly not in accord with Article 33 of the LOSC. The United States lodged a protest against this claim in 1992,[25] and has protested as well against the similar claims of some 17 other States. These include Bangladesh, Cambodia, Egypt, India, Iran, Pakistan, Saudi Arabia, Sri Lanka, Syria (to 40 M), the United Arab Emirates, Venezuela, and Vietnam.[26] The Chinese assertion of a right to enforce security laws against inbound and outbound vessels (and presumably aircraft as well) does not go as far as the Vietnamese Decree of 17 March 1980, which purports to require that foreign military vessels must obtain prior permission before entering Vietnam's contiguous zone.[27]

---

23. Center for Oceans Law and Policy, University of Virginia School of Law, *United Nations Convention on the Law of the Sea 1982—A Commentary*, II, volume eds. S. N. Nandan and S. Rosenne (Dordrecht: Martinus Nijhoff, 1993), p. 274 (33.8(d)); B. Kwiatkowska, *The 200 Mile Exclusive Economic Zone in the New Law of the Sea* (Dordrecht: Martinus Nijhoff, 1989), pp. 219–20.

24. *Limits in the Sea*, No. 117 (1996). See also Department of Defense, *Maritime Claims Reference Manual*. Accessed 19 August, 2002 on the World Wide Web: <http://web7.whs.osd.mil/html/20051m.htm>.

25. *Maritime Claims Reference Manual*, pp. 2–86.

26. J. A. Roach and R. W. Smith, *United States Responses to Excessive Maritime Claims*, 2d ed. (The Hague: Nijhoff, 1996), pp. 162–72.

27. Roach and Smith, p. 169.

MILITARY ACTIVITIES IN THE EXCLUSIVE ECONOMIC ZONE

The earlier discussion has been directed at showing how the attitudes of some States, and of China in particular, have been hostile to the presence without permission of warships of other nations in waters adjacent to their coasts. The question is whether these concerns are now driving a form of "creeping jurisdiction" to assert security rights even beyond coastal waters in the exclusive economic zone.

It is well accepted that the exclusive economic zone is a zone of special characteristics and that it owes its definition to the laborious processes of compromise and consensus at UNCLOS III.[28] The title to Article 55 of the LOSC is "Specific legal regime of the exclusive economic zone." It may be regarded as a zone *sui generis*.

Ostensibly, the waters of a state's EEZ are high seas so far as navigation and overflight by other States are concerned. However, the view is widely shared by commentators on the Convention that the provisions relating to the EEZ do not effect an automatic residue of high seas freedoms in so far as rights and jurisdiction are not specifically attributed to the coastal state by Article 56. Rather than residual freedoms, the high seas freedoms are incorporated freedoms and are to be interpreted and applied in the light of the EEZ regime as a whole. The issue in dispute among writers is whether these incorporated freedoms are subject only to restrictions based on the essentially economic character of the EEZ or are subject to restrictions of a broader character, including matters affecting security.[29]

The history of the negotiations at UNCLOS III shows that the final text of Articles 55 to 59 of the Convention was the outcome of a compromise between two positions or trends: the territorialist and the preferentialist. In broad terms the territorialist position was that taken by the Latin American and African States, working from the hypothesis of a "patrimonial" or territorial sea of 200 M with the concession of certain rights of navigation and use to other States. The preferentialist position was taken by the major naval and sea-trading nations that saw the proposed 200 M zone continuing—as before—to be high seas but with the concession of certain preferential resource rights to the coastal state. The outcome was a compromise that has been described as based on a third trend, the zonist, which combined elements of the two other trends and viewed the EEZ as a zone *sui generis*.[30]

---

28. Churchill and Lowe (n. 14 above), pp. 160–62; O'Connell (n. 14 above), pp. 553–62.

29. See, for example, the illuminating debate between Lowe and Kwiatkowska: A. V. Lowe, "Some legal problems arising from the use of the seas for military purposes," *Marine Policy* 10 (1986): 171–86, a reply by B. Kwiatkowska, *Marine Policy* 11 (1987): pp. 249–50, and the rejoinder by Lowe, *Marine Policy* 11 (1987): pp. 250–52.

30. This summary, based on the account by Ambassador Lupinacci of Uruguay, and published in F. Orrego Vicuña, ed., *The Exclusive Economic Zone: A Latin American*

The lack of a clear tilt in the Convention text in favor of either the "residualist" or "incorporationist" views of high seas freedoms in the EEZ, is seen in the wording of Article 58 of the Convention. On the face of it, there would appear to be an unambiguous preservation of high seas freedoms in the wording of Article 58(2). The fact, however, that the high seas freedoms are not referred to collectively but by reference to particular articles in Part VII of the Convention may be argued to be significant evidence for the incorporationist position.[31] Article 59 also seems to assume that there is no automatic application of the maxim *expressio unius est exclusio alterius* to the wording of Articles 56 and 58:

Article 59. *Basis for the resolution of conflicts regarding the attribution of rights and jurisdiction in the exclusive economic zone.*

In cases where this Convention does not attribute rights or jurisdiction to the coastal State or to other States within the exclusive economic zone, and a conflict arises between the interests of the coastal State and any other State or States, the conflict should be resolved on the basis of equity and in the light of all the relevant circumstances, taking into account the respective importance of the interests involved to the parties as well as to the international community as a whole.

Specifically in relation to military activities by other States in a state's EEZ, the travaux preparatoires of the Convention appear to reveal a concern by a few delegations for "exercises," "weapons practice," and "manoeuvres," rather than surveillance or reconnaissance, which were not overtly mentioned.[32] This view of the matter is confirmed by the subsequent statements of understanding, on ratification of the Convention by Brazil, Cape Verde, India, and Uruguay, that military exercises in the EEZ require the permission of the coastal state, and the statement in the opposite direction of Italy that the Convention does not give the right to coastal States to require authorization of foreign military exercises or manoeuvres in their EEZs.[33] The sole reference to intelligence gathering as a form of objectionable military activity is contained not in a published declaration on signature or ratification of the Convention, but in the legislation of Iran of 1993, which claims the right to prohibit "foreign military activities and practices, *collection of information* and any other activity inconsistent with the rights and interests

---

Perspective (Boulder, Colorado: Westview, 1984), pp. 75, 93–94, is reproduced in Nandan and Rosenne, (n. 23 above), p. 499.
    31. H. Djalal, *Indonesia and the Law of the Sea* (Jakarta: Centre for Strategic and International Studies, 1995), pp. 88–89.
    32. Nandan and Rosenne (n. 23 above), p. 564.
    33. Ibid.

of Iran."[34] The wording of the Convention, as finally adopted, gives no warrant for the view that military activities of any type by foreign States are prohibited in the EEZ. Efforts to include a prohibition of military activities in the EEZ were defeated.[35] The United Nations' study of state practice on entry into force of the 1982 Convention noted also the practice of the Soviet Union in concluding a number of agreements with other States that explicitly recognized the freedom to conduct naval operations outside the territorial sea.[36]

The specification in Article 58(1) of "internationally lawful uses of the sea" and the general injunction to parties of Article 301 to engage in only "peaceful uses of the sea" does not give rise to a prohibition of military activities, because these merely repeat the general obligations of States under the Charter of the United Nations, Article 2(4), to refrain from the threat or use of force against other States.[37]

THE VIEWS OF PUBLICISTS

Most publicists support the view that the Convention does not prohibit military activities in and over the exclusive economic zone, or allow coastal States to regulate such activities. Representative of this position is Kwiatkowska, who states the following:

> As regards military activities, in view of the legislative history of Article 58, uses such as naval manoeuvres, weapons practice, the employment of sensor arrays, aerial reconnaissance, or collection of military intelli-

---

34. Iran, Marine Areas Act, Article 16 1993; Roach and Smith (n. 26 above), pp. 413–14. The United States has protested against the Iranian legislation: Iran, Marine Areas Act.

35. Nandan and Rosenne (n. 23 above), p. 563 (proposal of Peru); D. J. Attard, *The Exclusive Economic Zone in International Law* (Oxford: Clarendon Press, 1987), p. 85.

36. UN Division for Ocean Affairs and the Law of the Sea, Office of Legal Affairs, *The Law of the Sea: Practice of States at the Time of Entry into Force of the United Nations Convention on the Law of the Sea* (New York: United Nations, 1994), p. 133. The agreements referred to were with the United Kingdom, Germany, Canada, Italy, France, the Netherlands, Norway, Spain, and Greece. See also the Agreement on the Prevention of Incidents on or over the High Seas (the INCSEA Agreement) of 1972 between the United States and the Soviet Union, 852 *United Nations Treaty Series* 151. See also Roach and Smith (n. 26 above), pp. 407–14.

37. H. B. Robertson, "Navigation in the exclusive economic zone," *Virginia Journal of International Law* 24 (1984): 865, 886–88; Oxman (n. 21 above), pp. 809, 831. F. Orrego Vicuña, however, would accord to article 301 some effect in the context of navigation rights in the EEZ: *The Exclusive Economic Zone: Regime and Legal Nature under International Law* (Cambridge: Cambridge University Press, 1989), p. 109.

gence regarding foreign activities at sea, are clearly internationally lawful uses of the sea 'associated with the operation of ships, aircraft and submarine cables and pipelines', which are related to the exercise and protection of the freedoms enjoyed by all States in the EEZ.[38]

A more nuanced view is taken by O'Connell, who, writing before UNCLOS III concluded its work, considered that the wording of what became Article 58 provided for freedom of navigation "in a rather awkward way." He concluded that, "since [the Convention] thus oscillates between alternative inferences as to the status of the EEZ, its actual status is likely to come to depend less upon textual exegesis than on the outcome of the free play of political forces over a span of time."[39]

A similarly cautious view is taken by Orrego Vicuña who states that the issue of the activity of naval and air forces in the EEZ is "not so simple." He continues,

> There is no doubt that the key principle of the 1982 Convention is that foreign war fleets have open access to the exclusive economic zone, since in effect it is an integral part of the freedom of navigation and over-flight. The exercise of this right, however, is subject to some restrictions in reference to that zone, such as the limitations of a political nature and those that are derived from economic rights, as previously indicated. These restrictions do not originate in the purpose of hindering the freedom of navigation, but in the necessary harmonization with other interests that exist in this area. In this sense, again, the *sui generis* nature of the exclusive economic zone imposes some limitations that are not found in the regime of the high seas and that determine the difference between one and the other.... Some States have interpreted the Convention in a restrictive manner in this regard, a point of view that is not the consequence of the *sui generis* nature indicated, as occasionally it has been thought, but is the outcome of a position closer to identifying the zone as an area of integral national jurisdiction.[40]

It seems that for both O'Connell and Orrego Vicuña the implications of the juridical nature of the exclusive economic zone for military operations have yet to be realized fully and will depend on the future interplay of the

---

38. Kwiatkowska (n. 23 above), p. 203. See also Attard (n. 35 above), p. 85; Oxman (n. 21 above), pp. 835–41; Robertson (n. 37 above), pp. 886–88; J.-P. Queneudec, "Zone Economique Exclusive et Forces Aeronavales," Colloque, 1981, Academie de Droit International, 319:-24.

39. D. P. O'Connell, *The International Law of the Sea*, vol. 1, ed. I. A. Shearer (Oxford: Clarendon Press, 1982), p. 578.

40. Orrego Vicuña, *The Exclusive Economic Zone: Regime and Legal Nature under International Law*, p. 120.

traditional forces constitutive of customary international law and on "political" factors. So far there has been little actual state practice restricting freedom of navigation and overflight in the EEZ but the legislation of some States has the potential to bring about conflict, if invoked.[41]

THE CHINESE POSITION

It may be too early yet to attribute to the government of China a final considered legal position on the status of airspace over its EEZ. On the evening of 2 April, one day after the incident, the Assistant Minister of Foreign Affairs of China, Mr. Zhou Wenzhong, was reported as saying to United States Ambassador Prueher that "the act of the U.S. side constitutes a violation of the United Nations Convention on the Law of the Sea, which provides that the sovereign rights and jurisdiction of a coastal state over its exclusive economic zone, *particularly its right to maintain peace, security and good order* in the waters of the zone, shall all be respected, and that a country shall conform to the UNCLOS and other rules of international law when exercising its freedom of the high seas."[42] At the talks held between representatives of China and the United States under the umbrella of the Military Maritime Consultative Agreement on 18–19 April, Mr. Lu Shumin, head of the Chinese delegation and Director-General of the Department of North American and Oceanian Affairs of the Foreign Ministry stated that,

> the activities of the U.S. side in the airspace over the waters close to China's coast have seriously harmed China's national security and national defense interests, and gone far beyond the limit of freedom of over-flight provided for in the UN Convention on the Law of the Sea . . . Such military activities of the United States during peacetime threaten China's national security, peace and order, constitute a provocation against China's national sovereignty, and violate the basic norms of international law on mutual respect for sovereignty and territorial integrity among nations.[43]

---

41. S. A. Rose, "Naval activity in the EEZ—Troubled waters ahead?," *Naval Law Review* 39 (1990): 67, 90.

42. (3 April 2001). China's solemn position on the US military reconnaissance plane ramming and destroying a Chinese military plane. Accessed 19 August 2002 on the World Wide Web: <http://www.channelnewsasia.com/uschina_crash/0304china.htm>.

43. Accessed 19 August 2002 on the World Wide Web: <http://www.chinese-embassy.org.uk/page-news/0420-china-us-talks.htm>.

An article was published in a Chinese newspaper on 24 April 2001 setting out the legal argument in more detail. The article,[44] by Liu Wenzong, Professor of International Law at the Law Research Institute of the Institute of Diplomacy, advances a view of the legal nature of the EEZ that includes the right to assert a security jurisdiction within it. The following arguments are advanced in the article:

1. The EEZ is not part of the high seas, as Article 86 of the United Nations Convention on the Law of the Sea, 1982, confirms. Thus the airspace over China's EEZ is not "international airspace."
2. Article 58 of the Convention does not give an unqualified freedom of navigation and overflight of the EEZ but only if the conditions of paragraph 3 of that Article are observed, [namely that States exercising that freedom "shall have due regard to the rights and duties of the coastal state and shall comply with the laws and regulations of the coastal state in accordance with the provisions of this Convention and other rules of international law in so far as they are not incompatible with this Part."]
3. Article 301 of the Convention declares that States, in exercising their rights and performing their duties under the Convention must refrain from any threat against the territorial integrity or political independence of any state. "In other words, a foreign aircraft shall be banned from violating a coastal country's sovereignty and national defense security, from engaging in all types of unlawful acts which do not comply with 'freedom of flight,' such as spying out a coastal country's military secrets, and from violating a coastal country's territorial integrity, peaceful order, or political independence as well."
4. China has legislated in accordance with the Convention for its EEZ. Article 11 of the Law of 1998 has stipulated that "all countries shall enjoy freedom of navigation by water and air within the PRC exclusive economic zone, provided that they abide by both the international law and the PRC laws and decrees."
5. The United States itself claims the right to regulate flight in areas beyond its territorial sea through the institution of Air Defense Identification Zones (ADIZ). This is evidence of a double standard.

These arguments are not convincing for the reasons already stated. Those numbered (1) through (4) are based on a flawed understanding of the meaning of Article 301 of the United Nations Convention on the Law

---

44. Excuses concocted by the United States for 'aircraft collision incident' untenable. *Guangzhou Ribao.* Accessed 19 August 2002 on the World Wide Web: <http://www.dayoo.com>.

of the Sea by which the writer inflates the meaning of Article 58 to include security concerns. It restates in another spatial context the argument against a right of innocent passage through the territorial sea: that warships (and military aircraft) by their very nature threaten through their mere presence. Taken to its logical conclusion, the argument would exclude the operations of warships and military aircraft anywhere, even on the high seas.[45]

So far as argument (5) is concerned, there is a misunderstanding of the United States' position. ADIZ regulations promulgated by the United States, requiring certain aircraft to file flight plans and periodic position reports, apply only to aircraft *bound for U.S. territorial airspace* and do not apply to aircraft not intending to land in the United States or to overfly the territorial sea of the United States.[46] However, it would be true to state that the United States recognizes that, in times of imminent or actual hostilities, the right of self-defence would justify a requirement of identification of surface vessels and aircraft approaching or in the vicinity of national waters. The domestic right of the President of the United States to establish "defensive sea areas" by Executive Order has been established since the 1930s.[47] It is a right intended to be invoked only in case of war or to a declared national emergency involving the outbreak of hostilities.[48] It is a right referable to the right of self-defence in international law and not to the law of the sea.

CONCLUSION

There is considerable evidence that the majority of delegations at UNCLOS III, and the ostensible consensus achieved, intended in the Convention to uphold an unrestricted right of international navigation and overflight of the EEZ for military and other noneconomic purposes. Consensus, however, has come to be questioned in later practice, based on the ambiguities inherent in the *sui generis* nature of the EEZ, or—as Orrego Vicuña views it—based on the identification of the zone as an area of integral national jurisdiction.[49] Hitherto, this contrary practice has been mostly confined to decla-

---

45. Robertson (n. 37 above), pp. 886–88.
46. *Commander's Handbook on the Law of Naval Operations* (n. 17 above) para 2.5.2.3.
47. Regulations Governing the Issuance of Entry Authorizations for Naval Defensive Sea Areas, Naval Airspace Reservations, Areas under Navy Administration, and the Trust Territory of the Pacific Islands, OPNAV INSTRUCTION 5500.11E.
48. *Commander's Handbook on the Law of Naval Operations* (n. 17 above) para 2.4.4.
49. Orrego Vicuña (n. 40 above), p. 120.

rations (such as that of Brazil) or legislation (such as that of Iran),[50] but with China's recent actions, active state practice is beginning to appear.[51]

It is not clear from Chinese statements that China regards its entire EEZ as a zone in which foreign warships and military aircraft are subject to restrictions. Emphasis appears to have been laid more on the security concerns that China has arising from the collection of military intelligence. In that respect the unambiguous mission and design of the EP-3 aircraft have created the triggering event. Passage through, or overflight of, the EEZ of China for military purposes by vessels or aircraft not exclusively or primarily designed for intelligence gathering may continue to proceed unchallenged by China.

There may in time develop a practice or doctrine of certain kinds of restriction of foreign military activity in the EEZ that is based not on the law of the sea, with which they sit uncomfortably, but on notions of self-preservation antecedent to the law of self-defense. At the same time the United States will continue assertions of right under its Freedom of Navigation Program.[52] We are now beginning to see the interplay of forces anticipated by O'Connell.

---

50. Roach and Smith (n. 26 above), pp. 409, 413; Rose (n. 41 above), p. 90, likens such legislation to "potential time-bombs ticking away" and that, if not yet "operationalized, they remain primed for use."

51. For actions by Ecuador and Peru asserting overflight restrictions above waters constituting an EEZ in international law but regarded by them as territorial seas of 200 M, see Roach and Smith (n. 26 above), pp. 370–75.

52. The Freedom of Navigation Program (FON) was instituted in 1979: Department of State, *Digest of United States Practice in International Law, 1979,* ed. M. L. Nash (Washington, D.C.: Department of State, 1980), pp. 997–98. It was reaffirmed by President Reagan in the U.S. Oceans Policy Statement of 10 March 1983: Roach and Smith (n. 26 above), pp. 5–11; *Commander's Handbook on the Law of Naval Operations* (n. 17 above), p. 143, para 2.6.

Climate Change and Arctic Shipping

Global Warming and Canada's Shipping Lanes: An Oceanographer's View[†]

Carl Anderson
Challenger Oceanography, Dartmouth, Nova Scotia, Canada

INTRODUCTION

Canada has the longest coastline of any country in the world, and has thousands of miles of sea lanes in subpolar and polar climates. Climatic factors such as extreme low temperatures, poor visibility, and the presence of ice severely affect shipping on Canada's east coast and in the north. Consequently, much of Canada's shipping has an Arctic character that it shares only with other polar nations.

The earth's climate warmed significantly in the 20th century, and the observed warming is expected to continue.[1] Forecasts indicate that climate warming in northern latitudes over the next 50 to 100 years may be significantly above the global average. Scientists addressing the Royal Society in 1995 said that if forecasts are correct, "it is in the Arctic that we can expect to observe global warming at its most powerful."[2]

The completion of the Third Assessment in 2000 by the Intergovernmental Panel on Climate Change (IPCC) is an appropriate time for Canada to examine the potential impact of global climate change on its shipping industry. "Global warming" suggests positive impacts for Canadian shipping, with the possibility of moderating the coldest winter temperatures and reducing the extent and thickness of sea ice. Summer navigating seasons might be longer, and year-round shipping might occur where it is only seasonal today. New shipping routes, such as the Northwest Passage, could open up for seasonal, or eventually, year-round navigation. Transpolar shipping

---

[†] EDITORS' NOTE.—An earlier version of this paper was presented to a seminar on "Global Warming and Canada's Shipping Lanes" at the Canadian Maritime Law Association Annual General Meeting, 15 June 2001, Montreal, Canada.
 1. IPCC (Intergovernmental Panel on Climate Change). Summary For Policymakers. A Report of Working Group I of the Intergovernmental Panel on Climate Change. Cambridge, UK: Cambridge University Press, January 2001.
 2. P. Wadhams, J. A. Dowdeswell and A. N. Schofield, eds., "The Arctic and Environmental Change," *Philosophical Transactions of the Royal Society, London*, A (1995) 352, pp. 197–385.

might even become possible. Some of the indirect effects of climate warming, however, coupled with the nature of Canada's northern geography, could lessen or delay these positive impacts.

The following is a brief examination of the potential impact of global warming on Canada's ocean shipping lanes. The article starts by identifying the important environmental factors in Canadian shipping and briefly describes Canada's shipping lanes in the late 20th century. Observed and forecast climate changes are presented and their potential impact on Canadian shipping is discussed. The article closes with a summary and conclusions.

ENVIRONMENTAL FACTORS IN CANADIAN SHIPPING

The principal environmental factors affecting navigation are sea state, wind, currents, visibility, and the presence of sea ice and icebergs. Air and sea temperatures affect navigation indirectly as factors in the formation of fog, and with wind, in the structural icing of ships. The extremes of wind, sea state, currents, and poor visibility found in Canadian waters are formidable, but are not unique in the world. Ships making voyages on the high seas regularly encounter similar or worse conditions. The factor that is unique in northern and polar shipping is the presence of ice. For part or all of the year, sea ice blocks or severely restricts navigation by filling channels, gulfs, bays, and inlets, and forming barriers along open coasts (Fig. 1). Icebergs constitute hazards to navigation that even powerful icebreakers must avoid absolutely. This article focuses on the possible impact of global warming on the occurrence of sea ice and icebergs in Canadian waters.

CANADA'S SHIPPING LANES

Canada's international maritime trade is conducted through ports in British Columbia, Atlantic Canada, the Great Lakes, the St. Lawrence River and Gulf, Hudson Bay, and the Arctic. Domestic shipping interconnects ports throughout these regions. The northern character of Canadian marine shipping is most evident in the Gulf of St. Lawrence, Newfoundland, and Labrador, and in the Arctic and sub-Arctic north of the 55th parallel. This section describes shipping and ice conditions in those regions. Shipping routes and place names can be found in *Water Transportation Infrastructure*.[3]

Dates given here for the appearance of various ice conditions are based on the 30-year (1969–1998) biweekly median ice concentration maps com-

---

3. Energy, Mines and Resources, *National Atlas of Canada 5th Ed., Water Transportation Infrastructure, Canada* (1:7 500,000 map), (Ottawa, Canada: Energy, Mines and Resources, 1986).

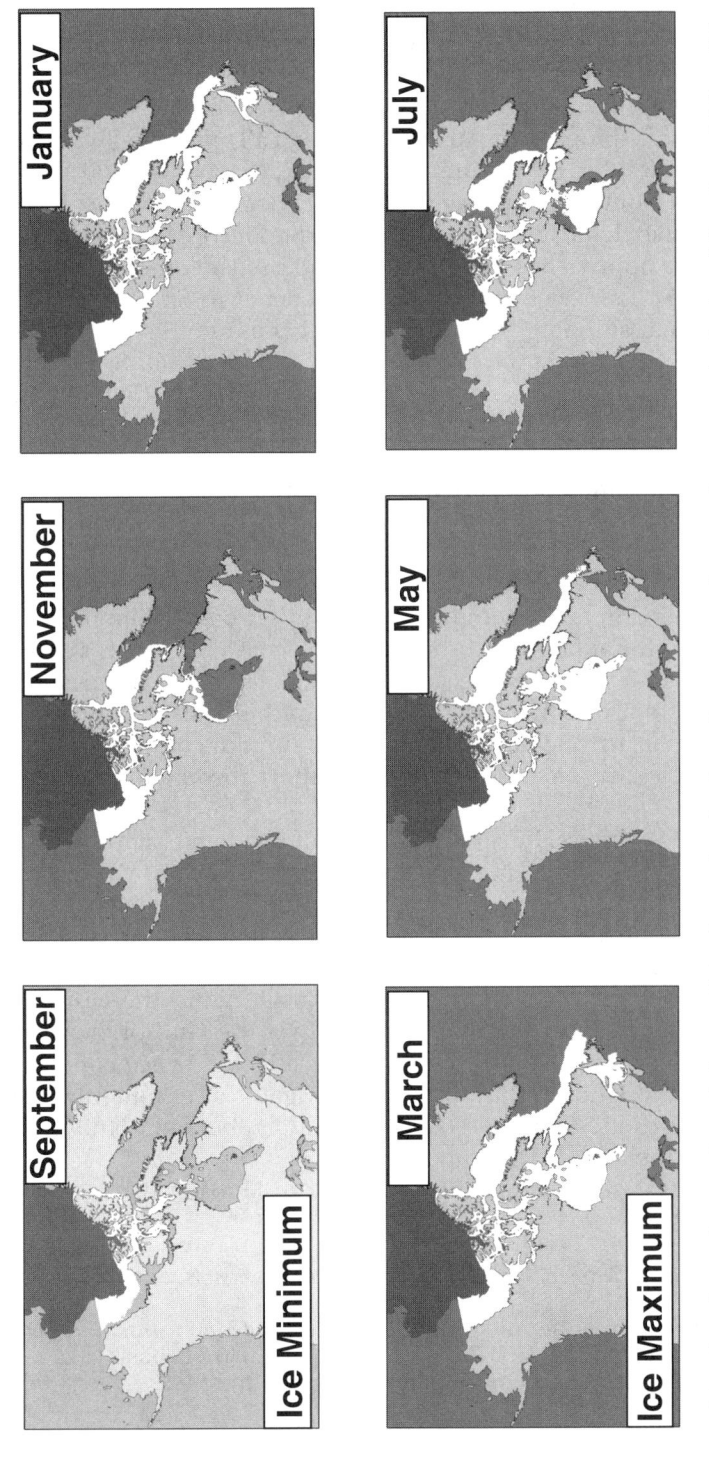

FIG. 1.—Annual growth and decay of Arctic sea ice in Canadian waters. Average ice extent is depicted for the middle of the months shown. Black represents the perennial sea ice cover in the Arctic Ocean. White is the annual sea ice extent. Minimum and maximum sea ice extent occurs in September and March, respectively (source: Adapted from Canadian Ice Service, Environment Canada, 2001).

piled by Environment Canada.[4] It should be kept in mind that *expected* ice concentrations are less than the median concentration half the time, and exceed the median half the time, and that the distribution about the median of heavy and light ice conditions varies from place to place.

International shipping in Atlantic Canada calls principally on the ice-free ports of Saint John, New Brunswick, Halifax, Nova Scotia and St. John's, Newfoundland. Domestic shipping and ferry services interconnect Nova Scotia, Îsles de la Madeleine, Newfoundland, and St. Pierre-Miquelon (France). Shipping in the approaches to Atlantic Canada must contend with sea ice in the Gulf of St. Lawrence and on the approaches to St. John's.

International shipping bound for inland ports enters the Gulf of St. Lawrence via either the Strait of Belle Isle or Cabot Strait. Sea ice significantly affects shipping and ferry traffic in these straits, in the Gulf, and in the St. Lawrence estuary. Sea ice begins to form in the estuary in the latter half of December, and reaches its maximum extent in the estuary and Gulf of St. Lawrence around mid-February. Cabot Strait and the Gulf are ice free by late April, but ice does not leave the Strait of Belle Isle until a month later.

Ships trading between Newfoundland and Nova Scotia encounter ice both in the Gulf of St. Lawrence and off St. John's. Coastal shipping between Lewisporte, Newfoundland, and the Labrador coast (Fig. 2) is strongly affected by Labrador coast sea ice and icebergs (Fig. 3). Ice starts to form along the Labrador coast in mid-December, and is at its maximum extent from mid-February through mid-March. The Labrador coast is free of sea ice by late July, but icebergs may be encountered there throughout the shipping season.

The Arctic Sealift to Hudson Bay and the Canadian eastern Arctic (Fig. 2) uses general cargo ships, tankers, and barges for the annual resupply of northern communities. Access to the communities of the north is restricted by the presence of sea ice, and is affected in the east by icebergs, as described later. Reduced visibility due to fog is also a problem.

Communities in Ungava Bay, Hudson Strait, along the east coast of Hudson Bay, and in Foxe Basin are supplied via the Gulf of St. Lawrence and Labrador Sea.[5] Shipping to these areas is affected by ice conditions in Hudson Strait and Hudson Bay, which are navigable after mid-July. New ice begins to fill Hudson Strait and Foxe Basin in October, and Hudson Bay is ice covered by the end of December.

Shore leads develop along the west coast of Hudson Bay in early July,

---

4. Canadian Ice Service, *Bi-weekly Arctic sea ice animation*. (Ottawa, Canada: Environment Canada, 2001).

5. K. J. Spears, *Arctic Marine Risks—The Interaction of Marine Insurance and Arctic Shipping*. Transportation Research Report No. 21, May 1986. Halifax, Canada: Canadian Marine Transportation Centre, Dalhousie University, 1986.

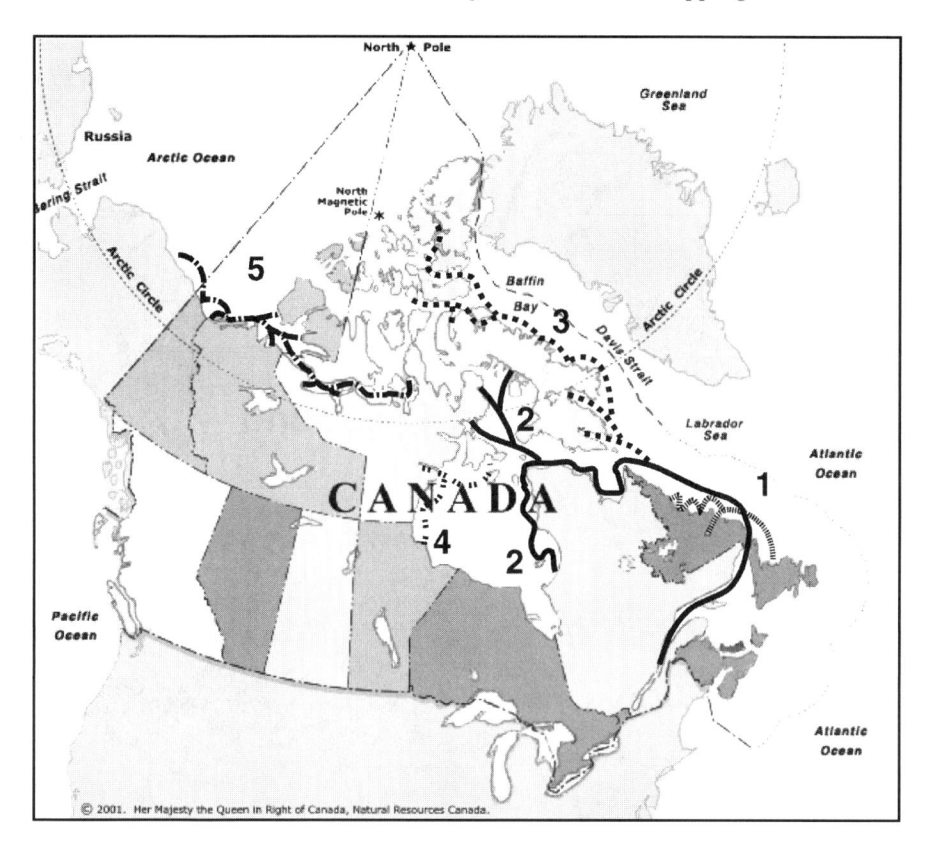

FIG. 2.—Coastal shipping and northern supply lines. (1) Labrador Coast, (2) Eastern Hudson Bay and Foxe Basin, (3) Baffin Island and High Arctic, (4) Western Hudson Bay, (5) Western Arctic.

about 2 weeks before the Bay is accessible from outside. This allows a resupply operation based in Churchill, Manitoba, to serve west coast Hudson Bay communities, Qamanittuaq (Baker Lake), and Salliit (Coral Harbour). Later in the season, grain from the Canadian midwest is shipped from Churchill via Hudson Bay and Hudson Strait (Fig. 4). From 1955 to 1962, ore was also shipped via this route from North Rankin Nickel Mines at Kangiqllinq (Rankin Inlet), north of Churchill.[6]

Baffin Island and the Canadian high Arctic are also supplied annually by sea as far north as Eureka (80° N) in the Canadian Archipelago (Fig. 2). These operations are subject to ice conditions in Davis Strait, Baffin Bay,

6. M. Soubilière, ed. *The Nunavut Handbook,* (Iqaluit, Canada: Nortext Media Inc., 1998).

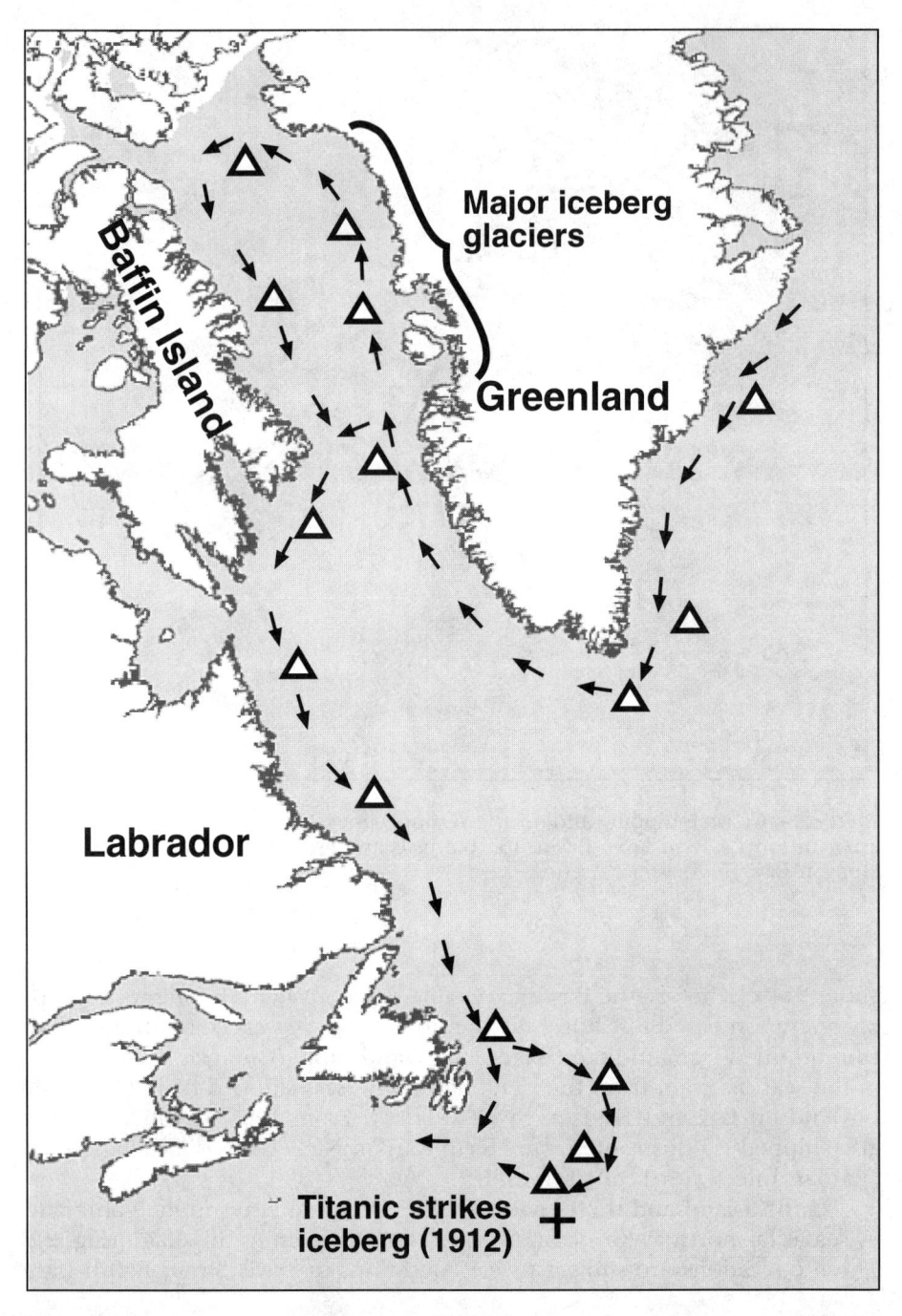

FIG. 3.—Typical summertime iceberg distribution in eastern Canadian waters. Arrows depict iceberg movement.

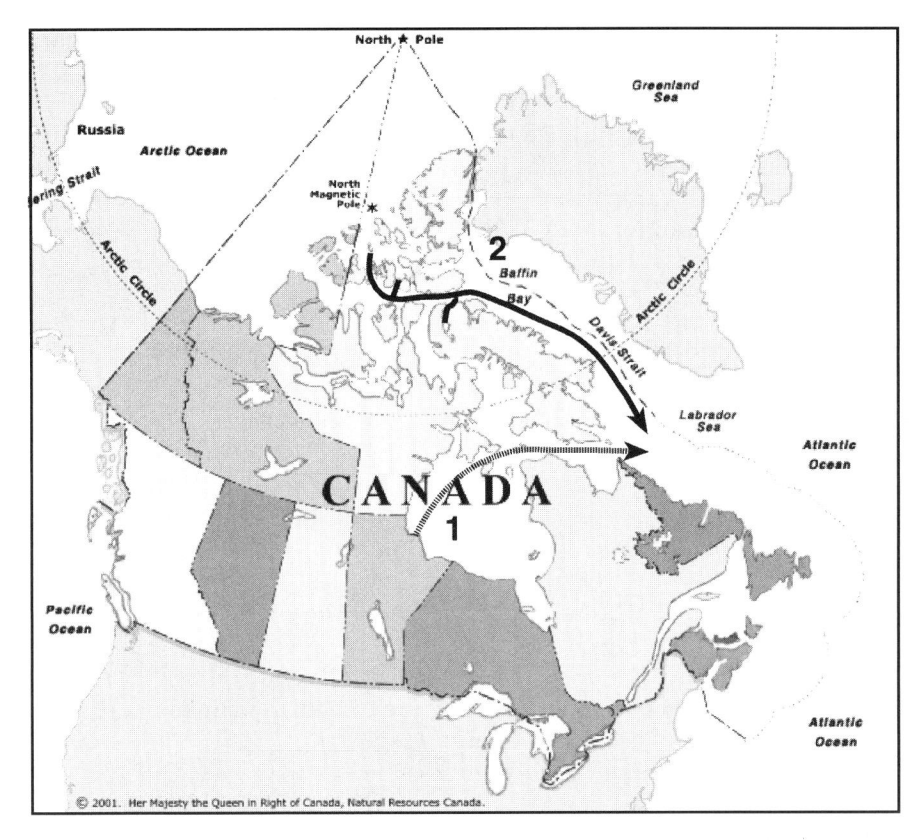

FIG. 4.—International raw materials shipments from northern ports. (1) From Churchill, Manitoba (grain), (2) from Nanasivik (lead, zinc), Cameron Island (oil) and Little Cornwallis Island (oil and gas).

Lancaster Sound, and in the waters north of Devon Island. High Arctic mineral and petroleum resources are shipped from the Polaris Mine on Little Cornwallis Island (lead, zinc), Nanasivik Mine on Baffin Island (lead, zinc), and the Bent Horn oil and gas field on Cameron Island[7] (Fig 4). Ice-free access to Lancaster Sound is possible about the end of July. Northern Baffin Island communities and the Nanasivik Mine (Strathcona Sound) are free of ice 1 week later in early August, but sea ice remains along the southern east coast of Baffin Island until the end of August. The most favourable ice

7. Geological Survey of Canada, Minerals and Metals Sector and National Energy Board, "Geological Survey of Canada Map 900A, scale 1:6 000 000," in *Principal Mineral Areas of Canada,* 46th ed. (Ottawa, Canada: Geological Survey of Canada, 1996).

conditions occur in the high Arctic in the first half of September, after which new ice begins to form.

In the western Arctic, a resupply operation originating at Inuvik extends westward along the Beaufort Sea coast to Alaska and eastward to communities on Banks Island, Victoria Island, and along the Canadian mainland from Amundsen Gulf to Talurruaq (Taloyoak) on Rae Strait (Fig. 2). The most favourable ice conditions along this supply route occur from late August to early October.

THE NORTHWEST PASSAGE AND NORTHERN SEA ROUTE

European maritime nations sought a northwest passage to Asia as early as the 16th century (Fig. 5). Although the Northwest Passage had been mapped piecemeal by 1859,[8] it was not transited in its entirety until the early 20th century. The Norwegian explorer Roald Amundsen made the voyage in 1903–1906, in the *Gjøa*, a 47-ton converted herring boat. It was not transited again until 1940–1942, when the Royal Canadian Mounted Police patrol vessel *St. Roche* made an eastward passage under the command of Sgt. Henry Larsen, RCMP. The *St. Roche* returned to the west coast via the Northwest Passage in 1944. Both Amundsen and Larsen followed routes that took advantage of the summer open water along the Canadian mainland, areas that are unsuited to large tankers and cargo vessels.

A succession of Canadian and American icebreakers and U.S. Navy submarines transited the Northwest Passage (NWP) starting in 1954, and in late summer 1969, the 155,000-ton U.S. icebreaking tanker *Manhattan* made the first (and only) "commercial" transit of the passage.[9] The westward route taken by *Manhattan* was from the east coast via Baffin Bay and Parry Channel to Viscount Melville Sound, thence via Prince of Wales Strait to Amundsen Gulf and the Beaufort Sea (Fig. 5). The most severe ice conditions on this route are in Viscount Melville Sound, where ice concentrations exceed 9/10 year-round. The most favourable ice conditions in the Sound occur about the first of September, when light ice appears for a brief time in shore leads along the southern shore of Melville Island.

At Prudhoe Bay, Alaska, the *Manhattan* loaded one symbolic barrel of crude oil and retraced her course eastward. The *Manhattan* voyage was conducted to test the feasibility of transporting North Slope Alaskan oil to east coast ports via the Northwest Passage. In 1970, however, the decision was made instead to construct the Trans-Alaska Pipeline System.

In the early 1980s consideration was given to the marine transport of

---

8. Canadian Hydrographic Service, *Pilot of Arctic Canada*, 2nd ed. (Ottawa, Canada: Department of Energy, Mines, and Resources, 1970) vol. 1.
9. W. D. Smith, *Northwest Passage*, (New York: American Heritage Press, 1970).

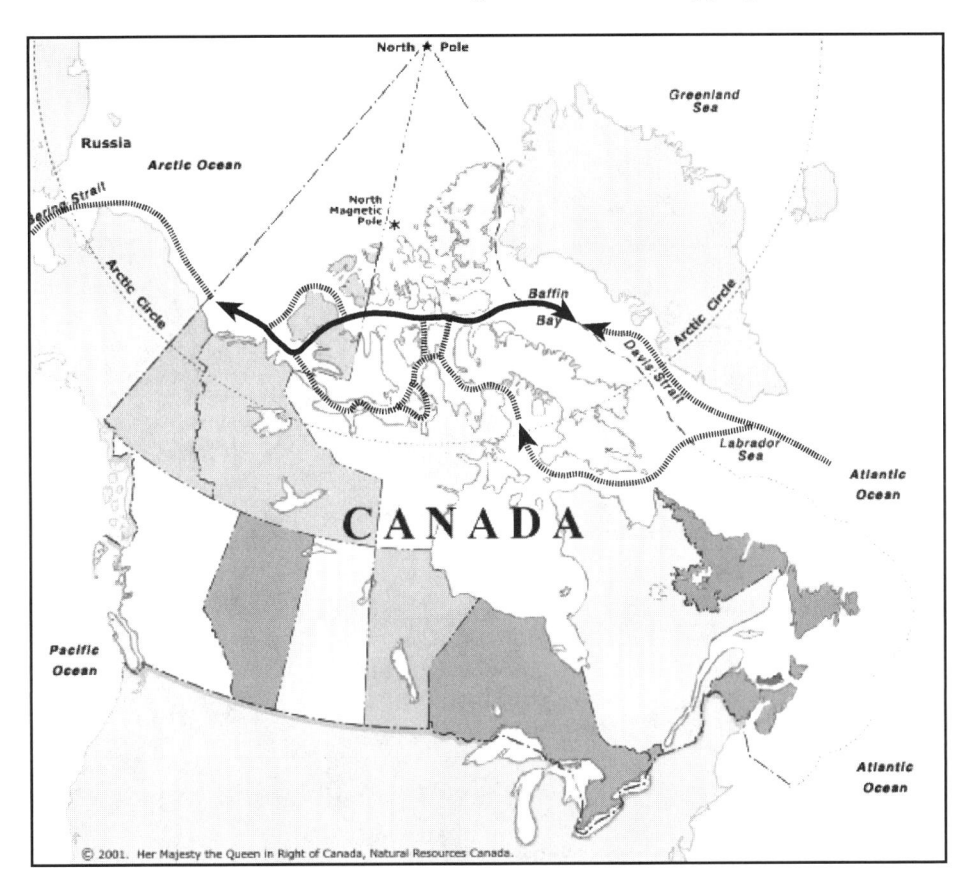

FIG. 5.—The Northwest Passage, showing possible transit routes. The route taken in 1969 by the U.S. icebreaking tanker *Manhattan* is indicated by the heavy arrow.

Canadian Arctic oil and gas via the NWP (Fig. 6). The Arctic Pilot Project proposed the transport of liquefied natural gas (LNG) in specially designed icebreaking LNG tankers from Melville Island to east coast markets.

Icebreaking tankers were also considered for transporting crude oil to east coast refineries from offshore production in the southern Beaufort Sea. The alternative of a land pipeline system was chosen instead for further study.

Commercial transits of the Northwest Passage had not begun by the end of the 20th century, with the exception of occasional voyages by chartered Russian icebreakers carrying tourists. Regular shipping in Canada's north is still confined to the eastern and western resupply operations, and the mineral, oil, and gas shipments described earlier.

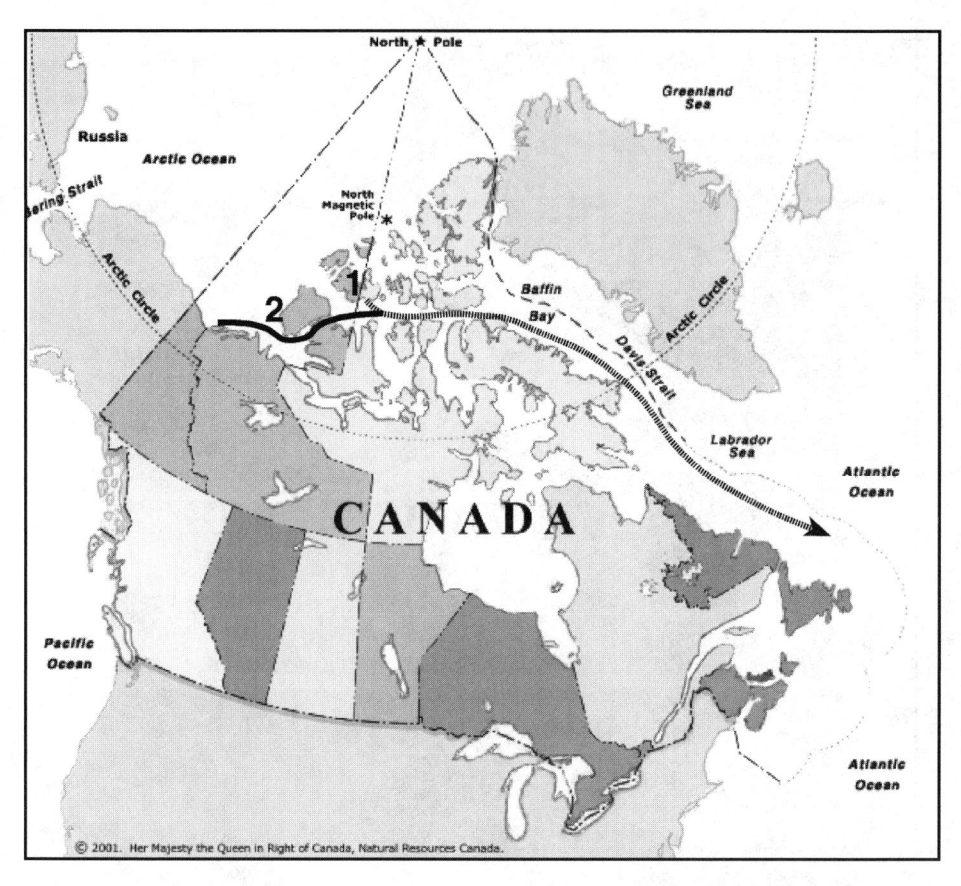

FIG. 6.—Arctic shipping proposals from the early 1980s. (1) From Melville Island (Arctic Pilot Project, liquefied natural gas), (2) from Beaufort Sea (offshore oil).

In contrast to Canada's Northwest Passage, the Russian Northern Sea Route (Fig. 7) has carried substantial volumes of commercial shipping annually since the 1950s. The NSR follows the Arctic coast of Russia from the Barents Sea to the Pacific Ocean, and was first navigated in its entirety in 1878–1879 by the Swedish explorer Otto Nordenskjöld. Vessels ply the shallow coastal waters of Russia from the Kara Sea to Bering Strait. In 1998, Russia declared 69 Arctic ports along the Northern Sea Route (NSR) open to foreign shipping.[10] The navigation season is about 140–150 days long. Several straits impose significant draft restrictions for modern vessels, and, as in the Canadian Arctic, sea ice is the overriding environmental factor.

10. W. Østreng, ed., *The Natural and Societal Challenge of the Northern Sea Route: A Reference Work* (Dordrecht: Kluwer Academic Publishers, 1999).

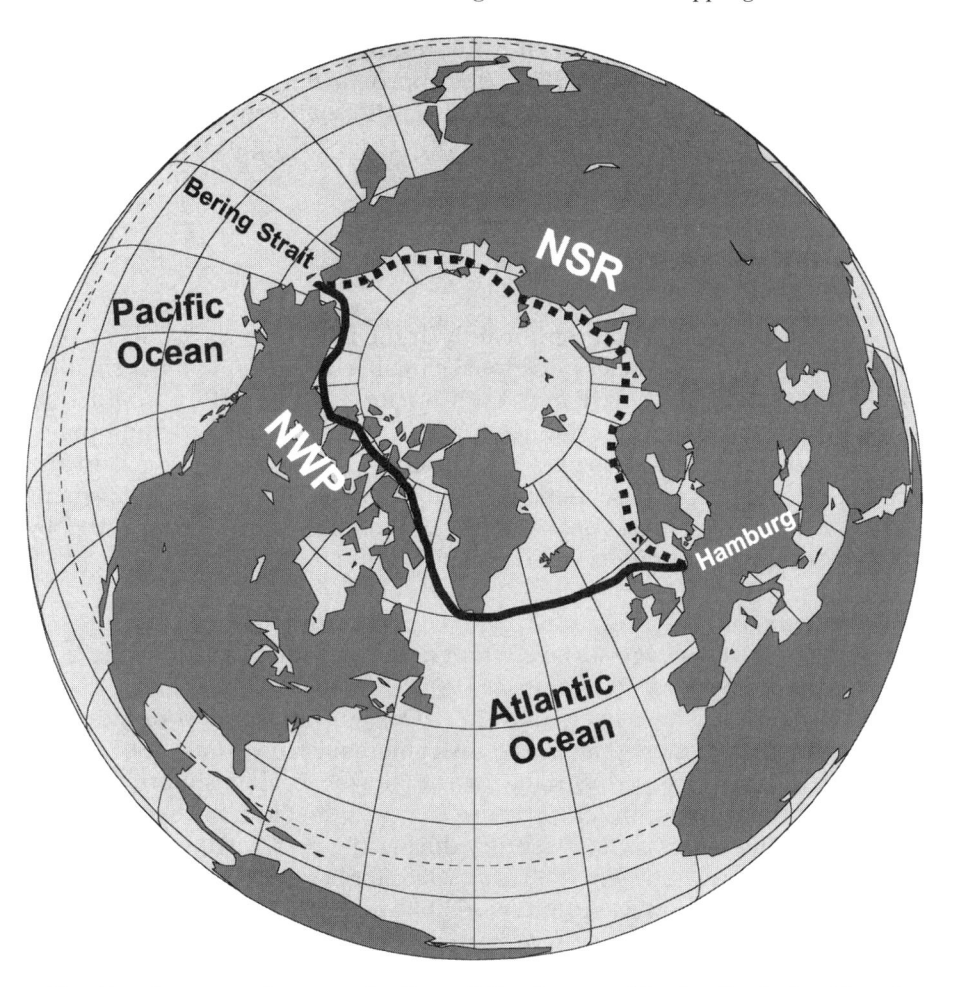

FIG. 7.—Sea routes between the Port of Hamburg and Bering Strait via either the Northwest Passage (NWP) or the Northern Sea Route (NSR).

Most Northern Sea Route traffic is coastal shipping and outward shipments of raw materials from the mines and forests of northern Russia. Traffic volume is highest on the western Kara Sea Route, with lesser volumes in the East Siberian and Chukchi Seas. Few complete transits are made between European and Pacific Ocean ports via the NSR. Northern Sea Route shipping uses special ice class vessels and depends heavily on icebreaker escort.

Under more favourable ice conditions than exist today, the Northwest Passage and Northern Sea Route represent two alternative shipping lanes between Europe and the Pacific Ocean. While the choice of route would not be made on distance alone, it is interesting to note that a voyage from Hamburg to the Bering Strait (Fig. 7) is approximately 4,910 nautical miles (M) via the NWP, compared to 4,610 M via the southern (winter) NSR, or 4,480 M via the northern (summer) NSR. The NSR figures use Northern Sea Route distances from Østreng.[11]

CLIMATE CHANGE IN THE 20TH CENTURY

The Arctic climate changed significantly in the 20th century. Concern over the possible effects of accelerated global climate change led to the formation of the Intergovernmental Panel on Climate Change (IPCC), which is charged with assessing the evidence for climate change and advising policy makers on its probable impacts. The observations in this section, and the predictions in the next section, are quoted from the Third Assessment (2000) *Summary for Policy Makers* of IPCC Working Group I, approved by IPCC member governments in Shanghai 17–20 January 2001:[12]

Global temperature
- The global average surface temperature has increased over the 20th century by about 0.6°C.
- Globally, it is very likely that the 1990s was the warmest decade and 1998 the warmest year in the instrumental record since 1861.
- New analyses of proxy data for the Northern Hemisphere indicate that the increase in temperature in the 20th century is likely to have been the largest of any century during the past 1,000 years.
- On average, between 1950 and 1993, night-time daily minimum air temperatures over land increased by about 0.2°C per decade. This is about twice the rate of increase in daytime daily maximum air temperatures (0.1°C per decade). This has lengthened the freeze-free season in many mid- and high-latitude regions. The increase in sea surface temperature over this period is about half that of the mean land surface air temperature.

Snow cover and ice extent have decreased
- Satellite data show that there are very likely to have been decreases of about 10 percent in the extent of snow cover since the late 1960s, and ground-based observations show that there is very likely to have been a reduction of about two weeks in the annual duration of lake and river ice cover in the mid- and high latitudes of the Northern Hemisphere, over the 20th century.

11. Ibid.
12. IPCC (n. 1 above).

- There has been a widespread retreat of mountain glaciers in non-polar regions during the 20th century.
- Northern Hemisphere spring and summer sea-ice extent has decreased by about 10 to 15 percent since the 1950s. It is likely that there has been about a 40 percent decline in Arctic sea-ice thickness during late summer to early autumn in recent decades and a considerably slower decline in winter sea-ice thickness.

Global average sea level has risen and ocean heat content has increased
- Tide gauge data show that global average sea level rose between 0.1 and 0.2 m during the 20th century.
- Global ocean heat content has increased since the late 1950s, the period for which adequate observations of subsurface ocean temperatures have been available.

Changes have also occurred in other important aspects of climate
- It is very likely that precipitation has increased by 0.5 to 1 percent per decade in the 20th century over most mid- and high latitudes of the Northern Hemisphere continents.
- In the mid- and high latitudes of the Northern Hemisphere over the latter half of the 20th century, it is likely that there has been a 2 to 4 percent increase in the frequency of heavy precipitation events.

CLIMATE CHANGE FORECASTS

Summary for Policy Makers[13] also observed that confidence in the ability of models to project future climate has increased in the 5 years since the IPCC Second Assessment.[14] The models in question are global, coupled atmosphere-ocean general circulation models that numerically simulate climate change in scenarios that assume different degrees of enhancement of the greenhouse effect. *Summary for Policy Makers* summarized climate forecasts based on output from these models, in part, as follows:

Global temperature
- The globally averaged surface temperature is projected to increase by 1.4°C to 5.8°C over the period 1990 to 2100.
- It is very likely, based on recent global model simulations, that nearly

13. IPCC (n. 1 above).
14. IPCC (Intergovernmental Panel on Climate Change), *Climate Change 1995: The Science of Climate Change. Contribution of Working Group I to the Second Assessment Report of the Intergovernmental Panel on Climate Change*, eds. J. T. Houghton, L. G. Meira Filho, B. A. Callander, N. Harris, A. Kattenberg and K. Maskell (Cambridge: Cambridge University Press, 1996).

all land areas will warm more rapidly than the global average, particularly those at northern high latitudes in the cold season. Most notable of these is the warming in the northern regions of North America, and northern and central Asia, which exceeds global mean warming in each model by more than 40 percent.

Snow and ice
- Northern Hemisphere snow cover and sea-ice extent are projected to decrease further.
- Glaciers and ice caps are projected to continue their widespread retreat during the 21st century.

Precipitation
- Based on global model simulations and for a wide range of scenarios, global average water vapour concentration and precipitation are projected to increase during the 21st century. By the second half of the 21st century, it is likely that precipitation will have increased over northern mid- to high latitudes and Antarctica in winter.

Long-term trends
- Global mean surface temperature increases and rising sea level from thermal expansion of the ocean are projected to continue for hundreds of years after stabilization of greenhouse gas concentrations (even at recent levels), owing to the long timescales on which the deep ocean adjusts to climate change.
- Ice sheets will continue to react to climate warming and contribute to sea level rise for thousands of years after the climate has been stabilized. Climate models indicate that the local warming over Greenland is likely to be 1 to 3 times the global average. Ice sheet models project that a local warming of larger than 3°C, if sustained for millennia, would lead to virtually a complete melting of the Greenland ice sheet.

IMPACT OF CLIMATE CHANGE ON CANADA'S SHIPPING LANES

These predicted climate changes would affect Canadian shipping primarily through reductions in sea ice extent. Warmer air temperatures and more open water may also increase the occurrence of fog, adversely affecting visibility at sea. The possible long-term (millennial) melting of the entire Greenland ice sheet would raise sea level by about 6 m, and the melting of all Antarctic land ice would raise sea level about ten times this amount. Thus, in the short term, the melting of land ice could increase sea level by a few meters, which would significantly affect port operations.

The impact of the forecast atmosphere and ocean warming at high latitudes would clearly be a decrease in sea ice *formation*, but the details of changes in ice extent, concentration, and thickness are not immediately apparent. Longer ice-free seasons and improved ice conditions would benefit existing shipping, resupply operations, and ferry services in Canada's ice-infested waters.

Substantial improvements in ice conditions in the Northwest Passage (Parry Channel, Viscount Melville Sound, Prince of Wales Strait, Beaufort Sea) could open up a new shipping route between Europe and the Pacific. Russia's Northern Sea Route would benefit similarly.

Predicting the timing and extent of these impacts in specific bodies of water, however, requires more reliable regional ice forecasts than are now available. The numerical general circulation models used in climate forecasting do not resolve sea ice conditions in sufficient detail to answer questions relating to when, for example, Northwest Passage navigation might become feasible without icebreaker escort.

Wadhams[15] summarises the findings of several investigators of recent hemispheric and regional reduction in sea ice extent. He points out that the observed decrease in Arctic sea ice extent of 3 percent per decade since 1978 (and more than 4 percent from 1987 to 1994) obscures regional trends in ice extent that are quite different over the same time period. For example, between 1979 and 1996, the length of the sea ice season (navigation restricted or impossible) decreased approximately 10 percent per decade in the Barents Sea, Kara Sea, Sea of Okhotsk, and central Arctic. If this trend were to continue, the sea ice season in those areas would be shortened by 50 percent in 70 years, and by 90 percent in 220 years. Weaker decreases, however, were observed in Greenland, Hudson Bay, and the Canadian Archipelago, and in the same period, the ice season lengthened perceptibly in the Bering Sea, Gulf of St. Lawrence, Baffin Bay, and Labrador Sea.

Regional differences in trends in ice extent are due to spatial variation in the factors that govern ice growth and destruction. The volume of sea ice at a given time and place is essentially the outcome of thermodynamic processes (the net atmosphere and ocean heat transfer to the ice or surface layer of water) and dynamical processes (the net ice transport to the place in question by ocean currents and wind). Sea ice forecast models (Fig. 8) simulate the formation, growth, and destruction of ice by evaluating the outcome of these processes at future times, using, as input, the output from a general circulation model. Differing balances between thermodynamic and dynamical processes occur in different situations, as can be seen by comparing the balance for fast ice and pack ice.

Fast ice is sea ice that remains stationary along a coast where it is attached to the shore, or is otherwise unable to move under the influence

---

15. P. Wadhams, *Ice in the Ocean* (Amsterdam: Gordon and Breach Science Publishers, 2000).

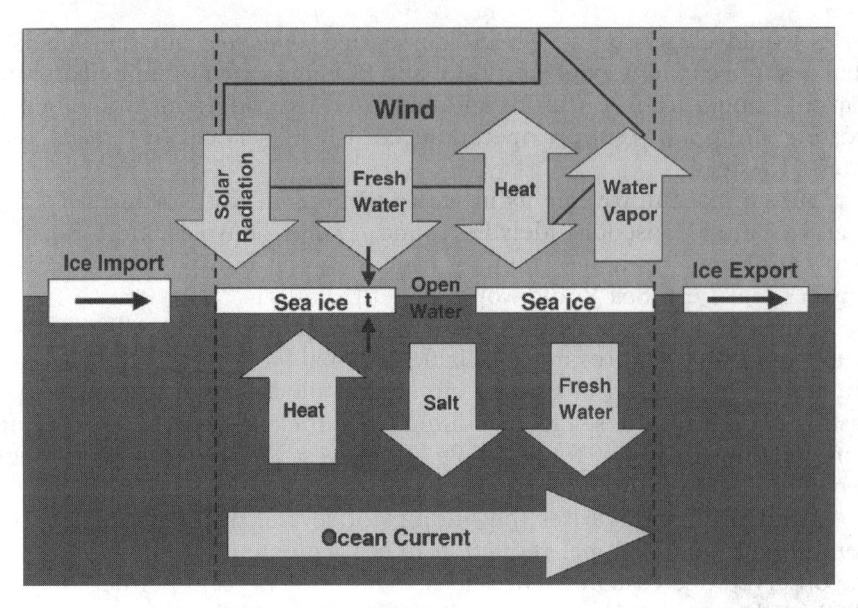

FIG. 8.—A sea ice forecast model for predicting the volume of ice within the area bounded by the dotted lines.

of wind and currents. In fast ice, therefore, the export and import of ice may be ignored, and the ice extent and thickness is almost entirely the result of heat transfer into the water or ice. In very shallow water, oceanic heat transfer is negligible relative to atmospheric heat transfer and solar radiation. The under-ice variables in Fig. 8 are therefore unimportant relative to those above the ice, with the result that the growth and destruction of fast ice is more predictable than that of pack ice.

In pack ice, which is free to move under the influence of currents and wind, both thermodynamic and dynamical processes are significant. Consequently, forecast models of pack ice must include all the variables shown in Fig. 8. In addition to solar radiation and atmospheric heat transfer, pack ice forecast models also require input data describing ocean temperature, salinity, currents, wind, and the strength properties of the ice. In areas such as Parry Channel and Viscount Melville Sound, ice forecasting is further complicated by the difficulty of predicting the southward transport of Arctic Basin multi-year ice southward through the channels of the Canadian Archipelago.

To reliably predict future ice conditions along the complex of shipping lanes in Canada's north and Arctic, will require sophisticated high-resolution regional climate and sea ice forecasting models, which are now being developed.

Apart from its influence on the occurrence of sea ice, global warming is also predicted to continue the retreat of ice caps and glaciers. A decrease in sea ice extent and concentration means an increase in the amount of open water, which, coupled with warmer temperatures, will increase evaporation. Thus there is the potential for more precipitation, which may lead to heavier snowfall, a net growth of West Greenland tidewater glaciers, and an increase in the production of icebergs. A warmer atmosphere and warmer ocean water would hasten iceberg melt, but available models are unable to predict the impact of climate warming on the number and size spectra of icebergs in Canada's shipping lanes (Fig. 3).

SUMMARY AND CONCLUSIONS

The most important environmental factor affecting navigation of Canada's shipping lanes is the presence of ice. Sea ice restricts shipping in the Gulf of St. Lawrence, Newfoundland and Labrador, Hudson Bay, and the Arctic.

The earth's climate warmed significantly in the 20th century, and forecasts predict the warming will continue. The rate of warming in high latitudes and the polar regions exceeds the global average. Sea ice extent and thickness have declined measurably in most parts of the Arctic since the 1970s.

Climate warming is likely to have positive impacts on northern and Arctic shipping in Canada and Russia. Navigation seasons are likely to become longer, but detailed forecasts of future ice conditions for particular locations are not possible with present-day climate and sea ice forecast models. Finer resolution models are needed to predict ice conditions in the complex web of waterways in Canada's north.

The numbers of icebergs encountered in the Canada's northern and east coast shipping lanes may increase if warming results in heavier snowfall and consequent growth of Greenland's tidewater glaciers. Reliable predictions of changes in iceberg numbers and sizes are not available with available glacier models.

Climate Change and Arctic Shipping

Climate Change and Canada's Shipping Lanes: The Background Science†

Henry G. Hengeveld
Meteorological Service of Canada, Environment Canada, Downsview, Ontario

INTRODUCTION

Less than a decade ago, few people outside of the scientific community had heard about climate change. Yet, today, climate change has become a term recognized and discussed daily in newspapers, households and boardrooms around the world. Internationally, politicians have agreed to a Framework Convention on Climate Change and signed an accompanying Kyoto Protocol, which if ratified, will commit the developed world to the first modest but concrete steps towards addressing the issue. Nationally, provinces, territories, and the federal government have developed an action strategy to help meet Canada's commitments under the Protocol. In keeping with these policy developments, many sectors of industry are now beginning to seriously address their responsibilities for reducing emissions of greenhouse gases, and to think about how they should prepare to adapt to those aspects of climate change that already appear to be inevitable.

There are, of course, still some who argue that climate change is poorly understood and may be a non-issue. Hence, they suggest we should first understand it better before we take mitigative action. Others argue that 'global warming' may actually be good for us. After all, we live in a rather cold climate! And then there are those who worry that action to reduce the risks of climate change may be too costly and that future technological development can save the day.

So what is the real story? What is the basis of the concern about climate change? Why has it become such an important issue so quickly? Why does it seem so controversial? Furthermore, how will it affect us here in Canada, where warmer weather sometimes seems so attractive? More importantly, for

†EDITORS' NOTE.—Presented to a seminar on "Global Warming and Canada's Shipping Lanes" at the Canadian Maritime Law Association Annual General Meeting, June 2001, Montreal, Canada.

© 2003 by The University of Chicago. All rights reserved.
0-226-06620-7/03/0017-0021$01.00 *Ocean Yearbook* 17:580–595

those involved in marine law, how will it impact those involved in the marine shipping industry?

The following sections of this article will first examine the process used within the international community for assessing and communicating the current state of scientific understanding of climate change and its risks. It will then consider what scientists can and cannot say about how and why the climate has been changing in recent centuries and decades, what changes are expected over the next century, and how such changes will affect global ecosystems and society. Some of the impacts on Canada's coastal and inland waters will be considered in greater detail. Finally, the article will briefly describe the history and current state of international efforts to mitigate climate change.

ASSESSING THE RISKS OF CLIMATE CHANGE: THE PROCESS

Scientists have recognized for more than a century that atmospheric greenhouse gases play an important role in the global climate system, and that changes in their concentrations can have significant effects on climate. However, the first serious concerns about the interfering role that humans might play in the natural balance of these concentrations, and hence in the global climate system, only emerged some 40 years ago.[1] Since then, systematic observations of changing atmospheric composition and a coordinated international climate research program (under the joint auspices of the United Nations' World Meteorological Organization (WMO) and the International Council of Scientific Unions (ICSU)) have helped to improve our understanding substantially. An international army of scientists from nations around the world, involved in almost every scientific discipline imaginable, is now publishing new results in literally thousands of peer-reviewed papers in scientific journals each year. The climate system is very complex and may never be fully understood because of the many nonlinear interactions between its geophysical components. However, slowly but incrementally, the mysteries of the climate system are being unraveled and our impacts upon our climate are becoming more and more apparent.

In 1985, some one hundred of these scientists met in Villach, Austria and issued the first formal warning to policy makers. They noted that "many important economic and social decisions are being made today on long-term projects . . . based on the assumption that past climate data . . . are

1. R. Revelle and H. E. Suess, "Carbon dioxide exchange between the atmosphere and ocean and the question of an increase of atmospheric CO_2 during the past decades," *Tellus* 9 (1957): 18–27.

582   *Climate Change and Arctic Shipping*

a reliable guide to the future. This is no longer a good assumption . . ."[2] Three years later, scientific experts, politicians, economists, engineers, environmentalists, and others met at the World Conference on Atmospheric Change, held in Toronto, to formally discuss the scientific basis for concern about climate change and to suggest possible courses of action to reduce related risks. Participants at that conference agreed that "humanity is conducting an unintended, uncontrolled, globally pervasive experiment [with the Earth's atmosphere] whose ultimate consequences could be second only to a global nuclear war."[3] That same year, the UN community demonstrated its concern by establishing the Intergovernmental Panel on Climate Change (IPCC) as a formal mechanism to undertake periodic comprehensive international assessments of the science of climate change and to advise appropriate UN bodies of its conclusions.

The mandate of the IPCC, as established by its UN organizers (the WMO and the United Nations Environment Programme), was to assess available scientific information on climate change and its environmental and socioeconomic impacts and to formulate appropriate response strategies.[4] Throughout its subsequent evolution, the IPCC Bureau and plenary bodies have attempted to keep their activities focused on science assessment, keeping the assessments independent of policy debate yet policy relevant. The IPCC has now completed three full comprehensive assessments of the state of climate change-related science and has released a number of supplementary special reports focused on specific concerns.

To ensure the scientific independence and integrity of the IPCC assessments, the preparation of the main chapters of the assessments is led by internationally recognized experts chosen on the basis of their publication record within the peer-reviewed scientific literature. They are assisted by contributing authors who bring additional specialized expertise to bear. These technical reports undergo a double peer-review process—first by the international community of experts, then, after revision, a second review by both the expert community and government representatives to the IPCC. The contents of these reports remain the responsibility of the lead authors, and are accepted (without further revision) by the IPCC for information. During the most recent assessment, this process

2. World Meteorological Organization. *Report of the International Conference on the Assessment of the Role of Carbon Dioxide and of Other Greenhouse Gases in Climate Variations and Associated Impacts.* WMO Report #661. Geneva: World Meteorological Organization, 1986, p. 1.

3. World Meteorological Organization. *Proceedings of the World Conference on the Changing Atmosphere: Implications for Global Security.* WMO Report #710. Geneva: World Meteorological Organization, 1988, p. 292.

4. J. T. Houghton, G. J. Jenkins and J. J. Ephraums, eds., *Climate Change: The IPCC Scientific Assessment* (Cambridge: Cambridge University Press, 1990), preface.

involved more than 2000 scientific experts from a broad range of scientific disciplines.[5]

These expert reports are highly technical scientific documents, replete with jargon and complex terminology often incomprehensible to lay audiences. Hence the IPCC Secretariat, with the assistance of the lead authors of the technical reports, also prepares a Summary for Policy Makers (SPM) for each report that seeks to synthesize its key findings in a more comprehensible manner. Each SPM is then debated extensively, often word for word, in IPCC plenary meetings to seek consensus from participating representatives of member countries on the final wording of the text. Lead authors of the technical reports attend these debates to ensure that the essence of their assessments is properly captured, and that they can agree with the final wording of the documents. Thus, the final text has the acceptance and ownership of both the science community and representatives of the participating countries. The entire process, from initial outline to final printing, takes about 2 years.

The IPCC process as outlined earlier is not without its critics. Some of these argue that the IPCC fails to adequately pay attention to those skeptical of the risks of climate change, or that it has a hidden agenda of stimulating increased funding for science research, not environmental protection. There are also warnings that, in its effort to seek consensus amongst scientists, it stifles the adversarial debate upon which the scientific process is based. Furthermore, the process of simplification from technical report to SPMs drops many of the caveats, and hence may inadequately retain the context for statements within it. Others, however, claim that it is far too conservative in its conclusions and must seek to make its results more policy relevant. It is also argued that the language of the documents continues to be couched in terms of uncertainties and caveats, rather than in terms of probabilities and risks more familiar to nonscientific audiences. Hence, the IPCC, while generally accepted as the most authoritative and credible voice on climate change science, also has its limitations as a means of communicating climate change science to lay audiences.

Despite the concerns of critics, both the international science and policy communities regard the IPCC reports as the most up-to-date, comprehensive, and extensively peer-reviewed assessment of the science of climate change currently available. That is in good measure due to the ability of the reports to not only address scientific debate on points of uncertainty but to also profile the good agreement between scientists on the fundamental

---

5. S. Agrawala, "Structural and process history of the Intergovernmental Panel on Climate Change," *Climatic Change* 39 (2000): 621–42; IPCC., 2001. *Third Assessment Report on Climate Change, Volumes I, II and III*. Cambridge: Cambridge University Press, 2001.

principles of physics and chemistry behind the climate change issue and the scientific basis for concern. Scientists now agree, for example, that the earth's climate is already changing; most of the global warming during the past 50 years is likely due to human activities; expected changes within the next century will be large and potentially problematic; and the related risks to humanity and ecosystems already justify mitigative actions beyond simple precautionary measures.

OUR CHANGING CLIMATE

Climatologists have been monitoring the Earth's climate, with the help of meteorological instruments, for more than a century. The resulting data, which are collected from many thousands of land and ocean weather and climate stations and corrected as carefully as possible for biases due to urbanization and other nonclimatic influences, suggest that the average surface temperatures around the world have warmed by about 0.6°C ± 0.2°C since 1860. Retreating alpine glaciers, thinning and retreating sea ice in the Arctic Ocean, and other forms of proxy data support the conclusion that this change is real. Furthermore, comparison with temperature records reconstructed from geophysical indicators of climate conditions, such as tree ring characteristics and chemical composition of polar ice cores and ocean corals, indicate that, at least in the Northern Hemisphere, the 20th century was the warmest of the past millennium, the 1990s the warmest decade, and 1998 the single warmest year (fig. 1).

Attributing the recent changes in climate to specific causes, however, is a more complex challenge. There is increasing evidence that internal variability within the climate system can cause significant oscillatory shifts in climate lasting at least several decades. Natural causes of climate change can also cause climates to vary with time. Such causes include the effects of variations in solar intensity over time scales of decades and centuries or longer and the large eruptions of volcanoes, which release sunlight-filtering aerosols into the atmosphere. Finally, there is increasing evidence of changes in the composition of the atmosphere due to human emissions of greenhouse gases (which heat the planet) and aerosols (which tend to cause surface cooling). For example, the atmospheric concentration of carbon dioxide (one of the key greenhouse gases) has increased by 31 percent since pre-industrial conditions, while that for methane (another important greenhouse gas) has more than doubled.

These possible causes of climate change, whether natural or induced by humans, are referred to as climate forcings. To help sort out this complicated interplay of different climate forcings, scientists turn to the combination of statistical analysis of observed climate data and computer simulations

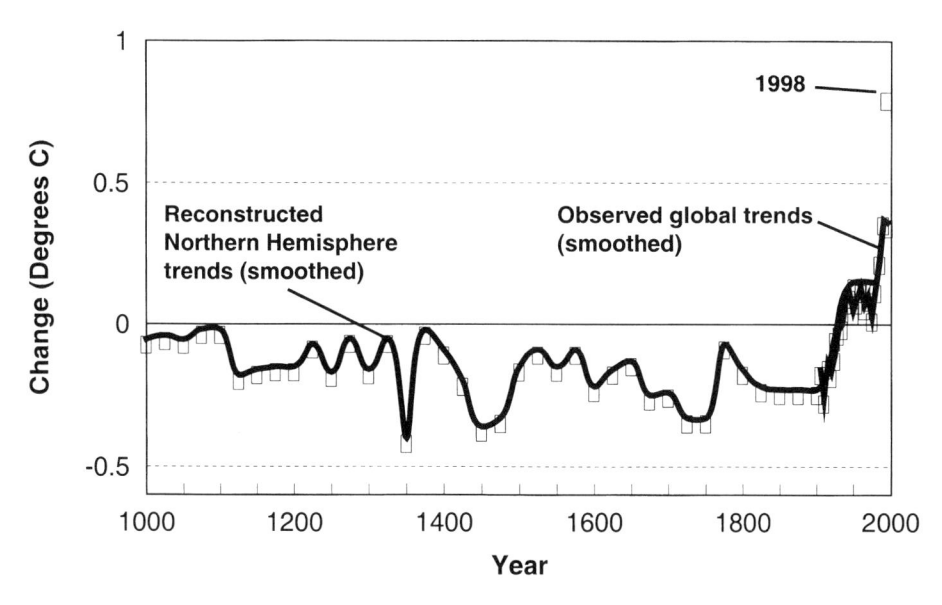

FIG. 1.—Comparison of observed trends in average surface temperatures in the Northern Hemisphere with that reconstructed from proxy data suggest that recent trends are unprecedented in at least the past millennium. Solid black line is smoothed proxy trend, while the line from 1900–2000 provides an estimate of uncertainty in the proxy data. Proxy data for the Southern Hemisphere are as yet too sparse to make similar comparisons in that region.

of the climate system. The latter use fundamental laws of physics and chemistry and advanced mathematics to replicate how the various components of the climate system interact and respond to changes. They are first tested against climate observations to assess their performance. Once the models can realistically reconstruct current climate behaviour, they are used to both study past climates and project how future climates might change due to changing climate forcings. The results of studies into the behaviour of the climate over the past century, using these tools, suggest that the changes in the early part of the 20th century were likely due to a combination of increasing solar intensity, increasing greenhouse gas concentrations, and possibly reduced volcanic aerosol concentrations (due to the paucity of major volcanic eruptions during that time period). However, for the past 50 years, the net forcing from solar and volcanic sources has been towards a cooling effect, in sharp contrast to the rapid rise in temperatures observed. For this period, the evidence clearly points towards human factors as the primary cause.

PROJECTING OUR FUTURE CLIMATE

Atmospheric scientists use the same computer models, together with other assessment tools, to project how the earth's climate may evolve over the next century or so in response to continued human interference with the climate system. To do so, they must address three key questions. How will human socioeconomic evolution affect greenhouse gas and aerosol emissions? How will these emissions alter the composition of the atmosphere? Finally, how will such atmospheric change alter the climate?

To address the first question, experts use integrated economic, social, and physical models to estimate future trends in greenhouse gas and aerosol emissions in response to a range of plausible changes (regional and global) in five key socioeconomic variables: population, economic growth, energy efficiency, technology, and land use activity. Out of 40 some possible futures that have emerged out of recent IPCC analyses (known as SRES scenarios), six representative scenarios are used for further scientific investigation and as references for policy discussions.

Global carbon budget models and similar models for other greenhouse gases and aerosols are then used to estimate how these six scenarios will alter the composition of the atmosphere. For CO_2, for example, they suggest that, without action to mitigate climate change, atmospheric concentrations by 2100 will almost certainly be twice pre-industrial levels, and could triple.

Finally, the projected changes in atmospheric composition can be applied to climate models to estimate how the climate may change over time. Results suggest that, within the next 100 years, global average temperatures are almost certain to increase by more than 1.4°C, and could rise by 5.8°C. Even the lower limit would be unprecedented in at least the past 10,000 years. At the upper limit, the magnitude of change would be similar to that between the peak of the last glacial maximum some 25,000 years ago and today, but at rates of change several orders of magnitude greater.

Projected climate changes during the next century involve much more than a simple increase in average global temperatures. While models disagree on the details of the regional characteristics of future climates, they agree on several key points: continents will warm more than oceans; high latitudes will warm more than low latitudes; and the hydrological cycle will become more active. The altered distribution of surface temperatures around the world, in turn, causes changes in wind currents (which are driven largely by temperature patterns) and hence in the distribution of rainfall. Interior continental regions of the Northern Hemisphere are expected to become drier in summer. Meanwhile, thermal expansion of ocean waters as they warm, together with melting glaciers, will add some 9 to 88 cm to ocean levels by 2100. Even greater increases in sea-level rise are expected in subsequent centuries as the oceans and ice sheets very slowly respond to warmer air temperatures, even if the climate at the earth's surface becomes stabilized.

Most models also predict major changes in ocean circulation. In the North Atlantic, for example, cold, more saline surface waters currently sink into the deep ocean under the forces of gravity, causing a conveyor-belt-like process that brings very warm tropical waters (often referred to as the Gulf Stream) northward to replace the sinking water. This advection northward of tropical waters, and the heat within them, keeps Western Europe some 10°C warmer than land areas in North America at similar latitudes. However, warmer climates are expected to cause enhanced precipitation at higher latitudes, thus increasing the flow of freshwater from land into the North Atlantic. This, in turn, makes the surface waters of the North Atlantic less saline and more buoyant, causing the conveyor-belt process, and hence the Gulf Stream, to slow down by as much as 50 percent. The decrease in flow of heat northward could therefore cause some surface areas of the North Atlantic Ocean to actually cool, significantly moderating the expected warming in Europe.

GLOBAL IMPACTS OF CLIMATE CHANGE

Climatologists use a variety of analysis tools to assess how such changes might affect people and ecosystems. The details of future climate change are as yet rather uncertain, particularly at the regional and local scales where the impacts of climate change are especially important. Hence, such studies can only give an approximate sense of possible impacts, and results must still be used with caution. These suggest, however, that the social and economic consequences of climate change will be unevenly distributed.

Wealthier countries in temperate to cool latitudes, for example, are expected to experience both the benefits of warmer temperatures and the hazards associated with drier interiors, possible increases in extreme weather events, and other related problems. The net effects in these countries are therefore unlikely to cause major economic catastrophes unless the change in climate exceeds 2°C to 3°C.

In contrast, nations located in tropical regions and most small island states surrounded by oceans will receive few benefits due to more heat but are much more likely to be faced with danger. A 1-m sea-level rise, for example, could threaten the very existence of small island nations such as the Maldives or Tuvalu, and displace many millions of people in countries with heavily populated coastal regions, such as Bangladesh, China, Japan, and Egypt. Projections also suggest decreased food production in many tropical countries, some of which already face food shortages and have low adaptive capabilities. For societies within these countries, many of which are among the poorest in the world, almost any significant change in climate can substantially increase the risk of disaster.

Canada belongs to the first group of countries. Many of the social and economic activities taking place within it are limited by cold temperatures, and for much of the year, snow-covered lands and ice-covered waters. Hence, climate change will clearly provide both benefits and liabilities. Warmer temperatures, for example, mean shorter and less harsh winters and longer summer growing seasons, with their attendant benefits. Offsetting these are the risks of drier summers, lower lake and river levels in the south, risks of more frequent intense rainfall events and summer heat waves (and related health implications), and the loss of cultural activities that rely on snow and ice. Furthermore, because trees cannot uproot themselves and move with the changes in climate, major portions of Canada's forest ecosystems will likely become increasingly stressed by the warmer temperatures and drier summers. In response, the risks of insect infestations, forest dieback, wild fire and loss of ecological habitats rise substantially. Hence, some regions and some economic sectors will benefit, and others will lose. Anticipation and preparation for these changes (that is, adaptation) can help to maximize the benefits and minimize the losses.

IMPLICATIONS FOR CANADA'S MARINE ENVIRONMENT

Most studies into the impacts of climate change are still based on climate model results that assess how equilibrium climate conditions might change in a world where atmospheric carbon dioxide has doubled over pre-industrial levels. Such conditions might be similar to those we could expect towards the end of the current century if we were able to stabilize carbon dioxide concentrations at that level by about 2050. The real world, however, will respond to a transient change in climate, where greenhouse gas concentrations and hence climate evolve with time in a complex and nonlinear manner. Such transient climate change scenarios are now possible with advanced coupled climate models that incorporate a circulating ocean fully interacting with a circulating atmosphere. Only a few impact studies have as yet used results from experiments with such models. Together with studies based on earlier model results, these help to better understand the sensitivities and vulnerabilities of ecosystems and society to climate change.

The results from such studies suggest that climate change, as discussed in the preceding overview, will have profound effects on the marine environment within and adjacent to Canada. Sea ice, for example, will become thinner, with shorter ice seasons. Sea levels along most of Canada's coasts will rise. In contrast, average water levels in interior lakes and rivers in southern Canada are likely to decrease with time. Shipping routes may also become stormier. While the implications for sea ice conditions in the maritime regions adjacent to Canada will be discussed elsewhere,[6] the following provides a brief assessment of some of the other environmental changes likely to occur as the climate warms, and the implications that these changes may have for marine shipping.

Water Levels in Canada's Freshwater Systems

As already noted, most climate models project that most of southern Canada is likely to become drier, particularly in summer. The primary reason is increased evaporation due to higher temperatures. In addition, total evaporation increases because longer snow and ice-free seasons expose soils and lake surfaces to longer periods of evaporative loss. In some regions, increased precipitation may partially or completely offset this loss due to evaporation. In others, decreases in precipitation, particularly in summer, will exacerbate it. However, when rain does occur on drier soils, more of it is soaked up to replenish water tables and less of the water is available for runoff into streams and rivers. Hence, changes in stream and river flows and in lake levels will likely be disproportionately larger than that expected solely from the direct changes in the regional hydrological budget.

For the Great Lakes, studies using a number of different equilibrium and transient climate model projections suggest that by the second half of this century water levels could drop significantly. Most scenarios, for example, project that Lake Michigan, Lake Huron, and Lake Erie could all drop by a meter or more. Drops in Lake Ontario and Lake Superior would be more modest, but also substantial. Net Great Lakes Basin outflow could decrease by about 25 to 50 percent. Hence, the St. Lawrence River levels at Montreal may also drop by more than a meter. However, not all study results agree. For example, a transient climate scenario recently developed by a British modelling group suggests that increased rainfall could completely offset evaporative losses, resulting in virtually no change in lake levels or runoff by 2050.[7] Hence, while most studies suggest reductions in water resources, such results must as yet be considered with caution.

An additional consideration in assessing such changes in Great Lakes water levels and runoff is natural variability. Records of lake levels over the past century suggest these can vary substantially from decade to decade. That for Lake Erie, for example, indicates very low levels in the early 1960s, and even lower conditions in the 1930s. In contrast, levels were well above nor-

6. C. Anderson, "Global Warming and Canada's Shipping Lanes: An Oceanographer's View," in *Ocean Yearbook* 17 (Chicago: University of Chicago Press, 2003), pp. 563–579.
7. L. Mortsch, H. Hengeveld, M. Lister, B. Lofgren, F. Quinn, M. Slivitzky and L. Wenger, "Climate change impacts on the hydrology of the Great Lakes-St. Lawrence system," *Canadian Water Resources Journal* 25 (2000): 153–80.

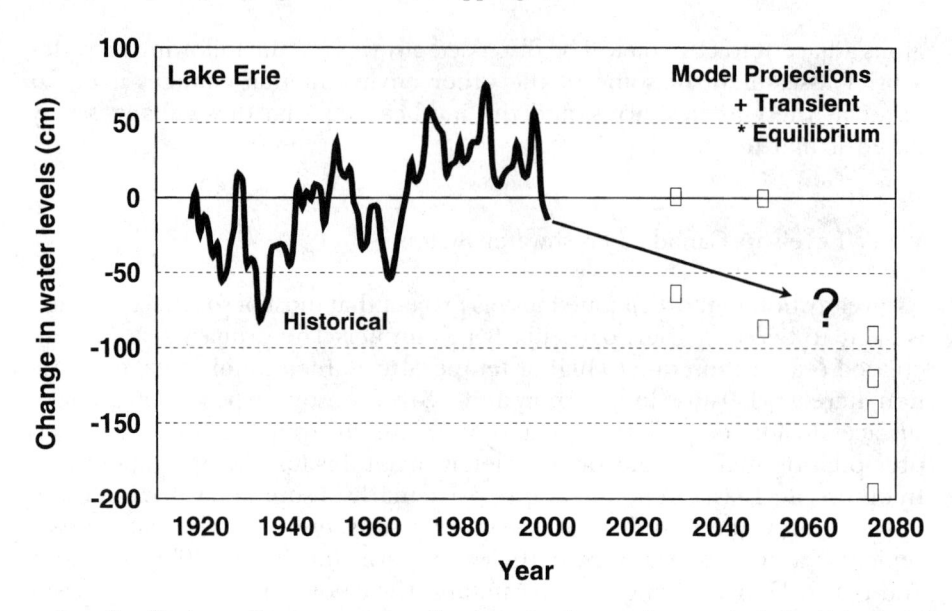

FIG. 2.—Various climate model projections for future changes in Lake Erie water levels suggest average levels by 2050 may be as low or lower than the lowest levels experienced during the past century.

mal throughout most of the past 3 decades (until the last few years). Such natural variability is likely to continue, and would be super-imposed upon any long-term trend that might be induced by climate change (fig. 2). Hence, it is likely that low water conditions like those of the 1960s and 1930s will occur more frequently in the coming decades, and may set new extreme values, while the risks of high lake levels decrease substantially. These prospects clearly suggest the likelihood of shallower ship bottom clearance in the interconnecting channels between the lakes. The negative impacts on commercial shipping, particularly grains and coal, may be quite large, and would only partially be offset by longer shipping seasons. Dredging operations to deepen such channels in turn may have unacceptable environmental risks related to disturbance of contaminated sediments.[8]

Analyses for other inland regions of interest to marine transportation also suggest similar results. In the MacKenzie Basin, for example, projected

8. F. Millerd, "The impact of water level changes on commercial navigation in the Great Lakes and St. Lawrence River," *Canadian Journal of Regional Science* 19 (1996): 119–30; J. Smith, B. Lavender, H. Auld, D. Broadhurst and T. Bullock, "Adapting to Climate Variability and Change in Ontario," in *The Canada Country Study: Climate Impacts and Adaptation, vol. IV* (Toronto: Environment Canada, 1998), p. 117.

reductions in net inflow into Great Slave and Great Bear Lakes suggest a potential lowering of average lake levels by a few decimeters in a world with doubled carbon dioxide concentrations. This, in turn, results in consistently lower outflows, increasing the risks of ice jams in the spring and other navigational concerns for the MacKenzie River system. Snow melt and hence peak run-off would also occur earlier in the year.[9]

Lake and River Ice

In the Great Lakes Basin, the thickness and seasonal duration of ice cover will both gradually decrease. By the end of the century, ice cover may pose shipping hazards for less than 1 month per year in most years.

In the MacKenzie Basin, studies also suggest shorter ice cover seasons and thinner winter ice conditions. Great Slave Lake for example, could break up 2 weeks earlier, and the ice-free season in the MacKenzie River could lengthen by up to a month by 2050.[10] Frequency and magnitude of major build-up of river frazil-ice (ice crystals in super-cooled waters that have not yet solidified into solid ice but can attach quickly to ships and other objects) are also likely to decrease.

Storms

Most climate models do not have adequate resolution to deal with storms with considerable confidence. Related studies, however, offer some clues as to how these may change in a warmer world. A Canadian study, for example, suggests that a warmer world may result in a decrease in the total number of synoptic storms in the Northern Hemisphere during the winter season, but that the number of severe storms will likely increase.[11] This is consistent with projections of increased northward transport of moisture in a warmer world, providing much of the latent energy for such storms. Dominant storm tracks over the North Atlantic and Pacific are also likely to shift with changing atmospheric circulation, although exactly how is still an important point of discussion.[12]

Likewise, there is still disagreement on how hurricane behaviour may

9. S. J. Cohen, ed. *MacKenzie Basin Impact Study (MBIS)*. Toronto: Environment Canada, 1997, p. 372.
10. Ibid.
11. S. Lambert, "The effect of enhanced greenhouse warming on winter cyclone frequency and strengths," *Journal of Climate* 8 (1995): 1447–52.
12. R. E. Carnell, C. A. Senior and J. F. B. Mitchell, "An assessment of measures of storminess: simulated changes in Northern Hemisphere winters due to increasing CO_2," *Climate Dynamics* 12 (1996): 467–76.

change, because there are a number of climatic factors that contribute to their evolution and properties. For example, if hurricanes move more slowly over warmer surface waters, they induce mixing of ocean surfaces with cooler waters below, thus decreasing their potential intensity. Conversely, fast moving hurricanes have increased potential for becoming super hurricanes. Warmer ocean surfaces may also increase the theoretical limit for their intensity.[13]

Sea-level Rise

Sea-level rises will have varied effects on coastal regions around the world, both because changing atmospheric pressure patterns can cause greater sea-level rise in one region than another, and because land masses in some regions of the world are slowly sinking while others are rising. In Canada, for example, land in the Hudson Bay region is still rebounding in response to deglaciation thousands of years ago. Such land rise may be enough to fully offset expected global sea-level rise. Conversely, like the opposite end of a teeter-totter, eastern Canada is slowly sinking, adding to the effects of sea-level rise. Hence, a local sea-level rise in Halifax or Saint John of a meter or more within the next century is quite realistic. While this may improve harbour access for shipping, it may also cause repeated flooding and damage to marine shipping infrastructure.[14]

SLOWING DOWN AND ADAPTING TO CLIMATE CHANGE

In 1992, nations around the world endorsed the United Nations Framework Convention on Climate Change (FCCC). The ultimate objective of the FCCC is to stabilize atmospheric concentrations of greenhouse gases at a level that would prevent dangerous human interference with the climate system. However, policy makers recognized the need for three key elements of a global strategy to address this challenge: research (improving our understanding of the risks to help us better define our response); mitigation (reducing the rate of climate change through reductions in global greenhouse gas emissions); and adaptation (reducing risks by anticipating and preparing

---

13. L. R. Schade and K. A. Emanuel, "The ocean's effect on the intensity of tropical cyclones: results from a simple coupled atmosphere-ocean model," *Journal of the Atmospheric Sciences* 56 (1999): 642–51; T. R. Knutson, R. E. Tuleya and Y. Kurihara, "Simulated increase of hurricane intensities in a CO_2-warmed climate," *Science* 279 (1998): 1018–20.

14. J. Shaw, R. B. Taylor, D. L. Forbes, S. M. Solomon and M.-H. Ruz. *Sensitivity of the Coasts of Canada to Sea Level Rise*. GSC Report #505. Ottawa: Geological Survey of Canada, 1998, pp. 1–79.

for the consequences). Canada, in developing its national action strategy for climate change, has also endorsed these three elements.

While some would still argue that more research is needed before we undertake mitigation efforts, there is an broad consensus within both the science and international policy communities that the risks of climate change are sufficiently significant to already justify mitigative action beyond simple precautionary measures (that is, those measures that already make sense for other socioeconomic reasons). In 1997, developed nations party to the Convention agreed under the Kyoto Protocol to take the first small step towards that FCCC's ultimate objective by collectively reducing their emissions by 5.2 percent below 1990 levels by 2010. Individual country targets vary, with pledged reductions of 6 percent for Canada, 7 percent for the United States, and 8 percent collectively for the member countries of the European Union. Some countries, such as Australia, successfully argued for increases in emissions. Most of these developed countries have not as yet ratified this agreement, nor have they as yet prepared detailed action plans on how they will meet their obligations under this agreement. Emissions in Canada, in fact, are already more than 10 percent above 1990 levels and, without concerted efforts to reduce emissions, are projected to increase by another 10 percent or more by 2010. Meanwhile, the new American administration has indicated its intent not to ratify the Protocol. Hence, the international resolve to mitigate climate change that was so evident in the early 1990s seems to be wavering. The Kyoto Protocol may never come into force!

However, even a successful and complete implementation of the Kyoto Protocol agreement would only delay attaining the various climate change thresholds of the future by a decade or so. In fact, studies suggest that, to achieve stabilization of greenhouse gas concentrations, greenhouse gas emissions at a global scale would eventually need to decrease to some 50 percent or more below 1990 levels. The developed country commitments under the Kyoto Protocol only achieve about 10 percent of their contribution to that target. More importantly, emissions from developing countries, which are expected to comprise the larger share of global emissions within a few decades, are as yet not addressed in the Protocol, and hence can continue to rise unabated.

Even if the international community were able to eventually reach agreement on how to fully stabilize concentrations at below a doubling of pre-industrial carbon dioxide concentrations (a major challenge, because this would require a major shift in energy technologies towards an almost totally renewable energy future), this is unlikely to eliminate the risks of danger due to climate change. That is because climate models predict that the earth would still likely warm by 2°C to 3°C under these conditions. In other words, while we can reduce the risks by reducing the rates of climate change through mitigation, it is too late to stop climate change. This is a reminder that adaptation to climate change must also be a key part of response strategies. That is, the development of major long-term socio-economic infrastructures, as they are being planned or renewed, must carefully consider the possible consequences of climate change. Often, simple changes in design can both deal with current reality and reduce the risks of hazards in a warmer world. The design of the bridge from New Brunswick to Prince Edward Island, for example, included allowances for a potential 1-m sea-level rise at modest incremental cost. Likewise, most other sectors of Canada's economy, whether land use planning, offshore development, forest management, emergency preparedness, or water resource management, need to carefully assess the possible implications of climate change and prepare for it. This can help to both capitalize on the benefits a warmer climate may provide and reduce the economic, social, and ecological losses that would otherwise occur. Collectively, such response also helps to raise the threshold at which climate change may become "dangerous."

Unfortunately, successful adaptation often demands accurate projections of future climates. Yet, climate models are as yet unreliable as prescriptive forecast tools. Hence, there is universal agreement on the need for more research, not as a prerequisite to action, but as a means to ensure that such actions are well designed and effective.

SOME CONCLUDING THOUGHTS

The scientific understanding of the climate system and the risks of future climate change are still constrained by inadequate understanding of the many complex processes involved and by the inherent uncertainty that will always be part of such a nonlinear, somewhat chaotic system. However, the available knowledge does allow for some tentative conclusions, as follows:

1. The global climate system is already changing, and changes in recent decades are likely due to humans;
2. The risks of future climate change at magnitudes unprecedented in human history are real and significant;
3. The potential rates and regional characteristics of such change are still poorly understood;
4. Climate change poses serious risks of harm to humans, particularly in poor and developing regions of the world;
5. The consequences for the Canadian marine environment will be significant, providing both benefits and problems;
6. International objectives for mitigating greenhouse gas emissions can reduce risks by slowing down the rate of climate change;
7. Some additional climate change is unavoidable, and adaptation is an important part of the response; and

8. Better understanding through enhanced research will improve response opportunities.

It is clear from recent international developments within the policy community that responding to the risks of climate change within a time frame that would not jeopardize the well being of future generations will require international statesmanship that places the interest of such generations at par with those of today. Policy makers must also recognize that in an increasingly interactive global village, a disaster anywhere in the world also affects the rest of the world, and that measures to protect the environmental well being of our planet can also be measures that make economic sense. Such statesmanship, in a democratic world, also demands a supportive electorate—something that still seems to be largely absent in North America. Perhaps it is indeed time for the science community to make their case more effectively!

Climate Change and Arctic Shipping

The Future of the Arctic Ocean: Competing Domains of International Public Policy

Douglas M. Johnston
Emeritus Professor of Law and Emeritus Chair in Asia-Pacific Legal Relations, University of Victoria

INTRODUCTION

The Arctic region has been peopled, albeit thinly, for thousands of years.[1] In the eyes of the outside world even the High Arctic has had a storied past for well over 400 years.[2] Yet the Arctic Ocean remains the most remote, most inaccessible, and most mysterious of all our seas.

During the first half of the 20th century, advances in vessel technology were just beginning to offer a glimpse of a more eventful future for the Arctic Ocean. However, the two-generational Cold War era (1946–89) had the effect of paralyzing almost any effort to achieve cooperative arrangements among the five littoral states adjacent to the Arctic Ocean: Canada, Denmark (for Greenland), Norway, Russia and the United States (the "Arctic Five").[3] Only in the early 1990s did hopes for significant international cooperation start to become realistic. Since then the Arctic Ocean has been "re-discovered" by the international community. The future of that frozen

---

1. Many ethnologists have found archeological evidence of a common Stone Age culture in the circumpolar North. Some believe that the general climate and a rich maritime environment could have sustained Aleut coastal communities for 12,000 years, and certainly for over 8,000 years. W. S. Laughlin and J. S. Aiguer, "Aleut Adaptation and Evolution," in *Prehistoric Maritime Adaptation of the Circumpolar Zone*, ed. W. Fitzhugh (The Hague: Mouton Publishers, 1975), pp. 181–201. But the Arctic Small Tool Tradition cannot be traced back as far as 3,000 B.C.
2. It is difficult to substantiate early accounts of explorations to "Thule," such as those of the Greek navigator Pytheas in the 4th century B.C. The Vikings certainly discovered Greenland in the 10th century A.D., and Iceland in the previous century. Exploration of the Northeast and Northwest Passages began in the 16th century, mainly by English and Dutch navigators looking for possible trade routes.
3. These five states all possess Arctic coastlines: that is, within the Arctic Circle (66 degrees 30' N). The three other "Arctic states" (Finland, Iceland and Sweden) are essentially sub-Arctic territories. Iceland has a tiny part of its coastline virtually on the Arctic Circle, but it is normally characterized as a North Atlantic coastal state.

sea, and of the circumpolar North in general, has become a candidate for inclusion on the global public policy agenda, despite the fact that most of the region consists of sovereign lands and national waters.

Most of the new global concern about the future of the Arctic arises from the mounting evidence of rapid climate change. More than ever before, the circumpolar North has become the focus of large-scale international research programs. But, unlike the de-territorialized Antarctic, the Arctic cannot be converted into a scientific laboratory. On the other hand, the Arctic Ocean—or part of it—might be regarded as a sort of international area. Fundamental questions arise about its future governance, and the potential role of the international community.

The initial public policy problem is that there are several "domains" of expertise relevant to international policy-making for the Arctic Ocean. Each domain has its own distinct set of assumptions, expectations, and perceptions, and each exhibits a diversity of interests and values. A reconciliation of all policy prescriptions across these domains is difficult to envisage. Yet it may be useful to sketch out the principal elements in the debate over the future of the Arctic Ocean.

On the face of things, it is necessary to take into account six important frames of reference: namely, those associated with (i) the law of the sea community, (ii) the environmental ethicists, (iii) the developmental ethicists, (iv) the Arctic science community, (v) the strategists, and (vi) the regime builders.

THE LAW OF THE SEA COMMUNITY

The negotiations before and during the Third United Nations Conference on the Law of the Sea (UNCLOS III) (1967–82) coincided with the final phase of the Cold War era. The shared interest in mutual deterrence on the part of the Soviet and U.S. superpowers involved an implicit agreement not to allow strategically sensitive Arctic Ocean issues to be placed on the agenda of this otherwise comprehensive law-making conference.[4] Yet both the process and the product of these negotiations, the 1982 United Nations Convention on the Law of the Sea, did have important repercussions for the Arctic Ocean in several ways. Three legal developments in particular deserve mention. Taken together, they constitute the international legal

4. The only specifically Arctic provision in the 1982 United Nations Convention on the Law of the Sea, the so-called "Arctic exception," is Article 234 on "ice-covered waters." The tacit understanding at UNCLOS III to leave the Arctic out of account has created uncertainties regarding the applicability of certain other provisions to these waters.

framework for public policy making in the Arctic Ocean, applicable even to states that are still nonparties such as Canada and the United States.

Most significantly, the extension of coastal state jurisdiction sanctioned at UNCLOS III, confirmed as a matter of state practice and established in customary international law, has allowed the five Arctic littoral states to extend their national jurisdiction in the Arctic Ocean. Under the new regime of functional jurisdiction for resource-related purposes, the exclusive economic zone (EEZ), each coastal Arctic state is permitted to exercise exclusive resource rights ("sovereign rights") as far as 200 nautical miles (M) beyond the baseline of their territorial sea.[5] Given the modest potentiality of fishery resources in the North,[6] the emergence of the EEZ regime is chiefly of Arctic importance because of its relevance to offshore oil and gas production.[7]

Admittedly, long before UNCLOS III the doctrine of the continental shelf had given rise to the sole entitlement of the adjacent state to the resources of the shelf.[8] But the creation of the EEZ regime eliminates the need for difficult Arctic research within 200-M limits that might have been necessary to justify jurisdictional claims to the seabed and subsoil under these waters on the basis of that doctrine.[9] On the Arctic shelf beyond the

---

5. 1982 United Nations Convention on the Law of the Sea, Articles 56–57.

6. It is a matter of controversy whether the renewable resources of the circumpolar North can be developed into a major sustainable industry. Most fish species are underexploited or unexploited, but their harvestable potentiality is limited. Most Arctic marine mammals, on the other hand, are exploited to the point of concern about their conservation, but scientists are divided on whether they are threatened or in danger of extinction. For a recent optimistic assessment, see L. A. Harwood and others, "The harvest of Beluga whales in Canada's Western Arctic: hunter-based monitoring of the size and composition of the catch," *Arctic* 55 (March 2002): 10. On current issues related to the harvesting of the living resources of the Arctic Ocean, see R. A. Caulfield, "Political Economy of Renewable Resources in the Arctic," in *The Arctic: Environment, People, Policy,* eds. M. Nuttall and T. V. Callaghan (Amsterdam: Harwood Academic Publishers, 2000), pp. 485–89.

7. Coastal state jurisdiction over offshore oil and gas reserves cannot be contested within undisputed EEZ areas, nor within undisputed areas of the seabed on the continental shelf beyond as defined in Article 76. The principal areas of current Arctic offshore oil and gas interest are in the Beaufort Sea and Mackenzie delta, off the coasts of Alaska, the Yukon, and the Northwest Territories.

8. Continental shelf doctrine prevailed in customary international law between the 1930s and 1970s, but was codified in the 1958 Geneva Convention on the Continental Shelf. Various difficulties, such as those related to the "exploitability" criterion, led to a fundamental reformulation at UNCLOS III. For a summary account of this shift, see D. M. Johnston, *The Theory and History of Ocean Boundary-Making* (Kingston and Montreal: McGill-Queen's University Press, 1988), pp. 85–94.

9. Obviously it would have been particularly difficult to satisfy the criterion of exploitability in ice-covered waters, and even in relatively ice-free Arctic waters questions might have arisen about exploitability in the sense of the economic feasibility of offshore production, given the exceptionally high costs of operation in remote frigid zones.

200-M limits, jurisdictional uncertainties still arise for Russia and Canada by virtue of the presence of submarine ridges in more distant waters,[10] although Article 76 of the 1982 Convention does provide guidance on how such claims beyond the EEZ should be addressed.[11]

The waters of the Arctic Ocean beyond 200 M, constituting the circumpolar "core" of that ocean, can be characterized as an "enclave" that might be analogized with other enclaves surrounded by the EEZs of opposite states.[12] Technically, it can be argued that such enclaves beyond national jurisdiction possess the status of high seas.[13] In practice, it seems clear that

10. In 2001, Russia made a claim to Arctic continental shelf areas beyond its EEZ based on geological evidence of submarine ridges that appear to be "non-oceanic" and therefore may be included under the regime of the continental shelf as qualified under Article 76 (3). On the general problem associated with ridges, see P. A. Symonds and others, "Ridge Issues," in P. J. Cook and C. M. Carlton, eds., *Continental Shelf Limits: The Scientific and Legal Interface* (Oxford: Oxford University Press, 2000), pp. 285–307.

11. Under Article 76(8), such claims must be submitted to a technical, non-judicial, international body, the Commission on the Limits of the Continental Shelf, which is authorized only to make "recommendations to the coastal States." The provision adds that the limits of the shelf "established by a coastal State on the basis of the recommendations shall be final and binding." For an interpretation of the Commission's mandate, see T. L. McDorman, "The role of the Commission on the limits of the continental shelf: A technical body in a political world," *International Journal of Marine and Coastal Law* (2002), in press. See also various contributions to Cook and Carleton (n. 10 above).

12. Two of these enclaves are in sub-Arctic seas: the Bering Sea and the Sea of Okhotsk. See D. A. Balton, "The Bering Sea doughnut hole convention: regional solution, global implications," in O. S. Stokke, ed., *Governing High Seas Fisheries: The Interplay of Global and Regional Regimes* (Oxford: Oxford University Press, 2001), pp. 143–77; and A. G. Oude Elferink, "The Sea of Okhotsk peanut hole: *de facto* extension of coastal state control," ibid., pp. 179–205. A third enclave, described as a "loophole," is located between the EEZs of Norway and Russia in the Barents Sea, which is in the High Arctic. See O. S. Stokke, "The loophole of the Barents Sea fisheries regime," ibid., pp. 273–301.

13. Part VII of the 1982 UN Convention on the Law of the Sea, dealing with "High Seas," is applicable, under Article 86, to "all parts of the sea that are not included in the exclusive economic zone, in the territorial sea or in the internal waters of the State, or in the archipelagic waters of an archipelagic State." This seems clear enough, but it can be argued on functionalist grounds that, since Part VII does not refer explicitly to enclaves, such "zone-locked" areas need to be brought under a negotiated set of arrangements at the initiative of the surrounding states. When the enclave consists of ice-covered waters, this functionalist argument is strengthened by common sense considerations of a kind that could not be entertained at UNCLOS III, since the Arctic Ocean was not in contemplation. But the orthodox legal argument may be that ice shelves, considered geophysically, are "extraterritorial glacial extensions floating on ocean space," and therefore "in situ high seas in a frozen state." C. C. Joyner, "The Status of Ice in International Law," in A. G. Oude Elferink and D. R. Rothwell, eds., *The Law of the Sea and Polar Maritime Delimitation and Jurisdiction* (The Hague: Martinus Nijhoff Publishers, 2001), pp. 23–48 at p. 33.

an anomaly of this kind can be resolved only through negotiation between or among the opposite littoral states, and arguably on the part of non-littoral states claiming a right of access to the enclave by reference to traditional high seas doctrine or modern common heritage doctrine.[14]

Second, the 1982 Convention allows the argument that the Arctic Ocean consists primarily of the territorial seas and exclusive economic zones of the five littoral states bordering the Arctic Ocean, even if the term "primarily" has to be interpreted in functional rather than spatial terms.[15] If this interpretation were conceded, as a matter of common sense in a largely frozen sea, then the Arctic Ocean as a whole could be characterized as a unit of cooperative ocean management along the lines envisaged in Article 123.[16] If the five littoral states were to acknowledge their common responsibilities based on this provision of the 1982 Convention, it would certainly strengthen their claim to exclusive entitlements beyond 200-M limits in the Arctic Ocean—the "Arctic Mediterranean"—vis-à-vis the rest of the international community.[17]

---

14. The high seas concept is linked historically to the traditional right of unrestricted navigation and other ocean-based freedoms that have never been exercised in the "core" waters of the Arctic Ocean. The same division between legal formalism and functionalism complicates efforts to designate parts of the Arctic Ocean as falling under Part XI on "The Area," where the common heritage doctrine would be applicable under Article 136.

15. Whether the Arctic Ocean consists "primarily" of national waters, in the spatial sense, depends, of course, on how one defines "Arctic Ocean." At its centre is the "North Polar Sea" or "Arctic Sea," much of which lies beyond 200-M limits, but most definitions include also adjacent seas (e.g., Greenland Basin, Norwegian Sea, Barents Sea, Kara Sea, Laptev Sea, Chukchi Sea, Beaufort Sea, and Baffin Bay). It is possible to include even nonadjacent waters such as the Bering Sea, Hudson Bay, and the Norwegian Basin, if the purpose is to encompass all waters that are ice-covered at certain times of the year. If broadly defined, the Arctic Ocean consists primarily of national waters. If narrowly defined (e.g., as the ocean area within the Arctic Circle), most of the Arctic Ocean extends outside EEZ limits, but is only minimally available for "international" purposes.

16. Article 123 provides that states bordering enclosed or semi-enclosed seas "should cooperate with one another in the exercise of their rights and in the performance of their duties" under the Convention. Explicit purposes of such cooperation include fishery use and management, which has limited applicability to the Arctic Ocean as narrowly defined, but also scientific research and environmental protection, which are very important in the region. Few would deny that the Arctic Ocean is "semi-enclosed": its only "wide mouth" is between Greenland and Norway in the Svalbard area, which permits only limited navigation most of the year.

17. Since the introduction of the Arctic Environmental Protection Strategy (AEPS) and the subsequent creation of the Arctic Council, these five states, along with Finland, Iceland, and Sweden, have engaged in numerous environmental and scientific activities that include those which would seem to qualify as the kind of cooperation envisaged in Article 123. On the evolution of AEPS in critical perspective, see D. VanderZwaag, R. Huebert and S. Ferrara, "The Arctic Environmental Protection Strategy, Arctic Council and Multilateral Environmental Initiatives: Tin-

Third, chiefly by virtue of strenuous Canadian diplomacy, the Conference accepted that in "ice-covered areas within the limits of the exclusive economic zone" coastal states have the right to "adopt and enforce non-discriminatory laws and regulations for the prevention, reduction and control of marine pollution from vessels."[18] This authority vested by Article 234, not otherwise permitted under the Convention, is restricted to those areas "where particularly severe climatic conditions and the presence of ice covering such areas for most of the year create obstructions or exceptional hazards to navigation, and pollution of the marine environment could cause major harm to or irreversible disturbance of the ecological balance."[19] Article 234 certainly confers special powers on the "Arctic Five."[20]

These three features of the 1982 Convention might be regarded as the primary components of the law of the sea framework for the Arctic Ocean. But in addition, of course, we can add many other provisions of the Convention that seem applicable to all oceans, in the absence of any provision excluding applicability to the Arctic Ocean. The provisions on "straight baselines"[21] and "straits used for international navigation"[22] are only two examples that come readily to mind because of the controversies they have generated in the Arctic context. Yet the argument for general applicability of the Convention to the Arctic Ocean has to contend with the cogent evi-

---

kering while the Arctic Marine Environment Totters," in Elferink and Rothwell (n. 13 above), pp. 225–48.

18. 1982 United Nations Convention on the Law of the Sea, Article 234.

19. Ibid., See D. M. McRae, "The Negotiation of Article 234," in F. Griffiths, ed., *Politics of the Northwest Passage* (Kingston, Ont: McGill-Queen's University Press, 1987), pp. 25–45. For recent re-evaluations, see R. Huebert, "Article 234 and Marine Pollution Jurisdiction in the Arctic," in Elferink and Rothwell (n. 13 above), pp. 249–67; and R. D. Brubaker, "Regulation of Navigation and Vessel-Source Pollution in the Northern Sea Route and State Practice," in D. Vidas, ed., *Protecting the Polar Marine Environment: Law and Policy for Pollution Prevention* (Cambridge: Cambridge University Press, 2000), pp. 221–43.

20. Due to its rather vague formulation, Article 234 might be considered applicable to EEZ waters beyond those of the five Arctic littoral states. D. M. McRae and D. Goundrey, "Environmental Jurisdiction in Arctic Waters: The Extent of Article 234," *University of British Columbia Law Review* 16 (1982): 197.

21. 1982 United Nations Convention on the Law of the Sea, Article 7. See T. Scovazzi, "The Baseline of the Territorial Sea: The Practice of Arctic States," in Elferink and Rothwell (n. 13 above), pp. 69–84. For historical background and analysis, see D. Pharand, *Canada's Arctic Waters in International Law* (Cambridge: Cambridge University Press, 1988), pp. 131–84.

22. 1982 United Nations Convention on the Law of the Sea, Articles 34–35. For a Russian view on the legal status of the Northeast Passage, see L. Tymshenko, "The Northern Sea Route: Russian Management and Jurisdiction over Navigation in Arctic Seas," in Elferink and Rothwell (n. 13 above), pp. 269–91. On the status of the Northwest Passage, see Pharand, op cit., pp. 185–248; and D. Pharand (in association with L. H. Legault), *The Northwest Passage: Arctic Straits* (Dordrecht: Kluwer Academic Publishers, 1984).

dence of its truly unique physical characteristics. The uniqueness counterargument may seem sufficiently compelling to make the case for special treatment in international law. However, it might be fair for the international community to suggest that international law abhors a vacuum in the age of global law-making. Further, they may claim the right to fill that vacuum in due course if the littoral states show no disposition to accept a collective responsibility to negotiate a special regional regime for the Arctic Ocean with due regard to the interests of the international community in general.[23]

THE ENVIRONMENTAL ETHICISTS

Nothing is more conspicuous in world diplomacy today than the use of the intergovernmental arena for negotiation of environmental instruments. Pressures on the natural/human environment have accumulated as burgeoning human populations, most sectors of industry (manufacturing, resource extraction, agriculture and transportation), and many kinds of services, combine to have increasingly severe impacts on our air, water, and lands, and on the species and ecosystems they sustain. Over the last three decades—since the first mega-conference on these concerns[24]—hundreds

---

23. It has been strongly argued that the Arctic Ocean presents too many unique difficulties to be brought within the global law of the sea framework, which was designed almost entirely without consideration of these difficulties. By this argument a sui generis approach is necessary. D. R. Rothwell and S. Kaye, "The Law of the Sea and the Polar Regions: reconsidering the traditional norms," *Marine Policy* 18 (1) (Jan. 1994): 41. The uniqueness of the Arctic Ocean is often invoked as the rationale for exceptional precautionary measures to protect the Arctic marine environment, and thus for an Arctic-specific environmental regime. Some argue that this should be the prerogative of the Arctic states—either the "Arctic Five" or the "Arctic Eight"—whereas others argue that it should be the responsibility of the larger international community. The latter view can rest on the fact that the most serious threats to the Arctic marine environment, contamination and climate change, originate largely outside the Arctic region. With the establishment of the Arctic Council as an informal forum, there seems to be no immediate prospect of a treaty-based approach to regime-building in the Arctic. Compare: D. L. VanderZwaag and others, "Towards regional ocean management in the Arctic: from coexistence to cooperation," *University of New Brunswick Law Journal* 38 (1988): 1; D. D. Caron, "Towards an Arctic environmental regime," *Ocean Development and International Law* 24 (1993): 377; R. Huebert, "The Canadian Arctic and the International Environmental Regime," in J. Oakes and R. Riewe, eds., *Issues in the North*, Vol. 2 (1997), pp. 45–53; and D. Vidas, "The Polar Marine Environment in Regional Cooperation," in Vidas (n. 19 above), pp. 78–103.

24. The UN Conference on the Human Environment, held at Stockholm in June 1972, marked the beginning of a series of UN mega-conferences devoted to highly complex problems. Arguably, it also represented a new approach to conference diplomacy, and the birth of the international environmental movement.

of environmental instruments of one sort of another have been concluded, and a surprisingly large proportion of them are global in origin.

At least seven primary features of international environmental law and diplomacy set it apart as a distinct domain. First, by most definitions, "environment" is the most active and most productive area of contemporary world diplomacy. Second, it is also the most intrusive, because most areas of treaty-making, or normative development, can be shown to have an environmental aspect. Third, many of the global environmental instruments that have been negotiated in the wake of the 1972 Stockholm Conference on the Human Environment are extremely ambitious and potentially costly, generating difficult commitments for many of the less affluent members of the world community and creating risks of "over-expectation."[25] Fourth, most environmental instruments serve ethical as well as problem-solving purposes, and thus attract deep-seated emotional allegiances on the part of the "transnational ethical community" outside the arena of traditional, interstate diplomacy. Fifth, more than any other field of international law and diplomacy, the environmental sector has given rise to an extraordinary diversity of formal and informal modes of instruments and norms.[26] Sixth, because of its ethical content, international environmental law has made exceptional use of "soft law" instruments and norms that may be binding in a technical sense, but are not sufficiently precise to be enforceable as a matter of legal obligation.[27] Finally, as a result of its volume, intrusiveness, ambitiousness, "ethicality," and confusing combination of formal/informal and hard law/soft law commitments, the world's environmental treaty system is perhaps the most controversial area of international law.

Because of the very recent emergence of the circumpolar North as a political region, and thus as a level of regional diplomacy and treaty development, there is not yet any officially negotiated stock of local, littoral, or indigenous norms or constructs designed specifically to govern activities in the Arctic Ocean. By default, it is necessary for the governments and peoples of the Arctic to fill the vacuum by drawing upon the general language of commitment found in the rapidly growing stock of globally negotiated environmental instruments, none of which focuses on the special character of the frozen or semifrozen waters of the North. In an area noted for its controversiality, applications of the existing global treaty system to the Arctic Ocean

---

25. E. L. Miles, "Preface," in E. L. Miles et al., *Environmental Regime Effectiveness: Confronting Theory with Evidence* (Cambridge, Mass.: MIT Press, 2002), pp. xi–xix.

26. On this general trend, see D. M. Johnston, *Consent and Commitment in the World Community: The Classification and Analysis of International Instruments* (Irvington-on-Hudson, N.Y.: Transnational Publishers, Inc., 1997).

27. For a comprehensive review of this pattern see D. Shelton, ed., *Commitment and Compliance: The Role of Non-Binding Norms in the International Legal System* (Oxford and New York: Oxford University Press, 2000).

can make matters worse by introducing an additional dimension of controversy.

It might appear that the law of the sea framework and the environmental treaty system are bound to be inherently compatible. Both are primarily the work of the global community of lawyer-diplomats; and there is certainly a measurable overlap between the corpus of UNCLOS III (a product of the 1970s) and the succession of accords through the 1980s and 1990s that provide the new foundation of the environmental law of the sea for the early 21st century. Yet there is also a discernible "mind-set gap" between the two succeeding generations of norm-setting elites in the diplomatic community that is sufficient to justify the perception that they represent distinct domains.

The new (post-UNCLOS III) law of the sea specialists seem to vary considerably in their comfort with the "environmentalization" of their field. Many of them accept the 1982 Convention on the Law of the Sea (as revised) as the "constitutional framework," as far as basic state entitlements and duties are concerned, but accept also the relevance of new environmental goals, concerns, and issues as expressed through the global diplomacy of the post-Brundtland era.[28] Others challenge the operational significance of those elements of the environmental law of the sea that have assumed only informal or soft law modes of expression. The difficulties arise when the "post-classical," soft law ethical constructs of international environmental law are seen to impinge on the classically balanced, constitutional, hard-law provisions of the 1982 Convention.

In the context of Arctic Ocean issues, the prospect of nonconvergence between the law of the sea community and the environmental ethicists seems clearest in future disputes over international rights of navigation in Arctic waters. In such disputes law of the sea specialists will wish to argue the matter mostly on the basis of the provisions of the 1982 Convention and relevant decisions of international tribunals, which on the whole have not yet displayed wholehearted commitment to the soft law ethos of international environmental law. International environmental lawyers, on the other hand, will certainly prefer to place the relatively new ethical principles of their discipline at the centre of the argument as the "governing constructs," whether to ban, limit, or merely regulate the intrusion of the world shipping industry into the vulnerable waters of the Arctic Ocean. The two sets of norms are not, of course, mutually incompatible, but they represent two different

28. On the impact of "soft" environmental law upon the law of the sea, see D. R. Rothwell, "The Influence of Non-Binding Norms upon the 1982 Convention," in D. M. Johnston and A. Sirivivatnanon, eds., *Ocean Governance and Sustainable Development in the Pacific Region: Selected Papers, Commentaries and Comments* (Bangkok: SEAPOL, 2002), pp. 157–87; and D. L. VanderZwaag, "The Precautionary Principle and Marine Environmental Protection: Slippery Shores, Rough Seas and Rising Normative Tides," D. M. Johnston and A. Sirivivatnanon, pp. 188–209.

points of departure, and perhaps suggest the need to "revise" the 1982 Convention.[29]

THE DEVELOPMENTAL ETHICISTS

Since the end of the Second World War one of the most salient and compelling features of the organized world community has been the priority given to the development of less advanced states, nations, peoples and territories. A broad concept of "development"—with social, institutional, and educational, as well as economic, dimensions—is a major motif of the United Nations Charter.[30] Reinforced by the post-imperial drive to decolonization and independence, the quest for development assumed the status of a moral imperative.[31]

From the 1960s to the 1990s the paramountcy of the developmental ethic became unchallengeable in virtually every sector of international politics, except in the event of crises that truly represent a threat to the peace. In recent years the developmental ethic of the UN system has often had to compete for priority status with other ethical concerns, especially human rights and environmental concerns. The simple distinction between "developed" and "developing" has become less easy to maintain as more and more states, nations, peoples, and territories rise to intermediate levels of development. In value-conflict situations, compromise concepts have emerged, such as the "right to development" as a construct of international human rights law[32] and "sustainable development" as a balance between economic growth and environmental responsibility.[33] In most contexts of

---

29. D. M. Johnston, "RUNCLOS: the case for and against revision of the UNCLOS III Convention," *Chuo Law Review* 2002 (in press).

30. One of the basic objectives of the United Nations, as stated in the Preamble to the UN Charter, is to promote "the economic and social advancement of all peoples."

31. Proposals for reform of the UN Charter have included the call for a "World Development Authority." G. Clark and L. B. Sohn, *World Peace Through World Law: Two Alternative Plans*, 3rd ed. enl. (Cambridge, Mass.: Harvard University Press, 1958, 1966), pp. 345–48. However, this proposal has been no more successful than other calls for Charter reform.

32. The concept of a "right to development" is derived from the language of responsibilities in the UN Charter and from the subsequent language of rights in the 1966 International Covenant on Economic, Social, and Global Rights. Since it focuses on states or peoples, not individuals, the right to development seems to live in the twilight between legal and moral rights.

33. "Sustainable development," articulated by the Brundtland Commission in the 1980s, has been challenged by many critics as an oxymoron, assuming that "development" carries the idea of growth. See, for example, W. Beckerman, "Sustainable Development: Is it a Useful Concept?" *Environmental Values* 3 (1994): 191. However, the Brundtland terminology is established, not least in government documents,

international development assistance, "capacity-building" has attained the highest level of priority, replacing "infrastructure."[34]

The circumpolar North poses an unusual challenge to developmental ethic. None of the eight Arctic states—Canada, Denmark (for Greenland), Finland, Iceland, Norway, Russia, Sweden, and the United States (the "Arctic Eight")—is a developing state. On the other hand, Greenland is surely a developing territory, and in most of these countries the northern residents have regional developmental problems much more severe than those of their fellow citizens to the south. These residents are developing peoples, occupying developing regions, and in most cases can be characterized ethnically as developing nations.

The concept of "developing nation" has become a particularly strong factor in the public policy of the circumpolar North, especially since the emergence of a collective sense of aboriginal solidarity on the part of the ethnic majorities in the Northern territories. Both under the auspices of the Arctic Council and independently, Northern ethnic organizations meet frequently, and combine their resources to present their case for increased autonomy and special consideration before the international community. Increasingly their demands reflect a balance between development, environment, and human rights that is usually encapsulated as "sustainable development."[35]

As the indigenous peoples of the North become more frequently exposed to Southern standards and lifestyles, the pressure for Northern development increases. However, it seems likely that Arctic versions of sustainable development will display special Northern attributes reflecting a degree of ambivalence between modernization and enrichment interests, on the one hand, and reverence for environmental and traditional lifestyle values, on the other.

---

and it now seems difficult to replace it with terms that are free of objections, such as "sustainable use" or "sustainable activities." It might be protested that the term "sustainable development" was of diplomatic origin, intended to create a bridge between developmental ethic and environmental ethic. But imprecision inherent in basic concepts tends to weaken the intellectual foundation of the field which they purport to serve.

34. The most recent reorientation of thinking about development assistance is a return to the problem of poverty. Those inclined to be skeptical about the potential effectiveness of this approach might wish to read or re-read a distinguished economist's work on this problem. G. Myrdal, *The Challenge of World Poverty: A World Anti-Poverty Program in Outline* (New York: Pantheon Books, 1970).

35. See Inuit Circumpolar Conference, *Circumpolar Sustainable Development* (Ottawa: I.C.C., 1994); Nuttall and Callaghan (n. 6 above); and M. M. R. Freeman, ed., *Endangered Peoples of the Arctic* (Westport, Conn.: Greenwood Press, 2000). In recent years the concept of "indigenous peoples" has become a powerful mobilizer at the global level. B. Kingsbury, " 'Indigenous peoples' in international law: a constructivist approach to the Asian controversy," *American Journal of International Law* 92 (1998): 414. Particularly since the early 1990s northern indigenous peoples have been represented in most global arenas where issues of relevance to the Arctic have been discussed and negotiated.

With a view to achieving a higher standard of economic development in the North, three industrial sectors in particular present themselves as channels of opportunity: (i) mineral resource extraction (e.g., zinc, lead, iron, copper, lithium, diamond, coal, gold, and silver); (ii) marine transportation of cargoes; and (iii) oil and gas production. All three present challenges to Northern goals of sustainable development and all have implications for the future use of Arctic waters.

Mineral resources have been mined in the North for many years.[36] Because of severe ice conditions throughout most of the year, shipping out the ores from the High Arctic is a strictly seasonal activity, which in the past has often been limited to three summer months. With the prospect of rapid warming over the next 20–30 years, it is realistic to project a significant lengthening of the Arctic summer that would transform the commercial attractiveness of investment in the still largely unexploited mineral deposits of the region. Those concerned with Arctic development are unlikely to overlook this opportunity to expand an industry that is well established in certain indigenous communities, even though it is capital, rather than labour, intensive. A larger scale of mineral extraction operations would encourage the risk of environmental damage in Arctic waters, but probably also increase the use of improved mining, vessel, and related technology.

The same scenario of higher Arctic summer temperatures and more extensive summer melting of ice introduces the prospect of increased commercial interest in the Arctic Ocean as a waterway for the transportation of cargoes between the Pacific and Atlantic Oceans. Two decades ago it seemed to some observers that the ice factor in the Arctic Ocean might limit such transits to bulk cargoes that could be pulled under the ice by underwater freighters still at the design stage of existence. Others argued that a bolder approach to cargo selection might be imagined within the framework of an enlightened "transit management" system of strict regulatory controls.[37] Today many Arctic development specialists believe it is appropriate now to

---

36. Mineral resource development in the circumpolar North has been most rapid in Russia, pre-dating the Soviet period, reaching its peak during the Stalinist years, but scarcely diminishing in the post-Soviet era despite mounting concern over the environmental damage it causes. In the Canadian High Arctic, a lead-zinc mine began production in Strathcona Sound in the early 1970s, and in the following decade a second lead-zinc operation started up on Little Cornwallis Island. C. L. Mitchell, "The Development of Northern Ocean Industries," in *Transit Management in the Northwest Passage: Problems and Prospects*, eds. C. L. Lamson and D. L. VanderZwaag, (Cambridge: Cambridge University Press, 1988), pp. 65–99, at 89–91.

37. There is still considerable disagreement on the future of Arctic shipping technology. Vessel designs and operational methods are under continuous re-evaluation. E. G. Frankel, "Arctic Marine Transport and Ancillary Technologies," in Lamson and VanderZwaag (n. 36 above), pp. 100–29. Obviously technical choices would have to be made before the design of a regulatory system for the Northern Sea Route and the Northwest Passage.

start planning for the future establishment and negotiation of such a regime. The planning process would have to grapple with a variety of problems and issues associated with the opening up of transit routes through the Northwest Passage in the Canadian sector[38] and the Northern Sea Route (including the Northeast Passage) in the Russian sector of the Arctic Ocean.[39]

The third kind of major industrial development of relevance to the Arctic Ocean is that of oil and gas production.[40] Some resource development specialists see in this industry a much larger, and economically more beneficial, opportunity for Northern residents than in the expansion of the indigenous mineral extraction industry; and certainly a much more immediate development than the opening up of general cargo waterways through Arctic waters.[41] The extraction of the major onshore oil reserves of the North Slope of Alaska has not significantly affected the Arctic marine environment, since the oil has been taken by overland pipeline to tanker terminals in the sub-Arctic waters of Prince William Sound. But the *Exxon Valdez* spill of 1984 aggravated long-lingering fears of the damage that tanker traffic could cause in ice-covered waters further north. Environmental objections can be raised both to tanker and pipeline systems of delivery of Arctic oil and gas to distant markets, whether the resource is onshore or offshore at the point of origin. Yet it seems certain that the U.S. oil industry is too important to the Alaskan economy for its further development to be stopped in its tracks, especially with rising concerns about the security of U.S. energy supplies.[42] Moreover,

38. For a recent overview of these issues, see D. M. Johnston, "The Northwest Passage revisited," *Ocean Development and International Law* 33 (2002): 145. For detailed studies of related problems, see D. L. VanderZwaag and C. L. Lamson, eds., *The Challenge of Arctic Shipping: Science, Environmental Assessment and Human Values* (Montreal and Kingston: McGill-Queen's University Press, 1990). For a challenge to the traditional statist framework, see F. Griffiths, "The Northwest Passage in transit," *International Journal* 54 (Spring 1999), pp. 189–202.

39. C. L. Ragner, ed., *The 21st Century—Turning Point for the Northern Sea Route?* (Dordrecht: Kluwer Academic Publishers, 2000); and W. Ostreng, ed., *The Natural and Societal Challenges of the Northern Sea Route* (Dordrecht: Kluwer Academic Publishers, 2000). See also Brubaker (n. 19 above).

40. Canada's North is particularly rich in oil and gas resources. The most immediate prospects lie in the Mackenzie Delta and the Beaufort Sea, but vast opportunities for petroleum development exist in Canada's High Arctic Islands, Lancaster Sound, and the Baffin Bay-Davis Strait region, if the recovery problems can be overcome. Mitchell (n. 36 above), pp. 83–89. In the meantime, Arctic oil and gas activities are concentrated in Alaska, both onshore and offshore.

41. On the other hand, the oil and gas industry, like the mining industry, is heavily capital-intensive. The shipping and related services industry is clearly the most labour-intensive of these three sectors, so that the opening up of waterways in the Arctic Ocean would have a much greater significance in terms of long-term social and economic development for the residents of the North.

42. At the time of writing (May 2002), the U.S. Congress was not willing to support the George W. Bush Administration's proposal for making drilling permits available in Alaska's Arctic National Wildlife Refuge, but this could change along with the political arithmetic in the House and Senate.

also in the Canadian Western Arctic territories (the Yukon and the Northwest Territories) there is now majority support for the introduction of the petroleum industry in the Beaufort Sea, where U.S. offshore drilling has been under way for several years, and also in the McKenzie delta and upriver areas. Indeed, these two Canadian territories seem to be competing with Alaska for a new pipeline system that is seen to confer economic benefits outweighing the admitted risks of environmental damage to their terrain.

Accordingly, it seems probable that the developmental ethic is prevailing as a matter of national public policy in the Western (North American) Arctic, but that pipelines rather than tankers will be used for the delivery of oil and gas. It is less clear how the international goal of sustainable development will affect developmental policy-making in other areas of the Arctic Ocean.

THE ARCTIC SCIENCE COMMUNITY

Scientists were usually not far behind the early explorers of remote and inaccessible places. However, the climate of the Arctic is so inhospitable to scientific observers that no systematic study of the Arctic Ocean was attempted until the 1820s, almost three centuries after the first Arctic explorations. Since the 1880s many international Arctic expeditions have been mounted to compile and compare data of importance to geographers, geologists, glaciologists, oceanographers, hydrographers, and biologists.[43] Yet, given the physical constraints on Arctic field research and the emergence of uncontested national sovereignty over virtually all Arctic islands, most Arctic science up to the end of the Cold War era was done by the Arctic countries themselves.[44]

During the Cold War many sectors of Arctic science in the United States and the Soviet Union were deemed to be of potential strategic sensitivity. Even Canada and other close allies of the United States were given relatively restricted access to research data in many sectors of American Arctic science, and especially to the results of the principal programmes of oceanographic research directed by the U.S. Navy. Because most of the field of Arctic science was occupied traditionally by Russian scientists, the American Arctic

---

43. Regional cooperation in Arctic science dates from 1879 with the founding of the International Polar Commission, but inter-governmental cooperation has been erratic over the years since then.
44. The tendency has been for national scientific activity to peak at a time of strategic excitement. It has been observed that Canadian Arctic science has received governmental priority only at a time of challenge to Canadian Arctic sovereignty. C. Pyc, "An Arctic science policy? All we need is a sovereignty crisis," *Arctic* 53 (March 2000): iii. On current Canadian Arctic sovereignty issues, see Johnston (n. 38 above), pp. 146–49.

science community had reason to feel at a strategic disadvantage. In recent years the U.S. Arctic science community has grown considerably.[45]

In the later years of the Soviet period, the U.S.S.R. and the United States began to pursue opportunities for exchange of data in certain, carefully defined areas of common interest.[46] Bilateral instruments for cooperation on Arctic science were also negotiated by other Arctic states.[47] However, until the 1990s the only treaty vehicle for multilateral cooperation was the Agreement on Conservation of Polar Bears.[48]

Now, however, the circumpolar North has become the focus of unprecedented efforts to conduct fundamental science with international cross-disciplinary teams of researchers supported by very substantial funds. This new era in Arctic science, which promises to match the sophistication of Antarctic science, is facilitated, of course, by the opening up of opportunities for genuine scientific collaboration after the end of strategic Cold War fears. It is due also to the mounting evidence of two major threats to the Arctic Ocean: contamination[49] and climate change.[50]

---

45. A Canadian scientist has estimated the annual U.S. Arctic scientific research budget as high as $257 million (Can), creating a "brain-drain" of Canadian specialists southward to affluent U.S. research institutions. Pyc (n. 44 above).

46. In December 1987, a Soviet-U.S. Joint Statement was adopted at a summit meeting, encouraging cooperation in "marine navigation, fishing, search and rescue at sea, interaction of radio-navigation systems, and delimiting the sea expanses of the Northern Arctic Ocean or the Chukchi and Bering Seas adjacent thereto." A. L. Kolodkin and M. E. Volosov, "The legal regime of the Soviet Arctic: Major issues," *Marine Policy* 14 (2) (March 1990): 158 at pp. 167–68. This political breakthrough, inaugurating a new age in international Arctic relations, was made possible by President Gorbachev's famous Murmansk speech of October 1, 1987, in which he promised to open up the Northern Sea route for international shipping. W. Ostreng, "The Northern Sea Route: A new era in Soviet policy?" *Ocean Development and International Law* 22 (3) (1991): 259. However, major differences between Russian and American (and Canadian and American) positions on Arctic Ocean legal issues remain unresolved.

47. See, for example, Agreement between the United States and Canada on Arctic Cooperation, 11 January 1988, Canada T.S. 29, *International Legal Materials* 28 (1989): 142. In the same year a similar instrument between Canada and the Soviet Union was concluded and was succeeded in 1992 by an agreement between Canada and the Russian Federation.

48. Agreement on Conservation of Polar Bears, 27 U.S.T. 3918, T.I.A.S. 8409. On the effectiveness of this international regime, see A. Fikkan et al., "Polar Bears: The Importance of Simplicity," in O. D. Young and G. Osherenko, eds., *Polar Politics: Creating International Environmental Regimes* (Ithaca, N.Y.: Cornell University Press, 1993), pp. 96–115.

49. For an overview, see R. W. MacDonald et al., "Contaminants in the Canadian Arctic: 5 years of progress in understanding sources, occurrence and pathways," *Science of the Total Environment* 254 (2–3), (2000), pp. 93–234.

50. Not surprisingly, climate change concerns have generated new fundamental research in the Arctic. As far back as the 1970s, the glaciologist, John Mercer, observed a striking geographical similarity between the Western Antarctic and the

Most research in these two areas is done by the "Arctic Eight": that is, the member states of the Arctic Council, which has assumed responsibility for the programs initiated under the Arctic Environmental Protection Strategy. However, scientists in several non-Arctic countries have had a traditional interest and involvement in Arctic science, notably the United Kingdom and Germany. In recent years they have been joined by colleagues from Japan, and to a lesser extent from China. Today almost all serious Arctic science is done by these 12 scientific communities, increasingly in a collaborative mode.

This constellation of national Arctic research institutions is comparable to its Antarctic counterpart[51] and might be regarded as representing the global science community. The sources of Arctic Ocean contamination are virtually worldwide, and climate change is, of course, a universal concern. The territoriality factor in the Arctic, in contrast with the Antarctic, makes it difficult to persuade the governments—or peoples—of the Arctic states to accept the concept of a world community entitlement in their region. Yet, from the perspective of the transnational arctic science community, a credible case might be made for designating the Arctic Ocean as a region of "common concern," if not as an area of "common heritage."

To this extent, the perspective of the global scientific community might seem compatible with that of the environmental ethicists, but difficult to

---

Eurasian Arctic: both possess a large continental shelf (in the geological sense) that is no more than a few hundred meters deep, but the Western Antarctic has a 2.5 km-thick ice-sheet resting on it, whereas the Eurasian Arctic is completely free of "grounded ice." Recent research under the PONAM and QUEEN projects, involving 50 scientists from seven European countries, shows that the Barents Sea used to have grounded ice, which gradually disintegrated after the height of the last ice age 14,000 years ago. It is believed probable that a similar meltdown of the Western Antarctic ice-sheet might occur causing sea-level rise and mass inundations in many regions. M. J. Siegert et al., "The Eurasian Arctic during the last Ice Age," *American Scientist* 90 (1) (Jan.–Feb. 2002): 32–39. Indeed, the ice-sheet of Western Antarctica is already disintegrating and causing some concern. For the most authoritative report on global climate change, see Intergovernmental Panel on Climate Change, Working Group I. J. T. Houghton et al., ed., *Climate Change 2001: The Scientific Basis: Contribution of Working Group I to the third assessment report of the Intergovernmental Panel on Climate Change* (Cambridge: Cambridge University Press, 2001). Whether global climate change is the cause of current variations in Arctic ice cover patterns is still in debate among Arctic scientists, but the majority view is that there is an annual trend to more prolonged melting in the Northwest Passage over part of the year, sufficiently to raise the prospect of sustained summer transit through these waters. R. Huebert, "Climate change and Canadian sovereignty in the Northwest Passage," *Isuma: Canadian Journal of Policy Research*, vol. 2, no. 4 (Winter 2001): no pages. ISSN 1492-0611, accessed 30 August 2002 on the World Wide Web: http://www.isuma.net/v02n04/huebert/huebert_e.shtml; and Johnston (n. 38 above).

51. Antarctic research is organized, or at least coordinated, among the countries that participate in the Antarctic treaty system. For a recent study of this regime, see C. C. Joyner, *Governing the Frozen Commons: The Antarctic Regime and Environmental Protection* (Columbia, S.C.: University of South Carolina Press, 1998).

reconcile with that of the law of the sea community and the developmental ethicists. As argued later,[52] a multipurpose regime-building initiative in the Arctic would be needed to accommodate the diverse assumptions, expectations, and perceptions of these four distinct "domains."

It might be added that in recent years the scientific community has had many occasions to be uncomfortable with transnational ethicists, especially those devoted to environmental issues. But the "sovereignty of science" sentiment is still a powerful force, not least in the context of international ocean governance. The prospect of some kind of "prerogative" for science in the Arctic, though not institutionalized in the Antarctic manner, would no doubt stir emotions among many scientists outside the littoral states of the Arctic Ocean.[53]

Meanwhile, Arctic scientists may have to find a way of integrating their research with the observations of local residents and their traditional knowledge of the Northern lands and waters, a challenge that has ethical as well as intellectual overtones.[54] Professional Arctic scientists differ on the value of local knowledge and indigenous oral history in the field of Arctic science, and specifically in the contexts of contamination and climate change. Perhaps the mainstream still resists the goal of integrating "natural" and "social" knowledge in the North, but in most Arctic countries scientists are obliged to cooperate with elders and other local residents by reason of public policy.[55]

THE STRATEGISTS

The Arctic Ocean has been an area of high strategic priority since the beginning of the Second World War, and of course throughout the Cold War. In many ways the region was ideal for submarine espionage, and its remoteness,

52. See section on "The Regime-Builders," below.
53. Many scientists retain the vision of the Arctic as a global laboratory, a vision that has had to compete with more utilitarian visions for over 100 years. "Global laboratory" advocates tend to be scornful of national boundaries as an irrational interference with the civilizing mission of the scientific community. C. Lamson, "In Pursuit of Knowledge: Arctic Shipping and Marine Science," in VanderZwaag and Lamson (n. 38 above), pp. 16–19.
54. See, for example, J. Cruikshank, "Glaciers and climate change: perspectives from oral tradition," *Arctic* 54 (4), (Dec. 2001): 389; P. Usher, "Traditional ecological knowledge in environmental assessment and management," *Arctic* 53 (2) (June 2000): 183; and G. W. Wenzel, "Traditional ecological knowledge and Inuit: reflections on TEK research and ethics," *Arctic* 52 (2) (June 1999): 113.
55. A problem in public policy arises when indigenous belief is diametrically opposed to orthodox scientific opinion. For example, the Inuit believe that "the more a species is hunted, the more abundant it will become," rationalized on the ground that "animal populations which are hunted regularly have less disease, reproduce faster, and have more to eat than animals which are not hunted." Wenzell (n. 54 above), p. 119.

vastness, and relative emptiness made it as politically safe as any area could be for the deployment of advanced nuclear capabilities. It might be said that the Arctic Ocean contributed to the success of Soviet-American mutual deterrence strategy during the potentially dangerous Cold War period.[56]

Since the demise of the Soviet Union in 1989 the water areas of the Arctic have not lost their strategic value in the eyes of long-range planners. Admittedly the threat of a blow-up between rival superpowers no longer exists. The old Soviet Navy is now virtually a spent force in most parts of the world. Only the United States possesses overwhelming military power throughout all oceans. Yet the same characteristics that made the Arctic Ocean so important as a theatre of covert military operations in the bipolar era of international politics continue to endow these waters with a special status for more modern security reasons.

First, to the extent that Russia remains a major power, with an uncertain political history to be revealed in the coming years, the United States continues to maintain a policy of vigilance.[57] By virtue of its sole northern presence in Alaska, the United States will never be able to compete effectively with

---

56. The relative privacy of Arctic waters allowed the two superpowers "breathing room" for developing and testing nuclear vessels and related technologies in a cat-and-mouse sport that proved to be bloodless, though certainly costly in other ways. Arguably, no other region offered so many advantages for this exercise in mutual deterrence. Because of its shortage of year-long, ice-free ports, the Soviet Union was forced to develop the northern port of Murmansk as the main base for its nuclear attack submarines (SSNs) and nuclear ballistic missile submarines (SSBNs). This in turn forced the United States to give a high priority to the Arctic in developing its own program of submarine operations. But the advent of shorter-range submarine-launched cruise missiles (SLCMs) armed with nuclear weapons forced both superpowers to deploy their submarines closer to the shores of their enemies, further increasing the strategic importance of Arctic waters. R. Huebert, "Canadian Arctic security issues: transformation in the post Cold War Era," *International Journal* 54 (2) (Spring 1999): 203 at 205–06.

57. The U.S. policy of vigilance is considered by some critics to be the product of an obsession with the illusory goal of total security: the "national security paradigm." One critic confronts "the historical assumptions that *complete* security is a natural state of affairs and somehow an American right." M. McGwire, "The paradigm that lost its way," *International Affairs* 77 (4) (October 2001): 777 at p. 796. These words were written before 11 September 2001. The depth of emotions stirred among the American people by that shocking assault on their centers of power and wealth would be understandable in any culture, but may be especially significant if seen also as an assault on the central myth of American foreign policy since 1945. Since that assault, U.S. policy, vis-à-vis Russia, has undergone further transformation because of the latter's central role in the U.S. strategy of coalition-building in the war against terrorism, moving from containment through constructive engagement to a policy of alliance favoring Russian membership of NATO. Meanwhile the newly pervasive sense of insecurity in the United States strengthens the hand of strategic planners who advocate a policy of "comprehensive security." The package of concerns that might be addressed together includes terrorism, illegal immigration, resource conflict, environmental degradation, and transnational crime (e.g., people smuggling, drug traffic, and arms trade). Huebert (n. 56 above), p. 204.

Russia for geopolitical ascendancy in extreme Northern latitudes. There will be no industrial center emerging in Alaska to compare with Nirsk, nor a naval base such as Archangel, nor even a seaport with the relatively modest potentiality of Murmansk. Within the foreseeable future, the Arctic will always be perceived in Washington, D.C., as an area of strategic weakness that requires the deployment of advanced military vessel technology in Northern waters.

Second, as the only global hegemon, the United States may be expected to entertain the notion that its unique status, unprecedented in the modern history of interstate relations, entitles it to special prerogatives commensurate with its special responsibilities.[58] Such a hegemonial situation, it might be argued, has not existed since the demise of the Roman Empire. It is not beyond the realm of probability that the first half of the 21st century may be witness to a gradual incorporation of additional Arctic space under U.S. control, if not as part of U.S. territory then perhaps initially through some kind of quasi-federational arrangement with Canada and Denmark/Greenland.

Third, the conceivability of such a scenario of expanding U.S. control or influence in the Arctic depends very largely on the degree of U.S. vulnerability to terrorist and related attacks in the coming years. The Northern Command system established in 2002 is regarded by many Canadians as raising fundamental questions about the future of Canadian sovereignty at a time in history when Canadian military autonomy seems to be becoming subservient to U.S. defence requirements.[59] A Fortress America mind-set could settle upon the U.S. political and bureaucratic systems, if it should prove impossible to guarantee the safety of the American homelands.[60] It is certainly possible in such circumstances that the northern perimeter would assume an

---

58. For a brief editorial essay on the legal implications of the "single superpower" hypothesis, see D. F. Vagts, "Hegemonic international law," *American Journal of International Law* 95 (4) (October 2001): 843.

59. At the time of writing (May 2002), it appeared that the events of September 2001 had forced Canada into an even closer relationship with the United States, especially in areas perceived by the U.S. government to be security-related. Even before these events, many Canadian observers of strategic realities felt that Canada's room for manoeuvre was greatly reduced if the national security of the United States was at stake. In one article, written in 2000, it was argued that "[i]n the realm of security and defence cooperation Canada's options are really limited to Hobson's choice, not necessarily because of huge advantages to be gained from being onside, but because of the incalculable disadvantages of refusing to take U.S. security concerns seriously." A. Macleod, S. Roussel and A. Van Mens, "Hobson's choice," *International Journal* 55 (3) (Summer 2000): 341 at p. 354.

60. On American concerns with "homeland defence" before the assaults of September 2001, see R. J. Larsen and R. A. David, "Homeland defense: assumptions first, strategy second," *Strategic Review* 28 (Fall 2000): 4; and J. Train, "Who will attack America?," ibid., 11.

important role within the framework of a comprehensive security system for the American people.

Fourth, at least in the absence of such threats of violence, the Arctic Ocean is likely to be opened up to international navigation, if the summer temperatures in the North continue to rise as rapidly as projected by many scientists. The availability of the Northern Sea Route and/or the Northwest Passage for four or more summer months each year would convert the Arctic Ocean into a major waterway for the carriage of general cargoes, so that commercial security would have to be safeguarded on behalf of the international trade community.[61]

For any or all of these reasons, it is not difficult to imagine future events that would re-convert the Arctic Ocean into a geopolitical or economic region where strategic considerations might prevail over legal, environmental, developmental, and scientific values. The possible future paramountcy of strategic concerns would presumably not eliminate, but perhaps relegate, these other goals of international public policy in the circumpolar North.

THE REGIME BUILDERS

Of all the "domains" that focus on the North and might determine the future of the Arctic Ocean, only one seems sufficiently flexible to accommodate diverging assumptions, expectations, and perceptions. Hopes for reconciliation of contending values and interests reside with the history and the future of experiments in international regime-building.[62]

Regime-making studies, both theoretical and empirical, have become a prominent feature of the literature on international relations. Precisely because it had not begun to evolve as a normal political region, the Arctic attracted considerable attention as a future arena for innovative regime building even as early as the mid-1980s. Since then the concept of an "international regime" has expanded to the point that it becomes necessary to distinguish several kinds of sets of settlements and arrangements that might be considered as alternative approaches to regime building for the circumpolar North in general or for the Arctic Ocean in particular. Each of the five domains identified earlier would have its own distinctive version or versions of an appropriate Arctic regime.

61. See, for example, Ragner (n. 39 above); Griffiths (n. 38 above); and Huebert (n. 50 and n. 56 above). See also Johnston (n. 38 above).
62. Much of the literature on Arctic regime-building has been written by Oran R. Young and his colleagues. See, for example, Young and Osherenko (n. 48 above). On the merits and prospects of the regime-building approach to Arctic regional issues, see D. Rothwell, *The Polar Regions and the Development of International Law* (Cambridge: Cambridge University Press, 1996), pp. 406–23.

Law of the Sea Arctic Regimes

For the law of the sea community, the regime would, of course, be limited to the Arctic Ocean. Variance of opinions within this community would turn on jurisdictional and navigational issues. On the face of things, there are three principal variants of the first kind and two of the second.

Arctic Exceptionalism

The Arctic Ocean can be considered so totally different from any other ocean area that it should be removed from the normal law of the sea framework.[63] Extremists may argue that, through negotiation on the part of the five Arctic littoral states, the entire, mostly frozen, ocean should be divided into five "sectors" of national jurisdiction. This would be shocking to many law of the sea specialists, who cling to the traditional principle of high seas freedoms beyond the limits of national jurisdiction re-defined at UNCLOS III; and equally shocking to advocates of the common heritage principle, who may wish to argue that some unmeasured part of the "core" of the Arctic Ocean falls under the jurisdiction of the International Seabed Authority by virtue of the definition of the continental shelf offered in Article 76. On the other hand, a "national lakes" division of the Arctic Ocean might result in more substantial benefits to the residents of these six developed littoral states than under any kind of internationally shared authority. Presumably such a system of national allocations would force the five littoral states into a normal practice of consultations under the Arctic Council, or otherwise.

Arctic-Specific Regional Cooperation

An alternative, and less radical, approach to regime-building in the Arctic Ocean, would key on Article 123 of the 1982 UN Convention on the Law of the Sea, but conceivably go beyond the purposes of fishery management, environmental protection, and scientific research prescribed in that provision for cooperation among states "bordering on enclosed or semi-enclosed seas."[64] It is difficult to imagine such an outcome that did not include Arctic-specific measures related to navigational concerns, particularly in the new age of rapid climate change.

Arctic Enclave

The waters beyond the limits of national jurisdiction in the Arctic Ocean could be characterized, theoretically, as an enclave possessing the status of high seas in the contemporary international law of the sea. Accordingly, the regime-building task could take the form of cooperative arrangements

---

63. See Rothwell and Kaye (n. 23 above).
64. 1982 United Nations Convention on the Law of the Sea, Article 123.

between the "surrounding states," Russia and Canada, by analogy with arrangements made for other high-seas enclaves. In practice, however, such a proposition would lack credibility, given the near-absence of fishery stocks in "extra-national" Arctic waters.[65] It is hard to believe that a great deal of diplomatic energy would be available to expend on such an exercise in legal formalism, and the realities of Arctic geography seem to rule out the possibility that an enclave regime in these waters would be seen as a guarantee of navigational rights in the Arctic Ocean.

Discretionary Transit Management in Arctic Waterways
The prospect of opening up the Arctic Ocean to international navigation ("sustainable navigation") in the lengthening Arctic summer might be so attractive from a developmental perspective as to give priority to the design, establishment, and maintenance of a discretionary transit management system for designated routes through Russian and Canadian Arctic waters. The International Northern Sea Route Project has clarified the paths that would have to be explored to facilitate such an initiative in Russian waters, but no Canadian project of a comparable magnitude has so far been launched.[66] A regime of this kind, controlled essentially by Russia and Canada as the "manager states," would have to be designed as a state-of-the-art regulatory system fundable mostly by transiting vessels on the model of the Panama Canal, with which an Arctic system would be competing in the summer months.[67]

  65. Enclave management arrangements elsewhere have been motivated by the need to resolve issues over fishing and fishery management rights and responsibilities. See, for example, Balton (n. 12 above), Elferink (n. 12 above), and Stokke (n. 12 above).
  66. Efforts have been made to persuade the Canadian government to fund a project on the Northwest Passage similar in purpose, if not scale, to the International Northern Sea Route Project, in conjunction with the Inuit Circumpolar Conference and other proponents, but they have been frustrated by the domestic politics of federal-Nunavut relations.
  67. Transit through the Panama Canal produces huge revenues: allegedly $100,000 or more for the largest vessels. Presumably economists find it very difficult to meaningfully compare the potential costs of operating a waterway through Canadian or Russian Arctic waters with the actual costs of operating the Panama Canal. Also the political, diplomatic and legal problems associated with opening up "international navigable waterways" of natural character such as the Northwest or Northeast Passage may be quite different from those experienced historically with man-made canals. See R. R. Baxter, *The Laws of International Waterways with Particular Regard to Interoceanic Canals* (Cambridge, Mass.:, Harvard University Press, 1964). But the proclivity to international controversy may be common to both categories. It might be noted that the history of canal-building is not over: in Southeast Asia much attention has been given in recent years to proposals for building the Kra Canal through Thai or Malaysian territory. In North America, several existing canals (e.g., Welland Canal in Canada) are in serious disrepair, requiring large-scale reinvestment, and the international waterway (the St. Lawrence Seaway) is said to have serious financial problems.

International Arctic Straits Regime

Finally, world shipping interests, nervous of a discretionary and strictly regulatory approach to transit management in the Arctic, might deny any special status to these Northern waters and argue that they qualify as "straits used for international navigation" under Part III of the 1982 Convention. The initial formulation of such a proposition would presumably include language similar to that of Article 38, on the "right of transit passage," which is designed to limit severely any "transit management" rights on the part of the "bordering states."[68] However, it is difficult to imagine either Russia or Canada yielding to such a proposition applied to Arctic waters, and it must be supposed that special Arctic-specific exceptions in favor of the "managing states" would have to be agreed to by the potential "transit states." So in the interest of compromise the final version of such a regime might not be different from a "discretionary transit management system."

Arctic Environmental Regimes

A growing proportion of international regimes negotiated over the last two or three decades have been "environmental" in orientation. Because environmental agreements tend to be ambitious, they give rise to concerns about their actual or potential effectiveness. A recent major study framed around these concerns concluded, on the basis of several case studies, that most environmental regimes seem not to have an impressive record from a functional ("environmental enhancement") perspective, but a fair record from a behavioral ("behavioral modification") perspective.[69]

Such studies suggest lessons that might be learned of relevance to future environmental regime-building in the Arctic.[70] First, given the relatively strong capabilities of the Arctic littoral states, it might be supposed that a regime along the lines of the Regional Seas Programme for the Arctic Ocean might prove more effective than most of those negotiated for developing regions under UNEP auspices. However, Russia embarked on an industrial development program in its Arctic territories many decades ago, and seems

---

68. Under Article 38 (1) of the 1982 UN Convention on the Law of the Sea, the right of transit passage "shall not be impeded." Under Article 38 (2), "transit passage means the exercise . . . of continuous and expeditious transit of the strait between one part of the high seas or an exclusive economic zone and another part of the high seas or an exclusive zone."

69. Miles et al. (n. 25 above), pp. 435–36.

70. Recently various maritime regimes have been evaluated, more impressionistically, from a Northeast Pacific perspective. M. J. Valencia, ed., *Maritime Regime Building: Lessons Learned and Their Relevance for Northeast Asia* (The Hague: Martinus Nijhoff Publishers, 2001). See also Vidas (n. 19 above).

to have difficulty complying with contemporary environmental and sustainable development standards in these remote areas, where an industrial development culture still flourishes. The introduction of strict international environmental standards into the Arctic Ocean through regime building would seem likely to create an anomaly as long as there are gaps in the land-based environmental standards of the five Arctic littoral states.

On the other hand, behavior modification in Russia and perhaps in other Arctic countries might become a realistic goal only with the addition of an international regime negotiated for the Arctic Ocean. Such a regime would create international consultative and other procedures that might be expected, in due course, to result in improved bureaucratic performance at the national level.

Third, a properly designed environmental regime for the Arctic Ocean would surely assist the training of Arctic residents in environmental stewardship both on land and ice-covered water. The indigenous peoples of the Arctic territories might come together more easily—with fewer political constraints—under a training program designed for oceanic rather than terrestrial areas, if the ocean is likely to generate less sovereignty sentiment than the land.

Integrated Arctic Sustainable Development Regimes

To the extent there is a trend toward sustainable development (as distinct from environmental preservation) goals in the circumpolar North, the social and political conditions for "sustainable-development regime-building" are improving. Integrated ocean development and management goals might be cooperatively developed by the five littoral states of the Arctic Ocean.

The chief purpose would be compare and reconcile general development goals for the Arctic Ocean macro-region. It is necessary to conform as much as possible with the goals of the Global Programme of Action to Protect the Marine Environment from Land-Based Activities (GPA) without sacrificing development opportunities for the benefit of the indigenous and other resident communities of the North. Instruments required for the building of such a regime would depend upon political acceptance of the need for sustained cooperation between environmental and developmental ethicists. Building such a regime simply on the basis of the GPA is unlikely to satisfy those ministries responsible for Northern and industrial development in the Arctic countries.

This kind of Arctic regime building may have to be attempted for the entire circumpolar region, including all eight of the Arctic Council states and all seven of the nongovernmental Permanent Observers. Indeed a truly "integrated" approach to sustainable development in the North might also involve certain non-Arctic as well as sub-Arctic countries, such as those states that contribute most significantly to industrial emissions that are transported northwards and deposited in the water and ice of the Arctic Ocean.[71] As a practical matter, however, it is difficult to envisage a successful effort on this scale to effect a reconciliation between environmentalists and developmentalists.

Arctic Science Regimes

The Arctic Council grew out of the earlier Arctic Environmental Protection Strategy (AEPS), which was essentially devoted to the tasks of scientific cooperation. Sound science was seen as the necessary foundation of sound regional environmental policies and practices for the Arctic countries. So, to the extent that the Council evolved out of the AEPS, science has been at the center of the Arctic regime building that has already occurred. Moreover, some of this research program development over the last decade has focussed specifically on research related to the Arctic marine environment.

Over the last few years, more programs have been launched, with very substantial funds, to advance research in the high-priority field of cross-disciplinary climate studies. The choice of the Arctic for priority attention, as a matter of scientific strategy, necessitates the involvement of many scientists from non-Arctic as well as sub-Arctic countries. Arguably, the need for the best possible research effort on global warming and related concerns is resulting in the globalization of the Arctic science community: a community that no longer consists entirely of the research institutions of the Arctic countries. Global institutions such as IUCN, ICSU, and IWGCC already provide a global dimension to Arctic science regime building.

This kind of scientific regime development might be encouraged as a bridge between environmental and developmental ethicists. It is already anticipated that these global research findings will be used to guide the planning of future development projects that might require the re-settlement of certain Arctic communities that would be the most severely threatened by rising temperatures and other radically different climate patterns in the North. So Arctic science regime building should be seen as compatible with Arctic environmental protection and sustainable development goals.

Strategic Regimes

Since the terrorist assaults on the United States in September 2001, homeland security has become the paramount public policy concern in that country, and a matter of priority in many other countries. It would be unrealistic

---

71. See MacDonald et al. (n. 49 above).

to suppose that the huge effort to shore up U.S. security will not have important implications for regime development in the Arctic. As argued earlier, the United States is particularly vulnerable in the North, and has no choice but to continue to strengthen the Alaskan economy through additional petroleum and other industrial developments.

Efforts to create a Northern Command are certain to result in the reinforcement of U.S. military installations in Alaska and the negotiation of closer security arrangements with Canada and perhaps other Arctic states. Indeed it is conceivable that security concerns will force the U.S. government to devote more time and energy to Arctic diplomacy in general, perhaps specifically to create an Arctic multilateral security organization with elements drawn down from NATO, NORAD, and the Arctic Council.

Such a strategic regime could certainly become a fundamental influence on how the Arctic Ocean would be governed, if terrorist and related pressures on the United States continue to mount. Far from becoming a "zone of peace," the Arctic Ocean could become, in a high-danger scenario, a "security zone" subject to special controls over vessel and aircraft movements in and out. Within such a frame of reference, U.S., Canadian, and Russian strategists would have to compare carefully the security costs and economic benefits of opening up the Northwest Passage and the Northern Sea Route against the continued use of the Panama Canal and other waterways for the transit of cargoes in and out of North America.[72]

THE ANTARCTIC ANALOGY

There are several possible futures for the Arctic as a whole. Because of the diversity of uses to which it can be put, the regime-building approach seems the most useful. The most obvious way to design an Arctic regime, from a world community perspective, would be by starting with the Antarctic treaty system that has been evolving since the Antarctic Treaty was negotiated in 1959.[73]

The similarities between the two polar regions are fairly obvious. Both are very extensive and cold places, distant and relatively inaccessible from a world community perspective. Both alternate annually between extremes of total daylight and total darkness. Both possess enormous mineral resources, but also ecosystems of unique interest and scientific value. Above all, they are both of critical importance for research on the planet's changing climate patterns.

On the other hand, there are major differences. The Arctic is mostly an ice-covered ocean, whereas the Antarctic consists very largely of a moun-

72. See n. 67 above.
73. Antarctic Treaty (1959) 402 U.N.T.S. 71, 12 U.S.T. 794, T.I.A.S. No. 4780.

tainous, continental landmass. The Arctic for thousands of years has been the home of widely scattered human communities and seminomadic peoples, whereas the Antarctic is host to only a few scientific stations of recent date. Moreover, much of the Arctic falls under the jurisdiction of nation-states with uncontested sovereignty over vast island and landmass territories and with exclusive rights over extensive ocean areas. The Antarctic is treated, by general agreement, as a territorial vacuum regulated under the functionalist jurisdiction of an international regime. So the legal idea of a common heritage, subject to *jus communis,* does not fit the statist reality of the circumpolar North.

Are there, nonetheless, certain elements of the famous Antarctic regime that might be useful and acceptable in the other polar region? Or are there reasons, at least, for drawing upon the Antarctic analogy for the more limited purpose of regime formation in the Arctic Ocean?

First, it should be recalled that the Antarctic treaty system consists of five principal components: (i) the 1959 Antarctic Treaty; (ii) the 1972 Convention on the Conservation of Antarctic Seals;[74] (iii) the 1980 Convention on the Conservation of Antarctic Living Marine Resources;[75] (iv) the Agreed Measures for the Conservation of Antarctic Flora and Fauna;[76] and (v) the Scientific Committee on Antarctic Research (SCAR). The Antarctic Treaty is designed above all to ensure the use of Antarctica for peaceful uses only, and to this end Article I prohibits all activities of a military nature. Article II ensures the freedom of scientific investigation and international cooperation. Article III guarantees the suspension, without detriment, of all territorial claims to any part of Antarctica. Article V prohibits nuclear explosions and the disposal of radioactive waste material. So the famous instrument essentially serves four fundamental purposes: establishment of a zone of peace; scientific investigation; de-territorialization; and protection from nuclear risks. It was not originally devoted to environmental preservation in the generic sense,[77] but three of the subsequent components of the system have conservation purposes. An effort was made to add an additional component that would have permitted mineral extraction, under strict regulatory controls, but this initiative was abandoned, or at least suspended indefinitely.[78]

74. Convention for the Conservation of Antarctic Seals (1972), 1080 U.N.T.S. 175, 29 U.S.T. 441, T.I.A.S. No. 8826.
75. Convention on the Conservation of Antarctic Living Resources (1980), 1829 U.N.T.S. 47.
76. This arrangement was first entered into in 1964.
77. Yet Article IX provides that the measures which might be considered by the Contracting Parties include those relating to ". . . f) preservation and conservation of living resources in Antarctica".
78. The Convention on the Regulation of Antarctic Mineral Resource Activities was concluded in 1988, but has not come into force due to a change of policy on the part of several Contracting Parties. See C. C. Joyner and S. K. Chopra, eds., *The Antarctic Legal Regime* (Dordrecht: Martinus Nijhoff Publishers, 1988), pp. 131–59. The text of this instrument was reproduced at 27 *International Legal Materials* 868 (1988).

Second, the concept of common heritage, with its developmental connotation of universally shared entitlement, is unlikely to be politically acceptable to the Arctic states, even if restricted to those areas of the Arctic Ocean beyond the law of the sea limits of national jurisdiction. Even an attempt to state the common heritage case in the largely frozen sea would require an unprecedented alliance between the environmental ethicists and the Arctic science community, whose credibility would be challenged by the strategists and the developmental ethicists as well as many in the law of the sea community.

Third, the argument for any kind of legitimate world community presence in the Arctic Ocean—a presence acceptable to the "Arctic Five" or the "Arctic Eight"—would have to rest on the notion of "common concern," not common heritage: on the principle of universal responsibility, not universal entitlement. In the abstract, the prerogatives associated with strict national sovereignty are now being challenged by international idealists of various kinds, including many environmentalists; and national boundaries have never been conceded the prestige of a "governing construct" within the scientific community. To internationalist idealists the Arctic Ocean might be seen as coming as close to the status of an area of common concern as the Antarctic landmass does to that of an area of common heritage.

But what kind of multipurpose, common-concern regime for the Arctic Ocean could attract even conditional support from the strategists and developmental ethicists within the "Arctic Eight"? The Arctic Council has neither the mandate nor the endowment at present to attempt such a reconciliation. It is unlikely that any sectoral UN agency such as UNESCO would be so presumptuous as to propose a multipurpose regime for the Arctic Ocean, though reference might be made to ICES and PICES as oceanic models for scientific cooperation on a macro-regional scale.[79]

Presumably the prospect of multipurpose regime formation in the Arctic Ocean cannot be seriously entertained until there is stronger evidence of the need for such a bold initiative on the basis of universally shared concern. The level of concern might rise quite swiftly, however, if median projections of rapid warming in the circumpolar North prove to have underestimated the rate of warming, or if the fears of environmental alarmists prove to be justified. Concern over the Arctic might also rise quickly because of some kind of security threat sufficiently critical to engage the attention of the UN Security Council. Moreover, given the efforts of indigenous peoples to secure a better life and a better future, it is possible to imagine the UN General Assembly acceding to suggestions for more generous and more innovative sustainable development regimes designed for their special benefit.

---

79. The International Council for the Exploration of the Sea, established in 1901, has been the principal intergovernmental organizer, coordinator, and disseminator in the field of oceanography for the North Atlantic Ocean; and PICES was established almost 100 years later as the counterpart organization for the North Pacific.

In such circumstances, the strategists and the developmental ethicists might accept the appropriateness of a UN leadership role.

In this futuristic future of the Arctic Ocean, almost all the elements of the multipurpose Antarctic regime might seem to come together as candidates for incorporation in the North. The Security Council would have to respond to the demonstrated willingness of a reckless state to use its newly developed nuclear strike capability against the Northern countries (though presumably not directed at their Arctic territories). In such a scenario the conditions would exist for designating the Arctic Ocean a nuclear-free zone and a zone of peace. In these circumstances, such an action by the Security Council would have greater credibility than similar designations already applied to other regions (such as the Indian Ocean), even more so perhaps than in the case of the militarily unthreatened Antarctic. The action envisaged for the General Assembly in response to a morally compelling demand by indigenous peoples is more feasible in the circumpolar North, where none of the developing members of the UN may cling to sovereign prerogatives. So this combination of imagined contingencies would make the case for a multipurpose Arctic regime as broadly based as the Antarctic treaty system, and indeed more so if it included a sustainable development component.

CONCLUSION

The future of the Arctic Ocean has not yet arrived. At present most sensible observers of the Arctic scene are likely to put their faith in the capability of the Arctic Eight to look after their own region without politically threatening intrusions by the larger international community. Yet most observers would certainly agree that few of the Arctic states seem to attach a high priority to their northern perimeter, especially in the Arctic Ocean. Climate change concerns may be expected to attract the attention of the larger international community.

Of the six domains of expertise reviewed, the regime builders seem to possess the greatest flexibility in accommodating diverse interests and values. Lessons of relevance to the Arctic Ocean can be drawn from experience gained in the evolution of the Antarctic treaty system as a multifunctional regime designed in the world community interest. But the framing of such a regime for the Arctic Ocean will require the best efforts of regime builders to project the future.

The case for futurism applied to the Arctic Ocean hinges on conceivable events that would necessitate a bold response by the world community. Such a future may not be suggested by our sense of history. But how reliable is history as a guide to the future of the frozen sea?

Appendix A

Annual Report of the International Ocean Institute

The report contained in this appendix describes the activities of the International Ocean Institute. Because of publication restraints, reports and abridged reports previously published in this appendix will not be continued. The editors feel that this information is more readily available in other publications, as well as on the Internet.

THE EDITORS

Annual Report of the International Ocean Institute

Report of the International Ocean Institute, 2000–2001†

The International Ocean Institute (IOI) corporate profile is also undergoing further development, increasing the coherency of its focus and extending system-wide projects beyond the training level. Of all of these system-wide projects, the IOI Virtual University (IOIVU) is the most important, involving nearly all the host institutions.

The IOI continues to address the most important issues in the area of ocean governance. It also continues delivery of its training programmes, actively participates in the UN Informal Open-ended Consultative Process on the Law of the Sea and proposes new initiatives in the area of comprehensive security, sustainable development, and environmental protection.

HIGHLIGHTS OF THE REPORTING PERIOD

The Meeting of the Executive Committee and Operational Centre Directors was held at the University of Malta, 5–8 June 2000. It reviewed the programme implementation since the IOI Governing Body Meetings and the XXVII Pacem in Maribus Conference, held at IOI-South Pacific (now Pacific Islands in Fiji) on 6–7 November and 8–12 November 1999, respectively. It also recalled the results of the IOI Consultation Meeting, which was held on 17 February 2000, in association with the Risk Management Workshop organized by the IOI at the Bermuda Biological Station for Research, Bermuda, 14–15 February 2000.

The Meeting in Malta discussed the participation of the IOI in the UN Informal Open-ended Consultative Process on the Law of the Sea, the development of the IOI Virtual University (IOIVU), the work of the "Women and the Sea" and IOI Youth Programmes, the UNEP-GPA News Forum, project activities in the area of marine insurance, issues related to IOI funding, means of strengthening the organization, and other matters.

A second meeting of the Executive Committee was held in Bremen, at the Centre for Marine Tropical Ecology, 4–6 September 2001. This meeting

---

†EDITORS' NOTE.—This report has been edited for publication in the *Ocean Yearbook* and is based on the "Annual Report of the International Ocean Institute (IOI) for 2000–2001. Prepared by the Executive Directors, IOI," and provided by IOI Headquarters in Malta.

reviewed the programme implementation since December 2000, the work plans for 2002 and 2003, preparations for the upcoming IOI meetings and PIM, and financial matters.

The Eleventh Meeting of the IOI Planning Council took place in Hamburg, Germany on 29 and 30 November and 2 December 2000. One of its sessions was conducted jointly with the IOI Governing Board, which had its thirty-ninth session on 1–2 December 2000. These meetings preceded the 28th Pacem in Maribus Conference (Hamburg, Germany, 3–6 December 2000) entitled "The European Challenge," which was hosted by the International Tribunal of the Law of the Sea (ITLOS).

One goal of the meetings was to consider the results of IOI activities since December 1999. Results were presented in the IOI Executive Director's report, which gave an extensive account of all activities. Descriptions by directors of the IOI Operational Centres, of their own activities, plans, problems, and approaches, complemented the Executive Director's report. A significant diversity of work can be observed in these reports. At the meetings, there were also discussions and approvals of work plans and budgets for 2001 and 2002 and projected potential activities for 2003. Considerable attention was devoted to the organization of coastal community work, regional co-operation and networking, further strengthening of the IOI, the current state of system-wide projects, and the preparations being made for future Pacem in Maribus Conferences.

An important outcome of the Hamburg Meetings was the confirmation by the Planning Council and the Governing Board Members of their determination to proceed with development of the IOIVU. It was stressed that the establishment of the IOIVU would streamline IOI training activities and strengthen the links to the host institutions. Successful implementation of these plans would make high quality education in ocean or marine governance accessible and affordable to many people in developing and developed countries and would lead the IOI into a new era in its development.

On 21–24 September 2000, the IOI conducted a Leadership Seminar in Malta on Mediterranean Basin-Wide Co-development and Security involving 25 invited participants from 16 countries. An innovative aspect of the discussions presented an expanded concept of security, not limited to the military concept, but extending to political, economic, ecological, and social issues, with human health, the environment, and people's means and livelihoods being of central concern.

The Seminar emphasized integration of all aspects of regional security. In efforts to solve environmental problems, regional co-operation has been fairly successful, leading to enhanced confidence and dialogue between States and among various sectors of society. However, no consensus exists regarding the inclusion of environmental threats under the security umbrella. At the same time, environmental security remains an evolving concept with no consensual definition. Integration of sustainable development and regional security may include, inter alia, such issues as: suppression of piracy, drug smuggling, and other crimes at sea; smuggling of illegal immigrants; and surveillance and enforcement of regulations governing fisheries, maritime transport, safety standards, and environmental management, and might include innovative, peaceful uses of coast guards and navies.

The Leadership Seminar focused on technological co-operation and co-development as well as the relation between sustainable development and comprehensive regional security. The overall aim was to identify and possibly initiate a process for pursuing integration of sustainable development and comprehensive regional security, focusing on the Mediterranean region. The following topics were discussed in detail:

- Sustainable Development and the Implementation of the GPA-LBA;
- Technology and Co-development;
- Sustainable Development of two Basin-Wide Activities: Tourism and Fisheries; and
- Integration of Sustainable Development and Regional Security.

Proceedings of the Leadership Seminar have been published under a separate cover.

The Pacem in Maribus Conferences continue to be flagship activities of the Institute. A CD-ROM containing photos of the 1999 Pacem in Maribus Conference and the full Conference Proceedings was produced by the IOI-Pacific Islands in 2000, while a hard cover, printed version was published by the IOI-Black Sea in mid-2001.

The Pacem in Maribus XXVIII Conference was conducted in Hamburg, Germany, from 3 to 6 December 2000, and was co-sponsored by the Lighthouse Foundation and Zeit-Stiftung. The plenary and four workshops, entitled "European Seas," "Subtropical to Tropical Seas, with Particular Consideration for the Needs of Developing Countries," "Legal Conflicts and Problems," and "The Emerging Institutional Framework for Ocean Governance" were led by prominent individuals and created great interest among the more than 150 participants. Professor Federico Mayor, the former Director-General of UNESCO, gave the Arvid Pardo Lecture on the theme of "The Oceans and the Culture of Peace." Topics of the intensive plenary and sessional discussions were reflected in the session summaries.

The IOI has continued its development of the IOIVU. IOI-Southern Africa has achieved considerable progress in developing the hardware and software platforms for the IOIVU. The proposed software platform is their new product, KEWL (Knowledge Environment for Web-based Learning), a suite of integrated software tools for developing, delivering, and managing on-line courses. In April–May 2001, the IOI-SA conducted a training course for IOIVU course developers. It was attended by IOI experts from Headquarters and from Canada, China, Costa Rica, Fiji, India, and South Africa. The participants in the course found it very useful and well organized. Following an in-depth discussion of KEWL, the IOIVU Technical Advisory Group recommended that KEWL be selected as the initial software platform for the IOIVU and that the servers available at the IOI-Southern Africa be chosen as the initial hosts for the IOIVU. At present, development of KEWL continues and its capabilities are evolving at a considerable rate. The development focus of the IOIVU must now be on preparing online courses and seeking support opportunities for the IOIVU as a whole.

In accordance with the decision of the IOI Governing Board in Hamburg, 2000, an external review of the IOI is now in preparation. A forward-looking strategy and a group of marketing experts have been established. The group will consider new means of development, financing, operation, and marketing of the IOI. Publication of the group's report is anticipated in 2002.

In 2001, a technical upgrade of the IOI Headquarters was initiated. Two new personal computers were purchased, a local area network and ADSL connection to the Internet were established, and other improvements were commenced, aimed at enhancing the technical capabilities of Headquarters.

Several new IOI Operational Centres have been or are currently being created, including:

- IOI-Indonesia at the Centre for Marine Studies, at the University of Indonesia, Depok, which focuses on research and training in support of efforts to protect the marine environment, to achieve sustainable use of marine resources, and to gain extended regional co-operation (MOU signed);
- IOI-Brazil, hosted by the Federal University of Parana, at its Centre for Marine Studies, Curitiba, which focuses on the south-west Atlantic region in marine research, education (particularly distance-learning contributions to the IOIVU), and promotion of coastal community-based sustainable development projects. (Negotiations ongoing);
- IOI-Chile, at the Catholic University of Valparaiso, focusing on disseminating results of research programs and "Pacem in Maribus" Conferences to decision-makers, public servants, non-governmental organizations, the private sector, and the public at large, and also on promoting joint research in the Latin American region. This centre is being established through the efforts of Professor Alejandro Gutierrez, director of IOI-Costa Rica. (Negotiations ongoing);
- IOI-Germany, at the Centre for Tropical Marine Ecology at the University of Bremen, focusing on co-operation in Europe, on studies of marine resources in the world oceans (especially in tropical areas), and on development of education and training courses, including courses of the IOIVU. (Negotiations ongoing);
- An IOI-Regional Centre for Australia and the Western Pacific at the International Marine Project Activities Centre (IMPAC) in Towns-

ville, Australia. The Regional Centre focuses on developing programs and activities relevant to Queensland and to Australia and also aims to co-ordinate, stimulate, and assist the development of programmes and activities for existing and planned IOI Operational Centres and Affiliates in Oceania and the Western Pacific. (MOU signed).

In 2001, the IOI published "the IOI Story" as a book, authored by Professor E. Mann Borgese, Dr. K Saigal and Dr. G. Kullenberg. The book's talented prose describes the three phases of the IOI's development from its inception in 1972. Publication was facilitated by Dreiviertel Verlag in Hamburg.

IOI's work in coastal communities continues. New surveys and plans are being developed in connection with a planned visit by IOI experts to coastal villages in Kenya. Activities of the IOI "Women and the Sea" and "Youth" Programmes are also being continued by several Centres. In addition, IOI took active part in the second session of the UN Informal Open-ended Consultative Process on the Law of the Sea. Preparatory work is now underway for the Conference "Oceans and Coasts at Rio+10," Paris, 3–7 December 2001, which is co-sponsored by the IOI.

A Seminar on "Ocean Resources, Products and Services," 7–9 September 2001, was organized jointly by the IOI and the Centre for Tropical Marine Ecology at the University of Bremen and was hosted by the HWK (Hanse Institute for Advanced Study) in Delmenhorst. The success of this joint activity formed the basis for the Memorandum of Understanding on the establishment of the Operational Centre, IOI-Germany.

Activities of the Headquarters and the IOI Centres, budgetary matters, preparations for the PIM 2001 Conference to be held in Dakar, Senegal, work plans and budgets for 2002–2004, the composition of the IOI Governing Board, and the work of the IOI strategy and marketing group were discussed at the Meeting of the IOI Executive Committee, which was held in Bremen, Germany on 4–6 September 2001, hosted by the Centre for Tropical Marine Ecology of the University of Bremen.

Overall financial support for IOI activities, including the operations of Headquarters and the provision of seed money, has become reliable, owing to the contributions of the Ocean Science and Research Foundation (OSRF). In addition, IOI continues to receive funding from CIDA, private donors, and other foundations (e.g., the JFGE).

This section of the Report has aimed to provide a brief overview of the most important, system-wide IOI activities. More detailed information on the work of IOI and its Centres follows in the body of the report.

In addition to the development of human resources, most of the Centres of the IOI system are involved with other development issues, including: poverty eradication; generation of self-reliant means of development for local coastal communities; resource management, development of ecologically friendly technologies and the use of traditional environmental knowledge; co-development and co-management focusing on integrated coastal area management; sustainable livelihoods; mitigation of and adaptation to natural hazards (e.g., cyclones and storm-surges); and empowering communities in developing countries to manage their coastal and EEZ resources.

Guided by the IOI Centres (in co-operation with local NGOs and in consultation with local and national authorities as required), community-driven projects address the problems of coastal communities in an integrated way, taking into account social, economic, environmental, and survival issues. The basic approach involves co-management and sustainable livelihood considerations. The innovative aspect lies in exploring the links between social, economic, environmental, and survival needs in a balanced fashion. Impacts on local communities have included improved living conditions, increased self-reliance and confidence, establishment of local community-driven enterprises (e.g., mariculture, food processing, tree-planting, and gardening), and increased interest in education.

Appendix B

Selected Documents and Proceedings

The documents and proceedings included in this appendix represent a selection of international agreements, proceedings of international conferences, and other documents bearing on important ocean-related developments. For the sake of consistency, certain minor alterations to punctuation and spelling have been made to the text of documents where necessary. Any more substantive editorial interventions, such as the addition of more complete citations, are enclosed in brackets.

THE EDITORS

Selected Documents and Proceedings

Oceans and the Law of the Sea: Report of the Secretary-General, 2001†

CONTENTS

	Paragraphs
I. OVERVIEW	1–16
II. THE UNITED NATIONS CONVENTION ON THE LAW OF THE SEA AND ITS IMPLEMENTING AGREEMENTS	17–38
A. Status of the Convention and its implementing Agreements	17–20
B. Declarations and statements under articles 310 and 287 of UNCLOS	21–24
C. Meeting of States Parties (Tenth Meeting)	25–38
III. MARITIME SPACE	39–90
A. Recent developments	39–48
B. Summary of national claims to maritime zones	49–51
C. Continental shelf beyond 200 nautical miles and the work of the Commission on the Limits of the Continental Shelf	52–82
D. Deposit of charts and/or lists of geographical coordinates and compliance with the obligation of due publicity	83–90
IV. SHIPPING AND NAVIGATION	91–168
A. Shipping industry	91–95
B. Navigation	96–168
1. Safety of ships	99–131
2. Transport of cargo	132–143
3. Safety of navigation	144–153
4. Flag State implementation	154–165
5. Port State control	166–168
V. CRIMES AT SEA	169–242
A. Piracy and armed robbery against ships	174–223
1. Extent of the problem—reports on incidents	175–182
2. Action at the global level	183–201
2. Action at the regional level	202–205
4. Recommended actions for Governments and the industry	206–223
B. Smuggling of migrants	224–234
C. Stowaways	235–237
D. Illicit traffic in narcotic drugs and psychotropic substances	238–242
VI. MARINE RESOURCES, MARINE ENVIRONMENT AND SUSTAINABLE DEVELOPMENT	243–427
A. Conservation and management of marine living resources	243–291
1. Actions to combat IUU fishing activities	245–271

†EDITORS' NOTE.—This document is provided by the United Nations Division for Ocean Affairs and the Law of the Sea. It is extracted from "Oceans and the Law of the Sea: Report of the Secretary-General," A/56/58, 9 March 2001, Fifty-sixth session. Available online: <http://www.un.org/depts/los/index.htm>.

634   Selected Documents and Proceedings

   2. Review of the role of regional fisheries management
    organizations in fishery conservation and management 272–278
   3. Conservation and management of marine mammals 279–282
   4. Marine and coastal biodiversity 283–291
  B. Nonliving marine resources 292–319
   1. Offshore hydrocarbons 292–298
   2. Nonfuel minerals 299–319
  C. Protection and preservation of the marine environment 320–392
   1. Reduction and control of pollution 320–372
    (a) Land-based activities: the Global Programme of
     Action 320–333
    (b) Pollution by dumping; waste management 334–345
    (c) Pollution from vessels 346–372
   2. Regional cooperation 373–392
    (a) Review of UNEP regional seas programme and action
     plans 373–385
    (b) Other regions 386–392
  D. Sustainable development of small island developing States 393–400
  E. Protection of specific marine areas 401–415
  F. Climate change and sea level rise 416–420
  G. Ten-year review of the implementation of Agenda 21 421–427
VII. SETTLEMENT OF DISPUTES 428–451
  A. Cases before the International Court of Justice 430–434
  B. Cases before the International Tribunal for the Law of the Sea 435–446
  C. Case decided by an arbitral tribunal 447–451
VIII. MARINE SCIENCE AND TECHNOLOGY 452–547
  A. Legal regime for marine science and technology 454–476
   1. Legal regime for marine scientific research 455–472
   2. Legal regime for the development and transfer of marine
    technology 473–476
  B. Programmes on marine science and technology in the United
   Nations system 477–540
   1. Marine science programmes in the United Nations system 477–539
   2. Marine technology programmes in the United Nations system   540
  C. Identified needs in marine science and technology 541–547
IX. CAPACITY-BUILDING 548–585
  A. Capacity-building activities within the organizations of the United
   Nations system 557–573
  B. Capacity-building activities of the United Nations Division for
   Ocean Affairs and the Law of the Sea 574–585
   1. Amerasinghe Memorial Fellowship Programme 574–578
   2. TRAIN-SEA-COAST Programme 579–585
X. INTERNATIONAL COOPERATION AND COORDINATION 586–606
  A. Subcommittee on Oceans and Coastal Areas of the
   Administrative Committee on Coordination (SOCA) 586–596
  B. Other mechanisms 597–606
   1. Joint Group of Experts on the Scientific Aspects of Marine
    Environmental Protection (GESAMP) 597–603

2. Inter-Secretariat Committee on Scientific Programmes
relating to Oceanography (ICSPRO) 604–606
IX. REVIEW BY THE GENERAL ASSEMBLY OF DEVELOPMENTS IN
OCEAN AFFAIRS: UNITED NATIONS OPEN-ENDED INFORMAL
CONSULTATIVE PROCESS ON OCEANS AND THE LAW OF THE
SEA 607–619

Annexes

I. Status of UNCLOS and the implementing Agreements—chronological listing
II. Summary of national claims to maritime zones
III. Deposits of charts and lists of coordinates, maritime zones notifications
IV. Regional fisheries management organizations and arrangements
V. UNEP regional seas conventions and protocols
VI. Small island development States and Territories

ABBREVIATIONS

ABLOS	Advisory Board on Geodetic, Hydrographic and Marine Geo-Scientific Aspects of the Law of the Sea (Advisory Board on the Technical Aspects of the Law of the Sea)
ACOPS	Advisory Committee for Protection of the Sea
APFIC	Asia-Pacific Fishery Commission
ARGO	Array for Real-time Geostrophic Oceanography
ASEAN	Association of South-East Asian Nations
CARICOM	Caribbean Community
CCAMLR	Commission for the Conservation of Antarctic Marine Living Resources
CCSBT	Commission for the Conservation of Southern Bluefin Tuna
CECAF	Fishery Committee for the Eastern Central Atlantic
CLIVAR	Climate Variability and Predictability Study
CMI	Comité Maritime International
COLREG	Convention on the International Regulations for Preventing Collisions at Sea, 1972
COREP	Comité Régional des Pêches du Golfe de Guinée
ECLAC	Economic Commission for Latin America and the Caribbean
FFA	Forum Fisheries Agency
GCOS	Global Climate Observing System
GCRMN	Global Coral Reef Monitoring Network
GEF	Global Environment Facility
GEOHAB	Global Ecology and Oceanography of Harmful Algal Blooms
GESAMP	Joint Group of Experts on the Scientific Aspects of Marine Environmental Protection
GFCM	General Fisheries Commission (formerly General Fisheries Council) for the Mediterranean
GIPME	Global Investigation of Pollution in the Marine Environment
GIS	Geographic Information System
GLOMARD	Global Marine Radioactivity Database

GMDSS	Global Maritime Distress and Safety System
GODAE	Global Ocean Data Assimilation Experiment
GODAR	Global Oceanographic Data Archaeology and Rescue Project
GOOS	Global Ocean Observing System
GPA	Global Programme of Action for the Protection of the Marine Environment from Land-based Activities
GTOS	Global Terrestrial Observing System
HELCOM	Helsinki Commission for Baltic Marine Environment Protection
IACSD	Inter-Agency Committee on Sustainable Development
IAEA	International Atomic Energy Agency
I-ATTC	Inter-American Tropical Tuna Commission
IBSFC	International Baltic Sea Fishery Commission
ICAM	integrated coastal area management
ICC	International Chamber of Commerce
ICCAT	International Commission for the Conservation of Atlantic Tunas
ICES	International Council for the Exploration of the Sea
ICRI	International Coral Reef Initiative
ICJ	International Court of Justice
ICSPRO	Inter-Secretariat Committee on Scientific Programmes relating to Oceanography
ICSU	International Council for Science
IGOS	Integrated Global Observing Strategy (GOOS, GTOS and GCOS)
IHO	International Hydrographic Organization
IMCAM	integrated marine and coastal area management
IMSO	International Mobile Satellite Organization (formerly Inmarsat)
INF Code	International Code for the Safe Carriage of Packaged Irradiated Nuclear Fuel, Plutonium and High-Level Radioactive Wastes on Board Ships
IOC	Intergovernmental Oceanographic Commission (UNESCO)
IODE	International Oceanographic Data and Information Exchange programme
IOPC Fund	International Oil Pollution Compensation Fund, established by the International Convention on the Establishment of an International Fund for Compensation for Oil Pollution Damage
IOTC	Indian Ocean Tuna Commission
IPCC	Intergovernmental Panel on Climate Change
IPHC	International Pacific Halibut Commission
ISM Code	International Safety Management Code
IUCN	The World Conservation Union
IUU fishing	illegal, unreported and unregulated fishing
IWC	International Whaling Commission
LL	International Convention on Load Lines

LMR-GOOS	Living Marine Resources module of GOOS
MARPOL 73/78	International Convention for the Prevention of Pollution from Ships, 1973, as modified by the Protocol of 1978 relating thereto
MEPC	IMO Marine Environment Protection Committee
MSC	IMO Maritime Safety Committee
NAFO	Northwest Atlantic Fisheries Organization
NAMMCO	North Atlantic Marine Mammal Commission
NASCO	North Atlantic Salmon Conservation Organization
NEAFC	North-East Atlantic Fisheries Commission
NOAA	National Oceanic and Atmospheric Administration (United States)
NODCs	National Oceanographic Data Centres
NPAFC	North Pacific Anadromous Fish Commission
OLDEPESCA	Latin American Fisheries Development Organization
OSPAR	Commission for the protection of the Marine Environment of the North-East Atlantic
PAME	Protection of the Arctic Marine Environment (Arctic Council)
PICES	North Pacific Marine Science Organization
PMOs	Port Meteorological Officers
POPs	persistent organic pollutants
PSC	Pacific Salmon Commission
ROPME	Regional Organization for the Protection of the Marine Environment
SCOR	Scientific Committee on Ocean Research (ICSU)
SEAPOL	South-east Asian Programme in Ocean Law, Policy and Management
SOCA	Subcommittee on Oceans and Coastal Areas of the Administrative Committee on Coordination
SOLAS	International Convention for the Safety of Life at Sea
SOPAC	South Pacific Applied Geoscience Commission
SPREP	South Pacific Regional Environment Programme
STCW Convention	1978 International Convention on Standards of Training, Certification and Watchkeeping for Seafarers
STCW-F	1995 International Convention on Standards of Training, Certification and Watchkeeping for Fishing Vessel Personnel
SUA Convention	1988 Convention for the Suppression of Unlawful Acts against the Safety of Maritime Navigation
TEMA	Training, Education and Mutual Assistance programme of IOC
TEU	twenty-foot equivalent unit
TSS	traffic separation schemes
UNCTAD	United Nations Conference on Trade and Development
UNDP	United Nations Development Programme
UNEP	United Nations Environment Programme
UNIDO	United Nations Industrial Development Organization
VDRs	voyage data recorders

VMS	vessel monitoring system
VOS	Voluntary Observing Ships scheme (WMO)
WECAFC	Western Central Atlantic Fishery Commission
WGS 84	World Geodetic System 84
WHO	World Health Organization
WOCE	World Ocean Circulation Experiment

I. OVERVIEW

1. "The state of the world's seas and oceans is deteriorating. Most of the problems identified decades ago still elude resolution, and many are worsening."[1]

2. The pollution of the seas and oceans, which has caused great concern but was overshadowed by other threats such as the exhaustion of stocks and the destruction of habitats, has returned to the forefront of international concern. Pollution generally enters the sea from coastal industries and sewage systems. It also comes from inland industries via rivers and the air. The sewage pollution of the seas has become a great health hazard through contamination of seafood and degradation of coastal water quality. Such pollution also has detrimental economic effects as it ruins large areas for recreation and tourism.

3. Among other activities giving rise to concern, which not only hinder the process of sustainable development but also endanger the delicate legal balance struck in the United Nations Convention on the Law of the Sea (UNCLOS),[2] are the fisheries, including the overexploitation of stocks, the by-catch and discards, as well as the major changes in the shipping industry, which is showing the effect of the globalization of trade.

4. More than a billion people, mainly in developing countries, depend on the world's fisheries for their primary source of protein. The decline of the worldwide catch, which is an outcome of over-fishing, has reached serious proportions. As the competition for scarce resources continues unabated, there is a significant risk of threatening the peaceful order of the oceans established under UNCLOS. The fishing fleets, which operate near the coast where fish stocks are increasingly overexploited, are now venturing out into deeper waters in search of new stocks. The deep-sea stocks are more vulnerable than those in shallow waters. Trawling may do grave damage. Other practices, such as fishing with explosives, poisons or drift-nets, have a major ecological impact.

5. The globalization of exchange and the increase of trade have changed the face of the shipping industry. In 1999, international shipping registered its fourteenth year of consecutive growth, with seaborne trade reaching a record high of 5.23 billion tons. The increase in shipping has placed a heavy burden on the traffic through important navigation routes, particularly international straits, increasing

---

1. GESAMP, *A Sea of Troubles,* GESAMP Reports and Studies No. 70, United Nations Environment Programme, 15 January 2001.

2. In recent years, it has become established practice to refer to the Convention as "UNCLOS," although originally this acronym, in the form of "UNCLOS III," was used to refer to the Third United Nations Conference on the Law of the Sea.

the risk for major catastrophes. The distribution of world tonnage ownership has changed considerably over the past 20 years. The total world fleet continued to expand in 1999 by 1.3% to 799 million tons. The globalization of trade has created a new shipping environment where the world merchant fleet is not registered in the countries of domicile of the parent enterprise, i.e., the countries where the controlling interest of the fleet is located. The world container ship fleet registered in major open-registry countries continued to expand in 1999 to 39.5% of the world TEU (twenty-foot equivalent unit) capacity, as compared to 38.1% in 1998. It is noteworthy that the ships of the 35 most important maritime countries are registered under a foreign flag. This has shifted the burden of control from flag States to port States and coastal States. The seven major open-registry countries (Panama, Liberia, Cyprus, Bahamas, Malta, Bermuda and Vanuatu) represent 75% of vessels registered under their flags.

6. Another major problem facing the shipping industry is the ageing of the world's fleet. A considerable number of vessels, in particular large bulk carriers and tankers, are at least 25 years old, which increases risks of accidents with serious consequences to the marine environment and coastal areas. It also raises the issues related to the disposal of those ships when they are decommissioned: problems of recycling and of scrapping.

7. With the globalization of shipping, a global labour market for seafarers has emerged that has transformed the shipping industry into the world's first truly global industry. Therefore, a global response is required, as well as a body of global standards applicable to the whole industry. The end result of the technological and trade changes will depend upon the training, skills and experience of the people involved.

8. Parallel to these developments regarding shipping and navigation, crimes committed at sea are on the rise. Piracy and armed robbery are costing the shipping industry millions, while at the same time endangering the lives of seafarers. The smuggling of migrants and stowaways continues to rise. There is therefore a need to strengthen international efforts to combat these crimes at sea and for more effective surveillance and law enforcement. Moreover, many of these illicit acts have developed in the last decade and are not defined as crimes under international law.

9. Marine science and technology remain prerequisites for an understanding of many complex issues, such as the ocean/atmosphere relation, and to facilitate sound decision-making by managers. This requires the creation of favourable conditions for the integration of the efforts of scientists in the study of processes occurring in the marine environment and the interrelations between them. In order to ensure that the regime envisaged in UNCLOS will not remain an empty shell, there is a need to adopt national rules, regulations, and procedures to promote and facilitate the conduct of marine scientific research, as well as to develop guidelines and criteria to assist States in ascertaining the nature and implications of marine scientific research.

10. The development of marine technology has pushed the frontier of access to resources into deep waters and remote areas. It has also permitted mankind to face its past through access to underwater cultural objects lying at the bottom of the oceans and seas. Negotiations are continuing at the United Nations Educational, Scientific and Cultural Organization (UNESCO), now entering the crucial phase for the determination of the regime applicable to the cultural heritage found in

deep-water areas beyond the zones referred to in UNCLOS (see A/54/429, paras. 510–515, and A/55/61, paras. 222–223).

11. To deal with these issues linked with major usages and activities at sea, a great number of ocean-related treaties have been adopted. Apart from UNCLOS, which sets out the general legal framework, more than 450 treaties at the global and regional levels regulate fisheries, pollution from all sources (vessels, land-based, dumping) and navigation. Unfortunately, the link between the normative level and the implementation level is clearly insufficient. The adaptation of the institutional framework has been very slow and States need to enhance their institutional capacity to implement not only UNCLOS but also all the other specialized agreements often adopted with a view to developing technical aspects of the rules contained in UNCLOS. This proliferation of treaties, which overlap in many cases, is not producing the needed synergy because of the lack of coordination between their enforcement mechanisms. To add to the confusion, new international policy mechanisms, such as programmes, action plans and codes of conduct, have been and are being put in place, prepared and negotiated with as much effort as binding agreements. This complex web of binding and nonbinding instruments has contributed to render the task of policy makers and managers at the national level more difficult. A great barrier exists between the international normative level and the national implementing level. There is a need to reorganize what is becoming an incoherent and highly complex architecture of ocean governance. The lack of knowledge-sharing within national administrations prevents the orderly adoption of necessary legislation and measures for the implementation of treaties as well as the necessary follow-up at the institutional level to execute and enforce them.

12. With regard to the impact of these issues on the environment, fisheries or navigation, serious efforts are under way to refocus energies towards achieving more concrete results and efficiency. The Global Programme of Action for the Protection of the Marine Environment from Land-based Activities (GPA) was launched in 1995. It has encountered difficulties and is to be reviewed at the end of 2001. Agenda 21 is also to be reviewed in 2002, 10 years after its adoption. The revitalization of the regional seas programmes of the United Nations Environment Programme (UNEP) has been a step in the direction of the promotion of the integrated management and sustainable development of coastal areas. The regional level, which reflects the geographic scale of most problems, is paramount for ocean governance.

13. In relation to fisheries, actions are being taken to curb illegal, unreported, and unregulated fishing (IUU fishing): at the global level, by the preparation of an international plan of action, and at the regional level, by strengthening the regional fisheries bodies and arrangements, which provide an efficient mechanism to ensure compliance with existing rules.

14. As far as navigation is concerned, the effects of the globalization of shipping are seen in the increase of flags of open registry, weakening the principle of flag State jurisdiction, one of the pillars of the enforcement tool under UNCLOS. New enforcement mechanisms have emerged: port State control and coastal State jurisdiction. The issue of reflagging or flag-hopping is seen as one of the major obstacles in combating IUU fishing.

15. Overall, the lack of coordination and cooperation in addressing ocean issues, which call for a cross-sectoral response at all levels, starting at the national level, has prevented the emergence of more efficient and results-orientated ocean

governance. A new attempt by the international community has been launched to attempt to refocus the political debate on issues that needed to be addressed as a matter of urgency and to do so by promoting better cooperation and coordination at all levels: at the international level, to ensure that all competent international organizations are coordinating their actions; and at the national level, to encourage States to adopt national policies and ensure that all treaties to which they have become party are implemented by ensuring the adoption of necessary legislation and measures.

16. In this spirit, the United Nations Open-ended Informal Consultative Process on Oceans and the Law of the Sea (the Consultative Process) was established in 1999 to deepen the debate in the General Assembly and to contribute to a broader understanding of the issues covered by the report of the Secretary-General on oceans and the law of the sea, as well as to further strengthen the coordination and cooperation in ocean affairs at the international and inter-agency levels. The first meeting of the Consultative Process in 2000 offered a new opportunity to seek solutions in a concerted manner and constituted a major milestone in ocean affairs.

II. THE UNITED NATIONS CONVENTION ON THE LAW OF THE SEA AND ITS IMPLEMENTING AGREEMENTS

A. Status of the Convention and its Implementing Agreements

17. In its resolution 55/7 of 30 October 2000, the General Assembly stressed the importance of increasing the number of States parties to the Convention and the Agreement relating to the implementation of Part XI of the Convention in order to achieve the goal of universal participation. The pace of deposit of instruments of ratification or accession has slowed down noticeably: since the last report (A/55/61) was issued, only three States have deposited their instruments of ratification (Nicaragua, Maldives, and Luxembourg). The total number of States parties, including one international organization, currently stands at 135 (see annex I). The General Assembly also reiterated the call upon all States that had not done so to become parties to these instruments. Of the coastal States, the following 32 are not yet parties to the Convention: 6 States in the African region (Congo, Eritrea, Liberia, Libyan Arab Jamahiriya, Madagascar, and Morocco); 13 States in the Asian and Pacific region (Bangladesh, Cambodia, Democratic People's Republic of Korea, Iran (Islamic Republic of), Israel, Kiribati, Niue, Qatar, Syrian Arab Republic, Thailand, Turkey, Tuvalu, and United Arab Emirates); 7 States in Europe and North America (Albania, Canada, Denmark, Estonia, Latvia, Lithuania, and United States of America) and 6 States in the Latin American and Caribbean region (Colombia, Dominican Republic, Ecuador, El Salvador, Peru, and Venezuela). As far as the landlocked States are concerned, 27 States should also consider responding to the call by the General Assembly, in view of the importance of the provisions of Part X of UNCLOS for them. These States are: Afghanistan, Andorra, Armenia, Azerbaijan, Belarus, Bhutan, Burkina Faso, Burundi, Central African Republic, Chad, Ethiopia, Holy See, Hungary, Kazakhstan, Kyrgyzstan, Lesotho, Liechtenstein, Malawi, Niger, Republic of Moldova, Rwanda, San Marino, Swaziland, Switzerland, Tajikistan, Turkmenistan, and Uzbekistan.

18. The Agreement relating to the implementation of Part XI of UNCLOS was adopted on 28 July 1994 (General Assembly resolution 48/263) and entered into force on 28 July 1996. The Agreement is to be interpreted and applied together with UNCLOS as a single instrument, and in the event of any inconsistency between the Agreement and Part XI of UNCLOS, the provisions of the Agreement shall prevail. After 28 July 1994, any ratification of or accession to UNCLOS represents consent to be bound by the Agreement as well. Furthermore, no State or entity can establish its consent to be bound by the Agreement unless it has previously established or establishes concurrently its consent to be bound by UNCLOS.

19. One hundred States parties to UNCLOS are parties to the Agreement relating to the implementation of Part XI, including those that ratified UNCLOS in 2000 (see annex I). That year, the Agreement was also ratified by Indonesia, already a State party to UNCLOS. A number of other such States that became States parties to the Convention prior to the adoption of the Agreement on Part XI have yet to express their consent to be bound by the Agreement. These States that continue to apply the Agreement de facto are: Angola, Antigua and Barbuda, Bahrain, Bosnia and Herzegovina, Botswana, Brazil, Cameroon, Cape Verde, Comoros, Costa Rica, Cuba, Democratic Republic of the Congo, Djibouti, Dominica, Egypt, Gambia, Ghana, Guinea-Bissau, Guyana, Honduras, Iraq, Kuwait, Mali, Marshall Islands, Mexico, Saint Kitts and Nevis, Saint Lucia, Saint Vincent and the Grenadines, Sao Tome and Principe, Somalia, Sudan, Tunisia, Uruguay, Viet Nam, and Yemen.

20. Regarding the 1995 Agreement for the implementation of the provisions of UNCLOS relating to the conservation and management of straddling fish stocks and highly migratory fish stocks (1995 Fish Stocks Agreement), 27 States have deposited their instruments of ratification or accession, most recently Brazil and Barbados (see Annex I). Only three more instruments are needed for the entry into force of the Agreement. Although the Agreement provides, in its article 41, for the possibility of its provisional application, no State or entity has notified the depositary of its wish to do so.

B. Declarations and Statements Under Articles 310 and 287 of UNCLOS

21. Among States that have ratified UNCLOS in 2000, Nicaragua made a declaration under article 310 of UNCLOS, stating, *inter alia*, that it did not consider itself bound by any of the declarations or statements made by other States with respect to UNCLOS and that it reserved the right to state its position on any of those declarations or statements at any time; and that ratification of UNCLOS does not imply recognition or acceptance of any territorial claim made by a State party to the Convention, nor automatic recognition of any land or sea border. Nicaragua further declared that, in accordance with article 287, paragraph 1, of UNCLOS, it accepted only recourse to the International Court of Justice (ICJ) as a means for the settlement of disputes concerning the interpretation or application of UNCLOS.

22. Thus, declarations upon ratification, accession or formal confirmation of UNCLOS have been made by 49 States and the European Community. All declarations and statements with respect to UNCLOS and to the Agreement relating to the implementation of Part XI of UNCLOS made before 31 December 1996 have been analysed and reproduced in a United Nations publication in the Law of the Sea

series;[3] full texts of those made after that date have been circulated to Member States in depositary notifications and have been published in *Law of the Sea Bulletins,* Nos. 36–44. They are also available at the web site of the Division for Ocean Affairs and the Law of the Sea of the United Nations Office of Legal Affairs (www.un.org/Depts/los) as well as that of the Treaty Section of the United Nations (www.un.org/Depts/Treaty). The information concerning the choice of procedure, as provided for in article 287, is reflected, among others, in Law of the Sea Information Circular No. 13.

23. In resolution 55/7, the General Assembly called again upon States to ensure that any declarations or statements that they had made or would make when signing, ratifying or acceding to UNCLOS were in conformity therewith and, otherwise, to withdraw any of their declarations or statements that were not in conformity. Categories of declarations and statements generally considered not to be in conformity with articles 309 (prohibiting reservations) and 310 are listed in paragraph 16 of the 1999 report on oceans and the law of the sea (A/54/429).

24. Since the most recent report was issued, no States have made a declaration or statement pursuant to article 43 of the 1995 Fish Stocks Agreement.

C. Meeting of States Parties (Tenth Meeting)

25. In accordance with UNCLOS article 319 (2) (e), the Secretary-General of the United Nations shall convene necessary meetings of States Parties to the Convention. A total of 10 such meetings have been convened thus far since the first meeting was held in November 1994 following the entry into force of UNCLOS. The issues dealt with by the meetings have primarily been the election of the judges of the International Tribunal for the Law of the Sea and of the members of the Commission on the Limits of the Continental Shelf; the consideration and approval of the budget of the Tribunal; and other administrative matters of the Tribunal. The Tenth Meeting was held from 22 to 26 May 2000.[4]

26. *Budget of the Tribunal for 2001.* The budget of the Tribunal approved by the Meeting of States Parties amounts to a total of $8,090,900 (see SPLOS/56).

27. *Financial Regulations.* Some hard-core issues on the Financial Regulations of the Tribunal still remain to be agreed upon, and in view of the many proposals and suggestions emanating from delegations it was decided that a draft revision of the Financial Regulations, taking into account the various proposals put forward by delegations and the outcome of the discussions during the Ninth and Tenth Meetings, is to be prepared by the Secretariat and the Tribunal for the Eleventh Meeting of States Parties. Among the proposals that generated considerable discussion was the presentation of the draft budget of the Tribunal under a "split-currency system" and the contributions to be made by the international organizations that are States parties to UNCLOS.

3. *The Law of the Sea: Declarations and Statements with respect to the United Nations Convention on the Law of the Sea and to the Agreement relating to the Implementation of Part XI of the United Nations Convention on the Law of the Sea* (United Nations publication, Sales No. E.97.V.3).

4. The report of the Tenth Meeting is contained in SPLOS/60. It is also available at the web site of the Division for Ocean Affairs and the Law of the Sea: <www.un.org/Depts/los>.

28. Discussions continued on rule 53 (Decisions on questions of substance (SPLOS/2/Rev.3)) and focused on the proposal that decisions on budgetary and financial matters should be taken by a three-fourths majority of States parties present and voting, provided that such majority included States parties contributing at least three fourths of the expenses of the Tribunal and a majority of the States parties participating in the Meeting. While some delegations supported the proposal, others were of the view that such a provision would amount to weighted voting in violation of the principle of equality followed by the United Nations and its organs. Since the Meeting failed to produce a generally acceptable solution on the issue, it was decided to pursue the matter further during the Eleventh Meeting.

29. *The 10-year deadline under article 4 of Annex II to UNCLOS.* Another of the issues discussed at the Tenth Meeting of States Parties related to UNCLOS article 76 and article 4 of Annex II. Article 4 of Annex II to UNCLOS places a 10-year deadline on a coastal State that intends, from the entry into force of the Convention for that State, to establish the outer limits of its continental shelf. However, owing to the difficulties faced by certain States, particularly developing States, in complying with the time limit, the Meeting decided that the topic was to be included in its agenda for the Eleventh Meeting and requested the Secretariat to prepare a background paper on the matter (see paras. 70–74).

30. *Role of the Meeting of States Parties with respect to the implementation of UNCLOS.* Various views were expressed on the proposal to include in the agenda of the Eleventh Meeting of States Parties the item "Implementation of UNCLOS" or "Issues of a general nature related to UNCLOS" (SPLOS/CRP.22). Suggestions were made to the effect that the Meeting of States Parties should receive an annual report from the Secretary-General of the United Nations on issues of a general nature that had arisen with respect to the Convention pursuant to its article 319. It was also suggested that the Meeting should be informed annually on the work of the Commission on the Limits of the Continental Shelf and of the International Seabed Authority.

31. In that connection, a number of delegations were of the view that the mandate of the Meeting of States Parties should not be expanded beyond the budgetary and administrative matters of the Tribunal. It was argued that the proposed report by the Secretary-General referred to in article 319 of UNCLOS was referred to in General Assembly resolution 49/28, in which the Secretary-General was requested to prepare a comprehensive report for the consideration of the General Assembly on developments relating to the law of the sea, which, *inter alia,* could also serve as a basis for report to all States Parties to the Convention.

32. In reply, the view was expressed that the Meeting of States Parties was the only competent body responsible for taking decisions on issues relating to the implementation of UNCLOS and its role should not be confined to dealing with the budgetary and administrative issues of the Tribunal and that therefore certain issues pertaining to the implementation of UNCLOS should be discussed by the Meeting. Some delegations alluded to the Consultative Process as a body where some of the concerns raised could be addressed. The relationship between the Consultative Process and the Meeting of States Parties was noted as being complementary in that the Meeting of States Parties could consider issues relating to the implementation of UNCLOS while the Consultative Process was meant to promote international cooperation and coordination within the framework of the Convention.

33. On account of the divergent opinions expressed, the Meeting decided to

include in its agenda for the Eleventh Meeting the topic "Matters related to article 319 of UNCLOS."

34. *Trust funds.* The Tenth Meeting of States Parties decided to recommend to the General Assembly the establishment of a voluntary trust fund, similar to the trust fund established for ICJ, to provide States with financial assistance in proceedings before the International Tribunal for the Law of the Sea. Accordingly, the General Assembly requested the Secretary-General to establish the trust fund.[5]

35. The Meeting also decided to recommend the establishment of two other trust funds relating to the work of the Commission on the Limits of the Continental Shelf. Therefore, the General Assembly requested the Secretary-General to establish such funds[6] (see paras. 65–69).

36. *Other matters.* The Meeting took note of an oral progress report presented by the Secretary-General of the International Seabed Authority on its work. In addition, a proposal was put forward that, in the light of the provisions of UNCLOS on the establishment of regional marine scientific and technological research centres, consideration should be given to the establishment of an African institute for the oceans. The Meeting decided to include this item in the agenda of its Eleventh Meeting.

37. The Eleventh Meeting of States Parties to UNCLOS will be held in New York from 14 to 18 May 2001. It will have on its Agenda, *inter alia,* the following items: (a) annual report of the International Tribunal for the Law of the Sea covering the calendar year 2000; (b) draft budget of the Tribunal for 2002; (c) draft Financial Regulations of the Tribunal; (d) external audit report and financial statement for 1999; (e) Rules of Procedure of the Meetings of States Parties, in particular, the rules dealing with decisions on questions of substance (rule 53), including the establishment of a finance committee; (f) matters related to article 319 of UNCLOS; and (g) issues with respect to article 4 of Annex II to the Convention.

38. The Eleventh Meeting will also deal with the election of one judge to fill the vacancy created by the demise of Judge Lihai Zhao (China), who passed away on 10 October 2000. The newly elected judge will serve the remainder of Judge Lihai Zhao's term of six years, which will expire in September 2002.

III. MARITIME SPACE

A. Recent Developments

39. The developments relating to State practice during the period under review were, generally, a positive reconfirmation of the wide degree of acceptance of UNCLOS by States. Several States have adopted new legislation or amended the existing laws, taking into account UNCLOS provisions. It seems, however, that a considerable number of other States, including States parties to UNCLOS, still need to address more efficiently the issue of harmonization of their national legislation with UNCLOS provisions and thus respond positively to the calls by the General Assembly, as contained in paragraph 3 of resolution 55/7. In this connection, the Secre-

---

5. General Assembly resolution 55/7, para. 9.
6. *Ibid.,* paras. 18 and 20.

tary-General wishes to invite States parties to UNCLOS to communicate the information concerning steps undertaken by them in this respect. An analysis of the information received would then appear in the next report as an overall assessment of the implementation of UNCLOS 20 years after its adoption.

40. During the reporting period, a number of important developments have been brought to the attention of the Division for Ocean Affairs and the Law of the Sea. Among them were, in Europe and North America, the establishment by Belgium and the Netherlands of their exclusive economic zones, by, respectively, the Act concerning the exclusive economic zone of Belgium in the North Sea, 22 April 1999, and the Act of 27 May 1999 establishing an exclusive economic zone of the Kingdom of Netherlands, together with the Decree of 13 March 2000 determining the outer limits of the exclusive economic zone of the Netherlands and effecting the entry into force of that Act. The Division also received copies of the following legislation: the Norwegian Act of 29 November 1996, No. 72, relating to petroleum activities; the Law on the internal waters, the territorial sea and the contiguous zone of the Russian Federation of 31 July 1998; the Law of the Russian Federation on the exclusive economic zone of 17 December 1998; the Act by Belgium on protection of the marine environment and ocean space under Belgian jurisdiction dated 20 January 1999; and the United States of America's Oceans Act of 2000. In the Latin American and Caribbean region, Honduras adopted the Maritime Areas of Honduras Act, by means of Legislative Decree 172-99, dated 30 October 1999, and enacted its baselines by Executive Decree No. PCM 007-2000 of 21 March 2000. Certain elements of this legislation of Honduras were protested bilaterally by Guatemala, Nicaragua, and El Salvador.[7] In the Asia and the Pacific region, Australia enacted, on 29 August 2000, a Proclamation under the Seas and Submerged Lands Act, 1973.

41. The delimitation of maritime boundaries has certainly become an important element of the practice of States in the modern law of the sea. The following agreements concerning the delimitation of maritime boundaries were received by the Division or adopted during the reporting period: (a) in Africa, the Agreement of 29 August 2000 between Nigeria and Sao Tome and Principe over the contending issue of delimitation of their common maritime boundary, the Treaty of 23 September 2000 between Nigeria and Equatorial Guinea concerning their maritime boundary; (b) in Asia and the Pacific, the Maritime Agreement between Oman and Pakistan (providing for the delimitation of maritime boundaries) signed on 11 June 2000, the Agreement between Kuwait and Saudi Arabia on the delimitation of the continental shelf, signed on 2 July 2000, the Treaty on the final and permanent international land and sea borders between the Kingdom of Saudi Arabia and the Republic of Yemen of 12 June 2000, and the Agreement between China and Viet Nam on delimitation of the territorial sea in the Gulf of Tonkin, concluded in December 2001; (c) in the European region, a Protocol between the Government of the Republic of Turkey and the Government of Georgia on the confirmation of the maritime boundaries between them in the Black Sea, concluded on 14 July 1997; Additional Protocol to the Agreement of 28 May 1980 between Norway and Iceland concerning fishery and continental shelf questions and the Agreement derived therefrom of 22 October 1981 on the continental shelf between Jan Mayen and

---

7. References to the respective letters containing the protests may be found in Law of the Sea Information Circular No. 12, p. 37.

Iceland of 11 November 1997; Additional Protocol to the Agreement of 18 December 1995 between the Kingdom of Norway and the Kingdom of Denmark concerning the Delimitation of the Continental Shelf in the Area between Jan Mayen and Greenland and the Boundary between the Fishery Zones in the Area, also of 11 November 1997; and (d) in North America, the Treaty between the Government of the United States of America and the Government of the United Mexican States on the Delimitation of the Continental Shelf in the Western Gulf of Mexico beyond 200 Nautical Miles, of 9 June 2000.

42. While an important number of maritime boundary delimitation agreements have already been concluded, providing a wealth of State practice, it is estimated that approximately 100 maritime boundary delimitations throughout the world still await some form of resolution by peaceful means. Some recent developments demonstrate that the delimitation of maritime boundaries remains in a number of instances one of the most sensitive issues in the relations between neighbouring States, with a potential impact on peace and security.

43. Among unresolved maritime boundary delimitations brought to the attention of the Division through the world media, the following could be mentioned: (a) in Africa: Morocco and Spain's Canary Islands; (b) in Asia: China and Japan, Iran (Islamic Republic of) and Kuwait; (c) in Latin America and the Caribbean: Barbados and Trinidad and Tobago, Cuba and Honduras, Guyana and Suriname, Guyana and Venezuela; and (d) in Europe: Romania and Ukraine in the Black Sea, and Russian Federation and Ukraine in the Strait of Kerch. It appears that in some of those cases, a certain degree of progress has been reached in the negotiations.

44. The Government of Iraq protested against the delimitations between Kuwait and Saudi Arabia,[8] stating that any agreement that did not take into consideration Iraq's legitimate rights, in accordance with international law and UNCLOS, could not be legally binding on Iraq, and Iraq would not recognize it. With respect to the delimitation between Kuwait and the Islamic Republic of Iran, Iraq added that its legal stand applied to any agreement that would be concluded and that any delineation of the continental shelf in the area should be reached through an agreement among all countries possessing sovereign rights on the continental shelf, including Iraq, to explore and invest their natural resources, with the aim of reaching a fair solution on the basis of article 83 of UNCLOS.

45. In another development related to delimitation, the Government of Malta informed the Division that it had received information that the authorities of the Libyan Arab Jamahiriya and Tunisia had announced the issuance of the offshore acreage for oil exploration in areas considered to be within Malta's continental shelf and under its jurisdiction for the purpose of oil exploration and exploitation. Malta had brought the issue to the attention of major oil companies.

46. In view of some of these latest developments, the Secretary-General wishes to emphasize that the delimitation of maritime boundaries shall be reached by agreement, preferably obtained through negotiations. The overall benefits of an agreement negotiated on the basis of international law and in a spirit of understanding and cooperation among States involved cannot be overstated.

47. To facilitate the negotiating process to which States with adjacent or opposite coasts will have to resort in case of overlapping claims, the Division for Ocean

8. S/2000/821.

Affairs and the Law of the Sea has prepared a *Handbook* on the delimitation of maritime boundaries.[9] The *Handbook* presents legal, technical and practical information deemed essential in negotiating maritime boundary delimitation agreements between coastal States. It also contains information concerning the peaceful settlement of disputes in case the negotiations are unsuccessful.

48. The Division continues to publish all newly obtained legislation and delimitation treaties in the *Law of the Sea Bulletin*, which appears periodically, three times per year.

B. Summary of National Claims to Maritime Zones

49. The statistics about national claims presented in the table entitled "Summary of national claims to maritime zones" (see Annex II) remain basically unchanged during the reporting period (see A/54/429, paras. 85–87), apart from a few adjustments. Those adjustments were made to take into account legislation and other relevant information communicated to the Division during the past year. The table of claims to maritime jurisdiction itself represents a review of information published in *Law of the Sea Bulletin* No. 39 in 1998. Despite extensive research, however, the table may not always reflect the latest developments, owing to the lack of regular updates from Governments.

50. Regarding claims with respect to the continental shelf, it should be noted that their status may appear in certain cases rather ambiguous, especially where the claims and legislation were initially based on the Convention on the Continental Shelf, adopted at Geneva on 29 April 1958, and where the State concerned subsequently became a State party to UNCLOS.

51. The table reflects the fact that the rights of a coastal State over the continental shelf do not depend on occupation, effective or notional, or on any express proclamation. It highlights the discrepancies that seem to exist between claims as reflected in national legislation and the entitlements under the 1982 Convention, which, pursuant to its article 311, paragraph 1, prevails, as between States parties, over the 1958 Geneva Conventions. Consequently, States parties to UNCLOS concerned may wish to review their legislation on the continental shelf and bring it into harmony with the provisions of current international law.

C. Continental Shelf Beyond 200 Nautical Miles and the Work of the Commission on the Limits of the Continental Shelf

52. *Work of the Commission on the Limits of the Continental Shelf.* The Commission has held eight sessions since it was established in June 1997. Those sessions were devoted both to preparing the Commission for the receipt of submissions from coastal States and to producing materials to assist States in the preparation of their submissions. The ninth session of the Commission will be held in New York from 21 to 25 May 2001.

9. *Handbook on the Delimitation of Maritime Boundaries* (United Nations publication, Sales No. E.01.V.2).

53. More detailed information regarding the work of the Commission can be found in the recent annual reports of the Secretary-General (A/55/61, paras. 25–29; A/54/429, paras. 55–69; A/53/456, paras. 55–69; A/52/487, paras. 43–53; and A/51/645, paras. 77–84).[10]

54. The Commission has produced three basic documents to date: its rules of procedure (CLCS/3/Rev.3), of which the provisions on confidentiality were extensively revised at the eighth session from the previous version of the rules; its modus operandi (CLCS/L.3); and its Scientific and Technical Guidelines (CLCS/11 and Add.1). The Guidelines are intended to provide assistance to coastal States with regard to the technical nature and scope of the data and information they are expected to submit to the Commission. The annexes to the Guidelines include, *inter alia*, flowcharts providing a simplified outline of the procedures described in the relevant parts of the Guidelines themselves.

55. The highly complex nature of the Guidelines, which deal with geodetic, geological, geophysical, and hydrographic methodologies stipulated in article 76 for the establishment of the outer limits of the continental shelf, using such criteria as determination of the foot of the continental slope, sediment thickness and types of sea floor highs, led the Commission to take two important steps to assist coastal States in applying them: the first was to hold an open meeting, since the Commission generally meets in private (closed) session owing to the nature of its mandate as a scientific and technical expert body; and the second, to design an outline for a five-day training course.

56. The seventh session of the Commission was held in New York from 1 to 5 May 2000. The first day of the session was devoted to an open meeting, aimed at flagging the most important and challenging issues related to the establishment of the continental shelf beyond 200 miles, in accordance with the legal and scientific requirements of article 76 of the Convention. The meeting was also intended to give a general indication to policy makers and legal advisers of the benefits that a coastal State might derive from the valuable resources of the extended continental shelf and to explain to experts in marine sciences involved in the preparation of submissions how the Commission considered that its Scientific and Technical Guidelines should be applied in practice.

57. At the open meeting, the Chairman of the Commission emphasized that the importance of the resources to be derived from the continental shelf was enormous and that in future the shelf area would be the main source of world oil and gas supplies. Offshore oil production in 2000 was estimated at 1.23 billion tons, and natural gas at 650 billion cubic metres. The effect of the provisions of the Convention on the continental shelf was that practically all seabed oil and natural gas resources would fall under the control of coastal States.

58. Approximately 100 government officials, members of intergovernmental organizations, legal advisers and experts in marine sciences related to the establishment of an extended continental shelf attended the meeting.

59. Several other activities were also undertaken at the seventh session in connection with the issue of training. A review of existing training projects and capacities within the United Nations system was presented to the Commission. Approaches

---

10. These reports can be found at the web site of the Division for Ocean Affairs and the Law of the Sea: <www.un.org/Depts/los>.

were also made to explore the relevance of certain programmes of the Intergovernmental Oceanographic Commission and the International Hydrographic Organization to the scientific provisions of article 76. The possibility that these organizations may also be in a position to address the training needs of developing States is still being explored.

60. Although no submissions have yet been received, the Commission is aware that the process of preparing a submission is at an advanced stage in some coastal States.

61. At its eighth session, held in New York from 31 August to 4 September 2000, the Commission concentrated primarily on the issue of training with a view to aiding States, especially developing States, to further develop the knowledge and skills for preparation of a submission in respect of the outer limits of the continental shelf provided for by the Convention. A basic flowchart for preparation of a submission to the Commission was designed (CLCS/22). In the context of its responsibilities to provide advice to coastal States, the Commission also prepared an outline for a training course of approximately five days' duration, aimed at practitioners who would take part in the preparation of the submission of a coastal State (CLCS/24). It is not part of the mandate of the Commission to conduct or organize training, though members may be involved in their personal capacity. However, the suggested course could be developed and delivered by interested Governments and/or international organizations and institutions possessing the necessary facilities and pedagogic and subject expertise.

62. The aim of the outline developed by the Commission is to facilitate the preparation of submissions in accordance with the letter and spirit of the Convention, as well as with the Guidelines of the Commission. It is expected that courses offered using a standard outline would help ensure a uniform and consistent practice in the preparation of submissions to the Commission.

63. The intended participants in such courses should be from among professionals in geophysics, geology, hydrography and geodesy, as well as others who would be involved in preparing a submission to the Commission; the minimum prerequisite for participants would be a bachelor's degree or the equivalent.

64. Courses could be adapted to the particular needs of coastal States at the regional level, which would have several practical advantages. First, offering courses to be held in, and designed for, specific regions would be cost-effective for developing countries in the region. Secondly, such courses may take into account the wide variety of types of continental margins in different areas of the oceans, as well as the ways of applying the criteria contained in the Convention.

65. *Establishment of voluntary trust funds.* Four voluntary trust funds were established by the General Assembly in its resolution 55/7 (paras. 9, 18, 20 and 45). Two are related to the establishment of an extended continental shelf in accordance with the provisions of article 76 of the Convention.

66. The first trust fund was established based upon a request by the Commission to the Tenth Meeting of States Parties, which decided in turn to recommend to the General Assembly the establishment of the fund so that members of the Commission from developing countries might participate more fully in the work of the Commission. The fund would cover travel expenses and provide a daily subsistence allowance for those members of the Commission nominated by developing States

that requested such assistance. This decision was taken notwithstanding the provision of Annex II to the Convention that requires the State party nominating a member of the Commission to defray the member's expenses while in performance of Commission duties (SPLOS/58).

67. The second fund was established by the General Assembly also upon the recommendation of the Tenth Meeting of States Parties. Its purpose is: (a) to provide assistance to States parties to meet their obligations under article 76 of the Convention, and (b) to provide training to countries, in particular, the least developed among them and small island developing States, for preparing submissions to the Commission with respect to the outer limits of the continental shelf beyond 200 nautical miles, as appropriate (SPLOS/59). During the most recent regular session of the General Assembly, Norway pledged US$ 1 million to the fund (see A/55/PV.42), and the United Nations has already taken steps required for its establishment. In resolution 55/7, not only States, but also intergovernmental organizations and agencies, national institutions, nongovernmental organizations, and international financial institutions, as well as natural and juridical persons are called upon to make voluntary financial or other contributions to the fund. The impending deadline for submissions to the Commission of November 2004 for many developing States has lent a sense of urgency to the establishment and use of this fund.

68. One of the uses of the fund may be to provide both training to the appropriate technical and administrative staff of the coastal State making a submission to enable them to perform initial desktop studies and project planning, and to prepare the final submission documents when the necessary data have been acquired. It may be used as well to provide for advisory assistance or consultancies, if needed. The data acquisition campaigns themselves, however, are not the object of the fund.

69. The submission documents must be prepared in conformity with the provisions of article 76 and Annex II to the Convention (and for some States, Annex II of the Final Act) and the Scientific and Technical Guidelines of the Commission. The training provided should take these requirements into account and should aim at enabling the submitting State's personnel to prepare most of the required documents themselves. The preparation of the submission may entail other costs that may also be met through the fund (e.g., software and hardware equipment, technical assistance, etc.)

70. *Deadline for submissions to the Commission.* At the Tenth Meeting of States parties, a discussion took place with regard to the issue of the 10-year time limit under article 4 of Annex II to the Convention (see para. 29). It was pointed out that certain countries, particularly developing countries, might have difficulties in complying with the 10-year time limit, especially in view of their limited technical expertise. General agreement was expressed with regard to the difficulty of complying with the 10-year time limit.

71. In fact, for 14 of the 30 States originally identified in 1978 as appearing to meet the legal and geographic requirements to take advantage of the provisions of article 76 regarding an extended continental shelf, the deadline will fall in November 2004. Those States are: Angola, Australia, Brazil, Fiji, Guinea, Guyana, Iceland, Indonesia, Mauritius (4 December 2004), Mexico, Micronesia (Federated States of), Namibia, Seychelles and Uruguay.

72. The Meeting of States parties decided to include on the agenda for the Eleventh Meeting (14–18 May 2001), an item entitled "Issues with respect to article 4 of Annex II to the United Nations Convention on the Law of the Sea" and requested the Secretariat to prepare a background paper for that discussion (SPLOS/60, para. 62) (see para. 29).

73. According to article 2, paragraph 2, of Annex II to the Convention, the initial election for the members of the Commission should have taken place "within 18 months after the date of entry into force of this Convention," that is, by 16 May 1996. However, at the Third Meeting of States Parties to the Convention, in 1995, it was decided that the election of members of the Commission would be postponed until March 1997, in order to give an opportunity for additional States to become parties to the Convention and to nominate candidates for the Commission. In fact, during the period of the postponement, 31 additional countries acceded to the Convention, and 8 among them nominated candidates who were elected and are currently serving. A proviso was agreed upon that, should any State that was already a party to the Convention by 16 May 1996 (i.e., 18 months after the entry into force of the Convention) be affected adversely in respect of its obligation to make its submission to the Commission *within 10 years after the entry into force of the Convention for that State* (Annex II, article 4, emphasis added), States parties to the Convention, at the request of such a State, would review the situation with a view to ameliorating the difficulty in respect of that obligation (SPLOS/5, para. 20). The election of the 21 members of the Commission was held on 13 March 1997. The Government of the Seychelles has already submitted a request to the Meeting of States Parties to postpone its deadline based on the above proviso.

74. Although the time period during which submissions should be made to the Commission will be under consideration at the Eleventh Meeting of States Parties, and may be extended, the existing cut-off date, in line with the existing rule, is still 10 years from the entry into force of the Convention for the submitting State.

75. *International activities regarding the extended continental shelf. (i) International Conference on technical aspects of maritime boundary delineation and delimitation, including the issues relevant to the provisions on the continental shelf contained in the United Nations Convention on the Law of the Sea (Monaco, 7, 9 and 10 September 1999).* The International Hydrographic Bureau hosted in Monaco in September 1999 the International Conference on technical aspects of maritime boundary delineation and delimitation, including UNCLOS article 76 issues, sponsored by ABLOS.[11]

76. Seventy-six participants from 29 countries attended the Conference. Several

---

11. ABLOS is the acronym for the Advisory Board on Geodetic, Hydrographic and Marine Geo-Scientific Aspects of the Law of the Sea (also referred to as the Advisory Board on the Technical Aspects of the Law of the Sea). It was formed in September 1994 by the International Hydrographic Organization and the International Association of Geodesy to provide advice and guidance and, where applicable, offer expert interpretation of the hydrographic, geodetic, and other technical aspects of the law of the sea to the parent organizations, their member States, or to other organizations on request. In 1999 the Intergovernmental Oceanographic Commission of UNESCO was invited to become associated with ABLOS. The Advisory Board is composed of three representatives from each organization and one additional member representing the United Nations Division for Ocean Affairs and the Law of the Sea of the Office of Legal Affairs in an ex-officio capacity.

members of the Commission on the Limits of the Continental Shelf also participated. The Conference Proceedings containing the 26 papers presented have been published by the International Hydrographic Bureau.[12]

77. The Conference was divided into four sessions over a period of two days. Topics related to the approach of the Commission on the Limits of the Continental Shelf to submissions made by coastal States were considered in contributions presented by several members of the Commission in their personal capacities during the first session. Discussions included the mandate and work of the Commission to date; a review of the continental margins of the world; preparation of desktop studies; uncertainties and errors in sediment thickness; an update of coastal States that might potentially be included in the category of wide continental margin States; and the elements for inclusion in submissions by coastal States.

78. The remainder of the sessions were devoted to: "geodetic issues, with emphasis on errors in maritime boundaries and how to reduce them," dealing specifically with geodetic problems in the delineation and delimitation of maritime boundaries; "tools needed for boundary delimitations," concerning the hardware and software that would be necessary to obtain the data to substantiate the establishment of an extended continental shelf; and "other issues and case studies," which discussed specific issues and presented case studies, only some of which were related to article 76.

79. *(ii) "Continental Shelf—Buenos Aires 2000" Workshop (Buenos Aires, 13–17 November 1999).* The issue of the establishment of an extended continental shelf was recently discussed at a workshop in Buenos Aires, which was attended by a number of well-known specialists on the subject of the continental shelf.

80. The purpose of the workshop was to exchange viewpoints, illustrate various methodologies, analyse resources and present relevant studies carried out to date. Papers were presented by several members of the Commission and its Secretary. Presentations were also made by several experts engaged in preparing for the establishment of the extended continental shelf in their own countries. The members of the Technical Subcommittee of the Argentine Commission on the determination of the outer limit of the continental shelf also participated in the Workshop.

81. *Workshops and symposiums to be held in 2001.* A five-day training course on delineation of the outer limits of the continental shelf beyond 200 nautical miles in accordance with UNCLOS, and on practical aspects of completing a submission to the Commission on the Limits of the Continental Shelf, was scheduled to be held in Southampton, United Kingdom, from 26 to 30 March 2001. It was to be offered jointly by the Southampton Oceanography Centre and the Hydrographic Office of the United Kingdom. The course represents a modification of the core training programme published by the Commission on the Limits of the Continental Shelf (CLCS/24).

82. A Symposium on Marine Geophysics is scheduled to take place during the next International Congress of the Brazilian Geophysical Society, to be held at Salvador de Bahia from 28 October to 1 November 2001. Among the subjects on which papers are to be presented are the deep-sea structures in the South Atlantic, the

---

12. Proceedings of the International Conference on Technical Aspects of Maritime Boundary Delineation and Delimitation (Including UNCLOS Article 76 Issues), International Hydrographic Bureau, Monaco, 9 and 10 September 1999.

continental/oceanic crust boundary, sedimentary processes in the South Atlantic Ocean basin, slope stability and studies on submarine hazards to offshore structures.

D. Deposit of Charts and/or Lists of Geographical Coordinates and Compliance with the Obligation of Due Publicity

83. Coastal States, under article 16, paragraph 2, article 47, paragraph 9, article 75, paragraph 2, and article 84, paragraph 2, of UNCLOS, are required to deposit with the Secretary-General of the United Nations charts showing straight baselines and archipelagic baselines as well as the outer limits of the territorial sea, the exclusive economic zone and the continental shelf; alternatively, the lists of geographical coordinates of points, specifying the geodetic datum, may be substituted. Coastal States are also required to give due publicity to all these charts and lists of geographical coordinates. Furthermore, under article 76, paragraph 9, coastal States are required to deposit with the Secretary-General charts and relevant information permanently describing the outer limits of the continental shelf extending beyond 200 nautical miles. In this case, due publicity is to be given by the Secretary-General. Together with the submission of their charts and/or lists of geographical coordinates, States parties are required to provide appropriate information regarding original geodetic datum.

84. In this connection, it should be noted that the deposit of charts or of lists of geographical coordinates of points with the Secretary-General of the United Nations is an international act by a State party to UNCLOS in order to conform with the deposit obligations referred to above, after the entry into force of UNCLOS. This act is addressed to the Secretary-General in the form of a note verbale or a letter by the Permanent Representative to the United Nations or other person considered as representing the State party. The mere existence or adoption of legislation or the conclusion of a maritime boundary delimitation treaty registered with the Secretariat, even if they contain charts or lists of coordinates, cannot be interpreted as an act of deposit with the Secretary-General under the Convention.

85. In resolution 55/7, the General Assembly once again encouraged States parties to the Convention to deposit with the Secretary-General such charts and lists of geographical coordinates. So far, only 24 States have fully or partially complied with their deposit obligations (see Annex III).

86. Acting upon the request contained in General Assembly resolution 49/28 of 6 December 1994, the Division for Ocean Affairs and the Law of the Sea, as the responsible substantive unit of the United Nations Secretariat, has established facilities for the custody of charts and lists of geographical coordinates deposited and for the dissemination of such information in order to assist States in complying with their due publicity obligations. In this connection, States parties are encouraged to provide all the necessary information for conversion of the submitted geographic coordinates from the original datum into the World Geodetic System 84 (WGS 84), a geodetic datum system that is increasingly being accepted as the standard and is used by the Division to produce its illustrative maps.

87. The Division has also established a Geographic Information System (GIS). GIS enables the Division to store and process geographic information and produce

custom-tailored cartographic outputs through the conversion of conventional maps, charts and lists of geographical coordinates in digital format. GIS also helps the Division to identify any inconsistencies in the information submitted. The GIS database is connected with the National Legislation/Delimitation Treaties database, which facilitates retrieval of relevant information on certain geographic features.

88. The Division has also sought to assist States in fulfilling their other obligations of due publicity established by UNCLOS. These obligations relate to all laws and regulations adopted by the coastal State relating to innocent passage through the territorial sea (article 21 (3)) and all laws and regulations adopted by States bordering straits relating to transit passage through straits used for international navigation (article 42 (3)). During the reporting period, Ukraine submitted a copy of the Regulations on the Customs Control over the Transit of Foreign-going Vessels through the Customs Border of Ukraine, adopted by Resolution No. 283 of 29 June 1995 of the State Customs Committee of Ukraine and registered under No. 217/783 of 12 July 1995 by the Ministry of Justice of Ukraine (published in *Law of the Sea Bulletin* No. 44).

89. The Division informs States parties to UNCLOS of the deposit of charts and geographical coordinates through a "maritime zone notification." The notifications are subsequently circulated to all States by means of the periodic publication entitled Law of the Sea Information Circular, together with other relevant information concerning the discharge by States of the due publicity obligation. The 13 issues of the Law of the Sea Information Circular that have already been issued give ample evidence of the practice of States in this respect. The texts of the relevant legislation together with illustrative maps are then published in the *Law of the Sea Bulletin*.

90. In addition, States continue to discharge their obligations of due publicity regarding sea lanes and traffic separation schemes under articles 22, 41 and 53 of UNCLOS, *inter alia*, through IMO, which provides for the adoption of ships' routing systems under SOLAS regulation V/8 and the adoption or amendment of traffic separation schemes (TSS) in rules 1 (d) and 10 of Convention on the International Regulations for Preventing Collisions at Sea, 1972 (COLREG). Guidelines and criteria developed by IMO for the adoption of routing measures are contained in the IMO General Provisions on Ship's Routeing (IMO Assembly resolution A.572 (14), as amended). These measures include traffic separation schemes (TSS), two-way routes, recommended tracks, areas to be avoided, inshore traffic zones, roundabouts, precautionary areas and deep-water routes. Information on recent new and amended traffic separation schemes and associated routing measures is contained in annex 18 to the report of the Maritime Safety Committee on its 73rd session (MSC 73/21/Add.3) (see paras. 153–155).

IV. SHIPPING AND NAVIGATION

A. Shipping Industry

91. The shipping industry has been undergoing significant changes. A technological revolution is taking place regarding the size and speed of ships. The average gross tonnage of passenger ships is now 71,140, with more than 3,100 people on

board at any one time. But cruise ships of 100,000 gross tonnage with a capacity of 5,000 people on board are already a reality, and plans are under way to build ships of 450,000 gross tonnage, capable of carrying 9,600 people.[13] The carrying capacity of container ships has also increased significantly. They can now carry 8,000 boxes, and proposals have been put forward to build vessels that can transport 18,000 units. It is also expected that there will be an increasing number of high-speed craft for the movement of both passengers and freight.

92. Other areas of shipping are also experiencing major technological changes, ranging from the introduction of electronic charts (see para. 105) to the emerging role of Internet-based transportation service providers. According to UNCTAD, e-trade facilitation, the newly developed Internet technology, when combined with the vast knowledge and expertise of the shipping world, may become the centralizing environment for the complex and dispersed global industry of shipping.[14]

93. The global economics of shipping has also continued to change. By the end of 1999, the world merchant fleet had reached 799 million dead weight (dwt). The major open-registry countries expanded their tonnage substantially to a record high of 348.7 million dwt. Approximately two thirds of these fleets are owned by developed market-economy countries, and the rest by developing countries. The latter's share has continued to increase. Tonnage registered in developing countries in 1999 increased substantially to 153.6 million dwt. This increase resulted from investments made by shipowners in Asian developing countries, whose fleets now account for 73% of the developing countries' total fleet. The fleets of other groups of developing countries were marginally reduced in 1999.[15]

94. A third change, which has occurred over time, is in the legal field. Most gaps in the international rules related to shipping have now been filled and the emphasis has therefore shifted to scrutinizing their implementation. For example, in the case of the International Convention on Standards of Training Certification and Watchkeeping for Seafarers (STW Convention), States have delegated to IMO the authority for assessing the implementation of the Convention.

95. *The global mandate of IMO in the field of the safety of navigation and the prevention of marine pollution from vessels*. IMO, in its contribution to the present report, highlighted the organization's mandate in the field of safety of navigation and the prevention of marine pollution from vessels, as follows:

"Although IMO is explicitly mentioned in only one of the articles of UNCLOS (article 2 of Annex VIII), several provisions in the Convention refer to the 'competent international organization' to adopt international shipping rules and standards in matters concerning maritime safety, efficiency of navigation and the prevention and control of marine pollution from vessels and by dumping. In such cases the expression 'competent international organization', when used in the singular in UNCLOS, applies exclusively to IMO, bearing in mind the

---

13. Extracted from the statement of the Secretary-General of IMO on World Maritime Day 2000: "IMO—Building Maritime Partnerships."

14. *UNCTAD Review of Maritime Transport 2000* (United Nations publication, Sales No. E.00.II.D.34), para. 124.

15. *Ibid.*

global mandate of the oganization as a specialized agency within the United Nations system established by the Convention on the International Maritime Organization (the 'IMO Convention').[16]

"The wide acceptance and uncontested legitimacy of IMO's universal mandate in accordance with international law is evidenced by the following facts: 158 sovereign States representing all regions of the world are members of IMO; all members may participate at meetings of IMO bodies in charge of the elaboration and adoption of recommendations containing safety and anti-pollution rules and standards. These rules and standards are normally adopted by consensus, and all States, irrespective of whether they are or are not members of IMO or the United Nations, are invited to participate at IMO conferences in charge of adopting new IMO conventions. All IMO treaty instruments have so far been adopted by consensus.

"At present, between 110 and 143 States (depending on the treaty) have become parties to the main IMO conventions. Since the general degree of acceptance of these shipping conventions is mainly related to their implementation by flag States, it is of paramount importance to note that States parties to these Conventions in all cases represent more than 90% of the world's merchant fleet.

"Adoption of new treaties, and amendments to existing ones, have been guided by adherence to the philosophy according to which rules and standards should be developed in order to prevent accidents at sea, and not in response to them. Accordingly, operational features are constantly under review in order to ensure that shipping activities conform to the highest possible safety and anti-pollution preventive regulations.

"IMO attaches the highest priority to the need to ensure that its numerous rules and standards contained in these treaties are properly implemented. In order to help ensure this implementation, IMO focuses on the continuous strengthening of regulations to ensure that flag States and port States and ship-owners develop their capacities and exert their responsibility to the fullest. Technical cooperation has been intensified by the operation of the Integrated Technical Cooperation Programme, aimed at ensuring that funds from different donor sources are properly channelled towards the execution of projects, under the supervision of IMO as executing agency, aimed at strengthening the maritime infrastructure of developing countries.

"Against this background, the ability of IMO to provide a prompt response to the consequences of a maritime accident was tested in 2000 in both the safety and the environmental fields by the sinking of the tanker *Erika* off the west coast of France. While the main IMO bodies considered improving existing rules and standards contained in IMO treaties, the issue was raised whether action to regulate international shipping might be taken regionally or unilaterally. In response, the Secretary-General reaffirmed IMO's global mandate by restating the firm position that IMO should, always and without exception, be regarded as the only forum where safety and pollution prevention standards affecting inter-

---

16. *Law of the Sea Bulletin*, No. 31, table listing the competent international organizations.

national shipping should be considered and adopted. Regional, let alone unilateral, application of national requirements to foreign flag ships that go beyond IMO standards would be detrimental to international shipping, the international regulatory regime, and to IMO itself, and should therefore be avoided."[17]

B. Navigation

96. UNCLOS sets out in article 94 the necessary measures that a flag State must take for its ships to ensure safety at sea. Ships must conform to generally accepted international regulations, procedures and practices governing construction, equipment and seaworthiness and be surveyed before registration and thereafter at appropriate intervals. The flag State must take into account the applicable international instruments governing the manning of ships, labour conditions and the training of crews. It is responsible for ensuring that the master, officers and crew on board observe the applicable international regulations concerning the safety of life at sea, the prevention of collisions, the prevention, reduction and control of pollution and the maintenance of communications.

97. There is a clear link between the observance of rules regarding the safety of ships, the transport of cargo, the safety of navigation and the prevention of pollution from ships. This was also emphasized at the first meeting of the Consultative Process, where the need was identified to keep under review ongoing work on different outstanding issues relating to pollution from ships (e.g., implementation of relevant international legal instruments, the transport of cargo, safety rules, routing rules, reflagging), given the importance of the social, economic, and environmental impacts of these issues (A/55/274, part A, para. 29).

98. Coastal States also have a responsibility with regard to ensuring that routes within their maritime zones are safe for navigation. A recent incident involving a vessel carrying a cargo of 29,500 tonnes of unleaded gasoline that developed a structural problem but was denied access to the ports of a number of States has raised the question of whether coastal States also have a duty to provide access to their ports to a vessel in distress. The IMO Working Group on Oil Tanker Safety and Environmental Matters, which met from 28 November to 1 December 2000, said that IMO should examine the need to establish principles for coastal States, acting either individually or on a regional basis, to review their contingency arrangements regarding the provision of ports of refuge, taking into consideration national sovereignty rights. The identified areas of refuge should have arrangements in place to allow ships in distress to take refuge.[18]

1. Safety of Ships

99. It is the responsibility of the flag State to ensure compliance by its vessels with the generally accepted international regulations, procedures and practices governing the safety of ships. Indeed article 217 (2) of UNCLOS provides that the flag State must ensure that vessels flying its flag or of its registry are prohibited from sailing until they can proceed to sea in compliance with the requirements of the

17. See report of the 72nd session of MSC, document MSC 72/23.
18. IMO document MEPC 46/12/3, para. 13.

international rules and standards for the prevention, reduction, and control of pollution, including requirements in respect of the design, construction, equipment, and manning of vessels.

(a) Ship construction, equipment, and seaworthiness

100. The generally accepted international regulations, procedures, and practices governing the construction, equipment and seaworthiness of ships, referred to in UNCLOS, are basically those contained in the International Convention for the Safety of Life at Sea (SOLAS), the International Convention on Load Lines (LL) and the International Convention for the Prevention of Pollution from Ships, 1973, as modified by the Protocol of 1978 relating thereto (MARPOL 73/78). In view of their importance, this section provides information on amendments to those instruments that entered or will enter into force in 2001, major amendments that were adopted in 2000 and major policy decisions relating to those instruments.

(i) Entry into force of amendments in 2001
101. Amendments to annex I, regulation 13 G, of MARPOL 73/78, which were adopted by the IMO Marine Environment Protection Committee in its resolution MEPC.78(43) in July 1999, entered into force on 1 January 2001. Existing oil tankers between 20,000 and 30,000 tons dwt carrying persistent oils, such as heavy diesel and fuel oil, are now subject to the same construction requirements as crude oil tankers.

(ii) Adoption of amendments in 2000
102. In 2000, IMO adopted, *inter alia,* the following new regulations concerning ship construction and equipment:

- A new high revised SOLAS Chapter V (Safety of navigation);
- A new High-Speed Craft Code 2000. The Code will enter into force on 1 July 2002 and is mandatory under SOLAS Chapter X (Safety measures for high-speed craft);
- A revised SOLAS Chapter II-2 (Construction—Fire protection, fire detection and fire extraction) and a new International Code for Fire Safety Systems (FSS Code), which is mandatory under revised Chapter II-2. Both will enter into force on 1 July 2002 under tacit acceptance;
- A new regulation 3-5 in SOLAS Chapter II-1 (Construction—Structure, subdivision, and stability, machinery and electrical installations), which prohibits the new installation of materials containing asbestos on all ships. It will enter into force on 1 July 2002.

103. *Adoption of revised SOLAS Chapter V.* IMO reported that a new revised SOLAS Chapter V dealing with several aspects of safety of ships and safety of navigation was adopted by the Maritime Safety Committee (MSC) at its 73rd session (27 November to 6 December 2000)[19] and will enter into force on 1 July 2002 under the system

19. See report of the 73rd session of MSC, document MSC 73/21.

of tacit acceptance of amendments regulated by SOLAS. Once it is in force, all new ships and existing passenger and ro-ro ships would have to be fitted with voyage data recorders (VDRs). A study would be carried out to examine the need for mandatory carriage of VDRs on existing cargo ships. Like the black boxes carried on aircraft, VDRs enable accident investigators to review procedures and instructions in the moments before an incident and help to identify the cause of any accident.

104. Another requirement, which would apply upon entry into force of revised SOLAS Chapter V, would be for all new ships of 300 gross tonnage or more engaged on international voyages, cargo ships of 500 gross tonnage or more not engaged on international voyages, and passenger ships irrespective of size, built on or after 1 July 2002, to be fitted with an automatic identification system (AIS) capable of providing information about the ship to other ships and to coastal authorities automatically.

105. A third new major change concerns the carriage requirements for shipborne navigational systems and equipment. New regulation 19 of Chapter V allows an electronic chart display and information system (ECDIS) to be accepted as meeting the chart carriage requirements of the regulation. The regulation requires all ships, irrespective of size, to carry nautical charts and nautical publications to plan and display the ship's route for the intended voyage and to plot and monitor positions throughout the voyage.

(iii) Major policy decisions in 2000
Elimination of sub-standard oil tankers
106. IMO reported that an MSC Working Group had developed a proposed list of measures to eliminate sub-standard ships, and the MSC had agreed to refer the list of measures to the organization's subcommittees and to the Marine Environment Protection Committee for general consideration. This work follows upon agreement at MEPC in October 2000 to accelerate the current phase-out schedule for single-hull oil tankers. The actual finalized revised phase-out schedule was expected to be adopted in April 2001 (see paras. 358–361).
Safety of large passenger ships
107. IMO reported that MSC, at its 72nd session, had considered a proposal by the Secretary-General of IMO to undertake a global consideration of safety issues pertaining to passenger ships, with particular emphasis on large cruise ships.[20] In response, the Committee established a Working Group on Enhancing the Safety of Large Passenger Ships, with the aim of identifying the extent to which current regulations should be reviewed, in the light of the sheer size of these vessels and the numbers of persons carried on board, and in particular with regard to emergency situations and seafarer training.

108. At the 73rd session of MSC, the Working Group reviewed the current safety regime as it relates to large passenger ships and identified areas of concern relating to: (a) the ship, including construction and equipment, evacuation, operation, and management; (b) the people, including crew, passengers, rescue personnel, training, crisis, and crowd management; and (c) the environment, including search and rescue services, operation in remote areas and weather conditions.

20. See n. 17 above.

109. MSC endorsed the Working Group's decision that future large passenger ships should be designed for improved survivability based on the time-honoured philosophy that "a ship is its own best lifeboat." The Committee endorsed a preliminary work plan as developed by the Working Group, which includes elements relating to the following areas of concern: collision and grounding; equipment failure; escape, evacuation and rescue; fire safety; medical emergency; operations and management; vessel surveys; search and rescue; ship survivability; and evacuation, life-saving systems, and arrangements.

(b) Training and certification of crew

110. It has been estimated that some 80% of marine casualties are attributable in some part to human error. Efforts within IMO have therefore continued to focus on improving the training and certification standards for crew, particularly on ensuring that the minimum requirements set out in the 1995 amendments to the 1978 STCW Convention are being implemented. According to the 1995 amendments, States are required to provide detailed information to IMO concerning administrative measures taken by them to ensure compliance with the Convention.

111. IMO in its submission reported that it had recently published a so-called "White List" of countries deemed to be giving "full and complete effect" to the 1995 amendments to the STCW Convention. At its 73rd session, MSC formally endorsed the findings of a working group established to examine a report presented by the Secretary-General to MSC, which revealed that 71 countries and one Associate Member of IMO had met the criteria for inclusion in the list. A position on the White List entitles other parties to accept, in principle, that certificates issued by or on behalf of the parties on the list are in compliance with the Convention.

112. In setting out unambiguously which countries are meeting the latest standards and requirements, according to IMO, the White List marks a significant step forward in the IMO global effort to rid the world of sub-standard ships and shipping. For the first time, it provides an IMO "seal of approval" for countries that have properly implemented the provisions of a Convention.

113. It is expected that port State control inspectors will increasingly target ships flying flags of countries that are not on the White List. A flag State party that is on the White List may, as a matter of policy, elect not to accept seafarers with certificates issued by non-White List countries for service on its ships. If it does accept such seafarers, they will be required by 1 February 2002 also to have an endorsement, issued by the flag State, to show that the flag State recognizes their certificate. By 1 February 2002, masters and officers should hold STCW 95 (STCW Convention as amended by the 1995 amendments) certificates or endorsements issued by the flag State. Certificates issued and endorsed under the provisions of the STCW Convention will be valid until their expiry date.

114. It was stressed at the MSC meeting that giving "full and complete effect" to the revised Convention might not be the same for all parties. Some may choose not to have any maritime training institutes at all and rely on recognition of certificates issued to seafarers by other States. Similarly, some parties may only provide a limited scope of training, such as for ratings only.

115. The fact that a party is not listed on the White List does not invalidate certificates or endorsements issued by that party. Nevertheless, the White List will

become one of several criteria, including inspection of facilities and procedures that can be applied in the selection of properly trained and qualified seafarers. Countries not initially included in the White List will be able to continue with the assessment process with a view to inclusion on the list at a later stage.

116. In the opinion of the Secretary-General of IMO, the fact that member States delegated the authority for assessing the implementation of STCW 95 to IMO indicates that the will to give the organization a greater role in implementation does exist. IMO is ready to respond to similar approaches in other areas where quality assurance needs to be reinforced and the name of IMO would lend credibility. In this way the STCW verification process points in the direction of a new and expanded role for IMO in the future.

Forgery of certificates of competence of seafarers
117. IMO in its submission recalled that the IMO Assembly, in its 1999 resolution A.892(21) on unlawful practices associated with certificates of competency and endorsements, had highlighted the problem of fraudulent certificates of competency issued in relation to the STCW Convention and urged Member States to take all possible steps to investigate cases and prosecute, or assist in the investigation and prosecution of, those found to be involved in the processing or obtaining of fraudulent certificates or endorsements, including the holders of such certificates or endorsements. An MSC circular on fraudulent certificates of competency (MSC/Circ.900), issued on 2 February 1999, also invited member States and parties to STCW to report to IMO and to the relevant administration any cases or suspected cases of fraudulent certificates, to intensify efforts to eliminate the problem and to act under the terms of the Convention, including prosecution of those involved, if seafarers on board were found to be holding fraudulent certificates; this could also involve detaining the ship.

118. Preliminary results of an IMO research study to establish the nature and extent of unlawful practices associated with certificates of competency has revealed 12,635 cases of forgery in certificates of competency and equivalent endorsements.[21] The study, being carried out by the Seafarers International Research Centre, Cardiff, United Kingdom, is in its final stages, having completed the data collection phase, and a final report is being produced.

119. *Training and certification of fishing vessel personnel.* The 1995 International Convention on Standards of Training, Certification and Watchkeeping for Fishing Vessel Personnel (STCW-F) is not yet in force. Efforts to improve the training, certification and watchkeeping standards of personnel on board fishing vessels have been adopted as recommendations in IMO Assembly resolutions and in the Document for Guidance on Fishermen's Training and Certification produced jointly by IMO, FAO and ILO. Amendments to the latter were adopted by MSC at its 72nd session.[22]

(c) Labour conditions

Review of ILO maritime instruments
120. ILO reported that the 29th session of the ILO Joint Maritime Commission in January 2001 constituted the first full session of the Commission since 1991. At the

21. See IMO document STW 32/6.
22. MSC 72/6/2, annex.

session the Commission adopted a historic Agreement, known as the Geneva Accord, designed to improve safety and working conditions in the maritime industry. Participants, including representatives of shipowners and seafarers, resolved that "the emergence of the global labour market for seafarers had effectively transformed the shipping industry into the world's first genuinely global industry, which required a global response with a body of global standards applicable to the whole industry." The Commission decided that the existing ILO maritime instruments should be consolidated and brought up to date by means of a new, single framework Convention on maritime labour standards. With a view to ensuring acceptable standards of working and living conditions for seafarers of all nationalities and in all merchant fleets, this approach envisages a more logical and flexible structure for maritime labour instruments and a more streamlined process for keeping them up to date. The ILO Governing Body has been requested by the Commission to authorize a programme of tripartite meetings (shipowners, seafarers and Governments) to prepare for an ILO Maritime Conference in 2005 to adopt the anticipated new framework Convention.

121. The Commission updated the minimum basic wage of able seamen. It also expressed deep concern about recent arrests of seafarers, in particular, ship captains, following maritime accidents, even before any investigation had taken place, and called upon the ILO Director-General to bring those concerns to the attention of all ILO member States.[23]

Provision of financial security for seafarers' claims
122. ILO reported that the Joint IMO/ILO Ad Hoc Expert Working Group on Liability and Compensation regarding Claims for Death, Personal Injury and Abandonment of Seafarers had held its second session from 30 October to 3 November 2000 (for the report of the first session, see A/55/61, paras. 201–203). The Working Group considered a document containing information collected by the IMO and ILO secretariats on the issues of abandonment and financial security for personal injury and the death of crew members. The document also includes information received from Governments regarding obstacles to the ratification of relevant ILO and IMO conventions, as well as UNCLOS.

123. ILO explained that the issues raised in the IMO/ILO document had led to the development by the Working Group of preliminary draft terms for inclusion in two resolutions and associated guidelines, one relating to abandonment and the other to death and injury.[24] The proposed possible draft resolution on guidelines on the provision of financial security in cases of abandonment of seafarers states that abandonment of seafarers is a serious problem, involving a human and social dimension and requiring urgent attention. It affirms that payment and remuneration and provision for repatriation should form part of the seafarer's contractual and/or statutory rights and are not affected by the failure or inability of the shipowner to perform its obligations.

124. The proposed possible draft resolution on guidelines on shipowners responsibilities in respect of contractual claims for personal injury to or death of seafarers notes that there is a need to recommend minimum international standards

---

23. ILO press release ILO/01/05 of 26 January 2001, available on the ILO web site at: <www.ilo.org/public/english/bureau/inf/pr/2001/05.htm>.
24. The report of the second session is contained in IMO document LEG 83/3.

for the responsibilities of shipowners in respect of contractual claims for personal injury and death of seafarers. It notes with concern that if shipowners do not have effective insurance cover, or other form of financial security, seafarers may not obtain prompt and adequate compensation, and adds that recommendatory guidelines are an appropriate interim means of establishing a framework to encourage all shipowners to take steps to ensure that seafarers receive contractual compensation for personal injury and death. The accompanying draft guidelines provide definitions for contractual claims, effective insurance, and set out shipowners' responsibilities to arrange for effective insurance cover.

125. The Working Group agreed to hold a third meeting from 30 April to 4 May 2001 to finalize the resolutions and guidelines before presenting them to the IMO Legal Committee at its 83rd session in October 2001 and to the ILO Governing Body at its 279th session in late 2001. Following a review by these bodies, the resolutions and guidelines could then be adopted by the IMO Assembly in November 2001.

Labour conditions of fishermen
126. With more than 70 fatalities per day, fishing at sea may be the most dangerous occupation in the world, according to the FAO report on the "State of World Fisheries and Aquaculture 2000."[25] According to the report, the ILO estimate of the worldwide death toll among fishers of 24,000 may be considerably lower than the true figure because only a limited number of countries keep accurate records on occupational fatalities in their fishing industries.

127. Also as stated in the report, which was to be presented to the FAO Committee on Fisheries at its meeting to be held from 26 February to 2 March 2001, more than 97% of the 15 million fishers employed in marine capture fisheries worldwide are working on vessels that are less than 24 metres in length, placing them beyond the scope of international conventions and guidelines. Where inshore resources have been overexploited, fishers must work farther away from shore, sometimes for extended periods, and frequently in fishing craft designed for inshore fishing, which do not comply with security regulations, according to FAO.

128. One of the main reasons for the occurrence of fatal accidents, according to the report, is the as yet unratified status of an international legal instrument on safety at sea, i.e., the 1993 Protocol to the Torremolinos International Convention for the Safety of Fishing Vessels, 1977, which superseded the Torremolinos Convention. It also cites lack of national regulations or, where they do exist, their lack of enforcement, a lack of experience of offshore fishing operations and a lack of knowledge about essential issues such as navigation, weather forecasting, communications and the vital culture of safety at sea. FAO believes that many of these situations can be rectified and is involved in a number of activities with this objective in the Caribbean, Asia and the Pacific.

129. In developing countries, poorly designed and poorly built fishing craft, lack of safety equipment and inappropriate, outdated and inadequately enforced regulations are the main causes of fatalities. In one night, in November 1996, during

---

25. The report is available on the FAO web site: <www.fao.org/DOCREP/003/X8002E/X8002E00.htm>.

a severe cyclone, more than 1,400 fishers perished in India owing to poorly designed trawlers and lack of awareness of the intensity of the danger.

130. In developed countries, rapid progress in vessel construction and fishing technologies and the application of more stringent regulations have not always led to a significant decrease in fatalities. As the report points out: "It seems that, as vessels are made safer, operators take greater risks in their ever increasing search for good catches." It should be noted that all of the Nordic countries have introduced obligatory safety courses for fishers.[26]

131. ILO reported on the outcome of the Joint FAO/ILO/IMO Meeting on Safety and Health in the Fishing Industry, held in December 1999. At the first meeting of the Joint FAO/IMO Ad Hoc Working Group on Illegal, Unreported and Unregulated (IUU) Fishing, in October 2000, it had drawn attention to the connection between IUU fishing and the human dimension of fishing, in particular expressing concern about cases of abuse of fishermen on certain vessels (see paras. 252–255). An ILO paper, annexed to the report of the Working Group, discussed the issues of flag State and port State control of labour conditions on fishing vessels.[27]

2. Transport of Cargo

132. At the first meeting of the Consultative Process, several delegations, in addressing the maritime transport of oil, hazardous substances and wastes, pointed out that the following issues merited attention: making use of the "vessel monitoring system" obligatory; revising current main routes of maritime transport in order to improve security standards and surveillance; implementing monitoring programmes to control environmental quality; and verifying the effective respect for safety rules for cargo, ships and crews, especially in the context of flags of convenience and the prevention of reflagging of vessels posing safety hazards (A/55/274, part B, para. 114).

133. IMO cited the organization's constant review of safety codes as an example of the degree to which it continuously updated the comprehensive set of safety regulations on board ships. The International Dangerous Goods (IMDG) Code, which was introduced by IMO in 1965 as a uniform international code for the transport of dangerous goods by sea covering such matters as packing, container traffic and stowage, with particular reference to the segregation of incompatible substances, had recently been revised and reformatted to make it more user-friendly and understandable.[28] At its 73rd session, MSC decided, in principle, to make the IMDG Code mandatory, aiming at an entry-into-force date of 1 January 2004, and instructed the subcommittee on Dangerous Goods, Solid Cargoes and Containers at its sixth session in July 2001 and the secretariat to prepare relevant documents such as draft amendments to SOLAS. MSC agreed that some chapters of the IMDG Code would remain recommendatory in nature.

134. In 2000, IMO also adopted amendments to the following codes: International Code for the Construction and Equipment of Ships Carrying Dangerous

26. FAO press release 01/02.
27. For the report of the Working Group, see IMO document FSI 9/15.
28. See report of the 72nd session of MSC, document MSC 72/23.

Chemicals in Bulk (IBC Code), which is mandatory under SOLAS and MARPOL 73/78 (resolutions MEPC.90(45) and MSC.102(73)); International Code for the Construction and Equipment of Ships Carrying Liquefied Gases in Bulk (IGC Code), which is mandatory under SOLAS (resolution MSC.103(73)); Code for the Construction and Equipment of Ships Carrying Dangerous Chemicals in Bulk (BCH Code), which is mandatory under MARPOL 73/78 (resolutions MEPC.91(45) and MSC.106(73)); and Code for the Construction and Equipment of Ships Carrying Liquefied Gases in Bulk (GC Code) (resolution MSC.107(73)). The amendments will enter into force on 1 July 2002 under tacit acceptance.

Transport of radioactive materials
135. Amendments to SOLAS Chapter VII adopted in 1999 by the Maritime Safety Committee in resolution MSC.87(71) entered into force on 1 January 2001 and provide for the mandatory application of the International Code for the Safe Carriage of Packaged Irradiated Nuclear Fuel, Plutonium and High-level Radioactive Wastes on Board Ships (INF Code). The Code applies to all ships, regardless of the date of construction and size, engaged in the carriage of INF cargo. Specific regulations in the Code cover a number of issues, including damage stability, fire protection, temperature control of cargo spaces, structural considerations, cargo-securing arrangements, electrical supplies, radiological protection equipment and management, training and shipboard emergency plans.

136. Carriage requirements for highly radioactive cargo, for example, design, fabrication, maintenance of packaging, handling, storage, and receipt, which are applicable to all modes of transport, are contained in the IAEA Regulations for the Safe Transport of Radioactive Material. In its resolutions on "Safety of Transport of Radioactive Materials," adopted over the past three years (resolutions GC(44)/RES/17, GC(43)/RES/11 and GC(42)/RES/13, adopted respectively in September 2000, 1999, and 1998), the IAEA General Conference has invited States shipping radioactive materials to provide, as appropriate, assurances to potentially affected States, upon their request, that their national regulations take into account the IAEA Regulations for the Safe Transport of Radioactive Material and to provide them with relevant information relating to shipments of such materials. The information provided should in no case be contradictory to the measures of physical security and safety. In the resolution adopted in 2000, GC(44)/RES/17, the General Conference noted the concerns of small island developing States and other coastal States about the transport of radioactive materials by sea and the importance of the protection of their populations and the environment. The General Conference called for efforts at the international, regional and bilateral levels to examine and further improve measures and international regulations relevant to the international maritime transport of radioactive material and spent fuel, consistent with international law, and stressed the importance of having effective liability mechanisms in place.

137. A similar call to examine and further improve measures was also made in the Final Document of the 2000 Review Conference of the Parties to the Treaty on the Non-Proliferation of Nuclear Weapons (May 2000).[29] The Conference urged States to ensure that the IAEA Regulations for the Safe Transport of Radioactive Material were maintained; affirmed that it was in the interests of all States that any

---

29. NPT/CONF.2000/28, Part I, pp. 10–11.

transportation of any radioactive materials should be conducted in compliance with the relevant international standards of nuclear safety and security and environmental protection, without prejudice to the freedoms, rights, and obligations of navigation provided for in international law; took note of the concerns of small island developing States and other coastal States with regard to the transportation of radioactive materials by sea; recalled the invitation to the shipping States in IAEA resolution GC(43)/RES/11; and called upon States parties to continue working bilaterally and through the relevant international organizations to examine and further improve measures and international regulations relevant to the international maritime transportation of radioactive material and spent fuel.

138. Most recently, the General Assembly in its resolution 55/49 of 29 November 2000, entitled "Zone of peace and cooperation of the South Atlantic," called upon Member States to continue their efforts towards the achievement of appropriate regulation of maritime transport of radioactive and toxic wastes, taking into account the interests of coastal States and in accordance with UNCLOS and the regulations of IMO and IAEA.

139. Pursuant to the request of the IMO Marine Environment Protection Committee and as a step in addressing the subject of the environmental impact of accidents involving materials subject to the INF Code, IMO and IAEA presented to MEPC at its 45th session (October 2000) a literature review on the potential hazards of radioactive material in the environment (MEPC 45/INF.2). A decision on how to proceed is to be taken at the next session of the Committee (MEPC 45/20, sect. 12).

140. Shipments of mixed oxide fuel between Europe and Japan continue to be of great concern to the coastal States along the routes currently being used for the shipments. Such concerns are heightened by the anticipation of more shipments past their coasts in the future, since Japan has a long-term contract with the United Kingdom and France for them to reprocess radioactive waste from Japanese nuclear power plants. Plutonium extracted from spent fuel is mixed with uranium oxides to produce so-called mixed oxide fuel (MOX); the remaining radioactive waste is embedded in glass for burial. The United Kingdom and France return radioactive fuel and waste to Japan by armed convoys, which go around Africa and South America or through the Panama Canal. According to a recent news report, the Russian Federation and Japan are exploring the possible shipment of MOX from Europe via the northern route off the Arctic coasts of the Russian Federation during the summer months because it would be shorter and safer from terrorist attack.[30]

141. In the past, some coastal States have either warned ships carrying MOX to stay out of their territorial seas and exclusive economic zones, for example, New Zealand[31]—or said that they preferred them not to enter those waters—for example, South Africa.[32] The Caribbean Community (CARICOM) has repeatedly called for a cessation of MOX shipments through the Caribbean Sea. At the 2000 Review Conference of the Parties to the Treaty on the Non-Proliferation of Nuclear Weapons, CARICOM expressed the view that the INF Code, while binding, did not protect en-route coastal States, and as a consequence they had no legal recourse to compen-

---

30. *ITAR-TASS* news agency, Moscow, 5 February 2001.
31. *Kyodo News Service*, Tokyo, 22 December 2000.
32. *Business Day* web site, Johannesburg, 22 January 2001.

sation for accidents, which were becoming more likely as shipments of radioactive nuclear wastes were increasing dramatically. CARICOM consequently called for consultations leading to the establishment of a comprehensive international regime for the protection of the populations and the marine environment of en-route coastal States from harm resulting from shipments of nuclear material (see also para. 404).[33]

142. Participants in the Workshop entitled "The Prevention of Marine Pollution in the Asia-Pacific Region" (Australia, 7–12 May 2000) said that, consistent with UNCLOS, IMO should liaise with IAEA on steps that might be taken to establish a monitoring and control system and a liability and compensation regime for maritime transport of radioactive materials. In the Workshop Statement, they recommended that the United Nations urgently address regional concerns about the issue of maritime transport of radioactive materials.[34]

143. The Pacific Islands Forum said it was engaged in a constructive dialogue with government and nuclear industry representatives from France, Japan, and the United Kingdom on a liability regime for compensating the region for economic losses incurred by the tourism, fishery and other industries affected as a result of an accident involving a shipment of radioactive materials and MOX fuel even if no actual environmental damage was caused. The Forum considered it necessary to focus on intermediate innovative arrangements or assurances to address its concerns, since amendments to existing international instruments, though under negotiation, would, when concluded, take some time to enter into force. It therefore called for a high-level commitment from the three shipping States to carry the process forward. It welcomed the offer by Japan to establish a "goodwill" trust fund for Forum countries, with an initial principal of US$ 10 million, which would be available to cover the costs of the initial response to incidents during shipment of radioactive materials and MOX fuel through the region. The Forum understood this Fund to be quite separate from the issue of compensation and liability, which it was currently pursuing with the three shipping States.[35]

3. Safety of Navigation

144. A shipping accident can result from a failure in the structure of the ship or because of a navigational error, such as a collision. Weather conditions can also affect a ship's navigation. The flag State not only has the duty to ensure the safety of the ship in terms of construction, equipment, manning, training, labour conditions of the crew and safe carriage of the cargo, but is also responsible for ensuring that the ship is navigated safely. Article 94 of UNCLOS requires, *inter alia,* that the master, officers and crew on board observe the applicable international regulations concerning the use of signals, the maintenance of communications and the prevention of collisions. Masters and officers are required to have appropriate qualifications, in particular in seamanship, navigation, communications, and marine engi-

33. NPT/CONF.2000/SR.11.
34. The report of the Workshop is contained in IMO document LC/SG 23/11, annex 6.
35. Communiqué of the 31st meeting of the Pacific Island Forum, 27–30 October 2000; A/55/536, paras. 28–33.

neering, and the crew must be of the appropriate qualifications and size for the type, size, machinery and equipment of the ship.

145. Ships are also required by UNCLOS to observe the applicable rights of passage in the various maritime zones, as well as, where appropriate, the measures that coastal States can take in regulating maritime traffic, for example, designated sea lanes and prescribed traffic separation schemes. Detailed rules regarding the safety of navigation are provided in SOLAS, Chapter V, and the Convention on the International Regulations for Preventing Collisions at Sea, 1972 (COLREG). In this regard, attention is drawn to the adoption of a new revised SOLAS Chapter V by the MSC, at its 73rd session (27 November–6 December 2000), and its approval of draft amendments to COLREG, for submission to the IMO Assembly at its 22nd session in November 2001 for final adoption.[36]

146. The new regulations in the revised SOLAS Chapter V, which take into account advances in technology, relate predominantly to the introduction of new requirements for ship-borne equipment (see paras. 103–105). The current regulations on ship routing, ship reporting and vessel traffic services were not revised, only renumbered. Amendments were adopted to the regulation dealing with the Ice Patrol Service and a new appendix was added to provide rules for the management, operation and financing of the North Atlantic ice patrol (see para. 152).

(a) Ship routing and reporting systems

147. IMO reported that MSC at its 73rd session had adopted amendments to the General Provisions on Ships' Routeing (resolution A.572(14), as amended) to incorporate "no-anchoring areas."

148. New and amended ship routing and reporting systems adopted by MSC at the 73rd session include: a new mandatory ship-reporting system "Off Les Casquets and the adjacent coastal area" (central English Channel, to supplement the existing mandatory ship-reporting systems already established at Ouessant and in the Pas de Calais); three mandatory no-anchoring areas on coral reef banks (Flower Garden Banks) in the north-western Gulf of Mexico; four new traffic separation schemes along the coast of Peru; new traffic separation schemes and associated routing measures in the approaches to the River Humber on the east coast of England; and amendments to the existing traffic separation scheme in Prince William Sound (United States). The new measures take effect as from 1 June 2001.

(b) Archipelagic sea lanes

149. Indonesia informed MSC at its 72nd and 73rd sessions of progress made in finalizing its draft national regulations concerning the designated archipelagic sea lanes and other basic rules and regulations on related passages. It pointed out that as a result of the creation of the newly independent State of East Timor, a new regime would have to be applied to one of the three archipelagic sea lanes designated by IMO in its resolution MSC.72(69) in 1998 (see A/53/456, para. 196), i.e., the one that crosses sea lanes III-A (in the Ombai Strait) and III-B (in the Leti Strait), since the latter two straits border East Timor and are no longer part of Indonesian archipelagic waters. An additional provision to that effect had been incorporated

36. For the texts, see the report of the 73rd session of MSC, document MSC 73/21.

in the draft national regulations, to the effect that they would no longer apply to the archipelagic sea lanes in the Ombai and Leti straits. The Government had recognized the need for further consultations with other maritime users of sea lanes III-A and III-B, for which a new regime had been proposed, before the draft national regulations were officially enacted.[37]

(c) Meteorological warnings and forecasts

150. The World Meteorological Organization (WMO) pointed out that the report of the reopened formal investigation into the loss of the *MV Derbyshire,* published in the United Kingdom in November 2000, dramatically highlighted once again the vulnerability of all shipping to extreme meteorological and oceanographic conditions, as well as the value to shipping of accurate and timely meteorological warnings and forecasts as part of maritime safety services. Meteorological and oceanographic observations made by ships at sea (under the WMO Voluntary Observing Ships (VOS) scheme) and transmitted to shore in real time are an essential component of the observational data used by national meteorological services in the preparation of such maritime safety services. The availability of such observations has, unfortunately, remained static, or actually decreased, for several years, for a number of reasons. The *MV Derbyshire* inquiry report also reiterated the importance of these VOS observations and urged more ships to participate in the VOS scheme.

151. WMO reported that it continued to make major efforts to enhance the VOS scheme in support of maritime safety. Specifically, in 2000, a descriptive brochure on the VOS had been prepared for distribution to shipping companies, ships' masters, maritime administrations, and national meteorological services. In addition, a series of international training workshops for Port Meteorological Officers (PMOs), South Africa, has continued, with a workshop in Cape Town for African countries. PMOs were crucial to the recruitment and maintenance of the VOS. Finally, WMO planned to collaborate with IMO in the rewriting and reissue of an IMO/MSC circular letter on the subject of the VOS. This rewriting would, in particular, highlight the findings in the *MV Derbyshire* inquiry report relating to the VOS.

(d) Provision of services/sharing of costs

152. *North Atlantic Ice Patrol.* Amended regulation 6 on the Ice Patrol Service and the Rules for the management, operation and financing of the North Atlantic Ice Patrol appended to revised SOLAS Chapter V[38] provide, *inter alia,* that each SOLAS Contracting Government specially interested in the ice patrol services whose ships pass through the region of icebergs during the ice season will undertake to contribute to the Government of the United States its proportionate share of the costs of the management and operation of the ice patrol service. Each contributing Government has the right to alter or discontinue its contribution, and other interested Governments may undertake to contribute to the expense of the service. The Rules provide for a voluntary contribution system, while also providing the United States, as the manager of the ice patrol, with the legal basis for implementing a new system

---

37. See MSC 72/23, para. 10.75, and MSC 73/21, para. 11.33.
38. See n. 19.

of calculation. Upon the entry into force of the Rules, the 1956 Agreement regarding Financial Support for the North Atlantic Ice Patrol will terminate and the parties to the 1956 Agreement will be deemed to be contributing Governments under the new Rules. MSC at its 73rd session, in adopting the new regulation and the Rules, reaffirmed its previous decision in 1999 that "the Ice Patrol financing system was unique and should not create a precedent for charging ships navigating in international waters for services provided by coastal States" (see also A/54/429, paras. 173–176).

153. *Straits used for international navigation: article 43 of UNCLOS.* At the Workshop entitled "The Prevention of Marine Pollution in the Asia-Pacific Region" (Australia, 7–12 May 2000),[39] the participants noted that the risks of ship-based marine pollution, both accidental and intentional, were higher in the major international shipping channels in the Asia-Pacific region with a high intensity of shipping traffic. The States bordering the straits were understandably concerned about the high cost of maintaining maritime safety and reducing the impact of marine pollution. It was important for user States to honour their obligations under article 43 of UNCLOS and assume a greater share of this burden. The Workshop recommended "that competent international organizations address the financial and resource burden of coastal States in implementing article 43 of UNCLOS on the development of safety of navigation and protection of the marine environment in straits used for international navigation."

4. Flag State Implementation

154. Flag States have the primary responsibility to have in place an adequate and effective system to exercise control over ships entitled to fly their flag and to ensure they comply with relevant international rules and regulations.

155. It has been reported that one of the greatest impediments to a genuine "quality culture" in shipping is the lack of a sufficient degree of transparency in the information on the quality of ships and their operators. While much relevant information has been collected and made available, it is scattered and often difficult to access. One of the main conclusions of the Quality Shipping Conference held at Lisbon in June 1998 was a unanimous call from the participants, representing the whole range of industry professionals (including shipowners, cargo owners, insurers, brokers, classification societies, agents and port and terminal operators), to make such information more accessible. In response, the Commission of the European Communities and the maritime authorities of a number of countries in 2001 inaugurated an information system know as EQUASIS, with the aim of collecting existing safety-related information from both public and private sources and making it available on the Internet. A given ship's history as presented on the EQUASIS web site, www.equasis.org, includes information on its registry, classification and Protection and Indemnity (P&I) cover, port State control details and any deficiencies discovered, manning information, etc.

156. Measures adopted by IMO to improve the effective implementation of international rules and standards have focused on strengthening the management

39. See n. 34.

of shipping companies and assisting flag States in assessing their performance. IMO also provides technical assistance to individual States upon request (see A/55/61, paras. 225–226) and has been very active in strengthening port State control. Recently, the organization has also been considering new measures to improve the effective implementation of international rules and standards (see paras. 161–164), including measures to enhance the implementation of flag State responsibility relating to fishing vessels (see paras. 251–255).

International Safety Management (ISM) Code
157. The ISM Code seeks to provide a framework for shipping companies' management and operation of their fleets. It requires that a safety management system be established by "the Company," which is defined as the shipowner or any person such as the manager or bareboat charterer who has assumed responsibility for operating the ship, and specifies the responsibilities regarding marine safety and environmental legislation (see A/53/456, paras. 221–222). The Code entered into force on 1 July 1998 for passenger ships (including high-speed passenger craft), oil tankers, chemical tankers, gas carriers, bulk carriers and high-speed cargo craft of 500 gross tonnage and above. The deadline for the remaining thousands of cargo ships trading internationally is 1 July 2002.

158. Amendments to the ISM Code were adopted by MSC at its 73rd session in resolution MSC.104(73). The amendments replace the existing chapter 13 (certification, verification and control) with a new chapter 13 (certification) and additional chapters 14 (interim certification), 15 (forms of certificate) and 16 (verification); as well as a new appendix giving forms of documents and certificates. The amendments will enter into force in 1 July 2002 under tacit acceptance.

159. It is too early to assess the full impact of ISM Code implementation on the first set of ships, which had to comply with the Code by 1998, but there are signs that it has already had an effect, especially in making the management of shipping companies more aware of their responsibilities. From the commercial standpoint, there are clear indications that ISM certification provides real value.

Self-assessment of flag State performance
160. IMO recalled that the IMO Assembly at its 21st session in November 1999 had adopted resolution A.881(21) on self-assessment of flag State performance, in which it urged member Governments to assess their capabilities and performance in giving full and complete effect to the various instruments to which they were party. The resolution includes a flag State performance self-assessment form (SAF), which is intended to establish a uniform set of internal and external criteria to be used by flag States on a voluntary basis to obtain a clear picture of how well their maritime administrations are functioning and to make their own assessment of their performance as flag States. Member Governments are also encouraged to use the SAF when seeking technical assistance from or through IMO. However, the submission of a completed form is voluntary and is not a prerequisite for receiving technical assistance. The IMO Assembly invited member Governments to submit a copy of their self-assessment report to enable the establishment of a database to assist IMO in its efforts to achieve consistent and effective implementation of IMO instruments. At its 73rd session, MSC discussed in depth the features of the SAF database to be maintained by the IMO secretariat.

Consideration of new measures
161. MSC at its 73rd session considered a joint submission by Australia, Denmark, Italy, Norway, Poland, Portugal, Singapore, Sweden, and the European Commission (MSC 73/8/3) (see also A/55/61, para. 88) explaining why IMO should address the invitation in paragraph 35 (a) of decision 7/1 adopted by the Commission on Sustainable Development in 1999 to develop binding measures to ensure that ships of all flag States met international rules and standards so as to give full and complete effect to UNCLOS, especially article 91, as well as the provisions of relevant IMO conventions. After considerable discussion, the Committee decided to instruct its Subcommittee on Flag State Implementation to consider the Commission's request under the following terms of reference: development of measures to ensure that flag States give full and complete effect to IMO and other relevant conventions to which they are party so that the ships of all flag States meet international rules and standards; consideration of the form such measures should take and how that form would relate to applicable IMO instruments.

162. The Subcommittee on Flag State Implementation, at its 9th session in February 2001, noting that no proposals had been submitted to it, invited members to submit comments and proposals at its 10th session in 2002 to enable it to consider the above-mentioned request of MSC.

163. When a ship transfers from one flag to another, the receiving flag State must have available to it all the necessary information to prevent the change of flag being used as a means of evading compliance with applicable regulations and standards.[40] In a document submitted to the Subcommittee at its 9th session, the United Kingdom proposed five principles for incorporation in an IMO Assembly resolution, against which the transfer of ships might be considered. The document suggests, *inter alia,* that prior to transfer, the "losing" State must advise the "gaining" State of any outstanding issues pertaining to the certificate the vessel has been issued or to any exemptions that may have been granted. The gaining State must then be satisfied, based on survey, that the vessel meets all relevant international standards. Once these conditions are satisfied, appropriate certificates, issued by or under the authority of the gaining State, can be provided to the vessel and the vessel may be deleted from one register and entered onto the other.[41] The Subcommittee agreed on the need to establish principles against which the transfer of ships might be considered and also agreed that some of the principles in the United Kingdom document could form the basis for developing such principles.[42]

164. The need for revision and improvement in the practices of registration of ships in order to avoid cases of double registration and the registration of so-called "phantom ships" (see paras. 180 and 201) was raised by Norway in its submissions to MSC at its 73rd session and to the Subcommittee at its 9th session.[43] MSC agreed to refer the matter to the Subcommittee for detailed consideration in the context

---

40. Introductory remarks by the Secretary-General of IMO to the 9th session of the Subcommittee on Flag State Implementation.
41. FSI 9/5/1.
42. Draft report of the 9th session of the Subcommittee, FSI 9/WP.7, paras. 3.5–3.7.
43. MSC 73/14/5. In its submission to the Subcommittee, Norway provided information on the procedure for registration of ships in Norway aimed at avoiding double registration; FSI 9/5.

of its review of the draft Assembly resolution it had prepared on measures to prevent the registration of phantom ships. At its 9th session, the Subcommittee considered the draft Assembly resolution prepared by MSC and agreed to restrict its scope to "phantom ships."[44]

165. In the context of fishing vessels, and in order to prevent dual registration, States members of the zone of peace and cooperation of the South Atlantic have undertaken the commitment to cooperate among themselves in exchanging information on the registry of fishing vessels flying their flags (see A/55/476, p. 2, para. 5).

5. Port State Control

166. IMO recalled that the IMO Assembly in 1999 had adopted resolution A.882(21) on amendments to the procedures for port State control, with a view to updating the comprehensive guidelines and recommendations on port State control procedures contained in resolution A.787(19) (see A/54/429, paras. 196–197).

167. During the reporting period, IMO continued its task of assisting in the implementation of the Memoranda of Understanding on Port State Control. A Workshop for Regional Port State Control Agreement Secretaries and Directors of Information Centres was held from 7 to 9 June 2000. Participants discussed harmonization and coordination of port State control procedures and exchange of information between regional Memoranda of Understanding agreements. Eight such regional agreements have been signed and are currently in operation. The Paris Memorandum of Understanding was the first to be adopted; others cover the following regions: Asia and the Pacific, Black Sea, Caribbean, Indian Ocean, Latin America, Mediterranean, and West and Central Africa (see A/54/429, paras. 199–207). The only region remaining to be covered by a Memorandum of Understanding is the Gulf region.

168. ILO reported that, as of December 2000, the Merchant Shipping (Minimum Standards) Convention, 1976 (No. 147) had been included as a relevant instrument in seven Memoranda of Understanding.

V. CRIMES AT SEA

169. Criminal activities at sea include piracy and armed robbery against ships, terrorism, smuggling of migrants, and illicit traffic in persons, narcotic drugs and small arms. They might also include violations of international rules dealing with the environment, such as illegal dumping, illegal discharge of pollutants from vessels or the violation of rules regulating the exploitation of the living marine resources, such as illegal fishing.

170. Most of the crimes that take place at sea, such as illicit traffic in narcotic drugs and psychotropic substances, smuggling of migrants, etc., are part of the broader, land-based problem of organized crime and the only way to effectively combat these crimes is for all States to cooperate at the global level. The recently adopted

---

44. FSI 9/WP.7, sect. 5.

United Nations Convention against Transnational Organized Crime, the Protocol against the Smuggling of Migrants by Land, Sea and Air, supplementing the above Convention (see para. 226) and the Protocol to Prevent, Suppress and Punish Trafficking in Persons, Especially Women and Children, also supplementing the above Convention,[45] represent major efforts by the international community to prevent and combat transnational organized crime.

171. In recognition of the importance of cooperation in the fight against crime not only at the global, but also at the regional and bilateral levels, some States have already concluded or are considering the conclusion of maritime cooperation agreements that address more than one crime.

172. At the national level, where efficient use must be made of limited resources in the area of enforcement, effective protection against all crimes at sea demands a multimission maritime expertise, since the kind of enforcement measures States can take to combat and suppress the various crimes at sea are differently regulated in various international instruments. Each case must be disposed of individually based on the complex humanitarian, diplomatic, environmental, and legal issues at stake. This poses a particular challenge to enforcement officers, who are often called upon to combat more than one crime at sea and who must therefore know what enforcement rights a State can exercise under international law for each crime and which of their national ministries must be involved.

173. As States are adopting a comprehensive and multidisciplinary approach to maritime security and are streamlining their enforcement capabilities at the national, bilateral and regional levels, they may wish to give particular attention to the importance of ensuring that all relevant national laws are in place; that there is a common understanding of what measures can be taken; that enforcement officers are trained and that the relevant ministries can work together rapidly and adopt appropriate and functional responses.

A. Piracy and Armed Robbery Against Ships

174. Acts of piracy and armed robbery against ships represent a serious threat to the lives of seafarers, the safety of navigation, the marine environment and the security of coastal States. They also impact negatively on the entire maritime transport industry, leading, for example, to increases in insurance rates and even the suspension of trade. For example, Royal Dutch/Shell operations suspended deliveries in January 2001 to an area in Papua New Guinea where armed robbers had attacked one of its oil tankers. It said it was seeking strong assurances from the authorities that such criminal actions would not recur.[46]

1. Extent of The Problem—Reports on Incidents
175. Reports on incidents of piracy and armed robbery against ships are received by the International Maritime Organization and the International Maritime Bureau of the International Chamber of Commerce and are issued periodically by those organizations.

45. The texts of the Convention and the Protocols are contained in A/55/383.
46. *The National* (Papua, New Guinea), 24 January 2001.

176. The IMO Secretariat stated that, based on the periodical reports and information it had provided, the Maritime Safety Committee at its 73rd session had expressed deep concern at the number of acts of piracy and armed robbery against ships reported to the organization during the first 10 months of 2000: a total of 314, representing an increase of 27% over the figure for the same period in 1999. The Committee also noted that the total number of reported incidents of piracy and armed robbery against ships from 1984 (when IMO began compiling relevant statistics) to the end of October 2000 had increased to 2017. From 1 January to 31 October 2000 the number of reported incidents had decreased from 32 to 23 in West Africa. In all other regions there had been an increase of incidents reported: in East Africa, from 14 to 15; in Latin America and the Caribbean, from 29 to 30; in the South China Sea, from 110 to 112; in the Indian Ocean, from 28 to 75; and in the Straits of Malacca, from 29 to 58.

177. Most of the attacks reported had occurred in territorial waters while the ships were at anchor or berthed. The Maritime Safety Committee was particularly concerned that, during the same period, 9 crew members had been killed, 5 had been reported missing and 22 had been injured; and that, in addition, one ship had sunk and two had been hijacked. Therefore, the Committee, endorsing the remark of the Secretary-General of IMO that this was a very alarming trend that needed to be addressed, once again invited Governments of flag States, port States and coastal States as well as the industry to intensify their efforts to eliminate these unlawful acts.[47]

178. According to the report of the International Maritime Bureau for 2000, the annual number of incidents of piracy and armed robbery against ships had risen by 57% as compared with 1999 and was nearly four and a half times as high as that of 1991. A total of 469 attacks on ships either at sea, at anchor or in port were reported to the Bureau during 2000; there were 307 instances of ship boardings and a total of 8 ship hijackings. The violence used in the attacks had also risen to new levels, with 72 seafarers killed and 99 injured, up from 3 killed and 24 injured the previous year. The Bureau believes that a large number of attacks remain unreported and that it expects to receive reports of additional incidents relating to 2000 in the coming months. More than a hundred incidents occurred in Indonesia. Elsewhere, the figures compiled by the Bureau show an alarming rise in incidents of piracy and armed robbery: in the Straits of Malacca, 75, compared with 2 in 1999; in Bangladesh, 55, compared with 25 in 1999; in India, 35, compared with 14 in 1999; in Ecuador, 13, compared with 2 in 1999; and in the Red Sea, 13, compared with none in 1999. One of the few areas to see a downturn in activity was the Singapore Straits (5 incidents, down from 14).[48] While the majority of the attacks had been carried out while the ships were berthed or anchored, all but one occurring in the Straits of Malacca had involved ships that were steaming, thus increasing the risk of a collision and possible pollution of the marine environment.

179. The International Maritime Bureau has identified four types of attacks carried out within the past decade, varying primarily with the region in which they occur. The first type occurs mainly in Asia, where ships are boarded with a minimum

---

47. MSC 73/21, sect. 14.

48. See article entitled "Piracy attacks rise to alarming new levels, ICC report reveals," dated 1 February 2001, on the web site of ICC at <www.iccwbo.org>.

of force unless resistance is offered and cash is taken from the ship's safe. India told MSC that 90% of the reported incidents along the Indian coast related to petty thefts. The second type occurs mainly in South America or in West Africa, where ships are attacked by armed gangs while berthed or at anchor. In these cases, there is a high degree of violence and the targets are cash, cargo, personal effects, ship's equipment, in fact anything that can be removed. The third type occurs mainly in South-East Asia, where ships are hijacked and the entire cargo and/or sometimes the vessel itself are stolen. The crew is occasionally set adrift in boats, thrown overboard or shot dead. The fourth type of attack is described as a type of maritime attack with military or political features.[49]

180. Hijackings, according to the Bureau, are the work of organized criminals since they require a degree of organization that only the international crime syndicates can muster. A hijacked ship is given a new name, repainted and given false registration papers and bills of lading, thereby creating a "phantom ship." The vessel is often put in to a port where the false identity of the vessel and cargo may escape detection. Even when identified, the hijacking gangs have been known to bribe local officials to allow them to sell the cargo and leave the port. Ships are sold and often end up in shipbreaking yards. The Bureau reports that there is evidence that organized crime is also backing some of the bands of pirates that prey on shipping in the coastal waters of Malaysia, Indonesia, the Philippines and other countries.[50]

181. The Bureau does not draw a clear distinction in its reports between an incident of petty theft, armed robbery or piracy. It defines piracy for statistical purposes as "an act of boarding or attempting to board any ship with the intent to commit theft or any other crime and with the intent or capability to use force in the furtherance of that act." The definition covers actual or attempted attacks, whether the ship is berthed, at anchor or at sea. IMO in its reports distinguishes between piracy and armed robbery, but not between petty theft and armed robbery (see para. 197).

182. The already alarmingly high number of acts of piracy and armed robbery against ships reported to IMO and the Bureau during 2000 probably does not even represent the true figure, as also noted by the Bureau in its annual report (see para. 178). In 1998, the Secretary-General of the United Nations, in his annual report to the General Assembly at its fifty-third session (see A/53/456, paras. 147 and 148), stated that the International Maritime Bureau and the International Transport Workers' Federation had expressed the view that official reports accounted only for 50% of the attacks, as shipowners were hesitant to report an incident for fear of having their ship immobilized during an enquiry (which could cost them up to $10,000 a day) and losing clients as a consequence. The insurance companies were said to settle cases discreetly and to simply increase premiums in high-risk regions. Reports of incidents would then be sent long after the incident had occurred, thus

---

49. Presentation by the International Maritime Bureau at the IMO Regional Seminar and Workshop on Piracy and Armed Robbery against Ships (India, March 2000); MSC 73/14/1, para. 28.

50. See article entitled "East Asian Governments must clamp down on piracy together," posted on the web page of ICC at: <www.iccwbo.org/home/news_archives/2000/piracy_east_asia.asp>.

frustrating the conduct of investigations by coastal States into incidents reported in their waters. While the situation has improved somewhat since 1998, under-reporting of incidents still remains a serious problem.

2. Action at The Global Level

183. The problem of piracy and armed robbery against ships has been brought to the attention of a number of fora, most notably the United Nations General Assembly and IMO, as well as the first meeting of the Consultative Process (see A/55/274, part A, issue K, paras. 45–47; part B, para. 37; and part C, para. 2 (b)), and the Meeting of States Parties to UNCLOS (SPLOS/31, para. 64).

(a) General Assembly

184. The Secretary-General of the United Nations drew attention to the problem of piracy and armed robbery against ships for the first time in his annual report on law of the sea to the General Assembly at its fortieth session in 1985 (A/40/923, para. 40) and has been including a separate section on the issue in his annual report on oceans and the law of the sea since 1993. The General Assembly addressed the problem of piracy and armed robbery against ships for the first time in its annual resolution on oceans and the law of the sea at its fifty-third session in 1998 (see resolution 53/32).

185. At its fifty-fifth session, the General Assembly had before it a note by the Secretary-General transmitting a copy of the letter addressed to him by the Secretary-General of IMO on 8 June 2000 (A/55/311, annex). The letter reported that MSC at its 72nd session, while acknowledging the positive action of the General Assembly and being appreciative of its support (as clearly demonstrated in resolution 54/31), was of the opinion that other bodies within the United Nations system might be able to provide additional assistance that would ensure that seafarers and ships could engage safely and peacefully in international maritime activities.

186. In its resolution on the item "Oceans and the law of the sea" adopted at its fifty-fifth session, the General Assembly noted the IMO Secretary-General's letter, and, as it had done the previous year in resolution 54/31, once again urged all States, in particular, coastal States, in affected regions to take all necessary and appropriate measures to prevent and combat incidents of piracy and armed robbery at sea, including through regional cooperation, and to investigate or cooperate in the investigation of such incidents wherever they occurred and bring the alleged perpetrators to justice in accordance with international law. The Assembly also repeated its call to States to cooperate fully with IMO, including by submitting reports on incidents to the organization and by implementing the IMO guidelines on preventing attacks of piracy and armed robbery. It furthermore once again urged States to become parties to the Convention for the Suppression of Unlawful Acts against the Safety of Maritime Navigation and its Protocol[51] and to ensure its effective implementation. The General Assembly moreover recommended that coordination and cooperation in combating piracy and armed robbery at sea should be one of the main areas of focus of consideration at the second meeting of the Consultative Process.

51. As at 31 January 2001, 52 States had ratified or acceded to the Convention and 48 States to the Protocol.

(b) Measures taken by IMO, as reported by the IMO Secretariat

187. The IMO Secretariat reported that MSC at its 73rd session had taken note of General Assembly resolution 55/7, in particular in connection with the need to take appropriate measures to prevent and combat incidents of piracy and armed robbery at sea. The aim of its contribution was to provide comprehensive information on the work being undertaken by IMO in this area.

Background
188. In 1993, the IMO Assembly, mindful of the duty of States to cooperate in the repression of piracy as stipulated in article 100 of UNCLOS, adopted resolution A.738(18) on measures to prevent and suppress piracy and armed robbery against ships. In the resolution, Governments were urged to recommend to vessels registered under their flags to take precautionary measures to avoid piratical attacks and to adopt procedures to be followed if they occurred, including in particular reporting immediately to the nearest or other appropriate rescue coordination centre and, if possible, to the coastal State as well as to the flag State concerned any such attacks or attempted attacks; and to establish and maintain close liaison with neighbouring States to facilitate the apprehension and conviction of all persons involved in piratical attacks.

189. The IMO Assembly also urged Governments of coastal States to make arrangements with coast earth stations to ensure prompt delivery of reports of piratical attacks to the authorities concerned. It invited Governments to consider using surveillance and detection techniques and acquiring the capability to prevent and respond to piratical attacks.

190. In the same resolution, IMO invited Governments to develop and continue cooperation agreements with neighbouring States, as appropriate, including the coordination of patrol activities and of the response by rescue coordination centres. Governments were requested to instruct national centres or other agencies involved, on receiving a report of an attack, to promptly inform the local security forces so that contingency plans might be implemented and to warn ships in the immediate area of the attack.

191. The IMO Assembly also requested the Secretary-General of IMO to seek means of providing support from donor countries and international financial institutions to Governments requesting technical and financial assistance in the prevention and suppression of piratical attacks. Finally, the Maritime Safety Committee was enjoined to keep the issue under continuous review and it has accordingly been included in IMO's Long-Term Work Plan.

192. IMO reported that, on the basis of resolution A.738(18), it had developed a comprehensive anti-piracy strategy consisting of compilation and distribution of periodical statistical reports, piracy seminars and field assessment missions to regions affected by piracy and the preparation of a code of practice for the investigation and prosecution of the crime of piracy and armed robbery against ships.

Periodic statistical reports
193. IMO compiles and distributes monthly, quarterly and annual reports on piracy and armed robbery against ships submitted by Governments and international organizations. Monthly reports list all incidents reported to the organization. Quarterly reports are composite reports accompanied by an analysis, on a regional basis, of

the situation and an indication as to whether the frequency of incidents is increasing or decreasing and advising on any new feature or pattern of significance. Information on the number of incidents reported to IMO during the first 10 months of 2000 is provided in paras. 176–177.

Seminars, workshops and missions
194. IMO arranges seminars and workshops to explain the problem of piracy and armed robbery and the organization's recommendations on how to deal with them. In addition, it carries out field missions to assess the actions Governments take to implement the inputs of the anti-piracy projects. Mission members examine, together with the responsible governmental representatives, what measures the national authorities responsible for anti-piracy activities have taken to implement the relevant IMO guidelines, where such measures have not been successful and what has impeded their implementation and, eventually, how IMO might assist in overcoming any difficulties encountered in the process.

195. Such missions also include advisory services and "tabletop" exercises at the national level to assess and evaluate the results of previous relevant IMO activities. IMO noted that seminars, workshops and missions could only be organized if Governments and governmental and nongovernmental organizations provided the necessary financial support for them.

Preparation of a code of practice for the investigation and prosecution of the crime of piracy and armed robbery against ships
196. MSC at its 73rd session approved the IMO draft Code of Practice for the Investigation of the Crimes of Piracy and Armed Robbery Against Ships.[52] The Code will be considered for adoption by the IMO Assembly at its 22nd session to be held from 19 to 30 November 2001.

197. The purpose of the draft Code of Practice is to provide IMO members with an aide mémoire to facilitate the investigation of the crimes of piracy and armed robbery against ships. The draft Code adopts the definition of piracy contained in article 101 of UNCLOS. Armed robbery against ships is defined as any unlawful act of violence or detention, or any act of depredation, or threat thereof, other than an act of piracy, directed against a ship or against persons or property on board, within a State's jurisdiction over such offences. The draft thus combines the geographical scope of jurisdiction over piracy, as laid down in UNCLOS with the jurisdiction over unlawful acts, as laid down in the Convention for the Suppression of Unlawful Acts against the Safety of Maritime Navigation (1988 SUA Convention) and its Protocol (SUA Protocol).

198. The draft Code includes a recommendation that States take the necessary measures to establish their jurisdiction over the offences of piracy and armed robbery against ships, including adjustment of their legislation, if necessary, to enable the apprehension and prosecution of persons committing such offences. States are explicitly encouraged to ratify, adopt, and implement UNCLOS and the SUA instruments.

199. In order to encourage masters to report all incidents of piracy and armed robbery against ships, the draft Code states that coastal and port States should make

---

52. Circulated as MSC/Circ.984.

every effort to ensure that the masters and their ships are not unduly delayed or burdened with additional costs related to such reporting. Coastal States are encouraged to enter into bilateral or multilateral agreements to facilitate the investigation of piracy and armed robbery against ships.

200. The draft Code also contains provisions for the specific training of investigators intervening in acts of piracy or armed robbery during or after the event. It further lays down the main principles for an investigative strategy and lists the responsibilities of the investigators in such matters as preservation of life, prevention of the escape of offenders, warnings to other ships, protection of crime scenes and the securing of evidence. A final chapter on the investigation lists measures to be taken to establish and record all relevant facts, record individual witness accounts, conduct detailed forensic examinations of scenes and, searches of intelligence databases and oversee the distribution of information and intelligence to appropriate agencies. In accordance with the principle of proportionality informing the draft, action to be pursued should be proportionate to the crime committed and consistent with the laws that were violated.

(c) "Phantom ships"

201. IMO also reported that, in order to reduce hijackings and the number of "phantom ships," i.e., ships with fraudulent registration, certification and identification, MSC has begun the consideration of a draft IMO Assembly resolution encouraging flag States to ensure that proper checks were made when registering a ship. The Subcommittee on Flag State Implementation at its 9th session (February 2001), prepared a draft resolution on "Measures to prevent the registration of 'phantom ships'," to be submitted for adoption by the IMO Assembly at its 22nd session (November 2001).

3. Action at the Regional Level

202. The strengthening of regional cooperation is imperative to prevent and respond effectively to incidents of piracy and armed robbery against ships. In this regard, the IMO regional seminars and workshops for the South and Central American and Caribbean region (Brazil, October 1998), the South-East Asian region (Singapore, February 1999), the West African region (Nigeria, October 1999), and for selected countries in the Indian Ocean region (India, March 2000) had proved very valuable not only in reviewing the effectiveness of any countermeasures the participating countries had put in place, but also for regional cooperation in general. Since the conclusion of the IMO seminars and workshops, efforts have continued, in particular, among States in South-East Asia, to advance regional cooperation. Two high-level international conferences on combating piracy and armed robbery were held in Tokyo in March and April 2000 and resulted in the endorsement of the Tokyo Appeal and the adoption of the "Asia Anti-Piracy Challenges 2000," and a Model Action Plan.[53] An ASEAN Regional Forum Workshop on Anti-Piracy was held in India in October 2000 and an Experts Meeting on Combating Piracy and Armed

---

53. For the texts, see IMO document MSC 73/INF.4.

Robbery against Ships was held in Malaysia in November 2000. The South-East Asian Programme in Ocean Law, Policy and Management (SEAPOL) Inter-Regional Conference on Ocean Governance and Sustainable Development in the East and Southeast Asian Seas: Challenges in the New Millennium (Thailand, 21–23 March 2001) was to devote one of its sessions to piracy and law enforcement, to discuss, *inter alia*, legal issues in piracy control, and piracy and the challenge of cooperative security and enforcement policy.

203. Cooperation in other regions is also being pursued. The High-level Meeting of five Coast Guard Agencies in the North-west Pacific Region was held in December 2000 to discuss ways of combating the illicit traffic in drugs and guns and piracy in the region. According to the Japanese news service Kyodo, national coast guard chiefs from Japan, the Republic of Korea, the Russian Federation and the United States attended the meeting.

204. Some of the main problem areas in dealing with pirates and armed robbers revealed as a result of the IMO expert missions[54] and the regional seminars and workshops in Brazil and Singapore were: the current economic situation in the regions concerned; certain resource constraints on law enforcement agencies; lack of communication and cooperation between the agencies involved; the length of the coastal State's response time following the affected ship's report of an incident; general problems of ship reporting; timely and proper investigation into reported incidents; the prosecution of pirates and armed robbers when apprehended; and lack of regional cooperation. The ASEAN Regional Forum (ARF) Workshop in October 2000 concluded that there was an urgent need for close coordination and cooperation among the maritime authorities and the law enforcement agencies of the States concerned to effectively curb piracy and armed robbery against ships; that piracy posed a transnational threat necessitating bilateral and regional arrangements among the ARF member States to unify measures to combat piracy; and that efficient exchange of information and intelligence was necessary for the successful conclusion of the investigation and prosecution of apprehended pirates.[55] The participants at the experts meeting in Malaysia stressed the importance of the last-named issue and agreed to pursue the matter further; they also considered that it was also necessary to standardize the format for ships' reporting to enforcement agencies, to enable immediate action by the enforcement agencies.[56]

205. Several of the regional meetings also discussed and agreed to further pursue the working definitions of piracy and armed robbery against ships. In this regard, it should be noted that piracy is defined in article 101 of UNCLOS and armed robbery against ships has recently been defined in the IMO draft Code of Practice (see para. 197).

4. Recommended Actions for Governments and the Industry

206. An act of piracy or armed robbery against ships affects different national interests: that of the flag State of the ship; the State in whose maritime zone the attack

54. In 1998, IMO conducted two expert missions: one to the Philippines, Malaysia and Indonesia, and another to Brazil.
55. For the oral report on the outcome of the meeting, see MSC 73/21, para. 14.7.
56. *Ibid.*, para 14.8.

took place; the State of suspected origin of the perpetrators; the State of nationality of persons on board; the State of ownership of cargo; and maybe also the State where the crime was prepared, planned, directed or controlled. In the Tokyo Appeal it was acknowledged that the issue of piracy and armed robbery against ships could not be resolved if the relevant authorities, the flag States and other substantially interested States and coastal States/port States, each took measures independently based on their individual positions; it could be tackled effectively only when such parties mutually coordinated and cooperated in a manner transcending their individual positions.

(a) Recommended actions for shipowners, ship operators, shipmasters and crews

207. Preparedness and action by shipping companies themselves is fundamental to the prevention of piracy and armed robbery against ships, for which IMO, the International Maritime Bureau, the Oil Companies International Marine Forum, the International Chamber of Shipping and the International Shipping Federation have all issued guidance materials.[57] The IMO Guidance to Shipowners and Ship Operators, Shipmasters and Crews on Preventing and Suppressing Acts of Piracy and Armed Robbery against Ships (MSC Circular 623/Rev.1, dated 16 June 1999) outlines steps that should be taken to reduce the risks of an attack and possible responses to them. It states that ships need to have a ship security plan or an action plan detailing the actions to be taken in case of an attack. The MSC Circular highlights, *inter alia,* the vital need to report attacks, successful as well as unsuccessful ones, to the authorities of the relevant coastal State, generally the Rescue Coordination Centres, and to the ships' own maritime administration. Such reports should be made as soon as possible to enable necessary action to be taken. The Model Action Plan adopted at the Tokyo Conference (see para. 202) also underlined the importance of filing both immediate and post-attack reports. It further proposed the re-enforcement of self-protection measures on board ships, including the examination of the use of ship-position reporting technology and enhanced defensive equipment. The International Maritime Bureau has said that shipowners should consider the installation of SHIPLOC, a low-cost vessel tracking system, which claims to be capable of instant location of a vessel.

(b) Recommended actions for Governments

208. Most attacks against ships occur in the territorial sea and therefore do not constitute piracy as defined in UNCLOS. According to the International Maritime Bureau, "what makes piracy a tempting crime is the difficulty of effective law enforcement, and the unwillingness of many countries to prosecute pirates caught in their own territorial waters for acts of piracy committed under another country's jurisdiction." The Bureau uses the expression "piracy" also to describe acts of armed robbery against ships.

---

57. International Maritime Bureau, Special Report on Piracy and Armed Robbery (March 1998); Oil Companies International Marine Forum, Piracy and Armed Robbery at Sea; International Chamber of Shipping/International Shipping Federation, Pirates and Armed Robbers—A Master's Guide (3rd ed., 1999).

(i) National action plans
209. The IMO Recommendations to Governments for Preventing and Suppressing Piracy and Armed Robbery against Ships (MSC Circular 622/Rev.1, dated 16 June 1999) set out the necessary actions to be implemented by Governments within areas identified as affected by acts of piracy and armed robbery. Some of these recommendations are also included in the draft Code of Practice. Coastal States/port States are recommended to develop action plans for preventing an attack as well as steps to take in the event of an attack. Because of the possibility of collisions or groundings as a result of an attack, coastal/port States are also recommended to develop plans to counter any subsequent oil spills or leakages of hazardous substances the ships may be carrying. This is especially important in areas of restricted navigation, for example, in straits used for international navigation, such as the straits of Malacca and Singapore. It should be noted that States may already have adopted measures for dealing with a pollution incident, either at the national level or in cooperation with other States, in implementation of the International Convention on Oil Preparedness, Response and Cooperation.

(ii) Rapid responses to reported incidents
210. The Model Action Plan adopted at the Tokyo Conference (see para. 202) proposed, *inter alia,* that each State should establish a system for communication and collaboration between relevant authorities within a Government to ensure that comprehensive and functional measures are taken in response to reports of incidents of piracy and armed robbery against ships.
    211. In order to ensure efficient communication and cooperation between various agencies and a rapid response after an incident has been reported to the coastal State, including the promulgation of threat warnings, MSC Circular 622/Rev.1 recommends that States adopt an incident command system and incorporate therein existing mechanisms for dealing with other maritime security matters, e.g., illicit traffic in narcotic drugs and terrorism, to enable the efficient use of limited resources. It is also recommended that States develop procedures for rapidly relaying alerts from the receiving communication centre to the entity responsible for taking action. The IMO regional seminar and workshop held in India in March 2000 highlighted the potential use of vessel traffic separation (VTS) information in piracy and armed robbery attack situations. It also recommended that IMO should develop harmonized procedures and guidelines on communication means for alerting other ships in the area.

(iii) Investigations of incidents and exchange of information/intelligence
212. MSC Circular 622/Rev.1 and the draft Code of Practice state that, to encourage masters to report all incidents, coastal States should make every effort to ensure that masters and their ships are not unduly delayed. They should clearly establish an entity responsible for conducting investigations into reported incidents. The IMO draft Code lays down the principles for an investigative strategy. Coastal States are encouraged, where appropriate, to enter into bilateral or multilateral agreements to facilitate the investigation of piracy and armed robbery.
    213. The draft Code also provides that it is important to involve relevant organizations (e.g., Interpol, International Maritime Bureau) at an early stage, where appropriate, to take account of the possibility that transnational organized crime may

be involved. Additionally, an important product of an effective investigation, even if it does not lead to any arrests, should be the generation of intelligence, and systems should be in place to ensure that potentially useful intelligence information is disseminated to all appropriate parties. The Model Action Plan adopted at the Tokyo Conference proposed the establishment of an international network for the exchange and analysis of information. It was deemed important that in both intelligence and operational terms piracy should not be viewed in isolation, as pirates are most likely involved in other crimes, such as smuggling of migrants and illicit traffic in narcotic drugs. Anti-crime measures should be linked to minimize duplication of efforts.

(iv) Bilateral/regional/multilateral cooperation
214. MSC Circular 622/Rev.1 recommends that States sharing borders in areas threatened by piracy and armed robbery establish bilateral/regional cooperation arrangements to provide, *inter alia*, for the coordination of patrol activities by both ships and aircraft. Further development of such cooperation may involve an agreement to facilitate coordinated response at the tactical as well as the operational level. Such an agreement would specify how information would be disseminated; establish a regional incident command system; set policies for joint operations and entry and pursuit into each others territorial seas; establish links between the entities involved in all maritime security matters, etc. An example of such a regional agreement is appended to the Circular.

215. Cooperative arrangements may also be established with States outside the region, either at the bilateral level (a recent example includes an offer by Sweden to lend Malaysia four assault boats to help strengthen security in its waters)[58] or at the multilateral level. For example, Japan's coast guard patrol vessels recently conducted joint exercises with the maritime authorities of India and Malaysia, and at the ASEAN workshop Japan offered to provide training to nonmilitary personnel.[59] According to a 3 January 2001 Kyodo News Service report, "Japan is mulling an anti-piracy pact with [ASEAN] to allow Japanese maritime authorities to join international patrols in piracy-prone waters in South-East Asia." Government sources were quoted as saying that they expected the Diet to pass laws to allow direct involvement by Japanese vessels in anti-piracy patrols and in measures against smugglers and illegal immigrants.

216. At the IMO regional seminar and workshop held in India, the IMO Secretariat reported that some Governments and shipowners had suggested that an international naval force should be established under the auspices of the United Nations to patrol danger areas, while others had urged coastal States to take more action (MSC 73/14/1, para. 30).

(v) Jurisdiction
217. Both the IMO recommendations in MSC Circular 622/Rev.1 and the IMO draft Code of Practice recommend to States to take the necessary measures to establish their jurisdiction over the offences of piracy and armed robbery against ships, including adjustment of their legislation, if necessary, to enable apprehension and

58. *Bernama* news agency web site, Kuala Lumpur, 23 October 2000.
59. MSC 73/21, para. 14.7.

prosecution of persons committing such offences. States are explicitly encouraged in the draft Code to ratify, adopt and implement the 1988 SUA Convention and its Protocol (SUA Protocol).

218. As States are adjusting their legislation and establishing their jurisdiction over the offences of piracy and armed robbery against ships, it is important for there to be a common understanding among States of the applicable enforcement rights they have under international law with respect to acts of piracy and armed robbery against ships. It is also important for States to establish effective penalties in their laws.

219. The basic enforcement rights with respect to acts of piracy are contained in UNCLOS. Piracy is defined in its article 101, and article 105 grants States universal jurisdiction on the high seas to seize a pirate ship or aircraft, or a ship or aircraft taken by piracy and under the control of pirates, and arrest the persons and seize the property on board. The same rights apply also in the exclusive economic zone by virtue of article 58, paragraph 2. The definition of piracy in UNCLOS excludes the territorial sea. However, article 25 permits the coastal State to take the necessary steps in its territorial sea to prevent passage which is not innocent. In accordance with article 27, the coastal State can exercise criminal jurisdiction against foreign ships that engage in acts that disturb the peace of the country or the good order of the territorial sea, or if the consequences of the crime extend to the coastal State.

220. Armed robbery against ships, as defined in the draft Code (see para. 197), constitutes an offence under article 3 of the 1988 SUA Convention, article 6 of which requires a State party to establish its jurisdiction when the offence is committed against or on board a ship flying its flag, in its territory, including its territorial sea, or by one of its nationals. The 1988 SUA Convention also permits a State, provided it has established its jurisdiction and notified IMO thereof, to exercise its jurisdiction if the offence has been committed by a stateless person whose habitual residence is in that State; or if one of its nationals is seized, threatened, injured or killed; or if the offence has been committed in an attempt to compel the State to do or abstain from doing any act. A State must establish its jurisdiction if the alleged offender is present in its territory and it has not extradited him to any of the States parties that have established their jurisdiction in accordance with the 1988 SUA Convention.

221. Unlike the 1988 SUA Convention, a State party is not required under the 2000 United Nations Convention against Organized Transnational Crime to establish its jurisdiction if the crime was committed against the ship or by one of its nationals. It is only required to take the necessary measures to establish its jurisdiction when the offence has been committed on board a ship flying its flag, or in the territory of that State (article 15). "Territory" is not defined, but is presumed to include the territorial sea. Article 4 of the Convention provides that nothing in the Convention entitles a State party to undertake in the territory of another State party the exercise of jurisdiction and performance of functions reserved exclusively for the authorities of the other State by its domestic law. States may wish to address the relationship between the provisions on jurisdiction in the 2000 United Nations Convention against Organized Transnational Crime and those in the 1988 SUA Convention and clarify which provisions they should use to suppress an act of armed robbery against ships, which also constitutes an organized crime.

222. In order to promote a common understanding of States' enforcement rights under international law and provide guidance to them in drafting their na-

tional laws, it may be useful to identify the elements that should be included in national legislation, or, alternatively, to develop model national laws. The Comité Maritime International has been working on the development of a model national law on piracy and maritime violence and presented the results of its deliberations to the CMI International Conference in February 2001.[60]

(vi) Technical assistance
223. Some States may have difficulties in effectively implementing the recommendations contained in MSC Circular 622/Rev.1 and the IMO draft Code of Practice because they lack the necessary equipment and trained personnel. Lack of equipment, such as patrol boats, radar and radio communications, and of trained personnel have been identified as major obstacles to the functioning of the effective machinery to combat piracy and armed robbery at sea (IMO regional seminar and workshop, India, March 2000). The assistance other States can provide to either individual States or to an entire region affected by acts of piracy and armed robbery can take various forms, such as training of personnel, provision of equipment or funds, etc. As noted in paragraph 222, assistance to States can also include legal advice in the drafting of national laws, either through seminars, or by preparing elements for inclusion in national laws.

B. Smuggling of Migrants

224. Global statistics show that the problem of smuggling of migrants is increasing. The demands for smugglers' services are increasing as potential emigrants become more desperate and less concerned with safety. As an example, on 16 February 2001, a Cambodian-registered vessel with a Syrian ship-owner ran aground and was abandoned off the coast of the Côte d'Azur of France with 800 Kurds on board. Poverty in the developing world and the tightening of legal immigration possibilities in many of the developed countries are among the root causes of much contemporary migration. Ever more smugglers are treating their cargo poorly, arranging transport that could under no circumstances be called safe and illegal migrants are themselves taking great risks. For example, it has been estimated that 120 people lost their lives attempting to cross the Strait of Gibraltar illegally in the first six months of 2000,[61] and as with all data, these numbers only represent those cases that have come to the attention of the authorities.[62] In response to the increased migration across the Strait, Spain has been working on a strategic plan for Sub-Saharan Africa, the main focus of which will be aid for development cooperation.[63]

225. The Commercial Crime Services of the International Chamber of Commerce have pointed out that the ingenious methods already in use for smuggling of migrants by sea are becoming more sophisticated. One of the preferred ways of

60. For the text of the model national law, see *CMI Yearbook 2000—Singapore I: Documents for the Conference,* Report of the Joint International Working Group, annex A.
61. *Migration News Sheet,* June 2000, p. 6.
62. Extract from *Quarterly Bulletin,* No. 21, summer 2000, published by the International Organization for Migration and available on its web site at <www.iom.int/iom>.
63. *RNE Radio 1,* Madrid, 3 October 2000.

circumventing normal barriers is by stowing away on a container ship; another is by signing on as a crew member. Another growing problem is the use of false documents by illegal immigrants. The problem has escalated as smuggling of migrants has become a big business. The International Group of P&I Clubs (protection and indemnity) currently spends approximately $10 million annually on fines and costs relating to illegal migrants. The actual cost is much higher, as shipowners pay substantial amounts themselves, due to higher deductibles on their protection/indemnity cover.[64]

226. IMO recalled the provisions of IMO Assembly resolution A.867(20) on combating unsafe practices associated with the trafficking or transport of migrants by sea, in which Governments were requested to detain all unsafe ships and report pertinent information to IMO (see A/55/61, para. 109). The Maritime Safety Committee at its 73rd session (November/December 2000) agreed to implement a reporting procedure similar to that for acts of piracy and armed robbery against ships to keep track of incidents of unsafe practices associated with the trafficking or transport of illegal migrants by sea. Governments and international organizations were urged to promptly report any such practices brought to their attention. The reports should include, where available, ship and shipowners' details, voyage details, date, time, and position of the incident, a description of the incident and measures taken, and information concerning the migrants, including number, nationality, break down by sex, and whether any were minors. Details of the reported incidents would be issued biannually in an IMO circular.

227. IMO also recalled that MSC at its 70th session (December 1998) had approved an MSC Circular (MSC/Circ.896) advising Governments what "Interim Measures for Combating Unsafe Practices Associated with the Trafficking or Transport of Migrants by Sea" they could take pending the entry into force of a convention against transnational organized crime and a protocol on smuggling of migrants (see also A/54/429, paras. 223–228, and A/55/61, paras. 111–113).

228. In this regard, attention is drawn to the adoption in 2000 of a legally binding instrument aimed at preventing and combating the smuggling of migrants by land, sea and air. The United Nations Centre for International Crime Prevention reported that the 2000 United Nations Convention against Organized Transnational Crime and the Protocol against the Smuggling of Migrants by Land, Sea and Air, supplementing the above Convention, had been finalized by the Ad Hoc Committee for the elaboration of the Convention at its tenth and eleventh sessions, in July and October 2000, respectively. Both instruments had subsequently been adopted by the General Assembly on 15 November 2000 (resolution 55/25), and opened for signature at Palermo, Italy, on 12 December 2000. At the Palermo Conference, 128 States had signed the Convention, and 78 States the Protocol. The Convention and Protocol contain comprehensive measures against all forms of transnational organized crime.[65] Part II (articles 7–9) of the Protocol deals specifically with issues arising from the smuggling of migrants by sea. The provisions deal with the powers and

---

64. See article on "Warning to ship agents against conspiracies to ship illegal immigrants," at the web site of ICC, at: <www.iccwbo.org/ccs.news_archives/2000/illegal_immigrants.asp>.

65. The texts of the Convention and the Protocol are contained in document A/55/383, annexes I and III.

procedures for dealing with vessels suspected of being engaged in the smuggling of migrants and the protection of the safety, security, rights, and other interests of the vessels, those on board, flag States and other interested States. The United Nations Centre for International Crime Prevention reported that those provisions had been based on the 1988 United Nations Convention against the Illicit Traffic in Narcotic Drugs and Psychotropic Substances, UNCLOS and IMO Circular MSC/Circ.896.

229. In addition to the information provided by the United Nations Centre for International Crime Prevention, it should be noted that article 7 of the Protocol, entitled "Cooperation," follows closely the wording of article 17, paragraph 1, of the 1988 Convention. Under article 7, States parties are called upon to cooperate to the fullest extent possible to prevent and suppress the smuggling of migrants by sea, in accordance with the international law of the sea, which according to the interpretative notes is to be understood as including UNCLOS as well as other relevant international instruments.[66]

230. The provisions in the Protocol are intended to cover vessels "engaged" both directly and indirectly in the smuggling of migrants. Of particular concern during the negotiations was the inclusion of vessels ("mother ships") that transport smuggled migrants on open ocean voyages, but which are sometimes not apprehended until after the migrants have been transferred to smaller local vessels.

231. Article 8 (Measures against the smuggling of migrants by sea) follows closely the wording of article 17, paragraphs 2, 3, 7 and 8, of the 1988 Convention. It permits a State party, *inter alia,* to board, search or take other appropriate action against a vessel suspected of being engaged in the smuggling of migrants by sea. What is new, *inter alia,* is the incorporation in article 8, paragraph 1, of the Protocol of the reference to the right of a State under article 110, paragraph 1(e), of UNCLOS to take measures against a ship that, though flying a foreign flag or refusing to show a flag, is in reality of the same nationality as the State party concerned. A new paragraph has also been included in article 8 that specifically addresses the rights of States to take measures against ships without nationality. In this connection, attention is drawn to past reports of the Secretary-General that highlighted the need for States to have national legislation in place granting enforcement jurisdiction over ships without nationality (see, e.g., A/54/429, para. 221).

232. Article 9 (Safeguard clauses) of the Protocol follows closely the wording of article 17, paragraphs 5, 10 and 11, of the 1988 Convention, as well as articles 94, paragraph 1, and 110, paragraph 3, of UNCLOS. What is new, *inter alia,* is the requirement for States parties, when taking measures against a vessel, to ensure the humane treatment of the persons on board, and to ensure within available means that any measure taken is environmentally sound.

233. The interpretative notes for the official records of the negotiation of the Protocol state that the *travaux préparatoires* should indicate that it is understood that the measures set forth in chapter II of the Protocol cannot be taken in the territorial sea of another State, except with the authorization of the coastal State concerned. During the negotiations it was felt that this principle did not need to be restated in the Protocol, since it was well enshrined in the law of the sea.

---

66. See the interpretative notes for the official records (*travaux préparatoires*) of the negotiation of the Convention and the Protocols, A/55/383/Add.1.

234. Even though the Protocol has not yet entered into force, the question may be raised as to whether it is advisable for States to continue using the earlier IMO circular (see para. 227) as a basis for their action to prevent and suppress smuggling of migrants by sea, as opposed to the Protocol. Not only is the Protocol a binding legal instrument, which was adopted by the United Nations General Assembly and has already received a large number of signatures, but it also is linked to the 2000 United Nations Convention against Transnational Organized Crime, as well as to the Protocol to Prevent, Suppress and Punish Trafficking in Persons, Especially Women and Children, supplementing that Convention.

C. Stowaways

235. According to the latest available report on stowaway incidents issued by IMO, 231 incidents were reported between 1 May to 30 September 2000 (FAL/Circ.61), bringing the total number reported since 1998 to 1,170.

236. At its 28th session (30 October–3 November 2000), the IMO Facilitation Committee noted that the continuing high number of stowaway incidents indicated that measures taken in ports and on board ships to prevent stowaways gaining access to ships ought to be strengthened. It agreed that the IMO Guidelines on the Allocation of Responsibilities to Seek the Successful Resolution of Stowaway Cases, adopted by the IMO Assembly in its resolution A.871(20) in 1997 (see A/55/61, paras. 115–116) were not strong enough to prevent stowaway cases and decided to incorporate stowaway regulations in the Convention on Facilitation of International Maritime Traffic, 1965. To that end, the Committee prepared and approved amendments to the Convention for adoption at its next session.[67] It also decided to scrutinize the Guidelines at a later stage to strengthen their content and invited Governments to submit proposals to the Committee at its next session.

237. IMO pointed out that the Guidelines advocate close cooperation between shipowners and port authorities and establish in detail the responsibilities of the master, the shipowner or operator, the country of the first scheduled port of call after the discovery of the stowaway (the port of disembarkation), the country where the stowaway first boarded the ship, the stowaway's apparent or claimed country of nationality, the flag State of the vessel and any countries of transit during repatriation. IMO also drew attention to the provisions of IMO Circular MSC/Circ.896 (see para. 227).

D. Illicit Traffic in Narcotic Drugs and Psychotropic Substances

238. The United Nations International Drug Control Programme (UNDCP) stated that fundamental to the full implementation of the cooperative provisions of article 17 of the 1988 Convention was the need for States parties to designate a competent national authority, or as appropriate authorities, having the legal power to grant or deny authorization to another State party to board, search or take other appropriate

---

67. See the report of the Facilitation Committee at its 28th session, document FAL 28/19, annex 4.

action against a vessel suspected of illicit drug trafficking. States must, at the time of becoming a party to the Convention, designate an authority or authorities to receive and respond to requests and must notify the Secretary-General of the designation. The competent national authority needs to be able to respond expeditiously both to the request for verification of registry and to the request for consent to take action. UNDCP pointed out that full implementation of the cooperative provisions of article 17 has been hampered by the inability to verify registry quickly or by the fact that competent authorities have not been identified or lack the necessary legal powers to quickly grant or deny consent. Often small pleasure craft or fishing boats are not registered or States lack a single central registry.

239. The Commission on Narcotic Drugs in its resolution 43/5, entitled "Enhancing multilateral cooperation in combating illicit trafficking by sea," adopted at its forty-third session (March 2000), recognized the increasing prevalence of illicit traffic by sea of narcotic drugs and psychotropic substances. It encouraged Governments to develop regional agreements where appropriate and requested UNDCP to support the negotiation of such agreements. The Commission supported efforts of UNDCP to facilitate the coordination of practical ways to ensure effective suppression of maritime drug trafficking and encouraged States to review regularly and to communicate the names of their competent authorities to the United Nations and to respond expeditiously to requests made pursuant to article 17 for verification of nationality and for consent to board, search, and, if evidence of involvement in illicit traffic was found, to take appropriate action with respect to the vessel, persons, and cargo on board.

240. An informal open-ended working group on maritime cooperation against illicit drug trafficking by sea was convened by UNDCP from 5 to 8 December 2000. The group examined current trends in illicit drug trafficking at sea, including the use of "go-fast" boats (speedboats), particularly in the Caribbean, and the use of containers to smuggle drugs. Both methods were seen as presenting special difficulties for law enforcement: "go-fast" boats because of their speed and the fact that they were rarely registered in any State, and containers because of the enormous volume of legitimate commerce by container and of the difficulty of searching them. The group also examined recent regional and subregional initiatives, including the consultations on a possible regional convention on maritime drug law enforcement for the Caribbean.

241. The working group identified a number of discussion points to bring to the attention of the Commission on Narcotic Drugs. Among the ideas proposed were the provision of equipment and training to developing countries, the exchange of information and intelligence, and cooperation between States in conducting joint operations and pooling equipment in appropriate circumstances. Consideration was given to the possibility of developing model reference forms and a user-friendly reference handbook for national competent authorities as a guide for receiving and making requests under article 17. The need to give special attention to the problem of the smuggling of drugs in containers and other commercial shipments was discussed, as well as the need to improve shore facilities. Finally, the group encouraged States to consider entering into agreements or arrangements on liability and compensation for any loss, damage or injury arising from actions taken pursuant to article 17.

242. IMO noted it had considered that the problem of drug trafficking in 1990

in the context of elaborating amendments to the Convention on Facilitation of International Maritime Traffic, 1965. Furthermore, in 1997, the IMO Assembly had adopted Guidelines for the prevention and suppression of the smuggling of drugs, psychotropic substances and precursor chemicals on ships engaged in international maritime traffic (see A/55/61, paras. 106 and 107).

VI. MARINE RESOURCES, MARINE ENVIRONMENT AND SUSTAINABLE DEVELOPMENT

A. Conservation and Management of Marine Living Resources

243. Overfishing in many parts of the world's oceans and seas has led the international community to take new actions over the past decade to restore sustainability in the use of fisheries resources and to commit itself to improved conservation and management of marine living resources for the sake of future generations. Such actions have included the adoption of the 1993 Agreement to Promote Compliance with International Conservation and Management Measures by Fishing Vessels on the High Seas (1993 Compliance Agreement), the 1995 Agreement for the Implementation of the Provisions of the United Nations Convention on the Law of the Sea of 10 December 1982 relating to Straddling Fish Stocks and Highly Migratory Fish Stocks (1995 Fish Stocks Agreement), the 1995 Code of Conduct for Responsible Fisheries,[68] and the three 1999 international plans of action for the Management of Fishing Capacity, for Reducing Incidental Catch of Seabirds in Longline Fisheries and for the Conservation and Management of Sharks.

244. However, efforts to improve the conservation and management of the world's fisheries have been confronted by the increase in illegal, unreported and unregulated fishing activities (IUU fishing) on the high seas, in contravention of conservation and management measures adopted by regional fisheries organizations and arrangements, and in areas under national jurisdiction in violation of coastal States' sovereign rights to conserve and manage marine living resources (see A/54/429, paras. 249–257; A/55/61, paras. 120–125). IUU fishing is perpetrated both by vessels of States members of regional fisheries management organizations, in some circumstances flying flags of convenience, as well as by vessels of States not members to these organizations. The problem is believed to have been aggravated by excess fleet capacity, the payment of government subsidies, strong market demand for particular fish products and ineffective monitoring, control and surveillance.[69] The adverse effects of IUU fishing on the good governance of world's fisheries, as well as on the economies and food security of coastal States, particularly developing coastal States, have prompted the international community to take measures at the national, regional and global levels to combat it.

---

68. For the text of the three instruments, see *International Fisheries Instruments with Index* (United Nations publication, Sales No. E.98.V.11).

69. Contribution of FAO to the present report.

1. Actions to Combat IUU Fishing Activities

(a) Actions at the global level

245. In the early 1990s, the United Nations General Assembly called upon States to take the responsibility, consistent with their obligations under international law, to take measures to ensure that no fishing vessels entitled to fly their national flag fished in zones under the national jurisdiction of other States unless duly authorized by the coastal States concerned.[70] It subsequently extended the prohibition to unauthorized fishing activities on the high seas, stipulating that no flag States should allow vessels flying their flag to fish on the high seas unless duly authorized by them and that no fishing activities should take place in contravention of applicable conservation and management measures.[71] Moreover, the FAO Ministerial Conference on Fisheries, which met in Rome in March 1999, adopted a declaration requesting FAO to develop a global plan of action to deal effectively with all forms of IUU fishing. That request was endorsed by the United Nations Commission on Sustainable Development, meeting at its seventh session, in April 1999,[72] which also invited IMO and FAO to urgently develop measures to ensure that ships of all flag States meet international rules and standards to give effect to the relevant provisions of UNCLOS, especially article 91 on the nationality of ships.[73] Furthermore, the Consultative Process at its first meeting (A/55/274, para. 10) and the General Assembly at its fifty-fifth session made urgent calls to States, *inter alia*, to continue the development of an international plan of action against IUU fishing and invited them, as well as competent United Nations specialized agencies, to continue their cooperation to that end.[74]

Joint FAO/IMO Ad Hoc Working Group[75]

246. In response to the above appeals, a Joint FAO/IMO Ad Hoc Working Group on IUU Fishing and Related Matters met in Rome in October 2000. The outcome of this first meeting among members of the two specialized agencies contained a number of recommendations aimed at enhancing flag State and port State control over fishing vessels, with a view to eliminating the roots of IUU fishing.

247. With respect to flag State responsibilities, the Working Group agreed that there was a need for States, *inter alia*: (a) to enhance implementation of flag State responsibility and focus on fisheries issues, including through regional fisheries management organizations; (b) to ensure that the flag State linked the registration of a fishing vessel with the authorization to fish in national administrations; (c) to establish cooperation between the flag State and the coastal State when a vessel was

---

70. General Assembly resolution 49/116, para. 1.
71. General Assembly resolutions 52/29, para. 7, and 53/33, para 7.
72. *Official Records of the Economic and Social Council, 1999, Supplement No. 9* (E/1999/29), chap. I. C, decision 7/1, para. 18.
73. *Ibid.*, para. 35 (a).
74. General Assembly resolutions 55/7, para. 24, and 55/8, paras. 16–18.
75. Illegal, Unregulated and Unreported (IUU) Fishing and Related Matters, Outcome of the first meeting of the Joint FAO/IMO Ad Hoc Working Group on IUU Fishing and Related Matters, Rome, 9–11 October 2000, Subcommittee on Flag State Implementation, 9th session, agenda item 15, IMO, 8 November 2000, document FSI 9/15.

fishing in areas under the jurisdiction of the coastal State, particularly to ensure that the flag State continued to exercise effective control over that vessel; (d) to avoid deregistering a vessel that failed to comply with the authorization to fish as the practice had the effect of "exporting" the problem; (e) to give effect to existing rights and obligations under international law and become parties to existing legal instruments relating to flag State control; and (f) to give consideration to the application to fishing vessels of the IMO number scheme to enable vessels to be traced regardless of changes in registration or name.

248. Concerning the need for flag States to continue exercising effective control over their vessels conducting fishing operations in the exclusive economic zones of other States, under UNCLOS, coastal States, as a corollary to their sovereign rights over natural resources, are entitled in their exclusive economic zones to take all measures against foreign fishing vessels, including boarding, inspection, arrest and judicial proceedings, as may be necessary to ensure compliance with the laws and regulations adopted by them in the area for the purpose of conserving and managing their marine living resources (article 73 (1)). However, enforcement powers of coastal States over the conduct of fishing operations in their exclusive economic zones, are without prejudice to the right of flag States to continue to exercise their jurisdiction and control in respect of administrative, technical, and social matters pertaining to vessels flying their flag. It is therefore hoped that flag State control, with regard to administrative, technical, and social matters, over their fishing vessels operating in the exclusive economic zones of coastal States will be directed towards giving effect to their obligations to ensure the compliance by those vessels with fisheries laws and regulations in areas under the national jurisdiction of coastal States, in accordance with General Assembly resolution 49/116.

249. The Working Group also expressed the view that a more appropriate approach would be to address the possible key issues constituting effective flag State control of a fishing vessel, rather than attempting to define the concept of the genuine link between a vessel and the State whose flag it is flying.[76] While the conclusion may be valid for merchant shipping, one may argue that, if a foreign fishing vessel seeks registration or reflagging in another State, with which it does not have real links (as in the case of flag of convenience), and the State involved either does not participate in the implementation of management measures established by regional fisheries management organizations in a particular subregion or region, or is known to lack the capacity to control fishing activities of vessels flying its flag, one may conclude that the main purpose of such registration or reflagging is to evade compliance with applicable fisheries conservation and management measures, which its flag State of origin would have otherwise enforced. Indeed, if by any chance such a State decided to require vessels flying its flag to abide by high seas conservation and management measures, it would be unlikely that it would be in a position to achieve prompt compliance, since it would lack the legal and economic leverage over the owners and operators of vessels flying its flag that allow it to compel such compliance.

250. Moreover, the Working Group concluded that port States, in the exercise of sovereignty over their ports in accordance with international law, were entitled, *inter alia,* to introduce domestic legislative measures to deal with foreign fishing

76. *Ibid.*, para. 24.

vessels entering or leaving their ports. Such measures may relate to the control of vessels engaged in the trans-shipment and transport of fish or the resupply of fishing vessels, as they are subject to port State control with respect to maritime safety, pollution prevention and living and working conditions. At the international level, the Working Group encouraged FAO, in cooperation with relevant international organizations, to consider developing measures for port State control, with emphasis on the management of fisheries resources and taking into account IMO port State control procedures. The Working Group also agreed that the mechanism of Memoranda of Understanding relating to port State control of fishing vessels could be used as an effective tool for enhancing fisheries management.[77]

251. In addition, the recent adoption by the FAO Committee on Fisheries[78] of the International Plan of Action (IPOA) to Prevent, Deter and Eliminate IUU fishing for the purpose of complementing the existing international instruments so as to counter their ineffectiveness in addressing the phenomenon of IUU fishing, is a further milestone in the fight against IUU fishing.

The International Plan of Action (IPOA)
252. IPOA is an instrument of a voluntary character aimed at addressing the legal and economic dimensions of IUU fishing in an integrated manner, whereby flag States, port States and coastal States are invited to take measures at the national, regional or global level to combat illegal fishing activities. The Plan of Action contains provisions that address: (a) the nature and scope of IPOA, (b) the objective and principles of the Plan, (c) the key actions in combating IUU fishing, (d) the special requirements of developing countries, (e) reporting by States and regional fisheries organizations and (f) the role of FAO in support of the Plan of Action.

253. Another important feature of IPOA is the inclusion in its provisions of a definition of IUU fishing, which identifies and describes the constitutive elements of IUU fishing; "illegal fishing," "unreported fishing," or "unregulated fishing,"[79] as undertaken either in areas under national jurisdiction or on the high seas.

254. The Plan of Action generally reaffirms the strengthening of the duties of flag States provided for in the Compliance Agreement (articles III and IV), the 1995 Fish Stocks Agreement (articles 18 and 19) and the Code of Conduct for Responsible Fisheries (article 8.2).

255. With regard to port State jurisdiction, IPOA provides that, in addition to the right of port States to conduct inspections and request information of foreign fishing vessels calling voluntarily at their ports or offshore terminals,[80] port States are entitled, prior to allowing fishing vessels access to their ports, to request a copy of their authorization to fish, details of their fishing trip and quantities of fish on board, with due regard to confidentiality requirements (IPOA, para. 45). With the

77. *Ibid.*, annex.
78. See Report of the Committee on Fisheries, Twenty-fourth session, Rome, Italy, 26 February–2 March 2001.
79. International Plan of Action to Prevent, Deter and Eliminate Illegal, Unreported and Unregulated Fishing, Nature and Scope of IUU Fishing and the International Plan of Action, II, 3.1–3.3.
80. 1995 Fish Stocks Agreement, article 23; Compliance Agreement, article V (2); Code of Conduct for Responsible Fisheries, article 8.3.2.

exception of *force majeure*, these provisions would allow a port State to deny a fishing vessel access to its port facilities if it has reasonable grounds to believe that the vessel is engaged in IUU fishing. Another important feature of the IPOA is its invitation to port States to cooperate bilaterally and multilaterally, as well as within relevant regional fisheries management organizations, to develop compatible measures for port State control over fishing vessels (IPOA, para. 49).

256. With respect to the duties of coastal States vis-à-vis IUU fishing, IPOA encourages those States to regulate fishing access in areas under their jurisdiction in a manner that will help to prevent, deter and eliminate IUU fishing. To that end, they should enforce measures they have adopted or by which they are otherwise bound for the conservation and management of fish stocks, through monitoring, control and surveillance of fishing activities, and through cooperation with other States and regional fisheries management organizations.

257. Moreover, IPOA provides that all States should adopt multilateral trade-related measures consistent with international law, including the provisions of the World Trade Organization (WTO), and implement them in a fair, transparent and nondiscriminatory manner, to prevent the trade or import into their territories of fish emanating from IUU fishing activities.

258. As to cooperation within regional fisheries management organizations, IPOA invites all States, members as well as nonmembers of those organizations, to enforce and ensure compliance with policies and measures on IUU fishing adopted by regional fisheries management organizations in conformity with international law. States are also called upon to take action to strengthen the role of these organizations in fisheries conservation and management in order to allow them to deter, prevent and eliminate IUU fishing. States are similarly encouraged to cooperate in the establishment of regional fisheries management organizations in regions where none exist.

259. The integrated approach adopted by the Plan of Action, if applied in good faith, may close the loopholes that in the past have hampered the international community in the prevention, deterrence and elimination of IUU fishing.

(b) Actions at the regional level

260. At the regional level, regional fisheries management organizations have developed measures to combat IUU fishing (see paras. 272–278), such as the adoption of enforcement and compliance schemes against the IUU fishing of noncontracting parties; market-related measures such as catch documentation schemes aimed at identifying the origin of the harvested fish, so as to regulate their sale; nondiscriminatory trade-restrictive measures; and port State measures to control the landings of fish. These measures have been adopted by, *inter alia*, the International Commission for the Conservation of Atlantic Tunas (ICCAT),[81] the Northwest Atlantic Fisheries Organization (NAFO),[82] the North-East Atlantic Fisheries Commission (NEAFC),[83] the Commission for the Conservation of Antarctic Marine Living Resources (CCAMLR)[84] and the Commission for the Conservation of Southern Bluefin

---

81. A/53/456, para. 267; A/54/429, para. 269; A/55/386, paras. 151–152.
82. A/53/456, paras. 268–271; A/53/473, para. 135; A/55/386, paras. 156–157.
83. A/53/473, para. 138; A/54/429, para. 275; A/55/386, para. 159–161.
84. A/55/386, para. 144.

Tuna (CCSBT).[85] Other organizations[86] are examining the possibility of adopting these measures to supplement the often ineffective diplomatic protests lodged with the flag States of IUU fishing vessels sighted conducting fishing operations in their respective regulatory areas. Moreover, in the South Pacific region, the Forum Fisheries Agency (FFA) has developed an information system known as the Violations and Prosecutions (VAP) Database, which contains information on vessels that have been involved in violations of the fisheries laws of FFA member countries. The VAP Database would allow the licensing authorities of the coastal States to verify the historical compliance records of fishing vessels before granting them fishing licences.[87]

(c) Actions at the national level

261. Several States have taken steps to strengthen national measures against IUU fishing by revising national fisheries laws and related legislation to close "loopholes" that permit such practices to take place. They have also taken measures to implement the relevant provisions of the 1993 Compliance Agreement, the 1995 Fish Stocks Agreement and the 1995 Code of Conduct for Responsible Fisheries.

262. Many flag States, recognizing that effective flag State control is fundamental to fisheries management, have taken steps to ensure that fishing vessels entitled to fly their flag do not engage in any activity that undermines the effectiveness of international conservation and management measures or in any activity that constitutes unauthorized fishing in areas under the national jurisdiction of other States. Measures have been introduced to prohibit fishing on the high seas without a proper authorization by the flag State. Such fishery regulations have been put into force by Denmark, Japan, Norway, Mauritius, the United States, Saudi Arabia, Panama, Uruguay, New Zealand, Mexico and Guyana (see A/53/473). Under some regulations, a fishing vessel may be registered or granted a licence only if sufficient links exist between the flag State and the vessel. Such requirements are found, for example, in the Mauritius Fisheries and Marine Resources Act of 1998, which provides that, in order to avoid the licensing of vessels of flag of convenience, only vessels wholly owned by the State of Mauritius, or owned by a corporation controlled at least 50% by the State or by Mauritius citizens, can be licensed to fish on the high seas. Similarly, the Mexico Fisheries Act, among other provisions, prohibits reflagging and stipulates that the national flag may be granted only to vessels that have surrendered their flag of origin (see A/55/386).

263. Moreover, fisheries regulations in a growing number of States, provide that conditions for granting fishing permits to vessels for high seas fishing require compliance by vessels with applicable conservation and management measures (New Zealand Fisheries Act 1996 (Amendment No. 2), Guyana Revised Fisheries Legislation and United States High Seas Fishing Compliance Act of 1995) (see A/55/386). Most such fisheries laws (e.g., those of Japan, Guyana, Norway, Mexico and the United States) require flag States to maintain a record of fishing vessels entitled to fly their flag and authorized by them to fish on the high seas (see A/53/473 and

---

85. *Ibid.*, para. 147.
86. *Ibid.*, para. 250.
87. Contribution by the South Pacific Applied Geoscience Commission (SOPAC) to the present report.

A/55/386). They also provide that fishing vessels must be marked in accordance with the FAO Standard Specifications for the Marking and Identification of Fishing Vessels. Some of them even stipulate that licence applications may be denied or withdrawn if the vessel or its owner had taken part in IUU fishing on the high seas (Norway, United States) or if a previous licence for high seas fishing granted to the vessel by a foreign State had been suspended or withdrawn because the vessel had undermined the effectiveness of international conservation and management measures (Guyana).

264. In addition, flag States have introduced provisions that make it mandatory for vessels flying their flag to submit catch reporting and other fishery data from their fishing operations or to have on board national observers or vessel monitoring systems as means of enhancing national monitoring, control and surveillance. Most of these fisheries regulations also provide for sanctions of sufficient severity for high seas fisheries violations, which may include suspension, withdrawal or cancellation of registration or fishing permits. Under other regulations fishing vessels that cease to be entitled to fly the national flag also lose their authorization to fish on the high seas. And some States (Japan, Norway) have placed restrictions on the export of vessels decommissioned from their national fishing fleets, to avoid exporting excess fishing capacity and reflagging (see A/55/386). While these laws and regulations significantly enhance flag States' control over vessels flying their flag, it is also believed that flag States' responsibilities must not be limited to ensuring compliance by fishing vessels flying their flag with agreed conservation and management measures, but should also encompass assistance in enforcement wherever allegations of violations occur.[88]

265. As part of their growing role in ensuring compliance with fisheries conservation and management, many port States exercise control in respect of foreign fishing vessels calling voluntarily at their ports or at offshore terminals through, *inter alia,* the monitoring of trans-shipments and landings and the collection of data on catch and effort (see A/52/555). They have also enacted national legislation establishing restrictions or prohibitions on landings or requiring the issuance of licences for fishing vessels to enter a port.[89] Under these laws, they have denied port access to vessels known to have engaged in IUU fishing or have closed off to their owners or operators access to markets by prohibiting landings to catches that have not been harvested in conformity with agreed regional conservation and management measures.[90] Of particular relevance are the Chilean Fisheries Law, which allows for the prohibition of landings, supplying of ships or other services in ports or in areas under national jurisdiction in respect of fishing vessels that have engaged in high seas activities that have an adverse impact on fishery conservation and management in the Chilean exclusive economic zone[91] (see cases before the International Tribu-

---

88. Contribution by the South Pacific Applied Geoscience Commission (SOPAC) to the present report.
89. FAO Fishery and Agricultural Legislation: United Kingdom, Fishery Limits Act 1976, sect. 3 (6), vol. 26, No. 2, 1977, p. 89; Sri Lanka, Fisheries Act No. 59 of 1979, Regulation of Foreign Fishing Boats, vol. 29, No. 1, 1980, p. 89; Trinidad and Tobago, Archipelagic Waters and Exclusive Economic Zone Act 1986, sect. 32, vol. 36, No. 2, 1987, p. 107.
90. Contribution by FAO to the present report.
91. Law No. 18.892 of 1989 and modifications, General Law on Fisheries and Aquaculture, art. 165, Library of the National Congress of Chile, Juridical-Legislative System, Most consulted law, web site at: <www.congreso.cl/biblioteca/leyes/otras/pesca13.htm>.

nal for the Law of the Sea, paras. 442–443), as well as coastal States' legislation in force in the South Pacific that prohibits the importation by a State of fish caught illegally in areas under the national jurisdiction of another State.[92]

266. For those who argue that port States restrictions may contravene the WTO Agreement on free trade, it is believed that such measures could qualify under the exceptions of the 1994 GATT rules under article XX (g), insofar as they are intended to promote the conservation of exhaustible natural resources.[93]

267. Many coastal States have taken measures to control foreign fishing operations in areas under their national jurisdiction through the adoption of laws and regulations governing fishing activities and the implementation of monitoring, control and surveillance systems for fishing operations in their exclusive economic zones. These include requirements for fishing authorization or fishing permits, types of gear, daily maintenance of logbooks, daily reportings of catch and vessel geographical positions, statistical data reporting, vessel monitoring system (VMS), prior authorization for trans-shipments at sea, obligation to land all or part of the catch, prohibition of discard of by-catches, and obligation to stow fishing gear when fishing vessels are in transit in areas under national jurisdiction.[94] They have also taken measures to enforce such fisheries laws and regulations in areas under their national jurisdiction.

268. However, for many developing coastal States, limited resources and the large size of the ocean space over which they exercise control have had an adverse impact on their ability to enforce their conservation and management measures against IUU fishing. For these countries, IUU fishing has been able to be carried out through, *inter alia*, the use of flags of convenience, illegal fishing on the ocean areas between the exclusive economic zones and the high seas, and misreporting of catch. Developing coastal States dependent on access fees for their economic development are particularly vulnerable because of distortions to fee levels, which are conditional upon the volume of catch.[95]

269. Consequently, in regions such as the South Pacific, coastal States have established a regional register of foreign vessels with a common database of all relevant information about vessels, updated annually, and containing information about their owners, operators and masters, call sign and port of registry. The regional register is used not only as a source of information on fishing vessels but also as a tool to ensure compliance with coastal States' laws and regulations. Under the scheme, no fishing vessel can be licensed unless it has good standing on the regional

---

92. Contribution by the South Pacific Applied Geoscience Commission (SOPAC) to the present report.

93. Article XX (g) of the 1994 General Agreement on Tariffs and Trade (GATT) reads as follows: "Subject to the requirement that such measures are not applied in a manner which would constitute a means of arbitrary or unjustifiable discrimination between countries where the same conditions prevail, or disguised restriction on international trade, nothing in this Agreement shall be construed to prevent the adoption or enforcement by any contracting party of measures:

". . . (g) Relating to the conservation of exhaustible natural resources if such measure are made effective in conjunction with restrictions on domestic production or consumption; . . ."

94. Coastal State Requirements for Foreign Fishing Database, FAO Legal Office, http://faolex.fao.org/cgi-bin/fishery.

95. Contribution by the South Pacific Applied Geoscience Commission (SOPAC) to the present report.

register. In addition, coastal States have taken measures to harmonize the terms and conditions of access, so that fishing vessels are not subject to different regulatory regimes in the exclusive economic zones of coastal States of the same region or subregion. These include provisions for licensing; prohibition of trans-shipment at sea; maintenance of catch logbook data and other information; access by authorized officers of the licensing State; regular catch reporting; use of observers; requirements for vessel marking and identification; and the requirement of the use of a satellite-based vessel monitoring system (VMS). Coastal States have also established cooperation within the framework of the Niue Treaty on Cooperation in Fisheries Surveillance and Law Enforcement, which permits a party, through a subsidiary agreement, to extend its fisheries surveillance and law enforcement activities to the territorial sea and archipelagic waters of another party, thereby allowing a cross-jurisdictional exercise of enforcement powers by each party's surveillance and enforcement officers.

270. Under UNCLOS, however, the exercise by coastal States of their sovereign rights in areas under their national jurisdiction shall not impede the freedoms of navigation and communications of all ships in the exclusive economic zones (article 58). Thus, the practice of some coastal States of requiring notification by foreign fishing vessels when transiting the exclusive economic zone[96] is considered to be inconsistent with the right of navigation recognized to all vessels in the zone, as provided in article 58 of UNCLOS. In addition, since the exercise by coastal States of their sovereign rights over marine resources is subject in UNCLOS to a *ratione loci* competence, i.e., limited to a clearly defined area, it is believed that the legislation of some States that implements their conservation and management measures with respect to straddling fish stocks and highly migratory fish stocks in the adjacent high seas areas beyond their exclusive economic zones,[97] is inconsistent with the relevant provisions of the Convention.

(d) Other Developments

271. The seriousness of IUU fishing is reflected in two judgements rendered by the International Tribunal for the Law of the Sea in 2000 in cases involving applications for the prompt release of vessels alleged to have fished illegally in the exclusive economic zone of a coastal State: The *Camouco* case (*Panama* v. *France*) (see A/55/61, paras. 250–257) and the *Monte Confurco* case (*Seychelles* v. *France*) (see paras. 435–441). In both cases, the Tribunal has "taken note of the gravity of the alleged offences"[98] as well as the "the general context of unlawful fishing in the region"[99]

---

96. France, Law No. 66-400 of 18 June 1966, as amended by Law of 18 November 1997.

97. Argentina, Act No. 23.968 of 14 August 1991, reprinted in *Law of the Sea Bulletin*, No. 20 (1992); Chile, Act 19.079 of 12 August 1991, amending Act 18.892, article 154, *Official Journal of the Republic of Chile*, 6 September 1991; and Peru, Ley General de Pesquarías, approved by Decree-Law 25977 of 7 December 1992, article 1, *Diario el Peruano-Normas Legales*, 22 December 1992.

98. International Tribunal for the Law of the Sea, 2000, *The "Camouco" Case (Panama v. France)*, Application for Prompt Release, Case No. 5, Judgment, para. 68.

99. International Tribunal for the Law of the Sea, *The "Monte Confurco" Case (Seychelles v. France)*, Application for Prompt Release, Case No. 6, Judgment, para. 79.

pointed out by France, as among the factors to be considered in the assessment of the reasonableness of bonds or other financial security.

2. Review of the Role of Regional Fisheries Management Organizations in Fishery Conservation and Management

272. UNCLOS imposes upon States a general obligation to cooperate in the implementation of the legal framework provided in the Convention. With respect to the conservation and management of marine living resources, it stresses the critical role played by regional fisheries management organizations (see Annex IV to the present report) in regional fisheries governance and as forums for cooperation in all aspects of fisheries conservation and management. In the exclusive economic zone, articles 61(2), (3) and (5); 63; 64(1); 65; and 66 of UNCLOS provide that the coastal State and competent international organizations shall ensure that marine living resources are not endangered by overexploitation. The Convention also invites States to cooperate within subregional, regional or global organizations, as the case may be, for the conservation and management of straddling stocks, shared stocks, anadromous stocks, highly migratory species and marine mammals. On the high seas, articles 118 and 119 of UNCLOS require States to cooperate within regional fisheries management organizations for the conservation of high seas marine living resources and, where no such organization exists in a particular subregion or region, to establish one as a forum for their cooperation.

273. Regional fisheries management organizations have only been variably successful in conserving the resources under their competence, owing, *inter alia*, to the inadequacies of their mandates and their inability to enforce their own management decisions. These organizations have also been confronted with such fisheries issues as overfishing; insufficiency of scientific advice; inadequacy of monitoring, control and surveillance; weakness of the decision-making process; noncompliance by members of management decisions; and IUU fishing by members and nonmembers. Nonetheless, a number of these fishery bodies and arrangements have started to work within the legal framework provided for in UNCLOS for the conservation and management of marine living resources, and have undertaken to strengthen their role in fisheries management, as required by developments[100] in international fisheries law. These developments require the international community, *inter alia*, to ensure the long-term sustainability of marine living resources, apply the precautionary approach, follow an ecosystem-based management approach, emphasize scientific advice, stress the importance of the collection and exchange of adequate data, implement effective monitoring, control and surveillance systems, agree on decision-making procedures that facilitate the timely adoption of conservation and management measures and evolve effective mechanisms for settlement of disputes.

274. To this end, a number of FAO regional fishery bodies have already strengthened their functions and responsibilities, strengthening their role from advisory to regulatory bodies. For instance, the General Fisheries Council for the Mediterranean (GFCM) has amended its establishing Agreement and its rules of proce-

---

100. Adoption of the 1995 Fish Stocks Agreement and the 1995 Code of Conduct for Responsible Fisheries.

dure, renamed itself a Commission and opted for an autonomous budget. It has established also a Scientific Advisory Committee to obtain scientific advice in the management of Mediterranean fisheries.[101] At its sessions in 1999 and 2000, the GFCM Scientific Advisory Committee addressed such issues as the definition of management units, definition of parameters for measuring fishing effort, identification of the actual state of resources and methodologies for determining such status, as well as the definition of required fleet data to be included in the regional register of vessels, as a necessary starting point to monitor fishing effort.[102] In addition, the Asia-Pacific Fishery Commission (APFIC) has also amended its constitution and updated its terms of reference to take into full account the recent fundamental changes in world fisheries and, in particular, to be equipped to play its role in the implementation of UNCLOS and in the promotion of the Code of Conduct for Responsible Fisheries. APFIC agreed that its future programmes should be more specific and pragmatic, assisting members to move closer towards self-reliance in sustainable fisheries.[103]

275. In the Indian Ocean region, the Indian Ocean Tuna Commission (IOTC) has also endeavoured to strengthen its role in the conservation and management of highly migratory fish stocks in its area of competence. On the advice of its Scientific Committee, it has implemented mandatory requirements for IOTC members to provide a timely, standardized statistical data for catch, effort and size for all species covered by the Commission,[104] as well as data for catches of nontarget species. It has also recommended a substantial reduction of the fishing capacity of distant longline tuna fleets operating in the region, as well as registration and exchange of information on vessels, including flag-of-convenience vessels fishing for tropical tunas in IOTC areas of competence.

276. Non-FAO regional fishery bodies and arrangements have also implemented or are in the process of implementing the new approaches to fisheries conservation and management. For example, ICCAT, NAFO and NEAFC have considered developing the precautionary approach as a tool for fisheries management.[105] The ecosystem-based approach[106] and precautionary total allowable catch[107] have already been integrated into the fisheries conservation and management programmes of CCAMLR and CCSBT.[108] Moreover, a number of other regional fisheries manage-

101. Progress Report on the Implementation of Conference Resolution 13/97 (Review of FAO Statutory Bodies) and the Strengthening of FAO Regional Fishery Bodies, Committee on Fisheries, Twenty-third Session, Rome, 15–19 February 1999, document COFI/99/4, para. 4.

102. Report of the General Fisheries Commission for the Mediterranean, Twenty-fourth session, Alicante, Spain, 12–15 July 1999 (GFCM Report 24), paras. 27–28.

103. COFI/99/4, para. 5.

104. Report of the Third Session of the Indian Ocean Tuna Commission, Mahe, Seychelles, 9–12 December 1998, document IOTC/03/98/R [E], appendix H.

105. A/54/461, paras. 33 and 44; Summary Report of the Eighteenth Annual Meeting of NEAFC, 22–25 November 1999, para. 7.

106. Report of the Fifteenth Meeting of the Commission for the Conservation of Antarctic Marine Living Resources, Hobart, Australia, 21 October–1 November 1996, document CCAMLR-XV, pp. 7–20.

107. *Ibid.*, Conservation measures adopted in 1996: Conservation measure 103/XV, p. 54.

108. A/54/429, para. 298.

ment organizations have adopted schemes to enforce their conservation and management measures.

277. As expected, these new trends have been incorporated in recent agreements establishing new regional fisheries management organizations or arrangements in various regions of the world. Especially noteworthy among these are the Convention on the Conservation and Management of Highly Migratory Fish Stocks in the Western and Central Pacific,[109] the Convention on the Conservation and Management of Fishery Resources in the South-East Atlantic Ocean[110] and the Framework Agreement for the Conservation of Living Marine Resources on the High Seas of the South-East Pacific ("Galapagos Agreement"),[111] which contain provisions underlining the requirement of long-term sustainability of fish stocks, compatibility of measures within and beyond areas under national jurisdiction, ecosystem-based management and application of the precautionary approach, as well the important role of scientific information in fisheries management. These new agreements also include strong provisions enhancing flag States' responsibilities as required in the FAO Compliance Agreement, as well as monitoring, control and surveillance and enforcement schemes involving reciprocal boarding and inspection by States parties and port State measures modelled along those in the 1995 Fish Stocks Agreement.

278. However, despite the strengthened role of regional fisheries management organizations in regional fisheries governance, progress is being hindered by, *inter alia*, the failure by States to accept and implement relevant international instruments, a lack of willingness by those States to delegate sufficient responsibility to regional bodies and the lack of the effective enforcement of management measures at both national and regional levels.[112] The performance of many regional fisheries management organizations is also adversely affected by inadequate financial resources, particularly in respect of FAO fishery bodies, ineffective decision-making procedures that allow noncompliance by members with management decisions, and IUU fishing by vessels flying flags of convenience.

3. Conservation and Management of Marine Mammals

279. Catch limits for commercial whaling, allocation of minke whales to Japan, elements of the Revised Management Scheme (RMS), catch limits for aboriginal subsistence whaling in the Bering-Chukchi-Beaufort Seas, the Eastern North Pacific, West Greenland, East Greenland, and in the Caribbean Sea, the status of whales, scientific permit catches by Japan and whale killing methods were among the main topics of discussion at the fifty-second annual meeting of the International Whaling Commission, held at Adelaide, Australia, from 3 to 6 July 2000.[113]

280. Concerning the contentious issue of commercial whaling, no consensus

---

109. Article 5 (a), (d) and (f); article 6; articles 12 and 13.
110. Article 2; article 3 (e) and (g); article 7 and 10; article 19.
111. Article 5 (a), (b), (c) and (e); article 7 (e), (f) and (g).
112. Report of the Meeting of FAO and Non-FAO Regional Fishery Bodies or Arrangements, Rome, 11–12 February 1999, document X1212/E, para. 27.
113. Final press release, 2000 Annual Meeting, Adelaide, Australia: http://ourworld.compuserve.com/homepages/iwcoffice/PRESSRELEASE2000.htm.

could be reached once again to break the deadlock between the States parties opposing the resumption of commercial whaling and those in favour of such resumption.[114]

281. In other developments, the North Atlantic Marine Mammal Commission (NAMMCO) held its tenth annual meeting in Sandefjord, Norway, from 25 to 28 September 2000. The meeting considered the report of Scientific Committee on the status of marine mammals under NAMMCO management.[115] Following a review of the state of whales under its competence, the Commission recommended that, in view of the depleted status of the West Greenland beluga, which required severe reductions in catch, closer links should be developed between NAMMCO and the Canada/Greenland Joint Commission on Conservation and Management of Narwhal and Beluga (JCNB) that has the competence to provide management advice for the stock. The Commission also requested the Scientific Committee to evaluate the migration patterns of West Greenland narwhal in Baffin Bay and Davis Strait and to monitor developments with regard to the Faeroese fin whales and dolphins.[116]

282. With respect to the economic aspects of marine mammal/fishery interactions, NAMMCO recommended that the Scientific Committee should proceed with its programme to develop multispecies economic models for candidate species and areas in the investigation of the problem, using for this purpose the Barents/Norwegian Seas and the area around Iceland. As to the collection of data on marine mammal by-catch, it endorsed the efforts undertaken by countries to establish mandatory logbook data collection systems and decided to initiate a system of by-catch reporting for NAMMCO member countries through the national progress reports, starting with data on numbers and species comprising marine mammal by-catch in fisheries.[117]

4. Marine and Coastal Biodiversity

283. The fifth meeting of the Conference of the Parties to the Convention on Biological Diversity was held at Nairobi from 15 to 26 May 2000 for the biennial review of the implementation of the Convention. In carrying out the marine component of its activities, the Conference adopted decision V/3 relating to its programme of work on marine and coastal biological diversity,[118] to assist in the implementation of the Jakarta Mandate on Marine and Coastal Biodiversity, in accordance with decision II/10.

284. It will be recalled that the Jakarta Mandate consisted originally of five thematic areas: integrated marine and coastal management (IMCAM); marine and coastal protected areas; marine and coastal living resources; mariculture; and alien species. The issue of coral reefs was added to the programme of work by the Conference of the Parties in its decision IV/5 adopted at its fourth meeting.

114. *Ibid.*
115. Final press release, 28 September, Tenth Meeting of NAMMCO, Sandefjord, Norway, 25–28 September 2000: <www.nammco.no/fi-pr-re.htm>.
116. *Ibid.*
117. *Ibid.*
118. UNEP/CBD/COP/5/23, pp. 74–80.

285. With respect to coral bleaching, the Conference of the Parties was of the view that since climate change was its primary cause, the United Nations Framework Convention on Climate Change ought to take actions to reduce the effects of changes in water temperatures and address the socio-economic impacts on the countries and communities most affected by coral bleaching. All States and relevant bodies were also urged to implement measures that would address the issue of coral bleaching, through information-gathering, policy development, financial assistance and capacity-building (see paras. 489–492).

286. In addition, the Conference requested the Secretariat of the Convention on Biological Diversity to integrate coral bleaching fully into its programme of work on the conservation and sustainable use of marine and coastal biological diversity and to develop and implement a specific work plan thereon. It also invited the Biodiversity Convention Secretariat's Subsidiary Body on Scientific, Technical and Technological Advice (SBSTTA) to include in its study of coral bleaching, the effects of the physical degradation and destruction of coral reefs as a threat to the biological diversity of coral reef ecosystems.

287. Experts have indicated in complementary findings that, in addition to climate change, destructive fishing practices, and the socio-economic conditions in coastal communities also contribute to coral reef degradation. They therefore suggested that activities that create wealth or add value in coastal areas, such as tourism, aquaculture and manufacturing, with due attention to environmental protection, could alleviate pressure on coral reefs.[119]

288. With regard to IMCAM, the Conference endorsed further work for the development of guidelines for coastal areas, taking into account the ecosystem approach, and encouraged SBSTTA to continue its work on ecosystem evaluation and assessment, *inter alia*, through the elaboration of guidelines on evaluation and indicators.

289. Regarding the management of marine and coastal living resources, the Conference requested the Biodiversity Convention Secretariat to gather information on approaches to the management of marine and coastal living resources currently in use by local and indigenous communities, and to disseminate that information through the clearing house mechanism. SBSTTA was invited to consider and prioritize, as appropriate, such issues as the use of unsustainable fishing practices; the failure to use marine and coastal protected areas for the management of living resources; the economic value of marine and coastal resources; and capacity-building for stock assessment and economic evaluation purposes. The subsidiary body was also requested to provide advice on scientific, technical, and technological matters related to the issue of marine and coastal genetic resources.

290. As to the issues of alien species and genotypes, the Conference of the Parties requested the Biodiversity Convention Secretariat to make use of existing information, expertise and best practices on alien species in the marine environment in the implementation of the work programme on alien species under decision IV/1 C of the fourth Conference of the Parties. The Conference also approved the terms of reference and duration of work specified for the ad hoc technical experts on marine and coastal protected areas and mariculture, as recommended by

---

119. Report of the thirtieth session of GESAMP, Monaco, 22–26 May 2000, GESAMP Reports and Studies, No. 69, para. 7.7.

SBSTTA,[120] and invited the Biodiversity Convention Secretariat to strengthen its cooperation with global organizations and coordinate with regional seas conventions and action plans in the implementation of the Jakarta Mandate.

291. On the subject of mariculture, the Conference requested the ad hoc technical experts to devise guidance on criteria, methods and techniques to avoid the adverse effects of mariculture on marine and coastal biodiversity. To this end, it might be suggested that due consideration should be given by the experts to the issue of the genetic interaction between farmed and wild fish stocks caused by mariculture escapees. Indeed, it is believed that such genetic interaction may enhance the risk of decreasing the natural genetic variability of one or more species, through the introduction in the wild of a great number of individuals presenting a higher inbreeding level resulting from domestication or from the practice of breeding programmes.[121]

B. Nonliving Marine Resources

1. Offshore Hydrocarbons

292. With the increasing demand for oil and gas, offshore exploration and development have been moving into deep-water areas and into the frontiers in remote and difficult places where little search and discovery activities have taken place in the past.

293. The move of the offshore oil and gas industry to deeper waters is reflected in the recent records set for water depth—6,079 feet (1,853 metres) for offshore production, and 8,016 feet (2,443 metres) for offshore exploration drilling—both by Petrobras of Brazil. Globally, the offshore oil and gas industry has three major deep-water "plays," areas where exploration and production are under way: the Gulf of Mexico, Brazil and West Africa. Beyond these "big three" areas, other countries are also intensifying their deep-water activities. Indonesia and Egypt have recently made substantial finds in their relatively unexplored deep-water areas. Israel, Malaysia and India have made their first deep-water discoveries. Other countries such as Turkey, Australia, New Zealand and Norway are continuing their deep-water search. Denmark (Faroe Islands), Greenland and Guyana may also have deep-water prospects.

294. With huge reserves of oil and natural gas beneath frigid seas, the Russian Arctic is one of the last frontiers for offshore hydrocarbons. The year 2000 witnessed initiatives on the part of oil and gas companies of the Russian Federation, the United States, France and Germany relating to the development of the Russian Arctic deposits.

295. While the delimitation of maritime boundaries remained in many cases a potential source of conflict with regard to offshore oil and gas development, during the reporting period there have been a number of constructive efforts among the parties involved to arrive at a mutually beneficial resolution (see paras. 41–47).

---

120. UNEP/CBD/COP/5/3, recommendation V/14, annex II.B.
121. GESAMP, Report of the thirtieth session, Monaco, 22–26 May 2000, GESAMP Reports and Studies, No. 69, para. 7.3.

Oil and gas installations
296. Ageing or damaged offshore facilities pose a number of challenges to the offshore oil and gas industry as well as to the government regulatory agencies (see A/55/61, paras. 145–146; A/54/429, paras. 345–360). An Asia-Pacific Economic Cooperation (APEC) workshop, co-sponsored by China and the United States, on assessing and maintaining the integrity of existing offshore facilities, was held at Beijing in October 2000, with the participation of about 150 representatives from China, the United States, Australia, Mexico, Malaysia and other countries. The aim of the workshop was to deepen understanding of the assessment and mitigation process for ageing and damaged offshore facilities in order to sustain safe operation, secure environmental quality and maintain efficient use of petroleum resources.

Methane hydrates
297. Methane hydrates, solid ice-like substances composed of water and natural gas (methane), occur in areas of the world's oceans where appropriate conditions of temperature and pressure cause water and methane to combine to form a solid. With time, as conventional oil and gas reserves decline, methane hydrates are expected to become an economically important source of hydrocarbons. Thought is currently being given to using methane as a starting material for more complex molecules to use as liquid fuels and lubricants, and for the manufacture of key chemicals.

298. Scientific studies of the economic geology of methane hydrates and preliminary studies on the economic feasibility of methane hydrate production indicate that deposits spread in a thin layer across large areas may be less economically productive than thick vertically stacked deposits limited to smaller areas. The methane hydrate deposits in the Blake Plateau off the Atlantic seaboard of the United States represent the former, while the deposits in the Gulf of Mexico and in deep-water off West Africa are an instance of the latter. Thick deposits form in leaky oil and gas basins where hydrocarbons seep to the sea floor, rapidly crystallizing as gas hydrates.

2. Nonfuel Minerals

299. In June 2000, an international workshop held under the auspices of the International Seabed Authority in Kingston, Jamaica, provided a unique opportunity for scientists, technologists, policy makers and representatives of the marine mining industry from both public and private sectors to review the global marine mineral situation and assess the challenges and prospects of marine mining. More than 60 participants from 34 countries, from developed and developing countries as well as countries in transition, attended the week-long workshop. Although the workshop focused on "Mineral resources of the international seabed area," since scientific, technical, economic, and environmental issues traverse boundaries, all marine minerals were addressed.[122]

300. The development of conventional marine minerals and our knowledge of new types of marine minerals are expanding rapidly, enhancing prospects for significant current as well as potential economic returns.

301. Conventional marine minerals comprise those minerals derived by mechanical and chemical erosion from rocks on continents and transported to the

---

122. Much of the following information is excerpted from the information notes prepared by the International Seabed Authority relating to the workshop.

ocean primarily by rivers. Minerals derived by mechanical erosion from continental rocks are concentrated into placer deposits, which are sorted by waves, tides, and currents by virtue of the relatively high density (mass per unit volume) of the constituent minerals. These minerals contain heavy metals (barium, chromium, gold, iron, rare-earth elements, tin, thorium, tungsten and zirconium) and nonmetals (diamonds, lime, siliceous sand, and gravel). Of the metals, gold is mined intermittently offshore Alaska depending on price (currently inactive) and tin continues to be mined at sites offshore Thailand, Myanmar, and Indonesia. Of the nonmetals, a growing diamond mining industry exists offshore (water depths to 200 metres; distance to about 100 kilometres) Namibia and the adjacent coast of South Africa, with recovery of 514,000 carats reported for 1999 by the principal producer (De Beers Marine). Sand and gravel are mined from beaches and shallow offshore accumulations at various sites around the world for construction material (concrete) and beach restoration as the marine material with the highest annual production value.

302. In November 2000, the Government of Namibia and De Beers signed four agreements aimed at substantially increasing the offshore diamond production of Namibia over the next decade, with a resulting growth in national income and government revenue. The long-term agreements involve the Namibian Diamond Corporation (NAMBEB), an equal partnership between the Namibian Government and De Beers, and are expected to secure the future of Namibia's diamond industry in an increasingly competitive international diamond market. Namibia's offshore reserves consist principally of high-quality diamonds. Moreover, industrial-scale marine diamond mining is relatively free of the perils associated with smuggling.

303. Materials dissolved from continental rocks by chemical weathering and transported to the ocean by rivers is considered to provide the source for several marine minerals of future use. One of these resources is phosphorite, which precipitates in the form of nodules and layers where sea water upwells from the deep ocean at the continental shelf within the trade wind belt (30° latitude north and south of the equator). Phosphorite is used as an agricultural fertilizer.

304. Two groups of metallic mineral resources of the deep seafloor incorporate dissolved metals from both continental and deep ocean sources. One such group is the golf-to-tennisball-sized polymetallic nodules (nickel, cobalt, iron, and manganese in varying concentrations). These nodules have precipitated from sea water over millions of years as sediment on vast expanses of the abyssal plains of the deep ocean (water depth 4 to 5 kilometres). The most promising of these deposits in terms of nodule abundance and metal concentration are found in the Clarion-Clipperton zone of the eastern equatorial Pacific between Hawaii and Central America, an area that has been licensed by pioneer investors; another prospective area lies in the Indian Ocean.

Cobalt-rich ferromanganese crusts

305. Cobalt-rich ferromanganese crusts are the second group of metallic mineral resources that incorporates metals from both land and sea sources. They precipitate from sea water as thin layers (up to 25 centimetres thick) on volcanic rocks of seamounts and submerged volcanic mountain ranges between water depths of 400 and 4,000 metres. The most favourable settings for the occurrence of these crusts lie within and beyond the 200-nautical-mile zones of the island nations of the Western Pacific. It is estimated that one seabed mine site could complement land production

to meet up to 25% of the annual global need for cobalt (used to make corrosion-resistant, light and strong metal alloys, and paints) contingent on development of mining and refining technology.

Polymetallic nodules

306. Polymetallic nodules, like cobalt-rich ferromanganese crusts and polymetallic sulphides (see para. 316), occur on the seabed and ocean floor within national jurisdiction as well as in the Area i.e., the seabed and ocean floor and subsoil thereof, beyond the limits of national jurisdiction (the international seabed area).

307. The International Seabed Authority is the organization through which States parties to UNCLOS shall, in accordance with Part XI of UNCLOS, the related annexes and the Agreement on Part XI of UNCLOS, organize and control activities in the Area, in particular with a view to administering the resources of the Area. The Authority came into existence upon the entry into force of the Convention in 1994. Its headquarters are located in Kingston, Jamaica.[123]

308. Details about the work of the Authority may be found in the annual reports of the Secretary-General of the Authority to the General Assembly and on the Authority's web site at www.isa.org.jm. The most recent such report was presented to the Authority at its sixth session in July 2000 (ISBA/6/A/9). Among the most significant achievements of the Authority are the approval, in 1997, of plans of work for exploration for polymetallic nodules of seven registered pioneer investors and the approval by the Assembly of the Authority during its sixth session in July 2000 of the Regulations for prospecting and exploration for polymetallic nodules in the Area (ISBA/6/A/18). The approval of the Regulations will enable the Authority to issue 15-year exploration contracts, in accordance with the regime established by the Convention and the Agreement, to each of the seven registered pioneer investors.[124] It is expected that such contracts will be issued during 2001.

309. During 2000, the Assembly of the Authority completed the first periodic review, pursuant to article 154 of UNCLOS, of the manner in which the international regime for the Area has operated in practice. In carrying out its review, the Assembly noted that the regime established by UNCLOS had been subjected to de facto review and modification both by the Preparatory Commission for the International Seabed Authority and for the International Tribunal for the Law of the Sea in its work, in particular relating to the elaboration of the rules of procedure for the various organs of the Authority and the registration of pioneer investors, and in the informal consultations of the Secretary-General leading to the adoption of the Agreement on Part XI of UNCLOS. The Assembly further noted that the first four years of operation of the Authority had been primarily devoted to consideration

---

123. The following paragraphs have been excerpted from the contribution of the International Seabed Authority to the present report.

124. The seven registered pioneer investors are the Government of India; Institut français de recherche pour l'exploitation de la mer (IFREMER)/Association française pour l'étude et la recherche des nodules (AFERNOD) (France); Deep Ocean Resources Development Co. Ltd. (DORD) (Japan); Yuzhmorgeologiya (Russian Federation); China Ocean Mineral Resources Research and Development Association (COMRA) (China); Interoceanmetal Joint Organization (IOM) (Bulgaria, Cuba, Czech Republic, Poland, Russian Federation and Slovakia); and the Government of the Republic of Korea.

of the organizational issues necessary for the proper functioning of the Authority as an autonomous international organization. While the Authority had commenced its operational and substantive activities, the Assembly considered that, in the light of the Authority's very short experience in implementing the regime, it would be premature to make any recommendations concerning measures to improve the operation of the regime (ISBA/6/A/19, para. 8).

310. The adoption by the Authority of the Regulations on prospecting and exploration for polymetallic nodules is an important milestone towards commercial exploitation of deep-sea minerals in the Area. UNCLOS stipulates that the Area and its resources are the common heritage of mankind. Exploration for and exploitation of the resources of the Area are to be carried out for the benefit of mankind as a whole; equitable sharing of financial and economic benefits derived from such activities is provided for in article 140 of UNCLOS. The Regulations are the first segment of a so-called "seabed mining code" that will eventually govern exploration for and exploitation of all deep-sea minerals in the Area. The regulations set out the provisions that prospective seabed miners, on the one hand, and the Authority, on the other, must follow in any work to locate and evaluate deposits of nodules rich in valuable metals such as nickel, copper, cobalt and manganese. Regulations governing the exploitation of nodules will be drawn up in due time.

311. The 40 regulations and four annexes were worked out over four years. Different rules are provided for prospecting and for exploration. Prospecting, defined as the search for deposits, including estimates of composition and value, confers no exclusive rights and requires little more than notification to the Authority of where the activity will take place. Exploration, defined to cover searching, analysis, tests of collecting and processing equipment and systems and commercial and other studies, does involve exclusive rights in a geographical area no other operator can work in. Exploration cannot take place until the Council of the Authority has approved a plan of work submitted by an operator and specified in a contract with the Authority. The Regulations also cover such matters as procedures to be followed by operators, fees, application for approval of plans of work for exploration, contracts for exploration and settlement of disputes.

312. In the drafting of the Regulations, most of the time was spent on fleshing out and reaching compromises on provisions in two areas: protection and preservation of the marine environment and safeguards for confidential data and information to be supplied to the Authority by the operators. Delegates sought to reconcile the need to safeguard the marine environment against potential threats of pollution from the activities of the operators and other damage with the need to encourage seabed investors by avoiding over-regulation. They also sought to ensure that the Authority obtained enough information from operators to evaluate and monitor their activities, while ensuring that commercially valuable data did not leak out to potential competitors.

313. The Authority will begin work at its forthcoming seventh session on regulations on prospecting and exploration for minerals that were just being discovered as UNCLOS was being drawn up (the seventh session of the Authority will be held at Kingston, Jamaica, from 2 to 13 July 2001). These are polymetallic sulphides (also referred to as polymetallic massive sulphides or sea-floor massive sulphides) and cobalt-rich ferromanganese crusts.

314. ECLAC reported that, in respect of nonliving marine resources, the focus of ECLAC in 2000 had been the regional follow-up to the work of the Authority and the provision of technical inputs for the negotiation and future implementation of the Regulations on prospecting and exploration for polymetallic nodules in the Area. In this context, ECLAC had published a study on the negotiation at the International Seabed Authority entitled: "A renewed opportunity for the contribution from the Latin American and the Caribbean Group."

315. National efforts to benefit from polymetallic nodules in the exclusive economic zone include those of the Cook Islands, which has recently established the National Research Institute. The main task of the Institute is to accelerate the development of polymetallic nodules in the exclusive economic zone of the Cook Islands. In mid-2000, the Government of the Cook Islands and the Norwegian Deep Seabed Mining Group signed a letter of intent that provided for the start-up of phase I of a polymetallic nodule project in August 2000. The phase lasted for four months and involved a business case study to determine the economic viability of mining the nodule deposits off the coast of the Cook Islands. The Cook Islands nodules are known to have a high abundance (estimated at 14 billion tons) and a high cobalt content, and are found in relatively obstacle-free ocean floor. These factors could make deep-sea mining in the Cook Islands attractive. Phase I of the project also evaluated the need to build up support institutions, formulate laws and regulations and study the environmental impacts of mining. The phase was completed in November 2000 and a team of researchers from the Norwegian Group was expected to visit the Cook Islands and report on the findings.

Polymetallic sulphides

316. Polymetallic massive sulphides are types of deposits discovered in the oceans in 1979; they contain copper, iron, zinc, silver and gold. They are deposited from sea-floor hot springs that are heated by molten rocks that upwell beneath a submerged volcanic mountain range that extends through all the ocean basins of the world (water depth 1 to 4 kilometres). At the current early stage, when only about 5% of the seabed has been systematically explored, about 100 such sites have been found, mostly associated with volcanic island chains that border the western margin of the Pacific Ocean. Polymetallic massive sulphide deposits constitute resources for the future. One such site actively forming on the floor of the Bismarck Sea within the 200-nautical-mile zone of Papua New Guinea was leased in 1997 from that Government by an Australian mining company and is under development for mining.

317. Sea-floor hot springs not only concentrate metals, but also provide chemical energy used by microbes to manufacture their food at the base of a food chain that supports an ecosystem of new life forms hosted in the metallic mineral deposits. This ecosystem is of scientific and commercial value in sustaining biodiversity, elucidating the early evolution of life and producing novel organic compounds valuable for industrial and pharmaceutical applications. The coincidence of nonliving and living resources poses the challenge to develop a regime that enables the sustainable development of both resources while protecting the ecosystems.

318. Scientists have recently discovered a field of hydrothermal vents with "chimneys" of carbonate and silica, potentially a new source of marine minerals. Their carbonate and silica composition differentiates them from the hydrothermal

vents associated with polymetallic sulphides, whose chimneys are formed from sulphur- and iron-based minerals. The new chimneys are also the tallest ever found, nearly 200 feet tall.

319. As pointed out by South Pacific Applied Geoscience Commission (SOPAC), major constraints to the development and exploitation of deep-sea mineral deposits continue to be the need to develop suitable and cost-effective mining technology and to resolve the legal and boundary issues related to the ownership of the resources. With regard to the issues of deep-sea mineral exploration licences, SOPAC indicates that support from the international community is urgently needed to assist national and regional efforts in the Pacific to assess resource information and to develop appropriate policies and legislative regimes for this activity. As it is a recent development, national capacity needs to be built to ensure that deep-sea mineral exploration is managed, regulated and monitored effectively.

C. Protection and Preservation of the Marine Environment

1. Reduction and Control of Pollution

(a) Land-based activities: the Global Programme of Action[125]

320. It is recalled that by its 1999 decision 20/19 B, the UNEP Governing Council decided to convene the first intergovernmental review of the implementation of the Global Programme of Action for the Protection of the Marine Environment from Land-Based Activities (GPA), which will be held in Montreal, Canada, from 19 to 23 November 2001, and requested the UNEP Executive Director to organize, in cooperation with Governments, United Nations bodies and agencies and other relevant organizations, an expert group meeting to facilitate the preparation for the review. In keeping with that request, an Expert Group Meeting to prepare for the first Intergovernmental Review Meeting on implementation of the GPA was held at The Hague, from 26 to 28 April 2000.

321. The experts noted that since the adoption of the GPA in November 1995 positive developments had taken place related to the protection of the marine and coastal environment in some regions, which had contributed to the implementation of the GPA.

322. The low level of participation in the Expert Group Meeting, which was attended by only two of the six United Nations agencies dealing with the implementation of GPA—UNESCO/IOC and United Nations Centre for Human Settlements (Habitat), was noted with concern.

323. In addition, the Expert Group Meeting also established a GPA Correspondence Group and recommended that the Executive Director of UNEP consider the establishment of a Steering Committee to advise UNEP on the intergovernmental review process and the 2001 Review Meeting. Accordingly, the GPA Coordination Office drafted two documents, which were circulated for comments to the GPA Correspondence Group and others: one outlining the proposed preparatory process

---

125. Excerpted from documents and contributions of UNEP and the GPA Coordination Office.

and the expected specific products of the Intergovernmental Review Meeting, and the other a draft GPA High-Level Statement.

324. The major goal of the Intergovernmental Review Meeting is to secure commitments from a full range of partners (including Governments, international and regional governmental and nongovernmental organizations, the private sector, international financing institutions, regional banks and commissions, civil society and other major groups) to advance GPA implementation, based on defined specific activities, targets and financial agreements. The Meeting also aims at mobilizing awareness and active participation and involvement of relevant stakeholders at the national, regional and global levels. The specific objectives of the Meeting, in keeping with paragraph 77 of the GPA, are: (a) to review progress on the implementation of the GPA at the national, regional and global levels; (b) to review the results of scientific assessments regarding land-based impacts upon the marine environment provided by relevant scientific organizations and institutions, including GESAMP (see paras. 597–603); (c) to consider reports on national plans to implement the GPA; (d) to review coordination and collaboration among organizations and institutions, regional and global, with relevant responsibilities and experience; (e) to promote the exchange of experience between regions; (f) to review progress in capacity-building and mobilization of resources to support the implementation of the GPA, in particular in countries in need of assistance, and where appropriate, to provide guidance; and (g) to consider the need for international rules, recommended practices and procedures to further the objectives of the GPA.

325. With respect to reporting by United Nations agencies to the Intergovernmental Review Meeting and, in particular, the Subcommittee on Oceans and Coastal Areas (SOCA) of the Administrative Committee on Coordination (see paras. 586–596), UNEP and SOCA agreed that each agency would submit by 31 March 2001 to the Coordination Office (a) a list of GPA-relevant projects, either using the United Nations Atlas of the Oceans or another vehicle, and (b) a report on their activities (including problems encountered, limitations and recommendations) in support of implementation of the GPA. The individual inputs of the agencies would be consolidated into a single report by the GPA Coordination Office and circulated to the agencies for comments. The final document would constitute the collective input of the members of SOCA to the GPA Intergovernmental Review Meeting and would be attached to the GPA Ministerial/High-Level Declaration, which would emanate from the Meeting.

326. The central node of the GPA clearing-house mechanism (www.gpa.unep.org) continues to be expanded with the addition of new content, the reorganization of some elements to improve ease of use and the enhancement and development of new functionality. Progress is also being achieved with other clearing-house initiatives, including the development of the pollutant-source category nodes by the relevant United Nations agencies, the development of regional prototype nodes and the acquisition of support and resources for additional activities. It is intended that the GPA clearing-house mechanism will be fully compliant with new UNEP-wide information management initiatives (UNEP.NET).

327. Pollutant source category nodes in various fields have been developed or are currently under development. These include the sewage clearing-house node, being developed with WHO and core partners; the nutrients and sediment mobilization clearing-house node, with FAO; the oils (hydrocarbons) and litter clearing-

714   *Selected Documents and Proceedings*

house node, with IMO (Government of Canada, OSPAR and the Swedish Environmental Protection Agency); the radioactive substances clearing-house node, with IAEA; the persistent organic pollutants (POPs), with UNEP Chemicals (Geneva); the physical alterations and destruction of habitats clearing-house node, being developed with the Biodiversity Convention Secretariat, the regional seas programmes and UNEP; and the heavy metals clearing-house node, being developed with UNEP Chemicals (Geneva).

328. The GPA Coordination Office is also initiating regional clearing-house activities in partnership with the regional seas programme. Two pilot projects have been ongoing since late 1999, one in collaboration with the South Pacific Regional Environment Programme (SPREP) and the other with the Caribbean Environment Programme. The needs evaluation and work plan for the South Pacific have almost been finalized. The next stage will entail developing a prototype node and obtaining the necessary funding and support to implement the work plan. The needs evaluation and work plan have been completed for the Caribbean Environment Programme and a prototype node has been developed. Assuming the availability of potential donor and partner support, funding, and support are being sought to initiate GPA clearing-house developments in other regional seas areas in early 2001.

329. Two recent regional assessments or overviews of land-based activities (available electronically through the GPA clearing-house (www.gpa.unep.org)), have been published: (a) Overview of land-based sources and activities affecting the marine environment in the East Asian seas (Regional Seas Report and Studies Series No. 173); and (b) Overview of land-based pollutant sources and activities affecting the marine, coastal and freshwater environment in the Pacific Islands region (Regional Seas Report and Studies Series No. 174). In addition, the Coordinating Office has established close collaborative links with the non-UNEP regional seas programme of the Helsinki Commission for Baltic Marine Environment Protection (HELCOM) (see paras. 386–387), the OSPAR Commission (for the North Atlantic) (see paras. 388–389) and the Protection of the Arctic Marine Environment (PAME) (see paras. 390–392) with the aim of exchanging information and experiences, receiving their contributions to the 2001 Intergovernmental Review and linking with their respective web sites and "twinning."

330. In the Pacific region, a recent assessment as part of the region's response to the Global Programme of Action for the Protection of the Marine Environment from Land-Based Activities identified the major source categories of marine pollution as domestic waste; agricultural; industrial; and physical alterations/habitat modifications, degradation, and destruction, including dredging, sand extraction, and seabed mining. In this connection, a work programme for 2002–2006 to further GPA implementation is being developed by the GPA Coordination Office together with the SPREP, which will act as the GPA regional focal point for the Programme. The same type of programme during the indicated period is being carried out for the region of the wider Caribbean, with Caribbean/Regional Coordination Unit, acting as the GPA regional focal point.

331. At the national level, a joint GEF/GPA project proposal for the development and implementation of national programmes of action on land-based activities, such as tourism, ports and harbours, bays and estuaries, in 20 countries (Brazil, Colombia, Costa Rica, Democratic People's Republic of Korea, Egypt, Georgia, India, Indonesia, Jamaica, Jordan, Panama, Saudi Arabia, Seychelles, Sri Lanka, Sudan, Philippines, Republic of Korea, United Republic of Tanzania, Vanuatu and Yemen),

within the framework of the regional seas programme, is being finalized for submission to GEF.

332. At the first Meeting of the Consultative Process (see paras. 612–615), the GPA Coordination Office was invited to give a presentation during Discussion Panel B on "Economic and Social Impacts of Marine Pollution and Degradation, especially in coastal areas: International aspects of combating them."

333. During the period under review, the GPA Coordination Office contributed to and participated in several forums in which consideration was given to the preparations for the GPA Intergovernmental Review Meeting:

- High-level Government-designated Expert Meeting of the Proposed Northeast Pacific Regional Seas Programme, Panama, 5–8 September 2000, which considered the draft of a regional Convention for the Protection and Sustainable Development of the Marine and Coastal Areas (a significant component of the draft Convention is pollution from land-based activities);
- Coastal Zone Canada Conference, Saint John, 17–22 September 2000, where two GPA-related sessions were organized, one on lessons learned and moving to GPA implementation within the context of the Intergovernmental Review Meeting, and another on municipal wastewater;
- International Ocean Institute Leadership Seminar on Mediterranean Basin-wide Co-development and Security, Malta, 21–22 September 2000, where, as a follow-up, the secretariat of the Mediterranean Action Plan (MAP) and the GPA Coordination Office have agreed on, *inter alia,* the input of the Mediterranean region into the GPA Intergovernmental Review, the participation of MAP in the Steering Committees of two GEF/GPA medium-size projects, forward "twinning" arrangements with other regional seas programmes to facilitate GPA implementation and implementation of the Jakarta Mandate with the Convention on Biological Diversity Secretariat;
- Third Global Meeting of Regional Seas Conventions and Action Plans, Monaco, 5–10 November 2000, in which secretariats of 17 regional seas programmes agreed to take the lead, together with the GPA Coordination Office, in the regional preparatory process leading to the GPA Intergovernmental Review Meeting, to strengthen the programmatic links with GPA activities and to work together with the GPA Coordinating Office and the Biodiversity Convention Secretariat in addressing GPA requirements on physical alteration and destruction of habitats (see paras. 378–382);
- Fifth Global Forum of the Water Supply and Sanitation Collaborative Council (WSSCC), Iguaçu, Brazil, November 2000, where the Recommendations for Decision-Making on Municipal Wastewater (developed jointly by the GPA Coordination Office, WHO, Habitat and WSSCC) were presented;
- North-West Pacific Intergovernmental Meeting, Tokyo, 3–4 December 2000, which agreed on the development of a regional programme of action on land-based activities.

(b) Pollution by dumping; waste management

334. It is estimated that dumping contributes to 10% of the potential pollutants in the oceans. Control of pollution of the marine environment by dumping is depen-

dent on finding solutions to problems engendered by land-based sources of marine pollution and proper waste management in general.

335. The United Nations General Assembly at its fifty-fifth regular session (2000) in its resolution on oceans and the law of the sea reiterated its concern about the degradation of the marine environment as a result of pollution by dumping of hazardous waste, including radioactive materials, nuclear waste, and dangerous chemicals, and urged States to take all practicable steps, in accordance with the 1972 Convention on the Prevention of Marine Pollution by Dumping of Wastes and Other Matter (London Convention), to prevent the pollution of the marine environment by dumping. The Assembly once again called upon States to become parties to and to implement the 1996 Protocol to the 1972 Convention.

336. As of January 2001, there were 78 Contracting Parties to the London Convention, and 13 States had ratified the 1996 Protocol. A number of countries have informed IMO that they will soon ratify or accede to the Protocol,[126] and it is thus likely that the Protocol will enter into force during 2002 (26 ratifications or accessions are required for its entry into force, of which 15 must come from Contracting Parties to the London Convention). IMO pointed out that once the Protocol comes into force, there would be a transitional period during which both the 1972 and the 1996 regimes would be in operation. Governments still party to the Convention should become parties to the Protocol as soon as possible to ensure that the 1972 Convention is entirely replaced by the 1996 Protocol as the sole international global regime regulating the dumping of wastes at sea.

337. *Relationship between the 1996 Protocol, the 1972 London Convention and UNCLOS.* The regime set out in 1996 Protocol is stricter than that of the 1972 London Convention. The 1996 Protocol, *inter alia*, prohibits the dumping of all wastes or other matter with the exception of certain materials listed in the annex, i.e., dredged material, sewage sludge, fish waste or material resulting from industrial fish processing operations, vessels, platforms or other man-made structures at sea, inert, inorganic geological material, organic material of natural origin, bulky items comprising iron, steel, concrete, etc.

338. States parties to UNCLOS, which are party to neither the 1996 Protocol nor the 1972 Convention are faced with the question of whether it is the 1996 Protocol or the 1972 London Convention that contains the global rules and standards referred to in articles 210 and 216 of UNCLOS, and therefore sets the minimum standard for the national laws and regulations that parties to UNCLOS must adopt and enforce, irrespective of whether they are also party to the 1972 Convention or the 1996 Protocol.

339. *Waste assessment guidance.* Guidelines for the assessment of each of the eight wastes permitted to be dumped under the 1996 Protocol (see para. 337), were adopted by the Contracting Parties to the London Convention at the twenty-second Consultative Meeting (September 2000). The Guidelines, which can also be applied to the wastes allowed to be dumped under the London Convention,[127] give a stepwise

---

126. See the report of the twenty-second Consultative Meeting of Contracting Parties to the London Convention (September 2000), IMO document LC 22/14, paras. 2.3–2.6, available on the web site of the Office of the London Convention, at: <www.londonconvention.org>.

127. *Ibid.*, annexes 3–10.

orientation to ensure that sufficient scientific and technical advice is collected to select appropriate waste management options and assess implementation of the chosen option.

340. The Contracting Parties agreed to keep the Guidelines under review and update them in five years, or earlier as warranted in the light of new technical developments and the results of scientific research. The adequacy of the existing international provisions for the disposal of vessels at sea would be reviewed in four years time, particularly in the light of the experience with implementing the waste-specific guidelines for the assessment of vessels as adopted by the Contracting Parties.

341. *Implementation of and compliance with the London Convention.* It is difficult to assess the current extent of dumping at sea by States. Information provided by the Secretariat to the Consultative Meeting indicates that only a small percentage of Contracting Parties to the London Convention have been meeting their notification and reporting requirements under article VI (4) of the Convention and sent reports to the Secretariat on their dumping activities from 1976 to 1998.[128] Indeed, the initiative taken in 2000 by the Contracting Parties to address the problem (see A/55/61, paras. 162–164), namely the circulation of a questionnaire (LC.2/Circ.403) to all Contracting Parties requesting information on areas possibly presenting barriers to compliance, was met with a low return. Only 15 out of 78 Contracting Parties sent a response, thus making it impossible to draw a firm conclusion on the views and needs of States with regard to compliance. At the twenty-second Consultative Meeting, the Contracting Parties requested the Secretariat to communicate with the States concerned once again to encourage them to respond to the questionnaire and to submit their reports as a matter of urgency.

342. IMO reported that other initiatives to improve compliance with the Convention included development of guidance for States on the implementation of the 1996 Protocol, including the establishment of a Correspondence Group to prepare a proposal for the adoption of such guidance at the next Consultative Meeting; the development of a proposal for a scheme to obtain funds for various projects to facilitate compliance with the London Convention; and reporting on "illegal dumping activities."

343. The Consultative Meeting viewed technical cooperation as a critical component in promoting the Convention and the Protocol and agreed, *inter alia*, to develop a long-term strategy for technical cooperation, to improve coordination with other international organizations, e.g., UNEP, and to intensify outreach to States wishing to join the Protocol.

344. *Radioactive waste management.* The London Convention prohibits the disposal at sea of radioactive wastes. The International Atomic Energy Agency stated that all materials, including those that can be disposed at sea in accordance with the Convention, contain radionuclides, of both natural and artificial origin. At the request of the London Convention Secretariat, IAEA developed definitions and criteria for making judgements on whether materials considered for dumping at sea could be treated as essentially "nonradioactive" for the purposes of the London Convention. Its advice on the matter was presented in IAEA-TECDOC-1068, published in March 1999 (see A/54/429, para. 392). At the twenty-second Consultative Meeting of the Contracting Parties (September 2000), IAEA presented another doc-

128. IMO document LC 22/3/2.

ument entitled "Guidance on Radiological Assessment Procedures to Determine if Materials for Disposal at Sea are within the Scope of the London Convention 1972," which further elaborates IAEA's advice on the subject. The final report, to be published in 2001, contains guidance on how to perform an assessment to determine if levels of radioactivity in materials to be disposed of at sea meet the exemption criteria established in IAEA-TECDOC-1068.

345. IAEA reported that it had been working for some years on assembling information on all inputs of radioactivity into the world oceans. A report on the disposal at sea of radioactive waste was published in August 1999 (IAEA-TECDOC-1105) (see A/55/61, para. 165). A second report on accidents and losses at sea resulting in actual or potential release of radioactive material into the marine environment as well as accidents and losses where the radioactive material had been recovered intact was presented at the twenty-second Consultative Meeting and is also to be published in 2001. The information gathered on the inputs of radioactive material into the oceans is incorporated into the IAEA Clearing House on Radioactive Substances, which will be linked to the GPA node (see paras. 326–327). A new database, the Global Marine Radioactivity Database (GLOMARD), has also been created. It covers the distribution of radionuclides in the Atlantic, Pacific, Indian and Southern oceans and has been extensively used for the development of time series of the worldwide distribution of radionuclides in seawater and sediment.

(c) Pollution from vessels

346. Some pollutants, such as oil, noxious liquid substances, sewage, garbage, antifouling paints or unwanted aquatic organisms, are released into the marine environment by ships in the course of their routine operations, either as a result of accidents, or illegally. However, most pollutants enter the marine environment as a result of routine operational discharges. As much as 92% of all oil spills involving tankers occur at the terminal during loading or unloading.

347. UNCLOS regulates pollution from ships by requiring States, acting through the competent international organization or a general diplomatic conference, to establish international rules and standards to prevent, reduce and control the pollution of the marine environment from vessels and to re-examine them from time to time as necessary. For the flag State such global rules and standards constitute the minimum standard that it must adopt for vessels flying its flag. Coastal States can adopt stricter rules and standards than the generally accepted global standards for application in their territorial sea, so long as such standards do not apply to the design, construction, manning or equipment of foreign ships or hamper innocent passage. In the exclusive economic zone, the generally accepted international rules and standards apply.

348. Apart from the IMO safety-related conventions, which are critical for the prevention of accidents (see sect. IV.B of the present report), the generally accepted international rules and standards for the prevention of pollution from vessels are mainly contained in MARPOL 73/78 (see para. 101). That Convention sets out where and under what conditions a vessel may discharge oil (Annex I), noxious liquid substances (Annex II), sewage (Annex IV) and garbage (Annex V). Annex III to MARPOL 73/78 contains regulations for the prevention of pollution by harmful

substances carried by sea in packaged form. Instruments on controlling the use of harmful anti-fouling systems and on ballast water management are currently being developed by IMO (see paras. 360–367).

349. At its fifty-fifth session (2000), the General Assembly in its resolution on oceans and the law of the sea reiterated its concern about the degradation of the marine environment as a result of pollution from ships, in particular through the illegal release of oil and other harmful substances, and urged States to take all practicable steps, in accordance with MARPOL 73/78, to prevent the pollution of the marine environment from ships.

(i) Developments in relation to the MARPOL Annexes
Entry into force of amendments in 2001
350. The amendments to Annexes I and II adopted by the IMO Marine Environment Protection Committee in July 1999 (resolution MEPC.78(43); see A/54/429, para. 401) entered into force on 1 January 2001. The amendments to Annex III adopted in resolution MEPC.84(44) in March 2000 will enter into force on 1 July 2001.[129]

New amendments adopted in 2000
351. Apart from the above-mentioned amendments to Annex III, new amendments to MARPOL Annex V were adopted by MEPC at its forty-fifth session, in October 2000 (resolution MEPC.89(45)) and are expected to enter into force on 1 March 2002 under the system of tacit acceptance of amendments. They include, *inter alia*, an update of the definition of "nearest land" and the addition of incinerator ashes as discharges from plastic products, which may contain toxic or heavy metal residues, to the list of materials, whose disposal is prohibited. And in its resolution MEPC.92(45),[130] MEPC amended the Revised Guidelines for the Implementation of Annex V.

Phasing-out of single-hull tankers
352. As a result of the sinking of the *Erika*, much of the attention of the shipping community during the reporting period has focused on examining the adequacy of existing global rules and standards, and in the context of MARPOL, the current timetable phasing out single-hull tankers. IMO reported that MEPC at its forty-fifth session had approved the first, formal step towards a global timetable for the accelerated phasing-out of single-hull oil tankers, thereby enabling the adoption of a revised regulation 13G of MARPOL at the forty-sixth session of MEPC, the dates of which had already been moved forward to April 2001 to permit the swiftest possible introduction of the new rules.[131]

353. IMO reported that the draft revised text of regulation 13G as developed by an MEPC working group set out two clear alternative schemes, A and B, for phasing out single-hull tankers. Both schemes would see category 1 vessels (oil tank-

---
129. For the text, see report of the 44th session of MEPC, document MEPC 44/20, annex 3.
130. For the text of the amendments, see report of the 45th session of MEPC, document MEPC 45/20, annexes 3 and 7.
131. *Ibid.*, paras. 7.18–7.107, and annex 9.

ers of more than 20,000 dwt, which do not comply with the requirements for protectively located segregated ballast tanks (commonly known as pre-MARPOL tankers)) phased out progressively between 1 January 2003 and 1 January 2007, depending on their year of delivery. Category 2 tankers (same size as category 1 tankers, but complying with the protectively located segregated ballast tank requirements (MARPOL tankers)), built in 1986 or earlier would be phased out after their 25th year of operation under both schemes, but category 2 ships built after 1986 would be phased out between 2012 and 2015 under alternative A and between 2012 and 2017 under alternative B. For category 3 tankers (oil tankers with less tonnage than category 1 and 2 tankers) built in or before 1987, both schemes entail progressive phasing-out of tankers between 2003 and 2013, but ships built after 1987 would be phased out between 2013 and 2015 for ships under scheme A and between 2013 and 2017 under scheme B. The continued operation of category 1 and 2 oil tankers beyond 2005 and 2010, respectively, would only be permitted for ships that had been subject to a Condition Assessment Scheme.

354. IMO stated that there was general agreement at the MEPC that the phasing-out of single-hull tankers should be seen as just one of several measures needed to help eliminate sub-standard tankers. The working group therefore drew up a preliminary list of topics to be considered in this regard. The Committee invited the Maritime Safety Committee to establish a working group at MSC 73 (November/December 2000), to examine fully all the measures listed and, initially, to separate the list into maritime safety and environmentally related issues. MSC would also request the technical subcommittees to develop relevant issues further and report to MEPC and MSC with a proposed implementation plan.

355. IMO reported that most delegations had cautiously welcomed the proposed revision of regulation 13G; many had expressed their approval of the constructive spirit in which the meeting had addressed the issue.

Pollution from fishing vessels and small craft
356. Fishing vessels have been identified in many countries and by many projects as a major source of marine pollution through the release of marine debris, discarded fishing nets and waste to the marine environment. Another difficulty is said to arise from unregulated carriage and refuelling at sea to support fishing activities. At the Workshop entitled "The Prevention of Marine Pollution in the Asia-Pacific Region" (Townsville, Australia, 7–12 May 2000),[132] it was noted that these problems were compounded by a relatively low level of awareness of the problems of marine pollution in many fishing communities and should be the targets of focused education and awareness-raising programmes. The Workshop also observed that small craft, including yachts and other recreational vessels, posed threats to the marine environment since they were sources of debris and waste and might carry marine pests.

357. The United Nations General Assembly in its resolution 55/8 of 30 October 2000, entitled "Large-scale pelagic drift-net fishing, unauthorized fishing in zones of national jurisdiction and on the high seas, fisheries by-catch and discards, and other developments," called upon FAO, IMO, regional and subregional fisheries management organizations and arrangements and other appropriate intergov-

---

132. See n. 34.

ernmental organizations to take up, as a matter of priority, the issue of marine debris as it relates to fisheries and, where appropriate, to promote better coordination and help States to fully implement relevant international agreements, including Annex V of MARPOL and the Guidelines for the Implementation of Annex V.

358. Other possible responses to the problem of marine debris, in particular from derelict fishing gear, which have been put forward include the establishment of an international plan of action to prevent the discard, minimize the loss and maximize the recovery of fishing gear[133] and the enhancement of the effectiveness of MARPOL Annex V by integrating into the annex itself the provisions of the Guidelines for the Implementation of Annex V relating to discarded or lost fishing gear, in particular those for reporting and recording discarded or lost fishing gear and shipboard operational waste.[134]

(ii) Progress in the drafting of new instruments
359. Two other major areas of focus by IMO during the period under review have been the control of harmful anti-fouling systems and ballast water management and their regulation at the global level.

Draft international convention on the control of harmful anti-fouling systems
360. IMO reported that MEPC had approved in principle the draft International Convention on the Control of Harmful Anti-fouling Systems,[135] which had been elaborated pursuant to IMO Assembly resolution A.895(21), "Anti-fouling systems used on ships," adopted in November 1999, which had called on MEPC to develop an instrument, legally binding throughout the world, to address the problem. The resolution had called for a global prohibition on the application of organotin compounds acting as biocides in anti-fouling systems on ships by 1 January 2003, and a complete prohibition on the presence of such compounds by 1 January 2008. A number of issues, including entry-into-force criteria, remained open for discussion before the Conference scheduled to be held in October 2001 to adopt the convention.

361. Under the terms of the proposed new Convention, parties would be required to prohibit and/or restrict the use of harmful anti-fouling systems on ships flying their flag. The Convention would apply to all ships; ships above a certain size (to be determined) would be required to have their anti-fouling systems surveyed and to carry an anti-fouling certificate. Anti-fouling systems to be prohibited or controlled would be listed in annex I to the Convention. Initially, the annex would include a reference to "organotin compounds that act as biocides in anti-fouling systems."

362. The Convention would allow for additional substances to be included in the annex and set out a procedure therefor: a proposal to prohibit or restrict a particular substance would be put before an expert group established by IMO that would assess the adverse effects of the particular anti-fouling system. The Conven-

---

133. Participants' Declaration of Resolve from the International Marine Debris Conference on Derelict Fishing Gear and the Ocean Environment, 6–11 August 2000, Hawaii, IMO document MEPC 46/INF.8, annex 2.
134. MEPC 46/INF.8, para. 22.
135. IMO document AFS/CONF/2.

tion would provide an agreed format for an international anti-fouling certificate and set out procedures for survey and certification.

363. Further to the information provided by IMO, attention is drawn to two of the articles in the draft Convention, which are currently within square brackets. If the existing wording of the draft article entitled "Dispute settlement" is adopted, parties can choose the means for the peaceful settlement of disputes, which include the dispute settlement procedures in UNCLOS. However, if the current wording of the draft article entitled "Relationship to international law and other agreements" were adopted, it would provide that nothing in the Convention shall prejudice the rights and obligations of any State under customary international law as reflected in UNCLOS or under any existing international agreement. The advisability of referring to UNCLOS as a mere reflection of customary international law is questionable, in view of the annual call by the General Assembly to all States that have not done so to become parties to UNCLOS in order to achieve the goal of universality. It is also not clear what is intended by the reference to "any existing international agreement," especially since UNCLOS also meets that description.

Harmful aquatic organisms in ballast water
364. It is estimated that about 10 billion tons of ballast water are transferred globally each year, potentially transferring from one location to another species of marine life that may prove ecologically harmful when released into a non-native environment. A new initiative to respond to this severe environmental problem is the Global Ballast Water Management Programme (GloBallast). This IMO/GEF/UNDP project entitled "Removal of Barriers to Effective Implementation of Ballast Water Control and Management Measures in Developing Countries" is intended to help countries implement effective measures to control the introduction of unwanted aquatic organisms (see para. 583, and also the web site of GloBallast at http://globallast.imo.org).

365. Another response to the problem is the development of mandatory regulations. IMO reported on the progress made by the MEPC Working Group (see also A/55/61, para. 189) in developing draft new regulations for ballast water and sediments management to prevent the transfer of harmful aquatic organisms in ballast water. A diplomatic conference is planned for 2002 or 2003 to adopt the new measures. The proposed new instrument is being developed on the basis of a two-tiered approach. Tier 1 requirements would apply to all ships and include mandatory requirements for a ballast water and sediments management plan, a ballast water record book and a requirement that new ships carry out ballast water and sediment management procedures to a given standard or range of standards. Existing ships would be required to carry out ballast water management procedures after a phase-in period, but these procedures may differ from those to be applied to new ships.

366. Tier 2 includes special requirements that may apply in certain areas, and would include procedures and criteria for the designation of such areas in which additional controls may be applied to the discharge and/or uptake of ballast water. It was noted at the forty-fifth session of MEPC that careful consideration should be given to the definition of zones for the discharge and/or uptake of ballast water in the light of the provisions of UNCLOS.[136] The Working Group has requested advice

---

136. MEPC 45/20, para. 2.22.

on "the implications/limitations under article 196 and other relevant articles of [UNCLOS] when establishing Ballast Water Management Areas beyond an exclusive economic zone."[137]

367. IMO reported that the Working Group had confirmed that ballast exchange on the high seas was the only widely used technique currently available to prevent the spread of unwanted aquatic organisms in ballast water and that its use should continue to be accepted. However, it was stressed that the technique had a number of limitations: it was of variable efficiency in removing organisms; the percentage removed depended upon the type of organism; the discharged water quality depended upon the original quality of the water taken up. It also had geographical limits. Furthermore, although existing ships might be subject to operational constraints, new ships might be designed to accommodate ballast exchange in a much wider range of circumstances. The Working Group concluded that development of alternative treatment technologies might produce techniques that were substantially more reliable and that ballast water exchange was an interim solution.

(iii) Liability and compensation for oil pollution damage
Increase in the limits of compensation for oil pollution damage
368. Under the 1992 Protocol to the International Convention on Civil Liability for Oil Pollution Damage (CLC Convention) the shipowner is strictly liable for damage suffered as a result of a pollution incident. If an accident at sea results in pollution damage of a value, which exceeds the compensation available under the CLC Convention, the IOPC Fund, created by the 1992 Protocol to the 1971 International Convention on the Establishment of an International Fund for Compensation for Oil Pollution Damage (the Fund Convention), which is made up of contributions from oil importers, will be available to make up the balance. The regime established by the two treaties thus ensures that the burden of compensation is spread more evenly between shipowner and cargo interests.

369. IMO reported that its Legal Committee at its eighty-second session (October 2000) had adopted amendments to the CLC Convention and to the IOPC Fund.[138] The amendments raise by 50% the limits of compensation payable to victims of pollution by oil from oil tankers. They are expected to enter into force on 1 November 2003, unless objections from one fourth of the Contracting States are received before then.

370. The increased limits were adopted in the wake of two major incidents, the *Nakhodka* in 1997 off Japan and the *Erika* disaster off the coast of France in December 1999. The amendments to the CLC Convention raise the limits payable to 89.77 million Special Drawing Rights (SDR) (approximately US$ 115 million) for a ship over 140,000 gross tonnage, up from 59.7 million SDR ($76.5 million) established in the 1992 Protocol. The amendments to the IOPC Fund raise the maximum amount of compensation payable from the IOPC Fund for a single incident, including the limit established under the CLC amendments, to 203 million SDR ($260 million), up from 135 million SDR ($173 million). However, if three States contributing to the Fund receive more than 600 million tons of oil per annum, the

---

137. MEPC 46/3, para. 3.2.2.
138. IMO document LEG 82/12, annexes 2 and 3.

maximum amount is raised to 300.74 million SDR ($386 million), up from 200 million SDR ($256 million).

371. Further to the information provided by IMO, it should be noted that the 1971 Fund Convention will cease to apply as of 27 March 2001, the date on which the Protocol of 2000 will enter into force. The 2000 Protocol to the Fund Convention was adopted by an International Conference to amend article 43, paragraph 1, of the Convention and facilitate the orderly termination of the Convention, while ensuring that the IOPC Fund was able to meet in full its obligations to pay compensation to victims of oil pollution damage covered by the Convention.[139] This need had arisen because most of the major Contracting States contributors to the 1971 Fund had left the 1971 Fund to join the 1992 Fund regime (see also A/54/429, para. 439). The 1971 Fund was therefore losing its financial viability. The Conference also adopted a resolution entitled "Resolution on the Termination of the 1971 Fund Convention and Accession to the 1992 Protocols."[140]

Draft international convention on civil liability for bunker oil pollution damage
372. IMO reported that a Diplomatic Conference, to be convened in March 2001, was expected to adopt an international convention on civil liability for bunker oil pollution damage.[141] The prospective convention would complete the task initiated by MEPC more than 30 years ago, namely, the adoption of a comprehensive set of unified international rules governing the award of prompt and effective compensation to all victims of pollution from ships.

2. *Regional Cooperation*

(a) Review of UNEP regional seas programme and action plans

373. The regional seas programme is currently undergoing a period of revitalization. Inaugurated in 1974, it is based on a periodic revision of action plans adopted by high-level intergovernmental meetings. There are currently 15 regions[142] covered by action plans, 11 of them supported by regional seas conventions (see annex V to the present report). Negotiations of the 12 regional seas conventions and action plans in the developing world were conducted under the auspices of UNEP. UNEP is also supporting the negotiations in the North-East Pacific and the Upper South-West Atlantic.

374. The main objectives of the regional seas programmes and action plans are the promotion of the integrated management and sustainable development of coastal areas and associated river basins and their living aquatic resources; promotion of the implementation of appropriate technical, institutional, administrative and legal measures for the improved protection of the coastal and marine environ-

---

139. For the text of the 2000 Protocol, see IMO document LEG/CONF.11/6.
140. For the text of the resolution, see IMO document LEG/CONF.11/8.
141. The draft text is contained in IMO document LEG/CONF.12/3.
142. Black Sea, Caribbean, East Africa, East Asia, ROPME Sea Area (Kuwait region), Mediterranean, North-West Pacific, Red Sea and Gulf of Aden, South Asia, South-East Pacific, South Pacific, West and Central Africa, Baltic, Arctic and North-East Atlantic.

ment; and facilitating the assessments of the coastal and marine environment, including their conditions and trends.

375. While not all of the 140 States[143] participating in at least one of the regional seas programmes and action plans are States parties to UNCLOS, the regional seas programme is an example of the realization of general obligations contained in Part XII of UNCLOS, which highlights the need to cooperate internationally and regionally on matters concerning the protection and preservation of the marine environment (articles 192 and 197).

376. UNEP initiated actions to revitalize the regional seas programme following the adoption of the Global Programme of Action for the Protection of the Marine Environment from Land-based Activities in 1995 (see paras. 320–333). In February 1999, the UNEP Governing Council, in its decision 20/19 A, stressed the need for UNEP to strengthen the regional seas programme as its central mechanism for the implementation of its activities relevant to chapter 17 of Agenda 21.

377. As a result of this revitalization movement the second Global Meeting of Regional Seas Conventions and Action Plans was convened at The Hague from 5 to 8 July 1999.

378. The third Global Meeting of Regional Seas Conventions and Action Plans was convened in Monaco from 6 to 11 November 2000. The four principle objectives of the Meeting were: (a) to promote and increase horizontal collaboration among regional seas conventions and action plans in addressing more effectively the protection and sustainable use of the marine environment; (b) to strengthen the linkages between the regional seas conventions and action plans and the global environment conventions and related agreements; (c) to strengthen the linkages between the regional seas conventions and action plans and the Global Programme of Action through agreed concrete actions; and (d) to continue to advance the revitalization of the regional seas conventions.

379. At the third Meeting a round-table discussion was held on the theme "Critical Problems and Issues Facing Regional Seas Conventions and Action Plans." The most commonly raised issue by the representatives of the regional seas conventions and action plans was the financial constraints hindering the implementation of the conventions and action plans. Other frequently raised concerns included: inadequate exchange of information; the need for the increased participation of civil society and the private sector; compliance and enforcement; marine pollution prevention and response; and improved monitoring.

380. The Meeting recommended that representatives of the shipping industry, the chemical industry and the tourism industry should be invited to participate in the fourth Global Meeting of Regional Seas Conventions and Action Plans to address the issue of closer collaboration in regional seas programmes, including the financing of activities.

381. The Meeting requested the UNEP Division of Environmental Conventions to prepare a document, for consideration by the Governing Council at its twenty-first session (5 to 9 February 2001), on strengthening the work of UNEP in the continued vitalization of the regional seas programmes.

382. The Meeting also adopted recommendations on: (a) "Innovative Financ-

---

143. United Nations Environment Programme, *Regional Seas: A Survival Strategy for our Oceans and Coasts* (Geneva, UNEP, 2000), p. 3.

ing Option for Regional Seas Conventions and Action Plans"; (b) "Exploring New Options for Horizontal Cooperation among Regional Seas Conventions and Action Plans"; "Implementation of the Global Programme of Action"; (c) "Assessment and Monitoring of Oceans"; (d) "Strengthening Linkages between the Regional Seas Conventions and Action Plans and the Chemical-Related Conventions"; and (e) "Strengthening Linkages between Regional Seas Conventions and Action Plans and Biodiversity-related Conventions and Agreements."

383. Overall, these recommendations called for closer cooperation between the regional seas conventions and action plans and various institutions and secretariats having mandates and objectives related to aspects of the marine environment. The recommendations pointed towards a new era of enhanced collaboration that, *inter alia,* will enable the exchange of information and experiences and encourage capacity-building on issues of concern that affect the marine ecosystem. The establishment of "twinning" relationships between regional seas conventions and action plans themselves were welcomed and encouraged.[144]

384. There was also a call for a more coherent and coordinated approach among international environment instruments by Ministers of the Environment and heads of delegation who met at Malmö, Sweden, from 29 to 31 May 2000. This First Global Ministerial Environment Forum adopted the Malmö Declaration that, *inter alia,* stated that the evolving framework of international environment law and the development of national law provided a sound basis for addressing current major environmental threats. They added that this must be underpinned by a more coherent and coordinated approach among international environment instruments.

385. The revitalization of the regional seas programme also included the following notable recent developments. In April 2000, UNEP launched the Regional Seas web site (www.unep.ch/seas). In October 2000, a monograph on UNEP and the world's 17 regional seas conventions and action plans was published. A joint UNEP/FAO initiative exploring possibilities for cooperation between the regional seas programme and the regional fisheries management organizations was initiated. On 9 February 2001, the UNEP Governing Council adopted decision 21/28 (d), in which it welcomed the initiative and requested the Executive Director of UNEP, in conjunction with FAO, to support actions for enhancing cooperation between regional fisheries bodies and regional seas conventions and action plans (see paras. 272–278). A draft paper entitled "Financing Regional Seas Conventions: Paying for a Regional and Public Good" was also prepared that examined alternative and innovative financial mechanisms for mobilizing resources to support the secretariats of the Conventions and the activities of the action plans for the North-East Pacific and the Wider Caribbean regions.[145]

---

144. The Meeting recognized the twinning arrangements between the Baltic Marine Environment Commission and UNEP as Secretariat of the Nairobi Convention and between the Jeddah Convention and Kuwait Convention.

145. Further information on the work undertaken by UNEP regarding the regional seas programme in 1999–2000 in general and in specific regions can be found in document UNEP (DEC) R/S 3.1.0, on the web site of the Division for Ocean Affairs and the Law of the Sea: <www.un.org/Depts/los>.

(b) Other regions

(i) Baltic Marine Environment Protection Commission (HELCOM)
386. The 1992 Convention on the Protection of the Marine Environment of the Baltic Sea Area (1992 Helsinki Convention) entered into force on 17 January 2000, thus superseding the 1974 Helsinki Convention. The parties are obliged to take all legislative, administrative or other relevant measures to prevent and eliminate pollution in order to promote the ecological restoration of the Baltic Sea area and the preservation of its ecological balance.[146] The work of the Commission is carried out by five subsidiary bodies and a Programme Implementation Task Force and complemented by various working groups and projects.[147]

387. The Helsinki Commission launched a new project to safeguard maritime transportation. According to figures presented at the second meeting of the Commission's Sea-based Pollution Group (Brussels, January 2001), the probability of occurrence of incidents of marine pollution from ship accidents is increasing. A study of ship accidents within the entire Baltic Sea over the period 1989–1999 reveals that of a total of 232 accidents, one fifth of them resulted in oil pollution. High-risk areas for accidents are concentrated around port areas and in narrow straits. The project involved the compilation of a reliable maritime transportation inventory in the entire Baltic Sea area and the pinpointing of probable areas at risk. It is envisaged that it may be possible to tailor precautionary measures for each risk zone. Based on the results of the project, the Helsinki Commission hopes to prioritize response actions to be taken in real-time accidents to protect sensitive sea areas such as breeding and spawning grounds for the benefit of the Baltic Sea and its people.

(ii) OSPAR Commission for the Protection of the Marine Environment of the North-East Atlantic
388. At its annual meeting, held at Copenhagen from 26 to 30 June 2000, the OSPAR Commission adopted two measures on implementation of its strategy on radioactive substances: (a) adoption of national plans and submission of a detailed forecast for the achievement of the elimination or reduction of radioactive substances from both nuclear and non-nuclear sources; and (b) a binding decision, as adopted by 12 States,[148] on the reduction and elimination of radioactive discharges, emissions and losses, especially from nuclear reprocessing.

389. The OSPAR Commission also adopted and launched the Quality Status Report on the entire North-East Atlantic, or "QSR 2000," work on which had initially been mandated by the 1992 Ministerial Meeting. To implement this commitment the Commission had decided in 1994 to undertake the preparation of QSRs for five regions of the North Atlantic: Arctic Waters; the Greater North Sea; the Celtic Sea; the Bay of Biscay and the Iberian Coast; and the wider Atlantic. These

---

146. Convention on the Protection of the Marine Environment of the Baltic Sea Area, 1992, article 3.
147. The subsidiary bodies are: Strategy Group; Monitoring and Assessment Group; Sea-based Pollution Group; Land-based Pollution Group; Nature Conservation and Coastal Zone Management Group.
148. France and the United Kingdom abstained and are therefore not bound. Luxembourg was absent.

regional QSRs, which have been published separately, form the basis of QSR 2000. The six chapters of QSR 2000 deal with, *inter alia,* geography, hydrography and climate, human activities, chemistry, biology, as well as overall assessment. The purpose of the conclusions and recommendations contained in QSR 2000 is to draw attention to problems and to identify priorities for consideration within appropriate forums as a basis for further work.

(iii) Arctic region: Programme for the Protection of the Arctic Marine Environment (PAME)
390. The Second Ministerial Meeting of PAME was held in Barrow, Alaska, United States, on 12 and 13 October 2000. The meeting set the Arctic Council's[149] agenda for the 2000–2002 period. During the United States chairmanship, the Council's accomplishments included, *inter alia,* continued strong progress in each of its four environmental working groups, including contributions to the development and implementation of the Russian National Programme of Action for the Protection of the Arctic Marine Environment from Land-based Activities and preparation of a map of resources at risk from oil spills in the Arctic.

391. At the Meeting, the Council adopted the Barrow Declaration, which endorsed the Council's Sustainable Development Framework Document, which will form a basis for continuing cooperation on sustainable development in the Arctic. It also noted with appreciation the work done by the PAME Working Group in the following areas: implementation of the Regional Programme of Action; offshore oil and gas; shipping; and review of international conventions and agreements. PAME's future activities, as outlined in a report to the Ministers, were endorsed by the Council, which recognized that the Regional Programme of Action should be used as a management framework for improved working group collaboration on the protection of the Arctic marine and coastal environment and that programme activities should also cover impacts on the coastal zone, which should be more fully addressed.

392. The Council also took note of the work being done by IMO with respect to the draft Guidelines for Ships Operating in Ice-covered Waters and welcomed further cooperation on them.

D. Sustainable Development of Small Island Developing States

393. The oceans and seas have an immense impact on small island developing States[150]—on their economies, their environment and their climate. Oceans continue to be the primary food source for the subsistence of many of the peoples of these States. Given their heavy reliance on the oceans, it is understandable that they have placed and continue to place such great importance on ocean affairs. In recognition of their economic vulnerabilities and environmental fragility, specific provisions catering to the special geographic characteristics and vulnerabilities of

---

149. The Arctic Council is a high-level intergovernmental forum that provides a mechanism to address the common concerns and challenges faced by the Governments and the peoples of the Arctic.
150. According to a list compiled by the United Nations Department of Economic and Social Affairs, there are 41 States and Territories listed as small island developing States; see annex VI to the present report.

small island developing States are embedded in international law and other major nonbinding instruments.[151] The United Nations General Assembly at its fifty-fifth session (2000) adopted four resolutions specifically relating to small island developing States and ocean affairs (resolutions 55/202, 55/203, 55/7 and 55/8). As of 31 January 2001, of the 41 SIDS (see annex VI to the present report) 34 have ratified UNCLOS; 3 have signed UNCLOS but have yet to express their consent to be bound; and 21 are parties to the Agreement relating to the implementation of Part XI of UNCLOS. Of the current total of 27 ratifications/accessions to the 1995 Fish Stocks Agreement, 15 are by small island developing States. However, efforts of these States towards the full implementation of UNCLOS and related agreements are hampered by the constraints on their national capacities, including the lack of trained and qualified manpower in technical fields, coupled with their limited financial resources. The Declaration and state of progress and initiatives for the future implementation of the Programme of Action for the Sustainable Development of Small Island Developing States, adopted by the General Assembly at its twenty-second special session in 1999 (see A/S-22/9/Rev.1, para. 22), together with the outcomes of the first meeting of the Consultative Process (see A/55/274), have highlighted the need for capacity-building among small island developing States to enable them to fully undertake their commitments contained in oceans-related international programmes of action and instruments. Capacity-building in the areas of training, research and technical skills that promote the sustainable management of the oceans and seas continues to be a priority for these States.

394. Cooperation between the international community and small island developing States remains a vital component in the efforts of the latter to achieve sustainable development, including ocean resources development. As emphasized in the Malmö Declaration,[152] there is an alarming discrepancy between commitments and action. Goals and targets for sustainable development agreed by the international community, such as the adoption of national sustainable development strategies and increased support to developing countries, must be implemented in a timely fashion. The Declaration also emphasized that the mobilization of domestic and international resources, including development assistance far beyond current levels, is vital to the success of this endeavour.[153] In the area of assistance in the sustainable management of fisheries resources, some assistance, including technical and financial resources, has been provided by regional fisheries bodies and other international organizations, including FAO, UNDP, UNEP, GEF,[154] and, in a few cases, regional

---

151. UNCLOS arts. 6, 7(1), 13, 47(1), 47(41, 121(2) re archipelagic States; arts. 46, 47(1), 53(5) re regime of islands; art. 121 re island States; Agreement relating to the Implementation of Part XI of the United Nations Convention on the Law of the Sea of 10 December 1982, annex, sect. 3, para. 15 (d); United Nations Framework Convention on Climate Change; Agenda 21, chap. 17.G; the 1994 Declaration of Barbados and Programme of Action for the Sustainable Development of Small Island Developing States, which includes chapter I, "Climate change and sea level rise," and chapter IV, "Coastal and marine resources."

152. See para. 384.

153. Malmö Declaration, para. 2.

154. The Global Environment Facility operates the financial mechanisms for the Convention on Biological Diversity and the United Nations Framework Convention on Climate Change.

banking institutions.[155] GEF and UNDP have been involved in projects in four focal areas[156] of GEF; the area of greatest relevance to ocean matters, and the sustainable development of small island developing States, especially as regards the integrated management of the coastal marine environment, is the International Waters Programme, which focuses, *inter alia,* on Africa, Asia and the Pacific, and Latin America and the Caribbean.[157] Furthermore, national and regional workshops aimed at capacity-building to assist recipient countries in fostering an ongoing two-way dialogue between GEF and the workshop participants were conducted by GEF-UNDP in the Organization of Eastern Caribbean States (OECS) subregion from 8 to 11 August 2000, in the Caribbean subregion from 5 to 8 December 2000 and in Cuba from 12 to 15 December 2000. Additional workshops are being scheduled for 2001, among them a series to be held for Comoros, Mauritius and Seychelles from 10 to 13 July 2001. Approximately 50 national and regional workshops are to be conducted over three years.

395. Fisheries activities within the exclusive economic zones of small island developing States continue to play a major role in the economic development of those States. However, IUU fishing remains a major threat to the sustainable harvesting of living marine resources (see paras. 245–259). Lack of capacity and resources to enforce international and regional agreements, as noted in General Assembly resolution 55/7, continues to be a concern to those small island developing States whose exclusive economic zones are often larger than their land areas. Continued cooperation with the international community in the area of monitoring of fishing activities and surveillance within the exclusive economic zones of small island developing States is important if the sustainable management of ocean resources within their exclusive economic zones is to be a reality.

396. Among major recent initiatives with regard to the preservation and conservation of the ocean resources and marine environment of small island developing States are the conclusion of negotiations on the Convention on the Conservation and Management of Highly Migratory Fish Stocks in the Western and Central Pacific Ocean[158] (see para. 77) and the adoption of General Assembly resolution 55/203 of 20 December 2000, entitled "Promoting an integrated management approach to the Caribbean Sea area in the context of sustainable development." The objective of the Convention is to ensure, through effective management, the long-term conservation and sustainable use of highly migratory fish stocks in the Western and Central Pacific in accordance with UNCLOS and the 1995 Fish Stocks Agreement.[159] In resolution 55/203, the General Assembly called upon the international community to provide international assistance and cooperation in the protection of the

---

155. See A/55/386.

156. Biodiversity; climate change; ozone layer; international waters.

157. For GEF/UNDP-funded projects affecting small island developing States (focal area: International Waters Programme), see the GEF-UNDP web site: <http://www.undp.org/gef/>.

158. Fifteen Pacific island States and Territories participated in the consultations in a series of multilateral high-level conferences over six years which was concluded on 4 September 2000.

159. Convention on the Conservation and Management of Highly Migratory Fish Stocks in the Western and Central Pacific Ocean, article 2.

Caribbean Sea. Other notable initiatives concerning small island developing States, the protection and preservation of their marine environment and the sustainable development of their marine resources have included: (a) in the Caribbean, the adoption of an Environmental Strategy at the sixth ordinary meeting of the Ministerial Council of the Association of Caribbean States (San Pedro Sula, Honduras, December 2000); and in the South Pacific, the endorsement of the South Pacific International Waters Programme at the eleventh biennial meeting of SPREP (Guam, October 2000). A Regional Pacific Oceans Policy is also being developed by the Pacific Islands Forum.[160]

397. The impact of climate change and its associated sea level rise continues to be a concern to small island developing States, particularly the low-lying island States and atolls. According to the Third Assessment Report of Working Group I of the Intergovernmental Panel on Climate Change (IPCC),[161] with regard to ocean temperatures, tide-gauge data show that the average sea level worldwide rose between 0.1 and 0.2 metres during the twentieth century. Global mean sea level is projected to rise by 0.09 to 0.88 metres between 1990 and 2100 (the full range of scenarios is presented by IPCC in Special Report on Emission Scenarios (SRES).[162] An IPCC report on the impacts of climate change on SIDS prepared for the sixth Conference of the Parties to the United Nations Framework Convention on Climate Change (The Hague, 13–24 November 2000), projected a warming of 1°C to 2°C for the Caribbean Sea and the Atlantic, Pacific, and Indian oceans in the future. And although much uncertainty in climate model projections of the distribution, frequency, and intensity of tropical cyclones and El Niño-Southern Oscillation events, the most significant climate-related projection for small islands is sea-level rise. While the level of vulnerability will vary from island to island, it is expected that practically all small island developing States will be adversely affected by sea-level rise (see paras. 416–420).[163]

398. The trans-shipment of radioactive materials through the territories and the exclusive economic zones of small island developing States continues to be a cause for concern to those States and their surrounding regions. In a communiqué released during the Thirty-First Pacific Islands Forum (Tarawa, October 2000), the Pacific Islands Forum drew attention to the continuing constructive dialogue between Forum members and France, Japan and the United Kingdom on developing a liability regime to compensate the region for damage or loss resulting from accidents involving trans-shipment of these materials.[164] Representatives of Caribbean States

---

160. Formerly called the South Pacific Forum, an intergovernmental organization with observer status in the United Nations General Assembly. It is made up of 16 member States, 14 of which are Pacific Small Island States.

161. Summary for Policy Makers, posted by IPCC on its web site <www.ipcc.ch>, 22 January 2001.

162. See IPCC report, pp. 7 and 10 and fig. 5, for illustrations.

163. See IPCC report entitled "The Regional Impacts of Climate Change," chap. 9; "Small Island States," at the IPCC web site: <www.ipcc.ch>.

164. Thirty-first Pacific Islands Forum, Tarawa, Kiribati, 27–30 October 2000, Forum Communiqué 2000, paras. 28–31. Paragraph 29 noted the continuation of constructive dialogue between Forum members and government and nuclear industry representatives from France, Japan and the United Kingdom on a liability regime for compensating the region for economic losses to tourism, fisheries and other industries affected as a result of an accident

also discussed the issue of trans-shipment of radioactive materials at the fourth meeting of the Special Committees for the Protection and Conservation of the Environment and the Caribbean Sea and Natural Resources (Port of Spain, 21–23 June 2000).[165]

399. Nine small island States were registered as having participated in the first meeting of the Consultative Process (see paras. 608–614).[166] Clearly, it is important for small island developing States to participate in such a process if it is to retain its integrity. A voluntary trust fund is being set up pursuant to General Assembly resolution 55/7 to assist developing States, including small island developing States, in participating in the Consultative Process.[167] Another trust fund was established in accordance with the same resolution to assist those States in the preparation of submissions pursuant to article 76 and annex II to UNCLOS (see paras. 65–69).[168]

400. Lack of capacity, coupled with limited resources, financial as well as technical, remain the major obstacles for small island developing States to implement the obligations they have undertaken under UNCLOS and other ocean-related agreements including international programmes of action. More concrete actions will need to be taken by the international community to assist those States in the regional and national implementation efforts of UNCLOS and other ocean-related agreements.

E. Protection of Specific Marine Areas

401. States may wish to protect a particular marine area for a variety of reasons, for example, because of its ecological, biogeographic, scientific, economic or social importance, and/or because of the vulnerability of its resources to certain activities. A number of global as well as regional instruments provide various types of measures aimed at the protection of marine areas and their resources. The kind of measures a State may wish to adopt to regulate certain activities in an area depends on the specific characteristics of the marine area, its species and the ecosystem the State seeks to protect.

402. UNCLOS, for example, permits a coastal State to take measures in its exclusive economic zone to regulate fishing seasons and areas to be fished (article 62 (4) (c)), or, subject to the approval of IMO, to protect an area from shipping (article

---

involving a shipment of radioactive materials and MOX fuel even if there were no actual environmental damage.

165. Fourth meeting of the Special Committees for the Protection and Conservation of the Environment and the Caribbean and Natural Resources, Port of Spain, 21–23 June 2000, Meeting records, "Transportation of Nuclear Waste in the Caribbean Sea." While acknowledging the legality of shipping under internationally accepted standards, participants raised concerns that even one accident would be devastating to the region, given its fragility and vulnerability.

166. Forms were distributed to delegates who were in attendance at the Consultative Process to allow them to register their participation. The number quoted is a reflection of the number of States that filled out the forms indicating their participation. The form also provided for a listing of participants in each delegation.

167. General Assembly resolution 55/7, para. 45.

168. *Ibid.*, para. 18 and annex II.

211 (6)). Agenda 21 in its chapter 17 calls upon States to undertake measures to maintain the biological diversity and productivity of marine species and habitats under national jurisdiction through, *inter alia,* the establishment and management of protected areas (para. 17.7). Under the Convention on Biological Diversity parties are required to establish a system of protected areas where special measures need to be taken to conserve biological diversity and to develop guidelines for the selection, establishment and management of such areas (article 8 (a) and (b)). At its second meeting, the Conference of the Parties to the Convention designated marine and coastal protected areas as one of the five thematic issues/spheres for action under the Jakarta Mandate on the Conservation and Sustainable Use of Marine and Coastal Biological Diversity (see paras. 283–291).

403. Other global measures available to States include the establishment of an area as a Special Area under MARPOL 73/78, as a Particularly Sensitive Sea Area; as a sanctuary under the International Convention for the Regulation of Whaling; as a biosphere reserve under the UNESCO Man and the Biosphere Programme; as a cultural or national heritage for inclusion in the World Heritage List under the Convention concerning the Protection of the World Cultural and Natural Heritage; or as a Wetland of International Importance under the Convention on Wetlands of International Importance Especially as Waterfowl Habitat. Regional measures include the establishment of an area as a specially protected area under the UNEP regional seas agreements, or other protective measures available under other regional agreements (see paras. 386–392). However, content, focus and the binding nature of possible measures vary greatly among the various regimes mentioned.

1. Marine Protected Areas

404. Marine protected areas have been identified as an essential tool for helping to conserve species and restore marine ecosystem health. They can be small or vast in size and can be established for a variety of objectives, ranging from strict protection to multiple uses. The Jakarta Mandate states that the provision of critical habitats for marine living resources should be an important criterion for selection.[169] States have so far mainly established marine protected areas in coastal vicinities. However, a number of States have adopted or are in the process of adopting national legislation providing for the establishment of such areas within their exclusive economic zones as well.

405. At the first meeting of the Consultative Process some delegations emphasized the need to give consideration to the use of marine protected areas as a tool for integrated ocean management. They also stressed that such areas could provide for a regime incorporating biodiversity conservation, fisheries, mineral exploration, tourism and scientific research in a sustainable manner. In this connection, mention was made of the need for identifying methods to establish and manage marine protected areas on the high seas. Some delegations expressed reservations about establishing and managing such areas on the high seas. The topic of marine protected areas was identified as an issue to be considered for possible inclusion in the agendas of future meetings.[170]

---

169. Second meeting of the Conference of the Parties to the Convention on Biological Diversity, decision II/10, annex I, para. (iv).

170. A/55/274, part B, para. 28, and part C, para. 2 (e).

406. Two workshops on marine protected areas on the high seas held during 2000 reflect the increasing interest of the international community in finding mechanisms for establishing high-seas protection zones. The expert workshop entitled "Marine Protected Areas on the High Seas: Scientific Requirements and Legal Aspects," organized by the German Federal Agency for Nature Conservation (27 February–4 March 2001), had as its aims: to identify conservation needs and priorities on the high seas; to review existing activities aimed at the conservation of valuable sites; and to develop ideas on achieving a sound protection regime for such sites on the high seas. The second workshop, entitled "Protection of the High Seas Marine Biodiversity in the South West Pacific: Role of Marine Protected Areas," to be hosted by Australia in April 2001, will build on the results of the German workshop and apply them to the specific circumstances of the South-West Pacific.

407. The possible establishment of marine protected areas beyond the limits of national jurisdiction was raised at the seventh session of the Commission on Sustainable Development in 1999 (see A/54/429, paras. 508–509) and at the first Meeting of Experts on Marine and Coastal Biological Diversity, convened by the Biodiversity Convention secretariat in March 1997. The experts highlighted the unique significance of certain high seas and deep seabed areas (such as identified spawning areas, deep ocean trenches and certain hydrothermal vents) beyond the limits of national jurisdiction and called for consideration to be given to the development of means and modalities for the establishment of marine protected areas in such areas (see A/52/487, para. 241).

2. Special Areas and Particularly Sensitive Sea Areas

408. IMO reported that its Marine Environment Protection Committee at its forty-fifth session had noted the amendments to the IMO Guidelines for the Designation of Special Areas and the Identification of Particularly Sensitive Sea Areas (PSSAs) contained in IMO Assembly resolutions A.720(17) and A.885(21), prepared by its Drafting Group. The Committee recognized the need for an Assembly resolution to revoke the existing Guidelines and requested the IMO secretariat to prepare a draft Assembly resolution together with a revised text of the Guidelines for discussion and approval at the forty-sixth session of MEPC, in April 2001.

409. MEPC also agreed that some guidance on selecting the most appropriate regime for a given area of the sea could be included in a separate document. The Committee requested the IMO secretariat to prepare a draft document for the forty-sixth session based on an outline prepared by the Drafting Group, as well as a flow chart to assist member States in deciding the most appropriate method of providing additional protection for sensitive sea areas.[171]

3. Developments at the Regional Level

410. All protocols relating to specially protected areas, adopted under the framework of the UNEP regional seas programme (see annex V to the present report), which cover the regions of East Africa, the wider Caribbean, the Mediterranean and the South-East Pacific, have entered into force. For the parties to those protocols the focus with regard to specially protected areas as a conservation and protection

---

171. See MEPC 45/20.

tool will now shift to the implementation and consolidation of the established rules, while in other regions legislative action might still be taken. Additional regions that have specific regimes in place providing protection and conservation measures for marine areas are Antarctica, the Baltic Sea area and the North-East Atlantic. In addition, other regional agreements provide protection to certain species and habitats more generally. These include the African Agreement on the Conservation of Nature and Natural Resources; the ASEAN Agreement on the Conservation of Nature and Natural Resources (not yet in force); and the Convention on the Conservation of Nature in the South Pacific. Furthermore, a large number of regional fisheries agreements provide for the establishment of conservation regimes applicable in specific marine areas to manage the resources in question.

411. The most recent UNEP instrument to enter into force is the 1990 Protocol concerning Specially Protected Areas and Wildlife to the Convention for the Protection and Development of the Marine Environment of the Wider Caribbean Region, in May 2000. The Protocol provides for the establishment of protected areas in zones over which the parties exercise sovereignty, sovereign rights or jurisdiction and envisages a variety of potential measures for those areas utilizing an ecosystem approach. The first meeting of the Scientific and Technical Advisory Committee and the first meeting of the parties to the Protocol will be held in Cuba, from 24 to 29 September 2001.

412. The regimes for the Mediterranean, the Baltic and the North-East Atlantic have been adjusted to take account of the recent shift in approaches to conservation and protection. They make use of such concepts as biodiversity protection and integrated management, as advocated by Agenda 21 and the Convention on Biological Diversity, and may serve as models for other regions in this respect.

413. The 1995 Barcelona Protocol concerning Specially Protected Areas and Biological Diversity in the Mediterranean entered into force in December 1999 and replaced the earlier 1982 Protocol on Specially Protected Areas. The 1995 Protocol introduces the concept of Specially Protected Areas of Mediterranean Interest (SPAMIs) and is applicable to the whole Mediterranean Sea, including its seabed, subsoil and the coastal areas including wetlands. The Protocol provides that it must be applied in a manner consistent with the relevant provisions of UNCLOS and other rules of international law.

414. The 1992 Convention on the Protection of the Marine Environment of the Baltic Sea Area entered into force in January 2000. HELCOM recommendation 15/5 is used by the parties as a basis for the designation of Baltic Sea Protected Areas (BSPAs). A number of such areas have already been designated and an even greater number are currently under consideration.

415. The 1998 amendments to the Convention for the Protection of the Marine Environment of the North-East Atlantic, resulting in the adoption of a new Annex V concerning the protection and conservation of the ecosystem and biological diversity of the maritime area covered by the Convention, together with a related appendix, entered into force in 2000 for Denmark, Finland, Luxembourg, Spain, Sweden, Switzerland, the United Kingdom and the European Community. Annex V provides, *inter alia*, for the development of protective and conservation measures related to specific areas, making use of the precautionary approach and other recent concepts, i.e., best environmental practice, best available techniques and clean technology.

F. Climate Change and Sea Level Rise

416. According to the Third Assessment Report of Working Group I of the Intergovernmental Panel on Climate Change, the average sea level worldwide has risen and ocean heat content has increased.[172] Moreover, the northern hemisphere volume of spring and summer sea ice has decreased by about 10 to 15% since the 1950s. It is likely that there has been about a 40% decline in Arctic sea-ice thickness in late summer and early autumn in recent decades as well as a decline in winter sea-ice thickness, although at a considerably slower rate.

417. However, some aspects of climate appear not to have changed. The few areas of the globe that have not warmed in recent decades include some parts of the southern hemisphere oceans and parts of Antarctica. No significant changes in volume of Antarctic sea ice have been apparent since 1978, when reliable satellite measurements became available.

418. The report stated that there is new and stronger evidence that most of the warming observed over the past 50 years is attributable to human activities. It further stated that it is very likely that the twentieth century warming has contributed significantly to the observed sea-level rise, through thermal expansion of sea water and widespread loss of land ice.

419. From the report it is clear that unless actions are taken soon to implement the commitments[173] undertaken by States in accordance to the 1997 Kyoto Protocol to the United Nations Framework Convention for Climate Change, climate change and sea level rise will continue to adversely affect the earth and human livelihood. The advent of sea-level rise could also have implications that affect the rights and obligations of some States parties to UNCLOS, especially in relation to the breadth of the territorial sea, the contiguous zone and the exclusive economic zone, all of which are measured from the baselines of States.[174] Although the effect of the advent of sea-level rise will vary from country to country, small island States, especially those with atolls and those that are low-lying, will be most affected, just through the possible loss of territorial integrity, but, what is far more sacred to them, the loss of their cultures and way of life (see para. 397).

420. During the sixth Conference of the Parties to the United Nations Framework Convention on Climate Change (The Hague, November 2000), the IPCC reminded delegates that if actions are not taken to reduce the projected increase in greenhouse gas emissions, the earth's climate is likely to change at a rate unprecedented in the last 10,000 years, with adverse consequences for societies and undermining the very foundation of sustainable development.[175] The Conference concluded without a finalization of the rules contained in the Kyoto Protocol.

---

172. See n. 161.
173. Not yet in force. The Protocol is based on the concept of "common but differentiated responsibilities" (art. 3), in which developed States (known as Annex 1 countries) take the lead in reducing national greenhouse gas emissions (although some have argued in favour of increasing their levels of emission) to agreed levels that would lead overall to decreased emission of greenhouse gases into the Earth's atmosphere.
174. See, in particular, UNCLOS, parts II and V.
175. Report to the Sixth Conference of the Parties to the United Nations Framework Convention on Climate Change, 20 November 2000.

G. Ten-Year Review of the Implementation of Agenda 21

421. The General Assembly, in its resolution 55/199 on 20 December 2000, sets out the timing and modalities for four sessions to be held by the Commission on Sustainable Development at its tenth session in 2002. The tenth session of the Commission is to serve as an open-ended intergovernmental preparatory committee for the 10-year review of progress achieved in the implementation of the outcome of the United Nations Conference on Environment and Development (UNCED) in 2002 at the summit level, the "World Summit on Sustainable Development," to be held in Johannesburg, South Africa. One of the aims of the Summit, including its preparatory process, is to ensure a balance between economic development, social development and environmental protection. The active participation of all major groups, as identified in Agenda 21, is encouraged.

422. The Preparatory Committee is charged with undertaking, among other things, a comprehensive review and assessment of the implementation of Agenda 21, identifying major constraints hindering its implementation and formulating ways to strengthen the institutional framework for sustainable development and evaluating the role and the programme of work of the Commission.

423. The Commission has invited the United Nations Secretariat, working in close cooperation with UNEP, the regional commissions, the secretariats of UNCED-related conventions, as well as other relevant organizations, agencies and programmes within and outside the United Nations system, including international and regional financial institutions, to support activities in preparation for the 2002 World Summit on Sustainable Development.

424. In dealing with chapter 17 of Agenda 21, on "Oceans and Seas," modalities for inter-agency preparations were agreed by the Inter-Agency Committee for Sustainable Development at its sixteenth session (Geneva, September 2000),[176] which decided that task managers would submit a short report to the Commission on Sustainable Development by 1 February 2001.

425. The Subcommittee on Oceans and Coastal Areas of the Administrative Committee on Coordination, at its meeting in Paris in January 2001, held a brainstorming session to consider the draft compilation of inputs from various heterogeneous standpoints with a view to providing a focus on common themes and constraints in the implementation of chapter 17.

426. With regard to preparations at the national level, the Commission invited all Governments to undertake national review processes as early as possible. It was agreed that national reports on the implementation of Agenda 21, which Governments had prepared since 1992 and to which major groups had contributed, could provide a reasonable basis for guiding national preparatory processes. In this connection, the Department of Economic and Social Affairs of the United Nations Secretariat has been preparing country profiles, which cover most of the main thematic areas of Agenda 21 and the Programme for the Further Implementation of Agenda 21 (General Assembly resolution S/19-2, annex). The profiles are based on information contained in national reports submitted to the Commission by Governments between 1997 and 2001. Moreover, in consultation with other parts of the Secretariat, the Department has elaborated a proposed framework for addressing key issues

176. See ACC/2000/12, paras. 24–29.

738   Selected Documents and Proceedings

in reviewing and assessing progress made in implementation of Agenda 21 at the national and regional levels. The proposed framework, in the form of a brief, user-friendly questionnaire, was communicated to all permanent missions in New York on 7 August 2000, with a suggested deadline for replies of 1 March 2001. The Department is also discussing with UNDP modalities for its effective involvement in the 2002 Summit process, in particular in support of national preparatory activities.

427. With regard to financial support for the preparatory process, the Commission at its eighth session recommended that steps be taken to establish a trust fund and urged international and bilateral donors to make voluntary contributions to the trust fund and to support the participation of representatives from developing countries in the regional and international preparatory process as well as the 2002 Summit.[177]

VII. SETTLEMENT OF DISPUTES[178]

428. Under Part XV, section 1, of UNCLOS States parties are required to settle their disputes concerning the interpretation or application of the Convention by peaceful means, in accordance with Article 2, paragraph 3, of the Charter of the United Nations. However, when States parties to UNCLOS involved in a dispute have not reached a settlement by peaceful means of their own choice, they are obliged to resort to the compulsory dispute settlement procedures provided for under the Convention (Part XV, section 2).

429. During 2000, the International Court of Justice, the International Tribunal for the Law of the Sea and an arbitral tribunal established under Annex VII to UNCLOS were seized of disputes relating to the law of the sea. (Further details on the cases concerned may be found at the web site of the Division for Ocean Affairs and the Law of the Sea of the United Nations Secretariat: www.un.org/Depts/los.)

A. Cases Before the International Court of Justice[179]

430. *Case concerning Oil Platforms (Islamic Republic of Iran v. United States of America)*. The case arose out of the attack on and destruction of three offshore oil production complexes, owned and operated for commercial purposes by the National Iranian

---

177. For further details on the 10-year review of Agenda 21 and the preparatory processes, see the following web site: <www.un.org/rio+10>.

178. The updated list of conciliators and arbitrators drawn up by the Secretary-General of the United Nations in accordance with article 2 of Annex V and article 2 of Annex VII, respectively, to UNCLOS, as well as the updated list of special arbitrators under article 2 of Annex VIII to UNCLOS received by the Secretary-General of the United Nations from FAO, UNEP, IOC and IMO, are periodically published in the *Law of the Sea Information Circular* by the Division for Ocean Affairs and the Law of the Sea.

179. Excerpted from the contribution of ICJ. See also the annual report of the Court to the General Assembly, *Official Records of the General Assembly, Fifty-fifth Session, Supplement No. 4* (A/55/4), and the ICJ web site: <www.icj-cij.org>.

Oil Company, by several warships of the United States Navy on 19 October 1987 and 18 April 1988. The acts, it was alleged, constituted a fundamental breach of international law and various provisions of the Treaty of Amity, Economic Relations and Consular Rights between the United States of America and Iran, signed at Tehran on 15 August 1955. At the request of the United States of America, the Court issued an Order dated 4 September 2000 extending from 23 November 2000 to 23 March 2001 the time limit for the filing of its Rejoinder. The Islamic Republic of Iran expressed no objection to the extension. However, Iran pointed out that the Court, in its Order of 10 March 1998, had reserved "the right of Iran to present its views in writing a second time on the United States counterclaim, in an additional pleading the filing of which may be the subject of a subsequent Order."

431. *Case concerning Maritime Delimitation and Territorial Questions between Qatar and Bahrain (Qatar v. Bahrain).* The case deals with disputes relating to sovereignty over the Hawar Islands, sovereign rights over the shoals of Dibal and Qit'al Jaradah, and the delimitation of the maritime areas of the two States. On 29 June 2000, the public hearings in the longest case in the history of the Court were concluded. (Qatar had filed its Application with the Court against Bahrain on 8 July 1991.) On 16 March 2001, the Court, in rendering its Judgment on the merits of the case, decided *inter alia,* that Qatar has sovereignty over Zubarah, Janan island, including Hadd Janan, and the low-tide elevation of Fasht ad Dibal; and that Bahrain has sovereignty over the Hawar Islands and the island of Qit'at Jaradah. Moreover, the Court recalled that vessels of Qatar enjoy in the territorial sea of Bahrain, which separates the Hawar Islands from the other Bahraini islands, the right of innocent passage accorded by customary international law. As regards the question of the maritime boundary, the Court also recalled that customary international law was applicable to the case and that the parties had requested it to draw a single maritime boundary (in the southern part, the Court drew a boundary delimiting the parties' territorial seas over which they enjoy territorial sovereignty, including the seabed, superjacent waters and superjacent aerial space; in the northern part, the Court had to carry out a delimitation between areas in which the parties have only sovereign rights and functional jurisdiction, i.e., over the continental shelf and in the exclusive economic zone). With respect to the territorial sea, the Court drew provisionally an equidistance line (a line every point of which is equidistant from the nearest points on the baselines from which the breadth of the territorial sea of each of the two States is measured) and then considered whether that line should be adjusted in the light of any special circumstances. The Court rejected Bahrain's argument that the existence of certain pearling banks situated to the north of Qatar, which had been predominantly exploited in the past by Bahraini fishermen, constituted a circumstance justifying a shifting of the equidistance line. It also rejected Qatar's argument that there is significant disparity between the coastal lengths of the parties calling for an appropriate correction. The Court further stated that considerations of equity required that the maritime formation of Fasht al Jarim should have no effect in determining the boundary line.

432. *Case concerning Sovereignty over Pulau Ligitan and Pulau Sipadan (Indonesia/ Malaysia).* The Court is requested to determine, on the basis of treaties, agreements and any other evidence furnished by the parties, whether sovereignty over Pulau Ligitan and Pulau Sipadan, two islands in the Celebes Seas, belong either to the

740   Selected Documents and Proceedings

Republic of Indonesia or to Malaysia. By an Order dated 11 May 2000, the President of the Court, at the request of the parties, further extended to 2 August 2000 the time limit for the filing of the Counter-Memorials, which were filed within the time limit as thus extended. In addition, by an Order dated 19 October 2000, the President of the Court fixed 2 March 2001 as the time limit for the filing of a Reply by each of the parties in the case. On 13 March 2001, the Philippines filed an Application for permission to intervene in the case, stating that it wished to "preserve and safeguard [its Government's] historical and legal rights arising from its claims to dominion and sovereignty over the territory of North Borneo, to the extent that those rights are affected, or may be affected, by a determination of the Court of the question of sovereignty over Pulau Ligitan and Pulau Sipadan."

433. *Case concerning Maritime Delimitation between Nicaragua and Honduras in the Caribbean Sea (Nicaragua v. Honduras)*. The dispute deals with the delimitation of the maritime zones appertaining to each State in the Caribbean Sea. Taking into account the agreement of the parties, the Court decided, by an Order dated 21 March 2000, that Nicaragua would file a Memorial by 21 March 2001 and that Honduras would file a Counter-Memorial by 21 March 2002.

434. *Case concerning the Land and Maritime Boundary between Cameroon and Nigeria (Cameroon v. Nigeria)*. The case deals with the question of sovereignty over the peninsula of Bakassi. The Court is also requested to determine the course of the maritime frontier between the two States beyond the line fixed by them in 1975 (Maroua Declaration of 1 June 1975). By an Order dated 21 October 1999, the Court authorized Equatorial Guinea to intervene in the case "to the extent, in manner and for the purposes set out in its Application for permission to intervene." In addition, the Court fixed 4 April 2001 as the time limit for the filing of a written statement by Equatorial Guinea and 4 July 2001 as the time limit for the filing of written observations by Cameroon and by Nigeria on that statement. Subsequently, by an Order of 20 February 2001, the Court authorized Cameroon to submit an additional pleading, which would relate solely to the counterclaims submitted previously by Nigeria, no later than 4 July 2001.

B. Cases Before the International Tribunal for the Law of the Sea[180]

1. Case Adjudicated
435. *The "Monte Confurco" Case (Seychelles v. France)*. The dispute concerns the arrest of the fishing vessel *Monte Confurco,* flying the flag of Seychelles, which was apprehended by the French frigate *Floréal* in the exclusive economic zone of the Kerguelen Islands on 8 November for alleged illegal fishing and failure to announce its presence in the exclusive economic zone of the Kerguelen Islands. The *Monte Confurco* was escorted by French naval authorities to Réunion.

436. The district court of Saint-Denis, Réunion, ordered that the vessel could be released upon the posting of a bond of 56.4 million French francs.

437. At the hearing at the Tribunal, the Agent for Seychelles stated that the Master of the ship had entered the exclusive economic zone of the Kerguelen Islands, heading in the direction of Williams Bank. However, since his fax machine

180. Excerpted from Tribunal documents.

was not functioning, the Master was unable to notify the French authorities of the vessel's presence in the exclusive economic zone, in keeping with articles 2 and 4 of French Law No. 66-400 of 18 June 1966, as amended. The Agent disputed the allegation that the *Monte Confurco* had been engaged in illegal fishing. He maintained that the fish on board the vessel had been caught in international waters. The Agent for Seychelles also requested the immediate release of the Master, who was being detained in Réunion, and the return of his passport as well as the release of the vessel upon the posting of a reasonable bond, arguing that the bond set by the French authorities was not reasonable.

438. The Agent for France contended that the *Monte Confurco* had been discovered in the exclusive economic zone without having given notification of its presence and its catch, even though the vessel was equipped with radio-telephone and an Inmarsat station. Also, it was alleged, *inter alia*, that the vessel did not stop when ordered to do so.

439. The Agent for France also referred to the increase in illegal fishing in the area and the means used by vessels to avoid detention or punishment. He also emphasized the environmental danger to the stock of toothfish in the waters of the southern Indian Ocean. The expert called by France stated that overexploitation of the species could have serious consequences for the stock, especially as it had a long maturation phase. He also expressed the opinion that it was not possible for the *Monte Confurco* to have been fishing where it claimed to have fished, owing to the great depths in the areas concerned. However, on cross-examination by the Agent for Seychelles, the expert asserted that Spanish fishermen had developed techniques that allowed fishing in waters up to a depth of 2,500 metres.

440. On 18 December 2000, the Tribunal rendered its judgement in the case concerning the Application for prompt release of the *Monte Confurco*. The Tribunal ordered the prompt release by France of the vessel and its Master, upon the provision by Seychelles, the flag State of the vessel, of a security of FF 18 million. The Tribunal decided that the bond set by the national court in Réunion of FF 56.4 million for the release of the *Monte Confurco* and its Master was not reasonable.

441. The Tribunal unanimously found that it had jurisdiction under article 292 of UNCLOS to entertain the Application made on behalf of Seychelles; that the claims of Seychelles that France had failed to comply with article 73, paragraphs 3 and 4, of UNCLOS were inadmissible; and that the Application with respect to the allegation of noncompliance with article 73, paragraph 2, of the Convention was admissible.

2. Case Settled By Agreement
442. *Case concerning the Conservation and Sustainable Exploitation of Swordfish Stocks in the South-eastern Pacific Ocean (Chile/European Community).* In February 2001, Chile and the European Union (EU) reached an agreement that settled their dispute on both access for EU fishing vessels to Chilean ports and bilateral and multilateral scientific and technical cooperation on the conservation of swordfish stocks.

443. Prior to that, on 25 January 2001, both parties to the dispute had reached a negotiated settlement as a result of which EU had requested a suspension of panel proceedings within the World Trade Organization (WTO) and Chile had suspended proceedings before the Tribunal.

444. Initially, at the request of Chile and the European Community, the Tribu-

742 Selected Documents and Proceedings

nal, in accordance with article 15 of its Statute, had by an Order dated 20 December 2000 formed a special chamber of five judges[181] to deal with their dispute concerning the conservation and sustainable exploitation of swordfish stocks in the south-eastern Pacific Ocean.

445. The special chamber was requested to decide the following issues, to the extent that they were subject to compulsory procedures entailing binding decisions under Part XV of UNCLOS:

(a) On behalf of Chile:

(i) Whether the European Community had complied with its obligations under UNCLOS, in particular articles 116 to 119, to ensure conservation of swordfish in the fishing activities undertaken by vessels flying the flag of any of its member States in the high seas adjacent to Chile's exclusive economic zone;
(ii) Whether the European Community had complied with its obligations under UNCLOS to cooperate directly with Chile as a coastal State for the conservation of swordfish in the high seas adjacent to Chile's exclusive economic zone and also to report its catches and other information relevant to that fishery to the competent international organization and to the coastal State;
(iii) Whether the European Community had challenged the sovereign right and duty of Chile, as a coastal State, to prescribe measures within its national jurisdiction for the conservation of swordfish and to ensure their implementation in its ports, in a nondiscriminatory manner, as well as the measures themselves, and whether such challenge would be compatible with UNCLOS;
(iv) Whether the obligations arising under articles 300 and 297, paragraph 1 (b), of UNCLOS, and the general thrust of the Convention, had been fulfilled in the present case by the European Community;

(b) On behalf of the European Community:

(i) Whether Chilean Decree No. 598, purporting to apply Chile's unilateral conservation measures relating to swordfish on the high seas, was in breach of articles 87, 89 and 116 to 119 of UNCLOS;
(ii) Whether the "Galapagos Agreement" of 14 August 2000 had been negotiated in keeping with UNCLOS, especially articles 64 and 116 to 119;
(iii) Whether Chile's actions concerning the conservation of swordfish were in conformity with article 300 of UNCLOS and whether Chile and the Euro-

---

181. The composition of the special chamber (P. Chandrasekhara Rao, President; Judges: Caminos, Yankov, Wolfrum; and Judge ad hoc: Orrego Vicuña) was determined by the Tribunal with the approval of the parties. In keeping with the Statute of the Tribunal, a judgment rendered by the special chamber will be considered as having been rendered by the full Tribunal (see annual report of the International Tribunal for the Law of the Sea for 2000, SPLOS/63).

pean Community remained under a duty to negotiate an agreement on cooperation under article 64 of UNCLOS;
(iv) Whether the jurisdiction of the special chamber extended to the issue referred to in point (a) (iii) above.

446. In parallel to the procedure before the Tribunal, on 10 November 2000, the European Commission had requested the establishment of a WTO panel against Chile in order to secure access for EU fishing vessels to Chilean ports, which had been closed to the European Community since 1991.

C. Case Decided by an Arbitral Tribunal

447. *Award of 4 August 2000 rendered by the arbitral tribunal in the Southern Bluefin Tuna Case (Australia and New Zealand v. Japan).* A five-member international arbitral tribunal (Judge Stephen M. Schwebel, President; Judge Florentino Feliciano, Justice Sir Kenneth Keith, Judge Per Tresselt and Professor Chusei Yamada) rendered its award on 4 August 2000 on jurisdiction and admissibility in the *Southern Bluefin Tuna* case. At the request of the parties and the arbitral tribunal, the International Centre for Settlement of Investment Disputes, one of the five organizations comprising the World Bank Group in Washington, administered the proceedings.

448. Australia and New Zealand had commenced arbitral proceedings against Japan under Annex VII of UNCLOS and, on 30 July 1999, pending the constitution of the arbitral tribunal, both countries had requested the Tribunal to prescribe provisional measures under article 290 (5) of UNCLOS.

449. The dispute among the three States had arisen over whether southern bluefin tuna, a valuable migratory species of tuna that is fished mainly in the southern Atlantic Ocean near the Antarctic and is highly prized in Japan as a delicacy, was recovering from a state of severe overfishing. Australia, Japan and New Zealand in 1993 had concluded the Convention on the Conservation of Southern Bluefin Tuna, which established a Commission responsible for setting a total allowable catch among the parties as well as for taking other measures to promote the recovery of the stock. In addition, the 1993 Convention contains a provision for the settlement of disputes arising under it, permitting the parties to choose whatever means of peaceful settlement of disputes they prefer (article 16). However, the three States concerned are also parties to UNCLOS, which itself also contains provisions for compulsory settlement of disputes arising under it, including arbitration (articles 286 et seq.). Moreover, UNCLOS contains provisions on the fishing of migratory fish species, such as the southern bluefin tuna.

450. One of the main issues before the arbitral tribunal was whether it had jurisdiction over the merits of the dispute. Japan argued that the dispute had arisen solely under the 1993 Convention and that therefore it could not be compelled to arbitrate the merits of the dispute under UNCLOS. Furthermore, Japan contended that under article 282 of UNCLOS parties could avoid compulsory dispute settlement if another treaty to which they were bound governed the case and excluded it.

451. The arbitral tribunal held that a dispute could arise under more than one treaty, and indeed did so in the present case, in keeping with article 30 (3) of the 1969 Vienna Convention on the Law of Treaties, thus rejecting the claim by Japan

that the dispute concerned only the 1993 Convention. Nonetheless, the arbitral tribunal sustained Japan's contention that a provision in the 1993 Convention excluded compulsory jurisdiction over disputes arising both under it and under UNCLOS and held that the parties were involved in a single dispute arising under both Conventions. In that connection, it held that the meaning and intent of the dispute settlement provision of the 1993 Convention was to exclude procedures for compulsory settlement under UNCLOS. As a result, it revoked, in accordance with article 290 (5) of UNCLOS, the provisional measures ordered by the International Tribunal for the Law of the Sea enjoining Japan from conducting an experimental fishing programme for southern bluefin tuna, while stating that the prospects for a successful settlement on the merits depended upon the parties' abstaining from unilateral action that could aggravate the dispute.[182]

VIII. MARINE SCIENCE AND TECHNOLOGY

452. At its first meeting, the Consultative Process emphasized the important role of marine science and technology in promoting the sustainable management and use of the oceans and seas as part of efforts to eradicate poverty, to ensure food security and to sustain economic prosperity and the well-being of present and future generations. It also underlined the importance of marine science in the assessment of fish stocks, their conservation, management, and sustainable use, including the consideration of ecosystem-based approaches, and, to that end, the improvement of status and trend reporting for fish. Finally, the Consultative Process pointed to the consequent need to ensure access for decision makers to advice and information on marine science and technology, the appropriate transfer of technology and support for the production and diffusion of factual information and knowledge for end-users.

453. The General Assembly, in its resolution 55/7 of 30 October 2000, stressed the need to consider, as a matter of priority, the issues of marine science and technology and to focus on the best ways to implement the many obligations of States and competent international organizations under Parts XIII and XIV of UNCLOS, and called upon States to adopt, as appropriate and in accordance with international law, the necessary national laws, regulations, policies and procedures to promote and facilitate marine scientific research and cooperation. The Assembly also recommended that, at its second meeting in May 2001, one of the areas of focus of the Consultative Process should be "marine science and development and transfer of marine technology as mutually agreed, including capacity-building in this regard."

A. Legal Regime for Marine Science and Technology

454. Since one of the basic concerns of the General Assembly, as reflected in resolution 55/7, is the implementation of the provisions of Parts XIII and XIV of UNCLOS, dealing with marine scientific research and the development and transfer

---

182. The full text of the Award and the dissenting opinion are posted on the ICSID web site: <www.worldbank.org/icsid>.

of marine technology, respectively, it is worthwhile to point out the salient features of the legal regime for marine science and technology as set forth in UNCLOS.

1. Legal Regime for Marine Scientific Research[183]

455. UNCLOS, principally in its Part XIII, lays down a comprehensive global regime under which States are required to promote and conduct marine scientific research and cooperate in such research. It has struck a balance and an important compromise between the rights of the coastal State to regulate and authorize the conduct of marine scientific research in the zones under its sovereignty and the rights of the researching States to carry out research as long as it does not have any bearing on exploration and exploitation of natural resources.

456. However, concerns have been expressed that the legal regime as set forth in Part XIII (the consent regime in article 246, in particular) and as implemented by States might in fact have damaging effects on the international marine science community.[184] At the same time, most of the developing States face substantial challenges in implementing the marine scientific research regime. One important objective of the description of the regime below is to re-ascertain that the provisions on marine scientific research as drafted in Part XIII, far from being inhibitive, promote the development of marine scientific research and should be actively implemented.

457. Section 1 (General provisions) of Part XIII establishes the general principles under which all States and competent international organizations shall conduct marine scientific research subject to the rights and duties of other States (articles 238–241).

458. In section 2, States and competent international organizations are called upon to promote international cooperation in marine scientific research, as well as to cooperate so as to create favourable conditions for the conduct of such research, and to publish and disseminate information on proposed major programmes and their objectives and knowledge resulting from that research. Cooperation should also entail the strengthening of research capabilities of developing States through, *inter alia*, programmes to provide adequate education and training (articles 242–244).

459. *The consent regime for the conduct of marine scientific research.* The consent regime as established in section 3 of Part XIII of UNCLOS represents a compromise between the coastal States' interests and those of the researching States. This compromise is reflected through the articles on tacit or implied consent and the right of the coastal State to withhold consent under specified conditions or to require the suspension or cessation of the research in progress in the exclusive economic zone and the continental shelf if the research does not comply with the information or the obligations required. In this regard, the provisions on settlement of disputes in section VI (articles 264–265) of Part XIII also stipulate that disputes concerning

---

183. For further details, see *Marine Scientific Research, a Guide to the Implementation of the Relevant Provisions of UNCLOS* (United Nations publication, Sales No. E.91.V.3); and ''Marine scientific research: report of the Secretary-General'' (A/45/563).

184. See David A. Ross and Judith Fenwick: *Marine scientific research: US perspective on jurisdiction and international cooperation;* in Lewis M. Alexander, Scott Allen, Lynne Carter Hanson (eds.), *Proceedings of the 22nd Annual Conference of the Law of the Sea Institute, June 12–16, 1988,* p. 217.

the rights of States to withhold consent for marine scientific research or to order its suspension and cessation are only, and to a limited extent, subject to the conciliation procedure under Annex V, section 2.

460. The basic consent provision is contained in article 246, paragraphs 1 and 2, whereby the coastal State in the exercise of its jurisdiction has the right to regulate, authorize and conduct marine scientific research in its exclusive economic zone and on its continental shelf in accordance with relevant provisions of the Convention. It is specified that marine scientific research in such maritime zones shall be conducted with the consent of the coastal State. However, the right of the coastal State is not absolute, in that UNCLOS, true to the balance struck between the coastal State's interests and those of the scientific community, differentiates between "normal circumstances" and situations where the discretionary powers may be exercised. It is emphasized in article 246, paragraph 3, that coastal States shall, in normal circumstances, grant their consent for marine scientific research projects. The granting of consent is thus established as the norm and not the exception. Furthermore, coastal States shall establish rules and procedures ensuring that such consent will not be subject to unreasonable delay or denial.

461. UNCLOS also identifies circumstances in which the coastal State can exercise its discretionary power to withhold consent. These are limited to four cases (article 246, paragraph 5): the research project (a) is of direct significance for the exploration and exploitation of the natural resources, whether living or nonliving; (b) involves drilling into the continental shelf, the use of explosives or the introduction of harmful substances into the marine environment; (c) involves the construction, operation or use of artificial islands, installations or structures referred to in articles 60 and 80; or (d) contains information communicated pursuant to article 248 regarding the nature and objectives of the project that is inaccurate or if the researching State or competent international organization has outstanding obligations to the coastal State from a prior research project.

462. In order to facilitate research, article 252 contains an implied consent rule that allows States or competent international organizations to proceed with a research project six months after the pertinent information has been supplied to the coastal State (article 248 lists the information to be supplied), unless within four months of receipt of the information the coastal State has informed the researching State or organization that it has not met certain conditions.

463. The coastal State will have the right to require suspension or cessation (article 253) of the research in progress in the exclusive economic zone or on the continental shelf if the research does not comply with the information or obligations required.

464. Section 4 of Part XIII contains provisions on the legal status of the installations and equipment, which must have identification markings and adequate warning signals to ensure safety at sea and the safety of air navigation. Their deployment should not interfere with international shipping routes, and safety zones of a reasonable breadth may be created around them. Section 5 deals with responsibility and liability, while section 6 establishes provisions for the settlement of disputes.

465. *Implementation of the consent regime for marine scientific research.* With some exceptions,[185] little is known about State practice with regard to the implementation

---

185. See Alfred H. A. Soons: *The developing regime of marine scientific research: recent European experience and State practice,* and David A. Ross and Judith Fenwick, op. cit., n. 184 above.

of the consent regime, and in particular, the provisions of article 246, paragraph 5. However, the Division for Ocean Affairs and the Law of the Sea has attempted to monitor developments in this field. Studies carried out by the Division include, *inter alia National Legislation, Regulations and Supplementary Documents in Areas under National Jurisdiction*,[186] the report of the Secretary-General on marine scientific research (A/45/563), *Marine scientific research: a guide to the implementation of the relevant provisions of the United Nations Convention on the Law of the Sea*,[187] *Marine Scientific Research: Legislative History of Article 246 of the United Nations Convention on the Law of the Sea*[188] and *Practice of States at the Time of Entry into Force of the Convention*.[189]

466. In the guide to the implementation of the above-mentioned provisions of UNCLOS on marine scientific research, a number of recommendations are made. They are worth bearing in mind because they correspond to some of the issues that the international community is still facing and the difficulties that still persist in the implementation of the legal regime for marine scientific research under the Convention.

467. With regard to the consent regime,[190] it has been revealed that in some instances coastal States have refused to give their consent to research projects. This may be attributable to: (a) difficulties coastal States may encounter in ascertaining the nature of the research proposal, arousing suspicion and resulting in rejection of the proposal, or (b) a lack of established internal administrative or legal structures for the coastal State to receive and process the research proposal, giving rise to delay or lack of response.

468. In this regard, one of the main recommendations in the Guide to marine scientific research issued by the Division is the proposal for the use of a standardized form[191] when making a request to the coastal State for marine scientific research in its maritime zone. Its aim is to reflect accurately the relevant provisions of UNCLOS and at the same time act as an assisting tool (or job aid) for the concerned authorities on both ends of the application. Coastal States are encouraged to agree on a standard form and to incorporate it into their rules and regulations and procedures.

469. International efforts are continuing, at the global and regional levels, to devise practical means for the efficient and effective functioning of the consent regime for marine scientific research. The Advisory Board of Experts on the Law of the Sea (ABE-LOS) of IOC intends to hold its first substantive meeting from 11 to 13 June 2001 (see para. 523). In preparation for the meeting, IOC circulated a questionnaire[192] seeking information on State practice with regard to the conduct of marine scientific research in accordance with UNCLOS. ABE-LOS will also base its agenda and selection of priority topics on a document prepared by the IOC

186. United Nations publication, Sales No. E.89.V.9.
187. See n. 183.
188. United Nations publication, Sales No. E.94.V.13.
189. United Nations publication, Sales No. E.94.V.9.
190. See also the discussions on article 246 in: *Marine Scientific Research: Legislative History of Article 246 of the United Nations Convention on the Law of the Sea* (see n. 189 above).
191. See United Nations publication *Marine Scientific Research: A guide to the relevant provisions of the United Nations Convention on the Law of the Sea* (United Nations publication, Sales No. E.91.V.3), annex I. International organizations are encouraged to use the same form, although they are not required to do so under article 247.
192. The questionnaire was prepared in collaboration with the Division for Ocean Affairs and the Law of the Sea. Fewer than 25 States responded.

secretariat entitled "A synthesis of IOC's possible role and responsibilities under the United Nations Convention on the Law of the Sea."[193]

470. At the regional level, issues related to the consent regime for marine scientific research are also being addressed. The South Pacific Geoscience Commission (SOPAC) indicated that most South Pacific island countries do not have a national system in place based on the UNCLOS regime. One of the few that do, where there is also a history of issues/problems between the coastal country and the researching countries, is Papua New Guinea. SOPAC was organizing a workshop to be held at Port Moresby from 28 February to 2 March 2001, with the aim of bringing all parties together to find solutions to the issues that have arisen in the case of Papua New Guinea and using it as a case study to develop practical solutions for other countries of the region. Constructive discussions were expected to be held on marine scientific research issues of concern to: coastal States, researching States, especially those active in the region (e.g., Japan, the Republic of Korea, Australia and France), and the region as a whole; together with a general synthesis of issues in marine scientific research. Special attention was to be devoted to the important marine resources in the region, such as minerals and living and nonliving resources associated with hydrothermal vents (see para. 318), and the related issue of the distinction between marine scientific research, prospecting and exploration for resources.

471. In this context, SOPAC pointed out that the potential of the South Pacific region, which comprises 22 Pacific island countries and territories for achieving prosperity, has largely been demonstrated through the numerous marine scientific research campaigns of researching States and international organizations. Given the nature of this research, costly collaboration and cooperation between national, regional and international organizations is imperative if small island developing States in the Pacific region are to succeed in collecting the necessary information for an understanding of their marine resource base. Consequently, according to SOPAC, these countries need to be receptive towards marine scientific research related requests for access to their waters. However, at the same time they need to develop and strengthen their internal procedures for handling such requests, to ensure that measures are taken to avoid the abuse of such access.

472. Since the granting in 1997 of exploration licences by the Government of Papua New Guinea to a private company for exploration of polymetallic sulphides, concerns have been raised by interested stakeholders, i.e., the Government, the tenement holder and the researchers who want to conduct marine scientific research within the tenement area. The issues that have been articulated include access to exploration sites, sampling rights, commercial alliances of the researchers, data-sharing, confidentiality of data and ship berthing rights. Some research institutes have indicated that, in the light of these emerging issues, they are reviewing their future commitment to continued marine scientific research activities within the SOPAC region. SOPAC is mindful of the fact that the activities of both researchers and industry are vital to the success of research, discovery, exploration and exploitation.

2. Legal Regime for Development and Transfer of Marine Technology

473. The provisions of UNCLOS dealing with the development and transfer of marine technology are contained mainly in Part XIV. UNCLOS provides for the promo-

---

193. See IOC/WG-LOS-I/6/Rev.1.

tion of the development and transfer of marine technology (article 266 (1)). More importantly, the Convention also provides for the promotion of the development of the marine scientific and technological capacity of States that may need and request technical assistance in this field, particularly developing States, including landlocked and geographically disadvantaged States, with regard to the exploration, exploitation, conservation and management of marine resources, the protection and preservation of the marine environment, marine scientific research and other activities in the marine environment compatible with the Convention, with a view to accelerating the social and economic development of the developing States (article 266 (2)). Considerable misrepresentation of UNCLOS has resulted from the tendency to overlook the fact that it provides that, in promoting marine science and technology, legitimate interests should be protected, including, *inter alia,* the rights and duties of "holders, suppliers and recipients of marine technology" (article 266 (3)).

474. UNCLOS places emphasis on international cooperation and coordination in the development and transfer of marine technology. International cooperation for the development and transfer of marine technology shall be carried out, where feasible and appropriate, through existing bilateral, regional or multilateral programmes, as well as through expanded and new programmes, in order to facilitate marine scientific research, the transfer of marine technology, particularly in new fields, and appropriate international funding for ocean research and development (article 270). States, directly or through competent international organizations, shall promote the establishment of generally accepted guidelines, criteria and standards for the transfer of marine technology on a bilateral basis or within the framework of international organizations and other forums, taking into account, in particular, the interests and needs of developing States (article 271). In the field of transfer of marine technology, States shall endeavour to ensure that competent international organizations coordinate their activities, including any regional or global programmes, taking into account the interests and needs of developing States, particularly landlocked and geographically disadvantaged States (article 272). The competent international organizations referred to in Part XIV as well as in Part XIII shall take all appropriate measures to ensure, either directly or in close cooperation among themselves, the effective discharge of their functions and responsibilities under Part XIV (article 278).

475. UNCLOS identifies the establishment of national and regional marine scientific and technological centres as an important measure of the development and transfer of marine technology. States, directly or through competent international organizations and the International Seabed Authority, shall promote the establishment, particularly in developing coastal States, of national marine scientific and technological research centres and the strengthening of existing national centres, in order to stimulate and advance the conduct of marine scientific research by developing coastal States and to enhance their national capabilities to utilize and preserve their marine resources for their economic benefit. States, through competent international organizations and the Authority, shall give adequate support to facilitate the establishment and strengthening of such national centres so as to provide for advanced training facilities and necessary equipment, skills and know-how as well as technical experts to such States that may need and request such assistance (article 275). States, in coordination with the competent international organizations, the Authority and national marine scientific and technological research institutions, shall promote the establishment of regional marine scientific

and technological research centres, particularly in developing States, in order to stimulate and advance the conduct of marine scientific research by developing States and foster the transfer of marine technology. All States of a region shall cooperate with the regional centres therein to ensure the more effective achievement of their objectives (article 276).

476. Finally, the Third United Nations Conference on the Law of the Sea adopted a resolution promoting the development of national marine science, technology and ocean service infrastructures.[194]

B. Programmes on Marine Science and Technology in the United Nations System

1. Marine Science Programmes in the United Nations System

(a) Intergovernmental Oceanographic Commission

477. The Intergovernmental Oceanographic Commission of the United Nations Educational, Scientific and Cultural Organization is recognized as the competent international organization with regard to Part XIII of UNCLOS on marine scientific research. The activities of IOC are channelled through three interrelated programmes: ocean science, ocean observation operational observing systems, and ocean services, all of which are related to or based on marine science. IOC also has a direct capacity-building programme, the Training, Education and Mutual Assistance (TEMA) programme. In addition, IOC has a number of regional subsidiary bodies that carry out marine science-related activities.

(i) Ocean science

478. Currently, the ocean science programme of IOC has four major areas of focus: (a) oceans and climate; (b) ocean science in relation to living marine resources; (c) marine pollution; and (d) marine science for integrated coastal area management.

479. In relation to ocean science, one of the major areas of focus of the work of IOC is oceans and climate. The purpose of the work carried out under the World Climate Research Programme (WCRP), co-sponsored by IOC, WMO and the International Council for Science (ICSU), is to foster activities in research and observing systems development leading to improved understanding of the ocean's role in climate in order to enhance government decision-making processes for dealing with global change.

480. IOC reported that WCRP was pressing ahead with its two main ocean climate research programmes: the World Ocean Circulation Experiment (WOCE) and the Climate Variability and Predictability (CLIVAR) Study.

---

194. Final Act of the Third United Nations Conference on the Law of the Sea, annex VI. In: *The Law of the Sea: Official Texts of the United Nations Convention on the Law of the Sea of 10 December 1982 and of the Agreement relating to the Implementation of Part XI of the United Nations Convention on the Law of the Sea of 10 December 1982, with Index and excerpts from the Final Act of the Third United Nations Conference on the Law of the Sea* (United Nations publication, Sales No. E.97.V.10), p. 206.

481. *World Ocean Circulation Experiment.* In 1998, WOCE completed nearly a decade of fieldwork and brought closure to the intensive observational phase initiated in 1990. The WOCE data set collected during this period serve as a much-needed benchmark with which to compare all past and future ocean observations in order to assess natural and anthropogenic change. The first set of CD-ROMs of WOCE data was made available in 1998, the second set in 2000.

482. The year 1998 also witnessed the finalization of the analysis, interpretation, modelling and synthesis phase of WOCE, which will enable the full benefits of all the investment to date by participating nations. The phase is expected to continue for at least five years. WOCE has also continued to expand its analysis phase during the period under review. The final regional workshop on the North Atlantic Ocean was held in August 1999 in Kiel, Germany. The report of the meeting highlights the tremendous progress made in the analysis of Atlantic data but also reveals how much more remains to be done in synthesizing these data and using them to improve models. To this end a Working Group on Ocean Model Development was established jointly with CLIVAR.

483. *Climate Variability and Predictability.* CLIVAR is the first scientific programme for the study of climate variability at the time scales of decades and centuries as well as to attribute causes to observed climate change. The enormous task of implementing CLIVAR moved forward by building on the statements of commitment and interest made at the International CLIVAR Conference held at UNESCO headquarters in Paris in December 1998. Important progress was made in defining priorities and developing plans for CLIVAR activities in Africa and South America. Planning started for implementation meetings to develop CLIVAR activities in the Atlantic, the Southern Ocean and the Pacific sectors. At a 1999 Conference co-sponsored by CLIVAR and the Global Ocean Observing System (GOOS) (see paras. 506–515) substantial progress was achieved in defining the global observational networks that would serve both operational and research needs. At a practical level, extensions to the Pacific and Atlantic moored-buoy arrays were undertaken, with installations in the Indian Ocean to follow shortly. These arrays, which have proved so valuable in the Pacific, will soon be providing real-time data on the state of all the tropical oceans to scientists operating global numerical models.

484. *Array for Real-time Geostrophic Oceanography (ARGO).* Firm commitments began to be received from nations intending to participate in the ARGO programme, which will deploy 3,000 free-drifting floats capable of regularly providing temperature and salinity profiles (see also paras. 513 and 524–526). These floats, together with the moored-buoy arrays, will provide the basic underpinning of the in situ upper-ocean observations required for pursuit of the year-to-year component of CLIVAR in each ocean basin. On the longer time scale CLIVAR scientists have been focusing on the forecast potential of subtle decadal climate signals in the oceanic mid-latitudes. These are commonly referred to as the North Atlantic Oscillation and the Pacific Decadal Oscillation.

485. The past decade was a turning point in gaining an understanding of the role of the oceans in climate and global change. Improvements in computer technology enabled the design and implementation of ocean-atmospheric physical integrated models with an unprecedented resolution power. Forecasting the rate of climate change and the regional expression of these changes require data and information previously unavailable. The collection of such data and information is

now being organized and executed through a concerted effort to monitor continuously the major planetary processes. The Global Ocean Observing System (GOOS), the Global Terrestrial Observing System (GTOS) and the Global Climate Observing System (GCOS) have been integrated into a single Integrated Global Observing Strategy (IGOS), at the same time developing a strong partnership with the space agencies.

486. A second major area of focus of the work of IOC is ocean science in relation to living marine resources (OSLR). This work, after two decades of evolution, has evolved into several research and observational components. These include: the Harmful Algal Blooms (HAB) Programme and a related new international initiative, Global Ecology and Oceanography of Harmful Algal Blooms (GEOHAB); the Global Coral Reef Monitoring Network (GCRMN); and the Living Marine Resources Module of GOOS (LMR-GOOS).

487. *Harmful Algal Blooms Programme.* The overall goal of HAB programme is to foster the effective management of and scientific research on harmful algal blooms to understand their causes, predict their occurrence, and mitigate their effects. The HAB programme comprises a number of global and regional working groups and offers databases, technical manuals and guides as well as an international HAB newsletter. A main activity of the programme is networking and capacity-building for improved research and routine monitoring.

488. A coordinated international programme on the ecology and oceanography of blooms was needed to gain an understanding of their causes and to be able to predict when they would occur. To that end, in 1998, IOC and the Scientific Committee on Ocean Research (SCOR) of ICSU established GEOHAB. An IOC-SCOR Scientific Steering Committee for GEOHAB has been established and a GEOHAB Science Plan has been prepared and recently accepted by SCOR.

489. *Global Coral Reef Monitoring Network.* GCRMN is a joint programme of IOC, UNEP, IUCN and the World Bank, aimed at (a) improving the conservation, management, and sustainable use of coral reefs and related coastal ecosystems by providing data and information on the trends in the biophysical status and the social, cultural, and economic values of these ecosystems; and (b) providing individuals, organizations and Governments with the capacity to assess the resources of coral reefs and related ecosystems and to collaborate within a global network to document and disseminate data and information on their status and trends. The programme is funded through contributions from coral reef countries. It functions through regional "nodes," which serve to fund and coordinate the coral reef monitoring activities of coral reef countries in several regions throughout the world.

490. GCRMN produced the "Status of the Coral Reefs of the World: 2000" report, which appeared two years after the "Status of the Coral Reefs of the World: 1998" report documented massive coral bleaching, particularly in the Indian Ocean and South-East and East Asia, with major shifts in population structure on many reefs. The 2000 status report documents some encouraging news. Recruitment of new corals has occurred in some reefs in the Indian Ocean and East Asia, suggesting that sufficient parent corals have survived to produce larvae. However, it may be years before it is known whether the reefs will fully recover, or if the structure of the reef community will be changed.

491. The degradation of coral reefs also affects the human communities that depend on them for their livelihood, through such activities as coral harvesting, fishing and tourism. The study of human communities and their social and eco-

nomic conditions and motivations associated with coral reef use is becoming a major focus within the coral reef monitoring community. In 2000, GCRMN, in association with the United States National Oceanic and Atmospheric Administration (NOAA), IUCN and the Australian Institute of Marine Science, published the *Socio-economic Manual for Coral Reef Management*. The manual is intended to familiarize reef managers with socio-economic assessment methodology and provide practical guidelines on conducting baseline socio-economic assessments of coral reef stakeholders. It will be used for gathering socio-economic information in parallel with the biophysical information already collected by GCRMN. The IOC-coordinated South Asia node of GCRMN, with funding from the United Kingdom Department for International Development, is conducting training using the manual and is establishing demonstration projects in India, Sri Lanka and the Maldives, where socio-economic monitoring will be conducted. The South Asia GCRMN node is also developing a regional GCRMN database to enable the management and exchange of socio-economic and biophysical data between participating countries and institutions.

492. In response to the severe coral bleaching event of 1998, IOC has established a Study Group on Indicators of Coral Bleaching and Subsequent Effects. Its major objectives are: (a) to develop possible molecular, cellular, physiological and community indicators of coral bleaching that can reliably detect early stress signals; (b) to examine potential reef coral mechanisms for adaptation/acclimatization to global environmental change; and (c) to investigate the long-term responses of reef corals to large-scale changes in environmental variables. The activities of the Study Group will involve consideration of current physiological research and promotion of molecular and biochemical techniques that may lead to the recognition of indicators of stress on corals and early detection of coral bleaching.

493. *Living Marine Resources Module of Global Ocean Observing System (LMR-GOOS)*. The sustainability of the oceans' living marine resources is threatened by a wide variety of factors. These issues can only be addressed successfully through improved information-gathering regarding the status of the world's living marine resources and the factors driving change. With this in mind, the goal of LMR-GOOS is to "provide operationally useful information on changes in the state of living marine resources and ecosystems. The objectives are to obtain from various sources relevant oceanographic and climatic data, along with biological, fisheries and other information on the marine ecosystems, to compile and analyse these data, to describe the varying state of the ecosystems and to predict future states of the ecosystems, including exploited species, on useful time scales. A consequence of these efforts should be the identification and development of the more powerful and cost-effective means for monitoring marine ecosystems required to meet the LMR-GOOS goal."

494. To address these needs, the LMR-GOOS strategic design plan utilizes a broad, ecosystem-based approach that considers living marine resources in relation to their physical, chemical and biological environment. Recognizing the increasing heterogeneity of marine ecosystems from the open ocean towards the shore, the approach is structured into three systems: open ocean, coastal ocean, and inshore.

495. Data and information management and the process of transforming data into useful products are an essential element of the LMR-GOOS approach. LMR data products, such as forecasts of ecosystem states, will be produced on an ecosystem scale, typically involving large ocean areas. Appropriate basin-scale regional analysis centres, which would serve to compile data and information on appropriate

ecosystem scales and to generate appropriate forecasts and other data products, should be the fundamental unit on which LMR-GOOS is developed. Existing regional marine science organizations such as the International Council for the Exploration of the Sea (ICES) could host the centres, as could existing regional ecosystem observing programmes.

496. A key linkage between the observing programme and useful predictions of system dynamics is process studies and modelling. Programmes such as the Global Ocean Ecosystem Dynamics Programme (GLOBEC) will provide critical information on physical-chemical-biological processes, develop advanced observing technologies and identify crucial variables and locations for long-time series analyses of climate variability and marine ecosystem response. In turn, LMR-GOOS will provide time series data for research programmes.

497. The first steps towards the implementation of a global LMR-GOOS must be the integration of existing observing systems into a more consistent, ecosystem-based approach utilizing regional design principles, together with a significant increase in capacity to enable full participation throughout the developing world. In many areas, ongoing observing programmes such as those identified as LMR components of the GOOS Initial Observing System are significant components of a regional system that need only minor augmentation and linkage through a regional analysis centre. In other areas, not even rudimentary monitoring capacities exist. The challenge to LMR-GOOS is to identify existing programmes and gaps and to find the resources to develop the programme on a global scale.

498. A third major area of the ocean science work of IOC relates to research on marine pollution, which is carried out under the Global Investigation of Pollution in the Marine Environment (GIPME) Programme.

499. *Global Investigation of Pollution in the Marine Environment.* GIPME is an international cooperative scientific investigation programme focused on marine contamination and pollution, co-sponsored by IOC, UNEP and IMO. In addition, the Marine Environment Laboratory of IAEA, through the Inter-agency Programme on Marine Pollution, is a partner in matters related to inter-comparison exercises and reference materials and methods. The overall objectives of GIPME have been: (a) to provide authoritative evaluations of the state of the marine environment at both global and regional levels, particularly in the identification of the nature and severity of the effects of marine contaminants; (b) to identify requirements for measures to prevent or correct marine pollution; and (c) to develop procedures for assessing/improving compliance monitoring and for surveillance of the marine environment, including risk assessments.

500. The GIPME programme, in collaboration with the Marine Environment Laboratory, has been engaged in the study of issues of contaminants in the marine environment. Through methodological development programmes, workshops and intercalibration exercises, techniques have been developed to assess contaminant concentrations in many matrices. Through the biological effects programme, measurement techniques have been developed to investigate the effects of contaminants on marine organisms. With the development of the GOOS project, scientists from within the GIPME programme have been involved in the development of the Health of the Ocean (HOTO) module of GOOS, specifically addressing the means of developing integrated mechanisms for observing and forecasting the effects of anthropogenic activities on the marine environment.

501. A fourth area of focus of the marine science—integrated coastal area management (ICAM)—is particularly useful for decision makers and managers. Established as a programme in 1998, the purpose of ICAM is to assist IOC member States in their efforts to build marine scientific and technological capabilities as a follow-up to Agenda 21. Fundamental to effective management of the coastal zone is the provision of scientific information to support the development of policies and coastal zone development options. ICAM provides a forum for identifying emerging issues and accessing and developing scientific information to underpin the work of the programme at regional and national levels. IOC has a range of scientific programmes of its own as well as access to other scientific programmes and skills that can be mobilized and focused to benefit coherent and relevant ICAM approaches. The ICAM programme focuses on interdisciplinary studies of coastal processes; scientific and technological information systems; methodological tools development; coastal monitoring; and training and education through symposia, workshops, seminars, and training courses.

502. The Coastal Regions and Small Islands Unit is an intersectoral programme within UNESCO devoted to the coastal sustainable development of small islands and coastal regions, focusing on socio-economic issues and following an integrated management approach. Jointly with the Advisory Committee for Protection of the Sea, the Unit will undertake a root-cause analysis of the status of the coastal and marine environment in the sub-Saharan countries, in the context of the follow-up to the Pan African Conference on Sustainable Integrated Coastal Management.

503. In October 1998, IOC, together with NOAA and the University of Delaware, United States, launched a web site on ICAM (http://www.nos.NOAA.gov/icm), which site provides practitioners with timely access to information on international guidelines on ICAM, descriptions of the ICAM programmes of other countries and ICAM approaches to specific problems (coastal erosion, coral reef management, beach replenishment, etc.). IOC and the other partners are contributing financially to the development and maintenance of the site.

504. IOC developed a strong training and education component for ICAM. Its general objective is to improve and promote training and education programmes at all levels relating to coastal and ocean management. The project covers a broad range of activities, including preparation of education materials, convening of specific workshops and courses and preparation of guidance documents for the facilitation of donor programmes. Some important components of the project are: training workshop on science policy in ICAM; national workshops on ICAM; specialized technical training courses; distance learning courses; and regional consortia of universities.

505. *Revitalization of the IOC Ocean Science Programme.* Following external reviews of OSLR, of GIPME and of the structure of the entire IOC science programme, and with the approval of the IOC governing bodies, the Ocean Science Programme has been undergoing a process of revitalization. As a result, IOC is expected to consolidate its current ocean science programme, divided into various sub-headings, into a single interdisciplinary programme, in recognition of the growing need to tackle complex environmental issues in an integrated and interdisciplinary way. The revitalized programme is expected to operate on two highly interacting main tracks: global and coastal ocean processes in the context of ocean ecosystems and climate variability; and integrated ocean and coastal area management.

(ii) Ocean observation

506. In the area of operational observing systems, the centrepiece of the work of IOC is the Global Ocean Observing System. Created in response to the need, also emphasized by Agenda 21, for an integrated and comprehensive global ocean observing and information system to provide the information required for oceanic and atmospheric forecasting, for ocean and coastal zone management by coastal nations and for research in global environmental change, GOOS is an operational system planned, established and coordinated by IOC, together with WMO, UNEP and ICSU. It is designed to provide real time descriptions of the current state of the sea and its contents, and forecasts of these for as far ahead as possible, for a wide range of users, and to meet the needs of the United Nations Framework Convention on Climate Change by underpinning forecasts of climate changes.

507. While the aims of GOOS are operational, it includes research to develop new operational approaches and tools. GOOS makes and integrates observations across all the disciplines and across all data-gathering media from ships and buoys to satellites and aircraft, covering the sea and its contents, sea ice and the air above the ocean. It is being designed to meet the needs of a broad user community for particular services or products. It will operate as an end-to-end, or production-line system, in which the data, and how they have been processed, are traceable from first observation to final product.

508. GOOS is already beginning to provide States Members of the United Nations with the ability to convert research results into useful products to meet societal needs. It is already influencing national thinking and planning. Many countries are now planning or executing their own coastal and ocean observations in line with the GOOS Strategic Plan and Principles. Many countries have created National GOOS Committees to develop contributions to GOOS at the national or regional level, by improving the way their methods of operational oceanography meet management needs and address policy issues.

509. Since the publication in 1998 of the GOOS Strategic Plan and an action plan for implementing the open ocean physical component of GOOS, the GOOS organization has been simplified into two implementation modules, one dealing with all aspects of coastal seas, and the other with the open ocean. Efforts have focused on two topics: (a) improving the design for open ocean observations in support of weather and climate forecasting, and (b) development of designs for the implementation of GOOS in coastal seas, which were made available by year-end 2000 on the GOOS web site (http://ioc.unesco.org/goos). The coastal observing system will detect and predict changes in coastal ecosystems and environments.

510. In Bonn in 1999 and in The Hague in 2000, GOOS continued to receive additional intergovernmental support from the Conference of the Parties to the United Nations Framework Convention on Climate Change, as the ocean component of the Global Climate Observing System (GCOS). The Conference is requiring the Parties to develop action plans to implement climate-monitoring systems, including ocean components that will form part of GOOS.

511. In Monaco, in November 2000, discussions were held between UNEP and IOC regarding the use of GOOS as a distributed tool for meeting the needs of the various UNEP regional seas conventions (which collectively form in effect a distributed convention on seas and oceans) (see annex V to the present report). Already there are plans for the Baltic component of GOOS to form the primary mechanism for gathering the data needed for the Helsinki Commission (see paras. 386–387).

512. The GOOS Initial Observing System (GOOS-IOS), created in 1998 to unite existing global ocean-observing sub-systems, incorporates measurements from voluntary ships, buoys, coastal stations, including tide gauges, and satellites, as well as data centres and means of communication. The system has continued to grow with the addition of components such as the Continuous Plankton Recorder survey and the California Cooperative Fisheries Investigations. Further development of GOOS-IOS will be facilitated by the development of the new Joint WMO/IOC Technical Commission for Oceanography and Marine Meteorology, merging previous bodies dealing with oceanography and marine meteorology, which will hold its first intergovernmental meeting in Iceland in June 2001. A significant problem facing GOOS-IOS continues to be the vandalism by fishing vessels of the weather and climate forecasting ocean buoys.

513. The main GOOS pilot project continues to be the Global Ocean Data Assimilation Experiment (GODAE), designed to demonstrate the power of integrating satellite and in situ data, the importance of model assimilation and the value of a global system capable of working in real time. GODAE requires global coverage of the temperature and salinity of the ocean interior that can be integrated with satellite data from the ocean surface so as to greatly improve the numerical models that forecast ocean behaviour, weather and climate. To obtain these data IOC and WMO have launched the ARGO Pilot Project to collect upper-ocean measurements every two weeks and radio the information back to shore stations via satellite (see also para. 484). Several countries have already made substantial financial commitments to ARGO and some of the floats are already in the water. An IOC Technical Coordinator will inform member States about the locations of the floats and what data may be obtained from them.

514. The implementation of GOOS depends ultimately on nations working individually or in groups. The two main regional GOOS programmes are EuroGOOS in Europe, and NEAR-GOOS in the North-East Asian region. EuroGOOS continues to be successful in attracting funds from the European Commission for pre-operational research projects to develop the skills and capabilities to implement GOOS. MedGOOS and IOCARIBE-GOOS have both developed secretariats and are developing work programmes and proposals to fund their future activities in the Mediterranean and the Caribbean respectively. PacificGOOS held a meeting in August 2000 to begin developing its work programme for the Pacific islands. Black-Sea-GOOS and GOOS-Africa are planning meetings in 2001 to develop GOOS in those areas. A new IOC regional programme office in Perth, Western Australia, is helping to develop GOOS in the Indian Ocean. Regional GOOS programmes are being developed around North America by the United States and Canada.

515. GOOS is part of an Integrated Global Observing Strategy (IGOS) developed by the United Nations sponsors of global observing systems, along with ICSU and the Committee on Earth Observation Satellites. IGOS involves the major space-based and in situ systems for global observation of the Earth, including in particular the climate and atmosphere, oceans, land surface, and Earth interior, in an integrated framework. It is expected to improve Governments' understanding of global observing plans, provide a framework for decisions on the continuity of observation of key variables, reduce duplication, help to improve resource allocation and assist the transition from research to operations. It is consistent with the drive towards increasing efficiency and effectiveness within the United Nations system. The IGOS Partners have agreed to focus initially on an oceans theme, which was presented by

the National Aeronautics and Space Agency (NASA) of the United States at the sixth IGOS Partners Meeting in November 2000.

(iii) Ocean services

516. In the area of ocean services, one of the core programmes of IOC during the past 40 years has been the International Oceanographic Data and Information Exchange (IODE) programme, which has as its aim to improve the knowledge and understanding of marine resources and the environment by providing a mechanism for the management and exchange of ocean data and information from which that knowledge can be generated. The IODE programme has assisted member States in establishing national oceanographic data centres, now numbering more than 60, which are linked with the ICSU world data centres (oceanography) and world data centres (marine geology and geophysics). This network has enabled the ocean community to build and access huge archives of oceanographic data and information, preserving these valuable resources for posterity.

517. At the global level, the IODE Committee at its 16th Session (Lisbon, October/November 2000) adopted an ambitious work plan focusing on, *inter alia:* (a) the establishment of a global ocean metadata management programme (following the successful completion of the pilot project in 2000); (b) strengthening of the IODE regional coordinator mechanism; (c) the establishment, maintenance and strengthening of cooperation between IODE and ocean research and monitoring programmes; (d) increased activities related to biological and chemical data management and exchange; (e) continuation (after the successful implementation of the first phase, ending in 1998) of the Global Oceanographic Data Archaeology and Rescue Project to safeguard data at risk of being lost owing to media decay or neglect; (f) establishment of the IODE Resource Kit project, an Internet-based tool for IODE capacity-building and distance learning (following the completion of the pilot project in 2000); (g) active collaboration within the Joint WMO/IOC Technical Commission for Oceanography and Marine Meteorology; and (h) IODE participation in a marine Extensible Markup Language (XML) consortium to develop an XML as a standard for data interchange on the Internet.

518. One of the cornerstones of the IODE programme is its capacity-building programme. Every year, IODE and its members organize national and regional training courses or workshops with the objective of building or strengthening national capacity. In the case of developing countries, the IODE programme links training with equipment and operational support within the framework of regional Ocean Data and Information Networks (ODINs). An excellent example of such a network is the Ocean Data and Information Network for Africa (ODINAFRICA), a pan-African network of 20 member States. Based on its regional predecessors in East Africa and in the Western Indian Ocean region and in the Central Eastern Atlantic, ODINAFRICA will: (a) provide assistance in the development and operation of national oceanographic data (and information) centres and establish their networking in Africa; (b) provide training opportunities in marine data and information management applying standard formats and methodologies as defined by IODE; (c) assist in the development and maintenance of national, regional and pan-African marine metadata, information and data-holding databases; and (d) assist in the development and dissemination of marine and coastal data and information products responding to the needs of a wide variety of user groups using national and regional

networks. In this regard emphasis is placed on data and information for coastal area management and data and information for development. An important component of the project is bringing together the ocean science/data management communities with managers/ decision makers.

(iv) Training, education and mutual assistance
519. The IOC Training, Education and Mutual Assistance (TEMA) capacity-building programme is central to the overall IOC role and supports the capacity-building efforts that are focused within the IOC scientific programmes. A strong TEMA policy ensures that the capacity-building process is linked to existing and planned national and regional programmes, thereby enhancing the success rate of capacity-building activities. IOC is developing principles and a programme to develop national capabilities in marine sciences and services. This programme for the building of capacity involves a wide range of activities, depending on the starting capacity (level of ability) of the countries concerned.

520. Developing and strengthening capacities in marine research, observation and effective use of services that organizations such as IOC can offer involves human resources, the necessary institutions and a framework that supports and sustains marine activities. These components must be integrated into a network, but implementation can be difficult because of the complexity of jurisdictions within and among nations and the large differences in ability and capacity among countries. Because of these differences, capacity-building activities must be tailor-made to the specific needs of a country or a region.

521. During 1999, IOC contributed to the implementation of 94 activities with strong capacity-building content. These activities were hosted in 29 member States and included 30 specific training events, 18 workshops and 1 beach-cleanup public awareness exercise. Thirty-six persons from 19 countries benefited from individual grants (24 travel grants and 12 research/study grants). More than 1,000 people from 102 member States participated in all activities. A large number of people (scientists and students) also benefited from access to scientific literature facilities and training tools. Marine science institutions of East Africa benefited from access to the Internet, acquisition of computer equipment and support for operational expenses. Funding for capacity-building amounted to more than $1.5 million, of which approximately 70% was derived from extrabudgetary sources. TEMA activities for 2000 included 20 workshops and training courses.

522. The entry into force of UNCLOS has prompted IOC to begin an examination of ways to expand its role and functions vis-à-vis UNCLOS. IOC is called upon to assume responsibilities such as the promotion of marine scientific research, the establishment of practical measures for the conduct of marine scientific research facilitating the implementation of the provisions of the Convention, the publication and dissemination of marine science information and knowledge, the coordination of international marine scientific research projects and the provision of basic scientific information towards the protection of the marine environment and transfer of technology.[195]

---

195. See IOC/INF-961 (1994).

523. In this regard, since 1994,[196] IOC has consistently placed on the agenda of its Executive Council and Assembly (19th and 20th sessions) the item entitled "IOC in relation to the United Nations Convention on the Law of the Sea." In response to its new responsibilities, e.g., the evolving expansion of its legal functions and other activities as a result of the entry into force of UNCLOS, IOC has established an Advisory Body of Experts on the Law of the Sea (ABE-LOS) (see para. 469),[197] an open-ended group composed of experts with expertise in the law of the sea and in marine science, to which each State member of IOC may nominate two experts.[198]

(b) World Meteorological Organization

524. WMO, observed that ocean data buoys, both freely drifting and moored, constitute valuable and sometimes unique sources of essential meteorological and oceanographic data from remote ocean areas. Such data, reported in real time via satellite, are distributed globally and made freely available on the Global Telecommunications System of the World Weather Watch of WMO. The data are inputted operationally into a variety of meteorological and oceanographic models, as well as being archived for delayed-mode applications. They directly support meteorological forecast and warning services (including for maritime safety), global climate and global change monitoring, research and prediction (including El Niño/La Niña), and meteorological and oceanographic research.

525. WMO, however, expressed the concern that these unattended, automatically operating ocean data collection platforms are sometimes the subject of vandalism, both deliberate and inadvertent, despite the value of their data to all maritime users. Often the vandalism results from ignorance of the purpose and value of the platform, and efforts have been made over many years to sensitize marine communities, in particular fishermen, to their purpose and value, but with little success. The problem was again highlighted at the annual session of the WMO/IOC Data Buoy Cooperation Panel in late 1999.

526. As an additional means of better sensitizing mariners of all types to the purpose and value of ocean data buoys, WMO and IOC have sought the assistance of the International Hydrographic Organization. Following the agreement of the IHO Commission for the Promulgation of Radio Navigational Warnings, an IHO circular letter containing an agreed text relating to buoy vandalism was distributed in August 2000 for promulgation in national Notices to Mariners. It is expected that this text will be promulgated in a similar way at regular intervals in the future.

527. WMO coordinates the operational delivery of meteorological and oceanographic data, analyses and forecasts for coastal areas, including maritime safety services under the Global Maritime Distress and Safety System (GMDSS), as well as the provision of comprehensive operational storm surge warning services. WMO

---

196. See Twenty-seventh Session of the IOC Executive Council, Paris, 5–12 July 1994, document IOC/EC-XXVII/3.
197. See IOC Assembly resolution XIX-19 and annex for terms of reference of ABE-LOS.
198. The Division for Ocean Affairs and the Law of the Sea is also an invited observer at ABE-LOS meetings.

also coordinates a global system to provide comprehensive marine climatological databases for all ocean areas, including coastal areas and exclusive economic zones.

528. A new paradigm in international cooperation is represented by the Joint Technical Commission for Oceanography and Marine Meteorology, established by the governing bodies of WMO and IOC, as an operative body responding simultaneously to both organizations in which both organizations have agreed to pool resources and expertise to address common challenges.

529. WMO not only stresses the need for capacity-building in developing countries, but also points to the detrimental effects of the lack of capacity-building in marine meteorology. As WMO noted, it is clear that many developing countries lack the capacity either to participate in and contribute to the major marine observation and services programmes of WMO and IOC, or to benefit from the data and products these programmes generate. This in turn means that the programmes themselves are deficient in data, product and service availability in many major ocean areas, which is to the detriment of all maritime users. This is particularly the case in large sections of the Indian Ocean. At the same time, isolated capacity-building efforts in individual countries in the marine area have not often been cost-effective and have had minimal overall impact. In an attempt to address capacity-building needs in a wider context, WMO, in collaboration with IOC, has developed a Western Indian Ocean Marine Applications Project (WIOMAP) for the enhancement of marine observing networks, data management and services in the Western Indian Ocean as a regional cooperative project involving both meteorological and oceanographic agencies and institutions.

(c) International Hydrographic Organization

530. The International Hydrographic Organization drew attention to a number of activities that should be carried out in the interest of the safety of navigation and the protection of the marine environment. Such activities include conducting, following IHO standards, hydrographic surveys (including bathymetry and measurements of oceanographic parameters) in ports, harbours and sensitive coastal areas as a first priority, and in the territorial sea, the exclusive economic zone and the continental shelf as a second priority; publishing and distributing the information derived through the hydrographic surveys in the form of nautical charts (electronic and paper) and nautical books for the safety of navigation of all ships; and making available the hydrographic and oceanographic survey information related to the sea areas under the coastal State's jurisdiction in the form of bathymetric maps and Geographic Information System (GIS) products, for the purposes of, for example, fishing, coastal zone management and scientific studies.

531. IHO pointed out that while in the developed countries there are well-established hydrographic services carrying out the above activities, many other countries also need to be assisted in this area. IHO has prepared a chart that broadly depicts the geographical regions where coordination and cooperation should be enhanced in the interest of the navigational safety and the protection of the marine environment; these include the West Pacific Islands, South Asia, the Persian Gulf, the Red Sea, southern Africa, Western and Central Africa, the southern Mediterranean, the Black Sea, the Baltic Sea, and Central America and the Caribbean. There

762   *Selected Documents and Proceedings*

are thus vast areas worldwide requiring robust intervention, most particularly in the African region.

532. It should be noted in this connection that the General Assembly, in paragraph 21 of its resolution 53/32 of 24 November 1998, invited States to cooperate in carrying out hydrographic surveys and nautical services for the purpose of ensuring safe navigation as well as to ensure the greatest uniformity in charts and nautical publications and to coordinate their activities so that hydrographic and nautical information is made available on a worldwide scale. Furthermore, in its resolution 54/31 of 24 November 1999, the Assembly specifically noted that developing countries, in particular small island developing States, might need assistance in the preparation and publication of charts under articles 16, 22, 47, 75, and 84 and annex II to UNCLOS, dealing with limits of national maritime zones, and in that context, urged the international community to assist such States (see paras. 83–87).

(d) Food and Agriculture Organization of the United Nations

533. FAO remarked that there are increasing demands for objective, neutral, verified and comprehensive information on fisheries, their resources and their environment. These demands stem from such concerns as the poor state of many fishery resources; the potential of unconventional fisheries resources; the collapse of some fisheries; the overcapacity and poor economic performance of many others; the threats of unabated environmental degradation; the risk of significant shifts in resources as a result of climate change; the requirement for a precautionary approach; and the threats and opportunities of globalization and free trade. Confronted with these problems, Governments, industry, NGOs, development banks, fisherfolk, and the public at large are facing increasing difficulty in understanding the situation. There is often a lack of the necessary information for a better understanding of the implications of agreed international instruments as well as the options available and their implications. The unavoidable conflicts of interests create a danger of misuse or misinterpretation of the information available and there is a huge demand for greater transparency and higher-quality information (e.g., the best scientific information available). The demand for information from FAO grew rapidly during the past decade.

534. FAO has undertaken to improve, and make fully available on the Internet, all of its information systems and databases, developing a Fisheries Global Information System (FIGIS) with financial support from France and Japan. A first partial version dealing with statistics and species has already been made available. Preparation of modules dealing with vessels, gears, fishery commissions and strategic issues will soon be completed. Modules dealing with stocks, fisheries products and trade are to be produced in the near future. In addition, a Fisheries Resources Monitoring System (FIRMS) has been conceptualized and will be developed within FIGIS to foster the development of an international network of regional fishery management organizations and centres of excellence collaborating in the maintenance of a global information system on the state of world resources. In order to facilitate access to this information by countries, institutions and people with insufficient access to the Internet, FAO is developing a World Fisheries Atlas on CD-ROM in which similar information will be made available. In order to formalize and establish an institutional basis for the necessary national, regional and global efforts, the FAO Commit-

tee on Fisheries was to consider at its meeting in February 2001 a proposal for an international plan of action for status and trends reporting on fisheries.

535. FAO pointed out that to fulfil their commitments with regard to agreed international instruments and initiatives, countries need to improve significantly the quality of their information on ocean fisheries. For this purpose, higher priority should be given to improve statistical systems that, in many developing countries, are deteriorating for lack of due recognition and resources. The efforts of FAO to improve information on IUU fishing, fishing capacity, trade, the state of stocks, employment, prices, etc., as required by its members, need to be supported, and the capacities of countries to contribute to and use the Internet should be improved.

(e) International Maritime Satellite Organization

536. The International Maritime Satellite Organization (Inmarsat) was established in 1979 to make provision for the space segment necessary for improved maritime communications and, in particular, for improved safety of life at sea communications and the Global Maritime Distress and Safety System (GMDSS). Its purpose was later extended to provide the space segment for land mobile and aeronautical communications, and the name of the organization was changed to the International Mobile Satellite Organization (IMSO) to reflect the amended purposes. In its contribution to the present report, IMSO states that after 20 years of successful operation, member States and signatories to the intergovernmental organization Inmarsat decided to challenge the rapidly growing competition from private providers of satellite communications services and pioneered the first-ever privatization of assets and business carried on by the intergovernmental organization while adhering to the continuous provision of its public service obligations and governmental oversight. At its twelfth session, in April 1998, the Inmarsat Assembly adopted amendments to the Inmarsat Convention and Operating Agreement that were intended to transform the organization's business into a privatized corporate structure while retaining intergovernmental oversight of certain public service obligations, in particular, GMDSS (see A/53/456, para. 215). In April 1999, Inmarsat was privatized, IMSO was created and a Public Service Agreement between IMSO and the privatized Inmarsat Ltd. was also executed.

537. IMSO also observed that the horizons of mobile satellite communications are expanding with ever-increasing speed and there are several different options for the design and capability of new services. The adoption by the IMO Assembly of resolution A.21/Res.888, "Criteria for the Provision of Mobile Satellite Communication Systems in the Global Maritime Distress and Safety System (GMDSS)," has provided a clear indication of the intention of IMSO to consider granting provision of GMDSS services in the future to any satellite operator whose system fits these criteria. This is most likely to happen in the context of a revision of Chapter IV (Radiocommunications) of SOLAS and will provide the opportunity for specifying more effective services in a way that permits the use of evolutionary capabilities and nongeostationary satellite constellations. At present, Inmarsat Ltd., with the satellite communications system it operates, is the sole global provider of these services and its position in the marketplace is, for the time being, unrivalled.

538. IMSO added that in recent years, the process of the liberalization and privatization of global and regional satellite communications services has become

a given fact. It is encouraging to note, in this context, that IMSO has not been able to detect any reduction or deterioration in the level and quality in the provision of GMDSS services by Inmarsat Ltd. under the new regime, compared with the situation prior to privatization. All other public service obligations were also fulfilled, or due attention has been given thereto by the Company. It may therefore be concluded that, after more than one year of distinct, but workable interface between IMSO and Inmarsat Ltd., the restructuring has paid off and the principles under which the process of restructuring took place have proved to be effective.

(f) Economic Commission for Latin America and the Caribbean

539. ECLAC reported that one of the areas of focus of its Division on Natural Resources and Infrastructure, which is responsible for ocean and the law of the sea affairs, is the role of marine scientific research in the sustainable development of oceans and coastal areas and in the most effective implementation of UNCLOS on the biological diversity of the marine environment. In 2000, a study was prepared by ECLAC on a preliminary approach to the constraints and opportunities for marine scientific research in Latin America and the Caribbean. In view of the current analysis it is undertaking in the field of marine scientific research and the incorporation of the subject as one of the areas of focus in the second meeting of the Consultative Process, ECLAC envisaged that it would carry out further work in the area and that the results thereof would be made available at the meeting.

2. *Marine Technology Programmes in the United Nations System*
540. UNIDO, together with its partner organizations, NOAA of the United States, the Centre for Coastal and Marine Sciences (Natural Environment Research Council) of the United Kingdom and the International Centre for Science and High Technology of UNIDO, in Italy, have at their disposal a broad range of the highest-quality expertise in integrated coastal zone management. Areas covered include fisheries management, environmental quality assessment (impact and risk assessment, eco-toxicology, monitoring tools, human health risks), risk factors related to political instability and international terrorism, image engineering for GIS, simulation modelling, climatology, expert systems, and software engineering. Additional areas of expertise include the use of biotechnology to combat environmental degradation, experience and databases on the application of cleaner technologies, technologies for municipal and industrial waste management, investment promotion for industrial modernization and environmental planning requirements for the development of "environmentally friendly" technologies (see A/55/61, para. 247).

C. Identified Needs in Marine Science and Technology

541. The General Assembly, in its resolution 55/7, annex II, paragraph 1, stated that promoting and developing the marine scientific and technological capacity of developing States, in particular the least developed countries and small island States, with a view to accelerating their social and economic development, is essential for the effective implementation of UNCLOS. There already exists a comprehensive inventory of needs built up over two decades; the Consultative Process can benefit

from revisiting the inventory so that it can focus on the most critical needs and the requisite measures amenable to international coordination and cooperation.

542. One of the earliest studies on the implications of the provisions of UNCLOS on marine science and technology was reviewed by the Third United Nations Conference on the Law of the Sea itself. One of the primary objectives of the study (document A/CONF.62/L.76 of 18 August 1981)[199] was to aid in the consideration of such questions as mechanisms for establishing policy, processes of preparing legislation and other measures, allocation of functions among existing agencies or departments, and coordination among all sectors concerned for these purposes.

543. With respect to marine scientific research and the development of associated technology, one of the requirements identified was the establishment of an administrative framework for marine scientific research activities and the development of marine scientific capabilities. This involves in particular: (a) the development of adequate arrangements to coordinate programmes and projects, particularly those having a complementary nature—arrangements might involve all the various entities conducting or sponsoring research, whether directed towards environmental protection, the provision of meteorological services, and the utilization and conservation of living resources and other offshore resources, or towards other marine uses; such coordination would serve the purposes of reviewing research needs and establishing priorities as well as ensuring the compliance of marine scientific research activities with environmental, navigational and other rules; (b) an examination of the interests and capabilities (qualified personnel, level of funding, facilities and equipment, etc.) of institutions and sectors concerned, taking into account the establishment or development of a national marine scientific and technical centre; (c) the establishment of priorities and the formulation of programmes and specific projects, taking account, *inter alia,* of opportunities provided by international programmes designed to strengthen the marine scientific capabilities of developing countries; (d) an examination of needs associated with participation in the development and execution of global, regional and subregional programmes of research, education and training, and data and information exchange; and (v) cooperation as regards the creation of favourable conditions for the conduct of marine scientific research.

544. The requirements identified in the study also included arrangements to deal with research projects that another State or an international organization intends to undertake in the exclusive economic zone or on the continental shelf of a coastal State. This involves in particular: (a) assessment of suitably qualified personnel for participation in each project and appropriate arrangements; (b) maintenance of information on the research activities (vessels used, areas studied, institutions involved, deployment and use of installations and equipment, participants, reports) and dissemination of research results nationally and internationally; (c) establishment of specific arrangements ensuring expeditious consideration of the projects, which may require devising methods to review projects in terms of their significance for resource development, taking into account general criteria and guidelines developed pursuant to article 251 of UNCLOS; (d) arrangements for

---

199. *Official Records of the Third United Nations Conference on the Law of the Sea,* vol. XV (United Nations publication, Sales No. E.83.V.4), Documents of the Conference, document A/CONF.62/L.76.

facilitating marine scientific research, access to harbours and assistance to research vessels (in coordination with port authorities); (e) special arrangements associated with the designation of areas on the continental shelf; (f) procedures in the event of the suspension or cessation of the research activities pursuant to article 253 of UNCLOS; (g) arrangements with respect to the participation of neighbouring landlocked and geographically disadvantaged States and provision of the relevant information and data; and (h) arrangements for dealing with disputes, including the requests of the coastal State for expert advice and assistance.

545. Additional requirements covered administrative aspects of the deployment and use of scientific research installations or equipment, including safety zones, identification markings and warning signals (in coordination with maritime and aviation authorities); and administrative arrangements (of the researching State) for the preparation and submission of information to the coastal State in compliance with its rules and procedures and with the conditions listed in articles 249 and 254 of UNCLOS.

546. With respect to the development and transfer of marine technology, the requirements included: (a) arrangements to take into account cooperative activities at all levels, including participation in forums dealing with economic and legal conditions for the transfer of technology and with policies and methods, or establishing guidelines, criteria and standards, and in programmes established to assist in the development of technical capacity in marine science and in marine resource development; (b) administrative measures associated with the development of infrastructure; manpower development (including education and training and exchange of scientists, technological and other experts); acquisition, evaluation and dissemination of scientific and technical information and data, including information on the marketing of technology, contracts and other arrangements; the development of appropriate technology, etc.; and (c) arrangements for establishing or developing national and regional marine science and technology centres, taking into account international programmes of technical cooperation, and outlining functions to be performed.

547. Three reports of the Secretary-General issued in 1990 and 1991 (A/45/563, A/45/712 and A/46/722) deal with marine scientific and technological needs and possible measures to address them. The difficulties faced by many States, especially developing States are several and varied. They include the following: (a) there is a lack of awareness of the overall development potential of the marine sectors, national capacity for development has been strained and capabilities in the ocean sectors are limited; (b) there has been a scarcity of available financing and external assistance; in the few cases where it has been available, the level of international financing has been limited; (c) the acquisition of new technologies is beyond the reach of most; for instance, several developing countries are inadequately equipped to deal with the environmental implications of marine development and other ocean uses; they cannot respond to catastrophes or threats to the ocean ecology; (d) the development of skilled manpower in the several disciplines required for the sustainable development and management of ocean resources, including the environmental implications of marine development, is yet a distant goal to many; (e) there is a lack of awareness of the types of data and information needed to secure resource jurisdiction and of the sustainable development and management of ocean resources; (f) there is a lack of access to such data and information, as well as data

and information resulting from marine scientific research; (g) there is a need for scientific information and technology development, whether applicable to traditional uses or to new avenues.

IX. CAPACITY-BUILDING

548. The issues of capacity-building and assistance to developing States were discussed during the first Meeting of the Consultative Process. The Meeting agreed to bring to the attention of the General Assembly the need for capacity-building to ensure that developing countries, and especially the least developed countries, and those that are landlocked, have the ability both to implement UNCLOS and to benefit from the sustainable use and development of seas and oceans and their resources, and the need to ensure the access of small island developing States to the full range of skills essential for these purposes.

549. In its resolution 55/7, the General Assembly underlined the essential need for capacity-building. In paragraph 23, it requested the Secretary-General, in cooperation with the competent international organizations and programmes as well as representatives of regional development banks and the donor community, to review the efforts taking place to build capacity as well as to identify the duplications that need to be avoided and the gaps that may need to be filled for ensuring consistent approaches, both nationally and regionally, with a view to implementing UNCLOS, and to include a section on this subject in his annual report on oceans and the law of the sea.

550. *Capacity-building: Suggested measures in chapter 17 of Agenda 21.* Agenda 21 itself, in its chapter 17, includes suggestions about capacity-building in matters relating to oceans and seas. It identifies programme areas and, for each programme area, provides suggestions for capacity-building of developing States.

551. With respect to the programme area of integrated management and sustainable development of coastal and marine areas, including exclusive economic zones, chapter 17 of Agenda 21 suggests that full cooperation should be extended, upon request, to coastal States in their capacity-building efforts and that, where appropriate, capacity-building should be included in bilateral and multilateral development cooperation. Coastal States may consider, *inter alia:* (a) ensuring capacity-building at the local level; (b) consulting on coastal and marine issues with local administrations, the business community, the academic sector, resource user groups and the general public; (c) coordinating sectoral programmes while building capacity; (d) supporting "centres of excellence" in integrated coastal and marine resource management; and (e) supporting pilot demonstration programmes and projects in integrated coastal and marine management. The suggested measures for human resource development include education and training in integrated coastal and marine management and sustainable development; and development of educational curricula and public awareness campaigns.

552. With respect to capacity-building for the programme area of marine environment protection, chapter 17 of Agenda 21 suggests that national planning and coordinating bodies should be given the capacity and authority to review all land-based activities and sources of pollution for their impacts on the marine environment and to propose appropriate control measures. An international funding mech-

anism should be created for the application of appropriate sewage treatment technologies and for building sewage treatment facilities, including grants or concessional loans from international agencies and appropriate regional funds, replenished at least in part on a revolving basis by user fees. The suggested measures for human resources development include provision of training based on training needs surveys; development of curricula for marine studies programmes; establishment of training courses for oil-spill and chemical-spill response personnel; conduct of workshops on environmental aspects of port operations and development; support for specialized international centres of professional maritime education; and supporting and supplementing national efforts as regards human resources development.

553. For the programme area on the sustainable use and conservation of marine living resources of the high seas, chapter 17 of Agenda 21 suggests that States, with the support, where appropriate, of relevant international organizations, whether subregional, regional or global, should cooperate to develop or upgrade systems and institutional structures for monitoring, control and surveillance, as well as the research capacity for the assessment of marine living resource populations. The suggested measures for human resources development include training in high seas fishing techniques and resource assessment; strengthening cadres of personnel to deal with high seas resource management and conservation; and training observers and inspectors to be placed on fishing vessels.

554. Regarding the sustainable use and conservation of marine living resources under national jurisdiction, chapter 17 of Agenda 21 suggests that coastal States, with the support of relevant subregional, regional and global agencies, where appropriate, should: (a) provide support to local fishing communities, in particular those that rely on fishing for subsistence, as well as to indigenous people and women; (b) establish sustainable aquaculture development strategies; and (c) develop and strengthen, where the need may arise, institutions capable of implementing the objectives and activities related to the conservation and management of marine living resources. The measures for human resources development include expanding multidisciplinary education, training and research on marine living resources, particularly in the social and economic sciences; creating training opportunities to support artisanal (including subsistence) fisheries; and introducing topics relating to the importance of marine living resources in educational curricula.

555. For the programme area entitled "Addressing critical uncertainties for the management of marine environment and climate change," chapter 17 of Agenda 21 recommends that States should strengthen or establish, as necessary, national scientific and technological oceanographic commissions or equivalent bodies to develop, support and coordinate marine science activities and work closely with international organizations. The suggested measures for human resources development include the development and implementation of comprehensive programmes for a broad and coherent approach to meet core human resource needs in the marine sciences.

556. In respect of the sustainable development of small islands, chapter 17 of Agenda 21 explains that the total capacity of small island developing States will always be limited. Existing capacity must therefore be restructured to meet efficiently the immediate needs for sustainable development and integrated management. At the same time, adequate and appropriate assistance from the international community must be directed towards strengthening the full range of human resources

needed on a continuous basis to implement sustainable development plans. New technologies that can increase the output and range of capability of the limited human resources should be employed to increase the capacity of very small populations to meet their needs. With regard to human resources development, since populations of small island developing States cannot maintain all necessary specializations, training for integrated coastal management and development should aim to produce cadres of managers or scientists, engineers and coastal planners able to integrate the many factors that need to be considered in integrated coastal management.

A. Capacity-Building Activities Within the Organizations of the United Nations System

557. Almost all the organizations of the United Nations system carry out capacity-building activities with respect to oceans and the law of the sea within their respective areas of competence. The range and the diversity of such activities are reflected in a recent survey carried out by UNDP (see paras. 568–572). That report, structured along subject areas, deals with various measures in the respective subject areas, many of which could be considered as capacity-building measures, following the definition used by the Organisation for Economic Cooperation and Development (OECD). Capacity development is defined by the OECD Development Assistance Committee (a definition adopted also by the Canadian International Development Agency and others), as "the process by which individuals, groups, organizations, institutions and societies increase their abilities to: (a) perform core functions, solve problems and achieve objectives, and (b) understand and deal with their development needs in a broad context and in a sustainable manner."[200] The present section, however, describes the direct capacity-building activities of various organizations within the United Nations system, based on the information provided by the organizations themselves in response to a request by the Secretary-General.

558. *Food and Agriculture Organization of the United Nations.* FAO provided information on its activities in respect of capacity-building in developing countries, in particular training, with special emphasis on monitoring, control and surveillance and enforcement of fishing regulations. Within the programmes managed from FAO headquarters during 2000, training was provided to a number of persons for various periods, the total being in the order of 1,500 person training days. Training has focused primarily upon two key areas for fisheries: training in support of improved post-harvest practices and training for the implementation of the Code of Conduct for Responsible Fisheries, particularly aspects associated with the monitoring, control and surveillance of fisheries. In addition, a number of training programmes are managed and delivered at the local level via the FAO regional offices. Besides direct training activities, the large number of manuals and publications produced by FAO may also be considered to contribute to capacity-building.

559. Another major area of activities of FAO in respect of capacity-building is the strengthening of regional fishery management organizations, as necessary,

---

200. Cited in "Capacity Assessment and Development," Technical Advisory Paper No. 3, Bureau for Development Policies, UNDP, January 1998, footnote 5.

through cooperation with such bodies (see annex III). The second meeting of FAO and non-FAO regional fisheries management organizations was scheduled to be held at FAO headquarters in Rome on 20 and 21 February 2001.

560. *International Maritime Organization.* Within the United Nations system, capacity-building in maritime infrastructure is the focus of IMO. IMO attaches the highest priority to the need to ensure that the numerous rules and standards contained in its body of treaties are properly implemented. To ensure this, it focuses on the continuous strengthening of regulations to enable flag States, port States and shipowners as well as all other industrial partners in the chain of responsibility to develop their capacities and exert their responsibilities to the fullest. Technical cooperation activities of IMO have been intensified by the operation of the Integrated Technical Cooperation Programme, aimed at ensuring that funds from different donor sources are properly channelled towards the execution of projects under the supervision of IMO as executing agency for strengthening the maritime infrastructure of developing countries.

561. The importance of technical assistance to developing countries and the form that such assistance will take are reflected in IMO Assembly resolution A.901(21), entitled "IMO and technical cooperation in the 2000s" (see A/55/61, paras. 245–246). The resolution states that capacity-building for safer shipping and cleaner oceans is the main objective of the IMO technical cooperation programme during the current decade. The development and implementation of ITCP should continue to be based on a number of key principles, including the following: ownership of the development and implementation process vested in the recipient countries themselves; integration of the IMO regulatory priorities in the programme-building process; development of human and institutional resources, on a sustainable basis, including the advancement of women; promotion of regional collaboration and technical cooperation among developing countries; promotion of partnerships with Governments, the shipping industry and international development aid agencies; mobilization of regional expertise and resources for technical assistance activities; coordination with other development aid programmes in the maritime sector; feedback from recipients on the effectiveness of the assistance being provided; and monitoring systems and impact assessments so that programme targets are met and lessons learned are transferred back to the programme-building process.

562. The resolution urges parties to IMO instruments containing provisions on technical cooperation to respond to their commitments and invites member States to use IMO as a coordination mechanism in relation to technical cooperation in the maritime sector. It also invites member States, the shipping industry and partner organizations to continue and, if possible, to increase their support for ITCP and affirms that ITCP can and does contribute to sustainable development.

563. The resolution highlights that the IMO mission statement, in relation to the organization's technical cooperation programme during the current decade, shall be to help developing countries improve their ability to comply with international rules and standards relating to maritime safety and the prevention and control of marine pollution, giving priority to technical assistance programmes that focus on human resources development, particularly through training and institutional capacity-building.

564. With respect to human resources development, the IMO capacity-building activities under ITCP are expected to produce an increased number of trained experts (both male and female) to develop and manage national programmes for maritime safety administration; marine environment protection; the development of maritime legislation; facilitation of maritime traffic; technical port operations; and training of seafarers and shore-based personnel. In the area of institutional capacity-building, activities under ITCP are expected to strengthen public sector departments capable of ensuring the effective exercise of flag State, port State and coastal State jurisdiction.

565. Priority global programmes within the ITCP framework include the development of administrative, legal and technical advisory services to Governments on the implementation of IMO safety and anti-pollution conventions, enhancement of training institutions and provision of fellowships, strengthening in the integration of women in the maritime sector, harmonization of the operation of various regional agreements on port State control, enhancement of maritime safety, prevention and control of illicit drug-trafficking and enhancement of port security. The main constituent programmes foreseen within the ITCP framework include regional programmes for Africa, Arab States/Mediterranean, Asia and the Pacific Islands, the Commonwealth of Independent States and Eastern Europe, and Latin America and the Caribbean.

566. The training of crews of ships is an essential capacity-building measure. The requirements regarding the training of crews that the flag State must implement under article 94 of UNCLOS are those contained in the International Convention on Standards of Training, Certification and Watchkeeping for Seafarers (STCW) and the STCW Code (see paras. 110–116).

567. In a broader sense, the various guidelines, assessment and verification procedures, and assessment and verification forms prepared by IMO are important tools that strengthen the capacity of officials of the national maritime administrations and related agencies to implement the complex provisions of the corresponding conventions. In the same vein, various workshops and seminars organized by IMO also help to strengthen the capacity of national officials through the exchange of information and experience and sharing of expertise. For example, during 2000 IMO continued its task of assisting member countries in the implementation of Memorandum of Understanding agreements on port State control. A Workshop for Regional Port State Control Agreement Secretaries and Directors of Information Centres was held at IMO headquarters in June 2000. Participants discussed the harmonization and coordination of port State control procedures and the exchange of information between regional agreements.

568. *United Nations Development Programme.* UNDP is one of the leading organizations in the area of capacity-building, not only in terms of developing and implementing measures of capacity-building, especially education, training, and field projects, but also in refining, promoting and giving practical effect to the concept of capacity-building.

569. In an attempt to facilitate the response by the Secretary-General to the request of the General Assembly, UNDP prepared a study for the tenth session of the Subcommittee on Oceans and Coastal Areas (SOCA) of the Administrative Com-

mittee on Coordination.[201] The study presented a listing of capacity-building activities of the organizations of the United Nations system focusing on education, training and field projects relating to oceans and seas. The range of activities is quite wide and the scope quite diverse. Fellowships in the field of ocean affairs are awarded by UNESCO, IMO and the United Nations. Training programmes are carried out by IOC, IMO, FAO, IAEA and the United Nations. Field projects are executed by FAO, IAEA, IMO and UNDP.

570. With respect to oceans and seas, the capacity-building activities of UNDP itself are carried out under the Strategic Initiative for Ocean and Coastal Management (SIOCAM), a global UNDP programme with the goal of enhancing the capabilities of existing and future ocean and coastal management projects through the systematic identification, documentation and sharing of best practices and lessons learned.

571. UNDP reported that during the past few years, SIOCAM has accomplished a number of objectives. With respect to assessing needs and resources of projects, two assessments were carried out: (a) an assessment of five UNDP coastal projects (1996, New York); and (b) an assessment of all UNDP, UNEP and World Bank GEF International Waters (IW) projects (2000, Budapest). With respect to establishing training and information networks, SIOCAM focused on two programmes: IW: LEARN (information exchange) and TRAIN-SEA-COAST (course development and sharing) (see paras. 579–585). With respect to identifying, documenting and disseminating best practices and resources, SIOCAM focuses on a number of studies and manuals that include case descriptions of best practices; a coastal management appraisal manual; and an integrated convention management matrix linking actions to six conventions/agreements.

572. One major emphasis of SIOCAM is enhancing UNDP support for donor coordination activities. In this context, cooperation has been strengthened with, *inter alia*, the Strategy for International Fisheries Research (SIFAR), the International Coral Reef Initiative (ICRI) and the World Commission on the Oceans. SIOCAM is also building up a number of support capabilities. These include the SIOCAM web site (siocam.sdnp.undp.org); the UNDP Coastal Group, combining IW and Marine and Freshwater Biodiversity; a Programme Advisory Note for project preparation; the UNDP-World Bank IW Partnership; and Strategic Global Projects on Ballast and Mercury from Gold Mining.

573. *United Nations University.* The United Nations University (UNU) Fisheries Training Programme is carried out as a formal cooperative venture of four institutions in Iceland. Since its inception in 1997, the programme has provided 29 fellowships to professionals in various areas of fisheries from 14 developing countries. The programme's primary focus is on developing countries that are dependent on fisheries or have major development potential in fisheries. The selection of UNU fellows also takes into account the regional context. The training covers diverse disciplines within fisheries and consists of two distinct parts: a common seven-week introductory course, followed by specialist training of four to five weeks. The fellows working under the programme are required to return to their home institutes after the training so that the capacity-building process continues even after the UNU training.

201. See the web site of the UNDP SIOCAM programme: siocam.sdnp.undp.org.

B. Capacity-Building Activities of the United Nations Division for Ocean Affairs and the Law of the Sea

1. Amerasinghe Memorial Fellowship Programme
574. The capacity-building measures of the United Nations system in relation to ocean law, policy and management are exemplified by the fellowship and training programmes of the Division for Ocean Affairs and the Law of the Sea.

575. Every year, the Division provides one or two fellowships under its Hamilton Shirley Amerasinghe Memorial Fellowship programme to qualified persons specializing in the law of the sea and ocean affairs who wish to broaden their knowledge and acquire additional skills, which in turn would benefit their countries. Although the fellowship is limited in quantity, the qualitative impact of the strengthened capabilities of the individual fellows can be enormous in view of the fact that in many cases, the fellows are the sole decision makers in their respective countries in the field of responsibility.

576. The Hamilton Shirley Amerasinghe Memorial Fellowship programme was established in 1982 in memory of the late President of the Third United Nations Conference on the Law of the Sea. It is part of the Programme of Assistance in the Teaching, Study, Dissemination and Wider Appreciation of International Law, which encompasses all training and fellowship programmes of the United Nations system in the field of international law. Under the fellowship programme, the fellows pursue a postgraduate level research/study programme at a participating university of their choice for a period of not less than six months. Thereafter they work as interns in the Division for Ocean Affairs and the Law of the Sea for a period of approximately three months.[202]

577. To date, the Legal Counsel of the United Nations, on the recommendation of a High-level Advisory Panel, has made 15 Annual Fellowship Awards and 3 Special Awards to fellows from 17 developing countries. Currently there are 15 universities and institutes in seven developed countries and one developing country participating in the fellowship programme.

578. In December 2000, Margaret N. Mwangi of Kenya, was awarded the fifteenth annual fellowship. Mrs. Mwangi is a Senior State Counsel in the Attorney-General's Office in Kenya and intends to utilize the fellowship award to pursue a programme of study in the control of marine pollution. (For participating universities and members of the High-level Advisory Panel, see press releases SEA/1654, SEA/1695 and SEA/1698/Rev 1. The information can also be obtained from the web site of the Division for Ocean Affairs and the Law of the Sea: www.un.org/Depts/los/HSA.htm.)

2. TRAIN-SEA-COAST Programme
579. The training activities under the Division's TRAIN-SEA-COAST (TSC) Programme encompass integrated coastal and ocean management. The Programme was developed as part of a system-wide training strategy that emphasizes: (a) building up permanent national capabilities for training; (b) sustainability of training efforts; (c) cost-effectiveness; (d) responsiveness to specific training priorities of the countries involved; (e) transfer of experience and sharing of training resources; and (f) long-term impact.

202. For a detailed description of the programme, see A/54/429, paras. 588–594, and A/55/61, paras. 267–268.

580. The basic objective of the TSC Programme is to create capacity at the local level to produce high-quality training courses to be shared among the TSC members, and at the same time to strengthen local institutions to become centres of excellence on training at the national or regional levels. In 2000, three major initiatives were undertaken in this regard. The Division for Ocean Affairs and the Law of the Sea conducted the Fourth Course Developers Workshop and Planning Meeting. Through intensive training, 16 individuals selected from existing as well as forthcoming TSC course development units (CDUs) learned how to apply the TSC pedagogic methodology in the course development process and made a tentative plan for their training activities to be initiated upon their return to their respective countries. This was the first workshop/meeting in the series with participants from CDUs from the first phase of the TSC Programme, CDUs associated with the GEF International Waters Projects together with new members of the network, such as TSC/Germany and the forthcoming TSC/Indonesia.

581. As part of a second national-level initiative, the Rockefeller Brothers Fund is providing support to the TSC Programme in the planning for a course development unit in Indonesia and the training of one course developer. This new CDU, together with TSC/Philippines and TSC/Thailand, will constitute a stronghold of training capacity in the South-East Asia region. And thirdly, in the South Pacific, the TSC Programme is about to commence activities for the sixth TSC course development unit associated with a GEF project. The unit is located at the University of the South Pacific.

582. The TSC Programme has grown and diversified. Over the first five years of its existence, it concentrated its efforts in building capacity at the national level. At the end of the last decade, with the support of UNDP/GEF, six additional CDUs were created at the national level, but were geared to attend to training needs at the regional level as well. Currently, the TSC Programme is expanding the breadth of training opportunities as well as its cooperating partners. This aspect of the TSC strategy encompasses the establishment of training initiatives in cooperation with various United Nations agencies. IMO and GPA, for example, both requested the TSC Programme to assist in the development of training courses tailored to their particular needs.

583. At the request of IMO, the TSC Central Support Unit at the Division for Ocean Affairs and the Law of the Sea prepared a project proposal for the development and delivery of a training course on "Control and Management of Ship's Ballast Water," which will be validated at two demonstration sites of the IMO Global Ballast Water Management (GloBallast) programme: Sepetiva, Brazil, and Saldahna, South Africa. The objective is to build capacity, through training at both local and regional levels, to implement the IMO voluntary Guidelines for the Control and Management of Ship's Ballast Water to minimize the transfer of harmful aquatic organisms and pathogens during ballasting, and thus to prepare for the IMO mandatory regulatory regime. The course will be developed by two TSC CDUs, namely, TSC/Brazil (at the University of Rio Grande) and TSC/Benguela Current (at the University of the Western Cape in South Africa) and delivered by teams of instructors from other TRAIN-X[203] sister programmes. The course will be adapted and deliv-

---

203. The TRAIN-X network is a UNDP-sponsored cooperative network of United Nations agency human resources development programmes, one of which is the TSC Programme.

ered at four other demonstration sites of the GloBallast programme, with the support of TSC/Brazil and TSC/Benguela Current. Once the project is approved and the course is ready for delivery, it is expected that up to 270 individuals will be trained at the site, in first deliveries alone.

584. Another key goal of the TSC Programme is the sharing of training courses and personnel among the members of the TSC network, thus utilizing the full capacity of the network as a sharing system and making training a more cost-effective endeavour. In 2000, TSC/Rio de la Plata requested the TSC Central Support Unit to organize the sharing and adaptation of the course entitled "Integrated Coastal Management: Exchange and Interrelationships between Coastal and Oceanic Systems" developed by TSC/Brazil. After several consultations between both course development units, the course was successfully adapted to the local conditions of the Rio de la Plata region and delivered jointly by TSC/Brazil and TSC/Rio de la Plata.

585. The Database on Education and Training in Integrated Coastal and Ocean Management, a cooperative project between the United Nations University, Institute of Advanced Studies (UNU/IAS), the Division for Ocean Affairs and the Law of the Sea, and the Bureau of Development Planning of UNDP, is a valuable tool for capacity-building in ocean policy and management. The Division and UNU organized an Expert Group Meeting on Training and Education at Sassari, Italy, in 1993. As a follow-up UNU, in cooperation with the Division and UNDP, initiated the Database, one of the most important and very few post-UNCED initiatives in capacity-building for integrated coastal and ocean management. The Database is an extremely useful tool for both developed and developing countries in that it provides immediate online access to information on academic programmes, extension courses, short training courses and on-the-job training opportunities. As a global inter-agency project, the Database draws upon the comparative advantage of the partnership among UNU, the host of the Database with its research and training mandate; the Division for Ocean Affairs and the Law of the Sea, with its long-standing experience in training and networking in integrated coastal and ocean management; and UNDP, a recognized institution in capacity development with a network of field offices. The Database is now fully operational and can be accessed at http://db.ias.unu.edu/published/icon/. An advisory body to be established will be responsible for data quality control and other important actions for the functioning of the Database. Keeping pace with the new information and communication technology, all activities related to the Database will be undertaken online, including the participation of the three partners in the annual conference.

X. INTERNATIONAL COOPERATION AND COORDINATION

A. Subcommittee on Oceans and Coastal Areas of the Administrative Committee on Coordination (SOCA)

586. In 1993, the Administrative Committee on Coordination (ACC), acting on a proposal of the Inter-Agency Committee on Sustainable Development, established the Subcommittee on Oceans and Coastal Areas with the purpose of meeting the coordination needs defined in chapter 17 of Agenda 21.

587. SOCA held its ninth session at IMO headquarters in London from 26–

28 July 2000 and its tenth session at the Intergovernmental Oceanographic Commission in Paris from 9 to 11 January 2001. At the tenth session, discussions focused on a number of subjects of ongoing concern and activity: the United Nations Atlas of the Oceans; status of implementation of the GPA; the 10-year review and appraisal of the implementation of Agenda 21: reporting and participation; coordination and cooperation in combating piracy and armed robbery at sea; matters related to the fifty-fifth session of the United Nations General Assembly and preparations for the second meeting of the Consultative Process, 7–11 May 2001, New York; making the Subcommittee more transparent, effective and responsive: follow-up to Commission on Sustainable Development decision 7/1 and General Assembly resolutions 54/33 and 55/7; and review of the Joint Group of Experts on the Scientific Aspects of Marine Environmental Protection (GESAMP) (see paras. 597–603).

588. The Subcommittee took a number of decisions to advance work on the United Nations Atlas of the Oceans relating, *inter alia*, to the allocation of funds to accelerate the inputting of information, arrangements with the publisher regarding a mutual understanding of responsibilities and the long-term management of the project, the long-term economic sustainability of the Atlas and the production of a CD-ROM and videos.

589. Regarding the status of implementation of the GPA and the 10-year review and appraisal of the implementation of Agenda 21, the Subcommittee focused on issues related to advancing the preparation of a coordinated input into the two review exercises (see para. 325).

590. The Subcommittee also considered a new item on its agenda, "Coordination and cooperation in combating piracy and armed robbery at sea," and agreed that it was not in a position to suggest any specific actions for consideration by the Consultative Process at its second meeting since IMO was the appropriate body to deal with the issue and that a section on "Crimes at sea" should be established for the United Nations Atlas of the Oceans with the Division for Ocean Affairs and the Law of the Sea taking the lead in the development of such a section (see paras. 169–226).

591. Regarding the preparations for the second meeting of the Consultative Process (see paras. 618–619), the Subcommittee decided that the United Nations Atlas of the Oceans would be the main topic of the SOCA presentation by its Chairman at the meeting and that the Chairman, in his capacity as Executive Secretary of IOC, would prepare an annotated draft paper on the topic of marine science for comments and inputs by other interested agencies before its submission to the Consultative Process meeting and that the Division for Ocean Affairs and the Law of the Sea and IMO would prepare a background paper on piracy and armed robbery at sea.

592. Under its item on GESAMP, the Subcommittee was briefed by the representative of IMO, who serves as the Administrative Secretary of GESAMP, on recent developments concerning the evaluation of GESAMP (see paras. 697–702). While recognizing the need for improving the functioning and effectiveness of GESAMP, the Subcommittee reiterated the value of independent scientific advice on ocean issues and expressed concern that transforming GESAMP into an intergovernmental panel could threaten the independence of its work and lead to a complicated and expensive process.

593. In the light of information on the ongoing restructuring of the ACC sys-

tem, the Subcommittee was informed that ACC at its October 2000 session had agreed to establish two new high-level committees with the immediate task of reviewing the functioning of all ACC subsidiary bodies. The review was to be "zero-based," i.e., to consider what needed to be done rather than what was currently being done. IACSD and its Subcommittees on Water Resources and SOCA would be reviewed by the new High-Level Committee on Programmes at the end of February 2001, at a meeting to be held in Vienna and chaired by the Director-General of UNIDO, and based on its recommendations, ACC is expected to take a final decision on the continued existence of its subsidiary machinery when it meets in Nairobi on 2 and 3 April 2001.

594. Based on the October 2000 decision of ACC, the Assistant Secretary-General for Policy Coordination and Inter-Agency Affairs of the United Nations Department of Economic and Social Affairs (who is also the Secretary of ACC) on 3 January 2001 addressed a letter to the Secretaries of all ACC subsidiary bodies advising, *inter alia,* that, pending the final review, care should be taken by those bodies to avoid any long-term decisions on work programmes and selection of officers that might prejudge the relevant ACC conclusions.

595. In that connection, SOCA unanimously agreed to recommend to the Interagency Committee on Sustainable Development the extension of the terms of the current Chairperson and Vice-Chairperson until such time as the status of the Subcommittee was clarified.

596. SOCA nevertheless reiterated its decision that the United Nations Atlas of the Oceans would be the main topic of its presentation at the second meeting of the Consultative Process, as it had the potential to best demonstrate cooperation and coordination by the United Nations system in working together on the oceans.

B. Other Mechanisms

1. Joint Group of Experts on the Scientific Aspects of Marine Environmental Protection (GESAMP)

597. Constituted in 1968 under an inter-agency Memorandum of Understanding, GESAMP is an expert scientific advisory body supported by organizations of the United Nations system. As of May 2001, the sponsoring agencies were: the United Nations, through its Division for Ocean Affairs and the Law of the Sea; UNEP; UNESCO/IOC; FAO; WHO; WMO; IMO; and IAEA. Each GESAMP sponsoring agency provides a technical secretary and supports the participation of experts at GESAMP meetings (plenary and working groups). IMO also provides the Administrative Secretariat for GESAMP in addition to a technical secretary. The principal task of GESAMP is to provide independent, multidisciplinary scientific advice to the sponsoring agencies concerning the prevention, reduction and control of the degradation of the marine environment with a view to sustaining its life support systems, resources and amenities. The annual reports of GESAMP and the reports of its working groups thus represent substantial contributions to the technical work of the sponsoring agencies under their respective mandates and programmes of work, including in relation to the implementation of UNCLOS and chapter 17, among others, of Agenda 21 and, through the agencies, to their governing bodies

and member States, to assist them in policy and decision-making for the marine environment, particularly coastal areas.

598. At the thirtieth session of GESAMP, hosted by IAEA at its Marine Environment Laboratory in Monaco,[204] from 22 to 26 May 2000, the Intersecretariat of GESAMP (comprising the GESAMP Administrative Secretary and the technical secretaries of the sponsoring agencies) reviewed a March 2000 proposal by the Executive Director of UNEP to the Administrative Secretary for an in-depth independent evaluation of GESAMP to make the advisory mechanism more effective and responsive. A subsequent, further elaborated proposal developed in consultation with IMO led to an agreement by the Intersecretariat to carry out the evaluation. The Intersecretariat also agreed on the terms of reference for the exercise and on the mechanism for its implementation, namely, the establishment of an evaluation team. The team, with a maximum of five members, would consist of two independent scientific experts who have so far not been involved in the GESAMP mechanism: one from a developed and another from a developing country; two Government-nominated experts: one representing a developed and another representing a developing country; and one scientific expert who has been a GESAMP member, to provide first-hand information and hands-on experience.

599. It was further agreed that the costs to carry out the evaluation, which were estimated at $60,000–$80,000, should be distributed among all the GESAMP sponsoring agencies, using to the extent possible an equally applied cost-sharing formula,[205] and that the Administrative Secretariat should make all the necessary arrangements.

600. At the May 2000 session, the GESAMP Chairman and the Vice-Chairman expressed their support for the exercise and the executive heads of the sponsoring agencies subsequently endorsed it, the first such review since GESAMP became operational in 1969.

601. The Evaluation Team held its first, largely organizational meeting at IMO headquarters in London on 29 and 30 January 2001, and is to meet again from 29 April to 1 May 2001 to analyse responses to questionnaires developed at the first meeting and the results of extensive interviews scheduled during the intersessional period. The IMO Administrative Secretariat will also prepare a status report on the evaluation for submission to the Consultative Process at its second meeting. At a final meeting planned for end June/early July 2001, the Evaluation Team will complete its report.

602. In addition to commenting on the proposed evaluation exercise, at its plenary meeting GESAMP reviewed and approved two draft reports prepared by its Working Group on Marine Environmental Assessments: "A Sea of Troubles" and "Protecting the Oceans from Land-based Activities." Both reports have been published under the GESAMP Reports and Studies Series as issues Nos. 70 and 71 respectively. The plenary meeting also addressed the progress of ongoing activities being carried out under the auspices of working groups and a number of issues of current and growing concern regarding the degradation of the marine environment, with

---

204. GESAMP 2000. Report of the Thirtieth Session, Monaco, 22–26 May 2000. GESAMP Reports and Studies, No. 69, 68 p.

205. WHO confirmed its support for the evaluation and its preparedness to be actively involved, but stated that it was not in a position to contribute to defraying the costs.

a view to bringing them to the attention of the marine environment community and, in some cases, to undertaking assessment activities intersessionally in order that a more in-depth evaluation of the issues might be considered in the future.

603. The thirty-first plenary session of GESAMP will be held at United Nations Headquarters in New York from 13 to 17 August 2001.

2. Inter-Secretariat Committee on Scientific Programmes Relating to Oceanography (ICSPRO)

604. During the first meeting of the Consultative Process, the Chairman of IOC brought to its attention the role of the Inter-Secretariat Committee on Scientific Programmes relating to Oceanography, an existing inter-agency coordinating mechanism whose enabling instrument is deposited with UNESCO. The members of ICSPRO are the executive heads of the relevant United Nations divisions, FAO, UNESCO, WMO, IMO and any other United Nations agencies wishing to cooperate. This level of participation provides for an executive management group that is invested with real decision-making power with regard to the implementation of oceans-related programmes capable of providing timely reaction and guidance for the United Nations system on emerging ocean issues. IOC provides the secretariat of ICSPRO.

605. At the time of its creation in 1969, ICSPRO was extremely effective in promoting and implementing the major ocean research programmes, resulting, for example, in an increased understanding of ocean/atmosphere interactions and the role of the ocean in climate change. However, there have been many changes since then. Today global needs have expanded to include the priorities of ocean and coastal management and other applications of marine science. In 1999, the mandate of IOC was modified to take into account these priorities. To be effective as a coordinating mechanism, ICSPRO would also need to be brought up to date in this regard, reflecting current needs and new trends on ocean issues.

606. It was pointed out that the implementation of actions agreed by the ACC/SOCA requires following the due process of endorsement of and financing within the governing structure of each agency. This process alone cannot solve the need for guidance on new, emerging cross-sectoral ocean issues. A revitalized ICSPRO can be an effective body with sufficient executive authority to address this task.

XI. REVIEW BY THE GENERAL ASSEMBLY OF DEVELOPMENTS IN OCEAN AFFAIRS: UNITED NATIONS OPEN-ENDED INFORMAL CONSULTATIVE PROCESS ON OCEANS AND THE LAW OF THE SEA

607. The General Assembly of the United Nations had been undertaking an annual review of all important developments in oceans and the law of the sea based on a comprehensive annual report prepared by the Secretary-General. However, it was felt that there was a need to broaden and deepen the debate in the General Assembly and to further enhance the coordination and cooperation in ocean affairs at the intergovernmental and inter-agency levels. In 1999, at its seventh session devoted to the review of progress in the area of sustainable development of oceans and seas, the Commission on Sustainable Development re-emphasized this need. Following the recommendation of the Commission, the General Assembly by its resolution

54/33 of 24 November 1999 decided to establish an annual open-ended informal consultative process in order to facilitate, in an effective and constructive manner, its own review of developments in ocean affairs.

608. The Consultative Process, consistent with the legal framework provided by UNCLOS and the goals of chapter 17 of Agenda 21, was established to discuss the annual report of the Secretary-General on oceans and the law of the sea and to suggest particular issues to be considered by the General Assembly, with an emphasis on identifying areas where coordination and cooperation at the intergovernmental and inter-agency levels should be enhanced. Moreover, the Consultative Process is intended to study overall developments in ocean affairs.

609. The two co-chairpersons of the consultative process, Ambassador Tuiloma Neroni Slade (Samoa) and Mr. Alan Simcock (United Kingdom), were appointed, after consultations with Member States, by the President of the General Assembly in accordance with paragraph 3 (e) of General Assembly resolution 54/33.

610. On the basis of consultations with delegations from 14 to 16 March 2000, deliberations in an informal meeting with delegations held on 12 April 2000 at United Nations Headquarters and comments subsequently submitted by delegations, the co-chairpersons proposed to the first Meeting of the Consultative Process a draft format for discussions and an annotated provisional agenda. The format for the first Meeting provided, among other things, the opportunity to receive input from major groups, as identified in Agenda 21, especially nongovernmental organizations. The first Meeting worked through plenary sessions and two discussion panels. On the basis of further consultations with delegations, the co-chairpersons also proposed areas of focus for the discussion panels: (a) responsible fisheries and illegal, unregulated and unreported fisheries: moving from principles to implementation; and (b) economic and social impacts of marine pollution and degradation, especially in coastal areas: international aspects of combating them. The first Meeting of the Consultative Process was held at United Nations Headquarters from 30 May to 2 June 2000.

611. The first Meeting adopted its format and agenda by consensus (A/AC.259/L.1). In addition, in the light of comments from some delegations, which sought to add a reference to the "law of the sea" to the designation of the Consultative Process in resolution 54/33, it was agreed to refer to the process as the "United Nations Open-ended Informal Consultative Process on Oceans and the Law of the Sea."

612. The first Meeting of the Consultative Process was opened by the Under-Secretary-General for Legal Affairs, the Legal Counsel, and the Under-Secretary-General for Economic and Social Affairs.

613. Discussions at the first and second plenary sessions of the first Meeting were based on annual reports of the Secretary-General on oceans and the law of the sea (A/54/429 and Corr.1 and A/55/61) as well as on other documents before the Meeting, including written submissions by States and international organizations (see A/AC.259/1 and A/AC.259/2).

614. The first Meeting focused on broadening and deepening the understanding of the issues discussed and the need to approach them in a cross-sectoral and integrated manner. Participation by the relevant intergovernmental organizations and representatives of major groups increased the value of the discussions. Consen-

sus was reached on 13 issues,[206] which merited attention by the General Assembly. In the area of international coordination and cooperation, there was an exchange of views with the Chairman and other members of the Subcommittee on Oceans and Coastal Areas of the Administrative Committee on Coordination. Lastly, many delegations, while avoiding fixing the issues to be discussed at subsequent meetings, put forward various possible issues for consideration by the Consultative Process at future meetings.[207]

615. By a letter dated 28 July 2000 addressed to the President of the General Assembly (A/55/274), the co-chairpersons submitted the report on the work of the first Meeting of the Consultative Process, proposing a number of nonexhaustive elements for the consideration of the General Assembly under the agenda item entitled "Oceans and the law of the sea" and for potential inclusion in the relevant General Assembly resolutions, in accordance with paragraph 3 (h) of resolution 54/33. The report was composed of three parts: (a) issues to be suggested and elements to be proposed to the General Assembly; (b) Co-Chairpersons' summary of discussions; and (c) issues for consideration for possible inclusion in the agendas of future meetings.

616. At the fifty-fifth session of the General Assembly, during the general debate on agenda item 34, entitled "Oceans and the law of the sea," delegations expressed appreciation for the work of the first Meeting of the Consultative Process and were in general agreement on its usefulness, especially as regards the informal consultations conducted in preparation of the General Assembly resolution.

617. The two resolutions adopted by the General Assembly on 30 October 2000 (resolutions 55/7 and 55/8), incorporate many of the issues discussed at the first Meeting of the Consultative Process. These issues were, *inter alia,* the need for capacity-building for the implementation of UNCLOS; the problems of illegal, unreported and unregulated (IUU) fishing; and the degradation of the marine environment, from both land-based sources and pollution from ships.

618. In paragraph 41 of resolution 55/7, the General Assembly recommended that the second meeting of the Consultative Process, to be held in New York from 7 to 11 May 2001, should organize its discussions around the following areas of focus: (a) marine science and the development and transfer of marine technology as mutually agreed, including capacity-building in this regard; and (b) coordination and cooperation in combating piracy and armed robbery at sea.

619. The two resolutions on oceans and the law of the sea adopted by the General Assembly at its fifty-fifth session show the usefulness of the discussions that took place at the first Meeting of the Consultative Process and make this process an invaluable tool for the effective and constructive review by the Assembly of developments in oceans and the law of the sea, and thus for the results-oriented stewardship of the world's oceans and seas by the General Assembly.

---

206. See also the report on the work of the Consultative Process at its first meeting, A/55/274, part A, paras. 1–50.

207. *Ibid.*, part C, paras. 1–4.

Selected Documents and Proceedings

Oceans and the Law of the Sea: Report of the Secretary-General, 2001†

Addendum

CONTENTS
I. Introduction
II. The United Nations Convention on the Law of the Sea and its implementing Agreements
    A. Status of the Convention and its implementing Agreements
    B. Declarations and statements under articles 310 and 287 of UNCLOS
    C. Meeting of States Parties (Eleventh Meeting)
III. Maritime space
    A. Recent developments
    B. Continental shelf beyond 200 nautical miles and the work of the Commission on the Limits of the Continental Shelf
    C. Deposit of charts and/or lists of geographical coordinates and compliance with the obligation of due publicity
IV. Shipping and navigation
    A. Safety of ships
        1. Ship construction, equipment and seaworthiness
        2. Training of crew
        3. Labour conditions
    B. Safety of navigation
    C. Marine casualties
    D. Flag State implementation
V. Crimes at sea
    A. Piracy and armed robbery against ships
    B. Smuggling of migrants
    C. Illicit traffic in narcotic drugs and psychotropic substances
VI. Marine resources, the marine environment and sustainable development
    A. Conservation and management of marine living resources
        1. Marine fisheries
        2. Conservation and management of marine mammals
    B. Non-living marine resources
    C. Protection and preservation of the marine environment
        1. Reduction and control of pollution
            (a) Land-based activities: the Global Programme of Action
            (b) Pollution by dumping; waste management
            (c) Pollution from vessels
        2. Regional cooperation
        Review of UNEP regional seas programme and action plans

†EDITORS' NOTE.—This document was provided by the United Nations Division for Ocean Affairs and the Law of the Sea (DOALOS). The source is "Oceans and the Law of the Sea: Report of the Secretary-General, Addendum, A56/58/Add.1, 5 October, 2001." Fifty-sixth session. Available online: <http://www.un.org/depts/los/index.htm>.

       D. Protection of specific marine areas
       E. Climate change and sea-level rise
       F. Ten-year review of the implementation of Agenda 21
  VII. Underwater cultural heritage
 VIII. Marine science and technology
   IX. Settlement of disputes
       A. Cases before the International Tribunal for the Law of the Sea
       B. Arbitration and conciliation
    X. International cooperation and coordination
       A. Subcommittee on Oceans and Coastal Areas of the Administrative Committee on Coordination
       B. Other mechanisms
   XI. Review by the General Assembly of developments in ocean affairs: United Nations Open-ended Informal Consultative Process established by the General Assembly in its resolution 54/33 in order to facilitate the annual review by the Assembly of developments in ocean affairs

I. INTRODUCTION

1. The importance of the oceans and seas for the earth's ecosystem and for providing the vital resources for food security and for sustaining economic prosperity and the well-being of present and future generations is reiterated by the General Assembly in its annual resolutions on "Oceans and the law of the sea" (see in particular General Assembly resolution 55/7).

2. In 1999, in its resolution 54/33 of 24 November 1999, the General Assembly, convinced of the importance of the annual consideration and review of ocean affairs and the law of the sea by the Assembly as the global institution having the competence to undertake such a review, established an open-ended informal consultative process in order to facilitate its annual review, in an effective and constructive manner, of developments in ocean affairs by considering the Secretary-General's report on oceans and the law of the sea. In the same resolution, the General Assembly requested the Secretary-General to make the report available at least six weeks in advance of the meeting of the Consultative Process. Accordingly, the Secretary-General's report on oceans and the law of the sea for the fifty-sixth session of the General Assembly was submitted to the second meeting of the Consultative Process, held from 7 to 11 May 2001 (A/56/58).

3. In the dynamic field of ocean affairs and the law of the sea, developments occur on a continuing basis. In that context, views were expressed during the deliberations on the agenda item entitled "Oceans and the law of the sea" at the fifty-fifth session of the General Assembly in 2000 that the Assembly, when it considers the item in the fourth quarter of the year, would benefit from a supplementary report which would cover the significant developments that had occurred after the preparation of the main annual report in the first quarter of the year, submitted to the meeting of the Consultative Process in May.

4. The present report has thus been prepared as an addendum to the main report to the General Assembly at its fifty-sixth session (A/56/58) and should be read in conjunction with the latter, as well as with the report on the work of the Consultative Process at its second meeting (A/56/121). The attention of the Gen-

eral Assembly is drawn to another report, entitled "Agreement for the Implementation of the Provisions of the United Nations Convention on the Law of the Sea of 10 December 1982 relating to the Conservation and Management of Straddling Fish Stocks and Highly Migratory Fish Stocks" (A/56/357), submitted to the General Assembly at the current session pursuant to General Assembly resolution 54/32 of 24 November 1999. All of the four above-mentioned reports are available to the General Assembly when it considers the item on "Oceans and the law of the sea" at the fifty-sixth session.

II. THE UNITED NATIONS CONVENTION ON THE LAW OF THE SEA AND ITS IMPLEMENTING AGREEMENTS

A. Status of the Convention and Its Implementing Agreements

5. Since the report of the Secretary-General (A/56/58) was issued, two further States have deposited their instruments of ratification of the United Nations Convention on the Law of the Sea (UNCLOS): Bangladesh and Madagascar. Thus, as at 30 September 2001, the total number of States parties, including one international organization, stood at 137.

6. Bangladesh and Madagascar also expressed their consent to be bound by the Agreement relating to the implementation of Part XI of UNCLOS of 28 July 1994. In addition, Costa Rica acceded to the Agreement in September 2001 and the number of parties to that Agreement has thus risen to 103.

7. As for the 1995 Agreement for the Implementation of the Provisions of the United Nations Convention on the Law of the Sea of 10 December 1982 relating to the Conservation and Management of Straddling Fish Stocks and Highly Migratory Fish Stocks 1995 (the Fish Stocks Agreement), instruments of ratification/accession have been deposited by New Zealand and Costa Rica since the issuance of the main annual report for the current session. The number of States parties to the Agreement currently stands at 29, so that only one more instrument is needed for the entry into force of the Agreement.

8. The entry into force of the Agreement in the near future would necessarily create a new situation with a number of implications, especially in respect of the exercise of rights of States parties and the fulfilment of their obligations, including the fulfilment of the enhanced duties of flag States regarding fishing vessels flying their flag on the high seas. The issues that would gain in significance include, *inter alia*, the establishment and implementation of conservation and management measures through existing or new, as appropriate, subregional or regional fisheries management organizations or arrangements, including the application of the precautionary approach, ecosystem-based management and ensuring compatibility of measures; the collection and provision of information and cooperation in scientific research; compliance and enforcement, including the implementation of cooperation schemes at the subregional and regional levels; and recognition of special requirements of developing States and cooperation with such States, including through the establishment of special funds to assist them in the implementation of the Agreement. (See also A/56/357 in this connection.)

B. Declarations and Statements Under Articles 310 and 287 of UNCLOS

9. Since the issuance of the main annual report, two additional States have made declarations. On 31 May 2001, Tunisia accepted, in its declaration under article 287 and in order of preference, the International Tribunal for the Law of the Sea and an arbitral tribunal established in accordance with Annex VII to UNCLOS as the means for the settlement of disputes relating to the interpretation or implementation of the Convention.

10. Bangladesh declared upon ratification of the Convention that, *inter alia,* it understood that the provisions of the Convention did not authorize other States to carry out in the exclusive economic zone and on the continental shelf military exercises, or manoeuvres, in particular those involving the use of weapons or explosives, without the consent of the coastal State. Bangladesh also declared that it was not bound by any domestic legislation or by any declaration issued by other States upon signature or ratification of the Convention and reserved its right to state its position concerning all such legislation or declarations at the appropriate time. In particular, Bangladesh stated that its ratification of the Convention in no way constituted recognition of the maritime claims of any other State which had signed or ratified the Convention where such claims were inconsistent with the relevant principles of international law and were prejudicial to the sovereign rights and jurisdiction of Bangladesh in its maritime areas.

11. Bangladesh reserved its right to adopt legislation regarding the exercise of the right of innocent passage of warships through its territorial sea and expressed the view that a notification was needed in respect of nuclear-powered ships or ships carrying nuclear or other inherently dangerous or noxious substances, stating that no such ships should be allowed within Bangladesh waters without the necessary authorization.

12. Other parts of Bangladesh's declaration related to the responsibility and liability in respect of damage caused by pollution of the marine environment by certain vessels or aircraft, objects of an archaeological and historical nature found within the maritime areas over which Bangladesh exercises sovereignty or jurisdiction which shall not be removed without its prior notification and consent, and matters relating to the settlement of disputes and harmonization of national legislation with the provisions of the Convention.[1]

13. In its resolution 55/7, the General Assembly once again called upon States to ensure that any declarations or statements that they had made or would make when signing, ratifying or acceding to UNCLOS were in conformity therewith and, otherwise, to withdraw any of their declarations or statements that were not in conformity (see also A/56/58, para. 23). No action by States parties in this connection has been reported.

14. Since the issuance of the main annual report, no additional States have made a declaration or statement pursuant to article 43 of the 1995 Fish Stocks Agreement.

---

1. Full texts of the declarations have been circulated through Depositary Notifications, and are available at the web site of the Division for Ocean Affairs and the Law of the Sea of the United Nations Office of Legal Affairs (www.un.org/Depts/los) as well as that of the Treaty Section of the United Nations Office of Legal Affairs (www.un.org/Depts/Treaty).

C. Meeting of States Parties (Eleventh Meeting)

15. The Eleventh Meeting of States Parties to UNCLOS was held in New York from 14 to 18 May 2001. Ambassador Cristián Maquieira (Chile) was elected President of the Eleventh Meeting by acclamation. The representatives of Australia, India, and Nigeria were elected as Vice-Presidents.

16. The Meeting of States Parties, *inter alia,* dealt with the budget of the International Tribunal for the Law of the Sea for 2002, the financial regulations of the Tribunal, matters related to the continental shelf and matters related to article 319 of the Convention. The Meeting also elected Mr. Xu Guangjian (China) to serve the remainder of the term of Judge Lihai Zhao, who passed away on 10 October 2000.

17. *Budget of the Tribunal for 2002.* The budget of the Tribunal, totalling $7,807,500, was approved by the Eleventh Meeting of States Parties for the financial year 2002. This included a recurrent expenditure of $6,522,400, a nonrecurrent expenditure of $340,800 essentially for the acquisition of furniture, equipment and special equipment, and $894,300 as a contingency fund to provide the necessary financial means to consider cases in 2002.

18. *Financial regulations.* The Secretariat, in consultation with the Registry, prepared a working paper on the financial regulations of the Tribunal (SPLOS/WP.14), taking into account the various proposals and the outcome of the discussions during the Ninth and Tenth Meetings. Progress was made in the Working Group chaired by the President regarding some of the pending issues. Tentative agreement was reached on most of the outstanding provisions in regulations 1 to 5. The proposals made in reference to the establishment of a Finance Committee were withdrawn in view of the decision taken by the Meeting regarding the establishment of an open-ended working group on financial and budgetary matters (see SPLOS/73, paras. 49–50).

19. *Matters relating to the continental shelf.* In support of the concerns raised by developing States regarding the difficulty of complying with the time limit laid down in article 4 of Annex II to the Convention, and in the light of the discussions and of proposals and amendments put forward by delegations, the Meeting of States Parties adopted a decision based on the draft prepared by an open-ended Working Group (see SPLOS/72). The decision provided that, for a State for which the Convention entered into force before 13 May 1999, the date of commencement of the 10-year time period for making submissions to the Commission on the Limits of the Continental Shelf is 13 May 1999. There was general agreement, however, that States that were in a position to do so should make every effort to make a submission within the time period established by the Convention. (See also paras. 40–43 below in this connection.)

20. *Matters related to article 319 of UNCLOS.* Divergent views were still held by delegations; some expressed their support for an expanded role for the Meeting of States Parties beyond budgetary and administrative matters while others maintained that the role of the Meeting of States Parties should not go beyond that laid down in the Convention (see A/56/58, paras. 30–33), according to which the General Assembly has the oversight role to review the overall implementation of the Convention. In that regard, the Assembly had established the Consultative Process in order to facilitate its annual review of developments in ocean affairs.

21. Some other delegations, while supporting an expanded role for the Meeting of States Parties, were of the view that the modalities of such a role should be defined and this should include legal issues regarding the implementation of the Convention.

22. In view of the divergent views still held by delegations, the Meeting decided to retain on the agenda for its next meeting the item entitled "Matters related to article 319 of the United Nations Convention on the Law of the Sea."

23. *Other matters.* The Twelfth Meeting of States Parties to UNCLOS will be held in New York from 13 to 24 May 2002.

III. MARITIME SPACE

A. Recent Developments

24. At the second meeting of the Consultative Process, the European Union noted the collection of information by the Division for Ocean Affairs and the Law of the Sea of the United Nations Office of Legal Affairs regarding legislative measures undertaken by States parties in implementing UNCLOS and welcomed the Secretary-General's idea that an analysis of the information received should appear in his next annual report, as part of an overall assessment of the implementation of UNCLOS 20 years after its adoption (see A/56/121, part B, para. 17). Although the Secretariat intends to send a note verbale requesting information concerning steps undertaken by States to harmonize their national legislation with UNCLOS as well as relevant texts of their legislative acts, as appropriate, the Secretary-General would also appreciate it if States wishing to contribute to this effort would communicate to the Division such information at their earliest convenience. The Division disseminates such information, especially regarding national legislation and delimitation treaties, through, inter alia, its web site, www.un.org/Depts/los.

25. Several developments relating to State practice have been brought to the attention of the Division. One of the notable developments was the adoption, in March 2001, of the comprehensive Maritime Code of Slovenia, which entered into force on 12 May 2001. Also in March 2001, Norway adopted "Regulations relating to foreign marine scientific research in Norway's internal waters, territorial sea and economic zone and on the continental shelf," which have been in force since 1 July 2001. The Regulations relating to the limits of the Norwegian territorial sea around Svalbard adopted in June 2001 entered into force on the same day (see also para. 50 below).

26. Concerning the deposit by Pakistan in June 1999 of the list of geographical coordinates of points for the drawing of straight baselines (see A/54/429, para. 90), India stated that, in its view, certain baseline points of Pakistan were inconsistent with international law and the relevant provisions of UNCLOS. India noted, *inter alia,* that Pakistan had employed straight baselines along its entire coastline, notwithstanding the fact that the Pakistani coastline was quite smooth and was rarely deeply indented or fringed by islands, and that the appropriate baseline for Pakistan's entire coast should be the normal baseline. India further objected against the use of certain basepoints and declared that it did not recognize the arbitrary method of drawing straight baselines and that any claim Pakistan would make on the basis of

the above notification to extend its sovereignty or jurisdiction on Indian waters or to extend its internal waters, territorial sea, exclusive economic zone and continental shelf would be rejected by India (see *Law of the Sea Bulletin* No. 46).

27. Regarding the delimitation of maritime boundaries, France and the Seychelles, on 19 February 2001, concluded an agreement on the delimitation of the maritime boundary of the exclusive economic zone and the continental shelf of France (around the territory of Île Glorieuse and Île du Lys) and of Seychelles (Assumption and Astove Islands).

28. On 9 January 2001, Peru issued a statement concerning the 18°21'00" parallel, which had been referred to by the Government of Chile as the maritime boundary between Chile and Peru in the charts that Chile had deposited with the Secretary-General on 21 September 2000. Peru stated that Peru and Chile had not concluded a specific maritime delimitation treaty pursuant to the relevant rules of international law and that Peru did not recognize the parallel as the maritime boundary between the two States (see *Law of the Sea Information Circular No. 13*).

29. The Division continues to publish all newly obtained legislation and delimitation treaties in the *Law of the Sea Bulletin*, which appears three times per year.

B. Continental Shelf Beyond 200 Nautical Miles and the Work of the Commission on the Limits of the Continental Shelf

30. *Commission on the Limits of the Continental Shelf.* The ninth session of the Commission was held in New York from 21 to 25 May 2001. At the session, the Commission followed up on decisions on training that had been approved at previous sessions. In addition, a discussion took place regarding the decision taken by the Eleventh Meeting of States Parties on the date of commencement of the 10-year period for making submissions to the Commission as well as other matters of relevance discussed by the Meeting. The issues of confidentiality in the work of the Commission were also extensively discussed (CLCS/29).

31. The Editorial Committee of the Commission prepared a document entitled Internal procedure of the subcommission of the Commission on the Limits of the Continental Shelf, which was subsequently adopted by the Commission (CLCS/L.12). During the discussion on the document several issues were raised which the Chairman of the Editorial Committee felt might eventually require amendments to the Modus Operandi of the Commission.

32. On the issue of training, the Commission requested the Secretariat to prepare a "Training manual on the preparation of a submission to the Commission on the Limits of the Continental Shelf" to facilitate the preparation of submission by States concerned, especially developing States (see CLCS/29, para. 15).

33. Following positive comments by the President of the Meeting regarding the benefits of a relationship between the Commission and the Meeting of States Parties, the Commission decided to seek observer status at the next Meeting.

34. The Commission decided not to hold its tenth session in August-September of 2001 but, rather, to convene the session in 2002 for three weeks' duration beginning with the week of 15 April, should there be a submission. If no submission was received, the session might be reduced to one week, or cancelled altogether, depending on the workload of the Commission. In view of the forthcoming election

of 21 members of the Commission at the next Meeting of States Parties in May 2002, the Commission proposed that the eleventh session of the Commission in its new composition should be held from 24 to 28 June 2002.

35. *Establishment of voluntary trust funds.* The General Assembly resolution 55/7 requested the Secretary-General to establish two trust funds, related respectively to the establishment of an extended continental shelf in accordance with the provisions of article 76 of the Convention, and to the work of the Commission.

36. The first trust fund, established pursuant to paragraph 18 of the resolution, is to provide assistance to States parties to meet their obligations under article 76 and annex II to the Convention, and to provide training to countries, in particular the least developed among them and small island developing States, for preparing submissions to the Commission with respect to the outer limits of the continental shelf beyond 200 nautical miles. Norway has donated $1 million to the trust fund and has also transferred to this trust fund the undisbursed portion ($9,220) of its contribution to the Voluntary Fund for Supporting Developing Countries Participating in the United Nations Conference on Straddling Fish Stocks and Highly Migratory Fish Stocks, which is now closed.

37. The uses to which the new trust fund are to be put are spelled out in very detailed terms of reference, which are contained in annex II to resolution 55/7. Applications for financial assistance from the fund may be submitted by any developing State, in particular the least developed countries and small island developing States, that is a Member of the United Nations and a party to the Convention. The intended recipients of monies from the fund are first and foremost coastal States wishing to prepare a submission to the Commission on the Limits of the Continental Shelf. The stated purpose of the fund is to provide, in accordance with the terms and conditions specified in the Financial Regulations and Rules of the United Nations: (a) training to the appropriate technical and administrative staff of the coastal State in question, in order to enable them to perform initial desktop studies and project planning, or at least to take full part in these activities; (b) funds for such studies and planning activities, including funds for advisory/consultancy assistance if needed.

38. This trust fund is not intended to be used to finance activities conducted by an international organization; however, reimbursements may be requested from the fund for airfare and per diem (presumably based on United Nations rates) for the participants from developing countries. Developing States interested in having their experts participate in any appropriate training course as trainees are asked to address their applications to the Division for Ocean Affairs and the Law of the Sea, Office of Legal Affairs, United Nations. All applications will be considered by the Division with the assistance of an independent panel of experts who will examine them on the basis of section 4 of the terms of reference (resolution 55/7, annex II) and recommend the amount of financial assistance to be given. The Division will be guided solely by the financial needs of the requesting developing State and the availability of funds, with priority given to least developed countries and small island developing States, taking into account the imminence of pending deadlines. The Secretary-General will provide financial assistance from the fund on the basis of the evaluation and recommendations of the Division. Payments will be made against receipts evidencing actual expenditures for approved costs.

39. The second trust fund, referred to in paragraph 20 of resolution 55/7, was created to enable members of the Commission from developing countries to

participate fully in the work of the Commission. Thus far, no contribution has been received for this trust fund, nor was there any request for assistance from the fund.

40. *Ten-year time limit for submissions to the Commission.* At the Tenth Meeting of States Parties, several States pointed out that certain countries, particularly developing countries, might have difficulties in complying with the 10-year time limit from the entry into force of the Convention for the countries in question to make the submission to the Commission regarding the outer limits of the continental shelf beyond 200 miles. The time limit was viewed as especially onerous for a large number of developing States in view of their limited technical expertise and lack of financial means. General support was expressed regarding the difficulty of complying with the 10-year time limit and the matter was placed on the agenda of the eleventh meeting.

41. At the Eleventh Meeting of States Parties (14–18 May 2001), an item entitled "Issues with respect to article 4 of Annex II to the United Nations Convention on the Law of the Sea" was discussed. At the request of the Tenth Meeting, the Secretariat had prepared a background paper for that discussion (SPLOS/64). Among other important considerations, in the background paper it was pointed out that, according to the provisions of the Convention, for 14 of the 30 States originally identified in 1978 as appearing to meet the legal and geographic requirements to take advantage of the provisions of article 76 regarding an extended continental shelf (12 of the 14 States being developing States), the deadline for submission of the outer limits would fall in November 2004. In the background paper the Secretariat also identified a number of possibilities for dealing with the issue of the 10-year time limit.

42. In addition to the background paper prepared by the Secretariat, the Meeting also had before it notes verbales from the Government of the Seychelles regarding the extension of the time period for submissions to the Commission (SPLOS/66) and a position paper (SPLOS/67) on the time frame for submissions put forward by all States members of the Pacific Island Forum which are also States parties to the Convention.

43. The Meeting adopted a decision (SPLOS/72) whereby in the case of a State party for which the Convention had entered into force before 13 May 1999, it was understood that the 10-year time period referred to in article 4 of Annex II to the Convention shall be taken to have commenced on 13 May 1999, and that the general issue of the ability of States, particularly developing States, to fulfil the requirements of article 4 of Annex II to the Convention would be kept under review (see also para. 19 above in this connection). It should be noted that 13 May 1999 was the date of the adoption of the Scientific and Technical Guidelines (CLCS/11 and Add.1) by the Commission; the Guidelines are intended, *inter alia,* to provide assistance to coastal States regarding the technical nature and scope of the data and information which they are expected to submit to the Commission according to the provisions of article 76 of UNCLOS.

44. *Workshops and symposia (2001–2002).* The General Assembly in its resolution 55/7 encouraged concerned coastal States and relevant international organizations and institutions to consider developing and making available training courses on the delineation of the outer limits of the continental shelf beyond 200 nautical miles and for the preparation of submissions to be presented to the Commission.

45. At its eighth session, which was held from 31 August to 4 September 2000,

the Commission concentrated primarily on the issue of training in order to assist States in further developing the knowledge and skills necessary to prepare a submission in respect of the outer limits of the continental shelf provided for by the Convention. Although it is not part of its mandate to conduct or organize training, the Commission decided to design an outline for a five-day training course for the delineation of the outer limits of the continental shelf beyond 200 nautical miles and for the preparation of a submission of a coastal State to the Commission (CLCS/24). The Commission undertook this work with a view to facilitating the preparation of submissions, especially by developing States, in accordance with the letter and spirit of the Convention, as well as with the Guidelines of the Commission; it was also felt that the use of the outline would ensure a uniform and consistent practice among the courses. Several regional training courses were conducted in 2001 and are scheduled for 2002 using this outline as the basis for the core curriculum. The practice of offering regional courses appears to be cost-effective for developing countries in the same region and allows the courses to take into account the wide variation in types of continental margins in different areas of the oceans.

46. In this context, a five-day training course was conducted jointly by the Southampton Oceanography Centre and the Hydrographic Office of the United Kingdom of Great Britain and Northern Ireland from 26 to 30 March 2001. The course emphasized both the delineation of the outer limits of the extended continental shelf and the practical aspects of completing a submission to the Commission, and represented a modification of the core outline for a five-day training course designed by the Commission. A similar course is being contemplated for 2002.

47. A regional course which was also a modified form of the Commission's training outline was given by the External Affairs Ministry of the Government of India in New Delhi from 3 to 7 September 2001. The course focused on the application of article 76 and the Statement of Understanding regarding the Bay of Bengal (see Final Act of the Third United Nations Conference on the Law of the Sea, annex II).

48. A Symposium on Marine Geophysics is scheduled to be held during the forthcoming International Congress of the Brazilian Geophysical Society, to be held from 28 October to 1 November 2001 in Salvador de Bahia, Brazil. The papers to be presented, *inter alia,* will deal with subjects related to the delineation of the continental shelf.

49. In addition, the Government of Brazil, as a result of the experience acquired in preparing its submission, has decided to develop and make available for interested coastal States a five-day regional training course, again based on the outline prepared by the Commission. The course will be held in Rio de Janeiro from 3 to 9 March 2002, under the sponsorship of the Brazilian Interministerial Commission on Sea Resources (CIRM), with the support of the Directorate of Hydrography and Navigation (the Brazilian Hydrographic Office) and Petrobras (the Brazilian State Oil Company), and with the assistance of the Division.

C. Deposit of Charts and/or Lists of Geographical Coordinates and Compliance with the Obligation of Due Publicity

50. Information concerning the obligation of coastal States parties to deposit charts and/or lists of geographical coordinates of points (specifying the geodetic datum),

regarding the baselines as well as the outer limits of various maritime zones, is contained in the main annual report for the current session (A/56/58, paras. 83–90). On 7 June 2001, Norway deposited with the Secretary-General, in accordance with article 16, paragraph 2, of the Convention, a list of geographical coordinates of points for drawing the baselines for measuring the breadth of the territorial sea around Svalbard, as contained in the Regulations of 1 June 2001 relating to the limit of the Norwegian territorial sea around Svalbard.

IV. SHIPPING AND NAVIGATION

51. During the period under review, the following main developments affecting the shipping industry and navigation can be highlighted.

A. Safety of Ships

1. Ship Construction, Equipment and Seaworthiness
52. The IMO Marine Environment Protection Committee (MEPC) at its forty-sixth session (23–27 April 2001) adopted amendments to regulation 13G of the International Convention for the Prevention of Pollution from Ships, 1973, as modified by the Protocol of 1978 relating thereto (MARPOL 73/78) to phase out single-hull oil tankers (see para. 87 below).

53. New amendments to the 1994 International Code of Safety for High-Speed Craft (HSC Code) were adopted by the IMO Maritime Safety Committee (MSC) at its seventy-fourth session (30 May–8 June 2001) in order to bring the provisions in line with the relevant provisions of the 2000 HSC Code, which will enter into force on 1 July 2002 for ships built after that date. The amendments relate in particular to the carriage of voyage data recorders and automatic identification systems (AIS).[2]

2. Training of Crew
54. An extraordinary session of MSC has been scheduled for two days in November 2001 for the evaluation of information on a number of parties to the International Convention on Standards of Training, Certification and Watchkeeping for Seafarers (STCW) so that they might be placed on the list of confirmed STCW parties before 1 February 2002, the deadline by which all seafarers must have been trained in compliance with the 1995 amendments to STCW and carry certificates to that effect.

3. Labour Conditions
55. The International Commission on Shipping, in its enquiry into ship safety published in March 2000, concluded that "for thousands of today's international seafarers life at sea is modern slavery and their workplace is a slave ship." The Commission made a number of recommendations for action, mainly on crew issues and port State control activities, directed at flag States, coastal States, shippers' councils, classification societies, the Government of the United States of America, the European Commission, the International Maritime Organization (IMO), the International La-

---

2. MSC 74/24, annex 4, resolution MSC.119 (74).

bour Organization (ILO), the Food and Agriculture Organization of the United Nations (FAO), the International Group of P&I Clubs, international shipping organizations and owners.[3]

56. The problems faced by seafarers, in particular the growing threat of pirate attacks, abandonment and the erosion of traditional seafarers' rights, were also highlighted at the Eleventh Meeting of States Parties to the United Nations Convention on the Law of the Sea (see SPLOS/73, paras. 97 and 98). In that connection, the Joint IMO/ILO Ad Hoc Expert Working Group regarding Claims for Death, Personal Injury and Abandonment of Seafarers, at its most recent meeting (30 April–4 May 2001), approved a draft resolution and guidelines on provision of financial security in case of abandonment of seafarers and a draft resolution and guidelines on shipowners' responsibilities in respect of contractual claims for personal injury or death of seafarers. These texts have been submitted by the Working Group for approval by the IMO Legal Committee and submission to the IMO Assembly for adoption.[4]

B. Safety of Navigation

57. The IMO Subcommittee on the Safety of Navigation (NAV) at its forty-seventh session in July 2001 approved a number of new ship routing measures and amendments to existing measures for submission to MSC at its seventy-fifth session in 2002 for adoption, among them the establishment of a precautionary area of 10 nautical miles around a floating production storage and offloading vessel (FPSO) located on the Grand Banks of Newfoundland in Canada. Initially an area to be avoided had been proposed instead of the precautionary area, but some delegations felt that such establishment restricted the freedom of navigation in contravention of UNCLOS, while others expressed concern regarding the excessive radius.[5] In this regard, it should be noted that UNCLOS provides that safety zones around artificial islands, installations and structures in the exclusive economic zone and on the continental shelf shall not exceed 500 metres in distance around them (see articles 60 and 80).

58. IMO, as the implementing agency of the project entitled "GEF/World Bank/IMO Development of a Regional Marine Electronic Highway (MEH) in the East Asian Seas," commenced the first phase of the project in March 2001 in the Straits of Malacca and Singapore for a duration of one year, at the end of which an action plan for implementing the regional MEH and a project brief for implementing the first phase of the regional project is to be developed. The MEH is intended to be a regional network of electronic navigational charts to enhance navigational safety and environmental management.[6]

3. The report of the Commission is available on its web site at www.icons.org.au.
4. LEG 83/4/1.
5. See NAV 47/13, paras. 3.63-3.66.
6. See IMO document MEPC 46/INF.35.

C. Marine Casualties

59. *Places of refuge.* IMO has decided to address, as a matter of priority, the issue of places of refuge, from the operational safety point of view, and will prepare guidelines on:

  (a) actions the master of a ship should take when in need of a place of refuge (including actions on board and actions required in seeking assistance from other ships in the vicinity, salvage operators, flag States, and coastal States);
  (b) the evaluation of risks associated with the provision of places of refuge and relevant operations in both a general and a case-by-case basis; and
  (c) actions expected of coastal States for the identification, designation and provision of such suitable places together with any relevant facilities.[7] The Legal Committee is to consider any matters relating to international law, jurisdiction, rights of coastal States, liability, insurance, bonds, etc.[8]

D. Flag State Implementation

60. The IMO Assembly at its twenty-second session in November 2001 will consider for adoption a number of draft resolutions aimed at strengthening flag State implementation, including, *inter alia,* those dealing with self-assessment of flag State performance, and revised guidelines on the implementation of the International Safety Management (ISM) Code by Administrations.[9]

61. As regards measures to strengthen flag State implementation in the area of fisheries, MSC at its seventy-fourth session noted the outcome of the discussions of the Subcommittee on Flag State Implementation (FSI) on illegal, unreported and unregulated (IUU) fishing and recognized that, although measures relating to fisheries management were outside the competence of IMO, there were many safety and environmental protection issues relating to IUU fishing which were within the purview of IMO, and the consideration of those issues would assist FAO. MSC also noted that IMO could cooperate with FAO to develop a port State control regime of its own through sharing of experience and expertise on the matter and that, in the context of the seventh session of the Commission on Sustainable Development, there was a need to establish principles against which the transfer of ships might be considered, as FSI had recognized that the transfer of ships was also a problem in relation to illegal fishing activities.

V. CRIMES AT SEA

62. Criminal activities at sea can range from acts of piracy and armed robbery to smuggling of migrants and illicit traffic in drugs or firearms, and often are the work of organized criminals. The Protocol against the Illicit Manufacturing of and Traf-

---

7. See NAV 47/13, paras. 12.28-12.33 and annexes 18 and 19.
8. See MSC 74/24, paras. 2.29 and 2.31.
9. For the text of the draft resolutions, see MSC 74/24, annexes 11 to 13.

ficking in Firearms, Their Parts and Components and Ammunition, supplementing the United Nations Convention against Transnational Organized Crime, which was adopted by the General Assembly on 31 May 2001,[10] is the most recent global instrument aimed at strengthening cooperation among States in preventing, combating, and eradicating transnational organized crime.

A. Piracy and Armed Robbery Against Ships

63. The number of incidents of piracy and armed robbery against ships continued to rise dramatically in the period under review and remains a cause of great concern to the shipping community and affected States, especially coastal States.

64. In recognition of the need to strengthen international cooperation and coordination in combating piracy and armed robbery at sea, the Consultative Process chose this issue as one of two areas of focus of its discussions at the second meeting, in May 2000. The outcome of the discussions and the suggested issues and elements with regard to the prevention of and response to incidents of piracy and armed robbery, which have been proposed to the General Assembly, are contained in the report of the Consultative Process at its second meeting (see A/56/121).

65. The IMO secretariat made an oral report to the Maritime Safety Committee (MSC) at its seventy-fourth session (30 May–8 June 2001) on the outcome of the discussions at the Consultative Process. The Committee requested the secretariat to submit the full report to its seventy-fifth session in 2002.

66. MSC at its seventy-fourth session expressed deep concern at the continuous upward trend in the number of incidents of piracy and armed robbery and once again invited all Governments (of flag, port, and coastal States) and the industry to intensify their efforts to eradicate these acts. It endorsed the outcome of the IMO evaluation and assessment missions to Jakarta and Singapore in March 2001.[11] It agreed that there should be a more precise distinction between the reporting of actual and of attempted attacks. The industry was urged to ensure that all incidents are reported to flag/coastal States. Flag States were urged to use the agreed format for reporting attacks and coastal States were urged to report on follow-up action taken when informed of such attacks and to put in place national legislation for dealing with incidents of piracy and armed robbery. The Committee also approved a draft resolution on the Code of Practice for the Investigation of the Crimes of Piracy and Armed Robbery against Ships and a draft resolution on measures to prevent the registration of "phantom" ships for submission to the IMO Assembly at its twenty-second session (19–30 November 2001) for adoption.[12]

B. Smuggling of Migrants

67. The number of people being smuggled by sea continues to increase. France, Greece and Italy informed IMO that they had detected around 3,375 illegal migrants

10. See General Assembly resolution 55/255.
11. The report of the evaluation and assessment missions conducted in Jakarta and Singapore is contained in IMO document MSC 74/17/1.
12. See MSC 74/24, sect. 17 and annexes 14 and 18.

being transported by sea between April 1999 and April 2001.[13] Spain, in its submission to IMO, reported that it had detected around 17,035 illegal migrants in waters under its sovereignty or jurisdiction during 2000 and that those figures did not even include the unknown number of migrants who could not be detected or detained by the authorities, including those who might have lost their lives.[14]

68. The increase in the smuggling of migrants by sea has been accompanied by an increase in the complexity of ways to deal with the problem. The recent rescue by the Norwegian vessel *Tampa* of more than 400 illegal migrants from a sinking Indonesian ferry and the refusal by Australia to permit them to disembark at Christmas Island demonstrate the potential for tensions between the rendering of humanitarian assistance and national sovereignty considerations. It is to be hoped that the *Tampa* case will not be viewed by shipmasters as a deterrent to rendering assistance to people in distress at sea, which is not only an obligation under article 98 of UNCLOS but also an enshrined tradition and principle of maritime law.

69. MSC at its seventy-fourth session revised the Interim Measures for Combating Unsafe Practices Associated with the Trafficking or Transport of Migrants by Sea contained in document MSC/Circ.896 on the basis of a proposal submitted jointly by France, Greece, Italy, and the United Kingdom.[15] The revised text, *inter alia*, specifies that the carriage of more than 12 persons on board a cargo ship constitutes an automatic infringement of the International Convention for the Safety of Life at Sea (SOLAS).

C. Illicit Traffic in Narcotic Drugs and Psychotropic Substances

70. The Commission on Narcotic Drugs in its resolution 44/6, entitled "Enhancing multilateral cooperation in combating illicit traffic by sea," adopted at its forty-fourth session in March 2001, considered the report of the informal open-ended working group on maritime cooperation against illicit trafficking by sea which met in December 2000 (see A/56/58, paras. 240–241). It requested the United Nations International Drug Control Programme (UNDCP) to, *inter alia*, develop a user-friendly reference training guide to assist parties making requests for verification of nationality and for consent to board, search and take appropriate action under article 17 of the 1988 United Nations Convention against Illicit Traffic in Narcotic Drugs and Psychotropic Substances and to assist competent authorities who have the responsibility to receive and respond to such requests.[16]

---

13. First biannual report on the trafficking or transport of illegal migrants by sea issued by the IMO secretariat on the basis of incidents reported to the organization. IMO Circular MSC.3/Circ.1, available on the web site of IMO at www.imo.org/HOME.html.

14. See MSC 74/23/4.

15. See MSC 74/23/8.

16. See *Official Records of the Economic and Social Council, 2001, Supplement No. 8* (E/2001/28-E/CN.7/2001/12), available on the web site of UNDCP at www.undcp.org/cnd_documents.html.

VI. MARINE RESOURCES, THE MARINE ENVIRONMENT AND SUSTAINABLE DEVELOPMENT

A. Conservation and Management of Marine Living Resources

1. Marine Fisheries
71. A Conference on Responsible Fisheries in the Marine Ecosystems organized jointly by FAO and the Government of Iceland, is to be held in Reykjavik from 1 to 4 October 2001. The objectives of the Conference are to gather and review the best available knowledge on marine ecosystem issues and identify means by which ecosystem considerations can be included in fisheries management. The Conference will also identify future challenges and strategies in ecosystem-based fisheries management.

2. Conservation and Management of Marine Mammals
72. The fifty-third annual meeting of the International Whaling Commission (IWC) was held in London from 23 to 27 July 2001 to consider recurrent issues pertaining to the conservation and management of marine mammals, such as the renewal of the zero catch limits for commercial whaling; continuation of the work on the Revised Management Procedure for commercial whaling, including specification of an inspection and observer system; catch limits for aboriginal subsistence whaling; status of whales; scientific permits issued by Japan for the taking of whales in the western North Pacific; whale killing methods; and environmental research, including the IWC Scientific Committee's plan to hold a workshop on interactions between fisheries and cetaceans.[17]

73. At the July meeting, members of the Commission defeated two proposals, one by Australia and New Zealand and the other by Brazil, to establish whaling sanctuaries in the South Pacific and in the South Atlantic respectively. The Commission also denied a request by Iceland to become a member of IWC with a reservation on the 1982 commercial whaling moratorium.[18]

B. Non-living Marine Resources

International Seabed Authority[19]

74. Following the adoption by the Assembly of the International Seabed Authority in July 2000 of the Regulations for Prospecting and Exploration for Polymetallic Nodules in the Area (ISBA/6/A/18), the Authority since 29 March 2001, has signed 15-year contracts for exploration with six of the seven registered pioneer investors,

---

17. IWC press release 2001, Final Press Release, http://ourworld.compuserve.com/homepages/iwcoffice/pressrelease2001.htm.
18. Ibid.
19. Excerpted from documents and press releases of the International Seabed Authority and information contained in its web site at www.isa.org.jm.

namely, Institut français de recherche pour l'exploitation de la mer (IFREMER)/ Association française pour l'étude et la recherche des nodules polymétalliques (AFERNOD) (France), Deep Ocean Resources Development Co. Ltd. (DORD) (Japan), Yuzhmorgeologiya (Russian Federation), China Ocean Mineral Resources Research and Development Association (COMRA) (China), Interoceanmetal Joint Organization (IOM) (Bulgaria, Cuba, Czech Republic, Poland, Russian Federation, and Slovakia), and the Government of the Republic of Korea. The contract between the Authority and the Government of India had not yet been signed at the time of the preparation of the present report.

75. The seventh session of the International Seabed Authority was held at Kingston, Jamaica, from 2 to 13 July 2001. A major item for the consideration of the Council of the Authority was the regulations and procedures for prospecting and exploration for polymetallic sulphides and cobalt-rich crusts in the international seabed area (see also A/54/429, para. 341). The Council held extensive discussions, on issues outlined in a paper prepared by the secretariat (ISBA/7/C/2), and decided to continue its consideration of the item at its next session. The Council also decided to request the secretariat of the Authority to collect and assemble the necessary information to facilitate further discussion in the Council on important considerations raised in the secretariat paper and to assist the Legal and Technical Commission in its work on the matter.

76. In accordance with regulation 38 of the Regulations for Prospecting and Exploration for Polymetallic Nodules in the area, the Legal and Technical Commission of the Authority had adopted and issued its recommendations for the guidance of contractors for the assessment of possible environmental impacts arising from exploration for polymetallic nodules in the Area (ISBA/7/LTC/1/Rev.1 and Corr.1). The Council took note of the recommendations and decided that further consideration should be given to them at its next session, as necessary.

77. During the session, the Council elected 24 members of the Legal and Technical Commission (ISBA/7/C/6) and the Assembly elected 15 members of the Finance Committee (ISBA/7/A/7, para. 5). The Assembly also approved the Staff Regulations of the Authority (ISBA/7/A/5).

78. Immediately preceding the session, a workshop to standardize the environmental data and information required by the Regulations and the recommendations of the Legal and Technical Commission for the guidance of contractors was convened by the Authority at its headquarters in Kingston from 25 to 29 June 2001. The workshop highlighted the fact that the environmental effects of seabed exploration are hard to predict, given the lack of experience in this area and the relative paucity of information about the deep ocean. It concentrated on identifying key types of data needed to assess the state of the deep ocean environment as a prerequisite to determining the effect of future mineral resource development on the environment and thereafter determining ways to shape such development so as to cause the least possible harm to the environment. The workshop's output included specific recommendations as to what should be collected and measured (in relation to benthic biology, chemical, and geological factors as well as the water column) and even, in many cases, what methods and procedures should be employed to ensure comparability of data and information.

C. Protection and Preservation of the Marine Environment

1. Reduction and Control of Pollution

(a) Land-based Activities: The Global Programme of Action[20]

79. The Global Programme of Action (GPA) Coordination Office, under the United Nations Environment Programme (UNEP), at The Hague, will sponsor the First GPA Intergovernmental Review Meeting, hosted by the Government of Canada, from 26 to 31 November 2001 in Montreal. The Meeting is expected to bring together senior representatives from more than 100 Governments, a large number of international organizations, global and regional nongovernmental organizations, the private sector, and other GPA stakeholders, which are the partners involved in both the current and future implementation of the GPA.

80. The main themes to be discussed during the meeting are: binding and nonbinding agreements at the national and regional levels; voluntary agreements and involvement of the private sector; capacity-building; innovative financing and use of economic instruments; and sharing experiences through reporting and the further development of the clearing-house mechanism.

81. One of the major tasks of the meeting is to increase awareness at all levels, especially at the national level, of the importance of addressing land-based activities as the major source of marine and coastal degradation. The meeting will also endeavour to highlight the lack of funding as a major impediment to dealing with land-based problems and will seek to increase private sector involvement. Other objectives of the meeting include developing a long-term work plan for a framework for a new long-term vision and preparing and adopting a high-level declaration to constitute the GPA input to the World Summit on Sustainable Development in Johannesburg in 2002. Further details on the First GPA Intergovernmental Review Meeting may be obtained by consulting the GPA web site at www.gpa.unep.org/igr.

(b) Pollution by Dumping; Waste Management

Disposal of wastes at sea
82. The Scientific Group of the Consultative Meeting of Contracting Parties to the Convention on the Prevention of Marine Pollution by Dumping of Wastes and Other Matter (London Convention) at its 24th meeting in May 2001 completed its work on the eight waste-specific guidelines for the assessment of wastes or other matter that may be considered for dumping.[21] It decided to refer the specific Guidelines for Assessment of Vessels to the Marine Environment Protection Committee (MEPC) for its consideration in view of the relationship between the discussions on recycling of ships and the recommendations in the Guidelines on the evaluation of

---

20. Excerpted from UNEP/GPA Coordination Office documents.
21. IMO document LC/SG 24/11, annexes 3–10.

alternatives to the disposal of vessels at sea and on the preparation of a decommissioned vessel in case disposal at sea is chosen.[22]

83. Australia, Japan, Norway and the United States informed the meeting of the Scientific Group that they were planning a joint research project involving the release of 15,000 gallons of liquid carbon dioxide at a depth of more than 800 metres to assess the feasibility of disposal of carbon dioxide at sea and that additional information on the project would be provided at future meetings.[23]

Management of radioactive wastes
84. The Joint Convention on the Safety of Spent Fuel Management and on the Safety of Radioactive Waste Management, which was adopted by the International Atomic Energy Agency (IAEA) in 1997, entered into force on 18 June 2001. The Convention is the first international instrument to address the safety of the management and storage of radioactive wastes and spent fuel in countries with or without nuclear programmes. One of its main objectives is to ensure that during all stages of spent fuel and radioactive waste management there are effective defences against potential hazards. The Convention contains requirements related to the transboundary movement of spent fuel and radioactive waste which are based on the 1990 IAEA Code of Practice on the International Transboundary Movement of Radioactive Waste. The State of origin must ensure that it obtains the prior notification and consent of the State of destination. The Convention provides that "transboundary movement through States of transit shall be subject to those international obligations which are relevant to the particular modes of transport utilized" (article 27, para. 1 (ii)).

85. The Convention establishes a mechanism whereby each Contracting Party is obliged to submit for review by meetings of Contracting Parties a report on the measures taken to implement each of the obligations under the Convention. This includes reporting on national inventories of radioactive wastes and spent fuel.

(c) Pollution from Vessels

86. The following major developments during the period under review in the regulation of pollution from ships can be highlighted:

(a) adoption of amendments to regulation 13G of MARPOL 73/78 to phase out single-hull oil tankers;
(b) adoption of the International Convention on Civil Liability for Bunker Oil Pollution Damage; and
(c) IMO Conference to consider and adopt the draft Convention on the Control of Harmful Anti-Fouling Systems.

Adoption of amendments to regulation 13G of MARPOL 73/78 to phase out single-hull oil tankers
87. The IMO Marine Environment Protection Committee at its forty-sixth session in March 2001 amended regulation 13G of annex I to the International Convention

22. Ibid., para. 2.15.
23. Ibid., paras. 9.10–9.11.

for the Prevention of Pollution from Ships, 1973, as modified by the Protocol of 1978 relating thereto (MARPOL 73/78) in order to expedite the phase-out of most single-hull oil tankers by 2015 or earlier (resolution MEPC.95[46] of 27 April 2001). According to revised regulation 13G, the phase-out period will depend on the category of oil tankers. The continued operation of oil tankers beyond 2015 or beyond the twenty-fifth anniversary of their delivery is only permitted for high-quality ships which had been subjected to a Condition Assessment Scheme (CAS).[24] However, any port State can deny entry to its ports or offshore terminals, to single-hull tankers, that are allowed to operate up until the twenty-fifth anniversary of their delivery.

Adoption of the International Convention on Civil Liability for Bunker Oil Pollution Damage
88. With the adoption of the Bunkers Convention on 23 March 2001, the last significant gap in the international regime for compensating victims of oil spills from ships has been closed. The Convention establishes a liability and compensation regime for damage caused by spills of oil when carried as fuel in ships' bunkers "in the territory, including the territorial sea of a State Party, and in the exclusive economic zone of a State Party, established in accordance with international law or, if a State Party has not established such a zone, in an area beyond and adjacent to the territorial sea of that State determined by that State in accordance with international law and extending not more than 200 nautical miles from the baselines from which the breadth of the territorial sea is measured."[25] Modelled on the International Convention on Civil Liability for Oil Pollution Damage, a key requirement in the new Bunkers Convention is for the registered owner of a vessel to maintain compulsory insurance coverage.

IMO Conference to consider and adopt the draft Convention on the Control of Harmful Anti-Fouling Systems
89. MEPC at its 46th session made further progress in resolving some of the outstanding issues in the draft Convention on the Control of Harmful Anti-Fouling Systems prior to its consideration and scheduled adoption at a conference in October 2001. The principal points for consideration by the conference will be the entry into force provisions, the removal of existing organotin tributyltin (TBT) paints versus overcoating with sealer paints, the proposed damage clause, the provisions on amendments, and other issues that might be raised before or during the Conference.[26]

2. Regional Cooperation
Review of UNEP Regional Seas Programme and Action Plans

Convention for the North-east Pacific
90. The third high-level Government-designated expert meeting of the proposed North-east Pacific regional seas programme, which was held in Panama from 6 to

---

24. The Condition Assessment Scheme was adopted on 27 April 2001 by resolution MEPC.94(46).
25. For the text of the Convention, see IMO document LEG/CONF.12/19.
26. For the discussions at the forty-sixth session of MEPC and the text of the draft convention to be submitted to the conference, see MEPC 46/23, sect. 5 and annex 5.

9 August 2001, approved the text of the draft Convention on Cooperation for the Protection and Sustainable Development of the Marine and Coastal Zones of the North-east Pacific as well as a plan of action and the programme of work for 2001–2006 of this new regional seas programme. It is expected that the plenipotentiaries of the eight coastal States of the region which participated in the negotiations (Colombia, Costa Rica, El Salvador, Guatemala, Honduras, Mexico, Nicaragua, and Panama) will meet during the first trimester of 2002 to sign the Convention.[27]

91. The Convention establishes the framework of operation of the plan of action. The programme of work for 2001–2006 also addresses the implementation of the GPA in the North-east Pacific. The Convention for the North-east Pacific is the first regional seas convention negotiated since the adoption of the GPA in 1995 that has integrated the implementation of the GPA within its framework.

D. Protection of Specific Marine Areas

Marine Protected Areas

92. As part of the preparations for the in-depth consideration by the Subsidiary Body on Scientific, Technical and Technological Advice (SBSTTA) at its eighth meeting in 2002 of the topic of protected areas and to assist in the implementation of the programme of work on marine and coastal protected areas,[28] the secretariat of the Convention on Biological Diversity is convening a Technical Expert Group Meeting on Marine and Coastal Protected Areas, in Leigh, New Zealand, from 22 to 26 October 2001. According to its terms of reference,[29] the Expert Group will, *inter alia*, focus on marine and coastal protected areas or similarly managed areas and their value for and effects on the sustainable use of marine and coastal living resources.

93. One of the conclusions reached at an expert workshop on the scientific requirements and legal aspects of marine protected areas on the high seas, organized by the German Federal Agency for Nature Conservation at Vilm from 27 February to 4 March 2001, was that UNCLOS provides the framework for all action to conserve biodiversity and other components of the marine environment of the high seas and that it was the bedrock on which all actions had to be based. The workshop suggested that an important subject for discussion within the Consultative Process in the very near future should be the management of risks to biodiversity and other

---

27. The report of the third meeting is contained in document UNEP(DEC)/NEP/EM.3/4. In order to facilitate the negotiating process, UNEP and the Division closely cooperated with respect to certain issues related to maritime jurisdiction. That cooperation was a good example of ensuring consistent approaches with a view to implementing UNCLOS, as called for by the General Assembly in resolution 55/7, para. 23.

28. The programme of work is contained in decision IV/5 adopted by the Conference of Parties to the Convention on Biological Diversity. For the text see the web site of the secretariat of the Convention on Biological Diversity at www.biodiv.org.

29. The terms of reference are contained in SBSTTA recommendation V/14, annex II, available on the web site of the secretariat of the Convention on Biological Diversity at www.biodiv.org.

components of the marine environment of the high seas.[30] In that connection, at the second meeting of the Consultative Process, in May 2001, one delegation proposed that the concept of marine protected areas should be applied to waters beyond the limits of national jurisdiction.[31]

Special Areas Under MARPOL 73/78 and Particularly Sensitive Sea Areas

94. New revised guidelines for the designation of special areas under MARPOL 73/78 and guidelines for the identification and designation of particularly sensitive sea areas (PSSAs) were approved by MEPC at its forty-sixth session in April 2001 and are scheduled to be adopted in the form of an IMO Assembly resolution at the twenty-second session in November 2001.[32] They will update and replace the 1991 IMO Guidelines, as amended in 1999 (IMO Assembly resolutions A.720(17) and A.885(21)).

95. At its forty-sixth session, MEPC also approved in principle the designation of the marine area around the Florida Keys of the United States and the Malpelo Islands off Colombia as PSSAs, subject to a review by the Subcommittee on Navigation of the proposed navigational measures. Such measures were approved by NAV at its forty-seventh session in July 2001 and its decision will be conveyed to MEPC at its next session in 2002.[33]

E. Climate Change and Sea Level Rise

96. Following its suspension after the November 2000 session at The Hague, at which negotiators had failed to reach agreement, the sixth Conference of the Parties to the United Nations Framework Convention on Climate Change resumed deliberations in Bonn, Germany, from 16 to 27 July 2001. A political agreement was reached in Bonn on some fundamental issues which, according to the Executive Secretary of the Conference, succeeded in enabling the ratification of the Kyoto Protocol.[34]

97. Agreement was reached on issues and concepts that are integral to the implementation of the Kyoto Protocol once it enters into force, including those concerning funding, reduction mechanisms, emissions trading, clean development, joint implementation, carbon sinks and compliance. The Agreement will be adopted formally at the seventh Conference of Parties, to be hosted by the Government of Morocco in Marrakech from 29 October to 9 November 2001. Several decisions still

---

30. The proceedings of the workshop are available on the web site of the German Federal Agency for Nature Conservation at www.bfn.de/06/060301_workshoptp.htm.
31. See A/56/121, part B, para. 84.
32. For draft text of the guidelines, see IMO document MEPC 46/23, annex 6.
33. See IMO document NAV 47/13, annex 4.
34. The Kyoto Protocol will enter into force after it has been ratified by at least 55 parties to the United Nations Framework Convention on Climate Change, including industrial countries accounting for 55 per cent of the total 1990 carbon dioxide emissions from the whole group of industrialized countries. As of the date of the present report, 37 countries have ratified, including one industrial country.

requiring some additional work are expected to be finalized at the seventh Conference of Parties and adopted together as a package with the decisions reached in Bonn.

F. Ten-Year Review of the Implementation of Agenda 21[35]

98. The Commission on Sustainable Development, acting as the preparatory committee for the World Summit on Sustainable Development, convened its tenth session from 30 April to 2 May 2001, to begin preparations for the Summit, to be held in Johannesburg, South Africa, in 2002. The 2002 Summit will assess the degree of progress made thus far in translating into practice the principles of sustainable development and the measures referred to in Agenda 21, agreed upon 10 years ago in Rio de Janeiro, at the United Nations Conference on Environment and Development. The three-day session was the first in a series of preparatory sessions of the Commission and was essentially an organizational session. The Commission, *inter alia*, considered the progress in preparatory activities at the local, national, subregional, regional and global levels as well as by major groups; focused on a process for setting the agenda and determining possible main themes for the Summit; and recommended to the General Assembly the adoption of the provisional rules of procedure of the Summit. A multistakeholder panel was also held during the session to allow representatives of major groups to bring their views to the organizational discussion.[36]

99. The Commission agreed on the timetable of regional and global preparatory meetings at which the details of the agenda for the Summit would be developed. Regional preparatory meetings, to be supported by national and subregional meetings, will be held from August to November 2001 and preparatory meetings at the global level will be held from January to June 2002.

100. The second and third preparatory sessions of the Commission are scheduled to be held in New York from 28 January to 8 February and from 25 March to 5 April 2002, respectively. The fourth and final preparatory session is scheduled to be held at the ministerial level in Indonesia from 27 May to 7 June 2002.

101. A number of regional round tables of eminent persons were also held from June to August 2001 to ensure that a wide range of views was brought into the preparatory process. Further details on the ongoing preparations for the Summit may be found at www.johannesburgsummit.org.

VII. Underwater Cultural Heritage

102. The fourth meeting of governmental experts on the draft convention on the protection of the underwater cultural heritage was held at UNESCO headquarters from 26 March to 6 April and from 2 to 7 July 2001. A text for submission to the

---

35. Excerpted from United Nations press releases and United Nations Department of Economic and Social Affairs documents.

36. See Official Records of the General Assembly, Fifty-sixth Session, Supplement No. 19 (A/56/19).

UNESCO General Conference at its thirty-first session, to be held in Paris from 15 October to 3 November 2001, was adopted by 49 votes in favour and 4 against, with eight abstentions. Among the sensitive issues were the protection of the underwater cultural heritage on the continental shelf and the inclusion of provisions concerning State vessels and aircraft. The text of the draft convention was recommended by the Director-General of UNESCO, the chairman of the fourth meeting and the co-chairmen of the Drafting Committee to the General Conference for adoption.

VIII. Marine Science and Technology

103. As recommended by the General Assembly in its resolution 55/7, one of the areas of focus of the second meeting of the Consultative Process was "marine science and development and transfer of marine technology as mutually agreed, including capacity-building." There were extensive discussions (see A/56/121, part B, paras. 18–19, 21, 23–24, 27–67) and the Consultative Process suggested a number of issues and a number of elements relating to each issue for consideration by the General Assembly (see A/56/121, part A, paras. 3–51). Delegations emphasized the fundamental importance of implementing the provisions of Parts XIII and XIV of UNCLOS on marine scientific research and development and transfer of marine technology, respectively, with the aim of making these important parts of the Convention operational in practical terms, as well as the marine science and technology provisions of chapter 17 of Agenda 21.

Intergovernmental Oceanographic Commission Advisory Body of Experts on the Law of the Sea (ABE-LOS)

104. The Intergovernmental Oceanographic Commission (IOC) of UNESCO is recognized in UNCLOS as the competent international organization for matters related to marine scientific research (MSR). The Advisory Body of Experts on the Law of the Sea (ABE-LOS) was established by IOC in its resolution XIX-19 with specific terms of reference whereby, upon request, it would give advice to the Assembly, the Executive Council and/or the Executive Secretary of IOC with regard to the possible role of the Commission in the implementation of UNCLOS. ABE-LOS held its first meeting in Paris on 11–13 June 2001 (ABE-LOS I). The meeting was attended by 29 member States and five representatives of institutions, including the Division, as observers. In this connection, attention is drawn to the low level of participation and States are encouraged to increase their participation in the upcoming meetings of ABE-LOS. The June meeting focused on two main topics on its agenda: "Matters pertaining to Part XIII of UNCLOS" and "Matters pertaining to Part XIV of UNCLOS."

105. *Matters pertaining to Part XIII of UNCLOS.* ABE-LOS I concentrated its discussions mainly on three articles of Part XIII under which the role of IOC could be further developed. During the consideration of UNCLOS article 251 on the establishment of general criteria and guidelines to assist States in ascertaining the nature and implications of marine scientific research, some members of ABE-LOS pointed

out the linkages between that article and article 246, paragraph 5 (a), on the "consent regime" for MSR, while others were of the view that article 251 should be viewed in conjunction with articles 248 and 249 on, respectively, the duty to provide information and to comply with certain conditions when carrying out MSR.

106. There were further discussions on article 246 and the "consent regime." States were encouraged to identify "a central MSR office" (or any equivalent body) at the national level, to facilitate the processing of applications for consent and for ensuring uniformity in the application and interpretation of the relevant provisions of UNCLOS.

107. ABE-LOS I also held discussions on UNCLOS article 247 dealing with MSR projects undertaken by or under the auspices of international organizations. It was stated that the article offered the benefit of a simplified procedure for consent for carrying out MSR projects when undertaken by or under the auspices of an international organization in the exclusive economic zone or on the continental shelf of one or more countries. In that regard, it was recognized that IOC had a major role to play in identifying specific rules and procedures to be followed to fully implement article 247. Preliminary work had been done on the issue, as reflected in document IOC/INF-1055, which nevertheless needed extensive revision.

108. *Matters pertaining to Part XIV of UNCLOS.* In view of the leading role of IOC in promoting the establishment of generally accepted guidelines, criteria and standards for the transfer of marine technology (UNCLOS article 271), Part XIV was considered a priority issue for implementation. In that connection, ABE-LOS I discussed the possible role of IOC as a clearing-house mechanism, based upon existing models (e.g., the GPA clearing house), with the purpose of meeting the needs of suppliers and recipients of marine technology. One component of such a mechanism would be an integrated database on the transfer of marine technology, which would at the same time accommodate the need for capacity-building.

109. It was recalled that the issue of a clearing house had been raised at the meeting of the IOC Inter-sessional Intergovernmental Working Group in Lisbon, on 29 and 30 March 2001. The Working Group had instructed the IOC Executive Secretary to initiate the development of a clearing-house mechanism for ocean sciences for facilitating access by member States to: relevant information derived from ongoing research; a list of global ocean science programmes and projects; opportunities for capacity-building in ocean science; and a list of sources of information on ocean science.

110. During the discussion at ABE-LOS I, it was recognized that document IOC/INF-1054, entitled "Draft IOC principles on transfer of marine technology," would constitute a good starting point for establishing accepted guidelines, criteria and standards for the transfer of marine technology and should be redrafted in close cooperation with the Division.

111. ABE-LOS I also discussed the question of the establishment and functions of regional marine scientific and technological research centres as envisaged in UNCLOS (articles 276 and 277). It was suggested that, through existing regional IOC mechanisms, regional bodies should be strengthened to carry out the functions spelled out in UNCLOS. The IOC subsidiary bodies could serve as effective platforms for the identification of needs and the implementation of marine science and technological transfers.

112. At the conclusion of the consideration of the topics on its agenda, ABE-

LOS I adopted three recommendations, which were submitted to the IOC Assembly at its twenty-first session for adoption.

113. *Recommendations adopted by ABE-LOS I.* The first two recommendations on Part XIV and Part XIII of UNCLOS, respectively, established two open-ended subgroups to work by correspondence, in close cooperation with the Division. The first subgroup would work on redrafting document IOC/INF-1054, taking into account the debate on the issues in ABE-LOS I. The second subgroup would assist IOC in establishing appropriate internal procedures related to the effective and appropriate use of UNCLOS article 247 on marine scientific research undertaken by or under the auspices of international organizations. The third recommendation concerned the continuation and completion of the collection and analysis of information from member States on their MSR practices, in close cooperation with the Division.

Twenty-First Session of the IOC Assembly

114. The IOC Assembly at its twenty-first session (Paris, 3–13 July 2001) adopted resolution XXI-2, entitled "IOC and UNCLOS," and its annex, entitled "First Meeting of the Advisory Body of Experts on the Law of the Sea (ABE-LOS I): Recommendations." In the resolution, the IOC Assembly noted with satisfaction the progress made by ABE-LOS I and instructed the Executive Secretary of IOC to take the necessary actions for the full implementation of the ABE-LOS I recommendations.

115. The IOC Assembly also adopted resolution XXI-11 on "African priorities," in which the IOC Executive Secretary was requested to assist African member States, without prejudice to the competence of the United Nations Commission on the Limits of the Continental Shelf, in developing their capacity within the context of UNCLOS article 76.

Workshop Organized by the South Pacific Geoscience Commission (SOPAC)[37]

116. Fifty-five participants representing coastal and researching States attended a three-day Regional Workshop on the Issues and Challenges of Marine Scientific Research in the Pacific Region, held at Port Moresby in February 2001 (see also A/56/58, paras. 470–472). Four key areas were discussed extensively, with recommendations proposed for each: relating to the legal framework for conduct of marine scientific research; capacity-building; transfer of marine science and technology, including data; and marine mineral exploration and marine scientific research as parallel activities.

117. SOPAC highlighted that, for Pacific States, UNCLOS provisions concerning obligations with regard to MSR incumbent upon the researching State in relation to participation and (post-cruise) data and information requirements remained a priority issue which demanded continued emphasis and attention.

---

37. Information provided by SOPAC.

IX. SETTLEMENT OF DISPUTES

118. During the period under review, the International Tribunal for the Law of the Sea was seized of the following cases: the *Grand Prince* case (*Belize v. France;* and the *Chaisiri Reefer 2* case (*Panama v. Yemen*). (Further details on these cases may be found at the web site of the Division for Ocean Affairs and the Law of the Sea: www.un.org/Depts/los.)

119. *Trust fund.* Pursuant to paragraphs 9 and annex I to General Assembly resolution 55/7, the Secretary-General established a trust fund for the purposes of assisting States in the settlement of disputes through the International Tribunal for the Law of the Sea. The United Kingdom made two contributions to the trust fund, amounting to $24,865. To date, no formal request was received by the Secretariat for assistance from the trust fund.

A. Cases Before the International Tribunal for the Law of the Sea

120. *Case concerning the Conservation and Sustainable Exploitation of Swordfish Stocks in the South-eastern Pacific Ocean* (*Chile v. European Community*) (see also A/56/58, paras. 442–446). In February 2001, Chile and the European Union (EU) reached an agreement by which they settled their dispute with regard to both access for EU fishing vessels to Chilean ports and bilateral and multilateral scientific and technical cooperation on the conservation of swordfish stocks. In view of the agreement, EU requested a suspension of panel proceedings within the World Trade Organization and Chile suspended proceedings before the Tribunal. However, each party reserved its right to revive the proceedings before the Tribunal at any time. By an Order of 15 March 2001, at the request of the parties, the President of the special chamber of the Tribunal formed to deal with the case extended the time limit for making preliminary objections. Accordingly, the time limit of 90 days for making preliminary objections would commence from 1 January 2004 and each party would have the right to request that the said time limit should begin to apply from any date prior to 1 January 2004.

121. *The "Grand Prince" Case (Belize v. France).* On 26 December 2000, the fishing trawler *Grand Prince,* flying the flag of Belize, was arrested by French authorities in the exclusive economic zone of the Kerguelen Islands in the French Southern and Antarctic Territories for allegedly engaging in illegal fishing. The court of first instance at Saint-Paul, Réunion, confirmed the seizure of the vessel, catch, and equipment on board by an order of 12 January 2001. The court also fixed a bond for the release of the vessel in the amount of 11.4 million French francs. On 23 January 2001, the criminal court at Saint-Denis, Réunion, ordered the confiscation of the vessel. On 21 March 2001, an Application was made to the Tribunal on behalf of Belize against France for the prompt release of the vessel in accordance with article 292 of UNCLOS. On 20 April 2001, the Tribunal found that it did not have jurisdiction under article 292, paragraph 2, of UNCLOS to hear the Application as there was not sufficient basis for holding that Belize was the flag State of the vessel. Therefore, the Tribunal was not called upon to deal with the remaining questions of jurisdiction, admissibility and merits of the Application.

22. *The "Chaisiri Reefer 2" Case (Panama v. Yemen).* In accordance with article

292 of UNCLOS, proceedings were instituted on 3 July 2001 before the Tribunal by an Application made on behalf of Panama against Yemen for the prompt release of the vessel *Chaisiri Reefer 2,* its crew, and cargo, which had been detained by Yemeni authorities. However, by a note verbale dated 12 July 2001, the Embassy of Yemen in Germany, on behalf of its Government, informed the Tribunal that the vessel, its cargo, and crew had been released and were free to sail from Mukalla Port, Yemen. In addition, the Government of Yemen guaranteed that the same load that had been unloaded previously from the vessel would be loaded back and that the case would therefore be withdrawn by Panama. Accordingly, the Agent of Panama informed the Tribunal that the parties had agreed to discontinue the proceedings as a result of having settled their dispute on the arrest of the vessel. Consequently, by an Order dated 13 July 2001, the President of the Tribunal recorded the discontinuance of the proceedings and directed the removal of the case from the Tribunal's List of Cases.

B. Arbitration and Conciliation

123. The following names have been added to the list of arbitrators in accordance with article 2, of Annex VII to UNCLOS: Prof. Dr. Hasjim Djalal, Dr. Etty Roesmaryati Agoes, Dr. Sudirman Saad, and Lieutenant Commander Kresno Bruntoro, nominated by Indonesia; and Mr. Walter Sá Leitão, nominated by Brazil.

124. The following names have been added to the list of conciliators in accordance with article 2, of Annex V to UNCLOS: Prof. Dr. Hasjim Djalal, Dr. Etty Roesmaryati Agoes, Dr. Sudirman Saad, and Lieutenant Commander Kresno Bruntoro, nominated by Indonesia; and Mr. Walter Sá Leitão, nominated by Brazil.

125. The full list of arbitrators and conciliators is available on the Division's web site at www.un.org/Depts/los. The list is also available in the Law of the Sea Information Circular published by the Division.

126. The list of special arbitrators under Annex VIII to UNCLOS is available on the web sites of the respective specialized agencies which have responsibility in the different fields concerned. The list has been drawn up in accordance with article 2 of Annex VIII to the Convention.

X. INTERNATIONAL COOPERATION AND COORDINATION

A. Subcommittee on Oceans and Coastal Areas of the Administrative Committee on Coordination

127. The Subcommittee on Oceans and Coastal Areas (SOCA) of the Administrative Committee on Coordination (ACC) held its eleventh session at United Nations Headquarters in New York on 3 and 4 May 2001. The session was hosted by the United Nations Development Programme (UNDP).

128. In considering the ongoing review of ACC machinery and its implications for SOCA, the Subcommittee welcomed the conclusions and approaches advocated by the newly established ACC High-level Committee on Programmes at its first session and noted that "international coordination and cooperation is of vital impor-

tance in addressing all aspects of oceans and coastal areas. The cooperation between the relevant parts of the United Nations Secretariat for the purpose of ensuring better coordination of United Nations work on oceans and seas is thus considered imperative. The existence of a mechanism such as ACC/SOCA is needed." The Subcommittee went on to express "its conviction [that the most productive course of future action lay in] building on existing mechanisms through innovative and more integrated approaches for effective coordination and cooperation."

129. The Subcommittee, *inter alia*, also reviewed the status of the preparation, under its auspices with FAO as the lead agency, of the United Nations Atlas of the Oceans; its role in the implementation of the GPA; and the preparations for the World Summit on Sustainable Development.

130. In addition, the Subcommittee discussed matters relating to progress in the independent evaluation of the Joint Group of Experts on the Scientific Aspects of Marine Environmental Protection (GESAMP) (see also paras. 132–133 below); and future directions for the IAEA Marine Environmental Studies Laboratory.

131. In regard to UNEP Governing Council decision 21/13 concerning a feasibility study for establishing a regular process for the assessment of the state of the marine environment, the Subcommittee, *inter alia*, expressed its willingness to participate in the consultative process for the study and stressed the need for the participation of Governments in the process (see also para. 134).

B. Other Mechanisms

Joint Group of Experts on the Scientific Aspects of Marine Environmental Protection (GESAMP)

132. At its thirty-first session, hosted by the United Nations through the Division for Ocean Affairs and the Law of the Sea, Office of Legal Affairs, at Headquarters in New York from 13 to 17 August 2001, GESAMP considered the final report of the independent evaluation team that it had established at its thirtieth session to recommend ways to make GESAMP more effective, more inclusive and more responsive to emerging problems and to the needs of policy makers and decision makers. Progress in this exercise, among other GESAMP matters, was reported by the chairperson of GESAMP and by its IMO Administrative Secretary to the Consultative Process at its second meeting in May 2001.

133. At the session, following considerable discussion, GESAMP responded positively and constructively to the evaluation team's recommendations, some of which contained substantial financial implications. Follow-up actions have been undertaken and will be reported on to the Consultative Process at its third meeting scheduled for May 2002.

134. GESAMP also addressed a UNEP Governing Council decision regarding a feasibility study for establishing a regular process for the assessment of the state of the marine environment (decision 21/13). That initiative warranted substantial attention by GESAMP in view of its own current and established role and competence in preparing global assessments of the state of the marine environment and of the need to define its role and position vis-à-vis the envisaged feasibility study. An informal consultative meeting to discuss the UNEP Governing Council decision

was convened in Reykjavik from 12 to 14 September 2001, jointly hosted by the Ministry of Environment of Iceland and UNEP. The report of the meeting is forthcoming. However, there appeared to be a consensus that the goal of a regular assessment of the state of the marine environment would best be served not by the establishment of new structures or institutions but rather by the adaptation of existing mechanisms, structures, and programmes and the optimization of cooperation and coordination among them.

XI. REVIEW BY THE GENERAL ASSEMBLY OF DEVELOPMENTS IN OCEAN AFFAIRS: UNITED NATIONS OPEN-ENDED INFORMAL CONSULTATIVE PROCESS ESTABLISHED BY THE GENERAL ASSEMBLY IN ITS RESOLUTION 54/33 IN ORDER TO FACILITATE THE ANNUAL REVIEW BY THE ASSEMBLY OF DEVELOPMENTS IN OCEAN AFFAIRS

135. The General Assembly, by its resolution 54/33 of 24 November 1999, decided to establish an open-ended informal consultative process in order to facilitate, in an effective and constructive manner, its own review of overall developments in ocean affairs.

136. Consistent with the legal framework provided by UNCLOS and the goals of chapter 17 of Agenda 21, the Consultative Process discusses the annual report of the Secretary-General on oceans and the law of the sea and suggests particular issues to be considered by the General Assembly, with an emphasis on identifying areas where coordination and cooperation at the intergovernmental and interagency levels should be enhanced.

137. The second meeting of the Consultative Process was held at United Nations Headquarters from 7 to 11 May 2001. Pursuant to paragraph 3 (e) of General Assembly resolution 54/33 and after consultations with Member States, the President of the General Assembly reappointed Ambassador Tuiloma Neroni Slade (Samoa) and Mr. Alan Simcock (United Kingdom) as Co-Chairpersons of the second meeting of the Consultative Process.

138. In the light of the results of informal consultations held by the Co-Chairpersons preceding the second meeting (three rounds of informal consultations were held, on 23 February, 23 March and 4 May 2001 respectively) and the comments from some delegations, Co-Chairperson Simcock proposed that the second meeting should adopt its format and annotated agenda (A/AC.259/L.2) with a number of amendments. The second meeting adopted by consensus the format and the annotated agenda, as amended (A/AC.259/5). In accordance with one of those amendments, the Consultative Process would henceforth be referred to as the "United Nations Open-ended Informal Consultative Process established by the General Assembly in its resolution 54/33 in order to facilitate the annual review by the Assembly of developments in ocean affairs." Some delegations would have wished to stress further the link between the Consultative Process and item 41 of the provisional agenda of the fifty-sixth session of the General Assembly entitled "Oceans and the law of the sea." Some other delegations did not share this view. Nevertheless, it was noted that, in resolution 54/33, the General Assembly, in establishing the Consultative Process, had recalled that the United Nations Convention on the Law of the Sea sets out the legal framework within which all activities in the oceans and

seas must be carried out and with which those activities should be consistent, as recognized also by the United Nations Conference on Environment and Development in chapter 17 of Agenda 21, and had also acknowledged the importance of maintaining the integrity of UNCLOS (see also A/56/121, letter from the Co-Chairpersons, fifth paragraph; and ibid., part B, para. 7).

139. There was an in-depth discussion on the two areas of focus selected for the second meeting of the Consultative Process, identified by the General Assembly in resolution 55/7:

(a) marine science and the development and transfer of marine technology as mutually agreed, including capacity-building in that regard; and

(b) coordination and cooperation in combating piracy and armed robbery at sea.

140. In the area of international cooperation and coordination, there was an exchange of views with the Chairman of the Subcommittee on Oceans and Coastal Areas (SOCA) of the Administrative Committee on Coordination. It was pointed out that SOCA was in a period of transition and was undergoing a period of reviewing its mechanism. However, it was stressed that, while the structure for coordination might undergo changes, the function and goal of coordination in ocean affairs would remain and would be carried out (see also para. 128 above).

141. *Trust funds.* Pursuant to paragraph 45 of General Assembly resolution 55/7, the Secretary-General established a trust fund for the purposes of assisting developing countries, in particular the least developed countries, small island developing States and landlocked developing States, in attending the meetings of the Consultative Process. Japan transferred to the fund the undisbursed portion ($17,130) of its contribution to the Voluntary Fund for Supporting Developing Countries Participating in the United Nations Conference on Straddling Fish Stocks and Highly Migratory Fish Stocks, which is now closed. Representatives from three developing countries were provided with travel expenses from the trust fund to attend the second meeting of the Consultative Process in May 2001.

142. By a letter dated 22 June 2001 addressed to the President of the General Assembly (A/56/121), the Co-Chairpersons submitted the report on the work of the second meeting of the Consultative Process, proposing a number of issues and elements for consideration by the General Assembly under the agenda item "Oceans and the law of the sea" and for potential inclusion in the relevant General Assembly resolutions, in accordance with paragraph 3 (h) of resolution 54/33. The report was composed of three parts:

(a) issues to be suggested, and elements to be proposed to the General Assembly;
(b) Co-Chairperson's summary of discussions; and
(c) issues for consideration for possible inclusion in the agenda of future meetings.

Selected Documents and Proceedings

United Nations Convention on the Law of the Sea Report of the Eleventh Meeting of States Parties New York, 14–18 May 2001†

CONTENTS

	Paragraphs
1. INTRODUCTION	1–4
2. ORGANIZATION OF WORK	5–17
A. Opening of the Eleventh Meeting of States Parties and election of officers	5–7
B. Introductory statement by the president	8–16
C. Adoption of the agenda and organization of work	17
3. REPORT OF THE CREDENTIALS COMMITTEE	18–19
4. MATTERS RELATED TO THE INTERNATIONAL TRIBUNAL FOR THE LAW OF THE SEA	20–44
A. Annual report of the Tribunal	20–30
B. Budget of the Tribunal for 2002	31–36
C. Financial Regulations of the Tribunal	37–40
D. Report of the External Auditors and financial statements of the Tribunal for 1999	41
E. Election of one member of the Tribunal	42–44
5. RULES OF PROCEDURE FOR MEETINGS OF STATES PARTIES	45–50
A. Proposed amendment to Rule 53 (Decisions on questions of substance)	45–46
B. Proposal to establish a finance committee	47–50
6. INFORMATION ON THE ACTIVITIES OF THE INTERNATIONAL SEABED AUTHORITY	51–59
7. MATTERS RELATED TO THE CONTINENTAL SHELF AND THE COMMISSION ON THE LIMITS OF THE CONTINENTAL SHELF	60–84
A. Statement by the Chairman of the Commission on the Limits of the Continental Shelf	60–66
B. Issues with respect to article 4 of annex II to the United Nations Convention on the Law of the Sea	67–80
8. MATTERS RELATED TO ARTICLE 319 OF THE UNITED NATIONS CONVENTION ON THE LAW OF THE SEA	85–92
9. OTHER MATTERS	93–109
A. Trust Funds	93–96
B. Statement by a representative of a non-governmental organization regarding seafarers	97–98

†EDITORS' NOTE.—This document was provided by the United Nations Division for Ocean Affairs and the Law of the Sea (DOALOS). The source is Report of the Eleventh Meeting of States Parties, New York, 14–18 May 2001, SPLOS/73, 14 June 2001. Available online: <http://www.un.org/depts/los/index.htm>.

C. Statement by the President at the closure of the eleventh Meeting of States Parties   97–107
D. Dates and programme of work for the twelfth Meeting of States Parties

1. INTRODUCTION

1. The eleventh Meeting of States Parties to the United Nations Convention on the Law of the Sea was convened at United Nations Headquarters from 14 to 18 May 2001, in accordance with article 319, paragraph 2 (e), of the Convention and the decision taken by the General Assembly at its fifty-fifth session (resolution 55/7, para. 6).

2. Although the tenth Meeting of States Parties had decided that the eleventh Meeting would be held from 7 to 11 May 2001 (SPLOS/60, para. 85), subsequent to that decision, with a view to accommodating the ninth session of the Commission on Sustainable Development to be followed by the second meeting of the Consultative Process, the General Assembly at its fifty-fifth session decided that the eleventh Meeting of States Parties would be convened from 14 to 18 May 2001.

3. Pursuant to that decision and in accordance with rule 5 of the Rules of Procedure for Meetings of States Parties (SPLOS/2/Rev.3), invitations to participate in the Meeting were addressed by the Secretary-General of the United Nations to all States Parties to the Convention. Invitations were also addressed to observers in conformity with rule 18 of the Rules of Procedure (SPLOS/2/Rev.3/Add.1), including to the President and the Registrar of the International Tribunal for the Law of the Sea and the Secretary-General of the International Seabed Authority.

4. In addition to a number of relevant documents from previous Meetings, the following documents were before the Meeting:

(a) Rules of Procedure for Meetings of States Parties (SPLOS/2/Rev.3 and SPLOS/2/Rev.3/Add.1);
(b) Report of the tenth Meeting of States Parties (SPLOS/60 and Corr.1);
(c) Provisional agenda (SPLOS/L.19);
(d) Annual report of the International Tribunal for the Law of the Sea for 2000 (SPLOS/63);
(e) Draft budget proposals of the International Tribunal for the Law of the Sea for 2002 (SPLOS/WP.13);
(f) Level of compensation for judges ad hoc (SPLOS/WP.15);
(g) Report of the External Auditors for the financial year 1999, with financial statements of the International Tribunal for the Law of the Sea as at 31 December 1999 (SPLOS/53);
(h) Issues with respect to article 4 of Annex II to the United Nations Convention on the Law of the Sea (SPLOS/64);
(i) Notes verbales from the Government of Seychelles regarding the extension of the time period for submission to the Commission on the Limits of the Continental Shelf (SPLOS/66);
(j) Position paper on the time frame for submissions to the Commission on the Limits of the Continental Shelf. Submitted by Australia, Fiji, Marshall

Islands, Micronesia (Federated States of), Nauru, New Zealand, Papua New Guinea, Samoa, Solomon Islands, Tonga, and Vanuatu (SPLOS/67);

(k) Letter dated 30 April 2001 from the Chairman of the Commission on the Limits of the Continental Shelf addressed to the President of the eleventh Meeting of States Parties (SPLOS/65);

(l) Financial Regulations of the Tribunal (SPLOS/WP.14 and Corr.1);

(m) Germany: Proposals relating to the Financial Regulations of the Tribunal (SPLOS/CRP.27);

(n) European Community, Germany and Japan: Proposal relating to the Financial Regulations of the Tribunal (SPLOS/CRP.28);

(o) United Kingdom of Great Britain, and Northern Ireland: Proposals relating to the Rules of Procedure for Meetings of States Parties (SPLOS/CRP.20/Rev.1);

(p) Germany: Proposal relating to the Rules of Procedure for Meetings of States Parties (SPLOS/CRP.26).

2. ORGANIZATION OF WORK

A. Opening of the Eleventh Meeting of States Parties and Election of Officers

5. The eleventh Meeting of States Parties was opened by the President of the tenth Meeting, Ambassador Peter D. Donigi (Papua New Guinea).

6. The Meeting elected by acclamation Ambassador Cristián Maquieira (Chile) as President of the eleventh Meeting of States Parties.

7. The Meeting also elected the representatives of Australia, India, and Nigeria as Vice-Presidents.

B. Introductory Statement by the President

8. In his opening statement, the President extended his welcome to all States Parties, particularly to Nicaragua, Maldives, and Luxembourg, which he noted had become Parties to the Convention since the last Meeting of States Parties, bringing the total number of Parties to 135. He asserted that States had to remain committed to reaching the common objective of universal participation in the Convention.

9. He noted that since the States Parties last met, three cases had been submitted to the International Tribunal for the Law of the Sea. Furthermore, the International Seabed Authority had signed 15-year contracts for exploration for polymetallic nodules with three of the seven registered pioneer investors, while the Commission on the Limits of the Continental Shelf was ready to receive submissions from coastal States on the delineation of their continental shelf beyond 200 nautical miles.

10. The President recalled that following the recommendations made at the tenth Meeting of States Parties, the General Assembly, at its fifty-fifth session, had approved the establishment of three voluntary trust funds. He noted that all the funds had been established by the Secretary-General and were now operational.

11. He outlined the programme of work of the eleventh Meeting. The Meeting would elect one member of the International Tribunal for the Law of the Sea to

serve the remainder of the term of Judge Lihai Zhao of China, who had passed away in October 2000. It would also examine the proposed budget of the Tribunal for 2002. In addition, it would consider the annual report of the Tribunal as well as the report of the External Auditors. Thereafter, the Meeting would consider the Financial Regulations of the Tribunal, as well as proposals to amend the Rules of Procedure for Meetings of States Parties, including a proposal providing for the establishment of a finance committee.

12. The Meeting would continue the consideration of the role of the Meeting of States Parties in the implementation of the United Nations Convention on the Law of the Sea. The President recalled that Chile had submitted a proposal in that respect.

13. The Meeting would also examine issues related to article 4 of Annex II to the United Nations Convention on the Law of the Sea. The President noted that at the tenth Meeting delegations had expressed general support for the concerns voiced regarding the difficulty experienced by States, particularly developing countries, in complying with the time limit outlined in that article.

14. The President stated that the Meeting would invite the Chairman of the Commission on the Limits of the Continental Shelf, Yuri Kazmin, to report on the progress of work in the Commission. In that context, he recalled that the Chairman of the Commission, in a letter to the President of the Meeting of States Parties, had addressed the question of training, particularly the ways training could be organized to assist developing States in preparing their submissions to the Commission.

15. The Secretary-General of the International Seabed Authority, Satya Nandan, would also be invited to report on the activities of the Authority, the President stated.

16. Following the statement by the President, one delegation made a general statement on matters related to the law of the sea. He observed that universal participation in the Convention remained the ultimate goal and stressed the need for States to enact the necessary legislation to ensure the effective and uniform implementation of the provisions of the Convention. He underlined the vital role of the institutions established by the Convention.

C. Adoption of the Agenda and Organization of Work

17. The Meeting considered the provisional agenda for the eleventh Meeting (SPLOS/L.19). The agenda as adopted is contained in document SPLOS/68.

3. REPORT OF THE CREDENTIALS COMMITTEE

18. The Meeting of States Parties appointed a Credentials Committee consisting of the following members: China, Indonesia, Monaco, Romania, Sierra Leone, Sudan, Trinidad and Tobago, United Kingdom of Great Britain and Northern Ireland, and Uruguay.

19. The Credentials Committee held two meetings, on 15 and 16 May 2001. The Committee elected Ferry Adamhar (Indonesia) as Chairman. At its meetings, the Committee examined the credentials of representatives to the eleventh Meeting of States Parties. It accepted the credentials submitted by the representatives of 94

States Parties to the Convention, including the European Community. On 16 May 2001, the Meeting of States Parties approved the report of the Committee (SPLOS/69 and Add.1).

4. MATTERS RELATED TO THE INTERNATIONAL TRIBUNAL FOR THE LAW OF THE SEA

A. Annual Report of the Tribunal

20. The annual report of the International Tribunal for the Law of the Sea, covering the calendar year 2000 (SPLOS/63), was submitted to the Meeting of States Parties under rule 6, paragraph 3 (d), of the Rules of Procedure for Meetings of States Parties.

21. In his introductory statement, the President of the Tribunal, Judge P. Chandrasekhara Rao, at the outset informed the Meeting that the Registrar of the Tribunal, Gritakumar Chitty, had tendered his resignation with effect from 1 July 2001.

22. Turning to the work of the Tribunal, he recalled that during 2000 the Tribunal had delivered judgments in two cases: the "Camouco" case between Panama and France; and the "Monte Confurco" case between Seychelles and France. More recently, on 20 April 2001, the Tribunal had delivered its judgment in the "Grand Prince" case between Belize and France. Further, at the request of Chile and the European Community, the Tribunal had formed a special chamber under article 15, paragraph 2, of its Statute (annex VI to the Convention) to hear a dispute concerning the conservation and sustainable exploitation of swordfish stocks in the South-eastern Pacific Ocean.

23. The President noted that judgments in all cases decided by the Tribunal had been delivered within remarkably short periods, adding that the Tribunal made special efforts to make that possible in view of the need to settle international disputes expeditiously. However, parties to prompt release proceedings under article 292 of the Convention had underlined the difficulties they faced in complying with the time limits fixed in the Rules of the Tribunal for the filing of written statements by both parties before the commencement of oral proceedings. In that context, the Tribunal had reviewed its Rules in the light of the experience gained, and on 15 March 2001, it had amended articles 111 and 112 of its Rules, extending the time period allocated for the disposal of an application made under article 292 of the Convention from 21 to 30 days.

24. During the two administrative sessions held in 2000, the Tribunal had discussed, among other things, issues that had a direct bearing on its judicial work, such as costs to be borne by parties in judicial proceedings, bonds or other financial securities to be furnished by parties and time factors in the handling of cases. The Tribunal also considered administrative matters such as budget proposals, budget performance, audit report, staff regulations and rules, recruitment of staff, instructions for the Registry, buildings and electronic systems, and library facilities.

25. The President of the Tribunal recalled that the official opening of the permanent headquarters of the Tribunal had taken place on 3 July 2000 in a ceremony attended by the Secretary-General of the United Nations, as well as by the President of the tenth Meeting of States Parties. On behalf of the Tribunal, he expressed its

deep appreciation to the Government of Germany for making the new building available to the Tribunal, which he noted had also served recently as a centre for several international conferences on matters concerning the law of the sea.

26. The President noted that the Tribunal and the Government of Germany had, on 18 October 2000, concluded an Agreement on the Occupancy and Use of the Premises of the Tribunal. With regard to the finalization of the Headquarters Agreement between the Tribunal and Germany, he hoped that the outstanding issues would soon be resolved in a spirit of good will and accommodation. He also noted that the Agreement on the Privileges and Immunities of the Tribunal had not yet entered into force, although it had been adopted nearly four years ago. The President recalled that the General Assembly had called upon States to consider ratifying or acceding to the Agreement.

27. He drew the attention of the States Parties to communications received with respect to the judgment of the Tribunal in the M/V "Saiga" (No. 2) case. In drawing attention to those communications, the Tribunal was not expressing any view with regard to their contents. In that context, he referred to General Assembly resolution 55/7 of 30 October 2000, in paragraph 8 of which the Assembly had recalled the obligations of parties to cases before a court or a tribunal referred to in article 287 of the Convention to ensure prompt compliance with the decisions rendered by such court or tribunal.

28. Delegations expressed their appreciation to the President and the Tribunal for the annual report. Some emphasized the vital role of the Tribunal in resolving disputes regarding the application and interpretation of the United Nations Convention on the Law of the Sea.

29. Many delegations expressed their regret at the resignation of the Registrar, Gritakumar Chitty, and expressed their appreciation for his valuable contribution to the law of the sea and in particular to the establishment and the commencement of functioning of the Tribunal. The need to have the next Registrar elected from a broad range of candidates, as well as the need for transparency in the election process, were emphasized by some delegations.

30. The Meeting took note, with appreciation, of the report of the Tribunal.

B. Budget of the Tribunal for 2002

31. The President of the Tribunal introduced the draft budget of the Tribunal for 2002 (SPLOS/WP.13). He emphasized that in making its budget proposals, the Tribunal had scrupulously followed the evolutionary approach. The proposals were based on the principle of zero growth of the overall budget. Moreover, compared with the approved budget for 2001, there was a decrease of about $0.28 million in the proposed budget for 2002, which had been made possible through the use of the latest version of the United Nations standard salary costs for calculating the budgetary estimates in respect of the staffing requirements.

32. The budget proposals were first considered in an open-ended Working Group under the chairmanship of the President of the Meeting. The Working Group deliberated on the overall budget proposals and also carried out an item-by-item examination. It agreed on the draft budget of the Tribunal for 2002, as proposed by the Tribunal in document SPLOS/WP.13. On the basis of the agreement

in the Working Group (SPLOS/L.20), the Meeting approved the budget of the Tribunal for 2002, which is contained in document SPLOS/70.

33. The approved budget amounted to a total of $7,807,500, including:

(a) A recurrent expenditure of $6,522,400, consisting of:
   (i) $1,808,100 for the remuneration, travel, and pension of judges;
   (ii) $2,916,900 for salaries and related costs of staff (15 posts at the Professional level and above and 21 posts at the General Service level);
   (iii) $252,600 for general temporary assistance, overtime, representation allowance, and official travel;
   (iv) $129,100 for temporary assistance for meetings;
   (v) $1,415,700 for other items, including communications, supplies and materials, printing and binding, maintenance of premises, rental and maintenance of equipment, hospitality, special services, library, training, and miscellaneous services;
(b) A nonrecurrent expenditure of $340,800, essentially for the acquisition of furniture, equipment and special equipment.

With a view to providing the Tribunal with the necessary financial means to consider cases in 2002, the Meeting of States Parties approved $894,300 as contingency funds of the Tribunal, which shall only be used in the event of cases being submitted to the Tribunal during that period. The contingency funds include an amount intended to meet the compensation of a judge *ad hoc* when required. The Meeting also approved an additional amount of $50,000 to be appropriated to the Working Capital Fund of the Tribunal in 2002 in order to build up the Fund to the recommended level of $650,000.

34. The budget of the Tribunal in 2002, including its contingency funds and the appropriations to its Working Capital Fund, is to be financed by all States and international organizations that are Parties to the United Nations Convention on the Law of the Sea. These contributions to be made by States Parties are to be based upon the scale of assessments for the regular budget of the United Nations for the corresponding financial year, adjusted to take account of participation in the Convention. The Meeting of States Parties decided that a floor rate of 0.01% and a ceiling rate of 25% would be used in establishing the rate of assessment for States Parties for the budget of the Tribunal in 2002. The European Community indicated that its contribution to the budget would be proportionate to the approved budget and would amount to $77,000.

35. One delegation proposed that contributions made by States Parties to the budget of the Tribunal should be subject to a floor rate of 0.01% and a ceiling rate of 22%. That would reflect a recently adopted change to the scale of assessments for the regular budget of the United Nations. Following a brief discussion on the proposal, it was decided that the issue would be taken up again at the twelfth Meeting of States Parties.

36. With respect to the level of compensation for judges ad hoc, the Meeting had before it a working paper prepared by the Tribunal (SPLOS/WP.15), in which it was proposed that the level of compensation for judges ad hoc should be consistent with the level of remuneration of elected members of the Tribunal. The Meeting adopted the proposal.

C. Financial Regulations of the Tribunal

37. The Financial Regulations of the Tribunal (SPLOS/36) had generated considerable discussions since the President of the Tribunal had introduced them during the ninth Meeting of States Parties. A number of oral and written proposals had been submitted by delegations during that Meeting and during the tenth Meeting in 2000. While some of the proposals had attracted broad support, further deliberations were required in respect of others. In that regard, the tenth Meeting had requested the Secretariat and the Registry of the Tribunal to prepare a revised version of the Financial Regulations, taking into account the various proposals and the outcome of the discussions during the ninth and tenth Meetings. In view of the number of outstanding issues, the Secretariat, in consultation with the Registry, decided that a working paper would better serve the discussions at the eleventh Meeting. Following the preparation of the working paper by the Secretariat (SPLOS/WP.14), the proposals attributed to the Tribunal in the working paper were withdrawn by the Tribunal. This is reflected in document SPLOS/WP.14/Corr.1.

38. The working paper was discussed in an open-ended Working Group, which was chaired by the President. The Working Group held three meetings. In considering the working paper, delegations took into account additional proposals submitted by Germany (SPLOS/CRP.27) and the European Community, Germany and Japan (SPLOS/CRP.28), and an informal proposal presented by Japan on regulations 5.2 and 5.3. The latter proposal was deferred for discussion until the next Meeting, since it was linked to the discussions on the scale of assessments for the budget of the Tribunal (see para. 35 above). The Working Group was able to reach a tentative agreement on most of the outstanding provisions in regulations 1 to 5. The proposals made in reference to a finance committee were withdrawn in view of the decision taken by the Meeting regarding the establishment of an open-ended working group on financial and budgetary matters (see paras. 49–50 below).

39. One of the pending issues relating to the Financial Regulations is a proposal put forward by Germany during the tenth Meeting that a "split currency system" should be used in the presentation of the budget, i.e., United States dollars and euros. While some delegations supported this view, others expressed a preference for the presentation of the budget in United States dollars, which was considered to be a more stable currency.

40. Owing to time constraints, the Meeting was not able to conclude the consideration of the working paper and will take up the item again at its twelfth Meeting. At the end of the Meeting, the President circulated an informal paper dated 18 May 2001, containing the regulations that had been tentatively agreed upon by the Working Group. In view of a number of references made by delegations to the Financial Regulations of the International Seabed Authority and of the United Nations, and in order to further facilitate the consideration of the Financial Regulations of the Tribunal at the next Meeting, the President suggested that the Secretariat prepare a comparative table of the Financial Regulations of the three institutions.

D. Report of the External Auditors and Financial Statements of the Tribunal for 1999

41. The report of the External Auditors for the financial year 1999 was initially made available to the tenth Meeting of States Parties (SPLOS/53) in 2000. Following an

introduction by the Registrar, the eleventh Meeting of States Parties considered and took note of the report.

E. Election of One Member of the Tribunal

42. As a result of the passing away on 10 October 2000 of Judge Lihai Zhao of China, whose term of office would have ended on 30 September 2002, a vacancy occurred in the Tribunal. In accordance with article 6, paragraph 1, of the Statute of the Tribunal, vacancies shall be filled by the same method as that laid down for the first election of the members of the Tribunal. Article 6, paragraph 2, of the Statute provides that a member of the Tribunal elected to replace a member whose term of office has not expired shall hold the office for the remainder of the predecessor's term.

43. An invitation calling for nominations was addressed to all States Parties in accordance with the provisions of the Statute. One candidate, Mr. Xu Guangjian, was nominated by China. The election was scheduled to take place on 16 May 2001 based on consultations carried out by the President of the Tribunal and the President of the tenth Meeting of States Parties.

44. There was only one round of balloting, during which the representatives of Belize, Croatia, Papua New Guinea, Senegal, and Sweden acted as tellers. Out of 94 delegations present and voting, a majority of 62 was required for election. Mr. Xu Guangjian obtained 92 votes, with 1 abstention and 1 invalid vote cast, and was elected to serve the remainder of the term of the late Judge Lihai Zhao. On behalf of the Meeting of States Parties, the President congratulated Mr. Xu Guangjian on his election.

5. RULES OF PROCEDURE FOR MEETINGS OF STATES PARTIES

A. Proposed Amendment to Rule 53 (Decisions on Questions of Substance)

45. The Meeting continued its discussion on a proposed amendment to rule 53 of the Rules of Procedure for Meetings of States Parties on the basis of a revised proposal submitted by the United Kingdom (SPLOS/CRP.20/Rev.1). The proposal provided for decisions on budgetary and financial matters to be taken by a three-fourths majority of States Parties present and voting, provided that such majority included a majority of States Parties participating in the Meeting.

46. Many delegations expressed reservations with respect to the proposed change to rule 53. In the light of the views expressed, the United Kingdom decided to withdraw its proposal.

B. Proposal to Establish a Finance Committee

47. The Meeting also considered a proposal by Germany (SPLOS/CRP.26) regarding the addition of a new rule 53bis providing for the establishment of a finance committee at each Meeting of States Parties at which financial and budgetary matters

would be discussed. The finance committee would serve as a subsidiary body to the Meeting to review the proposed budget of the Tribunal and make recommendations to the Meeting.

48. During the discussions, many delegations reiterated the views expressed during the tenth Meeting. While some delegations were of the view that a finance committee would expedite the work of the Meeting of States Parties, others maintained that there was no need for such a committee since the practice adopted so far for the consideration of the budget had worked very well, as exemplified by the timely manner in which the budget proposal of the Tribunal for 2002 had been approved.

49. The President decided to draft a compromise text reflecting the various views expressed (SPLOS/L.21). The text was adopted by consensus as a new rule 53bis of the Rules of Procedure for Meetings of States Parties (SPLOS/71).

50. Rule 53bis requires that an open-ended working group be established as a matter of priority during the Meetings of States Parties at which financial and budgetary matters will be discussed. The open-ended working group, to be chaired by the President of the Meeting, will review the proposed budget of the Tribunal and make recommendations to the Meeting. Decisions on budgetary and financial matters taken by the Meeting shall be based on those recommendations.

6. INFORMATION ON THE ACTIVITIES OF THE INTERNATIONAL SEABED AUTHORITY

51. At the ninth Meeting of States Parties, it had been agreed that the Secretary-General of the International Seabed Authority would be given an opportunity to address the Meetings of States Parties and provide information with respect to the activities of the Authority.

52. Pursuant to that decision and in accordance with rule 37 of the Rules of Procedure for Meetings of States Parties, the Secretary-General of the Authority, Satya Nandan, reported to the eleventh Meeting on recent developments with respect to the work of the Authority. He stated that the main achievement of the Assembly of the Authority during the sixth and the resumed sixth sessions in 2000 was the approval of the Regulations for Prospecting and Exploration for Polymetallic Nodules in the Area on the recommendations of the Council.

53. Following the adoption of the Regulations, draft contracts for exploration had been prepared in respect of each of the seven registered pioneer investors whose plans of work for exploration were considered to be approved by the Council on 27 August 1997. Fifteen-year contracts with three of the seven pioneer investors had already been signed, while another was scheduled for signature on 22 May 2001. The three other contracts would be signed in the near future.

54. Recalling that a request had been submitted to the Authority in August 1998 with respect to the adoption of regulations for exploration for polymetallic sulphides and cobalt-rich crusts, the Secretary-General of the Authority noted that, pursuant to article 162, paragraph 2 (o) (ii), of the Convention, such rules, regulations and procedures would need to be adopted within three years from the date of the request. In that regard, the secretariat of the Authority had commenced work in 1999 on a review of the status of knowledge and research on the resources concerned.

In June 2000, the Authority convened a workshop, the third in a series, the objective of which was to provide technical information to assist in drafting regulations for prospecting and exploration for these mineral deposits. The proceedings of the workshop would contain technical papers on the geology and mineralogy of polymetallic sulphides and cobalt-rich crusts, their distribution and resource potential, as well as the status of research on such resources and the technical requirements for their exploration and future mining.

55. He stated that the Legal and Technical Commission of the Authority had continued its consideration of draft recommendations for the assessment of the possible environmental impacts arising from exploration for polymetallic nodules. The recommendations would elaborate procedures to be followed in the acquisition of baseline data by contractors, including the monitoring to be performed during or after any activities with the potential to cause serious harm to the environment, and would also facilitate reporting by contractors.

56. With respect to the budget of the Authority, he noted that 34% of the assessed contributions to the 2001 budget and 97% of the contributions to the budget for 2000 had been received. However, he expressed concern that, as of 30 April 2001, 68 members of the Authority were in arrears of contributions for a period exceeding two years. In accordance with the Convention and the Rules of Procedure of the Assembly of the Authority, members whose arrears equalled or exceeded the amount of their assessed contributions for the preceding two full years would lose their vote. He urged all members of the Authority to pay their assessed contributions and arrears as soon as possible.

57. He encouraged States to become parties to the Protocol on the Privileges and Immunities of the International Seabed Authority, noting that only 4 of the 10 instruments of ratification or accession required for its entry into force had been deposited.

58. In conclusion, noting that the lack of a quorum could hamper the taking of decisions, he encouraged as many delegations as possible to participate in the upcoming seventh session of the Authority, where, among other things, elections would be held to the Legal and Technical Commission and to the Finance Committee.

59. The Meeting took note, with appreciation, of the report of the Secretary-General of the Authority.

7. MATTERS RELATED TO THE CONTINENTAL SHELF AND THE COMMISSION ON THE LIMITS OF THE CONTINENTAL SHELF

A. Statement by the Chairman of the Commission on the Limits of the Continental Shelf

60. The President invited the Chairman of the Commission on the Limits of the Continental Shelf, Yuri Kazmin, to provide any additional information on the matters contained in the letter dated 30 April 2001 (SPLOS/65) addressed to him and on the recent activities of the Commission. He pointed out that because the Commission had not been established at the time when the Meeting of States Parties had adopted its Rules of Procedure no formal relationship existed between the Meeting

of States Parties and the Commission as it did with the other two entities established by the Convention, namely the International Tribunal for the Law of the Sea and the International Seabed Authority, which enjoyed observer status. Since States Parties had displayed great interest in the activities of the Commission, the President was of the view that the Meeting of States Parties might wish to establish such a relationship and grant observer status to the Commission.

61. The Chairman of the Commission pointed out that the Commission was an autonomous body established by the Convention with no formal accountability to the Meeting of States Parties. However, he was of the view that that was a procedural issue that could be resolved in the course of time. He pointed out that in its own Rules of Procedure, the Commission had provided for consultations with the States Parties on certain issues. He noted, for example, that the States Parties had been consulted on the issue of submissions in cases of unresolved land or maritime disputes.

62. He was particularly grateful for the successful role played by the Meeting of States Parties in the establishment of two voluntary trust funds requested by the Commission. The first one had been established to provide training and technical and scientific advice, as well as personnel, to assist developing States, in particular the least developed and small island developing States, for the purpose of preparing submissions under article 76 and annex II to the Convention in accordance with the procedures of the Scientific and Technical Guidelines of the Commission. The second one had been created to defray the costs of participation of the members of the Commission from developing States to enable them to attend the meetings of the Commission.

63. The Chairman of the Commission called the attention of the Meeting of States Parties to the activities of the Commission as presented in his letter to the President of the Meeting. He drew attention to annex I of the Rules of Procedure of the Commission, concerning submissions in case of a dispute between States with opposite or adjacent coasts or in other cases of unresolved land or maritime disputes. The Commission had also adopted its *modus operandi* (CLCS/L.3) and, more significantly, its Scientific and Technical Guidelines (CLCS/11 and CLCS/11/Add.1 and Corr.1), the aim of which was to assist coastal States in dealing with the technical content and extent of information necessary to prepare submissions to the Commission.

64. He recalled that the Commission in May 2000 had held an open meeting to highlight the most important issues regarding the implementation of article 76 of the Convention. At the meeting, which had been attended by approximately 100 participants, representing Governments, intergovernmental organizations and other experts in marine science, members of the Commission had made presentations and exchanged views with the participants.

65. While no submissions had been made to date, it was the Chairman's understanding that some States had reached a fairly advanced stage in the preparation of their submissions. Referring to the 10-year time period established by the Convention for making submissions to the Commission, he appreciated that the determination of the outer limits of the continental shelf was a complicated task, particularly for developing States. In that respect, he emphasized the need to train the appropriate staff to enable States to carry out the activities entailed in preparing their submissions. In addition to the Scientific and Technical Guidelines, the Commission

had prepared an outline for a five-day training course (CLCS/24) designed to assist in the preparation of submissions. That outline, together with the relevant documents prepared by the Commission and the establishment of the trust fund, provided a good basis on which training could be organized. He emphasized, however, that the mandate of the Commission did not include the conduct of training. It was therefore up to States, international or regional organizations and any other institutions to take the initiative in that area. One such initiative had recently been taken by a scientific institution in the United Kingdom, which had held one seminar based on the outline for a five-day training course and on the Scientific and Technical Guidelines of the Commission. In conclusion, the Chairman emphasized the necessity of presenting submissions within the 10-year time frame set by the Convention.

66. The Meeting of States Parties took note, with appreciation, of the statement of the Chairman of the Commission.

B. Issues with Respect to Article 4 of Annex II to the United Nations Convention on the Law of the Sea

67. Under article 4 of annex II to the Convention, a coastal State intending to establish the outer limits to its continental shelf beyond 200 nautical miles is obligated to submit particulars of such limits to the Commission on the Limits of the Continental Shelf along with supporting scientific and technical data as soon as possible but in any case within 10 years of the entry into force of the Convention for that State.

68. At the tenth Meeting of States Parties, concerns had been voiced by developing States regarding the difficulty of complying with the time limit in article 4 of annex II to the Convention. The Meeting had expressed general support for the concerns raised and decided to include in the agenda for the eleventh Meeting the item "Issues with respect to article 4 of annex II to the United Nations Convention on the Law of the Sea." It also requested the Secretariat to prepare a background paper on the matter.

69. In addition to the document prepared by the Secretariat (SPLOS/64), the eleventh Meeting of States Parties also had before it notes verbales from the Government of Seychelles regarding the extension of the time period for submissions to the Commission (SPLOS/66) and a position paper (SPLOS/67) on the time frame for submissions put forward by the following States members of the Pacific Island Forum: Australia, Fiji, Marshall Islands, Micronesia (Federated States of), Nauru, New Zealand, Papua New Guinea, Samoa, Solomon Islands, Tonga, and Vanuatu.

70. The representative of the Federated States of Micronesia, introducing the position paper, emphasized the complexity of the task of preparing submissions to the Commission on the Limits of the Continental Shelf, which required significant resources, capacity and expertise to carry out the necessary activities such as the collection, collation and analysis of a large amount of bathymetric, seismic and geophysical data. He pointed out that a crucial theme of the Convention was that developing States should not, through lack of resources or capacity, be disadvantaged in respect of access to or use of their resources. Therefore, it would be inconsistent with the general approach of the Convention if developing States were unable to define the limits of their extended continental shelf owing to a lack of resources

or capacity. In that regard, he emphasized that the Convention contained important provisions on transfer of technology so as to ensure that developing States were able to exercise their rights and fulfil their obligations under the Convention.

71. He emphasized that many countries would not be able to make a submission within the 10-year time frame stipulated in the Convention for reasons of capacity, financial and technical resources; the lack of settlement of key jurisdictional boundaries and the complexity of the technical issues involved. Furthermore, States had had a clear idea of how to prepare their submissions only after the Commission had adopted its Scientific and Technical Guidelines on 13 May 1999. The representative recalled that the election of the members of the Commission had not taken place until May 1997, nearly three years after the entry into force of the Convention. In the light of the foregoing, the Pacific Island Forum States proposed the following:

(a) That the States Parties agree to extend the 10-year period prescribed in annex II, such an extension to be agreed through a decision of the Meeting of States Parties or through an understanding on the interpretation of annex II;
(b) Such an understanding would include an agreement that the 10-year period would not begin to run for any State Party, regardless of its date of ratification or accession, until the date of adoption of the Commission's Guidelines; and
(c) The time for making a submission would be further extended beyond 10 years where a State Party had been unable, for technical reasons, including lack of technical capacity, to comply in good faith with the time limitation (SPLOS/67, para. 8).

72. Many delegations agreed that the development and strengthening of the capabilities of developing States, including small island developing States, in order to enable them to benefit fully from the legal regime for the oceans as established by the Convention, was an issue of crucial importance. They supported the arguments put forward in the position paper of the Pacific Island Forum States that the Meeting of States Parties should consider issues with respect to article 4 of annex II and take such a decision on the starting date for the calculation of the 10-year time period for making submissions, which would ameliorate the difficulty in complying with the 10-year deadline envisaged in the Convention.

73. The Meeting generally supported a step-by-step approach to the issues raised with respect to article 4 of annex II to the Convention. The first step was to address the issue of selecting the date for calculating the 10-year time limit, which could be done at the present Meeting of States Parties. The second step was to deal with the issue of a possible extension of the 10-year time limit, which required a sound legal solution on the substance of the matter and on the procedures to be followed.

74. Many delegations agreed that the starting date should be 13 May 1999, the date of adoption of the Scientific and Technical Guidelines, which also marked the completion of the three basic documents of the Commission; the other two being its Rules of Procedure and its modus operandi. They pointed out that the Guidelines gave clear and detailed guidance to States as to the procedures to be followed in the preparation of submissions to the Commission and to the particulars that would be expected to be included in such submissions. One delegation emphasized that

the adoption of the Guidelines was not a prerequisite or a condition for making submissions by States, and that States should avoid taking on any additional obligations not included in the Convention.

75. Some delegations pointed out that there was no legal consequence stipulated by the Convention if a State did not make a submission to the Commission. Several delegations underscored the principle that the rights of the coastal State over its continental shelf were inherent, and that non-compliance with the 10-year time period specified in article 4 of annex II would not adversely affect those rights, which did not depend on occupation, effective or notional, or any express proclamation, as stated in article 77, paragraph 3, of the Convention.

76. On the issue of a possible further extension beyond 10 years of the time period for making submissions to the Commission, as proposed by the Pacific Island Forum States (SPLOS/67, para. 8 (c)), several delegations recognized that such an extension would accommodate the needs of developing countries, which lacked the requisite expertise and resources to fulfil the requirements of article 4 of annex II within the prescribed period. A number of other delegations were of the view that at the current stage the adoption of the decision that the 10-year period would not begin to run for any State Party, regardless of its date of ratification or accession, until the date of adoption of the Commission's Guidelines, would have already ameliorated substantially the situation for the first group of States by extending their deadline, in fact, for an additional five years. The delegations agreed that meanwhile further discussions were needed on the issue of the ability of States, particularly developing States, to fulfil the requirements of article 4 of annex II to the Convention.

77. Some delegations were of the view that a coastal State that for economic, financial, or technical reasons was able to make only a partial submission within the 10-year time period should be viewed as having complied with the requirements of article 4 of annex II to the Convention.

78. The procedural issue of how to give effect to any decision extending the 10-year time period was also discussed. Four possible procedures were put forward. They were similar to what was outlined in the background paper by the Secretariat (SPLOS/64, paras. 71–75):

(a) An amendment in accordance with article 312 of the Convention;
(b) An amendment by means of the simplified procedure provided for in article 313;
(c) An agreement relating to the implementation of article 4 of annex II to the Convention; and
(d) A decision by the Meeting of States Parties along the lines of the procedure used by the Meeting regarding the postponement of the election of the members of the Tribunal and of the members of the Commission on the Limits of the Continental Shelf.

79. Many delegations were of the view that it fell within the competence of the Meeting of States Parties to adopt by consensus a decision expressing general agreement on the starting date for calculating the 10-year time period. Such a decision, they stated, would be of a procedural nature similar to the ones the Meeting had taken with respect to the postponement of the election of members of the

International Tribunal for the Law of the Sea and that of the members of the Commission. However, one delegation was of the view that the issue of the starting date was of direct relevance to the rights and obligations of States Parties to the Convention and therefore could not be considered as simply procedural.

80. With respect to a possible decision, the Chairman of the Commission on the Limits of the Continental Shelf stated that the 10-year deadline was a matter that fell within the competence of States; the Commission would be guided by whatever deadline was decided upon by the States Parties on the condition that the decision was legally correct. In response, some delegations, while acknowledging the independent nature of the Commission, pointed out that the Commission had been established by the Convention, of which the States Parties were the custodians. In that regard, they underscored the importance of any decision taken by the Meeting of States Parties on the matter.

81. In the light of the discussions and of a proposal put forward by Papua New Guinea, an open-ended Working Group was convened by the President. A draft decision was prepared by the Group (SPLOS/L.22), which was subsequently adopted by the Meeting of States Parties (SPLOS/72). The decision provides that, for a State for which the Convention entered into force before 13 May 1999, the date of commencement of the 10-year time period for making submissions to the Commission is 13 May 1999.

82. There was general agreement that States that were in a position to do so should make every effort to make submissions within the time period established by the Convention. In that regard, it was pointed out that the deferral of the deadline should not place an undue burden on those States that were ready to make their submissions by requiring them to present new data at that time.

83. Many delegations pointed out that the issue of training and transfer of technology was closely linked to the discussions on the time period for making submissions to the Commission. Some stated that capacity-building was of vital importance irrespective of the decision on the starting date for the 10-year period and a possible decision on the extension of that period.

84. Many delegations noted with satisfaction the establishment of a trust fund and the contribution that had been made to it so far (see paras. 94 and 95 below). They expressed the hope that further contributions would be made to the fund. Referring to the statement by the Chairman of the Commission on the lack of a mandate on the part of the Commission to conduct training, some delegations emphasized the need for the relevant institutions to actively support training activities. It was suggested that cooperation between the Commission, regional centres of excellence, and the United Nations University should be pursued.

9. MATTERS RELATED TO ARTICLE 319 OF THE UNITED NATIONS CONVENTION ON THE LAW OF THE SEA

85. At the tenth Meeting, Chile had proposed that the Meeting of States Parties consider issues relating to the implementation of the Convention and that, to that end, the Meeting should receive a report every year from the Secretary-General on issues of a general nature that had arisen with respect to the Convention (see SPLOS/CRP.22 and SPLOS/60, paras. 73–78).

86. On the issue of reporting, the Under-Secretary-General for Legal Affairs, The Legal Counsel, Hans Corell, made a statement at the opening of the Meeting of States Parties on the mandate given to the Secretary-General by the General Assembly with regard to the preparation of comprehensive reports on developments relating to the implementation of the Convention, the law of the sea and ocean affairs, as reflected in the relevant General Assembly resolutions. He recalled that in paragraph 15 of its resolution 49/28 of 6 December 1994 the General Assembly had specified that the Secretary-General should prepare annually a comprehensive report, for the consideration of the General Assembly, on developments relating to the law of the sea which could also serve as a basis for reports to all States Parties to the Convention, the International Seabed Authority and competent international organizations, which the Secretary-General was required to provide under article 319 of the Convention. He highlighted the comprehensive nature of the issues addressed in the most recent report of the Secretary-General on oceans and the law of the sea (A/56/58), which included those that had arisen with respect to the Convention. He noted also that, in addition to the report of the Secretary-General, the three institutions established under the Convention, namely the International Tribunal for the Law of the Sea, the International Seabed Authority, and the Commission on the Limits of the Continental Shelf, also reported to the Meeting of States Parties.

87. Some delegations stated that the responsibility of the Secretary-General to report on matters of a general nature was clearly established in article 319, paragraph 2 (a), of the Convention and that the Secretary-General should submit a report to the Meeting of States Parties, as he had in 1996 (SPLOS/6). One delegation pointed out that in doing so, the Secretary-General should draw the attention of the States Parties to issues that had arisen with regard to the Convention, including issues of nonconformity with its provisions. However, some delegations were of the view that the role of the Secretary-General was not to raise issues of nonconformity with the Convention, particularly with respect to national laws; only States Parties should consider such matters.

88. Many delegations expressed their support for an expanded role for the Meeting of States Parties beyond budgetary and administrative matters. In their view, the Meeting had the competence to discuss issues of implementation of the Convention bearing in mind the need to avoid duplication with the work in other forums. The decision regarding the date of commencement of the 10-year period for making submissions to the Commission on the Limits of the Continental Shelf was cited by some as an example of the role that the Meeting of States Parties had already played in the implementation of the Convention. The Meeting also provided an opportunity to examine reports on the activities of the International Seabed Authority, the Commission on the Limits of the Continental Shelf and the International Tribunal for the Law of the Sea and, as one delegation pointed out, to make recommendations to the General Assembly. Another delegation expressed the view that arguments in favour of a more substantive role for the Meeting of States Parties did not represent an attempt to give the Meeting decision-making powers not provided for in the Convention.

89. Other delegations stated that the interpretation of article 319 of the Convention did not support an expanded role for the Meeting of States Parties. The mandate of the Secretary-General in article 319, paragraph 2 (e), to convene necessary

meetings of States Parties was qualified in two respects: first, it was limited to meetings that were "necessary"; and secondly, the mandate was linked to the provisions of the Convention, which clearly specified the matters to be considered by Meetings of States Parties, i.e., the election of the members of the Commission on the Limits of the Continental Shelf, the election of the members of the Tribunal and the consideration and approval of the budget of the Tribunal. In their view, no other provisions of the Convention either required action or acknowledged the possibility of action by the Meeting of States Parties. A strict reading of the text of article 319, paragraph 2 (e), therefore suggested that the provision should not be interpreted to mandate or authorize the Secretary-General to convene a Meeting of States Parties for the purpose of undertaking a far-reaching review of general matters related to the Convention. Moreover, the negotiating history of the article demonstrated that proposals to establish a mechanism for the periodic review of the Convention had failed to attract sufficient support. If the drafters had intended to do so, they would have, as in the case of other conventions, expressly provided for a monitoring and review role for the Parties. In addition, the implementation of the Convention involved a number of United Nations bodies and the General Assembly was the only forum with the overall competence to review the implementation of the Convention. Furthermore, the Assembly had established the Consultative Process in order to facilitate its annual review of developments in ocean affairs.

90. While recognizing the oversight role of the General Assembly, a number of delegations expressed the view that the Meeting of States Parties nevertheless had the right to discuss issues of implementation of the Convention, since it was, as one delegation stated, an autonomous body and the "supreme organ" for the implementation of the Convention. Another delegation expressed the view that, since the Consultative Process only dealt with ocean affairs, it would be necessary for the Meeting of States Parties in the future to decide on legal issues regarding the implementation of the Convention. On the relationship between the Meeting of States Parties and the Consultative Process, some delegations explained that issues of implementation of the Convention might be raised at the Process, following which they might need to be addressed by the Meeting.

91. With regard to the future work of the Meeting of States Parties, a number of delegations expressed their support for the proposal of Chile to include an agenda item entitled "Implementation of the United Nations Convention on the Law of the Sea." One delegation, noting that the Meeting of States Parties would not need to examine all issues of implementation, suggested instead the title "Issues which require consideration by the Meeting of States Parties." Another delegation proposed that any State Party wishing to include an item in the agenda of the Meeting should first circulate the proposal, through the Secretary-General, to all Parties before the next Meeting. The Meeting of States Parties would then decide whether or not to discuss the item. However, other delegations stated that they did not see the need for a special item on the implementation of the Convention and that States could raise any issue they deemed relevant under the agenda item entitled "Other matters."

92. In the light of the various views expressed, the Meeting of States Parties decided to retain the current agenda item entitled "Matters related to article 319 of the United Nations Convention on the Law of the Sea" for its next Meeting.

8. OTHER MATTERS

A. Trust funds

93. The tenth Meeting of States Parties had decided to recommend to the General Assembly at its fifty-fifth session, the establishment of three trust funds to be financed through voluntary contributions (SPLOS/60, paras. 47, 57, and 60).

94. The President informed the Meeting that the establishment of all three trust funds had been approved by the General Assembly at its fifty-fifth session (resolution 55/7, paras. 9, 18, and 20) and the funds were now established and operational. The trust funds are as follows:

(a) a voluntary trust fund to assist States in the settlement of disputes through the International Tribunal for the Law of the Sea;
(b) a voluntary trust fund to provide training for technical and administrative staff, and technical and scientific advice, as well as personnel, to assist developing States, in particular the least developed countries and small island developing States, for the purpose of desktop studies and project planning, and preparing and submitting information under article 76 and annex II to the Convention in accordance with the procedures of the Scientific and Technical Guidelines of the Commission on the Limits of the Continental Shelf; and
(c) a voluntary trust fund for the purpose of defraying the costs of participation of the members of the Commission on the Limits of the Continental Shelf from developing States in the meetings of the Commission.

95. The President also informed the Meeting that the United Kingdom had made two contributions to the trust fund for the International Tribunal for the Law of the Sea and that Norway had made a contribution to the trust fund to provide training for technical and administrative staff and to provide technical and scientific advice as well as personnel, to assist developing countries to prepare submissions and submit information under article 76 and annex II to the Convention.

96. On behalf of the Meeting of States Parties, the President thanked both Governments for their generous contributions and urged other States to make contributions to the trust funds.

B. Statement by a Representative of a Non-Governmental Organization Regarding Seafarers

97. In accordance with rule 18, paragraph 4, of the Rules of Procedure for Meetings of States Parties (SPLOS/2/Rev.3/Add.1), the Seamen's Church Institute was invited by the Meeting to participate as an observer. In his statement, the representative of the Institute drew attention to the problems currently faced by seafarers, in particular the growing threat of pirate attacks, abandonment, and the erosion of traditional seafarers' rights. With a particular focus on the latter two issues, he pointed out that crews were often abandoned by insolvent shipowners and that there had also been cases where the crews had been unfairly detained in response to

pollution incidents because the shipowner had not paid the coastal State concerned. Abandonment was devastating to crews, who in many cases could not afford to pay litigation costs and legal fees or support themselves during protracted legal procedures that would be required to avail themselves of the protection of the law. Many crews were abandoned in ports where there was insufficient community support to sustain them or where the legal system could not provide effective relief.

98. The representative of the Seamen's Church Institute also expressed concern about recent trends that attempted to dilute the traditional rights of seafarers to free medical care. In one case a mariner had been deprived of basic medical care and had been left to die under circumstances that appeared to be motivated by financial considerations. When the health, safety, or welfare of mariners was in jeopardy, the international community looked to the United Nations Convention on the Law of the Sea to protect them. The most fundamental function of the Convention was to provide order and predictability for people in the marine sector. When one flag State did not honour its obligations under the Convention, all States Parties were affected. In such situations, and especially those where persons' rights were involved, the community of nations, as well as individual States, had to step in to protect the seas' most valuable resource: the human beings who live and work on ships.

C. Statement by the President at the Closure of the Eleventh Meeting of States Parties

99. In his closing statement, the President reviewed the work that had been achieved during the Meeting. He noted that Judge Xu Guangjian from China had been elected to serve the remainder of the term of the late Judge Lihai Zhao. The budget of the Tribunal for 2002 had been adopted very expeditiously owing to the excellent proposal prepared by the Tribunal and the cooperation of all delegations. The Meeting had also approved the level of compensation for judges ad hoc of the Tribunal. The proposal by Japan concerning the adjustment of the scale of assessments for contributions to the budget of the Tribunal would be taken up at the next Meeting. The President drew attention to the need to ensure that the assessed contributions to the Tribunal were paid in full and in a timely fashion, so that the Tribunal could discharge its functions effectively and efficiently. The same also applied in respect of the payment of assessed contributions to the International Seabed Authority.

100. Good progress, he noted, had been made on outstanding issues relating to the Rules of Procedure for Meetings of States Parties. A new rule had been adopted providing for an open-ended working group on financial and budgetary matters, which would make recommendations to the Meeting. The Meeting also had made progress with regard to the Financial Regulations of the Tribunal. However, owing to time constraints, it would be necessary to take up the item again at the next Meeting.

101. The President observed that the Meeting had had an extremely interesting discussion on the 10-year time period for making submissions to the Commission on the Limits of the Continental Shelf. Noting the adoption of the Scientific and Technical Guidelines on 13 May 1999 and bearing in mind the difficulties encountered by some States, in particular developing States, in complying with their obliga-

tions under article 4 of annex II to the Convention, the Meeting had decided that in the case of a State Party for which the Convention entered into force before 13 May 1999, the 10-year period was taken to have commenced on 13 May 1999. He stressed that States that were in a position to do so should make every effort to make their submission to the Commission as soon as possible.

102. The discussions on the issue of the extension of the 10-year period had clearly indicated that a more in-depth consideration was necessary. The decision of the Meeting to keep under review the more general issue of the ability of States, particularly developing States, to fulfil the requirements of article 4 of annex II reflected the discussions on the matter.

103. The President highlighted the importance of training to enable States to prepare their submissions to the Commission. Recalling what the Chairman of the Commission had said regarding the need for interested Governments and relevant scientific organizations to provide training, the President noted that the Meeting of States Parties should explore the ways and means of organizing training.

104. He noted that an interesting discussion had taken place on matters related to article 319 of the Convention. Although there were opposing views, many delegations had supported an expanded role for the Meeting of States Parties.

105. He also noted the statement by the representative of the Seamen's Church Institute and thanked him for drawing the attention of the Meeting to the plight that seafarers often faced.

106. The President outlined the agenda items for the twelfth Meeting (see para. 109 below) and noted that the year 2002 would mark the twentieth anniversary of the signing of the Convention. He stated that he would consult on how the Meeting could commemorate the occasion.

107. In closing, he thanked all delegations for their cooperation and assistance. He also offered his best wishes to Mr. Gritakumar Chitty, the outgoing Registrar of the Tribunal, and to his family.

D. Dates and programme of Work for the Twelfth Meeting of States Parties

108. The twelfth Meeting of States Parties will be held in New York from 13 to 24 May 2002.

109. The twelfth Meeting will have on its agenda, inter alia, the following items:

(a) Report of the International Tribunal for the Law of the Sea to the Meeting of States Parties covering the calendar year 2001 (rule 6 of the Rules of Procedure for Meetings of States Parties);
(b) Draft budget of the International Tribunal for the Law of the Sea for 2003;
(c) Scale of assessments for the contribution of States Parties to the budget of the International Tribunal for the Law of the Sea;
(d) Consideration of the Financial Regulations of the International Tribunal for the Law of the Sea; Report of the External Auditors for the financial year 2000, with financial statements of the International Tribunal for the Law of the Sea as of 31 December 2000;
(f) Election of seven members of the International Tribunal for the Law of the Sea;

(g) Election of 21 members of the Commission on the Limits of the Continental Shelf;
(h) Issues with respect to article 4 of annex II to the United Nations Convention on the Law of the Sea;
(i) Matters related to article 319 of the United Nations Convention on the Law of the Sea; and
(j) Other matters.

Selected Documents and Proceedings

Report on the Work of the United Nations Open-ended Informal Consultative Process Established by the General Assembly in its Resolution 54/33 in Order to Facilitate the Annual Review by the Assembly of Developments in Ocean Afairs at its Second Meeting, Held at United Nations Headquarters from 7 to 11 May 2001†

Letter dated 22 June 2001 from the Co-Chairpersons of the Consultative Process addressed to the President of the General Assembly

Pursuant to General Assembly resolution 54/33 of 24 November 1999, you reappointed us as the Co-Chairpersons of the Open-ended Informal Consultative Process on ocean affairs established to facilitate the review by the General Assembly, in an effective and constructive manner, of developments in ocean affairs by considering the report of the Secretary-General on oceans and the law of the sea and by suggesting particular issues to be considered by the General Assembly, with an emphasis on identifying areas where coordination and cooperation at the intergovernmental and inter-agency levels should be enhanced.

We now have the honour to submit to you the attached report on the work of the Consultative Process at its second meeting, which was held at United Nations Headquarters from 7 to 11 May 2001.

The Consultative Process has suggested a number of issues for consideration by the General Assembly and, in accordance with paragraph 3 (h) of resolution 54/33 and bearing in mind General Assembly resolutions 55/7 and 55/8 of 30 October 2000, has proposed a number of elements for the consideration of the General Assembly in relation to its resolutions under the agenda item entitled "Oceans and the law of the sea."

These elements are, of course, not intended as an exhaustive list of material relevant to the General Assembly's consideration of the item "Oceans and the law of the sea."

---

†EDITORS' NOTE.—This document was provided by the United Nations Division for Ocean Affairs and the Law of the Sea (DOALOS). The source is Report on the work of the United Nations Open-ended Informal Consultative Process established by the General Assembly in its resolution 54/33 in order to facilitate the annual review by the Assembly of developments in ocean affairs at its second meeting, held at United Nations Headquarters from 7 to 11 May 2001, A/56/121, 22 June 2001. Fifty-sixth session. Available online: <http://www.un.org/depts/los/index.htm>.

836   United Nations Open-ended Informal Consultative Process

In the light of the terms in which the General Assembly referred to the Consultative Process in its resolution 55/7, this year the Consultative Process has been referred to as the "United Nations Open-ended Informal Consultative Process established by the General Assembly in its resolution 54/33 in order to facilitate the annual review by the Assembly of developments in ocean affairs." Some delegations wished, in addition, to stress the link between the Consultative Process and item 42 of the preliminary list of items to be included in the provisional agenda of the fifty-sixth session of the General Assembly: "Oceans and the law of the sea." Some other delegations did not share this view. Nevertheless, it was noted that the General Assembly, in establishing the Consultative Process, in its resolution 54/33, had recalled that the United Nations Convention on the Law of the Sea set out the legal framework within which all activities in the oceans and seas must be carried out, and with which those activities should be consistent, as recognized also by the United Nations Conference on Environment and Development in chapter 17 of Agenda 21, and had also recognized the importance of maintaining the integrity of the Convention.

(*Signed*) Tuiloma Neroni Slade and Alan Simcock, Co-Chairpersons

CONTENTS

Part A. Issues to be suggested, and elements to be proposed to the General Assembly
Part B. Co-Chairpersons' summary of discussions
Part C. Issues for consideration for possible inclusion in the agendas of future meetings

Annexes
1.   Statement by Mr. Hans Corell, Under-Secretary-General for Legal Affairs, The Legal Counsel
2.   Statement by Mr. Nitin Desai, Under-Secretary-General for Economic and Social Affairs

PART A

Issues to be suggested, and elements to be proposed to the General Assembly.

Issue A

Further progress on the prevention, deterrence, and elimination of illegal, unreported, and unregulated fishing
1.   It is proposed that the General Assembly should welcome the adoption by the Food and Agriculture Organization of the United Nations (FAO) Committee on Fisheries of the International Plan of Action to Prevent, Deter and Eliminate Illegal, Unreported and Unregulated Fishing and should invite States to take all necessary steps to implement it effectively.

Issue B

Protecting the marine environment from pollution and degradation from land-based activities
2.   It is proposed that the General Assembly should welcome the recent progress on the implementation of the Global Programme of Action for the Protection of the Marine Environment from Land-Based Activities (GPA), invite States to participate in the intergovernmental review of GPA which is to be held in Montreal, Canada, from 26 to 30 November 2001, and invite the relevant international and regional organizations and international financial institutions to make inputs to the review in order to overcome the obstacles to the full establishment of the clearing-house mechanism under the Global Programme of Action and the development of regional and national plans of action.

Issue C

"Science for sustainable development": the importance of marine scientific research for the objectives of sustainable development
3.   Marine science, and its supporting technologies, through improving knowledge and applying it to management and decision-making, can make a major contribution to eliminating poverty, to ensuring food security, to supporting human economic activity, to conserving the world's marine environment and to helping predict, mitigate the effects of and respond to natural events and disasters, and generally, to promoting the use of the oceans and their resources for the objective of sustainable development.
4.   Because of the wide range of different circumstances and characteristics in different marine regions, there needs to be, where appropriate, a strong regional focus on international cooperation, including the support of the international community, in promoting marine scientific research and deploying marine scientific knowledge and technology; this regional focus needs to reflect the linkages to large marine ecosystems.

5. Effective marine science does not consist simply of a series of one-off projects; sustained efforts are needed to monitor and understand the development of the highly dynamic marine environments and to apply that knowledge to prediction and to management decisions.

Issue D

Strengthening international cooperation at the regional level

6. To ensure an intersectoral research approach, there is a need to establish or strengthen, as appropriate, regional cooperation, including that between relevant regional fisheries organizations and arrangements, regional seas programmes and other regional marine environment bodies, including their scientific and technical advisory bodies, and the regional marine science organizations, including those under the aegis of the Intergovernmental Oceanographic Commission (IOC).

7. Such cooperation is also proposed to include working, where appropriate, with global organizations, such as the FAO, the International Maritime Organization (IMO) and the World Meteorological Organization (WMO) and with regional projects under the aegis of IMO. The aim of such cooperation should be both the most effective use of the available resources, particularly by the avoidance of duplication, and the achievement of a holistic approach to the scientific study of the oceans and their resources.

8. To achieve better dialogue and cooperation, regional fisheries, and environmental and scientific bodies could arrange meetings of their representatives.

9. States should be encouraged to fulfil their relevant obligations under international agreements. In particular, the regional centres foreseen by Part XIV of the United Nations Convention on the Law of the Sea (UNCLOS) (articles 276 and 277) should be established, with the technical assistance from the IOC and FAO, where they do not exist, and should be strengthened where they already exist.

10. To ensure a proper linkage between global and regional levels, the relevant bodies of the United Nations system should ensure appropriate interactions in marine science between them and the collaborative work of regional fisheries, environmental and scientific bodies, or regional centres; the Intergovernmental Oceanographic Commission should act as a focal point for those interactions.

Issue E

Establishing better links between marine scientists and policy makers and managers

11. It is essential to achieve an integrated approach to national marine policy by all the many public authorities that are necessarily involved in oceans management, in accordance with programme area A of chapter 17 of Agenda 21.

12. To achieve the effective application of marine scientific knowledge and technology, it is essential that national and regional institutions, systems and approaches are developed, with the support of relevant global bodies which can draw on their experience in this field, so as to ensure that the results of marine science can be understood, assimilated and used by decision makers and resource managers, and that decisions drawing on marine science take, where applicable, full account of socioeconomic factors and traditional ecological knowledge.

13. For these purposes, as part of the collaborative work of regional fisheries, environmental and scientific bodies, exchanges of experience among public officials from participating States should be organized, with the assistance of the Intergovernmental Oceanographic Commission, the Food and Agriculture Organization of the United Nations and other relevant international bodies, where appropriate.

Issue F

Proper planning of marine science projects and better implementation of the United Nations Convention on the Law of the Sea

14. The proper planning of marine science projects, whether basic or applied, should, among other things, be based upon the specific circumstances and needs of the local communities and national priorities and take account of the strategies developed by regional intergovernmental cooperation and the global context.

15. The consent regime under Part XIII of UNCLOS is the basis of all marine scientific research by third States in maritime areas under the national jurisdiction of coastal States. There is, however, a need to develop the general scientific criteria and guidelines referred to in article 251 of UNCLOS as well as national procedures based on a standard approach for seeking and granting consent as provided for in Part XIII, particularly in article 246.

16. There is an urgent need for cooperation at the international level to address the issue of the acquisition and transfer of marine scientific data to assist coastal developing States.

17. There is an urgent need to develop means to protect instruments and equipment deployed at sea for marine scientific research from vandalism and accidental damage.

18. The Intergovernmental Oceanographic Commission should be invited to request its Advisory Body of Experts on the Law of the Sea (ABE-

LOS) to work, in close cooperation with the Division for Ocean Affairs and the Law of the Sea, Office of Legal Affairs, of the United Nations Secretariat, on the development of procedures under Part XIII of UNCLOS. States could consider nominating a suitable regional intergovernmental cooperative body as their common focal point under this consent regime, where this helps their particular circumstances. When this is done, such information should be published in the Law of the Sea Bulletin of the Division for Ocean Affairs and the Law of the Sea.

19. The Intergovernmental Oceanographic Commission and the World Meteorological Organization should be invited to consider, with the assistance of International Hydrographic Organization (IHO), how protection might be provided for moored and drifting scientific instruments and equipment on the high seas.

Issue G

Exchange and flow of data

20. It is important that the knowledge derived from marine scientific research and monitoring is made available to those who need it, especially to developing countries. Where this information has been collected under the consent regime of Part XIII of UNCLOS, it is also essential that the rights of the coastal State under articles 248 and 249 are respected, in particular those under article 249, paragraph 1 (d).

21. It is equally important that such information is made available to those who need it, especially to developing countries, at the regional and global levels, in a consistent data format and by means of information on where the results of the research can be found.

22. The Intergovernmental Oceanographic Commission should be invited to request its Committee on International Oceanographic Data Exchange (IODE) to expand its work on data formats to include meta-data (information on where to find data).

23. Relevant international bodies should be invited to consider the questions of intellectual property rights in relation to the marine scientific research regime established by Part XIII of UNCLOS.

Issue H

Capacity-building for marine science and technology

24. Bearing in mind the importance of marine science for eliminating poverty, for ensuring food security, for supporting human economic activity, for conserving the world's marine environment, for helping predict, mitigate the effects of and respond to natural events and disasters, and for pro-

moting the use of the oceans and their resources with the objective of sustainable development, it is essential to build the capacities, in particular in developing countries, to conduct marine scientific research.

25. The development of human resources is the foundation to ensure a better understanding of marine science and technology and their potential. In developing countries, the fostering of these national capabilities presents special challenges, given the scarcity of financial resources and the reduced domestic awareness of the overall potential of marine resources. International cooperation, through bilateral, regional and international financial organizations and technical partnerships has played a key role in enhancing capacity-building activities for the transfer of environmentally sound technology associated with the sustainable development of marine resources, in particular in developing countries.

26. The general programmes for capacity-building may include, inter alia:
   (a) Sustaining efforts toward developing the necessary skilled personnel, both by encouraging individuals to engage in marine science and by providing the necessary training and experience, including under the possibility of serving as observers under the right referred to in article 249 of UNCLOS of the coastal State to participate or be represented on board research vessels;
   (b) Providing the necessary equipment, facilities and vessels, together with the essential infrastructure, such as electricity; to this end, the relevant international organizations, international financial institutions and the donor community should review their investment programmes to ensure that marine science is given adequate priority;
   (c) Ensuring the development of the necessary skills and techniques, both for the efficient and effective use of equipment, and for implementing the Part XIII regime and for adopting and enforcing the necessary implementing provisions, as well as for interpreting scientific results and for their publication and dissemination so that they can be applied by decision makers to be presented to the wider public; and
   (d) Transferring environmentally sound technologies, in accordance with Part XIV of UNCLOS and the future programme regarding the implementation of Agenda 21, together with the provision to developing countries of financial and technical assistance for this purpose.

Issue I

Strengthening global action to deliver effective marine science

27. Bearing in mind the importance of marine science for eliminating poverty, for ensuring food security, for supporting human economic activity,

for conserving the world's marine environment, for helping to predict and mitigate the effects of and respond to natural events and disasters, and for promoting the use of the oceans and their resources with the objective of attaining sustainable development, there is an essential need for a clear focal point for international cooperation on marine science. The extension of the mandate of the Intergovernmental Oceanographic Commission to cover ocean sciences and services, embodied in the 1999 revision of its statutes, should be welcomed and encouraged.

28.   The United Nations Atlas of the Oceans, a project being developed by the Administrative Committee on Coordination (ACC) Subcommittee on Oceans and Coastal Areas (SOCA) to bring together existing marine scientific knowledge, should be welcomed as providing a means to integrate the marine scientific knowledge held in the databases within the United Nations system and a basis for the further development of means to improve access to the world's marine scientific knowledge by those who need it.

29.   The United Nations Educational, Scientific and Cultural Organization (UNESCO) should be requested to strengthen the Intergovernmental Oceanographic Commission so that it has the resources needed to promote effective international cooperation on marine science and to carry out the tasks set out in the present conclusions.

30.   The relevant bodies of the United Nations system, with the Intergovernmental Oceanographic Commission as the focal point, should review the aspects of their programmes that are relevant to marine science to ensure that appropriate priority is given on a consistent basis.

Issue J

General policy on marine science

31.   The twenty-first century will be the era for the oceans. Humankind will need to devote ever greater effort to understanding, developing, and conserving the oceans, and the oceans will play an ever greater role in the development of human society and economy. Understanding the oceans thoroughly, protecting the marine environment effectively, and achieving the use of the oceans and their resources for the objective of sustainable development will become ever more important tasks for States.

32.   The approach to understanding the oceans needs to be integrated, interdisciplinary, and intersectoral. The ecosystem approach needs to be part of the global context of marine scientific research.

33.   If the information resulting from marine scientific research and monitoring is to play its proper role in ensuring that important decisions are properly informed, that information must be available and reliable. The need for availability implies that it should be accessible through appropriate data centres, such as that of the International Council for Science (ICSU).

The need for reliability implies equally the need for quality assurance of the data produced from any marine scientific research.

34.   The Global Ocean Observation System (GOOS), coordinated by the Intergovernmental Oceanographic Commission in collaboration with other agencies, should be developed in a balanced way through the implementation of its various modules dealing with ocean and climate (of which the Array for Real-time Geostrophic Oceanography (ARGO) project is an example), marine pollution, and coastal zones, so as to respond to the diversity of the requirements of Member States and other users.

35.   There should be dialogues at the national, regional and global levels, as appropriate, between those responsible for marine policy decisions and those responsible for organizing marine scientific research programmes, in order to establish, within each appropriate area, the issues on which scientific advice is needed and the best means to provide it, taking particular account, in international cooperation, of the issues important to coastal developing States and their needs for capacity-building and transfer of technology.

36.   When marine scientific research and monitoring projects are being set up, appropriate arrangements should be established for submitting the data to relevant national, regional or global data centres, and consideration should be given to the appropriate level of quality assurance for the data to be produced.

Issue K

Interactions between the atmosphere and the oceans

37.   The interactions between the atmosphere and the oceans are fundamental for life, both on land and in the sea. Understanding the interactions between the atmosphere and the oceans is a crucial step toward understanding the way in which the oceans work and, therefore, toward assessing what can be done.

38.   Scientific understanding of the interactions between oceans and the atmosphere is, however, not enough. It is also essential both that decision makers should be aware of the implications of that understanding, and that it is properly presented to the public in general, so that they can contribute appropriately to decision-making. It is equally important that the options for management decisions are clearly presented. This scientific understanding can also be translated into practical use to increase the adaptive capacity of the community, especially in developing countries.

39.   International action to promote marine scientific research, whether by bodies of the United Nations system or by other forms of intergovernmental cooperation, should aim to increase understanding of the ocean/atmosphere interface and its effects on living marine resources and the coastal

zone and its communities, together with the scientific understanding of the other factors needed for the integrated ecosystem-based approach to the management of oceans and coastal areas and for the safe execution of maritime operations. These categories are not exclusive but shade into one another. International actors should also seek to address the disparity and availability of data, particularly meteorological data, in the different regions of the world.

40. New innovative projects should be welcomed and encouraged where, through international cooperation, they can provide an understanding of the structure and mechanisms of the circulation systems of the oceans and result in prompt and transparent sharing of the resulting information by as wide a range of users as possible.

41. Equally, such projects should, from the start, aim at the effective use by all States of the information generated, and should therefore be designed in such a way, and be accompanied by such capacity-building and transfer of technology, as will enable developing countries to make effective use of that information.

Issue L

The needs for scientific understanding for the management of marine ecosystems

42. The management of marine ecosystems is driven, inter alia, by the needs to eradicate poverty, to support economic prosperity, to safeguard food security and to conserve biodiversity. It requires a knowledge of the dynamics of the ecosystems, in relation to both living marine resources and biogeochemical factors. This must involve understanding, on the one hand, the status and trends of stocks of living marine resources, their location, quantification and long-term sustainable yield, the methods of fisheries management, and, on the other hand, the factors affecting water quality, including eutrophication, waste dumping, and the source and fate of contaminants and their eco-toxicology. These factors are relevant to questions of pollution of both the seas and freshwater resources. The development of an ecosystem approach to ocean management should bring together monitoring and basic and applied research by both the fisheries science community and the ecological science community. At the global level, FAO should work together with relevant global and regional organizations to develop this concept.

43. Associated themes also requiring study include the scientific understandings needed for crisis management and for carrying out environmental impact assessments in relation to fragile marine environments, the introduction of non-native species, the impacts of pollutants from vessels and from land-based sources, the economic, environmental, and social impacts of sub-

sidies and their effects on fishing efforts and the role of coral reefs as a means of obtaining early warning of ecological modifications resulting from climate change and other pressures.

44.   Early action on these aspects may include, as appropriate, the further development of the concept of the ecosystem management approach and the completion of work on the draft International Plan of Action for Status and Trends Reporting on Fisheries.

45.   Since large parts of the marine biosphere are still unexplored, there should be a welcome and support for projects aimed at investigating the biological diversity of the high seas and the biota, biotopes, and habitats of the deep sea.

Issue M

The needs for scientific research for integrated management of oceans and coastal areas

46.   Integrated management is driven, inter alia, by the need to manage the development of human activities in a sustainable manner. It requires scientific inputs from many disciplines; in particular, in the coastal areas, it requires an understanding of the interactions of land and water, of the factors affecting water quality and of the basis for settling differences on the use of the coastal areas, in both their seaward and landward parts. It also must be based on scientific information for land and sea area planning decisions and the information needed to predict, mitigate the effects of and respond to natural events and disasters. In addition, work is needed to collect and maintain a local knowledge base.

47.   Early action is required to make progress on issues highlighted by the forthcoming intergovernmental review of the Global Programme of Action for the Protection of the Marine Environment from Land-based Activities, which should aim to identify areas where scientific research is needed and to investigate the problems that may arise from marine pollution by groundwater discharges to the sea from aquifers.

Issue N

The need for scientific research for maritime operations

48.   Marine scientific research and technological development for maritime operations are driven, inter alia, by the vital role of shipping in world trade. The fields that are particularly relevant are hydrography and meteorology (which is also relevant to the management of marine ecosystems and the integrated management of oceans and coastal areas) and the informa-

tion needed to predict, mitigate the effects of and respond to natural events and disasters.

49. There is a need to provide accurate and up-to-date charts of the world's oceans in order to promote maritime safety, and for assistance to build hydrographic capacity for those coastal States that do not yet have adequate hydrographic services.

50. The International Hydrographic Organization, in consultation with other relevant international organizations, provides the necessary assistance to States, in particular to developing countries, where lack of hydrographic capability undermines the safety of navigation, the protection of the marine environment or the enforcement of laws against piracy and armed robbery at sea.

51. The World Meteorological Organization and the Intergovernmental Oceanographic Commission should assist States, in particular developing countries, which do not have an adequate coverage of stations to monitor weather conditions and the sea state in waters under their jurisdiction to help overcome these problems, which can threaten maritime safety and undermine efforts to predict, mitigate the effects of and respond to extreme weather and sea events.

Issue O

General policy to promote cooperation and ensure coordination on combating piracy and armed robbery at sea

52. The recent rapid growth in incidents of piracy and armed robbery at sea, the harm that they cause to seafarers, and the threats that they pose to the safety of shipping and, consequently, to marine and coastal environments and to the trade carried by sea make it essential to give higher national and international priority to efforts to eradicate these crimes which are often the result of transnational crime.

53. States and relevant international organizations should therefore consider whether their policies and programmes give adequate emphasis to the needs to prevent piracy and armed robbery at sea, to provide a proper framework for response to these crimes and to ensure an effective response to such incidents as they occur.

54. Effective prevention of and response to piracy and armed robbery at sea will require the support of the international community by providing adequate support to developing countries, in particular to coastal and flag developing States, in the areas of transfer of technology and capacity-building in their efforts to prevent piracy and armed robbery at sea.

55. In this connection, international financial institutions and the donor community should review their programmes to determine whether adequate

provision is being made for investment in vessels and other equipment, including satellite tracking equipment.

56.   It is suggested that the General Assembly should reiterate the need for all States and relevant international bodies to work together to prevent and combat piracy and armed robbery at sea.

57.   The business sectors, such as chambers of shipping, maritime insurance industries and trade unions, can also play a useful role in support of the work led by the International Maritime Organization in combating piracy and armed robbery at sea.

Issue P

Prevention of piracy and armed robbery at sea

58.   Effective prevention will involve the flag States of ships sailing into areas where piracy and armed robbery at sea are known to be likely, the owners, masters, and crew of such ships, the coastal States in regions where incidents have occurred, and regional and international organizations concerned with shipping and crime prevention.

59.   The International Maritime Organization should be invited to consider requiring that seafarers in regions where incidents of piracy and armed robbery at sea are likely to occur receive training on precautions against incidents of piracy and armed robbery at sea under the International Convention on Standards of Training, Certification and Watchkeeping for Seafarers.

60.   Governments should ensure that their procedures for registering ships guard against fraudulent registrations, can give prompt and accurate responses about the details of ships which may be involved in incidents of piracy or armed robbery at sea and record details of such involvement. The International Maritime Organization should be invited to quickly complete its work on guidance on how this should be done. The work of IMO to require ships to be fitted with automatic identification systems is welcomed and any further relevant work should be encouraged.

61.   States should ensure that port authorities have appropriate measures in place to deter attempts at armed robbery within the ports, and that port staff have appropriate training in such measures. There should be a welcome and support for the work of the World Maritime University and of States in providing such training, or support for attendance at the World Maritime University, by way of capacity-building.

Issue Q

The framework for responses to piracy and armed robbery at sea

62.   Articles 100 to 107 and article 58, paragraph 2, of UNCLOS set out the proper framework for response to piracy. The Convention for the Sup-

pression of Unlawful Acts against the Safety of Maritime Navigation and its Protocol for the Suppression of Unlawful Acts against the Safety of Fixed Platforms located on the Continental Shelf ("the Rome Convention and Protocol") may also be used for the purpose of the prevention and suppression of armed robbery at sea.

63.   It is proposed that the General Assembly should reiterate its call for States that have not done so to become parties to the Rome Convention and Protocol. Where they have not already done so, coastal States should adopt legislation to ensure that there is a proper framework for responses to incidents of armed robbery at sea. It is suggested for convenience that the approach in such legislation should work together with the approaches adopted by other States in their region.

64.   All States should also ensure that the various public authorities, which are necessarily involved in dealing with incidents of piracy and armed robbery at sea, have a consistent approach to such incidents and are able to operate in an integrated manner.

Issue R

Response to incidents of piracy and armed robbery at sea

65.   Effective responses to incidents of piracy and armed robbery at sea must be based on measures for prevention, for reporting incidents and for enforcement, including the training of enforcement personnel and the provision of enforcement vessels and equipment. The ability of States to make such effective responses is substantially enhanced when regional cooperation arrangements are in place. The aim should be the creation of a network of contacts between the public authorities concerned, based on mutual trust, assistance and the fostering of a common approach to enforcement and capacity-building between States as to enforcement techniques, and to the investigation and prosecution of offences. Such regional cooperation arrangements may, in suitable cases, be strengthened by the conclusion of formal agreements. It is suggested that the General Assembly should welcome the initiatives of the International Maritime Organization and individual Governments to that effect.

66.   Since under-reporting of incidents of piracy and armed robbery leads to an underestimation of the seriousness of the problem, and consequently to enhanced risks, the owners and masters of ships should be encouraged to ensure that all incidents and threats of incidents are reported to the appropriate authorities and, through the flag State concerned, to the International Maritime Organization. The reporting procedures developed by IMO should be used to make it easy for reports to be submitted promptly.

67.   States in regions where incidents of piracy and armed robbery at sea are likely to occur should ensure that there are adequate arrangements in

place for receiving reports, communicating them without delay to all relevant authorities and alerting neighbouring States and ships in the area to incidents or threats of incidents. In this context, the cooperation of all States is essential.

68.   The States concerned should take measures that the personnel involved in all aspects of the response, including apprehension, investigation, prosecution and exchange of evidence, are properly trained. There should be a welcome and support for work by international organizations and States to provide such training or to support its provision by others. The International Maritime Organization should be invited to complete its work quickly on a code of practice for investigations. The International Law Enforcement Academy should be asked to consider what contribution it can make to the development of good practice and training in enforcement in this field. States that might have information about facts or circumstances which lead to a presumption of the possible occurrence of acts of piracy and armed robbery at sea should provide such information to the relevant States.

69.   Coastal States in regions where incidents of piracy and armed robbery are likely to occur should establish and keep up to date contingency plans for handling such incidents. In doing so, those States should, with the assistance of international and regional organizations, formulate arrangements to include handling of incidents which could result in major pollution of the marine environment.

Issue S

Coordination and cooperation within the United Nations system

70.   It is suggested that the General Assembly should continue to invite the Secretary-General to include in his annual report on oceans and the law of the sea material on the progress of the processes of collaboration and coordination between the relevant parts of the United Nations Secretariat and the United Nations system as a whole, as described in paragraph 8 of resolution 54/33 and paragraph 42 of resolution 55/7.

PART B

Co-Chairpersons' summary of discussions

Agenda item 1: Opening of the meeting
1.   The discussions at the first and the second plenary sessions of the second meeting of the United Nations Open-ended Informal Consultative

Process established by the General Assembly in its resolution 54/33 in order to facilitate the annual review by the Assembly of developments in ocean affairs were based on General Assembly resolutions 54/33, 55/7, and 55/8, the annual report of the Secretary-General on oceans and the law of the sea (A/56/58), as well as on other documents before the meeting, including written submissions by States and international organizations, in particular document A/AC.259/4 submitted by Norway.

2. The overall legal framework for the discussions was provided by the United Nations Convention on the Law of the Sea of 10 December 1982 and its two implementing Agreements,[1] while chapter 17 of Agenda 21 provided the programme of action for the sustainable development of oceans and seas, which was re-emphasized in decision 7/1 adopted by the Commission on Sustainable Development at its seventh session.

3. The discussions were opened, on behalf of the Secretary-General of the United Nations, by Mr. Hans Corell, Under-Secretary-General for Legal Affairs, The Legal Counsel, and Mr. Nitin Desai, Under-Secretary-General for Economic and Social Affairs.

4. In his introductory statement, Mr. Corell placed emphasis on the transition from the establishment of norms to their implementation, on challenges facing the developing States such as limited capacity, scarce resources and inadequate means of implementation and on the need for global responses and international coordination and cooperation to address problems of the oceans.

5. Mr. Desai focused in his introductory statement on the convergence of the legal and programmatic dimensions of international cooperation on matters relating to the oceans, on the shared interest of all nations in the future of the oceans and seas, and on the need to address global environmental issues. He also spoke about the connection of the Consultative Process to the World Summit on Sustainable Development to be held in September 2002 in Johannesburg, South Africa. (The texts of the statements by Mr. Corell and Mr. Desai are contained respectively in annexes I and II to the present report.)

6. In his opening statement, Ambassador Tuiloma Neroni Slade (Samoa), Co-Chairperson of the meeting, focused on marine science and technology as being fundamental for decision-making in all sectors. Capacity-building and development of information and skills to manage the oceans are integral to the issue of marine science and technology. He highlighted the need for clear and concrete ideas about how to obtain scientific information and, then, how to apply it.

---

1. Agreement relating to the implementation of Part XI of the Convention and Agreement for the Implementation of the Provisions of the Convention relating to the Conservation and Management of Straddling Fish Stocks and Highly Migratory Fish Stocks.

Agenda item 2: Approval of the format of the meeting and adoption of the agenda

7.   Mr. Alan Simcock (United Kingdom of Great Britain and Northern Ireland), Co-Chairperson of the meeting, presented the proposals of the Co-Chairpersons for the format and annotated agenda of the second meeting (A/AC.259/L.2). In the light of the results of informal consultations preceding the meeting[2] and the comments of some delegations, he proposed that the meeting adopt its format and annotated agenda with several amendments. The meeting then adopted by consensus the format and annotated agenda as amended (A/AC.259/5). In accordance with one of the amendments, the Consultative Process would henceforth be referred to as the "United Nations Open-ended Informal Consultative Process established by the General Assembly in its resolution 54/33 in order to facilitate the annual review by the Assembly of developments in ocean affairs."

Agenda item 3: Exchange of views on areas of concern and actions needed

The Consultative Process

8.   Delegations re-emphasized their support for the Consultative Process and expressed their readiness to contribute to its effectiveness and success. They highlighted the value of the integrated approach to all matters concerning oceans and seas and of intergovernmental and inter-agency cooperation and coordination. It was pointed out that strengthening coordination at all levels in matters related to the oceans and seas was the overriding purpose of the Consultative Process.

9.   Delegations noted with satisfaction the results of the first meeting of the Consultative Process, and the facts that General Assembly resolutions 55/7 and 55/8 had incorporated many elements resulting from it and that there had been some concrete progress in some of the areas discussed at the meeting. This, in their view, fully demonstrated the usefulness of the Consultative Process. In that connection, they expressed their appreciation to both chairpersons for their efforts and leadership.

10.   It was further noted that the Consultative Process represented a unique entity within the more formal and sectoral approach of the United Nations family. One delegation pointed out that the Process needed to embody a comprehensive approach based on shared goals, understanding, and information. One delegation stressed that the Process was a part of the consideration by the General Assembly of the agenda item "Oceans and the law of the sea" and emphasized its informal nature, stating that the Consultative Process should not be "institutionalized" in any way.

2. Three rounds of informal consultations were held, on 23 February, 23 March, and 4 May 2001.

11. Another delegation suggested that, in order to underscore the informal character of the Consultative Process and better record the nature and scope of the discussions, the consensus report emerging from the discussions should be restricted to broad elements and themes.
12. The forthcoming review by the General Assembly of the Consultative Process and of its effectiveness and utility in 2002 was mentioned as well.
13. One group of States expressed the view that it was important to avoid duplication of work and engaging in debates falling beyond the mandate of the Consultative Process. In that context, those States did not find it appropriate to consider the issues concerning the continental shelf and underwater cultural heritage.

Implementation of UNCLOS, the related Agreements and relevant international instruments

14. Many delegations reiterated that the United Nations Convention on the Law of the Sea was of strategic importance and provided the fundamental legal framework for all activities related to oceans and seas. The historic significance of the entry into force of UNCLOS and the needs to achieve universal participation in it, to preserve its integrity and to ensure its full implementation were noted as well. Together with chapter 17 of Agenda 21, UNCLOS was once again reconfirmed as the basis for the discussions on effective cooperation and coordination of matters relating to the oceans and seas.
15. Importance was attached by delegations to the need for cooperation and coordination at global and regional levels in implementing UNCLOS and to the necessity of enacting national legislation in order to implement the provisions of UNCLOS.
16. Some delegations also welcomed the recent progress in the pace of ratification of the Agreement for the Implementation of the Provisions of the Convention relating to the Conservation and Management of Straddling Fish Stocks and Highly Migratory Fish Stocks and noted that only two more ratifications were required for its entry into force.

Report of the Secretary-General
17. Many delegations expressed appreciation to the Secretary-General for the annual report on oceans and the law of the sea, highlighting its extensive and comprehensive nature and informational value. It was noted that the report was of central importance to the Consultative Process and its deliberations. The European Union noted the collection of information regarding legislative measures undertaken by States parties in implementing UNCLOS and welcomed the Secretary-General's idea that an analysis of the information received would appear in his next annual report, as part of an

overall assessment of the implementation of UNCLOS 20 years after its adoption.

Areas of focus
18.   Delegations expressed appreciation at the identification by the General Assembly of the two areas of focus for the second meeting of the Consultative Process and welcomed the fact that in the agenda of the meeting the Consultative Process had organized its discussions around those two areas, e.g., (a) marine science and the development and transfer of marine technology as mutually agreed, including capacity-building in this regard; and (b) coordination and cooperation in combating piracy and armed robbery at sea.
19.   It is to be noted that the annual report of the Secretary-General devotes a section (section VIII) and a subsection (subsection V.A), respectively, to the two topics. The report describes the legal regime for marine science and technology, as laid out in Parts XIII and XIV of UNCLOS, especially the "consent regime" for the conduct of marine scientific research in maritime areas under the sovereignty or jurisdiction of coastal States. The regime strikes a balance between the rights of coastal States to regulate and authorize the conduct of research in maritime zones under their jurisdiction and the rights of researching States to carry out research as long as it does not have any bearing on the exploration and exploitation of resources. The report describes the existing programmes on marine science and technology in the United Nations system. The report also addresses identified needs in marine science and technology, including the establishment of an administrative framework for marine scientific research activities and the development of national and regional marine science and technology centres.
20.   With respect to piracy and armed robbery against ships, the report expresses concern at the increasing number of incidents reported in recent years. Actions taken or envisaged at the global and the regional levels are described, especially those under the auspices of the International Maritime Organization. Recommended actions for Governments and for the shipping industry are also set forth in the report.
21.   The document by the delegation of Norway on "Marine science and the development and transfer of marine technology, including capacity-building" (A/AC.259/4) emphasizes that at the core of activating the marine science regime established by Part XIII of UNCLOS lies the adoption and implementation of national regulations relating to the conduct of foreign marine scientific research in waters under national jurisdiction and the identification of national focal points to coordinate such research activities. The document proposes a plan of action using Norwegian model legislation as an example to that end. The document also suggests a plan of action for

assisting developing countries in drawing up a scientifically based integrated ocean management regime.

22. In accordance with the format of the meeting, a discussion panel was to lead off the discussions in each area of focus by making short presentations on relevant questions. The Co-Chairpersons underlined that the proposed descriptions of each area of focus were intended to be the starting points for the discussions and aimed at identifying issues that the discussion panel might choose to consider. These descriptions are contained in appendices I and II to annexes I and II of the document entitled "Draft format and annotated provisional agenda" (A/AC.259/L.2). The description of the area of focus of marine science and technology is divided into two parts, part I dealing with improving structures and effectiveness and part II with priorities in marine science and technology.

23. In their opening statements, both Mr. Corell and Mr. Desai underscored the importance of the two areas for the effective implementation of UNCLOS and chapter 17 of Agenda 21.

24. Mr. Corell stated that the issues relating to marine science and technology that called for international coordination and cooperation included the unhampered conduct of marine scientific research, a better understanding of oceans and also of their interaction with the earth and the atmosphere, a more effective interface between scientific knowledge and decision-making, the development and transfer of marine technology, and the strengthening of marine science and technology capacity. Mr. Desai emphasized that while oceans were central to the problems of sustainable development, a better and shared understanding of oceans was central to the sustainable use and management of oceans. He pointed out that human knowledge about oceans was far more inadequate than about land. He identified certain key areas where human knowledge needed to be expanded, among them oceans and global climate change, biomass, fisheries and effects of marine pollution. He was of the view that marine science could be at the centre of international cooperation and coordination, including capacity-building.

25. Piracy and armed robbery at sea threaten the shipping industry and endanger the well-being of seafarers, Mr. Corell stated. He added that other crimes at sea, such as illicit traffic in drugs, smuggling of migrants and stowaways were continuing to rise. Parallel to these developments, the globalization of trade and the shipping industry was bringing newer issues to the fore: open-registry of ships and flags of convenience. Mr. Desai added that crimes at sea, for example, piracy and armed robbery against ships, jeopardized the very foundation of sustainable development. They could also constitute threats against the marine environment.

26. In addition, during the general exchange of views at the second meeting of the Consultative Process, delegations highlighted also the importance of follow-up of the areas of focus discussed at the first meeting, i.e.,

fisheries and the protection of the marine environment. They also expressed their wish to receive the most up-to-date information from the organizations and bodies concerned.

    (a) Marine science and the development and transfer of marine technology as mutually agreed, including capacity-building

27. During the exchange of views on areas of concern and actions needed, delegations addressed the issues of marine science and technology. Delegations emphasized the fundamental importance of implementing the provisions of Parts XIII and XIV of UNCLOS, on marine scientific research and development and transfer of marine technology, respectively, as well as the marine science and technology provisions of chapter 17 of Agenda 21.

28. Delegations reiterated that Part XIII of UNCLOS established an overall global regime for the promotion and conduct of marine scientific research.

29. Many delegations emphasized that the focus should be on identifying what is necessary to make this important part of the Convention operational in practical terms. They strongly supported calls for an "action plan" for this purpose, containing concrete policies and results-oriented initiatives.

30. Many delegations indicated that in addition to the importance of implementing Part XIII of UNCLOS, a number of multilateral treaties in the environmental field, such as the Convention on Biological Diversity and the United Nations Framework Convention on Climate Change, also had a bearing on marine scientific research.

31. Many delegations, in particular those of island States, identified marine science as an area of particular significance and focus in their region.

32. There was a consensus that knowledge about the oceans had to be expanded and that the promotion of marine science was essential for that purpose. There was also a consensus that the distribution of the existing knowledge was uneven. Developing countries, generally speaking, suffered from a lack of or insufficient access to the results of marine scientific research.

33. The provisions of articles 246, 248, and 249 of UNCLOS were recalled and their importance for developing coastal States underscored for the purposes of access to existing marine data from relevant databases, access to samples, obtaining assessment of data and research results and obtaining assistance in their assessment or interpretation. Recognizing that there might be a gap between the provisions of article 249 and practice in this regard, it was suggested that States could be encouraged to submit data to an international repository such as the IODE, and to participate in international oceanographic research projects.

34. The disparity in the availability of ocean data, particularly meteorological data, was mentioned.

35. Many delegations pointed out that it was in the interest of all that knowledge in the field of marine environment and sustainable use of the oceans and seas was developed and shared.

36. Beyond the expansion and distribution of the information base, there is also a need to develop a mechanism to ensure that scientific information is the "best possible," some delegations maintained. Mechanisms for vigorous peer review of scientific information can be useful for this purpose, they added.

37. A review of recent developments in marine science, especially oceanography and remote sensing, was offered by some delegations. Remote sensing and satellite-derived communications have been used to track the types and numbers of vessels in fishing areas or in specially protected areas. Those satellite tracking capabilities have also been used to track the amount of fish caught in specific fisheries and to track the migration patterns of protected species and of particularly threatened fish. Such data provide invaluable insight for national, regional, and global management and protection strategies based on the scientific application of remote sensing data.

38. Remotely sensed data have also been used in weather and severe storm forecasting. In combination with in situ data, national Governments have significantly improved warnings to communities and populations in coastal areas to evacuate low-lying areas that are prone to storm surge and flooding attributable to hurricanes and tsunamis. Coupled with land remote sensed data, coastal managers and urban planners had been able to use historical data to identify areas that should not be developed for housing or hotels owing to their vulnerability. These data are also useful for mapping the coastal area, to identify critical watersheds and habitats, current human and potential human uses, such as urbanization, industrialization, tourism development, and agriculture.

39. Coupled with other remotely sensed data, ocean circulation and primary productivity data can be acquired to map the best places to site installations such as sewage treatment plants, monitor primary productivity due to non-point source run-off and the potential for harmful algal blooms. With this information and utilizing GIS techniques, coastal managers and land use planners can develop special management regimes to address outstanding concerns and potential impacts.

40. For many years, oceanographers have used, among other tools, satellite-derived sea surface temperature (SST) data to model the onset and severity of the El Niño Southern Oscillation (ENSO) and La Niña events. Elevated SSTs correlate with the incidences of coral bleaching throughout the tropics. Increases in the severity and duration of coral bleaching episodes potentially harm the economies of small island States that depend entirely on coral reef ecosystems for their livelihood.

41. Over the past five years, researchers have used satellites to measure sea surface height, oceanic winds and oceanic circulation patterns to moni-

tor the formation and movement of giant ocean gyres. By monitoring and tracking these oceanic gyres, coastal managers have tracked the formation and movement of marine debris and derelict fishing gear that collect and float in the open ocean.

42. As a result of recent advances in information technologies in near real time, national Governments can now receive and process remotely sensed data and use them to support their decisions.

43. Passive underwater acoustics provide an ideal means to monitor ocean phenomena on a global basis. Significant discoveries have included the ability to monitor underwater seismic activity at levels far below the threshold of the land seismic networks; the detection of undersea volcanic activity associated with seafloor spreading and the discovery of the sub-seafloor microbial biosphere; and the distribution and migratory paths of large baleen whales, in particular the blue whale.

44. Hydrography and the need for hydrographic services were also discussed. Hydrographic services, which can carry out hydrographic survey, nautical charting and maritime safety information dissemination, are needed for marine navigation, coastal management, marine environment preservation, exploitation of marine resources, definition of marine boundaries and scientific studies connected to the sea and near-shore zone.

45. Many coastal States lack even the most elementary tools to carry out their own charting and surveying operations, even in the most elementary forms. IHO has a record of the countries that need assistance, e.g., in Africa, Central America, the South-West Pacific, East Asia, the Black Sea region, South America, and other areas.

46. Many delegations identified key areas where marine science can contribute on an urgent basis. These areas include: delineating ecosystem boundaries, identifying key ecosystem functions and components, integrating scientific, technical and socioeconomic information, developing predictive models and risk assessment, developing performance indicators, and assessing the state of ecosystem health, especially in the context of integrated management of ocean affairs; fisheries conservation and management; biodiversity and the environment of the deep oceans, in particular relating to seamounts; interaction of the oceans and the atmosphere and its implication for climate change; pollution in oceans and seas and its impacts on freshwater resources; impacts of pollution on fragile ecosystems, including closed and semi-closed seas; the role of fisheries in the socioeconomic welfare of developing countries; ways of controlling and preventing unsustainable fishery; ballast water and its impacts on the marine environment; dumping of wastes, hazardous wastes and radioactive and chemical wastes; dismantlement of ships; marine pollution in coastal areas and its effects on agriculture and freshwater; crisis management in emergency situations and environmental impact assessment for implementation of projects potentially considered dangerous in fragile marine environments; study of sustainable harvest

and the dynamic nature of exploited marine species and stocks; exploration of ecosystem impacts of ocean harvest, taking into account natural environmental fluctuations and the impact of pollutants on the marine ecosystem, its rational exploitation and other marine ecosystem services; and coral reef conservation and fisheries and the coral reef ecosystem.

47.   On the other hand, many delegations expressed their reservations with regard to taking up certain issues in the Consultative Process which by their very nature fell within the competence of specific forums, so as not to prejudice or duplicate efforts. Those issues included, according to them, questions relating to the continental shelf, the submarine cultural heritage and marine mammals.

48.   That marine science was fundamental to sound decision-making was underscored by delegations.

49.   The two-pronged approach to marine science—"science for science" and "science for development"—was highlighted by a number of delegations. Many delegations were of the view that while the "science for science" approach had its value and contributed to human knowledge, the "science for development" approach had not been pursued in the past to a desirable degree. More emphasis should therefore be placed on the latter approach.

50.   Many delegations, especially those of island States, emphasized the promotion of "science for development" for addressing their practical needs, especially for their immediate and medium-term sustainability, and also for exploring potential areas for positive cooperation. In that context, there was a need for countries undertaking marine scientific research not only to fulfil their obligations to share data under Part XIII of UNCLOS, but to do so in a manner that was meaningful for small island developing States and to provide assistance in relevant product development.

51.   In the same vein, many delegations emphasized "marine science and technology for sustainable development." They emphasized capacity-building and access to the means of implementation to that end, including capabilities for development planning and the incorporation of marine sectors therein, international financial resources and technological capabilities.

52.   The interrelated nature of ocean affairs calls for an integrated management approach and integrated management is especially dependent on information from marine science: this was asserted by many delegations.

53.   The need for a holistic and interdisciplinary approach to marine science was pointed out by many delegations. In that context, some delegations referred to the concept of "science within science," i.e., the ability to integrate observations from various sub-disciplines of the marine sciences. The needs, in that connection, were to ensure that ocean science programmes were balanced to cover all critical aspects of ocean systems, to possess the capacity to integrate data and information from a wide variety of sources, and for scientists to continue to work in multidisciplinary teams. Others asserted that the usual "piecemeal approach" to marine sciences

must give way to a more holistic approach that took into account the needs of the various sectors that required sound marine science for their operations. Still others offered a view that the large marine ecosystem (LME) approach, endorsed by many important institutions such as the Global Environment Facility (GEF), was the most appropriate one from a holistic perspective.

54.   Going beyond collaboration among marine scientists themselves, many delegations pointed to collaboration in wider circle of stakeholders, including marine scientists and social scientists, pure science and corporate science, and academic-based knowledge and traditional and customary knowledge and management practices.

55.   Moving from marine science to marine technology, many delegations pointed to the provisions of Part XIV of UNCLOS on the development and transfer of marine technology. In that context, appropriate international funding in research and development was an important aspect, in their view. The need of many States, especially developing States, for advice and assistance was also identified.

56.   Many delegations considered the issue of transfer of technology as a priority in the area of marine science and technology. Many others pointed to the need of developing countries for the acquisition of the most up-to-date technology.

57.   Some delegations mentioned an urgent challenge with respect to the development and transfer of marine technology for providing developing countries, including the least developed States and small island developing States, with adequate funding and technical assistance for the submission of technical and scientific data with respect to their extended continental shelf to the Commission on the Limits of the Continental Shelf, in accordance with article 76 of UNCLOS.

58.   Many others concurred that a critical example of the need to "operationalize" marine science and technology transfer could be found in the context of the continental shelf issue.

59.   Delegations underscored the needs of developing countries with respect to capacity-building in marine science and technology. It was acknowledged that there was a marine science and technology gap between developed and developing countries. Credible and practical ways were to be devised to encourage the exchange of information between developed and developing countries on marine science and technology.

60.   Many delegations were of the view that capacity-building was necessary to achieve the common goals of the preservation and sustainable use of the oceans and seas and that capacity-building went together with appropriate transfer of technology. Cooperation between developing and developed countries was essential in that regard. Many delegations believed that capacity-building should be strengthened within the existing institutions at the global, regional and national levels.

61. Capacity-building was considered by many delegations to be a priority in the area of marine science and technology. Others emphasized ensuring that any action plan to implement Part XIII of UNCLOS would profile capacity-building initiatives in a cross-sectoral manner and in a way that would guarantee the position of developing countries, particularly coastal States, as active participants and beneficiaries. Still others pointed to the challenges of developing national programmes on marine science and technology, such as organizational and institutional requirements, drawing in civil society and NGOs, the rational utilization of scarce resources to further national goals and mobilizing regional synergies and cooperation.

62. Capacity-building requires effective training of scientists and administrators, it was stressed by many delegations. In addition to training of scientists, the efficient use of equipment and calibration are necessary. Some delegations elaborated the effective design of ocean science programmes for developing countries, which would incorporate clearly defined objectives, clearly enumerated specific aspects and clearly postulated methodologies. Such programmes would have to involve all major stakeholders; should address social and economic goals and should be ongoing rather than one-shot.

63. Crucial areas of building capacity in marine science were identified by many delegations. These included marine fisheries, coastal ecosystems and sustainable coastal fisheries, coastal and marine biodiversity, marine non-living resources and the continental shelf, marine pollution, global climate change and linking national activities to regional systems and groupings.

64. Delegations urged the strengthening of international coordination and cooperation in marine science and technology, including at the intergovernmental and inter-agency levels. The necessary steps were encouraged at all levels for an effective and coordinated implementation of the provisions of UNCLOS and Agenda 21, including institutional adjustments and an improved coordination mechanism for chapter 17 of Agenda 21 to support action at the national and regional levels in developing countries and the provision of financial and technical assistance for the transfer of environmentally sound technologies. In that context, the international community was urged to promote, facilitate and finance access to and transfer of environmentally sound technologies and the corresponding know-how to developing countries on concessional and preferential terms. The importance of regional cooperation was stressed by delegations. It was added that successful regional cooperation needed to be supplemented by global cooperation. At the inter-agency level, IOC could serve as the focal point ensuring coordination, it was suggested by some delegations. Some other delegations suggested cooperation between IOC and the United Nations Environment Programme (UNEP). Still others suggested cooperation among IOC, UNEP, FAO and the regional organizations of Regional Seas Programme and regional fisheries management organizations.

65. Many delegations pointed to national and multilateral measures, existing or under development, which in their view addressed the issues of marine science and technology and of international coordination and cooperation therein in an effective manner. Such measures included: the IOC programme on the international exchange of data and information; GOOS, a cooperative programme of States and the organizations of the United Nations system and the related ARGO project; the Global International Waters Assessment (GIWA); the efforts of FAO relating to information on status and trends with respect to fisheries and marine living resources, including the development of an international plan of action (IPOA) and assistance in national capacity-building in fishery statistics; the IOC-WMO Joint Technical Commission on Oceanography and Marine Meteorology; the development and implementation jointly by the United Nations Industrial Development Organization (UNIDO) and the United States of America of GEF-supported ecosystem-based international waters projects involving 16 countries in Africa; the United Nations University's Fisheries Training Programme for practising professionals from the fisheries sectors in developing countries; the European Union (EU) Programme for Scientific and Technological Cooperation with Developing Countries and, within its framework, research on oceans and seas by the Research and Development Programme of EU (INCODEV); the multilateral programme, Census of Marine Life, to assess and explain the diversity, distribution and abundance of marine life in the world's oceans, and its component, Ocean Biogeographic Information System (OBIS), designed to be an online, worldwide atlas of marine life; existing regional and global mechanisms to promote the access of developing countries to science and technology; regional cooperation along the lines of active scientific cooperation in the North-East Atlantic within the International Council for the Exploration of the Sea (ICES); training and technical assistance available in developed States, for example, the United States, including educational and training programmes, fellowships and scholarships, clearing houses, databases and web sites; the Canadian International Development Agency's "Strategy for Ocean Management and Development"; Norway's programme of assistance in developing national regulations relating to the conduct of marine scientific research in waters under national jurisdiction and its contribution to the trust fund for facilitating the preparation of submissions to the Commission on the Limits of the Continental Shelf by developing States; and the International Marine Projects Activity Centre (IMPAC) of the Cooperative Research Centre for the Great Barrier Reef World Heritage Area of Australia, facilitating cooperation in the areas of fisheries management, coastal planning, management and research, and policy development for oceans governance.

66. A number of concrete suggestions were offered for the improvement of international coordination and cooperation. These included: the establishment of a clearing-house mechanism for marine science similar to the

existing GPA clearing-house mechanism; establishment of focal points for marine science and linking them up with relevant actors, such as the Joint Group of Experts on Scientific Aspects of Marine Environmental Protection (GESAMP) and GOOS, with regional organizations playing an important role in this respect; development of programmes for the competent international organizations that have implementing responsibilities under Part XIII of UNCLOS; strengthening of the regional organizations of the regional seas programme of UNEP through further cooperation of relevant international organizations with them; establishment of centres for the dissemination of information on marine scientific research and technology; strengthening of GEF and other financial institutions, enabling them to actively finance capacity-building projects in developing countries; identification of the existing intergovernmental centres of excellence on marine science and technology with a view to disseminating information on them and exploring the possibility of cooperation among them; development of regional marine science and technology centres, and financing for such centres; exploring the feasibility of establishing a regular process for the assessment of the state of the marine environment, with greater cooperation between regional seas organizations and regional fisheries organizations; workshops and joint technical meetings among regional organizations on subjects of mutual interest, for example marine science and its impact on fisheries, habitat destruction and pollution; joint programmes that may result from such joint meetings; transfer of technology through training; transfer of data through the use of the Internet; development of a single, comprehensive web site on international ocean affairs to facilitate exchange of information; and hydrography to be included in the appropriate development projects proposed by the United Nations funding agencies, the European Commission and other participant donor agencies, national as well as international, in order to achieve an adequate hydrographic data coverage by means of the creation of national hydrographic services.

67. Some delegations pointed out that the two areas of focus for the meeting, marine scientific research and piracy, while seemingly far apart in scope, were related at one level. Research vessels operating around the world were beset by the problems of piracy and armed robbery at sea. In addition, there was an increasing problem of vandalism of floating high-tech research equipment as well as moored oceanographic instruments. In view of the apparent correlation between mooring's data return and fishing activities in the oceans, it was suggested that efforts to combat vandalism could include the distribution of information brochures to national fishing agencies, fishing boats in ports and industry representatives.

(b) Piracy and armed robbery at sea

68. Delegations stressed the importance of discussing the matters of piracy and armed robbery in the context of the Consultative Process. It was pointed out that all nations needed to be actively engaged in combating

these growing threats that seriously affected navigation, the security of the crews of ships, as well as international maritime trade. It was further pointed out that, owing to the global nature of the threats, there was a need to consider countermeasures on a global level. Particular concern was expressed in connection with the recent increase in piracy and armed robbery in the seas of South-East Asia.

69.   Delegations also commended and endorsed efforts by the International Maritime Organization in this respect. A number of endeavours were mentioned, such as the correspondence group on "Code of Practice for the Investigation of the Crime of Piracy and Armed Robbery against Ships," the IMO Regional Expert Meeting on Combating Piracy and Armed Robbery against Ships, held in Singapore in March 2001; various seminars organized by IMO; as well as other actions, including resolutions of the IMO Assembly encouraging member States to cooperate to combat piracy. It was pointed out that IMO should be further reinforced in order to be the institution for coordination in the suppression of crimes at sea.

70.   In addition to the cooperation within the framework of IMO, delegations mentioned several regional initiatives, such as the Regional Conference on Combating Piracy and Armed Robbery against Ships, held at Tokyo in March/April 2000, and the planned Asian cooperation conference on combating piracy and armed robbery, to be held at Tokyo in the latter part of 2001. It was stressed that regional cooperation should be strengthened to develop an efficient information-exchange system among the States concerned with crimes at sea.

71.   It was reiterated that the Consultative Process should address issues of piracy mainly from the perspective of cooperation and coordination and that the relevant organizations should deal with specifics, in the light of the duty of all States to combat piracy.

Conservation and management of marine living resources; illegal, unreported and unregulated fisheries

72.   Many delegations noted with appreciation the adoption in March 2001 by the FAO Committee on Fisheries of the International Plan of Action to Prevent, Deter and Eliminate Illegal, Unreported and Unregulated Fishing and stressed the overall importance of the plan. A number of delegations called upon all States and fishing entities to implement it, as a matter of urgency, together with other FAO plans and instruments, such as the International Plan of Action for the Management of Fishing Capacity and the Code of Conduct for Responsible Fisheries. They suggested that implementation could be achieved through national legislation, international and regional organizations and fisheries management bodies and that FAO should provide coordination and assistance in that regard.

73.   Several delegations expressed their continued concern about illegal, unreported, and unregulated fishing (IUU fishing) and in particular about

the use of flags of convenience. They called upon all flag States to take measures, in accordance with international law, with a view to solving the problem and encouraged cooperation through regional fisheries management organizations. They further noted the progress made by FAO and IMO in identifying the possibilities of more effective actions against IUU fishing by flag States and port States.

74. In that connection, the central role of FAO as a coordinator for those regional fisheries management organizations was reiterated. It was also suggested that the General Assembly of the United Nations and the Consultative Process should monitor closely the implementation of the Plan of Action.

75. Among other initiatives, delegations recalled the concept of the Forum for Sustainable Fisheries, a worldwide coalition of multilateral agencies, Governments, organizations, the private sector and banks, other components of civil society as well as fishermen, which was being formed to assist developing States in achieving sustainable management of their living marine resources. They also mentioned the convening of the Conference on Responsible Fisheries in the Marine Ecosystem in Reykjavik in October 2001, which should highlight the application of marine science to ecosystem-based fisheries management.

76. Regarding various achievements in the cooperative approach by both coastal States and distant-water fishing States, the adoption of the Convention on the Conservation and Management of Highly Migratory Fish Stocks in the Central and Western Pacific in 2000 was highlighted, as well as efforts that were under way to bring that instrument into force as soon as possible.

77. The representative of FAO discussed issues related to fisheries, including fisheries trends and status reporting, successful efforts to improve coordination, and technical consultations and seminars. Regarding IUU fishing, he highlighted the adoption of the International Plan of Action and referred to parallel efforts of FAO and IMO to address the issues of implementation by flag States and by port States.

Marine environment and marine pollution

78. With respect to the marine environment and marine pollution, a number of delegations referred to the role of UNEP in the field of the protection and preservation of the marine environment, especially from land-based sources. Mention was made of the development of frameworks, such as Global Programme of Action for the Protection of the Marine Environment from Land-based Activities or the regional seas programme, together with examples of regional cooperation, such as North-West Pacific Action Plan promoted by Japan.

79. Delegations welcomed the efforts deployed by UNEP to prepare the forthcoming Intergovernmental Review Meeting on the Implementation of the GPA, to be held at Montreal in November 2001. Some observed that

there had been a low level of participation from the United Nations agencies dealing with the implementation of the GPA. Also, the representative of the GPA Coordination Office informed delegations about the focus and organization of the Review Meeting. In that context, the European Union underscored, in view of the still incomplete donor base, the need for adequate financing.

80. A view was expressed that UNEP should act as a focal point in the field of the protection and preservation of the marine environment and provide for coordination among different regions. Accordingly, the regional organizations of the UNEP regional seas programme should be strengthened and their cooperation with regional fisheries organizations should be improved. It was also suggested that the cooperation between UNEP and IOC should be enhanced to develop the scientific methodology necessary for both coastal management and the protection of marine environment.

81. It was further proposed that GEF and other financial institutions should be strengthened to enable them to actively finance the capacity-building projects in developing States. In that respect, it was also proposed to invite international financial institutions, including GEF, to support the implementation of projects in developing States in a number of areas, such as the control and reduction of pollution, waste management and recycling projects, prevention of dumping of wastes and hazardous substances, environmental impact assessment for projects potentially harmful to the marine environment, etc.

82. It was noted that it was important for the multilateral treaties in the environmental field, such as the Convention on Biological Diversity and the United Nations Framework Convention on Climate Change, to be understood also in the context of UNCLOS. In that context, one delegation recalled the proposal it had made at the first meeting, to conduct a review of the national, regional and global implementation of Part XII of UNCLOS.

83. One group of States reiterated its concern at the transit of radioactive material and hazardous wastes along coastal routes or navigable waterways, given the risk of harm which the practice carried for marine ecosystems, and called for strict compliance with the security norms and standards applicable to the transport of such material and wastes established by the International Atomic Energy Agency (IAEA) and the International Maritime Organization. It further reiterated its commitment to strengthen the international regime on the security of the transport of radioactive material.

84. Referring to the risk of serious and irreversible damage to the marine environment, in particular to sensitive habitats, from unsustainable development and practices, which were not confined to the exclusive economic zones of coastal States, one delegation proposed that the concept of marine protected areas should be applied to waters beyond the limits of national jurisdiction. The delegation expressed the view that such "international marine protected areas" might serve as a tool for integrated conservation and

management, without prejudice to the rights and obligations of States under UNCLOS.

85. The representative of the Baltic Marine Environment Protection Commission (HELCOM) spoke about the achievements of regional cooperation in reducing marine pollution and about the state of the marine environment in the Baltic Sea.

Capacity-building and assistance to developing States
86. It was noted that capacity-building, together with appropriate transfer of technology, was necessary to achieve the common goals of the preservation and sustainable use of the oceans and seas and that cooperation between developing and developed States was essential in that regard. It was pointed out that capacity-building should be strengthened within the existing institutions on global, regional and national levels and that the efforts which were being undertaken in the United Nations system and at the regional level should continue to be supported.
87. A number of delegations welcomed the inclusion of a section on capacity-building in the Secretary-General's report. Some expressed the wish for further analysis of the gaps and overlaps in the capacity-building activities. Several delegations expressed their appreciation to donor States which had pledged or made contributions to the trust funds established pursuant to resolution 55/7.
88. In addition to the important discussion on capacity-building which took place under the area of focus on marine science and technology, it should be noted that the United States representative encouraged delegations to investigate the numerous possibilities for training and technical assistance available in the United States and announced that information on many of those programmes would be posted on the Department of State web site in June (http:www.state.gov).

International coordination and cooperation
89. It was noted by many delegations that, while the discussions of the second meeting would focus on marine science and piracy, the mandate of the Consultative Process included areas in which cooperation and coordination could be enhanced among international bodies.
90. A number of delegations concurred with the Secretary-General's assessment that there was an overall lack of coordination and cooperation in addressing ocean issues, which prevented more efficient and results-oriented ocean governance. Those delegations stressed the need for cross-sectoral responses at all levels, starting at the national level, and urged the Secretary-General to take further measures aimed at ensuring more effective collaboration and coordination between the relevant parts of the Secretariat and the United Nations system with respect to ocean affairs and the law of the sea, with the aim, inter alia, of avoiding duplication and streamlining the

activities in different forums. In that regard, collaboration between UNEP and FAO on sustainable fisheries was highlighted as a positive example.

91.   Several delegations expressed the view that full advantage should be taken of existing organizations and bodies, such as IMO, FAO, UNEP, IOC–UNESCO, which had played an important role in dealing with the relevant issues of oceans and had expertise and knowledge in the area of coordination. They proposed that each of those organizations or bodies should act as a focal point within their respective areas of competence and that they should coordinate all other organizations or bodies concerned. In addition, it was suggested that the International Seabed Authority should undertake overall responsibility with respect to the development and management of the non-living marine resources of the international seabed area.

92.   It was pointed out that, at the present time, the main focus should be to strengthen the functions of relevant organizations and mechanisms involved in ocean management, to enhance coordination and cooperation among them and to strengthen assistance to developing States for their capacity-building.

93.   On the international level, the clearing-house mechanism established under the Global Programme of Action was highlighted as a successful approach to improve coordination and cooperation.

94.   It was further proposed that, with respect to the dissemination of information on ocean affairs, all international organizations concerned should consider jointly setting up a single comprehensive web site.

95.   Regarding cooperation at the regional level, the Pacific Islands Forum States reported that they were developing a regional integrated ocean policy which would, in part, examine ways to improve coordination and cooperation among their regional organizations and provide a more coherent framework for addressing the priority needs of their region.

96.   As far as the national level was concerned, examples of comprehensive legislation devoted exclusively to oceans, such as the Oceans Act of Canada, were mentioned as an illustration of blueprints for the integrated management of ocean activities. It was noted that the shift toward such integrated management had gained momentum also on the international level, presenting challenges in relation to planning process as well as to governance.

Panel discussions: Areas of focus
    (a) Discussion Panel A: Marine science and the development and transfer of marine technology as mutually agreed, including capacity-building

Part I

Improving structures and effectiveness
97.   The discussions in Part I of Panel A on improving structures and effectiveness with respect to marine science and technology were led off

by presentations from the following representatives: Mr. Patricio A. Bernal, Executive Secretary, Intergovernmental Oceanographic Commission (IOC) of the United Nations Educational, Scientific and Cultural Organization (UNESCO); Ms. Lene N. Lind, alternate head of delegation, Norway; Mr. Robert Duce, Chairman, Joint Group of Experts on Scientific Aspects of Marine Environmental Protection (GESAMP); Mr. Jorge E. Illueca, Assistant Executive Director, United Nations Environment Programme (UNEP); and Mr. Alfred Simpson, Director, South Pacific Applied Geoscience Commission (SOPAC).

98. Mr. Bernal provided an overview of ocean sciences, in particular from the perspective of IOC, which was task manager for science in the context of chapter 17, Oceans and seas, of Agenda 21. Ocean science can be pursued from two approaches: science for understanding (or "science for science," as described in earlier sections of the present report) and science for development. Science for understanding currently focused on climate change, biogeochemical cycles, and regulation of climate. The focus of science for development was on the sustainable use of resources, the protection and preservation of the marine environment and integrated coastal area management (ICAM).

99. Mr. Bernal then described three strands of ocean sciences: ocean sciences 1, 2, and 3. Ocean sciences 1 are aimed at preserving the integrity of the natural services provided by the oceans based on an understanding of the oceans. There are many natural cycles and processes relating to the oceans and the unique life-support system of the earth depends on a balanced interplay of these cycles and processes. Ocean sciences 1 attempts to deepen the scientific understanding of the interplays. Because of the intensity of human use of the oceans in the recent period, the maintenance of the balance is an issue of increasing concern. The central objective of ocean sciences 2 is to provide a sound basis for policy formulation, akin to that of science for development. Many coastal and marine resources are affected by over-exploitation and unsound practices. These concerns are at the centre of ocean sciences 2. Loss of life and property attributable to ocean-generated natural disasters, such as storms, hurricanes and tsunamis are the concerns that are at the centre of ocean sciences 3, the basic objective of which is forecasting future states of the oceans.

100. Mr. Bernal then discussed international ocean science, in which he identified four main streams: climate change, ocean-atmosphere interaction, human dimension of global change, and ocean observation at the global level. The programmes of international ocean science are planned and coordinated internationally, implemented jointly by organizations of the United Nations system and NGOs, with the active participation of the international scientific community and the engagement of governmental agencies and institutions.

101. International scientific work on climate change is carried out under

the World Climate Research Programme (WCRP), which uses a multidisciplinary strategy for the investigation of the physical aspects of climate and climate change. The main projects under the programme are Climate Variability and Predictability (CLIVAR), World Ocean Circulation Experiment (WOCE), and Global Energy Water Cycle Experiment (GEWEX).

102. The International Geosphere Biosphere Programme (IGBP) and its eight core projects address the issues of ocean-atmosphere interaction. Two of the important projects are the Joint Global Ocean Flux Study (JGOFS), which investigates carbon cycles and the role of the oceans, and the Global Ocean Ecosystem Dynamics (GLOBEC), which studies large marine ecosystems and large-scale shifts in ocean regimes.

103. The human dimension is studied under the International Programme on the Human Dimension of Global Change (IHDP), which is an emerging programme.

104. Ocean observations at the global level are carried out by the Global Ocean Observing System (GOOS). GOOS is a sustained and coordinated international system for gathering data about the world's oceans and seas, sponsored jointly by IOC/WMO/UNEP/ICSU with the active involvement of States. Its initial observing system is operational, and substantially increased observations are planned during the period 2002–2005, especially through the deployment of about 3,000 ARGO floats. GOOS constitutes a single system in which all ocean data, from both remotely sensed sources and in situ sources, would be collected, combined and processed. Its goals are universal participation by developing and developed States and the provision of services for use by end-users for the purposes of, inter alia, the protection and preservation of the marine environment and international ocean governance.

105. Ms. Lind's presentation focused on the implementation of Part XIII of UNCLOS, especially the consent regime for the conduct of marine scientific research. In her view, Part XIII tries to strike a balance between the principle of full freedom of research and the coastal State's interest in controlling activities in maritime areas under its sovereignty and jurisdiction. On the one hand, marine scientific research may only take place with the consent of the coastal State. On the other hand, the coastal State must exercise its powers in a predictable and reasonable way, and with a view to promoting and encouraging the conduct of scientific research as much as possible.

106. Under article 255 of the Convention, States are encouraged to adopt reasonable rules, regulations and procedures to promote and facilitate marine scientific research beyond their territorial waters. The adoption of such rules and regulations based upon a common understanding of the rules of Part XIII will provide clarity and predictability for scientists involved in planning research projects facilitate the standardization of requests for research projects, and ensure an improved flow of information through authorized organizations and channels.

107. For coastal States, the provisions of section 3 of Part XIII are particularly important. Section 3 lays down the balance of interests between the coastal State in relation to research activities of other States and international organizations, in its territorial sea and exclusive economic zone, and on the continental shelf.

108. The consent regime applies in the exclusive economic zone and on the continental shelf (article 246). In the territorial sea, the coastal State exercises full sovereignty and jurisdiction (article 245). The conduct of the marine scientific research shall be with its express consent.

109. Ms. Lind informed the meeting that Norway had found it most practical to adopt unified and coherent regulations on marine scientific research covering all areas under Norwegian sovereignty and jurisdiction. Research cruises often cover areas of both the territorial sea and the exclusive economic zone, and requests for cruises inside the territorial sea might occur as often as requests for cruises outside it.

110. In section 3, the core of the compromise between the coastal State's interests and those of the researching States is shown through the articles on tacit or implied consent and the right of the coastal State to withhold consent under specified conditions, or to require the suspension or cessation of the research in progress in the exclusive economic zone and the continental shelf if the research does not comply with the information or the obligations required.

111. For an effective and efficient implementation of the Part XIII regime, it is desirable that all States designate national focal points to coordinate research activities and respond to applications. Ideally, the designated body should be part of the government organization involved in marine matters, particularly marine scientific research activities. An important function of such office would be to ensure that all relevant government agencies are notified of the research project and to coordinate the reply to the researching State. The office should also be responsible for informing all relevant agencies and authorities, such as the coast guard and port authorities, of the decision to grant consent.

112. Marine scientific research may be conducted freely in the water column beyond the limits of the exclusive economic zone, according to articles 257 and 87 of UNCLOS. The same is the case, according to articles 143 and 256, for the area defined in the Convention as the seabed and ocean floor and subsoil thereof beyond the limits of national jurisdiction. In Ms. Lind's view, these provisions are particularly relevant in relation to compliance with article 76 and article 4 of annex II to the Convention. Research institutions and organizations conducting studies of the continental margin will collect data of the same type to be acquired for the purpose of mapping the limits of the continental shelf. Similarly, all the bathymetric and geophysical data acquired on the outer edge of the continental margin and adjacent deep sea by the world's marine research institutions and organizations are

highly relevant for any State that intends to establish the outer limits of its continental shelf beyond 200 nautical miles. Ms. Lind suggested that within the United Nations, the so-called GRID-system (Global Resource Information Database) of UNEP might be a suitable candidate to host and develop a centre for research data from the outer continental margin intended to serve the needs of coastal States and developing countries in particular.

113. Mr. Duce discussed the work and role of GESAMP, sponsored by IMO, FAO, UNESCO, WMO, the World Health Organization (WHO), IAEA, the United Nations and UNEP. GESAMP was established to provide advice relating to the scientific aspect of marine environmental protection to the sponsoring agencies and, through them, to their member States. The other purpose of GESAMP is to prepare a periodic review and assessment of the state of the marine environment and to identify problem areas requiring special attention.

114. The unique characteristics of GESAMP are that it is the only inter-agency mechanism designed to provide independent and cross-sectoral analysis and advice, based on marine science, concerning the prevention, reduction and control of the degradation of the marine environment.

115. GESAMP works through its working groups formed to address a particular issue or problem that has been identified by the agencies or member States. Working group members work inter-sessionally. GESAMP itself meets once a year, when it reviews working group reports, decides on new issues that require working groups and evaluates emerging issues for further consideration.

116. Over the past 30 years, 140 scientists have served as members of GESAMP and over 340 have participated in GESAMP working groups. These scientists are unpaid independent experts from over 50 countries, both developing and developed, selected by the sponsoring agencies based on their scientific expertise. GESAMP has produced 41 specific reports on issues related to the protection of the marine environment. Topics of recent reports include the safe and effective use of chemicals in coastal aquaculture, the global input of pollutants from the atmosphere to the oceans, marine biodiversity: patterns, threats and conservation methods, and the contribution of science to integrated coastal management.

117. Mr. Duce then highlighted the two most recent reports: A Sea of Troubles and Protecting the Oceans from Land-based Activities. The former is a state-of-the-marine-environment report and was often quoted by participants in the meeting. The latter assesses the problems relating to the protection of the marine environment from land-based activities, identifies emerging problems and new perspectives, and highlights the regional perspective. It then develops certain strategies and measures and concludes with priorities for action.

118. Based on the work of GESAMP, Mr. Duce identified a number of priority problems for the global marine and coastal environment. These in-

clude alteration and deterioration of habitats and ecosystems; the effects of sewage on human health and the environment; widespread and increasing eutrophication of coastal waters; and the decline of fish stocks and other renewable resources.

119. Mr. Duce concluded by stating that GESAMP is currently undergoing a comprehensive review, the first in its 30 years of existence, by an independent group of peers to make it more effective and more responsive.

120. Mr. Illueca focused on the work of UNEP in the area of the marine and coastal environment, which was a central issue of the twenty-first session of the Governing Council of UNEP in February 2001. One-fourth of the 31 programmatic decisions of the Governing Council were related to the work of UNEP in oceans and coastal areas, touching on issues such as the strengthening of the regional seas programmes, coral reefs, the GPA, the establishment of a secretariat for the Northwest Pacific Action Plan (NOWPAP), the finalization of negotiations on a new regional seas convention for the North-East Pacific extending from Colombia to Mexico, interlinkages in the work programmes of regional seas programmes and global conventions such as the Convention on Biological Diversity and chemicals-related conventions, and the assessment of the state of the marine environment. Most of these decisions contained elements related to marine science.

121. Three important decisions are particularly relevant to the issue under consideration by the meeting: Governing Council decision 21/13 on global assessment of the state of the marine environment; decision 21/28, entitled "Further development and strengthening of regional seas programmes: promoting the conservation and sustainable use of the marine and coastal environment, building partnerships and establishing linkages with multilateral environment agreements"; and decision 21/12 on coral reefs.

122. The work of UNEP relevant to marine science is basically focused on eight areas: the assessment programmes of the regional seas conventions and action plans; the GPA; the Global International Water Assessment (GIWA); work on marine and coastal biodiversity of the UNEP World Conservation Monitoring Centre; the Millennium Ecosystem Assessment; the International Coral Reef Action Network (ICRAN) and the International Coral Reef Initiative; the Joint Group of Experts on the Scientific Aspects of Marine Environmental Protection (GESAMP); and the Global Environment Outlook (GEO) reports.

123. There are currently 17 regional seas programmes currently in operation. Fourteen of them were facilitated by UNEP and three were developed independently but today are working closely with UNEP and its regional seas programmes as partners. Environmental assessment is an essential element of the action plans of regional seas programmes. Monitoring and assessment activities provide a scientific basis for setting regional priorities and policies, particularly for issues such as integrated coastal area management.

Assessments are also made of the social and economic factors that relate to environmental degradation and the status and effectiveness of national environmental legislation.

124. Under the GPA, regional seas secretariats are undertaking a number of regional assessment activities, including the preparation of regional diagnostic studies of marine degradation from land-based activities. The GPA clearing-house mechanism is developing as a useful tool for disseminating and exchanging information. Subject to availability of resources, future emphasis will be placed on developing web-based geographic information capabilities to better support decision makers.

125. GIWA is focusing on the root causes of environmental degradation in 66 international marine, freshwater and groundwater areas around the world. Financed by the Global Environment Facility, GIWA aims to provide the most objective and comprehensive assessment of transboundary water problems ever made.

126. Through ICRI and the ICRAN project, assessment and management activities for the protection and sustainability of coral reefs are being promoted worldwide.

127. With regard to GESAMP, for the recent publications A Sea of Troubles and Protecting the Oceans from Land-based Activities, UNEP provided the Technical Secretary of the GESAMP Working Group on Marine Environmental Assessments, which produced both of these important reports. The preparation of the latter report was initiated by UNEP as a contribution to the first intergovernmental review meeting on the progress in the implementation of the GPA, to be held at Montreal in November of 2001.

128. Through the Global Environment Outlook (GEO), several thematic areas, including the coastal and marine environment, are periodically assessed on a regular basis. GEO aims to provide policy-relevant assessments. GEO-2000 is the latest in this flagship series of assessment reports, with the next scheduled for 2002 as a contribution to the World Summit on Sustainable Development.

129. Mr. Illueca then focused on Governing Council decision 21/28, which, in his view, was of particular relevance to the deliberations of the Consultative Process, particularly in section (d), entitled ''Partnerships with international organizations.'' Largely as the result of the report of the first meeting of the Consultative Process, UNEP and FAO had embarked on a joint initiative resulting in the preparation of a report entitled ''Ecosystem-based Management of Fisheries: Opportunities and Challenges for Coordination between Marine Regional Fisheries Bodies and Regional Seas Conventions.'' At the Third Global Meeting of Regional Seas Conventions and Action Plans, held in Monaco in November 2000, the recommendations of the report were endorsed by the 17 regional seas programmes. Subsequently, the Governing Council in its decision 21/28 endorsed the recommendations of the Monaco meeting, as well as the following actions for en-

hanced cooperation: the formalization of the observer status of the regional seas conventions and action plans at the meetings of the governing bodies of regional fisheries bodies and their technical subsidiary organs, and vice versa; the exchange of data and information available at the levels of regional fisheries bodies and regional seas conventions and action plans that may be of mutual interest; and the design and implementation of joint programmes between regional fisheries bodies and regional seas conventions and action plans, taking fully into account the respective mandates, objectives and scope of the regional seas programmes.

130. In addition to welcoming the joint initiative between FAO and UNEP for enhanced cooperation between regional fisheries bodies and regional seas conventions and action plans on issues relevant to ecosystem-based management of fisheries, the UNEP Governing Council in decision 21/28 invited the IOC/UNESCO, through its Global Ocean Observing System, given the complementary work that it was undertaking, to participate in the UNEP/FAO initiative. As in the case of the close UNEP partnership with IMO in supporting the regional seas programmes in the area of emergency response to oil spills and accidents from other ship-borne hazardous substances, UNEP would like to strike a similar partnership with IOC in support of the assessment activities of the regional seas programmes.

131. Mr. Illueca concluded by drawing the attention of the meeting to Governing Council decision 21/13 on the global assessment of the state of the marine environment. In that important decision, inter alia, the Council requested the Executive Director, in cooperation with IOC/UNESCO and other appropriate United Nations agencies, the secretariat of the Convention on Biological Diversity and in consultation with the regional seas programmes to explore the feasibility of establishing a regular process for the assessment of the state of the marine environment, with the active involvement of Governments and regional agreements, building on ongoing assessment programmes, such as GESAMP.

132. Mr. Simpson began his presentation by underscoring the status quo in dealing with ocean matters. In his view, the attitudes have not changed: a perspective of ownership rather than stewardship is still applied in ocean matters, and a terrestrial mindset rather than an oceanic one is still prevalent. He demonstrated his point by providing certain statistics about the island countries of the Pacific region. For example, with the exclusion of Papua New Guinea, 2.5 million people in 500 islands in the Pacific region are responsible for about 27.1 million square kilometres of the Earth's surface, composed of 27 million sq km of exclusive economic zone and 93,500 sq km of land territory. This translates into an ocean:land ratio of 290:1, which can be viewed as an index of stewardship of oceans that has to be provided by the island countries in the region, which in Mr. Simpson's view should be referred to as "large-ocean island developing States" rather than "small island developing States."

133. He then dealt with the challenges to the capacity of such States: essentially the limitations of financial, technical and human resources. The institutions are weak, which means that the legal and policy arrangements are not complete. There are few specialists in legal and ocean governance. The region does not have any research vessels and has only a few research institutions and few regional marine scientists. The occurrence of a number of marine minerals, such as manganese nodules, cobalt-rich crusts, and polymetallic sulphides has been established, but their quantity is too small for commercial exploitation. Only one licence has been issued to date, for the exploration of polymetallic sulphides. The region is rich in fishery resources, but less than 4% of the catch value is returned to the region as access fees from distant water fishing nations. To compound the problems, the region has the highest concentration of natural hazards in the world. The region is also flanked by the countries in the Pacific rim, with an estimated coastal population of 2 billion, who generate considerable amounts of waste, part of which ends up in the oceans.

134. According to Mr. Simpson, although there are 19 States parties to UNCLOS in the region and only 3 non-States parties, the high rate of participation in UNCLOS does not appear to have made much material difference.

135. The countries in the region support participation in the international ocean observing systems. They are planning to participate in the ARGO programme, with the initial ARGO floats scheduled to be deployed in the region later in 2001. There have been a number of marine scientific research cruises since 1953, 196 of them since 1990 involving nine researching States, in the exclusive economic zones of 16 Pacific island States. However, only 25% of the known collected data are at the disposal of the island States; according to Mr. Simpson, this means that the exchange of data envisaged in Part XIII of UNCLOS has not materialized. SOPAC is proposing a process whereby the data collected from marine scientific research by foreign countries in the exclusive economic zones of Pacific island States will be exchanged to a fuller extent through the SOPAC Cruise Database, Oceanographic Databank and Seismic Databank.

136. Pacific island States have already taken a regional approach and are planning to enhance the applications of that approach. On the issue of the implementation of UNCLOS, two regional workshops were held, in 1998 and 1999 respectively. One regional workshop in 2001 dealt with marine scientific research and the implementation of the marine scientific research regime established by Part XIII of UNCLOS. The region is envisaging the development of a regional oceans policy. Such a policy was recommended by a regional workshop in 1999, was subsequently endorsed by the Leaders of the Pacific islands Forum and is currently being developed by the Council of Regional Organizations in the Pacific.

137. The regional workshop on marine scientific research recommended the development of regional marine scientific research guidelines,

to focus on capacity-building involving effective participation in marine scientific research, and not merely representation. The guidelines would comprise the following key components: a standardized form; identification of contact points and establishment of national arrangements; development of a data protocol to unify the various formats currently in use, many of them unreadable; development of a regional data standard to set norms in the face of the diversity and consequent loss of utility of data produced and shared by current marine scientific research activities; and the development of a regional meta-data database. In recognition of the fact that marine scientific research and exploration were parallel activities, the workshop also recommended the development of policy and guidelines in this context.

138. Mr. Simpson concluded by enumerating certain enabling factors that would translate the concept of sustainable development into reality, including: the development of baseline data and information; enhancement of carrying capacity; ensuring sustainable yield, effective monitoring and review; development of policy and legislation; participation of trained personnel; achievement of economies of scale through regional cooperation; and application of transparent and clearly formulated guidelines.

139. The discussions that took place after the presentations focused on a substantial number of issues. Delegations placed in context the importance of the area of focus by recalling that marine scientific research and the knowledge resulting from it contributed to the eradication of poverty, addressed food security issues, sustained economic development and the well-being of present and future generations, and in general, provided for the effective protection of the marine environment.

140. Delegations focused on their national experience with regard to marine scientific research, transfer of technology and capacity-building. Cooperative activities and programmes were suggested by many delegations. The regional approach was also emphasized.

Legal framework for the conduct of marine scientific research (MSR)
141. Some delegations suggested that the establishment of a focal point for dealing with requests for marine scientific research could be useful. It was also suggested that States could forward the names of the focal points to the United Nations Secretariat to be published in the Law of the Sea Bulletin of the Division for Ocean Affairs and the Law of the Sea.

142. While some delegations pointed to the need to establish reasonable conditions in accordance with Part XIII for the granting of consent, others shared their positive experience in that regard. They indicated that the infrastructure in place in their countries had allowed for consent to be given to all requests under Part XIII and within the time limit of four months established by article 252. Other delegations pointed out that even in the case of research having direct significance for the exploration and exploitation of natural resources or research taking place in their territorial sea, they

were not aware of any denials to requests under Part XIII. They explained that the coastal State might have an economic interest in taking advantage of research cruises in areas under their national jurisdiction. This point was emphasized by Mr. Simpson, who stressed the importance and interest of most developing countries, in the South Pacific in particular, in science for development or science for management.

143. With reference to the issue of consent, Mr. Simpson stated that even though there might be some cases where consent was denied because of suspicion or lack of internal structure to handle the marine scientific research-related request, in his region, however, what was more important was the "track record" of the requesting State or institution. In that connection, he reiterated that his organization had in return received barely 25% of all collected data and information from foreign marine scientific research cruises. Many delegations concurred and expressed concerns about the fact that despite all the marine scientific research projects that were taking place globally, UNCLOS provisions about data and information exchange in particular might not be fulfilled. Other delegations added that that was true not only for cruise data but also for post-cruise data.

144. Many delegations stated that in view of all the marine scientific research projects currently being carried out, full and effective implementation of the Convention should be ensured, particularly with regard to equitable sharing of information and with respect to transfer of technology and environmentally sound technology.

145. Some delegations pointed out that compliance with the duty to have the coastal State's representatives on board the research vessel, when requested, had allowed for some marine scientific research cruises to contribute to capacity-building in those countries by involving research scientists from the coastal State in the research programme being carried out. Whenever possible, that should be encouraged, they added.

146. Also with reference to the issue of consent, a question was raised as to the practice of States when research requests would likely be interfering with other legitimate uses of the sea.

Exchange of data

147. With reference to the lack of exchange of data with and communication of data to the coastal States concerned, some delegations attributed this state of affairs partly to the fact that many States did not have the necessary internal structure or the capability to handle the data obtained.

148. In some circumstances, particularly for most developing countries, the data provided could not be interpreted and put to use because of unreadable format. Some delegations called for data to be transmitted in an appropriate manner and format. In that connection, some States stressed the need to adopt a data protocol.

149. As to the question of the provision of cruise data, some delegations

suggested the use of the ROSCOP (Report of Observations/Samples Collected by Oceanographic Programmes, also known as Cruise Summary Report) format, a meta-data recording format hosted by ICES that could allow coastal States to keep track of the data collected, the instruments which had been used, and the site of storage of the data. Questions were also raised regarding the issue of intellectual property rights and patents, and in this connection suggestions were made that States needed to clarify the issue from the legal point of view.

150. Some delegations called for a more transparent and systematic system for the exchange of data and information in order to allow States, inter alia, to better coordinate the communication of the information to their public. This would help avoid negative perceptions among concerned communities.

151. Also with respect to the exchange of data, mention was made of the example of JAMSTEC, the Japanese marine science and technology centre, which provided their data and outcomes through its web site. It was also observed that data were exchanged through the framework of the International Ocean Data Exchange (IODE), promoted by IOC.

152. It was pointed out that in the case of the SOPAC countries, with the exception of Papua New Guinea, which had established its own structure to handle marine scientific research requests, SOPAC, as the extension of national competence, was the focal point for all marine scientific research activities and maintained a cruise database for the region. The issue was deemed to be related to that of capacity-building since, in the particular case of SOPAC countries, the lack of trained personnel had led to the transfer of most UNCLOS responsibilities to the regional organization.

153. Some delegations, although recognizing the importance of the establishment of regional mechanisms, suggested that in the cases of many countries, that arrangement might not be effective since in those cases the Ministries of Foreign Affairs were generally considered to be the appropriate channel for marine scientific research activities.

154. Delegations pointed to the importance of establishing national marine scientific research centres to process the requests, advise the appropriate stakeholders of decisions to grant or deny consent and deal with cruise-related and post-cruise issues. The centres could also assist in the establishment of priorities and guidelines for research activities, which would in turn assist researching countries. They would be faced with known conditions and practices that would allow them to adjust their research project requests.

Transfer of technology and capacity-building

155. Some delegations pointed out that a substantial portion of the world's oceans and seas fell under the national jurisdiction of the developing countries. Furthermore, a sizeable part of the ocean was under the national jurisdiction of small island developing States. Capacity-building, therefore,

would be central to delivering concrete results in marine scientific research and activities based on such research, along with the appropriate transfer of technology. In that regard, some delegations also emphasized, as another aspect of capacity-building, the necessity for developing countries to be actually involved in all the relevant programmes and organizations.

156. Some delegations stated that the transfer of appropriate technology and know-how was essential for building effective marine scientific research capacity in the developing countries. The importance of direct investment and bilateral aid was stressed for assisting developing countries in building the scientific and administrative basis of their fisheries management systems, in view of the central role of fisheries for developing countries.

157. Some delegations pointed out that it was important if not imperative for the developing countries to have access to reliable technical advice and information on effective management practices and the experience from such practices. Access to such information and advice would assist in the improvement of fisheries management arrangements in a manner befitting domestic circumstances and would ensure conservation and optimal sustainable yield of living marine resources.

158. Many delegations agreed with the panellists that initiatives such as the Forum for Sustainable Fisheries should be encouraged and should be given a new momentum.

159. Many delegations emphasized that the lack of technical, financial, technological and institutional capacity in the developing countries to effectively tackle the catastrophes and threats to the ecology of the oceans and seas was among the main constraints they were facing in establishing integrated ecosystem-based approaches.

160. The representative of the International Ocean Institute (IOI) noted that the international law dealing with ocean affairs consisted of many conventions, protocols, codes of conduct and action plans. The international institutional framework that had emerged was therefore fragmented, inadequately coordinated and difficult to manage. This posed still further difficulties in particular for small and poor States in their efforts to keep abreast of the current situation. In that regard, there was a need to build capacities at both national and regional levels and to develop the existing international institutions and programmes to take into account the special needs of those States. It was especially important to create first of all, at the national level, some of the most important tools of implementation of the established legal framework, namely: enforcement capability, scientific/technological capacity and financial capacity. Without national capacity, there could be no effective international cooperation.

Protection of the marine environment

161. Several delegations maintained that a thorough and comprehensive knowledge of the state of the oceans and seas was essential for protecting the marine environment. The two recent GESAMP reports had concluded

that despite improvements in managing some of the pressures, the overall state of the world's seas and oceans was deteriorating. This was attributed by some delegations to the fact that despite the wealth of information on the marine environment and the availability of new information, there was a lack of an overview particularly on the links between the state of the marine environment and cross-cutting issues of human health, seafood safety and sustainable use of living marine resources. Decision makers thus needed to have at their disposal regular assessments of the impact of human activity on the state of the marine environment, including its socioeconomic consequences, at the national, regional and—with regard to pollution—global levels.

162. Delegations stressed that it was important for all studies to be based on a holistic approach which would take into account both the living and the non-living parts of the marine environment. Ecosystem models based on such an approach could constitute an important tool in furthering the understanding of interactions of marine ecosystem components and identifying specific gaps in knowledge and for defining research priorities. Delegations were of the view that such models should be encouraged.

Marine pollution

163. Delegations pointed to the impact of marine pollution on the sustainable use of living marine resources and on other marine ecosystems. They stressed that the current process of assessment of marine pollution needed to be strengthened. It was observed that the process of making scientific results policy-relevant was just as important as the process of collecting the data. There was a lack of coherence in the follow-up at the international level and the development of policy recommendations based on the assessment reports.

164. Some delegations suggested that consideration should be given to undertaking a global assessment of the marine pollution. In this context, reference was made to UNEP Governing Council decision 21/13. The comprehensive assessment envisaged in that decision would, inter alia, focus on the impact of marine pollution and physical alteration and destruction of habitats in relation to public health, food security, biodiversity and marine ecosystem health, including the marine ecosystem services. The other appropriate agencies to be involved in such an assessment would include WHO, IMO, FAO, IAEA, UNIDO and WMO.

165. Some delegations were of the view that such an assessment, in which an effort would be made to involve all stakeholders, should not only identify improved end-uses of the assessment but should also identify ways to improve communication with decision makers.

166. The representative of Greenpeace urged the cessation of the maritime transport of nuclear material because of the threat of accidents, which could have a major impact on the environment and human health conse-

quences and potentially bring about important economic losses. He also cautioned against using the ocean as a carbon sink.

167.   Some delegations called attention to what was in their view, the least studied phenomenon, namely submarine groundwater discharge and its impact on the coastal zone. While the magnitude of such discharge might be relatively minor, in areas dominated by river flow, recent studies had indicated that groundwater might occasionally account for a significant fraction of the freshwater inflow. The problem was to develop from both a scientific and a management standpoint a method for assessing the ways in which this phenomenon altered coastal ecosystems, with its effects on the water level and fluxes, caused withdrawal or alterations in the recharge patterns and the groundwater water quality, as well as its potential impact on coral reefs. Such major interventions in the coastal zone management system required a sound scientific justification and a degree of technical understanding not currently available. In that connection, the National Oceanographic Committee of the Russian Federation had started a study on SGD at the international level and undertaken to conduct a collaborative project on assessment and management implications of submarine groundwater discharge. Such a programme would need financing to carry out studies at the five sites selected.

Living resources

168.   Many delegations stated that the global monitoring of stocks of marine living resources was an area in need of strengthened cooperation and coordination. It was necessary to ensure that information supplied was up to date, comprehensive and reliable, particularly when it was being used for policy purposes. While FAO had a central role to play, the submission of basic biological information from member States as well as cooperation with regional fisheries organizations were also essential factors in successfully addressing such problems.

169.   Some delegations stated that there was a need to improve the understanding of ways in which ecosystems worked. This would allow for better multi-species management of living resources. Research in this area should be conducted primarily at the local or regional level as the characteristics of ecosystems varied greatly among the different areas of the world. Long-term monitoring and detailed investigations of different species and their interactions were the only safe means of obtaining the necessary level of understanding to ensure sustainable development.

170.   In that connection, delegations stressed the importance of the establishment of precautionary reference points as a basis for decisions on fisheries and marine ecosystems management. This was a necessary precondition for the application of the precautionary principle envisaged in the 1995 Fish Stocks Agreement. Cooperation between research institutions, regional fisheries organizations and FAO needed to be improved to establish

such reference points, particularly for the large number of stocks where such data were still insufficient.

171. In addition, the representative of Greenpeace cautioned against overfishing and the threats to many ecosystems that were suffering increasing degradation as a result of a multiplicity of activities. He called attention to the phenomenon of genetically engineered fish, characterized by increased size and accelerated growth. Such fish had the potential of becoming invasive species that could cause irreversible damage to wild fish stocks as well as to the wider marine environment. Scientific experiments had suggested that the introduction of a few transgenic individuals could wipe out entire populations within just a few generations. The international community needed to address this newly emerging threat.

Decision-making: Science for management, science for development

172. Many delegations, in particular those from the South Pacific region, made observations regarding the necessity of applied research for the benefit of developing countries. In that regard, it was suggested that the concept of science for development should be further articulated.

173. The representative of ICES also emphasized that the success of all the international agreements and other arrangements formulated to address the issue of the sustainable management of the marine environment and living resources depended heavily on the quality of the scientific advice available to decision makers. Decision makers needed advice that was unbiased, sound and credible. In response to the newly developing trends, ICES had modified its organizational structure to foster interdisciplinary collaborative science and had established a strategic planning process to better position itself in addressing emerging challenges. In that regard, it has also developed a close partnership with decision makers and management organizations. Through a Memorandum of Understanding which spells out the type and timing of advice to be provided by the Council, and makes provision for dealing with extraordinary requests, ICES is attempting to respond to the need for scientific advice to support more integrated management of marine ecosystems. This interactive process could be used as a model, which might be pursued at both the national and the international level.

174. Other delegations recognized that, with regard to fisheries management, there was a need to improve on the structure and effectiveness of marine science. In some regional fisheries management organizations there was a lack of clearly defined and agreed management objectives, with consequences for the science/stock assessment process that had left those organizations with no agreed basis for management responses to stock assessment. To increase the organizations' effectiveness, there was a need to develop stronger links between science and management through the development of, inter alia, clearly defined management objectives incorporating the precautionary approach, to promote effective communication between scien-

tists and managers so that the scientific assessment objectives were aligned with management issues, and to ensure that managers comprehended the likely impacts of advances in science on stock assessments and any associated management decision practices and strategies.

175. The representative of the International Hydrographic Organization pointed out that IHO was an intergovernmental and consultative organization involved in systematic surveys of the sea bottom with the aim of producing electronic nautical charts that can use the GIS. The bathymetric aspects of the surveys were carried out in collaboration with IOC. The data gathered could be used for many different applications, including the identification of the outer limits of the continental shelf, fisheries monitoring and assessment, investigation of water level changes, monitoring of ocean dump sites, and so forth. In view of the necessity to improve the knowledge of the sea bottom, many coastal States, with inadequate hydrographic services, needed to build their capacity. Investment was necessary in that regard. IHO, for its part, offered training programmes through national and international centres.

176. Some representatives of international agencies and organizations cautioned that scientific organizations needed to focus on the production line of science, since the necessity of "packaging" the scientific information together with social and economic information might distract those organizations from their central mandate. The demand for science per se needed to be clearly defined. In that context cooperation among organizations with different mandates, purposes, and goals within specific programmes would respond to the specific needs and concerns of States with regard to sustainable development. Attention was drawn to the United Nations Atlas of the Oceans as an endeavour of several United Nations organizations and agencies to offer comprehensive information on the oceans.

International cooperation and coordination

177. Several delegations reaffirmed that the responsibility for the state of the oceans and seas was a matter for both national Governments and international bodies. To respond adequately to the needs and problems of the marine environment, it was crucial to provide for coordination and cooperation at the national, regional, and global levels.

178. Some delegations emphasized the importance of improved coordination and cooperation between agencies at the international and regional levels in view of the reliance of most developing countries on those organizations for marine scientific research and transfer of technology. Those organizations were called upon to develop their technical cooperation programmes so as to foster capacity-building in developing countries which would enable them to comply with international standards and obligations. In that regard, more training and scholarships were needed. Many delegations pointed out that research programmes, particularly those of interna-

tional institutions, should take into account the specific needs of developing coastal States. In addition, there should be more synergy between the developed countries, donor countries, and the United Nations system in addressing issues related to the oceans.

179. Many delegations recognized the need for an integrated management of the oceans and coastal zones to be effected through the establishment of intersectoral and interdisciplinary approaches. In that regard, coordination might also require institutional adjustments. It was pointed out that coordination and cooperation between existing research efforts needed to be improved and strengthened at all levels.

180. Some delegations considered the global monitoring of marine living resources to be one of the areas in need of strengthened cooperation and coordination.

181. Many delegations pointed out that it was encouraging to note that the needs of developing countries relating to their lack of technical, financial, technological, and institutional capacity were being addressed through international cooperation and international programmes.

182. In that regard, some delegations cited various specific international programmes with a mandate in science and technology as good examples of cooperation and coordination. It was also pointed out that certain international programmes were in need of improvement.

Regional cooperation

183. Many delegations expressed their support for regional initiatives; particular mention was made of regional fisheries organizations and the regional seas programme of UNEP. Delegations recognized that activities at the regional level had often proved effective.

184. The representative of IOI pointed to the example of the regional seas programmes of UNEP. It was in such regional seas, which were closely associated with most of the identified large marine ecosystems that pollution control as well as management of living resources and other uses of the common ocean space could be facilitated. From the economic standpoint such ocean spaces offered opportunities for economies of scale; and from the cultural and historical points of view, there were often a commonality of interests. Coastal States, especially small or poor States, could do together what none of them could do alone. The representative stressed that the international community needed to be aware that when dealing with tools of implementation in a regional context, it had become necessary to coordinate and integrate the various convention regimes.

185. She added that the legal basis for cooperation in the development and transfer of marine technology was found in articles 276 and 277 of UNCLOS, which provided for the establishment of regional marine scientific and technological research centres. It was suggested that, through adaptation to all the new conventions adopted subsequently, provisions of those

articles had been reinforced and should be implemented. The implementation of those provisions could be viewed in the context of the UNEP regional seas programme.

International organizations and agencies
186.   The representative of IMO, the organization in charge of organizing the review of GESAMP, highlighted GESAMP as a good model for coordination and cooperation among United Nations agencies that needed to be preserved. In that regard, the review of GESAMP could be considered as an effort to assess the effectiveness of such a mechanism in adequately addressing the emerging problems and priorities of the international community. The review was expected to lead to the conclusion that there was no need for the establishment of additional scientific bodies for oceans assessment. What was needed, as had also been pointed out by other delegations, was to improve communication among United Nations agencies and Governments. It was necessary to create an environment needed in which scientific advice could be properly and more rigorously used in the decision-making process.
187.   The representative of WMO highlighted the establishment of the Joint Technical Commission on Oceanography and Marine Meteorology (JCOMM) as one response to the need for an interdisciplinary approach in ocean matters. JCOMM was a coordinating body for all current and future marine activities of WMO and IOC. A major initial priority was the implementation of an ocean observing system for climate, which will require the equal engagement of both meteorologists and oceanographers. The JCOMM sought to pool the expertise and resources of the meteorological and oceanographic communities, both nationally and internationally, through WMO and IOC. An outreach programme would be conducted to enhance the capacity of all maritime countries both to contribute to JCOMM and to benefit to the maximum extent from the outcome of its activities.

Role of the Intergovernmental Oceanographic Commission
188.   Many delegations welcomed and supported the role of IOC in coordinating marine scientific activities, ocean services and related capacity-building. The Commission was encouraged to continue developing its role as the focal point for marine scientific research. The potential functions of the IOC Advisory Body of Experts on the Law of the Sea (ABE-LOS) were also highlighted. Other delegations called for more capacity to an enhanced level for IOC to enable it to fulfil its role. The IOC regional bodies could play a central role in regional scientific cooperation and monitoring and their cooperation with regional seas arrangements and regional fisheries organizations and arrangements should be strongly encouraged. Such regional cooperation could provide a means of fulfilling the obligation under

UNCLOS regarding the establishment of regional centres for marine science and technology.

189.   The IOC was strongly encouraged to increase its cooperation and forge partnerships in particular with the regional seas programmes, of UNEP, with other agencies and programmes, and, in the area of scientific programmes, even with organizations outside the United Nations system. The Commission was also urged to promote the open exchange of oceanographic information and data, using internationally accepted formats that could be utilized and managed by its members. Finally, IOC should encourage the integration of coastal zone management policies in the development of marine scientific research programmes.

Consultative Process

190.   Some States reaffirmed that the establishment of the Consultative Process constituted an appropriate answer to States' concerns about coordination and that it contributed to an integrated approach to ocean issues. That had been achieved partly because of the different dimensions of the Process where lawyers, managers, scientists, custodians of the marine environment, etc., gathered in a single revenue, and where law could be checked against implementation. In that connection, the improved ocean affairs culture at the agency and the United Nations level could be attributed to the Process.

Part II

Priorities in marine science and technology

191.   The discussions in Part II of Panel A on priorities in marine science and technology were led off by presentations from the following representatives: Mr. Patricio A. Bernal, Executive Secretary, IOC/UNESCO; Mr. Hein Rune Skjoldal, Institute of Marine Research, Bergen, Norway; Dr. Li Jingguang, Director General, State Oceanic Administration, China; Dr. Norman P. Neureiter, Science and Technology Adviser to the Secretary, United States Department of State, jointly with Dr. W. Stanley Wilson, Director, International Ocean Programmes, Oceanic and Atmospheric Research, United States National Oceanic and Atmospheric Administration (NOAA); and Dr. Sian Pullen, World Wide Fund for Nature (WWF).

192.   Mr. Bernal addressed the issue of priorities in marine science and identified ocean and climate, ocean ecosystem science and marine science for integrated coastal area management (ICAM) as priority areas. He stressed that in marine science supply was serving demand and that the current and new priorities were defined by the supply-demand nexus.

193.   The driving forces behind science relating to ocean and climate are the needs to understand climate change and to mitigate the effects of

climate change. In understanding climate change, the important issues are seasonal and inter-annual forecasting, studying long-term effects on marine ecosystems and investigating the integrity of the life-support system on earth, of which oceans are an integral part. Climate change, manifested in sea-level rise, El Niño and La Niña type oscillations and increased occurrence of extreme events, can have devastating effects on society and the economy, mitigation of which is an urgent need.

194. Mr. Bernal gave examples of societal impacts of the 1997/98 El Niño, which included human fatalities, health risks, property damage, damage to crops, food shortages, water shortages and disruption in a number of sectors such as energy, transportation and tourism.

195. Mr. Bernal then stated that marine science was in the verge of moving from the current priorities focusing on the physics of the oceans to the new priorities which emphasized the chemistry and the biology of the oceans.

196. Ocean ecosystem science is driven by a number of factors, including dependence of a significant part of the world's population on the ocean ecosystem for its livelihood and food security. Other important engines of ocean ecosystem sciences include the need to study the effects of intensive and extensive exploitation as well as the accumulated effects over time and combined effects over sectors.

197. The current priorities of ocean ecosystem science include ecosystem-based fisheries management, land/ocean interface and coral reefs and other critical habitats. One example of this is the Global Ocean Ecosystem Dynamics (GLOBEC) project, which studies the effects of large-scale shifts in ocean regimes. In fisheries, ecosystem-wide multiple-species changes in population are found which appear to be attributable to certain environmental phenomena, in addition to the direct fishing activities of man. Mr. Bernal identified the following new priorities of ocean ecosystem science: the control and regulation of ecosystems; the identification and quantification of structural ecosystem changes, including valuation of ecosystem services; and ecotoxicology.

198. Marine science for ICAM is driven by management-oriented needs. It is estimated that by 2020 75% of the world's population will live near the coasts. Sixteen out of the world's 23 mega-cities are situated on the coast and the uses of the coasts by a number of industries, especially tourism, are growing. The priorities in marine science for ICAM are to increase the knowledge base at the localized level and to enhance local capacities. The local knowledge base includes information about the typology of coasts, the dynamics of sediment movements, including erosion, local current systems and local and regional bioproductivity regimes.

199. Mr. Bernal concluded his presentation by providing information about a global web service on ICAM (http://www.nos.noaa.gov/icm), a co-operative effort of IOC, UNESCO, the World Bank, the NOAA National

Ocean Service and the Center for Marine Policy of the University of Delaware, in conjunction with a number of other partners around the world. The service is aimed at providing timely and accurate information on developments and advances in ICAM at the global, regional and national levels.

200.   Mr. Skjoldal discussed marine ecosystems and the appropriate approach to their management. Marine ecosystems are open and subject to weather and climate patterns; their components are interlinked; and they face impacts of multiple human activities.

201.   Mr. Skjoldal provided examples of both long-term and short-term effects of ocean climate on fish populations and demonstrated that certain distinctive patterns emerge in different marine ecosystems. Presenting statistics on the populations of various species in a given marine ecosystem, he showed how the levels and fluctuations in populations of different species, especially in the food chain, are interlinked and how the distribution of various species is disturbed, implying a loss of integrity of the ecosystem.

202.   The interlinkages, combined with the multiplicity of human activities which have varying impacts on the interlinkages, necessarily call for an integrated approach in studying the marine environment. In the view of Mr. Skjoldal, an ecosystem approach can achieve the desired integration in management. He provided the definition used by ICES for an ecosystem approach to ocean management: integrated management of human activities based on knowledge of ecosystem dynamics to achieve sustainable use of ecosystem goods and services and maintenance of ecosystem integrity. He then provided a framework for an ecosystem approach to ocean management: ecosystem objectives are to be defined; monitoring and research, and thereafter integrated assessment of the findings are to be carried out, advice is to be provided based on such assessment; and such advice can be used to adapt management practice geared to the achievement of the ecosystem objectives. Science is the basis of the whole framework. All major stakeholders are to be involved in the management exercise. He cautioned, however, that a major challenge was to maintain the objectivity and integrity of science in the face of demands from various stakeholders.

203.   Mr. Skjoldal then considered the issue of environmental assessment, the main challenges of which are, first, to separate anthropogenic influence from natural variability, and then to distinguish the effects of different human activities. The recent Quality Status Report (QSR) of the OSPAR Commission was an excellent example of environmental assessment meeting the above challenges.

204.   After a demonstration of fish stocks' close connections with and adaptations to ocean circulation, he stressed the importance of ecosystem monitoring and assessment. An ecosystem is defined as a dynamic complex of plant, animal and organism communities and their non-living environment interacting as a functional unit. A large marine ecosystem (LME) is

an extensive region, typically larger than 200,000 sq km, with a unique hydrographic regime, submarine topography, productivity and trophically dependent population. On the global scale, the crucial factor is climate variability and change; on the LME scale, resources, especially biological resources, and physical environmental aspects are most important; on the local scale which is also relevant for ICAM, it is the land/ocean interaction and habitats, including effects of contaminants, eutrophication, microorganisms, mariculture, spatial use and physical disturbances, that are of primary importance.

205. Mr. Skjoldal concluded by providing elements of a plan of action for an ecosystem approach to management. The prerequisite of international coordination and cooperation was national coordination and cooperation. The elements of a plan of action included: a stronger international cooperation; use of GOOS as a core element; coordination between GOOS and international research programmes; and execution of selected LME test cases. Such test cases would involve combined monitoring and research, transferable experience and results, and training and capacity-building.

206. Dr. Li addressed the issue of developing marine science and technology to promote sustainable development. The twenty-first century was described as an era for the oceans, when man will devote greater effort to understand, develop and protect the oceans and the oceans will play a more important role in the development of human society and the economy. Marine science and technology will play an essential role in enhancing man's knowledge of the natural processes of the ocean; it can provide a rational basis for decision-making on sustainable development, can help to improve integrated coastal management, can improve the utilization of marine resources and can provide effective means for the protection of the marine environment and for the conservation of marine resources.

207. Dr. Li enumerated the achievements of the United Nations system in promoting marine science and technology; among them were UNCLOS, chapter 17 of Agenda 21, the Commission on Sustainable Development and its decision 7/1; the programmes of the organizations of the United Nations system and the work of the Consultative Process.

208. He then offered certain suggestions for the work of the United Nations system in the field of marine science and technology:

1) formulating guidelines for the development of marine science and technology geared to the social, economic and environmental goals at the global level;
2) encouraging States to formulate laws and regulations, compiling collections of existing laws, regulations and policies to that end, and providing training in formulating laws and regulations;
3) intensifying the role of the relevant organizations within the United Nations system responsible for marine science and technology mat-

ters in planning, guiding and coordinating global, regional and national marine scientific research projects—the functions of IOC should be further strengthened in this context;

4) improving the coordination between United Nations organizations responsible for marine science and technology matters and other ocean-related organizations, programmes and projects within the United Nations so as to avoid unnecessary overlapping and duplication and to improve utilization of available financial, human and material resources;

5) encouraging bilateral cooperation on the basis of equality and mutual benefit, especially cooperation between developed and developing countries; in regions with favourable conditions, encouraging multilateral cooperation on a regional basis; encouraging the establishment of joint research centres to study issues of common interest; and building joint virtual laboratories;

6) strengthening the Training, Education and Mutual Assistance (TEMA) programme of IOC;

7) developing practical and feasible plans for capacity-building to help developing countries: at present, capacities most urgently needed by most developing countries are those for marine scientific research, marine environmental observation and monitoring, marine resources survey and exploitation, and marine environmental protection; capacity-building in developing countries may be improved by the creation of marine scientific research centres provided with the necessary equipment, skills and expertise; demonstration centres may be established in countries where favourable conditions exist; and

8) promoting the transfer of marine science and technology, especially from the developed countries to the developing countries; preparing marine technology transfer plans and programmes and coordinating global and regional marine technology transfer activities in the spirit of UNCLOS; regular seminars or workshops should be held at the global level to provide a forum for discussing issues related to marine science and technology transfer and for exchanging experiences.

209. Dr. Li then described the marine science and technology activities of China with a view to stimulating the exchange of information and experiences among States. Since the 1980s, China's marine science and technology has developed rapidly, and major achievements have been made in the fields of coastal and ocean survey, oceanographic research, research and development and application of new and high ocean technologies.

210. In the new century, China's social and economic development will rely increasingly on the ocean; this will result in a growing level of activities of developing and utilizing marine resources and will exert great pressure

on the marine environment. In order to rationally exploit marine resources and protect the marine environment, an important task will be to vigorously develop marine science and technology. To achieve sustainable development and to enhance the contribution of marine science and technology to its social and economic development, China will continue to implement various programmes. Efforts will be made to promote the development of marine high technologies, ocean-related appropriate technologies, basic oceanographic research and applied research so as to expedite the application of research results to marine and ocean-related operations and industries, to serve the rational exploitation of marine resources and effective protection of the marine environment and to ensure the safety of offshore operations.

211.   Dr. Li concluded by enumerating the priority areas for China that will be closely connected with its economic and social development in the near future. These areas include observation, research and prediction of coastal natural hazards; integrated coastal area management; marine environment protection; mariculture and fishery; and utilization of seawater and desalination.

212.   Dr. Neureiter and Dr. Wilson gave a joint presentation, Dr. Neureiter concentrating on marine science priorities from a developed country perspective, and Dr. Wilson on ARGO, operational oceanography and marine scientific research.

213.   Dr. Neureiter stressed the importance of the concept of "science for development" for developing countries and of capacity-building in this context. Another emerging theme was the need for Governments to cooperate at the regional level to improve regional coordination in marine sciences and to assure that political decisions are based on sound science. For this, decision makers must be given the best available scientific information when making policy decisions. This presents a challenge, because scientific results are often interpreted by different groups in different ways. Also, often when decisions need to be made the supporting science is often incomplete.

214.   The paradigm is now shifting from managing single species and maximizing yields of every species to the sustainable management of marine ecosystems. This requires the integration of scientific information from many disciplines, ranging from species abundance studies to physical and biological oceanography to the study of changes in habitat with the introduction of land-based pollutants. In that context, Dr. Neureiter stressed the importance of the draft International Plan of Action to Improve Status and Trends Reporting developed by the FAO Advisory Committee on Fisheries Research.

215.   With regard to ocean observation systems, Dr. Neureiter stated that oceanography was maturing from the simple collection and description of observations to a real understanding of ocean processes which is close to achieving the ability to forecast events. Man currently has the capability to

implement long-term, operational observing systems for the global ocean, comparable to those in operation for the atmosphere for the past 30 years.
216.   A key element of marine scientific research is the ability to draw on a broad-based range of tools, including linking space-based and in situ observations. A common framework is needed to link these two techniques. In addition, recent innovations in ocean engineering and information technology are broadening man's ability to study the oceans and to use multiple layers of information to understand marine ecosystems. Cooperation is indispensable to these efforts, especially as one moves from physical oceanography to biological and chemical oceanography and the multidisciplinary approaches required to understand marine ecosystems. In this connection, Dr. Neureiter gave the example of the operational capability to collect in situ observations across the Equatorial Pacific, the El Niño/Southern Oscillation (ENSO) Observing System, a legacy of the decade-long Tropical Ocean and Global Atmosphere (TOGA) research programme. Together with data from satellites, the observations obtained have enabled an understanding of ENSO events, which allows one to forecast these events and anticipate their impacts. The TOGA/ENSO experience demonstrates how almost two decades of international cooperation in physical oceanography and meteorology have resulted in a forecasting capability of great societal and economic importance.
217.   Turning to chemical and biological oceanography, Dr. Neureiter stated that the observation of chemical and biological characteristics presented a greater challenge to the science community than the observation of physical ones.
218.   Many programmes have been initiated to investigate biological and chemical problems. For example, the Global Coral Reef Monitoring Network assesses the health of coral reefs and has become a key tool in understanding the diverse effects of the human activities that are causing the global decline of coral ecosystems. The Harmful Algal Bloom programme, which studies eutrophication and plankton blooms, is critical to human health and local economies. Dr. Neureiter also cited the examples of GPA, the work of GEF on LMEs, and the Census of Marine Life.
219.   He then emphasized that all of these programmes shared a common need for research, data collection, assessments, monitoring and the development of operational observations in the coastal oceans. The need for cooperation was underscored. Done independently, or without complementary approaches, each programme could implement components of its own observing system, resulting in a situation where the whole is less than the sum of its parts. Member States acting through the United Nations and its specialized agencies have a critical role to play: to facilitate, coordinate and set consensus standards for operational ocean observing systems. WMO/IOC is providing an organizational focus within the United Nations system for basin-scale physical observations. However, a similar organiza-

tional focus is needed for the inclusion of biological and chemical observations to address ecosystem issues, especially in the coastal regions. Cooperation could be facilitated through joint meetings, web sites, the publication of directories of specialists and regular regional reporting of priorities for incorporation in the Secretary-General's report.

220. Dr. Neureiter concluded by addressing the issues of capacity-building in developing countries. He emphasized that continuous consideration should be given about local capacity-building in every programme, every project and every organization. Only in this way will sustainable development be truly achieved on a global scale.

221. Dr. Wilson explained that drifting buoys currently collect global in situ observations at the sea surface, and surface and sub-surface observations are taken by vessels of opportunity along major shipping lanes. While satellites observe conditions at the sea surface globally, there is no comparable long-duration basin-scale capability, beyond the ENSO Observing System in the Equatorial Pacific, to observe sub-surface conditions spanning ocean basins.

222. IOC and WMO, working with UNEP and ICSU, have been leading the Global Ocean Observing System (GOOS). Their work is motivated by the premise that if everyone who had a need for ocean observations had his own independent observing system there would be duplication and gaps and there would be no means of integrating the resulting observations. GOOS is an effort to implement, by international consensus, complementary observing systems capable of meeting multiple needs, both in real time for operational users and in delayed mode for research. Complementary systems will facilitate the integration of observations, avoiding duplication and filling gaps, where the whole will be greater than the sum of the parts.

223. ARGO, the international programme to use 3,000 profiling floats to observe the upper ocean in real time, is one key in situ element of GOOS. These floats are oceanic analogues to the radiosondes used by meteorologists to profile the atmosphere. ARGO floats are programmed to drift at depths of 2,000 metres, rising to the surface every 10 days to observe temperature and salinity profiles. While briefly at the surface, they report their position and data to a satellite for relay to shore, and then sink to begin another 10-day cycle. They have a design life of about four years. ARGO has grown from 55 floats funded in 1999 to 525 in 2001. Global coverage by 2005 is anticipated, with a spacing between floats of 300 kilometres. The newly established Joint Technical Commission on Oceanography and Marine Meteorology (JCOMM) of IOC and WMO is to develop a consensus approach for the collection, distribution and archiving of marine observations, both atmospheric and oceanic. By providing an organizational focus within the United Nations system, JCOMM will help ensure the availability of consistent sets of observations to support research.

224. ARGO features a full and open data policy, a policy also in place for surface drifting buoys, volunteer observing ships and the ENSO Observing

System. Under it, there will be no period of exclusive use, and all data will be available to meet the needs of both operational agencies and the research community, thus bringing potential benefit to all. For example, the sharing of ARGO float data will facilitate new global-scale research into an understanding of the coupled ocean/atmosphere system, extending well beyond the discipline of oceanography. At the same time, the availability of real-time ARGO data will lead to improvements in operational climate forecasts by national meteorological services. Finally, the development of new sensors by the research community will enable the collection of chemical and biological observations from ARGO profiling floats.

225. Countries can be involved in the ARGO project in a number of ways: by helping in the deployment of ARGO floats; helping to implement complementary in situ observing systems; using ARGO data for research and operational demonstrations; and deriving benefit from improved operational forecasts.

226. Looking to the future, ARGO is one of a number of systems that collect routine, long-term observations of the oceans. Another such system is a set of time-series stations for collecting integrated observations at fixed sites. With their full and open data policy, these observing systems are changing the way oceanography is practised, enabling the development of a broader understanding of the interplay between the physical, chemical and biological components, and of how the oceans function as a system.

227. Dr. Pullen addressed the issue of targeting marine science and technology to develop an ecosystem-based approach for the protection and sustainable development of the marine environment. She highlighted the recommendations of WWF to the parties to UNCLOS and explained the rationale behind them. The recommendations are as follows:

(1) Adopt an integrated and multidisciplinary ecosystem-based approach to the management of the seas and oceans;
(2) Manage activities and demands, by using marine science and technology to assess the resources, make decisions about the management of the use of resources and apply and enforce the management tools;
(3) Promote regional cooperation in applying the ecosystem approach across national borders and establish political frameworks (e.g., joint declarations) to facilitate such cooperation;
(4) Integrate the ecosystem approach across sectoral and intersectoral policies, plans and programmes, including national biodiversity strategies and action plans and national strategies for sustainable development;
(5) Promote integrated, international monitoring and assessment programmes; the urgency with which these are required has accelerated owing to the rate of change in the environment and associated socioeconomic factors as a result of climate change;

(6) Target research and technical development to improve the management of marine resources, especially in fields where there are linkages between science, technology, social welfare and economics;
(7) Invest effort and resources to restore the marine environment, both as a means of protecting biodiversity in its broad sense and as an investment in the future sustainable economics of a region;
(8) Facilitate inter-agency coordination and support on a regional basis to provide sufficient information and appropriate technology to enable management measures to be implemented in a timely fashion and adequately enforced; in particular, mechanisms should be examined which would afford protection to threatened high-seas areas outside of exclusive economic zones;
(9) Apply the precautionary principle, as agreed at the United Nations Conference on Environment and Development, to focus scientific research and technology development on aspects and regions to prevent environmental and social degradation before it occurs;
(10) Adopt a strategic, applied programme of research that responds promptly to the needs of decision makers;
(11) Encourage forums in which specialists in fields including natural resources, social sciences, economics and legislation can interact and forums in which participants from developing and developed countries can share their perspectives and priorities for future research and development; and
(12) Establish a task force that includes members from United Nations agencies, Governments, intergovernmental organizations, non-governmental organizations, and academic institutions to develop specific proposals for research and development in line with recommendations from the current meeting of the Consultative Process; such a task force would need to include specialists in the fields of natural resources, social sciences, economics and legislation.

Scope of marine science programmes
228. The discussion that took place after the presentations focused on marine science programmes, priority in marine science and the linkage among different fields. Many delegations recalled the importance of marine science for sustainable marine development in order to ensure food security, alleviate poverty, foster economic prosperity and provide disaster prediction, prevention and mitigation.
229. With regard to future marine science and technological programmes, it was suggested that rigorous review processes of existing programmes, such as the Intergovernmental Panel on Climate Change, should include questions related to the extent to which the objectives of the pro-

grammes or projects were being achieved. In that connection, simple considerations such as how many people were trained and whether they were usefully deployed would allow measurement of the effectiveness of capacity-building.

230.   Many delegations underlined the fact that ocean management decisions should draw upon well-documented scientific and technical information. In that regard, it was important to ensure quality control and quality assurance of data across programmes so that data might be safely integrated for better management decisions. The delegations also highlighted the need to strengthen the link between marine scientific research and policy-making institutions. There was a need for sustained long-term marine environment observation and monitoring programmes, which were essential for an enhanced understanding of global changes, thus leading to the improvement of the scientific basis for policy-making.

231.   Other delegations emphasized that the protection of the marine environment, as well as an integrated approach to coastal management, were important elements that must be part of any marine scientific research programme and objective.

232.   Delegations cited a number of existing programmes which were effective and encouraged wider participation in them. Those programmes could also be models for programmes to be developed in the future. In that connection mention was made of GOOS and also its ARGO project; SEACAMP (South-East Asian Centre for Atmospheric and Marine Protection) and the WIOMAP (Western Indian Ocean Marine Applications Project), two major WMO-IOC regional cooperative programmes, currently under development, the aims of which would be to coordinate the enhancement of marine observing systems, modelling capabilities and services based on cooperation among interested agencies and institutions; and the GEF Regional Baltic Sea Project, a case study for the large marine ecosystem approach using the riparian countries of the Baltic Sea and a joint project by HELCOM, ICES and the International Baltic Sea Fishery Commission (IBSFC).

233.   Several delegations pointed out that, with respect to climate change, GOOS should be implemented in a balanced manner. To that end, opportunities must be created to enable developing countries to participate fully in scientific research and monitoring programmes such as GOOS.

234.   Other delegations still had questions about how the developing countries in particular could benefit from GOOS and all its mechanisms of observation of the oceans. Many developing countries were still facing basic problems such as limited electricity resources which would impede their use of sophisticated technology, including, inter alia, their access to data via computers, their ability to advance beyond the former system of VMS, which basically relied on the integrity of the ship captain and the owner of the vessel and their access to satellite imagery systems, which provided informa-

tion directly. It was suggested that specific ocean programmes for development might be needed. It was also pointed out that without an analysis of training needs, there could be no effective and relevant programme of capacity-building.

235.   With regard to capacity-building, some delegations stated that there was a need to pay greater attention to investing in people, training, development of the appropriate skills and providing means for the retention of trained and skilled people in developing countries.

Priorities in marine science

236.   With regard to priorities in marine science, the representative of WMO suggested that it was important to develop modalities for the close interaction of various marine disciplines. The Conference on Oceans and Coasts, to be organized by IOC and other organizations in Paris in December 2001, in the run-up to the World Summit on Sustainable Development, would provide a good opportunity for such interaction between various marine scientific disciplines. Many of the presentations would focus on different aspects of marine scientific research.

237.   Many delegations pointed to the importance of adopting an integrated ecosystem approach for the management of ecosystems and for the marine environment in general. Such an approach would include the involvement of various sectoral users of a specific ecosystem to identify and put in place specific arrangements for the sustainable use and protection of the ecosystem. There was a need for reliable, relevant and available scientific information to support that approach. Another important consideration would be to specifically address the role of science in risk-based decision-making and the operational application of the precautionary approach.

238.   Many delegations called for a better understanding of the interaction between the oceans and the atmosphere and its implications for climate change. Such knowledge would be aimed at enhancing and adapting the capacity of countries to handle information and respond to the negative impacts of climate change.

239.   Many delegations listed as priorities the following issues: ways of controlling and preventing unsustainable patterns of fisheries; necessity of carrying out environmental impact assessments in fragile marine environments for the implementation of potentially dangerous projects; study of pollution in oceans and seas and its impact on freshwater resources; impacts of pollution on fragile ecosystems, in particular closed and semi-enclosed seas; impact of ballast water, dumping of wastes, hazardous wastes, radioactive and chemical wastes in oceans and seas on marine living and non-living resources; marine pollution in coastal areas and its effects on agriculture; and crisis management in emergency situations.

240.   Several delegations pointed to the potentials of the increased utilization of non-living resources of the seabed. It could be considered crucial

for its future international strategies and international as well as national coordination programmes of marine scientific research to be developed not only with an interdisciplinary focus but also with a focus on the integrated goals of the sustainable use of the common heritage of mankind.

241.   In the view of some delegations, the high seas contained a significant biodiversity which was as yet poorly known. For example, while 40% of the species from seamounts were known, specialists had indicated that there were a significantly greater number of species yet to be discovered. The ecological dependencies and the role of those species and systems were even more poorly understood. In that context, support should be given to the Census of Marine Life to be undertaken by the United States and others. In addition, there was a need for an improved and coordinated scientific focus on identifying and managing risks to biodiversity and the environment of the high seas, which would lead also to the adoption of improved management mechanisms, including the use of the precautionary approach.

242.   Many delegations recognized the danger posed by marine pests introduced into the marine ecosystems, both in terms of their productivity for human use and their intrinsic integrity. Recent estimates indicated that over 3,000 non-native species were being moved around the world daily in shipping and other means. A key requirement for the international management of ballast water introduction was an international framework that would include the following considerations: sharing of information on pest distribution and impacts; the scientific vetting of proposed ballast water treatment options; and helping to set internationally acceptable standards for ballast water cleanliness.

243.   Many delegations were of the view that the consideration of issues of the underwater cultural heritage and the continental shelf belonged in different forums and therefore they were not deemed relevant to the discussions of the Consultative Process.

Linkages among different fields

244.   In response to the questions posed in the annotated agenda and format (A/AC.259/L.2, appendix I, para. 31) regarding the strengthening of the linkages between different fields of marine scientific study and linkages between the study of the marine environment and the study of social and economic factors, many delegations proposed the following:

- Strengthening coordination at the international level, as well as the inter-agency level, with the aim of avoiding duplication and streamlining the activities in different forums;
- Strengthening UNEP regional seas programmes through further cooperation with relevant international organizations;
- Establishment of centres for dissemination of information on marine scientific research and technology;
- Strengthening GEF and other financial institutions with a view to en-

abling them to actively finance capacity-building projects in developing countries, in particular in the areas of: (a) controlling and reducing pollution in oceans and seas, especially in fragile ecosystems like closed and semi-closed seas; (b) coastal cities waste management and recycling projects; (c) controlling and reducing the pollution from shipping, dumping of hazardous and radioactive wastes, as well as chemical wastes and other harmful substances; and (d) carrying out environmental impact assessments for projects which are potentially harmful to the marine environment;
- Implementation of joint projects between regional organizations of the UNEP regional seas programme and other relevant international organizations;
- Encouraging, at all levels, the steps necessary for an effective and coordinated implementation of UNCLOS and Agenda 21.

(b) Discussion Panel B: Coordination and cooperation in combating piracy and armed robbery at sea

245.   The discussions in Panel B on coordination and cooperation in combating piracy and armed robbery at sea were led off by presentations from the following representatives: Mr. E. E. Mitropoulos, Assistant Secretary-General/Director, Maritime Safety Division, International Maritime Organization; Mr. J. Abhayankar, Deputy Director, International Maritime Bureau, International Chamber of Commerce; and Mr. H. Sato, Director of the Ocean Division, Ministry of Foreign Affairs of Japan.

246.   Mr. Mitropoulos in his presentation pointed out that in addition to piracy and armed robbery, other unlawful acts under consideration in IMO included unlawful seizures of cargo and other forms of maritime fraud, terrorism at sea, illicit drug trafficking, stowaway cases and illegal transport of migrants by sea.

247.   He said that piracy and armed robbery had consistently figured on the agenda of the IMO Maritime Safety Committee (MSC) since 1984. A number of IMO Assembly resolutions and MSC circulars had been adopted, but as statistical information received demonstrated, there had been a considerable deterioration of the situation in the Malacca Strait, the South China Sea, the Western/Central African region, parts of the northern area of Latin America and the Caribbean, and parts of the Indian Ocean.

248.   Mr. Mitropoulos provided information on past efforts by IMO to assist countries most affected by acts of piracy and armed robbery against ships, for example, the dispatch of an IMO working group to Malaysia, Singapore and Indonesia in 1993 to report on the situation in the Malacca Strait; and an IMO mission to China, Hong Kong and the Philippines in 1994 to study the situation in the South China Sea. Following those missions, a sig-

nificant improvement had been experienced, albeit temporarily. In 1998, MSC had launched an anti-piracy project comprising a number of missions of experts and seminars/workshops in countries in the South China Sea and the Malacca Strait and in Brazil; followed by a regional seminar and workshop for the Latin American and Caribbean region, held in Brasilia in October 1998; a regional seminar and workshop for the South-East Asia region, held in Singapore in February 1999; a mission of experts to Abuja, followed by a regional seminar and workshop for the West and Central African region, held in Lagos in October 1999; and a regional seminar and workshop for the Indian Ocean region, held in Mumbai, India, in March 2000.

249. Unfortunately, he noted, the completion of the 1998 anti-piracy project had not coincided with any significant improvement of the situation; on the contrary, the situation had worsened and this had caused grave concern to MSC, which, at its seventy-second session in May 2000, had decided, subject to the availability of funds, that a number of assessment missions should be undertaken to countries bordering waters where pirates and armed robbers continued to operate unabated.

250. The purpose of this new IMO effort (phase 2 of the anti-piracy project) was: (a) to evaluate the actions taken by the invited Governments to implement the IMO recommendations to prevent and suppress acts of piracy and armed robbery against ships within areas of their jurisdiction; (b) to receive information on the measures the national authorities of the participating countries had put in place for the purpose of implementing at the national level the recommendations of the IMO regional seminars and the workshops held within the 1998 anti-piracy project as well as those contained in revised MSC circulars 622 and 623; (c) to identify where such measures had not been successful and what had impeded their implementation; (d) to explain the reasons behind any total or partial inability to implement the measures; (e) to seek information on any ideas/proposals the participating Governments might have with respect to regional cooperation for the purpose of combating piracy and armed robbery against ships (for example, joint or coordinated exercises, patrolling of particular vulnerable sea areas, exchange of intelligence on moves of pirates/armed robbers); and (f) to specify ways in which IMO could assist in overcoming any difficulties the participating countries had encountered in the process.

251. The first such mission had been dispatched to Jakarta from 13 to 14 March 2001, and a regional meeting had been held in Singapore from 15 to 16 March 2001, which was attended by representatives from countries that were experiencing extensive piracy in waters off their coasts. The States participating in the Singapore meeting either: (a) could play a substantial role in addressing the problem by virtue of their strategic location vis-à-vis the most affected areas, stretching from the South China Sea to the Malacca Strait to the Eastern Indian Ocean; or (b) had a genuine interest in seeing

the problem effectively addressed because of the large number of ships under their national flag using the waters concerned.

252.   The mission to Jakarta and the meeting in Singapore had been undertaken against a backdrop of a deteriorating situation with respect to piracy and armed robbery. In the Malacca Strait, the situation had dramatically worsened, with 75 attacks reported during 2000, as opposed to 2 incidents in 1999.

253.   The number of acts of piracy and armed robbery against ships in 2000, as reported to IMO, was 471, an increase of 52% over the figure for 1999. The total number of incidents of piracy and armed robbery against ships reported to have occurred from 1984 to the end of April 2001 was 2,289. From March 2000 to March 2001, as compared with 1999, in the Mediterranean Sea the number of reported incidents had decreased from 4 to 2, while in West Africa there had been a decrease from 36 to 33. On the other hand, the number of incidents had increased from 37 to 112 in the Malacca Strait, from 136 to 140 in the South China Sea, from 51 to 109 in the Indian Ocean, from 16 to 29 in East Africa and from 29 to 41 in Latin America and the Caribbean. Most of the attacks worldwide were reported to have occurred in the territorial waters of coastal States while the ships were at anchor or berthed. During the same period, 72 crew members had been killed, 129 had been wounded and five reported missing. One ship had been destroyed, two ships had been hijacked and three ships had been reported missing.

254.   The Singapore meeting identified the following main problem areas (not applicable in all participating countries): the continuing adverse economic situation prevailing in certain parts of the region; the geographical configuration of certain countries; the resource constraints on law-enforcement agencies; the lack of communication and cooperation among the various national agencies involved; the delayed response time after an incident had been reported to the coastal State concerned by affected ships; general problems of incident reporting, such as alerting the nearest coastal States and other ships in the area of a ship under attack or threat of attack; the prosecution of pirates and armed robbers when apprehended; and the lack of regional cooperation.

255.   Mr. Mitropoulos said that the Singapore meeting had agreed upon a number of recommendations, which he believed IMO would consider when MSC met in May 2001. The meeting had recommended that participating Governments identify, on the basis of experience and statistical information, vulnerable areas off their coasts and in their ports and direct their resources to cope with the increased risks to safe navigation and environmental protection in such areas, with particular emphasis on areas used by international shipping; and provide specific advice for ships on protective measures and local reporting procedures. The participants identified focal points in their respective administrations for the exchange of information

and coordination of efforts in the fight against piracy and armed robbery in the region. The effectiveness of coordinated patrols and joint exercises, where appropriate, to test existing anti-piracy systems and strengthen cooperation among neighbouring countries in their efforts to eradicate piracy and armed robbery against ships in their waters had been recognized and encouraged. Governments were encouraged to continue and further strengthen regional initiatives, such as the 2000 Tokyo conferences and the follow-up meeting in Kuala Lumpur.

256. Governments in the region which had not yet done so were encouraged to ratify the 1988 Convention for the Suppression of Unlawful Acts against the Safety of Navigation (SUA Convention) and the 1988 Protocol for the Suppression of Unlawful Acts against the Safety of Fixed Platforms Located on the Continental Shelf (the SUA Protocol) to consider doing so. Currently there were 52 States parties to the 1988 Convention, representing 48% of the world tonnage; and 48 States parties to the Protocol.

257. Furthermore, the industry was encouraged to ensure that all attacks or attempted attacks were reported promptly to the nearest Rescue Coordination Centre as well as the designated focal points of the coastal and flag State concerned. Concern was expressed at the lack of reporting to IMO by flag States on most instances of attacks or attempted attacks on their ships and MSC was invited to urge all flag States to make such reports in accordance with the relevant IMO instruments.

258. The meeting also invited the Secretary-General of IMO to undertake consultations with Governments in the region for the purpose of convening, at an appropriate time, a meeting to consider concluding a regional agreement on cooperation against piracy and armed robbery against ships.

259. Participating Governments lacking the necessary expertise and associated resources were encouraged to seek technical assistance from IMO in order to improve their capabilities to prevent and suppress piracy and armed robbery against ships in their waters.

260. Mr. Mitropoulos explained that other IMO activities regarding the combating of piracy and armed robbery included the preparation of a draft Code of Practice for the Investigation of the Crimes of Piracy and Armed Robbery against Ships for submission to the IMO Assembly at its twenty-second session in November 2001.

261. Member Governments were invited to use the Code when arranging for investigations into the crimes of piracy and armed robbery against ships to be conducted under their jurisdiction. He noted that the draft Code provided a definition of "armed robbery." IMO was also working on the prevention of the registration of "phantom ships."

262. In conclusion, he stated that IMO would like to see the Consultative Process increase awareness of the problem of piracy and armed robbery; motivate the political will to act; build a consensus position and shape a

uniform policy to prevent and suppress acts of piracy and armed robbery; and sensitize countries in areas affected to act and others in a position to assist or provide resources to do so.

263. Mr. Abhayankar in his presentation said that pirates today generally fell into two categories: poor, opportunistic people; and professional pirates. Both types were usually armed and difficult to apprehend. He pointed out that pirates did not hesitate to murder the crew, often with extreme brutality, as had been the case with the crew of the MV Erria Inge.

264. He said that in 2000, the highest number of acts of piracy had been recorded. The trend was continuing in 2001: as at the end of April, the number of reported incidents had approached 100. At the same time, it was to be noted that only one in three attacks was actually reported. There had been an increase in the number of incidents, especially in the Malacca Strait, and also in the Red Sea, where there had been 13 incidents so far, compared with none in 1999. A comparison of statistics of incidents reported to the International Maritime Bureau during 1999 and 2000 showed a total number of attacks reported in 2000 of 469, representing a 56% increase over 1999; 15 crew members killed in 2000, as opposed to 3 in 1999; and 8 ships hijacked in 2000, as against 10 in 1999.

265. Mr. Abhayankar described some common features of what he termed "maritime mugging," which accounted for 85% of the incidents: the target of most attacks would be cash and valuables. Vessels were boarded in port, at anchorage or steaming, and attacks lasted from 30 to 60 minutes, during which time the ship usually was not under command. Pirates were armed, but not necessarily organized. Violence was employed mainly if the crew resisted. Regions or particular countries most affected were South Asia, South-East Asia, the Far East, West Africa and Brazil.

266. The common scenario of a cargo or ship hijacking was described by Mr. Abhayankar as follows: the target of the attack would be the entire cargo or the ship; the ship was boarded at sea by heavily armed pirates; the attacks lasted several days; violence was used and the crew was locked up or killed; and organized criminal syndicates were involved. Such incidents, of which there were about 12 annually, were predominant in South-East Asia or the Far East.

267. He then enumerated the major hijackings that had taken place between 1998 and 2000 and provided a detailed account of the hijacking of the Alondra Rainbow in October 1999 and its subsequent recovery by the Indian authorities. One of the highlights of that incident had been the name changes that the vessel had undergone in attempts to conceal its identity. The pirates had painted new names over the original name of the vessel, but when the vessel was brought into port by the Indian authorities and the paint was removed, the true identity of the vessel was revealed. To thwart such activities in future, the International Maritime Bureau had proposed that an IMO identification number be welded onto vessels.

268. He outlined some of the legal issues arising from the typical hijacking of a vessel. Problems of jurisdiction might arise in situations where UNCLOS had not been ratified or was not applicable, the State concerned was not a party to the SUA Convention or there was no provision in the national legislation on piracy. In such cases one possibility might be to try the offenders under jure gentium. In the area of investigations, possible issues might centre on the number of countries involved or the expertise and the costs involved—which in the case of the Alondra Rainbow had been borne by India, the intercepting State. The possibility of such a burden might discourage States from becoming involved in preventing and suppressing acts of piracy and armed robbery at sea. With regard to the prosecution of the offenders, he pointed out that in these types of criminal matters the degree of proof required was high. Mr. Abhayankar underscored the paramount importance of enacting appropriate national legislation.

269. He emphasized that piracy was responsible for bringing about physical and psychological trauma for the crew, monetary loss and threats to the marine environment. He made the following recommendations for combating piracy:

   (1) preventive measures by the crew—the IMO circulars provided excellent guidance in that respect;
   (2) industry initiatives—establishment of a piracy reporting centre in Kuala Lumpur financed by shipowners and the Protection and Indemnity (P&I) Clubs;
   (3) application of the model law developed by the Comité Maritime International (CMI), which might be of assistance in answering some of the problems;
   (4) use of technology-tracking devices cost less than US$ 300 per month and could be hidden on board a ship;
   (5) a proactive approach by coastal and flag States—for example, if a coastal State were unable or unwilling to prosecute pirates, the flag State should have a role to play;
   (6) regional cooperation—joint patrols had proved to be a deterrent; and
   (7) intergovernmental involvement—IMO had done an excellent job, but it might be useful to have some law enforcement involvement at the intergovernmental level. He did not recommend the use of arms by the crew or the use of armed guards providing security. In conclusion, he raised the question of whether an international task force was needed.

270. Mr. Sato, in his presentation, noted that Asian waters, particularly the South China Sea and the Malacca Strait, had been affected severely by piracy and armed robbery against ships. Over the past two years, Japan had been doing its utmost to promote global as well as regional cooperation in dealing with the problem of piracy.

271. With regard to cooperation at the global level, he said that, firstly, Japan felt it was taking effective steps to combat piracy by taking up the problem at international forums, such as the United Nations General Assembly and the Consultative Process. In doing so, it was demonstrating the political will to fight piracy as well as heightening public awareness of the issue. Secondly, his country appreciated and supported the efforts undertaken by IMO, as demonstrated, for example, in Japan's participation in the correspondence group on the IMO draft Code of Practice for the Investigation of the Crimes of Piracy and Armed Robbery against Ships as well as in the IMO Regional Expert Meeting on Combating Piracy and Armed Robbery against Ships held at Singapore in March 2001. Thirdly, he said, Japan would also continue to make efforts to urge States to become parties to the SUA Convention and its Protocol, and to ensure its effective implementation.

272. In connection with regional cooperation, Mr. Sato said that in response to the proposal made by former Japanese Prime Minister Keizo Obuchi at the Summit Meeting of Japan and the Association of South-East Asian Nations (ASEAN) in November 1999, Japan had held the Regional Conference on Combating Piracy and Armed Robbery against Ships at Tokyo, in April 2000, in which 17 countries had participated. The Conference had adopted three documents, namely the Tokyo Appeal, a Model Action Plan and "Asia Anti-Piracy Challenges 2000."

273. The Tokyo Appeal proposed the establishment of contact points for all maritime-related concerns as well as the elaboration of action plans for combating piracy and armed robbery against ships, in particular a plan for strengthening the self-defence capability of private ships. The Model Action Plan contained more concrete measures based upon the proposals put forward in the Tokyo Appeal.

274. The paper entitled "Asia Anti-Piracy Challenges 2000" provided guidelines for facilitating regional cooperation on combating piracy and armed robbery against ships and proposed such measures as: information exchange among coast guard authorities; mutual cooperation in dealing with unlawful activities; technical cooperation to enhance the individual capability of coast guard authorities; and the continuous holding of expert meetings.

275. As a follow-up to the Conference, Japan had dispatched a mission to the Philippines, Malaysia, Singapore, and Indonesia in September 2000 to consult with the Governments concerned about concrete measures aimed at implementing the proposals enunciated at the Tokyo Regional Conference. As a result of those consultations and bearing in mind the mutual cooperation proposed in "Asia Anti-Piracy Challenges 2000," the Japanese Coast Guard had conducted joint exercises with India and Malaysia. The exercises covered areas of communication, search and rescue, interception, and boarding. A further joint exercise with the Indian Coast Guard was to be held in May 2001 in Japan.

276. At the Regional Experts Meeting on Combating Piracy and Armed Robbery against Ships, held at Kuala Lumpur in November 2000, Japan had expressed its willingness to accept students from the Asian region at the Japan Coast Guard Academy starting in April 2001. Mr. Sato stated that students from Thailand, Viet Nam, Malaysia, Indonesia, and the Philippines had already enrolled at the school.

277. Japan would also be holding a Maritime Law Enforcement Seminar in 2001, where participants would be provided with the knowledge, skills and techniques for planning, conducting and supervising maritime law-enforcement activities.

278. Mr. Sato recalled that at the "ASEAN + 3" (Japan, China, Republic of Korea) Summit Meeting in November 2000, Prime Minister Mori of Japan had proposed holding an Asian Cooperation Conference on Combating Piracy and Armed Robbery against Ships in 2001, again to be held in Japan. Issues proposed for the conference included information exchange, the future direction of regional cooperation and capacity-building.

279. He summarized the main problem areas with respect to piracy and armed robbery at sea as: (a) the lack of communication and cooperation among the various national agencies involved within individual countries; (b) the response time after an incident has been reported to the coastal State concerned by affected ships; and (c) general problems of incident reporting. It was the view of Japan that the solution to these problems lay in the improvement of the system of information exchange.

280. In order to address other problems, including the timely and proper investigation of reported incidents, the prosecution of pirates and armed robbers when apprehended and the lack of regional cooperation, it was necessary to strengthen regional cooperation among the maritime law-enforcement authorities of the countries concerned.

281. A third set of problems included: the continuing adverse economic situation prevailing in certain parts of the region; the geographical configuration of certain countries; and the resource constraints on law-enforcement agencies. Mr. Sato pointed to capacity-building as one possible solution to those issues.

282. In the discussions that followed the presentations by the three speakers, the following points were raised.

283. Delegations, including those not currently affected by the problem of piracy and armed robbery, expressed concern at the recent dramatic increase in incidents of piracy and armed robbery at sea and the associated level of violence, particularly in South-East Asia and the Malacca Straits.

284. It was recognized that acts of piracy and armed robbery represented a serious threat to the lives of seafarers, the safety of navigation, the marine environment and the security of coastal States. They also had a negative impact on the entire maritime transport industry, leading, for example, to increases in insurance rates and even the suspension of trade.

285. Several delegations noted that armed robbery, particularly repeated acts committed in the territorial seas of coastal States and in international straits, could threaten the rights of innocent passage and transit passage and, in archipelagic waters, the passage through archipelagic sea lanes, as enjoyed by all States under the United Nations Convention on the Law of the Sea. It was further noted that armed robbery also affected the management of ports.

286. Delegations highlighted the potential of an environmental disaster from an attack on a ship, particularly if a ship carrying hazardous cargo was left steaming with no one in command in a high-traffic area and/or a narrow waterway.

287. Delegations agreed that piracy and armed robbery at sea was a global problem, which in order to be combated effectively required action and cooperation at all levels. Acts of piracy and armed robbery against ships were characterized as international crimes that were often part of organized transnational crimes, which no State could combat on its own. This was particularly true in cases which involved so-called "phantom ships."

288. It was also noted that piracy and armed robbery should be seen in the larger context of illegal activities at sea, such as illicit traffic in narcotic drugs and psychotropic substances, the illegal transport of migrants, and organized crime, which constituted a threat to international peace and security.

289. It was pointed out by several delegations that the flag State, the coastal State, the port State and the State of which the criminals were nationals had a particular responsibility to combat piracy and armed robbery. The importance of consistent and uniform measures in combating piracy and armed robbery at the regional and global levels was underscored.

290. Delegations recalled the duty of all States under article 100 of UNCLOS to cooperate to the fullest extent possible in the repression of piracy on the high seas and in any other place outside the jurisdiction of any State. In that connection, it was pointed out that the Convention only applied to piracy on the high seas or in areas outside the jurisdiction of States, while the majority of acts of violence against ships occurred in the territorial waters or ports of States while the ships were at anchor or berthed.

291. The importance of cooperation at the international level was raised by a number of delegations. They noted with appreciation the activities of IMO in order to prevent and combat piracy and armed robbery against ships. It was observed that while regional initiatives and activities should be strongly encouraged and supported, IMO should be recognized as the international organization with the primary mandate to deal with the problem of piracy and armed robbery at sea at the global level. Several delegations, concerned at the current under-reporting of acts of piracy and armed robbery and stressing the importance of reporting all incidents, proposed that IMO should be a focal point for receiving such reports.

292.   Several delegations highlighted the importance of implementing the IMO guidelines on preventing attacks of piracy and armed robbery contained in the MSC circulars, as also called for by the United Nations General Assembly in its resolutions 54/31 and 55/7 on oceans and the law of the sea. Reference was also made to the recent circulation of the IMO draft Code of Practice for the Investigation of the Crimes of Piracy and Armed Robbery against Ships and the work of the IMO Subcommittee on Flag State Implementation in preparing a draft resolution to prevent registration of so-called "phantom ships." The draft texts were to be submitted to the IMO Assembly in November 2001 for adoption.

293.   A number of delegations highlighted the importance of developing preventive measures with respect to phantom ships. Reference was made to a proposal from Hong Kong, China, to the forthcoming meeting of the IMO Maritime Safety Committee, that the IMO Ship Identification Number should be visibly welded onto the stern of all vessels required to possess an IMO number. It was proposed that States should transmit to IMO and flag States information they have on phantom ships and that IMO should establish a database of this information, which could be accessed by shipowners. It was also suggested that consideration should be given to increasing the onus upon flag States not to register stolen vessels. Furthermore, it observed that article 110 of UNCLOS provided a basis for boarding vessels flying questionable flags.

294.   Several delegations expressed the view that more attention should be paid to the applications of current technology to vessel-tracking, e.g., the role that an Automatic Identification System could play in curbing criminality. It was suggested that home port verification should also be considered.

295.   A number of delegations noted the valuable work carried out by the International Maritime Bureau of the International Chamber of Commerce in combating piracy and armed robbery and also the work of other organizations, such as Interpol, the International Chamber of Shipping, the Baltic and International Maritime Council, the International Transport Workers' Federation, the International Union of Marine Insurance, and the International Group of P&I Clubs.

296.   In a letter it had addressed to the Secretariat, the International Labour Organization emphasized the need to protect the lives of seafarers and to ensure that they were not deprived of the elementary and universal rights to freedom, security, and dignity.

297.   Reference was made to the work of the Comité Maritime International, particularly its elaboration of a model national law on piracy and maritime violence. Several delegations said that the work of CMI should be honoured. Piracy was truly a global problem which cut across all sectors of society. However, those delegations pointed out that, given the number of forums in which the problem was under discussion, there was a risk that it was being dealt with in disparate ways. Also, attempts to define criminal acts

against ships not included in the definition of piracy in UNCLOS represented one area that needed further discussion.

298. Encouraging cooperation between States and relevant international bodies, several delegations pointed to the need for a global management regime, as well as to the need to ensure that measures taken by individual States were consistently enforced within the framework of international law. The World Bank, together with regional bodies and States, should support those measures, it was suggested.

299. Several delegations proposed that the World Maritime University should serve as the focal educational point at the global level and begin a more organized education campaign on the issue of piracy and armed robbery against ships. It was noted that the University already provided lectures on piracy, and that it could develop a seminar on piracy in which representatives of United Nations bodies could participate. Alternatively, a "professional development course," with the participation of maritime administrators from around the world, might be developed. Such courses would include the training of investigators of piracy and armed robbery against ships and might also serve as a contact point for representatives from those regions where the problem of piracy was most serious.

300. Also with respect to training, the United States stated that its Federal Bureau of Investigation was providing training, which could enhance regional efforts to combat piracy and armed robbery.

301. Many delegations underlined the importance of cooperation at the regional level, in particular among States in regions most affected by acts of piracy and armed robbery. The value of the IMO regional workshops and seminars and their follow-up was emphasized in that regard.

302. With a particular focus on the problem in South-East Asia, many delegations commended the holding of recent regional conferences and other cooperative initiatives among the States in the region. Reference was made to the 1976 Declaration of the ASEAN Concord and the 1997 ASEAN Plan of Action to Combat Transnational Organized Crime. The latter was described as constituting a milestone for combating piracy in the South-East Asian region. It put in place a cohesive regional strategy and facilitated information exchange among ASEAN member States, cooperation in legal and law-enforcement matters, institutional capacity-building, training, and extra-regional cooperation. Other ongoing efforts in the South-East Asian region included the two high-level international conferences on combating piracy held at Tokyo in March and April 2000, the meeting of the ASEAN Regional Forum, held in India in October 2000, the Experts Meeting on Combating Piracy and Armed Robbery against Ships, held in Malaysia in November 2000, and the South-East Asian Programme in Ocean Law, Policy and Management (SEAPOL) Inter-Regional Conference on "Ocean Governance and Sustainable Development in the East and Southeast Asian Seas: Challenges in the New Millennium," held in Thailand from 21 to 23

March 2001, which devoted one of its sessions to piracy and law enforcement.

303. Several delegations pointed out that tangible progress had been made in the South-East Asian region in terms of both deterrence and enforcement; cooperation had been intensified with neighbouring countries by increasing maritime patrol and equipment with greater use of satellite tracking vessel and monitoring systems. However, the States of the region still required the assistance and expertise of developed nations and relevant international organizations.

304. Several delegations were encouraged by the initiatives taken by ASEAN to combat acts of piracy and armed robbery and hoped that the outcome of the Consultative Process would help further consolidate such regional efforts. One delegation proposed that the issue of piracy and armed robbery could also perhaps be addressed within the framework of the organization for Asia Pacific Economic Cooperation.

305. Several delegations pointed out that serious consideration should be given by States sharing borders in areas threatened by piracy to establishing—preferably formal—bilateral/regional cooperation arrangements. Regional agreements would provide a legal framework for cooperation. In that regard, reference was made to the example of a draft regional agreement appended to IMO Circular 622/Rev.1. It was noted, however, that in drafting regional agreements, attention should be paid to the different characteristics of the various regions, as well as to their political environments. The ideal, therefore, would be to forge a consensus before calling on IMO's expertise in the elaboration of regional agreements.

306. One delegation noted that it was important for regional cooperative arrangements or agreements to be open not only to the States of the region, but also to other States with a substantial interest in navigation in the region. Capacity-building in law enforcement and sharing of expenses, etc., could be addressed within the framework of such arrangements or agreements.

307. Since acts of piracy and armed robbery could result in collisions or groundings, the importance of regional emergency plans in the event of a pollution incident was emphasized by several delegations.

308. It was noted by several delegations that flag States whose ships were sailing in waters affected by crimes at sea and were the targets of piracy attacks or armed robbery should make an increased effort to advise their ships on how to take precautions against those attacks. The importance of the IMO Guidance to Shipowners and Ship Operators, Shipmasters and Crews on Preventing and Suppressing Acts of Piracy and Armed Robbery against Ships (MSC Circular 623/Rev.1) and several other guidelines developed by other organizations or Governments were referred in this regard.

309. Several delegations expressed the view that IMO and other organizations should strongly discourage the carrying and use of firearms on board merchant vessels.

310. The importance of alerting other ships in the vicinity of an attack was highlighted as an important tool in combating piracy and armed robbery at sea. It was pointed out by IHO that the appropriate coordinator of the navigational warning service needed also to be informed of all actual and attempted attacks which took place in the area for which he or she was responsible under the IMO/IHO worldwide navigational warning service.

311. One delegation stressed that States which had information about facts or circumstances leading to a presumption that acts of piracy or acts against the safety of navigation might occur should provide information to the relevant States.

312. One delegation raised the issue of the effects of acts of piracy and armed robbery against ships on the level of insurance premiums that shipowners might be charged and the subsequent impact on the costs of transportation of goods and their delivery.

313. Delegations recognized the importance of action at the national level to combat acts of piracy and armed robbery. Many reported on steps they had undertaken to increase port security and strengthen their maritime enforcement capabilities, including through increased coordination among various national administrations and departments.

314. Some delegations stated it was imperative for Governments to consider taking the actions identified in paragraphs 209 to 223 of the report of the Secretary-General on oceans and the law of the sea (A/56/58), including: the development of national action plans for preventing an attack as well as steps to take in the event of an attack; augmenting surveillance; and enhancing port security.

315. Reference was made to the need for full cooperation between the coastal State and the flag State, including the role of the State in warning ships in the area where the attacks were likely to take place, in particular in cases of repeated attacks.

316. The importance of adequate charting of waters was highlighted in order, inter alia, to locate the hide-outs of pirates and armed robbers. In that connection, IHO pointed out that hydrographic surveying was very expensive.

317. Delegations agreed that the strengthening of the capacity of developing countries was intrinsically linked to improved efforts in the suppression of piracy and armed robbery against ships.

318. The importance of training for port personnel was stressed. The United States, referring to the current two-year course offered by the World Maritime University, said it would be in a position to fund requests for attendance by personnel from developing countries at the course if such requests were identified as a priority. In addition, the United States Coast Guard was available for instruction on law-enforcement tactics and port security measures and enhancements to member States. Member States were encouraged to send their personnel to the United States for training or, alternatively, the

United States Coast Guard could send international training detachments to other States.

319. It was suggested that the World Bank and other donor agencies such as the United Nations Development Programme should give priority to requests for assistance from developing countries in addressing two major issues with respect to piracy and armed robbery at sea, namely enhancing enforcement capability and implementation of port security measures. One delegation stated that donor institutions should be encouraged to engage in a dialogue with developing countries to assess the needs identified to address piracy and report their findings to the Secretary-General.

320. Delegations noted that acts of piracy were by definition confined to the high seas or the exclusive economic zone and that not all of the attacks that took place in these maritime zones could be classified as traditional acts of piracy over which all States might exercise jurisdiction under the provisions of the United Nations Convention on the Law of the Sea.

321. It was noted that when attacks against ships occurred in port or at anchorage they were most likely proscribed and should be punishable by local criminal law. When acts endangered the safety of navigation and occurred on board foreign flag ships while under way in the territorial sea, in international straits or in waters beyond the limits of the territorial sea, those acts were frequently not proscribed or punishable by the criminal law of the coastal State. It was noted in that regard that the SUA Convention and its Protocol could fill many of the jurisdictional gaps. The SUA Convention required States parties to criminalize such acts under national law and to cooperate in the investigation and prosecution of their perpetrators.

322. One delegation recalled the draft ocean space treaty proposed by Malta in 1971 to the Committee on the Peaceful Uses of the Seabed and the Ocean Floor beyond the Limits of National Jurisdiction, which contained provisions on combating piracy also in the territorial sea.

323. Many delegations underlined the importance for States to ratify or accede to UNCLOS and the SUA Convention and its Protocol, as well as the United Nations Convention against Transnational Organized Crime. It was suggested that the Consultative Process could endorse the United Nations General Assembly's call for States that had not done so to consider adhering to the SUA Convention and its Protocol and to implement its provisions.

324. Several delegations stated that States should be encouraged to enact and enforce national legislation for effective implementation and enforcement of the above Conventions and that all States should review their national legislation and practice to see if they fully reflected the rights and duties embedded in the Conventions. One delegation suggested that elements of legislation necessary to implement the obligations under the SUA Convention should be identified, following the approach taken for the 1988 United Nations Convention against Illicit Traffic in Narcotic Drugs and Psychotropic Substances.

325. The importance of ensuring that measures by individual States were consistently enforced within the framework of international law was highlighted by several delegations.

326. They underlined the importance of a common understanding of States' enforcement rights under international law. One delegation emphasized that, more than reaching a common understanding on existing rules, there should be a direct reference to and application of such rules.

327. In conclusion, several delegations said that they looked forward to a strong statement in the report of the meeting to the United Nations General Assembly on the importance of preventing and combating piracy, suggesting measures and decisions that could be reviewed at future meetings of the Consultative Process and by the General Assembly.

Agenda item 4: Exchange of views with the Subcommittee on Oceans and Coastal Areas of the Administrative Committee on Coordination

328. Mr. Patricio Bernal, Executive Secretary of IOC-UNESCO and Chairman of the Subcommittee on Oceans and Coastal Areas (SOCA) of the Administrative Committee on Coordination (ACC), presented an overview of the structure and functions of ACC, an internal body of the United Nations system on coordination, and outlined current activities of SOCA.

329. He pointed out that ACC was undergoing a phase of reviewing its mechanism in order to improve coordination and that, in that respect, SOCA itself was in a period of transition. In that connection, he stressed that while the structure for coordination might undergo changes, the function and goal of coordination in ocean affairs would remain and would be carried out.

330. He informed delegations that at its two most recent meetings, in January and May 2001, SOCA had focused on reporting tasks, in particular with regard to chapter 17 of Agenda 21 and the forthcoming World Summit on Sustainable Development to be held in September 2002. The Subcommittee had also discussed its assistance in the coordination and cooperation in the implementation of the GPA, although for resource-related and practical reasons it had been obliged to relinquish its initial function of a GPA steering committee. Furthermore, SOCA had devoted considerable attention to its new project, the United Nations Atlas of the Oceans, for which it acted as coordinator and manager. That project was aimed at integrating dispersed databases and poorly catalogued information available at the United Nations agencies and presenting the material on a single web site or on a compact disk. Despite management problems, the project, which was partly financed and supported by the United Nations Foundation, was expected to be completed by November 2001.

331. In the ensuing dialogue, a number of delegations made comments and suggestions and asked questions.

332. Delegations recognized the importance of the participation of SOCA in the meetings of the Consultative Process and acknowledged with appreciation information provided by Mr. Bernal. Many of them reiterated that the improvement of inter-agency coordination and cooperation on ocean affairs was one of the main purposes of the Consultative Process.

333. Delegations further noted the importance of ensuring the effectiveness, transparency and responsiveness of SOCA and the need for enhanced cross-sectoral cooperation and coordination, not only at the inter-agency level but also at the intergovernmental level and at the regional level, e.g., between the regional fisheries organizations and UNEP regional seas programmes.

334. Delegations also took note of the ongoing restructuring of the ACC system. In that context, they reiterated the importance of SOCA and unanimously called for the strengthening of its role and for the provision of adequate resources for the Subcommittee. The role of IOC as the SOCA secretariat should be maintained, some delegations suggested.

335. A number of suggestions were made regarding the functions of SOCA. Among the proposed functions were:
- To review ocean-related activities and problems encountered by United Nations agencies and programmes so as to achieve coordination and cooperation and avoid duplication of effort;
- To exercise a strong role with regard to the GPA and the 2002 World Summit for Sustainable Development;
- To coordinate inter-agency responses regarding the sustainable use of living resources and the protection of biological diversity on the high seas;
- To provide advice on subjects before the Consultative Process;
- To enhance coordination among agencies so as to ensure an integrated approach to implementation and financing of programmes;
- To increase the cooperation between agencies and bodies and the World Bank in linking the needs for resources for projects to adequate funding;
- To review budget proposals from various United Nations bodies in order to coordinate responses to needs;
- To provide assistance and coordination with regard to training and technical assistance programmes.

336. In answering questions and responding to comments and suggestions, Mr. Bernal pointed to the difficulties of a practical nature facing SOCA and its secretariat. These included inadequate resources, lack of permanent support staff, technical coordination by SOCA without the executive power to make decisions, rigidity of the administrative and budgetary procedures of the United Nations agencies and bodies, as well as the need to respect their hierarchical structure. Despite those difficulties, he assured delega-

tions of the commitment of the members of SOCA to continued and enhanced cooperation and coordination.

337. Furthermore, during the discussion under agenda item 4, it was suggested that the analytical content of the chapter on international coordination and cooperation in the Secretary-General's report could be enhanced and that the Division for Ocean Affairs and the Law of the Sea could regularly brief delegations in New York on the work of SOCA.

338. With respect to the participation of various parts of the United Nations system in the Consultative Process, delegations once again stressed the importance of the presence of all relevant United Nations organizations and bodies, including funding institutions, such as the World Bank and GEF. It was further suggested that the United Nations Office for Outer Space Affairs should be invited to the meetings of the Process in view of its potential contribution in the field of observation of oceans by satellites.

339. One delegation recalled its plea made at the first meeting of the Consultative Process concerning better coordination among UNDP, UNEP, FAO, IMO, and UNESCO through negotiations of memorandums of understanding with respect to particular programmes.

340. During the discussion on Part A of the draft report on the work of the Consultative Process at its second meeting, a number of delegations made additional suggestions. It was proposed, inter alia, that ACC should organize the preparation of a report, identifying the full range of United Nations organizations, agencies, programmes and funds engaged in ocean affairs, their mandates and the relationship between them, including the description of their current activities, and that open briefings on the work of SOCA should be held at United Nations Headquarters. Some other delegations, while concurring with the opinion that there was a need for detailed information on the internal functioning of the United Nations system, considered that the current format of the report of the Secretary-General on oceans and the law of the sea was sufficient to cover that issue. Yet another delegation expressed the opinion that paragraph 42 of General Assembly resolution 55/7 already contained a similar request for a study and that it was for the General Assembly and not for the Consultative Process to consider the need for further reports, taking into account the already sufficiently complex and comprehensive nature of the report of the Secretary-General.

Agenda item 5: Identification of issues for possible consideration at the third meeting of the Consultative Process in 2002

341. Co-chairperson Slade opened the discussion on agenda item 5 with reference to paragraphs 10 (c) and 11 of the format of the second meeting of the Consultative Process. Pursuant to those paragraphs and to the annotated agenda, the second meeting was to discuss additions or amendments to the list in the report of the first meeting (A/55/574, Part C) entitled

"Issues for considerations for possible inclusion in the agendas of future meetings." A note on those additions or amendments would then be contained in the draft report of the second meeting and be open for comments during a plenary session.

342. During the ensuing debate, a substantial number of possible issues were put forward by delegations. Among them were the following:
- Capacity-building and regional cooperation;
- Capacity-building for developing States;
- Regional approach in oceans management and development;
- Development and transfer of marine technology;
- Evaluation of the progress achieved under the issues discussed at the first and second meetings of the Consultative Process;
- Marine protected areas;
- Review of the national, regional and global implementation of Part XII of UNCLOS;
- Ecosystem-based integrated management of the marine environment;
- Potential and new uses of the oceans;
- Oceans stewardship;
- Food security and mariculture;
- Cooperation and coordination between regional fisheries organizations and regional seas programmes of UNEP;
- Impact of the activities in the international seabed area as a source of contamination of the marine environment;
- Fishery subsidies and their clear and negative effect on the conservation of marine living resources;
- Marine debris;
- Integration of the applicable legal provisions and programme issues;
- Navigation in ecologically sensitive areas;
- Protection of coastal areas from introduction of non-native species.

343. The suggestions received varying degrees of support; it was felt, however, that priorities should be established by the General Assembly.

344. There was a consensus among delegations that the theme of capacity-building had been recurrent during the first and the second meetings of the Consultative Process. This had resulted in an overwhelming support for its future consideration. Some delegations were of the opinion, however, that the theme of capacity-building was too vast to be considered as a separate issue and suggested focusing the capacity-building item on the needs of developing States.

345. Several delegations expressed the view that areas of focus for the meetings of the Consultative Process should be as concrete as possible. Some of them felt that the theme of capacity-building would necessarily need to be addressed under each specific issue, as had been the case with issues already discussed, and that that element could be reflected, in accordance

with the practice of the Consultative Process, in the annexes to the format and annotated agenda which contained issues for consideration.

346. Many delegations supported consideration of the issue of the development and transfer of marine technology. A number of delegations also concurred with the view that the next meeting of the Consultative Process should devote some time to the assessment of achievements in the areas of focus discussed at the first and the second meetings.

347. Regarding oceans stewardship, it was noted that it included assuming responsibility and taking actions for a greater improvement of the marine environment and ensuring the stability of development. It embodied many different activities and initiatives and was the responsibility of all actors of the international community.

348. Many delegations felt that the next meeting of the Consultative Process should also devote time to the review of its effectiveness and utility and recalled that such a review was due, pursuant to General Assembly resolution 54/33, in 2002. Some delegations considered that such self-evaluation would be, for both formal and practical reasons, inappropriate and that, according to the same resolution, that was clearly a prerogative of the General Assembly.

349. Many delegations also deemed that the third meeting of the Consultative Process could make a valid contribution and provide input to the forthcoming World Summit on Sustainable Development, to be held in Johannesburg, South Africa, in September 2002. In that regard, the issues of food security and mariculture were mentioned as deserving particular attention. One delegation pointed out that, since the General Assembly would not be in a position to consider the output from the third meeting in time for the summit, it would not be appropriate to include its consideration in the agenda of the third meeting.

350. In addition, during the general exchange of views, one delegation expressed the hope that future meetings of the Consultative Process would accord high priority to discussions on the management of risks to biodiversity and other components of the marine environment beyond the limits of national jurisdiction

PART C

Issues for consideration for possible inclusion in the agendas of future meetings

1. There was broad support for including capacity-building and the regional approach in oceans management and development as areas of focus for the third meeting of the Consultative Process.

2. Other suggestions put forward included:

(a) Marine protected areas;
(b) Review of the national, regional and global implementation of Part XII of the United Nations Convention on the Law of the Sea;
(c) Potential and new uses of the oceans;
(d) Development and transfer of marine technology;
(e) Oceans stewardship/ecosystem-based integrated management of the marine environment;
(f) Food security and mariculture;
(g) Cooperation and coordination between regional fisheries organizations and regional seas programmes of the United Nations Environment Programme;
(h) Impact of the activities in the international seabed area as a source of contamination of the marine environment;
(i) Effect of fishery subsidies on the conservation of marine living resources;
(j) Marine debris;
(k) Convergence of the legal and programmatic dimensions of international cooperation;
(l) Navigation in ecologically sensitive areas; and
(m) Protection of coastal areas from the introduction of non-native species.

3.   Support was expressed for evaluation of the progress achieved under the four areas of focus at the first and the second meetings: "responsible fisheries and illegal, unreported and unregulated fisheries: moving from principles to implementation"; "economic and social impacts of marine pollution and degradation, especially in coastal areas"; "marine science and the development and transfer of marine technology as mutually agreed, including capacity-building"; and "coordination and cooperation in combating piracy and armed robbery at sea."

ANNEX I

Statement by Mr. Hans Corell, Under-Secretary-General for Legal Affairs, The Legal Counsel

1.   On behalf of the Secretary-General, Mr. Nitin Desai, Under-Secretary-General for Economic and Social Affairs, and I would like to welcome you to the 2nd meeting of the United Nations Open-ended Informal Consultative Process established by the General Assembly in its resolution 54/33 in order to facilitate the annual review by the Assembly of developments in ocean affairs. I am pleased to note that, in keeping with the inclusive nature of the Consultative Process, delegations present here today represent States Members of the United Nations, States members of the specialized agencies

and parties to the United Nations Convention on the Law of the Sea. They also represent entities that have received a standing invitation to participate as observers in the work of the General Assembly pursuant to its relevant resolutions, intergovernmental organizations as well as major groups as identified in Agenda 21.

2. It is almost two decades since a constitution for the oceans was adopted in the form of the United Nations Convention on the Law of the Sea. It is almost a decade since a programme of action for the world's oceans and seas was adopted in the form of chapter 17 of Agenda 21. Thus, the international norms for the world's oceans and seas are in place. The legal and the programmatic frameworks for effective action in the marine sector are set.

3. The efforts of the international community in the past few years focused on the transition from the establishment of the norms to their implementation. This is a daunting challenge. It is therefore not a mere coincidence that a need was felt for establishing a Consultative Process. At the core of this Process is the attempt to find ways and means to meet the challenges of implementing the law and the programme of action for the world's oceans and seas.

4. The challenges are many and diverse. Developing countries, in particular the least developed countries and small island developing States, are finding their capacity to be limited, their resources scarce and their means of implementation inadequate. International cooperation is therefore essential in building their capacity, in enhancing their resources and in strengthening their means of implementation.

5. International coordination is necessary for identifying and filling the gaps so that weak links in the common effort do not undermine the whole structure of ocean governance. Coordination is also essential for eliminating duplications and overlaps so that the outcomes of a given amount of effort can be maximized.

6. The challenges for developed countries, on the other hand, are of a different nature. The challenges they face emanate from a multiplicity of activities, in many cases carried out in isolation from one another: one sector may have little interaction with another; one discipline may have little interface with another. Once again, coordination and cooperation at the national and international levels become essential so that actions taken will benefit from synergy effects.

7. Above all, there is the need for international coordination and cooperation between developing and developed countries. The conventional dichotomy between the North and the South is literally washed away by the waters of the world's oceans and seas. The political and economic boundaries do not match the ecological boundaries of the oceans, or the boundaries of marine resource occurrence. The marine environment does not distinguish between the North and the South, defined in political terms.

8. "The problems of ocean space are closely interrelated and need to be considered as a whole": this was a fundamental principle for the framers of the present legal regime and also of the programme of action. At the implementation level, this principle takes on immense dimensions. International coordination and cooperation is the most effective means of considering the problems of ocean space as a whole; in fact, it is the only means. The Convention and Agenda 21 themselves prescribe international coordination and cooperation in almost every area. The Convention is hailed as the most comprehensive framework for such coordination and cooperation in ocean affairs.

9. That international coordination and cooperation is beneficial in dealing with ocean issues is easy to see. Let me give you an example of the opposite. In the Secretary-General's report this year on oceans and the law of the sea (A/56/58), the World Meteorological Organization provides an alarming example of the dangers of the lack of international coordination and cooperation. Developing countries are unable to provide sufficient meteorological data and services in the ocean areas within their jurisdiction. Such inability in turn causes a deficiency in the availability of data, products and services in many major ocean areas, especially those required for meteorological forecasting. Such deficiencies can put all maritime users in peril.

10. The main purpose of the Consultative Process is to suggest particular issues to be considered by the General Assembly. The emphasis should be on identifying areas where coordination and cooperation at the intergovernmental and inter-agency levels should be enhanced.

11. The Secretary-General's annual reports on oceans and the law of the sea, in particular the one before you for your consideration and to which I just referred, chronicle a number of persistent problems as well as emerging issues in ocean affairs. Each of these problems calls for international action.

12. In 2000, the Consultative Process identified two areas of focus and suggested ways and means of enhancing international coordination and cooperation. These areas were: responsible fisheries and illegal, unreported and unregulated (IUU) fisheries; and economic and social impacts of marine pollution and degradation, especially in coastal areas. Over-exploitation of fish stocks, damaging fishing equipment and practices, and by-catch and discards compound the problems of irresponsible fisheries. Marine environmental problems are accentuated by the impacts of land-based activities, especially sewage, and effects of dumping and vessel-source pollution. The ocean-atmosphere interaction, through a rise in the sea level and the occurrence of periodic oscillations, is raising concerns about the well-being of present and future generations and highlighting the vulnerability of many coastal States, especially developing States and small island States.

13. This year, the Consultative Process is going to focus on two other areas: marine science and technology; and combating piracy and armed robbery at sea.

14. The unhampered conduct of marine scientific research, a better understanding of oceans and also of their interaction with the Earth and the atmosphere, a more effective interface between scientific knowledge and decision-making, the development and transfer of marine technology, and the strengthening of marine science and technology capacity: these are issues that urgently call for international coordination and cooperation. The international community will be looking to the Consultative Process for concrete measures.

15. Piracy and armed robbery threaten the shipping industry and endanger the well-being of seafarers. Other crimes at sea, such as the illicit traffic in drugs, smuggling of migrants and stowaways continue to rise. Parallel to these developments, the globalization of trade and the shipping industry brings newer issues to the fore: open registry of ships and flags of convenience and global labour markets for seafarers. The ageing shipping industry gives rise to the problems of safety and of environment-friendly decommissioning of a large number of ships. Global responses are required to address these problems of the global industry. International coordination and cooperation in formulating and executing those responses is essential.

16. As the Head of the Department of the United Nations devoted to the promotion of the rule of law in international affairs, I would like to conclude by emphasizing two points.

17. First, problems of implementation in many cases may lead to the undercutting of the very norms of the rule of law. Such is the case with the field of marine science and technology, where the discrepancy between the norm and the implementation is so glaring that many fear the norm itself will be relegated merely to "an empty shell." Such is also the case with the field of fisheries and shipping, where the balance among the rights and duties of coastal States and those of flag States and of other States achieved in the Convention may be in jeopardy.

18. Secondly, while the general norms are in place, in the formulation of norms in specific areas, international coordination and cooperation is becoming imperative. This is not only to ensure that norms that have been developed or are under development in a wide variety of areas relating to oceans are complementary to one another, but also to safeguard that such norms conform to the unifying and coordinating framework of the Convention. Currently, my Department, in particular through its Division for Ocean Affairs and the Law of the Sea, is fostering international coordination and cooperation by providing advice and assistance in the development of legal regimes in conformity with the Convention in the specific fields of, for example, underwater cultural heritage, marine protected areas and marine genetic resources.

19. Within the United Nations Secretariat, we endeavour to achieve an efficient interdepartmental coordination and cooperation, especially be-

tween the Office of Legal Affairs and the Department of Economic and Social Affairs. The idea is to integrate the legal aspects of ocean affairs and the programmatic aspects of an economic and social nature. The servicing of the Consultative Process, as mandated by the General Assembly in resolution 54/33, is a cooperative endeavour between the two departments.

20. The inter-agency coordination and cooperation among the funds, programmes and organizations of the United Nations system is achieved essentially by two means: through direct communications, contacts and liaison among the various entities themselves; and through the system-wide inter-agency coordination and cooperation mechanism of the Subcommittee on Oceans and Coastal Areas of the Administrative Committee on Coordination. The report of the Secretary-General before you (A/56/58), exemplifies inter-agency coordination and cooperation in that it incorporates the contributions of relevant agencies within the coordinating framework of the United Nations Convention on the Law of the Sea and Agenda 21.

21. I am confident that the Co-Chairpersons will navigate the Consultative Process so that it can meet effectively the challenges of transition from establishing the required norms to their implementation. At the same time, the Process must stave off the centrifugal forces that may lead us away from the existing rule of law. I wish the 2nd meeting of the Consultative Process all success.

ANNEX II

Statement by Mr. Nitin Desai, Under-Secretary-General for Economic and Social Affairs

1. It is a pleasure for me to join my colleague, Mr. Hans Corell, at the opening of this second session of the informal consultative process. The fact that the two of us are here together reflects the intentions behind this Informal Consultative Process, which is to bring together the legal dimension and the programmatic dimension of international cooperation on matters relating to the oceans.

2. As my colleague, Mr. Corell, emphasized, partly this objective is based on the growing recognition that the establishment of norms has to be accompanied by a systematic effort at cooperation in implementation. And once one gets into issues of implementation, then one has to get into substantive programmatic areas such as fisheries, marine pollution, ocean science, coastal zone management, regional seas and so on. And that was the motivation that led to the five-year review process of the Rio Conference and then to the proposal of the Commission on Sustainable Development contained in its decision 7/1, that the Consultative Process could and should be set up.

3.   The proposal was finally endorsed when the recommendations of the Commission on Sustainable Development at its seventh session on oceans and seas were discussed by the General Assembly in 1999 and, as you know, the consultative process was launched. In many ways it is an experiment. It was scheduled to work for three years, and next year when you meet for the third time you will also have to review your experiences with the Consultative Process so that decisions can be taken on its future and how we continue to see best reflected the type of integration of the legal and the programmatic dimension that is required.

4.   Let me also underline another point which my colleague, Mr. Corell, made. The ocean is the one area where the case for international cooperation is absolutely clear. As he emphasized, ecosystems do not know political boundaries, and this is even more so in the case of the oceans, where a very substantial part of what we are talking about lies beyond national jurisdiction. And that is why for so many reasons the development of a legal regime specifying the rights and obligations of States has probably moved further in this area than in almost any other area involving natural resources.

5.   Let me also stress that it is not just a matter of the physical characteristics of the ocean itself. If you look—and I now speak as a social scientist, as an economist and as an historian—it is also the fact that economic zones are defined as much by the ocean as by land. I come from Western India, in Gujarat. And I can tell you that my part of India has had historically a stronger connection with the littoral States of the Indian Ocean in Arabia than maybe with other parts of India. After all, it was the Indian Ocean that defined the economic zones of the area. This has been true for the South China Sea, the Mediterranean and the Pacific Rim. So economic zones have been defined very much by the interaction of nations across oceans. And the great historian Fernand Braudel said that even civilizations are defined by the sea that they are surrounded by, when he wrote about the Mediterranean.

6.   So, I would say that for historical reasons, for economic reasons, besides ecological reasons, the reality is that the nations of the world share a common interest in the oceans. In terms of economic resources, the way the ocean affects all territories and the management of pollution, it is understandable that the issue of oceans figured very prominently in the 1992 Rio Conference.

7.   In fact, chapter 17 of Agenda 21 is perhaps still the only place where we have an integrated programme on oceans involving almost every aspect: fisheries, pollution, coastal zone management, scientific research, coordination and cooperation. And that chapter has provided a very useful way of bringing together the different parts of the work of the United Nations system on ocean affairs, at the Secretariat, inter-agency and intergovernmental levels.

8. At the Secretariat level, one of the important products of the Rio Conference was the establishment of the ACC Subcommittee on Oceans and Coastal Areas (SOCA), the inter-agency coordination process on oceans. This is something whose need had been felt for a long time and in many ways was a catalyst for this unified chapter 17 on oceans, which forms part of Agenda 21. And it was based on the earlier decisions that we are addressing in the follow-up of the Rio Conference. The very fact that the area reserved here in this room for the agencies to sit is almost as crowded as the one for the delegates is a reflection of the widespread interest in ocean affairs in the United Nations system. SOCA has been an important instrument in bringing all of these agencies and organizations together.

9. The primary focus of coordination is, of course, on exchanging information and launching joint programmes and initiatives, such as the United Nations Atlas of the Oceans. The basis of this hard work has been provided by the legislative process and now, since last year, by this Consultative Process that you have launched.

10. Some questions have been raised with regard to the future of SOCA in the context of the discussions that are going on in relation to ACC reforms. I would like to assure you that the focus of ACC reform is more in terms of ACC itself and not on the basis of the needs of its own work. It has been recognized that the arrangements which are in place at the working level, particularly in order to strengthen cooperation and coordination among agencies and organizations, have to be justified in terms of their own objectives and concerns, which are not all related to servicing ACC. A significant part of SOCA's concern is to be of service to the Consultative Process on oceans. Part of its concern is to ensure cooperation and coordination among agencies in terms of programmatic work. And therefore I would say that it has been recognized that these bodies will continue in a form that will be defined by their purpose. As of now, what ACC has asked is that each inter-agency process examine its objectives, its purposes and define its own rationale and its own method of work, which, besides supporting ACC, involve many other dimensions of cooperation and coordination that are relevant. I would assure you that I personally, particularly as the Head of the Department of Economic and Social Affairs, and within it the Commission on Sustainable Development, place a very high value on the mechanisms of coordination that have been established through SOCA.

11. The first specific issue before you at the current session deals with marine science. This work has been traditionally coordinated through the Joint Group of Experts on the Scientific Aspects of Marine Environmental Protection (GESAMP). I have before me GESAMP's latest report entitled "A Sea of Troubles." At first glance, it shows what valuable work GESAMP has been doing. I am aware that there are many questions and issues that have been raised with regard to the adequacy of the efforts for cooperation

in marine science and research. And I am sure that this is something that you will discuss. This is clearly one of the key areas that we need to strengthen. If we look at what we know about the oceans, we find that it is far, far less than what we know about the land. And we are increasingly aware of the fact that much of what we are talking of in terms of issues—global environment issues—have a strong connection with oceans.

12.   My colleague, Mr. Corell, alluded to the most important example of this: the global climate system. It is increasingly clear that much of our understanding of global change depends strongly on the atmosphere/ocean interaction, of which, frankly, we know very little—and much more research is required. But this is not the only area of research. There are key areas of research regarding biomass in the oceans. In view of the amount of biomass connected with the oceans in one way or another and how it has been threatened, as shown in the GESAMP report, I would say again that it remains a very important area of concern. An example of this of course is the coral reef initiative, but that is not the only one.

13.   The area of fisheries has of course been traditionally of great importance in the context of ocean affairs. That too is an area where the strengthening of research and understanding is crucial because much of international cooperation depends on scientific assessments of what is happening to fish stocks, which ultimately depends on information, research and analysis.

14.   The area that you are tackling today is absolutely central to the mechanisms of international cooperation. It is also an area where the agencies and organizations of the United Nations system have a strong interest and involvement in capacity-building. In fact, it is an important part of the work of the United Nations in relation to oceans. We look forward to seeing what your recommendations and suggestions will be in this area. I do not really have much to say on the second issue of your agenda: robbery and piracy on the seas. It is a legal issue and I really do not have much to add or contribute in that area.

15.   Let me turn finally to the connection with the Johannesburg Summit on Sustainable Development, the 10-year review of the Rio Conference, which will take place in September of 2002. We just completed the first preparatory meeting for the Johannesburg Summit, a very successful one, which has basically mandated a flexible process with a lot of interaction with civil society, major groups, stakeholders, in the Conference itself and in the preparation for the Conference. The focus of the Conference will be on trying to operationalize sustainable development. It will also focus a great deal of attention on the impact of issues and trends of globalization, risk management issues, finance for sustainable development, technology transfer and, I believe, also in reasserting, reinforcing and reinvigorating the sense of responsibility for environmental and sustainable development at the global level. The issue of the oceans is very central to this focus. I would certainly

invite this Consultative Process to consider how it can best contribute to the consideration of this issue at the World Summit on Sustainable Development in September 2002.

16.   The main preparatory process will begin in January 2002. A second preparatory meeting will be held in March 2002. A major Ministerial Preparatory Meeting, which will look at what will be done at Johannesburg, will take place in Indonesia, possibly on the island of Bali, in late May in 2002, and then we will go on to the World Summit in September 2002 in Johannesburg. In the first six months of this year the focus of the preparations has been on thematic round tables, regional round tables, creating a number of activities, on a very decentralized basis, which would all be brought together at the first preparatory meeting, which would be held in New York in January 2002. I mention this timetable so that you, as a Process, and individually, working through your national preparatory processes, would find ways of contributing to this exercise.

17.   I spoke of responsibility. Let me just conclude with something a little more philosophical if I might. Very close to Johannesburg there is a site called "the cradle of humanity." It is a world heritage site, where the oldest known fossils of hominids have been found. They are between 3 and 1/2 and 4 million years old. I hope that in some way we can connect what we are going to talk about in Johannesburg to that site. I mention this because in some ways the ocean is an even earlier cradle of humanity. Without the ocean we cannot sustain life on Earth. There are many ways in which oceans are absolutely central to the evolution of humanity. Therefore, the ocean is an issue that is central to the problématique of sustainable development. I look forward to your deliberations not only at the current session but also next year and to the contributions you can make to reinvigorate and reinforce sustainable development in Johannesburg next year.

Selected Documents and Proceedings

Report of the Fourth Global Meeting of Regional Seas Conventions and Action Plans†

INTRODUCTION

1. The Fourth Global Meeting of Regional Seas Conventions and Action Plans was held at the Hotel Delta Centre Ville, Montreal, Canada, from 21 to 23 November 2001 at the invitation of the Government of Canada.

I. OPENING OF THE MEETING

A. Opening Statements

1. Opening statements were made by Mr. Jorge Illueca, Director of the UNEP Division of Environmental Conventions, and Ms. Elisabeth Mann Borgese, Honorary Chair of the International Ocean Institute (IOI).

2. Mr. Illueca, in his capacity as Chair, welcomed participants to the Meeting on behalf of the Executive Director of UNEP, Mr. Klaus Töpfer, and introduced the provisional agenda (UNEP(DEC)/RS.4.0.1). The agenda of the meeting as adopted is given in annex I to the present report.

4. He reiterated the objectives of the Fourth Meeting as set out in the provisional agenda: to streamline the ways in which UNEP provided programme support to the regional seas conventions and action plans, in accordance with the blueprint provided by the decisions of the first three Meetings, in areas complementary to its own programme of work; to boost horizontal cooperation between the various regional seas conventions and action plans so that the longer-established organizations would be twinned with less developed organizations with a view to sharing experience and providing technical cooperation; to build bridges and form links with the secretariats of multilateral environmental agreements, such as the conventions on biodiversity, migratory species, and international trade in endangered species, and other secretariats such as those of the conventions dealing with hazardous chemicals and wastes, in pursuit of the objectives of the Global Programme of Action for the Protection of the Marine Environment from Land-based Activities; and to review the follow-up to the Second and Third Meetings.

5. The Fourth Meeting differed from previous Meetings in that the private sector, particularly representatives of the shipping, chemicals, and petroleum industries, had been invited to attend. Their input, both financial and in terms of exper-

---

†EDITORS' NOTE.—This document has been extracted from the Report of the Fourth Meeting of Regional Seas Conventions and Action Plans, held at Montreal, Canada, 21–23 November 2001, UNEP(DEC)/RS.4/6, 20 December 2001. It is published without Annex II, the List of Participants.

tise, would be critical for the future of international environmental governance, and it therefore behoved the Meeting to begin building bridges in that area too.

6. The role of the regional seas programmes was evolving from pollution abatement to addressing sustainable development issues. Multisectoral as those issues were, the need for dialogue with the private sector was both implicit and evident.

7. The cooperation between UNEP and the Food and Agriculture Organization of the United Nations (FAO) on fisheries was an example of a sustainable development issue which had been subsumed within an ecosystem-based management approach.

8. He pointed to Governing Council decision 21/13, which mandated cooperation with various other bodies including the International Maritime Organization (IMO), the International Oceanographic Commission of the United Nations Educational, Scientific and Cultural Organization (IOC-UNESCO) and the Convention on Biological Diversity and in consultation with regional seas programmes in exploring the feasibility of establishing a regular process for the assessment of the marine environment with active involvement by Governments and regional agreements, building on ongoing assessment programmes.

9. He then introduced Ms. Elisabeth Mann Borgese, Honorary Chair of the International Ocean Institute (IOI), who had been invited by decision of the Third Meeting to address the Fourth Meeting as a special guest.

10. Ms. Mann Borgese gave an overview of the history of her organization and drew attention to a presentation paper which included identification of areas of mutual interest between IOI and the regional seas conventions and action plans, and suggestions for cooperation (UNEP(DEC)/RS.4.0.2). IOI had been founded with a view to cooperating with the regional seas programmes because it considered them fundamental to the whole question of governance of the oceans, her organization's raison d'être. IOI's operational training centres, which had been established primarily to build the capacity of small and developing countries so that they could argue their case in the global forums on the law of the sea, now offered a virtual university, which enabled its students to obtain internationally recognized masters' degrees in ocean governance, published the *Ocean Yearbook* and carried out policy research.

11. The IOI "Echo Villages" project in Tamil Nadu, India, built capacity by empowering poor people, particularly women in poor, coastal villages, with microcredit and training to help themselves through environmentally friendly projects of direct usefulness. Such projects were necessary because they built a constituency for sustainable development in a way that top-down development models could never do, by making people's lives better.

12. Under paragraphs 276 and 277 of the Convention on the Law of the Sea, technology transfer centres were supposed to have been set up, but never had been. However, one of the outcomes of the General Assembly consultation process on ocean affairs had been a consensus recommendation that they should be. She pointed out that technology transfer was not the same as it had been even 20 years before: in a high-technology world, technology could not be transferred or bought, it had to be learned, and therefore cooperation in research and development was the only sustainable way to effect such transfers. Such research and development ventures should be 50 per cent funded by Governments, bilateral sources or the Global Environment Facility (GEF) and 50 per cent by the proposing private agency,

with a view to building synergy between public and private funding. Nor did such technology transfer centres need to be bricks and mortar; they could be Web-based and thus more cost effective.

13. The number of programmes, protocols, and agreements was becoming overwhelming, and there was a growing need to integrate them all properly into the United Nations Convention on the Law of the Sea and United Nations Conference on Environment and Development process at the operational level within the regional seas framework. Otherwise, the system would break down.

14. She pointed out that tourism was by some counts the planet's largest industry. The overwhelming majority of tourists wished to give something to assist the development of the places they visited and compensate their inhabitants for the social and environmental burden imposed by tourism. At even $1 per tourist, that was a major potential source of funding. However, participants pointed out that total collaboration would have to be obtained from national ministries of finance, and the opinion was expressed that the World Bank and the International Monetary Fund were no great lovers of environmental protection funds.

15. The insurance industry, she said, was a major stakeholder in the regional seas/Global Programme of Action process, and was deeply interested in integrated coastal area management from the point of view of risk reduction, where its interests coincided with those of environmental protection and sustainable development. Also, microinsurance could be made available by insurance companies in a manner analogous to the way in which financial institutions made finance available for microcredit. To comments from some participants that in developing countries people were poor and could not afford insurance, and that as a result developing countries' insurance industries were also underdeveloped, she replied that for large facilities such as hydroelectric dams and port facilities, insurance was in place and the insurers were the same big companies as in the developed world. The point was not to milk the insurance industry for funds but rather to draw on their expertise in areas such as disaster risk assessment where the industry's interests coincided with those of environmental protection and sustainable development.

B. Attendance

16. The Meeting was attended by representatives of the following organizations:
   (a) *Regional seas conventions and action plans:* Convention for Cooperation in the Protection and Development of the Marine and Coastal Environment of the West and Central African Region (Abidjan Convention); Convention for the Protection, Management and Development of the Marine and Coastal Environment of the Eastern African Region (Nairobi Convention); Convention for the Protection of the Mediterranean Sea against Pollution (Barcelona Convention); South Asian Seas Environment Programme (SACEP); East Asian Seas Action Plan; Convention for the Protection of the Natural Resources and Environment of the South Pacific Region (Noumea Convention), Convention for the Protection of the Marine Environment and Coastal Area of the South-East Pacific (Lima Convention); Convention for the Protection and Development of the Marine Environment of the Wider Caribbean Region (Cartagena Convention); Commission for the

Protection of the Marine Environment of the North-East Atlantic (OSPAR Convention); Programme for the Protection of the Arctic Marine Environment (PAME); Caspian Environment Programme (CEP); North-West Pacific Action Plan (NOWPAP);

(b) *Global and international agreements:* Basel Convention on the Control of Transboundary Hazardous Wastes and their Disposal; Convention on Biological Diversity;

(c) *Intergovernmental organizations:* International Oceanographic Commission of the United Nations Educational, Scientific and Cultural Organization (IOC-UNESCO); International Atomic Energy Agency (IAEA); Food and Agriculture Organization of the United Nations (FAO); Division for Oceans Affairs and the Law of the Sea (DOALOS), United Nations; Global Programme of Action for the Protection of the Marine Environment from Land-based Activities Coordination Office; Marine Environment Laboratory of the International Atomic Energy Agency (IAEA);

(c) *Non-governmental organizations:* International Ocean Institute (IOI); World Conservation Union (IUCN); International Oil Pollution Compensation Funds 1971 and 1992 (IOPC Funds); International Tanker Owners Pollution Federation Limited (ITOPF); International Petroleum Industry Environmental Conservation Association (IPIECA); Commission for Environmental Cooperation (CCEMTL); EnviroLaw Solutions.

17. The list of participants is reproduced as annex II to the present report.

II. PROGRESS REPORT ON FOLLOW-UP TO THE DECISIONS OF THE SECOND AND THIRD GLOBAL MEETINGS OF REGIONAL SEAS CONVENTIONS AND ACTION PLANS

18. The Chair introduced a report entitled "Status of implementation of the decisions of the Third Global Meeting of Regional Seas Conventions and Action Plans" (UNEP(DEC)/RS.4.1.1) and briefed the participants on follow-up to the decisions of the Third Meeting calling for closer cooperation between UNEP and IOC-UNESCO on the Global Ocean Observing System (GOOS), the development by IMO of an international forum on response to oil spills, the inventory on chemicals work undertaken by regional seas programmes (UNEP)DEC)/RS.4.12), the International Coral Reef Action Network (ICRAN) and the development by UNEP of a financial strategy for mobilizing additional resources for regional seas programmes. The problem of the cutbacks in GEF funding for activities in international waters emerged as a key concern. Presentations were given by Mr. Tim Turner of the Caspian Environment Programme on the situation of the Caspian sturgeon (UNEP(DEC)/RS.4/INF/2); by Mr. Benedict Satia of FAO on the ecosystem-based approach to fisheries management (UNEP(DEC)/RS.4.1.3 and 4.1.4; UNEP Regional Seas Reports and Studies No. 175); by Ms. Sachiko Kuwabara Yumamoto of the Basel Convention secretariat on opportunities for cooperation and coordination between the Basel Convention and the regional seas conventions and action plans (UNEP(DEC)/RS.4.1.5); and by Mr. Robert Droop of the Coordination Office for the Global Programme of Action on the state of preparations for the 2001 intergov-

ernmental review of the Global Programme of Action, including the role of the regional seas programmes.

19. A wide-ranging discussion of what should be the way forward ensued, with particular reference to the Meeting's input to the impending review of the Global Programme of Action. It was generally felt that a holistic approach was needed to the problems of proliferating agreements and protocols; of relations between regional seas and regional fisheries bodies; of the proliferation of activity and training centres of one kind and another with duplication of effort; of the evolution from pollution abatement to a sustainable development approach in which ecosystem-based management was key, and not just in the area of fisheries; of the interrelated issues of training and capacity building; and of the legal basis for any action, which should not run counter to the United Nations Convention on the Law of the Sea. Integrated coastal area management should be viewed as such a holistic approach, in the service not only of the environment but also of sustainable development.

20. In the context of the ecosystem-based management of fisheries, attention was drawn to a paper on the geographical overlapping between regional seas conventions and action plans and marine regional fisheries bodies (UNEP(DEC)/RS.4/INF/3). The question of overlap, both geographical and otherwise, with fisheries and other bodies was retained as an issue for reflection. There was wide-ranging discussion on the way forward, including the promotion of closer collaboration between regional seas programmes and regional fisheries bodies and the development of guidelines for the ecosystem-based management of fisheries as a joint initiative by FAO, UNEP, and IOC-UNESCO.

21. In the discussion on closer collaboration between the Basel Convention and the regional seas programmes, concern was expressed about obsolete agricultural chemicals, particularly in Africa. As hazardous chemicals and/or pesticides and/or persistent organic pollutants, such chemicals were a problem under the Basel, Rotterdam, and Stockholm Conventions, but also the context of the Global Programme of Action and regional seas programmes, the more so in that they were for the most part stored and deteriorating in seaports.

22. Regarding the Global Programme of Action, the discussion turned to methods of persuading Governments to take action and disburse funding. All present were of the view that the case for action in the context of the Global Programme of Action had been made and that there was no need to restate it.

23. Much of the discussion on the Global Programme of Action turned on the issue of how to effectively assess progress in its implementation. It was generally felt that to obtain a sound scientific basis for action in terms of useful statistical and other data would often cost more than taking the action itself and might even result in action being taken too late: the precautionary principle should be paramount. The use of indicators such as indicator fish species was generally held to be poor science even if it made good public relations. Other problems could arise from too simplistic an approach: an example was given of rising levels of polychlorinated biphenyls (PCBs) detected in harbour porpoises in the United Kingdom. A simplistic analysis would have resulted in further legislative clampdowns on PCBs and a huge and expensive effort to identify sources. The correct analysis, taking a wider view of the web of life, showed that the cause was overfishing of the preferred food species, forcing the animals to eat more benthic organisms, and PCB levels were naturally higher in bottom feeders.

24. Several participants made the point that Governments were unlikely to respond to proposals that to them smacked more of satisfying scientific curiosity than providing useful results. It was agreed that the needs of clients must be taken into account when providing them with information. It was also accepted that some Governments were in dire financial straits. Even so, the point must be made with all necessary forcefulness that money was needed and that commitments had been made to provide it.

25. One indicator that was generally felt to be of use for purposes of persuasion was the economic costs of action and inaction, such as expressing the cost of the destruction of a mangrove swamp in terms of thousands of dollars per metre of beach eroded as a consequence. Integrated coastal area management was held up as the holistic response to the challenge of ensuring that the environment was protected in the service of sustainable development. Also, for all the problems involved in quantifying such costs, and for all the problems involved in getting agreement from national ministries of finance, users such as desalination plants, power generators that used coastal waters for cooling, paper mills, and tourists should pay the costs of their activities.

Recommendations

International Ocean Institute (IOI)
26. In the light of the role of IOI in the fields of environmental management, public awareness and education, and bearing in mind the discussion that took place following the presentation by the keynote speaker from IOI before the consideration of agenda item 1, the Meeting recommended:
 (a) That UNEP and IOI should develop joint programmes of environmental management and environmental education relevant to the sustainable development of the marine and coastal environment, and that those programmes should be implemented through the regional seas programmes as appropriate;
 (b) That intergovernmental organizations, such as the European Union, IMO, IOC-UNESCO, the United Nations Development Programme (UNDP) and UNEP, and other nongovernmental organizations should be invited to consider and implement an innovative approach to cooperation by creating tripartite and multipartite projects and by seeking IOI involvement in the development of such projects; and
 (c) That the recommendations of the General Assembly consultative process on ocean affairs for the establishment of regional centres should be considered by regional seas programmes for implementation in the form of operational training and technology transfer centres, and in the form of virtual centres where appropriate.

International Maritime Organization (IMO)
27. Recognizing the progress made by IMO and UNEP and several regional seas programmes in strengthening their collaborative approach to issues of oil spill preparedness and response, the Meeting recommended:
 (a) That further collaboration should be developed, in particular in the estab-

lishment of regional systems for cooperation in preparedness for and response to oil spills. Such collaboration should take the form of developing and amending existing relevant protocols, establishing dedicated regional activity centres, developing regional contingency plans, and investing in regional training and exercises;
(b) That linkages with potential partners, such as the oil and shipping industry, and twinning arrangements with more developed regional seas programmes, should be explored and implemented, in particular for younger and less developed regional seas programmes; and
(c) That further joint activities in the various regions should be developed along the lines indicated at the Third Global Meeting of Regional Seas Conventions and Actions Plans (Monaco, 6–10 November 2000).

Ecosystem-Based Approach to Management of Fisheries and the Marine and Coastal Environment
28. Recognizing that the ecosystem-based approach to management of fisheries and the marine and coastal environment is a goal to be pursued, and recognizing also the impact of fisheries activities on the ecosystem, the Meeting recommended:
(a) That the regional seas programmes should consider the necessary steps to be taken towards the adoption of an ecosystem-based approach to the management of the marine and coastal environment. In that connection, the issues of integrated coastal area management and the ecosystem-based approach to fisheries management are of particular importance;
(b) That regional seas programmes should follow up the recommendations for closer cooperation with regional fisheries bodies contacted in the report on the ecosystem-based approach to fisheries management (UNEP Regional Seas Reports and Studies, No.175). To that end, it was agreed that the questionnaire on the status and planned development of cooperation between regional seas programmes and relevant fisheries bodies (UNEP(DEC)/RS.4.14) should be completed and returned to UNEP;
(c) That regional seas programme coordinators should explore opportunities for cooperation with the United Nations Division for Ocean Affairs and the Law of the Sea and with relevant institutions and organizations in their regions and should raise the innovative, ecosystem-based approach to the management of marine and coastal resources, including fisheries for adoption at forthcoming meetings of their contracting parties; and
(d) That IOC-UNESCO should be fully integrated into the joint UNEP/FAO initiative on ecosystem-based management of fisheries. As a first step, a joint programme of work should be developed by UNEP, FAO, and IOC-UNESCO in which special attention should be given to the preparation of technical guidelines for best practices in introducing ecosystem considerations into fisheries management.

Basel Convention on the Control of Transboundary Movements of Hazardous Wastes and their Disposal
29. Recognizing the potential benefit that could be derived from closer cooperation between the regional seas programmes and the Basel Convention, in particular

through the Convention's regional centres for training and for transfer of technology, the Meeting endorsed the actions recommended to enhance cooperation by developing and implementing cost-effective joint actions in such areas as:
- (a) Training in waste management principles, procedures, and technologies;
- (b) Public awareness-raising;
- (c) Assistance in developing national legislation and regulatory measures relating to waste management,
- (d) Harmonization of reporting requirements under the Basel Convention and the related instruments of the regional seas programmes; and
- (e) Development of joint waste management programmes in relation to the protection of the marine and coastal environment.

30. The Meeting also recommended that interested regional seas conventions and action plans and the Basel Convention secretariat should consider negotiating and implementing a memorandum of understanding covering specific actions such as:
- (a) Formalizing, on a reciprocal basis, the observer status of the regional seas programmes at meetings of the parties to the Basel Convention and of the relevant subsidiary bodies and structures coordinating the work of the Basel Convention regional centres;
- (b) Exchanging data and information of mutual interest and relevance available within the secretariats of the regional seas programmes, the secretariat of the Basel Convention, and at the Basel Convention regional centres;
- (c) Establishing joint advisory panels comprising the regional seas programmes and the Basel Convention, including the Basel Convention regional centres, and organizing joint technical meetings on subjects of mutual interest;
- (d) Seeking support for jointly agreed activities by the regional seas programmes and the Basel Convention through multilateral associations and through cooperation with the relevant global conventions, such as the Stockholm Convention on Persistent Organic Pollutants, the Rotterdam Convention on the Prior Informed Consent Procedure for Certain Hazardous Chemicals and Pesticides in International Trade, the International Convention for the Prevention of Pollution from Ships (MARPOL), and the London Convention on the Prevention of Marine Pollution by Dumping of Wastes and Other Matter, and the regional components of global programmes such as the Global Programme of Action for the Protection of the Marine Environment from Land-based Activities and the IOC Global Ocean Observing System (GOOS);
- (e) Designing and implementing joint programmes between the regional seas programmes and the Basel Convention, and by the Basel Convention regional centres as appropriate, taking fully into account the respective mandates, objectives and scope of those conventions; and
- (f) Coordinating with the Basel Convention the implementation of existing and future protocols to the regional seas conventions for the control of transboundary movements of wastes and their disposal by adopting a com-

mon approach to their development and implementation and harmonizing the reporting requirements.

31. The Meeting recommended further that UNEP should facilitate the development of cooperative arrangements between the Basel Convention and the regional seas programmes.

Global Programme of Action for the Protection of the Marine Environment from Land-based Activities

32. Recognition was given to the work of the Global Programme of Action Coordination in the preparation for the first Intergovernmental review of the Global Programme of Action scheduled for the following week. Final arrangements regarding participation by regional seas programmes were reviewed and agreed upon, including the final designation of the 12 regional seas programmes that would present progress reports on the implementation of the Global Programme of Action during informal session on the margins of the plenary sessions of the review meeting.

33. It was also agreed that further thought needed to be given to the development of indicators for measuring progress in the implementation of the Global Programme of Action at the global and regional levels for future intergovernmental reviews of the Global Programme of Action.

III. GLOBAL ASSESSMENT OF THE STATE OF THE MARINE ENVIRONMENT

34. The Chair drew attention to documents UNEP(DEC)/RS.4.2.1 and 4.2.2 concerning the feasibility of establishing a regular process for the assessment of the state of the marine environment and recalled UNEP Governing Council decision 21/13 calling for the feasibility of establishing a regular process for the assessment of the state of the marine environment to be explored and requesting the Executive Director to submit a progress report on the issue to the Council at its twenty-second session.

35. Mr. Geoffrey L. Holland of IOC-UNESCO gave a presentation on the decisions and actions taken by the IOC Assembly in response to UNEP Governing Council decisions 21/13 and 21/28. The Assembly had met in Paris from 3 to 13 July 2001, and the question of potential collaboration between the Commission and UNEP had been raised in the specific context of the regional seas programmes. Discussions on cooperation had also taken place on related programme areas such as integrated coastal area management, the monitoring of coral reefs, large marine ecosystems, and possibilities for an assessment of the marine environment.

36. The IOC Assembly had strongly endorsed the development of close links between the Commission and the regional seas programmes, and had instructed the Executive Secretary:
   (a) To enhance mutual awareness, transparency and knowledge concerning the activities being carried out by IOC and the regional seas programmes;
   (b) To complete the development of a memorandum of understanding as a formal instrument between the regional seas programmes and IOC-GOOS;
   (c) To coordinate projects of mutual interest;

(d) To develop further cooperation between IOC regional bodies and the relevant regional seas programmes;
(e) To examine possibilities for establishing joint regional activity centres;
(f) To provide input from GOOS to the UNEP/FAO initiative on ecosystem-based approaches to fisheries management;
(g) To ensure that GOOS contributes to the feasibility study on the Global Ocean Assessment process; and
(h) To inform the respective governing bodies concerning programmes of mutual interest.

37. IOC also had several programmes in the area of ocean science in relation to living resources, many of which, including the coral reef monitoring programme, were of direct relevance to the ecosystem-based approach. The Assembly had instructed the IOC secretariat to work closely with FAO and UNEP on issues related to the ecosystem-based approach to fisheries management in order to further the ecosystem-based approach within the United Nations system.

38. The discussion brought out many of the difficulties that would have to be resolved and due benefits that would have to be evaluated in future discussions and decisions on the question of the assessment of the global marine environment. Much of the necessary data would be available from programmes and projects carried out to satisfy national and regional priorities other than marine environmental assessment, but their data would be both necessary and useful in any such assessment. Governments and regional seas programmes should, however, ensure that their observation programmes were both compatible and consistent. Existing organizations could assist by facilitating the cooperation and interactions necessary to assemble from available national sources the expertise and the data required to produce answers to help solve the regional and global problems of the marine environment. Organizations such as the IAEA Marine Environment Laboratory in Monaco could play an important role in the global assessment.

Recommendations

International Oceanographic Commission of the United Nations Educational, Scientific and Cultural Organization (IOC-UNESCO)
39. In view of the progress made in enhancing collaboration between IOC-UNESCO and regional seas programmes, the Meeting recommended:
(a) That cooperation between the organizations should be further strengthened and developed along the lines of the joint plan of work presented to the Meeting;
(b) That an umbrella memorandum of understanding between the organizations should be developed. To that end, the Meeting welcomed the ongoing preparations for the signature of a memorandum of cooperation between the Caribbean Environment Programme and IOCaribe, the regional IOC body for Latin America and the Caribbean, which would identify scope for collaboration for GOOS and data exchanges.

(c) That practical measures aimed at the establishment of joint regional seas programme/IOC-GOOS regional office regional activity centres should be explored and implemented on a cost/benefit basis.

Governing Council Decision 21/13
40. In view of the potential benefit to Governments and to regional seas programmes of having direct access to comprehensive and broadly based global and regional assessment reports on the state of the marine environment as tools to support their decision making, the Meeting recommended:
  (a) That the conduct of the feasibility study for establishing a regular process for the assessment of the state of the marine environment mandated by the Governing Council at its twenty-first session should be facilitated, and that the regional seas programmes should become full participants in the consultations carried out in the context of that study;
  (2) That UNEP should establish a regular process of global coordination of ongoing assessment activities in full cooperation with IOC, the Convention on Biological Diversity and other relevant international and scientific organizations; and
  (3) That the regional seas programmes should play a major role in that global coordination process.

IAEA Marine Environment Laboratory, Monaco
41. Recognizing the benefit to the regional seas programmes of using the services of and being assisted by the Marine Environment Laboratory in developing national and regional monitoring programmes and in capacity building, the Meeting recommended:
  (a) That UNEP and IAEA should finalize the agreement on the modalities of co-sponsorship and co-ownership of the Laboratory by IAEA and UNEP;
  (b) That regional seas programmes should participate in the Laboratory's activities and use it as a principal tool to assist them in establishing and maintaining monitoring of the marine environment in their regions;
  (c) That the Laboratory should continue to play a part in capacity building and as a scientific advisory body to regional seas secretariats and member States that so request;
  (d) That UNEP should endeavour to incorporate the Laboratory's expertise and capabilities into projects and activities under the Global Programme of Action;
  (e) That UNEP should participate in funding the Laboratory's budget by financing additional regional seas programmes which would in turn purchase services from the Laboratory;
  (f) That the regional seas programmes which should be selected first for capacity building under this scheme are the Nairobi Convention (Eastern Africa), the Abidjan Convention (West Africa), SACEP (South Asia), and the Lima Convention (south-east Pacific); and
  (g) That IOC-UNESCO should remain as a partner in and a member of the

inter-agency advisory group on the Laboratory together with IAEA and UNEP, and should resume its contribution to the Laboratory's activities.

IV. PANEL DISCUSSION ON COOPERATION BETWEEN THE PRIVATE SECTOR AND THE REGIONAL SEAS CONVENTIONS AND ACTION PLANS

42. Mr. Ian White of the International Tanker Owners' Pollution Federation Ltd. (ITOPF), Mr. Masamichi Hasebe of the International Oil Pollution Compensation (IOPC) Funds 1971 and 1992, Mr. Eric Calonne of the International Petroleum Industry Environmental Conservation Association (IPIECA) and Ms. Wanda Hoskin of the UNEP Division of Technology, Industry and Economics gave presentations and then formed a panel with Mr. Lucien Chabason, Coordinator of the Coordinating Unit for the Mediterranean Action Plan, acting as Moderator.

43. Mr. White informed the Meeting that shipping was probably the most highly regulated international industry: shipowners and associated groups had to comply not only with conventions but with a plethora of national and regional regulations. Each major accident tended to result in further controls, adding to the high cost of compliance. Nevertheless, relatively few international associations looked after the interests of the shipping industry: in the fields of safety and pollution prevention the main associations were his own organization, the International Association of Independent Tanker Owners (INTERTANKO), the Oil Companies International Marine Forum (OCIMF) and the International Chamber of Shipping (ICS).

44. ITOPF focused its attention on spill preparedness and response. On matters of mutual interest there was a high level of coordination between his own and the aforementioned bodies and other industry associations such as IPIECA to ensure that limited resources were used to maximum benefit. That all those organizations had secretariats based in London made the task of coordination easier. He stressed that whilst all the international shipping associations cooperated extensively with private-sector and public-sector players at the national, regional and international levels, none was able to finance major projects. They did, however, provide expertise to assist capacity building. The numerous calls for such assistance meant that priorities had to be established at a global level which might not always coincide with regional priorities. He summarized the memberships and main objectives of INTERTANKO, OCIMF and ITOPF and drew attention to their various publications.

45. Tanker shipping was now safer and of a higher quality than ever before, as shown by the dramatic reduction in the annual incidence of major spills over the past three decades. Nevertheless, the world expected zero accidents. Both INTERTANKO and OCIMF were working hard to achieve that goal, but tanker owners and charterers were only two of the links in the chain of responsibility, and the continued improvement in the quality of international tanker shipping required among other things the active participation of Governments by imposing port controls and uniformly enforcing conventions and regulations.

46. The principal role of ITOPF was to respond to marine oil spills on behalf of its tanker owner members and shipowner associates, their third-party-liability insurers and the IOPC Funds. At the site of a spill staff gave objective, non-partisan advice to those in charge of the response operations on the most appropriate clean-

up techniques to maximize effectiveness and reduce the damage to the environment and economic resources. The practical experience that ITOPF had gained worldwide was now used to enhance preparedness among maritime States by providing assistance with contingency planning and training and disseminating technical information. Realistic contingency plans for various levels of risk and good organization and management of response operations were crucial. However, despite enormous efforts over many years, realistic plans still did not exist in many high-risk countries. Plans were best developed hands-on by the people who would carry them out, not by employing consultants, because the development process was more important than the final plan. A sustained commitment on the part of responsible government officials was required, which was occasionally lacking until they were confronted by a major spill.

47. The ITOPF series of country profiles summarized oil spill risk, response arrangements, and states of preparedness in some 160 maritime States, which may also be found on the organization's Web site www.itopf.org. He requested those present to assist ITOPF in keeping the information up to date, noting that on the basis of information in the country profiles, ITOPF had in 1996 produced a report on the risk of oil spills and the state of preparedness in 13 regional seas areas. ITOPF was interested in working with UNEP and other interested parties to extend and update the report as a starting point for enhanced cooperation. Regional seas conventions and action plans, moreover, should consider urging their member countries to accede to the IMO civil liability conventions, which would give them access to the IOPC Funds, which he described as "free insurance," as an additional source of funding for post-spill clean-up operations.

48. Mr. Hasebe of the IOPC Funds explained the conditions attached to disbursements from the Funds, the sums available and how they were spent. The purpose of the Funds was to provide supplementary compensation to persons who suffered damage caused by pollution resulting from the escape of discharge of oil from ships, and to ensure that the oil cargo interests shouldered a part of the economic consequences of such damage, to the relief of the shipping industry. The sums were raised not from Governments but from petroleum customers that purchased over 150,000 tonnes per year. The size of the funds had been increased within the past decade by about 50% to 203 million SDR,[1] although over 95% of spills had not reached the pay-out limit even before the increase. Moves were being made to add a third tier of coverage, but only the most costly spill imaginable would breach the 203 million SDR ceiling. However, payments from the Funds for environmental damage per se were not made.

49. Mr. Calonne of IPIECA explained that the oil and gas industry was interested in risk reduction and accident prevention and that his organization assisted that process by circulating information on best practices—in which connection he mentioned his organization's Web site www.ipieca.org—and by developing cost-effective, science-based and socially and economically acceptable solutions for when the inevitable did happen. Given the constraints, its approach was to identify hot spots and work on them. In that connection he pointed to the interactive sensitivity maps available on the Web site www.unep-wcmc.org/imaps.

1. On 31 December 2000, ISDR was worth US$1.30736.

50. For many oil spills the clean-up cure proved worse than the disease. The regional seas conventions and action plans could assist by ensuring that their regional activity centres passed on the educative message to the extent possible so that clean-up operations were handled optimally from the environmental point of view, using the "reasonable restoration" criterion. IPIECA would do what it could to place its members' expertise at the disposal of the regional seas programmes, which included vetting contingency plans on the basis of net environmental benefit analysis. He noted that clean-up operations mostly used local equipment and manpower: cleaning up oil spills was not a high-technology exercise, although a huge amount of cooperation of all kinds was involved.

51. He drew attention to the joint IMO/IPIECA global initiative to boost developing countries' contingency planning and preparedness for oil spills and encourage them to ratify and implement the International Convention on Oil Pollution Preparedness, Response and Cooperation and the conventions relating to oil spill compensation from the IOPC Funds.

52. There was a shared perception that the regional seas programmes could usefully encourage their member countries to ratify the various conventions and compensation arrangements described by the industry spokesmen. However, it was also noted that the IOPC Funds did not pay out for environmental damage per se. The exclusion clauses on pay-outs from the IOPC Funds were noted. Participants expressed the view that distinctions between sources of oil spills, such as that between bunker oil and oil carried as cargo, were otiose.

53. Ms. Hoskin of the UNEP Division of Technology, Industry and Economics gave a presentation on the experience of UNEP with voluntary initiatives by the financial, tourism, advertising and telecommunications industries in the area of protection of the marine environment as examples of what could be achieved. The concept of partnership was crucial, not only in the interests of enhancing horizontal cooperation and operationalizing programmes but also to proper recognition of the economic impacts involved. Even if the approach taken to industry and the private sector was regulatory rather than cooperative, they needed to be persuaded of the advantages they would derive from acting in a particular way, and only in partnership with them was it possible to find out how they perceived such advantages. In that connection, a report entitled *Carrots for the Sea* (GPA Report Series No. 2) was circulated.

54. It was pointed out that entering into a partnership with the private sector did not mean being in anyone's pocket. Many private enterprises resisted giving money to ministries responsible for the environment because, unlike the regional seas programmes, they were not directly accountable. Passed through the regional seas programmes, the money would for the most part be spent in the target country or countries for the desired purpose and in a manner which would satisfy an auditor or meeting of shareholders. Also, such programmes did not adopt legislation or regulations, so there could be no allegations of influence-peddling if they were given money. In the scientific area, industries were often happy to contribute to projects when they could share data of interest to them. The feeling was general that, with the private sector, there should be an appropriate mix of regulatory and voluntary approaches, "sticks" and "carrots," and that in more environmentally sensitive societies, "carrots" were more effective than elsewhere.

Recommendations

Environmental Damage
55. The Meeting recommended that the issue of compensation for environmental damage should be further explored by UNEP and the IOPC Funds in consultation with ITOPF and that the clearer picture of compensation schemes for environmental damage which emerged should be presented to Governments in order to increase their awareness of how to recover expenses incurred in reasonable restoration activities carried out on environmental goods which were affected by oil spills.

Strengthening Relations Between Regional Seas Programmes and the Oil and Shipping Industries
56. In view of the potential mutual benefits of enhancing collaboration between regional seas programmes and global and regional oil and shipping industry organizations, the Meeting recommended:
   (a) That UNEP and the regional seas programmes should encourage their coastal States members to ratify the Civil Liability and IOPC Fund Conventions;
   (b) That UNEP, regional seas secretariats and the IOPC Funds, in collaboration with IMO, should organize, where and when required, regional workshops to facilitate the process of ratification of those conventions;
   (c) That UNEP and regional seas secretariats and the IOPC Funds should participate in future meetings of mutual relevance such as general assemblies and conferences of parties, in order to enhance awareness and play an active role in capacity building in oil spill preparedness and response and in increasing the membership of global compensation schemes for oil spill-related damage;
   (d) That UNEP, the regional seas programmes, IPIECA and ITOPF should cooperate in developing joint training activities and producing relevant educational material such as paper publications, video productions, material for posting on Web sites and the like to assist countries in capacity building and in establishing oil spill preparedness and response systems;
   (e) That the UNEP Division of Environmental Conventions should serve as a link between the regional seas programmes and the private sector, with particular attention to the insurance industry and to liability and compensation issues; and
   (f) That the Regional Seas Unit and ITOPF should collaborate in updating the document "An assessment of the risk of oil spills and the state of preparedness in 13 UNEP regional seas areas."

V. THE ONGOING DISCUSSIONS ON INTERNATIONAL ENVIRONMENTAL GOVERNANCE: THE ROLE OF REGIONAL SEAS CONVENTIONS AND ACTION PLANS

57. The Coordinator of the Global Programme of Action Coordination Office stressed the need to strengthen cooperation between regional seas programmes and the Global Programme of Action as the former were the principal implementation

platforms for projects and activities carried out under the Global Programme of Action at the regional level. Such cooperation should proceed from an identification of problems and setting of priorities by the regional seas programmes, while the Global Programme of Action could serve as the principal instrument for identifying partners and donors and forming links with them. The Global Programme of Action could also assist the regional seas programmes in developing national and regional plans of action and protocols on land-based activities and in drafting project proposals and overcoming financial bottlenecks. In that connection, GEF was only one potential donor, and other sources should be identified. Joint approaches by UNEP, the Global Programme of Action and a particular regional seas programme were more likely to receive funding.

58. She expressed that hope that the regional seas programmes would be active participants in the first intergovernmental review of the Global Programme of Action, which was to be conducted the following week, noting that they had been allocated special time slots during the review meeting to present their achievements and plans for the future and to attract potential funding partners.

59. The representative of the Convention on Biological Diversity emphasized that the Convention secretariat wished to work closely with the regional seas programmes and promised collaboration in projects and activities of mutual interest.

60. The representative of the United Nations Division for Ocean Affairs and the Law of the Sea noted the importance of the regional seas programmes in the context of the United Nations Convention on the Law of the Sea and promised his Division's continuing interest in and collaboration with them.

61. The Chair described the review of international environmental governance taking place within UNEP as requested by the Secretary General, reporting in depth on the current situation of the multilateral environmental agreements and on how they were or could be clustered. The difficulties of ensuring proper coordination and collaboration between so many independent secretariats were compounded by their geographical dispersion. However, proposals for colocation of secretariats had encountered fierce opposition, as secretariats differed in their views of which cluster they should be in and host Governments were unwilling to lose secretariats. The point was made that whereas the Ozone Secretariat, which looked after the multilateral environmental agreement which was generally accepted as the most successful, was based in Nairobi, there was no interest in colocating other secretariats there.

62. On the issue of proliferating protocols and agreements, there was consensus that any new legal instrument must represent added value. However, the point was made that although much talk had been heard about the "fragmentation" of international law, a close reading of the various multilateral environmental agreements and related instruments showed a degree of overlap but very few contradictions. Even overlap was not necessarily a bad thing: while duplication of effort was to be deplored, mutual reinforcement was not. Nevertheless, no new instrument should be introduced unless it could be shown to be more specific, more detailed and tougher than the applicable global legislation in the field. It was felt, however, that there should be no vacuums in the global commons of the ocean: loopholes and lacunae in applicable international law and legislation were invitations to trouble.

63. On the proliferation issue in general, it emerged that the causes of problems tended to lie more at the national than the international level. The regional bodies

already cooperated and coordinated amongst themselves well enough that they were unlikely to try to reinvent the wheel. Similarly, it was unlikely that any regional body of interest to the Meeting would actively wish to be out of step with analogous bodies elsewhere in the world, which also operated within the framework of applicable international law; however, it was Governments that made the agreements founding them and made the decisions in their governing bodies. Also, it was a common experience that focal points for the various agreements could be based in different ministries in a country and sometimes appeared to be in disharmony. Equally, Governments' perceptions of national interest lay behind the proliferation of "centres" in some areas and had even led to rival "centres" being set up within the same country. Even getting representatives of some Governments to talk together was no mean achievement in some cases.

Recommendations

International Ocean Governance
64. The Meeting recognized the mutual benefits of speaking with a single voice in discussions of international environmental governance. Also, if clustered together, the horizontal linkages and cooperation between contiguous regional seas programmes would be facilitated. It was stressed that the regional seas programmes could serve as platforms for implementing relevant global conventions on a regional basis and that clustering would facilitate regional coordination in so doing. Clustering would also enable the regional seas programmes to serve better as foci for collaborating with the regional divisions of other organizations in the United Nations system.

65. The Meeting recommended that the regional seas conventions should be grouped in a single cluster addressing oceans in close collaboration with relevant multilateral environmental agreements such as the Convention on Wetlands of International Importance, especially as Waterfowl Habitat, the Convention on Biological Diversity, Convention on International Trade in Endangered Species of Wild Fauna and Flora, the Global Programme of Action and the Basel Convention and key ocean partners such as the United Nations Division of Ocean Affairs and Law of the Sea, IMO, IOC-UNESCO, IAEA, FAO, and relevant regional organizations.

66. The Meeting recommended that GEF/UNEP and GEF/UNDP, UNDP, the World Bank, other multilateral and bilateral donors, and partners from industry should be invited to send representatives to the next Meeting of Regional Seas Conventions and Action Plans in order to develop a dialogue and to examine possibilities for partnerships and project development and for finding innovative and realistic forms of funding with the involvement of regional-level resource providers.

67. The Meeting agreed that ocean governance would be strengthened through the following actions, bearing in mind that regional seas conventions and action plans are the best-placed platforms for promoting the protection and sustainable use of the marine and coastal environment:
   (a) Because of their multisectoral nature, regional seas programmes could and should provide complementary regional frameworks for the implementation of global multilateral environmental agreements and global conventions relevant to the environment, including the biodiversity-related Con-

ventions, the hazardous chemicals and wastes Conventions, the Global Programme of Action, the Rio Conventions, the IMO marine pollution Conventions and Protocols and the United Nations Convention on the Law of the Sea;

(b) Horizontal cooperation between regional seas conventions and action plans on issues of common concern, including the provision of technical cooperation by the more developed regional seas programmes to those that are less developed, should be promoted further;

(c) Cooperation between the regional bodies of international organizations, including UNEP (specifically the regional seas programmes), FAO, IOC-UNESCO, IMO and IAEA, among others, must be increased in order to enhance governance and management of the marine and coastal environment;

(d) Regional clustering of activities carried out by global multilateral environmental agreements regional seas conventions and action plans, international organizations and other regional bodies should be implemented in order to carry out activities in a more coordinated and cost-effective manner, particularly in areas such as capacity building, technology transfer, development of supportive national legislation, assessment and monitoring, and public awareness and information exchange; and

(e) In particular, greater efforts should be made to pool resources for developing collective regional technology transfer centres for the protection and sustainable use of the marine and coastal environment in support of regional seas programmes, global multilateral environmental agreements, and other international initiatives.

VI. ROUND-TABLE DISCUSSION WITH SECRETARIATS OF REGIONAL SEAS CONVENTIONS AND ACTION PLANS

A. Preparations for the World Summit on Sustainable Development

68. Following a discussion in which concern was expressed that the regional seas programmes might not be sufficiently involved in the preparations for the World Summit on Sustainable Development in Johannesburg in 2002, the Chair undertook to present the issues of relevance to the regional seas programmes at the coming "Oceans and Coasts" meeting organized in Paris IOC-UNESCO and the Center for Marine Policy of the University of Delaware and to ensure that the regional seas programmes were well represented in the preparation process for the Summit taking place within UNEP and its regional offices. There would also be an opportunity to provide input to the Summit from the regional seas programmes through the forthcoming Global Ministerial Environment Forum to be held in Cartagena, Colombia, in February 2002, at which a major side event on oceans was being organized. He also undertook to clarify in due time the question of participation in the Summit by the regional seas programmes and to ascertain the source of funding for such participation.

69. Regional seas coordinators agreed on the need to develop a new and improved booklet on regional seas based on the current one entitled "Regional Seas:

A Survival Strategy For Our Oceans and Coasts." The new booklet should aim to enhance the visibility of the regional seas programmes to potential funding partners and should present the strategic goals of each programme by setting out achievable and practical operational targets for the coming 5 to 10 years. The booklet should also set forth the commitments and planned inputs of Governments, as donors wanted to see a matching effort on the part of States. To the extent possible, the booklet should also include descriptions of the links between the regional seas programmes and international organizations and global initiatives.

Recommendation

70. The Meeting recommended:
   (a) That a second edition of the booklet on the regional seas conventions and action plans should be developed, setting forth future strategies and agreed targets for the next 5 to 10 years based on firm and clear ideas and taking into account the limitations, and informing potential donors of the commitments made by Governments;
   (b) That the booklet should contain approximately 500 words on each of the regional seas and should reflect also the linkages between those programmes, other international organizations, nongovernmental organizations and regional marine fisheries bodies.

Regional Activity Centres

71. The different status of regional activity centres in the various regional seas programmes was discussed. The Meeting also discussed the problems, benefits, advantages and disadvantages of decentralizing the work of the regional coordinating units and secretariats and the various political and financial problems surrounding the operation of regional activity centres that were not integral parts of the system but were instead operated by host Governments. The Meeting concluded that no one model could fit all regions or all functions. Some participants felt that the regional seas programmes should revisit the role and definition of the regional activity centres in order to adjust to the changing environmental and socio-economic contexts of their constituencies.

72. The proliferation of "centres" other than regional activity centres with the evidently useful function of responding to oil spills and other pollution emergencies was viewed in an analogous light. For every "centre" which did excellent work there were more whose output was less than ideal. It was generally felt that serious rationalization of the system in the broadest sense was required, not least in order to make the most of limited resources. Any "centre" must represent added value, and serious thought must be given to the possibility of generating synergies by having "centres" for various purposes, agreements and bodies under one, possibly virtual, roof. As a general principle also, no new bureaucracies should be established.

Recommendation

73. The Meeting recommended that further discussions and exchanges of knowledge and experience concerning the issue should take place. It was also recommended that thought should be given to establishing regional coordinating units along the lines of the UNEP World Conservation Monitoring Centre, whose director is a UNEP staff member. In such an arrangement, the director of the regional activity centre would be a staff member reporting directly to the Director of the regional coordinating unit of the regional seas programme, thus ensuring close coordination with the programme's secretariat and the programme of work adopted by member States.

C. Other Issues of Concern to Regional Seas Conventions and Action Plans

74. Examples were given of fruitful horizontal cooperation between regional seas programmes, such as that between the Mediterranean and north-west Pacific action plans and the ongoing development of cooperation between the Baltic Marine Environment Protection Commission and the Nairobi Convention. Coordinators of other regional seas programmes expressed a commitment to develop cooperation between their programmes.

75. Concern was expressed regarding the lack of a legal framework for cooperation between conventions and with action plans. However, it was accepted that Governing Council decision 21/28 called for just such horizontal cooperation.

76. The problem was raised of the very difficult financial situation facing UNEP in the coming biennium. A substantial cut would be required in the activities of all UNEP divisions, which would severely affect the ability of the Division of Environmental Conventions to offer support to the regional seas programmes.

77. Following a discussion, the regional seas coordinators expressed regret that potential funding partners had not been invited to attend the Meeting.

Recommendations

78. The Meeting recommended that the regional seas programmes should endeavour to build up their trust funds to the extent possible and engage in resource mobilization from their member States and other sources of funding.

79. The Meeting also recommended that potential funding partners and donor organizations should be invited to future Meetings of the regional seas conventions and action plans.

80. The Meeting recommended further that existing "twinning" arrangements between regional seas programmes should be developed and that horizontal linkages should be forged or reinforced, particularly between programmes covering contiguous areas.

D. Venue of the Fifth Global Meeting of the Regional Seas Conventions and Action Plans

81. The Meeting gratefully accepted the offer by the Executive Director of the North-West Pacific Region Environmental Cooperation Center to host the Fifth

Global Meeting of Regional Seas Conventions and Action Plans in Toyama, Japan in 2002.

ADOPTION OF THE REPORT OF THE MEETING

82. The present report was adopted on the basis of the draft that had been prepared by the secretariat, taking into account written corrections provided in writing by the participants and on the understanding that finalization of the report would be entrusted to the secretariat, subject to review.

CLOSURE OF THE MEETING

83. In their closing remarks, participants noted that the calendar of meetings which it would be desirable for them to attend in the next year and beyond was extremely full. Indeed, the absence of representatives of regional seas programmes from the current Meeting was attributable to that heavy load of meetings. No coping strategy could be completely successful. The Global Meetings remained, however, of great importance to their programmes. It had been very helpful that the agenda of the Meeting had not been over heavy.

84. The Chair concluded that the future of international environmental governance nevertheless remained positive in that the channels for dialogue remained open, whether or not they lay in formal meetings. In that connection, he was confident that the critical issues for the regional seas programmes would be successfully laid before the World Summit on Sustainable Development in 2002. He noted that the ecosystem-based approach to the management of the marine and coastal resources held out great promise for the future. The future looked positive also in that closer cooperation with IOC-UNESCO in particular was assured and that partnership was growing also between the regional seas programmes, UNEP, FAO and other bodies such as the oil and shipping industry associations represented at the current Meeting. In the light of earlier comments, he took it that representatives of the insurance industry should be amongst those invited to attend the Fifth Global Meeting.

85. The Chair thanked participants for their valuable contributions and expressed his wish that representatives of all regional seas programmes would attend the next Meeting.

86. After the customary exchange of courtesies, the Chair declared the Meeting closed at 4.45 p.m. on Friday, 23 November 2001.

ANNEX I

Agenda of the Meeting

1. Opening of the Meeting
    (a) Introductory statement by the representative of the Executive Director of the United Nations Environment Programme (UNEP);
    (b) Address by Professor Elisabeth Mann Borgese, International Ocean Institute (IOI).

2. Progress report on follow-up to the decisions of the Second and Third Global Meetings of Regional Seas Conventions and Action Plans.

3. Global assessment of the state of the marine environment.

4. Panel discussion on cooperation between the private sector and the regional seas conventions and action plans.

5. The ongoing discussions on international environmental governance: the role of regional seas conventions and action plans.

6. Round-table discussion with secretariats of Regional Seas Conventions and Action Plans
   (a) Preparations of Regional Seas Programmes for the World Summit on Sustainable Development;
   (b) Regional activity centres;
   (c) Other issues of concern to regional seas conventions and action plans.

7. Adoption of the report of the Meeting.

8. Closure of the Meeting.

Selected Documents and Proceedings

Review of Accomplishments in the Implementation of the Global Programme of Action, 1995–2001†

I. INTRODUCTION

1. Pursuant to paragraph 77 of the Global Programme of Action for the Protection of the Marine Environment from Land-based Activities, the United Nations Environment Programme (UNEP) should, *inter alia*, convene periodic intergovernmental meetings to review progress on implementation of the Programme of Action.

2. In response to UNEP Governing Council decision 20/19B, the Executive Director of UNEP convened the Expert Group Meeting[1] to Prepare the First Intergovernmental Review Meeting on Implementation of the Global Programme of Action for the Protection of the Marine Environment from Land-based Activities. The Expert Group Meeting was held in The Hague from 26 to 28 April 2000.

3. The Expert Group Meeting recommended, *inter alia*, that the reporting process should be an integral component of the implementation of the Global Programme of Action; that emphasis should be placed on information exchange and questions should be posed in such a way as to elicit information rather than to produce a comprehensive assessment report; that the reporting process should assist Governments and the international community in advancing the implementation of the Global Programme of Action; and that national and regional reports should incorporate indicative examples reflecting both successes achieved and barriers encountered in implementing the Programme.

4. The Expert Group Meeting also supported the proposal of the Global Programme of Action Coordination Office that the review process should focus on five main areas: binding and nonbinding agreements at the national and regional levels; voluntary agreements involving the private and public sectors; capacity-building; innovative financing and use of economic instruments; and sharing experiences through reporting and through the further development of the Clearing-House Mechanism.

5. To facilitate the review process, the Global Programme of Action Coordination Office developed a reporting format, in line with the above recommendations, which requested Governments, regions and stakeholders to report on initiatives taken, barriers encountered, and opportunities identified in implementing the Global Programme of Action.

6. The Global Programme of Action underlines the importance for States of providing regular progress reports on their efforts to implement the Programme.

---

†EDITORS' NOTE.—This document is extracted from UNEP/GPA/IGR.1/2, 12 September 2001, Intergovernmental Review Meeting on the Implementation of the Global Programme of Action for the Protection of the Marine Environment from Land-Based Activities, Montreal, Canada, 26–30 November 2001. It is available online at <http://www.gpa.unep.org/igr/>.

1. See document UNEP/GPA/EG.1/8.

As it is part of a nonbinding agreement, it contains no mandatory reporting requirements. Governments, regions and stakeholders that have submitted reports have done so on a voluntary basis to share experiences and expertise with others. In keeping with the recommendations of the Expert Group Meeting, the reports do not seek to provide comprehensive overviews of the implementation of the Global Programme of Action at the national, regional, and global levels. Rather, the reports highlight innovative and interesting examples that could be replicated elsewhere.

7. This review identifies some major achievements, some barriers encountered, and opportunities distilled from the reports received up to 15 August 2001. It highlights a few representative examples of successful initiatives undertaken by Governments and other stakeholders at the local, national, regional, and global levels. The accomplishments reported give an indication of how much progress has been made in implementing the Global Programme of Action at the various levels. Together with document UNEP/GPA/IGR.1/3, which reports on the activities of the Global Programme of Action Coordination Office during the period 1996–2001, this document provides an account on the progress achieved in the implementation of the Global Programme of Action since its adoption in 1995.

8. A comprehensive compilation and analysis of the reports received, including those received after 15 August 2001, will be made available at the Intergovernmental Review Meeting as a background document.[2] This will be an important source of information on experiences in implementing the Global Programme of Action. Also, all reports are available on the Global Programme of Action Clearing-House Website at www.gpa.unep.org. Governments, regions, stakeholders, and international organisations are invited to continue to share their experiences and expertise in implementing the Global Programme of Action and also to submit reports, should they not have done so, preferably before the Intergovernmental Review Meeting.

9. The Review Meeting is urged to consider the reported barriers, opportunities, and accomplishments in the implementation of the Global Programme of Action. Specifically, it is urged to consider what action needs to be taken at the local, national, regional and global levels in order to address the barriers and opportunities identified in the reports and also how the Global Programme of Action Coordination Office can help facilitate such action.

II. MAIN CONCLUSIONS

10. Based on the reports received so far, we can conclude that the Global Programme of Action has gained considerable impetus over the last five years, particularly since the establishment and full staffing of the Global Plan of Action Coordination Office in The Hague. Progress has been achieved to varying degrees in the various areas. If we are to make a fair assessment of the progress achieved, we should note that implementing the Global Programme of Action is an iterative process in which each step builds upon the one before and in which the guidance provided by the Global Programme of Action is continuously revisited and refined, leading to incremental action to protect coastal and marine environments from land-based sources of pollution and resource degradation.

2. UNEP/GPA/IGR.1/INF/10.

11. In the implementation of the Global Programme of Action, particular progress has been achieved in identifying problems and the action required to address them at both the national and regional levels and in furthering the objectives of regional cooperation. These objectives include the identification and assessment of problems; establishment of priorities for action; identification of management approaches; and identification of strategies to mitigate and remedy adverse impacts of land-based activities. Also, considerable progress has been achieved in developing legally binding agreements on land-based activities at the regional level in two regions.

12. There has been somewhat less progress in mobilizing financial resources, and in capacity building at the national level. Progress was weaker yet in the areas of mobilizing activities, exchanging experience and expertise at the national and regional levels, and in developing the necessary institutional arrangements, particularly arrangements for coordination between sectors and sectoral institutions at the national level.

13. In ranking the priorities assigned to the various Global Programme of Action marine pollution source categories, it is obvious that most reporting countries give top priority to the source category "sewage," followed by "nutrients," "oils," "heavy metals," and "litter," in that order. This is in accordance with the priorities identified in regional workshops of Government-designated experts held between 1996 and 1999 under the auspices of the Coordination Office and within the context of the UNEP Regional Seas Programme. The reports contained little information about achievements in the area of physical alterations to and destruction of habitats, though this source category was singled out for priority action at the regional workshops.

14. Many of the challenges which the Global Programme of Action has met since its inception have been faced by almost all Governments that reported on barriers to effective implementation. Such barriers include limited public and political awareness of the degradation of the marine environment attributable to land-based activities; a lack of appropriate legislation and enforcement mechanisms; inadequate capacity at all levels of government; and a lack of financial resources.

15. All countries reported on the development of new instruments for environmental protection. These varied from general environmental legislation to specific regulations controlling discharges and introducing environmental taxes, environmental quality criteria and emission standards. An increase in the use of environmental impact assessments and in reliance on coastal management practices is also evident from the reports. Practically all land-based activities are targeted to varying degrees by these new instruments. Among the most targeted sectors of land-based activities are "chemical industry" and "water management," followed by "urban development" and "agriculture." Tourism, aquaculture, harbours, mining, and road transport also received attention in some countries or regions. However, very little information has been reported as to the effectiveness and the efficiency with which land-based activities were addressed by those various instruments.

16. Generally speaking, the success and the degree of implementation of the Global Programme of Action in a country depends on the availability of a strong and coordinated institutional structure, including a viable national programme of action on land-based activities, and on the availability of resources both human and financial. In some cases, over and above the traditional approaches to addressing

land-based activities, innovative initiatives that altered normal practice were successfully introduced.

17. A number of experiences in the area of municipal wastewater management showed that public-private partnerships and voluntary agreements involving the private sector can improve the quality of sanitation services while protecting the coastal and marine environment from pollution from domestic and urban wastewater discharges. Public-private partnerships also proved to be useful, in some cases, in effectively mobilizing new and additional resources and in advancing government action in the field of policy formulation, including regulation and legislation and the setting of goals and targets.

18. Funding for projects in all areas pertaining to the Global Programme of Action remains the main barrier to implementing it. Very little progress was reported on new and additional funding or on the use of innovative or non-conventional ways and means to fund implementation at the national and regional levels. Financial arrangements with international financing institutions for protecting the marine environment from land-based activities have been used effectively in a number of cases to enhance Governments' efforts in implementing the Global Programme of Action at the national and regional levels. However, support for the Global Programme of Action has not been mainstreamed in the programme of work in the World Bank portfolio, nor has it been made an explicit part of the funding requirements: the objectives and approaches of the Global Programme of Action have not been taken into consideration in the allocation of funds. This may be indicative of a lack of mainstreaming of the objectives of the Global Programme of Action in the work programmes of other financial institutions also.

III. BARRIERS ENCOUNTERED

19. The barriers to effective implementation of the Global Programme of Action reported to the Coordination Office in the national and regional reports submitted by Governments and the regional bodies concerned can be grouped into four main clusters: financial and economic barriers; technical barriers; managerial and institutional barriers; and legal and policy barriers. Limited capacities at the local, national and regional levels are important in all of these. Also, the reports show that the more traditional "command and control" approaches were often predominant in Governments' views on how to deal with the barriers to addressing land-based activities.

20. The financial and economic barriers are principally: inadequate or nonexistent investment in marine environmental protection; inadequate or ineffective allocation of funds for municipal wastewater treatment; and Best Available Technology (BAT) that is either very or even prohibitively expensive for developing countries.

21. The most commonly reported technical barriers are a lack of technical capability and trained personnel; a lack of awareness of and scientific knowledge about problems resulting from land-based activities among policy-makers and stakeholders; a lack of understanding or awareness among the public concerning the impact of their activities on the marine environment; the use of outdated production techniques that result in excessive discharges of wastewater that do not meet environmental standards; and a lack of suitable monitoring equipment.

22. Most of the managerial and institutional barriers reported by Governments involve a lack of coordination and integration, at the national level, in the development and management of the coastal zone and its resources. The high concentrations of population, industry, and transport links along narrow coastal strips and the intense interactions between competing and often conflicting development activities make renewable resource systems in the marine environment and coastal areas particularly vulnerable to degradation. Here, the general lack of comprehensive land-use planning or enforcement is a fundamental barrier. A lack of information on which to base decision-making, generally low managerial skill levels, and a lack of specifically environmental management skills in private enterprises have also been reported as important barriers. Also, low or nonexistent levels of public participation in marine environmental management were also reported as a barrier to making such management effective.

23. The legal and policy barriers are attributable mainly to the relatively low priority given by some countries to environmental conventions, treaties, or framework agreements: Governments are either not convinced of the need for legally binding instruments to address land-based activities, or their commitment to tackling pollution from those activities is more or less inadequate. In some other countries, even where environmental legislation and regulations exist, a major barrier is poor or absent enforcement, which makes them largely ineffective. A few reports indicated that the legislation on land-based activities was either too general or too outdated to address specific or current problems from those activities.

24. Several reports mentioned lack of coordination between agencies and sectoral authorities at various levels of government as a policy barrier because legislation covering environmental issues may involve various organizations, for example, in relation to enforcement and compliance. Another important barrier—which was pointed to in some reports as a key barrier—is a lack of political will, determination, or initiative on the part of Government, mostly but not only in developing countries. Many developing countries, for legitimate reasons, pay much more attention to economic development issues, with environmental protection taking a back seat.

IV. OPPORTUNITIES IDENTIFIED

25. Several opportunities were identified with the potential to instigate and support effective action to address negative impacts of land-based activities and enhance the effectiveness of measures taken in that respect. A few examples of the opportunities that were highlighted in the reports received so far are given below.

26. The introduction and adoption of Integrated Coastal Area Management (ICAM) was viewed by several of the reporting countries as an important policy step towards implementing the Global Programme of Action at the national level. ICAM covers all activities and discharges along the coastal zone and it includes a participatory planning process, establishes priorities for action and supports all stages of developing and implementing national programmes of action. It was noted also that establishing linkages between management of river basins and management of the marine environment offers further opportunities.

27. The Environmental Impact Assessment (EIA) process also represents an opportunity, particularly in areas where planning is lacking. Environmental ap-

praisal committees, particularly when their membership includes representation from a wide cross-section of agencies which evaluate EIAs, can offer a tremendous opportunity for increasing awareness in key government agencies about environmental issues.

28. The initiation and implementation of major national projects on municipal wastewater management, and the participation of the private sector through public-private partnerships, have proved to be significant opportunities that can support effective implementation of the Global Programme of Action. Involving the private sector in water supply and sanitation and developing long-term concessions for providing these services have shown encouraging results in several countries in several regions.

29. Increased recreational use of water and nature is in itself an opportunity for increasing awareness of the importance of clean water and for providing business opportunities (particularly in tourism). By swaying opinion amongst policy makers and investors it increases the likelihood of investments being allocated to the restoration and protection of bodies of water and the natural environment, and to sustainable waste management.

30. It was mentioned also that the forthcoming World Summit on Sustainable Development, to be held in South Africa in 2002, could offer a unique opportunity to raise political awareness about all environmental issues, including the impacts of land-based sources and activities on the marine environment.

V. ACCOMPLISHMENTS AS INDICATORS OF PROGRESS

A. Binding and Nonbinding Agreements at the National and Regional Levels

31. The development and adoption of legally binding agreements at both the regional (regional conventions and protocols) and the national levels (national legislation, administrative rules and standards and so on) are important elements in the process of environmental protection. Binding agreements reflect the commitment of Governments to adopting policies and taking the necessary steps to address the issue of land-based activities, and increase the likelihood of enforcement and compliance.

32. In this connection, the Global Programme of Action Coordination Office has prepared an overview[3] of binding ("hard law") and nonbinding ("soft law") agreements in support of the implementation of the Global Programme of Action in all 17 Regional Seas regions.

33. The introduction of nonbinding agreements at the national level (such as national strategies or national programmes of action) or at the regional level (for example, through regional programmes of action) is as important as binding agreements for promoting regulatory policies and protective measures, including those taken by local Governments and communities, and for enhancing the capacity to carry out such policies and sustain such measures. Progress has been achieved in applying both binding and nonbinding agreements in order to implement the Global Plan of Action.

3. UNEP/GPA/IGR.1/INF/2.

1. Binding Agreements

34. Almost all reporting countries have provided details on one or more legally binding agreements at the national level, such as national legislation or regulations, or standards used as management tools. These instruments deal with land-based activities and control emissions or effluents that impact the marine environment and associated water bodies in a particular country. Some countries have also passed Coastal Area Management legislation to control future or ongoing development activities in coastal areas and to ensure the sustainable and wise use of coastal areas and resources.

35. In some countries, the reports showed that the general framework for environmental management is very comprehensive, covering every conceivable aspect of the environment, such as the geosphere, the hydrosphere, the atmosphere and material and immaterial values such as social and cultural components.

36. At the regional level, an important achievement in connection with the Global Programme of Action was the successful development and negotiation of three legally binding agreements. The first of these, the revised Protocol for the Protection of the Mediterranean Sea against Pollution from Land-based Sources and Activities to the Barcelona Convention for the Protection of the Mediterranean Sea, was adopted in 1996. The second, the Protocol Concerning Pollution from Land-based Sources and Activities to the Cartagena Convention for the Protection and Development of the Marine Environment in the Wider Caribbean Region, was adopted in 1999. The third, the draft Convention for the Protection and Sustainable Management of the Marine and Coastal Environment of the Northeast Pacific, was endorsed by high-level, Government-designated experts in August 2001 and is expected to be adopted at a Conference of Plenipotentiaries in late 2001 or early 2002 (see also document UNEP/GPA/IGR.1/3).

37. Seven Regional Seas are now covered by legally binding protocols or detailed regional legal regimes on land-based sources or activities: Wider Caribbean (CAR/RCU), Mediterranean (MEDU), South East Asia Pacific (CPPS), Black Sea (BSEP), Kuwait (ROPME), North East Atlantic (OSPAR) and Baltic (HELCOM).

38. The Stockholm Convention on Persistent Organic Pollutants (POPs), adopted by the Conference of Plenipotentiaries in May 2001, is a positive development for the Global Programme of Action. The POPs Convention directly addresses one of the nine source categories by seeking to prevent the adverse effects of the various POPs at all stages of their life cycle. Similarly, the 1998 Rotterdam Convention on Prior Informed Consent Procedures for Certain Hazardous Chemicals and Pesticides in International Trade is an important step toward implementing the actions at the global level recommended in Chapter IV of the Global Programme of Action.

2. Nonbinding Agreements

39. Since the adoption of the Global Programme of Action in 1995, UNEP and its Global Programme of Action Coordination Office, in association with other partners, have been supporting or initiating regional efforts to implement the Global Programme of Action at the regional level. In the beginning those support activities

consisted in convening and following-up a series of regional workshops of Government-designated experts and in preparing regional programmes of action, in the form of nonbinding agreements, to address land-based activities. Varying degrees of progress have been achieved through these regional efforts, which are detailed in document UNEP/GPA/IGR.1/INF/2.

40. The Global Programme of Action requests States, in accordance with their policies, priorities and resources, to develop or review national programmes of action within a few years. To date, at least 13 countries (including countries in Africa, Asia, North America, South America, and West Asia) have developed or are in the course of developing national programmes of action (further details are given in document UNEP/GPA/IGR.1/3). The available information shows that additional cross-sectoral and intersectoral action in the coming years will be needed to fulfil the objectives of Chapter II of the Global Programme of Action, which require action at the national level.

41. At the regional level, regional programmes of action are important implementation tools, supporting countries' compliance with and fulfilment of their obligations under regional agreements or protocols on land-based activities. Details of the development and implementation of regional programmes of action in 12 out of 17 regions are given in document UNEP/GPA/IGR.1/3. For the remaining five Regional Seas, States have still to decide whether they wish to establish regional programmes of action or to proceed with addressing land-based activities on the basis of existing legal and institutional arrangements. Further support action will be required to facilitate the preparation of regional programmes of action in regions that wish to establish and adopt them.

B. Voluntary Agreements and the Involvement of the Private Sector

42. Voluntary action may take the form of commitments by individual companies or groups of private entities, particularly in the industrial sector. Examples include codes of conduct adopted unilaterally at the national or international level, agreements between stakeholders on environmental performance targets and the establishment of effective self-regulatory mechanisms. Voluntary initiatives of this kind support existing regulatory measures and environmental policy instruments; they do not replace them. Voluntary initiatives by the private sector have proven effective in facilitating the implementation of environmental policies and management practices.

43. Several national and regional reports submitted to the Coordination Office showed that noteworthy progress had been achieved. In the East Asian Seas Region, the role of the private sector in the area of municipal wastewater treatment has been growing, with some indications of success. Several countries have chosen to transfer the provision of sanitation services to private operators. For Governments, this is an alternative to a State-managed system and a response to the problems of meeting urgent needs and keeping up with the rapid pace of urban, industrial and commercial development. In many East Asian countries, private enterprises are obliged to build facilities to treat effluent to a required standard before discharging it into public sewers. Industries with similar needs are encouraged to build common facili-

ties for wastewater treatment. The State is required to monitor the performance of these enterprises.

44. Similarly, in the South Asian Seas Region, a new partnership, the Public-Private Infrastructure Advisory Facility, has attracted wide support from the public, and also financial support from the Asian Development Bank. It is one of the largest regional initiatives for promoting public-private partnerships.

45. Within the context of the Mediterranean Action Plan, the Mediterranean Commission on Sustainable Development has set up a working group on industry in an effort to develop a dialogue with key industrial associations in the region. The aim is to encourage industries in the Mediterranean countries to adopt pollution prevention and eco-efficiency approaches and to circulate information to their members in support of the Mediterranean Strategic Action Programme to Address Pollution from Land-Based Activities.

46. In the Arctic Region, following the development and adoption of the Russian National Programme of Action for the Arctic, good prospects exist for private sector/business involvement through a Partnership Conference. The implementation of this National Programme of Action is being supported by the Arctic council's programme for the Protection of the Arctic Marine Environment (PAME) and the Advisory Committee on Protection of the Seas (ACOPS) through the provision of technical, scientific and financial assistance. The Global Environment Facility (GEF) is financing the first phase of implementation and partnership building.

47. At the national level, a project in Sri Lanka for the relocation and modernization of tanneries is a unique socioeconomic partnership with the private sector promoted by Government (Ministry of Industries). The Government, the private sector, and the donor community are jointly funding the project. Amongst the results of the project will be the construction of treatment facilities in the form of a common effluent plant that meets all discharge standards; the re-use of the treated effluent; and the establishment of a safe landfill for the solid waste.

48. Several projects to implement the Global Programme of Action at the regional level have been reviewed and the lessons to be learned have been distilled, particularly with regard to political structure or form of convention; the methodological approaches required, such as strategic action planning, setting regional emission standards, and identifying hot spots; and the need for stakeholder involvement. In addition, these regional projects were assessed on how well they instigated effective national action to address land-based activities. Regional approaches to implementing the objectives of the Global Programme of Action and the lessons learned from those regional projects are described in document UNEP/GPA/IGR.1/INF/5.

C. Capacity-Building

49. Building national and regional capacities is crucial to the successful, effective implementation of the Global Programme of Action. Several of the regional reports, and some reports by partner organizations, give specific examples of useful and promising initiatives in this area.

50. The initiatives reported varied in objective and scope, from initiatives focusing on a specific source category or specific target group to initiatives addressing a

wider range of substances or target audience. Also, most of the capacity-building activities were undertaken in the context of a regional capacity-building initiative, or a project or programme with a specific component or components aimed at enhancing the technical and/or institutional capacity to address one or more of the Global Programme of Action source categories.

51. The capacity-building initiatives at regional level, undertaken in the interests of the Strategic Action Programme of the Mediterranean Action Plan (MAP), are exemplary, providing for regional "Training of Trainers" activities in the area of technical information and advice on the environmentally sound operation of sewage treatment facilities. In these training sessions, modern training techniques are employed and a training package is given to the trainees at the end of each session. The experience gained from the first series of sessions will be used in a second regional training course for practitioners from Mediterranean countries. Also, a number of national training courses for operators of sewage treatment plants are planned for 2001–2003. These national training courses are to be given by staff trained at the regional courses.

52. Activities using the same "training of trainers" approach are planned in the areas of best environmental practices and clean production techniques for priority target industries in the region. Through the Clean Production Regional Activity Centre, based in Barcelona, Spain, the Mediterranean Action Plan is currently assisting businesses in applying cleaner production techniques, with priority to pollution prevention at source and the minimization of waste flows.

53. Through the Mediterranean Action Plan, UNEP is also pursuing an innovative initiative to build regional capacities in the area of compliance with and enforcement of legislation for the control of land-based pollution. This is being undertaken in cooperation with the World Health Organization and the International Network for Environmental Compliance and Enforcement with the aim of establishing an informal regional network for exchanging information on regional environmental protection and on networks of professionals involved in compliance issues.

54. In the West and Central Africa Region, the work undertaken under the Gulf of Guinea Large Marine Ecosystem Project (GOG-LME) has contributed considerably to building the capacities of the participating countries in areas directly related to the Global Programme of Action, particularly with respect to waste minimization and management. For example, the marine debris/solid waste monitoring activities on Cameroon's beaches have increased the monitoring capacities of that country and yielded information on the types and quantities of waste relating to major activities such as tourism and fisheries.

55. Another innovative capacity-building initiative has been taken by the Global Programme of Action Coalition for the Gulf of Maine (GPAC) in the United States of America, and aims to provide networking and capacity-building amongst communities involved in environmental monitoring in the Gulf of Maine watershed through the Coastal Network of the Gulf of Maine. The initiative's products include a database of monitoring activities, additional monitoring protocols and several other capacity-building, networking activities.

56. The experience of the Baltic Marine Environment Protection Commission (HELCOM) in promoting and building national and regional capacities in the Baltic Sea Area is also worth mentioning. The work is being undertaken mainly through the HELCOM Programme Implementation Task Force and the Monitoring and As-

sessment Group. The activities include technical workshops, training programmes, guideline development, and intercalibration and quality assurance procedures for stakeholders at the national, municipal, and local levels.

57. The capacity-building initiatives implemented in the East Asian Seas Region by UNEP and the International Maritime Organization (IMO) through the GEF-funded Partnerships in Environmental Management for the Seas of East Asia (PEMSEA) have been particularly successful. One such initiative is the Malacca Straits project, which has been reported as successful in providing a transparent and reliable mechanism to bring together scientists, geographers, engineers, economists and decision-makers from the three littoral States of the Malacca Straits in partnership on the issues, priorities and required actions to manage and protect that subregional sea area. The initiative was also successful in putting together a multidisciplinary, multisectoral team of stakeholders from the States, which are now well equipped to proceed with the further development and implementation of the action plans for the Straits. Other capacity-building projects within this same PEMSEA framework include demonstration sites for developing, testing, and implementing Integrated Coastal Management and several other projects on providing the authorities in the area with sustainable mechanisms for preventing and managing marine pollution off their coasts.

58. The full involvement of nongovernmental organizations in capacity building for the Global Programme of Action is essential. The International Ocean Institute (IOI) is offering several opportunities for capacity building relevant to the Global Programme of Action. Several IOI Operational Centres give courses for local and regional authorities, on Integrated Coastal Management and on land-based sources and activities and economic and legal frameworks for addressing them. The World Wide Fund for Nature (WWF) has developed a guidance document for nongovernmental organizations on the implementation of the Global Programme of Action, whereas Greenpeace is assisting local nongovernmental organizations in analysing the environmental and social risks posed by certain aquaculture developments, and in building local capacities in support of sustainable alternatives to monoculture activities.

59. Further information on capacity-building efforts in support of the Global Programme of Action is given in documents UNEP/GPA/IGR.1/3 and UNEP/GPA/IGR.1/INF/10.

D. Innovative Financing and Use of Economic Instruments

60. In the sections dealing with financial resources, the Global Programme of Action recognizes that mobilizing financial resources is an indispensable foundation for the development and implementation of national and regional programmes to protect the marine environment from land-based activities. It is essential, therefore, for innovative financing mechanisms to be identified and explored. Opportunities include direct international financing under bilateral or multilateral agreements and loans from regional development banks and other financial institutions.

61. Reports from several countries in the Caribbean, Latin America, Europe, Africa, and Asia contain information about methods of raising funding, such as user charges, polluter charges, local and national taxes imposed on the use of certain

products, and fees for certain tourism activities and also on financial incentives such as tax relief on certain favourable activities. Funding by development banks and grants from foundations, in addition to national funding is commonly used in those regions to support Global Programme of Action activities. Apart from State funding, other forms of financial arrangements such as grants and concessionary assistance, multilateral loans and export credits were used. The information from the Baltic Sea Region exemplified the way that the countries in a region use various economic instruments, both State and non-State.

62. Nevertheless, the reports received by the Global Programme of Action Coordination Office provide very little information on the use of innovative financing. One country in Latin America reported on the introduction of a tax for municipal wastewater treatment that is set as a function of household income: households with the lowest incomes enjoy free service while higher-income households pay a greater share of the costs.

63. For the South Asian Seas region, it was reported that financial support for projects to protect the coastal and marine environment from land-based activities comes mainly as grants from bilateral or multilateral donors. Such funding was provided both to governmental and nongovernmental sectors in the member countries in the region and was helpful in implementing projects relevant to the Global Programme of Action and also the Regional Programme of Action.

64. At the global level, the World Bank, together with its International Finance Corporation, emerges as a potential key player in activities of relevance to the Global Programme of Action because of its wide experience of innovations in financing. A study commissioned by the Global Programme of Action Coordination Office showed that in more than 165 projects worldwide in the World Bank portfolio, various types of innovative financial arrangements were used. Some of these innovative approaches were used in several projects, including projects concerning the Aral Sea, the Baltic, the Black Sea, the Caspian, the Wider Caribbean, the Gulf of Aqaba, the Indian Ocean, the Mediterranean, the Mesoamerican Reefs, the Nile Basin, the Red Sea, and the Senegal River Basin. The study was prepared for a joint World Bank/UNEP workshop on promoting sustainable financing for the protection of the marine environment from land-based activities (The Hague, 9–11 July 2001). The study, and the recommendations from the workshop concerning innovative financial mechanisms, can be found in document UNEP/GPA/IGR.1/INF/7.

65. Several regional and national projects implementing the Global Programme of Action receive financial assistance from GEF for their incremental costs. Document UNEP/GPA/IGR.1/INF/6 gives details of these projects.

E. Sharing of Experiences Through Reporting and the Further Development of the Clearing-House Mechanism

66. Mobilizing and exchanging experience and expertise of relevance to the Global Programme of Action, which includes facilitating effective scientific, technical and financial cooperation, as well as capacity building, is an essential component of the Programme itself. States have therefore agreed to cooperate in the development of the Global Programme of Action Clearing-House Mechanism as the key to facilitating such exchanges and cooperation and hence to furthering the objectives of the

Programme. A summary of the progress achieved and possible ways forward can be found in document UNEP/GPA/IGR.1/INF/9.

67. The submission by Governments, regional bodies and partner organizations of reports on their implementation of the Global Programme of Action for the First Intergovernmental Review Meeting is an important contribution to the further development of the Global Programme of Action Clearing-House Mechanism.

68. One related initiative that was reported is the European regional network on water, EUROWATERNET, established by the Finnish Environmental Institute in cooperation with 13 regional environment centres in Europe. EUROWATERNET operates in support of certain issues of relevance to the Global Programme of Action and is based on the current national monitoring networks of the European countries. It is designed to provide information on water quality and quantity at the European level, and will be developed in the future to meet the requirements of the European Union Water Framework Directive.

A. Concluding Remarks

69. The information in this document is designed to give a bird's-eye view of the national, regional and stakeholder reports received during the preparation of the First Intergovernmental Review Meeting on the Implementation of the Global Programme of Action for the Protection of the Marine Environment from Land-based Activities. The full reports are available on the Global Programme of Action Clearing-House Website and an extended summary will be made available.

70. UNEP takes this opportunity to thank all those that submitted contributions.

Selected Documents and Proceedings

Montreal Declaration on the Protection of the Marine Environment from Land-Based Activities†

1. We, the representatives of 76 Governments, with the valued support and concurrence of delegates from international financial institutions, international and regional organizations, the private sector, non-governmental organizations, other stakeholders and major groups, meeting in Montreal, Canada, from 26 to 30 November 201, for the first Intergovernmental Review Meeting on the Implementation of the Global Programme of Action for the Protection of the Marine Environment from Land-based Activities, agree as follows:

2. We are concerned that:
   (a) The marine environment is being increasingly degraded by pollution from sewage, persistent organic pollutants, radioactive substances, heavy metals, oils, litter, the physical alteration and destruction of habitats, and the alteration of timing, volume, and quality of freshwater inflows with resulting changes to nutrient and sediment budgets and salinity regimes;
   (b) The significant negative implications for human health, poverty alleviation, food security and safety, and for affected industries are of major global importance;
   (c) The social, environmental, and economic costs are escalating as a result of the harmful effects of land-based activities on human health and coastal and marine ecosystems and that certain types of damage are serious and may be irreversible;
   (d) The impacts of climate change on marine environments are a threat to low-lying coastal areas and small island States due to the increased degradation of the protective coastal and marine ecosystems; and
   (e) Greater urgency is not accorded to taking action at the national and regional levels for meeting the objectives of the Global Programme of Action.

3. *We are concerned* also about the widespread poverty, particularly in coastal communities of developing countries, and the contribution that the conditions of poverty make to marine pollution through, for example, lack of even basic sanitation; and how marine degradation generates poverty by depleting the very basics for social and economic development.

4. *We acknowledge* that the United Nations Convention on the Law of the Sea and Agenda 21 provide the key framework for implementing the Global Programme of Action.

---

†EDITORS' NOTE.—This declaration from the first Intergovernmental Review Meeting on the Implementation of the Global Programme of Action for the Protection of the Marine Environment from Land-Based Activities, Montreal, 26–30 November 2001, is available online: <http://www.gpa.unep.org/igr/> (accessed 28 December 2001). Adopted by the Intergovernmental Review Meeting on the Implementation of the Global Programme of Action for the Protection of the Marine Environment from Land-based Activities at its first meeting on Friday, 30 November 2001.

5. *We declare* that implementation of the Global Programme of Action is primarily the task of national Governments. Regional seas programmes also play an important role in implementation and both should include the active involvement of all stakeholders.

6. *We shall cooperate* to improve coastal and ocean governance for the purpose of accelerating the implementation of the Global Programme of Action, by mainstreaming, integrating coastal area and watershed management, and enhancing global, regional, and national governance processes.

7. *We shall also cooperate* to identify new and additional financial resources to accelerate the implementation of the Global Programme of Action, by building capacity for effective partnerships among Governments, industry, civil society, international organizations, and financial institutions, and by making better use of domestic and international resources.

Mainstreaming of the Global Programme of Action

8. *We commit* ourselves to improve and accelerate the implementation of the Global Programme of Action by:
   (a) Incorporating the aims, objectives and guidance of the Global Programme of Action into new and existing activities, action programmes, strategies and plans at the local, national, regional, and global levels and into sectoral policies within our respective jurisdictions;
   (b) Strengthening the capacity of regional seas organizations for multistakeholder cooperation and action, including through participation in partnership meetings focused on concrete problem identification and solution;
   (c) Supporting the ratification of existing regional seas agreements and development of additional ones, as appropriate, and promoting collaboration between existing regional seas organizations, including through twinning mechanisms;
   (d) Calling on the United Nations agencies and programmes and international financial institutions to incorporate, where appropriate, the objectives of the Global Programme of Action into their respective work programmes, giving priority in the period 2002–2006 to addressing the impacts of sewage, physical alteration and destruction of habitats and nutrients on the marine environment, human health, poverty alleviation, food security and safety, water resources, biodiversity, and affected industries; and
   (e) Calling upon regional seas programmes in light of assessments of their marine environment to:

      (i) Identify priorities with particular regard to those set out in paragraph 8 (d) above;
      (ii) Prepare action plans to address the implementation of those priorities and work, as appropriate, with national authorities to implement those plans; and
      (iii) Produce interim reports on the carrying out of these action plans with a view to completing full reports at the time of the next Global Programme of Action review.

Oceans and Coastal Governance

9.   *We further commit* ourselves to improve and accelerate the implementation of the Global Programme of Action by:
   (a) Taking appropriate action at the national and regional levels to strengthen institutional cooperation between, *inter alia,* river-basin authorities, port authorities, and coastal zone managers, and to incorporate coastal management considerations into relevant legislation and regulations pertaining to watershed management in particular transboundary watersheds;
   (b) Strengthening the capacity of local and national authorities to obtain and utilize sound scientific information to engage in integrated decision-making, with stakeholder participation, and to apply effective institutional and legal frameworks for sustainable coastal management;
   (c) Strengthening regional seas programmes to play a role in, as appropriate, coordination and cooperation:

   (i) In the implementation of the Global Programme of Action;
   (ii) With other relevant regional organizations;
   (iii) In regional development and watershed management plans; and
   (iv) With global organizations and programmes relating to implementation of global and regional conventions;

   (d) Supporting this new integrated management model for oceans and coastal governance as an important new element of international environmental governance;
   (e) Improving scientific assessment of the anthropogenic impacts on the marine environment, including, *inter alia,* the socio-economic impacts;
   (f) Enhancing the state-of-the-oceans reporting to better measure progress towards sustainable development goals, informing decision making (such as setting management objectives), improving public awareness, and helping assess performance; and
   (g) Improving technology development and transfer, in accordance with the recommendations of the United Nations General Assembly.

Financing of the Global Programme of Action

10.   *We commit* ourselves to improve and accelerate the implementation of the Global Programme of Action by:
   (a) Strengthening the capacity of local and national authorities with relevant financial and other resources to identify and assess needs and alternative solutions to specific land-based sources of pollution; and to formulate, negotiate and implement contracts and other arrangements in partnership with the private sector;
   (b) Calling on international financial institutions and regional development banks and other international financial mechanisms in particular the World Bank and the Global Environment Facility, consistent with its operational strategy and policies, to facilitate and expeditiously finance activities

related to the implementation of the Global Programme of Action at regional and national levels;
(c) Giving due consideration to the positive and negative impacts of domestic legislation and policies, including, *inter alia,* fiscal measures, such as taxation and subsidies, on land-based activities degrading the marine and coastal environment; and
(d) Taking appropriate action at the national level including, *inter alia,* institutional and financial reforms, greater transparency and accountability, the development of multiyear investment programmes and providing an enabling environment for investment.

Other Provisions

11. *We welcome* the Strategic Action Plan on Municipal Wastewater and urge the United Nations Environment Programme to finalize this document as a tool for implementing the objectives of the Global Programme of Action.

12. *We call upon* Governments to ratify the Stockholm Convention on Persistent Organic Pollutants, the 1996 Protocol to the London Convention on the Prevention of Marine Pollution by Dumping of Wastes and Other Matter and other relevant agreements in particular regional conventions, such as the 1999 Aruba Protocol to the Cartagena Convention for the Protection and Development of the Marine Environment of the Wider Caribbean Region and protocols dealing with the prevention of pollution of the marine environment as a means of implementing the Global Programme of Action. We also stress the need for increased international cooperation on chemicals management.

13. *We welcome* also the work done by the Global Programme Coordination Office, commend its 2002–2006 work programme to the Governing Council of the United Nations Environment Programme and encourage it to implement the programme at a strengthened level, subject to availability of resources.

14. *We note* the outcome of the first Intergovernmental Review of the Global Programme of Action as a valuable contribution to the implementation of Agenda 21. We request that the next Global Ministerial Environment Forum endorse this outcome. We commend the outcome to the attention of the Monterey International Conference on Financing for Development, as well as of the Third World Water Forum to be held in Kyoto, Japan, in 2003. We request the preparatory process of the World Summit on Sustainable Development to take full account of the outcome of this meeting and the objective of the Global Programme of Action as it considers measures on protection of the marine environment.

15. *We request* the Executive Director of the United Nations Environment Programme to convene the second Intergovernmental Review Meeting in 2006 and seek support for organizing the meeting.

Selected Documents and Proceedings

Conclusions of the Co-Chairs from the First Intergovernmental Review Meeting on the Implementation of the Global Programme of Action for the Protection of the Marine Environment from Land-Based Activities, Montreal, 26–30 November 2001†

INTRODUCTION

1. In pursuance of decision 21/10 of February 2001 of the Governing Council of the United Nations Environment Programme (UNEP), government representatives, international financial institutions, international organizations, the private sector, nongovernmental organizations, other stakeholders and major groups, have met from 26 to 30 November 2001, in Montreal, Canada, for the first Intergovernmental Review Meeting on the Implementation of the Global Programme of Action for the Protection of the Marine Environment from Land-based Activities.

2. We are honoured to co-chair this important event and have prepared these conclusions as part of the proceedings of the meeting. We are pleased to recommend the following conclusions as an accompaniment to the Montreal Declaration and commend them for the consideration of Governments in preparation for the World Summit on Sustainable Development to be held in Johannesburg, South Africa, in September 2002 and all other forums at which activities relating to the goals of the Global Programme of Action are dealt with.

3. The 2001 report prepared by the Joint Group of Experts on the Scientific Aspects of Marine Environmental Protection—Protecting the Oceans from Land-based activities: Land-based sources and activities affecting the quality and uses of the marine, coastal, and associated freshwater environment—highlighted the alarming conclusion that "on a global scale marine environmental degradation has continued and in many places even intensified."

4. The productive capacity and ecological integrity of the marine environment, including estuaries and near-shore coastal waters, continue to be degraded for a variety of reasons, including pollution from sewage, nonpoint source runoff from agricultural and urban areas, the physical alteration and destruction of habitat nutrients, sediment mobilization and chemicals. The negative implications for coastal and marine industry, human health, poverty alleviation, food security, and safety are continuing, in many cases, unabated.

5. The social, environmental and economic costs to society are escalating as a result of disproportionately low levels of action to mitigate the harmful effects of land-based activities on coastal and marine environments and associated freshwater

---

†EDITORS' NOTE.—This Concluding Statement by the Conference Chairs is available online at <http://www.gpa.unep.org/igr/> (accessed 28 December 2001).

systems. Some types of damage are serious and irreversible. Indeed, the massive negative implications for human health, particularly as a result of pathogen laden sewage pollution of bathing beaches and shellfish harvesting areas have been seriously underestimated and neglected by the world community. A study by the World Health Organization has shown that such pollution results in millions of cases of disease and thousands of deaths annually.

A. Accomplishments of the Global Programme of Action

6. Since the inception of the Global Programme of Action, its implementation has witnessed considerable progress, and there has been continuing progress in integrated coastal zone management and oceans governance.

7. Many countries have prepared national programmes of action or have integrated the goals of the Global Programme of Action into their national strategies, policies, programmes, and legislation.

8. Many regions have cooperatively prepared regional programmes of action, both binding and nonbinding. Many of these provide excellent examples of coordination and cooperation and demonstrate the capacity of the regional seas programmes to serve as a central platform for improving coastal and oceans governance.

9. A good example of a regional approach to the Global Programme of Action and its emphasis upon developing partnerships in financing implementation is the Russian National Programme of Action for the Arctic. Similarly, a good example of multilateral partnership is the Africa Process on Cooperation for the Development and Protection of the Coastal and Marine Environment, particularly in sub-Saharan Africa.

10. The continuing development of the Global Programme of Action Clearinghouse Mechanism by UNEP, in collaboration with respective United Nations organizations has proved to be a major achievement for the implementation of the Global Programme of Action. It will prove to be a valuable tool for use by local, national, regional, and global stakeholders in implementing the Global Programme of Action.

11. Many Governments have made considerable contributions in support of the Global Programme of Action Coordination Office. Special recognition should be given to the Kingdom of the Netherlands, host of the Coordination Office for the very generous continued support for the Office, and the Government of Belgium for donations allowing the development of national programmes of action in several countries. Many donor countries have also contributed significant funds in support of projects related to the Global Programme of Action in developing nations and regional programmes.

12. With regard to multilateral financing, the Global Environment Facility (GEF) has allocated substantial resources to projects relevant to the objectives of the Global Programme of Action. The World Bank has also provided substantial support for projects that address objectives of the Global Programme of Action.

13. The Stockholm Convention on Persistent Organic Pollutants, adopted at the Conference of Plenipotentiaries in May 2001, is a major binding instrument which directly addresses one of the pollutant source categories identified in the Global Programme of Action.

B. Opportunities and Barriers

14. The first Intergovernmental Review Meeting provided Governments and other stakeholders an opportunity to consider the barriers and opportunities associated with the implementation of the Global Programme of Action. The Global Programme of Action is a suitable means of improving governance under ocean-related conventions, including strengthening the regional seas conventions and protocols. It can serve as an effective global harmonizing mechanism to improve coordination and cooperation among these regional conventions and relevant global conventions.

15. The need for international cooperation and for a coordinated approach at the national level to address the problems of freshwater as well as coastal and marine pollution from land-based activities is stressed. Bringing together the many different economic sectors contributes invaluably to poverty alleviation, food security, and peace.

16. Globally, the impact of sewage, physical alteration of coastal and marine ecosystems and high nutrient levels merit the highest priority for action. Addressing these priorities cannot be achieved in isolation of the broader objectives of sustainable development. The causative relationship between poverty, human health, unsustainable consumption and production patterns, poorly managed social and economic development, and environmental degradation must be emphasized when implementing the Global Programme of Action.

17. There is an urgent need to integrate coastal resource management and the requirements of coastal zone protection with river basin management. In this regard, the potential of institutional partnerships to ensure an integrated and holistic approach to coastal zone management, catchment or watershed management, and land-use planning is recognized.

C. The Strategic Action Plan on Municipal Wastewater

18. The Strategic Action Plan on Municipal Wastewater expands on what is provided in the Global Programme of Action with the aim of seeking consensus, promoting alternative solutions, and facilitating partnerships and regional cooperation. The three-pronged functional approach outlined in the Strategic Action Plan is widely supported but a number of issues could be expanded upon. Examples include:

(a) Provision of guidance on implementing new financial mechanisms;
(b) Giving adequate attention to alternatives to large and costly treatment facilities;
(c) Consideration of the impact of small industry on sewage systems;
(d) Role of water conservation measures in reducing demand for water treatment; and
(e) Monitoring and evaluation.

19. The Draft Guidelines on Municipal Wastewater, developed by the Coordination Office as a critical element of the Strategic Action Plan, provide valuable

guidance to manage urban wastewater worldwide, in accordance with national policies and plans.

20. The transfer of technology and expertise is critical to the global implementation of the Global Programme of Action, and in particular, with regard to management of municipal wastewater. A shortage of adequately trained personnel with technical skills to manage new facilities, or administrative skills to develop management schemes, is holding back the implementation of the Global Programme of Action in some parts of the world.

21. Initiatives concerning technology transfer should be compatible with local environmental and cultural circumstances. In this context, it is noted that a high percentage of coastal communities in developing countries suffer from a lack of basic sanitation services. There is no doubt that initiatives related to the Global Programme of Action in such communities can contribute toward efforts to address this situation.

22. Capacity-building initiatives related to the Global Programme of Action require consistent attention at the local and national levels, and deserve attention within the framework of national development plans.

23. The "polluter pays" principle provides a significant catalyst for changing attitudes and facilitating the wise use of water. It is being used successfully in a number of countries and has the combined effect of raising revenue and discouraging pollution. In implementing this principle, however, there is a need to appropriately consider the social costs and its impact on the poorest members of society. There may also be considerable costs associated with identifying the polluters and establishing a payment scheme. The "polluter pays" principle may also discourage some development and should therefore be balanced with positive economic incentives for reducing pollution.

24. Finally, it would be valuable to further develop the Strategic Action Plan on Municipal Wastewater in cooperation with international financial institutions.

D. The Work Programme of the Global Programme of Action Coordination Office for the period 2002–2006

25. The focus of the programme of work is to move the implementation of the Global Programme of Action from the planning to the action phase by developing toolkits, facilitating partnerships, and initiating demonstration and capacity-building projects. In this regard, it aims to:

(a) Facilitate the mobilization of financial resources;
(b) Further involve the private sector and civil society;
(c) Establish stronger working links with the freshwater community;
(d) Expand capacity building by enhancing the Global Programme of Action Clearing-house Mechanism; and
(e) Strengthen cooperation with United Nations agencies.

26. The programme of work could be further enhanced through the development of performance indicators, specific targets, and the incorporation of monitoring and assessment. These activities should build upon existing and ongoing pro-

grammes and efforts should be made to link the programme of work with those of other United Nations agencies, especially at the regional level, while avoiding duplication and overlapping. The cost-effectiveness of initiatives within the programme of work should also be analysed.

27. Opportunities also exist for achieving efficiencies by combining the efforts of United Nations agencies in cross-cutting issues, such as clearing-house mechanisms, capacity building, technology transfer, indicators, and monitoring. Specifically, in relation to the clearing-house mechanism, stronger links could be made with the non-governmental organizations community and academia. Furthermore, the meeting was reminded that in many developing countries, access to the Internet is severely limited, especially for local practitioners.

28. Expanding the links with the freshwater community to also incorporate land-use planning would also significantly enhance the programme of work. In all aspects of the programme of work, however, the central role of Governments in setting priorities and ensuring compliance must be emphasized.

29. Many United Nations agencies and other international organizations have initiated activities that complement the proposed programme of work. Significant examples include the regional virtual centers for technology transfer being developed by the International Ocean Institute, and the Coastal Cities Network being developed by the International Council of Local Environmental Initiatives.

E. Oceans and Coastal Governance

30. Recognizing the central authority of the United Nations Convention on the Law of the Sea, and the guidance of Agenda 21, the implementation of the Global Programme of Action can be both a catalyst for, and a beneficiary of, improved coastal and oceans governance. It provides an excellent framework for harmonizing the activities of coastal and marine institutions and mechanisms at the local, national, regional and global levels, and for producing efficiencies by bringing stakeholders together from different sectors, both public and private, to address common objectives. For example, at the international level, the Global Programme of Action could serve as a harmonizing mechanism for the Rotterdam Convention on the Prior Informed Consent Procedure for Certain Hazardous Chemicals and Pesticides in International Trade, the Stockholm Convention on Persistent Organic Pollutants and the Convention on Biological Diversity. Its active implementation at the local, national, and regional levels will contribute to the protection of human health, food security, economic development, and environmental protection.

31. Improvements in coastal and oceans governance should be at a level commensurate with the problem of coastal and marine degradation. However, the harmonizing capacity of the Global Programme of Action is especially relevant at the regional level and the regional seas programmes provide an excellent and existing vehicle for implementing the Global Programme of Action. They are a fundamental pillar for improved coastal and oceans governance.

32. The utilization of twinning arrangements involving information sharing, capacity building, and technology transfer between selected or contiguous regional seas programmes, can also strengthen coastal and oceans governance.

33. The objectives of the Global Programme of Action are complementary to

many other multilateral environmental agreements, such as the Convention on Biological Diversity, and institutional mechanisms such as the respective regional fisheries management organizations. Consequently, efforts should be made to integrate the Global Programme of Action into these initiatives in a more systemic manner. In this context, coordinating capacity building amongst multilateral environmental agreements will improve efficiency and expand their reach and positive impacts. To facilitate this process, the Global Programme of Action Coordination Office should take active steps in collaboration with the regional seas programmes to reach out to other United Nations agencies. Consideration should be given to organizing a meeting of all regional seas programmes to coordinate a strategic approach to this effect, and to consider the possible role of the regional seas programmes as a platform for multistakeholder participation.

34. At the global level the Coordination Office should explore the potential for memoranda of understanding with multilateral environmental agreements, such as the Convention on Wetlands of International Importance especially as Waterfowl Habitat, to coordinate their initiatives. Similarly, there is a need to ensure the currency of the Global Programme of Action in the United Nations Oceans Consultative Process.

35. The importance of regional and global efforts to implement the Global Programme of Action should not undermine the importance of national action. Indeed, a bottom-up approach to improving global oceans governance is also needed. In this regard, in many countries there is a need for capacity building and institutional strengthening to improve the governance of coastal and ocean resources at the national level. Similarly, there is a need to better understand the oceans while supporting the economic development of the oceans.

F. Financing the Global Programme of Action

36. Financing appropriate action to implement the Global Programme of Action should, in the first place, come from a country's own resources. It is therefore important to engineer a country-driven demand for implementing the Global Programme of Action amongst decision makers, industry, academia, and the community.

37. The lack of adequate resources is a major impediment to the implementation of the Global Programme of Action. Innovative approaches must be adopted to attract new finances for implementation of the Global Programme of Action. Such approaches should be tailored to national and local needs, including the needs of municipalities and local government entities, and solutions must encapsulate appropriate lower cost alternatives. Lower cost solutions should, however, be assessed for their total economic, social, and environmental costs and impacts that may not be immediately apparent in some cases.

38. It is essential to integrate Global Programme of Action related activities into national development strategies and development assistance frameworks in order to facilitate interventions by international financial institutions, regional development banks, and the donor community.

39. In implementing the Global Programme of Action increased emphasis should be given to the issues of poverty alleviation, human health and food security. Emphasizing the effect of projects related to the Global Programme of Action on

these issues will attract political will, media attention, and the interest of international financial institutions. In this context, the goals of the Global Programme of Action should be incorporated into national development programmes and sustainable development strategies. Similarly, efforts should be directed toward building the capacity of Governments to assess the economic value of coastal and marine resources, and to fully engage the private sector and community groups in the implementation of the Global Programme of Action.

40. The development of financial partnerships, including public-private partnerships, will benefit the Global Programme of Action by increasing the level of participation in and awareness of, the Global Programme of Action and by opening new financial opportunities. For example, Governments could take action to facilitate wider application of microfinancing and enterprise financing mechanisms, involving the private sector and financial institutions. Similarly, stakeholders of the Global Programme of Action could contribute to national, regional or global studies related to the development of economic instruments, such as water markets and pollution reduction trading mechanisms, and to studies on the need and feasibility of multistakeholder water funds.

41. Learning partnerships with organizations such as the World Bank Institute should also be developed by the Coordination Office as an avenue to build national and regional capacity.

42. Finances for the Global Programme of Action can also be obtained indirectly. For example, by requiring the best available techniques in both existing industries and new investment in potentially polluting industries, Government can stem the increasing demand for spending related to the Global Programme of Action. Similarly, the introduction of the "polluter pays" principle will provide both economic disincentives for pollution, and economic incentives for cleaner production. Appropriate debt relief is yet another option for freeing much needed financial resources so that they can be directed toward the Global Programme of Action.

43. Finally, it is imperative that the Global Environment Facility continue to address the priorities and objectives of the Global Programme of Action, especially in relation to the current replenishment process and within established rules and modalities.

Mr. Tuiloma Neroni Slade
Ambassador/Permanent Representative
Permanent Mission of Samoa
to the United Nations

Mr. Magnús Jóhannesson
Secretary-General
Ministry for the Environment
Iceland

Selected Documents and Proceedings

Oceans and Coasts at Rio+10: Toward the 2002 World Summit on Sustainable Development, Johannesburg: Concluding Statement by Conference Co-Chairs, Paris, 3–7 December 2001†

The goals of next year's World Summit on Sustainable Development can only be met if effective action to protect ocean and coastal areas is implemented—and soon. This conclusion was reached at a meeting held in UNESCO headquarters, Paris, France, of more than 400 coastal and ocean experts assembled for *The Global Conference on Oceans and Coasts at Rio+10: Toward the 2002 World Summit on Sustainable Development, Johannesburg*.[1]

"Sustainable development and poverty reduction cannot be achieved without healthy oceans and coasts," said the meeting's Co-Chairs, Dr. Patricio Bernal, Executive Secretary of the Intergovernmental Oceanographic Commission (IOC) of UNESCO, and Dr. Biliana Cicin-Sain, Director of the Center for the Study of Marine Policy of the University of Delaware (USA). "The key question is how to sustain the natural resource base and the integrity of coastal and ocean ecosystem services, while continuing to expand economically. We strongly recommend that the United Nations put sustainable development of oceans—comprising 70% of the Earth's surface—as a central feature of the World Summit."

The United Nations will convene heads of state for the World Summit on Sustainable Development in Johannesburg, South Africa, in September 2002, the tenth anniversary of the 1992 Earth Summit held in Rio de Janeiro.

"We have great reason for concern about the health of our oceans and coastal areas. Participants at the Conference generally agreed that we are in a critical situation of declining trends that requires immediate action by nations and governing bodies worldwide." This sense of urgency and priority was corroborated in ministerial statements, as well as by nongovernmental, intergovernmental experts, scientists, commercial fishing, and industrial representatives attending the meeting. "It is sig-

---

†EDITORS' NOTE.—This Concluding Statement by the Conference Co-Chairs is available online: <http://www.udel.edu/CMS/csmp/rio+10/> (accessed 28 December 2001). Reprinted with permission.

1. For further information, please contact: Julian Barbiere at the Intergovernmental Oceanographic Commission in Paris (Tel: +33 1 4568-3983; Fax: +33 1 4568-5810; E-mail: j.barbiere@unesco.org), or Catherine Johnston at the Center for the Study of Marine Policy (Tel: +1 302 831-8086; Fax: +1 302 831-3668; E-mail: johnston@udel.edu). Conference reports (Co-Chairs' Summary of the Conference, Ministerial Perspectives Presented at the Conference, and Working Group Reports) will be posted at the conference's website <http://www.udel.edu/CMS/csmp/rio+10>. A detailed summary of the conference, prepared by the Earth Negotiations Bulletin, is available at <http://www.iisd.ca/linkages/sd/ocrio+10/>.

nificant that this broad array of ocean and coastal experts agrees with this statement," said Drs. Bernal and Cicin-Sain.

THE HIGHLIGHTS OF THE CONFERENCE CO-CHAIRS' REPORT

1. Poverty Reduction During the Coming Decade Requires More Access to Sustainable Economic Livelihoods and Wealth Derived from the Ocean, and Development of Safer, Healthy Coastal Communities

The UN Millennium Declaration notes the need to halve, by 2015, the proportion of very poor people in the world, and to reduce the scourge of diseases like malaria and water-borne infections (250 million clinical cases of gastroenteritis and upper respiratory diseases are caused annually by bathing in contaminated sea water). This is a key concern, and perhaps the most difficult challenge facing our use of the oceans.

Meeting these needs requires a new commitment to making the benefits of trade and globalization available to coastal communities, participatory management of resources, programs specifically targeted to reducing vulnerability of coastal people and infrastructure, and commitments to full participation of women and youth in decision-making and activities related to locally based coastal and ocean decisions.

2. Full Implementation and Effective Compliance with International Agreements Is Needed

The significant number of international agreements that have come into effect since 1992 now need to be properly implemented and enforced, and their implications for national level action must be more fully addressed. There is an urgent need for better cooperation and coordination among regional and international bodies governing oceans and fisheries to ensure harmonized and efficient implementation. For example, the implementation of the fishing instruments concluded in recent years (UN Straddling Fish Stocks Agreement, Food and Agriculture Organization (FAO) Code of Conduct for Responsible Fishing, and the FAO Compliance Agreement) is an essential element in putting fisheries on a sustainable development path that could address existing overcapacity and subsidized fishing fleets.

3. Capacity Building for Good Governance of Coastal and Ocean Use Is Necessary

Scientific advances and technology development will continue to open untapped potential for use of coastal, offshore, and Exclusive Economic Zones, and deep ocean areas. Yet our understanding of the role and vulnerability of these resources and habitats is still limited. And all countries, rich and poor, lack the needed capacity to manage even the existing level of development in a well-integrated way.

Thus the capacity of local and national governments to apply effective institutional and legal frameworks for integrated coastal and ocean management must be strengthened. This will enable them to pursue opportunities for economic development in the coasts and oceans while protecting their ecological integrity and biodiversity. It will require, among other things, raising public awareness of coastal and ocean issues, the retargeting of financial assistance to take into account lessons learned from experience, and building of the capacity of the educational institutions of coastal nations. Capacity building is required within governments, local communities, and NGOs, as well as to enable effective involvement of the private sector.

4. The Health of the Oceans and Coasts Is Directly Linked to the Proper Management of River Basins, Including Freshwater Flows to the Marine Environment

Eighty percent of marine pollution comes from land-based sources. In the developing world, more than 90% of sewage and 70% of industrial wastes are dumped untreated into surface waters where they pollute water supplies and coastal waters. Ecosystem approaches that link management of river basins to marine ecosystems, such as the Global Programme of Action for the Protection of the Marine Environment from Land-Based Activities, must be effectively implemented. This is especially important in the context of the coastal megacities (70% of cities with over 8 million people are coastal), such as Lagos, Nigeria—where 65% of the estimated 13.4 million population live in poverty.

5. Protecting Coastal and Marine Areas and Biodiversity Takes an Ecosystem Approach

The very significant shift from a sectoral to an ecosystem-based approach that recognizes precaution and linkages among activities is an important achievement of the past decade. The Convention on Biological Diversity provides an international framework for an ecosystem-based approach that will depend upon protection of marine habitats at regional and national levels. A global representative system of marine-protected areas is now needed as one essential component for ecosystem understanding and management and biodiversity protection.

6. Strengthening Science-Based Monitoring and Assessment of the Oceans Is Essential for Managing the Long-Term Sustainability of Marine Ecosystems

Effective international coordination needs to be developed to support an integrated assessment of the status of oceans and coasts, and their use. A periodic, comprehensive global report on the State of Oceans and Development that builds upon existing regional and sectoral efforts is needed. It could be complemented by similar reports at the national level. This report should anticipate and plan for emerging ocean and coastal issues, such as offshore aquaculture and bioprospecting of marine genetic resources.

7. The Special Problems and Issues of Small Island Developing States Must Be Addressed

Small island developing states have special problems and opportunities related to the oceans that need to be recognized and addressed. These nations, small in land area, typically have control and stewardship responsibilities over huge Exclusive Economic Zones. As Ambassador Tuiloma Neroni Slade, Chair of the Alliance of Small Island Developing States, put it, "Small island states are a special case since they are most vulnerable to the effects of climate change, especially sea-level rise. We are responsible for the stewardship of our islands and vast areas of the oceans, containing high biological diversity, the most extensive coral reef systems in the world, and significant seabed minerals. We have a critical role to play in the future of the oceans."

Fishing remains the most widespread economic activity in the ocean. "The future integrity of our coastal communities and of the world's food security is directly linked to sustaining our fisheries and their related ecosystems. The 400 million fishing men and women of the world are a testimony to one of the richest heritages of mankind. Fishing brings us one of the last sources of wild food—let us not take it for granted," said Mr. Pietro Parravano, World Fisheries Forum of Fish Workers and Fish Harvesters, and a participant at the Conference.

A substantial body of scientific evidence supports the urgent call by the conference to place coastal and ocean issues squarely on the World Summit's agenda. More than half of the world's population currently lives within 100 km of the coast, and by 2025 it is estimated that 75% of the world's population, or 6.3 billion people, will live in the coastal zone, concentrated in coastal megacities and many living in poverty on less than two dollars a day. FAO says that, in the last 40 years, the demand for fish has been growing at twice the rate of population growth. Over 500 million people depend on coral reefs for food and income, yet 70% of reefs worldwide are threatened. Eighty-eight out of 126 species of marine mammals are threatened, and several are extinct or close to extinction.

"It is essential that we link economic development, social welfare and resource conservation in order to achieve sustainability. Governments worldwide must realize that resource conservation and protection must be an integral part of the development process and cannot be considered in isolation," said Drs. Bernal and Cicin-Sain.

The Paris discussions did report some good news about the decade since the Rio Earth Summit: significant progress has been made in laying the groundwork toward sustainable development of the oceans—a new cluster of some eight global agreements provide the direction for good governance of coastal and ocean use; many countries, both developing and developed, have experimented with various approaches to ocean and coastal management; significant funding, by both national and international donors has taken place; and a significant body of knowledge and practical experience on ocean and coastal management has been accumulated.

Ocean resources and environmental conditions have continued to decline, however, and, unless oceans and coasts are given high priority by the world's governments, "under present trends and circumstances, the outlook for our oceans and coasts in the year 2020 leaves little room for optimism. It is obvious that action is required now to correct our present course," said Drs. Bernal and Cicin-Sain. "As the world's population continues to grow and if current development and social trends continue, there will be even greater pressures on our coastal resources. We have an alternative vision for the future—one of healthy and productive seas, clean coastal waters, and prosperous coastal communities. We have an obligation at the World Summit to look at the root causes of many of the world's economic and social crises, and nearly all of these are affected by the way we care for our oceans and coastal areas.''

Directory

DIRECTORY OF OCEAN-RELATED ORGANIZATIONS

This directory was updated in January 2003. Additions and amendments are welcomed and should be sent by e-mail to ocean.yearbook@dal.ca.

THE EDITORS

Global Intergovernmental Organizations	979
Global Nongovernmental Organizations	984
Global Academic Organizations	1002
Regional Intergovernmental Organizations	1003
Africa	1003
The Americas	1007
Asia	1011
Europe	1015
Australia, New Zealand and Oceania	1021
Polar Regions	1022
Regional Nongovernmental Organizations	1023
Africa	1023
The Americas	1024
Asia	1038
Europe	1040
Australia and Oceania	1051
Polar Regions	1053
Regional Academic Organizations	1053
Africa	1053
The Americas	1057
Asia	1069
Europe	1075
Australia, New Zealand and Oceania	1088
Polar Regions	1091
National Government Organizations	1091

1. GLOBAL INTERGOVERNMENTAL ORGANIZATIONS

Alliance of Small Island States (AOSIS)
Permanent Representative of Samoa to the United Nations
800 Second Avenue, 4th floor
New York, NY, 10017, USA
Contact: Chairman
Tel: 1 212 599-6196
Fax: 1 212 599-0797
E-mail: samoa@un.int
http://www.sidsnet.org/aosis/

Basel Convention on the Control of Transboundary Movements of Hazardous Wastes and Their Disposal (Basel Convention)
International Environment House
13-15, chemin des Anémones
CH-1219 Châtelaine, Geneva, Switzerland
Contact: Secretariat
Tel: 41 22 917 8218
Fax: 41 22 797 3454
E-mail: sbc@unep.ch
http://www.basel.int/

Commission for the Geological Map of the World (CGMW)
77, rue Claude Bernard
F-75005 Paris, France
Contact: Secretary General
Tel: 33 1 4707 2294
Fax: 33 1 4336 9518
E-mail: ccgm@club-internet.fr
http://perso.club-internet.fr/ccgm/index.html

Commission on Science and Technology for Development
Division on Investment, Technology and Enterprise Development
United Nations Conference on Trade and Development (UNCTAD)
Palais des Nations, Building E, 9th Floor
CH-1211 Geneva 10, Switzerland
Contact: Chief, Technology for Development Section
Tel: 41 22 917 5069
Fax: 41 22 907 0197
E-mail: stdev@unctad.org
http://www.unctad.org/stdev/

Commission on Sustainable Development (CSD)
United Nations Two UN Plaza, Room DC2-2220
New York, NY, 10017, USA
Contact: Secretariat
Tel: 1 212 963-3170
Fax: 1 212 963-4260
E-mail: dsd@un.org
http://www.un.org/esa/sustdev/csd.htm

Commission on the Limits of the Continental Shelf
Division for Ocean Affairs and the Law of the Sea
Office of Legal Affairs
Room DC2-0450, United Nations
New York, NY, 10017, USA
Contact: Secretariat
Tel: 1 212 963-3966
Fax: 1 212 963-5847
E-mail: doalos@un.org
http://www.un.org/Depts/los/clcs_new/clcs_home.htm

Committee on International Oceanographic Data and Information Exchange (IODE)
IOC-UNESCO
1, rue Miollis
F-75732 Paris Cedex 15, France
Contact: Head, Ocean Science
Tel: 33 1 4568 4046
Fax: 33 1 4568 5812
E-mail: p.pissierssens@unesco.org
http://ioc.unesco.org/iode/

Convention on Biological Diversity
World Trade Centre
393 St Jacques Street, Office 300
Montréal, QC, H2Y 1N9, Canada
Contact: Secretary General
Tel: 1 514 288-2220
Fax: 1 514 288-6588
E-mail: secretariat@biodiv.org
http://www.biodiv.org/

Convention on International Trade in Endangered Species of Wild Fauna and Flora (CITES)
International Environment House
13-15, chemin des Anémones
CH-1219 Châtelaine, Geneva, Switzerland
Contact: Secretariat
Tel: 41 22 917 8139/40
Fax: 41 22 797 3417
E-mail: cites@unep.ch
http://www.cites.org/

Division for Ocean Affairs and the Law of the Sea (DOALOS)
Office of Legal Affairs
United Nations
Room DC2-0450
New York, NY, 10017, USA
Contact: Director
Tel: 1 212 963-3950
Fax: 1 212 963-5847
E-mail: doalos@un.org
http://www.un.org/depts/los/

Food and Agriculture Organisation of the United Nations (FAO)
Viale delle Terme di Caracalla
I-00100 Rome, Italy
Contact: Director General
Tel: 39 06 5705 1
Fax: 39 06 5705 3152
E-mail: FAO-HQ@fao.org
http://www.fao.org/

Global Biodiversity Information Facility (GBIF)
Universitetsparken 15
DK-2100 Copenhagen Ø, Denmark
Contact: Secretariat
Tel: 45 35 32 14 70
Fax: 45 35 32 14 80
E-mail: gbif@gbif.org
http://www.gbif.org/

Global Environment Facility (GEF)
1818 H St. NW
Washington, DC, 20433, USA
Contact: Secretariat
Tel: 1 202 473-0508
Fax: 1 202 522-3240 or 522-3245
E-mail: secretariatofgef@world-bank.org
http://www.gefweb.org/

Global Programme of Action (GPA)
GPA Coordinating Office, UNEP
PO Box 16227
NL-2500 BE, The Hague, The Netherlands
Contact: Coordinator
Tel: 31 70 311 4460
Fax: 31 70 345 6648
E-mail: gpa@unep.nl
http://www.gpa.unep.org/

GLOBEFISH
FAO, Fishery Industries Division
Viale delle Terme di Caracalla
I-00100 Rome, Italy
Tel: 39 06 5705 6313/5594
Fax: 39 06 5705 5188
E-mail: globefish-web@fao.org
http://www.globefish.org/

Intergovernmental Oceanographic Commission (IOC)
UNESCO
1, rue Miollis
F-75732 Paris Cedex 15, France
Contact: Head, Ocean Science
Tel: 33 1 4568 4046
Fax: 33 1 4568 5812
E-mail: p.pissierssens@unesco.org
http://ioc.unesco.org/iocweb/default.htm

Intergovernmental Panel on Climate Change (IPCC)
c/o World Meteorological Organisation
7, bis Avenue de la Paix
CP 2300
CH-1211 Geneva 2, Switzerland
Contact: Secretariat
Tel: 41 22 730 8208 184
Fax: 41 22 730 8025 113
E-mail: ipcc_sec@gateway.wmo.ch
http://www.ipcc.ch/

International Continental Scientific Drilling Program
GeoForschungsZentrum Potsdam
ICDP-OSG
Telegrafenberg
D-14473 Potsdam, Germany
Contact: Research Coordinator
Tel: 49 331 288 1805
Fax: 49 331 288 1002
E-mail: icdp@gfz-potsdam.de
http://www.icdp-online.org/

International Court of Justice (ICJ)
Peace Palace
NL-2517 KJ, The Hague, The Netherlands
Tel: 31 70 302 2323
Fax: 31 70 364 9928
E-mail: information@icj-cij.org
http://www.icj-cij.org/

International Hydrographic Organisation (IHO)
4, Quai Antoine 1er
BP 445
MC-98011 Monaco Cedex, Principality of Monaco
Contact: IH Bureau
Tel: 377 93 108100
Fax: 377 93 108140
E-mail: info@ihb.mc
http://www.iho.shom.fr/

International Labour Organization (ILO)
4, route des Morillons
CH-7211, Geneva 22, Switzerland
Contact: International Labour Office
Tel: 41 22 799 6111
Fax: 41 22 798 8685
E-mail: ilo@ilo.org
http://www.ilo.org/

International Maritime Organisation (IMO)
4 Albert Embankment
London, SE1 7SR, United Kingdom
Contact: Secretary General
Tel: 44 20 7735 7611
Fax: 44 20 7587 3210
E-mail: info@imo.org
http://www.imo.org/

International Mobile Satellite Organisation (IMSO)
99 City Road
London, EC1Y 1AX, United Kingdom
Contact: Customer Care
Tel: 44 20 7728 1000
Fax: 44 20 7728 1044
E-mail: information@inmarsat.org
http://www.inmarsat.org/

International Oil Pollution Compensation Funds (IOPC)
Portland House, Stag Place
London, SW1 5PN, United Kingdom
Tel: 44 20 75927100
Tel: 44 1223 233971
Fax: 44 20 75927111
E-mail: info@iopcfund.org
http://www.iopcfund.org/

International Satellite System for Search and Rescue
c/o INMARSAT
99 City Road
London, EC1Y 1AX , United Kingdom
Contact: Cospas-Sarsat Secretariat
Tel: 44 20 7728 1391

Fax: 44 20 7728 1170
E-mail: cospas_sarsat@imso.org
http://www.cospas-sarsat.org/

International Sea-Bed Authority
 (ISA)
14-20 Port Royal Street, 2nd floor
Kingston, Jamaica, West Indies
Contact: Secretary General
Tel: 1 876 922-9105
Fax. 1 876 922-0195
E-mail: snandan@isa.org.jm
http://www.isa.org.jm/

International Tribunal for the Law
 of the Sea
Am Internationalen Seegerichtshot
 #1
D-22609 Hamburg, Germany
Contact: President
Tel: 49 40 356070
Fax: 49 40 35607245
E-mail: itlos@itlos.org
http://www.itlos.org/

International Tsunami Information
 Center
Grosvenor Center, Mauka Tower
737 Bishop Street, Suite 2200
Honolulu, HI, 96813-3213, USA
Contact: Director
Tel: 1 808 532-6423
Fax: 1 808 532-5576
E-mail: laura.kong@noaa.gov
http://www.shoa.cl/oceano/itic/
 frontpage.html

International Whaling Commission
The Red House
135 Station Road, Impington
Cambridge, CB4 9NP, United
 Kingdom
Contact: Secretary
Tel: 44 1223 233971
Fax: 44 1233 232876
E-mail: iwc@iwcoffice.org
http://www.iwcoffice.org/

IOC Science and Communication
 Centre on Harmful Algae
University of Copenhagen
Øster Farimagsgade 2 D
DK-1353 Copenhagen K, Denmark
Contact: Technical Secretary
Tel: 45 3313 4446
Fax: 45 3313 4447
E-mail: hab@bot.ku.dk
http://ioc.unesco.org/hab/
 default.htm

IOC-IEO Science and Communica-
 tion Centre on Harmful Algae
Instituto Español de Oceanografía
Centro Oceanografico de Vigo,
 Cabo Estay-Canido
36390 Vigo, Spain
Tel: 34 986 49 21 11
Fax: 34 986 49 20 03
E-mail: vigohab@vi.ieo.es
http://ioc.unesco.org/hab/act4.htm

IOC-WMO-UNEP Committee for
 Global Ocean Observing System
 (I-GOOS)
c/o GOOS Project Office
IOC-UNESCO
1, rue Miollis
F-75732 Paris Cedex 15, France
Contact: Director
Tel: 33 1 4568 4042
Fax: 33 1 4568 5812
E-mail: c.summerhayes@unesco.org
http://ioc.unesco.org/goos/

Joint Group of Experts on the Scientific Aspects of Marine Environmental Protection (GESAMP)
c/o International Maritime Organisation
4 Albert Embankment
London, SE1 7SR, United Kingdom
Contact: Administrative Secretary
Tel: 44 20 7587 3119
Fax: 44 20 7587 3210
E-mail: ksekimizu@imo.org
http://gesamp.imo.org/

Joint IOC/IHO Guiding Committee for the General Bathymetric Chart of the Oceans (GEBCO)
c/o Challenger Division for Seafloor Processes
Southampton Oceanography Centre
Empress Dock
Southampton, SO14 3ZH, United Kingdom
Contact: Permanent Secretariat
Tel: 44 23 80 593052
Fax: 44 23 80 596564
E-mail: bob.whitmarsh@soc.soton.ac.uk
http://www.ngdc.noaa.gov/mgg/gebco/gebco.html

Joint WMO/IOC Technical Commission for Oceanography and Marine Meteorology (JCOMM)
c/o WMO Marine Programme
7, bis Avenue de la Paix
CP 2300
CH-1211 Geneva 2, Switzerland
Tel: 41 22 730 8473
Fax: 41 22 730 8021
E-mail: mermaid@gateway.wmo.ch
http://ioc.unesco.org/jcomm/

Paris Memorandum on Port State Control
Ministry of Transport, Public Works and Water Management
Transport and Water Management Directorate
PO Box 90653
NL-2509 LR, The Hague, The Netherlands
Contact: Secretariat
Tel: 31 70 351 1508
Fax: 31 70 351 1599
E-mail: office@parismou.org
http://www.parismou.org/

Ramsar Convention on Wetlands
28, rue Mauverney
CH-1196 Gland, Switzerland
Contact: Ramsar Convention Bureau
Tel: 41 22 999 0170
Fax: 41 22 999 0169
E-mail: Ramsar@ramsar.org
http://www.ramsar.org/

Small Island Developing States Network (SIDSNET)
Small Islands Developing States Unit
Division for Sustainable Development
United Nations Department of Economic and Social Affairs (UN/DESA)
Two United Nations Plaza, DC2-2020
New York, NY, 10017, USA
Contact: SIDS Unit Team
Tel: 1 212 963-4135
Fax: 1 917 367-3391
E-mail: sidsnet@sdnhq.undp.org
http://www.sidsnet.org/

Stockholm Convention on Persistent Organic Pollutants (POPs)
UNEP Chemicals
11-13, chemin des Anémones
CH-1219 Chatelaine, Geneva, Switzerland
Tel: 41 22 917 8193
Fax: 41 22 797 3460
E-mail: pops@unep.ch
http://www.unep.ch/pops/

Tokyo MOU on Port State Control
Tomoecho Annex Building 6F, 3-8-26, Toranomon Minato-ku
Tokyo 105-0001, Japan
Contact: Secretariat
Tel: 81 3 3433 0621
Fax: 81 3 3433 0624
E-mail: secretariat@tokyo-mou.org
http://www.tokyo-mou.org/

United Nations Commission on International Trade Law (UNCITRAL)
Vienna International Centre
PO Box 500
A-1400 Vienna, Austria

Contact: Secretariat
Tel: 43 1 26060 4061
Fax: 431 26060 5813
E-mail: uncitral@uncitral.org
http://www.uncitral.org/

United Nations Conference on Trade and Development (UNCTAD)
8-14, avenue de la Paix
Palais des Nations
CH-1211 Geneva 10, Switzerland
Contact: External Relations and Communications
Tel: 41 22 907 1234
Fax: 41 22 907 0043
E-mail: info@unctad.org
http://www.unctad.org/

United Nations Development Programme (UNDP)
One United Nations Plaza
New York, NY, 10017, USA
Tel: 1 212 906-5558
Fax: 1 212 906-5364
E-mail: aboutundp@undp.org
http://www.undp.org/

United Nations Environment Programme (UNEP)
UN Avenue, Gigiri
PO Box 30552
Nairobi, Kenya
Contact: Executive Director
Tel: 254 2 621234
Fax: 254 2 624489/90
E-mail: eisinfo@unep.org
http://www.unep.org/

United Nations Industrial Development Organization (UNIDO)
Vienna International Centre
PO Box 300
A-1400 Vienna, Austria
Tel: 43 1 26026
Fax: 43 1 269 2669
E-mail: unido@unido.org
http://www.unido.org/

United Nations Framework Convention on Climate Change (UNFCCC)
PO Box 260124
D-53153 Bonn, Germany
Contact: Secretariat
Tel: 49 228 815 1000
Fax: 49 228 815 1999
E-mail: secretariat@unfccc.int
http://unfccc.int/

World Meteorology Organisation (WMO)
WMO Building
7, bis Avenue de la Paix
CP 2300
CH-1211 Geneva 2, Switzerland
Tel: 41 22 730 8111
Fax: 41 22 733 8181
E-mail: ipa@www.wmo.ch
http://www.wmo.ch/

World Tourism Organization
Capitán Haya, 42
E-28020 Madrid, Spain
Tel: 34 91 5678100
Fax: 34 91 5713733
E-mail: omt@world-tourism.org
http://www.world-tourism.org/

World Trade Organization (WTO)
Centre William Rappard
154, rue de Lausanne
CH-1211 Geneva 21, Switzerland
Contact: Committee on Trade and Environment
Tel: 41 22 739 5111
Fax: 41 22 731 4206
E-mail: enquiries@wto.org
http://www.wto.org/

2. GLOBAL NONGOVERNMENTAL ORGANIZATIONS

Advisory Committee on Protection of the Sea (ACOPS)
11 Dartmouth St.

London, SW1H 9BN, United Kingdom
Contact: Director
Tel: 44 20 7799 3033
Fax: 44 20 7799 2933
E-mail: info@acops.org
http://www.acops.org/

Association Internationale Villes et Ports
(International Association Cities and Ports–AVIP/IACP)
45, rue Lord Kitchener
F-76600 Le Havre, France
Tel: 33 2 3542 7884
Fax: 33 2 3542 2194
E-mail: bureau@aivp.com
http://www.aivp.com/

Baltic and International Maritime Council (BIMCO)
Bagsvaerdvej 161
DK-2880 Bagsvaerd, Denmark
Contact: Secretary General
Tel: 45 44 36 6800
Fax: 45 44 36 6868
E-mail: mailbox@bimco.dk
http://www.bimco.org/

Charles Darwin Foundation for the Galapagos Islands
Av. 6 de Diciembre N 36-109 y Pasaje California
PO Box 17-01-3891
Quito, Ecuador
Tel: 593 5526 147
E-mail: galapagosinfo@darwinfoundation.org
http://www.darwinfoundation.org/

Comité International Radio-Maritime (CIRM)
(International Association for Marine Electronics Companies)
Southbank House
Black Prince Road
London, SE1 7SJ, United Kingdom
Contact: Secretary General
Tel: 44 20 7587 1245
Fax: 44 20 7587 1436
E-mail: secgen@cirm.org
http://www.cirm.org/

Comité Maritime International (CMI)
(International Maritime Committee)
Mechelsesteenweg, 196
B-2018 Antwerp, Belgium
Tel: 32 3 227 3526
Fax: 32 3 227 3528
E-mail: admini@cmi-imc.org
http://www.comitemaritime.org/

Commission for Cooperation with Developing Countries
c/o Mariculture and Fisheries Division
Kuwait Institute of Scientific Research
PO Box 1638
22017 Salmiya, Kuwait
Contact: President
Tel: 965 575 1984
Fax: 965 571 1293
E-mail: sdurva@kisr.edu.kw
http://www.olympus.net/IAPSO/comm.html

Commission on Ecosystem Management (CEM)
IUCN–The World Conservation Union
28, rue Mauvernay
CH-1196 Gland, Switzerland
Contact: Chair
E-mail: cem@iucn.org
http://www.iucn.org/

Commission on Education and Communications (CEC)
IUCN–The World Conservation Union
28, rue Mauvernay
CH-1196 Gland, Switzerland
Contact: Chair
Tel: 41 22 999 0282
Fax: 41 22 999 0025
E-mail: wjg@hq.iucn.org
http://info.iucn.org/cec/

Commission on Environmental, Economic and Social Policy (CEESP)
c/o CENESTA (Centre for Sustainable Development)
5 Lakpour Lane, Suite 24
IR-16936 Tehran, Iran
Contact: Executive Officer
Tel: 98 21 295 4217
Fax: 98 21 332 8599
E-mail: ceesp@iucn.org
http://ceesp.cenesta.org/

Commission on Environmental Law (CEL)
IUCN–The World Conservation Union
Environmental Law Centre
Godesberger Allee 108-112
D-53175 Bonn, Germany
Contact: Secretariat
Tel: 49 228 2692 231
Fax: 49 228 2692 250
E-mail: secretariat@elc.iucn.org
http://www.iucn.org/themes/law/elp_home.html/

Commission on Mean Sea Level and Tides
Director de Recherche
Laboratoire d'Oceanographie et de Geophysique Spatiale
GRGS/ Observatoire Midi Pyrenees
14, avenue Edouard Belin
F-31400 Toulouse, France
Tel: 33 5 61 33 29 23
Fax: 33 5 61 25 32 05
E-mail: leprovos@pontos.cst.cnes.fr
http://www.olympus.net/IAPSO/comm.html

Commission on Sea Ice
Scott Polar Research Institute
Lensfield Road
Cambridge, CB2 1ER, United Kingdom
Contact: President
E-mail: pw11@cus.cam.ac.uk

http://www.olympus.net/IAPSO/comm.html

Coral Reef Alliance (CORAL)
417 Montgomery Street, Suite 205
San Francisco, CA, 94105, USA
Tel: 1 415 834-0900
Fax: 1 415 834-0999
E-mail: info@coral.org
http://www.coral.org/

Cruise Lines International Association (CLIA)
500 Fifth Avenue, Suite 1407
New York, NY, 10110, USA
Tel: 1 212 921-0066
Fax: 1 212 921-0549
E-mail: info@cruising.org
http://www.cruising.org/

Det Norske Veritas
Veritasveien 1
N-1322 Høvik, Norway
Contact Head Office
Tel: 47 67 57 9900
Fax: 47 57 67 9911
http://www.dnv.no/

Earth Action Network
17 The Green
Wye, Kent, TN25 5AJ, United Kingdom
Contact: Executive Director
Tel: 44 12 3381 3796
Fax: 44 12 3381 3795
E-mail: wye@earthaction.org.uk
http://www.oneworld.org/earthaction/

Earth Council
PO Box 319-6100
San José, Costa Rica
Contact: Secretariat
Tel: 1 506 205-1600
Fax: 1 506 249-3500
E-mail: eci@ecouncil.ac.cr
http://www.ecouncil.ac.cr/

Engineering Committee on Oceanic Resources
c/o Royal Institution of Naval Architects
10 Upper Belgrave Street
London, SW1X 8BQ, United Kingdom
Contact: Executive Director
Tel: 44 20 7235 4622
Fax: 44 20 7259 5912
E-mail: tblakeley@rina.org.uk
http://www.engr.mun.ca/ECOR/

Estuarine Research Federation
PO Box 510
Port Republic, MD, 20676, USA
Tel: 1 410 586-0997
Fax: 1 410 586-9226
E-mail: webmaster@erf.org
http://www.erf.org/

Federation of National Associations of Ship Brokers and Agents (FONASBA)
3 St. Helen's Place
London, EC3A 6EJ, United Kingdom
Contact: Secretariat
Tel: 44 20 7628 5559
Fax: 44 20 7628 5445
E-mail: fonasba@ics.org.uk
http://www.fonasba.com/

Fish for All
PO Box 500, GPO
10670 Penang, Malaysia
Tel: 604 6261606
Fax: 604 6265530
E-mail: fishforall@cgiar.org
http://www.fishforall.org/

Friends of the Earth–International
PO Box 19199
NL-1000 GD Amsterdam, Netherlands
Contact: International Secretariat
Tel: 31 20 622 1369
Fax: 31 20 639 2181
E-mail: foei@foei.org
http://www.foei.org/

Global Aquaculture Alliance
5661 Telegraph Road, Suite 3A
St. Louis, MO, 63129, USA
Tel: 1 314 293-5500
Fax: 1 314 293-5525
E-mail: homeoffice@gaalliance.org
http://www.gaalliance.org/

Global Coral Reef Monitoring Network (GCRMN)
c/o Australian Institute of Marine Science
PMB No. 3
Townsville, MC 4810, Australia
Contact: Coordinator
Tel: 61 7 4772 4314
Fax: 61 7 4772 2808
E-mail: c.wilkinson@aims.gov.au
http://www.coral.noaa.gov/gcrmn/

Global Ocean Ecosystem Dynamics (GLOBEC)
Plymouth Marine Laboratory
Prospect Place
Plymouth, PL1 3DH, United Kingdom
Contact: International Project Office
Tel/Fax: 44 17 5263 3160
E-mail: globec@pml.ac.uk
http://www.pml.ac.uk/globec/

Greenpeace International
Keizergracht 176
NL-1016 DW, Amsterdam, The Netherlands
Contact: Executive Director
Tel: 31 20 523 6222
Fax: 31 20 523 6200
E-mail: supporter.services@ams.-greenpeace.org
http://www.greenpeace.org/

Hydrographic Society
PO Box 103

Plymouth, PL4 7YP, United Kingdom
Contact: Manager
Tel: 44 17 5222 3512
E-mail: helen@hydrographicsociety.org
http://www.hydrographicsociety.org/

ICC Commission on Maritime Transport
c/o International Chamber of Commerce
38 Cours Albert 1er
F-75008 Paris, France
Contact: Policy Manager
Tel: 33 1 4953 2895
Fax: 33 1 4953 2859
E-mail: viviane.schiavi@iccwbo.org
http://iccwbo.org/home/transport/maritme-commission.asp

IGU Commission on Coastal Systems
Department of Marine and Coastal Sciences
71 Dudley Road
Cook College
Rutgers University
New Brunswick, NJ, 08901, USA
Contact: Executive Secretary
Tel: 1 732 932-6555, ext. 506/500
Fax: 1 732 932-1820
E-mail: psuty@imcs.rutgers.edu
http://igu-ccs.ucc.ie/

IGU Commission on Marine Geography
Department of Geographical Sciences
University of Plymouth
Plymouth, Devon, PL4 8AA, United Kingdom
Contact: Chairman
Tel/Fax: 44 17 5223 3051
E-mail: dapinder@plymouth.ac.uk
http://www.geog.plymouth.ac.uk/marine/

ILA Operations Center
741 Cathedral Pointe Ln
Santa Barbara, CA, 93111, USA
Tel: 1 805 967-8649
Fax: 1 805 967-8471
E-mail: ila@loran.org
http://www.loran.org/

Industrial Shrimp Action Network
14420 Duryea Lane
Tacoma, WA, 98444, USA
Tel: 1 253 539-5272
Fax: 1 253 539-5054
E-mail: isanet@shrimpaction.org
http://www.shrimpaction.org/

Institute for Fisheries Management and Coastal Communities Development (IFM)
North Sea Center
PO Box 104
DK-9850 Hirtshals, Denmark
Tel: 45 9894 2855
Fax: 45 9894 4268
E-mail: ifm@ifm.dk
http://www.ifm.dk/

Institute of Chartered Shipbrokers (ICS)
3 St Helen's Place
London, EC3A 6EJ, United Kingdom
Tel: 44 20 7628 5559
Fax: 44 20 7628 5445
E-mail: info@ics.org.uk
http://www.ics.org.uk/

Institute of International Container Lessors (IICL)
555 Pleasantville Road, Suite 140 South
Briarcliff Manor, NY, 10510, USA
Tel: 1 914 747-9100
Fax: 1 914 747-4600
E-mail: info@iicl.org
http://www.iicl.org/

Inter-Association Tsunami Commission (IACT)
Institute of Computational Mathematics and Mathematical Geophysics, Siberian Division
6, Pr. Lavrentieva
Novosibirsk 630090, Russia
Contact: Chairman
Tel: 7 3832 342070
Fax: 7 3832 343783
E-mail: gvk@omzg.sscc.ru
http://omzg.sscc.ru/tsulab/

International Arctic Social Sciences Association (IASSA)
University of Alaska Fairbanks
PO Box 757730
Fairbanks, AK, 99775-7730, USA
Contact: Secretariat
Tel: 1 907 474-6367
Fax: 1 907 474-6370
E-mail: fyiassa@uaf.edu
http://www.uaf.edu/anthro/iassa/

International Association for the Study of Common Property (IASCP)
Indiana University
PO Box 2355
Gary, IN, 47409, USA
Contact: Executive Director
Tel: 1 219 980-1433
Fax: 1 219 980-2801
E-mail: iascp@indiana.edu
http://www.iascp.org/

International Association for the Physical Sciences of the Ocean (IAPSO)
PO Box 820440
Vicksburg, MS, 39182-0440, USA
Contact: Secretary General
Tel: 1 601 636-1363
Fax: 1 601 629-9640
E-mail: camfield@vicksburg.com
http://www.olympus.net/IAPSO/

International Association for the Study of Maritime Mission (IASMM)
School of Theology
The College of Ripon and York St. John
Lord Mayor's Walk
York, YO31 7EX, United Kingdom
Contact: Secretariat
Tel: 44 1904 716861
Fax: 44 1904 612512
E-mail: s.friend@ucrysj.ac.uk
http://www.ucc.uconn.edu/shpark/iasmm.html

International Association of Aquaculture Economics and Management (IAAEM)
Department of Economics
University of Queensland
Brisbane, QLD 4072, Australia
Contact: Secretariat
Tel: 61 7 3365 6570
Fax: 61 7 3365 7299
E-mail: c.tisdell@economics.uq.edu.au
http://www.uq.edu.au/aem/

International Association of Aquatic and Marine Science Libraries and Information Centers (IAMSLIC)
c/o Harbor Branch Oceanographic Institution Library
5600 US 1 North
Fort Pierce, FL, 34946, USA
Tel: 1 561 465-2400, ext. 201
Fax: 1 561 465-2446
E-mail: metzger@hboi.edu
http://www.iamslic.org/

International Association of Biological Oceanography (IABO)
c/o Scottish Association for Marine Science
PO Box 3
Argyll, PA34 4AD, United Kingdom
Contact: President

Tel: 44 1631 567853
Fax: 44 1631 565518
E-mail: dmlsec@dml.ac.uk

International Association of Classification Societies (IACS)
5 Old Queen Street
London, SW1H 9JA, United Kingdom
Contact: Permanent Secretariat
Tel: 44 20 7976 0660
Fax: 44 20 7976 0440
E-mail: permsec@iacs.org.uk
http://www.iacs.org.uk/

International Association of Dredging Companies (IADC)
Duinweg 21
NL-2585 JV, The Hague, The Netherlands
Tel: 31 70 352 3334
Fax: 31 70 351 2654
E-mail: info@iadc-dredging.com
http://www.iadc-dredging.com/

International Association of Dry Cargo Shipowners (INTERCARGO)
4 London Wall Buildings, 2nd Floor
Blomfield Street
London, EC2M 5NT, United Kingdom
Contact: Secretary General
Tel: 44 20 7638 3989
Fax: 44 20 7638 3943
E-mail: info@intercargo.org
http://www.intercargo.org/

International Association of Hydrogeologists (IAH)
PO Box 9
Kenilworth, CV8 1JG, United Kingdom
Contact: Secretary General
Tel: 44 1926 450677
Fax: 44 1926 856561
E-mail: iah@iah.org
http://www.iah.org/

International Association of Independent Tanker Owners (INTERTANKO)
Bogstadveien 27B
PO Box 5804
Majorstua
N-0308 Oslo, Norway
Tel: 47 22 122640
Fax: 47 22 122641
E-mail: postmaster@intertanko.com
http://www.intertanko.com/

International Association of Institutes of Navigation (IAIN)
The Royal Institute of Navigation
1 Kensington Gore
London, SW7 2AT, United Kingdom
Contact: Secretary General
E-mail: prentpage@aol.com
http://www.iainav.org/

International Association of Marine Aids to Navigation and Lighthouse Authorities (IALA/AISM)
20 ter rue Schnapper
F-78100 St. Germain-en-Laye, France
Contact: Secretary General
Tel: 33 1 3451 7001
Fax: 33 1 3451 8205
E-mail: aismiala@easynet.fr
http://www.iala-aism.org/

International Association of Maritime Economists (IAME)
IAME Permanent Secretariat (H12-11)
Erasmus University
PO Box 1738
NL-3000 DR, Rotterdam, The Netherlands
Tel: 31 10 408 1490
Fax: 31 10 408 9156
E-mail: rhv-iame@few.eur.nl
http://www.iame.nl/

International Association of Maritime Institutions
Faculty of Nautical Sciences
South Tyneside College
St. Georges Avenue
South Shields, NE34 6ET, United Kingdom
Contact: Honorary Secretary
Tel: 44 191 427 3696
Fax: 44 191 427 3653
E-mail: nautical@stc.ac.uk

International Association of Maritime Universities
c/o Kobe University of Mercantile Marine
5-1-1, Fukae-minami, Higashi-nada
Kobe 658-0022, Japan
Contact: Secretariat
E-mail: k-inoue@cc.kshosen.ac.jp
http://www.iamu-edu.org/

International Association of Ports and Harbours (IAPH)
5th Floor, North Tower
New Pier Takeshiba
1-11-1 Kaigan
Minato-Ku, Tokyo, 105-0022, Japan
Contact: Secretary General
Tel: 81 3 5403 2770
Fax: 81 3 5403 7651
E-mail: info@iaphworldports.org
http://www.iaphworldports.org/

International Association of Meiobenthologists
c/o Department of Zoology
The Natural History Museum
London, SW7 5BD, United Kingdom
Contact: Chair
http://www.meiofauna.org/

International Cable Protection Committee (ICPC)
PO Box 150
Lymington, SO41 6WA, United Kingdom
Contact: Secretary
Tel: 44 15 9068 1673
Fax: 44 87 0052 6049
E-mail: secretary@iscpc.org
http://www.iscpc.org/

International Cargo Handling Coordination Association (ICHCA)
85 Western Road, Suite 2
Romford
Essex, RM1 3LS, United Kingdom
Contact: International Secretariat
Tel: 44 1 70873 4787
Fax: 44 1 70873 4877
E-mail: info@ichca.org.uk
http://www.ichca.org.uk/

International Cartographic Association (ICA)
PO Box 80115
NL-3508TC Utrecht, The Netherlands
Contact: Secretary General
Tel: 31 30 253 1373
Fax: 31 30 254 0604
E-mail: f.ormeling@geog.uu.nl
http://www.icaic.org/

International Chamber of Shipping (ICS)
Carthusian Court
12 Carthusian Street
London, EC1M 6EZ, United Kingdom
Contact: Secretary General
Tel: 44 20 7417 8844
Fax: 44 20 7417 8877
E-mail: ics@marisec.org
http://www.marisec.org/

International Christian Maritime Association (ICMA)
's-Gravendijkwal 64
NL-3014 EG, Rotterdam, The Netherlands
Contact: Secretariat

Tel: 31 10 225 1799
Fax: 31 10 225 0692
E-mail: icma@wanadoo.nl
http://www.icma.as/

International Collective in Support of Fishworkers (ICSF)
27 College Road
600006 Chennai, India
Contact: Executive Secretary
Tel: 91 44 2827 5303
Fax: 91 44 2825 1145
E-mail: icsf@vsnl.com
http://www.icsf.net/

International Commission on Polar Meteorology
British Antarctic Survey
High Cross, Madingley Road
Cambridge, CB3 0ET, United Kingdom
Contact: President
E-mail: J.Turner@bas.ac.uk
http://www.nerc-bas.ac.uk/public/icd/icpm/

International Commission on Remote Sensing (ICRS)
c/o 11 Innovation Blvd
National Water Research Institute
Saskatoon, SK, S7N 3H5, Canada
Contact: Secretary ICRS
E-mail: al.pietroniro@ec.gc.ca
http://www.cig.ensmp.fr/iahs/

International Commission on Snow and Ice (ICSI)
c/o INRS EAM
CP 7500
2700 Einstein
Ste-Foy, QC, G1V 4C7, Canada
Contact: President
Tel: 1 418 654-2533
Fax: 1 418 654-2600
E-mail: jones@inrs-eau.uquebec.ca
http://geowww.uibk.ac.at/research/icsi/

International Committee on Seafarers' Welfare (ICSW)
11 Lancaster Mews
London, W2 3QE, United Kingdom
Contact: Executive Secretary
Tel: 44 20 7402 1554
Fax: 44 20 7402 1522
E-mail: icsw@icsw.org.uk
http://www.icsw.org.uk/

International Committee on the Underwater Cultural Heritage
Western Australian Maritime Museum
Cliff Street
Fremantle, WA 6160, Australia
Contact: Secretariat
Tel: 61 08 9431 8488
http://www.mm.wa.gov.au/Museum/public/ICOMOS.html

International Congress of Maritime Museums (ICMM)
PO 326
Mystic, CT, 06355, USA
Contact: Secretary General
E-mail: thomas@marinpro.com
http://www.icmmonline.org/

International Cooperative Fisheries Organization
National Federation of Fisheries Co-operative Association (ZENGYOREN)
Co-operative building, 7th floor
1-1-12 Uchikanda, Chiyoda-ku
Tokyo 101, Japan
Contact: Chairman
Tel: 81 3 3294 9617
Fax: 81 3 3294 9602
E-mail: icfo@zengyoren.jf-net.ne.jp
http://www.coop.org/ica/ica/sb/fish.html

International Coral Reef Action Network (ICRAN)
c/o UNEP-World Conservation Monitoring Centre

219 Huntingdon Road
Cambridge, CB3 0DL, United
  Kingdom
Contact: ICRAN Assistant Director
Tel: 44 12 2327 7314
Fax: 44 12 2327 7136
E-mail: icran@icran.org
http://www.icran.org/

International Coral Reef Initiative
c/o Robert Jara
Department of Environment and
  Natural Resources (DENR)
2nd Floor, FASPO Building
DENR Compound
Visayas Avenue
Diliman, Quezon City, The Philippines
Tel: 632 928 1215
Fax: 632 928 1225
E-mail: icri_secretariat@hotmail.com
http://www.icriforum.org/

International Council of Cruise
  Lines (ICCL)
2111 Wilson Boulevard, 8th Floor
Arlington, VA, 22201, USA
Contact: President
Tel: 1 703 522-8463
Fax: 1 703 522-3811
E-mail: info@iccl.org
http://www.iccl.org/

International Council of Marine Industry Associations (ICOMIA)
Marine House
Thorpe Lea Road
Egham, Surrey, TW20 8BF, United
  Kingdom
Contact: Secretariat
Tel: 44 1784 22 3700
Fax: 44 1784 22 3705
http://www.icomia.com/

International Council of Scientific
  Unions (ICSU)
51, boulevard de Montmorency
F-75016 Paris, France
Contact: Secretariat

Tel: 33 1 4525 0329
Fax: 33 1 4288 9431
E-mail: secretariat@icsu.org
http://www.icsu.org/

International Council on Monuments and Sites (ICOMOS)
49-51, rue de la Fédération
F-75015 Paris, France
Contact: International Secretariat
Tel: 33 1 4567 6770
Fax: 33 1 4566 0622
E-mail: secretariat@icomos.org
http://www.icoms.org/

International Desalination Association
PO Box 387
Topsfield, MA, 01983, USA
Contact: Secretary General
Tel: 1 978 887-0410
Fax: 1 978 887-0411
E-mail: ida1pab@ix.netcom.com
http://www.ida.bm/

International Dolphin Watch (IDW)
10 Melton Road
North Ferriby
East Yorkshire, HU14 3ET, United
  Kingdom
Contact: Secretary
Tel: 44 14 8264 5789
Fax: 44 14 8263 4914
E-mail: info@idw.org
http://www.idw.org/

International Ecotourism Society
PO Box 668
Burlington, VT, 05402, USA
Tel: 1 802 651-9818
Fax: 1 802 651-9819
email: ecomail@ecotourism.org
http://www.ihei.org/

International Federation of Shipmasters' Associations (IFSMA)
202 Lambeth Road
London, SE1 7JY, United Kingdom
Contact: General Secretary

Tel: 44 20 7261 0450
Fax: 44 20 7928 9030
E-mail: hq@ifsma.org
http://www.ifsma.org/

International Fishmeal and Fish Oil Organization (IFFO)
2 College Yard
Lower Dagnall Street
St. Albans, Hertfordshire, AL3 4PA, United Kingdom
Contact: Director General
Tel: 44 17 2784 2844
Fax: 44 17 2784 2866
E-mail: secretariat@iffo.org.uk
http://www.iffo.org.uk/

International Game Fish Association (IGFA)
300 Gulf Stream Way
Dania Beach, FL, 33004, USA
Contact: President
Tel: 1 954 927-2628
Fax: 1 954 924-4299
E-mail: igfahq@aol.com
http://www.igfa.org/

International Hydrofoil Society
PO Box 51
Cabin John, MD, 20818, USA
Contact: President
E-mail: president@foils.org
http://www.foils.org/

International Institute for Sustainable Development (IISD)
161 Portage Avenue East, 6th Floor
Winnipeg, MB, R3B 0Y4, Canada
Contact: President & CEO
Tel: 1 204 958-7700
Fax: 1 204 958-7710
E-mail: info@iisd.ca
http://www.iisd.org/

International Marine Contractors Association (IMCA)
Carlyle House
235 Vauxhall Bridge Road
London, SW1V 1EJ, United Kingdom
Contact: Secretary
Tel: 44 20 7931 8171
Fax: 44 20 7931 8935
E-mail: imca@imca-int.com
http://www.imca-int.com/

International Marine Mammal Association Inc.
1474 Gordon St.
Guelph, ON, N1L 1C8, Canada
Tel: 1 519 767-1948
Fax: 1 519 767-0284
E-mail: ccosgrove@imma.org
http://www.imma.org/

International Marine Minerals Society (IMMS)
c/o Marine Minerals Technology Center
University of Hawaii
1000 Pope Road, MSB 303
Honolulu, HI, 96822, USA
Contact: Administrative Office
Tel: 1 808 956-6036
Fax: 1 808 956-9772
E-mail: karynnem@hawaii.edu
http://www.ngdc.noaa.gov/mgg/imms/

International Marinelife Alliance
Stangenwald Building, Suite 610
119 Merchant Street
Honolulu, HI, 96813, USA
Contact: International Headquarters
Tel: 1 808 523-0143
Fax: 1 808 523-0140
E-mail: info@marine.org
http://www.marine.org/

International Maritime Arbitration Organisation (IMAO)
38 Cours Albert 1er
F-75008 Paris, France
Contact: Secretariat of the ICC/CMI Standing Committee of the IMAO

Tel: 33 1 4953 2828
Fax: 33 1 4953 2933
E-mail: arb@iccwbo.org
http://www.iccwbo.org/court/
english/maritime/all_topics.asp

International Maritime Bureau (IMB)
Maritime House
1 Linton Road
Barking, Essex, 1G11 8HG, United Kingdom
Contact: Director
Tel: 44 20 8591 3000
Fax: 44 20 8594 2833
E-mail: imb@icc-ccs.org.uk
http://www.icc-ccs.org/

International Maritime Economic History Association
Maritime History Publications
Memorial University of Newfoundland
St. John's, NL, A1C 5S7, Canada
Contact: President
Tel: 1 709 737-2602
Fax: 1 709 737-8427
E-mail: mhp@mun.ca
http://www.mun.ca/mhp/imeha.htm

International Maritime Pilots' Association (IMPA)
HQS Wellington
Temple Stairs
Victoria Embankment
London, WC2R 2PN, United Kingdom
Contact: Secretary General
Tel: 44 20 7240 3973
Fax: 44 20 7240 3518
E-mail: sec@impahq.org
http://www.impahq.org/

International Navigation Association (PIANC/AICPN)
Graaf de Ferraris Bldg, 11th Floor
Boulevard du Roi Albert II, 20–Box 3
B-1000 Brussels, Belgium
Contact: General Secretariat
Tel: 32 2 553 7160
Fax: 32 2 553 7155
E-mail: info@pianc-aipcn.org
http://www.pianc-aipcn.org/

International Network of Basin Organisations
c/o Office international de l'eau/ International Office for Water
21, rue de Madrid
F-75008 Paris, France
Contact: Secretariat
Tel: 33 1 4490 8860
Fax: 33 1 4008 0145
E-mail: riobs@wanadool.fr
http://www.inbo-news.org/

International Ocean Institute (IOI)
University of Malta
PO Box 3
MT-Gzira, GZR 01, Malta
Contact: Executive Director
Tel: 356 346529
Fax: 356 346502
E-mail: ioihq@ioihq.org.mt
http://www.ioinst.org/

International Ombudsman Centre for Environment and Development (OmCED)
PO Box 319-6100
San José, Costa Rica
Tel: 1 506 205-1600
Fax: 1 506 249-3500
E-mail: omced@ecouncil.ac.cr
http://www.omced.org/

International Organization of Masters, Mates and Pilots (IOMMP)
700 Maritime Blvd
Linthicum Heights, MD, 21090-1941, USA
Contact: Secretary-Treasurer
Tel: 1 410 850-8700, ext. 21
Fax: 1 410 850-0973
E-mail: info@bridgedeck.org
http://www.iommp.org/

International Organization for Standardization (ISO)
1, rue de Varembé, Case postale 56
CH-1211 Geneva 20, Switzerland
Contact: Secretariat
Tel: 41 22 749 01 11
Fax: 41 22 733 34 30
E-mail: central@iso.org
http://www.iso.org/

International Petroleum Industry Environmental Conservation Association (IPIECA)
Monmouth House, 2nd Floor
87-93 Westbourne Grove
London, W2 4UL, United Kingdom
Contact: Executive Secretary
Tel: 44 20 7221 2026
Fax: 44 20 7229 4948
E-mail: info@ipieca.org
http://www.ipieca.org/

International Salvage Union (ISU)
PO Box 32293
London, W5 1WZ, United Kingdom
Contact: Secretary General
Tel: 44 20 7345 5122
Fax: 44 20 7345 5722
E-mail: isu@randell.fsnet.co.uk
http://www.marine-salvage.com/

International Seaweed Association
Instituto de Biociências
Universidade de São Paulo
CP 11461
São Paulo, SP 05422-970, Brazil
Contact: Secretary
Tel: 55 11 3818 7630
Fax: 55 11 3818 7547
E-mail: euricodo@usp.br
http://www.intseaweedassoc.org/

International Ship Managers' Association (ISMA)
Suite 202, Eastlands Court
St. Peter's Road
Rugby, CB21 3QP, United Kingdom
Tel: 44 20 7417 8844
Fax: 44 20 7417 8877
E-mail: alan.ward@isma-london.org
http://www.isma-london.org/

International Ship Suppliers Association (ISSA)
The Baltic Exchange
St. Mary Avenue
London, EC3A 8BH, United Kingdom
Contact: Secretariat
Tel: 44 20 7626 6236
Fax: 44 20 7626 6234
E-mail: issa@dial.pipex.co.uk
http://www.shipsupply.org/

International Shipping Federation (ISF)
Carthusian Court
12 Carthusian Street
London, EC1M 6EZ, United Kingdom
Contact: Secretary General
Tel: 44 20 7417 8844
Fax: 44 20 7417 8877
E-mail: isf@marisec.org
http://www.marisec.org/

International Society for Environmental Protection (ISEP)
Marxergasse 3/20
A-1030 Vienna, Austria
Tel: 43 1 715 28280
Fax: 43 1 715 282819
E-mail: office@isep.at
http://www.isep.at/

International Society for Microbial Ecology (ISME)
c/o Center for Microbial Ecology
Michigan State University
East Lansing, MI, 48824, USA
Tel: 1 517 353-9021
Fax: 1 517 353-2917
E-mail: isme@msu.edu
http://microbes.org/

International Society for Reef Studies (ISRS)
c/o School of Biological Sciences, Hatherly Laboratory
University of Exeter, Prince of Wales Road
Exeter, EX4 4PS, United Kingdom
Contact: Secretary
Tel: 44 13 9226 3798
Fax: 44 13 9226 3700
E-mail: p.j.mumby@exeter.ac.uk
http://www.uncwil.edu/isrs/

International Society for the Prevention of Water Pollution
Little Orchard, Bentworth
Alton, Hampshire, GU34 5RB, United Kingdom
Contact: Chairman
Tel: 44 14 2056 2225

International Society of Acoustic Remote Sensing of the Atmosphere and Oceans (ISARS)
Physics Department
University of Auckland
Private Bag 92019
Auckland, New Zealand
Contact: Chairperson
Tel: 64 9 3737999
Fax: 64 9 3737445
E-mail: s.bradley@auckland.ac.nz
http://www.boku.ac.at/imp/isars/

International Society of Offshore and Polar Engineers (ISOPE)
PO Box 189
Cupertine, CA, 95015-0189, USA
Tel: 1 408 980-1784
Fax: 1 408 980-1787
Contact: Executive Director
E-mail: info@isope.org
http://www.isope.org/

International Support Vessel Owners' Association (ISOA)
Carthusian Court
12 Carthusian Street
London, EC1M 6EZ, United Kingdom
Tel: 44 20 7417 8844
Fax: 44 20 7417 8877
E-mail: isoa@marisec.org
http://www.marisec.org/isoa/

International Tanker Owners Pollution Federation (ITOPF)
Staple Hall
Stonehouse Court
87-90 Houndsditch
London, EC3A 7AX, United Kingdom
Contact: Managing Director
Tel: 44 20 7621 1225
Fax: 44 20 7621 1783
E-mail: central@itopf.com
http://www.itopf.com/

International Transport Workers' Federation (ITF)
ITF House
49-60 Borough Road
London, SE1 1DR, United Kingdom
Tel: 44 20 7403 2733
Fax: 44 20 7357 7871
E-mail: mail@itf.org.uk
http://www.itf.org.uk/

International Union of Biological Sciences (IUBS)
51, boulevard de Montmorency
F-75016 Paris, France
Contact: Executive Director
Tel: 33 1 4525 0009
Fax: 33 1 4525 2029
E-mail: secretariat@iubs.org
http://www.iubs.org/

International Union of Geodesy and Geophysics (IUGG)
c/o CIRES
Campus Box 216
University of Colorado
Boulder, CO, 80309, USA
Contact: Secretary General
Tel: 1 303 497-5147

Fax: 1 303 497-3645
E-mail: jjoselyn@cires.colorado.edu
http://www.iugg.org/

International Union of Geological Sciences (IUGS)
Geological Survey of Austria
Rasumofskygasse 23
PO Box 127
A-1031 Vienna, Austria
Contact: Secretary General
Tel: 43 1 712 56 74 180
Fax: 43 1 712 56 74 56
E-mail: wjanoschek@cc.geolba.ac.at
http://www.iugs.org/

International Union of Marine Insurance (IUMI)
Swiss Re
Mythenquai 50/60
CH-8022 Zurich, Switzerland
Contact: General Secretary
Tel: 41 1 285 2121
Fax: 41 1 282 2126
E-mail: Fritz_Stabinger@swissre.com
http://www.iumi.com/

International Water Association
Alliance House
12 Caxton Street
London, SW1H 0QS, United Kingdom
Contact: Executive Director
Tel: 44 20 7654 5500
Fax: 44 20 7654 5555
E-mail: water@iwahq.org.uk
http://www.iawq.org.uk/

Islamic Shipowners' Association (ISHA)
PO Box 14900
Jeddah 21434, Saudi Arabia
Contact: Secretary General
Tel: 966 2 667 0288

Lloyd's Register of Shipping
71 Fenchurch Street
London, EC3M 4BS, United Kingdom
Contact: Enquiries
Tel: 44 20 7709 9166
Fax: 44 20 7488 4796
E-mail: lloydsreg@lr.org
http://www.lr.org/

Marine Aquarium Council
923 Nu'uanu Avenue
Honolulu, HI, 96817, USA
Tel: 1 808 550-8217
Fax: 1 808 550-8317
E-mail: info@aquariumcouncil.org
http://www.aquariumcouncil.org/

Marine Technology Society
5565 Sterrett Place, Suite 108
Columbia, MD, 21044, USA
Contact: Executive Director
Tel: 1 410 884-5330
Fax: 1 410 884-9060
E-mail: mtspubs@aol.com
http://www.mtsociety.org/

Multiport Ship Agencies Network
City Reach
14 Greenwich Quay, 2nd Floor
Clarence Road
London, SE8 3EY, United Kingdom
Contact: Secretary General
Tel: 44 20 8469 9188
Fax: 44 20 8469 9189
E-mail: multiport@dial.pipex.com
http://www.multiport.org/

National Shellfisheries Association, Inc.
Patrick Center for Environmental Research
Academy of Natural Sciences
1900 Ben Franklin Parkway
Philadelphia, PA 19103 USA
Contact: President
Tel: 1 215 299-1184
Fax: 1 215 299-1079
E-mail: kreeger@say.acnatsci.org
http://www.shellfish.org/

Ocean Drilling Program (ODP)
Joint Oceanographic Institutions
1755 Massachusetts Avenue, NW, Suite 700
Washington, DC, 20036-2102, USA
Contact: Program Manager
Tel: 1 202 232-3900
Fax: 1 202 462-8754
E-mail: joi@joiscience.org
http://www.oceandrilling.org/

Ocean, Offshore and Arctic Engineering Division (OOAE)
American Society of Mechanical Engineers
ASME International
Three Park Avenue
New York, NY, 10016-5990, USA
Contact: Chair
Tel: 1 973 882-1167
E-mail: chair@ooae.org; infocentral@asme.org
http://www.ooae.org/

Oil Companies International Marine Forum (OCIMF)
27 Queen Anne's Gate
London, SW1H 9BU, United Kingdom
Tel: 44 20 7654 1200
Fax: 44 20 7654 1205
E-mail: enquiries@ocimf.com
http://www.ocimf.com/

Partnership for Observation of the Global Oceans (POGO)
Bedford Institute of Oceanography
1 Challenger Drive
Dartmouth, NS, B2Y 4A2, Canada
Contact: Executive Director
Tel: 1 902 426-8044
Fax: 1 902 426-9388
E-mail: pogo@sio.ucsd.edu
http://www.oceanpartners.org/

Passenger Shipping Association (PSA)
Walmar House
288-292 Regent Street
London, W1R 5HE, United Kingdom
Tel: 44 20 7436 2449
Fax: 44 20 7636 9206
Email: admin@psa-psara.org
http://www.psa-psara.org/

Permanent Service for Mean Sea Level
Bidston Observatory
Birkenhead, Merseyside, CH43 7RA, United Kingdom
Contact: Director
Tel: 44 15 1653 8633
Fax: 44 15 1653 6269
E-mail: psmsl@pol.ac.uk
http://www.nbi.ac.uk/psmsl/index.html/

Pew Center on Global Climate Change
2101 Wilson Boulevard, Suite 550
Arlington, VA, 22201, USA
Tel: 1 703 516-4146
Fax: 1 703 841-1422
http://www.pewclimate.org/

Pew Oceans Commission
2101 Wilson Boulevard, Suite 550
Arlington, VA, 22201, USA
Contact: Executive Director
Tel: 1 703 516-0624
Fax: 1 703 516-9551
E-mail: tulouc@pewoceans.org
http://www.pewoceans.org/

Reef Relief
PO Box 430
Key West, FL, 33041, USA
Tel: 1 305 294-3100
Fax: 1 305 293-9515
E-mail: info@reefrelief.org
http://www.reefrelief.org./

Scientific Committee on Antarctic Research (SCAR)
Scott Polar Research Institute

Lensfield Road
Cambridge, CB2 1ER, United
  Kingdom
Contact: Executive Secretary
Tel: 44 12 2336 2061
Fax: 44 12 2333 6549
E-mail: info@scar.org
http://www.scar.org/

Scientific Committee on Oceanic Research (SCOR)
Dept. Earth and Planetary Sciences
Olin Hall, San Martin Drive
Johns Hopkins University
Baltimore, MD, 21218, USA
Contact: Executive Director
Tel: 1 410 516-4070
Fax: 1 410 516-4019
E-mail: scor@jhu.edu
http://www.jhu.edu/scor/

Scientific Committee on Problems of the Environment (SCOPE)
51, boulevard de Montmorency
F-75016 Paris, France
Contact: Executive Director
Tel: 33 1 4525 0578
Fax: 33 1 4288 1466
E-mail: secretariat@icsu-scope.org
http://www.icsu-scope.org/

Sea Shepherd Conservation International
22774 Pacific Coast Highway
Malibu, CA, 90265, USA
Tel: 1 310 456-1141
Fax: 1 310 456-2488
E-mail: seashepherd@seashepherd.org
http://www.seashepherd.org/

Seacology
2009 Hopkins Street
Berkeley, CA, 94707, USA
Tel: 1 510 599-3505
Fax: 1 510 599-3506
E-mail: islands@seacology.org
http://www.seacology.org/

Seas at Risk Federation
Drieharingstraat 25
NL-3511 BH Utrecht, Netherlands
Contact: Secretariat
Tel: 31 30 670 1291
Fax: 31 30 670 1292
E-mail: secretariat@seas-at-risk.org
http://www.seas-at-risk.org/

Sir Alister Hardy Foundation for Ocean Science
The Laboratory
Citadel Hill
Plymouth, PL1 2PB, United
  Kingdom
Tel: 44 17 5260 0016
Fax: 44 17 5260 0015
E-mail: sahfos@mail.pml.ac.uk
http://www.npm.ac.uk/sahfos/sahfos2.html

Society for Underwater Technology (SUT)
80 Coleman Street
London, EC2R 5BJ, United Kingdom
Contact: Executive Secretary
Tel: 44 20 7382 2601
Fax: 44 20 7382 2684
E-mail: ian.gallett@sut.org.uk
http://www.sut.org.uk/

Society of Environmental Toxicology and Chemistry
SETAC North America
1010 North 12th Avenue
Pensacola, FL, 32501-3370, USA
Tel: 1 850 469-1500
Fax: 1 850 469-9778
E-mail: setac@setac.org
http://www.setac.org/

Society of International Gas Tanker and Terminal Operators (SIGTTO)
17 St. Helen's Place
London, EC3A 6DG, United Kingdom
Contact: Secretariat
Tel: 44 20 7628 1124

Fax: 44 20 7628 3163
E-mail: secretariat@sigtto.org
http://www.sigtto.org/

Society of Naval Architects and Marine Engineers (SNAME)
601 Pavonia Avenue
Jersey City, NJ, 07306, USA
Contact: Executive Director
Tel: 1 201 798-4800
Fax: 1 201 798-4975
E-mail: ccali-poutre@sname.org
http://www.sname.org/

Species Survival Commission (SSC)
IUCN–The World Conservation Union
28, rue Mauvernay
CH-1196 Gland, Switzerland
Contact: Chair
Tel: 41 22 999 0152
Fax: 41 22 999 0015
E-mail: ssc@iucn.org
http://www.iucn.org/themes/ssc/index.html/

The Nature Conservancy
4245 North Fairfax Drive, Suite 100
Arlington, VA, 22203-1606, USA
E-mail: comment@tnc.org
http://www.nature.org/

The Nautical Institute
202 Lambeth Road
London, SE1 7LQ, United Kingdom
Contact: Secretary
Tel: 44 20 7928 1351
Fax: 44 20 7401 2817
E-mail: sec@nautinst.org
http://www.nautinst.org/

The Reef-World Foundation (UK)
Linton House
Danesbury Park Road,
Welwyn, Hertfordshire, AL6 9SH, United Kingdom
Contact: General Manager
Tel: 44 14 3871 8881
Fax: 44 14 3871 8880
E-mail: amiller@reef-world.org
http://www.reef-world.org/

The World Conservation Union – International Union for the Conservation of Nature (IUCN)
28, rue Mauverney
CH-1196 Gland, Switzerland
Contact: Director
Tel: 41 22 999 0001
Fax: 41 22 999 0002
E-mail: mail@hq.iucn.org
http://www.iucn.org/

UNEP World Conservation Monitoring Centre
219 Huntingdon Road
Cambridge, CB3 0DL, United Kingdom
Tel: 44 12 2327 7722
Fax: 44 12 2327 7136
E-mail: info@unep-wcmc.org
http://www.unep-wcmc.org/

Water Environment Federation (WEF)
601 Wythe Street
Alexandria, VA, 22314-1994, USA
Contact: Executive Director
Tel: 1 703 684-2452
Fax: 1 703 684-2492
E-mail: csc@wef.org
http://www.wef.org/

Wetlands International
International Coordination Unit
PO Box 471
NL-6700 AL, Wageningen, The Netherlands
Tel: 31 317 478854
Fax: 31 317 478850
E-mail: post@wetlands.agro.nl
http://www.wetlands.org/

World Aquaculture Society (WAS)
Louisiana State University
143 J.M. Parker Coliseum
Baton Rouge, LA, 70803, USA

Contact: Secretary
Tel: 1 225 388-3137
Fax: 1 225 388-3493
E-mail: wasmas@aol.com
http://www.was.org/

World Commission on Protected
  Areas (WCPA)
IUCN–The World Conservation
  Union
28, rue Mauverney
CH-1196 Gland, Switzerland
Contact: Head, Programme on Protected Areas
Tel: 44 22 999 0162
Fax: 44 22 999 0115
E-mail: das@iucn.org
http://wcpa.iucn.org/

World Fish Center (ex-ICLARM)
PO Box 500, GPO 10670
Penang, Malaysia
Contact: Director General
Tel: 60 4 626 1606
Fax: 60 4 626 5530
E-mail: iclarm@cgiar.org
http://www.worldfishcenter.org/

World Fisheries Trust
#204 1208 Wharf Street
Victoria, BC, V8W 3B9, Canada
Contact: President
Tel: 1 250 380-7585
Fax: 1 250 380-2621
E-mail: bharvey@worldfish.org
http://www.worldfish.org/

World Glacier Monitoring Service
  (WGMS)
Department of Geography
University of Zurich-Irchel
Winterthurerstrasse 190
CH-8057 Zurich, Switzerland
Contact: Director
Tel: 41 1 635 5120
Fax: 41 1 635 6848
E-mail: haeberli@geo.unizh.ch
http://www.geo.unizh.ch/wgms/

World Resources Institute
10 G Street, NE, Suite 800
Washington, DC, 20002, USA
Tel: 1 202 729-7600
Fax: 1 202 729-7610
E-mail: front@wri.org
http://www.wri.org/

World Travel and Tourism Council
1-2 Queen Victoria Terrace
Sovereign Court
London, E1W 3HA, United
  Kingdom
Tel: 44 87 0727 9882
Fax: 44 87 0728 9882
E-mail: enquiries@wttc.org
http://www.wttc.org/

World Wide Fund for Nature
  (WWF)
Avenue du Mont-Blanc
CH-1196, Gland, Switzerland
Contact: Director General
Tel: 41 22 364 9111
Fax: 41 22 364 5358
E-mail: members@wwfint.org
http://www.panda.org/

3. GLOBAL ACADEMIC
ORGANIZATIONS

Chartered Institute of Logistics and
  Transport (CILT)
11/12 Buckingham Gate
London, SW1 6LB, United Kingdom
http://www.citrans.org.uk/

Circumpolar Universities Association
The Roald Amundsen Centre of Arctic Research
University of Tromsø
N-9037 Tromsø, Norway
Tel: 47 77 64 52 40
Fax: 47 77 67 66 72
E-mail: frits.jensen@arctic.uit.no
http://arctic.uit.no/cua_secr/

Institute of Marine Engineering, Science and Technology (IMarEst)
80 Coleman Street
London, EC2R 5BJ, United Kingdom
Contact: Secretariat
Tel: 44 20 7382 2600
Fax: 44 20 7382 2670
E-mail: imare@imare.org.uk
http://www.imare.org.uk/

International Maritime Law Institute (IMLI)
PO Box 31
MT-Msida MSD 01, Malta
Contact: Director
Tel: 356 21 319343
Fax: 356 21 343092
E-mail: info@imli.org
http://www.imli.org/

International Maritime Lecturers' Association (IMLA)
c/o World Maritime University
PO Box 500
Citadellsvagen 29
SE-201 24 Malmö, Sweden
Contact: President
Tel: 46 40 35 63 67
Fax: 46 40 12 48 27
E-mail: rajendra.prasad@wmu.se
http://www.wmu.se/imla/

United Nations University (UNU)
53-70 Jingumae 5-chome Shibuya-ku
Tokyo 150-8925, Japan
Tel: 81 3 3499 2811
Fax: 81 3 3499 2828
E-mail: mbox@hq.unu.edu
http://www.unu.edu/

World Maritime University (WMU)
PO Box 500
Citadellsvagen 29
SE-201 24 Malmö, Sweden
Contact: President
Tel: 46 40 35 63 00
Fax: 46 40 12 84 42
E-mail: info@wmu.se
http://www.wmu.se/

4. REGIONAL INTERGOVERNMENTAL ORGANIZATIONS

4.1 Africa

Association of African Maritime Training Institutes (AAMTI)
Gamal Abdel Nasser
PO Box 1029
Alexandria, Egypt
Contact: Secretary General
Tel: 20 3 560 0245
Fax: 20 3 560 2144
E-mail: aaen0038@aast.egnet.net

Atlantic Africa Fisheries Conference (Conference Ministerielle Halieutique des Etats Africains Riverains de l'Ocean Atlantique– Ministerial Conference on Fisheries Cooperation among African States Bordering the Atlantic Ocean)
BP 476 Nouvelle Cite Administrative
Rue Mohammed Bel Hassan El Ouazzani
Haut-Agdal-Rabat, Morocco
Contact: Permanent Secretary of the Conference
Tel: 212 37 688303
Fax: 212 37 688329
E-mail: alahlou@mp3m.gov.ma

Centre de Recherce Scientifique de Conakry-Rogbane (CERESCOR)
BP 1615
Conakry, Guinea
Contact: ODIN Africa Regional Coordinator
Fax: 224 413811
E-mail: s.cisse@unesco.org

FAO Regional Office for the Near
    East (RNO)
11 El Eslah El Zerai Street
Dokki, Cairo
PO Box 2223
Cairo, Egypt
Tel: 20 2 337 2229
Fax: 20 2 349 5981
E-mail: fao-rne@fao.org

FAO Regional Office for Africa
PO Box 1628
Accra, Ghana
Tel: 233 21 666851
Fax: 233 21 668427
E-mail: fao-raf@fao.org
http://www.fao.org/regional/
    africa/default.htm

FAO Sub-Regional Office for North
    Africa
PO Box 300
Tunis, Tunisia
Tel: 216 1 847 553
Fax: 216 1 791 859
E-mail: fao-snea@fao.org

FAO Subregional Office for South-
    ern and East Africa
PO Box 3730
Harare, Zimbabwe
Tel: 263 4 791407
Fax: 263 4 703497
E-mail: fao-safr-registry@field.fao.org

Fishery Committee for Eastern Cen-
    tral Atlantic (CECAF)
c/o FAO Regional Office for Africa
PO Box 1628
Accra, Ghana
Contact: Assistant Director-General
Tel: 233 21 675000
Fax: 233 21 668427
E-mail: fao-raf@fao.org
http://www.fao.org/fi/body/rfb/
    CECAF/cecaf_home.html

Indian Ocean Commission
Q4, avenue Sir Guy Forget
Quatre Bornes, BP 7, Mauritius
Contact: Secretary General
Tel: 230 425 1652
Fax: 230 425 2709
E-mail: coi7@intnet.mu
http://www.coi-info.org/

Indian Ocean Rim Association for
    Regional Cooperation (IOR-ARC)
c/o Indian Ocean Rim Business Net-
    work (IORNET)
Federation House, Tansen Marg
New Delhi 110001, India
Tel: 91 11 3325110
Fax: 91 11 3721504
E-mail: iornet@ficci.com
http://www.iornet.com/newiornet/
    index.htm

INFOPECHE (Intergovernmental
    Organization for Marketing In-
    formation and Technical Advi-
    sory Services for Fishery Prod-
    ucts in Africa)
Tour C, 19éme étage, Cité Administ-
    trative
BP 1747
Abidjan 01, Côte d'Ivoire
Contact: Director
Tel: 225 2022 8980
Fax: 225 2021 8054
E-mail: infopech@africaonline.co.ci
http://www.globefish.org/
    entry_infopeche.htm

INFOSAMAK (Centre for Marketing
    Information & Advisory Services
    For Fishery Products in the
    Arab Region)
71, boulevard Rahal El Meskini
16243, Casablanca, Morocco
Tel: 212 22 540856
Fax: 212 22 540855
E-mail: info@infosamak.org
http://www.infosamak.org/
    english/

INFOYU
Rm. 203 Building 18 Maizidian
    Street
Chaoyang District
Beijing 100026, People's Republic of
    China
Tel: 86 10 64195140
Fax: 86 10 64195141
Email: infoyu@agri.gov.cn
http://www.globefish.org/
    globefishpartners/
    fishinfonetworks/

Inter-African Committee on Oceanography, Inland and Sea Fisheries
Scientific, Technical and Research
    Commission of the African
    Union (AU/STRC)
26-28 Marina, Port Authority Bldg.
    4th Floor
PMB 2359, Marina
101231 Lagos, Nigeria
Contact: Executive Secretary
Tel: 234 1 2633752
Fax: 234 1 2636093
E-mail: oau/strcl@rcl.nig.com

Inter-governmental Standing
    Committee on Shipping
    (ISCOS)
Bima Tower, 9th floor
Digo Road, PO Box 89112
Mombasa, Kenya
Tel: 254 11 312196
Fax: 254 11 316387

IOC Regional Committee for the
    Central Eastern Atlantic
    (IOCEA)
Nigerian Institute for Oceanography
    and Marine Research
PMB12729
Victoria Island, Lagos, Nigeria
Tel/Fax: 234 1 2619517
Mobile: 234 8033123967
http://ioc.unesco.org/iocweb/
    activities/regions/iocea.htm

IOC Regional Committee for the Cooperative Investigation in the
    North and Central Western Indian Ocean (IOCINCWIO)
PO Box 95832
Mombasa, Kenya
Contact: IOCINCWIO Project Office
Tel: 254 11 472527
Fax: 254 11 475157
E-mail: m.odido@unesco.org
http://ioc.unesco.org/iocincwio/

La Commission sous-régionale des
    pêches (CSRP)
(Sub-Regional Commission on Fisheries–SRCF)
Km 10 Boulevard de Centenaire de
    la Commune de Dakar-Senegal
BP 20505
Dakar, Senegal
Contact: Secretariat
Tel: 221 345580
Fax: 221 238720
E-mail: sp_csrp@metissacana.sn
http://www.csrp-afrique.org/

Maritime Organization for West and
    Central Africa
BP V 257
Abidjan, Côte d'Ivoire
Contact: Secretary General
Tel: 225 217115
Fax: 225 224532
E-mail: minconmar@africaonline.co.ci

Niger Basin Authority (NBA)
BP 729
Niamey, Niger
Contact: Executive Secretary
Tel: 227 723102
Fax: 227 724208
E-mail: abnsec@intnet.ne
http://www.abn.ne/

Pan-African Association for Port Cooperation (PAPC)
12 Park Lane
PO Box 1113

Apapa Lagos, Nigeria
Tel/Fax: 234 1 587 4108
E-mail: papc@afriports.org
http://www.afriports.org/

Port Management Association of Eastern and Southern Africa (PMAESA/AGPAEA)
PO Box 99209
Mombassa, Kenya
Contact: Secretary General
Tel: 254 11 223245
Fax: 254 11 228344
E-mail: pmaesa@africaonline.co.ke
http://www.pmaesa-agpaea.org/

Port Management Association of West and Central Africa (PMAWCA/AGPAOC)
Box 1113
Apapa, Lagos, Nigeria
Contact: Secretary General
Tel: 234 1 587 4108
Fax: 234 1 587 1278
E-mail: secgen@pmawac-agpaoc.org
http://www.pmawca-agpaoc.com/

Regional Activity Centre for Specially Protected Areas (RAC/SPA)
Boulevard de l'Environnement
PB 337 Cedex
1080 Tunis, Tunisia
Contact: Director
Tel: 216 1 795760
Fax: 216 1 797349
E-mail: car-asp@rac-spa.org.tn
http://www.rac-spa.org.tn/

Regional Co-ordinating Unit for Eastern African Action Region/Seychelles
UNEP Secretariat of the Nairobi Convention & Related Action Plan
PO Box 487, Str. Anne Island
Mahe, Seychelles
Tel: 248 32 52 45
Fax: 248 32 45 73
E-mail: uneprcu@seychelles.net
http://www.unep.org/eaf/

Regional Fisheries Committee for the Gulf of Guinea (COREP)
BP 161
Libreville, Gabon
Contact: Secretariat
Fax: 241 73 7149
http://www.fao.org/fi/body/rfb/COREP/corep_home.html

Secretariat for Eastern African Coastal Area Management (SEACAM)
874, Av. Amílcar Cabral, 1st Floor
Caixa Postal 4220
Maputo, Mozambique
Contact: Secretariat
Tel: 258 1 300641/2
Fax: 258 1 300638
E-mail: seacam@virconn.com
http://www.seacam.mz/

Southeast Atlantic Fisheries Organisation (SEAFO)
Ministry of Fisheries and Marine Resources
Private Bag 13355
Windhoek, Namibia
Contact: Interim Secretariat
Tel: 264 6220 53121
Fax: 264 6120 53041
E-mail: pmwatile@mfmr.gov.na
http://www.mfmr.gov.na/seafo/seafo.htm

Southern African Transport and Communications Commission (SATCC)
Avenida Martires de Inhaminga 170, 2nd floor
SATCC-TU CP 2677
Maputo, Mozambique
Contact: Director
Tel: 258 1 420214
Fax: 258 1 420213
E-mail: director@satcc.org
http://www.satcc.org/

UNEP Regional Office for Africa
PO Box 30552
Nairobi, Kenya
Contact: Director
Tel: 254 2 624279
Fax: 254 2 623928
E-mail: sekou.touve@unep.org
http://www.unep.org/roa/

West and Central African Action
   Plan Regional Coordinating
   Unit (WACAF/RCU)
Abidjan Convention
c/o Ministry of Environment, Water
   and Forest
BP 153
Abidjan 20, Côte d'Ivoire
Contact: Coordinator
Tel: 225 2021 0323
Fax: 225 2021 0495
E-mail: kaba@cro.orstom.ci
http://www.unep.org/water/
   regseas/wacaf.htm

4. REGIONAL
INTERGOVERNMENTAL
ORGANIZATIONS

4.2 The Americas

Caribbean Community (CARICOM)
PO Box 10827
Georgetown, Guyana
Contact: Secretariat
Tel: 592 226 9280
Fax: 592 226 7816
E-mail: webmaster@caricom.org
http://www.caricom.org/

CARICOM Fisheries
PO Box 642
Princess Marquaret Dr.
Belize City, Belize
Tel: 501 234 444
Fax: 501 234 446
E-mail: cframp@btl.net
http://www.caricom-fisheries.com/

Caribbean Environment Programme
   Regional Coordinating Unit
   (CAR/RCU)
(Cartagena Convention)
14-20 Port Royal Street
Kingston, Jamaica
Contact: Coordinator
Tel: 1 876 922-9267
Fax: 1 876 922-9292
E-mail: uneprcuja@cwjamaica.com
http://www.cep.unep.org/

Central American Commission on
   Maritime Transport
(Comision Centroamericana de
   Transporte Maritimo–COCA
   TRAM)
Contiguo a Mansíon de Teodolinda
Apartado Postal 2423
Managua, Nicaragua
Contact: Executive Director
Tel: 505 222 3560
Fax: 505 222 2759
E-mail: cocatram@cocatram.org.ni
http://www.cocatram.org.ni/

Centro del Agua del Trópico
   Húmedo para América Latina y
   el Caribe (CATHALAC)
(Water Center for the Humid Trop-
   ics of Latin America and the Ca-
   ribbean)
PO Box 873372
Panamá, Republic of Panama
Contact: Director
Tel: 507 317 0125
Fax: 507 317 0127
E-mail: cathalac@cathalac.org
http://www.cathalac.org/

Comisión de Pesca Continental para
   América Latina (COPESCAL)
c/o FAO Regional Office for Latin
   America and the Caribbean
Avenida Dag Hammarskjold 3241
Vitacura
Casilla 10095
Santiago, Chile

Tel: 56 2 337 2100
Fax: 56 2 337 2101
E-mail: francisco.pereira@fao.org
http://www.rlc.fao.org/organos/
copescal/copescal.htm

Comisión Permanente del Pacífico
   Sur (CPPS)
Edificio Inmaral, ler piso
Av. Carlos Julio Arosemena, Km. 3
Guayaquil, Ecuador
Contact: Secretariat
Tel: 593 4 2221 202
Fax: 593 4 2221 201
E-mail: cpps@ecuanex.net.ec
http://www.cpps-int.org/

Comisión Técnica Mixta del Frente
   Marítimo (COFREMAR)
1355 Juncal, oficina 604
11000 Montevideo, Uruguay
Contact: President
Tel: 598 2 916 1973
Fax: 598 2 916 1578
http://www.cofremar.org/

Commission for the Conservation
   and Management of Highly Migratory
   Fish Stocks in the Western
   and Central Pacific
Contact: Interim Secretariat
E-mail: interimsec@isa.org.jm
http://www.ocean-affairs.com/

FAO Regional Office for Latin
   America and the Caribbean
Avenida Dag Hammarskjold 3241
Vitacura
Casilla 10095
Santiago, Chile
Tel: 56 2 337 2100
Fax: 56 2 337 2101
E-mail: german.rojas@fao.org
http://www.fao.org/Regional/
   LAmerica/

FAO Subregional Office for the Caribbean
   (SLAC)

PO Box 631-C
Bridgetown, Barbados
Tel: 854 426 7110
Fax: 854 427 6075
E-mail: fao-slac@fao.org

Gulf of Maine Council on the Marine
   Environment
c/o New Hampshire Department of
   Environmental Services
PO Box 95
Concord, NH, 03302-0095, USA
Contact: Council Coordinator
Tel: 1 603 225-5544
Fax: 1 603 225-5582
E-mail: info@gulfofmaine.org
http://gulfofmaine.org/

INFOPESCA (Centro para los servicios
   de información y asesoramiento
   sobre la comercialización
   de los productos pesqueros
   en América Latina y el Caribe)
Julio Herrera y Obes 1296
Casilla de Correo 7086
CP 11200
Montevideo, Uruguay
Tel: 598 2 9028701
Fax: 598 2 9030501
E-mail: infopesc@infopesca.org
http://www.infopesca.org/

Inter-American Institute for Global
   Change Research (IAI)
Av. dos Astronautas, 1758
CEP 12227-010 São José dos Campos
São Paulo, Brazil
Contact: IAI Directorate–Brazil
Tel: 55 12 3945 6855
Fax: 55 12 3941 4410
E-mail: iaibr@dir.iai.int
http:// www.iai.int/

Inter-American Tropical Tuna Commission
   (IATTC)
8604 La Jolla Shores Drive
La Jolla, CA, 92037-1508, USA
Contact: Secretariat

Tel: 1 858 546-7100
Fax: 1 858 546-7133
E-mail: info@iattc.org
http://www.iattc.org/

International Coordination Group for the Tsunami Warning System in the Pacific (ICG/ITSU) International Tsunami Information Centre (ITIC)
737 Bishop Street, Suite 2200
Honolulu, HI, 96813-3213, USA
Contact: Chair
Tel: 1 808 532-6423
Fax: 1 808 532-5576
E-mail: laura.kong@moana.itic.noaa.gov
http://www.shoa.cl/oceano/itic/itsu.html

International Pacific Halibut Commission (IPHC)
PO Box 95009
Seattle, WA, 98145-2009, USA
Contact: Director
Tel: 1 206 634-1838
Fax: 1 206 632-2983
E-mail: info@iphc.washington.edu
http://www.iphc.washington.edu/

IOC Sub-Commission for the Caribbean and Adjacent Regions (IOCARIBE)
Casa del Marqués de Valdehoyos
Centro, Calle de la Factoría No. 36-57
Apartado Aéreo 1108
Cartagena de Indias, Colombia
Contact: Senior Assistant
Tel: 57 5 664 6399
Fax: 57 5 660 0407
E-mail: iocaribe@col3.telecom.com.co; iocaribe@cartagena.cetcol.net.co
http://ioc.unesco.org/iocaribe/

North American Commission for Environmental Cooperation (CEC)
393 rue St-Jacques Ouest, Suite 200
Montréal, QC, H2Y 1N9, Canada
Contact: Executive Director
Tel: 1 514 350-4300
Fax: 1 514 350-4314
E-mail: info@ccemtl.org
http://www.cec.org/

North Atlantic Treaty Organization (NATO) Maritime Headquarters
RHQ EASTLANT/HQ NAVNORTH
Northwood Headquarters
Northwood
Middlesex, HA6 3HP, United Kingdom
Tel: 44 1923 843746
Fax: 44 1923 843762
E-mail: pio@eastlant.nato.int
http://www.eastlant.nato.int/

North East Pacific Action Plan
Regional Seas Unit & Coral Reef Unit
Division of Environmental Conventions
United Nations Environment Programme
PO Box 30552
Nairobi, Kenya
Contact: Mr. Jorge E. Illueca, Assistant Executive Director UNEP and Director of DEC
Tel: 254 2 624 052
Fax: 254 2 624 618
E-mail: Jorge.Illueca@unep.org
http://www.unep.ch/seas/rshome.html

North Pacific Anadromous Fish Commission (NPAFC)
889 West Pender Street, Suite 502
Vancouver, BC, V6C 3B2, Canada
Contact: Executive Director
Tel: 1 604 775-5550
Fax: 1 604 775-5577
E-mail: secretariat@npafc.org
http://www.npafc.org/

North Pacific Marine Science Organization (PICES)
c/o Institute of Ocean Sciences
PO Box 6000
Sidney, BC, V6L 4B2, Canada
Contact: PICES Secretariat
Tel: 1 250 363-6364
Fax: 1 250 363-6827
E-mail: bychkov@pices.int
http://www.pices.int/

Northwest Atlantic Fisheries Organization (NAFO)
2 Morris Drive
PO Box 638
Dartmouth, NS, B2Y 3Y9, Canada
Contact: Executive Secretary
Tel: 1 902 468-5590
Fax: 1 902 468-5538
E-mail: info@nafo.ca
http://www.nafo.ca/

Organizacion Latino Americana de Desarrollo Pesquero (OLDPESCA)
(Latin American Organization for Fisheries Development)
Las Palomas 422
URB, Limatambo
Apartado 10168
Lima 34, Peru
Contact: Executive Director
Tel: 51 14 427655
Fax: 51 14 429925
E-mail: oldepesc@bellnet.com.pe
http:// fish.com/oldepesca/

Pacific Salmon Commission (PSC)
1155 Robson St., Suite 600
Vancouver, BC, V6E 1B5, Canada
Contact: Executive Secretary
Tel: 1 604 684-8081
Fax: 1 604 666-8707
E-mail: kowal@psc.org
http://www.psc.org/

Programa Hidrológico Internacional (PHI)
Dr. Luis Piera 1992, 2o.piso
Avda. Brasil 2697
CP 11200
Montevideo, Uruguay
Tel: 598 2 413 2075
Fax: 598 2 413 2094
E-mail: phi@unesco.org.uy
http://www.unesco.org.uy/phi/

South East Pacific Action Plan
Comisión Permanente del Pacífico Sur (CPPS)
Edificio Inmaral, lerpsiso
Av. Carlos Julio Arosemena, Km. 3
Guayaquil, Ecuador
Contact: Secretariat
Tel: 593 4 2221 202
Fax: 593 4 2221 201
E-mail: cpps@ecuanex.net.ec
http://www.plandeaccion.cpps-int.org/

South West Atlantic Action Plan
Regional Seas Unit & Coral Reef Unit
Division of Environmental Conventions
United Nations Environment Programme
PO Box 30552
Nairobi, Kenya
Contact: Mr. Jorge E. Illueca, Assistant Executive Director UNEP and Director of DEC
Tel: 254 2 624 052
Fax: 254 2 624 618
E-mail: Jorge.Illueca@unep.org
http://www.unep.ch/seas/rshome.html

UNEP Regional Office for North America (RONA)
1707 H Street, NW, Suite 300
Washington, DC, 20006, USA
Tel: 1 202 785-0465
Fax: 1 202 785-2096
E-mail: info@runa.unep.org
htttp://www.rona.unep.org/

UNEP Regional Office for Latin America and Carribean
Boulevard de los Virreyes No. 155
Lomas de Virreyes
CP 11000
México, D.F. Mexico
Contact: Regional Representative
Tel: 52 5 202 6394
Fax: 52 5 202 0950
E-mail: unepnet@rolac.unep.mx
http://www.rolac.unep.mx/

United States Coast Guard International Ice Patrol
1082 Shennecossett Road
Groton, CT, 06340-6095, USA
Contact: Commander
Tel: 1 860 441-2626
Fax: 1 860 441-2773
E-mail: iipcomms@rdc.uscg.mil
http://www.uscg.mil/lantarea/iip/home.html

Western Central Atlantic Fisheries Commission (WECAFC)
FAO Sub-Regional Office for Latin American and the Caribbean
PO Box 631C
Bridgetown, Barbados
Tel: 1 246 426-7110
Fax: 1 246 426-7111
E-mail: bisessar.chakalall@fao.org
http://www.fao.org/fi/body/rfb/wecafc/wecafc home.html

4. REGIONAL INTERGOVERNMENTAL ORGANIZATIONS

4.3 Asia

Arab Federation of Shipping
PO Box 1161
Baghdad, Iraq
Tel: 964 1 719 3947
Fax: 964 1 719 7243

Arab Fisheries Company
PO Box 17604
Jeddah 21494, Saudi Arabia
Contact: Chairman and Director-General
Tel: 966 2 651 9140
Fax: 966 2 651 0528

ASEAN Council on Petroleum (ASCOPE)
Petronas International Business Ventures
Level 45, Tower 1, Petronas Twin Towers
50088 Kuala Lumpur, Malaysia
Contact: Secretariat
Tel: 60 3 5814804
Fax: 60 3 584203
E-mail: ascopesec@petronas.com.my
http://www.ascope.com.my/

ASEAN Fisheries Federation (AFF)
Association of Southeast Asian Nations (ASEAN)
ASEAN Secretariat Building
Jalan Sisingamangaraja 70A
Jakarta 12110, Indonesia
Tel: 62 21 726 2991
Fax: 62 21 725 7916
E-mail: aff@asean.or.id
http://www.aff.or.id/

ASEAN Ports Association
c/o Philippine Ports Authority
Marsman Building South Harbour
Port Area, Manila, The Philippines
Contact: Permanent Secretariat
Tel: 632 527 4746
Fax: 632 527 4749
E-mail: aida@ppa.gov.ph
http://www.psa.com.sg/apa/

Asia-Pacific Economic Cooperation (APEC)
APEC Secretariat
438 Alexandra Road
#14-00 Alexandra Point
Singapore 119958

Tel: 65 6276 1880
Fax: 65 6276 1775
E-mail: tth@mail.apecsec.org.sg
http://www.apecsec.org.sg/

Asia-Pacific Fisheries Commission (APFIC)
FAO Regional Office for Asia and the Pacific
39 Phra Atit Road
Bangkok 10200, Thailand
Contact: Secretariat
Tel: 66 2 280 7844
Fax: 66 2 280 0445
E-mail: veravat.hongskul@fao.org
http://www.fao.org/fi/body/rfb/apfic/apfic_home.html

Asian-African Legal Consultative Organization (AALCO)
E-66 Vasant Marg
Vasant Vihar
New Delhi, 110057, India
Tel: 91 11 6152 251
Fax: 91 11 6152 041
E-mail: aalcc@vsnl.com
http://www.aalco.org/

Coordinating Committee for Coastal and Offshore Geoscience Programmes in East and Southeast Asia (CCOP)
Thai CC Tower, 24th Floor, Room 244-5
889 Sathorn Road
Satho Tai
Bangkok 10120, Thailand
Contact: Director
Tel: 66 2 672 3080-1
Fax: 66 2 672 3082
E-mail: ccopts@ccop.or.th
http://www.ccop.or.th/

East Asia Seas Regional Co-ordinating Unit (EAS/RCU)
United Nations Building
Rajdamnern Nok Avenue
Bangkok 10200, Thailand

Tel: 66 2 288 1870-4
Fax: 66 2 280 3829
E-mail: andrewsni@un.org
http://www.roap.unep.org/easrcu/index.htm/

FAO Regional Office for Asian and the Pacific (RAP)
Maliwan Mansion
39 Phra Atit Road
Bangkok 10200, Thailand
Tel: 66 2 697 4000
Fax: 66 2 697 4445
E-mail: fao-rap@fao.org
http://www.fao.or.th/

Fisheries Working Group
Asia Pacific Economic Cooperation (APEC)
35 Heng Mui Keng Terrace
Singapore 119616
Contact: APEC Secretariat
Tel: 65 6 276 1880
Fax: 65 6 276 1775
E-mail: tth@mail.apecsec.org.sg
http://www.apecsec.org.sg/workgroup/fish.html

Indian Ocean Tuna Commission (IOTC)
PO Box 1011
Victoria, Mahé, Seychelles
Contact: Secretariat
Tel: 248 225494
Fax: 248 224364
E-mail: iotcsecr@iotc.org
http://www.iotc.org/

INFOFISH (Intergovernmental Organization for Marketing Information and Technical Advisory Services for Fishery Products in the Asia and Pacific Region)
1st. Floor, Wisma PKNS
Jalan Raja Laut
PO Box 10899
50728 Kuala Lumpur, Malaysia
Tel: 60 3 291 4466

Fax: 60 3 291 6804
E-mail: infish@po.jaring.my
http://www.infofish.org/

IOC Sub-Commission for the Western Pacific (WESTPAC)
c/o National Research Council Thailand
196 Phaholyothin Road, Chatujak
Bangkok 10900, Thailand
Tel: 66 2 561 5118
Fax: 66 2 561 5119
E-mail: westpac@samart.co.th
http://ioc.unesco.org/westpac/index.html

Marine Mutual Aid Centre (MEMAC)
PO Box 10112
Manama, Bahrain
Contact: Director
Tel: 973 274554
Fax: 973 277551
E-mail: memac@batelco.com.bh
http://www.computec.com.bh/memac/

Marine Resource Conservation Working Group
Asia Pacific Economic Cooperation (APEC)
35 Heng Mui Keng Terrace
Singapore 119616
Contact: APEC Secretariat
Tel: 65 276 1880
Fax: 65 276 1775
E-mail: tth@mail.apecsec.org.sg
http://www.apecsec.org.sg/workgroup/marine.html

Mekong River Commission
364 M.V. Preah Monivong Blvd
PO Box 1112
Phnom Penh, Cambodia
Contact: Secretariat
Tel: 855 23 720 979
Fax: 855 23 720 972

E-mail: mrcs@mrcmekong.org
http://www.mrcmekong.org/

Network of Aquaculture Centres in Asia-Pacific (NACA)
Kasetsart Post Office Box 1040
Bangkok 10903, Thailand
Contact: Secretariat
Tel: 66 2 561 1728
Fax: 66 2 561 1727
E-mail: naca@enaca.org
http://www.enaca.org/

North West Pacific Action Plan (NOWPAP)
Special Monitoring and Coastal Environmental Assessment
Regional Activity Centre NOWPAP
Toyama City, Japan
Contact: Director
Tel: 81 76 445 1571
Fax: 81 76 445 1581
E-mail: oritani@npec.or.jp

NOWPAP Marine Environmental Emergency and Preparedness Response Regional Activity Centre (MER/RAC)
PO Box 23
Yusung, Taejon 305-600, Republic of Korea
Contact: Director
Tel: 82 42 868 7738
Fax: 82 42 868 7281
E-mail: nowpap@kriso.re.kr
http://merrac.nowpap.org/

Partnership in Environmental Management for the Seas of East Asia (PEMSEA)
PO Box 2502
Quezon City 1165
Metro-Manila, The Philippines
Tel: 632 920 2211
Fax: 632 926 9712
E-mail: info@pemsea.org
http://www.pemsea.org/

Regional Organisation for the Protection of the Marine Environment (ROPME)
c/o MEMAC
PO Box 10112
Manama, Bahrain
Contact: Coordinator
Tel: 973 274 554
Fax: 973 274 551
E-mail: memac@batelco.com.bh
http://www.computec.com.bh/memac/

Regional Organization for the Conservation of the Environment of the Red Sea and Gulf of Aden (PERGSGA)
PO Box 53662
Jeddah, 21583, Saudi Arabia
Contact: Deputy Secretary-General
Tel: 966 2 657 3324
Fax: 966 2 652 1901
E-mail: information@persga.org
http://www.unep.ch/seas/main/PERSGA/index.html

South Asia Co-operative Environment Programme (SACEP)
No. 10 Anderson Road
Colombo 5, Sri Lanka
Contact: Director
Tel: 941 589787
Fax: 941 589369
E-mail: info@sacep.org
http://www.sacep.org/

Southeast Asian Fisheries Development Center (SEAFDEC)
Suraswadi Building
PO Box 1046, Kasetsart Post Office
Bangkok 10903, Thailand
Tel: 66 2 940 6326
Fax: 66 2 940 6336
Contact: Secretariat
E-mail: secretariat@seafdec.org
http://www.seafdec.org/

Transportation Working Group
Asia Pacific Economic Commission (APEC)
35 Heng Mui Keng Terrace
Singapore 119616
Contact: APEC Secretariat
Tel: 65 276 1880
Fax: 65 276 1775
E-mail: sbs@mail.apecsec.org.sg
http://www.apecsec.org.sg/workgroup/transportation.html

UNEP Regional Office for West Asia (UNEP/ROWA)
PO Box 10880
Manama, Bahrain
Contact: Regional Director
Tel: 973 826600
Fax: 973 82510
E-mail: myunrowa@unep.org.bh
http://www.unep.org.bh/

UNEP Regional Office for Asia and the Pacific (UNEP/ROAP)
United Nations Bldg.
Rajadamnern Ave.
Bangkok 10200, Thailand
Contact: Regional Director
Tel: 66 2 288 1870-4
Fax: 66 2 280 3829
E-mail: andrewsni@un.org
http://www.roap.unep.org/

United Nations Economic and Social Commission for Asia and the Pacific (ESCAP)
United Nations Building
Rajadamnern Nok Avenue
Bangkok 10200, Thailand
Tel: 66 2 288 1234
Fax: 66 2 288 1000
E-mail: webmaster@unescap.org
http://www.unescap.org/

Western Indian Ocean Tuna Organization (WIOTC)
c/o Seychelles Fishing Authority
PO Box 449

Victoria, Mahé, Seychelles
Tel: 248 224 597
Fax: 248 224508
E-mail: sfasez@seychelles.net
http://www.fao.org/fi/body/rfb/wiotc/wiotc_home.htm

4. REGIONAL INTERGOVERNMENTAL ORGANIZATIONS

4.4 Europe

Arc Manche
c/o West Sussex County Council
County Hall
Chichester, PO19 1RQ, United Kingdom
Contact: English Group Coordinator
Tel: 44 1243 777 927
Fax: 44 1243 777 697
E-mail: rachel.gapp@westsussex.gov.uk
http://www.arcmanche.com/

Baltic 21
Strömborg
PO Box 2010
SE-103 11 Stockholm, Sweden
Contact: Secretariat
Tel: 46 84 40 19 38/41
Fax: 46 84 40 19 44
E-mail: marek.maciejowski@cbss.st
http://www.ee/baltic21/

Baltic Environment Forum (BEF)
26/28 Peldu Street, Room 505
LV-1050, Riga, Latvia
Tel: 371 7 214 477
Fax: 371 7 214 448
E-mail: bef@latnet.lv
http://www.bef.lv/index.htm

Baltic Marine Biologists
c/o Coastal Research and Planning Institute
Klaipeda University
H. Manto gat. 84
LT-5808 Klaipeda, Lithuania
Contact: President
Tel: 370 6 398 847
Fax: 370 6 398 802
E-mail: s.olenin@samc.ku.lt
http://www.smf.su.se/bmb/

Black Sea Environmental Programme
Black Sea Commission
Domabahce Sarayi
II. Harekat Kosku
80680 Besiktas
Istanbul, Turkey
Contact: Coordinator
Tel: 90 212 259 6716
Fax: 90 212 227 9933
E-mail: semaacar@blacksea-environment.org
http://www.blacksea-environment.org/

Blue Plan Regional Activity Centre (BP/RAC)
15, rue Beethoven
Sophia Antipolis
F-06560 Valbonne, France
Contact: Director
Tel: 33 4 9238 7130
Fax: 33 4 9238 7131
E-mail: planbleu@planbleu.org
http://www.planbleu.org/

Bonn Agreement
New Court
48 Carey Street
London, WC2A 2JQ, United Kingdom
Contact: Secretary
Tel: 44 20 7430 5200
Fax: 44 20 7430 5225
E-mail: secretariat@bonnagreement.org
http://www.bonnagreement.org/

Central Commission for the Navigation of the Rhine (CCNR)

2, place de la République
Palais du Rhin
F-67082 Strasbourg Cedex, France
Contact: President
Tel: 33 3 8852 2010
Fax: 33 3 8832 1072
E-mail: ccnr@ccr-zkr.org
http://www.ccr-zkr.org/

Centre for Environment and Development for the Arab Region and Europe (CEDARE)
2 El Hegaze Street, CEDARE Building,
Heliopolis
PO Box 1057
Heliopolis Bahary
Cairo, Egypt
Tel: 20 2 4513921
Fax: 20 2 4513918
E-mail: email@cedare.org.eg
http://www.cedare.org.eg/

Cephalopod International Advisory Council (CIAC)
c/o FRS Marine Laboratory Aberdeen PO Box 101, Victoria Road
Aberdeen, Scotland, AB11 9DB, United Kingdom Contact: Executive Secretary
Tel: 44 12 2429 5507
Fax: 44 12 2429 5511
E-mail: hatfielde@marlab.ac.uk
http://www.nbs.ac.uk/public/mlsd/ciac/index.html

Committee on the Challenges of Modern Society (CCMS)
North Atlantic Treaty Organization (NATO)
Scientific Affairs Division
B-1110 Brussels, Belgium
Contact: CCMS Secretariat
Tel: 32 2 707 4846
Fax: 32 2 707 4232
http://www.nato.int/ccms/index.html

Commonwealth Secretariat
Marlborough House
Pall Mall
London, SW1Y 5HX, United Kingdom
Tel: 44 20 7747 6500
Fax: 44 20 7930 0827
E-mail: info@commonwealth.int
http://www.thecommonwealth.org/

Conference of Peripheral Maritime Regions (CPMR)
6 Rue Saint-Martin
F-35700 Rennes, France
Contact: Secretary General
Tel: 33 2 9935 4050
Fax: 33 2 9935 0919
E-mail: secretariat@crpm.org
http://www.crpm.org/

Council of Europe
F-67075 Strasbourg Cedex, France
Tel: 33 3 8841 2033
Fax: 33 3 8841 2745
E-mail: infopoint@coe.fr
http://www.coe.fr/

Council of the Baltic Sea States (CBSS)
PO Box 2010
SE-103 11 Stockholm, Sweden
Contact: Secretariat
Tel: 46 84 40 19 20
Fax: 46 84 40 19 44
E-mail: cbss@cbss.st
http://www.baltinfo.org/

Danube Commission (Donaukommission)
Benczúr utca 25
H-1068 Budapest, Hungary
Contact: General Director
Tel: 36 1 461 8023
Fax: 36 1 352 1839
E-mail: e.schulze-r@danubecom-intern.org
http://www.danubecom-intern.org/

Emergency Response Activity Centre (ERAC)
1, Slavejkov Sq., Port Complex, IVT
Varna 9000, Bulgaria
Contact: Coordinator
Tel: 359 52 221407
Fax: 359 52 602594
E-mail: riseco@mbox.digsys.bg

Environment Remote Sensing Regional Activity Centre (ERS/RAC)
2 Via G. Giusti
I-90144 Palermo, Italy
Contact: Managing Director
Tel: 39 091 342368
Fax: 39 091 308512
E-mail: ctmrac@tin.it; ctm.ersrac@ctmnet.it
http://www.ctmnet.it/

EUROFISH (ex-EASTFISH)
UN Centre
Midtermolen 3
DK-2100 Copenhagen Ø, Denmark
Tel: 45 35 46 71 87
Fax: 45 35 46 71 81
E-mail: info@eurofish.dk
http://www.eastfish.org/eurofish/

European Commission
200, rue de la Loi
B-1049 Brussels, Belgium
Tel: 32 2 2999 1111
E-mail: europawebmaster@cec.eu.int
http://europa.eu.int/comm/index_en.htm

European Conference of Ministers of Transport (ECMT)
2, rue André Pascal
F-75775 Paris Cedex 16, France
Tel: 33 1 4524 9710
Fax: 33 1 4524 9742
E-mail: ecmt.contact@oecd.org
http://www.oecd.org/cem/index.htm

European Environment Agency (EEA)
Kongens Nytorv 6
DK-1050 Copenhagen K, Denmark
Tel: 45 33 36 71 00
Fax: 45 33 36 71 99
E-mail: eea@eea.eu.int
http://www.eea.eu.int/

European Group on Ocean Stations (EGOS)
Christian Michelsen Research A/S
PO Box 6031 Postterminalen
N-5892 Bergen, Norway
Contact: Technical Secretariat
Tel: 47 55 57 42 64
Fax: 47 55 57 40 41
E-mail: anne.hageberg@cmr.no
http://www.meteo.shom.fr/egos/

European Inland Fisheries Commission (EIFAC)
Viale delle Terme di Caracalla
I-00100 Rome, Italy
Contact: Chair
Tel: 39 06 5705 2944
Fax: 39 06 5705 3020
E-mail: rudolf.mueller@eawag.ch
http://www.fao.org/fi/body/eifac/eifac.asp

FAO Regional Office for Europe
Viale delle Terme di Caracalla
I-00100 Rome, Italy
Tel: 39 06 5705 4963
Fax: 39 06 5705 5634
http://www.fao.org/regional/Europe/

FAO Subregional Office for Central and Eastern Europe (SEUR)
Benczur utca 34
H-1068 Budapest, Hungary
Tel: 361 461 2000
Fax: 361 351 7029
E-mail: fao-seur@fao.org
http://www.fao.org/regional/seur/

General Fisheries Commission for
the Mediterranean (GFCM)
FAO, Room F-211
Viale delle Terme di Caracalla
I-00100 Rome, Italy
Contact: Secretary
Tel: 39 6 5705 6435
Fax: 39 6 5705 6500
E-mail: alain.bonzon@fao.org
http://www.fao.org/fi/body/rfb/
gfcm/gfcm home.htm

Helsinki Commission (HELCOM)
Baltic Marine Environment Protection Commission
Katajanokanlaituri 6 B
FIN-00160 Helsinki, Finland
Contact: Executive Secretary
Tel: 358 9 6220 220
Fax: 358 9 6220 2239
E-mail: helcom@helcom.fi
http://www.helcom.fi/

Institute for Environment and Sustainability
Joint Research Centre of the European Commission
José Jimenez
Via Fermi
PO Box TP 290
I-21020 Ispra (VA), Italy
Contact: Inland and Marine Waters Unit
Tel: 39 0332 789834
Fax: 39 0332 789222
E-mail: ies-contact@jrc.it
http://ies.jrc.cec.eu.int/

International Baltic Sea Fishery Commission (IBSFC)
Hozastrasse 20
PO-00-528 Warsaw, Poland
Contact: Secretariat
Tel: 48 22 628 86 47
Fax: 48 22 625 33 72
E-mail: ibsfc@polbox.pl
http://www.ibsfc.org/

International Commission for Scientific Exploration of the Mediterranean Sea (ICSEM)
16 boulevard de Suisse
MC-98000 Monte Carlo, Monaco
Contact: Director General
Tel: 377 93 30 38 79
Fax: 377 92 16 11 95
E-mail: comments@ciesm.org
http://www.ciesm.org/

International Commission for the Protection of the Rhine (ICPR)
Postfach 200253
D-56002 Koblenz, Germany
Tel: 49 261 12495
Fax: 49 261 36572
E-mail: sekretariat@iksr.de
http://iksr.firmen-netz.de/icpr/

International Commission for the Protection of the Danube River (ICPDR)
Vienna International Center
D0412 PO Box 500
A-1400 Vienna, Austria
Contact: Permanent Secretariat
Tel: 431 260 60 5738
Fax: 431 260 60 5895
E-mail: icpdr@unvienna.org
http://www.icpdr.org/

International Commission for the Conservation of Atlantic Tunas (ICCAT)
Calle Corazón de María, 8, 6th Floor
E-28002 Madrid, Spain
Contact: Executive Secretary
Tel: 34 91 416 5600
Fax: 34 91 415 2612
E-mail: info@iccat.es
http://www.iccat.es/

International Council for the Exploration of the Sea (ICES)
Palaegade 2-4
DK-1261 Copenhagen K, Denmark
Contact: General Secretary

Tel: 45 33 15 42 25
Fax: 45 33 93 42 15
E-mail: info@ices.dk
http://www.ices.dk/

Interreg North Sea Region Programme Secretariat
Transnational Cooperation on Spatial Planning
c/o Viborg Amt
Skottenborg 26
DK-8800 Viborg, Denmark
Tel: 45 87 27 19 99
Fax: 45 86 62 68 62
E-mail: crbbj@vibamt.dk
http://www.interregnorthsea.org/

Joint Research Centre of the European Commission (JRC)
SDME 10/78
B-1049 Brussels, Belgium
Tel: 32 2 295 7624
Fax: 32 2 295 6322
E-mail: jrc-info@cec.eu.int
http://www.jrc.org/

Local Authorities International Environmental Organisation
Kommunenes Internasjonale Miljøorganisasjon (KIMO)
Shetland Islands Council
Information Services
Grantfield
Lerwick, Shetland, ZE1 0NT, United Kingdom
Contact: Secretariat
Tel: 44 01 5957 44800
Fax: 44 01 5956 95887
E-mail: kimo@zetnet.co.uk
http://www.zetnet.co.uk/coms/kimo/

Marine Environment Laboratory
International Atomic Energy Agency
4, Quai Antoine 1er
BP 800
MC-98012 Monaco
Contact: Director

Tel: 377 9797 7272
Fax: 377 9797 7273
E-mail: mel@iaea.org
http://www.iaea.org/monaco/

Mediterranean Action Plan Regional Coordinating Unit (MEDU)
UNEP/MEDU
48 Vassileos Konstantiou Avenue
PO Box 18019
GR-116 35 Athens, Greece
Tel: 30 1 727 3100
Fax: 30 1 725 3196
E-mail: unepmedu@unepmap.gr
http://www.unepmap.org/

NATO SACLANT Undersea Research Centre
Viale San Bartolomeo 400
I-19138 La Spezia, Italy
Tel: 39 0187 527 (1 or extension)
Fax: 39 0187 527 700
E-mail: registry@saclantc.nato.int
http://www.saclantc.nato.int/

Nordic Council of Ministers
Store Strandstraede 18
DK-1255 Copenhagen K, Denmark
Contact: Senior Advisor for Fisheries Affairs
Tel: 45 33 96 02 55
Fax: 45 33 93 20 47
E-mail: jh@nmr.dk
http://www.norden.org/fisk/

North Atlantic Marine Mammal Commission (NAMMCO)
Polar Environmental Centre
N-99296 Tromsø, Norway
Contact: General Secretary
Tel: 47 77 75 01 78
Fax: 47 77 75 01 81
E-mail: nammco-sec@nammco.no
http://www.nammco.no/

North Atlantic Salmon Conservation Organisation (NASCO)
11 Rutland Square

Edinburgh, Scotland, EH1 2AS,
   United Kingdom
Tel: 44 13 1228 2551
Fax: 44 13 1228 4384
E-mail: hq@nasco.int
http://www.nasco.int/

North Atlantic Treaty Organization
   (NATO)
Scientific and Environmental Affairs
   Division
B-1110 Brussels, Belgium
Tel: 32 2 707 4111
Fax: 32 2 707 4232
E-mail: science@hq.nato.int
http://www.nato.int/science/

North-East Atlantic Fisheries Commission (NEAFC)
22 Berners Street
London, W1T 3DY, United
   Kingdom
Tel: 44 20 7631 0016
Fax: 44 20 7636 9225
E-mail: info@neafc.org
http://www.neafc.org/

Organisation for Economic Cooperation and Development (OECD)
2, rue Andre Pascal
F-75775 Paris Cedex 16, France
Tel: 33 1 4524 8200
Fax: 33 1 4524 8500
E-mail: webmaster@oecd.org
http://www.oecd.org/

OSPAR Commission for the Protection
   of the Marine Environment of the
   North-East Atlantic (OSPAR)
New Court
48 Carey Street
London, WC2A 2JQ, United
   Kingdom
Contact: Secretariat
Tel: 44 20 7430 5200
Fax: 44 20 7430 5225
E-mail: secretariat@ospar.org
http://www.ospar.org/

Priority Actions Programme Regional
   Activity Centre (PAP/RCU)
Kraj sv. Ivana 11
HR-21000 Split, Croatia
Contact: Director
Tel: 385 21 343 499
Fax: 385 21 361 677
E-mail: pap@gradst.hr
http://www.papthecoastcentre.
   org/

Regional Activity Centre for Specially
   Protected Areas (RAC/SPA)
Boulevard de l'Environnement
PB 337
Cedex 1080 Tunis, Tunisia
Contact: Director
Tel: 216 1 795 760
Fax: 216 1 797 349
E-mail: car-asp@rac-spa.org.tn
http://www.rac-spa.org.tn/

Regional Activity Centre for Cleaner
   Production (CP/RAC)
C/París, 184–3ª planta
E-08036 Barcelona, Spain
Contact: Director
Tel: 34 93 415 1112
Fax: 34 93 237 0286
E-mail: cleanpro@cema-sa.org
http://www.cipn.org/

Regional Marine Pollution Emergency Response Centre for the
   Mediterranean Sea (REMPEC)
Manoel Island
MT-Gzira GZR 03, Malta
Contact: Director
Tel: 356 21 337296
Fax: 356 21 339951
E-mail: rempec@rempec.org
http://www.rempec.org/

The North Sea Conference
Ministry of Environment
Myntgata 2
PO Box 8013 Dep
N-0030 Oslo, Norway

Contact: North Sea Secretariat
Tel: 47 22 24 58 03
Fax: 47 22 24 60 64
E-mail: postmottak@md.dep.no
http://www.dep.no/md/nsc/

UNEP Regional Office for Europe (ROE)
International Environment House
15, chemin des Anémones
CH-1219 Chatelaine, Geneva, Switzerland
Tel: 41 22 917 8279
Fax: 41 22 917 8024
E-mail: roe@unep.ch
http://www.unep.ch/roe/

Vision and Strategies Around the Baltic Sea 2010 (VASAB 2010)
Dlugi Targ Str. 24
PO-80 828 Gdansk, Poland
Tel: 48 58 301 8255
Fax: 48 58 305 4005
E-mail: infov@vasab.org.pl
http://www.vasab.org.pl/

Western European Union
15, rue de l'Association
B-1000 Brussels, Belgium
Tel: 32 2 500 4412
Fax: 32 2 500 4470
E-mail: ueo.secretariatgeneral@skynet.be
http://www.weu.int/

4. REGIONAL INTERGOVERNMENTAL ORGANIZATIONS

4.5 Australia, New Zealand and Oceania

Commission for the Conservation of Southern Bluefin Tuna (CCSBT)
PO Box 37
Deakin West, ACT 2600, Australia
Tel: 61 2 6282 8396
Fax: 61 2 6282 8407
E-mail: bmacdonald@ccsbt.org
http://www.ccsbt.org/

FAO Subregional Office for Pacific Islands (SAPA)
FAO Private Mail Bag
Apia, Samoa
Tel: 685 22127
Fax: 685 22126
E-mail: fao-sapa@fao.org
http://www.fao.or.th/

Forum Fisheries Agency (FFA)
PO Box 629
Honiara, Solomon Islands
Contact: Director
Tel: 677 21124
Fax: 677 23995
E-mail: info@ffa.int
http://www.ffa.int/

Marine Resources Pacific Consortium
c/o University of Guam Marine Laboratory
UOG Station
Mangilao, GU, 96923, USA
Contact: Secretariat
Tel: 671 735 2175/2176
Fax: 671 734 6767
http://www.uog.edu/marepac/

Pacific Islands Development Program
East-West Center
1601 East-West Road
Honolulu, HI, 96848-1601, USA
Tel: 1 808 944-7778
Fax: 1 808 944-7670
E-mail: pidp@EastWestCenter.org
http://www.eastwestcenter.org/pidp-ab.asp

Pacific Islands Marine Resources Information System (PIMRIS)
Pacific Information Centre
USP Library
PO Box 1168
Suva, Fiji

Contact: Coordinator
Tel: 679 313900
Fax: 679 300830
E-mail: mamtora_j@usp.ac.fj
http://www.usp.ac.fj/library/

Secretariat of the Pacific Community (SPC)
BP D5
98848 Noumea Cedex, New Caledonia
Contact: Secretary General
Tel: 687 262000
Fax: 687 263818
E-mail: spc@spc.int
http://www.spc.org.nc/

South Pacific Applied Geoscience Commission (SOPAC)
Private Mail Bag GPO
Suva, Fiji
Contact: Director
Tel: 679 338 1377
Fax: 679 337 0040
E-mail: director@sopac.org
http://www.sopac.org.fj/index.html

South Pacific Forum Secretariat (ForumSec)
Private Mail Bag
Suva, Fiji
Contact: Secretary General
Tel: 679 312600
Fax: 679 305573
E-mail: info@forumsec.org.fj
http://www.sidsnet.org/pacific/forumsec/

South Pacific Regional Environment Programme (SPREP)
PO Box 240
Vaitele, Apia, Western Samoa
Contact: Director
Tel: 685 21 929
Fax: 685 20 231
E-mail: sprep@sprep.org.ws
http://www.sidsnet.org/pacific/sprep/

Treaty of Waitangi Fisheries Commission (Te Ohu Kai Moana)
48 Mulgrave Street
Thorndon
PO Box 3277
Wellington, New Zealand
Tel: 64 4 499 5199
Fax: 64 4 499 5190
E-mail: glenn.inwood@tokm.co.nz
http://www.tokm.co.nz/

4. REGIONAL INTERGOVERNMENTAL ORGANIZATIONS

4.6 Polar Regions

Arctic Council
c/o Ministry for Foreign Affairs of Iceland
Raudararstigur 25
IS-150 Reykjavik, Iceland
Tel: 354 545 9900
Fax: 354 562 2373
E-mail: arctic.council@utn.stjr.is
http://www.arctic-council.org/

Arctic Monitoring and Assessment Programme (AMAP)
Stromsveien 96
PO Box 8100 Dep.
N-0032 Oslo, Norway
Contact: Secretariat
Tel: 47 23 24 16 30
Fax: 47 22 67 67 06
E-mail: amap@amap.no
http://www.amap.no/

Barents Euro-Arctic Council
c/o Ministry of Foreign Affairs
Central and Eastern Europe Department
Fredsgatan 6
SE-103 39 Stockholm, Sweden
Contact: Secretariat
Tel: 46 84 05 37 94
Fax: 46 87 23 11 76

E-mail: helena.odmark@foreign.
  ministry.se
http://www.utrikes.regeringen.se/
  inenglish/policy/barents/

Commission for the Conservation of
  Antarctic Marine Living Re-
  sources (CCAMLR)
PO Box 213
North Hobart, TAS 7002, Australia
Tel: 61 3 6231 0366
Fax: 61 3 6234 9965
E-mail: ccamlr@ccamlr.org
http://www.ccamlr.org/

International Arctic Science Commit-
  tee (IASC)
Middelthunsgate 29
PO Box 5156 Majorstua
N-0302 Oslo, Norway
Contact: Secretariat
Tel: 47 22 95 99 00
Fax: 47 22 95 99 01
E-mail: iasc@iasc.no
http://www.iasc.no/

The Northern Forum
4101 University Drive
Carr-Gottstein Center, Suite 221
Anchorage, AK, 99508 USA
Contact: Office of the Secretariat
Tel: 1 907 561-3280
Fax: 1 907 561 6645
E-mail: secretariat@northernforum.org
http://www.northernforum.org/

5. REGIONAL NONGOVERNMENTAL ORGANIZATIONS

5.1 Africa

African Association of Remote Sensing of the Environment
  (AARSE)
c/o Laboratory of Cartography and
  Remote Sensing

Geological Survey Division, ITC
PO Box 6
NL-500AA Enschede, The Nether-
  lands
Contact: President
Tel: 31 53 4874 279
Fax: 31 53 4874 336
E-mail: worldai@itc.nl
http://www.itc.nl/aarse/

Arab Institute of Navigation
El Sebaei 45th Street
Zahret El Sebaei Building
Miami, Alexandria, Egypt
Contact: President
Tel: 203 550 9824
Fax: 203 550 9686
E-mail: ain@aast.edu
http://www.aast.edu/ain/

Empowerment for African Sustain-
  able Development
PO Box 165
Green Point
Cape Town 8051, South Africa
Tel: 27 83 306 0030
Fax: 27 21 434 6134
E-mail: dmacdev@icon.co.za
http://easd.org.za/

International Ocean Institute–East-
  ern Africa
Kenya Marine and Fisheries Re-
  search Institute (KMFRI)
Box 81651
Mombasa, Kenya
Contact: Director
Tel: 254 11 475 527
Fax: 254 11 475 157
E-mail: ioi-ea@recoscix.com;
  kmfri@recoscix.com
http://www.ioinst.org/

International Ocean Institute–Senegal
Centre de Recherches Oceanographiques de Dakar-Thiaroye
  (CRODT)

BP 2241
Dakar, Senegal
Contact: Director
Tel: 221 8 34 8041
Fax: 221 8 34 2792
E-mail: dtoure@crodt.isra.sn

International Ocean Institute–Southern Africa
Botany Department
University of the Western Cape
P Bag X17
Bellville 7535, South Africa
Contact: Director
Tel: 27 21 959 3088
Fax: 27 21 959 1213
E-mail: t.potts@uwc.ac.za
http://www.ioinst.org/ioisa/

IOI West African Operational Centre of NIOMAR
Wilmat Point Road
Big Beach, Victoria Island
PMB 12729
Lagos, Nigeria
Fax: 234 1 261 9517
E-mail: niomar@linkserve.com.ug; niomar@hyperia.com

South African Association for Marine Biological Research
PO Box 10712
Marine Parade
Durban 4056, South Africa
Contact: Director
Tel: 27 31 337 3536
Fax: 27 31 337 2132
E-mail: seaworld@dbn.lia.net
http://www.seaworld.org.za/

South African Network for Coastal and Oceanic Research (SANCOR)
c/o National Research Foundation
PO Box 2600
Pretoria 0001, South Africa
Contact: Secretariat
Tel: 27 12 481 4107
Fax: 27 12 481 4005
E-mail: annette@nrf.ac.za
http://www.botany.uwc.ac.za/sancor/

Western Indian Ocean Marine Science Association (WIOMSA)
PO Box 3298
Zanzibar, Tanzania
Tel: 255 24 223 3472
Fax: 255 24 223 3852
E-mail: wiomsa@yahoo.com; secretary@wiomsa.org
http://www.wiomsa.org/

5. REGIONAL NONGOVERNMENTAL ORGANIZATIONS

5.2 The Americas

Alliance for Marine Remote Sensing Association
NSCC, 5685 Leeds Street
PO Box 1153
Halifax, NS, B3J 2X1, Canada
Contact: Systems Manager
Tel: 1 902 491-2160
Fax: 1 902 491-2162
E-mail: amrsadmin@waterobserver.org
http://www.waterobserver.org/

American Association of Petroleum Geologists
Box 979
Tulsa, OK, 74101-0979, USA
Contact: Executive Director
Tel: 1 918 584-2555
Fax: 1 918 560-2665
E-mail: postmaster@aapg.org
http://www.aapg.org/

American Association of Port Authorities
1010 Duke Street
Alexandria, VA, 22314-3589, USA
Tel: 1 703 684-5700

Fax: 1 703 684-6231
E-mail: info@aapa-ports.org
http://www.aapa-ports.org/

American Boat and Yacht Council (ABYC)
3069 Solomons Island Road
Edgewater, MD, 21037, USA
Contact: President
Tel: 1 410 956-1050
Fax: 1 410 956-2737
E-mail: info@abycinc.org
http://www.abycinc.org/

American Bureau of Shipping (ABS)
ABS Plaza
16855 Northchase Drive
Houston, TX, 77060, USA
Contact: President
Tel: 1 281 877-5800
Fax: 1 281 877-5803
E-mail: abs-worldhq@eagle.org
http://www.eagle.org/

American Cetacean Society
PO Box 1391
San Pedro, CA, 90733-1391, USA
Tel: 1 310 548-6279
Fax: 1 310-548-6950
E-mail: info@acsonline.org
http://www.acsonline.org/

American Coastal Coalition
5460 Beaujolais Lane
Fort Myers, FL, 33919, USA
Contact: President
Tel: 1 941 489-2616
Fax: 1 941 489-9917
E-mail: info@coastalcoalition.org
http://www.coastalcoalition.org/

American Fisheries Society
5410 Grosvenor Lane, Suite 110
Bethesda, MD, 20814, USA
Contact: Executive Director
Tel: 1 301 897-8616
Fax: 1 301 897-8096

E-mail: main@fisheries.org
http://www.fisheries.org/

American Geological Institute (AGI)
4220 King Street
Alexandria, VA, 22302-1502, USA
Tel: 1 703 379-2480
Fax: 1 703 379-7563
E-mail: agi@agiweb.org
http://www.agiweb.org/

American Geophysical Union (AGU)
2000 Florida Ave. NW
Washington, DC, 20009, USA
Contact: President
Tel: 1 202 462-6900
Fax: 1 202 328-0566
E-mail: service@agu.org
http://www.agu.org/

American Institute of Fishery Research Biologists (AIFRB)
c/o 6211 Madawaska Road
Bethesda, MD, 20816, USA
Contact: President
Tel: 1 301 320-5202
E-mail: DickSchaef@aol.com
http://www.aifrb.org/

American Littoral Society
Building 18
Sandy Hook
Highlands, NJ, 07732, USA
Contact: Executive Director
Tel: 1 732 291-0055
Fax: 1 732 291-3551
E-mail: als@netlabs.net
http://www.americanlittoralsoc.org/

American Marinelife Dealers Association
c/o Aquascapes/The Reef Shop
6527½ Naomi
Portage, MI, 49002, USA
Contact: Secretary

E-mail: mark.schreffler@amdareef.com
http://www.amdareef.com/

American Maritime Congress
1300 Eye St. NW, Suite 250 W
Washington, DC, 20005, USA
Contact: Executive Director
Tel: 1 202 842-4900
Fax: 1 202 842-3492
E-mail: info@us-flag.org
http://www.us-flag.org/

American Maritime Officers
2 West Dixie Highway
Dania Beach, FL, 33004, USA
Contact: National President
Tel: 1 954 921-2221
Fax: 1 954 926-5112
http://www.amo-union.org/

American Maritime Safety, Inc.
Westchester Financial Center
11 Martine Avenue, Suite 1450
White Plains, NY, 10606-4025, USA
Tel: 1 914 997-2916
Fax: 1 914 997-7125
E-mail: ams@maritimesafety.org
http://www.maritimesafety.org/

American Meteorological Society
45 Beacon Street
Boston, MA, 02108-3693, USA
Contact: Executive Director
Tel: 1 617 227-2425
Fax: 1 617 742-8718
E-mail: amsinfo@ametsoc.org
http://www.ametsoc.org/AMS/

American Petroleum Institute
1220 L Street, NW
Washington, DC 20005-4070
Tel: 1 202 682-8000
http://www.api.org/

American Polar Society
4816 Huron Drive
Pensacola, FL, 32507, USA
Contact: Capt. Frank Stokes
Tel: 1 850 497-0759
E-mail: ampolars@aol.com
http://www.oaedks.net/amerpolr.htm

American Shore and Beach Preservation Association (ASBPA)
1724 Indian Way
Oakland, CA, 94611, USA
Contact: Business Office
Tel: 1 510 339-2818
Fax: 1 510 339-6710
E-mail: business@asbpa.org
http://www.asbpa.org/

American Society of International Law (ASIL)
2223 Massachusetts Avenue, NW
Washington, DC, 20008, USA
Tel: 1 202 939-6000
Fax: 1 202 797-7133
E-mail: info@asil.org
http://www.asil.org/

American Society of Limnology and Oceanography
5400 Bosque Blvd., Suite 680
Waco, TX, 76710-4446, USA
Contact: Business Office
Tel: 1 254 399-9635
Fax: 1 254 776-3767
E-mail: business@aslo.org
http://www.aslo.org/

American Society of Naval Engineers (ASNE)
1452 Duke Street
Alexandria, VA, 22314-3458, USA
Tel: 1 703 836-6727
Fax: 1 703 836-7491
E-mail: asnehq@navalengineers.org
http://www.navalengineers.org/

American Sportfishing Association
225 Rainekers Lane, Suite 420
Alexandria, VA, 22314, USA
Contact: President

Tel: 1 703 519-9691
Fax: 1 703 519-1872
E-mail: info@asafishing.org
http://www.asafishing.org/

American Waterways Operators
801 North Quincy Street, Suite 200
Arlington, VA, 22203, USA
Tel: 1 703 841-9300
Fax: 1 703 841-0389
E-mail: aburns@vesselalliance.com
http://www.americanwaterways.com/

Aquacultural Engineering Society
c/o Freshwater Institute
PO Box 1889
Shepherdstown, WV, 25443, USA
Contact: Secretary/Treasurer
Tel: 1 304 876-2815
Fax: 1 304 870-2208
E-mail: b.vinci@freshwaterinstitute.org
http://www.aesweb.org/

Aquaculture Association of Canada
16 Lobster Lane
St. Andrews, NB, E5B 3T6, Canada
Tel: 1 506 529-4766
Fax: 1 506 529-4609
E-mail: aac@mar.dfo-mpo.gc.ca
http://www.aquaculture.ca/

Association of Canadian Port Authorities
85 Albert Street, Suite 1502
Ottawa, ON, K1P 6A4, Canada
Contact: Executive Director
Tel: 1 613 232-2036
Fax: 1 613 232-9554
E-mail: leroux@acpa-ports.net
http://www.acpa-ports.net/

Association of Coastal Engineers
c/o Applied Technology and Management Inc.
2770 NW 43rd Street, Suite B
Gainesville, FL, 32606, USA
Contact: Executive Secretary

E-mail: pehrman@appliedtm.com
http://www.coastalengineers.org/

Association of Environmental and Resource Economists (AERE)
1616 P St. NW, Suite 410, Room 400
Washington, DC, 20036, USA
Contact: Secretary
Tel: 1 202 328-5077
Fax: 1 202 939-3460
E-mail: voigt@rff.org
http://www.aere.org/

Association of Pacific Ports
1402 Eastview Court, NW
Olympia, WA, 98502, USA
Contact: Executive Director
Tel: 1 360 352-5346
Fax: 1 360 754-7460
E-mail: PortsWest@aol.com
http://www.associationofpacificports.com/

Association of Ship Brokers and Agents-USA, Inc.
510 Sylvan Avenue, Suite 201
Englewood Cliffs, NJ, 07632, USA
Contact: Executive Director
Tel: 1 201 569-2882
Fax: 1 201 569-9082
E-mail: asba@asba.org
http://www.asba.org/

Atlantic Coastal Zone Information Steering Committee (ACZISC)
International Oceans Institute of Canada
Dalhousie University
1226 LeMarchant Street
Halifax, NS, B3H 3P7, Canada
Contact: Secretariat
Tel: 1 902 494-3879
Fax: 1 902 494-1334
E-mail: aczisc@is.dal.ca
http://www.dal.ca/aczisc/

Atlantic Salmon Federation
PO Box 5200

St. Andrews, NB, E5B 3S8, Canada
Contact: President
Tel: 1 506 529-1033
Fax: 1 506 529-4438
E-mail: asfweb@nbnet.nb.ca
http://www.asf.ca/

Bay of Fundy Ecosystem Partnership (BoFEP)
Acadia Centre for Estuarine Research
Acadia University
PO Box 115
Wolfville, NS, B0P 1X0, Canada
Contact: Secretariat
Tel: 1 902 585-1054
E-mail: nancy.roscoe-huntley@acadiau.ca
http://www.auracom.com/bofep/

Canadian Aquaculture Industry Alliance (CAIA)
907-75 Albert Street
Ottawa, ON, K1P 5E7, Canada
Tel: 1 613 239-0612
Fax: 1 613 239-0619
E-mail: caiaoffice@aquaculture.ca
http://www.aquaculture.ca/

Canadian Centre for Fisheries Innovation
PO Box 4920
St. John's, NL, A1C 5R3, Canada
Tel: 1 709 778-0517
Fax: 1 709 778-0516
E-mail: ccfi@mi.mun.ca
http://www.ifmt.nf.ca/ccfi/

Canadian Centre for Marine Communications
155 Ridge Road
PO Box 8454
St. John's, NL, A1B 3N9 Canada
Tel: 1 709 579-4872
Fax: 1 709 579-0495
E-mail: ccmc@ccmc.nf.ca
http://www.ccmc.nf.ca/

Canadian Coastal Science and Engineering Association (CCSEA)
c/o National Water Research Institute
Canada Centre for Inland Waters
PO Box 5050
865 Lakeshore Road
Burlington, ON, L7R 4A6, Canada
Tel: 1 905 335-4981
Fax: 1 905 336-6444
E-mail: michael.skafel@ec.gc.ca
http://www.cciw.ca/ccsea/intro.html

Canadian Council of Professional Fish Harvesters (CCPFH/CCPP)
102 Bank Street, Suite 202
Ottawa, ON, K1P 5N4, Canada
Contact: Executive Director
Tel: 1 613 235-3474
Fax: 1 613 231-4313
E-mail: fish@ccpfh-ccpp.org
http://www.ccpfh-ccpp.org/

Canadian International Freight Forwarders' Association (CIFFA)
1243 Islington Avenue, Suite 706
Toronto, ON, M8X 1Y9, Canada
Tel/Fax: 1 416 234-5100
E-mail: info@ciffa.com
http://www.ciffa.com/

Canadian Maritime Law Association
240 Rue St. Jacques West, Suite 300
Montreal, QC, H2Y 1L9, Canada
Tel: 1 514 849-4161
Fax: 1 514 849-4167
E-mail: cmla@cmla.org
http://www.cmla.org/

Canadian Nature Federation
1 Nicholas Street, Suite 606
Ottawa, ON, K1N 7B7, Canada
Tel: 1 613 562-3447
Fax: 1 613 562-3371
E-mail: cnf@cnf.ca
http://www.cnf.ca/

Canadian Nautical Research Society (CNRS)
PO Box 511
Kingston, ON, K7L 4W5, Canada
Contact: President
E-mail: william@cae.com
http://www.marmus.ca/cnrs/

Canadian Shipowners Association
350 Sparks Street, Suite 705
Ottawa, ON, K1R 7S8, Canada
Tel: 1 613 232-3539
Fax: 1 613 232-6211
E-mail: csa@shipowners.ca
http://www.shipowners.ca/

Canadian Water and Wastewater Association
5330 Canotek Road, 2nd Floor, Unit 20
Ottawa, ON, K1J 9C3, Canada
Contact: Executive Director
Tel: 1 613 747-0524
Fax: 1 613 747-0523
E-mail: admin@cwwa.ca
http://cwwa.ca/

Canadian Wildlife Federation
350 Michael Cowpland Drive
Kanata, ON, K2M 2W1, Canada
Tel: 1 613 599-9594
Fax: 1 613 599-4428
E-mail: info@cwf-fcf.org
http://www.cwf-fcf.org/

Caribbean Alliance for Sustainable Tourism
1000 Ponce de Leon Ave., 5th Floor
San Juan, Puerto Rico, 00907, USA
Tel: 1 787 725-9139
Fax: 1 787 725-9108
E-mail: cast@caribbeanhotels.org
http://www.cha-cast.com/

Caribbean Community Ocean Sciences Network (CCOSNET)
c/o Institute of Marine Affairs
PO Box 3160
Carenage Post Office
Trinidad and Tobago
Contact: Coordinator
Tel: 1 868 634-4291/4
Fax: 1 868 634-4433
E-mail: director@ima.gov.tt
http://www.ima.gov.tt/ccosnet.htm

Caribbean Conservation Association (CCA)
"Chelsford"
The Garrison
St. Michael, Barbados
Tel: 1 246 426-5373
Fax: 1 246 429-8483
E-mail: admin@ccanet.net
http://www.caribbeanconservation.org/

Caribbean Natural Resources Institute (CANARI)
Fernandes Industrial Centre
Administrative Building
Eastern Main Road
Laventille, Trinidad
Tel: 1 868 626-6062
Fax: 1 868 626-1788
E-mail: info@canari.org
http://www.canari.org/

CEDAM International
1 Fox Road
Croton-on-Hudson, NY, 10520, USA
Contact: President
Tel: 1 914 271-5365
Fax: 1 914 271-4723
E-mail: cedamint@aol.com
http://www.cedam.org/

Center for Coastal Studies
59 Commercial Street
PO Box 1036
Provincetown, MA, 02657, USA
Contact: Executive Director
Tel: 1 508 487-3622
Fax: 1 508 487-4495
E-mail: ccs@coastalstudies.org
http://www.coastalstudies.org/

Center for International Environmental Law (CIEL)
1367 Connecticut Avenue, NW, Suite 300
Washington, DC, 20036, USA
Contact: Executive Director
Tel: 1 202 785-8700
Fax: 1 202 785-8701
E-mail: info@ciel.org
http://www.ciel.org/

Centro Ecuatoriano de Derecho Ambiental (CEDA)
(Center for Environmental Law)
Eloy Alfaro 1770 y Rusia, 3rd Floor
Quito, Ecuador
Tel: 593 2 223 1410
Fax: 593 2 223 8609
E-mail: comercio-e@ceda.org.ec
http://www.ceda.org.ec/

Chamber of Maritime Commerce
350 Sparks Street, Suite 704A
Ottawa, ON, K1R 7S8, Canada
Tel: 1 613 233-8779
Fax: 1 613 233-3743
E-mail: info@cmc-ccm.com
http://www.cmc-ccm.com/

Chamber of Shipping
100–1111 West Hastings Street
Vancouver, BC, V6E 2J3, Canada
Tel: 1 604 681-2351
Fax: 1 604 681-4364
E-mail: csbc@chamber-of-shipping.com
http://www.chamber-of-shipping.com/

Chilean Maritime League
(Liga Marítima de Chile)
Arda Errázuriz 471, 2º piso
Casilla Postal 117-V
Valparaíso, Chile
Tel: 56 32 235280
Fax: 56 32 255179
E-mail: ligamar@armada.cl
http://www.ligamar.cl/

Circum-Pacific Council on Energy and Mineral Resources (CPCEMR)
Halbouty Center
5100 Westheimer Road, Suite 500
Houston, TX, 77056, USA
Contact: Chairman
Tel: 1 713 622-1130
Fax: 1 713 622-5360
http://www.circum-pacificcouncil.org/index.html

Clean Islands International
8219 Elvaton Drive
Pasadena, MD, 21122-3903, USA
Tel: 1 410 647-2500
Fax: 1 410 647-4554
E-mail: info@islands.org
http://islands.org/

Coast Alliance
600 Pennsylvania Ave., SE, Suite 340
Washington, DC, 20003, USA
Contact: Executive Director
Tel: 1 202 546-9554
Fax: 1 202 546-9609
E-mail: coast@coastalliance.org
http://www.coastalliance.org/

Coastal Communities Network
PO Box 1613
Pictou, NS, B0K 1H0, Canada
Contact: Coordinator
Tel: 1 902 485-4754
Fax: 1 902 445-7134
E-mail: coastalnet@ns.sympatico.com
http://www.coastalcommunities.ns.ca/

Coastal Community Network
B-3019 4th Avenue
Port Alberni, BC, V9Y 2B8, Canada
Tel: 1 250 812-5652
Fax: 1 250 383-1903
E-mail: coastcom@island.net
http://www.coastalcommunity.bc.ca/

Coastal Conservation Association (CCA)
6919 Portwest, Suite 100
Houston, TX, 77024, USA
Contact: Chairman
Tel: 1 713 626-4234
Fax: 1 713 626-5852
E-mail: ccantl@joincca.org
http://www.joincca.org/

Coastal Research and Education Society of Long Island, Inc. (CRESLI)
Division of Natural Sciences and Mathematics
Kramer Science Center
Dowling College
Oakdale, NY, 11769-1999, USA
Tel: 1 631 244-3352
E-mail: information@cresli.org
http://www.cresli.org/

Consortium for Oceanographic Research and Education
1755 Massachusetts Avenue, NW, Suite 800
Washington, DC, 20036-2102, USA
Tel: 1 202 332-0063
Fax: 1 202 332-8887
E-mail: core@coreocean.org
http://www.coreocean.org/

Council on Ocean Law (COL)
3314 P Street NW
Washington, DC, 20007, USA
Contact: Chair
Tel: 1 202 338-2629
http://www.oceanlaw.org/

David Suzuki Foundation
2211 West 4th Avenue, Suite 219
Vancouver, BC, V6K 4S2, Canada
Tel: 1 604 732-4228
Fax: 1 604 732-0752
E-mail: solutions@davidsuzuki.org
http://www.davidsuzuki.org/

Dredging Contractors of America
643 South Washington St
Alexandria, VA, 22314, USA
Contact: Executive Director
Tel: 1 703 518-8408
Fax: 1 703 518-8490
E-mail: mark@dredgingcontractors.org
http://www.dredgingcontractors.org/

Earth Island Institute
300 Broadway, Suite 28
San Francisco, CA, 94133, USA
Tel: 1 415 788-3666
Fax: 1 415 788-7324
http://www.earthisland.org/

Ecological Society of America
1707 H St., NW, Suite 400
Washington, DC, 20006, USA
Contact: Executive Director
Tel: 1 202 833-8773
Fax: 1 202 833-8775
E-mail: esahq@esa.org
http://www.esa.org/

Ecology Action Centre
1568 Argyle Street, Suite 31
Halifax, NS, B3J 2B3, Canada
Contact: Marine Issues Committee
Tel: 1 902 429-2202
Fax: 1 902 422-6410
E-mail: eac_hfx@istar.ca
http://www.ecologyaction.ca/

Fisheries Council of Canada
38 Antares Drive, Suite 110
Ottawa, ON, K2E 7V2, Canada
Contact: President
Tel: 1 613 727-7450
Fax: 1 613 727-7453
E-mail: info@fisheriescouncil.org
http://www.fisheriescouncil.org/

Friends of the Earth (FOE)
1025 Vermont Ave NW, Suite 300
Washington, DC, 20005, USA
Contact: President
Tel: 1 202 783-7400
Fax: 1 202 783-0444

E-mail: foe@foe.org
http://www.foe.org/

Geoscience Information Society
c/o Geological and Planetary Sciences Library
100-23
Caltech
Pasadena, CA, 91125, USA
Contact: Secretary
Tel: 1 626 395-2199
Fax: 1 626 568-0935
E-mail: jimodo@caltech.edu
http://www.geoinfo.org/

Global Programme of Action Coalition for the Gulf of Maine (GPAC)
c/o Atlantic Coastal Action Program Saint John
PO Box 6878, Station A
Saint John, NB, E2L 4S3, Canada
Contact: Coordinator
Tel: 1 506 652-2227
Fax: 1 506 633-2184
E-mail: acapsj@fundy.net
http://www.gpac-gom.org/

Huntsman Marine Science Centre
1 Lower Campus Road
St. Andrews, NB, E5B 2L7, Canada
Tel: 1 506 529-1200
Fax: 1 506 529-1212
E-mail: huntsman@huntsmanmarine.ca
http://www.huntsmanmarine.ca/

IEEE Geoscience and Remote Sensing Society
c/o IEEE Operations Centre
445 Hoes Lane
PO Box 1331
Piscataway, NJ, 08855-1331, USA
Tel: 1 732 981-0060
Fax: 1 732 981-1721
E-mail: grss@ieee.org
http://ewh.ieee.org/soc/grss/

IEEE Oceanic Engineering Society
2403 Lisbon Lane
Alexandria, VA, 22306-2516, USA
Contact: President
Tel: 1 703 768-9522
E-mail: t.wiener@ieee.org
http://www.oceanicengineering.org/

Inland Seas Education Association (ISEA)
PO Box 218
Suttons Bay, MI, 49682, USA
Contact: Executive Director
Tel: 1 231 271-3077
Fax: 1 231 271-3088
E-mail: info@greatlakeseducation.org
http://www.schoolship.org/

Institute of Navigation
3975 University Drive, Suite 390
Fairfax, VA, 22030, USA
Tel: 1 703 383-9688
Fax: 1 703 383-9689
E-mail: membership@ion.org
http://www.ion.org/

International Association of Fish and Wildlife Agencies (IAFWA)
444 North Capitol St. NW, Suite 544
Washington, DC, 20001, USA
Contact: Executive VP
Tel: 1 202 624-7890
Fax: 1 202 624-7891
E-mail: iafwa@sso.org
http://www.iafwa.org/

International Marina Institute
PO Box 7197
Jupiter, FL, 33468, USA
Contact: President
Tel: 1 561 741-0626
Fax: 1 561 741-0676
E-mail: info@imimarina.org
http://www.imimarina.com/

International Ocean Institute–Costa Rica
Universidad Nacional
PO Box 86
Heredia 3000, Costa Rica
Contact: Director
Tel: 506 277 3594
Fax: 506 260 2546
E-mail: ioicos@una.ac.cr
http://www.una.ac.cr/ioi/

Inuit Tapiriit Kanatami
170 Laurier Avenue West, Suite 510
Ottawa, ON, K1P 5V5, Canada
Tel: 1 613 238-8181
Fax: 1 613 234-1991
E-mail: info@itk.ca
http://www.itk.ca/

Island Resources Foundation
6292 Estate Nazareth No. 100
St. Thomas, US Virgin Islands, 00802-1104, USA
Tel: 1 340 775-6225
Fax: 1 340 779-2022
E-mail: info@irf.org
http://www.irf.org/

Lake Carriers' Association
Suite 915, Rockefeller Bldg.
614 Superior Avenue, W
Cleveland, OH, 44113-1383, USA
Contact: President
Tel: 1 216 621-1107
Fax: 1 216 241-8262
E-mail: ggn@lcaships.com
http://www.lcaships.com/

Latin American Shipowners' Association (LASA)
(Asociación Latinoamericana de Armadores)
Blanco 869, Piso 3
Valparaíso, Chile
Contact: Executive Secretary
Tel: 56 32 212057
Fax: 56 32 212017
E-mail: armadore@entelchile.net

Liberian Shipowners' Council Ltd.
99 Park Avenue, Suite 1700
New York, NY, 10016, USA
Contact: General Secretary
Tel: 1 212 973-3896
Fax: 1 212 994-6763
E-mail: jl@liberianshipowners.com
http://www.liberianshipowners.com/

Marina Operators Association of America (MOAA)
1819 L Street NW, Suite 700
Washington, DC, 20036, USA
Contact: Executive Director
Tel: 1 866 367-6622
Fax: 1 202 861-1181
E-mail: moaa@nmma.org
http://www.moaa.org/

Marine Conservation Biology Institute
15806 NE 47th Court
Redmond, WA, 98052, USA
Tel: 1 425 883-8914
Fax: 1 425 883-3017
E-mail: mcbiweb@mcbi.org
http://www.mcbi.org/

Marine Retailers Association of America (MRAA)
150 E Huron St., No. 802
Chicago, IL, 60611, USA
Contact: President
Tel: 1 312 944-5080
Fax: 1 312 944-2716
E-mail: mraa@mraa.com
http://www.mraa.com/

Maritime Affairs
Naval Officers' Association of Canada
PO Box 33078
Halifax, NS, B3L 4T6, Canada
Fax: 1 902 835-5994
E-mail: maritime@naval.ca
http://www.naval.ca/maritimeaffairs/

Maritime Security Council
3471 N. Federal Highway, Suite 611
Fort Lauderdale, FL, 33306, USA
Tel: 1 954 567-2536
Fax: 1 954 567-2511
E-mail: mailbox@maritimesecurity.
   org
http://www.maritimesecurity.org/

Monterey Bay Aquarium Research
   Institute
7700 Sandholdt Road
Moss Landing, CA, 95039-9644, USA
Contact: President
Tel: 1 831 775-1700
Fax: 1 831 775-1620
E-mail: mcnutt@mbari.org
http://www.mbari.org/

National Aquaculture Association
111 W. Washington Street, Suite 1
Charles Town, WV, 25414-1529,
   USA
Tel: 1 304 728-2167
Fax: 1 304 728-2196
E-mail: naa@intrepid.net
http://www.natlaquaculture.org/

National Association for the Marine
   Assistance Industries (C-PORT)
1600 Duke Street, Suite 400
Alexandria, VA, 22314, USA
Contact: Executive Director
Tel: 1 703 519-1713
Fax: 1 703 519-1716
E-mail: c-port@wpa.org
http://www.c-port.org/

National Association of Charterboat
   Operators (NACO)
1600 Duke Street, Suite 400
Alexandria, VA, 22314, USA
Contact: Executive Director
Tel: 1 703 519-1714
Fax: 1 703 519-1716
E-mail: naco@wpa.org
http://www.charterboat.org/

National Association of Marine Sur-
   veyors, Inc. (NAMS)
PO Box 9306
Chesapeake, VA, 23321-9306, USA
Contact: Office Manager
Tel: 1 757 488-9538
Fax: 1 757 488-0584
E-mail: office@nams-cms.org
http://www.namsurveyors.org/

National Association of Marine Services (NAMS)
5458 Wagonmaster Drive
Colorado Springs, CO, 80917, USA
Contact: Executive Director
Tel: 1 719 573-5946
Fax: 1 719 573-5952
E-mail: nams@namsshipchandler.
   com
http://www.namsshipchand
   ler.com/

National Coalition for Marine Con-
   servation (NCMC)
3 North King St.
Leesburg, VA, 20176, USA
Contact: Executive Director
Tel: 1 703 777-0037
Fax: 1 703 777-1107
E-mail: christine@savethefish.org
http://www.savethefish.org/

National Fisheries Institute
1901 North Fort Myer Drive, Suite
   700
Arlington, VA, 22209, USA
Contact: President
Tel: 1 703 524-8880
Fax: 1 703 524-4619
E-mail: tressler@nfi.org
http://www.nfi.org/

National Marine Distributors Associa-
   tion (NMDA)
37 Pratt Street
Essex, CT, 06426-1159, USA
Contact: Executive Director
Tel: 1 860 767-7898

Fax: 1 860 767-7932
E-mail: info@nmdaonline.com
http://www.nmdaonline.com/

National Marine Educators Association
PO Box 1470
Ocean Springs, MS, 39566-1470, USA
Contact: Secretary
Tel: 1 228 374-7557
Fax: 1 228 374-5559
E-mail: johnette.bosarge@usm.edu
http://www.marine-ed.org/

National Marine Electronics Association (NMEA)
7 Riggs Avenue
Severna Park, MD, 21146, USA
Contact: Executive Director
Tel: 1 410 975-9425
Fax: 1 410 975-9450
E-mail: info@nmea.org
http://www.nmea.org/

National Marine Manufacturers Association (NMMA)
200 East Randolph Drive, Suite 5100
Chicago, IL, 60601, USA
Tel: 1 312 946-6200
Fax: 1 312 946-0388
E-mail: webmaster@nmma.org
http://www.nmma.org/

National Maritime Historical Society
5 John Walsh Blvd.
PO Box 68
Peekskill, NY, 10566, USA
Tel: 1 914 737-7878
Fax: 1 914 737-7816
E-mail: david@seahistory.org
http://www.seahistory.org/

National Ocean Industries Association (NOIA)
1120 G St. NW, Suite 900
Washington, DC, 20005, USA
Contact: President
Tel: 1 202 347-6900
Fax: 1 202 347-8650
E-mail: tom@noia.org
http://www.noia.org/

National Shellfisheries Association
c/o PO Box 350
VIMS Eastern Shorelab
Wachapreague, VA, 23480, USA
E-mail: nlewis@vims.edu
http://www.shellfish.org/

Newfoundland Ocean Industries Association
Atlantic Place, Suite 602
215 Water Street
St. John's, NL, A1C 6C9, Canada
Contact: President
Tel: 1 709 758-6610
Fax: 1 709 758-6611
E-mail: noia@noianet.com
http://www.noianet.com/

North American Native Fishes Association, Inc. (NANFA)
1107 Argonne Drive
Baltimore, MD, 21218, USA
Contact: Secretary
Tel: 1 410 243-9050
E-mail: nanfa@att.net
http://www.nanfa.org/

Northwest Marine Trade Association (NMTA)
1900 N. Northlake Way, Suite 233
Seattle, WA, 98103-9087, USA
Contact: Executive Director
Tel: 1 206 634-0911
Fax: 1 206 632-0078
E-mail: info@nmta.net
http://www.nmta.net/

Ocean Voice International
PO Box 37026
3332 McCarthy Road
Ottawa, ON, K1V 0W0, Canada
Tel: 1 613 721-4541
Fax: 1 613 721-4562

E-mail: oceans@superaje.com
http://www.ovi.ca/

Oceana
2501 M Street, NW
Washington, DC, 20037-1311, USA
Contact: Executive Director
Tel: 1 202 833-3900
Fax: 1 202 833-2070
E-mail: info@oceana.org
http://www.oceana.org/

Oceanic Society
Fort Mason Center, Building E,
San Francisco, CA, 94123, USA
Tel: 1 415 441-1106
Fax: 1 415 474-3395
E-mail: office@oceanic-society.org
http://www.oceanic-society.org/

Oceans Blue Foundation
405-134 Abbott Water Street
Vancouver, BC, V6B 2K4, Canada
Contact: Executive Director
Tel: 1 604 684-2583
Fax: 1 604 684-6942
E-mail: sails@oceansblue.org
http://www.oceansblue.org/

Offshore Marine Service Association (OMSA)
990 N. Corporate Dr., Suite 210
Harahan, LA, 70123, USA
Contact: President
Tel: 1 504 734-7622
Fax: 1 504 734-7134
E-mail: robert@offshoremarine.org
http://www.offshoremarine.org/

Pacific Coast Federation of Fishermen's Associations (PCFFA)
PO Box 29370
San Francisco, CA, 94129-0370, USA
Contact: Executive Director
Tel: 1 415 561-5080
Fax: 1 415 561-5464
E-mail: fish4ifr@aol.com
http://www.pcffa.org/

Pacific Maritime Association
PO Box 7861
San Francisco, CA, 94120-7861, USA
Tel: 1 415 576-3200
Fax: 1 415 989-1425
http://www.pmanet.org/

Passenger Vessel Association
801 North Quincy Street, Suite 200
Arlington, VA, 22203, USA
Contact: Executive Director
Tel: 1 703 807-0100
Fax: 1 703 807-0103
E-mail: pva@vesselalliance.com
http://www.passengervessel.com/

Protected Areas Conservation Trust
2 Mango Street
Belmopan, Cayo, Belize
Tel: 501 822-3637
Fax: 501 822-3759
E-mail: info@pactbelize.org
http://www.pactbelize.org/

Reef Care Curaçao
PO Box 676
Curaçao, Netherlands Antilles
Contact: President
Tel: 1 599 869-5285
E-mail: reefcare@cura.net
http://www.willemstad.net/reefcare

Shipbuilders Council of America (SCA)
1455 F Street, Suite 225
Washington, DC, 20005, USA
Contact: President
Tel: 1 202 347-5462
Fax: 1 202 347-5464
E-mail: jmccluer@de.bjllp.com
http://www.shipbuilders.org/

Shipping Federation of Canada
300 rue de Sacrement, Suite 326
Montreal, QC, H2Y 1X4, Canada
Tel: 1 514 849-2325
Fax: 1 514 849-6992
E-Mail: info@shipfed.ca
http://www.shipfed.ca/

Sierra Club
85 Second St., 2nd Floor
San Francisco, CA, 94105, USA
Contact: Executive Director
Tel: 1 415 977-5500
Fax: 1 415 977-5799
E-mail: information@sierraclub.org
http://www.sierraclub.org/

Sierra Club of Canada
412-1 Nicholas Street
Ottawa, ON, K1N 7B7, Canada
Tel: 1 613 241-4611
Fax: 1 613 241-2292
E-mail: sierra@web.ca
http://www.sierraclub.org/canada/

Society of Exploration Geophysicists
8801 South Yale
PO Box 702740
Tulsa, OK, 74170-2740, USA
Contact: Executive Director
Tel: 1 918 497-5500
Fax: 1 918 497-5557
E-mail: web@seg.org
http://seg.org/

Sustainable Development Institute
3121 South Street, NW
Washington, DC, 20007, USA
Tel: 1 202 338-1017
E-mail: susdev@igc.org
http://www.susdev.org/

The Coastal Society
PO Box 25408
Alexandria, VA, 22313-5408, USA
Contact: President
Tel: 1 703 768-1599
Fax: 1 703 768-1598
E-mail: info@thecoastalsociety.org
http://www.thecoastalsociety.org/

The Cousteau Society
870 Greenbrier Circle, Suite 402
Chesapeake, VA, 23320, USA
Contact: President
E-mail: cousteau@cousteausociety.org
http://www.cousteau.org/

The Ecological Society of America
1707 H St. NW, Suite 400
Washington, DC, 20006, USA
Tel: 1 202 833-8773
Fax: 1 202 833-8775
E-mail: esahq@esa.org
http://www.esa.org/

The Ocean Conservancy
1725 DeSales St., NW, Suite 600
Washington, DC, 20036, USA
Tel: 1 202 429-5609
Fax: 1 202 872-0619
E-mail: info@oceanconservancy.org
http://www.oceanconservancy.org/

The Oceanography Society (TOS)
PO Box 1931
Rockville, MD, 20849-1931
Contact: Executive Director
Tel: 1 301 251-7708
Fax: 1 301 251-7709
E-mail: info@tos.org
http://www.tos.org/

West Gulf Maritime Association
1717 East Loop, Suite 200
Houston, TX, 77029, USA
Contact: President
Tel: 1 713 678-7655
Fax: 1 713 672-7452
E-mail: barbara@wgma.org
http://www.wgma.org/

Western Dredging Association
PO Box 5797
Vancouver, WA, 98668-5737, USA
Contact: Executive Director
Tel: 1 360 750-0209
Fax: 1 360 750-1445
E-mail: weda@juno.com
http://www.westerndredging.org/

Wildlife Conservation Society
2300 Southern Blvd.
Bronx, NY, 10460, USA
Contact: President
Tel: 1 718 220-5100

E-mail: nyageneral@wcs.org
http://www.wcs.org/

Women's Aquatic Network, Inc.
PO Box 4993
Washington, DC, 20008, USA
Contact: Secretariat
E-mail: info@womensaquatic.net
http://orgs.womensaquatic.net/

World Shipping Council
1015 15th Street NW, Suite 450
Washington, DC, 20005, USA
Tel: 1 202 589-1230
Fax: 1 202 589-1231
E-mail: info@worldshipping.org
http://www.worldshipping.org/

Yacht Brokers Association of America
105 Eastern Avenue, Suite 104
Annapolis, MD, 21403, USA
Contact: Executive Director
Tel: 1 410 263-1014
Fax: 1 410 263-1659
E-mail: jmt@ybaa.com
http://yachtworld.com/ybaa/

5. REGIONAL NONGOVERNMENTAL ORGANIZATIONS

5.3 Asia

Asian Fisheries Society
PO Box 2725
Quezon City Central Post Office
1167 Quezon City, Philippines
Tel: 63 2 921 1914
Fax: 63 2 920 2757
E-mail: afs@compass.com.ph
http://www.compass.com.ph/afs/

China Society for Fisheries
Building 22, Maizidian Street
Chaoyang District
Beijing, People's Republic of China
Contact: President
Tel: 86 10 6419 4233
Fax: 86 10 6419 4231
E-mail: cnscfish2public.bta.net.cn
http://www.fisheries.moa.gov.cn/

Chinese Ports and Harbours Association
12 Zhong Shan Rd. (E2)
Shanghai 200002, People's Republic of China
Contact: Secretary General
Tel: 86 21 311 5654
Fax: 86 21 329 0202

Chinese Society of Naval Architects and Marine Engineers (CSNAME)
5 Yuetan Beijie St.
Beijing 100861, People's Republic of China
Contact: President
Tel: 86 10 6803 8833
Fax: 86 10 6803 1560
E-mail: b24@csname.org
http://www.csname.org/

Coastal Development Planning Centre
25 Vasundhara Colony, Gulbai Jkrd
Ahmedabad
Gujurat, 380006, India
Tel: 91 79 656 8421
Tel: 91 79 642 0056
E-mail: mihir@adl.vsnl.net.in

Environmental Engineers Association of Thailand
122/4 Soi Rawadee, Rama VI Road
Samsen Nai, Phayathai
Bangkok 10400, Thailand
Contact: President
Tel: 66 2 617 1530
Fax: 66 2 279 9720
E-mail: info@eeat.or.th
http://www.eeat.or.th/

Federation of ASEAN Shipowners' Associations (FASA)
c/o Singapore Shipping Association
456 Alexandra Road #09-02, NOL Bldg
Singapore 119962
Tel: 65 278 3464
Fax: 65 274 5079
E-mail: fasa@pacific.net

Haribon Foundation
41F Fil Garcia Tower
140 Kalayaan Avenue
Cor Mayaman St.
Diliman, Quezon City, The Philippines
Tel: 632 920 7430
Fax: 632 924 8976
E-mail: info@haribon.org.ph
http://www.haribon.org.ph/

Institute of Chartered Shipbrokers
GPO Box 12292
Hong Kong
Contact: Secretary
Tel: 852 2866 1488
Fax: 852 8106 1480
E-mail: icshk@netvigator.com
http://www.ics.org.hk/

International Cetacean Education Research Centre–Japan
3-37-12, Nishihara
Shibuya-ku
Tokyo, 151, Japan
Tel: 81 3 5712 3480
Fax: 81 3 5712 3481
E-mail: info@icerc.org
http://www.icerc.org/

International Ocean Institute–China
c/o National Marine Data and Information Service (NMDIS)
State Oceanic Administration
93 Liuwei Road, Hedong District
Tianjin 300171, People's Republic of China
Contact: Director
Tel/Fax: 86 22 2401 0859
E-mail: houwf@netra.nmdis.gov.cn

International Ocean Institute–India
Indian Institute of Technology (IIT Madras)
IC & SR Building
Chennai 600036, India
Contact: Director
Tel: 91 44 445 8805
Fax: 91 44 220 0559
E-mail: ioimas@md2.vsnl.net.in; ioi@vsnl.com

International Ocean Institute–Japan
Intercom Inc.
#302 1-26-10 Komaba
Meguro-ku
Tokyo 153-0041, Japan
Contact: Director
Tel: 81 3 5454 0231
Fax: 81 3 5454 0232
E-mail: intercom@qb3.so-net.ne.jp

International Society for Mangrove Ecosystems (ISME)
c/o Faculty of Agriculture
University of Ryukyus
Okinawa 903-0129, Japan
Contact: Secretariat
Tel: 81 98 895 6601
Fax: 81 98 895 6602
E-mail: mangrove@ryukyu.ne.jp
http://www.mangrove.or.jp/

Japan Society on Water Environment
201 Green-Plaza-Fukagawa-Tokiwa
2-9-7 Tokiwa, Koto-ku,
Tokyo 135-0006, Japan
Tel: 81 3 3632 5351
Fax. 81 3 3632 5352
E-mail: info@jswe.or.jp
http://www.jswe.or.jp/

Jordan Royal Ecological Diving Society
PO Box 831051
Amman 11183, Jordan

Tel/Fax: 962 6 06 5676183
E-mail: information@jreds.org
http://www.jreds.org/

Malaysian Shipowners' Association (MASA)
Menara Dayabuni, 17th Floor
Jalan Sultan Hisamudi
PO Box 10371
50712 Kuala Lumpur, Malaysia
Tel: 603 2275 2136
Fax: 603 2260 2575
E-mail: masa kl@tm.net.my
http://www.malaysianshipowners.org/

Nippon Kaiji Kyokai (Class NK)
4-7 Kioi-cho
Chiyoda-ku
Tokyo 102-8567, Japan
Contact: Chairman and President
Tel: 81 3 3230 1201
Fax: 81 3 5226 2012
E-mail: bnd@classnk.or.jp
http://www.classnk.or.jp/nonjava/index.html

Small Fishers Federation of Sri Lanka
Pambala, Kakkapalliya
PO Box 01
Chilaw, Sri Lanka
Contact: Director
Tel: 94 032 48707
Fax: 94 032 47960
E-mail: sffl@sri.lanka.net
http://www.shrimpaction.com/titlessffl.html

Southeast Asia START Regional Committee (SARCS)
National Central University
Chung-Li 320, Taiwan
Contact: Secretariat
Tel: 886 3 426 2726
Fax: 886 3 426 2640
E-mail: sarcs@sarcs.org.tw
http://www.sarcs.org.tw/

Telapak Indonesia
Jl. Sempur Kaler No.16
Bogor West Java, Indonesia
Tel: 62 251 320 792
Fax: 62 251 351 069
E-mail: info@telapak.org
http://www.telapak.org/

Yadfon Association
16/4 Rakchan Road
Tambon Tabtieng, Amphur Muang
Trang 92000, Thailand
Tel: 66 75 219 737
Fax: 66 75 219 327
E-mail: yadfon@loxinfo.co.th
http://www.shrimpaction.com/yadfon.html

5. REGIONAL NONGOVERNMENTAL ORGANIZATIONS

5.4 Europe

Alliance of Maritime Regional Interests in Europe (AMRIE)
68, avenue Michel Ange
B-1000 Brussels, Belgium
Contact: Secretariat
Tel: 32 2 736 1755
Fax: 32 2 735 2298
E-mail: info@amrie.org
http://www.amrie.org/

Association for the Preservation of Underwater Historical Heritage (Association pour la Mise en Valeur du Patrimoine Historique Sous-Marin)
20, rue de Verdun
F-29000 Quimper, France
Contact: President
Tel/Fax: 33 98 952007

Association of European Shipbuilders and Shiprepairers (AWES)

Rue Marie de Bourgogne 52-54, 3rd floor
B-1000 Brussels, Belgium
Contact: Director
Tel: 32 2 230 2791
Fax: 32 2 230 4332
E-mail: info@awes-shipbuilding.org
http://www.awes-shipbuilding.org/

Association of Port Health Authorities (APHA)
Dutton House
46 Church St.
Runcorn, Cheshire, WA7 1LL, United Kingdom
Contact: Executive Secretary
Tel: 44 87 0744 4505
Fax: 44 19 2858 1596
E-mail: apha@cieh.org.uk
http://www.apha.org.uk/

Association of River Waterworks (Vereniging van Rivier Waterbedrijven – RIWA)
PO Box 402
NL-3430 AK Nieuwegein, The Netherlands
Tel: 31 30 600 9030
Fax: 31 30 600 9039
E-mail: riwa@riwa.org; iawr@riwa.org
http://www.iawr.org/

Atlantic Salmon Trust
Moulin
Pitlochry
Perthshire, Scotland, PH16 5JQ, United Kingdom
Contact: Director
Tel: 44 17 9647 3439
Fax: 44 17 9647 3554
E-mail: salmontrust@aol.com
http://www.atlanticsalmontrust.org/

Atlantic Whale Foundation
St Martins House
59 St Martins Lane
Covent Garden
London, WC2N 4JS, United Kingdom
Tel: 44 20 7240 6604
Fax: 44 20 7240 5795
E-mail: edb@huron.ac.uk
http://www.whalefoundation.f2s.com/

Baltic Ports Organization (BPO)
c/o Ports of Stockholm
PO Box 27314
SE-102 54 Stockholm, Sweden
Contact: Secretariat
Tel: 46 86 70 26 00
Fax: 46 86 70 26 45
E-mail: bpo@stoports.com
http://www.bpoports.com/

British Marine Life Study Society
Glaucus House
14 Corbyn Crescent
Shoreham-by-Sea, BN43 6PQ, United Kingdom
Tel/Fax: 44 12 7346 5433
E-mail: glaucus@hotmail.com
http://ourworld.compuserve.com/homepages/BMLSS/homepage.htm

British Maritime Law Association
c/o Ince & Co.
Knolly's House
11 Byward Street
London, EC3R 5EN, United Kingdom
Contact: Secretary & Treasurer
Tel: 44 20 7623 2011
Fax: 44 20 7623 3225
E-mail: patrick.griggs@ince.co.uk
http://www.bmla.org.uk/

British Ports Association
Africa House
64-78 Kingsway
London, WC2B 6AH, United Kingdom
Contact: Director

Tel: 44 20 7242 1200
Fax: 44 20 7405 1069
E-mail: info@britishports.org.uk
http://www.britishports.org.uk/

Central Dredging Association
 (CEDA)
PO Box 488
NL-2600 AL, Delft, The Netherlands
Contact: General Manager
Tel: 31 15 278 3145
Fax: 31 15 278 7104
E-mail: ceda@dredging.org/
http://www.dredging.org/

Centre for Documentation, Research and Experimentation on Water Pollution Accidents (CEDRE)
rue Alain Colas
BP 20413
F-29604 Brest Cedex, France
Contact: Director
Tel: 33 2 9833 1010
Fax: 33 2 9844 9138
E-mail: contact@le-cedre.fr
http://www.le-cedre.fr/

Cetacea Defence
PO Box 78
Shaftesbury
Dorset, SP7 8ST, United Kingdom
E-mail: cetaceadefence@hotmail.com
http://free.freespeech.org/cetaceadefence/

Chamber of Shipping
Carthusian Court
12 Carthusian St.
London, EC1M 6EB, United Kingdom
Contact: Director General
Tel: 44 20 7417 2800
Fax: 44 20 7600 1534
E-mail: postmaster@british-shipping.org
http://www.british-shipping.org/

Coalition Clean Baltic
Östra Ågatan 53
SE-753 22 Uppsala, Sweden
Contact: Secretariat
Tel: 46 18 71 11 70
Fax: 46 18 71 11 55
E-mail: secretariat@ccb.se
http://www.ccb.se/

Coastwatch Europe International
Civil and Environmental Engineering
Trinity College, Dublin
Dublin 2, Ireland
Tel: 353 55 25843
Fax: 353 55 25842
E-mail: dubsky@iol.ie
http://www.coastwatcheurope.org/

Committee of European Shipowners' Associations (CESA)
52-54, rue Marie de Bourgogne, 3rd floor
B-1000 Brussels, Belgium
Contact: Secretary General
Tel: 32 2 230 2791
Fax: 32 2 230 4332
E-mail: info@cesa-shipbuilding.org
http://www.cesa-shipbuilding.org/

Council of European and Japanese National Shipowners' Associations (CENSA)
Carthusian Court
12 Carthusian St.
London, EC1M 6EB, United Kingdom
Contact: Secretary General
Tel: 44 17 1600 5405
Fax: 44 17 1600 5398
E-mail: censa@marisec.org

Deltalinqs
PO Box 54200
NL-3008 JE Rotterdam, The Netherlands
Tel: 31 10 4020 399
Fax: 31 10 4120 687

E-mail: info@deltalinqs.nl
http://www.deltalinqs.nl/

ESF Consortium for Ocean Drilling (ECOD)
ESF Management Committee for the ODP (EMCO)
European Science Foundation
1, quai Lezay-Marnésia
F-67080 Strasbourg Cedex, France
Contact: Scientific Secretary
Tel: 33 3 8876 7122
Fax: 33 3 8837 0532
E-mail: jdalton@esf.org
http://www.esf.org/ecod/

ESF Marine Board
European Science Foundation
1, quai Lezay Marnésia
F-67080 Strasbourg, France
Contact: Secretariat
Tel: 33 3 8876 7141
Fax: 33 3 8825 1954
E-mail: marineboard@esf.org
http://www.esf.org/marineboard/

Estuarine and Coastal Sciences Association (ECSA)
Department of Biological Science
University of Hull
Hull, HU6 7RX, United Kingdom
Contact: Secretary
http://www.ecsa.ac.uk/

EUCC—The Coastal Union
PO Box 11232
NL-2301 EE Leiden, The Netherlands
Tel: 31 71 512 2900
Fax: 31 71 512 4069
E-mail: admin@eucc.nl
http://www.eucc.nl/home/index.htm

European Aquaculture Society (EAS)
Slijkensesteenweg 4
B-8400 Oostende, Belgium
Tel: 32 59 323 859
Fax: 32 59 321 005
E-mail: eas@aquaculture.cc
http://www.easonline.org/

European Association for Aquatic Mammals
c/o Särkänniemi
FIN-33230 Tampere, Finland
Contact: Secretary/Treasurer
Tel: 358 3 2488 111
Fax: 358 3 2121 279
E-mail: info@eaam.org
http://www.eaam.org/

European Association of Fisheries Economists
c/o FOI
Rolighedsvej 25
DK-1958 Frederiksberg, Denmark
Contact: President
Tel: 45 35 28 68 00
Fax: 45 35 28 68 01
E-mail: jl@foi.dk
http://www.eafe-fish.org/

European Association of Remote Sensing Laboratories (EARSeL)
2 avenue Rapp
F-75340 Paris Cedex 07, France
Contact: Secretariat
Tel: 33 1 4556 7360
Fax: 33 1 4556 7361
E-mail: earsel@meteo.fr
http://www.earsel.org/

European Cetacean Society
c/o Department of Marine Fisheries
Danish Institute for Fisheries Research
Charlottenlund Slot
DK-2920 Charlottenlund, Denmark
Contact: Chairman
Tel: 45 33 96 33 00
Fax: 45 33 96 33 33
http://web.inter.NL.net/users/J.W.Broekema/ecs/

European Coastal Association for Science and Technology (EUROCOAST)
Department of Earth Sciences
Cardiff University
PO Box 914
Cardiff, Wales, CF10 3YE, United Kingdom
Contact: Secretariat
Tel: 44 29 2087 4830
Fax: 44 29 2087 4326
E-mail: eurocoast@cardiff.ac.uk
http://www.eurocoast.org/

European Community Shipowners' Associations (ECSA)
45 rue Ducale
B-1000 Brussels, Belgium
Contact: Secretariat
Tel: 32 2 511 3940
Fax: 32 2 511 8092
E-mail: mail@ecsa.be
http://www.ecsa.be/

European Desalination Association
Science Park of Abruzzo
via Antica Arischia, 1
I-67100 L'Aquila, Italy
Contact: Secretariat
Tel: 39 0862 3475 308
Fax: 39 0862 3475 213
E-mail: balaban@desline.com
http://www.edsoc.com/

European Dredging Association (EuDA)
2-4 rue de Praetere
B-1000 Brussels, Belgium
Contact: Secretariat
Tel: 32 2 646 8183
Fax: 32 2 646 6063
E-mail: info@euda.be
http://www.european-dredging.info/

European Federation of Sea Anglers (EFSA)
Inglewood, Braal Road
Halkirk, Caithness, KW12 6XE, United Kingdom
Contact: Secretary
Tel/Fax: 44 18 4783 1985
E-mail: enquiries@efsa.co.uk
http://www.efsa.co.uk/

European Geophysical Society (EGS)
Max Planck Strasse 13
D-37191 Katlenburg-Lindau, Germany
Contact: EGS Office
Tel: 49 5556 1440
Fax: 49 5556 4709
E-mail: egs@copernicus.org
http://www.copernicus.org/egs/egs.html/

European Maritime Pilots' Association (EMPA)
St. Aldegondiskaai 36-38
B-2000 Antwerp, Belgium
Contact: Secretary General
Tel: 32 3 205 9436
Fax: 32 3 205 9437
E-mail: empa@skynet.be
http://www.empa-pilots.org/

European Polar Board (EPB)
European Science Foundation
1, quai Lezay-Marnésia
F-67080 Strasbourg, France
Contact: Secretariat
Tel: 33 3 8876 7159
Fax: 33 3 8876 7180
E-mail: pegerton@esf.org
http://www.esf.org/epb/

European Science Foundation (ESF)
1, quai Lezay-Marnésia
F-67080 Strasbourg, France
Contact: Secretary General
Tel: 33 3 8876 7100
Fax: 33 3 8837 0532
E-mail: esf@esf.org
http://www.esf.org/

European Sea Ports Organisation (ESPO)
68 Michelangelolaan Avenue
B-1000 Brussels, Belgium
Contact: Secretary General
Tel: 32 2 736 3463
Fax: 32 2 736 6325
E-mail: mail@espo.be
http://www.espo.be/

European Shippers' Council (ESC)
Park Leopold
Rue Wiertz 50
B-1050 Brussels, Belgium
Contact: Secretary General
Tel: 32 2 230 2113
Fax: 32 2 230 4140
E-mail: enquiries@europeanshippers.com
http://www.europeanshippers.com/

European Society for Marine Biotechnology
c/o Burgess Laboratory
Centre for Marine Biodiversity and Biotechnology
Division of Life Sciences
Heriot-Watt University
Edinburgh, EH14 4AS, United Kingdom
E-mail: M.Searle@hw.ac.uk
http://www.esmb.org/

European Tugowners' Association (ETA)
Docklands Business Centre
10-16 Tiller Road
London, E14 8PX, United Kingdom
Tel: 44 020 7345 5122
Fax: 44 020 7345 5722
E-mail: associationservices1@compuserve.com

European Union Fish Processors Association (Association des Industries du Poisson de la CE–AIPCEE)
c/o SIA
Avenue de Roodebeek, 30
B-1030 Brussels, Belgium
Contact: Secretary General
Tel: 32 2 743 8730
Fax: 32 2 736 8175
E-mail: sia01@sia-dvi.be

European Union for Coastal Conservation (EUCC)
PO Box 11232
NL-2301 EE Leiden, Netherlands
Tel: 31 71 512 2900
Fax: 31 71 512 4069
E-mail: admin@eucc.nl
http://www.eucc.nl/

European Union of Geosciences (EOST)
5 rue René Descartes
F-67084 Strasbourg Cedex, France
Contact: Executive Secretary
Tel: 33 3 9024 0353
Fax: 33 3 9024 0125
E-mail: webmaster@eost.u-strasbg.fr
http://eost.u-strasbg.fr/

European Water Association
Theodor-Heuss-Allee 17
D-53773 Hennef, Germany
Contact: Chairman
Tel: 49 22 42 872189
Fax: 49 22 42 872135
E-mail: ewa@atv.de
http://www.atv.de/ewpca/

EUROPECHE
(Association of National Organisations of Fishing Enterprises of the EC)
rue de la Science 23-25, bte. 15
B-1040 Brussels, Belgium
Contact: Secretary General
Tel: 32 2 230 4848
Fax: 32 2 230 2680
E-mail: europeche@free.way.net

EuroTurtle–Mediterranean Sea Turtle Conservation Database
Biology Department
King's College
Taunton, Somerset, TA1 3DX, United Kingdom
Tel: 44 18 2327 5092
Fax: 44 18 2333 4236
E-mail: roger@kings-tauton.co.uk
http://www.euroturtle.org/

Federation of British Port Wholesale Fish Merchants Association
Wharncliffe Road
Fish Docks
Grimsby, Humberside, DN31 3QJ, United Kingdom
Contact: Chair
Tel: 44 14 7235 0022
Fax: 44 14 7224 0838

Federation of European Aquaculture Producers (FEAP)
30, rue Vivaldi
B-4100 Boncelles, Belgium
Contact: Secretary General
Tel: 32 4 338 2995
Fax: 32 4 337 9846
E-mail: secretariat@feap.info
http://www.feap.info/feap/

Federation of European Microbiological Societies
Poortlandplein 6
NL-2628 BM Delft, The Netherlands
Contact: Central Office
Tel: 31 15 278 5604
Fax: 31 15 278 5696
E-mail: fems@fems-microbiology.org
http://www.fems-microbiology.org/

Fisheries Society of the British Isles
c/o Granta Information Systems
82A High Street
Sawston, Cambridge CB2 4HJ, United Kingdom
Contact: President
Tel: 44 12 2383 0665

Fax: 44 12 2383 9804
E-mail: fsbi@grantais.demon.uk
http://www.le.ac.uk/biology/fsbi/index.html

Fridtjof Nansen Institute
PO Box 326
N-1326 Lysaker, Norway
Contact: Director
Tel: 47 67 11 19 00
Fax: 47 67 11 19 10
E-mail: post@fni.no
http://www.fni.no/

Geographical Information Systems (GISIG) International Group
Via Piacenza 54
16138 Genova, Italy
Tel: 39 010 8355 588
Fax: 39 010 8357 190
E-mail: gisig@gisig.it
http://www.gisig.it/

German Institute of Navigation (DGON)
Kölnstrasse 70
D-5311 Bonn, Germany
Tel: 49 228 201970
Fax: 49 228 2019719
E-mail: dgon.bonn@t-online.de
http://www.dgon.de/

German Shipbuilding and Ocean Industries Association
(Verband für Schiffbau und Meerestechnik–VSM)
An der Alster 1
D-20099 Hamburg, Germany
Tel: 49 40 28 01520
Fax: 49 40 28 015230
E-mail: info@vsm.de
http://www.vsm.de/

High North Alliance
PO Box 123
N-8390 Reine i Lofoten, Norway
Contact: Chair
Tel: 47 76 09 24 14

Fax: 47 76 09 24 50
E-mail: hna@highnorth.no
http://www.highnorth.no/

Icelandic Fisheries Laboratories
Skulagata 4
PO Box 1405
IS-101 Reykjavik, Iceland
Tel: 354 562 0240
Fax: 354 562 0740
E-mail: info@rf.is
http://www.rfisk.is/home.htm

Institute for Fisheries Management and Coastal Community Development
PO Box 104
DK-9850 Hirtshals, Denmark
Tel: 45 98 94 28 55
Fax: 45 98 94 42 68
E-mail: ifm@ifm.dk
http://www.ifm.dk/

Institute of Fisheries Management
3 Chase Plain Cottage
Portsmouth Road
Hind Head, Surrey, GU26 6BZ, United Kingdom
Tel: 44 19 3362 3206
Fax: 44 11 5982 6150
E-mail: admin@ifm.org.uk
http://www.ifm.org.uk/

International Association for the Rhine Vessels Register (IVR)
Vasteland 12E
NL-3011 BL, Rotterdam, Netherlands
Tel: 31 10 411 6070
Fax: 31 10 412 9091
E-mail: info@ivr.nl
http://www.ivr.nl/

International Association of Waterworks in the Rhine Basin (IAWR)
(Internationale Arbeitsgemeinschaft der Wasserwerke im Rheineinzugsgebiet)
PO Box 402
NL-3430 AK Nieuwegein, The Netherlands
Tel: 31 30 600 9030
Fax: 31 30 600 9039
E-mail: riwa@riwa.org; iawr@riwa.org
http://www.iawr.org/

International Commission for Protection of the Rhine Against Pollution (ICPR)
(Internationale Kommission zum Schutze des Rheins Gegon Verunreinizung–IKSR)
Hohenzollernstrasse 18
Postfach 200253
D-56002 Koblenz, Germany
Tel: 49 261 12495
Fax: 49 261 36572
E-mail: sekretariat@iksr.de
http://www.iksr.de/

International Marine Certification Institute (IMCI)
rue Abbé Cuypers 3
B-1040 Brussels, Belgium
Tel: 32 2 741 6836
Fax: 32 2 741 2418
E-mail: imcibrux@aol.com
http://members.aol.com/imcibrux/

International Ocean Institute–Black Sea
National Institute for Marine Research and Development "Grigore Antipa"
Blvd Mamaia 300
RO-8700 Constanta 3, Romania
Contact: Director
Tel: 40 41 543 288
Fax: 40 41 831 274
E-mail: abologa@alpha.rmri.ro

International Ocean Institute–Malta
The Farmhouse–Mediterranean Institute
University of Malta
MT-Msida MSD 06, Malta

Contact: Director
Tel: 356 3290 2430
Fax: 356 320 717
E-mail: mzammit@arts.um.edu.mt;
   mzamz@um.edu.mt

Marine Biological Association of the United Kingdom
The Laboratory
Citadel Hill
Plymouth, Devon PL1 2PB, United Kingdom
Contact: Director
Tel: 44 17 5263 3207
Fax: 44 17 5263 3102
E-mail: sec@mba.ac.uk
http://www.mba.ac.uk/

Marine Conservation Society
9 Gloucester Rd.
Ross-on-Wye,
Herefordshire, HR9 5BU, United Kingdom
Contact: Secretary
Tel: 44 19 8956 6017
Fax: 44 19 8956 7815
E-mail: info@mcsuk.org
http://www.mcsuk.org/

Marine Mammal Council
Nachimovskiy Avenue 36
117218 Moscow, Russia
Tel/Fax: 7 095 124 7290
E-mail: support@2mn.org
http://www.2mn.org/

Marine Society
202 Lambeth Rd.
London, SE1 7JW, United Kingdom
Contact: General Secretary
Tel: 44 20 7261 9535
Fax: 44 20 7401 2537
E-mail: enq@marine-society.org
http://www.marine-society.org.uk/

Marine Studies Group
The Geological Society of London
c/o British Antarctic Survey
High Cross
Madingley Rise
Cambridge, CB3 0ET, United Kingdom
Contact: Secretary
Tel: 44 12 2322 1573
Fax: 44 12 2336 2616
E-mail: rdla@pcmail.nerc-bas.ac.uk
   http://www.ocean.cf.ac.uk/
   people/neil/MSG/

Maritime Development Center of Europe
Amaliegade 33
DK-1256 Copenhagen K, Denmark
Tel: 45 33 33 74 88
Fax: 45 33 33 75 88
E-mail: info@maritimecenter.dk
http://www.maritimecenter.dk/

MEDCOAST
Middle East Technical University
06531, Ankara, Turkey
Tel: 90 312 210 5429
Fax: 90 312 210 1412
E-mail: medcoast@metu.edu.tr
http://www.medcoast.org.tr/

Mediterranean Association to Save the Sea Turtles (MEDASSET/GR)
c/o 24 Park Towers
2 Brick Street
London W1J 7DD, United Kingdom
Contact: President
Tel: 44 20 7629 0654
E-mail: medasset@medasset.org
http://www.euroturtle.org/

Mediterranean Marine Bird Association (MEDMARAVIS)
PO Box 2
F-83470 Saint Maximin, France
Tel: 33 4 9459 4069
Fax: 33 4 9459 4738
E-mail: medmaraxm@aix.pacwan.net
http://www.geocities.com/
   med avis/

National Federation of Fishermen's Organisations (NFFO)
Marsden Road
Fish Docks
Grimsby, DN31 3SG, United Kingdom
Contact: Secretary
Tel: 44 14 7235 2141
Fax: 44 14 7224 2486
E-mail: nffo@nffo.org.uk
http://www.nffo.org.uk/

National Union of Marine, Aviation and Shipping Transport Officers (NUMAST)
Oceanair House
750-760 High Road Leytonstone
London, E11 3BB, United Kingdom
Contact: General Secretary
Tel: 44 20 8989 6677
Fax: 44 20 8530 1015
E-mail: enquiries@numast.org
http://www.numast.org/

Netherlands Institute of Navigation
Seattleweg 7 (Haven 2801)
NL-3195 ND Pernis, The Netherlands
Contact: Director
Tel: 31 010 498 7518
Fax: 31 010 452 3457
E-mail: r.vanrhee@loodswezen.nl
http://www.nlr.nl/nin/

Nordic Marine Insurance Pool (NMIP)
(Nordisk Sjoforsikringspool–NSP)
c/o Swedish Association of Marine Underwriters
Klara Norra Kyrkogata 33
SE-111 22 Stockholm, Sweden
Contact: Secretary
Tel: 46 87 83 98 50
Fax: 46 87 83 98 95
E-mail: sten.gothberg@sjoass.se

Northern Shipowners' Defence Club
(Nordisk Skipsrederforening)
Kristinelundveien 22
Postboks 3033 El.
N-0207 Oslo 9, Norway
Contact: General Manager
Tel: 47 22 13 56 00
Fax: 47 22 43 00 35
E-mail: post@nordisk.no
http://www.nordisk.no/

Offshore Pollution Liability Association
Bank Chambers
29 High Street
Ewell, Surrey, KT17 1SB, United Kingdom
Contact: Managing Director
Tel: 44 20 8786 3640
Fax: 44 20 8786 3641
E-mail: opol@compuserve.com
http://www.opol.org.uk/

Royal Institute of Navigation
1 Kensington Gore
London, SW7 2AT, United Kingdom
Contact: Director
Tel: 44 20 7591 3130
Fax: 44 20 7591 3131
E-mail: info@rin.org.uk
http://www.rin.org.uk/

Royal Institution of Naval Architects
10 Upper Belgrave St.
London, SW1X 8BQ, United Kingdom
Contact: Secretary
Tel: 44 20 7235 4622
Fax: 44 20 7259 5912
E-mail: hq@rina.org.uk
http://www.rina.org.uk/

Royal Meteorological Society
104 Oxford Road
Reading, Berkshire, RG1 7LL, United Kingdom
Contact: Executive Director
Tel: 44 11 8956 8500
Fax: 44 11 8956 8571

E-mail: execdir@royalmetsoc.org
http://www.royal-met-soc.org.uk/

Scottish Fishermen's Federation
14 Regent Quay
Aberdeen, AB11 5AE, United Kingdom
Tel: 44 12 2458 2583
Fax: 44 12 2457 4958
E-mail: ssf@ssf.co.uk
http://www.sff.co.uk/

Sea Watch Foundation
36 Windmill Road
Headington, Oxford, OX3 7BX, United Kingdom
Contact: Director
Tel: 44 18 6576 4794
Fax: 44 18 6576 4757
E-mail: info@seawatchfoundation.org.uk
http://www.seawatchfoundation.org.uk/

Shetland Fish Processors' Association
Shetland Seafood Centre
Stewart Building
Lerwick, Shetland, ZE1 0LL, United Kingdom
Tel: 44 15 9569 3644
Fax: 44 15 9569 6126
E-mail: sfpa@fishuk.net
http://www.fishuk.net/sfpa/index.htm

Shetland Salmon Farmers' Association
Shetland Seafood Centre
Stewart Building
Lerwick, Shetland, ZE1 0LL, United Kingdom
Contact: Chairman
Tel: 44 15 9569 5579
Fax: 44 15 9569 4494
E-mail: ssfa@fishuk.net
http://www.fishuk.net/ssfa/index.htm

Ship Suppliers Organization of the European Community (OCEAN)
Van Stolkweg 31
NL-2585 JN, The Hague, Netherlands
Contact: Secretariat
Tel: 31 70 358 9137
Fax: 31 70 358 4538
E-mail: secretariaat@vanstolkweg.nl

Shipbuilders and Shiprepairers Association (SSA)
Marine House
Thorpe Lea Road
Egham, Surrey, TW20 8BF, United Kingdom
Contact: Director
Tel: 44 17 8422 3770
Fax: 44 17 8422 3775
E-mail: office@ssa.org.uk
http://www.ssa.org.uk/

Society for Nautical Research (SNR)
National Maritime Museum
Greenwich, London, SE10 9NF, United Kingdom
Contact: Honorary Secretary
Tel: 44 20 8312 6772
Fax: 44 20 8312 6661
http://www.snr.org/

Society of Consulting Marine Engineers and Ship Surveyors (SCMES)
202 Lambeth Road
London, SE1 7JW, United Kingdom
Contact: Secretary
Tel: 44 20 7261 0869
Fax: 44 20 7261 0871
E-mail: sec@scmshq.org
http://www.scmshq.org/

Society of Maritime Industries
30 Great Guildford Street, 4th Floor
London SE1 0HS, United Kingdom
Tel: 44 20 7928 9199
Fax: 44 20 7928 6599

E-mail: info@maritimeindustries.org
http://www.maritimeindustries.org/

Stockholm Environment Institute
Lilla Nygatan 1
Box 2142
SE-103 14 Stockholm, Sweden
Contact: Director
Tel: 46 84 12 14 00
Fax: 46 87 23 03 48
E-mail: postmaster@sei.se
http://www.sei.se/

Swedish Ports and Stevedores Association
Box 1621
SE-111 86 Stockholm, Sweden
Tel: 46 87 62 71 00
Fax: 46 86 11 12 18
E-mail: ports@transportgruppen.se
http://www.transportgruppen.se/

Tethys Research Institute
c/o Acquario Civico
Viale G.B. Gadio 2,
I-20121 Milano, Italy
Tel: 39 02 7200 1947
Fax: 39 02 7200 1946
E-mail: tethys@tethys.org
http://www.tethys.org/

The Netherlands' Shipbuilding Industry Association (VNSI)
PO Box 138
NL-2700 AC, Zoetermeer, The Netherlands
Tel: 31 79 353 11 65
Fax: 31 79 353 11 55
E-mail: info@vnsi.nl
http://www.vnsi.nl/

Whale and Dolphin Conservation Society
Brookfield
38 St. Paul Street
Chippenham, Wiltshire, SN15 1LY, United Kingdom
Tel: 44 12 4944 9500
Fax: 44 12 4944 9501
E-mail: info@wdcs.org
http://www.wdcs.org/

5. REGIONAL NONGOVERNMENTAL ORGANIZATIONS

5.5 Australia and Oceania

Australian Conservation Foundation
60 Leicester Street, Floor 1
Carlton, VIC 3053, Australia
Tel: 61 03 9345 1111
Fax: 61 03 9345 1166
http://www.acfonline.org.au/

Australian Coral Reef Society
c/o Centre for Marine Sciences
University of Queensland
St. Lucia, QLD 4072, Australia
Contact: Secretary
Tel: 61 7 3346 9576
Fax: 61 7 3365 4755
E-mail: acrs@jcu.edu.au
http://www.australiacoralreefsociety.org/

Australian Marine Conservation Society
PO Box 3139
Yeronga, QLD 4104, Australia
Tel: 61 07 3848 5235
Fax: 61 07 3892 5814
E-mail: amcs@amcs.org.au
http://www.amcs.org.au/

Australian Marine Sciences Association (AMSA)
PO Box 902
Toowong, QLD 4066, Australia
E-mail: amsa@amsa.asn.au
http://www.amsa.asn.au/

Australian Water Association
PO Box 388
Artaman, NSW 1570, Australia

Tel: 61 02 9413 1288
Fax: 61 02 0413 1047
E-mail: info@awa.asn.au
http://www.awa.asn.au/

Cawthron Institute
Private Bag 2
98 Halifax Street East
Nelson, New Zealand
Tel: 64 3 548 2319
Fax: 64 3 546 9464
E-mail: info@cawthron.org.nz
http://www.cawthron.org.nz/

Eastern Dredging Association (EADA)
c/o Port Klang Association
Mail Bag Service 202
42009 Port Klang, Malaysia
Contact: Secretary General
Fax: 603 3157 0211
E-mail: david@pka.gov.my

International Ocean Institute – Pacific Islands
c/o Marine Studies Programme
University of the South Pacific
PO Box 1168
Suva, Fiji
Contact: Director
Tel: 679 305 446
Fax: 679 301 490
E-mail: veitayaki_j@usp.ac.fj
http://www.usp.ac.fj/marine/

Marine and Coastal Community Network (MCCN)
PO Box 3139
Yeronga, QLD 4104, Australia
Contact: National Coordinator
Tel: 61 7 3848 5360
Fax: 61 7 3892 5814
E-mail: aus@mccn.org.au
http://www.mccn.org.au/

Maritime Union of Australia
365 Sussex Street, Level 2
Sydney, NSW 2000, Australia
Contact: General Secretary

Tel: 61 2 9267 9134
Fax: 61 2 9261 5897
E-mail: muano@mua.org.au
http://www.mua.org.au/

New Zealand Marine Sciences Society
c/o Carlyon Rd, RD1
Upper Moutere
Nelson, New Zealand
Contact: Secretary
Tel: 643 543 2109
E-mail: nzmss@hotmail.com
http://nzmss.rsnz.org/

Oceania Project
PO Box 646
Byron Bay, NSW 2481, Australia
Tel: 61 2 6685 8128
Fax: 61 2 6685 8998
E-mail: trish.wally@oceania.org.au
http://oceania.org.au/

Pacific Ocean Research Foundation
PO Box 4800
74-381 Kaeakehe Pkwy, Suite C
Kailua-Kona, HI, 96740, USA
Contact: Chairman
Tel: 1 808 329-6105
Fax: 1 808 329-1148
E-mail: porf@aloha.net
http://holoholo.org/porf/

PACON International
PO Box 11568
Honolulu, HI, 96828-0568, USA
Tel: 1 808 956-6163
Fax: 1 808 956-2580
E-mail: pacon@hawaii.edu
http://www.hawaii.edu/pacon/

Seafriends Marine Conservation and Education Centre
7 Goat Island Road
Leigh R.D.5., New Zealand
Tel: 64 9422 6212
Fax: 64 9422 6245
E-mail: floor@seafriends.org.nz
http://www.seafriends.org.nz/

South Pacific Tourism Organization (SPTO)
PO Box 131119
Suva, Fiji
Tel: 679 330 4177
Fax: 679 330 1995
E-mail: info@spto.org
http://www.tcsp.com/

5. REGIONAL NONGOVERNMENTAL ORGANIZATIONS

5.6 Polar Regions

Antarctic and Southern Ocean Coalition (ASOC)
The Antarctica Project
1630 Connecticut Ave., NW, 3rd Floor
Washington, DC, 20009, USA
Contact: Director
Tel: 1 202 234-2480
Fax: 1 202 234-2482
E-mail: antarctica@igc.org
http://www.asoc.org/

Antarctic Institute of Canada
PO Box 1223, Main Post Office
Edmonton, AB, T5J 2M4, Canada
Contact: Executive Officer
Tel/Fax: 1 780 452-7392
E-mail: mardon@freenet.edmonton.ab.ca

Arctic Ocean Science Board (AOSB)
National Science Foundation, Room 1070
4201 Wilson Blvd
Arlington, VA, 22230, USA
Contact: Secretariat
Tel: 1 703 292-7856
Fax: 1 703 292-9152
E-mail: info@aosb.org
http://www.aosb.org/

Canadian Arctic Resources Committee (CARC)
7 Hinton Ave. N., Suite 200
Ottawa, ON, K1Y 4P1, Canada
Contact: Executive Director
Tel: 1 613 759-4284
Fax: 1 613 722-3318
E-mail: info@carc.org
http://www.carc.org/

International Polar Heritage Committee
c/o Directorate for Cultural Heritage
PO Box 8196 Dep
N-0034 Oslo, Norway
Contact: Secretary General
Tel: 47 22 94 04 00
E-mail: pchaplin@online.no
http://www.polarhertiage.com/

Inuit Circumpolar Conference (ICC)
c/o ICC (Greenland)
PO Box 204
3900 Nuuk, Greenland
Contact: President
Tel: 299 3 23632
Fax: 299 3 23001
E-mail: aqqaluk@inuit.org
http://www.inuit.org/

Institute for Environmental Monitoring and Research
PO Box 1859, Stn. "B"
Happy Valley-Goose Bay, NL, A0P 1E0, Canada
Tel: 1 709 896-3266
Fax: 1 709 896-3076
E-mail: iemr@iemr.org
http://www.iemr.org/

6. REGIONAL ACADEMIC ORGANIZATIONS

6.1 Africa

Académie régionale des Sciences et Techniques de la Mer
(Regional Maritime Academy)

BP 158
Abidjan, Côte d'Ivoire
Contact: Director General
Tel: 225 371823

Arab Academy for Science and Technology and Maritime Transport
Gamal Abdel Naser Street
PO Box 1029
Alexandria, Egypt
Contact: Director General
Tel: 20 3 556 1497
Fax: 20 3 548 7786
http://www.aast.edu/

Benguela Environment Fisheries Interaction and Training Programme (BENEFIT)
PO Box 912
Swakopmund, Namibia
Contact: Secretariat
Tel: 264 64 4101164
Fax: 264 64 405913
E-mail: chocutt@mfmr.gov.na
http://www.benefit.org.na/

Centre de Recherches Océanolographiques (CRO)
Département des Ressources Halieutiques Vivantes
29, rue des Pêcheurs BP V18
Abidjan, Côte d'Ivoire
Tel: 225 21 35 5014
Fax: 225 21 35 1155
E-mail: postmaster@cro.orstom.ci
http://www.ci.refer.org/ivoir_ct/rec/cdr/cro/accueil.htm

Centre for Documentation, Research and Training on the South West Indian Ocean (CEDREFI)
PO Box 91
Rose-Hill, Republic of Mauritius
Contact: Director
Tel: 230 465 5036
Fax: 230 465 1422

Centre for Environment and Development in Africa (CEDA)
Université Nationale du Bénin
BP 7060
081 Cotonou, Bénin
Tel: 229 33 1917
Fax: 229 33 1981
E-mail: ceda@bow.intnet.bj

Centre for Marine Studies
University of Cape Town
Oceanography Annexe
Residence Road
Private Bag
Rondebosch 7701, South Africa
Contact: Director
Tel: 27 21 650 3277
Fax: 27 21 650 3282
E-mail: cms@physci.uct.ac.za
http://www.sea.uct.ac.za/cms/

Departement de Biologie et Physiologie Animale
Faculté des Sciences
Université de Yaoundé I
BP 337
Yaoundé, Cameroon
Contact: Chef de Departement
Tel: 237 22 0744
Fax: 237 22 1320

Departement de Biologie et de Physiologie Animale
Faculté des Biosciences
Université Cocody
BP V 34
Abidjan 01, Côte d'Ivoire
Contact: Chef de Departement
Tel: 225 449000
Fax: 224 441407

Departement de Zoologie
Faculté des sciences et techniques
Université Nationale du Bénin (UNB)
BP 526
Cotonou, Bénin
Contact: Head of Department

Tel: 229 360074
Fax: 229 301638

Departemento de Biologia
Universidade Agostinho Neto
PO Box 815 BT
Avenida 4 de Fevereiro
Luanda, Angola
Contact: Chef de Departement
Tel: 244 330089
Fax: 244 330520

Department of Environmental and
    Geographical Science
University of Cape Town
Private Bag
Rondebosch 7701, South Africa
Tel: 27 21 650 2873/4
Fax: 27 21 650 3791
E-mail: admin@enviro.uct.ac.za
http://www.egs.uct.ac.za/

Department of Natural Resources
    and Conservation
Faculty of Agriculture and Natural
    Resources
University of Namibia
Private Bag 13301
Windhoek, Namibia
Tel: 264 61 206 3890
Fax: 264 61 206 3013
E-mail: fanr@unam.na
http://www.unam.na/faculties/
    agriculture/nrc.htm

Department of Oceanography
University of Cape Town
Private Bag
Rondebosch 7701, South Africa
Tel: 27 21 650 3278
Fax: 27 21 650 3979
E-mail: scifac@science.uct.ac.za
http://www.uct.ac.za/

Environmental Study and Research
    Centre
Université Gamal Abdel Nasser de
    Conakry
BP 1147
Conakry, Guinea
Contact: Director
Tel: 224 46 4689
Fax: 224 46 4808

Faculte des Sciences
Université Omar Bongo
131 Blvd Léon M'Ba
BP 13131
Libreville, Gabon
Tel: 241 72 6910
Fax: 241 73 0417

Faculty of Law
University of Namibia
Private Bag 13301
Windhoek, Namibia
Contact: Dean
Tel: 264 61 206 3998
Fax: 264 61 242 644
E-mail: smbalwe@unam.na
http://www.unam.na/faculties/
    law/

Henties Bay Marine and Coastal Resources Research Centre
University of Namibia
Private Bag 13301
Windhoek, Namibia
E-mail: fmolloy@unam.na
http://www.unam.na/research/
    henties/

Institut des Sciences de la Mer et de
    l'Aménagement du Littoral
    (ISMAL)
BP 54
Sidi-Fredj
42321 d'Alger, Algeria
Tel: 213 2 377076
Fax: 213 2 376806
E-mail: dir-ismal@hotmail.com
http://www.ismal.net/

Institut Scientifique
Université Mohammed V
Ave Ibn Batouta

BP 703 Rabat-Agdal
10106 Rabat, Morocco
Contact: Directeur
Tel: 212 3777 4548
Fax: 212 3777 4540
http://www.israbat.ac.ma/

Institute of Marine and Environmental Law
University of Cape Town
Private Bag
Rondebosch 7701, South Africa
Contact: Director
Tel: 27 21 650 5642
Fax: 27 21 650 5607
E-mail: lawmar@law.uct.ac.za
http://www.uct.ac.za/depts/pbl/imel/intro.htm

Institute of Marine Biology and Oceanography
Fourah Bay College
University of Sierra Leone
Mount Aureol
Freetown, Sierra Leone
Contact: Director
Tel: 232 22 227 924
Fax: 232 22 224 260
E-mail: fbcadmin@sierratel.sl
http://fbcusl.8k.com/index.html

Institute of Marine Sciences
University of Dar es Salaam
PO Box 668
Zanzibar, Tanzania
Contact: Director
Tel: 255 24 223 2128
E-mail: director@ims.udsm.ac.tz
http://www.ims.udms.ac.tz/

Institute of Maritime Technology
PO Box 181
Simon's Town 7995, South Africa
Tel: 27 21 786 1092
Fax: 27 21 786 3634
E-mail: jwjvw@msi.imt.za
http://www.imt.za/imt/

Institute of Oceanography
University of Calabar
PMB 1115
Calabar, Cross River State, Nigeria
Contact: Coordinator
Tel: 234 222 855
Fax: 234 224 996

Laboratoire Biologie Marine
Institut Fondamental d'Afrique Noire (IFAN)
Cheikh Anta Diop Université
BP 206
Dakar, Senegal
Contact: Chef de Departement
Tel: 221 825 0090
Fax: 221 244 918
E-mail: bifan@telecomplus.sn
http://www.refer.sn/ifan/

Nigerian Institute for Oceanography and Marine Research (NIOMR)
Wilmot Point Road
Bar Beach
PO Box 12729
Lagos, Nigeria
Tel/Fax: 234 1 2617 530
E-mail: niomr@linkserve.com.ng

Organisation Africaine de Cartographie et Télédétection (OACT)
African Organisation of Cartography and Remote Sensing (AOCRS)
BP 102-5 route de Badjarah
16040 Hussein Dey
Algiers, Algeria
Contact: Chairman
Tel: 213 2 77 7938
Fax: 213 2 77 7934
E-mail: oact@wissal.dz
http://oact.citeglobe.com/oact1.htm

Regional Maritime Academy
PO Box 1115
Accra, Ghana
Contact: Director General

Tel: 233 21 712775
Fax: 233 21 712047
E-mail: rma@africaonline.com.gh
http://www.rma-edu.com/

School of Life and Environmental Sciences
University of Natal
Durban 4041, South Africa
Tel: 27 31 260 3192
Fax: 27 31 260 2029
E-mail: info@biology.und.ac.za
http://www.nu.ac.za/sles/

School of Maritime Studies
University of Natal
Durban, 4041, South Africa
Tel: 27 31 260 2994
Fax: 27 31 260 3163
E-mail: maritime@nu.ac.za
http://www.maritime.und.ac.za/

Shipping Law Unit
University of Cape Town
Private Bag
Rondebosch 7701, South Africa
Contact: Director
Tel: 27 21 650 3065
Fax: 27 21 650 5607
E-mail: shiplaw@shiplaw.uct.ac.za
http://www.uctshiplaw.com/

UNESCO Chair in Integrated Coastal Management and Sustainable Development
Departement de Geographie
Faculté des Lettres et des Sciences Humaines
Cheikh Anta Diop Université
BP 5005
Dakar-Fann, Senegal
Contact: Chef de Departement
Tel: 221 825 7528
Fax: 221 825 3724
E-mail: info@ucad.sn
http://www.ucad.sn/

6. REGIONAL ACADEMIC ORGANIZATIONS

6.2 The Americas

Acadia Centre for Estuarine Research
Acadia University
Campus Box 115
Wolfville, NS, B0P 1X0, Canada
Contact: Director
Tel: 1 902 585-1113
Fax: 1 902 585-1054
E-mail: nancy.roscoe-huntley@acadiau.ca
http://ace.acadiau.ca/science/cer/home.htm

AquaNet
c/o Ocean Sciences Centre
Memorial University of Newfoundland
St. John's, NL, A1C 5S7, Canada
Contact: Administrative Centre
Tel: 1 709 737-3245
Fax: 1 709 737-3500
E-mail: info@aquanet.ca
http://www.aquanet.mun.ca/

Association of Marine Laboratories of the Caribbean (AMLC)
Mote Marine Laboratory
1600 Ken Thompson Parkway
Sarasota, FL, 34236, USA
Contact: Executive Director
Tel: 1 941 388-4441
Fax: 1 941 388-4312
E-mail: slegore@mindspring.com
http://amlc.uvi.edu/

Atmospheric and Oceanic Sciences
McGill University
805 Sherbrooke St. West
Montreal, QC, H3A 2K6, Canada
Contact: Chair
Tel: 1 514 398-3764
Fax: 1 514 398-6115

E-mail: ornella@zephyr.meteo.
  mcgill.ca
http://www.mcgill.ca/meteo/

Bamfield Marine Station
Bamfield, BC, V0R 1B0, Canada
Contact: Director
Tel: 1 250 728-3301
Fax: 1 250 728-3452
E-mail: info@bms.bc.ca
http://www.bms.bc.ca/

Bellairs Research Institute
St. James, Barbados
Tel: 1 246 422-2087
Fax: 1 246 422-0692
E-mail: bellairs@sunbeach.net
http://www.mcgill.ca/bellairs/

Belle W. Baruch Institute for Marine Biology and Coastal Research
University of South Carolina
607 EWS Building
Columbia, SC, 29208, USA
Contact: Director
Tel: 1 803 777-5288
Fax: 1 803 777-3935
E-mail: fletcher@biol.sc.edu
http://inlet.geol.sc.edu/

Bermuda Biological Station for Research
17 Biological Station Lane
Ferry Reach
St. George's, GE01, Bermuda
Contact: Director
Tel: 1 441 297-1880
Fax: 1 441 297-8143
E-mail: webmaster@bbsr.edu
http://www.bbsr.edu/

Bodega Marine Laboratory
University of California at Davis
2099 Westside Road
PO Box 247
Bodega Bay, CA, 94923-0247, USA
Tel: 1 707 875-2211
Fax: 1 707 875-2009

E-mail: ucdbml@ucdavis.edu
http://www-bml.ucdavis.edu/

Canadian Coast Guard College
PO Box 4500
Sydney, NS, B1P 6L1, Canada
Tel: 1 902 567-3208
Fax: 1 902 567-3233
E-mail: registrar@cgc.ns.ca
http://www.cgc.ns.ca/

Canadian Institute for Climate Studies
University of Victoria
130 Saunders Annex
PO Box 1700 Sta CSC
Victoria, BC, V8W 2Y2, Canada
Tel: 1 250 721-6236
Fax: 1 250 721-7217
E-mail: climate@uvic.ca
http://www.cics.uvic.ca/

Canadian Institute of Fisheries Technology (CIFT)
Faculty of Engineering
Dalhousie University
PO Box 1000
1360 Barrington St.
Halifax, NS, B3J 2X4, Canada
Tel: 1 902 494-6030
Fax: 1 902 420-0219
E-mail: alex.speers@dal.ca
http://www.dal.ca/cift/

Caribbean Marine Research Center
Perry Institute for Marine Science
100N US Highway 1, Suite 202
Jupiter, FL, 33477, USA
Tel: 1 561 741-0192
Fax: 1 561 741-0193
E-mail: cmrc@cmrc.org
http://www.cmrc.org/

Center for Coastal and Land-Margin Research (CCALMR)
Department of Environmental Science and Engineering

OGI School of Science and Engineering
Oregon Health & Science University
20000 NW Walker Road
Beaverton, OR, 97006, USA
Contact: Director
Tel: 1 503 690-1147
Fax: 1 503 690-1273
E-mail: baptista@ccalmr.ogi.edu
http://www.ccalmr.ogi.edu/

Center for Coastal Studies
Scripps Institution of Oceanography
University of California, San Diego
La Jolla, CA, 92093-0209, USA
Contact: Director
Tel: 1 858 534-4333
Fax: 1 858 534-0300
E-mail: rguza@ucsd.edu
http://www-ccs.ucsd.edu/

Center for Oceans Law and Policy
University of Virginia School of Law
580 Massie Road
Charlottesville, VA, 22903, USA
Contact: Director
Tel: 1 434 924-7441
Fax: 1 434 924-7362
E-mail: colp@virginia.edu
http://www.virginia.edu/colp/

Center for the Study of Marine Policy
University of Delaware
301 Robinson Hall
Newark, DE, 19716, USA
Contact: Director
Tel: 1 302 831-8086
Fax: 1 302 831-3668
E-mail: bcs@udel.edu
http://www.udel.edu/CMS/cmsp/

Centre for Asian Legal Studies
Faculty of Law
University of British Columbia
1822 East Mall
Vancouver, BC, V6T 1Z1, Canada
Tel: 1 604 822-3151

Fax: 1 604 822-8108
E-mail: borthwick@law.ubc.ca
http://www.law.ubc.ca/centres/index.htm

Centre for Coastal Management and Coastal Ecology
Universidad de Guadalajara
Gomez Farias #82
San Patricio Melaque
CP 48980
Jalisco, Mexico
http://www.udg.mx/

Centre for Community-based Management
St. Francis Xavier University
PO Box 5000
Antigonish, NS, B2G 2W5, Canada
Tel: 1 902 867-2433
E-mail: ccbm@stfx.ca
http://www.stfx.ca/institutes/ccbm/

Centre for Earth and Ocean Research (CEOR)
University of Victoria
Petch Building, Room 169
PO Box 3055, Stn CSC
Victoria, BC, V8W 3P6, Canada
Contact: Director
Tel: 1 250 721-8848
Fax: 1 250 472-4100
E-mail: ceor@uvic.ca
http://ceor.seos.uvic.ca/

Centre for Indigenous Peoples' Nutrition and Environment (CINE)
Macdonald Campus of McGill University
21,111 Lakeshore Road
Ste-Anne-de-Bellevue, QC, H9X 3V9, Canada
Tel: 1 514 398-7544
Fax: 1 514 398-1020
E-mail: cine@cine.mcgill.ca
http://cine.mcgill.ca/

Centre for Marine Sciences
University of the West Indies, Mona Campus
Mona, Kingston 7, Jamaica
Tel: 1 876 927-0262
Fax: 1 876 927-2765
http://www.uwimona.edu.jm/

Centre for Tourism Policy and Research
Simon Fraser University
8888 University Drive
Burnaby, BC, V5A 1S6, Canada
Contact: Director
Tel: 1 604 291-3074
Fax: 1 604 291-4968
E-mail: peter_williams@sfu.ca
http://www.sfu.ca/dossa/

Centro de Investigaciones Marinas
Universidad de la Habana
Calle 16 # 114, entre 1era y 3era, Miramar
Playa, Ciudad de la Habana, Cuba
E-mail: cim@nova.uh.cu
Tel: 537 23 0617
Fax: 537 24 2087
http://www.uh.cu/centros/cim/index.htm

Charles Darwin Research Station
Puerto Ayora, Santa Cruz
Galapagos Islands, Ecuador
Contact: Director
Tel: 593 5526 146/7
E-mail: info@darwinfoundation.org
http://www.darwinfoundation.org/

Chesapeake Biological Laboratory
Center for Environmental Science
University of Maryland
PO Box 38
1 Williams Street
Solomons, MD, 20688, USA
Contact: Director
Tel: 1 410 326-4281
Fax: 1 410 326-7264
E-mail: webmaster@cbl.umces.edu
http://www.cbl.umces.edu/

Coastal and Oceanic Geological Research Centre
(Centro de Estudos de Geologia Costeira e Oceânica – CECO)
Universidade Federal do Rio Grande do Sul
Av. Bento Gonçalves, 9.500 Prédio 43125
CEP 91.509-900, Campus do Vale
Porto Alegre, RS, Brazil
Tel: 55 51 3166388
Fax: 55 51 3191811
E-mail: ceco@if.ufrgs.br
http://www.ufrgs.br/ceco/

Coastal Fisheries Institute
Louisiana State University
204 Wetland Resource Building
Baton Rouge, LA, 70803, USA
Tel: 1 255 578-6455
Fax: 1 255 578-6513
E-mail: rshaw@lsu.edu
http://www.cfi.lsu.edu/

Coastal Institute
Narragansett Bay Campus
University of Rhode Island
Narragansett, RI, 02882, USA
Tel: 1 401 874-6513
Fax: 1 401 874-6869
E-mail: ci@edc.uri.edu
http://www.ci.uri.edu/

Coastal Resources Center
Narragansett Bay Campus
University of Rhode Island
Narragansett, RI, 02882, USA
Tel: 1 401 874-6224
Fax: 1 401 789-4670
E-mail: communications@crc.uri.edu
http://www.crc.uri.edu/

Coastal Resources Institute
ECMAR-UNA
Universidad Nacional
Facultad de Ciencias Exactas y Naturales
Heredia, Costa Rica

Tel: 506 277 3313
Fax: 506 277 3485
Tel/Fax: 506 661 2394
E-mail: ecmar@una.ac.cr
http://www.una.ac.cr/ecmar/

College of Ocean and Fishery Sciences
University of Washington
1492 NE Boat Street
PO Box 355350
Seattle, WA, 98195-5350, USA
Contact: Dean
Tel: 1 206 543-6605
Fax: 1 206 543-6393
E-mail: webmaster@cofs.washington.edu
http://www.cofs.washington.edu/

College of Oceanic and Atmospheric Sciences
Oregon State University
104 Ocean Administration Bldg
Corvallis, OR, 97331-5503, USA
Tel: 1 541 737-3504
Fax: 1 541 737-2064
E-mail: www@coas.oregonstate.edu
http://www.oce.orst.edu/

Cooperative Institute for Arctic Research (CIFAR)
University of Alaska Fairbanks
PO Box 757740
Fairbanks, AK, 99775-7740, USA
Contact: Administrator
Tel: 1 907 474-5698
Fax: 1 907 474-6722
E-mail: slynch@iarc.uaf.edu
http://www.cifar.uaf.edu/

Darling Marine Center
University of Maine
193 Clark's Cove Road
Walpole, ME, 04573, USA
Contact: Director
Tel: 1 207 563-3146
Fax: 1 207 563-3119
E-mail: darling@maine.edu
http://www.dmc.maine.edu/

Department of Earth and Ocean Sciences
University of British Columbia
6339 Stores Road
Vancouver, BC, V6T 1Z4, Canada
Contact: Department Head
Tel: 1 604 822-2449
Fax: 1 604 822-6088
E-mail: webmaster@eos.ubc.ca
http://www.eos.ubc.ca/

Department of Marine Science
University of Texas at Austin
Marine Science Institute
750 Channel View Drive
Port Aransas, TX, 78373-5015, USA
Contact: Chair
Tel: 1 361 749-6711
Fax: 1 361 749-6777
E-mail: gardner@utmsi.utexas.edu
http://www.utmsi.utexas.edu/

Department of Marine Sciences
Marine Sciences Bldg
University of Georgia
Athens, GA, 30602-3636, USA
Contact: Director
Tel: 1 706 542-7671
Fax: 1 706 542-5888
E-mail: aquadoc@arches.uga.edu
http://alpha.marsci.uga.edu/

Department of Oceanography
0102 OSB, West Call Street
Florida State University
Tallahassee, FL, 32306-4320, USA
Tel: 1 850 644-6700
Fax: 1 850 644-2581
E-mail: admissions@ocean.fsu.edu
http://www.ocean.fsu.edu/

Department of Oceanography
Texas A&M University
College Station, TX, 77843-3146, USA
Tel: 1 979 845-7211
Fax: 1 979 845-6331
E-mail: web@ocean.tamu.edu
http://www-ocean.tamu.edu/

Department of Oceanography
Fundação Universidade do Rio
   Grande
Caixa Postal, 474
96201-900, Rio Grande, RS, Brazil
Tel: 55 532 336534
Fax: 55 532 33 6601
E-mail: docosta@super.furg.br
http://www.furg.br/ciencias.marinhas

Department of Oceanography
Dalhousie University
1355 Oxford Street
Halifax, NS, B3H 4J1, Canada
Tel: 1 902 494-3557
Fax: 1 902 494-3877
E-mail: oceanography@dal.ca
http://www.dal.ca/wwwocean/
   index.html

Duke University Wetland Center
Nicholas School of the Environment
   and Earth Sciences
Box 90333
Durham, NC, 27708-0333, USA
Contact: Director
Tel: 1 919 613-8009
Fax: 1 919 613-8101
E-mail: randyn@duke.edu
http://www.env.duke.edu/wetland/

Escuela Nacional de Marina Mer-
   cante (ENAMM)
Av. Progreso 632
Chucuito-Callao, Peru
Tel: 511 4298210
Fax: 511 4298218
E-mail: informes@enamm.edu.pe
http://www.enamm.edu.pe/

Escuela Superior Politécnica del Li-
   toral
Facultad de Ingenieria Maritima y
   Ciencias del Mar
Campus Gustavo Galindo Velasco
Km. 30,5 Via Perimetral
Apartado: 09-01-5863
Guayaquil, Ecuador

Tel: 593 4 2 269494
Fax: 593 4 2 269492
E-mail: ecervantes@goliat.espol.edu.ec
http://www.espol.edu.ec/

Facultad de Ciencias del Mar
Universidad de Valparaíso
Av. Borgoño s/n, Montemar
Viña de Mar
Valparaíso, Chile
Tel: 56 32 507829
E-mail: montemar@uv.cl
http://www.uv.cl/

Facultad de Ciencias del Mar
(Faculty of Marine Sciences)
Universidad Católica del Norte
Campus Guayacán
Casilla 117
Coquimbo, Chile
Tel: 56 51 209736
Fax: 56 51 209750
E-mail: faculmar@ucn.cl
http://www.faculmar.ucn.cl/

Faculty of Law
University of Victoria
PO Box 2400, STN CSC
Victoria, BC, V8W 3H7, Canada
Tel: 1 250 721-8150
Fax: 1 250 721-6390
E-mail: lawrecep@uvic.ca
http://www.law.uvic.ca/

Fisheries Centre
University of British Columbia
2204 Main Mall
Vancouver, BC, V6T 1Z4, Canada
Contact: Administrative Clerk
Tel: 1 604 822-2731
Fax: 1 604 822-8934
E-mail: office@fisheries.ubc.ca
http://www.fisheries.ubc.ca/

Five College Coastal and Marine Sci-
   ences
Smith College
Northampton, MA, 01063, USA

Tel: 1 413 585-3799
E-mail: marinesci@smith.edu
http://websci.smith.edu/departments/MARINE/

Florida State University Marine Laboratory
034 Montgomery Bldg
Tallahassee, FL, 32306-2300, USA
Tel: 1 850 644-8436
Fax: 1 850 644-8297
E-mail: buclp@mailer.fsu.edu
http://www.marinelab.fsu.edu/

Graduate College of Marine Studies
University of Delaware
111 Robinson Hall
Newark, DE, 19716-3501, USA
Contact: Director
Tel: 1 302 831-2841
Fax: 1 302 831-4389
E-mail: marinecom@udel.edu
http://www.ocean.udel.edu/

Graduate School of Oceanography
University of Rhode Island
South Ferry Road
URI Bay Campus Box 52
Narragansett, RI, 02882-1197, USA
Contact: Dean
Tel: 1 401 874-6246
Fax: 1 401 874-6889
E-mail: TheDean@gso.uri.edu
http://www.gso.uri.edu/

Gulf and Caribbean Fisheries Institute
c/o Harbor Branch Oceanographic Institution, Inc.
5600 US 1 North
Fort Pierce, FL, 34946, USA
Contact: Executive Secretary
Tel: 1 561 462-1660
Fax: 1 561 462-1510
E-mail: leroy.creswell@gcfi.org
http://www.gcfi.org/

H. John Heinz III Center for Science, Economics and the Environment
1001 Pennsylvania Ave, NW, Suite 735 South
Washington, DC, 20004, USA
Contact: Chairman
Tel: 1 202 737-6307
Fax: 1 202 737-6410
E-mail: info@heinzctr.org
http://www.heinzctr.org/

Harbor Branch Oceanographic Institution
5600 US 1 North
Fort Pierce, FL 34946, USA
Tel: 1 561 465-2400
Fax: 1 561 465-2446
E-mail: education@hboi.edu
http://www.hboi.edu/

Hopkins Marine Station of Stanford University
Oceanview Boulevard
Pacific Grove, CA, 93950-3094, USA
Tel: 1 831 655-6200
Fax: 1 831 375-0793
E-mail: info@marine.stanford.edu
http://www-marine.stanford.edu/

Institut des sciences de la mer de Rimouski (ISMER)
Université du Québec à Rimouski
300, allée des Ursulines
Rimouski, QC, G5L 3A1, Canada
Tel: 1 418 724-1650
Fax: 1 418 724-1842
E-mail: ismer@uqar.uquebec.ca
http://www.uqar.uquebec.ca/ismer/

Institut maritime du Québec
53, rue Saint-Germain Ouest
Rimouski, QC, G5L 4B4, Canada
Tel: 1 418 724-2822
Fax: 1 418 724-0606
E-mail: jlavigne@imq.qc.ca
http://www.imq.qc.ca/

Institute of Marine Sciences
University of North Carolina at Chapel Hill

3431 Arendell Street, CB#3100
Morehead City, NC, 28557, USA
Contact: Director
Tel: 1 252 726-6841
Fax: 1 252 726-2426
E-mail: john_wells@unc.edu
http://www.marine.unc.edu/organization/IMS.html

Instituto de Technologías Ciencias Marinas
Universidad de Simón Bolívar
Apdo. 89.000
Caracas 1080, Venezuela
Tel: 58 2 906 3111
Fax: 58 2 962 1615
http://www.usb.ve/

Interdisciplinary Studies in Aquatic Resources (ISAR)
St. Francis Xavier University
PO Box 5000
Nicholson Hall
Antigonish, NS, B2G 2W5, Canada
Contact: Program Officer
Tel: 1 902 867-3905
Fax: 1 902 867-2448
E-mail: lpatters@stfx.ca
http://iago.stfx.ca/people/aqua_res/

International Institute of Fisheries Economics and Trade (IIFET)
Department of Agricultural and Resource Economics
Oregon State University
Corvallis, OR, 97331-3601, USA
Contact: Executive Director
Tel: 1 503 737-1416
Fax: 1 503 737-2563
E-mail: Ann.L.Shriver@orst.edu
http://www.orst.edu/Dept/IIFET/

International Law Institute (ILI)
1615 New Hampshire Ave., NW
Washington, DC, 20009, USA
Contact: Executive Director
Tel: 1 202 483-3036
Fax: 1 202 483-3029

E-mail: training@ili.org
http://www.ili.org/

International Marine Simulator Forum (IMSF)
Marine Institute
Memorial University of Newfoundland 155 Ridge Road
PO Box 8454
St. John's, NL, A1B 3N9, Canada
Contact: Chairman
Tel: 1 709 579-4872, ext. 217
Fax: 1 709 579-0495
E-mail: imsf@oarmo.nf.ca
http://www.imsf.org/

Joint Oceanographic Institutions for Deep Earth Sampling (JOIDES)
University of Miami
RSMAS
600 Rickenbacker Causeway
Miami, FL, 33149, USA
Contact: JOIDES Office
Tel: 1 305 361-4668
Fax: 1 305 361-4632
E-mail: joides@rsmas.miami.edu
http://joides.rsmas.miami.edu/

Marine Affairs Programme
Dalhousie University
1234 Seymour Street
Halifax, NS, B3H 3J5, Canada
Contact: Coordinator
Tel: 1 902 494-3555
Fax: 1 902 494-1001
E-mail: marine.affairs@dal.ca
http://www.dal.ca/mmm/

Maine Maritime Academy
Castine, ME, 04420-5000, USA
Tel: 1 207 326-2206
E-mail: admissions@mma.edu
http://www.mainemaritime.edu/

Marine and Environmental Law Programme
Dalhousie Law School
6061 University Avenue

Halifax, NS, B3H 4H9, Canada
Contact: Director
Tel: 1 902 494-3495
Fax: 1 902 494-1316
E-mail: melp@dal.ca
http://www.dal.ca/wwwlaw/melp/

Marine and Environmental Programs
Woodward Hall
University of Rhode Island
Kingston, RI, 02881, USA
Tel: 1 401 874-2957
Fax: 1 401 874-4017
E-mail: omp@gso.uri.edu
http://omp.gso.uri.edu/urime/

Marine Ecology Station
Port Sydney Marina
9835 Seaport Place
Sydney, BC, V8L 4X3, Canada
Tel: 1 250 655-1555
Fax: 1 250 655-1573
E-mail: info@mareco.org
http://www.mareco.org/

Marine Environment Research Institute (MERI)
Center for Marine Studies
55 Main Street
PO Box 1652
Blue Hill, ME, 04616, USA
Tel: 1 207 374-2135
Fax: 1 207 374-2931
E-mail: meri@downeast.net
http://www.meriresearch.org/

Marine Institute
Fisheries and Marine Institute of Memorial University of Newfoundland
PO Box 4920
St. John's, NL, A1C 5R3, Canada
Tel: 1 709 778-0200
Fax: 1 709 778-0346
E-mail: coordinator_advanced_programs@mi.mun.ca
http://www.mi.mun.ca/

Marine Invasion Research Laboratory
Smithsonian Environmental Research Center
PO Box 28
Edgewater, MD, 21037, USA
Tel: 1 443 482-2414
Fax: 1 443 482-2380
E-mail: noblem@si.edu
http://invasions.si.edu/

Marine Law Institute
University of Maine
School of Law
246 Deering Avenue
Portland, ME, 04102, USA
Contact: Director
Tel: 1 207 780-4474
Fax: 1 207 780-4239
E-mail: jduff@usm.maine.edu
http://www.law.usm.maine.edu/mli/

Marine Policy Center
Woods Hole Oceanographic Institution
Crowell House, Mail Stop # 41
Woods Hole, MA, 02543-1138, USA
Contact: Director
Tel: 1 508 289-2449
Fax: 1 508 457-2184
E-mail: mpc@whoi.edu
http://www.whoi.edu/mpcweb/

Marine Resource Management Program
College of Oceanic and Atmospheric Sciences
Oregon State University
104 Ocean Administration Bldg
Corvallis, OR, 97331-5503, USA
Contact: Coordinator
Tel: 1 541 737-1339
Fax: 1 541 737-2064
E-mail: good@coas.oregonstate.edu
http://www.oce.orst.edu/

Marine Science Center
Northeastern University
430 Nahant Road

Nahant, MA, 01908, USA
Contact: Director
Tel: 1 781 581-7370
Fax: 1 781 581-6076
E-mail: e.maney@neu.edu
http://www.marinescience.neu.edu/

Marine Science Institute
Department of Marine Science
University of Texas at Austin
750 Channel View Drive
Port Aransas, TX, 78373-5015, USA
Contact: Director
Tel: 1 361 749-6711
Fax: 1 361 749-6777
E-mail: gardner@utmsi.utexas.edu
http://www.utmsi.utexas.edu/

Marine Sciences Program
University of North Carolina
12-7 Venable Hall, CB# 3300, UNC-CH
Chapel Hill, NC, 27599, USA
Contact: Director
Tel: 1 919 962-1252
Fax: 1 919 962-1254
E-mail: cisco@unc.edu
http://www.marine.unc.edu/
   organization/MASC.html

Marine Sciences Research Center
Stony Brook University
Endeavour Hall, Room 145
Stony Brook, NY, 11794-5000, USA
Tel: 1 631 632-8700
Fax: 1 631 632-8915
E-mail: nglover@notes.cc.sunysb.edu
http://www.msrc.sunysb.edu/

Mote Marine Laboratory
1600 Ken Thompson Parkway
Sarasota, FL, 34236, USA
Tel: 1 941 388-4441
Fax: 1 941 388-4312
E-mail: info@mote.org
http://www.mote.org/

Naval Postgraduate School
1 University Circle
Monterey, CA, 93943-5001, USA
Tel: 1 831 656-2441/2
Fax: 1 831 656-2337
http://www.nps.navy.mil/

Naval War College
686 Cushing Avenue
Newport, RI, 02841, USA
Tel: 1 401 841-3089
E-mail: qdeck@nwc.navy.mil
http://www.nwc.navy.mil/

Nicholas School of the Environment
   and Earth Sciences
Levine Science Research Center
PO Box 90328
Beaufort, NC, 27708, USA
Tel: 1 919 613-8000
Fax: 1 919 684-8741
http://www.env.duke.edu/

Ocean and Coastal Law Program
University of Miami School of Law
PO Box 248087
Coral Gables, FL, 33124-8087, USA
Contact: Director, International and
   Foreign Law Programs
Tel: 1 305 284-5402
Fax: 1 305 284-5497
E-mail: intl-llm@law.miami.edu
http://www.law.miami.edu/ifp/

Ocean and Coastal Law Center
School of Law
University of Oregon
1221 University of Oregon
Eugene, OR, 97403-1221, USA
Contact: Director
Tel: 1 541 346-3845
Fax: 1 541 346-1564
E-mail: rghildre@law.uoregon.edu
http://oceanlaw.uoregon.edu/

Ocean Engineering Program
Department of Civil Engineering
Texas A&M University
College Station, TX, 77843-3136,
   USA

Tel: 1 979 845-4515
E-mail: lori@civil.tamu.edu
http://edge.tamu.edu/

Ocean Engineering Research Centre (OERC)
Faculty of Engineering and Applied Science
Memorial University of Newfoundland
St. John's, NL, A1B 3X5, Canada
Contact: Director
Tel: 1 709 737-8805
Fax: 1 709 737-2116
E-mail: oerc@engr.mun.ca
http://www.engr.mun.ca/OERC/

Ocean Sciences Centre
Memorial University of Newfoundland
St. John's, NL, A1C 5S7 Canada
Contact: Director
Tel: 1 709 737-3708
Fax: 1 709 737-3220
E-mail: osc@mun.ca
http://www.osc.mun.ca/

Office of Marine Programs
University of Rhode Island, North Bay Campus
Narragansett, RI, 02882-1197, USA
Tel: 1 401 874-6211
Fax: 1 401 874-6486
E-mail: omp@gso.uri.edu
http://omp.gso.uri.edu/

Pan-American Institute of Naval Engineering (IPEN)
Instituto Panamericano de Ingenieria Naval
Avenida Presidente Vargas 542–S913
20071-000 Rio de Janeiro, Brazil
Tel/Fax: 55 21 2536263
E-mail: ipin@openlink.com.br
http://www.ipen.org.br/

Rosenstiel School of Marine and Atmospheric Science (RSMAS)
University of Miami
4600 Rickenbacker Causeway
Miami, FL, 33149-1098, USA
Contact: Dean
Tel: 1 305 361-4000
Fax: 1 305 361-4711
E-mail: dean@rsmas.miami.edu
http://www.rsmas.miami.edu/

School for Field Studies
16 Broadway
Beverly, MA, 01915-4435, USA
Contact: President
Tel: 1-800-989-4435
Fax: 1 978 927-5127
E-mail: admissions@fieldstudies.org
http://www.fieldstudies.org/

School of Aquatic and Fisheries Sciences
University of Washington
Box 355020
Seattle, WA, 98195-5020, USA
Tel: 1 206 543-4270
Fax: 1 206 685-7471
E-mail: frontdesk@fish.washington.edu
http://www.fish.washington.edu/

School of Fisheries and Ocean Sciences
University of Alaska Fairbanks
245 O'Neill Building
Fairbanks, AK, 99775-7220, USA
Tel: 1 907 474-7824
Fax: 1 907 474-7204
E-mail: fysfos@uaf.edu
http://www.sfos.uaf.edu/

School of Marine Affairs
University of Washington
3707 Brooklyn Avenue, NE
Seattle, WA, 98105-6715, USA
Tel: 1 206 543-7004
Fax: 1 206 543-1417
E-mail: uwsma@u.washington.edu
http://www.sma.washington.edu/

School of Marine Sciences
University of Maine
214 Libby Hall
Orono, ME, 04469-5741, USA
Contact: Director
Tel: 1 207 581-4388
E-mail: davidt@maine.edu
http://www.marine.maine.edu/

School of Oceanography
University of Washington
Box 357940
Seattle, WA 98195-7940, USA
Contact: Director
Tel: 1 206 543-5060
E-mail: webmaster@ocean.washington.edu
http://oceanweb.ocean.washington.edu/ocean_web/

Scotiabank Marine Geology Research Laboratory
Department of Geology
University of Toronto
22 Russell Street
Toronto, ON, M5S 3B1, Canada
Tel: 1 416 978-5424
Fax: 1 416 978-3938
E-mail: scottsd@geology.utoronto.ca
http://www.geology.utoronto.ca/marinelab/

Scripps Institution of Oceanography
University of California, San Diego
8602 La Jolla Shores Drive
La Jolla, CA, 92037, USA
Contact: Director
Tel: 1 619 534-2826
Fax: 1 619 453-0167
E-mail: siodir@sio.ucsd.edu
http://sio.ucsd.edu/

Sea Education Association
PO Box 6
Woods Hole, MA, 02543, USA
Contact: President
Tel: 1 508 540-3954

Fax: 1 508 457-4673
E-mail: admission@sea.edu
http://www.sea.edu/

Sea Grant Law Center
University of Mississippi
Kinard Hall, Wing E, Room 262
PO Box 1848
University, MS, 38677-1848, USA
Tel: 1 662 915-7775
Fax: 1 662 915-5267
E-mail: sealaw@olemiss.edu
http://www.olemiss.edu/orgs/SGLC/

Shoals Marine Laboratory
Cornell University
G-14 Stimson Hall
Ithaca, NY, 14853-3717, USA
Contact: Director
Tel: 1 607 255-3717
Fax: 1 607 255-0742
E-mail: shoals-lab@cornell.edu
http://www.sml.cornell.edu/

Smithsonian Tropical Research Institute
Roosevelt Avenue, Ancon
PO Box 2072
Balboa, Panama
Contact: Director
Tel: 507 212 8000
Fax: 507 232 6197
E-mail: webmaster@stri.org
http://www.stri.org/

Tulane Law School
Weinmann Hall
6329 Freret Street
New Orleans, LA, 70118-6231, USA
Tel: 1 504 865-5939
Fax: 1 504 865-6710
E-mail: admissions@law.tulane.edu
http://www.law.tulane.edu/

Urban Harbors Institute
University of Massachusetts
100 Morrissey Blvd

Boston, MA, 02125, USA
Tel: 1 617 287-5570
Fax: 1 617 287-5575
E-mail: urban.harbors@umb.edu
http://www.uhi.umb.edu/

US Naval Institute (USNI)
291 Wood Road
Annapolis, MD, 21402, USA
Contact: CEO
Tel: 1 410 268-6110
Fax: 1 410 269-7940
E-mail: tmarfiak@usni.org
http://www.usni.org/

Virginia Institute of Marine Science
College of William and Mary
PO Box 1346
Gloucester Point, VA, 23062, USA
Contact: Director
Tel: 1 804 684-7000
Fax: 1 804 684-7097
E-mail: wright@vims.org
http://www.vims.edu/

Whale Research Lab
Department of Geography
University of Victoria
Box 3050, Stn CSC
Victoria, BC, V8W 3P5, Canada
Tel: 1 604 721-7327
Fax: 1 604 721-6216
E-mail: whalelab@office.geog.uvic.ca
http://office.geog.uvic.ca/dept/whale/wrlmp.html

Woods Hole Oceanographic Institution
Co-op Building, MS #16
Woods Hole, MA, 02543-1050, USA
Contact: Information Office
Tel: 1 508 289-2252
Fax: 1 508 457-2180
E-mail: information@whoi.edu
http://www.whoi.edu/

6. REGIONAL ACADEMIC ORGANIZATIONS

6.3 Asia

Akademi Laut Malaysia (ALAM)
(Malaysian Maritime Academy)
PO Box 31, Masjid Tanah
783000 Melaka, Malaysia
Tel: 606 387 6201
Fax: 606 387 6700
E-mail: mma@alam.edu.my
http://www.alam.edu.my/

Aquaculture and Aquatic Resources Management Program (AARM)
School of Environment, Resources and Development (SERD)
Asian Institute of Technology
PO Box 4, Klong Luang
Pathumthani 12120, Thailand
Tel: 662 524 5456
Fax: 662 524 6200
E-mail: amara@ait.ac.th
http://www.serd.ait.ac.th/field02.htm

Aquatic Resources Research Institute
9th floor, Institute Building No.3
Chulalongkorn University
Phayathai Rd., Pathumwan,
Bangkok 10330, Thailand
Tel: 662 2188 160
Fax: 662 2544 259
E-mail: arri@chula.ac.th
http://www.arri.chula.ac.th/

Asia-Pacific Centre for Environmental Law
Faculty of Law
National University of Singapore
13 Law Link
117590 Singapore
Tel: 65 874 6246
Fax: 65 872 1937
E-mail: lawapcel@nus.edu.sg
http://law.nus.edu.sg/apcel/

Centre for Coastal Pollution and
 Conservation
City University of Hong Kong
83 Tat Chee Avenue
Kowloon, Hong Kong
Tel: 852 2788 7404
Fax: 852 2788 1167
E-mail: bhrolin@cityu.edu.hk
http://www.cityu.edu.hk/

Centre for Transportation Research
Faculty of Engineering
National University of Singapore
10 Kent Ridge Crescent
119260 Singapore
Tel: 65 777 0170
Fax: 65 777 0994
E-mail: cvefwatf@nus.edu.sg
http://www.eng.nus.edu.sg/civil/
 C_CTR/

Centre for Transportation Studies
c/o School of Civil and Environ-
 mental Engineering
Nanyang Technical University
Block N1# 1a-29
Singapore 639798
Tel: 65 6790 5328
Fax: 65 6792 3886
E-mail: wwwcts@ntu.edu.sg
http://www.ntu.edu.sg/home/
 cshlam/cts/centre.html

Centre of Excellence in Marine Bi-
 ology
University of Karachi
Karachi 75270, Pakistan
Contact: Director
Tel: 92 43131 2256
E-mail: huss@cyber.net.pk
http://www.ku.edu.pk/research/
 cemb/intro.html

Coastal Resources Institute
 (CORIN)
Prince of Songkhla University
Hat Yai
Songkhla 90112 Thailand

Tel: 66 74 212800
Fax: 66 74 212782
E-mail: corin@ratree.psu.ac.th
http://www.psu.ac.th/corin

Coastal Resources Management
Department of Biology
Silliman University
Dumaguete City, 6200, The Philip-
 pines
Tel: 63 35 422 8880
Fax: 63 35 422 4776
E-mail: biology@su.edu.ph
http://www.su.edu.ph/cas/
 biology/biology.html

College of Fisheries and Ocean Sci-
 ences
University of the Philippines in the
 Visayas
Miag-ao, Iloilo, The Philippines
Contact: Dean
Tel: 632 920 5301
E-mail: upv_oc@up.edu.ph
http://www.upv.edu.ph/
 academicinfo/cfos/cfos.htm

College of Fisheries Sciences
Pukong National University
599-1 Daeyundong, Namgu
Pusan 608-737, Republic of Korea
Contact: Dean
Tel: 82 51 620 6100
Fax: 82 51 628 7430
E-mail: dsbyun@pknu.ac.kr
http://www.pknu.ac.kr/

College of Fisheries
Central Luzon State University
Nueva Ecija 3120, The Philippines
Tel/Fax: 63 44 456 0107
E-mail: clsu@mozcom.com
http://www2.mozcom.com/clsu/

College of Ocean Science and Technology
Kunsan National University
68 Miryong-dong

Kunsan Chollapuk-do 573-701, Republic of Korea
Tel: (0654)469-4114
E-mail: yujinhee@ks.kunsan.ac.kr
http://www.kunsan.ac.kr/old-www/

Dalian Maritime University
1 Linghai Road
Dalian 116026, Liaoning Province, People's Republic of China
Tel: 86 411 4727874
Fax: 86 411 4727395
E-mail: faodmu@dlmu.edu.cn
http://www.dlmu.edu.cn/

Department of Marine Science
Science Faculty
Chulalongkorn University
Bangkok 10330, Thailand
Contact: President
Tel: 662 215 0871
Fax: 662 215 4804
http://www.chula.ac.th/

Faculty of Fisheries
Bangladesh Agricultural University
Mymensingh 2202, Bangladesh
Tel: 880 91 52236
Fax: 880 91 55810
E-mail: ffbau@mymensingh.net
http://agri-varsity.tripod.com/fish/fis-fac.html

Faculty of Marine Science
University of Karachi
Karachi 75270, Pakistan
Contact: Acting Director
Tel: 92 43131 2378
E-mail: mfatima@ims.ku.edu.pk
http://www.ku.edu.pk/research/marinesc.html

Faculty of Marine Sciences
Cochin University of Science and Technology
"Aathira" Justice Lane
Janatha Road
Kochi 682 019, India
Contact: Dean
Tel: 91 484 351957
Fax: 91 484 577595
E-mail: damu@cusat.ac.in
http://www.cusat.ac.in/

Faculty of Marine Sciences and Fisheries
Universitas Hasanuddin
Jl. Perintis Kemerdekaan
Kampus Tamalanrea
Ujung Pandang 90245, Indonesia
Tel: 411 510102
Fax: 411 510088
E-mail: cio@unhas.ac.id
http://www.unhas.ac.id/

Faculty of Ocean Technology
Sepuluh Nopember Institute of Technology
Jl. Arief Rahman Hakim
Kampus ITS Sukolilo
Surabaya 60111, Indonesia
Tel. 62 31 5927939
Fax. 62 31 5923411
E-mail: pr_4_its@its.ac.id
http://www.its.ac.id/

Fisheries University
2 Nguyen Dinh Chieu
Nha Trang City, Khanh Hoa Province, Vietnam
Tel: 84 58 831145
Fax: 84 58 831147
E-mail: dhtsnt@dng.vnn.vn

Hue University of Agriculture and Forestry
24 Phung Hung
Hue, Viet Nam
Tel: 84 54 525049
Fax: 84 54 524923
E-mail: huaf@dng.vnn.vn
http://www.ire.ubc.ca/hong_ha/university/university.htm

Indonesian Centre for the Law of the Sea and Marine Policy (ICLOS)

University of Padjadjaran
Bandung, Indonesia
Contact: Director
E-mail: webleged@lawunpad.
elga.net.id
http://www2.elga.net.id/webleged/
iclos.htm

Institut Kelautan Malaysia (IKMAL)
(Malaysian Maritime Institute)
47A, Lorong Cungah
42000 Port Klang
Selangor Darul Ehsan, Malaysia
Tel: 60 3 3166 2472
Fax: 60 3 3166 2870
E-mail: president@ikmal.org
http://www.ikmal.org/

Institute for Marine Aquaculture
Can Tho University
3/2 Street
Can Tho City, Vietnam
Tel: 84 71 838237
Fax: 84 71 838474
E-mail: webmaster@ctu.edu.vn
http://www.ctu.edu.vn/
index_e.htm

Institute of International Maritime
Affairs
Korea Maritime University
1 Dongsam-dong,
Yeongdo-ku Pusan 606-791, Korea
Tel: 82 051 410 4089
Fax: 82 051 404 3985
E-mail: imirc@hanara.
kmaritime.ac.kr
http://hanara.kmaritime.ac.kr/
imirc

Institute of Marine Science
University of Chittagong
University Post Office
Chittagong, Bangladesh
Tel: 880 31 282031
Fax: 880 31 726310
E-mail: cu@ctgu.edu
http://www.ctgu.edu/

Institute of Sea Culture
Pukong National University
599-1 Daeyundong, Namgu
Pusan 608-737, Republic of Korea
Tel: 82 51 620 6410
Fax: 82 51 622 9248
http://www.pknu.ac.kr/

Integrated Tropical Coastal Zone
Management (ITCZM)
Asian Institute of Technology (AIT)
PO Box 4, Klong Luang
Pathumthani 12120, Thailand
Tel/Fax: 662 524 5442
E-mail: itczm@ait.ac.th
http://www.itczm.ait.ac.th/

Japan Institute of Navigation (JIN)
c/o Tokyo University of Mercantile
Marine
2-1-6 Etchujima, Koto-ku
Tokyo, 135-8533, Japan
Fax: 81 3 36303093
E-mail: navigation@nifty.com
http://homepage2.nifty.com/
navigation/english.html

Kobe University of Mercantile Marine
5-1-1, Fukae-minami, Higashi-nada
Kobe 658-0022, Japan
Contact: President
Tel: 81 78 431 6200
Fax: 81 78 431 6355
E-mail: gakutyou@office.kshosen.ac.jp
http://www.kshosen.ac.jp/

Korea Maritime University
1 Dongsam-dong Yengdo-Gu
Busan 606791, Republic of Korea
E-mail: webmaster@hhu.ac.kr
http://www.kmaritime.ac.kr/

Law of the Sea Program
Institute of International Legal
Studies
3/F Bocobo Hall
U.P. Law Center

University of the Philippines
Diliman, Quezon City 1100, The Philippines
Contact: Dean
Tel: 632 920 5301
Fax: 632 927 7180
E-mail: iils.claw@pacific.net.ph
http://www.upd.edu.ph/iils/

Marine Science Institute
College of Science
University of the Philippines, Diliman
1101 Quezon City, Philippines
Tel: 632 922 3962
http://www.msi.upd.edu.ph/

Mokpo National Maritime University
571-2 Chukyo-dong
Mokpo, Jeonnam, 530-729, Republic of Korea
Tel: 82 61 240 7114
Fax: 82 61 242 5176
E-mail: webinfo@mmu.ac.kr
http://www.mmu.ac.kr/english/

National Fisheries University
2-7-1 Nagata-Honmachi
Shimonoseki 759-6595, Japan
Tel: 81 832 86 5111
E-mail: zenpan@fish-u.ac.jp
http://www.fish-u.ac.jp/e-index.html

National Institute of Ocean Technology
Velacherry-Tambaram Main Road
Narayanapuram
Chennai, Tamil Nadu, 601 302, India
Tel: 91 44 2460063
Fax: 91 44 2460645
E-mail: webmaster@niot.ernet.in
http://www.niot.ernet.in/

National Research Institute of Fisheries Science
2-12-4, Fukuura
Kanazawa

Yokohama
Kanagawa 236-8648, Japan
Tel: 81 45 788 7615
Fax: 81 45 788 5001
E-mail: www@nrifs.affrc.go.jp
http://www.nrifs.affrc.go.jp/index-e.html

National Taiwan Ocean University
2 Pei-Ning Road
Keelung 20224, Taiwan, ROC
Tel: 886 2 2462 2192
Fax: 886 2 2462 0724
http://www.ntou.edu.tw/

Ocean Research Institute
University of Tokyo
1-15-1 Minamidai,Nakano-ku
Tokyo 164-8639, Japan
Contact: Director
Tel: 81 3 5351 6342
Fax: 81 3 5351 6836
E-mail: webmasters@ori.u-tokyo.ac.jp
http://www.ori.u-tokyo.ac.jp/

Ocean University of China
5 Yushan Road
266003 Qingdao, People's Republic of China
Tel: 86 0532 2032730
Fax: 86 0532 2032799
E-mail: xiaoban@mail.ouqd.edu.cn
http://www.ouqd.edu.cn/

Research and Development Centre of Oceanology
(Pusat Penelitian dan Pengembangan Oseanologi)
Indonesian Institute of Sciences (LIPI)
Jalan Pasir Putih 1
Ancol Timur
PO Box 4801 JKTF
11001 Jakarta, Indonesia
Tel: 62 21 683850
Fax: 62 21 681948
E-mail: ppolipi@jakarta.wasantara.net.id
http://www.oseanologi.lipi.go.id/

School of Environmental Science
and Management
University of the Philippines Los
Baños College
Laguna 4031, The Philippines
E-mail: dkv@mudspring.uplb.edu.ph
http://www.uplb.edu.ph/
academics/schools/sesam/
index.htm

School of Law
Soochow University
#56 Sec 1, Kuei-Yant Street
100 Taipei, Taiwan, ROC
Tel: 886 2 2311-1531, ext. 2521
Fax: 886 2 2375-1067
E-mail: law@mail.scu.edu.tw
http://www.scu.edu.tw/english/law/

Shanghai Fisheries University
334 Jungong Road
200090 Shanghai, People's Republic
of China
E-mail: master@shfu.edu.cn
http://www.shfu.edu.cn/

Shanghai Maritime University
1550 Road Pudong Shanghai
200135, People's Republic of China
Tel: 86 02158855200
E-mail: smuico@shmtu.edu.cn
http://www.shmtu.edu.cn/

South China Sea Institute of Oceanology
164 West Xingang Road
Guangzhou 510301, People's Republic of China
Tel: 86 20 89023132
Fax: 86 20 84451947
E-mail: sli@scsio.ac.cn
http://www.scsio.ac.cn/

Southeast Asian Programme on
Ocean Law, Policy and Management (SEAPOL)
Sukhothai Thammathirat Open University

Academic Building 2, 3rd Floor,
Room 2320
Pakkred, Nonthabur 11120, Thailand
Contact: Executive Director
Tel: 662 503 3858
Fax: 662 503 3608
E-mail: seapol@asianet.co.th
http://www.seapol.org/

Tokyo University of Fisheries
5-7, Konan 4
Minato-ku
Tokyo 108-8477, Japan
Tel: 81 3 5463 0400
Fax: 81 3 5463 0359
E-mail: www-master@tokyo-u-fish.ac.jp
http://www.tokyo-u-fish.ac.jp/

Transportation Institute (Merchant
Marine)
6th floor, Prajadhipok-Rambhaib-
arni Building
Chulalongkorn University
Phayathai Road
Bangkok 10330, Thailand
Tel: 662 218 7449
Fax: 662 214 2417
E-mail: tri_web@chula.com
http://www.tri.chula.ac.th/

Tropical Marine Science Institute
(TMSI)
National University of Singapore
14 Kent Ridge Road
Singapore 119223
Contact: Director
Tel: 65 774 9656
Fax: 65 774 9654
E-mail: tmsdir@nus.edu.sg
http://www.tmsi.nus.edu.sg/

Viet Nam Maritime University
484 Lach Tray Street
Haiphong, Vietnam
Tel: 84 31 845931
Fax: 84 31 843282
E-mail: tdsuu@hn.vnn.vn
http://www.vimaru.vnn.vn/

6. REGIONAL ACADEMIC ORGANIZATIONS

6.4 Europe

Archipelago Research Institute
University of Turku
Luonnontieteidentalo II
FIN-20014 Turku, Finland
Tel: 358 2 333 5933
Fax: 358 2 333 6592
E-mail: paula.rasanen@utu.fi
http://www.utu.fi/erill/saarmeri/en/

British Polarological Research Society
6 Beechvale
Hillview Road
Woking, Surrey GU22 7NS, United Kingdom
Contact: President

Cambridge Coastal Research Unit (CCRU)
Department of Geography
University of Cambridge
Downing Place
Cambridge, CB2 3EN, United Kingdom
Tel: 44 12 2333 9775
Fax: 44 12 2335 5674
E-mail: geog-CCRU@lists.cam.ac.uk
http://ccru.geog.cam.ac.uk/

Center for Coastal and Marine Environmental Studies
(Centro de Investigação dos Ambientes Costeiros e Marinhos – CIACOMAR)
Universidade do Algarve
Avenida das Forças Armadas
PT-8700 Olhão, Portugal
Tel: 351 289 707 087
Fax: 351 289 706 972
E-mail: webmaster.ciacomar@ualg.pt
http://www.ualg.pt/ciacomar/

Center for Marine and Environmental Research
(Centro de Investigação Marinha e Ambiental – CIMA)
Faculdade de Ciências do Mar e do Ambiente
Universidade do Algarve
Campus das Gambelas
PT-8000 810 Faro, Portugal
Tel: 351 289 800 900
Fax: 351 289 818 353
E-mail: webmaster.cima@ualg.pt
http://www.ualg.pt/cima/Indexing01.htm

Center for Maritime Research (MARE)
Plantage Muidergracht 4
NL-1018 TV, Amsterdam, The Netherlands
Tel: 31 20 527 0624
Fax: 31 20 622 9430
E-mail: info@marecentre.nl
http://www.marecentre.nl/

Centre d'Océanologie de Marseille (COM)
(Marseille Oceanology Institute)
Université de la Méditerranée Marine d'Endoume
Rue de la Batterie des Lions
F-13007 Marseille, France
Tel: 33 4 9104 1600
Fax: 33 4 9104 1608
E-mail: dekeyser@com.univ-mrs.fr
http://www.com.univ-mrs.fr/

Centre de Formation et de Recherche sur l'Environnement Marin (CEFREM)
CNRS-UMR 5110
Université de Perpignan
52 avenue de Villeneuve
F-66860 Perpignan Cedex, France
Contact: Director
Tel: 33 4 6866 2090
Fax: 33 4 6866 2096
E-mail: cefrem@univ-perp.fr

http://www.univ-perp.fr/see/rch/lsgm/index.htm

Centre for Coastal Conservation and Education
School of Conservation Sciences
Bournemouth University
Poole, Dorset, BH12 5BB, United Kingdom
Tel: 44 12 0259 5444
Fax: 44 12 0259 5255
E-mail: consci@bmth.ac.uk
http://csweb.bournemouth.ac.uk/consci/marinecentre/marinecentre.htm

Centre for Coastal Management
Ridley Building
University of Newcastle upon Tyne
Newcastle Upon Tyne, NE1 7RU, United Kingdom
Tel: 44 19 1222 5607
Fax: 44 19 1222 5095
E-mail: enquiries@sustainablecoasts.com
http://www.ncl.ac.uk/ccm/

Centre for Development and the Environment
University of Oslo
Box 1116 Blindern
N-0317 Oslo, Norway
Tel: 47 22 85 89 00
Fax: 47 22 85 89 20
E-mail: info@sum.uio.no
http://www.sum.uio.no/

Centre for Energy, Petroleum and Mineral Law and Policy (CEPMLP)
University of Dundee
Dundee, Scotland, DD1 4HN, United Kingdom
Contact: Director
Tel: 44 1382 344300
Fax: 44 1382 322578
E-mail: cepmlp@dundee.ac.uk
http://www.dundee.ac.uk/cepmlp/

Centre for Fisheries, Aquaculture Management, and Economics
Faculty of Social Sciences
University of Southern Denmark (Syddansk Universitet)
Campusvej 55
DK-5230 Odense M, Denmark
Tel: 45 65 50 10 00
Fax: 45 65 93 56 92
E-mail: fame@sam.sdu.dk
http://www.sam.sdu.dk/fame/

Centre for Fisheries Economics
Institute for Research in Economics and Business Administration (SNF)
Norwegian School of Economics and Business Administration
Breiviksveinen 40
N-5045 Bergen, Norway
Contact: Research Director
Tel: 47 55 95 92 60
Fax: 47 55 95 95 43
E-mail: rognvaldur.hannesson@nhh.no
http://www.snf.no/

Centre for International Economics and Shipping
(Senter for Internasjonal Økonomi og Skipsfart)
Institute for Research in Economics and Business Administration (SNF)
Norwegian School of Economics and Business Administration
Breiviksveinen 40
N-5045 Bergen, Norway
Contact: Research Director
Tel: 47 55 95 95 00
Fax: 47 55 95 93 50
E-mail: karenhelene.midelfart@snf.no
http://www.snf.no/

Centre for Marine and Climate Research
(Zentrum für Meeres-und-Klimaforschung–ZMK)
University of Hamburg

Bundesstrasse 55
D-20146 Hamburg, Germany
Tel: 49 40 428 38 4523
Fax: 49 40 428 38 5235
E-mail: walter.lenz@dkrz.de
http://www.uni-hamburg.de/wiss/
se/zmk/

Centre for Marine Environmental
Sciences
(Zentrum für Marine Umweltwissenschaften–MARUM)
Der Universität Bremen
Klagenfurter Strasse
D-28359 Bremen, Germany
Tel: 49 421 218 3389
Fax: 49 421 218 3116
E-mail: gwefer@marum.de
http://www.marum.de/

Centre for Maritime and Oceanic Law
(Centre de droit maritime et
océanique–CDMO)
University of Nantes
Faculty of Law
Chemin de la Censive du Tertre
BP 81307
F-44313 Nantes, Cedex 3, France
Contact: Director
Tel: 33 2 4014 1534
Fax: 33 2 4014 1500
E-mail: jean-pierre.beurier@
droit.univ-nantes.fr
http://www.univ-nantes.fr/

Centre for Maritime Studies
University of Turku
WTC-Building, Veistämönaukio 1-3
FIN-20100 Turku, Finland
Contact: Director
Tel: 358 2 281 3300
Fax: 358 2 281 3311
E-mail: pavi.soderholm@utu.fi
http://www.utu.fi/erill/mkk/

Centre for Ships and Ocean Structures
Department of Marine Technology
Norwegian University of Science
and Technology
(Nordisk Teknisk-Naturvitenskapelige Universitet)
Marine Technology Centre
Otto Nielsens vei 10
N-7491 Trondheim, Norway
Contact: Director
Tel: 47 73 59 55 01
Fax: 47 73 59 55 28
E-mail: coe@marin.ntnu.or
http://www.ntnu.no/cesos/

Centre for the Economics and Management of Aquatic Resources
(CEMARE)
Department of Economics
University of Portsmouth
Milton Campus
Locksway Road
Portsmouth, PO4 8JF, United
Kingdom
Contact: Director
Tel: 44 23 9284 4082
Fax: 44 23 9284 4037
E-mail: cemare.library@port.ac.uk
http://www.pbs.port.ac.uk/econ/
cemare/

Centre for Tropical Marine Ecology
(Zentrum für Marine Troppenökologie - ZMT)
University of Bremen
Fahrenheitstrasse 6
D-28359 Bremen, Germany
Contact: Director
Tel: 49 421 23800 21
Fax: 49 421 23800 30
E-mail: contact@zmt-bremen.de
http://www.zmt.uni-bremen.de/

Challenger Society for Marine Science
Room 251/20
Southampton Oceanography Centre
Waterfront Campus
Southampton SO14 3ZH, United
Kingdom

Contact: Executive Secretary
Tel: 44 17 0359 6097
Fax: 44 17 0359 6149
E-mail: jxj@soc.soton.ac.uk
http://www.soc.soton.ac.uk/
OTHERS/CSMS/

Chalmers/Göteborg Universitet Centre for Environment and Sustainability
(Göteborgs Miljö Vetenskapliga Centrum – GMV)
Vera Sandbergs Allé 513
SE-412 96 Göteborg, Sweden
Tel: 46 31 772 49 51
Fax: 46 31 772 49 58
E-mail: office@miljo.chalmers.se
http://www.miljo.chalmers.se/

Coastal Research and Planning Institute (CORPI)
Klaipëda University
H. Mantog. 84
LT-5808 Klaipeda, Lithuania
Tel: 370 6 398844
Fax: 370 6 398845
E-mail: simona@corpi.ku.lt
http://www.corpi.ku.lt/

Coastal Research Laboratory
Institute of Geosciences
Christian Albrechts University
Otto-Hahn-Platz 3
D-24118 Kiel, Germany
Tel: 49 431 880 2851
Fax: 49 431 880 7303
E-mail: info@corelab.uni-kiel.de
http://www.corelab.uni-kiel.de/

Coastal Studies Research Group
School of Environmental Studies
University of Ulster at Coleraine
Coleraine, Co. Londonderry, Northern Ireland, BT52 1SA, United Kingdom
Contact: Head
Tel: 44 28 7032 4428
Fax: 44 28 7032 4911

E-mail: jag.cooper@ulst.ac.uk
http://www.science.ulst.ac.uk/crg/

DEA européen en modélisation de l'environnement marin
(European DEA in Modeling of the Marine Environment)
Université de Liège
Sart Tilman, B5
B-4000 Liège, Belgium
Contact: Director
Tel: 32 4 366 3350
Fax: 32 4 366 2355
E-mail: J.Nihoul@ulg.ac.be
http://www.ulg.ac.be/deamodel/

Department of Geological, Environmental and Marine Sciences
Università delgi Studi di Trieste
Via E. Weiss 2
I-34127 Trieste, Italy
Tel: 39 40 558 2047
Fax: 39 40 558 2048
E-mail: disgam@units.it
http://www.units.it/disgam/

Department of Marine Biology
Institute of Ecology and Conservation Biology
University of Vienna
Althanstrasse 14
A-1090 Vienna, Austria
Contact: Head
Tel: 43 1 4277 54202
Fax: 43 1 4277 9542
E-mail: oekologie@univie.ac.at
http://www.univie.ac.at/marine-biology/

Department of Mechanical Engineering
Technical University of Denmark
Studentertorvet, Building 101E
DK-2800 Kgs. Lyngby, Denmark
Contact: Maritime Engineering Program
Tel: 45 45 25 13 60
Fax: 45 45 88 43 25
http://www.ish.dtu.dk/

Department of Oceanography
Göteborg University
PO Box 460
SE-405 30 Göteborg, Sweden
Contact: Chairman
Tel: 46 31 77 32 870
Fax: 46 31 77 32 888
E-mail: majo@oce.gu.se
http://www.oce.gu.se/

Department of Oceanography and Fisheries
University of the Azores
PT-9901 892 Horta, Portugal
Tel: 351 292 200400
Fax: 351 292 200411
E-mail: hisidra@horta.uac.pt
http://www.horta.uac.pt/

Estonian Marine Institute (EMI)
(Eesti Mereinstituut)
Mäealuse 10a
EE-12618 Tallinn, Estonia
Tel: 372 626 7402
Fax: 372 626 7417
E-mail: meri@sea.ee
http://www.sea.ee/

Euro-Mediterranean Centre on Insular Coastal Dynamics (ICoD)
Foundation for International Studies
University of Malta
St. Paul Street
MT-Valletta VLT 07, Malta
Contact: Director
Tel: 356 21 240 746
Fax: 356 21 230 551
E-mail: icod@maltanet.net
http://www.icod.org.mt/

European Institute–Denmark
Kvästhusgade 5C
DK-1252 Copenhagen K, Denmark
Tel: 45 33 33 91 00
Fax: 45 33 33 91 20
E-mail: euroinst@euroinst.dk
http://www.euroinst.dk/

European Institute for Marine Studies
(Institut Universitaire Européen de la mer – IUEM)
Université de Bretagne Occidentale
Technopôle Brest-Iroise
Place Nicolas Copernic
F-29280 Plouzane, France
Contact: Director
Tel: 33 02 9849 8600
Fax: 33 02 9849 8609
E-mail: Direction.iuem@univ-brest.fr
http://www.univ-brest.fr/IUEM/

Faculty of Nautical Science
(Facoltà de Scienze Nautiche)
Instituto Universitario Navale
Via Acton 38
I-80133 Naples, Italy
Tel: 39 081 552 4342
Fax: 39 081 552 7126
E-mail: presidesn@uninav.it
http://www.uninav.it/

Finnish Institute of Marine Research
(Merentutkimuslaitos)
Lyypekinkuja 3 A
PO Box 33
FIN-00931 Helsinki, Finland
Tel: 358 9 6139 41
Fax: 358 9 6139 4494
E-mail: info@fimr.fi
http://www2.fimr.fi/

Fiskeriforskning
Muninbakken 9-13
N-9291 Breivika Tromsø, Norway
Tel: 47 77 62 90 00
Fax: 47 77 62 91 00
E-mail: fiskforsk@norut.no
http://www.fiskforsk.norut.no/indexe.htm

Flanders Marine Institute (VLIZ)
Victorialaan 3
B-8400 Oostende, Belgium
Contact: Director
Tel.: 32 5934 2130

Fax: 32 5934 2131
E-mail: info@vliz.be
http://www.vliz.be/

Force Technology
Park Allé 345
DK-2605 Brøndby, Denmark
Contact: Main Office
Tel: 45 43 26 70 00
Fax: 45 43 26 70 11
E-mail: force@force.dk
http://www.force.dk/

Foundation for International
 Studies
University of Malta
Old University Building
St. Paul Street
MT-Valletta VLT 07, Malta
Contact: Director General
Tel: 356 21 234 121
Fax: 356 21 230 551
E-mail: intoff@um.edu.mt
http://www.um.edu.mt/intoff/

Gdynia Maritime University
Morska 81-87
PO-81 225 Gdynia, Poland
Tel: 48 58 621 7041
Fax: 48 58 620 6701
http://www.wsm.gdynia.pl/

Geohydrodynamics and Environmental Research (GHER)
Université de Liège
Sart Tilman, B5
B-4000 Liège, Belgium
Contact: Director
Tel: 32 4 366 3350
Fax: 32 4 366 2355
E-mail: J.Nihoul@ulg.ac.be
http://modb.oce.ulg.ac.be/GHER/

Graduate Institute of International
 Studies
(Institut universitaire de hautés
 études internationales, Genève)
132, rue de Lausanne

CH-1211 Geneva, Switzerland
Tel: 41 22 908 57 00
Fax: 41 22 908 57 10
E-mail: info@hei.unige.ch
http://heiwww.unige.ch/

IMAR–Center for Ecological Modelling
(IMAR–Centro de Modelação Ecológica)
Faculty of Sciences and Technology
New University of Lisbon
Quinta da Torre
PT-2829 516 Monte de Caparica,
 Portugal
Tel: 351 21 2948300
Fax: 351 21 2948554
E-mail: frg@mail.fct.unl.pt
http://tejo.fct.unl.pt/

Institute for Fisheries and Marine Biology
(Institutt for Fisheri og Marinbiologi–IFM)
University of Bergen
PO Box 7800
N-5020 Bergen, Norway
Tel: 47 55 58 44 00
Fax: 47 55 58 44 50
E-mail: epost@ifm.uib.no
http://www.ifm.uib.no/

Institute for Marine and Atmospheric Research Utrecht
 (IMAU)
University of Utrecht
PO Box 80005
NL-3508 TA, Utrecht, The Netherlands
Tel: 31 030 253 3275
Fax: 31 030 254 3163
E-mail: imau@phys.uu.nl
http://www.fys.ruu.nl/wwwimau/

Institute for Marine Research
(Institut für Meereskunde an der
 Universitaet Kiel)
Düstenbrooker Weg 20

D-24105 Kiel, Germany
Contact: Directorate
Tel: 49 431 600 1530
Fax: 49 431 600 1515
E-mail: ifm@ifm.uni-kiel.de
http://www.ifm.uni-kiel.de/

Institute of Aquaculture
University of Stirling
Pathfoot Building
Stirling, FK9 4LA, United Kingdom
Contact: Director
Tel: 44 1786 467750
Fax: 44 1786 466896
E-mail: natsci@stir.ac.uk
http://www.stir.ac.uk/aqua/

Institute of Estuarine and Coastal
    Studies (IECS)
University of Hull
Cottingham Road
Hull, HU6 7RX, United Kingdom
Contact: Director
Tel: 44 14 8246 5667
Fax: 44 14 8246 5001
E-mail: iecs@hull.ac.uk
http://www.hull.ac.uk/iecs/

Institute of Hydrobiology and Fish-
    ery Science
(Institut für Hydrobiologie und
    Fischereiwissenschaft)
University of Hamburg
Olbersweg 24
D-22767 Hamburg, Germany
Tel: 49 40 428 38 6601
Fax: 49 40 428 38 6618
E-mail: ihf-office@uni-hamburg.de
http://www.biologie.uni-
    hamburg.de/ihf/

Institute of Marine Biology of Crete
    (IMBC)
PO Box 2214
Iraklio
GR-71003 Crete, Greece
Tel: 30 281 346860
Fax: 30 281 337822

E-mail: imbc@imbc.gr
http://www.imbc.gr/

Institute of Marine Research (IMAR)
(Centro Interdisciplinar de
    Coimbra)
c/o Department of Zoology
University of Coimbra
PT-3000 Coimbra, Portugal
Tel: 351 239 836386
Fax: 351 239 823603
E-mail: imar@ci.uc.pt
http://www.imar.pt/

Institute of Marine Sciences and
    Management
University of Istanbul
34470 Vefa Istanbul, Turkey
Tel: 90 212 528 2539
Fax: 90 212 526 8433
E-mail: cemga@istanbul.edu.tr
http://www.istanbul.edu.tr/
    enstituler/denizbilimleri/
    denizbilimleri.htm

Institute of Marine Studies
University of Plymouth
Drake Circus
Plymouth, PL4 8AA, United
    Kingdom
Tel: 44 17 5223 2400
Fax: 44 17 5223 2406
E-mail: ims@plymouth.ac.uk
http://
    hydrography.ims.plym.ac.uk/

Institute of Maritime Law and the
    Law of the Sea
(Institut für Seerecht und Seehan-
    delsrecht–ISSR)
University of Hamburg
Heimhuder Strasse 71
D-20148 Hamburg, Germany
Tel: 49 40 42838 2240
Fax: 49 40 42838 6271
E-mail: seercht@jura.uni-hamburg.de
http://www.jura.uni-hamburg.de/
    issr/

Institute of Maritime Law
University of Southampton
Highfield, Southampton, SO17 1BJ,
    United Kingdom
Contact: Institute Secretary
Tel: 44 23 8059 3449
Fax: 44 23 8059 3789
Email: iml@soton.ac.uk
http://www.soton.ac.uk/iml/

Institute of Oceanography
(Institut für Meereskunde)
University of Hamburg
Troplowitzstrasse 7
D-22529 Hamburg, Germany
Tel: 49 40 42838 2605
Fax: 49 40 42560 5926
E-mail: suendermann@ifm.
    uni-hamburg.de
http://www.ifm.uni-hamburg.de/
    index.html

Institute of Oceanography and Fisheries
Setaliste Iva Mestrovica 63
HR-21000 Split, Croatia
Tel: 385 21 358688
Fax: 385 21 358650
E-mail: office@izor.hr
http://www.izor.hr/eng/intro.html

Institute of Oceanology
Polish Academy of Sciences
PO Box 68
Powstancow Warszawy 55
PO-81-712 Sopot, Poland
Tel: 48 58 551 7281
Fax: 48 58 551 2130
E-mail: office@iopan.gda.pl
http://www.iopan.gda.pl/

Institute of Shipping Economics and
    Logistics
(Institut für Seeverkhvswirtschaft
    und Logistik–ISL)
Universitätallee GW 1, Block A
D-28359 Bremen, Germany
Tel: 49 421 220960
Fax: 49 421 2209655
E-mail: info@isl.org
http://www.isl.uni-bremen.de/

International Boundaries Research
    Unit (IBRU)
Department of Geography
University of Durham
Durham, DH1 3UR, United
    Kingdom
Tel: 44 19 1374 7701
Fax: 44 19 1374 7702
E-mail: ibru@dur.ac.uk
http://www-ibru.dur.ac.uk/

International Centre for Coastal Resources Research
(Centre Internacional d'Investigació
    dels Recursos Costaners–CIIRC)
Campus Nord–Univsitat Politecnica
    de Catalunya
Jordi Girona, 1-3, Edif. D-1
E-08034 Barcelona, Spain
Tel: 34 93 280 6400
Fax: 34 93 280 6019
E-mail: ciirc@upc.es
http://lim-ciirc.upc.es/

International Centre for Coastal and
    Ocean Policy Studies (ICCOPS)
Via Piacenza, 54
16138 Genova, Italy
Contact: Secretary-General
Tel: 39 10 846 8526
Fax: 39 10 835 7190
E-mail: info@iccops.it
http://www.iccops.it

International Centre for Earth Tides
    (ICET)
Observatoire Royal de Belgique
Avenue Circulaire 3
B-1180 Brussels, Belgium
Contact: Executive Officer
Tel: 32 2 373 0248
Fax: 32 2 374 9822
E-mail: b.ducarme@oma.be
http://www.astro.oma.be/icet/
    index.html

International Fisheries Institute (HIFI)
University of Hull
Hull HU6 7RX, United Kingdom
Tel: 44 14 8246 6422
Fax: 44 14 8247 0129
E-mail: hifi@hull.ac.uk
http://www.hull.ac.uk/hifi/

International Geosphere-Biosphere Programme (IGBP)
Royal Swedish Academy of Sciences
Box 50005
SE-104 05 Stockholm, Sweden
Contact: Secretariat
Tel: 46 8 16 64 48
Fax: 46 8 16 64 05
E-mail: sec@igbp.kva.se
http://www.igbp.kva.se/

International Studies in Aquatic Tropical Ecology
Zentrum für Marine Tropenökologie (ZMT)
Fahrenheitstrasse 6
D-28359 Bremen, Germany
Tel.: 49 421 2380 042
Fax: 49 421 2380 030
E-mail: isatec@uni-bremen.de
http://www.isatec.uni-bremen.de/

Islands and Small States Institute
Foundation for International Studies
Old University Building
St. Paul Street
MT-Valletta VLT 07, Malta
Tel: 356 21 248 218
Fax: 356 21 230 551
E-mail: islands@um.edu.mt
http://www.comnet.mt/issi/

Laboratoire d'Océanologie
Université de Liège
Sart Tilman B6,
B-4000 Liège, Belgium
Tel: 32 4 366 3324
Fax: 32 4 366 3325
E-mail: claire.daemers@ulg.ac.be
http://www.ulg.ac.be/oceanbio/laboeng.htm

Laboratoire de Sédimentologie Marines
Université de Perpignan
52 avenue de Villeneuve
F-66 860 Perpignan Cedex, France
Tel: 33 4 6866 2000
Fax: 33 4 6866 2019
http://www.univ-perp.fr/

Laboratoire Environnement Marin Littoral
Université de Nice-Sophia Antipolis
Faculté des Sciences
Parc Valrose
F-06108 Nice Cedex 2, France
Tel: 33 4 9207 6846
Fax: 33 4 9207 6849
E-mail: meinesz@unice.fr
http://www.unice.fr/LEML/

Laboratori d'Enginyeria Marítima (LIM/UPC)
Campus Nord–Universitat Politècnica de Catalunya
Jordi Girona, 1-3, Edif. D-1
E-08034 Barcelona, Spain
Contact: Head of LIM/UPC
Tel: 34 93 401 6468
Fax: 34 93 401 1861
E-mail: info.lim@upc.es
http://www.upc.es/lim/

Legal Aspects of Marine Affairs Program
Cardiff Law School
Cardiff University
PO Box 427
Cardiff, Wales, CF10 3XJ, United Kingdom
Contact: Director
Tel: 44 29 2087 4353
Fax: 44 29 2087 4982
E-mail: law-pg@cf.ac.uk
http://www.cf.ac.uk/claws/

Marine and Coastal Environment
    Group
Department of Earth Sciences
Cardiff University
PO Box 914
Cardiff, Wales, CF10 3YE, United
    Kingdom
Contact: International Secretariat
Tel: 44 29 2087 4713
Fax: 44 29 2087 4326
E-mail: earth@cardiff.ac.uk
http://www.cf.ac.uk/masts/
    index.html

Marine Research Centre
(Marina Forskningscentrum)
Earth Sciences Centre
PO Box 460
Göteborg University
SE-405 30 Göteborg, Sweden
Contact: Director
Tel: 46 03 17 72 22 95
Fax: 46 03 17 72 27 85
E-mail: henrik.tallgren@
    science.gu.se
http://www.gmf.gu.se/

Maritime Institute
(Maritiem Instituut)
Universiteitstraat 6
B-9000 Gent, Belgium
Tel: 32 9 264 6894
Fax: 32 9 264 6989
E-mail: gwendoline.gonsaeles@
    rug.ac.be
http://www.maritieminstituut.be/

Max Planck Institute for Marine Mi-
    crobiology Bremen
(Max-Planck-Institut für Marine Mik-
    robiologie Bremen)
Celsiusstrasse 1
D-28359 Bremen, Germany
Contact: Director
Tel: 49 421 202850
Fax: 49 421 2028580
E-mail: contact@mpi-bremen.de
http://www.mpi-bremen.de/

Max Planck Institute for Meteo-
    rology
(Max-Planck-Institut für Metero-
    logie)
Bundesstrasse 55
D-20146 Hamburg, Germany
Contact: Executive Director
Tel: 49 40 41173 421
Fax: 49 40 41173 298
http://www.mpimet.mpg.de/

Mediterranean Programme for Inter-
    national Environmental Law
    and Negotiation (MEPIELAN)
Panteion University of Athens
136 Syngroy Avenue
GR-17671 Athens, Greece
Tel: 30 10 920 1841
Fax: 30 10 961 0591
E-mail: info@mepielan.gr
http://www.mepielan.gr/

Nansen Environmental and Remote
    Sensing Center (NERSC)
Edvard Grieg Vei 3a
N-5037 Bergen, Norway
Tel: 47 55 29 72 88
Fax: 47 55 20 00 50
E-mail: admin@nersc.no
http://www.nersc.no/

Netherlands Institute for Fisheries
    Research
Haringkade 1, IJmuiden
PO Box 68
NL-1970 AB IJmuiden, The Nether-
    lands
Tel: 31 25 556 4646
Fax.: 31 25 556 4644
E-mail: postmaster@rivo.wag-ur.nl
http://www.rivo.wag-ur.nl/

Netherlands Institute for the Law of
    the Sea (NILOS)
Utrecht University
Faculty of Law
Achter Sint Pieter 200

NL-3512 HT Utrecht, The Netherlands
Contact: Director
Tel: 31 30 253 7033
Fax: 31 30 253 7073
E-mail: nilos@law.uu.nl
http://www.law.uu.nl/english/isep/framenilos.asp

North Atlantic Fisheries College
Port Arthur
Scalloway, Shetlands, ZE1 0UN, United Kingdom
Tel: 44 15 9577 2000
Fax: 44 15 9577 2001
E-mail: admin@nafc.ac.uk
http://www.nafc.ac.uk/

Norwegian College of Fishery Science (Norges Fiskerihøgskole)
Universitetet i Tromsø
N-9037 Tromsø, Norway
Tel: 47 77 64 60 00
E-mail: webmaster@nfh.uit.no
http://www.nfh.uit.no/

Norwegian Marine Technology Research Institute (MARINTEK)
Otto Nielsens vei 10
PO Box 4125 Valentinlyst
N-7450 Trondheim, Norway
Contact: Main Office
Tel: 47 73 59 55 00
Fax: 47 73 59 57 76
E-mail: marintek@marintek.sintef.no
http://www.marintek.sintef.no/

Oceanographic Institute
(Institut océanographique)
195, rue Saint Jacques
F-75005 Paris, France
Tel: 33 1 44 321070
Fax: 33 1 40 517316
http://www.oceano.org/

Oceanographic Institute of the Western Mediterranean Sea (IOMO)
San Juan, 9
Vespella de Gaia
E-43763 Tarragona, Spain
Contact: President
Tel: 34 977 655214
Fax: 34 977 655584
E-mail: iomo@tinet.fut.es
http://www.fut.es/iomo/

Oceanography Laboratories
Earth Sciences Department
University of Liverpool
Matat Oceanography Building, Peach Street
Liverpool, L69 7ZL, United Kingdom Tel: 44 15 1794 4090
Fax: 44 15 1794 4099
E-mail: sn12@liv.ac.uk
http://pcwww.liv.ac.uk/ocean/oceanography/

Oceanology and Environmental Geophysics (OGA)
Instituto Nationale di Oceanografia
Borgo Grotta Gigante 42/c
I-34010 Sgonico (TS), Italy
Tel: 39 040 2140 1
Fax: 39 040 2140 266
E-mail: mailbox@ogs.trieste.it
http://doga.ogs.trieste.it/index.html

Plymouth Marine Laboratory
Prospect Place, The Hoe
Plymouth PL1 3DH, United Kingdom
Tel: 44 17 5263 3100
Fax: 44 17 5263 3101
E-mail: forinfo@pml.ac.uk
http://www.pml.ac.uk/

Proudman Oceanographic Laboratory
Bidston Observatory
Birkenhead, Merseyside, CH43 7RA, United Kingdom
Tel: 44 15 1653 8633
Fax: 44 15 1653 6269
http://www.pol.ac.uk/

Research Unit of Maritime Medicine
    (RUMM)
University of Southern Denmark
Niels Bohrs Vej 9
DK-6700 Esbjerg, Denmark
Contact: Senior Research Associate
Tel: 45 65 50 41 11
Fax: 45 65 50 10 91
E-mail: ocj@fmm.sdu.dk
http://www.esb.sdu.dk/vny/
    uksmiforside.htm

Rhodes Academy of Oceans Law
    and Policy
c/o Center for Oceans Law and Policy
University of Virginia School of Law
580 Massie Road
Charlottesville, VA, 22903, USA
Contact: Director
Tel: 1 434 924-7441
Fax: 1 434 924-7362
E-mail: colp@virginia.edu
http://www.virginia.edu/colp/
    COLP.html

SAR International Center for Marine Molecular Biology
Bergen High Technology Center
Thormøhlensgate 55
N-5008 Bergen, Norway
Tel: 47 55 58 43 44
Fax: 47 55 58 43 05
E-mail: trygve.serck-hanssen@sars.
    uib.no
http://www.uib.no/fa/sars/

Scandinavian Institute of Maritime
    Law
(Nordisk Institutt for Sjørett)
University of Oslo
Karl Johansgate 47
Domus Media Vest
N-0162 Oslo, Norway
Contact: Director
Tel: 47 22 85 97 48
Fax: 47 22 85 97 50
E-mail: sjorett-adm@jus.uio.no
http://www.jus.uio.no/nifs/

Scarborough Centre for Coastal
    Studies
University of Hull
Scarborough Campus
Filey Road
Scarborough, YO11 3AZ, United
    Kingdom
Tel: 44 17 2336 2392
Fax: 44 17 2337 0815
E-mail: sccs@hull.ac.uk
http://www.ccs.hull.ac.uk/

School of Environmental Sciences
University of East Anglia
Norwich, NR4 7TJ, United Kingdom
Tel: 44 16 0359 2542
Fax: 44 16 0350 7719
E-Mail: env@uea.ac.uk
http://www.uea.ac.uk/env/

School of Geography and
    Geology
University of Plymouth
Drake Circus
Plymouth, Devon, PL4 8AA, United
    Kingdom
Tel: 44 17 5223 3093
Fax: 44 17 5223 3095
E-mail: scienqs@plymouth.ac.uk
http://www.geog.plym.ac.uk/
    contact.htm

School of Marine Science and Technology
Armstrong Building
University of Newcastle upon Tyne
Newcastle Upon Tyne, NE1 7RU,
    United Kingdom
Tel: 44 19 1222 6718
Fax: 44 19 1222 5491
E-mail: atilla.incecik@ncl.ac.uk
http://www.ncl.ac.uk/marine/

School of Maritime and Coastal
    Studies
Southampton Institute, East Park
    Terrace

Southampton, Hants, SO14 0RD,
 United Kingdom
Tel: 44 23 8031 9755
Fax: 44 23 8033 4441
E-mail: tf@solent.ac.uk
http://www.solent.ac.uk/maritime/
 index.htm

School of Ocean Sciences
University of Wales, Bangor
Menai Bridge
Anglesey, North Wales, LL59 5EY,
 United Kingdom
Tel: 44 12 4838 2846
Fax: 44 12 4871 6367
E-mail: enquiries@sos.bangor.ac.uk
 http://www.sos.bangor.ac.uk/

Scottish Association for Marine Science (SAMS)
Dunstaffnage Marine Laboratory
Oban, Argyll, Scotland, PA34 4AD,
 United Kingdom
Tel: 44 16 3155 9000
Fax: 44 16 3155 9001
E-mail: mail@dml.ac.uk
http://www.sams.ac.uk/

Sea Mammal Research Unit
Gatty Marine Laboratory
University of St Andrews
St Andrews, Fife, Scotland, KY16
 8LB, United Kingdom
Tel: 44 13 3446 2630
Fax: 44 13 3446 2632
E-mail: pw18@st-andrews.ac.uk
http://smub.st-and.ac.uk/index.htm

Seafarers International Research
 Centre (SIRC)
Cardiff University
PO Box 907
Cardiff, Wales, CF10 3YP, United
 Kingdom
Contact: Secretariat
Tel: 44 29 2087 4620
Fax: 44 29 2087 4619
E-mail: sirc@cardiff.ac.uk
http://www.sirc.cf.ac.uk/

Sophia Antipolis Centre d'Etudes et
 de Recherches sur le Droit des
 Activités Maritimes
(CERDAM)
Université de Nice
7, Avenue Robert Schuman
F-06050 Nice, Cédex 1, France
Tel: 334 92 15 7043
Fax: 334 93 96 0131
E-mail: cerdam@unice.fr
http://www.unice.fr/cerdam/

Southampton Oceanography Centre
University of Southampton
Waterfront Campus
European Way
Southampton, SO14 3ZH, United
 Kingdom
Contact: Director
Tel: 44 23 8059 6666
Fax: 44 23 8059 6667
http://www.soc.soton.ac.uk/

Stazione Zoologica Anton Dohrn
Villa Comunale
I-80121 Naples, Italy
Contact: Secretary General
Tel: 39 81 583 3111
Fax: 39 81 764 1355
E-mail: pidigi@alpha.szn.it
http://www.szn.it/

Stockholm Marine Research Centre
(Stockholms Marina Forskningcentrum)
Stockholm University
SE-106 91 Stockholm, Sweden
Contact: Director
Tel: 46 8 16 37 18
Fax: 46 8 15 79 56
E-mail: smf@smf.su.se
http://www.smf.su.se/

Transport and Shipping Research
 Group
Cardiff Business School
Cardiff University
Aberconway Building, Colum Drive

Cardiff, Wales, CF10 3EU, United Kingdom
Tel: 44 29 2087 6081
Fax: 44 29 2087 4301
E-mail: gardner@cardiff.ac.uk;
marlow@cardiff.ac.uk
http://www.cf.ac.uk/carbs/index.html

Umeå Marine Sciences Centre
(Umeå Marina Forskningscentrum–UMF)
Umeå Universitet
Norrbyn
SE-910 20 Hörnefors, Sweden
Contact: Secretariat
Tel: 46 09 07 86 79 74
Fax: 46 09 07 86 79 95
E-mail: kristina.wiklund@umf.umu.se
http://www.umf.umu.se/

UNESCO–Cousteau Ecotechnie Programme (UCEP)
Division of Ecological Sciences
UNESCO
1, rue Miollis
75732 Paris Cedex 15, France
Fax: 33 1 4568 5804
E-mail: p.dogse@unesco.org
http://www.unesco.org/mab/capacity/ucep/ucephome.htm

Université du Littoral Côte d'Opale
Services Centraux
1, place de l'Yser–B.P. 1022
F-59375 Dunkerque CEDEX 1, France
Tel: 33 3 2823 7373
Fax: 33 3 2823 7313
http://www.univ-littoral.fr/

World Ship Trust
202 Lambeth Road
London, SE1 7JW, United Kingdom
Contact: Chairman
Tel: 44 20 7385 4267
E-mail: worldship@lynnmallet.demon.co.uk
http://www.worldshiptrust.org/

6. REGIONAL ACADEMIC ORGANIZATIONS

6.5 Australia, New Zealand and Oceania

Australasian Legal Information Institute
Faculty of Law
University of Technology Sydney
PO Box 123
Broadway, NSW 2007, Australia
Tel: 61 2 9514 3174
Fax: 61 2 9514 3168
http://www.austlii.edu.au/

Australian Defense Studies Centre
University College
Australian Defence Force Academy
Northcott Drive
Campbell, ACT 2600, Australia
Tel: 61 2 6268 8849
Fax: 61 2 6268 8840
E-mail: b.hall@adfa.edu.au
http://idun.itsc.adfa.edu.au/ADSC/

Australian Institute of Navigation
GPO Box 2250
Sydney, NSW 2001, Australia
Tel: 61 2 9264 6413
Fax: 61 2 9267 1682
E-mail: c.rizos@unsw.edu.au
http://www.gmat.unsw.edu.au/aion/aion.html

Australian Maritime College
PO Box 986
Launceston, TAS 7250, Australia
Tel: 61 3 6335 4711
Fax: 61 3 6326 6493
E-mail: amcinfo@amc.edu.au
http://www.amc.edu.au/

Centre for Coral Reef Biodiversity
School of Marine Biology and Aquaculture
James Cook University
Townsville, QLD 4811, Australia

Tel: 61 7 4781 4222
Fax: 61 7 4725 1570
E-mail: ccrbio@jcu.edu.au
http://www.jcu.edu.au/school/
    mbiolaq/ccrbio/

Centre for Marine Studies
University of Queensland
Brisbane, QLD 4072, Australia
Tel: 61 7 3365 4333
Fax: 61 7 3365 4755
E-mail: cms@uq.edu.au
http://www.marine.uq.edu.au/

Centre for Maritime Law
T. C. Beirne School of Law
University of Queensland
Brisbane, QLD 4072, Australia
Tel: 61 7 3365 1444
Fax: 61 7 3365 1454
E-mail: s.derrington@law.uq.edu.au
http://www.law.uq.edu.au/

Centre for Maritime Policy
University of Wollongong
Wollongong, NSW 2522, Australia
Tel: 61 2 4221 4883
Fax: 61 2 4226 5544
E-mail: sali_bache@uow.edu.au
http://www.uow.edu.au/law/cmp/

Coastal Marine Research Unit
Victoria University of Wellington
PO Box 600
Wellington, New Zealand
Contact: Dept of Biological Sciences
Tel: 64 4 463 5339
Fax: 64 4 463 5331
E-mail: biosci@vuw.ac.nz
http://www.vuw.ac.nz/sbs/

East-West Center
1601 East-West Road
Honolulu, HI, 96848, USA
Tel: 1 808 944-7111
Fax: 1 808 944-7376
E-mail: webhelp@eastwestcenter.org
http://www.eastwestcenter.org/

Indian Ocean Centre
John Curtin International Institute
Curtin University of Technology
GPO Box U 1987
Perth, WA 6001, Australia
Tel: 61 9 220 7033
Fax: 61 9 220 7020
E-mail: jcii@cc.curtin.edu.au
http://www.curtin.edu.au/curtin/
    centre/jc/jcii/

Institute of Marine Resources
University of the South Pacific
PO Box 1168
Suva, Fiji
Contact: Director
Tel: 679 313900
Fax: 679 301305
http://www.usp.ac.fj/imr/

Lincoln Marine Science Centre
PO Box 2023
Port Lincoln, SA 5606, Australia
Tel: 61 8 8683 2500
Fax: 61 8 8683 2525
E-mail: lmsc@flinders.edu.au
http://www.scieng.flinders.edu.au/
    biology/lmsc/

Marine Affairs Programme
University of the South Pacific
PO Box 1168
Suva, Fiji
Contact: Coordinator
Tel: 679 212890
Fax: 679 301490
E-mail: veitayaki_j@usp.ac.fj
http://www.usp.ac.fj/marineaf/

Marine Laboratory
University of Guam
UOG Station
Mangilao, GU, 96923, USA
Tel: 671 735 2175
Fax: 671 734 6767
http://www.uog.edu/marinelab/

Marine Studies Programme
University of the South Pacific
PO Box 1168
Suva, Fiji
Tel: 679 305272
Fax: 679 301490
E-mail: hay_c@usp.ac.fj
http://www.usp.ac.fj/marine/

New Zealand Centre for Environmental Law
Faculty of Law
University of Auckland
Private Bag 92019
Auckland Central, New Zealand
Tel: 64 9 373 7599, ext. 7827
Fax: 64 9 373 7473
E-mail: k.bosselmann@auckland.ac.nz
http://www.law.auckland.ac.nz/groups/cel/index.html

Oceans and Coastal Research Centre
University of Wollongong
Wollongong, NSW 2522, Australia
Tel: 61 2 4221 4134
Fax: 61 2 4221 4665
E-mail: john_morrison@uow.edu.au
http://www.uow.edu.au/science/research/ocrc/

School of Environmental Science
University of Auckland, Tamaki Campus
Private Bag 92019
Auckland, New Zealand
Tel: 64 9 373 7599, ext. 6815
Fax: 64 9 373 7042
E-mail: sems@auckland.ac.nz
http://sems.auckland.ac.nz/

School of Marine Biology and Aquaculture
James Cook University
Townsville, QLD 4811, Australia
Tel: 61 7 4781 4345
Fax: 61 7 4781 5511
E-mail: mbiolaq@jcu.edu.au

http://www.jcu.edu.au/school/mbiolaq/

School of Ocean and Earth Science and Technology (SOEST)
University of Hawaii
1680 East-West Road, POST 802
Honolulu, HI, 96822, USA
Contact: Office of the Dean
Tel: 1 808 956-6182
Fax: 1 808 956-9152
E-mail: raleigh@soest.hawaii.edu
http://www.soest.hawaii.edu/

School of Tropical Environment Studies and Geography
James Cook University
Townsville, QLD 4811, Australia
Tel: 61 7 4781 4325
Fax: 61 7 781 4020
E-mail: emma.gyuris@jcu.edu.au
htttp://www.tesag.jcu.edu.au/

Tasmanian Aquaculture and Fisheries Institute
University of Tasmania
PO Box 252-49
Hobart, TAS 7001, Australia
Tel: 61 3 6227 7277
Fax: 61 3 6227 8035
E-mail: tafi@utas.edu.au
http://www.utas.edu.au/docs/tafi/TAFI_Homepage.html

Tourism Program
School of Business
James Cook University
Townsville, QLD 4811, Australia
Tel: 61 7 4781 5134
Fax: 61 7 4781 4019
E-mail: robyn.yesberg@jcu.edu.au
http://www.jcu.edu.au/flbca/business/tourism/

William S. Richardson School of Law
University of Hawaii
2515 Dole Street
Honolulu, HI, 96822, USA

Contact: Dean
Tel: 1 808 956-7966
Fax: 1 808 956-6402
E-mail: lawadm@hawaii.edu
http://www.hawaii.edu/law/

6. REGIONAL ACADEMIC ORGANIZATIONS

6.6 Polar Regions

Arctic Centre
University of Lapland
PO Box 122
FIN-96101 Rovaniemi, Finland
Tel: 358 16 341 2758
Fax: 358 16 341 2777
E-mail: arctic.centre@urova.fi
http://www.arcticcentre.org/

Arctic Centre, Groningen
Rijksuniversiteit Groningen
Oude kijk in't Jatstraat 26
PO Box 716
NL-9700 AS, Groningen, The Netherlands
Tel: 31 50 363 6834
Fax: 31 50 363 4900
E-mail: arctisch@let.rug.nl
http://odin.let.rug.nl/arctic/

Arctic Institute of North America
University of Calgary
2500 University Dr. NW
Calgary, AB, T2N 1N4, Canada
Contact: Executive Director
Tel: 1 403 220-7516
Fax: 1 403 282-4609
E-mail: wkjessen@ucalgary.ca
http://www.ucalgary.ca/aina/

Canadian Circumpolar Institute
8625 112 Street, Suite 308, Campus Tower
University of Alberta
Edmonton, AB, T6G 0H1, Canada
Contact: Director
Tel: 1 780 492-4512
Fax: 1 780 492-1153
E-mail: ccinst@gpu.srv.ualberta.ca
http://www.ualberta.ca/ccinst/

Gateway Antarctica
University of Canterbury
Private Bag 4800
Christchurch
New Zealand
Tel: 64 3 364 2136
Fax: 64 3 364 2197
E-mail: gateway@canterbury.ac.nz
http://www.anta.canterbury.ac.nz/

Institute of Antarctic and Southern Ocean Studies
University of Tasmania
GPO Box 252-77
Hobart, TAS 7001, Australia
Contact: Director
Tel: 61 3 6226 2971
Fax: 61 3 6226 2973
E-mail: Kelvin.Michael@utas.edu.au
http://www.iasos.utas.edu.au/

Scott Polar Research Institute
University of Cambridge
Lensfield Road
Cambridge CB2 1ER, United Kingdom
Tel: 44 12 2333 6540
Fax: 44 12 2333 6549
E-mail: enquiries@spri.cam.ac.uk
http://www.spri.cam.ac.uk/

7.0 NATIONAL GOVERNMENT ORGANIZATIONS

Agriculture, Fisheries and Forestry–Australia
PO Box 858
Canberra, ACT 2601, Australia
Contact: Fisheries and Marine Sciences
Tel: 61 2 6272 5777
E-mail: fisheries@affa.gov.au
http://www.affa.gov.au/

Alfred Wegener Institute for Polar and Marine Research
Postfach 12 0161
D-27515 Bremerhaven, Germany
Tel: 49 471 4831 0
Fax: 49 471 4831 1149
http://www.awi-bremerhaven.de/

Asociacíon Argentina de Ingeniería Sanitaria y Ciencias del Ambiente (AIDIS Argentina)
Av. Belgrano 1580 Piso 3
1093 Buenos Aires, Argentina
Contact: Chairman
Tel/Fax: 54 11 4381 5832/5093
E-mail: aidisar@aidisar.org
http://www.aidisar.org/

Association of Sea Fisheries Committees of England and Wales
24 Wykeham
Scarborough
Yorkshire, YO13 9QP, United Kingdom
Contact: Chief Executive
Tel/Fax: 44 17 2386 3169

Atlantic Coastal Action Program
Community Programs
Environment Canada
16th Floor, Queen Square
45 Alderney Drive
Dartmouth, NS, B2Y 2N6, Canada
Tel: 1 902 426-5777
Fax: 1 902 426-6348
E-mail: colleen.mcneil@ec.gc.ca
http://www.ns.ec.gc.ca/community/acap/index_e.html

Atlantic States Marine Fisheries Commission (ASMFC)
1444 Eye St., NW, 6th floor
Washington, DC, 20005, USA
Contact: Executive Director
Tel: 1 202 289-6400
Fax: 1 202 289-6051
E-mail: comments@asmfc.org
http://www.asmfc.org/

Australian Department of Agriculture, Fisheries and Forestry
Edmund Barton Building
Broughton Street, Barton, GPO Box 858
Canberra, ACT 2601, Australia
Tel: 61 2 6272 5777
http://www.affa.gov.au/

Australian Institute of Marine Science
PMB No. 3
Townsville MC, QLD 4810, Australia
Contact: Chief Scientist
Tel: 61 7 4753 4444
Fax: 61 7 4772 5852
E-mail: reception@aims.gov.au
http://www.aims.gov.au/

Australian Maritime Safety Authority
GPO Box 2181
Canberra, ACT 2601, Australia
E-mail: generalenquiries@amsa.gov.au
http://www.amsa.gov.au/

Bangladesh Fisheries Research Institute
Mymensingh 2201, Bangladesh
Tel: 880 91 54874
Fax: 880 91 55259
E-mail: frihq@bdmail.net
http://www.bangladeshgov.org/mofl/fri/fritop.htm

Bedford Institute of Oceanography
PO Box 1006
Dartmouth, NS, B2Y 4A2, Canada
Contact: Director
Tel: 1 902 426-2373
http://www.bio.gc.ca/

British Geological Survey
Kingsley Dunham Centre
Keyworth
Nottingham, NG12 5GG, United Kingdom
Contact: Headquarters
Tel: 44 11 5936 3100

Fax: 44 11 5936 3200
E-mail: enquiries@bgs.ac.uk
http://www.bgs.ac.uk/

Bureau of Fisheries and Aquatic Resources
4th Floor Arcadia Bldg., 860 Quezon Avenue
3008 Quezon City, Metro Manila, The Philippines
Tel: 632 372 505
Fax: 632 372 5048
E-mail: bfarnmfd@info.com.ph
http://www.bfar.gov.ph/

Bureau of Ocean and International Environmental and Scientific Affairs
US Department of State
2201 C Street, NW, Room 7831
Washington, DC, 20520, USA
Contact: Assistant Secretary of State
Tel: 1 202 647-1554
Fax: 1 202 647-1554
http://www.state.gov/g/oes/

Canada-Newfoundland Offshore Petroleum Board
5th Floor TD Place
140 Water Street
St. John's, NL, A1C 6H6, Canada
Tel: 1 709 778-1400
E-mail: postmaster@cnopb.nf.ca
http://www.cnopb.nfnet.com/

Canada-Nova Scotia Offshore Petroleum Board
6th Floor TD Centre, 1791 Barrington Street
Halifax, NS, B3J 3K9, Canada
Tel: 1 902 422-5588
Fax: 1 902 422-1799
E-mail: postmaster@cnsopb.ns.ca
http://www.cnsopb.ns.ca/

Canadian Environmental Assessment Agency
200 Sacré-Coeur Blvd., 14th floor
Hull, QC, K1A 0H3, Canada
Tel: 1 819 997-1000
Fax: 1 819 994-1469
E-mail: info@ceea.gc.ca
http://www.ceaa.gc.ca/

Canadian Polar Commission
Suite 1710, Constitution Square
360 Albert Street
Ottawa, ON, K1R 7X7, Canada
Tel: 1 613 943-8605
Fax: 1 613 943-8607
E-mail: mail@polarcom.gc.ca
http://www.polarcom.gc.ca/

Caribbean Fishery Management Council
US Department of Commerce
268 Muñoz Rivera Avenue, Suite 1108
San Juan, Puerto Rico, 00918-2577, USA
Contact: Executive Director
Tel.: 1 787 766-5926
Fax: 1 787 766-6239
E-mail: miguel.a.rolon@noaa.gov
http://www.caribbeanfmc.com/headquar.htm

Centre for Environment, Fisheries and Aquaculture Science (CEFAS)
Lowestoft Laboratory
Pakefield Road
Lowestoft, Suffolk, NR33 0HT, United Kingdom
Tel: 44 15 0256 2244
Fax: 44 15 0251 3865
E-mail: marketing@cefas.co.uk
http://www.cefas.co.uk/homepage.htm

Center of Marine Research
Lithuanian Ministry of the Environment
Taikos pr.26
LT-5802 Klaipeda, Lithuania
Tel: 370 4641 0450

Fax: 370 4641 0460
E mail: CMR@klaipeda.omnitel.net
http://www1.omnitel.net/
 juriniai_tyrimai/

Centre de Recherches Océanographiques de Dakar-Thiaroye (CRODT)
Km 10, Route des Rufisque
BP 2241
Dakar, Senegal
Contact: Directeur
Tel: 221 834 3383
Fax: 221 834 2792
E-mail: tdiouf@crodt.isra.sn; asamba@crodt.isra.sn
http://www.waf-peche.crodt.isra.sn/

Centre de Recherches Oceanologiques (CRO)
Ministere de la Recherche Scientifique
29, rue des Pêcheurs
BP V18
Abidjan, Côte d'Ivoire
Contact: Directeur
Tel: 225 355014
Fax: 225 351155
E-mail: postmaster@cro.orstom.ci
http://www.refer.org/ivoir_ct/rec/cdr/cro/

Centre for Ocean Research and Information
Continental Shelf Committee
Cabinet of Ministers
103 Quan Thanh Street
Hanoi, Vietnam
Tel: 844 8455150
Fax: 844 8236920
E-mail: cori@cori.ac.vn; vinh@cori.ac.vn
http://www.cori.gov.vn/

Centre National de la Recherche Scientifique (CNRS)
3 rue Michel-Ange
F-75794 Paris Cedex 16, France
Contact: Sécretariat général
Tel: 33 1 44 96 4000
Fax: 33 1 44 96 5390
E-mail: websg@cnrs-dir.fr
http://www.sg.cnrs.fr/

Centre Océanologique du Pacifique, IFREMER
BP 7004
Taravao, Tahiti, French Polynesia
Tel: 689 546000
Fax: 689 546099
E-mail: emmanuel.thouard@ifremer.fr
http://www.ifremer.fr/cop/

China Institute for Marine Development Strategy
State Oceanic Administration
1 Fuxingmenwai Avenue
Beijing 100860, People's Republic of China
Tel: 86 10 6803 0767
Fax: 86 10 6804 7756
http://www.coi.gov.cn/eindex.html

China Institute for Marine Affairs
State Oceanic Administration
1 Fuxingmenwai Ave.
Beijing 100860, People's Republic of China
Contact: Executive Director
Tel: 86 10 6802 2137
E-mail: zgao@public.bta.net.cn
http://www.soa.gov.cn/

Coastal America
300 7th Street, SW, Suite 680
Washington, DC, 20024, USA
Contact: Director
Tel: 1 202 401-9928
Fax: 1 202 401-9821
E-mail: virginia.tippie@usda.gov
http://www.coastalamerica.gov/

Coastal Research Institute (CoRI)
Alexandria, Egypt
Contact: Director

Tel: 202 484 4614
Fax: 202 444 6761
E-mail: nwrc@idsc.gov.eg
http://www.nwrc.gov.eg/

Coastal States Organization
Hall of States, Suite 322
444 N. Capitol Street, NW
Washington, DC, 20001, USA
Contact: Executive Director
Tel: 1 202 508-3860
Fax: 1 202 508-3843
E-mail: cso@sso.org
http://www.sso.org/cso/

Coastal Zone Management Centre (CZMC)
PO Box 20907
NL-2500 EX, The Hague, Netherlands
Tel: 31 70 311 4311
Fax: 31 70 311 4380
E-mail: czmc@rikz.rws.minvenw.nl
http://www.netcoast.nl/

Coastal Zone Management Unit
Bay Street
St. Michael, Barbados, West Indies
Contact: Director
Tel: 1 246 228-5951
Fax: 1 246 228-5956
E-mail: info@coastal.gov.bb
http://www.coastal.gov.bb/

Commonwealth Scientific Industrial Research Organization (CSIRO)
CSIRO Marine Research
GPO Box 1538
Hobart, TAS 7001, Australia
Tel: 61 3 6232 5222
Fax: 61 3 6232 5000
E-mail: reception@marine.csiro.au
http://www.marine.csiro.au/

Cooperative Research Centre (CRC) for Antarctica and the Southern Ocean
GPO Box 252-80
Hobart, TAS 7001, Australia
Contact: Chief Executive Officer
Tel: 61 3 6226 7888
Fax: 61 3 6226 2973
E-mail: secretary@antcrc.utas.edu.au
http://www.antcrc.utas.edu.au/

CRC for Coastal Zone, Estuary and Waterway Management
Indooroopilly Sciences Centre
80 Meiers Road
Indooroopilly, QLD 4068, Australia
Contact: Chief Executive Officer
Tel: 61 7 3362 9398
Fax: 61 7 3362 9372
E-mail: roger.shaw@dnr.qld.gov.au
http://www.coastal.crc.org.au/

CRC for Sustainable Aquaculture of Finfish
PO Box 120
Henley Beach, SA 5022, Australia
Contact: Chief Executive Officer
Tel: 61 8 8290 2302
Fax: 61 8 8200 2481
E-mail: montague.peter@saugov.sa.gov.au

CRC for Sustainable Tourism
Griffith University Gold Coast
PMB 50
Gold Coast Mall Centre, QLD 9726, Australia
Tel: 61 7 5552 8116
Fax: 61 7 5552 8171
E-mail: info@crctourism.com.au
http://www.crctourism.com.au/

CRC Reef Research Centre
Northtown Towers, 6th Floor
PO Box 772
Townsville, QLD 4810, Australia
Contact: Secretariat
Tel: 61 7 4729 8400
Fax: 61 7 4729 8499
E-mail: info@crcreef.com
http://www.reef.crc.org.au/

Council for Scientific and Industrial
  Research (CSIR)
PO Box 395
Pretoria 0001, South Africa
Contact: Program Manager
Tel: 27 12 841 2911
Fax: 27 12 349 1153
E-mail: webmaster@csir.co.za
http://www.csir.co.za/

Countryside Council for Wales
Maes Ffynnon
Penrosgarnedd
Bangor, Gwynedd, Wales, LL57
  2LQ, United Kingdom
Contact: Head Office
Tel: 44 0845 1306 229
http://www.ccw.gov.uk/

Danish Directorate of Fisheries
Stormgade 2
DK-1470 Copenhagen K, Denmark
Tel: 45 33 96 30 00
Fax: 45 33 96 39 03
E-mail: fd@fd.dk
http://www.fd.dk/

Danish Institute for Fisheries Research
  (Danmarks Fiskeriundersøgelser)
Jägersborg vej 64-66
DK-2800 Lyngby, Denmark
Tel: 45 33 96 33 00
Fax: 45 33 96 33 49
E-mail: info@dfu.min.dk
http://www.dfu.min.dk/

Danish Institute of Agricultural and
  Fisheries Economics
(Fødevareøkonomisk Institut)
Rolighedsvej 25
1958 Frederiksberg C, Denmark
Tel: 45 35 28 68 00
Fax: 45 35 28 68 01
E-mail: foi@foi.dk
http://www.sjfi.dk/

Danish Maritime Authority (DMA)
Ministry of Economics and Business
  Affairs
PO Box 2605
DK-2100 Copenhagen Ø, Denmark
Tel: 45 39 17 44 00
Fax: 45 39 17 44 01
E-mail: dma@dma.dk
http://www.dma.dk/

Danish Polar Center
Strandgade 100H
DK-1401 Copenhagen K, Denmark
Tel: 45 32 88 01 00
Fax: 45 32 88 01 01
E-mail: dpc@dpc.dk
http://www.dpc.dk/

Department for Environment,
  Food & Rural Affairs
Nobel House
17 Smith Square
London, SW1P 3JR, United Kingdom
Tel: 44 20 7238 6951
Fax: 44 20 7238 3329
E-mail: helpline@defra.gsi.gov.uk
http://www.defra.gov.uk/fish/

Department of Communications,
  Marine and Natural Resources
Leeson Lane
Dublin 2, Ireland
Tel: 353 1 6782000
Fax: 353 1 6618214
E-mail: webmaster@dcmnr.gov.ie
http://www.marine.gov.ie/

Department of Conservation
PO Box 10-420
59 Boulcott Street
Wellington, New Zealand
Contact: Director-General
Tel: 64 04 471 0391
http://www.doc.govt.nz/
  Conservation/Marine-and-
  Coastal/

Department of Fisheries
Kasetlang, Chatuchak
Bangkok 10900, Thailand
Tel: 66 2 562 0600

Fax: 66 2 562 9406 203
E-mail: webmaster@fisheries.go.th
http://www.fisheries.go.th/

Department of Marine and Coastal
    Resources
Ministry of Natural Resources and
    Environment
92 Soi Paholyothin 7
Bangkok 10400, Thailand
Tel: 66 2298 2000
Fax: 66 2298 2002
E-mail: monre@environnet.in.th
http://www.monre.go.th/

Department of Ocean Development
Block No. 9 & 12, CGO Complex
Lodhi Road
New Delhi 110003, India
Fax: 91 11 4360336
E-mail: ocean@dod.delhi.nic.in
http://dod.nic.in/

Department of Transportation and
    Communications
PPL Building
1000 UN Avenue cor. San Marcelino Street
Manila, The Philippines
Tel: 63 2 523 8651
E-mail: feedback@marina.gov.ph
http://www.marina.gov.ph/

Dirección Nacional de Recursos Acuáticos (DINARA)
Constituyente 1497
CP 11200
PO Box 1612
Montevideo, Uruguay
Tel: 598 2 409 2964
http://www.dinara.gub.uy/

Directorate-General for Fisheries
Communication and Information
    Unit
European Commission
Rue de la Loi 200
B-1049 Brussels

Fax: 32 2 299 3040
E-mail: fisheries-info@cec.eu.int
http://europa.eu.int/pol/fish/
    index_en.htm

Dunstaffnage Marine Laboratory
Oban, Argyll, Scotland, PA34 4AD,
    United Kingdom
Tel: 44 16 3155 9000
Fax: 44 16 3155 9001
E-mail: mail@dml.ac.uk
http://www.sams.ac.uk/dml/

English Nature
Northminster House
Peterborough, PE1 1UA, United
    Kingdom
Tel: 44 17 3345 5000
Fax: 44 17 3356 8834
E-mail: enquiries@english-
    nature.org.uk
http://www.english-nature.org.uk/

Environment Agency
Rio House Waterside Drive, Aztec
    West
Almondsbury, Bristol, BS32 4UD,
    United Kingdom
Tel: 44 84 5433 3111
E-mail: enquiries@environment-
    agency.gov.uk
http://www.environment-
    agency.gov.uk/

Environment Australia
Marine and Water Division
GPO Box 787
Canberra, ACT 2601, Australia
Tel: 61 2 6274 1967
Fax: 61 2 6274 1006
http://www.ea.gov.au/coasts/

Marine Environment Branch
Environment Canada
351 St. Joseph Boulevard, 12th Floor
Hull, QC, K1A 0H3, Canada
Tel: 1 819 997-2800
Fax: 1 819 953-2225

E-mail: enviroinfo@ec.gc.ca
http://www.ec.gc.ca/

Environment Canada, Atlantic Region
45 Alderney Drive
Dartmouth, NS, B2Y 2N6, Canada
Tel: 1 902 426-7231
Fax: 1 902 426-6348
E-mail: 15th.reception@ec.gc.ca
http://www.ns.ec.gc.ca/

Federal Research Centre for Fisheries
(Bundesforschungsanstalt für Fischerei)
Palmaille 9
D-22767 Hamburg, Germany
Tel: 49 40 38905-145
Fax: 49 40 38905-200
E-mail: kfraft.ver@bfa-fisch.de
http://www.bfa-fisch.de/

Finnish Environment Institute
(SYKE)
PO Box 140
FIN-00251 Helsinki, Finland
Tel: 358 9 403 000
Fax: 358 9 4030 0190
E-mail: neuvonta.syke@ymparisto.fi
http://www.ymparisto.fi/eng/syke/syke.htm

Finnish Maritime Administration
PO Box 171
FIN-00181 Helsinki, Finland
Tel: 358 204 481
Fax: 358 204 48 4355
http://www.fma.fi/

Fisheries and Oceans Canada
(DFO)
200 Kent Street, 13th Floor Station 13228
Ottawa, ON, K1A 0E6, Canada
Contact: Communications Branch
Tel: 1 613 993-0999
Fax: 1 613 990-1866

E-mail: info@dfo-mpo.gc.ca
http://www.dfo-mpo.gc.ca/;
http://www.dfo-mpo.gc.ca/canwaters-eauxcan/index.html

Fisheries Information Center
(FICen)
Ministry of Fisheries
10 Nguyen Cong Hoan Street
Ba Dinh District, Hanoi, Vietnam
Tel/Fax: 84 4 771 6578
E-mail: ttam.bts@hn.vnn.vn
http://www.fistenet.gov.vn/index En.asp

Fisheries Resource Conservation Council
PO Box 2001 Station D
Ottawa, ON, K1P 5W3, Canada
Contact: Secretariat
Tel: 1 613 998-0433
Fax: 1 613 998-1146
E-mail: SheehanT@dfo-mpo.gc.ca
http://www.ncr.dfo.ca/frcc/

Florida Marine Research Institute
100 Eighth Avenue SE
St. Petersburg, FL, 33701-5095, USA
Tel: 1 727 896-8626
http://www.floridamarine.org/

Geological Survey of Canada (Atlantic)
Bedford Institute of Oceanography
1 Challenger Drive
PO Box 1006
Dartmouth, NS, B2Y 4A2, Canada
Contact: Director
Tel: 1 902 426-3448
Fax: 1 902 426-1466
E-mail: webmaster@gsca.nrcan.gc.ca
http://agcwww.bio.ns.ca/

German Oceanographic Museum
Museum for Marine Research and Fishery–Aquarium
Katharinenberg 14/20
D-18439 Stralsund, Germany

Tel: 49 3831 265010
Fax: 49 3831 265060
E-mail: info@meeresmuseum.de
http://www.meeresmuseum.de/

Great Barrier Reef Marine Park Authority
PO Box 1379
Townsville, QLD 4810, Australia
Tel: 61 7 4750 0700
Fax: 61 7 4772 6093
E-mail: info@gbrmpa.gov.au
http://www.gbrmpa.gov.au/

Gulf of Guinea Large Marine Ecosystem Project (GOGLME)
Regional Coordination Centre
c/o CRO
B.P. V18
Abidjan, Côte d'Ivoire
Tel: 225 35 50 14
Fax: 225 35 11 55
E-mail: gog-lme@africaonline.co.ci
http://www.africaonline.co.ci/AfricaOnline/societes/goglme/goglme.html

Gulf of Mexico Fishery Management Council
The Commons at Rivergate
3018 U.S. Highway 301 North, Suite 1000
Tampa, FL 33619-2266, USA
Tel: 1 813 228-2815
Fax: 1 813 225-7015
E-mail: gulfcouncil@gulfcouncil.org
http://www.gulfcouncil.org/contact.htm

Icelandic Marine Research Institute (Hafrannsóknastofnunin–HAFRO)
Skulagata 4
PO Box 1390
IS-121 Reykjavik, Iceland
Tel: 354 552 0240
Fax: 354 562 3790
E-mail: hafro@hafro.is
http://www.hafro.is/sjalf.e.html

Institut Agronomique et Vétérinaire Hassan II
Ministere de l'Agriculture et de la Reforme Agraire
BP 6202, Instituts
10101 Rabat, Morocco
Tel: 212 37 77145
Fax: 212 37 778135
E-mail: webmaster@iav.ac.mc
http://www.iav.ac.mc/

Institut de Recherche pour le Développement (IRD)
213, rue la Fayette
F-75480 Paris Cedex 10, France
Contact: Headquarters
Tel: 331 1 4803 7777
Fax: 331 1 4803 0829
E-mail: corlay@paris.ird.fr
http://www.ird.fr/

Institut de Technologie Alimentaire
Route des Peres Maristes
Dakar, Senegal
Contact: Directeur
Tel: 221 832 0070
Fax: 221 832 8295
E-mail: ita@ita.sn
http://www.obs-industrie.sn/

Institut Français de Recherche pour l'Exploitation de la Mer (IFREMER)
155, rue Jean-Jacques Rousseau
F-92138 Issy-les-Moiulineaux Cedex, France
Contact: CEO
Tel: 33 1 4648 2100
Fax: 33 1 4648 2121
http://www.ifremer.fr/

Institut für Ostseeforschung Warnemuende (IOW)
Baltic Sea Research Institute Warnemuende
Seestrasse 15
D-18119 Rostock, Germany
Tel: 49 381 5197 100

Fax: 49 381 5197 105
E-mail: barbara.hentzsch@
io-warnemuende.de
http://www.io-warnemuende.de/

Institut National de Recherche Halieutique
2 rue de Tiznit
Casablanca, Morocco
Contact: Directeur
Tel: 212 22 20 90
Fax: 212 22 69 67

Institute for Baltic Sea Fisheries
(Institut für Ostseeforschung–IOR)
An der Jägerbäk 2
D-18069 Rostock, Germany
Tel: 49 381 810344
Fax: 49 381 810445
E-mail: chammer@ior.bfa-fisch.de
http://www.bfa-fisch.de/

Institute for Marine Biosciences
National Research Council of Canada
1411 Oxford Street
Halifax, NS, B3H 3Z1, Canada
Contact: Director
Tel: 1 902 426-8332
Fax: 1 902 426-9413
E-mail: crm.imb@nrc.ca
http://www.imb.nrc.ca/

Institute for Marine Dynamics
PO Box 12093
St. John's, NL, A1B 3T5, Canada
Tel: 1 709 772-4939
Fax: 1 709 772-2462
E-mail: noel.murphy@nrc.ca
http://imd-idm.nrc-cnrc.gc.ca/

Institute for Sea Fisheries
(Institut für Seefischerei–ISH)
Palmaille 9
D-22767 Hamburg, Germany
Tel: 49 40 38905 177
Fax: 49 40 38905 263
E-mail: ish@bfa-fisch.de
http://www.bfa-fisch.de/ish/

Institute for Coastal Research
GKSS-Research Centre
System Analysis and Modelling
Max-Planck-Straβe
D-21502 Geesthacht, Germany
Tel: 49 41 5287 1831
Fax: 49 41 5287 2832
E-mail: storch@gkss.de
http://coast.gkss.de/

Institute of Coastal Resources
Swedish National Board of Fisheries
Gamla Slipvägen 19
SE-740 71 Öregrund, Sweden
Tel: 46 17 34 64 60
E-mail: fiskeriverket@fiskeriverket.se
http://www.fiskeriverket.se/

Institute of Marine Affairs (IMA)
Hilltop Lane
PO Box 3160
Carenage Post Office
Chaguaramas, Trinidad and Tobago
Tel: 1 868 634-4291
Fax: 1 868 634-4433
E-mail: director@ima.gov.tt
http://www.ima.gov.tt/

Institute of Marine Research
(IMRS)
Swedish National Board of Fisheries
PO Box 4
SE-45321 Lysekil, Sweden
Contact: Secretary General
Tel: 46 52 31 87 00
E-mail: fiskeriverket@fiskeriverket.se
http://www.fiskeriverket.se/

Institute of Ocean Sciences
9860 West Saanich Road
PO Box 6000
Sidney, BC, V8L 4B2, Canada
Tel: 1 250 363-6517
E-mail: olsonp@pac.dfo-mpo.gc.ca
http://www.pac.dfo-mpo.gc.ca/sci/
english/ facilities/ios.htm

Instituto de Fomento Pesquero
(IFOP)
(Fishery Research Institute)
Hulto 374, Casilla 8-V
Valparaíso, Chile
Tel: 56 32 2047858
Fax: 56 32 2254362
E-mail: info@ifop.cl
http://www.ifop.cl/

Instituto de Investigaciones Marinas
Eduardo Cabello 6
ES-36208 Vigo, Spain
E-mail: admin@iim.csic.es
http://www.iim.csic.es/

Instituto de Mar de Perú (IMARPE)
(Peruvian Marine Institute)
Esquina de Gamarra y General Valle
    S/N Chuciuto
Callao, Peru
Tel: 51 420 2000
E-mail: webmaster@imarpe.gob.pe
http://www.imarpe.gob.pe/

Instituto de Oceanología de Cuba
(Institute of Oceanology)
Calle 1ra No. 18406 esquina 184 y
    186
Reparto Flores, Playa
Havana 12100, Cuba
Contact: Director
Tel: 537 21 1380
Fax: 537 33 9112
E-mail: jperez@unepnet.inf.cu
http://www.cuba.cu/ciencia/
    citma/ama/oceanologia/
    inicio.html

Instituto Nacional de Pesca (INP)
Letamendi 102 y La Ría
PO Box 09-04-15131
Guayaquil, Ecuador
Contact: Director
Tel: 593 4 2401 773
Fax: 593 4 2402 304
E-mail: inp1@ecua.net.ec
http://www.inp.gov.ec/

Instituto Oceanográfico de la Armada
(Naval Oceanographic Institute)
Avda 25 de Julio (Base Naval Sur)
via al Puerto Marítimo
Guayaquil, Ecuador
Tel: 593 4 2481 300
Fax: 593 4 2485 166
E-mail: inocar@inocar.mil.ec
http://www.inocar.mil.ec/

Inter-Agency Committee on Marine
    Science and Technology
Southampton Oceanography Centre
European Way
Empress Dock
Southampton SO14 3ZH, United
    Kingdom
Tel: 44 23 8059 6611
Fax: 44 23 8095 6395
E-mail: enquiries@marine.gov.uk
http://www.marine.gov.uk/

International Association of Fish
    and Wildlife Agencies
444 Capitol Street, NW, Suite 544
Washington, DC, 20001, USA
Contact: Executive Vice President
Tel: 1 202 624-7890
Fax: 1 202 624-7891
E-mail: iafwa@sso.org
http://www.iafwa.org/

Irish Marine Institute
80 Harcourt Street
Dublin 2, Ireland
Tel: 353 1 476 6500
Fax: 353 1 478 4988
E-mail: institute.mail@marine.ie
http://www.marine.ie/

Japan Marine Science and Technology Center (JAMSTEC)
Natsushima-cho 2-15, Yokosuka-shi
Kanagawa Prefecture, 237-0061, Japan
Tel: 81 0468 66 3811
E-mail: www-admin@jamstec.go.jp
http://www.jamstec.go.jp/

Japan Sea National Fisheries Research Institute (JSNFRI)
5939-22, Suido-cho 1 chome,
Niigata-shi
Niigata 951, Japan
Tel: 81 025 228 0451
Fax: 81 025 224 0950
E-mail: www@jsnf.affrc.go.jp
http://www.affrc.go.jp/mirror/jsnf/

Korea Institute of Ships and Ocean Engineering
PO Box 23
Yusong, Taejon 305-600, Republic of Korea
Tel: 82 42 868 7701
Fax: 82 42 868 7711
http://www.kordi.re.kr/

Korea Ocean Research and Development Institute (KORDI)
Ansan PO Box 29
425-600, Republic of Korea
Tel: 82 31 400 6000
Fax: 82 31 408 5820
http://www.kordi.re.kr/

Le Centre national océanographique du CBRST (CNO)
BP 03-1665
Cotonou, Benin
Tel: 229 32 12 63
Fax: 229 32 36 71
http://www.bj.refer.org/benin_ct/rec/cbrst/cno.htm

Maine Atlantic Salmon Commission
172 State House Station
Augusta, ME, 04333-0172, USA
Tel: 1 207 287-9972
Fax: 1 207 287-9975
E-mail: joan.trial@maine.gov
http://www.state.me.us/asa/

Marine and Coastal Management
Department of Environmental Affairs and Tourism
PO Box X2
Roggebaai 8012, South Africa
Contact: Director
Tel: 27 21 4023111
Fax: 27 21 252920
E-mail: mlouw@sfri.wcape.gov.za
http://www.environment.gov.za/mcm/

Marine Board
Transportation Research Board
500 Fifth St. NW
Washington, DC, 20001, USA
Contact: Director
Tel: 1 202 334-3119
Fax: 1 202 334-2030
E-mail: jcambrid@nas.edu
http://www4.nationalacademies.org/trb/homepage.nsf/web/marine board

Marine Environment Branch
Environment Canada
351 St. Joseph Boulevard, 12th Floor
Hull, QC, K1A 0H3, Canada
Tel: 1 819 997-2800
Fax: 1 819 953-2225
E-mail: sarah.kennedy@ec.gc.ca
http://www.ec.gc.ca/marine/npa-pan/index e.htm

Maritime Administration
Office of Statistical and Economic Analysis, MAR-450
400 Seventh Street, SW
Washington, DC, 20590, USA
Tel: 1 202 366-2267
Fax: 1 202 366-8886
E-mail: data.marad@marad.dot.gov
http://www.marad.dot.gov/statistics/

Maritime and Coastguard Agency
Spring Place
105 Commercial Road
Southampton, SO15 1EG, United Kingdom
Tel: 44 23 8032 9100
Fax: 44 23 8032 9105

E-mail: infoline@mcga.gov.uk
http://www.mcagency.org.uk/

Maritime Institute of Malaysia (MIMA)
Units B-06-08 to B-06-11
Megan Phileo Avenue
No. 12 Jalan Yap Kwan Seng
50450 Kuala Lumpur, Malaysia
Contact: Director General
Tel: 603 2161 2960
Fax: 603 2161 4035
E-mail: mima@mima.gov.my
http://www.mima.gov.my/

Maurice Lamontagne Institute
Fisheries and Oceans Canada
850 route de la Mer
PO Box 1000
Mont-Joli, QC, G5H 3Z4, Canada
Tel: 1 418 775-0500
Fax: 1 418 775-0542
E-mail: info@dfo-mpo.gc.ca
http://www.qc.dfo-mpo.gc.ca/iml/

Mid-Atlantic Fisheries Management Council
Room 2115 Federal Building
300 S. New Street
Dover, DE, 19904, USA
Contact: Executive Director
Tel: 1 302 674-2331
E-mail: dfurlong@mafmc.org
http://www.mafmc.org/

Ministry of Fisheries and Ocean Resources
(Ministère des Pêches et des Ressources Halieutiques)
Route des 04 Canons
Alger, Algeria
Tel: 213 2143 3181
Fax: 213 2143 3168
E-mail: mprh@wissal.dz
http://www.mprh-dz.com/

Ministry of Marine Fisheries
(Ministère des Pêches Maritimes)
BP 475 Quartier Administratif Rabat, Agdal, Morocco
Tel: 212 37 688000
Fax: 212 37 688135
E-mail: webmaster@mpm.gov.ma
http://www.mpm.gov.ma/

Ministry of Maritime Affairs and Fisheries (MOMAF)
139 Chungjong-No 3, Seodaemun-Gu
Seoul 120-715, Korea
Tel: 82 2 3148 6114
E-mail: momweb@momaf.go.kr
http://www.momaf.go.kr/

National Center for Marine Research
Agios Kosmas
Hellinikon
GR-16604 Athens, Greece
Tel: 30 1 982 0214
Fax: 30 1 983 3095
E-mail: info@ncmr.gr
http://www.ncmr.gr/

National Environment and Planning Agency
John McIntosh Building
10 Caledonia Avenue
Kingston 5, Jamaica
Tel: 876 754 7546 51
Fax: 876 754 7595 6
E-mail: pubed@nepa.gov.jm
http://www.nrca.org/

National Environmental Research Institute, Denmark (NERI)
PO Box 358
Frederiksborgvej 399
DK-4000 Roskilde, Denmark
Tel: 45 46 30 12 00
Fax: 45 46 30 11 14
E-mail: dmu@dmu.dk
http://www.dmu.dk/

National Institute for Marine Research and Development "Grigore Antipa"
Blvd. Mamaia 300

RO-8700 Constanta 3, Romania
Tel: 40 41 543 288
Fax: 40 41 831 274
E-mail: rmri@alpha.rmri.ro
http://www.rmri.ro/

National Institute of Oceanography
    and Fisheries (NIOF)
Kayet Bay, Anfoushy
Alexandria, Egypt
Tel: 20 3 480 1553
Fax: 20 3 480 1174

National Institute of Water and Atmospheric Research
Private Bag 999 40
269 Khyber Pass Road
Newmarket, Auckland, New Zealand
Tel: 64 9 375 2090
Fax: 64 9 375 2091
E-mail: webmaster@niwa.co.nz
http://www.niwa.cri.nz/

National Marine Data & Information Service
93 Qiwei Road,
Hedong District, Tianjin 300171,
    People's Republic of China
Contact: Director
Tel: 022 24300881
E-mail: hwang@mail.nmdis.gov.cn
http://www.nmdis.gov.cn/

National Marine Fisheries Service
    (NMFS)
National Oceanic and Atmospheric
    Administration (NOAA)
1315 East-West Highway SSMC3
Silver Spring, MD, 20910, USA
E-mail: cyber.fish@noaa.gov
http://www.nmfs.noaa.gov/

National Ocean Service
National Oceanic and Atmospheric
    Administration (NOAA)
1305 East-West Highway
Silver Spring, MD, 20910, USA
Tel: 1 301 713-3074

E-mail: nos.info@nos.noaa.gov
http://www.nos.noaa.gov/

National Sea Grant College Program
National Oceanic and Atmospheric Administration 1315
    East-West Highway
Silver Spring, MD, 20910, USA
Tel: 1 301 713-2431
Fax: 1 301 713-0799
E-mail: Amy.Painter@noaa.gov
http://www.nsgo.seagrant.org/

Natural Environment Research
    Council (NERC)
Polaris House
North Star Avenue
Swindon, SN2 1EU, United
    Kingdom
Tel: 44 17 9341 1500
Fax: 44 17 9341 1501
E-mail: webmaster@nerc.ac.uk
http://www.nerc.ac.uk/

Netherlands Institute for Sea Research
(Nederlands Instituut voor Onderzoek der Zee – NIOZ)
PO Box 59
NL-1790 AB, Den Burg, Texel, The
    Netherlands
Tel: 31 222 369300
Fax: 31 222 319674
E-mail: webmaster@nioz.nl
http://www.nioz.nl/

New England Fishery Management
    Council
50 Water Street, Mill 2
Newburyport, MA, 01950, USA
Tel: 1 978 465-0492
Fax: 1 978 465-3116
E-mail: pfiorelli@nefmc.org
http://www.nefmc.org/

New Zealand Ministry of Fisheries
ASB Bank House
101–103 The Terrace

PO Box 1020
Wellington, New Zealand
Tel: 64 4 470 2600
Fax: 64 4 470 2601
E-mail: info@fish.govt.nz
http://www.fish.govt.nz/

Nigerian Institute of Oceanography
 and Marine Research (NIOMR)
Ministry of Science and Technology
Private Mail Bag 12729
Lagos, Nigeria
Contact: Director
Tel/Fax: 234 1 261 9517
E-mail: niomr@linkserve.com.ng

North Pacific Fishery Management
 Council
605 West 4th, Suite 306
Anchorage, AK, 99501-2252, USA
Contact: Executive Director
Tel: 1 907 271-2809
Fax: 1 907 271-2817
E-mail: Chris.Oliver@noaa.gov
http://www.fakr.noaa.gov/npfmc/

Northwest Atlantic Fisheries Centre
Science, Oceans, and Environment
 Branch
Newfoundland Region
Fisheries and Oceans
PO Box 5667
East White Hills Road Extension
St. John's, NL, A1C 5X1, Canada
Tel: 1 709 772-2027
E-mail: soe_feedback@nwafc.nf.ca
http://www.nwafc.nf.ca/

Norwegian Institute for Water Re-
 search
(Norsk Institutt for Vannforskning)
Brekkeveien 19
Postboks 173 Kjelsås
N-0411 Oslo, Norway Tel: 47 22 18
 51 00
Fax: 47 22 18 52 00
E-mail: niva@niva.no

http://www.niva.no/engelsk/
 welcome.htm

Norwegian Institute of Marine Re-
 search (IMR)
(Havforskningsinstituttet)
PO Box 1870 Nordnes
N-5817 Bergen, Norway
Tel: 47 55 23 8500
Fax: 47 55 23 8531
E-mail: havforskningsinstituttet@
 imr.no
http://www.imr.no/

Norwegian Polar Institute
(Norsk Polarinstitutt–NPI)
Polar Environmental Centre
N-9296 Tromsø, Norway
Tel: 47 77 75 05 00
Fax: 47 77 75 05 01
E-mail: postmottak@npolar.no
http://www.npolar.no/

Observatoire Océanologique Ville-
 franche-sur-Mer
BP 28
F-06238 Villefranche Sur Mer
 Cedex, France
Tel: 33 4 9376 3803
Fax: 33 4 9376 3834
E-mail: webmaster@www.obs-vlfr.fr
http://www.obs-vlfr.fr/

Oceans Policy Secretariat
Ministry for the Environment
PO Box 10362
Wellington Grand Avenue
84 Boulcott Street
Wellington, New Zealand
Tel: 64 4 917 7400
Fax: 64 4 917 7523
E-mail: oceans@mfe.govt.nz
http://www.oceans.govt.nz/

Ocean Studies Board
The National Academies
500 5th Street, NW, NA-752
Washington, DC, 20001, USA

Tel: 1 202 334-2714
Fax: 1 202 334-2885
E-mail: osb_feedback@nas.edu
http://www7.nationalacademies.org/osb/

Office of Naval Research
Ballston Centre Tower One
800 North Quincy Street
Arlington, VA, 22217-5660, USA
Tel: 1 703 696-5031
Fax: 1 703 696-5940
E-mail: onrpao@onr.navy.mil
http://www.onr.navy.mil/

Office of the Commissioner for Aquaculture Development (OCAD)
427 Laurier Avenue West, Suite 1210
Ottawa, ON, K1A OE6, Canada
Contact: Commissioner
Tel: 1 613 993-8603
Fax: 1 613 993-8607
E-mail: BastienY@dfo-mpo.gc.ca
http://ocad-bcda.gc.ca/

Office of Wetlands, Oceans and Watersheds
Environmental Protection Agency
Ariel Rios Building
1200 Pennsylvania Ave, NW
Washington, DC, 20460, USA
Contact: Administrator
Tel: 1 202 566-1300
E-mail: ow-owow-internet-comments@epa.gov
http://www.epa.gov/owow/org.html

Pacific Biological Station
3190 Hammond Bay Road
Nanaimo, BC, V9T 6N7, Canada
Tel: 1 250 756-7000
Fax: 1 250 756-7053
E-mail: ronalda@pac.dfo-mpo.gc.ca
http://www.pac.dfo-mpo.gc-ca/sci/pbs/

Pacific Fisheries Research Center (TINRO-Center)
4 Shevchenko Alley
Vladivostok, 690950, Russia
Contact: Director
Tel: 7 4232 400921
Fax: 7 4232 300752
E-mail: tinro@tinro.ru
http://www.tinro.ru/

Pacific Fishery Management Council
7700 NE Ambassador Place, Suite 200
Portland, OR, 97220-1384, USA
Contact: Executive Director
Tel: 1 503 820-2280
Fax: 1 503 820-2299
E-mail: pmc.comments@noaa.gov
http://www.pcouncil.org/

Pacific States Marine Fisheries Commission (PSMFC)
45 SE 82nd Dr., Suite 100
Gladstone, OR, 97027-2522, USA
Contact: Executive Director
Tel: 1 503 650-5400
Fax: 1 503 650-5426
E-mail: webmaster@psmfc.org
http://www.psmfc.org/

Philippine Council for Aquatic and Marine Research and Development (PCAMRD)
Eusebio Building, BPI Economic Garden
Los Baños, Laguna, 4030, The Philippines
Tel: 63 49 536 1574
Fax: 63 49 536 1566
E-mail: pcamrd@laguna.net
http://www.laguna.net/pcamrd/

Research Institute of Marine Products (RIMP)
170 Le Lai
Ngo Quyen District, Hai Phong, Vietnam
Tel: 84 31 836664

Fax: 84 31 836812
E-mail: dokhuong@hn.vnn.vn
http://www.fistenet.gov.vn/
   index_En.asp

Research Institute of Marine Products (RIMP)
170 Le Lai
Ngo Quyen District
Hai Phong, Vietnam
Tel: 84 31 836664
Fax: 84 31 836812
E-mail: dokhuong@hn.vnn.vn

Scottish Natural Heritage
12 Hope Terrace
Edinburgh, Scotland, EH9 2AS, United Kingdom
Tel: 44 131 447 4784
Fax: 44 131 446 2277
E-mail: www.enquiries@snh.gov.uk
http://www.snh.org.uk/

Servicio Hidrográfico y Oceanográfico de la Armade de Chile
(Hydrographic and Oceanographic Service of the Chilean Navy)
Calle Errázuriz 254, Cerro Playa Ancha
Casilla: 324
Valparaíso, Chile
Tel: 56 32 266666
Fax: 56 32 266542
E-mail: shoa@shoa.cl
http://www.shoa.cl/

South Atlantic Fishery Management Council
One Southpark Circle, Suite 306
Charleston, SC, 29407-4699, USA
Tel: 1 843 571-4366
Fax: 1 843 769-4520
E-mail: safmc@safmc.net
http://www.safmc.net/

South Australian Research and Development Institute (SARDI)
GPO Box 397
Adelaide, SA 5001, Australia
Contact: Aquatic Sciences Research Programme
Tel: 61 8 8200 2400
Fax: 61 8 8200 2481
E-mail: pirsa.sardi@saugov.sa.gov.au
http://www.sardi.sa.gov.au/

Southeastern Association of Fish and Wildlife Agencies
8005 Freshwater Farms Road
Tallahassee, FL, 32308, USA
Contact: Executive Secretary
Tel: 1 850 893-1204
Fax: 1 850 893-6204
E-mail: seafwa@aol.com
http://www.seafwa.org/

Spanish Institute of Oceanography
(Instituto Español de Oceanografía–IEO)
Avenida de Brasil, No. 31
ES-28020 Madrid, Spain
Tel: 34 915 974 443
Fax: 34 915 974 770
E-mail: leo@md.ieo.es
http://www.ieo.es/

State Oceanic Administration
93 Livwei Road
Hedong District
Tianjin 300171, People's Republic of China
Contact: Chief
Tel: 86 1068532211
Fax: 86 1068533515
E-mail: peibs@nmdis.go.cn
http://www.soa.gov.cn/

Swedish National Board of Fisheries
Ekelundsgatan 1
Box 423
SE-401 26 Göteborg, Sweden
Contact: Head Office
Tel: 46 31 743 03 00
Fax: 46 31 743 04 44
E-mail: fiskeriverket@fiskeriverket.se
http://www.fiskeriverket.se/

Swedish Polar Research Secretariat
PO Box 50 003
SE-104 05 Stockholm, Sweden
Tel: 46 8 673 96 00
Fax: 46 8 15 20 57
E-mail: office@polar.se
http://www.polar.se/

Thailand Department of Fisheries
Kasetklang, Chatuchak
Bangkok 10900, Thailand
Tel: 66 2 562 0600 15
Fax: 66 2 940 6203
E-mail: webmaster@fisheries.go.th
http://www.fisheries.go.th/

Transport Canada–Marine Safety
Tower C, Place de Ville, 330 Sparks Street
Ottawa, ON, K1A 0N5, Canada
Tel: 1 613 990-2309
Fax: 1 613 954-4731
E-mail: webfeedback@tc.gc.ca
http://www.tc.gc.ca/marinesafety/menu.htm

Ukrainian Scientific Center of Marine Ecology
89, Frantsuzki Blvd.
UA-Odessa 270009, Ukraine
Contact: Director
Tel: 380 482 636622
Fax: 380 482 636673
E-mail: root@accem.odessa.ua

United States Army Corps of Engineers
Cold Regions Research and Engineering Laboratory
CECRL-PP
72 Lyme Road
Hanover, NH, 03755-1290, USA
Tel: 1 603 646-4100
Fax: 1 603 646-4448
E-mail: info@crrel.usace.army.mil
http://www.crrel.usace.army.mil/

United States Coast Guard
2100 Second Street, SW
Washington, DC, 20593, USA
Contact: Headquarters
Tel: 1 202 267-2229
http://www.uscg.mil/

United States Coast Guard Auxiliary
Commandant (G-OCX)
US Coast Guard Headquarters
2100 2nd St. SW, Room 3501
Washington, DC, 20593-0001, USA
Contact: Chief Director, Auxiliary
Tel: 1 202 267-1001
Fax: 1 202 267-4409
E-mail: info@cgaux.org
http://www.cgaux.org/

United States Commission on Ocean Policy
1120 20th Street, NW, Suite 200 North
Washington, DC, 20036, USA
Tel: 1 202 418-3442
Fax: 1 202 418-3475
E-mail: mail@oceancommission.gov
http://oceancommission.gov/

Western Pacific Regional Fishery Management Council (WPRFMC)
1164 Bishop Street, Suite 1400
Honolulu, HI, 96813, USA
Tel: 1 808 522-8220
Fax: 1 808 522-8226
E-mail: info.wpcouncil@noaa.gov
http://www.wpcouncil.org/

Woods Hole Oceanographic Institution (WHOI)
Sea Grant Program
193 Oyster Pond Road, MS #2,
Woods Hole, MA, 02543-1525, USA
E-mail: seagrant@whoi.edu
http://www.whoi.edu/seagrant/

Contributors

William (Bill) Aalbersberg is professor of natural products chemistry and director of the Institute of Applied Sciences at the University of the South Pacific (USP) in Suva, Fiji. He was born and raised in San Clemente California and did his undergraduate studies at Cornell University in New York. He went to Fiji in 1970 as a Peace Corps volunteer in secondary education. Professor Aalbersberg continued his education at the University of California at Berkeley where he earned his Ph.D. in chemistry and taught there for several years before returning to Fiji in 1984 at USP. His research interests range beyond natural products chemistry (both plant and marine) to food chemistry and nutrition and environment chemistry. He believes that scientists working in developing countries should assist communities in meeting their development needs while at the same time ensuring that the environment base that supports much of this development remains intact. He is especially interested in using participatory methods to encourage communities to meet their sustainable development needs by blending residents' traditional knowledge with academic techniques. Professor Aalbersberg has also been chair of a regional environmental NGO called the South Pacific Action Committee on Human Ecology and the Environment and is on the executive committees of several NGOs.

R.P. Anand is professor (emeritus) of international law at the School of International Studies, Jawaharlal Nehru University, India. Professor Anand is a well-known scholar in the field of international law and widely recognized as a spokesman of the Third World views on the subject. The only elected member of the Institut de Droit International at present, Professor Anand has been the recipient of a number of awards and honors in the field, visiting scholar in several universities and institutes of higher learning in the United States and Europe, lecturer at the Hague Academy of International Law, U.G.C. National Lecturer in Law in India, and has served as legal consultant to the United Nations Secretary-General on law of the sea. Author or editor of eighteen books, Professor Anand has published nearly one hundred articles in professional journals in India, Europe, and the United States.

Carl Anderson is a self-employed physical oceanographer in Halifax, Nova Scotia. He earned a Ph.D. from McGill University in 1987, and has since been associated with the Bedford Institute of Oceanography and Dalhousie University. Anderson has taken part in studies of marine storms, sea ice, marine fisheries, sediment transport, and climate change.

François N. Bailet, B.Sc., Diplôme d'Étude Approfondies (D.E.A.), Ph.D. is the deputy executive director of the International Ocean Institute—Headquarters and a research fellow of the Centre for Foreign Policy Studies, Department of Political Science, Dalhousie University. His current academic research and lecture topics are in the areas of ocean governance and the implementation of the LOS Convention and the UNCED process, through all levels of governance, with a focus on maritime security and regionalization. As deputy executive director of the IOI, he is also involved in training and capacity building for developing nations and the development and coordination of IOI network-wide programs. He also works closely with the various "ocean" bodies of the UN system and serves as an IOI representative to numerous UN meetings and academic conferences.

Awni Behnam began his career at an early age in the merchant marine and naval services. Before joining UNCTAD in 1977 he lectured at his alma mata, the University of Wales, Department of Maritime Studies. He held executive posts in the shipping division of UNCTAD before assuming responsibility as the chief of liaison with developing countries for the Secretary-General of UNCTAD. He was promoted to the rank of director, and in 1999 he was appointed as secretary of the Tenth United Nations Conference on Trade and Development, and the Third United Nations Conference on the Least Developed Countries. Dr. Benham currently holds the position of senior adviser to the Secretary-General. He holds a first degree in business administration (B.A. Mustensiriya University) and a master's and a doctorate in development and shipping economics from the University of Wales. Since 1977, Dr. Behnam has been involved in international multilateral affairs. He was instrumental in the negotiations of several international conventions and multilateral instruments in the field of maritime transport.

Jeff Benoit has over 20 years of professional experience and leadership in coastal management and marine conservation in both national and international affairs. He is the proprietor of J.R. Benoit Consulting, offering expertise in coastal management, marine conservation, policy analysis, program assessment, coastal hazard mitigation and strategic planning. He served in the Clinton Administration for 7 years as director of the National Oceanic and Atmospheric Administration's (NOAA) Office of Ocean and Coastal Resource Management (OCRM) from 1993 to 2001. As the OCRM director, he led three national programs for coastal and marine stewardship including, Coastal Zone Management, Estuarine Research Reserves, and Marine Sanctuaries. He was chair of NOAA's Sustain Healthy Coasts strategic planning team from 1994 to 1997. He also chaired one of five working groups established by the U.S. Coral Reef Task Force and coordinated drafting the *National Action Plan to Conserve Coral Reefs*. Prior to accepting his position at OCRM, he served as coastal geologist and later was assistant secretary of

Environmental Affairs and director of the Massachusetts Coastal Zone Management Program. He earned a B.Sc. degree in marine geology from Southampton College and a master's degree in geophysical science from the Georgia Institute of Technology.

Jan Boon studied biology at the Free University of Amsterdam from 1974 to 1981. In 1986, he obtained a Ph.D. in environmental chemistry and toxicology from Groningen University for his thesis on the bioaccumulation of polychlorinated biphenyls (PCBs) in marine food chains. The practical work for this thesis was carried out at the Royal Netherlands Institute for Sea Research (NIOZ), located at the island of Texel. From 1984 to 1988, he worked as a toxicologist in the Department of Chemistry of the Tidal Waters Division of the Ministry of Transport, Public Works and Water Management. In 1988, he returned to NIOZ, where he became head of the Department of Marine Biogeochemistry and Toxicology in 1997. His main research interests are the fate and effects in marine ecosystems of persistent organic pollutants (POPs), such as PCBs, polyaromatic hydrocarbons (PAHs), organotin compounds, and most recently brominated flame retardants (BFRs). He also coordinates the Marine Environmental Awareness Course, which NIOZ organizes jointly with the educational institute 'Ecomare' for students of the Dutch higher maritime schools and staff of shipping companies. This course won the 'Clean Shipping Award' of the Dutch Ministry of Transport in 1999.

Elisabeth Mann Borgese was professor of political science and law at Dalhousie University, Halifax, Canada, Founder and Honorary Chair of the International Ocean Institute, headquartered in Malta, and former vice chair of the Independent World Commission on the Oceans. Professor Borgese served as an advisor to the Austrian delegation at the Third United Nations Conference on the Law of the Sea. For some years she was senior fellow at the Center for the Study of Democratic Institutions, Santa Barbara, California. In 1970, she initiated the *Pacem in Maribus* conference series on peaceful uses of the sea. She wrote numerous books, monographs, and essays on international ocean affairs and marine resource management and was the founding editor of the *Ocean Yearbook*. Her pioneering research and teaching in ocean affairs educated at least two generations of young professionals, primarily from developing countries.

Marcela Campa graduated with honors from UABC, Tijuana, Mexico, in 1996 and obtained a Master of Laws degree in international legal studies, with a concentration in international business, from the Washington College of Law at American University, Washington D.C., in 1997. She is currently a foreign legal consultant in the Business Practice Group at Luce, Forward, Hamilton & Scripps, LLP., where she focuses her practice on assisting clients with U.S.-Mexico cross-border business transactions, cross-border investment, formation of entities in Mexico, and intellectual property issues in Mexico. From 1998 to 2000, she was a legal assistant for the Inter-American Tropical Tuna Commission. She taught public international law at the Universidad Autonoma de Baja California in Tijuana, Mexico, during the 1998 spring semester.

Aldo Chircop is director of the Marine and Environmental Law Programme at Dalhousie Law School where he teaches law of the sea and maritime law. In addition to the law school, Professor Chircop holds faculty positions at the Marine Affairs Programme and International Development Studies Programme at Dalhousie University. His work experience has included directorships of the International Ocean Institute and the Mediterranean Institute in Malta, and a brief term with the United Nations Industrial Development Organization in Vienna. Dr. Chircop has extensive capacity-building experience in his areas of expertise in Southeast Asia, Caribbean, Latin America, the Mediterranean, and the South Pacific. He is counsel to Patterson Palmer, an Atlantic Canada-wide law firm. He has published widely in collective works and periodicals. He is also co-editor of the *Ocean Yearbook*.

Steven Chown is a professor in the Department of Zoology, University of Stellenbosch, South Africa. His research concerns, amongst other topics, the terrestrial ecology and biogeography of Southern Ocean island biotas, and the physiology of the invertebrate component of the fauna.

Peter Faust is Head, Trade Logistics Branch in the Division for Services Infrastructure for Development and Trade Efficiency, UNCTAD, Geneva. He is in charge of UNCTAD's work on transport, trade facilitation and cross-sectoral legal issues. From 1969 to 1973, he studied economics at California State University and the Universities of Bochum and Hamburg, Germany. In 1981, he completed a Ph.D. in economics at the University of Münster, Germany. Before joining UNCTAD in 1980, he worked with the Institute of Shipping Economics in Bremen.

Martha Gómez was born and raised in Bogotá, Colombia. She attended the Universidad de los Andes, from which she received a Bachelor's degree in business administration in 1994. In January 1999, after holding several jobs in Colombia, she moved to San Diego, where she is currently working as assistant for the Tuna-Dolphin Program of the Inter-American Tropical Tuna Commission.

José Luis Gómez-Ariza received his Ph.D. in analytical chemistry at the University of Sevilla, Spain in 1976 where he attained the rank of associate professor in 1977. He remained in this position until 1995 when he moved to the University of Huelva, Spain as full professor in analytical chemistry in the Department of Chemistry and Material Sciences. His research interests include topics related to environmental issues, especially the development of analytical speciation methods based on hyphenated techniques. Special studies have been focused on the survey of TBT and related compounds in the offshore southwest Spain and more recently in the snail imposex assessment of Spain and Portugal. His other interests include issues related to metal mobility and bioavailability and he has contributed to the survey of the Aznalcollar (Spain) mining environmental disaster in 1998. Recently, he has been appointed as director of the Environmental and Ecological International Institute, which is dependent on the Andalusia Research Organisation and the Huelva University, in studies related to Doñana National Park and the southwest coast of Spain.

Martin Hall was awarded the degree of Licenciado from the University of Buenos Aires, Argentina in 1973. He worked at the Centro Nacional Patagonico, Puerto Madryn, Argentina, from 1973 to 1977 on various aspects of the management and conservation of marine resources, including abundance, growth, and population dynamics of giant kelp, ecological studies of seabirds, and studies of aquatic communities. In 1977 he received a Fulbright Scholarship to study at the School of Fisheries, University of Washington, Seattle. He obtained his Ph.D. in 1982 with a dissertation on spatial models for clam populations. Since 1984, he has been in charge of the Tuna-Dolphin Program of the Inter-American Tropical Tuna Commission. He was a member of the National Academy of Sciences Committee on Reducing Porpoise Mortality from Tuna Fishing. He is currently a member of the Board of Directors of the National Fisheries Conservation Center and Adjunct Professor of the University of British Columbia, Vancouver, Canada. His more recent papers address the ecological significance of bycatches in fisheries and strategies to mitigate the impacts.

Matthew Heemskerk completed his Bachelor of Arts in history at the University of Victoria and graduated from Dalhousie Law School with a Bachelor of Laws. Recently, he has accepted a position in Vancouver at Alexander, Holburn, Beaudin & Lang in their Aviation Practice Group.

Henry G. Hengeveld is Environment Canada's Senior Science Advisor on climate change. In this role, he undertakes regular assessments of international and domestic scientific literature related to climate change and communicates related information to a broad range of scientific and lay audiences, including policy makers, industry groups, and the general public. Throughout the past decade, he has been actively involved in a variety of international meetings dealing with both climate change science assessment and the development of related global agreements on mitigative action.

Chad Hewitt joined CSIRO in 1995 to lead the Invasion Processes Program at the Centre for Research on Introduced Marine Pests in Hobart. His background in the field of marine bio-invasions both in the United States and Australia extends over a period of seventeen years. He holds a B.A. from UC Berkeley and a Ph.D. from the University of Oregon in the ecology of marine biological invasions under the supervision of Dr. Jim Carlton. He was recipient of the Department of Energy Global Change Postdoctoral Award in 1993 for work with Dr. Jim Drake (University of Tennessee) and Dr. Mike Huston (Oak Ridge National Labs) to model the process of biological invasions. Hewitt serves on the Australian Ballast Water Management Advisory Council—Research Advisory Group, the Australian Quarantine and Inspection Service Decision Support System Steering Committee, and represents CSIRO at the International Council for the Exploration of the Seas. He has led the development of a world-first National Introduced Species Ports Survey Program and the Biological Risk Assessment (in collaboration with Dr. Keith Hayes) that underpins the Australian Ballast Water Management Decision Support System.

Douglas M. Johnston received his university education at St. Andrews, McGill and Yale (M.A., LL.B., MCL, LL.M., and J.S.D.) and has taught at various universities in Canada, United States, and Singapore. In 1995, he retired from the Chair in Asia-Pacific Legal Relations at the University of Victoria, which he held from its establishment in 1987. During his career at Dalhousie Law School (1972–87) he founded the Marine and Environmental Law Programme and co-founded the Dalhousie Ocean Studies Programme. In 1981, he was instrumental, along with Phiphat Tangsubkul and Edgar Gold, in setting up the Southeast Asian Programme in Ocean Law, Policy and Management (SEAPOL), based in Bangkok, and he still serves as SEAPOL's Co-Director. His numerous publications span various fields: international and comparative law, theory of international law, law of the sea, international environmental law, marine and coastal policy, Asian affairs, and Arctic relations. He is now emeritus professor of law and emeritus chair in Asia-Pacific Legal Relations at the University of Victoria and adjunct professor at Dalhousie University, and is currently working on a history of international law.

John Kemp was a sailor in the UK merchant marine for 10 years before swallowing the anchor and becoming an academic. He is professor emeritus at the London (Guildhall) University and was, until 2001, a special professor at the University of Nottingham. He is a fellow of the Royal Institute of Navigation and of the Nautical Institute, and he edited the *Journal of Navigation* for 11 years. In retirement, he maintains many interests concerning safety at sea and protection of the marine environment, and he continues to represent Friends of the Earth International at the International Maritime Organization.

Maurice Knight, B.A., MURP, University of New Orleans, is a senior environmental policy advisor with extensive experience in the United States and numerous countries around the world. He specializes in evaluating, developing, and implementing environmental natural resources management programs (especially coastal and protected areas management), organizational engineering and conflict management related to reducing the rate of natural resources loss and degradation from rapid urban and economic development in developing countries. He is currently a senior program advisor with the Coastal Resources Center at the University of Rhode Island and the project director for the United States Agency for International Development (USAID) sponsored Indonesia Coastal Resources Management Project (CRMP). The CRMP is a multi-year, multi-million dollar project focused on supporting decentralized and strengthened coastal resources management in Indonesia. The CRMP is implemented via the Coastal Resources Management II Cooperative Agreement between USAID (G/ENV Bureau) and the Coastal Resources Center.

Christine LeBlanc is completing a master's degree in law at the Osgoode Hall Law School of York University in Toronto. Her thesis work explores the international trade implications of ecolabelling schemes in the sustainable fisheries sector. LeBlanc received both her B.Sc. (Hon. Biology) and LL.B. from the Université de Moncton, in Moncton, New Brunswick. She has worked many seasons as a naturalist in national parks across Canada. She is currently an articling student with the Legal Department of Ontario's Ministry of the Environment and will be called to the Bar of Ontario in the fall of 2002. She also does translation work for the Osgoode Hall Law Journal.

Moira L. McConnell, B.A., LL.B., Ph.D., is a professor of law at Dalhousie University, Canada. Between June 2000 and June 2002 she was on secondment to the World Maritime University, a postgraduate university created by the International Maritime Organization, located in Malmö, Sweden. She is also admitted as a barrister and solicitor in Nova Scotia, Canada. In her 12 years at Dalhousie, Professor McConnell has been a co-director of its Marine Affairs Programme, a director of the Marine and Environmental Law Programme and a founder and facilitator for its Negotiation and Conflict Management Programme. McConnell is a member of a number of organisations including the International Commission of Jurists (Vice-President (Atlantic), Canadian Council), Lawyers for Social Responsibility, and the IUCN Commission on Environmental Law. Her teaching and research interests are in the fields of marine environmental protection, biosecurity, integrated coastal and ocean management, business and environmental law/management, public and corporate governance, regulatory design, dispute resolution processes, international trade and environment, law of the sea, and feminist theory. Most recently she was lead consultant and coordinator for a six country Legislative Review Project under the GEF/ UNDP/IMO Global Ballast Water Management Programme. She is a co-editor of the international, interdisciplinary *Ocean Yearbook* and has published widely in the fields of international law, environmental law, maritime law and policy, social justice and human rights.

Bevon Morrison is president of Call Associates Consultancy Limited. She has a M.Phil. in environmental chemistry and a B.Sc. in marine biology and chemistry. Morrison has a diploma in international management from the Swedish Institute of Management. She has over 13 years of experience in environmental management, development policy, energy, and project management. Morrison is the former environmental advisor to the Ministry of Public Utilities, Mining and Energy, technical advisor to the National Water Commission and is the former chairperson of the Natural Resources Conservation Authority's Air Quality Standards Committee. Morrison is the director general-environment, housing and water of an international non-governmental organisation and their main representative to the United Nations Office in Vienna. She has worked extensively in Latin America and the Caribbean (Mexico, Venezuela, Columbia, Puerto Rico, Cuba, Bahamas, Jamaica, Trinidad, British Virgin Islands, St. Vincent, St. Lucia, Grenada and Barbados).

Adam Newman graduated from Dalhousie University with a LL.B. and a Master of Library and Information Studies in May 2001. At the time of finalizing his paper, he was clerking at the Federal Court of Canada, Trial Division. Newman has worked as a volunteer in the Solomon Islands (with Youth Challenge International) and in Costa Rica (with Canada World Youth). He spent one year at a university in Mexico City, studying Spanish.

Evgeny Pakhomov is an associate professor at the Department of Zoology of the University of Fort Hare, Alice, South Africa. He has been involved in Antarctic research since 1985 and participated in numerous research expeditions to the Southern Ocean. His research interests include oceanic biochemical processes, biogeography of waters surrounding African and Antarctic continents, sub-Antarctic and Antarctic islands and seamounts, ecophysiology of marine fish and invertebrates, and fishery ecology.

Jason M. Patlis conducted research for this article as a Fulbright Senior Scholar 2000–2001. As the founding director for the Environmental Law and Law Development Associates, he is collaborating with non-governmental organizations and multilateral donor agencies on drafting natural resource laws and developing legal institutions to address changing governance regimes in developing countries. From 1997 to 2000, he served as majority counsel for the Environment and Public Works Committee of the

U.S. Senate, chaired by the late Senator John H. Chafee, handling all natural resource issues for the Committee, including the Endangered Species Act. From 1993 to 1997, he served in the Office of the General Counsel for the National Oceanic and Atmospheric Administration of the U.S. Department of Commerce. He graduated with a B.A. from Haverford College in 1985 and from Cornell University School of Law with a J.D. in 1992.

Timothy Pickering is a Lecturer in aquaculture at the Marine Studies Programme of the University of the South Pacific (USP), based at the Laucala Bay campus in Suva, Fiji. He joined USP in 1995 as a lecturer in fish biology and aquaculture, then worked for a time as a lecturer in marine affairs before taking up his current post. Pickering is originally from New Zealand, where he has worked as an Aquaculture Policy Analyst in the Ministry of Fisheries involved in review of aquaculture legislation, and as a Ministry research scientist on seaweed, abalone, and spiny lobster aquaculture. Current research interests are diverse and span both natural science and social science aspects of tropical aquaculture, such as seaweed, coral, finfish and shellfish farming.

Donald R. Rothwell is an associate professor at the Faculty of Law, University of Sydney where he has taught since 1988. His major research interest is international law, with a specific focus on international environmental law, law of the sea, law of the polar regions, and implementation of international law within Australia. He has taught a range of courses including constitutional law, law of the sea, international environmental law, international law and use of armed force, and public international law. Major publications amongst nine books and over one hundred book chapters and articles include *The Polar Regions and the Development of International Law* (1996), and *International Environmental Law in the Asia Pacific* (1998) coauthored with Ben Boer and Ross Ramsay. His latest publication is *The Law of the Sea and Polar Maritime Delimitation and Jurisdiction* (2001) co-edited with Alex Oude Elferink. He is currently working on projects assessing international dispute resolution and the law of the sea, navigational rights and freedoms, and the globalisation of Australian law. Since 2001, he has been president of the Australian New Zealand Society of International Law (ANZSIL).

Ivan Shearer is Challis Professor of International Law at the University of Sydney, Australia. Prior to his election to that chair in 1993, he was professor of law and dean of the Faculty of Law at the University of New South Wales. In 2000–2001, Professor Shearer held the Charles H. Stockton Chair of International Law at the United States Naval War College, Newport, Rhode Island. In 2000, he was elected at a meeting of States Parties to the International Covenant on Civil and Political Rights to a four-year term as a member of the Human Rights Committee of the United Nations. His special fields of interest in international law are the law of the sea, international humanitarian law, the law of armed conflict, human rights, and international criminal law. In 1999, Professor Shearer served as a judge ad hoc of the International Tribunal for the Law of the Sea on the nomination of Australia and New Zealand in the Southern Bluefin Tuna Case. Professor Shearer is a captain (ret'd) in the Royal Australian Navy. For his work in promoting the teaching of international law in the Australian Defence Force, he was appointed by the Queen as a member of the Order of Australia in 1995.

Alifereti Tawake is a scientific officer with the Institute of Applied Science at the University of the South Pacific (USP). He did all of his undergraduate and graduate work at USP, receiving a B.Sc. degree in biology and chemistry in 1997, a Postgraduate Diploma in chemistry in 1999, and recently completed his M.Sc. research degree in chemistry and marine science. Tawake grew up in a traditional Fijian rural community. Since 1997, he has been working with local communities throughout Fiji and other Indo-Pacific Island countries in collaboration with NGOs and governments on community-based marine biodiversity management and monitoring projects, aspects of community-based conservation, ways to revive traditional management practices, and empowerment of local communities in managing their resources. He was one of three Pacific Islanders chosen to attend the 2000 Summer Institute in Coastal Management at the Coastal Resources Center at the University of Rhode Island. He is also the project officer for the Pacific Region on the Locally Managed Marine Area network, a Foundation of Success Learning Portfolio, and the coordinator of the Fiji Locally Managed Marine Area network.

Cato ten Hallers-Tjabbes is a marine biologist with a long-standing interest in policies to safeguard the marine environment. As a scientist in the Netherlands and in the United States she saw the marine environment being threatened; in the Netherlands government she learned about policy processes. In the late 1970s she joined a group of voluntary scientists, lawyers, and physical planners to develop ideas to protect the North Sea environment. While the group gained momentum and inspired government policies and activities in other North Sea countries, she contributed by editing the *North Sea Monitor* and organising conferences where scientists, policies makers and interest organisations joined in discussing current marine issues. Since the 1980s she employs her expertise in science and in decision-processes to try and connect both worlds and stimulate mutual understanding. Since a decade she co-operates with the Netherlands Institute for Sea Research, along the path from the first signs of impact of TBT in offshore seas to the adoption of the Convention to ban such compounds. She now seeks to generalize the experience from this case study.

Joeli Veitayaki is the Coordinator of the University of the South Pacific's (USP) Marine Affairs Programme and Director of the International Ocean Institute-Pacific Islands operational centre. Veitayaki is from Fiji and has a background in human geography and taught geography at USP before he joined the Marine Affairs Programme (Ocean Resources Management Programme) in 1992. His research interests are in human ecology, with M.A. research at USP on village-level fisheries and Ph.D. research at Australian National University on the reasons for the poor performance of fisheries development projects. He is collaborating with the Institute of Applied Science to set up community-based marine protected areas in Fiji. He has published widely in the areas of customary marine tenure, capacity building, marine resources management, and regional cooperation. He is currently writing his second book based on rural development and fisheries development projects.

Wolfgang Graf Vitzthum is professor of public and international law at the University of Tübingen, Germany. He has served as dean of the law faculty and vice-president of the university. After gaining his LL.M. at Columbia University Law School, he assisted Elisabeth Mann Borgese in the *Pacem in Maribus* project from the beginning and continuing through the 1970s. He earned his Doctorate in Law at Freiburg University with a dissertation on the legal status of the deep seabed. Having obtained his habilitation with a book on the German parliamentary system he became full professor at the University of Tübingen in 1981. He has been a visiting professor at UCLA and at the University of Aix-en-Provence in France. He has served as expert advisor to the German delegation to the Third United Nations Conference on the Law of the Sea. He is currently preparing a book on the identity of Europe and is working with colleagues on a handbook of the law of the sea.

Index for Ocean Yearbook Volume 17

Abalone, 209
Adriatic Sea, 6, 57
Africa, 5, 11, 14–15, 483, 519, 555
Agenda 21, 147, 191, 216, 238–239, 258, 421, 495
Algeria, 52
Antarctic, 576, 597, 611, 621–622, 624 (South Africa, Prince Edward Archipelago)
Antarctic Treaty, 621–622 (see also CCAMLR)
Aquaculture, 142, 151–156, 196, 203, 209, 222, 403, 447; eco-labelling, 94 (see also Tuna, ranching)
Arbitration, 30, 32–36, 37, 42–48, 50 (see also, ITLOS)
Archipelagos: (see South Africa, Prince Edward archipelago)
Arctic: and climate change, 563, 574–579, 584, 588, 590–591, 597, 607, 610–611, 616, 624; Agreement on Conservation of Polar Bears, 610; Arctic Council, 600, 602, 606, 611, 616, 619, 620, 621, 623; Arctic Environmental Protection Strategy, 600, 611, 620; "Arctic Mediterranean", 600; Arctic Ocean, 596–604, 607–619, 621, 623–624; indigenous peoples, 606, 619, 623, 624; mineral resources, 607; nuclear-free zone, 624; scientific research, 609–612, 620; strategic position, 612–615, 620–621; transit management, 617–618; "zone of peace", 621, 624 (see also, Canada, Northwest Passage; Maritime transport, and sea ice; Russian Federation, Northern Sea Route)
Argentina, 13, 52, 205, 223, 246
Asia, 4–8, 11, 14–15, 425, 426, 444, 453, 476; maritime traditions, 7, 9

Assyria, 2
Atlantic Ocean, 591
Gulf Stream, 587
Australia, 20, 40–45, 52, 93, 97, 118–121, 122, 143, 149, 154, 186, 195–196, 198–199, 201–202, 204–207, 209–210, 211, 220, 222–223, 228, 241, 246, 249, 251–252, 260, 379, 428, 520, 538, 547, 593; ANZECC, 198, 206, 210; Ballast Water Management Strategy, 199, 204; oceans policy, 206
Austria, 52
Azores, 5

Baffin Bay, 567–569, 570, 577
Bahamas, 170
Baltic Sea, 6, 196
Bangladesh, 483, 554, 587
Barents Sea, 572, 577
Baselines, 14, 549–551, 601
Beaufort Sea, 570–571, 609
Bêche–de–mer, 46, 150, 161
Belarus, 52
Belgium, 53, 541, 542
Bering Sea, 577
Bering Strait, 572–574
Bermuda, 170
Biodiversity, 50, 142, 215, 219, 221–222, 223, 228, 376, 479 (see also Convention on Biological Diversity)
Bioinvasions (see Marine biosecurity)
Biotechnology, 50
Black Sea, 196, 198, 222, 227
Bohai, Gulf of, 549
Boundaries (see Maritime boundaries)
Brazil, 205, 226, 241, 246, 418, 437, 469, 556, 562

Cadiz, Bay of, 433
Cambodia, 554

1121

Canada, 20, 30, 93, 97, 185, 205, 225, 232, 240, 246, 249, 256, 260, 263, 269, 273, 293, 300, 344, 400, 424, 517, 522–526, 538–539, 541, 542, 563–572, 576–581, 588–595, 596, 598, 599, 601, 606, 608–609, 614–615, 617–618, 621; Atlantic Coastal Action Program, 285; aboriginal rights, 275–285, 287–289; coastal management, 285–289; Oceans Act, 285–287, 289; Northwest Passage, 563, 570–572, 577, 608, 615, 621
Canadian International Development Agency (CIDA), 331
Capacity–building (see Education and training)
Cape Verde Islands, 5, 53, 556
Caribbean, 290, 307–310, 342, 344
Caribbean Sea, 476
Carp, Asian, 147
Caspian Sea, 222
CCAMLR (see Convention on the Conservation of Antarctic Marine Living Resources)
Center for Marine Conservation, 88
Chile, 13, 53, 205, 246
China, People's Republic, 2, 176, 178, 183, 205, 223, 241, 244, 469, 548–555, 559–562, 587, 611
Chukchi Seas, 573
Clams (giant), 144, 147, 150, 152, 161, 163
Climate change, 580–582, 594–595; and polar ecosystems, 366–371, 374; and sea ice, 563–564, 574–579, 584, 586, 588, 607, 615; and seabirds, 369–370; and seals, 369; climate forcings, 584; food chains, 36, 371; icebergs, 564, 579; modelling, 577–579, 584–594; ocean circulation, 369, 587; temperature, 574, 575–576, 584–588, 607 (See also, Extreme events; Intergovernmental Panel on Climate Change; Sea level change)

Climate processes: and freshwater systems, 589–591; and invasive species, 349, 372–375; El Niño, 64; frontal systems, 350–351, 363–371, 374; island mass effect, 352, 357–358; "life support system", 357–365, 371; marine-terrestrial interactions, 349, 357–371; mineral cycling, 353–356; oceanographic forcing, 365; trophic structure, 364–365 (see also Extreme events; Sea level change)
Coastal management, 297–298; and decentralized government, 380–390, 395–400, 415–416; capacity development, 386, 393–394, 400, 409; community–based, 266–268, 338–339, 407–408, 409; competing uses, 267, 390, 392, 406; development setbacks, 313–315, 338, 343; enforcement, 399, 406, 408–409, 411; evaluation, 394; ICZM, 257–259, 264–274, 279, 280, 281, 285–289, 298, 317, 337–338, 343, 394, 397–398, 400–401, 409–415, 442–444; indigenous peoples, 256–289; mapping, 313; participatory approaches, 269–273, 279–285, 450, 452–463; standards and guidelines, 408, 410–411, 413; voluntary incentive-based model, 381–382, 390–395, 398–400, 405–416
Coastal processes, 307, 316–317, 392 (see also Coastal management, development setbacks; Extreme events)
Coastal Society, 21
Coastal Zone Canada Association, 21
Colombia, 63, 86, 87
Commission on Sustainable Development, 216
Commission on the Limits of the Continental Shelf, 27, 599
Common heritage doctrine, 16, 18,

Index   1123

58, 467, 476, 479, 480, 600, 611, 616, 623
Conciliation, 35
Conservation, 349, 372–379, 381, 382, 388, 392, 402–403, 405, 407, 415 (see also Convention on Biological Diversity)
Contiguous zone, 3, 9, 13, 18, 286, 550, 553–554
Continental shelf, 13–15, 58, 286, 428, 425, 431, 550, 598–600, 616
Convention on the Conservation of Antarctic Marine Living Resources, 372, 376, 622
Convention for the Conservation of Antarctic Seals, 622
Convention for the Conservation of Southern Bluefin Tuna (CCSBT), 40–45
Convention for the Protection of the North East Atlantic (OSPAR), 45, 47
Convention on Biological Diversity, 50, 199–200, 203, 212, 220, 221, 227, 239, 242, 246, 250, 253, 254, 477, 478 (see also, Marine biosecurity)
Convention on Facilitation of International Maritime Traffic (FAL), 245, 254
Convention on Fishing and Conservation of Living Resources of the High Seas, 14
Convention on the Continental Shelf, 14, 29
Convention on the High Seas, 3, 14, 170, 173–174; hot pursuit, 520–524, 535
Convention on the Prevention of Marine Pollution by Dumping of Wastes and Other Matter (see London Convention)
Convention on the Territorial Sea and Contiguous Zone, 14, 29, 551, 552–553
Cook Islands, 142, 149, 165
Coral, 153, 163, 317, 406, 461
Costa Rica, 13, 20, 87, 91

Crabs, 208, 459
Crustaceans (see Lobster and Crabs)
Cuba, 20, 53, 469
Cyprus, 170

Denmark, 6, 10, 57, 144, 183, 205, 422, 576, 577, 596, 601, 606, 614, 617
Developing countries, 11, 15–17, 137, 142–143, 165, 253, 290–291, 305, 469, 470–478, 480, 484, 492, 494, 587–588, 593, 594, 603, 605–606; and community-based tourism, 328–335, 344–347; and decentralized governments, 380–389, 395–408, 415; eco-labelling, 137; fisheries trade, 138–139; "Pacific Paradox", 450; shipping, 167–192
Dispute settlement, 26–55; limits on jurisdiction, 34–36 (see also, United Nations Convention on the Law of the Sea, dispute settlement; International Court of Justice; International Tribunal for the Law of the Sea)
Dolphins: and tuna, 60–92; Agreement for the Conservation of Dolphins, 75–76 (see also, International Dolphin Conservation Program)

Earth Island Institute, 86
Ecological security (see Marine biosecurity)
Ecosystem management, 111, 136, 286, 375–378
Ecuador, 13, 63, 64, 86, 87, 91, 397
Education and training, 20–23, 70–74, 80–81, 98–100, 150, 151, 231, 244, 252, 253, 328, 333–334, 418, 423, 425–431, 432, 437, 449–463, 468–471, 477, 480, 619; curriculum standardization, 22–23; marine affairs professions, 19–20; professional associations, 21–25; EEC (see European Union)

EEZ (see Exclusive Economic Zone)
Egypt, 2, 53, 554, 587
El Salvador, 13
Environmental Defense Fund, 88
Environmental impact assessment, 148, 162, 377
Environmental protection (see Marine environmental protection)
Erosion (see Coastal processes)
Europe, 1–2, 6–8, 10–11, 15, 18, 173, 185, 300, 293, 577, 587; and sea power, 7, 12–13
European Union, 97, 215, 241, 250, 254, 422, 424, 426, 428, 432, 436, 441, 447–448, 469, 473, 493, 533–534, 545, 593
Exclusive Economic Zone, 18, 27, 65, 286, 402–403, 443–444, 548–551, 598–601; and military activities, 550–553, 555–562; freedom of navigation, 553–556, 558, 559–562; freedom of overflight, 548–549, 551, 555–556, 559–562 (see also United Nations Convention on the Law of the Sea)
Extreme events, 290–292, 295, 307–309, 337–339, 343–344, 393, 591–592; storm surges, 307, 313, 315, 317, 319 (see also, Climate change; Coastal processes)

Faeroe Islands (see Denmark)
FAO (see United Nations Food and Agricultural Organization)
Federated States of Micronesia, 149
Fiji, 142–151, 154, 156–166, 454–463
Finland, 53, 97, 205, 600, 606
Fisheries, 440, 436, 443, 447, 598; accidental capture, 60–92, 102, 120, 124, 125, 128, 132–134, 135, 375; capacity, 458–460; data needs, 113, 120, 124–125, 144; destructive techniques, 113, 120, 136, 138, 405, 457–461; dispute settlement, 27, 29; drift-nets, 121–127; eco-labelling, 95–104, 138–139; education and training, 70–74; embargoes, 74, 83, 84, 86–87, 91; enforcement, 112, 458, 460, 462, 463; fish aggregating devices (FADs), 62, 74, 77; habitat, 215, 406; high seas, 12, 18, 59; illegal, unreported and unregulated (IUU), 186, 190, 375, 459, 467, 560; observer programs, 65–67, 70, 76, 83–86, 89, 91; ornamental, 460; purse seines, 127; technology, 12, 61–62, 68–70, 77–79, 91–92; trawling, 122–123, 127, 131–132 (see also Marine resources management; Marine Stewardship Council)
Fisheries products: international trade, 138–139; marketing, 98–102 (See also Fisheries, eco-labelling; Marine Stewardship Council)
Flags of convenience (see Maritime transportation, flag state)
France, 11, 53, 177, 379, 424, 527, 566
Freedom of the seas doctrine, 2–18, 57–59, 167, 599–600, 616–617; and colonialism, 6–7, 9–12, 14–16, 57; and Roman law, 3–7, 57–58; codification, 11, 14–15; fisheries, 9–10, 12–14, 18, 59; laying submarine cables and pipelines, 35, 477; military activities, 9–11, 551–553, 555–562; navigation, 3, 5–6, 9–10, 18, 35, 553–556, 558, 559–562; overflight, 35, 548–549, 551, 555–556, 559–562 (see also, Innocent passage; International law; *Mare liberum*; Rhodian sea law)

General Agreement on Tariffs and Trade (GATT), 98, 138–139
Geographic information systems (GIS), 411
Germany, Federal Republic of, 11, 53, 176, 205, 611; Blue Angel Programme, 96, 97, 139, 140

Index   1125

Global Aquaculture Alliance, 94
Global Environment Facility (GEF), 200, 475, 479
Global Plan of Action for the Protection of the Marine Environment from Land-based Activities (GPA), (incl Washington Declaration), 466, 475, 494, 619
Great Britain (see United Kingdom)
Great Lakes (North America), 589–591
Greece, 53, 176, 205
Greenhouse gases (see Climate change)
Greenland (see Denmark)
Greenpeace, 88
Grotius, Hugo (see *Mare liberum*)
Guinea-Bissau, 53

Hainan Strait, 549
Hake, 93
Harvard Draft Convention on Jurisdiction with Respect to Crime, 537, 544
Herring, 121–126
High seas, 3, 12–13, 27, 482, 561; and abortions, 512–547; "enclaves", 599–600, 616–617 (see also, International law, constructive presence; United Nations Convention on the Law of the Sea)
Hoki, 114, 128, 130–134
Honduras, 13, 523
Hong Kong, 176, 205
Hudson Bay, 564–567, 577, 579, 592
Human security (see Marine biosecurity; Security issues)
Hydrocarbons (see Offshore oil and gas)

Iceland, 10, 30, 54, 600, 606
ICES (see International Council for the Exploration of the Sea)
Illicit drug trafficking, 325–326, 466, 490, 491, 523–524
IMO (see International Maritime Organization)

India, 2, 4, 6–7, 223, 241, 246, 469, 471, 554, 556
Indian Ocean, 4–7, 624
Indonesia, 20, 382, 387, 397, 400–416, 427, 483
Innocent passage, 186, 238, 530–531, 536, 547, 551–553, 561
Inter-American Development Bank (IDB), 332
Inter-American Tropical Tuna Commission, 63–65, 69–74, 82–84, 89
Intergovernmental Oceanographic Commission, 465, 470
Intergovernmental Panel on Climate Change (IPCC), 563, 574–576, 582–584, 586
International Bank for Reconstruction and Development (see World Bank)
International Bureau of Commerce:
International Maritime Bureau, 483, 488
International Center for Living Aquatic Resources Management (ICLARM), 149–150, 151
International Convention for the Prevention of Pollution of the Sea by Oil (see MARPOL)
International Convention for the Prevention of Pollution from Ships (MARPOL), 242–243, 245, 252, 254
International Convention for the Safety of Life at Sea (SOLAS), 184, 186, 219, 233
International Convention on the Control of Harmful Anti-fouling Systems on Ships, 246, 418–419, 434
International Council for the Exploration of the Sea (ICES), 623
International Council of Scientific Unions (ICSU), 581, 620
International Court of Justice, 14, 26, 30–31, 32, 34, 36–38, 39, 44, 541
International Criminal Court, 542

1126   Index

International Dolphin Conservation Program, 76–77, 87–91; La Jolla Agreement, 87–91; Declaration of Panama, 76, 88
International environmental law, 28, 48–52, 602–603 (See also, Multilateral environmental agreements)
International Labor Organization, 173, 175, 178
International law, 1–2, 5, 33–34, 47, 56; and coastal state jurisdiction, 16, 18, 58–59, 186, 227–228, 238–240, 253–254, 443, 551–557, 598–600; and criminal jurisdiction, 512–547; and EEZ, 533; active personality, 537–540, 543, 547; competing claims, 544–545; constructive presence, 519–524, 535–536, 544, 546, 547; customary, 6; double criminality, 543, 547; enforcement jurisdiction, 545–546; flag state jurisdiction, 527–528, 539–540, 546, 547; "floating island" theory, 546; passive personality, 537, 540, 541–543, 544; protective principle, 540; right of self-defence, 561; territoriality, 516–529, 535–536, 543–544, 547, 611; universality principle, 536–537, 543 (see also, Freedom of the seas doctrine; Innocent passage; Multilateral environmental agreements; United Nations Law of the Sea Convention)
International Maritime Organization, 49, 168, 173, 175–176, 178, 199–200, 203, 204, 216, 219, 222, 225, 229, 233, 235, 239–242, 244, 245, 444, 447, 465, 470, 484, 491, 493; Draft Ships' Ballast Water Convention, 242–246; Marine Environmental Protection Committee, 418, 424, 426, 436, 441, 443, 446, 447; Maritime Safety Committee, 466, 486–488, 491, 493

(see also, World Maritime University)
International Ocean Institute, 466, 473–475, 477–479
International Organization for Standardization (ISO), 24, 95, 187
International regimes, 421, 615–624 (see also, Marine biosecurity)
International Sea-Bed Authority, 27, 476–479, 616
International Tribunal for the Law of the Sea (ITLOS), 26–28, 32–34, 36–48, 59, 477; Sea-Bed Disputes Chamber, 33, 38; Special Chambers, 38, 49–50, 51
Introduced species (see Marine biosecurity)
IOI (see International Ocean Institute)
Iran, 241, 246, 554, 556–557, 562
Ireland, 45–48, 512, 515, 545 (see also, High seas, and abortions)
Israel, 205, 246
Italy, 6, 54, 57, 205, 429, 431, 433–436, 447, 520, 524, 556
IUCN (see World Conservation Union)

Jamaica, 290–291; coastal management, 297–298, 312–319, 327, 329, 332–333, 335, 342–344; disaster management, 310–312, 335–339, 343–344; economy, 302–307, 345–347, 339–340; tourism, 292–294, 298–302, 306–307, 315–347
Jan Mayen (see Norway)
Japan, 40–45, 97, 139, 144, 149, 154, 173, 176, 205, 223, 244, 246, 249, 293, 427, 489, 587, 611

Kara Sea, 572–573, 577
Kiribati, 149
Korea, Republic of, 41, 183
Kyoto Protocol (see United Nations Framework Convention on Climate Change)

Labrador Sea, 566, 577, 579
Latin America, 11, 14, 293, 326, 438, 447, 519, 555
League of Nations, 11, 494
Liberia, 170, 177
Ligurian Sea, 6
Lobster, 93, 118–121, 122
London Convention, 437, 438, 441, 446–447
LOS Convention (see United Nations Convention on the Law of the Sea)

Mackerel, 79, 94
Malacca, Straits of, 7, 483
Maldives, 587
Malta, 15–16, 474
Mangroves, 317, 406, 459, 461
Mare clausum, 8, 9, 57–58
Mare liberum, 3–8, 17, 57–58, 167
Mare nostrum, 57
Marine Affairs and Policy Association (MAPA), 20, 21
Marine biosecurity, 193–198, 200–201, 203, 213–216; and ballast water, 197, 199–207, 211, 213–255, 418, 437, 446–447; and hull fouling, 196, 201, 203–204, 206, 220, 224, 246; and ship design, 233–234, 247–248, 251; control strategies, 209–212, 224, 235–255; GloBallast, 200, 215, 225, 241, 248, 254; Global Introduced Species Program, 211; information needs, 197–200, 228; open sea exchange, 229–234, 244, 248; prediction, 207–209; remediation, 199, 223–224
Marine environmental protection, 30, 45–52, 184–186, 212, 264–267, 618–619 (See also, Multilateral environmental agreements)
Marine mammals, 406, 598 (see also Dolphins; Seals; Whales)
Marine pests (see Marine biosecurity)
Marine pollution: agricultural runoff, 457; dredging spoils, 418, 437; land-based, 51, 215, 218, 317–319, 338, 457, 459, 460, 461–462; ocean dumping, 437, 446; radioactive waste, 45–48; sediment contamination, 433–435, 441, 446; sewage, 218, 319, 459, 461; tributyltin (TBT), 203, 216, 246, 418, 423–426, 432–438, 442, 446–447; vessel source, 171, 181, 184–186, 189, 218, 437, 601 (see also Marine biosecurity)
Marine protected areas, 136, 163, 349, 372, 374, 375–378, 388, 392, 458, 461, 463; historical sites, 377, 378, 392
Marine resources management, 30, 35, 58, 108–113, 127–129, 136, 215, 264–265, 375, 378, 382, 403–404; access rights, 146, 153, 155–166, 273–285, 288–289; community-based, 449–463; customary, 449, 451–452; resource surveys, 457–458 (see also Coastal management; Fisheries; Fisheries products; Ocean mining)
Marine science and technology, 12, 15, 35, 36, 203, 351–353, 374, 376, 379, 396–397, 464–469; and policy making, 417–423, 425–432, 435–448, 471; and the public, 418, 423, 425–431, 432, 437, 468–469, 471, 583; "eco-technologies", 471; human resource development, 469, 470, 471, 477, 480; cooperation and transfer, 428–432, 435–437, 444–446, 465–481, 494–502; joint ventures, 468–469, 473, 479, 495; regional centres, 471–476, 495–502; traditional technological experience, 471 (see also, Intergovernmental Panel on Climate Change)
Marine stewardship, 619
Marine Stewardship Council, 93–95, 97; and accreditation bodies,

1128 Index

104–108, 137; certification scheme, 108, 113–118, 137, 141; certified fisheries, 118–137; sustainable fisheries principles, 108–113 (see also Fisheries, eco-labelling)
Marion Island (see South Africa, Prince Edward Archipelago)
Maritime boundaries, 27, 30, 34, 39, (see also International Court of Justice; United Nations Law of the Sea Convention, dispute settlement procedures)
Maritime fraud, 170, 182, 191 (see also, Maritime transport, flag state)
Maritime labour (see Seafarers)
Maritime law, 2, 56–57 (see also Freedom of the seas doctrine)
Maritime safety, 168–169, 171, 181–186, 187, 189, 190, 225, 228, 229–234, 245, 247; Maritime transport, 36, 197, 199, 201, 216–217, 429, 431, 432–433, 436, 440, 442: and sea ice, 563–574, 576–579, 616; anti-fouling systems, 447; classification societies, 182, 185, 189, 224, 243, 247–248, 252; climate change impacts, 588–592; flag state, 167–192, 227, 241, 243, 252–253, 488; International Safety Management Code, 186–189; marine paints, 418, 423–428, 431–436, 440; port state control, 168–169, 173, 184–186, 190, 225–227, 240, 243, 253–254; shipbuilding and repair, 169, 440 (see also Marine biosecurity, and ballast water; and Ports)
MARPOL (see International Convention for the Prevention of Pollution from Ships)
Marshall Islands, 149
Mediterranean Sea, 196, 223, 227, 427, 433, 473, 475, 476, 495; regional coast guard, 466

Mexico, 20, 63, 64, 67, 83–84, 86, 87, 91, 93, 137, 223, 418, 435, 469, 522
Milkfish, 146, 147, 152
Minerals (see Ocean mining)
Molluscs, 134–137; cockles, 116, 134–137; mussels, 135, 143, 209, 227; oysters, 143, 298, 424; pearl oyster, 143, 144–145, 147, 150, 152, 160–161; scallops, 154; snails, 418, 423–425, 433–435; whelks, 424, 433–435 (see also, Abalone; Clams)
Monitoring, control and surveillance (MCS) (see Fisheries, enforcement)
Multilateral environmental agreements, 27, 47–48, 51, 217, 221–222, 235–236, 250–251, 254, 444, 602–605, 618–619

National Wildlife Federation, 88
Nauru, 149
Navies (see Security issues)
Navigation, 9, 18, 30, 173, 180, 190, 601, 604, 607–608, 615–618; "sustainable navigation", 617 (see also, Maritime transport, and sea ice)
Negotiation, 29, 32, 44
Netherlands, 4, 10, 54, 57, 205, 435, 447, 512–516, 519, 529–530, 532, 536, 543–544, 546–547
Netherlands Institute for Sea Research, 426
New Caledonia, 150
New Zealand, 40–45, 93, 97, 114, 128, 130–134, 143, 149, 154, 155, 160, 155, 196, 198, 200, 204, 205, 207, 220, 246, 249, 256, 284, 289, 379; Environment Court, 271; Fisheries Claims Settlement Act, 273–274; Maori rights, 260–264, 265, 271–274; resource management, 264–274; Treaty of Waitangi, 261–264, 266, 269, 271, 272
Nicaragua, 54

Nippon Foundation, 486, 488–489
Niue, 149
Non-Governmental Organizations (NGOs), 328, 333–334, 341–342, 343, 422, 435, 436, 440, 449, 450
North America, 12, 15, 595
North American Aerospace Defense Command (NORAD), 621
North American Free Trade Agreement (NAFTA), 254
North Atlantic Treaty Organization (NATO), 621
North Pacific Marine Science Organization (PICES), 623
North Sea, 79, 219, 418, 424, 426, 429, 431, 433–434, 442, 444
North Sea Conferences, 422, 440
Northern Sea Route (see Russian Federation)
Northwest Passage (see Canada)
Norway, 10, 14, 30, 54, 97, 183, 205, 596, 601, 606
Nuclear waste (see Marine pollution, radioactive waste)

Ocean dumping (see Marine pollution)
Ocean governance (see Ocean management)
Ocean management, 28, 37, 40, 42, 51, 600, 619
Ocean mining, 403; deep sea, 15, 476–479; hydrates, 477, 478; JEFERAD, 477–478; sulphides, 477
"Oceans Senate", 192
Offshore oil and gas, 12, 598, 608–609
Okhotsk, Sea of, 577
Oman, 54
Organization of African Unity, 493
Organization of American States, 260, 296, 491, 493
Orkney Islands, 246

Pacific Island States, 449, 453, 476, 495; "Pacific Way", 455

Pacific Ocean, 572–573, 577, 591
Pakistan, 554
Palau, 149
Panama, 63, 86, 87, 90, 170, 177
Panama Canal, 617, 621
Papua New Guinea, 149
Paracel Islands, 549
Paralytic shellfish poisoning, 222–223
Patagonia (see Argentina)
Patagonian toothfish, 93, 375
Pathogen transfer (see Marine biosecurity, and ballast water)
Peaceful uses, 10, 476, 473
Peru, 13
Petroleum (see Offshore oil and gas)
Philippines, 20, 387–388, 397
PICES (see North Pacific Marine Science Organization)
Piracy and armed robbery at sea, 464–467, 481–495
Pollock, 93
Pollution (see Marine pollution)
Population growth, 381, 393, 397, 459, 461
Ports, 199, 215, 218, 225–227, 232, 240, 252–254, 326, 393, 423, 427, 428, 435, 440, 566, 576
Portugal, 4–8, 54, 57, 223, 429, 431–436, 447
Prawns (see Shrimp)
Precautionary approach, 49–51, 112, 228, 240, 242, 254, 286, 418, 421–423, 430, 436, 439, 440, 444–445, 447–448, 449, 593
Prince Edward Islands (see South Africa, Prince Edward Archipelago)

Quality Status Reports, 422, 440, 441

Radioactive waste (see Marine pollution)
"Red Tide" (see Paralytic shellfish poisoning)
Regional cooperation, 185, 191, 378–379, 428, 465–466, 471–

476, 485, 490–492; Regional Seas Programme, 466, 473, 475, 495, 618 (see also, Arctic, Arctic Council)
Rhodian sea law, 5–6, 56–57
Rio Declaration (see United Nations Conference on Environment and Development)
Risk management, 291, 295–296, 335–338, 342–343; hazard mapping, 311, 312–313, 317, 337; insurance, 291, 309–310, 342; mitigation measures, 309, 310–317, 326–339, 342–343; social risks, 291–292, 296, 300, 317–335, 339–342, 344–347 (see also, Coastal management, development setbacks; Extreme events)
Romania, 525–526
Russian Federation, 11, 14, 54, 308, 557, 571–574, 577, 579, 596, 597, 599, 601, 606, 608–611, 613–615, 617, 619, 621; Northern Sea Route, 572–574, 577, 608, 615, 617, 621

Saithe, 79
Salmon, 115, 116, 122, 126–130, 143
Saudi Arabia, 183, 554
Scientific Committee on Antarctic Research, 379, 622
Sea cucumbers, 406
Sea level change, 308–309, 575, 576, 586, 588, 592 (see also, Climate change; Intergovernmental Panel on Climate Change)
Sea turtles, 406
Seabed mining (see Ocean mining)
Seabed resources, 477, 479 (see also, Ocean mining)
Seabirds, 132–135, 353–357, 363, 369, 372, 374–375; albatrosses, 362, 375, 377; penguins, 362, 356–357
Seafarers, 171, 175, 180, 181, 183, 190, 232–233, 244, 249, 251, 252

Seals, 132–134, 222, 353, 354, 355–357, 363, 369, 372, 374
Seaweed, 144–145, 146, 147, 227
Secretariat for the Pacific Community (SPC), 149, 151
Security issues, 6–12, 215, 320–328, 344–347, 482, 492, 541–543; and information technology, 489, 490; cooperation, 466–467, 485–486, 490–493; illegal transport of persons and goods, 466, 482, 490, 491, 529; interdiction and enforcement, 466–467, 485, 491–492, 495; offshore gambling, 518, 539; regional coast guard, 466, 491–492, 502, 511; "sea police force", 493–494 (see also, Illicit drug trafficking; Piracy)
Sharks, 78–79
Shipping (see Maritime transport)
Shoreline protection (see Coastal management, development setbacks)
Shrimp, 144, 147, 360
penaeid, 144–145, 147, 153, 156
Singapore, 183, 205, 249
Slovenia, 54
Small Island Developing States (SIDS), 587 (see also, Caribbean and Pacific Island States)
Solomon Islands, 149, 150
South Africa, 93, 196, 241, 246, 397
Crozet Islands, 349, 362
Prince Edward Archipelago, 348–379
South America, 227, 326, 483
South China Sea, 483, 548–549
South Pacific Community (see Secretariat for the Pacific Community)
Southern ocean (see Antarctic Region)
Soviet Union (see Russian Federation)
Spain, 4–5, 30, 55, 57, 223, 428, 429, 431–436, 447, 448
Spratly Islands, 549

Sri Lanka, 397, 554
St. Lawrence, Gulf of, 564–567, 577, 579
Stockholm Declaration (see United Nations Conference on the Human Environment)
Straits, 30, 484, 488, 494, 601, 618
Subsidiarity principle, 396
Sustainable development, 49, 51, 142, 143, 215, 228, 258–259, 265, 268, 271, 286, 449, 605–606, 609, 619–620 (see also, Marine Stewardship Council, sustainable fisheries principles)
Sweden, 30, 55, 57, 97, 205, 432, 435, 600, 606
Syria, 554

Taiwan, 144, 183, 553, 550–551
Tanzania, 55, 397
Territorial sea, 3, 11, 13–15, 58, 286, 549–554, 600
Thailand, 20
Tilapia, 144–145, 147, 148, 156
Tonga, 149, 156
Tonkin, Gulf of, 550
Tourism, 158, 290–292, 315, 373–374, 377; and development planning, 295, 312, 335–339; community-based, 328–335, 344–347; heritage-based, 331–332; "sustainable tourism", 301–302, 326 (see also, Jamaica, tourism)
Traditional knowledge, 257, 259, 451–452, 471, 612
Transit passage, 618
Trochus, 147
Tuna: and dolphins, 62–78, 80–81, 90, 91–92; bigeye, 61, 62, 77; bluefin, 40–45, 50, 51, 79; dolphin-safe label, 74–75, 77, 85–86, 88–89, 91; purse seines, 60–63, 68–70, 87, 91–92; ranching, 79, 152; skipjack, 61, 62; yellow fin, 61, 62, 70, 77, 84–86
Tunisia, 55
Turkey, 527
Turtles (see Sea turtles)

Tuvalu, 149, 587
Tyrrhenian Sea, 433

Ukraine, 55, 241, 246
UNCED (see United Nations Conference on Environment and Development)
UNCLOS (see United Nations Conference on the Law of the Sea)
UNCTAD (see United Nations Conference on Trade and Development)
UNDP (see United Nations Development Programme)
UNEP (see United Nations Environmental Programme)
UNESCO (see United Nations Educational, Scientific and Cultural Organization)
United Arab Emirates, 554
United Kingdom, 3, 4, 6, 8–11, 20, 45–48, 55, 57–58, 93, 116, 121–126, 134–137, 170, 196, 205, 344, 348, 424, 432, 435, 522, 524, 533, 539–540, 611
United Nations, 14, 260, 465, 492–493, 623; Charter, 26, 28, 31, 35, 557, 605; consultative process, 216, 464–468, 492, 493, 495; DOALOS, 242; General Assembly, 16, 192, 464, 492, 623, 624; Secretary-General, 190, 464, 465, 493; Security Council, 35, 493, 623–624
United Nations Agreement on Straddling Fish Stocks and High Migratory Fish Stocks, 59
United Nations Conference on Environment and Development, 147, 421, 465; Rio Declaration, 259, 484
United Nations Conference on Trade and Development, 169, 170–171, 178; Committee on Shipping, 174, 176–177; Conference on Conditions for Registration of Ships, 177, 179–180; Convention on Conditions for

Registration of Ships, 169, 180–183, 184, 188
United Nations Conference on the Human Environment, 216
United Nations Conference on the Law of the Sea, 16–18, 58, 59, 175, 471–472, 533, 553, 555, 561, 597, 598
United Nations Development Programme, 200, 455
United Nations Disaster Relief Organization, 308
United Nations Draft Universal Declaration on the Rights of Indigenous Peoples, 259
United Nations Economic Commission for Latin America and the Caribbean (ECLAC), 308
United Nations Educational, Scientific and Cultural Organization, 378, 623
United Nations Environmental Program, 221, 222, 242, 465, 470, 474, 582 (see also, Regional cooperation, Regional Seas Programme)
United Nations Food and Agricultural Organization, 139, 143, 149, 153–154, 222, 465, 470
United Nations Framework Convention on Climate Change (UNFCC), 477, 478, 580, 592–593; Kyoto Protocol, 580, 593
United Nations Industrial Development Organisation (UNIDO), 465, 474
United Nations Informal Consultative Process on the Oceans (UNICPO) (see, United Nations consultative process)
United Nations Informal Consultative Process on the Oceans and the Law of the Sea (UNICPOLOS) (see, United Nations, consultative process)
United Nations Law of the Sea Convention, 26, 216, 465, 467, 514, 529, 546–547, 597–602, 604–605; alien species, 236–238, 242, 250, 253; contiguous zone, 553–554; continental shelf, 598–599, 616; criminal jurisdiction, 530, 531; deepsea bed, 59, 476–479; dispute settlement procedures, 26–36, 59; environmental protection, 48–51, 59, 186, 236–238, 421; exclusive economic zone, 243, 482–483, 531–532, 553–562, 598–601; flag state jurisdiction, 532, 533, 546; high seas, 482–483, 527, 532–536, 546, 547; hot pursuit, 524, 530, 535–536, 547; innocent passage, 238, 530–531, 536, 547, 552–553; marine science and technology, 464–466, 469–473, 476; peaceful uses of the sea, 476, 557; polar regions, 597–602; semi-enclosed seas, 600, 616; shipping, 168, 170, 174–175, 184, 186, 190, 191, 225, 227, 238; straits, 601, 618; territorial sea, 513, 517, 530, 536, 550, 551–554, 555, 598, 561; transit passage, 618; unauthorized broadcasting, 532–536, 547
United Nations Programme of Action for Sustainable Development (see Agenda 21)
United States, 11, 18, 63, 64–66, 81–91, 93, 99, 115, 116, 122, 126–130, 144, 176, 195–196, 198, 200–201, 204, 205, 209, 219–220, 227, 246, 249, 260, 293, 300, 308, 326, 333, 424, 468, 515, 518, 519, 520, 522, 527–528, 539, 540, 545, 547, 548–549, 552, 554, 559–562, 570, 593, 596, 597, 598, 601, 606, 608–611, 613–615, 620–621; Coastal Zone Management Act, 267, 381–382, 389–400, 409, 416; Magnuson-Stevens Act, 83, 389, 395; Marine Mammals Protection Act, 63, 65, 69,

74, 82–91, 389; NOAA, 390, 392, 394; National Marine Fisheries Service, 68, 82–83, 86, 91; Truman Proclamation, 13
University of the South Pacific, 150, 451
Urbanization (see Coastal management)
Uruguay, 55, 556

Vanuatu, 63, 86, 149, 170
Venezuela, 63, 64, 86, 87, 91, 223, 554
Vietnam, 554, 549

Water levels (see Sea level change)
Western Samoa, 149
Wetlands, 393 (see also Mangroves and Coral)
Whales, 353

Women on Waves Foundation, 512–516, 530, 536, 541–543
World Bank: Noordwijk Guidelines, 258
World Conservation Union (IUCN), 133, 349, 620
World Health Organization, 242
World Maritime University, 435
World Meteorological Organization, 581, 582
World Trade Organization, 217, 254–255, 295, 301–302; Agreement on Sanitary and Phytosanitary Measures, 255; Agreement on Technical Barriers to Trade, 138–139; Committee on Trade and Environment, 138–139
World Wildlife Fund, 88, 104

Yachting, 427, 423